# HUMAN

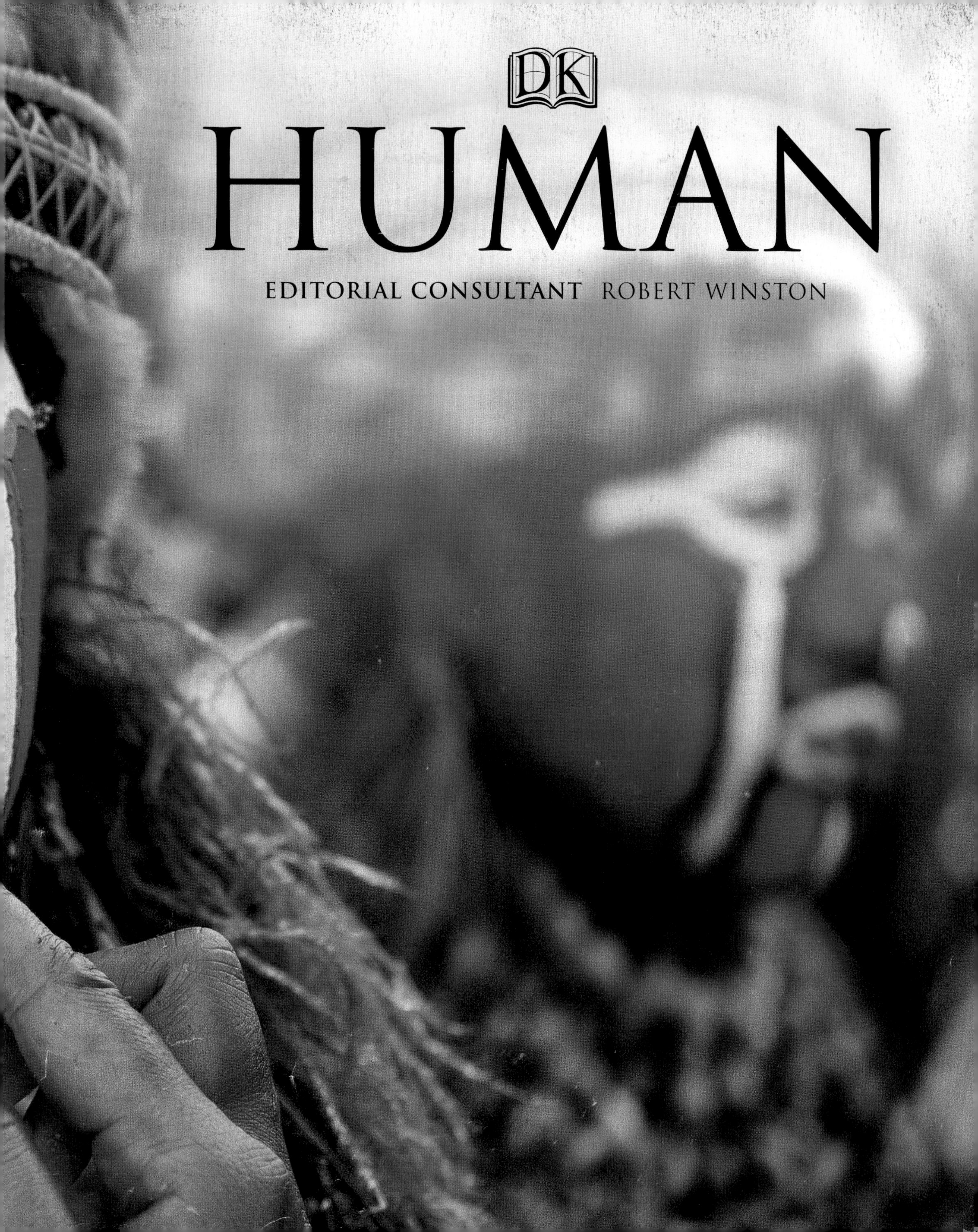


# HUMAN

EDITORIAL CONSULTANT ROBERT WINSTON

LONDON, NEW YORK, MELBOURNE, MUNICH AND DELHI

**Senior Editor** Janet Mohun
**Project Editors** Ann Baggaley, Joanna Benwell, Dawn Henderson, Rob Houston, Katie John
**Indexer** Hilary Bird

**Editorial Manager** Andrea Bagg
**Publishing Director** Jonathan Metcalf

**Senior Designer** Liz Sephton
**Project Art Editors** Sara Kimmins, Peter Laws, Maxine Lea, Mark Lloyd, Shahid Mahmood, Dan Newman
**Design Assistants** Iona Hoyle, Francis Wong
**DTP Designer** Julian Dams
**Picture Researchers** Gwen Campbell, Helen Stallion, Rob Nunn
**DK Picture Library** Romaine Werblow
**Illustrators** Joanna Cameron-Rutherford, Combustion Design and Advertising, Jane Fallows, Adam Howard, Kevin Jones Associates, Joern Kroeger, Debbie Maizels, Mirashade, Primal Pictures Ltd, 4site Visuals
**Photographer** John Davis
**Cartographers** Kenny Grant, Rob Stokes
**Production Controllers** Heather Hughes, Melanie Dowland

**Managing Art Editor** Marianne Markham
**Art Director** Bryn Walls

**Smithsonian Project Co-ordinators**
Ellen Nanney, Katie Mann

First published in Great Britain in 2004
Second edition published in 2005 by
Dorling Kindersley Limited
80 Strand, London WC2R ORL

A Penguin Company





2 4 6 8 10 9 7 5 3 1





A CIP catalogue record for this book is available from the British Library

ISBN 140531155X

Colour reproduction by Colourscan, Singapore
Printed and bound in China by Leo Paper Products

see our complete catalogue at
**www.dk.com**


# CONTENTS

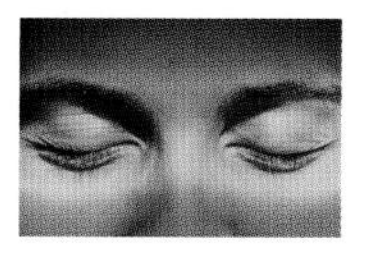

## MIND

## LIFE CYCLE

## SOCIETY

## CULTURE

## PEOPLES

## FUTURE

# FOREWORD

This book is about what it means to be human. Humans are relative newcomers on the planet. Most animals have been on Earth for much longer than the 150,000 years or so that *Homo sapiens* has existed. Our ancestors were hominids, walking on two legs, coming from the open valleys of East Africa. We were an endangered species, and few in number. We were among the weakest of animals on the savanna – not particularly fleet of foot, no large teeth or claws as weapons, and with vulnerable babies needing constant nurture and protection.

But our species had, among other attributes, two remarkable advantages over other animals. One was our large brain, and with it, a keen intelligence and aptitude to adapt to hostile environments. The other was our sociability. Like many other apes we could not have survived without the ability to work in groups, and to cooperate in hunting and protecting each other. And human society was enhanced by our ability to communicate. Our hominid ancestors probably evolved from chimpanzeelike creatures between 5 and 15 million years ago. So it is not entirely surprising that modern humans are relatively similar to modern-day chimps. We share much of our DNA with them, and we probably have very similar genes. Yet the tiny genetic difference makes for a remarkable dissimilarity. We have a totally different physique, more powerful mental abilities, and an inbuilt skill at language.

Once the Earth's climate grew warmer and more stable around 10,000 years ago, humans were able to establish settlements. Human civilization went on to develop complex symbols and a written language. Now we have buildings and machines that mean we can live almost anywhere on Earth's surface – at high altitude, at extremes of temperature, even under water or in space. The story of evolution has largely been about how animals adapt to their environment by survival of the fittest. In one sense, we have become masters of our environment, and we now escape many evolutionary pressures.

Humans are phenomenally inquisitive – hence the development of science. We investigate and experiment, theorize about the origins of existence and the nature of the universe, and have a powerful spiritual sense. This aspect of the human mind, together with our aesthetic sense, led humans to develop pleasure in painting and music, poetry and theatre.

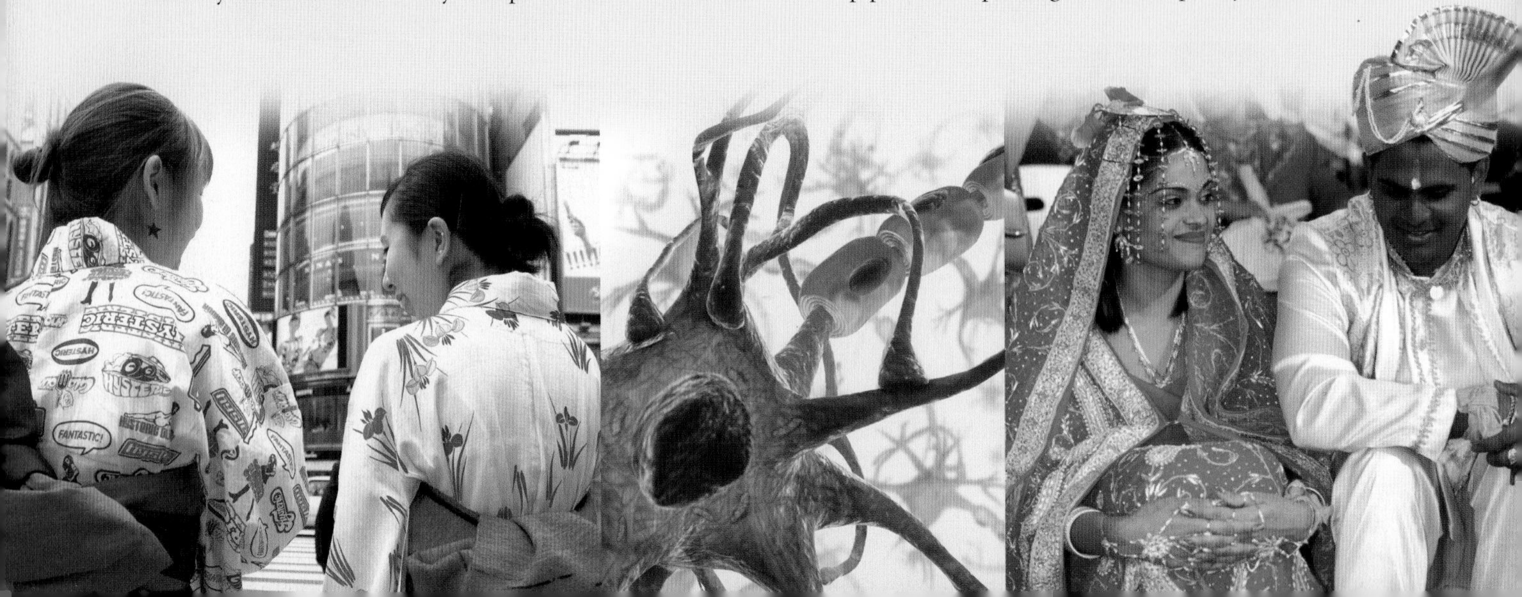

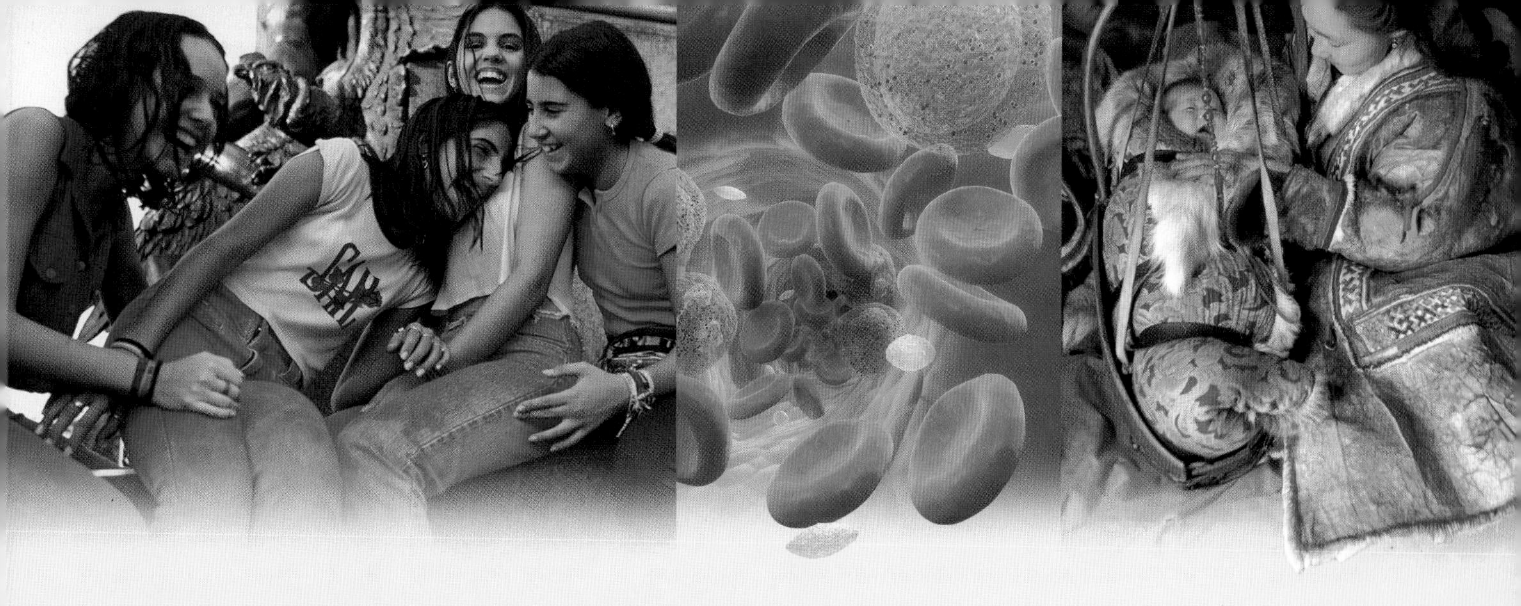

This book is one expression of humankind's natural inquisitiveness. No other species seems interested in how the body works. One great modern advance was the technology to make images of the brain at work, helping us understand the human mind – what makes each of us truly human.

We may all be from the same species, but humanity is very diverse. This book is testament to this amazing variety. Human development has proceeded in many ways in separate parts of the world. Each society found different ways to organize life – to survive, to educate, to dress, to govern itself, to communicate with the rest of the world, and to generate wealth. With rapid changes in technology, the internet, and increasing world trade and easy travel, we now live in a changing global society – where the consequences of our actions can have profound effects on people living at great distances.

So what of the future of the human species? It took 100,000 years to gain the ability to build crude dwellings, and even longer to learn to make stone tools. But in the last 200 years or so we have fashioned steam engines, harnessed this planet's stores of energy, ventured to the Moon, and learned the workings of the genome. Humankind's achievements are proceeding exponentially so it is impossible to predict our future. No doubt we shall improve our understanding of diseases and live longer. Maybe we shall even manipulate our own genes, changing the course of evolution. But fundamentally humans remain the fragile creatures they once were on the savanna. Even a modest change in Planet Earth could threaten our existence; and our use of technology, which has wrought good in the past, could see us lurch towards self-destruction.

It has been a privilege to be part of the team telling this remarkable story. I remain convinced that, while human existence is unlikely to continue untroubled, the nature of the human spirit and our moral sense means that we can be confident of human progress.

Robert Winston.

Robert Winston EDITORIAL CONSULTANT

# ABOUT THIS BOOK

*Human* is divided into seven main sections plus a main introduction and a section on the future. Each of these seven sections is represented on the right. The first section, Origins, introduces human evolution and history. Next are sections on Body and Mind, which are followed by Life Cycle, the story of human life from birth to death. There are also sections on Society, Culture (including belief and language), and on Peoples around the world.

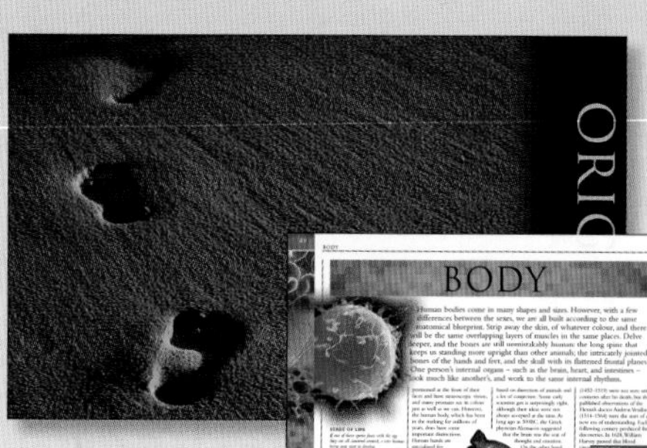

**"ORIGINS" OPENER**
*Each of the eight main sections of the book is identified by a stunning image, this one from the Origins section.*

**"BODY" INTRODUCTION**
*Introductions to each section give a historical background, cover our current state of knowledge, and put the topic in the context of human life.*

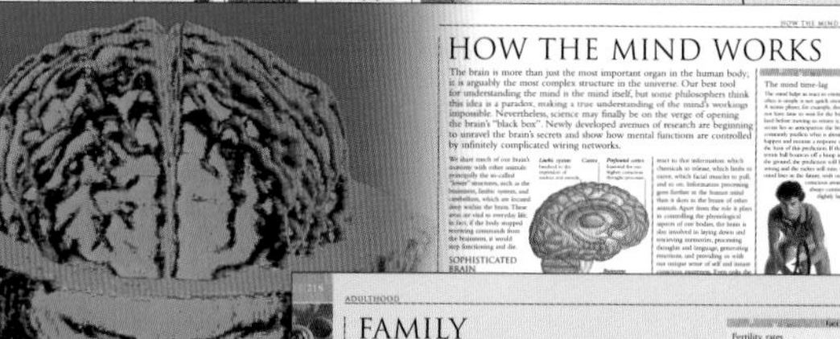

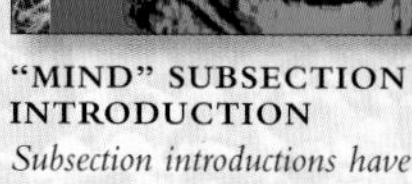

**"MIND" SUBSECTION INTRODUCTION**
*Subsection introductions have an eye-catching image on the lefthand page. The text gives an overview of the topics to follow. For example, this one for Mind introduces how we think, learn, and remember events.*

**"LIFE CYCLE" EXPLANATORY PAGES**
*These pages on Family describe the general features of family life. They set the background to the pages that follow, which show individual types of family.*

## INTRODUCTION

The book begins with an introduction to humans as a species. The gulf between humans and the rest of the animal kingdom seems vast. In fact, from a biological viewpoint, the difference is quite small. This introduction examines the features we share with animals and those that make us uniquely human, such as our large brains and capacity for language.

## FUTURE

Based on what we are currently able to achieve, this section realistically examines human prospects for the next 50–100 years. It addresses such issues as potential new medical technology, increased longevity, the limits of the human mind, and the changes we may see in societies.

Panel styles

### Thematic panels

Throughout the book, five types of thematic panels appear, highlighting topics of special interest. These colour-coded panels are headed Fact, Health, History, Issue, or Profile. Fact panels include intriguing facts on subjects such as Tibetan "sky burials". Health panels describe disorders or aspects of lifestyle that can affect our health. History panels give a perspective on subjects such as ancient calendar systems. Profile panels include brief biographies of influential individuals such as Chinese poet Li Bo. Issue panels discuss some of the dilemmas we face in the 21st century.

*Health*
Positive health issues as well as some common disorders are covered in Health panels

*Profile*
Biographies of notable individuals give key dates and achievements

*Fact*
Interesting or unusual facts are picked out in Fact panels

*History*
These panels give the background to a topic with key dates

*Issue*
Topics of contention or debate are discussed in Issue panels

Health

A sedentary childhood

In the West the time that children spend watching television, surfing the internet, and playing video games is increasing. The average American child spends 1,023 hours a year watching television, and 6½ hours a day on all media-related activities. The amount of time spent on physical activity is falling. This is partly responsible for the rise in childhood obesity.

Profi

Sigmund Freud

Austrian psychoanalyst Sigmund Freu

Fact

The meaning of the veil

The origins of the wedding veil

History

Discovering germs

The discovery that certain microorganisms cause diseases was

Issue

Where do we come from?

Some cultures believe in a single deity that created the world suddenly: many Christians believe God created Adam

## Consultants and contributors

### EDITORIAL CONSULTANT

**Robert Winston** Professor of Fertility Studies, Imperial College School of Medicine, London, UK

### CONSULTANT FOR THE SMITHSONIAN INSTITUTION

**Dr Don E Wilson** National Museum of Natural History, Smithsonian Institution, Washington, DC, US

### SECTION CONSULTANTS

**Professor Frances Ashcroft** Professor of Physiology, University of Oxford **Dr Sue Davidson** London **Dr Dylan Evans** Researcher and lecturer in Biomimetics, University of Bath **Dr Dena Freeman** Anthropologist, London **Professor Frank Furedi** Professor of Sociology, University of Kent at Canterbury **Professor Nick Humphrey** Psychologist and neuroscientist, Centre for Philosophy of Natural and Social Science, London School of Economics **Professor Tom Kirkwood** Professor of Gerontology, Institute for Aging and Health, Newcastle University **Dr Graham Ogg** Institute of Molecular Medicine, Oxford **Dr Max Steuer** Economist, London School of Economics **Professor Christopher Stringer** Department of Palaeontology, Natural History Museum, London

### CONTRIBUTORS

Susan Aldridge, Jo de Berry, Susan Blackmore, Rita Carter, David Brake, Liam D'Arcy Brown, Adam Burgess, Andrew Chapman, John Coggins, Carol Cooper, Ludovic Coupaye, Andrew Dalby, Robert Dinwiddie, Roger East, Dorothy Einon, Dena Freeman, Peggy Froerer, Ken Gilhooley, Amra Hewitt, Ben Hoare, Alan Hudson, Mark Jamieson, Dawn Marley, Magnus Marsden, Wim Mellaerts, Rory Miller, Zoran Milutinovic, Ben Morgan, Phil Mullan, Brendan O'Neill, Katie Parsons, Mukul Patel, Dr Penny Preston, Hazel Richardson, Sanna Rimpilainen, Nigel Ritchie, D J Sagar, Martin Shervington, John Skoyles, Max Steuer, Richard J Thomas, Edmund Waite, Helen Watson, Philip Wilkinson, James Woudhuysen, Eve Zucker, Cambridge International Reference on Current Affairs (CIRCA)

**"SOCIETY" PROFILE PAGES**
*Profile entries provide information on individual aspects of the subject covered. For example, within the Society section, types of social hierarchy, power structure, and conflict are covered.*

**MAP AND FACT FILE**
*Each entry in the Peoples section is headed by a map, pinpointing the main areas of distribution, along with details of population, languages spoken, and main belief systems.*

**Map**
Location of peoples is pinpointed on specially prepared relief maps

***Location***
Profiles are arranged on the page by geographic location

***Fact file***
Each profile is headed by a fact file listing location, population, languages spoken, and beliefs

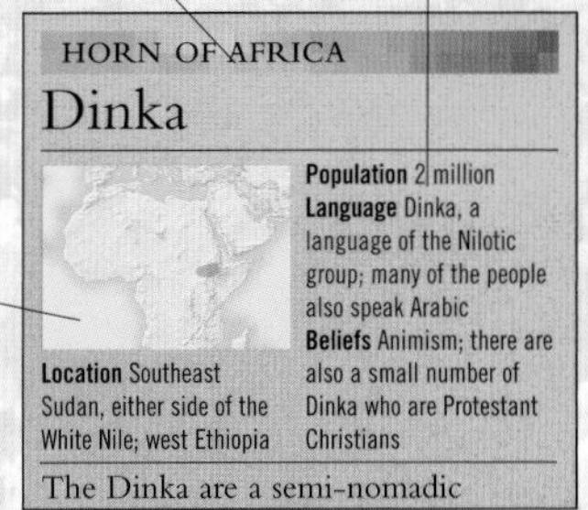

HORN OF AFRICA
Dinka
Population 2 million
Language Dinka, a language of the Nilotic group; many of the people also speak Arabic
Beliefs Animism; there are also a small number of Dinka who are Protestant Christians
Location Southeast Sudan, either side of the White Nile; west Ethiopia
The Dinka are a semi-nomadic


**"PEOPLES" PROFILE PAGES**
*More than 250 profiles of peoples from every part of the world are presented. Each profile begins with a location map and fact file.*

**"CULTURE" PROFILE PAGES**
*This section looks at cultural variations, including religions, clothing, and languages. The illustrated profiles here are from pages on types of clothing and adornment. Most profiles are illustrated with photographs.*

# INTRODUCTION

# INTRODUCTION

One of the curious things that makes our species different from others is that we can recognize ourselves in a mirror. To scientists and philosophers, our capacity to understand a reflection is a sign of one of our most important distinguishing features: self-awareness. Only the most intelligent animals, including chimps and gorillas, show hints of this very peculiar ability. Self-awareness not only defines us, it also drives our ongoing efforts to understand our very nature. Since the beginning of history, people have struggled to unravel the mystery of human nature and find out exactly what makes us so special.

**HUMAN BRAIN**
*The human brain is about three times bigger than it should be for an average primate of our body size. Our large brain underpins many of our unusual mental abilities.*

Humanity has not always seen itself as part of the animal kingdom. For centuries, humans thought of themselves as higher beings with souls, free will, and consciousness, and saw animals as mindless creatures driven by instinct. As recently as the 19th century, Charles Darwin caused uproar by saying we were descended from apes, and even today the word "animal" retains its historic meaning: something base, violent, and inhuman.

For biologists, however, "human" and "animal" are far from opposites. Our species, *Homo sapiens*, is unequivocally an ape. Peel away the distracting layer of clothes and the peculiar, naked skin, and we have exactly the same complement of organs and tissues as our ape cousins. To a geneticist, the difference is even smaller: human and chimp DNA differs by only 1–2 per cent. So the evidence from science says that humans are unquestionably animals, but common sense and tradition tell us there is a gulf between us and the rest of the animal world. This apparent contradiction lies at the heart of the mystery of human nature, and only by resolving it can we begin to understand what it means to be human. To do that, we need to understand our place in the animal kingdom.

## THE ANIMAL KINGDOM

We are just one of at least 1.5 million species that make up the kingdom Animalia, one of the five kingdoms of life (*see* The tree of life, below). In the past, zoologists arranged all these animal species into a family tree with two major branches: vertebrates (all the animals with backbones, including us) and invertebrates (worms, insects, spiders, and so on). Over the years, as more information was amassed about the evolutionary history of the animals, our branch of the tree shrank in importance and the classification changed. Today, vertebrates make up a mere twig on one of 30 or so main branches, or "phyla", in the animal kingdom. Despite this apparent relegation, vertebrates are the dominant animals. Squeezed onto that small twig in the tree of life are all the world's mammals, birds, reptiles, amphibians, and fish, and among their number are the biggest and most spectacular creatures ever to have lived.

## A TYPICAL VERTEBRATE

The first vertebrates, our distant ancestors of 400 million years ago, were fish. The legacy of those aquatic ancestors is still with us, including the defining vertebrate feature: a bony spine, flanked by pairs of muscles. This feature gave the first fish superb mobility in water, making them the oceans' top predators. Today, it

**MAN ON THE MOON**
*The first moon landing in 1969 is just one example of the human drive to explore and understand the universe. A total of 12 astronauts walked on the Moon during the six Apollo missions.*

### The tree of life

All living things can be classified into a tree of life that reflects their evolutionary history. At the base of the tree are the five kingdoms of life, each of which can be subdivided into a branching network of many different species. This chart shows in a very simplified form how our own species fits into the tree. Working from left to right, the species become increasingly closely related to each other. Like all species, humans have a scientific name made up of two parts: a genus and a species. We are the sole surviving members of the genus *Homo*, which also includes our direct ancestor *Homo erectus* and close cousin *Homo neanderthalensis*. Humans are also the only surviving members of a subfamily of apes now known as hominins (family Homininae), made up of species of ape that lived on the ground and moved around on two feet rather than "knuckle-walking", as gorillas and chimpanzees do. Hominins are believed to have split from the chimpanzee branch of the ape family tree about 5–6 million years ago.

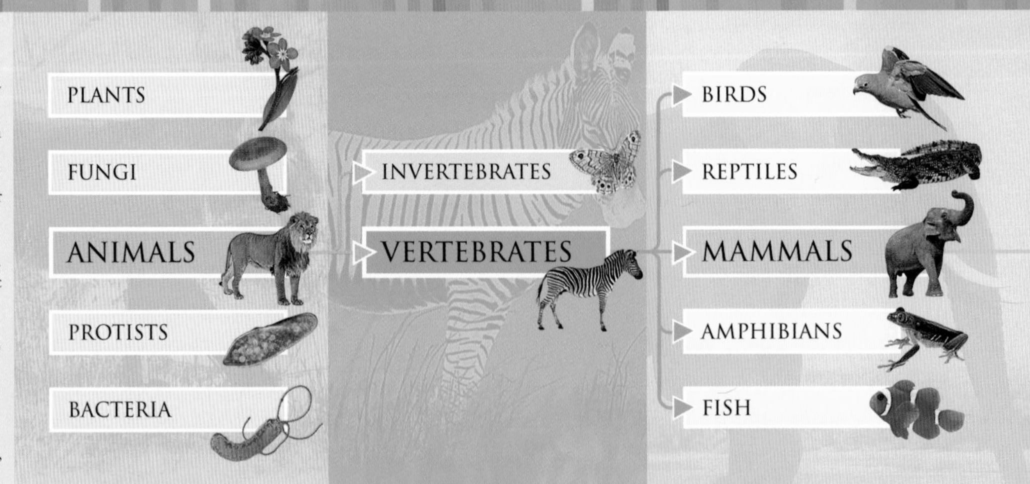


**KINGDOMS OF LIFE**
*Living things can be divided into five main kingdoms. Our species belongs to the animal kingdom, along with about 1.5 million other known species.*

**ANIMAL DIVISIONS**
*Animals have traditionally been divided into those with a backbone (vertebrates) and those without (invertebrates). All vertebrates have a similar skeletal plan.*

**VERTEBRATE CLASSES**
*Vertebrates are divided into five classes. We belong to the mammals, which are defined by the ability to produce milk. Most mammals also have hair or fur.*

provides the main supporting strut in the human skeleton, allowing us to stand and move. Beneath our skin, our body's tissues are still bathed in a salty solution, the chemical composition of which is surprisingly similar to that of seawater. This is yet another reminder of an aquatic past.

Except for our unusual upright stance, human anatomy is typical for a land vertebrate. We have the same skeletal arrangement as most other land-living vertebrates, with four jointed limbs that each end in a spread of five digits, the typical number for vertebrates. Our digits form our hands and feet, but those of other vertebrates have taken many different forms. A bat's digits, for example, form the bony struts of its wings, and a horse's single digit (all the others have withered away) ends in a grotesquely enlarged toenail – its hoof.

Humans have the typical sense organs of a vertebrate, including a single pair of eyes and a single pair of ears. The stream of data that comes from these organs is processed by another feature that is typical of vertebrates: a brain enclosed in a strong, bony case.

The tail that is characteristic of the majority of vertebrates appears and disappears before humans are born, and the gill slits present in aquatic vertebrates appear only fleetingly when we exist as embryos – one has evolved into the eustachian tube, a narrow airway that connects the middle ear to the throat.

## HUMANS AS MAMMALS

Further along the tree of life, the vertebrate twig splits into a spread of smaller twigs, which are known as classes. We belong to the class Mammalia – the mammals. There are several defining features that set mammals apart from other vertebrates, the most important of which is that mammals produce milk to nourish their young (the word "mammal" comes from the Latin *mamma*, which means "breast"). Like other mammals, we are warm-blooded, and our bodies are covered by hair. Our species belongs to a category of mammals known as primates. Nearly all primates are tree-dwellers, and most are confined to the belt of warm forest that girdles our planet between the tropics. Many aspects of human anatomy – for example, our grasping hands, forward-facing eyes, and colour vision – are typical primate features, vestiges of millions of generations spent clambering through trees. And we are unmistakably primates in our social lives and behaviour: like many monkeys and apes, we live in complex, hierarchical societies where survival depends on navigating an ever-changing web of social relationships.

Fact

### Ninety-nine per cent chimp?

Scientists have used a plethora of techniques to compare human genes to those of chimpanzees and other apes. The latest experiments suggest we differ from chimps in just 1.2 per cent of our active genes. Indeed, some scientists say humans, chimps, and the closely related bonobo should all be placed in the same genus, *Homo*. However, although the genetic distance between humans and apes is tiny, the differences in appearance, behaviour, and intelligence are profound. The few genes that make humans different probably include key controller genes that affect the activity of many other genes during our development, with far-reaching consequences for our bodies and brains.

**THE DNA CONNECTION**
*Deciphering the genetic code of living creatures enables scientists to understand the links between species. Shared DNA suggests shared ancestors.*

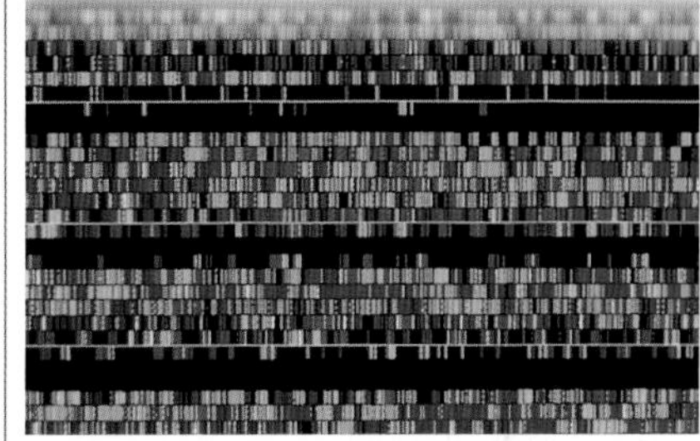

## BEING HUMAN

It is common knowledge that we share nearly all our genes with chimpanzees. Yet the tiny fraction that makes us different has a profound effect on our anatomy, so much so that, for centuries, scientists had difficulty believing we had evolved from apes at all. Unlike other apes, we lack the opposable (flexible) big toes needed for climbing, and we get about by walking on outsized hind limbs with our short forelimbs dangling in the air. Compared to apes, our skin is almost naked, our skeleton is bent out of shape, our head is swollen like a balloon, and our breasts and penis are bizarrely large. Even so, if alien scientists were to land on Earth and study *Homo sapiens*, they would have

**MAMMALIAN ORDERS**
*There are around 21 "orders" of mammals. We belong to the primates, most of which are tree-dwellers with grasping hands and large brains.*

**PRIMATE CATEGORIES**
*Primates are divided into three main groups: apes, monkeys, and prosimians. Apes and monkeys are active by day but many prosimians are nocturnal.*

**THE APES**
*Apes are characterized by their large, muscular arms, their mobile shoulders, and the lack of a tail. All but humans are restricted to areas of tropical forest.*

very little difficulty in recognizing us as apes. Physically unusual we may be, but the mountain of similarities between us and the other apes would soon become apparent.

If those visitors were to study the human mind, however, they might think that we too had arrived from another planet. In intellectual terms, a yawning chasm separates *Homo sapiens* from every other species on Earth. We have language of astonishing complexity. We build cities, cars, spaceships. We invented morality, religion, trade, science, world wars. We can think symbolically, create art, plan for the future, and solve problems using our imagination. And, perhaps even better than other primates, we can read each other's minds and intentions from only the slightest inflection of the voice or dart of the eyes.

## FLEXIBLE SHOULDERS

Many of the features that make humans special are unique to us, but others are shared with our close relatives. Like other apes, we have flatter chests and longer, more powerful arms than monkeys. These differences evolved because of the way that apes move through the trees. Instead of scurrying along branches on all fours as monkeys do, apes haul themselves up trees by grasping the trunk in their arms. The shoulder blades of apes are at the back of the chest rather than the sides, which frees up the shoulder joint and allows the arm to reach overhead and swing around. In humans, the flexible shoulders that evolved for climbing are put to other uses. Uniquely, we can throw objects with force and accuracy – an ability that proved enormously useful when our hunting ancestors invented throwable weapons. Flexible shoulders also make our grasping hands more useful. If our shoulders were less mobile, we would find it much harder to carry and manipulate objects.

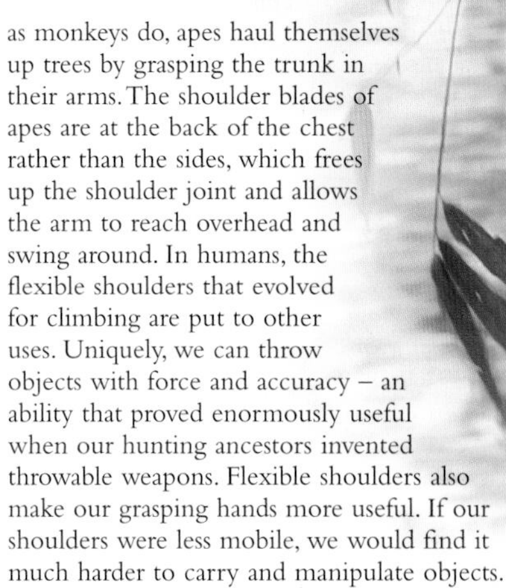

**SHOULDER ACTION**
*Humans and other apes have highly flexible shoulders that give us free overarm movement. As a result, humans can throw a javelin with accuracy, and orangutans are able to swing between branches, hanging from their hands.*

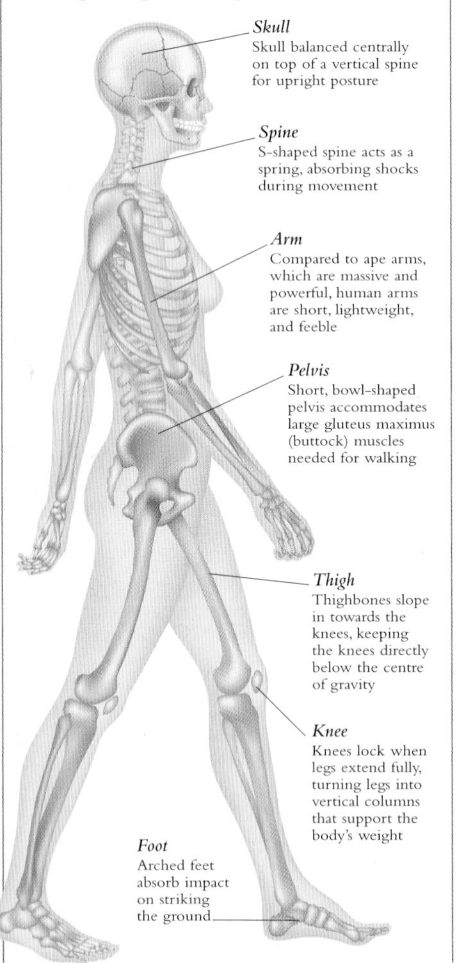

*Skull*
Skull balanced centrally on top of a vertical spine for upright posture
*Spine*
S-shaped spine acts as a spring, absorbing shocks during movement
*Arm*
Compared to ape arms, which are massive and powerful, human arms are short, lightweight, and feeble
*Pelvis*
Short, bowl-shaped pelvis accommodates large gluteus maximus (buttock) muscles needed for walking
*Thigh*
Thighbones slope in towards the knees, keeping the knees directly below the centre of gravity
*Knee*
Knees lock when legs extend fully, turning legs into vertical columns that support the body's weight
*Foot*
Arched feet absorb impact on striking the ground


**BUILT FOR WALKING**
*Evolution has dramatically re-engineered the human skeleton since we split from the rest of the apes, changing a body plan adapted for climbing into one adapted for standing, walking, and running.*

## ON TWO FEET

The ability to walk on two feet not only sets humans apart from other apes but marks a major divide between us and the rest of the mammals. No other mammal can match our ability to stand, walk, and run on two legs. Of course, many other animals are capable of two-legged – or "bipedal" – movement, including ostriches, kangaroos (although they hop), and penguins. However, unlike ostriches and kangaroos, which are counterweighted by long necks and tails that act like a tightrope walker's pole, humans stay vertical largely because our nervous system is so beautifully coordinated. Our sense of balance also enables us to learn how to ice-skate, ski, and even perform feats such as walking on our hands. One price we pay for moving in such a strange manner is that walking takes time for us to master – human babies cannot walk for about a year.

Although other apes cannot walk or run as we can, they show hints of our ability. Apes have a more upright posture than monkeys. In trees, they frequently stand on their hind limbs while holding branches with their arms. Chimps and gibbons can even walk – or waddle – a short distance on the ground on two feet, an ability that comes in handy when they need to wade across rivers, but one that requires great effort.

Bipedal motion is easy for humans as a result of changes in the skeleton of our ancestors. We can extend our legs fully to form a vertical column that supports the body's weight, and the knees lock to prevent overextension of the lower leg. In contrast, apes can only partially extend the lower leg, forcing them to stand with their knees bent and rely on muscle power alone to stay upright, which is tiring.

Seen from the front, the human femur (thighbone) slopes inward from the hip to the knee, ensuring that the knees and feet are directly below the body's centre of gravity. Apes'

Health

### An evolutionary compromise

During childbirth, a baby must pass through the space in the middle of the mother's pelvis. In most mammals this is a straightforward process, but in humans it is exceptionally painful, dangerous, and slow. Because we evolved a two-legged posture, we have a much narrower pelvis and a smaller pelvic outlet than other apes. However, human babies also have an unusually large head to accommodate the larger human brain. As a result, humans are born at a relatively early stage of their development, at which time they are physically helpless and therefore entirely dependent on their parents. Another evolutionary compromise is that women need a wider pelvis than men, which makes them slower runners, and slightly less athletic.

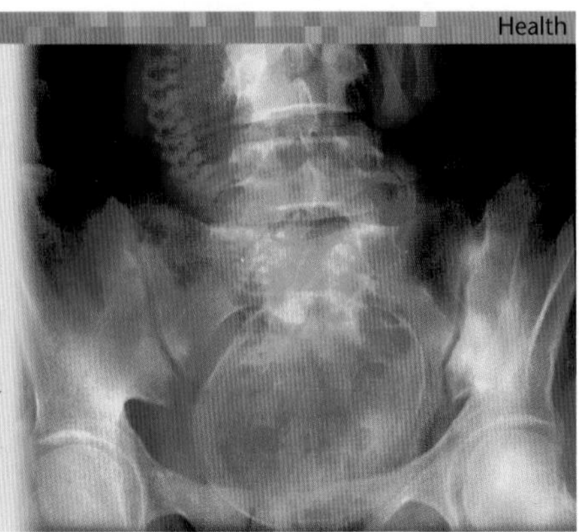

**X-RAY OF CHILDBIRTH**

legs are more splayed, resulting in an awkward, waddling gait. Human legs are much larger than our arms, giving our body a relatively low centre of gravity that aids stability. Our centre of gravity falls between the two hips, giving us a stable, vertical posture. In contrast, apes have large, muscular arms, short legs, and a high centre of gravity in front of its hips. An ape's high centre of gravity results in an unstable, crooked posture when upright.

Our feet are arched so that the heel and the ball of the foot carry our weight as we move. In contrast, apes stand with the whole length of the foot on the ground. Apes have opposable big toes for climbing, but in humans the big toe is aligned with the others. When we walk, weight is transmitted from the heel to the ball of the foot and on to the big toe, which is the last point of contact as the foot pushes off the ground. This is a more efficient way of walking.

The human skull is balanced on top of a vertical spine rather than being held by muscles in front of a horizontal spine. As a result, the hole through which our spinal cord passes (the foramen magnum) is shifted forwards relative to that of apes, so that it lies directly under the brain. The spine is curved into an undulating S-shape that acts as a spring, absorbing shocks during movement.

To accommodate the enormous muscles that we need for walking – especially the gluteus maximus (buttock) – the human pelvis is much shorter and broader than that of other apes. It is also bowl-shaped to support the abdominal organs cradled above it.

There are drawbacks to being bipedal. One is that childbirth is more painful and protracted in humans than in other mammals (*see* An evolutionary compromise, p14). Another is that lower back pain and injury are more common in humans. The vertebrae of the lower back have to carry all the weight of the upper body, yet they must be small to preserve the spine's flexibility. When we bend over to lift a heavy weight, the lower back is subjected to a force much greater than our body weight, and this force can rupture one of the shock-absorbing discs between the bones (called a slipped disc).

## HANDS FREE

One of the great advantages of being upright is that it frees the hands. Like other primates, we have opposable thumbs – digits that move in the opposite direction to the rest. This ability turns hands into pincers that can grasp and manipulate objects. Gorillas use their hands to pick apart thorny plants and uproot edible material, and chimps use theirs to handle simple tools, such as twigs with which to fish termites or rocks to crack nuts. Unlike humans, though, these apes also have to use their hands as feet, which limits how delicate and nimble the fingers can become. Human hands, being completely free, have evolved into precision instruments of amazing dexterity. Our hands can tie shoelaces, play pianos, thread needles, grasp hammers, and count coins in a pocket.

Human fingertips are packed with receptors sensitive to pressure, and they are backed by nails instead of claws, improving the sense of touch. The palms of our hands are among the only parts of our skin that are truly hairless (along with the soles of our feet and our lips), which improves our grip. To improve grip still further, the skin of our fingertips is roughened by tiny ridges bearing oil and sweat glands that keep the skin supple and damp, allowing us to pick up and grip the tiniest objects. At the same time, the part of the brain controlling hand–eye coordination is far more highly developed than in other primates. Manual dexterity is crucial to our species. Without it, we would be unable to make fire, throw spears, build houses, or invent endless varieties of tools.

Fact

### Slow but economical

Walking on two legs may have evolved to save us energy – we use less energy walking on two legs than do chimps or gorillas when they "knuckle-walk" on all fours. However, compared to most four-legged mammals, we are very poor runners. We burn through calories at an extravagant rate while running, and our top speed is unimpressive. Over a distance of about a mile, the best athletes average about 15mph (24km/h). Horses, dogs, and antelope can easily exceed 30mph (48km/h) and can sustain their speed for much longer.

## SCENTS AND SMELLS

Most mammals live in a world of smells. They use scent glands to annoint landmarks in their territory, leaving long-lasting signals that attract mates or repel rivals. Scents contain a wealth of information, telling visitors the age, sex, identity, and reproductive status of the marker.

For primates – and especially for us – the sense of smell is less important. Our olfactory bulbs (the parts of the brain that process smell) are a fraction of the normal mammalian size. Even so, we do produce scents that have social and sexual significance. Our scents are made in the hairiest parts of the adult body: the armpits and groin. Special sweat glands in these areas exude a viscous, strong-smelling type of sweat called apocrine sweat that contains chemicals known as pheromones. The composition of apocrine sweat varies according to the menstrual cycle and mood. Before the era of deodorants and perfume, apocrine sweat probably helped people choose their sexual partners. Scientists think that the complex odours contained in the sweat somehow convey information about whether a potential partner's immune system will be compatible with our own.

## VISION

Although smell and hearing are the primary senses for most mammals, vision is the dominant sense in humans. As primates, we have excellent colour vision and the ability to perceive tiny details, owing to the way light-detecting cells

**A SUPERB SENSE OF BALANCE**
*Our skeletal structure and sophisticated nervous system enable us to keep our balance even when carrying objects that are bigger than ourselves.*

are packed into the back of the human eye. In the centre of the retina (the light-sensitive membrane at the back of the eye) is a small pit called the fovea, where light-sensitive cells are crammed together very densely to create the detailed central point of the visual field. Only animals with a fovea have very sharp vision; these include birds of prey and primates. The human fovea is packed with colour-detecting cells called cone cells, of which there are three types: one for each primary colour, giving us full colour vision. Most mammals have only one or two types of cone cell, making them colour blind in the medical sense.

Excellent colour vision comes at a price: cone cells work only in daylight, and we are almost blind at night. Poor night vision is one of the reasons that humans and other apes spend the night asleep indoors or in treetops, far from nocturnal predators that can see in the dark.

Like all primates, we can see in 3-D. Our primate ancestors evolved three-dimensional vision partly because it helped them move in the treetops. For us, 3-D vision is more important for manipulating objects. This type of vision requires overlapping fields of view, with both eyes facing in roughly the same direction (quite unlike the sideways-facing eyes of animals such as rabbits, which need all-round vision to spot approaching predators). Humans are typical primates in this respect. Our eyes face forward, giving us a wide area of 3-D vision, and our nose is short to minimize obstruction of the field of view. To compensate for our lack of

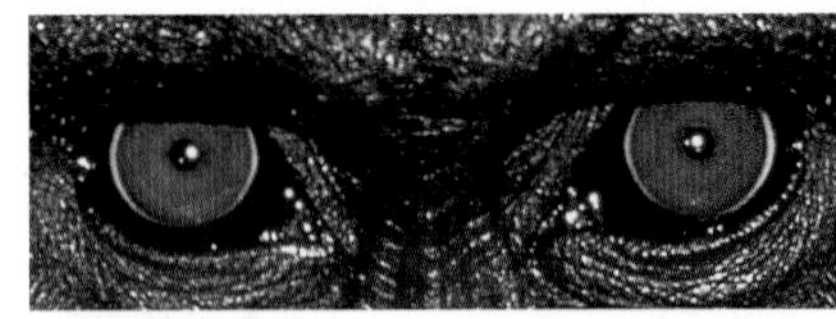

GORILLA

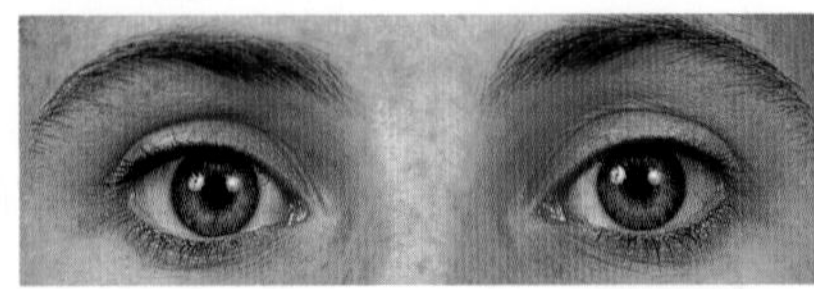

HUMAN

**WHITE EYES**

*Compared to apes, human eyes have a larger white zone, making direction of gaze obvious to observers and so aiding communication involving eye contact.*

all-round vision, we have mobile eyeballs and necks, allowing us to look around without turning the body. In one respect, our eyes are different from those of other primates: we have a large white zone around the iris that makes our eyes conspicuous and allows us to communicate direction of gaze. Most apes have largely brown eyes that camouflage their gaze.

## KEEPING WARM

Like all mammals, humans are "warm-blooded", which means that our bodies can maintain a consistently warm internal temperature. Reptiles and amphibians, in contrast, are "cold-blooded", which is a somewhat misleading term since these animals sometimes have warmer blood than mammals do. The crucial difference is that cold-blooded animals cannot maintain a constant body temperature independent of the conditions around them; they are therefore at the mercy of their environment. Desert lizards, for example, become sluggish at night as the temperature plummets. In the mornings, they need to sun themselves for some time in order to become active again. Mammals, however, can stay active even in Arctic winters. Of all the mammals, humans have made the most of the adaptability that warm-bloodedness brings.

Fact

### Our love of water

Our desire to be near water may have evolutionary roots. Early humans lived on the African savanna, where they came to tolerate the heat by evolving bare skin and producing huge amounts of sweat. This allowed them to be active during the hottest hours of the day, when lions and other predators usually rest. However, our ancestors had to stay near water to replenish the fluid lost through sweat. Until they invented flasks made from animal skins and ostrich eggs, early humans probably never strayed far from waterholes or rivers. Water sources were also a magnet when people first began building settlements.

## NEARLY NAKED

For most mammals, hair is a vital part of the warm-blooded heat-retention system. A dense layer of hairs (fur) traps warm air next to the body like a blanket, slowing down the loss of precious body heat. Only in the largest mammals, such as whales and elephants, does hair become redundant, because their huge bulk and resulting low surface area slows down the loss of heat. Although human skin appears hairless, we are not truly naked – we have just as many hairs as chimpanzees. The difference is that our hairs are much smaller and finer. With the exception of the patch of thick hair on the head, our body hair is of little use as insulation.

Why, then, are humans almost naked? The answer probably lies in our historical need to stay cool. When our ancestors left the forests for more open savannas, they were exposed to the full force of the stifling African sun. Only the most heat-tolerant animals are able to stay active during the hours of daylight in the African savanna. We could tolerate such conditions thanks to our upright stance (which minimizes exposure to the sun), our bare skin, and a sweat system more advanced than that of any other mammal.

Our skin is covered with millions of glands that produce a watery secretion called eccrine sweat when our body temperature rises. This sweat evaporates quickly from the skin, drawing heat away from the body. A layer of fur would not only make us overheat but would also block the flow of air across the surface of our skin, thereby slowing down the evaporation of sweat.

Unlike humans, other primates stay dry in hot conditions, and their eccrine glands are mainly restricted to the bare skin on their palms and soles, where the skin needs to be kept moist to improve grip and sensitivity. Our palms also have high numbers of eccrine glands that, just as in other primates, become active in times of stress – a vestige of our tree-dwelling past when good grip was vital for making a quick escape.

When our ancestors left Africa and spread to other parts of the world, they remained tropical in their biology and had to adapt to the cold by using their ingenuity. With help from clothes, fire, and shelter, they conquered almost every environment on the planet. Evolution did not stop altogether, however. In bitterly cold parts of the world, such as the Arctic, people seem to have evolved a shorter, more compact build that helps retain heat. People native to much hotter, drier climates, such as the Masai and the Arabs, typically have a more slender, long-limbed build that increases the body's surface area and so helps shed heat.

## TEETH AND DIET

To fuel their on-board central heating systems, mammals have to consume around 10 times more calories than do cold-blooded animals, and humans are no exception. Much like other mammals, we have a highly efficient digestive system and sophisticated teeth that mesh together precisely to grind up food like

INUIT

MASAI

**CLIMATE CONTROL**

*The Masai have a slender build that sheds heat; the Inuit have a compact frame that retains heat.*

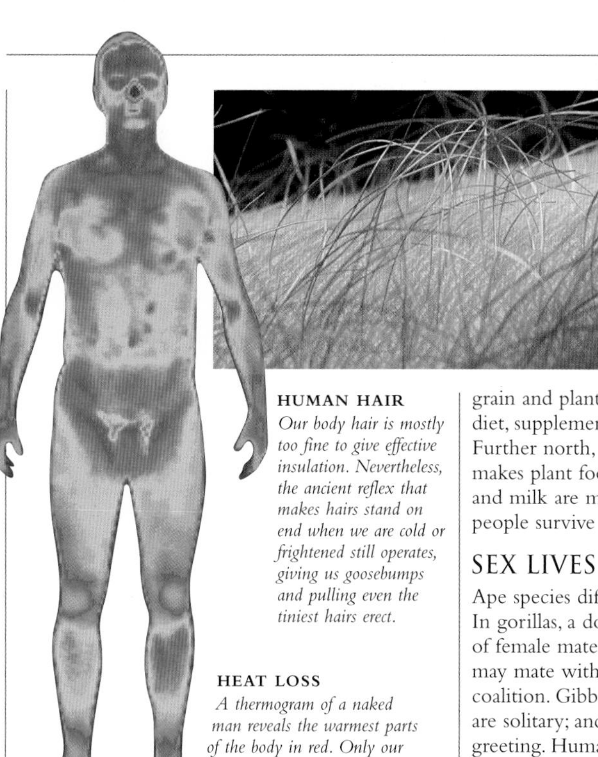

**HUMAN HAIR**
*Our body hair is mostly too fine to give effective insulation. Nevertheless, the ancient reflex that makes hairs stand on end when we are cold or frightened still operates, giving us goosebumps and pulling even the tiniest hairs erect.*

**HEAT LOSS**
*A thermogram of a naked man reveals the warmest parts of the body in red. Only our extremities can tolerate the cold; in cool climates, the rest of the body must be covered.*

the stones of a flourmill. Whereas reptiles shed and regrow their crude, peglike teeth throughout life, mammals typically grow just two sets of more finely-tuned teeth: a set of deciduous (milk) teeth and an adult set.

Some primates specialize in eating leaves, insects, or fruit; others are generalists, able to exploit a diverse range of foods, including fruit, leaves, seeds, insects, and small animals. Generalists are among the most inquisitive and intelligent of the primates. They are able to adapt to new situations, discover new types of food, and use their cunning and ingenuity to overcome the biological defences of the animals and plants they eat. Humans belong to this generalist category. Our teeth are small and relatively unspecialized, indicating that our natural diet includes a diverse range of foods rather than a single staple. We have the same dental formula as other apes: 8 incisors, 4 canines, 8 premolars, and 12 molars. However, our teeth are smaller because we can use our hands, tools, or cooking to deal with tough plants and meat. Some experts think that early humans were scavengers rather than hunters, but most agree that animal foods have long been an important part of our diet. Today, the amount of meat in the human diet depends mainly on where people live. In tropical countries, where plant foods are available all year round, starchy foods like grain and plant roots make up the bulk of the diet, supplemented by small amounts of meat. Further north, where the seasonal climate makes plant foods hard to find in winter, meat and milk are more important. And some Arctic people survive on almost nothing but meat.

Issue

### Is it natural to eat meat?

There is great debate about whether it is "natural" for humans to eat meat. Some people say that since meat-eating is rare in primates, the amount of meat in the human diet is unnatural. Moreover, we lack the large canine teeth and claws of natural carnivores. However, chimpanzees, our closest relatives, eat meat, and our ancestors probably ate even more meat than we do when they moved from the forests to game-rich savannas. Our teeth may be small because early humans learned to use stone blades and fire to soften tough foods. Paradoxically, herbivorous primates have much larger canines than we do because their large teeth are used to signal aggression, a function that became unnecessary when we invented weapons.

## SEX LIVES

Ape species differ enormously in their sex lives. In gorillas, a dominant male lives with a harem of female mates, but in chimpanzees, one female may mate with every single member of a male coalition. Gibbons are monogamous; orangutans are solitary; and bonobos have casual sex as a greeting. Humans follow none of these patterns. Biologists describe the human mating system as "mildly polygynous", which means that we are nearly monogamous but sometimes have a tendency towards promiscuity.

Compared to other apes, our sexual anatomy is unusual. The human penis – the longest and thickest of any primate's – is about four times larger than a gorilla's when erect, yet human testes are smaller than those of chimpanzees. Women's breasts are not only larger than those of other female apes, but are even larger than is necessary to produce milk. Why these differences exist is not entirely clear, but many scientists think they are caused by "sexual selection", an evolutionary process in which animals (usually males) acquire sexual adornments, such as the peacock's tail or the stag's antlers.

Like all higher primates, we have sex for pleasure, even when women are not ready to conceive. Chimpanzees and many other primates have an area of bare genital tissue that swells and colours during ovulation, advertising a female's fertility. In our species, however, a woman's readiness to conceive is hidden even from herself. Some biologists think that this "concealed ovulation" evolved in women to make men more faithful – if ovulation was signposted, a man would safely be able to neglect his partner during her infertile period in the knowledge that any offspring would be his. According to this theory, concealed ovulation and sex for pleasure are among a suite of adaptations that cement the bond between couples and encourage fathers to share in the childrearing process. However, another possible explanation for concealed ovulation is our upright posture. When humans became upright, the external female genitals changed position and became concealed between the legs. Upright posture may also explain the human male's long penis.

## LIFE COURSE

As mammals, we are typical in looking after our offspring and feeding them milk (many other animals simply abandon their eggs or offspring to fate), but we care for our offspring for much longer than other mammals do. Humans are not only born helpless but develop slowly and stay dependent on parents for an extraordinarily long period. Gorillas and chimps are able to feed themselves as soon as they are weaned, but human infants continue to be fed for years after weaning. Our period of dependency is longer because our brains need more time to grow, develop, and learn the complex social and technical skills needed for survival. Because young children need so much care, we usually live in monogamous families, with both parents sharing the burden of providing food.

As well as growing up slowly, humans live longer than other apes. Our life expectancy is about 80 years, compared to 40–50 years in chimps. Human females lose their fertility at around age 50, yet survive for another 3 decades or so – much longer than any other mammals survive after the end of the fertile period. The menopause seems to play an important role in our species' life course and family structure. By looking after their grown-up daughters and their grandchildren, older women may contribute more to their families' success than would be the case if they continued to bear

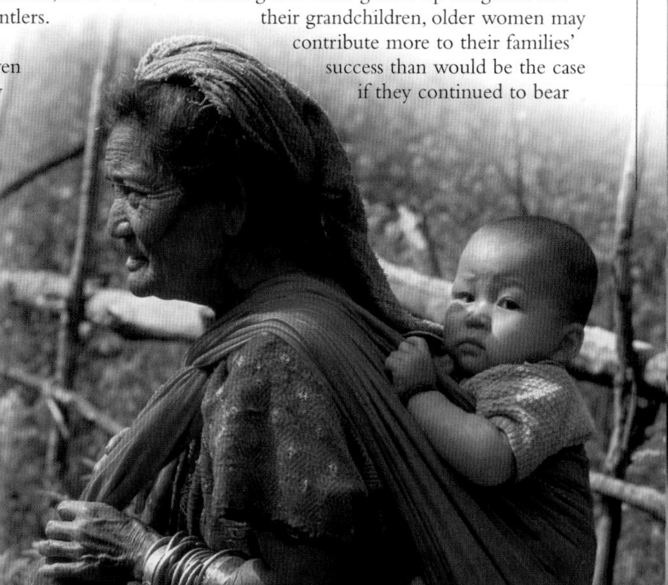

**THE ROLE OF GRANDMOTHERS**
*Grandmothers play a crucial role in the human life cycle by helping their offspring to raise families. The importance of grandmothering may have led to the evolution of the long human lifespan and early menopause.*

children. Some scientists think it was the evolution of this system of "grandmothering" that extended our life expectancy and slowed the entire human life course as a side effect, delaying puberty well into the teens.

## COMPLEX SOCIETY

Monkeys and apes have complex social lives and live in troops (tight-knit social groups). Many features common to monkey and ape societies are also true of human society. For instance, to get on in life, other primates have to climb a social ladder, exploiting a network of allegiances to get to the top. High-status animals attract the most mates and get first choice of the most important resources, such as food. Hierarchies and status are clearly also important in human society, and they provide us with similar benefits.

Understanding the complex relationships within a social group requires a certain type of intelligence, known to psychologists as Machiavellian intelligence. Primates are thought to use Machiavellian intelligence to keep track of their friends and enemies in the constantly changing social environment. The cleverest primates, the apes, even seem able to trick and deceive each other. To climb the social ladder, primates spend much of their time grooming each other's fur, an activity that seems to give them great pleasure. Humans are unusual among primates in that we do very little grooming, but then we have almost no hair to groom. According to one theory, conversation may serve the same purpose as grooming.

The need for Machiavellian intelligence may be one of the factors that led to what is arguably our most distinctive and important feature: a large brain.

## BIG BRAINS

The organ that underlies most of our special abilities is the brain. Exactly what is so special about the human brain is largely a mystery. In terms of hardware (the cells and tissues of which it is made), it is much the same as any mammal's brain. In terms of size, though, it is unusually large.

Humans do not have the largest brains in the animal world; those of elephants and whales are much bigger. However, if the scaling effect of body size is taken into consideration, humans have the largest brains of any mammal. Our brains are not only huge, they are also expensive to run in that they consume a disproportionate amount of our energy: an adult's brain accounts for 2 per cent of body weight yet uses 20 per cent of calorie intake; the brain of a newborn baby uses 60 per cent of its energy intake.

The part of the brain that is thought to be responsible for our higher mental faculties is the cerebral cortex – the "grey matter" in the outer part of the cerebrum (the main area of the brain). In both humans and apes, the cerebral cortex is thrown into a series of deep folds that give the brain a very large surface area. However, humans have an especially large area of cortex: about four times as much as chimps. Neuroscientists know that damage to specific areas of this cortex can impair our higher mental functions, including language, decision making, memory, and emotional control – but as yet we have very little idea of how a healthy cortex manages to conjure up any of our unusual mental abilities.

Scientists are not sure why our large brains evolved. The benefits seem obvious now, given the wealth of things we invent to make our lives easy. However, evolution cannot plan ahead, and our ancestors' brains increased in size hundreds of thousands of years ago. Since big brains consume much energy, there must have been an advantage to having them when we were living as foragers on the African savanna, but what was it?

One idea is that the evolution of the human brain was driven by the development of language. Language seems to have changed the structure of our brain: the speech areas make it asymmetrical because they are only found on one side, which has led to the human tendency to be right- or left-handed. The brains of other animals tend to be symmetrical, and it is likely that our early ancestors' brains were too. However, most experts think language actually appeared later than big brains – perhaps as recently as 50,000 years ago.

A more radical idea is that large brains are merely a side effect of the evolutionary process that increased our lifespan. Since brain tissue cannot regenerate, we need large brains just to provide the spare capacity that a longer life demands. This theory fails to account for the brain's voracious demand for energy, however. Our brains consume lots of energy because they are active; unused areas of "spare capacity" would be likely to lie dormant.

Another theory is that large brains evolved to make us sexually attractive. Charles Darwin was the first person to propose this idea. He suspected that sexual selection may have driven the evolution of intelligence and large brains in men in the same way it drove the evolution of gigantic tails in peacocks. One problem with this idea is that it accounts only for the intelligence of males, but men and women actually have the same average intelligence levels.

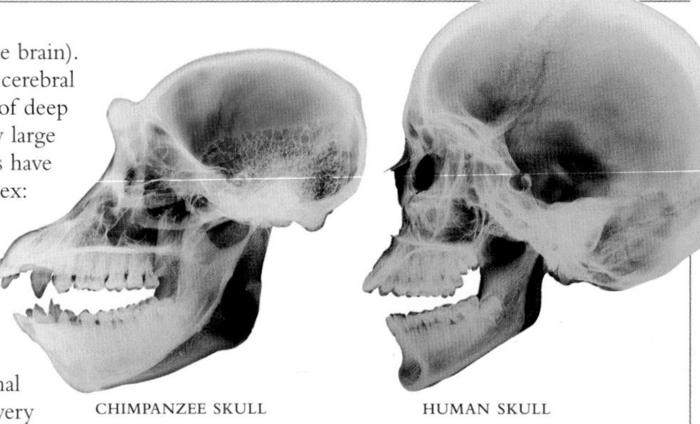


**SKULLS AND BRAINS**

*As humans evolved, brains got bigger and jaws got smaller. Our jaws and teeth are smaller than those of chimpanzees because we use tools rather than teeth to prepare food. Our brains are three times larger, making our skull appear grotesquely swollen in comparison.*

**HIGH-DENSITY SOCIETY**

*Complex society may have been a driving force behind human intelligence. Today we live in cities of many millions, such as Shanghai in China.*

It takes some ingenuity to visualize the shape of a blade in a lump of stone, so perhaps our ancestors' brains expanded as they honed their toolmaking skills. Surprisingly, evidence suggests this is not the case: prehistoric humans seem to have spent millions of years using the same old stone handaxes without innovation (although they may perhaps have been inventing ingenious tools from more perishable materials).

Then again, large brains may be related to our carnivorous diet. Animals that hunt for a living have large brains because they need to outwit their prey. Yet, although we eat more meat than other primates, other predators manage with much smaller brains than us, so hunting cannot be the only reason for human brain size.

An idea that has gained ground over recent years is that our large brains evolved for social reasons. Primates need cunning to understand the complexities of their social lives. Indeed, scientists have discovered a strong correlation between brain size in primates and the size of their social group. Our social groups are among the largest and most complex of any primate.

Perhaps the most convincing explanation for our intelligence, however, is that we evolved to fill what psychologists call a "cognitive niche". All species fill a niche that defines their way of life in an ecosystem. Our foraging ancestors were pioneers in using thought and knowledge to attain goals in the face of obstacles. Intelligence allowed them to think up countless new ways of finding food – by catching prey using traps, digging up plant roots with sticks, cracking open bones with rocks, and so on. As psychologist Steven Pinker has pointed out, "life for our ancestors was a camping trip that never ends, without the space blankets, Swiss army knives, and freeze-dried pasta".

## TOOL USERS AND MAKERS

Many animals use tools (among them dolphins, sea otters, and vultures), and there are a few that make them (chimps and orangutans), but humans are unsurpassed in mastery of tools. Opposable (flexible) thumbs and superb manual dexterity all contribute to making us expert toolmakers. Toolmaking also demands a certain amount of brain power, especially insight. With insight and imagination, we are able to visualize designs in our head and solve problems before we set to work with our hands.

Our technical intelligence has also given us fire for cooking, buildings, transport, spacecraft, medicines, and many other wonders of the modern world. However, there is a dark side to our mastery of tools. The first stone tools made by our ancestors were stone blades, and they may have been used for violence against people as much as for hunting. Violence is one of the hallmarks of *Homo sapiens*; with the possible exception of chimpanzees, we are the only species known to engage in warfare.

## THEORY OF MIND

Most animals see a stranger when they look in a mirror, but humans are among the few exceptions that recognize themselves. This rare ability hints at something special about our species: we each have a personal identity; a concept of the self. With this comes the ability to reflect on our own thoughts and feelings, and to visualize ourselves in imaginary situations. In turn, we can imagine what is going on in other people's minds by mentally putting ourselves in their shoes. Psychologists often refer to this phenomenon as a "theory of mind".

A theory of mind is very useful for a species such as ours, which lives in complex, dynamic social groups. Without it, we would not be able to deceive one another effectively, work out the devious motivations of our enemies, or predict how our friends and enemies might behave.

Chimpanzees may also be able to recognize themselves in mirrors, but does this mean they too have a theory of mind and the same kind of self-awareness as we do? Tantalizing anecdotes suggest they sometimes deceive each other, but as yet there is no unequivocal evidence for a theory of mind in any species besides our own.

Issue

### Weapons of war

Humans are the only species that regularly goes to war, commits genocide, and has the capacity to self-destruct. War has probably always been an important driving force in the invention of new technology: some of the first human tools were weapons, and we probably used them against each other as much as for hunting. Our passion for war was also the driving force behind some of our greatest scientific achievements, from splitting the atom to landing men on the moon.

## A TALKING APE

Language is arguably the greatest of human inventions. Other animals can communicate, but their efforts and abilities pale in comparison to our own. Vervet monkeys and meerkats have a vocabulary of alarm calls to warn each other of approaching predators, and even bees use a type of sign language to tell each other the way to nectar or pollen. However, other animal languages seem to lack one vital ingredient: grammar – a set of rules that enables words to be formed and strung together in a variety of different combinations, each with a distinct and unambiguous meaning.

The speed at which human language is both generated and interpreted is no less impressive than its infinite complexity. During normal speech, the human voicebox and mouth can generate up to around 25 separate speech sounds (roughly equivalent to letters of the alphabet) per second. Listeners can absorb this flood of data and effortlessly decode its meaning in an instant.

Language depends on the human ability to think symbolically. Words are essentially symbols for objects or concepts that need not be present or even physically real, for example people, food, emotions, and spirits. Symbolic thought also underlies another unique human creation: art. For tens of thousands of years, people have drawn paintings on rock walls, carved precious stones into ornaments, and adorned their bodies with tattoos and coloured dyes.

## HUMAN CULTURE

Thanks largely to language, humans possess culture – the complex patterns of behaviour and knowledge that vary among societies and are handed down through the generations. Culture can adapt and learn; in doing so, it helped our species discover new ways of living as we spread around the globe. Through culture, society itself becomes a body of knowledge, obviating the need for each generation to relearn survival skills from scratch. It is through language that knowledge can be quickly shared and survival skills passed on.

Common to all the world's human cultures is the ability to form contracts and make deals. We remember social commitments and feel a sense of moral outrage and a desire for vengeance when betrayed. These are uniquely human traits; other animals have no concept of commitment or moral obligation, and as such, their ability to cooperate is limited. Humans build complex societies that are based on a division of labour. Social rules allow us to specialize in separate careers, yet live together and support each other by exchanging the fruits of our labour.

Key to the human ability to form contracts is our ability to remember the past and plan for the future. Mentally, we can travel forwards or back in time. Although other animals may have sharper memories than ourselves (squirrels and jays can remember the hundreds of sites in which they have buried food), it seems that humans are unique in that their memories are so clearly etched in time.

EARLY CAVE PAINTING

**THE CREATION OF ART**

*Art and symbolic thinking may have existed for as long as our species. Even the oldest cave paintings show a sense of perspective comparable with that of the greatest modern works of art.*

LEONARDO DA VINCI'S *MONA LISA*

# ORIGINS

# ORIGINS

We must look to the past to see how far the human body, mind, and culture have evolved. The story winds through 6 million years, most of which are shrouded in mystery. Historians look back a few thousand years, to the dawn of writing and recorded history. Archaeologists peer a little farther back, digging through the litter of ancient civilizations and disinterring the dead. Beyond that, we have just faint glimpses of our past: the occasional bone or tooth, a scattering of stone tools, and a few clues from our DNA.

Some 6 million years ago, Africa's forests were home to an ape that probably looked quite similar to a chimpanzee. Like a chimp, it was an agile climber and spent a lot of time in the trees, but could perhaps take a few steps on two feet. It may also have mastered a few tools. By approximately 5 million years ago, the species had split in two. One group stayed in the tropical African forests, giving rise to the chimp and its close cousin the bonobo. The other adapted to life on land, gradually starting to walk upright and spreading to the savanna.

## PLANET HOMINID

This ground-dweller was the first member of a branch of the ape family that has been traditionally known as hominids, a group that ultimately came to include ourselves. However, classifications are being amended to refer to all great apes as "hominids". As a result, our ancestors from after the time of the split with the chimp line are now frequently termed "hominins". Here, the term "hominids" will continue to be used when referring specifically to our walking ancestors, and not to the other great apes.

Palaeontologists have now laid to rest the theory that big brains were responsible for driving us to walk upright in order to leave our hands free for making tools. In fact, our ancestors were walking on two feet by 4 or 5 million years ago, when their brains were no bigger than those of chimps. Even after their brains expanded, our ancestors continued to spend millennia hammering out the same crude tools with little sign of intelligence. The full impact of the human brain did not make itself apparent until around 40,000 years ago. By that time, though, a full 99 per cent of our history was behind us.

## SOLE SURVIVOR

Many years ago, scientists saw each species as an ancestor of our own. It seemed logical to organize them in a linear sequence, each one slightly more "human" than the last. With the discovery of more fossils, a very different picture has emerged. Human evolution looks like a bush made up of a maze of dead ends, and working out how all the species relate to each other is very difficult.

One of the most sobering discoveries is that our story is one of failure and extinction. All of our ancestors' contemporaries perished, leaving *Homo sapiens* as the sole survivor. Evidence suggests that our species very nearly met the same fate when population crashes significantly reduced its numbers. It may therefore be a matter of luck that we are here at all. Fossils reveal that our "cousins" survived

Issue

### Where do we come from?

Some cultures believe in a single deity that created the world suddenly: many Christians believe God created Adam and Eve just 6,000 years ago. Other societies think the world developed organically, like a fetus in the womb, or from "parents" symbolized by the sky and earth. Aboriginal Australians believe the world was created by mythological ancestors during the "dreamtime". In Judaeo-Christian mythology, special powers made us superior to plants and other animals; in some cultures, though, these entities embody the spirits of our ancestors.

ABORIGINAL DREAMTIME FIGURES

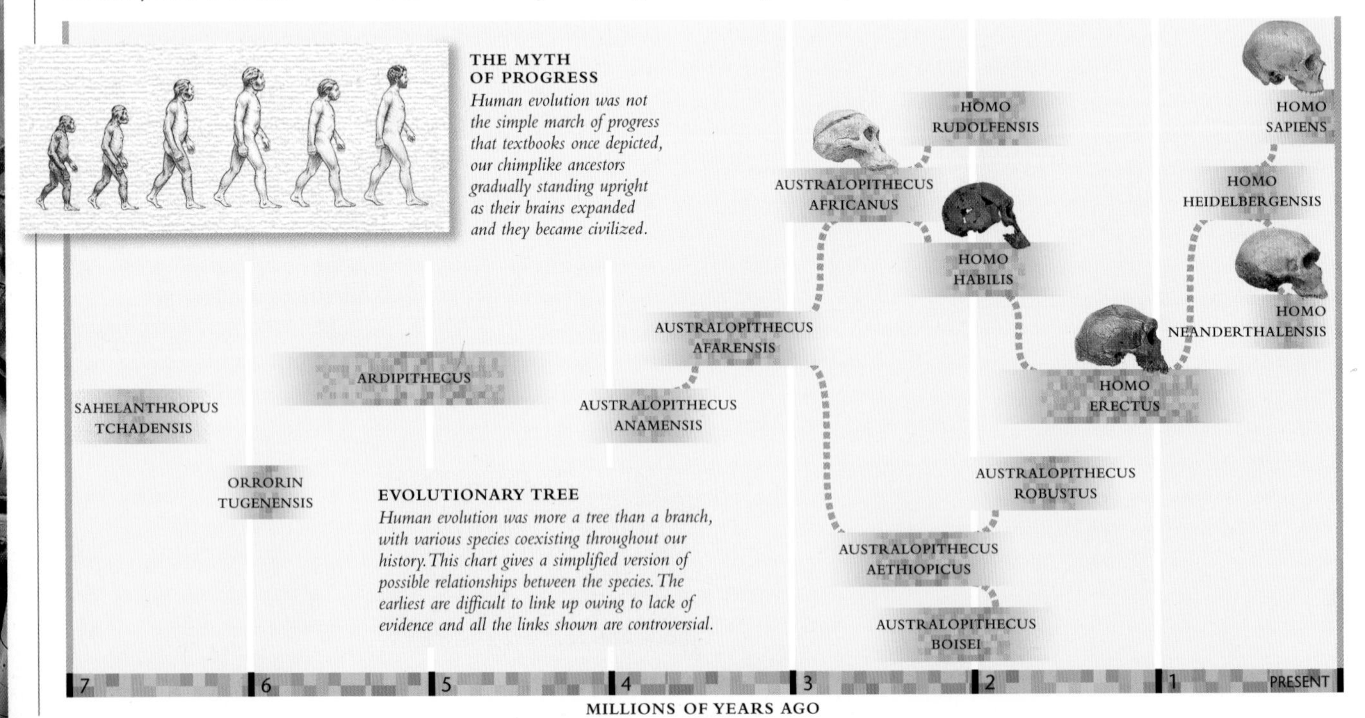


**THE MYTH OF PROGRESS** *Human evolution was not the simple march of progress that textbooks once depicted, our chimplike ancestors gradually standing upright as their brains expanded and they became civilized.*

**EVOLUTIONARY TREE** *Human evolution was more a tree than a branch, with various species coexisting throughout our history. This chart gives a simplified version of possible relationships between the species. The earliest are difficult to link up owing to lack of evidence and all the links shown are controversial.*

until as recently as 30,000 years ago, and that small pockets may in fact have clung on even longer. Almost every culture seems to have stories about mythical apemen, from the Sasquatch in North America to the Yeti in the Himalayas, the Alux in Central America, and the Orang Pendek in Sumatra. Perhaps these myths originated as Chinese whispers from a distant past when our hominid cousins still lived.

## SO WHY BIG BRAINS?

A traditional explanation for why we evolved such large brains focuses on technology: perhaps they evolved to help us devise complex weapons. With these, we unlocked a valuable new source of food – meat – which provided the protein and calories to make our brains even bigger. Yet perhaps stone tools dominate our thinking simply because so many examples have been found.

The probable benefits of big brains were complex language and social behaviour. Recently, the role of social interaction in evolution has been given more emphasis. In human society, success depends on the ability to exploit a web of social relationships. This skill may have been just as important as the ability to sharpen a rock or throw a spear.

## LEAPING FORWARD

Human history is all about quantum leaps. For thousands, even millions, of years, nothing much happens; then some discovery or twist of fate propels humanity forward.

One such leap showed itself around 40,000 years ago, when culture, art, and sophisticated tools made an appearance. In Europe, this cultural revolution is associated with the Cro-Magnons, whose paintings adorn the caves of southern France and northern Spain. Some experts claim the trigger was the origin of language and consciousness, but this remains an area of debate. The hunter-gatherer lifestyle remained our species' only way of life until around 10,000 years ago, when another leap occurred.

## CIVILIZATION DAWNS

As the last ice age drew to a close, people all over the world learned to domesticate and rear animals, and to cultivate plants. The invention of agriculture set in motion events that were to transform human society. Instead of roaming, people settled down. A good food supply led to a growth in populations and the emergence of towns. City life led to a greater division of labour and increased innovation.

**RECORDED HISTORY**
*Written language, such as this Sabaean script that dates from around 500BC, sprang from the emergence of civilization.*

However, evidence suggests the first farmers were less healthy than their nomadic ancestors. They lived in crowded, insanitary conditions and their diet was dominated by just a few staple grains. Social divisions also appeared: a small number of landowners lived in luxury, while most people toiled in the fields.

## THE AGE OF REASON

Writing is the sacred text of our human history. Written language was initially used only to draw up transactions and contracts. Later, however, it began to be used to record sacred knowledge and mythology, and by the time of ancient Greece, it was being used to store and communicate knowledge for its own sake.

This pure love of knowledge has continued to feature in intellectual movements, such as the Renaissance. This began in Italy in the 14th century and continued in Europe until around the middle of the 17th century. It was a time when action and reason started to challenge religious and contemplative life. The Renaissance was an era of exploration and discovery in every sense. It pushed back the boundaries of art, geography, music, science, and thought. Writing became a tool for recording and sharing new knowledge, be this scientific, artistic, or political.

**REASON BRINGS REVOLUTION**
*Increased interest in philosophy and politics not only led to cultural advances but gave the populace the tools with which to challenge its masters, as occurred in the French Revolution.*

## DIVISION AND UNITY

For most of history, groups of *Homo sapiens* did not share knowledge; it was passed from one generation to the next or discovered anew. When groups met it was often to fight – something humans seem predisposed to do. As technology develops, this seemingly innate urge could actually prove to be a threat to humanity. The more we learn about our world, the more we seem to be capable of destroying it. Today we live in a global village of 6.4 billion people and rising; all our knowledge is pooled; many of us speak the same language; and every major town is connected to the internet. The cultures that once divided us may be breaking down and merging. If the past is anything to go by, society will continue to change at breakneck speed. Only time will tell where this will take us.

Fact

### A day in the life

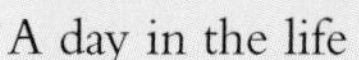

Imagine human history as a 24-hour clock, starting with the split from the other apes. We lived as hunter-gatherers until very nearly midnight. Suddenly, at 23:58, agriculture and towns appeared. We invented the wheel and built the pyramids at just over a minute to 12; the first Olympics took place at 40 seconds to; man walked on the moon at 3 seconds to; and the internet came about at a tenth of a second to. All of our greatest achievements have only just happened and the pace is accelerating.

Fact

### History repeating itself

The cradle of human evolution was the savanna of tropical Africa, where our ancestors survived for millions of years by gathering food and hunting. The legacy of this period is with us today. Despite thousands of years spent living all over the world, we remain tropical animals, dependent on warm clothes and heating in cold areas. Our food also reflects grassland origins: we choose grass seeds, in the form of wheat and rice, as our staple. And just as our ancestors survived best in small bands, we still sort ourselves into small communities, even in the most densely populated of cities.

**THE GLOBAL VILLAGE**
*Communications technology has enabled us to understand and even share in each other's lives. However, although the globe is shrinking, not all the "villagers" live in harmony.*

# FIRST STEPS

Around 6 million years ago, a momentous event occurred in the forests of Africa. An ancient species of ape left the safety of the trees and began to live on the ground. Exactly how, when, and why this species made the transition remain shrouded in mystery. Perhaps droughts dried out the rainforest, turning it into savanna. Or perhaps the land flooded, forming food-rich lagoons that tempted animals out of the trees. We may never know. One thing, however, we do know for sure: one pioneering ape adapted well to life on the ground. Instead of walking on its knuckles, it took to balancing precariously on its hind limbs, with its forelimbs dangling at its sides. It was the first step in an extraordinary process that was to transform apes into human beings.

**AN UNUSUAL SOLUTION** *Most big apes cope with life on the ground by knuckle-walking. Our ancestors were different: they stood up and walked.*

## A FAMILY IN DECLINE

Apart from *Homo sapiens*, the apes are a family in decline. Apes probably reached their peak around 20 million years ago, when at least 60 species inhabited the tropical forests of Africa and southern Asia. Around 10 million years ago, the apes started to disappear, and monkeys began taking their place. For millions of years, the planet has been getting cooler and drier, with the vast belt of rainforest once found in the tropics turning into sparser woodland or open grassland. The apes were fruit specialists, dependent on rainforest for their main source of food. So, as the forests dwindled, the apes disappeared. Monkeys learned to exploit the nuts and seeds of their new home. They had a secret weapon: the ability to digest unripe fruit, allowing them to steal the apes' food before it was edible. Around 5 million years ago, the Earth went through an especially severe bout of cooling and drying. In East Africa, the effect was heightened by slowly drifting continental plates generating immense pressure deep underground. The African continent tore open, producing a gash of monumental proportions: the Great Rift Valley. Into this landscape walked a new species with a secret weapon of its own: the ability to walk on two feet.

**THE GREAT RIFT VALLEY** *About 5 million years ago, the East African landscape began to pull apart, replacing dense jungle with the mosaic of grassland, desert, woodland, and rivers that exists today.*

Issue

### Why did we walk?

Perhaps standing enabled our ancestors to see over tall savanna grass to spot predators and prey, or perhaps they needed to be able to carry things. One idea is that an upright stance reduced exposure to the sun, allowing them to stay active during the midday heat. A controversial theory is that we became two-legged during an aquatic phase in our evolution. Animals that could wade in shallow lagoons would have found a rich supply of food.

## ON TWO FEET

Whatever tipped the balance towards a bipedal stance, becoming upright had its benefits. Walking on two legs uses less energy than knuckle-walking; more importantly, it leaves the hands free for many other purposes, including making tools, carrying things, gathering food, and throwing weapons. *Ardipithecus* is the oldest-known certain hominid (upright ape). Two distinct species have been found: *Ardipithecus kadabba* existed around 5.8 million years ago and *Ardipithecus ramidus* around 4.4 million years ago. Based on the few fossils found, *Ardipithecus* had a mixture of hominid and ape characteristics. It had the short canine teeth typical of hominids, but its molars were apelike in shape and its incisors large – more like a chimp than a human. The skull bones found suggest the skull rested on top of the spine, making *Ardipithecus* bipedal, but this has not been confirmed. Its arms appear large and powerful, so it was probably still a good climber. Maybe it spent its days on the ground but returned to the treetops to sleep, like chimps and gorillas do. Evidence suggests that *Ardipithecus* inhabited a woodland environment, not the grassy savanna previously believed to be home to the earliest hominids.

EARLY HOMINID

ANCIENT APE

**BIPEDAL STANCE** *On leaving the trees, our ancestors had to adapt to life on the ground. They gradually became proficient at walking on two legs.*

**7 million years ago** *Sahelanthropus* (*see* The Chad skull, opposite page) in existence

**5.8 million years ago** Earliest evidence of *Ardipithecus kadabba*, the first known hominid

**5–6 million years ago** Human branch had separated from chimp branch of ape family tree according to some genetic estimates

7,000,000

6,000,000

5,000,000

**6 million years ago** Leg bones dated to this time claimed to be the earliest evidence of upright walking on two legs

**5 million years ago** The Earth begins to go through severe bouts of cooling and drying, with rifting in northeast Africa

ORIGINS

# EARLY HUMANS

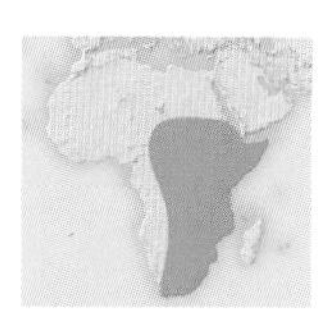

**WHERE THEY LIVED**
Australopithecus *is known to have inhabited much of eastern, southern, and central Africa.*

*Australopithecus* was a common and successful type of early hominid judging by the many fossils found. Several species spanned 3 million years, and these species divided into two main types: gracile and robust. Gracile australopithecines are likely to have been our ancestors and died out around 2.5 million years ago; robust australopithecines were an evolutionary dead end but survived later than the graciles. Both types of australopithecine would have looked a bit like upright chimps: they were only about 1–1.5m (3–5ft) tall, lacked the athletic build of later hominids, and had flat nasal openings. The shape of fossilized spinal, pelvic, and leg bones shows they could walk upright, as do the tracks of footprints found in a stretch of 3.65-million-year-old rock in Laetoli, Tanzania (*see below*). However, their short legs and long arms suggest they may only have been occasional walkers who spent much of their time sitting or climbing. Gracile species had smaller back teeth, and therefore probably subsisted on a varied diet of fruit, insects, seeds, roots, and maybe meat. The bigger jaws and back teeth of robust australopithecines point to a diet of coarse grassland vegetation. There is no real evidence that australopithecines could make stone tools. They perhaps made simple tools out of twigs, but may have lacked the intelligence to create anything more sophisticated.

**AUSTRALOPITHECINE SKULL**
*The skull of* Australopithecus *was shaped quite differently from ours, and its brain was only about the size of a chimp's. Shown here is the skull of* Australopithecus africanus, *a gracile australopithecine.*

*Jawbone*
*Upper arm bone*
*Ribcage*
*Pelvis*
*Thighbone*
*Shinbone*

**FAMOUS SKELETON**
*Almost half of this female* Australopithecus afarensis *(gracile) skeleton survived its 3.2 million years of burial. "Lucy" was adult, but she stood only 1.1m (3ft) tall.*

**EARLY FOOTPRINTS**
*These tracks, found in fossilized volcanic ash in Tanzania, may have been made by a family of three.*

**A SUCCESSFUL HOMINID**
*Judging from the many fossils found,* Australopithecus *was a common hominid, its several species spanning around 3 million years.*

Issue

## The Chad skull

From the front, this skull found in Chad in 2001 (named *Sahelanthropus*) is said to look more advanced than *Ardipithecus* or early *Australopithecus*. Yet it is 6–7 million years old, predating all known hominids and the hominid–chimp split. If it is a hominid, *Ardipithecus* and *Australopithecus* may be evolutionary dead ends; the split would also have to be pushed back or chimps redefined as hominids. We still do not know where *Sahelanthropus* sits on the ape evolutionary tree.

**4.4 million years ago**
*Ardipithecus ramidus* believed to be in existence

**4,000,000**

**4.2 million years ago**
Limited remains of bipedal hominid (*Australopithecus anamensis*) dated to this time found by Lake Turkana, Kenya

**3.6 million years ago**
First clear evidence of bipedalism: *Australopithecus afarensis* footprints found at Laetoli in Tanzania

**3.2 million years ago**
"Lucy" (*Australopithecus afarensis*) in existence; skeleton found at Hadar in Ethiopia

**3 million years ago**
*Australopithecus africanus*, notable for the powerful build of its upper body, in existence

**3,000,000**

**2.6 million years ago**
*Australopithecus aethiopicus* in existence; the earliest finds of stone tools also date to this time

# TOOLMAKERS

The arrival of *Homo*, the human genus (group of species), between 2 and 2.5 million years ago marks a major turning point in the story of our evolution. Apelike hominids such as *Australopithecus* had flourished, and continued to flourish, for millions of years with little change. With *Homo* there was a sudden leap in brain size, a dramatic change in anatomy, and the beginnings of stone-age technology. Perhaps the emergence of *Homo* also marked the first glimmerings of language, culture, and a social structure based on monogamous families. The origins of *Homo* are unclear. It may have evolved from one of the gracile australopithecines in Africa, but exactly where and when remain shrouded in mystery.

## THE HANDYMAN

*Homo habilis* ("handyman") marks a turning point in human evolution: it could make stone tools. Humans have a lack of natural weapons such as sharp claws or teeth, but stone tools allowed our ancestors to become more carnivorous. The tools of *Homo habilis* were simple and included small, sharp flakes of rock, probably used as blades or scrapers for cutting hides and butchering carcasses. Some experts think *Homo habilis* scavenged the leftovers of other predators rather than hunting live prey. Many of Africa's herbivores would have been too fast and strong to take on, but a gang of rock-throwing hominids could have driven cheetahs from their meals. The brain of *Homo habilis* probably consumed a disproportionate amount of energy. Stone tools unlocked the rich source of calories in meat necessary to power this hungry organ and sustain its expansion over the next 2 million years. A more versatile diet may also have helped to liberate our ancestors from their habitat.

**WHERE AND WHEN?**
Homo habilis *is thought to have lived in East Africa between 2.3 and 1.6 million years ago.*

**SMALL SKULL**
*The* Homo habilis *skull and brain were bigger than those of* Australopithecus, *but its brain was only half as big as a modern human's.*

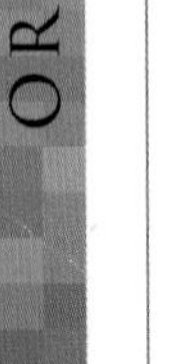

**FIRST TOOLS**
*Sharp flakes of rock were used as blades or scrapers. They were struck from cobbles that may also have been used for cracking nuts or releasing marrow from bones.*

**HUMAN GENUS**
Homo habilis *is one of the earliest species assigned to the human genus, and one of the first species known to have made stone tools.*

Profile

### Louis Leakey

Born in Kenya, Louis Leakey (1903–1972) always believed Africa was the cradle of human evolution, although many disagreed. He began fossil-hunting in East Africa's Great Rift Valley, and in 1959, he and his wife Mary discovered the oldest-known stone tools. Leakey's conviction that humans originated in East Africa led to a string of major fossil finds, including those of ancient apes. Leakey's son Jonathan found *Homo habilis*, his son Richard the Turkana boy (*Homo erectus*), and his wife the Laetoli footprints (*see* p25).

**2,500,000**

**2.3 million years ago** — *Homo habilis* appears

**2 million years ago** — *Australopithecus robustus* comes into existence

**2,000,000**

**1.8 million years ago** — *Homo erectus*, the first hominid to leave Africa, first emerges

**1.7 million years ago** — Earliest evidence of hominids in Asia

**1.6 million years ago** — *Homo habilis* dies out

# THE FIRST TO LEAVE AFRICA

*Homo erectus* ("upright man") first appeared in East Africa around 1.8 million years ago. *Homo erectus* was the first hominid to leave Africa and seems to have spread across the Old World with astonishing speed: by 1.7–1.8 million years ago it had reached Georgia, and by 1.6 million years ago it was in Java. Much of what we know about *Homo erectus* comes from a single individual: the Turkana boy. His 1.5-million-year-old, almost-complete skeleton was found in 1984 near Lake Turkana in Kenya. The Turkana boy was aged 8–11 when he died face down in a marsh that buried him before scavengers could destroy the carcass. He looked radically different from *Australopithecus*. Despite his age, he stood 1.6m (5ft 3in) tall; an adult may have been well over 1.8m (6ft). His long legs and narrow pelvis show he was as upright and athletic as us, but his sturdier bones suggest a more muscular build. His face projected less than that of *Australopithecus*, and he had a nose instead of flat nasal openings. Yet a prominent bony ridge jutted out above the eyes and would have given him a glowering expression. His brain was still far smaller than the modern average, but other physical measurements suggest the late puberty, or long childhood, characteristic of modern humans had started to evolve. Our long childhood is linked to the fact that we are born earlier in our development than other animals. Human infants are thus helpless and depend entirely on their parents, who tend to form long-term couples in which to look after them. The apparent late puberty of *Homo erectus* suggests this social feature had already begun to emerge.

**UPRIGHT MAN**
Homo erectus *was tall with long limbs, helping the body to shed heat. It is likely that he had bare skin to aid sweating, and he would have been black for sun protection.*

**ON THE MOVE**
Homo erectus *spread out of Africa around the globe, possibly surviving until as late as 100,000 years ago in China and Java.*

**TURKANA BOY**
*His barrel-shaped ribcage indicates a smaller abdomen and shorter intestine than* Australopithecus, *a sign he ate more meat.*

# INTELLECT AND INNOVATION

The still-small brain of *Homo erectus* may reflect a lack of intellect or simply a short lifespan. The shape of its skull base suggests it had a lower larynx than an ape, so could talk. But the Turkana boy's spinal cord was much narrower in the chest than ours and his chest muscles were unlikely to have had the nerve connections needed for true language. The hand axes made by *Homo erectus* were multifunctional tools probably used to skin and butcher animals. They were made with skill, yet hardly changed in over a million years, which suggests a mind very different from our own. It is as if this species was driven to make the same tools again and again. In 1997, 840,000-year-old stone tools, possibly those of *Homo erectus*, were found on an Indonesian island that may never have been joined to the mainland. Some therefore believe *Homo erectus* could build rafts from wood or bamboo and sail across the sea. Palaeontologists have also found evidence of a new type of sociality: one individual survived for months after being crippled, so someone may have been caring for her. A darker picture of early humans comes from Europe 800,000 years ago. *Erectus*-like creatures left behind hominid bones bearing scratch marks from being defleshed using tools.

**EVER-INCREASING SIZE**
*The brain of* Homo erectus *was somewhat bigger than that of* Homo habilis; *a more complex social life could be the explanation.*

**MORE ADVANCED TOOLS**
*More sophisticated than the cobbles and flakes of* Homo habilis, *the hand axe of* Homo erectus *was worked into a teardrop-shaped cutting implement by chipping it on both faces.*

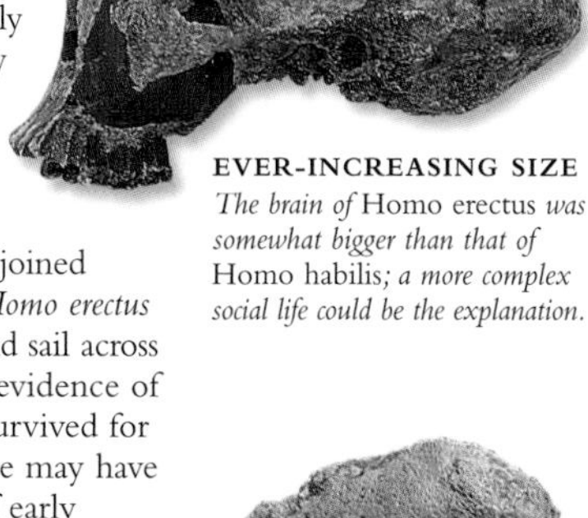


**SIGNS OF CANNIBALISM?**
*Hominid bones found at Burgos in northern Spain bear scratch marks made when the flesh was removed – possibly a sign of cannibalism or ritual defleshing of the dead.*

Issue

## Discovery of fire

Hearths first appear in the fossil record 250,000 years ago, and ash deposits in China are seen by some as the remains of 400,000-year-old hearths. Yet *Homo erectus* may actually have used fire 1.5 million years ago. Cooking turned indigestible plant matter into energy-rich brain fuel, so could explain the species' small teeth and intestines. It may even have transformed society: food now had to be gathered, carried, and prepared, and a female would have been at risk from thieves. The need for male protection could be a possible origin of the male–female bond.

**1.5 million years ago**
"Turkana Boy" (*Homo erectus*) in existence; hand axe developed; earliest evidence of fire, southern Africa

**800,000 years ago**
Oldest evidence of human life in Europe, near Burgos in northern Spain

1,500,000 — 1,000,000

**1.2 million years ago**
Extinction of *Australopithecus robustus* in Africa

# THE HUNTER-GATHERER

There is evidence of some European life before 600,000 years ago: fossils of *Homo antecessor*, a creature similar to *Homo erectus* that has been dated at 800,000 years old, have been found in northern Spain. However, we only have clear evidence that Europe was well populated from 600,000 years ago. Early Europeans were without doubt specialized hunters: both *Homo heidelbergensis* and *Homo neanderthalensis* are known to have crafted hunting tools, and some evidence suggests that they may even have hunted large animals at very close range.

**A WAY OF LIFE**
*It is likely that both* Homo heidelbergensis *and* Homo neanderthalensis *hunted deer and other large animals.*

> Issue
>
> **The Pit of Bones**
>
> The 400,000-year-old bones of about 32 *Homo heidelbergensis* bodies were found in a deep pothole in northern Spain. The site is thought by some to be a burial chamber with spiritual meaning. Ritualistic burial implies the dawn of sophisticated consciousness, so this would give the site huge significance. Others are less convinced, suspecting the bones slid into the pit or were dragged there by animals.

## MAN THE HUNTER

*Homo heidelbergensis* came into existence around 600,000 years ago. Although experts are unsure whether *Homo erectus* was more of a hunter or a scavenger, there is little doubt that *Homo heidelbergensis* was a skilled hunter. At Boxgrove, England, animal bones and stone tools have been found at what seems to be a prehistoric butchery site. Most of the animals were large game, such as deer and rhinos, and their bones bear scratch marks made by stone tools. These marks lie under the tooth marks of scavenging animals, showing that the hominids got to the prey first. More evidence of hunting comes from Shöningen, Germany, where 400,000-year-old hardwood spears have been found. The spear ends appear to have been split, possibly to hold stone spearheads.

**INHABITED EUROPE**
Homo heidelbergensis *lived in Europe, as well as in parts of Asia and Africa.*

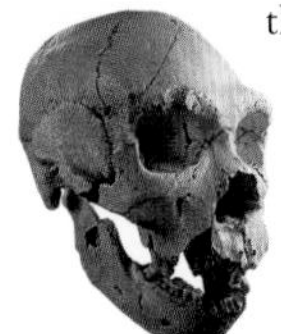

SKULL

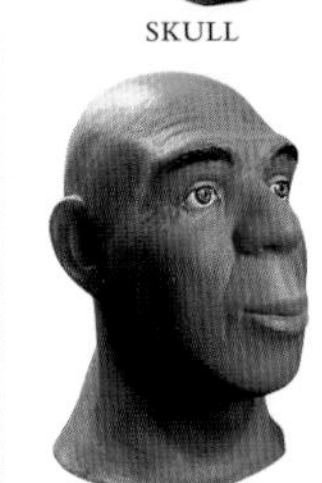

RECONSTRUCTION

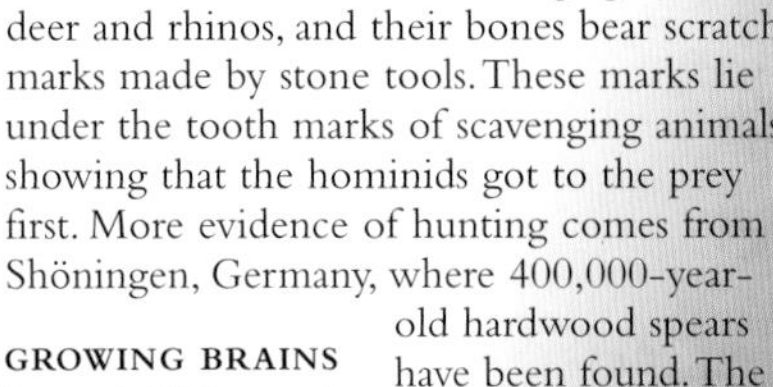

**GROWING BRAINS**
Homo heidelbergensis *had the glowering brow ridge and low forehead of* Homo erectus, *but its brain was bigger.*

## ICE-AGE SURVIVORS

Few hominids have caused as much controversy as *Homo neanderthalensis*. Wearing a hat and scarf, a Neanderthal might attract little attention in the street today. Yet with his or her face uncovered, it would be different. A bulbous nose made the centre of the face jut out and, instead of a forehead, there was a prominent brow ridge overshadowing deep-set eyes. The Neanderthal chin was receding, wide cheekbones flared back on each side of the face, and the front teeth were huge. People tend to think of Neanderthals as primitive cavemen, stooped and as hairy as apes. This distorted view of our close cousin is largely the result of the Old Man of Chapelle, a fossil found in 1908. Scientists emphasized his curved spine, stooped posture, apelike feet, and limited intellect. In the 1950s, however, the Old Man of Chapelle's curved spine was found to be the result of arthritis and his brain bigger than the modern average. Neanderthals were well adapted for surviving the cold Ice-Age winters: their short, stocky build helped to conserve heat.

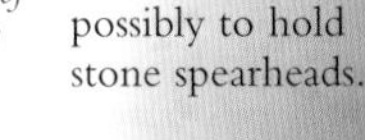

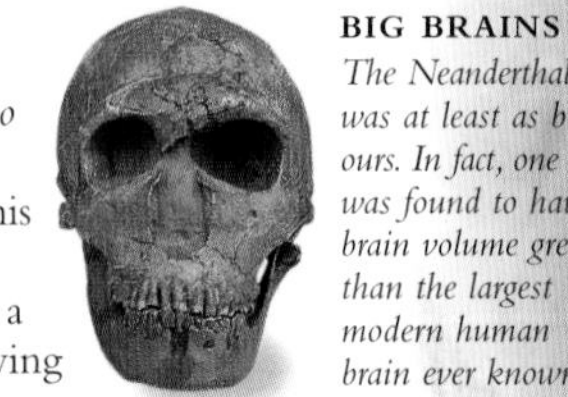

**BIG BRAINS**
*The Neanderthal brain was at least as big as ours. In fact, one skull was found to have a brain volume greater than the largest modern human brain ever known.*

**SKIN VARIATION**
*Most experts now believe that Neanderthal skin was hairless and varied from pale to dark.*

**COLD CLIMES**
*Neanderthals lived in Ice-Age Europe and western Asia between 250,000 and 30,000 years ago.*

**NEANDERTHAL MAN**
*Perceptions of our close cousin* Homo neanderthalensis *alternate between hairy savage and sophisticated hunter-gatherer.*

**600,000 years ago**
Emergence of *Homo heidelbergensis* in Europe, northern Africa, and western Asia

**400,000 years ago**
Oldest preserved wooden spears found in Shöningen, Germany, and Clacton, UK

**400,000 years ago**
The Pit of Bones ("*La Sima de los Huesos*") in the Atapuerca region of northern Spain dated to this time

600,000 | 500,000 | 400,000

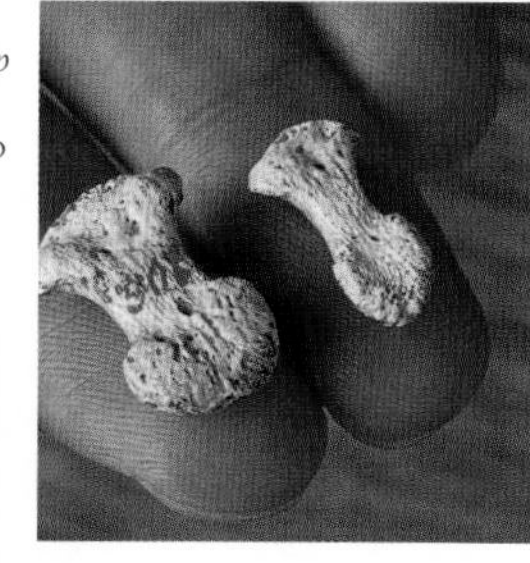

**GRIP COMPARISON**
*Neanderthal hands (see the fingertip bone near right) may have given a less subtle grip than those of* Homo sapiens *(fingertip bone, far right).*

**BROKEN BONES**
*Fractures and other traumatic injuries are almost universal among Neanderthal fossils, especially on upper body bones such as the ribs shown left.*

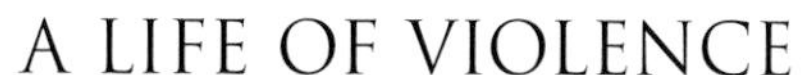

## A LIFE OF VIOLENCE

Neanderthals clearly took care of the elderly or infirm. At Shanidar Cave in Iraq, archaeologists found the skeleton of a 40-year-old man who had survived for several years with a catalogue of injuries and deformities. One side of his face was so badly crushed that he must have been partially blind, yet the wound had healed over. He had crippling arthritis in his right leg, healed fractures in a foot, and a withered arm. Any one of these afflictions could have killed him if he had lived alone, but with a social group to support him, he survived. The Shanidar skeleton's injuries bear witness to another feature of the Neanderthals – their violent lives. A likely explanation for the fractures commonplace among Neanderthal fossils is that they hunted big game at close range, using brute force and hand-held weapons to overpower animals as big as bison. Some experts think they may even have leaped on their prey. The price the Neanderthals paid for this physically demanding way of life was frequent injury and a life expectancy of only 30–40 years. Neanderthal tools were more advanced than those of earlier hominids and, judging from the pattern of wear on the Neanderthals' large front teeth, they used their mouths to hold meat while hacking at it with knives – much as Arctic people eat reindeer meat today. The Neanderthals also had fire: they built crude hearths and left layers of compressed ash on the floors of their caves. For all their skill with stone blades and fire, some experts think that Neanderthals were somewhat ham-fisted compared to modern humans, which may be why they rarely made wooden shafts for spears and knives.

**BURIAL CEREMONY**
*Some Neanderthal graves contain animal bones, possibly left as part of burial rituals. This reconstruction shows a Neanderthal burial being performed.*

**MORE ADVANCED TOOLS**
*Neanderthals replaced bulky hand axes with smaller implements, such as flakes of rock retouched along the edges to make scrapers or saw-tooth cutters.*

## WHAT HAPPENED TO THE NEANDERTHALS?

The greatest controversy around the Neanderthals is why they disappeared. They flourished for more than 200,000 years, endured countless Ice-Age winters, and became skilled hunters of some of the biggest animals on the planet. Yet, in the end, something drove them to extinction; it was probably us. Modern humans reached Europe about 40,000 years ago – just 10,000 years before the last Neanderthals perished. It seems the two could not coexist, but exactly why remains a mystery. Perhaps our ancestors waged war on the Neanderthals or hunted them for food. Or maybe the Neanderthals succumbed to a disease brought by the invaders. In recent years, however, another theory has been gaining ground: that modern humans had some kind of cultural or intellectual advantage over the Neanderthals that gave them a competitive edge, especially when facing increasingly unstable climates. Our arrival in Europe is accompanied by the emergence of symbolic thought, art, sophisticated new tools made from bone and antler, and more effective clothing to keep out the cold. We brought with us a complex culture and formed networks of social contacts with neighbouring groups, fostering trade and the flow of information. All of this points to the final emergence of the modern human mind and, according to some experts, the flowering of language. The Neanderthals simply could not compete for Europe's diminishing resources in the unstable climates of the time.

Fact

### Neanderthal DNA

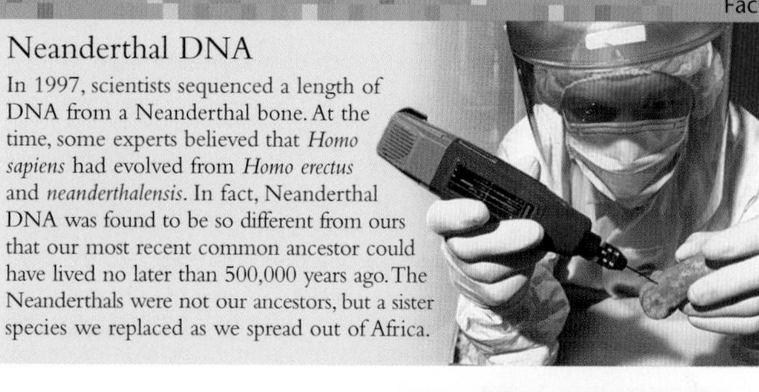

In 1997, scientists sequenced a length of DNA from a Neanderthal bone. At the time, some experts believed that *Homo sapiens* had evolved from *Homo erectus* and *neanderthalensis*. In fact, Neanderthal DNA was found to be so different from ours that our most recent common ancestor could have lived no later than 500,000 years ago. The Neanderthals were not our ancestors, but a sister species we replaced as we spread out of Africa.

**UNPICKING THE PAST**
*Many thousands of people around the world are involved in the search for evidence of where, when, and how the Neanderthals lived and what happened to them.*

**BURIED EVIDENCE**
*So much is known about the Neanderthals because they buried their dead, leaving many skeletons like this one for palaeontologists to unearth.*

**300,000**

**260,000 years ago**
Possible earliest *Homo sapiens* in Africa

**250,000 years ago**
*Homo neanderthalensis* emerges and exists in ice-age Europe and western Asia for the next 220,000 years

**200,000**

**160,000 years ago**
Definite earliest *Homo sapiens* in Africa

# MODERN HUMANS

*Homo sapiens* evolved 150,000 to 200,000 years ago in Africa. At first, our species led a similar life to other hominids, but modern behaviour began to evolve. From around 60,000 years ago, *Homo sapiens* swept out of Africa in waves, spreading across the planet and replacing the natives to become the only hominid on Earth. A cultural revolution occurred 40,000 years ago, with the appearance of Cro-Magnons (a type of *Homo sapiens*) in Europe. They were fully modern in their anatomy, society, and behaviour, and had a culture so sophisticated that some call this stage the "Great Leap Forward". In a few thousand years, there was more innovation than in the previous 6 million – a spectacular flowering of art, music, religion, mythology, trade, clothes, houses, and tools.

**CRO-MAGNON SKELETON**
*This amazingly intact specimen shows how similar the anatomy of Cro-Magnons was to our own. They were also fully modern in their society and behaviour.*

## COMMON ANCESTORS

Palaeontologists usually rely on fossils to find out about our ancestors. With modern humans, the study of DNA is a powerful new tool. During development, chromosomes pair up and swap bits of DNA, making each sperm or egg cell unique. This process could make it difficult to trace our genetic history, but it does not occur in mitochondria (tiny energy pumps in our cells, with their own genes) or in Y chromosomes. We inherit mitochondrial genes from our mothers and Y chromosomes from our fathers. Like any genes, they undergo mutations in their DNA at a roughly defined rate. By studying the mutation pattern in people around the world, scientists have worked out a family tree of the world's races. At its base is "Mitochondrial Eve", about 150,000 years ago. She was not the first *Homo sapiens*, but was the most recent common female ancestor we know about. Y chromosomes contain more DNA, so yield a more detailed picture. By studying the pattern of their mutations in the world's indigenous peoples, scientists found "Nuclear Adam", the most recent known common ancestor of all the world's men, who probably lived between 90,000 and 60,000 years ago; they also mapped our spread across the globe.

**"MITOCHONDRIAL EVE"**
*By studying the genes in mitochondria, (energy pumps in our cells; one shown above), we think our most recent common female ancestor lived 150,000 years ago.*

**"NUCLEAR ADAM"**
*Studies of the Y chromosome (right, alongside an X) provide a detailed picture of our ancestry.*

## PHYSICAL FEATURES

The early days of *Homo sapiens* are obscure. Very few early modern fossils have been found and there is no clue as to what would trigger the dramatic emergence of culture. Early modern humans had the tall foreheads, domed skulls, and flat faces typical of *Homo sapiens*, and their brains were bigger than those of the majority of their predecessors. Their limb proportions reveal a slender, tropical build unlike that of Neanderthals, so it seems likely they had black skin. Like Neanderthals, however, they would have lived as hunter-gatherers, collecting wild plants, eggs, and honey, and hunting or trapping animals. The hunter-gatherer way of life was to dominate our existence until at least 10,000 years ago, when the Earth's climate became more settled and agriculture was invented. As *Homo sapiens* spread around the globe, the species became increasingly physically diverse. When Cro-Magnons appeared in Europe, for example, they were lighter-skinned than earlier *Homo sapiens* in Africa, probably because they were less exposed to harmful rays of sunlight and their skin therefore needed less protection.

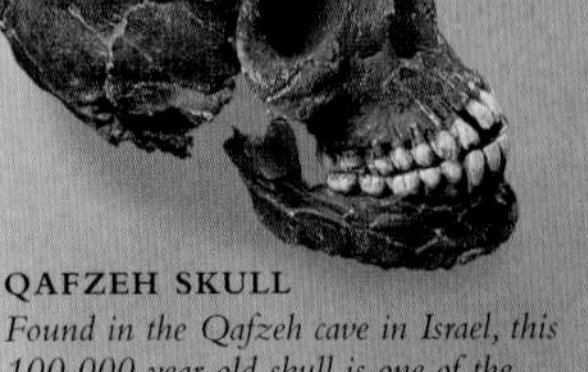

**QAFZEH SKULL**
*Found in the Qafzeh cave in Israel, this 100,000-year-old skull is one of the earliest known examples of* Homo sapiens.

**CRO-MAGNON MAN**
*Cro-Magnon people are likely to have looked physically distinct from early* Homo sapiens *in Africa, principally owing to their fairer skin.*

**EXODUS FROM AFRICA**
*When* Homo sapiens *first left Africa, they may have looked physically similar to these modern San Bushmen.*

**150,000 years ago**
"Mitochondrial Eve" alive; she was the most recent known common female ancestor of all the world's peoples

150,000

125,000

**120,000 years ago**
*Homo sapiens* starts to spread across Africa

**100,000 years ago**
*Homo sapiens* reaches southern Africa and Israel

100,000

Profile

## Charles Darwin

The observations made by naturalist Charles Darwin (1809–1882) while on board HMS *Beagle* during its world voyage led to his ideas on evolution. In 1859, he defined his theory of natural selection in *The Origin of Species*. Darwin was one of the first to suggest that Africa was the homeland of the human race and, when ancient hominid fossils began to turn up, Africa was where they were found. Not a single bone older than 1.8 million years has been found elsewhere.

# OUT OF AFRICA

"Mitochondrial Eve", our most recent common female ancestor, is believed to have lived in Africa because the mitochondria of African people have more genetic diversity than those of non-Africans. The results of studies of the Y chromosome also support the idea that *Homo sapiens* originated in Africa. The various non-African populations in the world today seem to have evolved from small subsets of migrating Africans. The oldest modern human bones found come from Herto in Ethiopia and date to 160,000 years ago. By 100,000 years ago, modern humans had evidently spread to southern Africa and Israel (graves dated to that time have been found there), and by 90,000 years ago they were in central Africa, judging from the discovery of intricately carved bone harpoons. As far as can be told, these early people mostly led similar lives to the Neanderthals and used the same limited range of stone tools. According to studies of the Y chromosome, the first wave of migrants began to leave Africa for the farther reaches of the globe around 60,000 years ago, or even a little earlier, spreading along the southern coast of Asia and reaching Australia at least 55,000 years ago. Aboriginal Australians are their descendants, as are the aboriginal peoples of New Guinea, the Andaman Islands, and Sri Lanka. A second wave of African migrants travelled to the steppes of Central Asia perhaps 40,000 years ago, from where they spread into India, Europe, eastern Asia, and Siberia.

**GLOBAL SPREAD**
*Modern humans left Africa in two main waves. The first moved through southern Asia to Australia and the second populated India, the rest of Asia, and Europe. The Americas were one of the last places to be reached.*

# GENETIC BOTTLENECKS

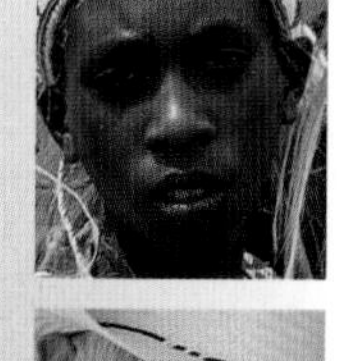

Compared to other primates, humans have very little genetic diversity. One reason for this is that *Homo sapiens* recently passed through a series of "genetic bottlenecks". A bottleneck occurs when a population falls to a dangerously low level, causing the gene pool to shrink. If the population expands again, all descendants carry copies of the same limited set of genes, preserving a record of the bottleneck for future generations. Drought, famine, and epidemics can all cause genetic bottlenecks. Owing to one or several environmental catastrophes, the number of modern humans may at one point have fallen to as few as 10,000 people. An even more common cause of genetic bottlenecks is migration, and one occurred when modern humans spread out of Africa. In fact, all the world's non-Africans could have descended from a founding population of only 50 people.

**GENE POOLS**
*Africans have the most genetic diversity; Native Americans have a particularly low diversity.*

**GENETICALLY SUPERIOR?**
*A single troop of chimpanzees has more variation in certain genes, such as those of mitochondria and Y chromosomes, than do all of the world's people.*

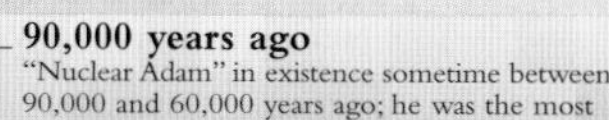

**90,000 years ago**
"Nuclear Adam" in existence sometime between 90,000 and 60,000 years ago; he was the most recent known common ancestor of today's men

75,000

**60,000 years ago**
*Homo sapiens* reaches China

**55,000 years ago**
*Homo sapiens* reaches Australia

50,000

**40,000 years ago**
Cro-Magnons appear in Europe

# THE GREAT LEAP FORWARD

There are many theories about what triggered the "Great Leap Forward", the jump in creativity that occurred 40,000 years ago in Europe. Because language enables knowledge to be shared, vastly aiding innovation, some think the trigger was a change in the brain or vocal tract that enabled complex language to emerge. Others think it was prompted by the evolution of a sense of self (shown by a belief in the soul and the afterlife), which helped us to know our own feelings and predict those of others – vital to the success of a social primate. A controversial theory is that there was a change in the architecture of the human mind. Some psychologists think that early hominid minds were divided into social, technical, and natural history modules that remained strictly separate. *Homo erectus*, for example, may have had language, but only as part of social intelligence, so could not talk creatively about toolmaking (part of technical intelligence). In *Homo sapiens*, these mental barriers broke down, causing a leap in imagination.

**INNOVATIVE LEAP**
*Cro-Magnons pioneered the use of man-made shelters, such as this one made from mammoth bones.*

**SEWING IMPLEMENTS**
*These eyed needles were used some 15,000 years ago to make clothes from animal hides – vital to survive the punishing winters of Ice-Age Europe.*

# A NEW TOOLKIT

The Cro-Magnons were much more inventive toolmakers than the Neanderthals. They made a wider range of stone tools by striking microliths (thin blades) of flint from a larger core and then adapting these "blanks" for a multitude of different purposes. Cro-Magnons also made innovative use of bone, antler, and wood and, for the first time in history, different materials were combined to make multipart tools. Palaeontologists have found Cro-Magnon fishhooks, net-sinkers, and rope (probably used for snares), all of which reveal the ingenious ways in which they caught and trapped fish and small game. Cro-Magnons are also known to have hunted large animals but, unlike Neanderthals, they killed their prey by throwing spears from a distance rather than fighting them at close range. This saved them from the serious injuries sustained by their Neanderthal cousins.

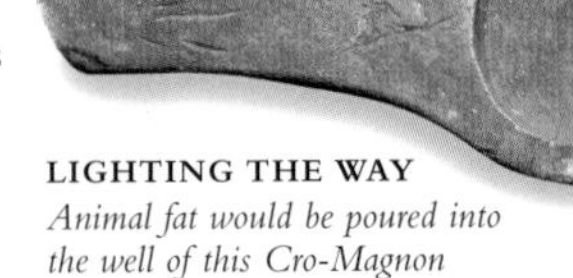

**LIGHTING THE WAY**
*Animal fat would be poured into the well of this Cro-Magnon stone lamp and set alight.*

# ART AND SYMBOLISM

Even more impressive than the explosion in technology was the appearance of prehistoric art, which reached its zenith 15–30,000 years ago with the famous cave paintings at Chauvet and Lascaux in France, Altamira in Spain, and other caves in the area. The paintings are not primitive; they are strikingly modern and show a fully developed sense of perspective. When Pablo Picasso saw the paintings at Lascaux, he compared the images to modern art, saying "we have discovered nothing". The Cro-Magnons' aesthetic sense is also evident in their love of necklaces, precious stones, musical instruments, and sculptures. Precious stones, such as amber, have been found far from their sites of manufacture – a sign that Cro-Magnon societies traded them with their neighbours. Many of the Cro-Magnons' artistic creations are recognizable as animals or people, but some are more abstract. Among the European cave paintings are geometric patterns of dots and lines similar to Aboriginal Australian rock art, as well as fabulous creatures that are part human and part animal. These abstract images seem rich with symbolic meaning and suggest the existence of an ancient mythology. If the people who created these paintings actually had mythology, then they must also have had articulate language.

**MAKING MUSIC**
*This 25,000-year-old bird-bone flute is an early example of a musical instrument.*

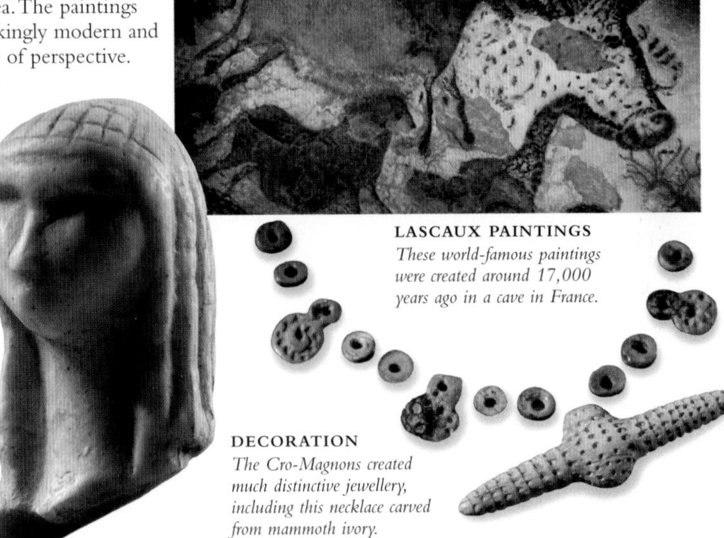

**LASCAUX PAINTINGS**
*These world-famous paintings were created around 17,000 years ago in a cave in France.*

**LADY OF BRASSEMPOUY**
*This 25,000-year-old masterpiece was exquisitely carved out of mammoth ivory.*

**DECORATION**
*The Cro-Magnons created much distinctive jewellery, including this necklace carved from mammoth ivory.*

35,000

30,000

**35,000 years ago**
Simple baboon-bone counting device, found in South Africa, dates from this time

**30,000 years ago**
*Homo neanderthalensis* disappears

**30,000 years ago**
Chauvet cave paintings created by Cro-Magnons, France

**28,000 years ago**
Sungir bodies buried, Russia

## BELIEF IN THE AFTERLIFE

Grave goods reveal that Cro-Magnons must have believed in the afterlife and perhaps had religion. A spectacular example comes from Sungir in Russia: an elderly man and two children were found covered in thousands of ivory beads that were once sewn into clothing. One child had a belt made from 250 fox teeth, an ivory sculpture, a mammoth-tusk lance, and a polished human thighbone packed with red ochre. These people were clearly important and had been buried with grave goods either to protect, entertain, or placate them in the next life – it is hard to know which. Cro-Magnon graves reveal something else: an increased lifespan. Neanderthals were lucky to survive to 40, but Cro-Magnons often lived until well after 60, perhaps reflecting the extra care older people received, and so the extra status they may have had.

Issue

### Myths and magic

Early rock art did not always depict the everyday. Some paintings hint of mythical or spiritual beliefs or show visions seen during altered states. This Tanzanian example (below) has been interpreted by some as a shamanistic trance dance. One possible explanation for the human figure with the head of an antelope (second from right) is that dancers may have seen fellow dancers as animals; alternatively, this figure may have been a vision of the spirit world seen at the height of a trance.

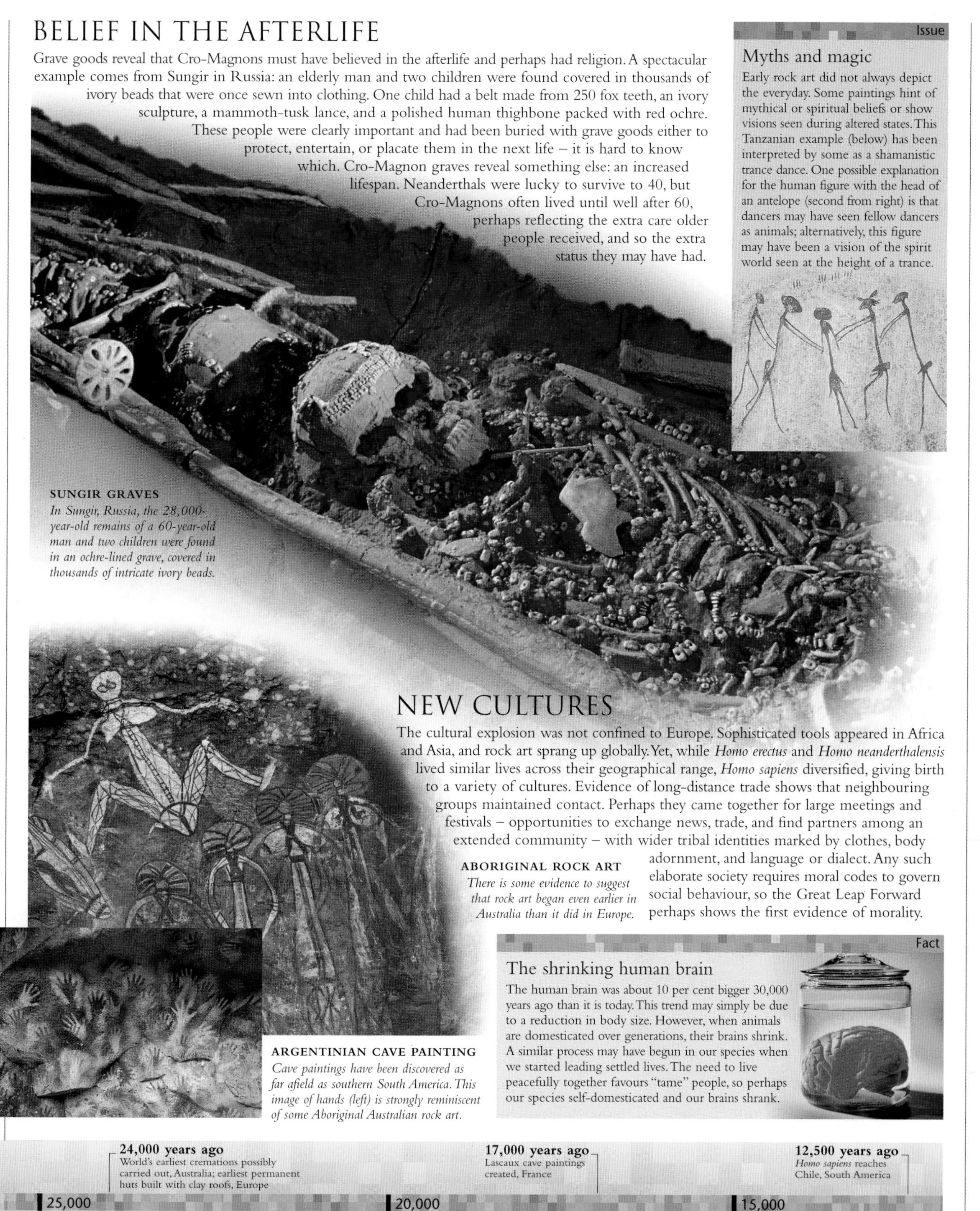

**SUNGIR GRAVES**
*In Sungir, Russia, the 28,000-year-old remains of a 60-year-old man and two children were found in an ochre-lined grave, covered in thousands of intricate ivory beads.*

## NEW CULTURES

The cultural explosion was not confined to Europe. Sophisticated tools appeared in Africa and Asia, and rock art sprang up globally. Yet, while *Homo erectus* and *Homo neanderthalensis* lived similar lives across their geographical range, *Homo sapiens* diversified, giving birth to a variety of cultures. Evidence of long-distance trade shows that neighbouring groups maintained contact. Perhaps they came together for large meetings and festivals – opportunities to exchange news, trade, and find partners among an extended community – with wider tribal identities marked by clothes, body adornment, and language or dialect. Any such elaborate society requires moral codes to govern social behaviour, so the Great Leap Forward perhaps shows the first evidence of morality.

**ABORIGINAL ROCK ART**
*There is some evidence to suggest that rock art began even earlier in Australia than it did in Europe.*

**ARGENTINIAN CAVE PAINTING**
*Cave paintings have been discovered as far afield as southern South America. This image of hands (left) is strongly reminiscent of some Aboriginal Australian rock art.*

Fact

### The shrinking human brain

The human brain was about 10 per cent bigger 30,000 years ago than it is today. This trend may simply be due to a reduction in body size. However, when animals are domesticated over generations, their brains shrink. A similar process may have begun in our species when we started leading settled lives. The need to live peacefully together favours "tame" people, so perhaps our species self-domesticated and our brains shrank.

**24,000 years ago** World's earliest cremations possibly carried out, Australia; earliest permanent huts built with clay roofs, Europe

**25,000**

**23,000 years ago** Lady of Brassempouy sculpted, France

**20,000**

**17,000 years ago** Lascaux cave paintings created, France

**16,000 years ago** First huts built with mammoth-bone roofs, western Russia

**15,000**

**13,500 years ago** *Homo sapiens* crosses into the American continent

**12,500 years ago** *Homo sapiens* reaches Chile, South America

# SETTLED SOCIETIES

Between 10,000 and 1000BC, humanity began to leave a bigger impression on the natural world. The changing climate and environment at the end of the last ice age offered new possibilities for human life. Agriculture emerged and was sustained by a new type of cereal and irrigation technology. Agricultural surplus gave rise to settled agrarian communities, then to cities with a wider division of labour. The need to store this surplus against bad times produced pottery and recording systems, and there is also evidence of elaborate burial rituals. Humanity did all these things in river valleys across the world, and even when specific cultures were shortlived, the struggle to sustain a settled life was uninterrupted.

**SETTING BOUNDARIES**
*This Babylonian boundary stone constitutes an early legal document. Written evidence of this kind allows us to read the story of our past.*

## THE WRITTEN WORD

Without written records, there would be no human story or "history". Written symbols are sophisticated tools that allowed humanity to refine skills, ideas, and understanding over thousands of generations. The first evidence of this capacity to record things in writing comes in the form of Sumerian clay counting tokens and written symbols (cuneiform) from around 3400BC. The wide variety of types of symbol indicated they were used across a range of activities, from accounting to recording creation stories and cosmologies, and writing soon became a necessity in societies that stored and traded goods. Early symbols were simple pictorial representations of common sources of livelihood, such as a horned head for an ox; more abstract systems, such as hieroglyphics and early alphabets, developed to express more complex ideas. Those who possessed the jealously guarded ability to use and interpret written language were able to establish the law.

**MAKING A MARK**
*Writing systems made up of abstract signs, such as the cuneiform script, enabled the expression of complex ideas.*

**ORACLE BONES**
*Questions about the future engraved on bone are the earliest known form of Chinese writing.*

## EARLY AGRICULTURE

Around 12,000 years ago, the cultivation of plants for food began in the Fertile Crescent in Mesopotamia (modern-day Iraq) when humans witnessed and then repeated accidents of nature. A chance crossing of wild grasses had produced a new hybrid strain with a much fuller head of seed. Scattered by the wind, the hybrid eventually crossed with another grass to create "bread wheat", an even richer grain ideal for human cultivation. As a result, agriculture sprang up in fertile regions: near lakes, rivers, the sea, or in areas with sufficient rainfall at the crucial times of year. However, as farming methods spread and larger populations became dependent on crops, people were forced to attempt an agricultural lifestyle in less fertile areas. Irrigation techniques were vitally needed in arid, unpredictable climates. The methods were basic at first – a case of digging a well and carrying water to the crops – but they soon became increasingly sophisticated. Networks of canals were built to store water from seasonal downpours and to channel it to crops, or even to divert water from lakes and rivers farther afield. Similar methods were used to drain waterlogged regions.

**SEEDS OF CHANGE**
*A new hybrid bread wheat required human cultivation. Tools such as sickles (left) and hand mills (far left) made the task easier.*

**DOMESTICATION**
*The domestication of plants and animals formed the basis of farming, which was transformed by inventions like the plough. This Egyptian fresco dates to c1300BC.*

**THE WHEEL**
*The invention of the wheel transformed agriculture, and also trade and military conquest; this example dates from c3500BC.*

### Fact

### The significance of pottery

Earthenware storage vessels are a classic feature of early Asian, Middle Eastern, and American civilizations. Within a few hundred years of the advent of agriculture, pots emerged to store the surplus food produced. The earliest vessels were found not in the Fertile Crescent, but in Japan's Honshu district. Similarly dated pots found in river valleys all over the world show that civilizations developed at the same time around the globe.

EARLY JAPANESE POT

**10,000BC** First known use of pottery vessels, Honshu, Japan

**9000BC** Settled agriculture established; the first selected grasses produce an early form of wheat

**8000BC** Sheep and goats domesticated in the Middle East and pigs in China

**7500BC** The world's first walled town established, Jericho, Israel

**6500BC** Cattle successfully domesticated in Sahara region, North Africa

**6000BC** Corn (maize) first cultivated, Ecuador

10,000BC | 9000BC | 8000BC | 7000BC | 6000BC

# THE FIRST SETTLEMENTS

The possibility of growing food almost anywhere, and so living almost anywhere, transformed human society. In fact, the advent of agriculture led to the simultaneous rise of civilizations all over the world. In the past, staying in one place had probably meant death; now it tended to mean security. Jericho, a farming settlement just north of the Dead Sea, seems to have been the first walled human community, dating from around 7500BC. Initially, such settlements were agricultural, comprising the homes of those who farmed the land in the immediate vicinity and came together for protection. These communities could expand as it became technically possible to build bigger settlements, and because larger communities could grow and store a greater surplus of food, they were able to expand even further. After a certain level of growth, some community members could be freed to undertake more specialized roles, such as those of priests, potters, or soldiers. Hierarchies began to form, with the interpretation of knowledge and physical protection of a settlement becoming privileged roles that were respected by those who produced and traded food.

**TOWER OF JERICHO**
*To give a vantage point, a community, such as that of Jericho, would construct a watchtower at the highest point of the wall surrounding its settlement.*

**CATAL HUYUK**
*In early cities, such as this Turkish settlement from around 7000BC, many people could live together carrying out complementary tasks.*

# BURIAL

The high status of certain individuals in these new societies is reflected in the manners of their burial. Across Asia, the "elite", who may have been members of the military caste or seen as gods or gods' messengers on Earth, were buried under huge mounds; in Egypt and areas of Mesoamerica, leaders were laid to rest within elaborate pyramids or mausoleums. In some areas, most bodies were mummified for preservation; in others, they were simply buried, cremated, or left exposed for birds to pick at the bones. Bodies of the social elites were invariably surrounded by a range of grave goods, which sometimes even included sacrificed servants, concubines, soldiers, or animals – although over time these were replaced with artificial representations, such as model armies. Choice of grave goods can tell us much about the religious beliefs of a society, the position a deceased person held, and which materials were considered valuable at that time. Burial rituals were intended to ease the passage of the dead to the afterlife; to celebrate their wealth, distinction, and achievements; and to give a sense of continuity to those who remained.

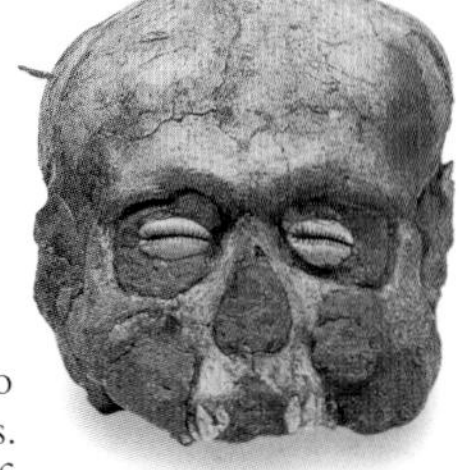

**BURIAL RITUALS**
*Shells in the eye sockets of this Jericho woman's skull indicate that a burial ritual may have been performed.*

**CREMATION URN**
*The Harappan civilization from the Indus Valley (3000–1500BC) cremated bodies and stored the ashes in funerary urns.*

**A NEW SOCIAL ORDER**
*As societies formed, some people gained status and had to protect their position, even after death. This model army guards the tomb of an Egyptian ruler.*

Fact

## Building the pyramids

According to the Greek historian Herodotus, it took 400,000 men a total of 20 years to construct the Great Pyramid at Giza in Egypt. This awesome human achievement stands 147m (482ft) tall and comprises some 2.5 million blocks of stone, each weighing an average of 2.5 tonnes (2.8 tons). The blocks were manually rolled up earth ramps to form "stepped" layers. Finally, the step effect was hidden by applying a smooth layer of limestone.

**5500BC** First irrigation system developed, Mesopotamia

**5000BC** First use of copper, Mesopotamia

5000BC

**4500BC** First large cemeteries, Europe

**3400BC** Use of clay counting tokens and first written symbols, Sumeria

4000BC

**3200BC** First wheeled transportation, Asia

**2750BC** Start of great period of pyramid building, Egypt

3000BC

**3000BC** Silk first produced, China

**1500BC** First evidence of metal-working, Peru

2000BC

**1360BC** Assyrian empire begins to build up, western Asia

ORIGINS

# THE CLASSICAL WORLD

From 1000BC to AD400, civilizations remained susceptible to decay and attack, although most nomadic invaders formed new, settled populations. As societies achieved greater levels of prosperity, they found new scope for reflection and, in some places, science and technology emerged. The greatest development was that of a political or civil society, with the emergence of concepts such as citizenship and democracy. Cultural evidence, from Homer's poetry to the Olympics, shows how people understood and celebrated their world. The forms of religion and philosophy we are familiar with also arose at this time: Buddhism and Confucianism were established and Jewish monotheism gained a new form with the birth of Christianity.

**THE GREAT WALL**
*China's Han empire extended this already-existing defensive wall to guard itself from attack.*

## AN AGE OF EMPIRES

At the start of the first millennium BC, the Assyrian empire, covering western Asia, was probably the world's greatest. It was destroyed by the Babylonians around 606BC, who were in turn conquered by the expanding Persian empire in 539BC. Persia dominated the world until about 325BC, when Alexander the Great conquered lands from Greece to the borders of India. Yet within 100 years of his death in 323BC, his empire had divided and disappeared. By the beginning of the 2nd century AD, four great empires reigned. Rome ruled the Mediterranean, parts of northern Europe, the Middle East, and Africa's north coast. In the east was the Han Chinese empire. In between, the Parthian empire covered much of western Asia and the Middle East, while the Kushan stretched from the Aral Sea through present-day Uzbekistan, Afghanistan, and Pakistan, and into northern India. By AD400, the demise of Rome and Han China had led to a power vacuum. Much of the world entered a state of chaos, with the recurring destruction of cities by barbarian attack.

**ALEXANDER THE GREAT**
*Alexander, King of Macedonia in northern Greece, rose to power in 330BC. In little more than 10 years, his armies had conquered much of the world known to him.*

## ECONOMIC EXPANSION

Coinage was invented in the 7th century BC in Greek Asia Minor as a convenient method of payment. India and China saw similar developments from 500BC and, by 200BC, coins had spread to western Europe. Coins are not only important for local payment: long-distance trade is more efficient if conducted through a medium of exchange rather than by the direct trade of goods, as long as the medium is portable and of mutually acknowledged value. The world's empires were linked by trade routes that skills and goods travelled along. From Spain in the west, trade routes skirted the Red Sea and went down the African coast. Others went from Africa to India, crisscrossed the Mediterranean and the Middle East, and went across the Asian steppes to China. Some even extended to the Pacific. Trade and military power were often linked, and wars were fought over raw materials and strategic routes.

**ROMAN HAND ABACUS**
*This method of calculating complex payment facilitated trade between individual community producers and also between empires.*

**SIGNS OF WEALTH**
*Early coins, such as these Chinese and Greek ones, were made from precious alloys and often carried a ruler's image to mark their value.*

**EARLY ROMAN SHIP**
*Trade benefited from better ships and navigational aids, which were in turn the result of increased wealth.*

**CLASSICAL CULTURE**
*The "golden age" of Greek classical culture is celebrated in Raphael's painting "School of Athens", which features Aristotle, Plato, Socrates, and Ptolemy, amongst others.*

1000BC | 900BC | 800BC | 700BC | 600BC | 500BC | 400BC

- **850BC** Homer believed to have created his *Iliad* and his *Odyssey*, Greece
- **776BC** First Olympic Games, Greece
- **753BC** Rome founded
- **650BC** Coinage invented in Greek Asia Minor
- **606BC** Assyrian empire destroyed by Babylonians
- **560BC** Buddha born, Nepal
- **539BC** Babylonian empire conquered by Persians
- **490BC** Battle of Marathon during the Persian–Greek wars
- **432BC** Parthenon completed, Athens
- **334BC** Alexander the Great, controller of Greece, invades Asia Minor

# EMERGENCE OF POLITICAL SOCIETY

The idea of a citizen's rights, responsibilities, freedoms, and duties developed in the city state of classical Athens under the stewardship of Pericles (about 495–429BC). Greek political thinkers, such as Plato (427–347BC) and Aristotle (384–322BC), went on to define not only how society was, but also how it ought to be. In *The Republic*, Plato set out his ideas for a utopian society. In *Politics*, Aristotle took a more scientific approach, comparing existing political states so as to formulate arguments for the best balance in the administration of public life. Together, they created and disseminated the rich language of modern politics, including concepts such as the citizen, the assembly, the republic, democracy, oligarchy, tyranny, and dictatorship. The vocabulary and framework of modern philosophy were similarly set out. In fact, the Greeks made no real distinction between political and moral philosophy, determining that political action and institutions and ethical conduct were all integral to life. At the same time in China, Confucius and his followers were also searching to understand the relationship between the individual and society.

**EASTERN PHILOSOPHY**
*Confucius (551–479BC) developed an influential view of how humanity should relate to the world, which was centred on duty to the state.*

**PERICLES**
*From the relatively short rule of this statesman, we derive many of our ideas about politics and public life.*

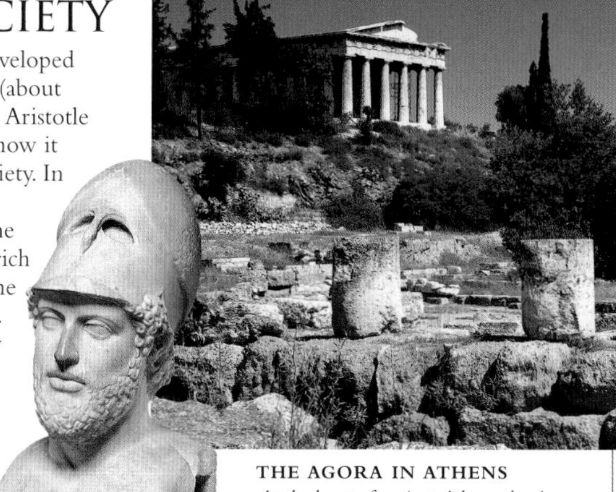

**THE AGORA IN ATHENS**
*At the heart of ancient Athens, the Agora (or "market") was the centre of the state justice system, political and commercial life, administration, and religious activity.*

# KNOWLEDGE AND CULTURE

A cultural revival around 800BC accompanied the emergence of city states across Greece. Greek culture gradually spread far beyond the boundaries of ancient Greece, influencing cultures all around the globe. Worldwide, scientific and technological developments comprised both abstract systems of knowledge and practical tools to facilitate life, agriculture, and trade – from Pythagoras' theorem, which defined the relationship of objects in space, to Archimedes' irrigation device and Chinese paper-making techniques. The desire for order in this period was also evident in an astronomy that sought not only harmony in the stars, but to provide for seafarers and travellers. Perhaps even more important for human history was the emergence of a new concern for humans themselves, a feature of classical Greece that Renaissance scholars would later seize upon. Although ancient civilizations intimately linked individual human life to a cosmic order, fate, destiny, and the will of the gods, the classical world came to believe that society was directed by humans. In-depth consideration of the individual and the consequences of his or her actions is at the very heart of what we understand as "modern".

**HUMAN FOCUS**
*Greek sculpture took the individual as an object of study, focusing on and celebrating the human form.*

**PRACTICAL INNOVATION**
*The remains of Roman underfloor heating systems (above) and those of piping systems show how advanced classical society was.*

Profile

## Claudius Ptolemy

The astronomer, mathematician, and geographer Ptolemy (AD87–150) lived in Alexandria, Egypt, a principal centre of Greek classical culture. Ptolemy rationalized the order and apparent movements of the planets known at the time. His elaborate patterns of astronomy were based on Aristotle's belief that the Earth was fixed, with the Sun, stars, and planets revolving around it. The Ptolemaic universe is now often derided because it works on the supposition that the Sun revolves around the Earth. However, it is incredible to think such a sophisticated idea came so early.

**PAPER-MAKING**
*The paper-making technique, which we could not live without today, originated in classical China in around AD105.*

**221BC** Shih Huang-ti of the Ch'in dynasty unites China
**30BC** Antony and Cleopatra die; Egypt annexed to Roman empire
**AD43** Start of Roman conquest of Britain
**AD105** Paper is invented in Han China
**AD228** End of Parthian empire
**AD300** Abacus in use, China

300BC | 200BC | 100BC | 0 | AD100 | AD200 | AD300

**247BC** Parthian empire emerges
**146BC** Rome conquers Greece and Carthage, becoming master of western Mediterranean
**4BC** Jesus is born
**AD79** Eruption of Vesuvius destroys the cities of Herculaneum and Pompeii
**AD230** End of Han dynasty in China: China first divides into three states, then fragments
**AD330** Constantinople (Byzantium) becomes capital of Roman empire

# WORLD AT A CROSSROADS

The period between the fall of Rome and the Renaissance is often termed the "Middle Ages". After a time of uncertainty and cultural loss, a period of creativity set in motion long-lasting developments based on the achievements of the classical world. This era saw the last great nomadic incursion against civilization, that of the Mongols, and the bubonic plague. There was a rapid expansion of Islamic culture, which cut the Chinese off from their western trade routes and forced Europe to turn west to reestablish trading links with the East. It was also through Islamic culture that the science, philosophy, and investigative outlook of the ancient world were renewed, developed, and dispersed, providing the foundations for the Renaissance.

**UNIVERSITY LIFE**
*Universities were founded in the new European centres of culture; this manuscript image shows a lecturer at Bologna University, one of the first.*

## THE SPREAD OF ISLAM

The scale and pace of Islamic expansion during the Middle Ages were unprecedented. When the prophet Muhammad died in 632, his authority extended only around Mecca and Medina, and perhaps to central and southern Arabia. Over the next century, Arab armies carried Islam as far west as Spain, and as far east as northern India and the Chinese border. The Arab armies emulated and surpassed Alexander the Great's achievements, and their influence lasted much longer. Their religious conviction swept aside the crumbling Byzantium (late Roman) and Persian empires. Yet the Arab expansion was not merely a military adventure underpinned by religious enthusiasm. It helped to sustain civilization, with learning and culture supported at a far more advanced level than in any other contemporary civilization, even that of the monasteries of Europe. Although the Islamic world soon lost its unity, it had an enormous influence on modern culture.

**A PROPHET'S BIRTH**
*This illustration depicts the birth of the prophet Muhammad in about 570. The religion he expounded was to have enormous global influence.*

**DIFFUSION OF CULTURE**
*The extent of Islam's spread is evident in southern European Islamic architecture, such as the Alhambra, Spain. Engineering, science, philosophy, and art flourished under Islam.*

## THE MONGOL INCURSION

Emerging from the depths of Asia, the Mongols were the last pastoralist and nomadic people to make a major and successful incursion against the settled civilizations of the world. In 1206, Genghis Khan (then a tribal chief) succeeded in uniting the Mongol tribes, gaining control of an army powerful enough to destroy the Ch'in empire in China. His successors went on to recast the map of the world. They advanced into Russia, swept through Persia's central Asian territories, inflicted the first significant defeats on the centres of Islam, and later invaded India. In 1241, Christian Europe was saved from invasion only by the death of the Great Khan, Maijke. Mongol power lasted until about 1405, but its social and cultural legacy is nothing compared with the achievements of the Arab diaspora. Despite their enormous military impact, Mongol influence survives only through their assimilation into the settled societies they conquered – in a dynasty of Chinese rulers and in the Moghul empire in the Indian subcontinent.

**GENGHIS KHAN**
*The Mongols' greatest leader was born around 1162. His death in 1227 did not halt the Mongol advance.*

**BRIDGING WORLDS**
*The period between the end of the Roman empire and the Renaissance was a bridge between the ancient and the modern, in which classical achievements were built on.*

| | |
|---|---|
| 400 | |
| 404 | Latin version of the Bible completed |
| 410 | Visigoths sack Rome, leading to collapse of Roman empire |
| 500 | |
| 600 | Windmills in use, Persia |
| 650 | Koran written, 18 years after death of the prophet Muhammad |
| 661 | Start of first Muslim dynasty, the Umayyads |
| 700 | |
| 730 | Paper-making spreads from China to Muslim world |
| 750 | Gunpowder invented, China |
| 760 | Arabs develop algebra and trigonometry |
| 800 | Printing with blocks developed, Japan |
| 850 | Collapse of classic Mayan culture in Mesoamerica |
| 900 | |

# THE IMPACT OF PRINTING TECHNOLOGY

The precursor to modern printing was xylography, the art of printing from carved wooden blocks. This process originated in China and began to appear in Europe during the last quarter of the 14th century. Printing technology developed very rapidly in 15th-century Europe. The advent of movable metal type was a revolutionary advance that marked the start of mass communications because it allowed texts to be set in relief much more quickly than when carved wooden blocks were used. The invention of typography, based on the use of movable type, is credited to Johannes Gutenberg in approximately 1450, and he produced the first printed bible approximately 6 years later. Mechanical printing was the gateway for an unprecedented diffusion of knowledge. Access to information was not only wider in terms of a greater quantity of printed materials, but also in the sense that books and pamphlets were now frequently presented in vernacular language. As the technology spread, so the price of books fell, with the result that more and more people learned to read. In a very real sense, the world expanded – or became more intelligible – for a larger number of people.

**GUTENBERG PRESS**
*This is the printing press on which the first typographic bible (left) was produced. Speed of printing made books cheaper, which led to the start of mass communications.*

**EASTERN PRINTING**
*This early Japanese wooden printing block shows that printing technology was developed the world over.*

**FIRST PRINTED BIBLE**
*The Gutenberg Bible was printed on the first typographic printing press (right) around 1456. From this time onwards, there was an avalanche of printed material.*

## GUNPOWDER "REVOLUTION"

The fierceness and stamina of the Mongol army had inspired shock and awe, but such primitive energy ceased to be a threat to better-established societies armed with gunpowder and firearms. Discovered in China, the formula for gunpowder was used to produce explosives at least from the time of the Sung dynasty (960–1279). The formula had clearly reached Europe by the second half of the 13th century because it is referred to several times in the work of the English scientist Roger Bacon. The first recorded evidence of using gunpowder in handguns is at the Italian town of Forli in 1284, and there are frequent references to siege guns employing the new form of explosive from the early 14th century. Such early devices did not make for much of a revolution in military tactics and strategy but, within a century, guns became much more effective. In fact, the technological superiority of the European armies was a decisive factor in the establishment of European domination and colonial power in the Americas.

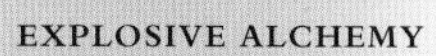

**EXPLOSIVE ALCHEMY**
*A 16th-century woodcut shows German monk and alchemist Berthold Scharz at work producing gunpowder in the early 14th century.*

Fact

### The Black Death

Carried by rats from China, bubonic plague caused the death of between a third and a half of Europe's population in the last half of the 14th century. It is likely to have arrived through a military incursion on Europe or on merchant ships. Surprisingly, the massive loss of life probably contributed to the rapid development of European domination in the world. As land became more abundant and labour scarcer, old feudal hierarchies became difficult to sustain, making cities more dynamic. It was easier for survivors to gain an economic surplus, and society in general became more at ease with change.

**EARLY GUNS**
*This type of cannon revolutionized military tactics once Europe had acquired the gunpowder formula.*

**1000** Camera obscura invented, Arabia; expansion of Inca empire, Peru; Vikings reach America

**984** Single or "flash" lock-gates used on canals, China

**1080** Magnetic compass invented, China

**1126** Arabian philosopher Averroes born

**1150** Completion of temple of Angkor Wat, Cambodia; paper-making reaches Europe

**1175** First Muslim empire in India

**1206** The Mongols under Genghis Khan begin their conquest of Asia

**1271** Marco Polo travels the Silk Road between Europe and China (until 1295)

**1341** The Black Death begins in Asia; goes on to ravage Europe in 1348

**1453** Ottoman Turks capture Constantinople and bring Byzantium empire to an end

**1492** Muslim rule ends in Spain

1000 | 1100 | 1200 | 1300 | 1400

# NEW HORIZONS

The period known as the Renaissance, which had begun around 1400, was a journey of discovery and debate that pushed back all known limits. From the late 15th century, scholars were looking at the world very differently, believing that humanity determined its own fate. Experimentation and observation became essential in all fields: artists observed their subjects and developed a sense of perspective; seafarers watched the stars and mapped the Earth; engineers solved life's practical problems; philosophers developed a new world order. Italy's city states played a critical role in this cultural revolution, but by the late 1500s, as European attention moved towards transatlantic expansion, Spain and France were leading powers; Britain was waiting in the wings.

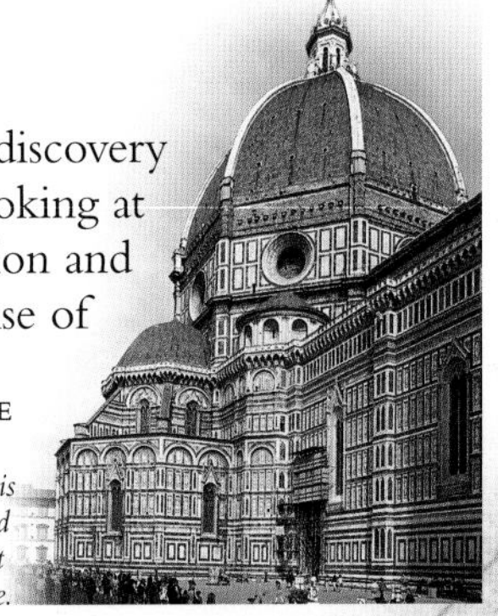

**AHEAD OF ITS TIME**
*Brunelleschi's dome (built 1419–1436) in Florence, is a miracle of engineering and artistic imagination that set the tone for the Renaissance.*

## SCIENTIFIC ADVANCEMENT

Like the Greeks, Renaissance scientists delighted in knowledge, but they sought to apply it more systematically. Arabic science provided the tools and methods to understand the world, and mathematical measurement and calculation became the basis of all science. Scientists moulded a relationship between mathematics, astronomy, and navigation, and used their knowledge of the planets and mapping methods to chart new lands. Wealth from this "new world" helped the scientific revolution to spread more widely. Astronomical research was severely restricted, however, because the Catholic Church declared it heretical to question the Earth's place at the centre of the universe. Yet both Copernicus (1473–1543) and Galileo did just that, showing that the Earth was spherical and in uniform motion around the Sun. They removed science from the sphere of Church authority.

**COPERNICAN WORLD SYSTEM**
*Copernicus revolutionized astronomy, showing that the Earth rotated daily and moved around the Sun once a year.*

**GALILEO'S TELESCOPES**
*Observation of the planets led Galileo (1564–1642) to argue for separating theology and science.*

## EXPLORATION AND DISCOVERY

The voyages of discovery for which the Renaissance is renowned were inspired more by the pursuit of trade and wealth than knowledge or adventure. Turkish domination of the Middle East had cut off European access to East Indian spices, so European monarchs hired navigators to find new routes to reestablish this trade. These voyagers discovered lands that fired European greed and imagination. Initially the voyages were tentative because sailors had trained in the Mediterranean where they were never far from a landfall: early navigators such as Vasco da Gama set out south for the Indies, skirting Africa and the Cape of Good Hope. Although this route proved valuable, more intrepid crews (including those of Christopher Columbus and Ferdinand Magellan) steered ships west. Sailors navigated using the stars and a compass, and charted their passage on a quadrant so they could follow it again. Such devices were the raw material of map-making. Europe's future lay to the west, but its exploration had a detrimental impact on the native populations of the "New World".

**COLUMBUS**
*Italian sailor Christopher Columbus (1451–1506) believed he could reach China by sailing across the Atlantic Ocean. He persuaded the rulers of Spain to back him.*

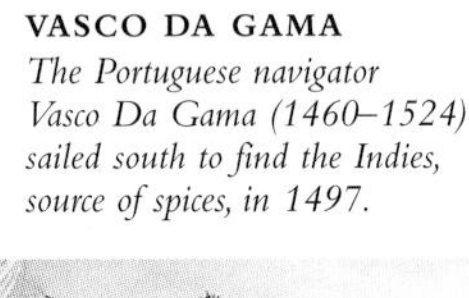

**VASCO DA GAMA**
*The Portuguese navigator Vasco Da Gama (1460–1524) sailed south to find the Indies, source of spices, in 1497.*

**PILGRIM FATHERS**
*After the discovery of new territory, many Europeans moved to colonize it. This image shows passengers travelling to America on board the* Mayflower.

Fact

### The Golden Triangle

The transatlantic slave trade, known as the Golden Triangle, began in 1441 when a Portuguese sailor seized 10 Africans to take to the Americas. In 1454, Pope Nicolas V officially started the trade. Merchant ships would leave Europe for West Africa laden with manufactured goods. In West Africa, these were exchanged for slaves from the interior who were transported to the Americas. Those who survived the journey were then sold to plantation owners in exchange for valuable produce, such as cotton, tobacco, molasses, and rum.

**1493** Columbus reaches America

**1497** Vasco da Gama sets off for the Indies from Lisbon

**1500**

**1519** Cortés arrives in Mexico; Aztec empire collapses

**1520** Chocolate introduced to Spain from the Americas

**1528** Mombasa revolts against Portuguese rule

**1530** Coal mining begins in Europe

**1540**

**1545** Discovery of silver at Potosí, Bolivia

**1550** Beijing is besieged by the Mongols for a week

**1560** Reunification begins in Japan, a land of warring nobles

**1570** Potato introduced into Europe from the Americas

**1580**

**1588** Famine and pestilence sweep China

**THE FIRST "UTOPIA"**
*Thomas More coined the term "utopia", in his 1516 book, to refer to a desirable yet unattainable place. The book linked the voyages of discovery to journeys of the mind.*

# A CHANGING PHILOSOPHY

Renaissance thinkers, scientists, and artists went on an even greater voyage of discovery than that of the explorers – to learn about mankind and push back the limits of humanity. A new feeling of self-determinism arose, with philosophers stating that humans were not simply pawns in a preordained cosmic order. This humanist stance became increasingly widespread and led to a new urgency and quality of artistic production, and to increased interest in scientific and geographical exploration. To the Greeks and Romans, chance had referred to the will of the gods. The Renaissance concept of chance, in contrast, held that a person made his or her own destiny by seizing opportunities and taking decisive action, which brought about a new sense of optimism and hope for the future. It is difficult to date this shift in perspective but, in the space of 350 years between 1350 and 1700, an enormous transformation in social outlook and philosophy did come about. Geographical exploration had opened up new vistas in the physical world and corresponding horizons in the human mind. The presence of God was still important and remained beyond the limits of human understanding, but religion was increasingly held open to enquiry and analysis.

**DESIDERIUS ERASMUS**
*Contemporary and confidant of Thomas More, Erasmus (c1469–1536) is considered prince of the humanists. He expounded the view that man was his own maker.*

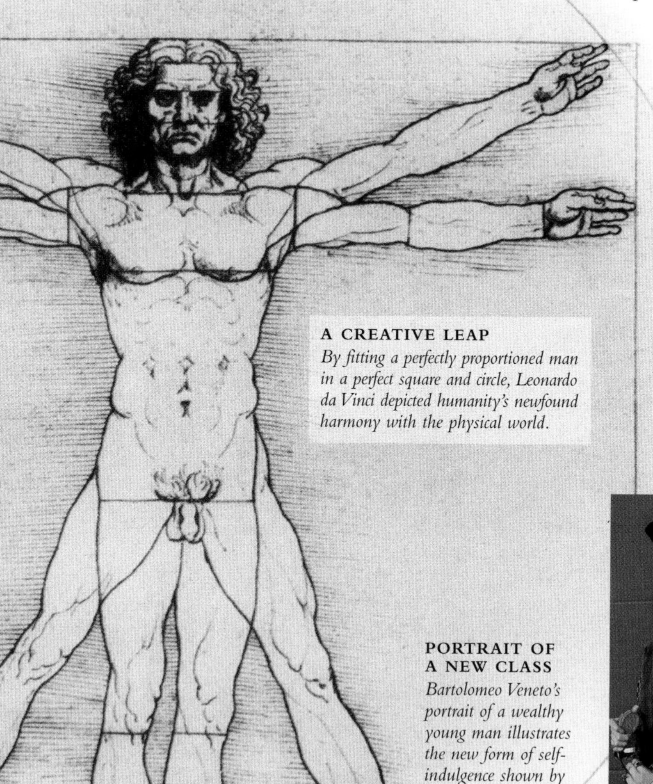

**A CREATIVE LEAP**
*By fitting a perfectly proportioned man in a perfect square and circle, Leonardo da Vinci depicted humanity's newfound harmony with the physical world.*

# THE ARTS

Although the social order was still prescribed and social mobility remained extremely limited, advances in trade and economic production, the growth of towns, and the decay of feudal hierarchies allowed a new concern for the individual to surface. A new urban elite developed, and these wealthy people not only prospered but reflected upon themselves and were proud of their own achievements. Therefore, as patrons of the arts and celebrators of their own success, members of the new elite were just as likely to commission a portrait of themselves as yet another one of the Virgin Mary. Such elites also owned their own books and, although they were certainly proud of their classical learning, these books were frequently written and published in common, everyday language rather than in Latin. Thanks in part to the earlier influences of Dante, Chaucer, and the printing press, and now – especially in England – to Shakespeare, the rich variety of human experience began to be communicated in the concentrated forms of language that we continue to use today all around the world. At the same time, theatre-going became an outlet for the hopes and fears of generations of Europeans from a wide variety of socially diverse backgrounds.

**PORTRAIT OF A NEW CLASS**
*Bartolomeo Veneto's portrait of a wealthy young man illustrates the new form of self-indulgence shown by moneyed individuals.*

Profile

## William Shakespeare

The playwright Shakespeare (1564–1616) was born in Stratford-upon-Avon, England. He wrote his plays, such as *Hamlet* and *Romeo and Juliet*, during the Elizabethan age, when English society was unusually dynamic. Trade, discovery, and innovation had transformed a static medieval world and, in Shakespeare's work, there is a sense both of a vanished past and the exhilaration of discovery. The form and content of his drama reached out to and touched a socially disparate audience as never before. Shakespeare possessed a wide vocabulary and even invented words now in common usage in the English language.

**SISTINE CHAPEL**
*Pope Julius II was a liberal patron of the arts and in 1508 asked Michelangelo to paint the ceiling of the Vatican's Sistine Chapel.*

1600 | 1620 | 1640 | 1660 | 1680

- **1600** British East India Company founded
- **1602** Dutch East India Company founded
- **1608** Telescope invented by Hans Lippershey, Holland
- **1630** Circulation of blood discovered by English physician William Harvey
- **1644** Qing dynasty, last of Chinese imperial dynasties, established
- **1648** Construction of Taj Mahal completed after 16 years
- **1655** Pendulum clock invented by Christian Huygens, Holland
- **1660** Gujaratis make earliest known Indian nautical charts
- **1682** English astronomer Edmund Halley observes Halley's comet
- **1688** William Dampier is the first Englishman to visit Australia

# A CHANGING WORLD

Agricultural and industrial revolutions brought rapid developments in human technical capacity, productivity, and the mass availability of goods, yet such technological advances were outstripped by developments in the political and social imagination. By 1815, the UK (formed 1801) had taken centre stage in the world scene, and optimism and a sense of progress prevailed in Europe. Within 100 years, Germany and the US had challenged British hegemony, and the world had changed immeasurably. Millions died in two world wars, empires collapsed, and the new power centres (the US and USSR) lived in ideological conflict. Political turmoil, economic instability, and human devastation continued through the 20th century, and there is, today, a renewed sense of unease.

Fact

### The Lisbon earthquake

On 1 November 1755, the populous city of Lisbon was destroyed by an earthquake. The event stimulated philosophical debate about God, fate, and human intervention. Jesuits saw it as God's retribution for human sins; others, as a powerful argument against His existence. Some said that people could avoid such disasters by moving from large cities. Other consequences were also significant: the Portuguese dictator rejected the Jesuits' argument and cut them off from influence, a precursor to the rapid secularization of European governments after the French Revolution.

## ENLIGHTENMENT AND UNCERTAINTY

The 18th-century "Enlightenment" was founded on the belief that by understanding life, the world, and its social institutions, the human condition could be improved. Until this era, people had either looked back to the Golden Ages of classical Greece and Rome or settled for a better life in the next world. Now, greater material prosperity and an escape from the worst ravages of disease brought new optimism. This quest for knowledge and scientific order continued into the 19th century, but the findings began to unsettle society. For example, Charles Darwin (1809–1882) revolutionized the way humanity thought about itself and challenged long-held beliefs in the measured hand of a divine creator. Psychologist Sigmund Freud (1856–1939) responded to the social and spiritual upheavals unleashed by scientists such as Darwin and political philosophers like Karl Marx (1818–1883). Freud noted that nothing escaped humankind's criticism or resentment, and he believed that the human capacity for destruction outweighed its creativity. After the Enlightenment's optimism, a sense of uncertainty and the need to "play it safe" had set in – despite humanity's ever-increasing ability to understand and shape the world.

**DIDEROT'S ENCYCLOPEDIE**
*French philosopher Denis Diderot compiled this vast record of the Enlightenment between 1745 and 1772. It aimed to represent all knowledge and, through this, combat fear.*

## POLITICAL REVOLUTION

The 1776 US Declaration of Independence stated that people had the right to abolish any government destructive of life, liberty, and the pursuit of happiness. The French Revolution of 1789 expounded an even more radical vision of human rights, based on a belief that people are inherently free and equal. Everyone should have the right to participate in the legislative process, to make leaders accountable, and to live without fear of arbitrary arrest and punishment. They should also enjoy freedom of speech, opinion, and religion, and equality of taxation. A new political language and framework was born.

In February 1792, the revolutionary French government issued a general emancipation of slaves, but ruthless ordinances against vagabondage kept labour on the plantations. In social and economic terms, therefore, little had changed. The French never espoused equal rights for women, but the issue could not be kept off the new political agenda. In 1792, Mary Wollstonecraft's *Rights of Women* articulated the place of the forgotten 50 per cent of society in the struggle for emancipation. In *The Communist Manifesto* of 1858, Karl Marx had already set out the possibility of humanity seizing control of its own destiny and building a new, liberated social order.

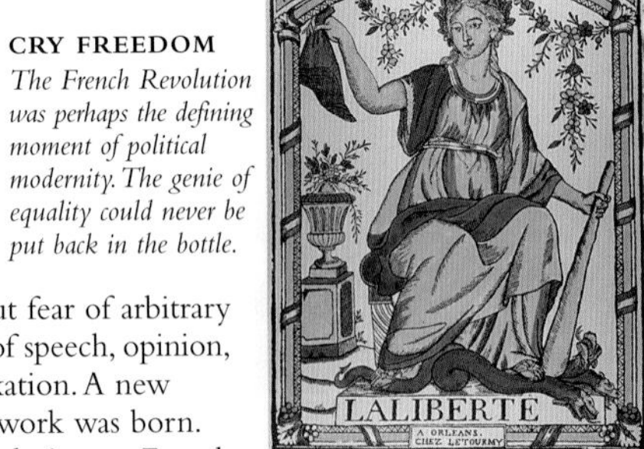
LALIBERTE

**CRY FREEDOM**
*The French Revolution was perhaps the defining moment of political modernity. The genie of equality could never be put back in the bottle.*

VOTES FOR WOMEN ON THE SAME TERMS AS MEN
VOTES FOR WOMEN ON THE SAME TERMS AS MEN

**EQUAL RIGHTS**
*Towards the end of the 19th century, suffragettes famously led by Emmeline Pankhurst took up the cause of freedom for women with a passion.*

- **1700**
- **1712** Early slave revolt in North America
- **1717** Britain, France, and the Netherlands form an alliance to contain expansionist plans of Spain
- **1720**
- **1724** German philosopher Emmanuel Kant born
- **1736** Rubber introduced to Europe from Central America
- **1740**
- **1760**
- **1764** James Hargreaves invents "spinning jenny", England
- **1770** Luigi Galvani and Alessandro Volta make electricity from chemicals, Italy
- **1776** US Declaration of Independence
- **1780**
- **1783** Russia annexes Crimea
- **1800**
- **1803** US acquires all French territory between Mississippi and the Rockies

# ECONOMIC REVOLUTION

In the late 17th and early 18th century, technological changes in farm machinery and experimentation in crop rotation and plant breeding stimulated agricultural productivity. Innovators such as Jethro Tull (1674–1741), famous for his seed drill, were practical people whose passion for innovation owed as much to a desire to commercialize food production as to the new spirit of enquiry. In the late 18th century, the same trend occurred in manufacturing. The Industrial Revolution began in England and spread to Europe, the US, and Japan. New technology and power sources were used in the textile industry. New spinning machines, and the invention of the steam engine, transformed Britain into the workshop of the world. Such technical changes were just part of a much wider transformation of trade, economic relations, and division of labour. In 1851, the Great Exhibition was held in London to show off the newly formed United Kingdom's industrial, military, and economic superiority. Yet the world's leading force soon faced competition. All the great powers were establishing strong financial institutions, and traders around the world were gaining access to the best markets and enjoying the security of operating through contracts.

**STEPHENSON'S ROCKET**
*This groundbreaking design was developed in 1829, 4 years after the first passenger steam train was introduced. Rail travel transformed society and made goods far easier to transport.*

**IRON BRIDGE**
*Built in 1779 in Shropshire, the world's first iron bridge dominates an English region that was transformed by iron, steel, and pottery works.*

Profile

## Simón Bolívar

Born in Venezuela, Simón Bolívar (1783–1830) liberated much of South and Central America from Spanish rule. "El Libertador" conquered the armies of Panama, Peru, Colombia, Ecuador, Bolivia, and Venezuela, and established presidential rule. Bolívar's dream was to unite Spanish America into one republic, but his autocratic rule led to dissent and his "empire" soon crumbled. Bolivia, the country, and the Venezuelan currency (the Bolivar) are still named after him today.

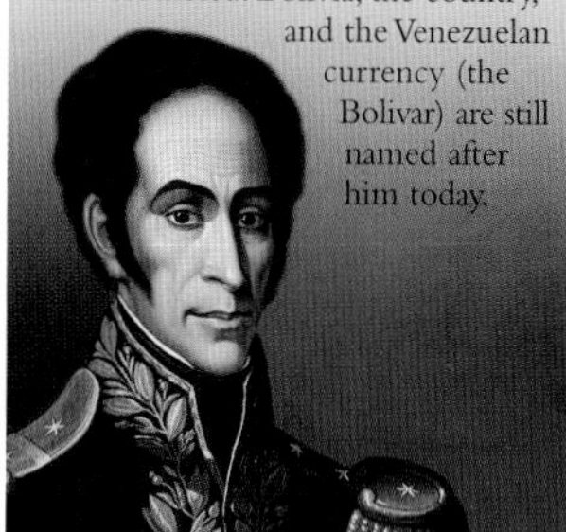

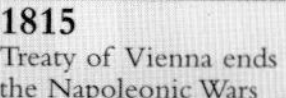

**HEIGHT OF PROGRESS**
*Technological innovation and changes in trade, economic relationships, and division of labour made possible the UK's Great Exhibition of 1851.*

# SHIFTING POWER

The reassertion of European monarchies at the 1815 Congress of Vienna quelled aspirations towards national liberty. The war for Greek independence (1820s) was an exception that succeeded in stirring the liberal conscience in Europe – probably because it did not threaten the new European order. A year of revolutions in 1848 again crushed ambitions for national autonomy, but the Italian struggle for independence attracted sympathy and unity was achieved in 1870. Chancellor Bismarck put Prussia at the core of a German empire that covered central Europe by the late 19th century. The major European powers fought for empires to guarantee their raw materials, economic markets, and strategic routes. The "Scramble for Africa" began in earnest in 1885, and within 15 years, the map of Africa was redrawn. The UK no longer dominated global trade and investment: the US had its own growing capitalist class and zone of influence, and Germany started to outpace the UK in iron and steel production. The search for a competitive edge soon took a more military form. Russia, the UK, and Germany each dominated an area of China but were, towards the end of the 19th century, threatened by the presence of the US and a modernized Japan. By 1918, the US brooked no interference from war-devastated Europe, but was not yet powerful enough to stabilize and rebuild the continent. A power vacuum was left at the centre of Europe after Germany had been defeated and Tsarist Russia (guardian of the old order) had become the Soviet Union.

**CARVING IT UP**
*The UK and France divide up the world in this political cartoon. The scale of colonial expansion and the rivalry between the world's leading powers are the crucial features of this era.*

**WORLD WAR I**
*The "Great War" was totally catastrophic for Europe, not only in terms of lives lost but also economies destroyed. It left a patchwork of small nation states beset with ethnic rivalries.*

**1810** Start of revolutions in Spanish America; by 1826, all Spain's American colonies are independent

**1815** Treaty of Vienna ends the Napoleonic Wars

**1820**

**1834** End of Spanish Inquisition (began 1478)

**1839** Opium wars begin in China, when Hong Kong is ceded to the UK

**1840**

**1858** Darwin publishes *The Origin of Species*, UK

**1860**

**1861** American Civil War begins (ends 1865)

**1877** Famine starts in northern China; kills 10 million in 2 years

**1880**

**1890** First underground railway opens, UK

**1896** First modern Olympics, Greece

**1900**

**1904** Russo–Japanese war reinforces Japan's dominance

**1914** World War I starts (ends 1918)

**1915** Einstein's Theory of Relativity formulated, UK

# THE GREAT DEPRESSION

A brief economic boom in the 1920s had little impact worldwide. Then, stock market prices collapsed in the Wall Street Crash of October 1929, and world trade actually shrank. In the resulting chaos, investors withdrew money, companies went bankrupt, and trade crumbled. Economic stagnation and mass unemployment brought hardship and suffering not just to the developed economies, but across the globe. In the US (the world's largest economy), 13 million were unemployed by 1931. Yet the impact was most disruptive in Germany, a society not yet recovered from military defeat in 1918 and revolutionary upheaval in the early 1920s. German economic meltdown provided the perfect context for the National Socialist (Nazi) Party not only to gain power but also to maintain it. Hitler's ruthless desire to carve out a preeminent role for his Third Reich in a new world order was the catalyst for barbarism on a scale previously unknown.

**SOUP KITCHENS**
*Germany suffered a cataclysmic economic collapse following the Wall Street Crash, reducing many people to poverty.*

**A FATEFUL DAY**
*Americans took to the streets in panic after the 1929 Wall Street Crash. They withdrew money from investments, companies went bankrupt, and unemployment skyrocketed.*

# WORLD WAR II

Within months of becoming German Chancellor in 1933, Hitler opened the first concentration camp. Initially intended for political opponents, these camps became repositories for non-Aryans, especially Jews. The Nuremberg Race Laws formalized longstanding anti-Semitism in 1935, and Jewish persecution culminated in the agreement of the Final Solution in 1942. Genocide was executed with bureaucratic efficiency. The 1939 German invasion of Poland had reopened hostilities between the Great Powers. Unlike in 1914, Germany brushed aside the French army and expelled Britain from mainland Europe. Despite failing in the Battle of Britain (1940), the German High Command risked a Soviet invasion in June 1941. That December, the war became global when Japan bombed Pearl Harbor naval base and the US joined forces with the USSR, the UK, and the colonies. The Soviet Red Army and US economic power crushed Hitler from both sides. War ended in August 1945, with the atomic annihilation of two Japanese cities.

**GENOCIDE**
*In 1941, Jews were told to wear yellow stars. By 1945, 6 million had been killed in Europe.*

**BATTLE OF STALINGRAD**
*The heroic efforts of the Red Army and the Soviet people helped to ensure German defeat. This battle was the turning point of the war.*

**ATOMIC BOMBS**
*The Hiroshima bomb killed tens of thousands instantaneously; radiation accounted for 140,000 lives. The iconic mushroom cloud remains a potent sign of what humanity can do to itself and the planet.*

Profile

## Nelson Mandela

From 1948, South Africa's Afrikaner government enforced a policy of total discrimination against South African blacks. Nelson Mandela (1918–) was a leading figure in the African National Congress (ANC) during its long-term campaign against the government. Imprisoned for over 27 years, Mandela became a role model for the oppressed. He was eventually released in 1990 and became President of the first ANC government in 1994.

# ALTERNATIVES TO CAPITALISM?

A new world order came about with the UN in October 1945. Cold War rivalry between the US, the dominant capitalist power, and the USSR, the dominant communist power, formed the framework of international relations for 40 years to come. Mao's China, outside the main world economy and increasingly distant from the USSR, offered another model for the developing world. The framework of US–Soviet relations regulated international relations. Despite major political crises, such as the Cuban missile crisis in 1962, direct conflict was avoided, and the world economy saw unprecedented expansion and success. Previous capitalist failings, such as economic depression and war, gave credibility to the Soviet model, and many colonial people struggling to build independent states adopted its structures and language. In the 1960s and '70s, it looked like this cultural conflict might lead to a total overhaul of Western society. Although this did not happen, hierarchies were eroded and the role of women changed. Communism fell with the Berlin Wall in 1989, but far from bringing security, the end of the Cold War has made many people feel more insecure.

**BREAKING BARRIERS**
*The tearing down of the Berlin Wall and collapse of the USSR have not brought about the prosperity or political stability anticipated.*

1920 | 1930 | 1940 | 1950 | 1960

- **1925** Frozen foods developed, US
- **1928** First antibiotic, penicillin, discovered by Alexander Fleming, UK
- **1929** Wall Street Crash and start of World Depression
- **1934** Long March begins, China
- **1939** World War II begins (ends 1945)
- **1945** Atomic bombs dropped on Hiroshima and Nagasaki, Japan
- **1947** Mahatma Gandhi achieves Indian independence
- **1949** Apartheid begins in South Africa; Mao declares People's Republic of China
- **1960** First female prime minister elected, Ceylon (now Sri Lanka)
- **1961** Berlin Wall erected

Issue

## Global insecurity

Throughout the history of different groups living in close proximity, there has always been a sense of insecurity, however vague. Since the Cold War, many of the more obvious global threats have disappeared, but insecurity remains. Conflict is increasingly defined in terms of terrorism, although this may be no more common today than in the past. Why terrorism has had a huge impact on our collective psyche is debatable – perhaps because it is an ill-defined threat and is carried out by ordinary people.

# A GLOBAL ECONOMY

The increasing tendency towards greater economic, social, and technological exchange between countries is known as globalization. Through our technical abilities, we have reduced the size of the world, and this has resulted in wide-ranging economic and cultural changes. The resulting "global village" is unevenly developed and unevenly welcomed. In protest at the increasing emphasis on commerce and profits, many people have adopted new forms of ideology, such as environmentalism, and some question the need to pursue global economic models. Paradoxically, these very environmental concerns have become a new market opportunity for global companies. Globalizers and antiglobalizers do not conform to the old ideological divisions of left and right. Globalizers may simply support capitalist growth or may believe that Western intervention is necessary to bring about global democracy. Equally, they may be rural workers in the developing world who want a road, a dam, or a better strain of wheat. Some antiglobalizers strive to reform or abolish capitalism and US dominance, preserve the planet, or help its more deprived inhabitants. Others are conservatives, wanting to preserve tradition or their own embattled economic and political status. Whether the trend towards globalization is inevitable or desirable will be debated for some time to come.

**ONE WORLD?**
*Some hope and others fear that the spread of Western culture will result in a homogenous global life experience.*

**WORLD OF CONTRASTS**
*Global markets tend to mean abundant choice in Western countries, where supersized meals cause obesity. At the same time, developing countries, such as the Sudan, are unable to feed their populations.*

**ENVIRONMENTAL DEVASTATION**
*Many claim deforestation occurs as a direct result of globalization, for example from logging by timber companies. Yet some local farmers also clear land.*

Fact

## Movement of peoples

Increasing prosperity in the developed world and economic and political problems across the developing world have led many, such as the African immigrants below, to seek new lives in the West. This has led to a new trade in "people smuggling" and growing resentment in developed countries. The perception, however false, that immigrants and asylum seekers gain more benefits than some of the less-advantaged sections of Western society has increased support for rightwing groups through the developed world.

**MODIFYING NATURE**
*We now have the coveted power to alter nature. We can even genetically modify plants and animals – for example, this GM cotton.*

# TECHNOLOGICAL LEAPS

The 1990s opened with the hope that a new period of stability and progress had begun, but it did not materialize. Debate about humanity's future centres on globalization, US power, the communication opportunities of the internet, and on the scientific possibilities of human intervention in genetic structure. The internet is a recent innovation: in 1989, Tim Berners-Lee produced an internal paper for his European company and a year later received the go-ahead from his boss to write a global hypertext system. Within 6 months he had created the *WorldWideWeb*. The use and size of the Web has grown exponentially. It contains a wealth of accessible data, although people in less-developed regions cannot afford to tap into it. Cheaper travel and mobile phones have also helped to shrink the world. Cultivation of a random mutation of grass led to the advent of bread wheat around 10,000 years ago, a technological leap that seems insignificant now that we are in the process of trying to identify all of our 30,000 genes and determine the sequence of 3 billion chemical base pairs that make up human DNA. We now have a more developed capacity to transform nature than ever before. However, in some ways we may have come full circle in our evolution – whether we allow our newfound power to bring progress or self-destruction is crucial to our future as a species.

**SPACE SHUTTLE**
*Cheap, accessible travel has transformed life for the rich; travel to other planets is even a theoretical possibility.*

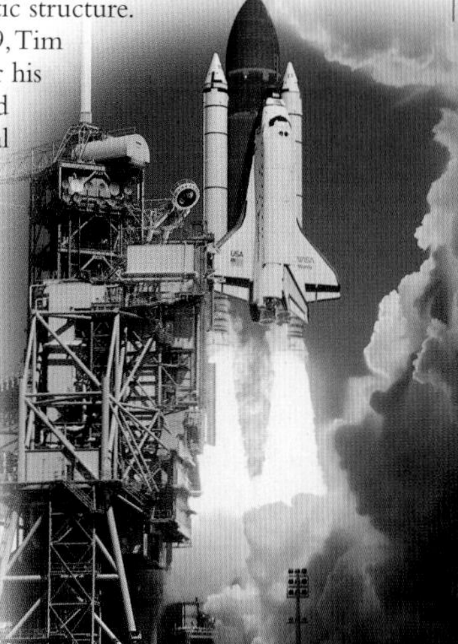

**1963** Martin Luther King delivers his "I have a dream" speech, US; Nelson Mandela imprisoned, South Africa

**1964** *Shinkansen* high-speed train invented, Japan

**1969** First moon landing

1970

**1976** Concorde's first flights from London and Paris

**1978** First test-tube baby born, UK

1980

**1981** First AIDS cases, US

**1986** Soviet nuclear reactor explodes, Chernobyl

**1989** Tiananmen Square demonstrations; fall of Berlin Wall

1990

**1990** Black political parties legalized, South Africa

**1997** Hong Kong returned to China

2000

**2003** Human Genome Project completed after 3 years

# BODY

# BODY

Human bodies come in many shapes and sizes. However, with a few differences between the sexes, we are all built according to the same anatomical blueprint. Strip away the skin, of whatever colour, and there will be the same overlapping layers of muscles in the same places. Delve deeper, and the bones are still unmistakably human: the long spine that keeps us standing more upright than other animals; the intricately jointed bones of the hands and feet, and the skull with its flattened frontal planes. One person's internal organs – such as the brain, heart, and intestines – look much like another's, and work to the same internal rhythms.

**START OF LIFE**
*Sperm cluster around an egg cell, trying to penetrate it. If one succeeds, it will fuse with the egg, possibly creating a new human.*



We are part of the animal kingdom and belong to a group of animals known as primates, which includes apes (our closest relatives) and monkeys. Primates in turn belong to a larger group of animals – the mammals. In order to understand how humans work, it is useful to see what characteristics we share with other group members and in what important ways we differ.

## HUMAN TRAITS

Humans possess the common mammalian characteristics of warm blood and body hair. Together with nearly all mammals, we give birth to live young and produce milk for them. As primates, we are not alone in being able to hold our upper bodies erect (although our firm balance on two legs is unrivalled). The structure and arrangement of the bones in our limbs are similar to those in many of our fellow primates. Like us, other primates have eyes at the front of their faces and stereoscopic vision, and many primates see in colour just as well as we can. However, the human body, which took a few million years to evolve once we branched away from our closest relatives, does have a number of important distinctions. Human hands are specialized for many tasks, such as the skilled use of sophisticated tools; and human feet are modified to bear the entire weight of the moving body. Most significant of all is the human brain, which went on growing during its evolution and is much bigger in proportion to body size than in any comparable species.

**HEART FUNCTION**
*Unlike the ancient physicians, we now know how blood is pumped through the heart.*

## UNDERSTANDING HUMAN ANATOMY

From the earliest days of civilized society, humans have been curious about what makes their bodies work. For some thousands of years, cutting up bodies to see what goes on inside was severely restricted by the taboos of religion and culture. Theories on human anatomy were, for the most part, based on the dissection of animals and a lot of conjecture. Some early scientists got it surprisingly right, although their ideas were not necessarily accepted at the time. As long ago as 500BC, the Greek physician Alcmaeon suggested that the brain was the seat of our thoughts and emotions. On the other hand, mistaken ideas were rife, most famously those of Galen, a Greek-born doctor who practised in Rome during the 2nd century AD. Galen's beliefs that blood is made in the liver and that a flame burning in the heart keeps us alive were just two of the theories that stayed unchallenged for centuries. Not until scientific knowledge took a leap forward in the Renaissance were some of the truths about the human body revealed. The observations of the Flemish doctor Andreas Vesalius (1514–1564), who published the first accurate illustrations of human anatomy, were the start of a new age of understanding. With each century, fresh discoveries were made. In 1628, William Harvey proved that blood circulates around the body; in 1775, the French chemist Antoine Lavoisier found out that cells use oxygen for fuel; British surgeon William Bowman first described the function and structure of the kidney in 1842; hormones were discovered in the early 1900s. So, gradually, the pieces of the human jigsaw were slotted into place. The growth of knowledge about the body accelerated as the 20th century passed, and is continuing unabated in the 21st century.

## THE MODERN BODY

In countries where there is an adequate food supply and good standards of basic hygiene, the average human body is taller and longer-lasting than it was in previous centuries. With a

**TESTING THE BODY**
*Like this young break-dancer, humans are constantly testing their strength and agility, pushing the body to the limits of its capabilities.*

History

### Anatomy lesson

Dissections of dead bodies were once carried out more as elaborate stage performances than as anatomy lessons. Many of the audience were merely curious onlookers rather than medical students. Until the 19th century, the bodies used were nearly always those of executed criminals. In this 1751 engraving, the English artist William Hogarth caricatures a public demonstration of the anatomist's craft.

**DEFENCE CELLS**
*These cells are lymphocytes, which belong to the body's disease-fighting system and are found mainly in the blood. They have just formed by division of a single cell.*

better understanding of health and body functioning, people are learning how to keep themselves in good order over a lifetime. Although humans in modern society tend to lead sedentary lives, we have thought up ways of meeting the body's need for regular physical activity. Many communities offer a wide choice of sports and exercise facilities; and some people enjoy the challenge of exploring the amazing capacity the human body has for physical improvement. The limits to which the body can be pushed are being extended all the time. Every year, new records of speed, strength, and stamina are set by athletes, gymnasts, and sportspeople. These achievements demonstrate the body's potential for increased lung power, more efficient use of the heart, stronger muscles, and greater flexibility. The modern body is under our control in other ways. If we are not satisfied with our appearance or shape, there are huge industries devoted to making the body thinner, more attractive, or younger-looking. We can even limit our fertility, making informed choices about whether or not to produce children. At the same time, we can also improve our chances of creating new life.

## VULNERABLE BODIES

Although the body has an efficient inbuilt disease-fighting system, from time to time it needs help. The enormous range of drugs available in the developed world provides invaluable back-up protection and can often cure diseases that get past our natural defences.

Modern medicine has wiped out or greatly reduced diseases such as smallpox or polio that once killed people in their millions worldwide. It is possible, too, to replace some body parts, such as the heart, liver, and joints, when the originals are diseased, damaged or worn out. Despite these great advances, the human body is still vulnerable. Old threats may have been overcome, but we constantly face new ones, some of our own making. In all affluent societies, illnesses caused by overeating and inactivity are increasing. We create chemicals to destroy diseases in the food-chain and to boost food production, but some of these chemicals could be gradually poisoning our bodies. In some areas, environmental pollution is thought to be a contributing factor in the increase of cancers and birth defects. Infectious organisms are becoming resistant to drugs, and are more dangerous than before. New diseases, such as HIV/AIDS and SARS (severe acute respiratory syndrome), are emerging to which we have no natural resistance. The scale and speed of international travel means that once-localized diseases are carried to new areas within a few hours.

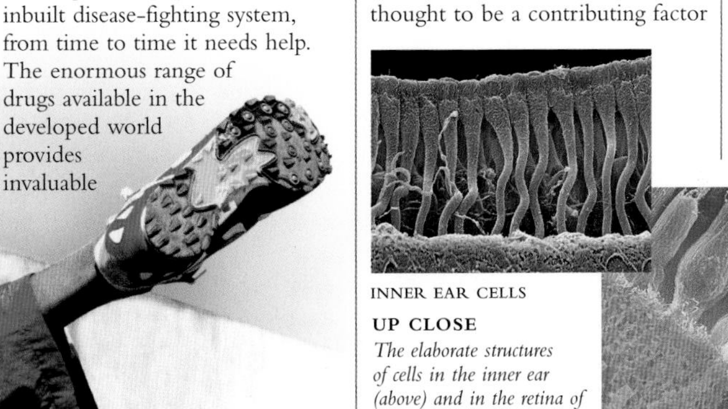

INNER EAR CELLS

RETINAL CELLS

**UP CLOSE**
*The elaborate structures of cells in the inner ear (above) and in the retina of the eye (right) are revealed by an electron microscope.*

## NEW DISCOVERIES

Doctors can now look at the innermost recesses of the living body without cutting it open. There are techniques for viewing the body that were unimaginable a few decades ago. For many years X-rays, discovered in 1895, were the only way to see inside the body without surgery – and only hard tissue, mainly bone, was shown. Newer imaging techniques enable soft tissues such as muscle to be seen, and can highlight hollow organs. It is also possible to look at tissue activity, to find out how well an organ is functioning (*see* Looking inside the body, below). Another modern technique is ultrasound, which uses sound waves and is a safe method for monitoring babies in the womb. Major advances have produced 4-D ultrasound, which constructs moving images. The examination of hollow organs is often carried out by endoscopy, in which a tube-like optical instrument is inserted through a body opening, such as the mouth or anus. One of the most exciting inventions is the electron microscope, which is used for examining samples of body tissue in minute detail. With a magnifying power thousands of times greater than an optical microscope, it facilitates the study of the internal structure of any cell in the body.

Fact

### Looking inside the body

Modern imaging techniques provide reliable pictorial information about the inner body. In computerized tomography (CT) scanning a series of X-rays passes through the body, producing cross-sectional images ("slices"), which are displayed on a monitor. MRI scanning uses radio waves combined with magnets to pick up tiny signals from atoms in the body. These signals are built into images by a computer. Positron emission tomography (PET) scanning is a technique that detects the body's uptake of injected radioactive substances. PET produces colour-coded images showing cell activity.

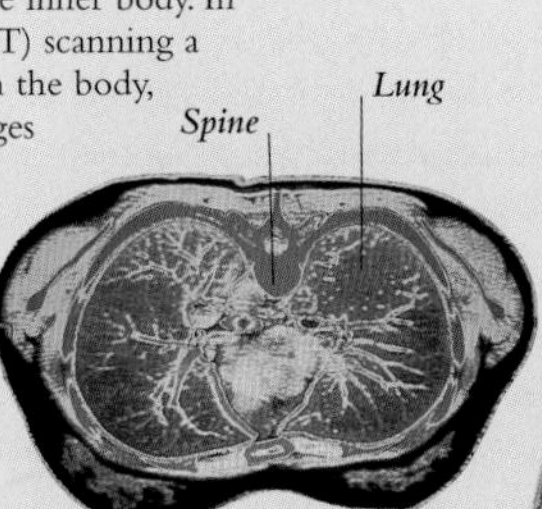


**CT SCAN**
*This "slice" is a cross-section of the chest, which shows tissues of different densities, including the lungs (red) and bones (green).*

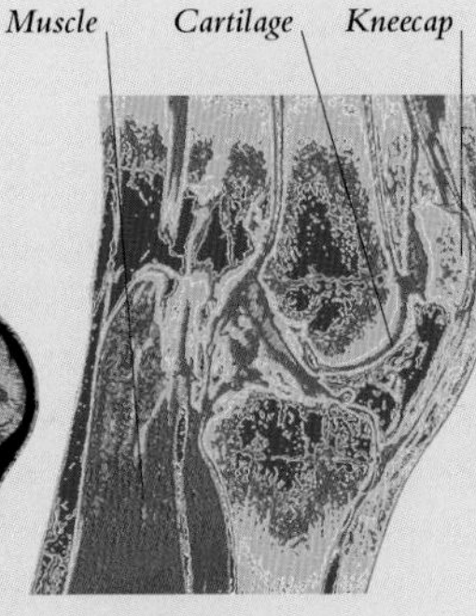


**MRI SCAN**
*An MRI scan of the knee gives a detailed picture of the different structures in the joint, including the bones, muscles, and cartilage.*

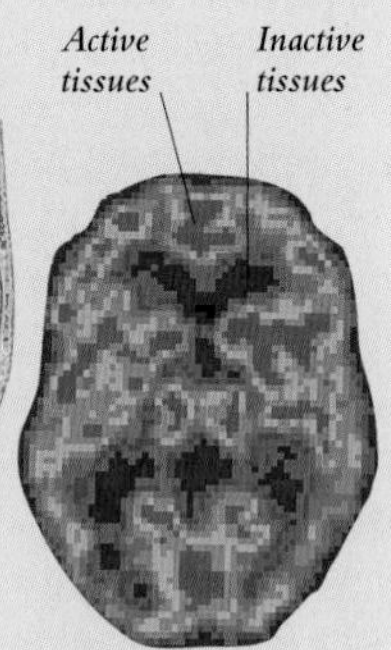


**PET SCAN**
*Radiation detected in this PET scan of the brain shows areas of active (red) and inactive (blue) tissues.*

BODY

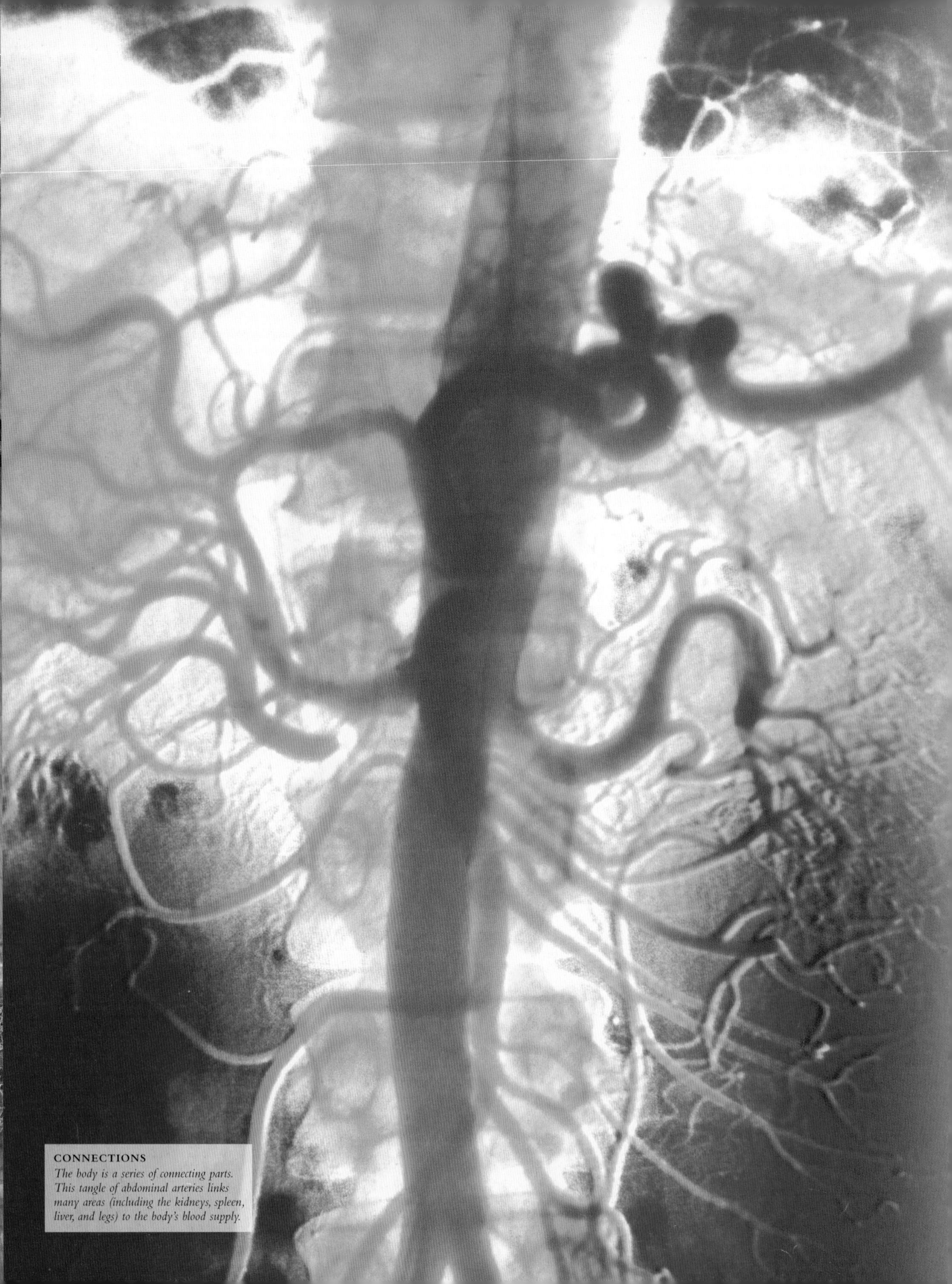

**CONNECTIONS**

*The body is a series of connecting parts. This tangle of abdominal arteries links many areas (including the kidneys, spleen, liver, and legs) to the body's blood supply.*

# BUILDING A BODY

Construction of the human body starts at microscopic level with cells, the basic units of all living things. Cells of the same type combine to make tissues, the body's materials; and collections of tissues make organs, the working parts of the human machine. A series of organs performing one of the body's major processes or functions, such as digestion or circulation of the blood, makes a system. To function properly, each system is reliant on the others and all of them interconnect to make the complete body.

The body's basic ingredients are thousands of different types of chemical substances that are similar to the substances found in food: for example, fats, proteins, and minerals. Each chemical consists of units known as molecules, and each molecule is constructed from combinations of tiny particles called atoms. All our body cells contain specialized working parts, known as organelles, that are made up from chemical molecules.

## THE DOUBLE HELIX

The whole of the body, from the organelles inside cells to the tissues, organs, and systems that keep us alive and functioning, is created under the instructions of a single chemical: DNA (deoxyribonucleic acid). Molecules of DNA are coiled tightly into tiny threadlike structures called chromosomes, which are packed into the nucleus (central unit) of each cell. The long DNA molecule is shaped like a twisted ladder with the rungs made of building blocks called nucleotide bases, which are arranged in specific pairs. Sections of DNA make up genes and these have two functions. Genes provide cells with the instructions they require for making proteins and other molecules that control cellular processes – elements that are needed for the development and growth of all the body's organs and structures. Genes are also the means by which physical and some mental characteristics are passed on from generation to generation. For instance, there are combinations of genes that are responsible for creating the colour of our eyes and hair or the shape of our noses. When a body cell divides in order to make a copy of itself – in the process of growth and repair that occurs naturally all over the body – the DNA in the cell is duplicated as well. This ensures that every new cell that is created contains a complete set of instructions for the human body.

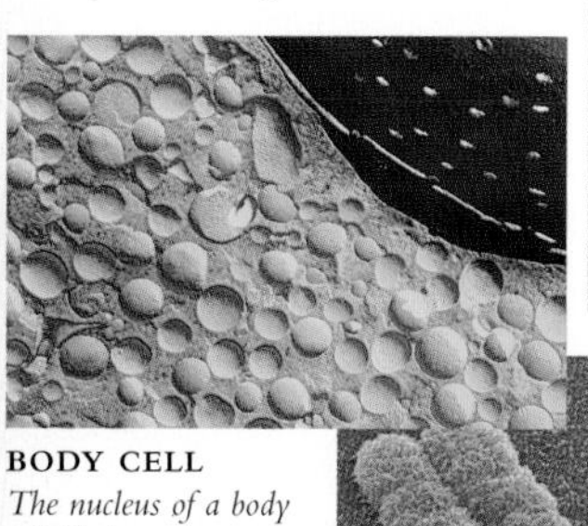

**BODY CELL**
*The nucleus of a body cell (dark area) contains genetic instructions. The cytoplasm (green) that surrounds the nucleus contains structures needed for cell function.*

**DNA**
*These structures are chromosomes, coils of DNA (genetic material) found in cells. Normally in single strands, chromosomes join in an X-shape when cells divide.*

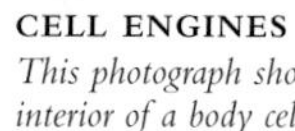

**CELL ENGINES**
*This photograph shows the interior of a body cell. The red structures are mitochondria, the driving forces of the cell, which turn food into energy.*

## LIVING CELLS

We have trillions of cells of various types in our bodies, and each one is a world in its own right. Cells are too minute to be seen with the naked eye but they can be studied under a microscope. The degree of magnification possible with modern instruments reveals the astonishing inner world of cells. These tiny living organisms are crammed with working parts that keep the body functioning. The cell nucleus houses our genetic material, DNA, while precisely arranged in the fluid that fills the rest of the cell are other essential components such as mitochondria, where biochemical energy is created. Although cells have many features in common – for example, all cells break down glucose to use for energy – they also have specialized roles and are often highly distinctive in appearance. For example, there are certain cells, neurons, that carry electrical signals to and from the brain. Neurons are easily identified by their long "wires", axons, down which the signals travel. The function of red blood cells is to transport oxygen around the body. The red coloration and dimpled shape of these blood cells make them unmistakable. The lifespans of cells vary. Some kinds of cell survive for a few hours, others last a lifetime.

## BODY BUILDING

Isolated cells are fragile and can build a body only when they join together to form tissues. There are various types of tissue in the body, – including muscle, the nervous tissue found only in the brain and spinal cord, and the connective tissues that bind everything together. These tissues all have different roles in the maintenance of our body structures and the functioning of our organs. The constant replacement of individual cells in any one mass of tissue keeps everything in good repair. For tissues to function efficiently, they need a support system that provides a supply of nutrients, and a communication system that keeps the tissue in touch with the rest of the body. Tissues have a direct blood supply that provides nutrients through a network of tiny vessels, and also removes waste products generated by cell action.

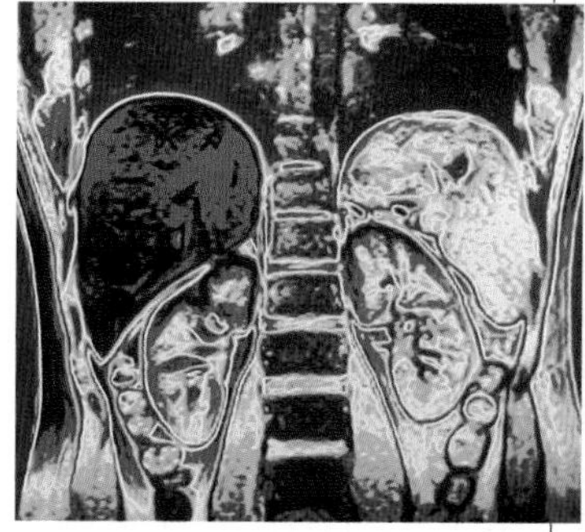

**VITAL ORGANS**
*The liver (dark structure on left) and the kidneys (seen on either side of the spine) are among the most vital parts of the body's working machinery.*

If the blood supply is interrupted, the tissue may die. Most tissue is also threaded by nerve fibres so that messages pass to and from the brain (which is how we feel pain).

Tissues do not perform their functions in isolation. It takes various types of tissue to make a working body organ, such as the heart, stomach or kidneys. For example, muscle tissue may provide the organ's movement (as in the intestines), while other types of tissue produce mucus to provide lubrication and protection.

Like cells and tissues, organs cannot operate alone but must instead be integrated with other organs to build body systems. Examples of organ groups are the respiratory system, which brings oxygen into the body, and the urinary system, which eliminates waste products from the body.

Fact

### Tissue engineering

It is possible to create tissues in the laboratory for replacing damaged parts of the body. Some replacement tissues are entirely synthetic. Others are created by "seeding" a chemical compound with human cells that are then stimulated into growth.

**NEW TISSUE**
*This bubble is replacement tissue for the cornea (front of the eyeball); it was cultured in the laboratory from cells.*

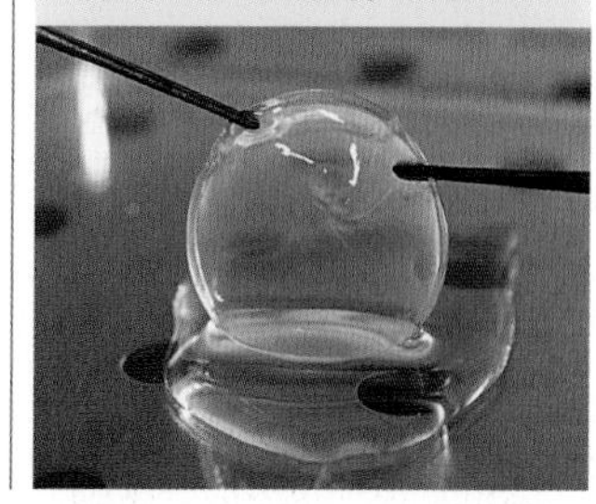

# DNA

The chemical DNA (deoxyribonucleic acid) is found within the nucleus of every cell, where it is spiralled up tightly into structures called chromosomes. DNA is the master chemical in the body – the "key" to all life. It contains the recipe for making proteins, which are needed for the development and growth of organs and structures. DNA is the basis for inheritance: the information that enables characteristics to be passed down the generations is carried by genes, which are made of DNA.

## MAKING PROTEINS

Some proteins in the body make up structures such as skin; others are hormones or enzymes that control cell activities. A major function of DNA is to provide the instructions for making proteins. DNA is made of molecules called bases (*see right*) whose arrangement provides the template for assembling proteins. Proteins are made from amino acids. Instructions held by DNA for assembling amino acids are relayed by a chemical called messenger ribonucleic acid (mRNA) in stages called transcription and translation.

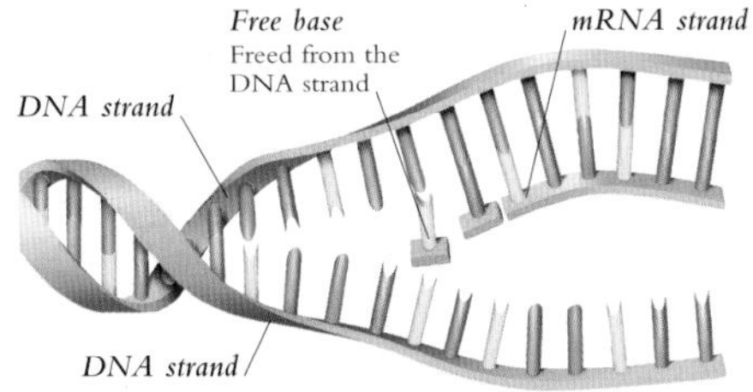


**TRANSCRIPTION**
*Strands of DNA separate along a portion of their length. Free bases attach to bases on one DNA strand to make mRNA. The mRNA carries the instruction for making a protein and moves into the cytoplasm surrounding the nucleus of the cell.*

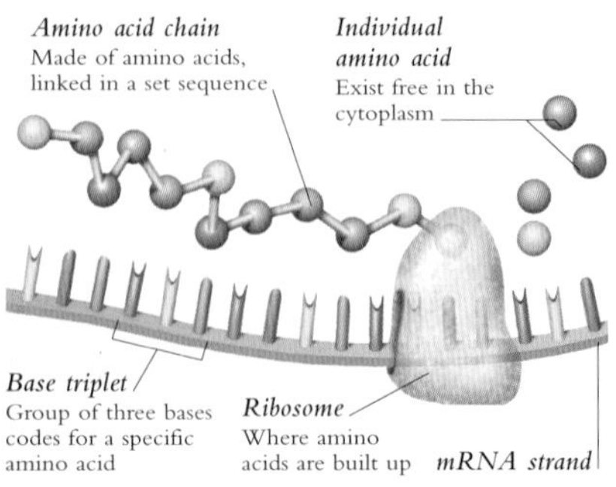


**TRANSLATION**
*A structure called a ribosome moves along the strand of mRNA three bases at a time. The ribosome attaches specific amino acids in place according to the sequence of bases in the mRNA triplets.*

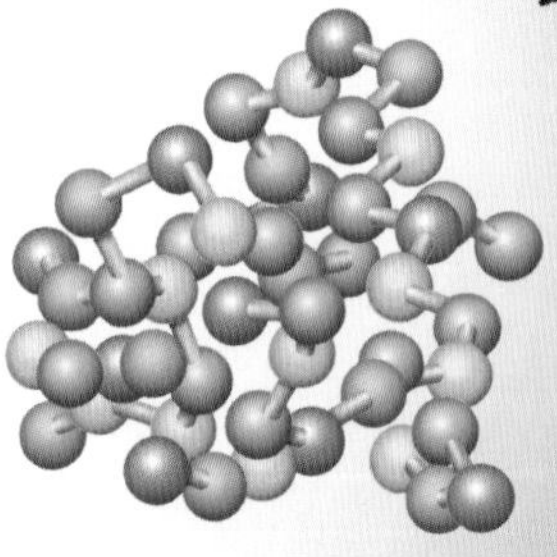

**COMPLETED PROTEIN**
*When the ribosome reaches the end of the mRNA strand, it detaches itself from the assembled chain of amino acids. The chain then folds up to form the newly completed protein.*

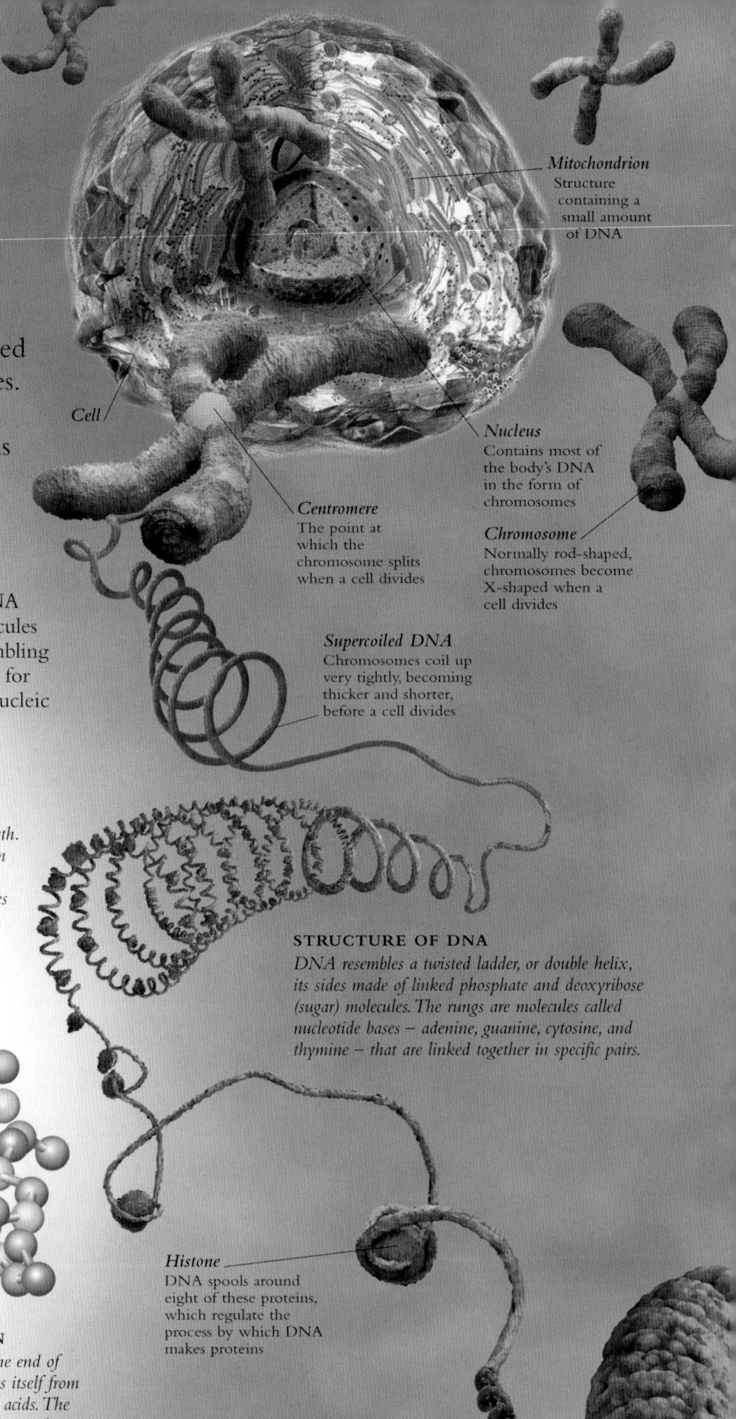

*Mitochondrion*
Structure containing a small amount of DNA

*Cell*

*Nucleus*
Contains most of the body's DNA in the form of chromosomes

*Centromere*
The point at which the chromosome splits when a cell divides

*Chromosome*
Normally rod-shaped, chromosomes become X-shaped when a cell divides

*Supercoiled DNA*
Chromosomes coil up very tightly, becoming thicker and shorter, before a cell divides

**STRUCTURE OF DNA**
*DNA resembles a twisted ladder, or double helix, its sides made of linked phosphate and deoxyribose (sugar) molecules. The rungs are molecules called nucleotide bases – adenine, guanine, cytosine, and thymine – that are linked together in specific pairs.*

*Histone*
DNA spools around eight of these proteins, which regulate the process by which DNA makes proteins

*DNA backbone*
The two rungs that form the backbone of DNA are made from sugar and phosphate molecules

History

### Watson and Crick

In 1953, scientists James Watson and Francis Crick – who were working at a laboratory in Cambridge, Britain – discovered the structure of DNA, using Chargaff's ratios of the bases of DNA and the X-ray crystallography of fellow scientists Maurice Wilkins and Rosalind Franklin. They built a model of the DNA molecule, showing the form of a double helix, with which they explained their findings. Watson, Crick, and Wilkins were awarded the Nobel prize in 1962, Rosalind Franklin having died of cancer in 1958.

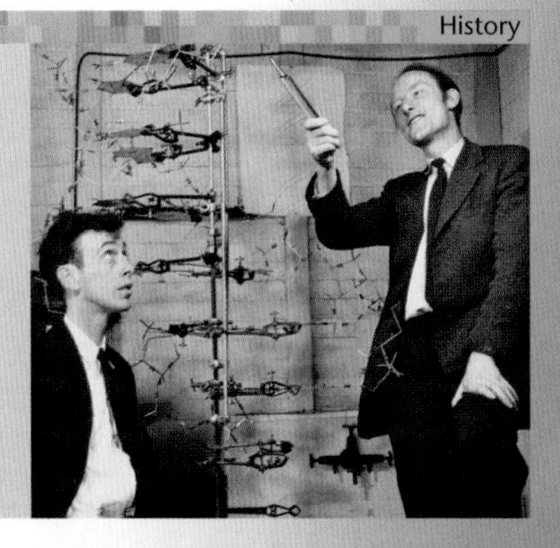

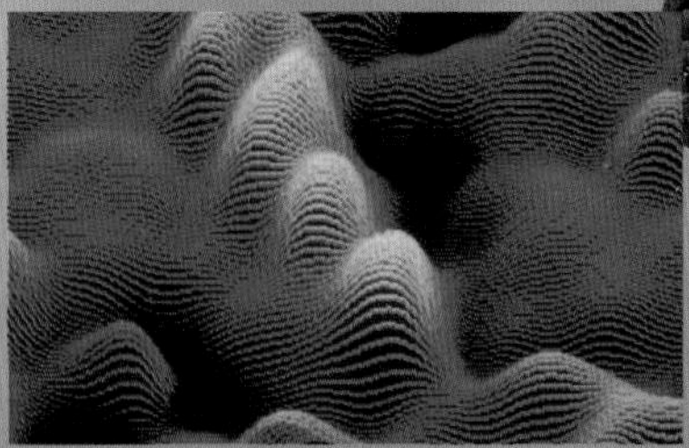

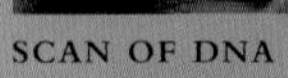

**SCAN OF DNA**
*The yellow "peaks" on this specialized microscope picture of DNA correspond to the coiled ridges of the DNA's two intertwined strands: its double helix.*

BODY

# REPLICATION OF DNA

Body cells are dividing continuously, during periods of growth and to compensate for cell damage. Before a cell can divide to make new body cells (a process called mitosis, *see right*) or egg and sperm cells (a process called meiosis, *see* p138) the DNA contained in the cell must be replicated, or copied. This process is possible because strands of DNA are able to "unzip" themselves along their length and separate. Each of the two strands in the original DNA acts as a template against which two new strands are built.

**STAGE ONE**

*The original DNA double helix splits open at several points along its length. This process produces areas where there are two separate single strands.*

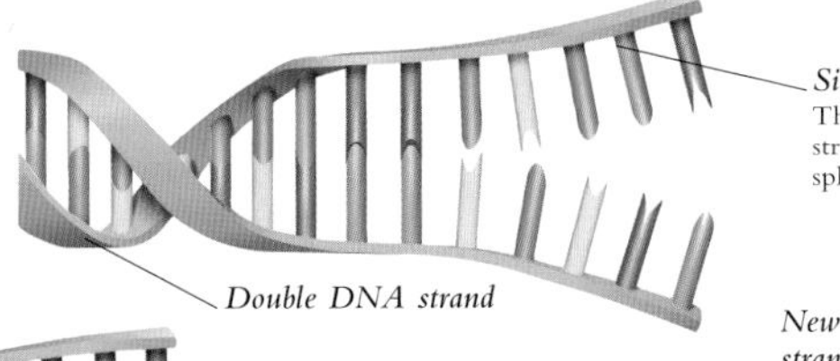


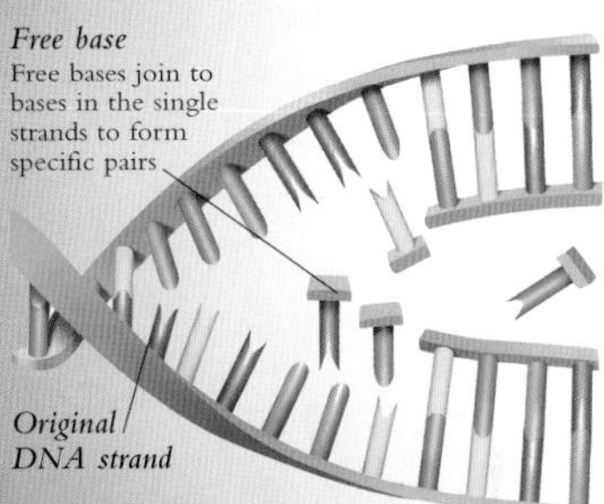


**STAGE TWO**

*Free bases (from the DNA strand) are attached to both of the single strands of DNA. The order in which free bases join to the single DNA strands is determined by the DNA bases that are already present on the single strand.*

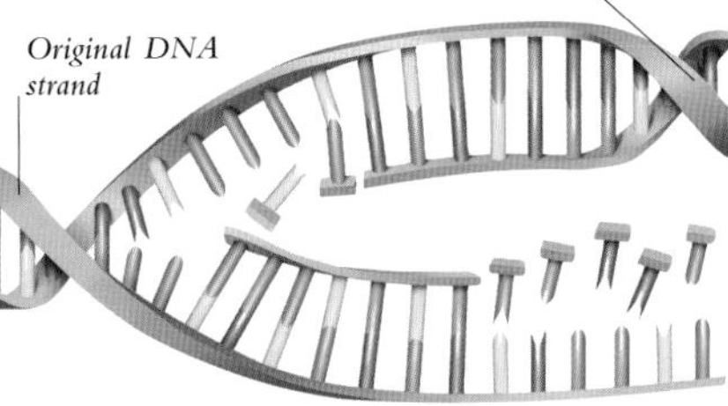


**STAGE THREE**

*While the bases attach to the strand, each of the two newly formed double strands starts to twist. The process continues along the whole length of the DNA, eventually producing two identical double DNA strands.*

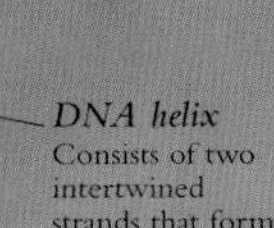
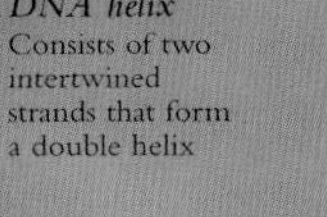


# MAKING NEW BODY CELLS

Before new body cells can be made, in a process called mitosis, DNA must first be replicated, or copied (*see left*). The double chromosomes of duplicated DNA can then line up and split, forming identical single chromosomes. The resulting cells are "daughter" cells, identical to the original cell, and these new cells contain the full complement of chromosomes. This process happens to the full set of 46 chromosomes that exist in each cell, but, for simplicity, only four chromosomes are shown in this representation of the process (*see below*).

**STAGE ONE**

*The DNA in the chromosomes is copied to form two identical strands joined in the centre by a structure called a centromere.*

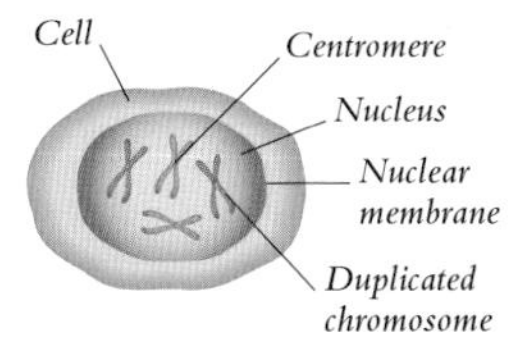


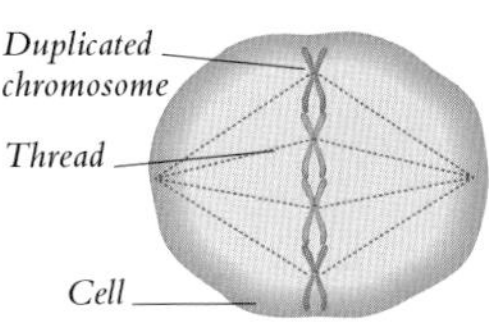


**STAGE TWO**

*The membrane around the nucleus breaks down and threads form across the cell. The chromosomes line up on the threads.*

**STAGE THREE**

*The duplicated chromosomes are pulled apart by the threads. The single chromosomes move to opposite sides of the cell.*

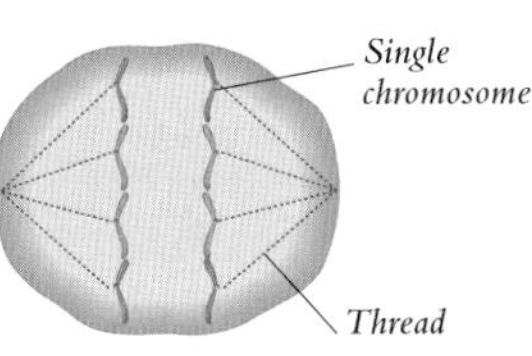


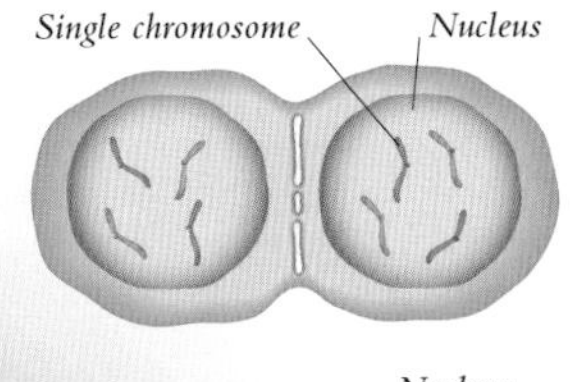


**STAGE FOUR**

*A nuclear membrane forms around each set of single chromosomes. The cell begins to divide into two cells.*

**STAGE FIVE**

*Two new cells form. Each cell has a nucleus containing an identical set of chromosomes.*

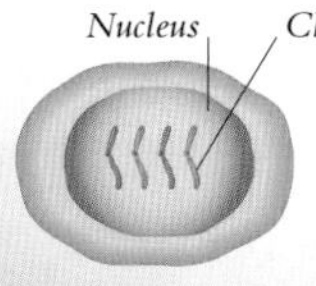
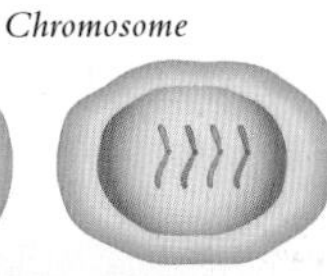

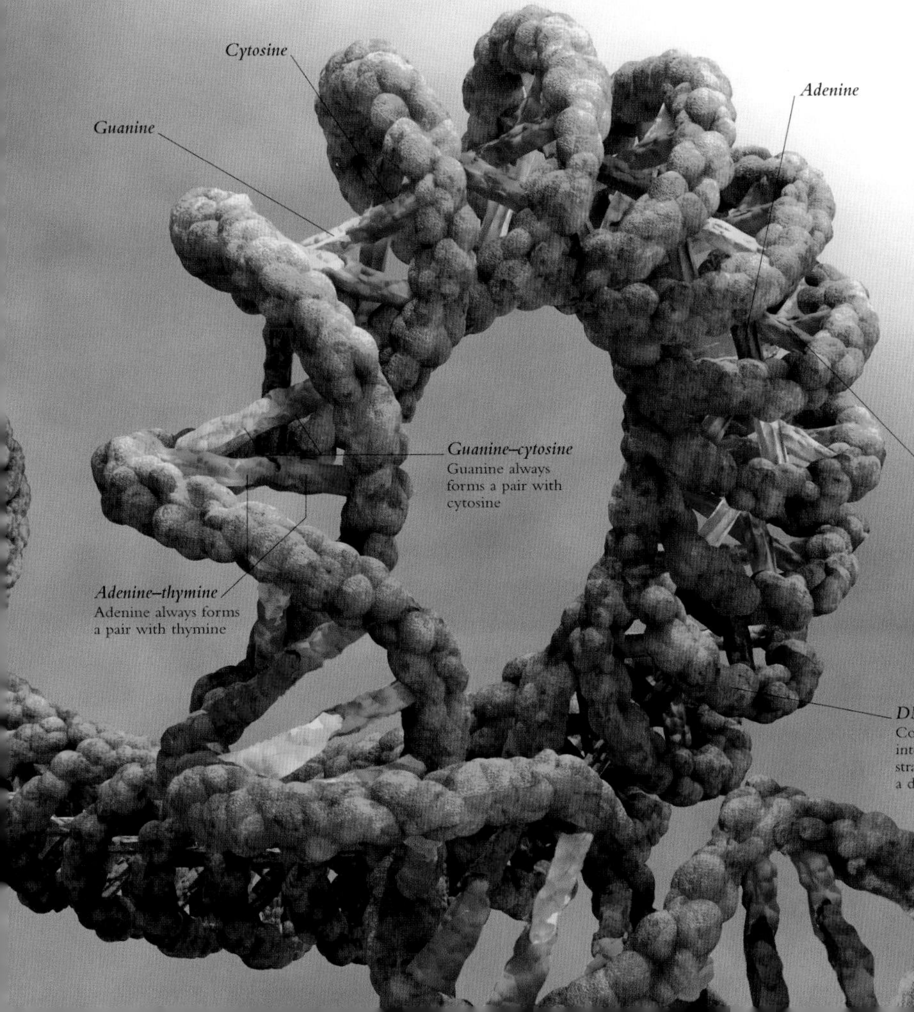

Fact

## Mitochondrial DNA

Mitochondria (energy-producing structures in cells) contain a small amount of DNA (shown in red, below). Unlike DNA in the cell nucleus, which is inherited from both parents, mitochondrial DNA is inherited only from the mother. Mistakes in copying mitochondrial DNA is the cause of a small number of genetic disorders.

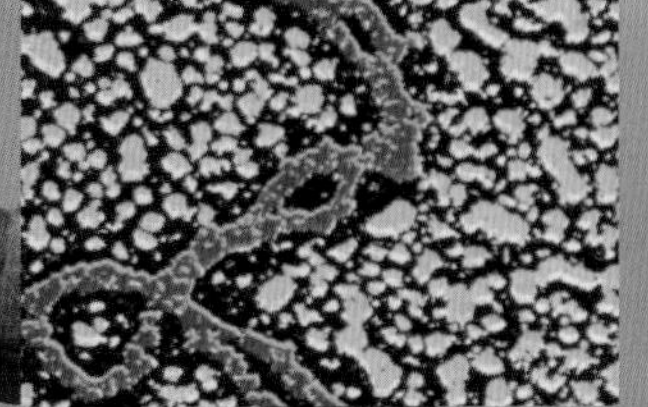

# CELLS TO SYSTEMS

The living structure of the human body is built from different types of cells. Cells mass together in clumps or layers to form tissues, which are specialized for different roles within the body. Some tissues form linings – for organs, for example – while others provide a framework, bind body parts together, act as insulation, or allow movement. Two or more tissues make up organs like the heart or the stomach, the working parts of the human machine. When a series of organs functions as a team to facilitate one of the body's major processes, they comprise a system, such as the digestive or nervous system. These systems themselves cannot function alone, but link up to and interact with other systems to form a living, working body.

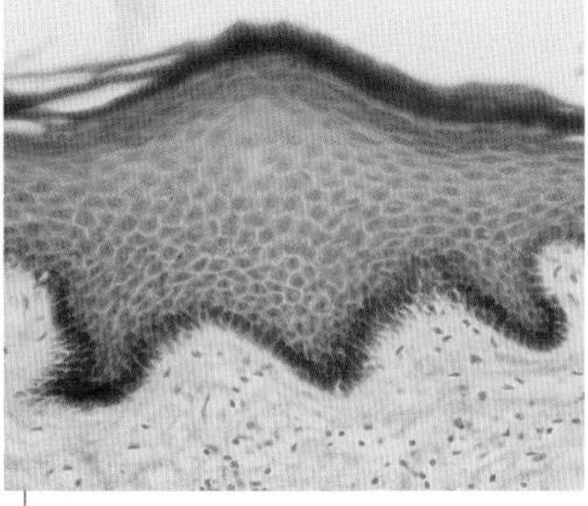

**EPITHELIAL TISSUE**
*The top layer of skin (coloured area above) is composed of a type of tissue (made of epithelial cells) whose main function is to act as a protective outer layer for the body.*

## UNITS OF LIFE

The cell is the fundamental unit of life. We all come into existence as a single fertilized cell that rapidly multiplies. By the time we reach adulthood, however, the cells that make up the body number approximately 75 trillion. Each of these tiny structures, visible separately only under a microscope, is packed with working components. Cells also contain DNA (*see* p52), the genetic material that is responsible for our development and our individual characteristics. Some cells last a lifetime; others wear out after a day or so, or fall victim to damage or disease. Every day, the body manufactures replacements (*see* Repair and replacement, p104). Although each cell is self-contained, it does not work in isolation. Cells act together in an organized way in order to function effectively – they communicate with one another via chemical messages.

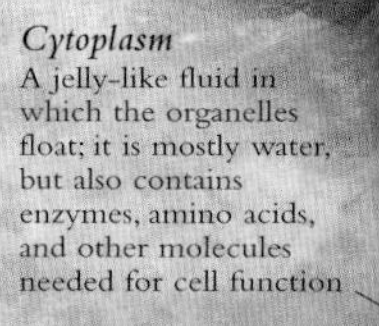

*Cytoplasm*
A jelly-like fluid in which the organelles float; it is mostly water, but also contains enzymes, amino acids, and other molecules needed for cell function

*Cytoskeleton*
The internal framework of the cell, made up of thread-like filaments

*Peroxisome*
Sac where some enzymes are produced and some cell substances oxidized

*Lysosome*
Contains powerful enzymes that break down harmful materials and dispose of any unwanted substances

*Ribosomes*
Small granular structures, here attached to the endoplasmic reticulum, that play a key role in the assembly of proteins

## THE VARIETY OF CELLS

Around 200 different types of cell have been identified in the human body. Cells begin to differentiate, becoming specialized for particular roles, at a very early stage in our development. Under the influence of growth chemicals, the primitive stem cells of an embryo pass rapidly through various stages before reaching their destinies, for example, as red blood cells, nerve cells, or muscle cells (*see below*). Cells may differ widely in their external appearance and in their activities, but almost all of them share the same internal components (although red blood cells are exceptional as they do not have a nucleus). They also have many basic functions in common. For example, most cells play a part in generating energy from food, a process that keeps our organs in full working order and, ultimately, keeps us alive.

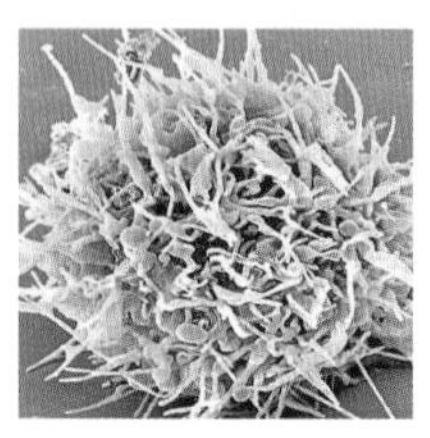

**STEM CELL**
*It is from the stem cells in the bone marrow that all blood cells originate.*

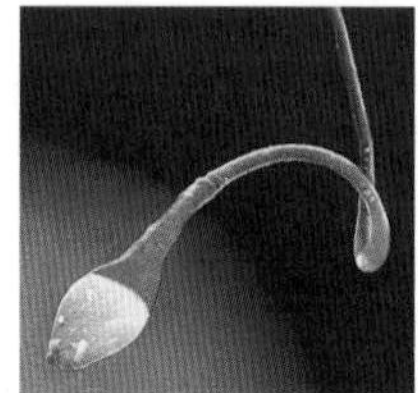

**SPERM**
*The whip-like tails of sperm cells propel them up the female genital tract.*

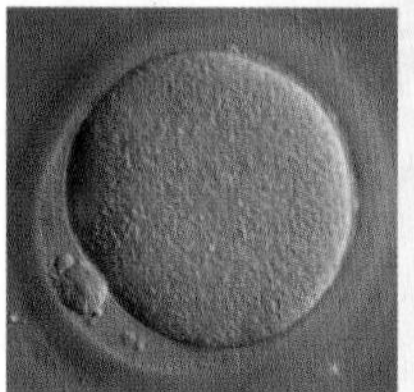

**OVUM (EGG CELL)**
*Egg cells have a protective membrane, which grows thicker after fertilization.*

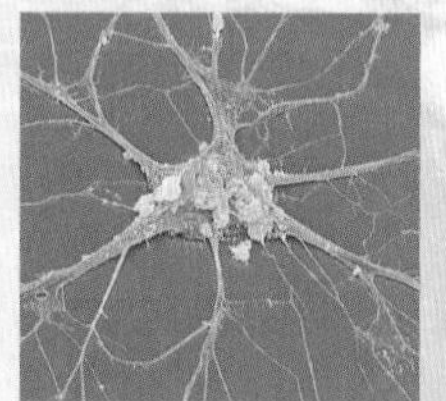

**NERVE CELL**
*Signals between nerve cells pass between long filaments called axons.*

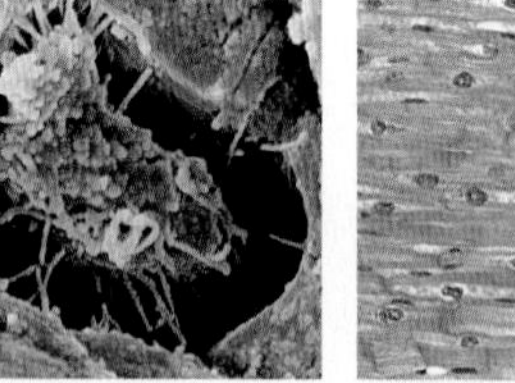

**RED BLOOD CELL**
*These pigmented cells give blood its colour. They carry oxygen around the body.*

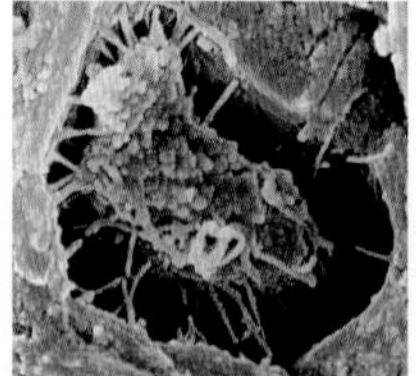

**OSTEOBLAST**
*This type of bone cell lays down calcium to maintain bone strength.*

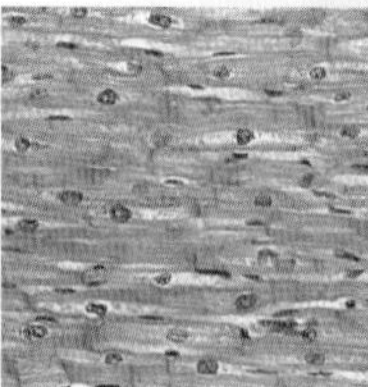

**MUSCLE CELL**
*These cells can alter their length, which varies the force of contraction.*

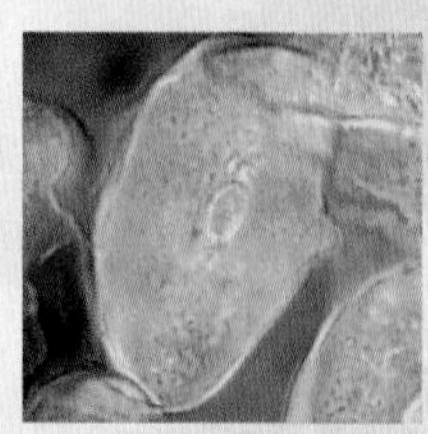

**EPITHELIAL CELL**
*Various types of epithelial cell form skin and tissues that line or cover organs.*

Fact

### Red blood cells

An average man has approximately 25 trillion red blood cells in his body – a third of the estimated total number of cells in the human body. These curved, disc-like cells have a large surface area, maximizing oxygen absorption from the lungs, but are flexible enough to squeeze into small blood vessels to transport the oxygen all over the body.

*Centriole*
Structure located near the centre of the cell that play a key role in cell division

*Vacuole*
Sac that transports and stores ingested materials, waste products, and water

*Mitochondrion*
Contains a small amount of DNA and produces adenosine triphosphate (ATP), which provides energy needed for many cell functions

*Nucleus*
Contains most of the cell's DNA (genetic material), in the form of chromosomes

*Pore*
Allows substances to pass in and out

*Smooth endoplasmic reticulum*
Helps to transport materials through the cell; also breaks down toxins and is the main site of fat metabolism

*Rough endoplasmic reticulum*
Helps to transport materials through the cell; also site of attachment for ribosomes, which are important for the manufacture of proteins

*Golgi complex*
A stack of flattened sacs that receive protein from the rough endoplasmic reticulum and repackage it for release at the cell membrane

*Vesicle*
Sac containing substances such as hormones or enzymes, which are secreted at the cell membrane

*Nucleolus*
Region at the heart of the nucleus; plays an important role in ribosome production

*Cell membrane*
Outer membrane enclosing the cell and regulating the entrance and exit of substances

History

## Identifying the cell

English physicist Robert Hooke coined the term "cell" in his ground-breaking book *Micrographia*, published in 1665. The book outlined his observations of a variety of organisms under a microscope that he devised. The box-like cells visible in this slice of cork reminded him of the cells of a monastery.

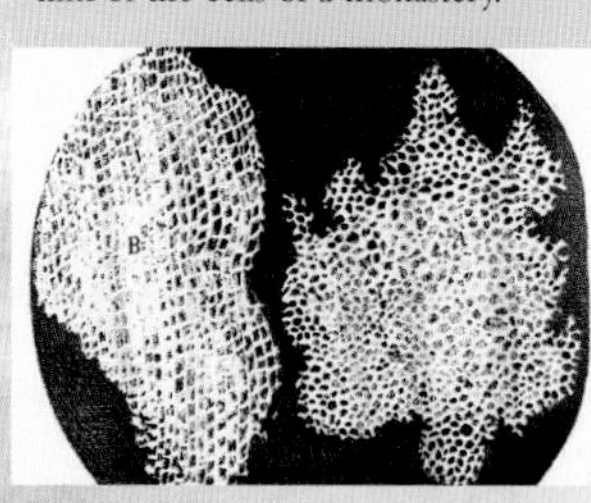

# CELL FUNCTION

Cells are busy factories, carrying out several thousand different tasks in an orderly, integrated way. Each type of cell has a specialized role, but most cells are involved in the breakdown of glucose, creating energy to drive their respective activities. Small structures called organelles within each cell carry out the cell's vital activities, coordinated by the cell's control centre, the nucleus. One important task of organelles is the production of proteins, which are needed to carry out vital biochemical reactions in the body as well as for development and growth. Some organelles are involved in digestion processes while others can destroy dangerous chemicals that may potentially harm the cell.

**FEATURES OF CELLS**
*Most of the cells in the human body contain smaller substructures called organelles ("little organs"), each of which performs a highly specialized task. Enclosing the cell contents is a membrane that regulates the passage of substances into and out of the cell.*

**MITOCHONDRION**
*The cell's powerhouse, this organelle is the site of respiration and the breakdown of fats and sugars for energy. Inner folds contain energy-producing enzymes.*

# TYPES OF TISSUE

Collections of similar cells, and the substances around them, form tissue. There are five main types of tissue: epithelial, connective, skeletal, nervous, and blood. Epithelial tissue lines and protects body organs. It is fairly closely packed and occurs in sheets that may be several layers thick. Connective tissue is the supporting tissue of the body, and includes bone, cartilage, and fat. It is more loosely packed than epithelial tissue. The protein collagen is very important in giving connective tissue its toughness. There are three types of muscle tissue. Skeletal muscle is attached to bone and allows us to make voluntary movements; smooth muscle tissue is involved in involuntary movements; and cardiac muscle is a specialized tissue allowing the coordinated beating of the heart. Nervous tissue consists of nerve cells (also known as neurons) that conduct electrical impulses, and supporting cells, called glial cells, which supply the working nerve cells with nutrients and oxygen.

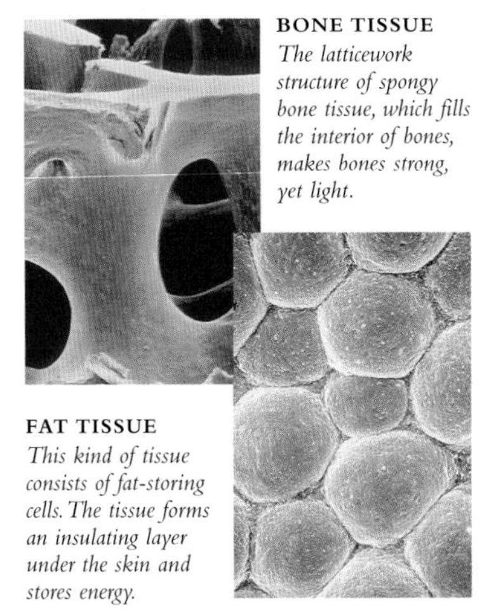

**BONE TISSUE**
*The latticework structure of spongy bone tissue, which fills the interior of bones, makes bones strong, yet light.*

**FAT TISSUE**
*This kind of tissue consists of fat-storing cells. The tissue forms an insulating layer under the skin and stores energy.*

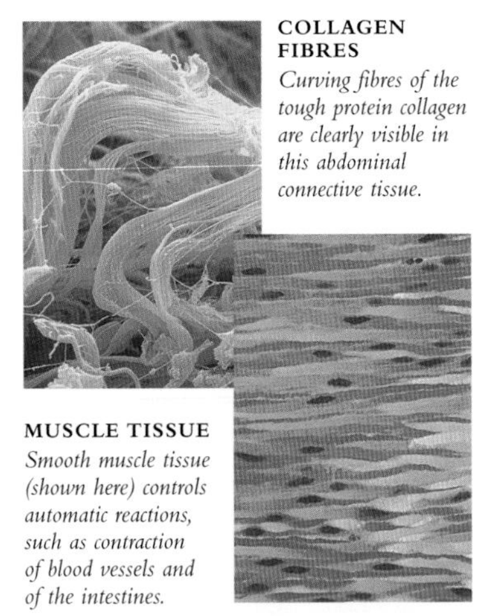

**COLLAGEN FIBRES**
*Curving fibres of the tough protein collagen are clearly visible in this abdominal connective tissue.*

**MUSCLE TISSUE**
*Smooth muscle tissue (shown here) controls automatic reactions, such as contraction of blood vessels and of the intestines.*

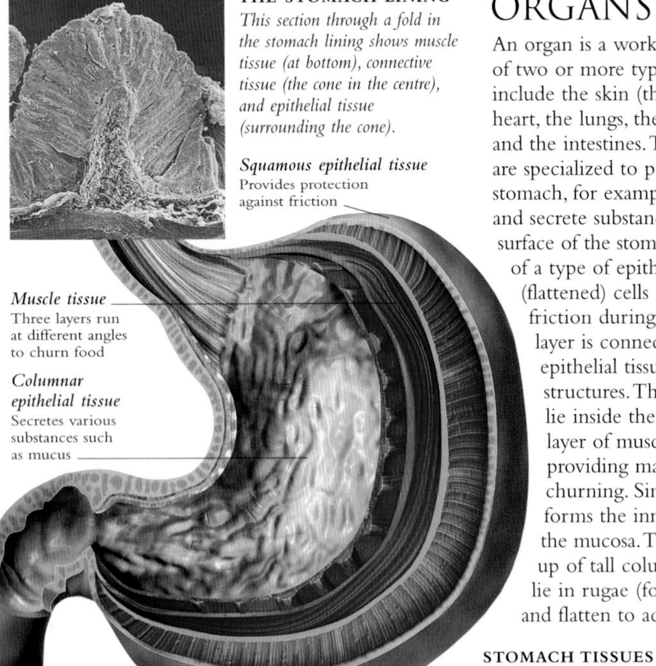


**THE STOMACH LINING**
*This section through a fold in the stomach lining shows muscle tissue (at bottom), connective tissue (the cone in the centre), and epithelial tissue (surrounding the cone).*

**STOMACH TISSUES**
*The stomach is made up of several different types of tissue lying in layers. Each of the layers of tissue has a different role to play in the structure and function of the stomach.*

# ORGANS OF THE BODY

An organ is a working part of the body, consisting of two or more types of tissue. Human organs include the skin (the largest organ in the body), the heart, the lungs, the stomach, the liver, the kidneys, and the intestines. The tissues that make up organs are specialized to perform specific functions. The stomach, for example, needs to be able to churn food and secrete substances that aid digestion. The exterior surface of the stomach's outer layer, the serosa, consists of a type of epithelial tissue made up of squamous (flattened) cells that protect the stomach against friction during churning. The serosa's interior layer is connective tissue that supports the epithelial tissue and nourishes the surrounding structures. Three layers of smooth muscle tissue lie inside the connective tissue layer. Each layer of muscle tissue runs at a different angle, providing maximum directional movement for churning. Simple columnar epithelial tissue forms the inner lining of the stomach, called the mucosa. This type of epithelial tissue is made up of tall columnar cells, so that the tissue can lie in rugae (folds) when the stomach is empty and flatten to accommodate incoming food when necessary. The epithelial tissue in the lining of the stomach also houses glands that secrete enzymes and acid to break down food as well as cells that produce protective mucus.

Health

## Organ transplants

Diseased or damaged organs in the body can be replaced with organs donated from another person. However, molecules in tissue vary chemically from person to person, so a "match" (a chemical agreement) must first be found to ensure that the donated organ will be accepted by the recipient body. Most matches are found between members of the same family. Commonly transplanted organs include the heart, liver, kidneys, and lungs. Tissue itself can also be transplanted.

**LIVER TRANSPLANT**
*A donor liver is placed on a sterile cloth over the clamped opening of a patient's abdomen, ready to be transplanted.*

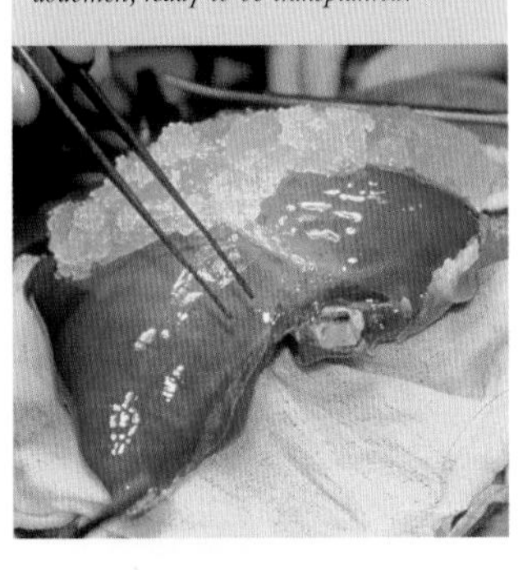

# BODY SYSTEMS

Systems of the body are made up of a group of tissues and organs that work together to carry out a specific function or set of functions. For example, the musculoskeletal system consists of bone, muscle, cartilage, and tendons, which together provide support for the body and enable us to move. The main functions of each body system are listed in the table at right and their components are described on p57. Systems cannot work alone: each system is dependent on the others to function. For example, all systems in the body are reliant on the cardiovascular system to bring them nutrient- and oxygen-rich blood that provides them with the energy they require to function. The nervous system and the endocrine system are the body's control systems: they continuously monitor body activities and adjust them appropriately.

**THE BODY SYSTEMS**
*Each of the body's systems has its own set of functions. They must not only fulfil these, but also must work together as a complementary unit to ensure the smooth functioning of the entire body.*

**BODY SYSTEMS AND THEIR FUNCTIONS**

| | |
|---|---|
| **Musculoskeletal** Provides the framework on which the body is built and facilitates movement. | **Integumentary** Provides protection from the environment through the skin, hair, and nails. |
| **Respiratory** Through breathing, supplies fresh oxygen to body tissues and expels carbon dioxide. | **Lymphatic and immune** Defend and protect the body from infection and some cancers. |
| **Cardiovascular** Circulates blood to deliver nutrients and oxygen to all body tissues. | **Nervous** Senses the environment; through nerve impulses, monitors and controls body activities. |
| **Digestive** Fuels the body by food breakdown and processing of nutrients; also removes waste. | **Endocrine** Controls the body through the action of hormones secreted by glands and tissues. |
| **Urinary** Forms urine to rid the body of waste and help maintain its chemical balance. | **Reproductive** Makes new bodies through the production of hormones, sperm, and eggs. |

# THE COMPLETE BODY

It is astonishing to think that our complex bodies are built up of simple building blocks – cells. Individual cells of similar size and shape build up into different types of tissue, each of which has a specific function. Organs, such as the stomach, are the working parts of the human machine. They are made up of two or more types of tissue that work together to carry out a particular function: for example, the stomach churns food and the ovaries produce eggs. A system is a group of tissues and organs that perform a body function, such as digestion or reproduction. Systems are dependent on each other to create a healthy, fully functioning human body.

**CELLS**
*The building blocks of the body, cells are the smallest units of living material capable of carrying out all of the activities that are necessary for life.*

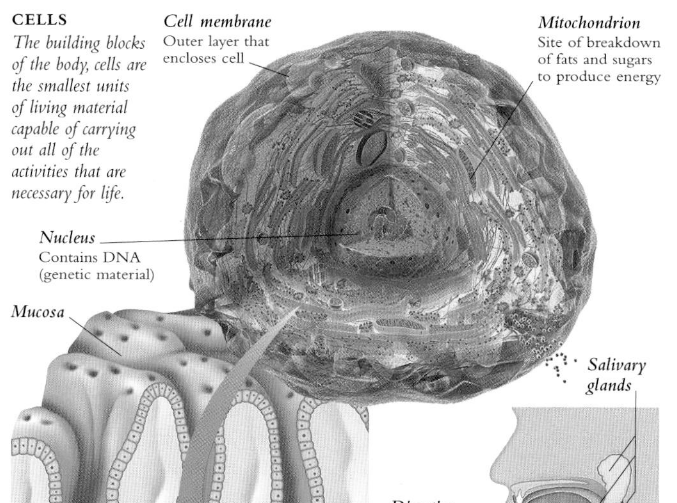

*Cell membrane*
Outer layer that encloses cell

*Mitochondrion*
Site of breakdown of fats and sugars to produce energy

*Nucleus*
Contains DNA (genetic material)

*Mucosa*

*Digestive gland*

*Mucus-secreting cell*

*Salivary glands*

*Mouth*

*Oesophagus*

**TISSUES**
*Collections of cells that make up tissues each have a distinct task. For example, the lining of the stomach (mucosa) is made of protective epithelial tissue that houses glands and mucus-secreting cells.*

*Liver*

*Gallbladder*

*Stomach*

*Pancreas*

*Oesophagus*

*Duodenum*

*Mucosa*

*Muscular stomach wall*

*Large intestine*

*Small intestine*

*Rectum*

**ORGANS**
*Made up of several different types of tissue, organs have key roles in each body system. The stomach's role in the digestive system is to churn, store, and partially digest food.*

**SYSTEMS**
*Each body system carries out an important function. The organs of the digestive system, assisted by accessory organs such as the liver, break down food into nutrients and process waste.*

**THE BODY**
*In a healthy body, all of the body systems work efficiently and in synchrony, so that the human body can function and reproduce.*

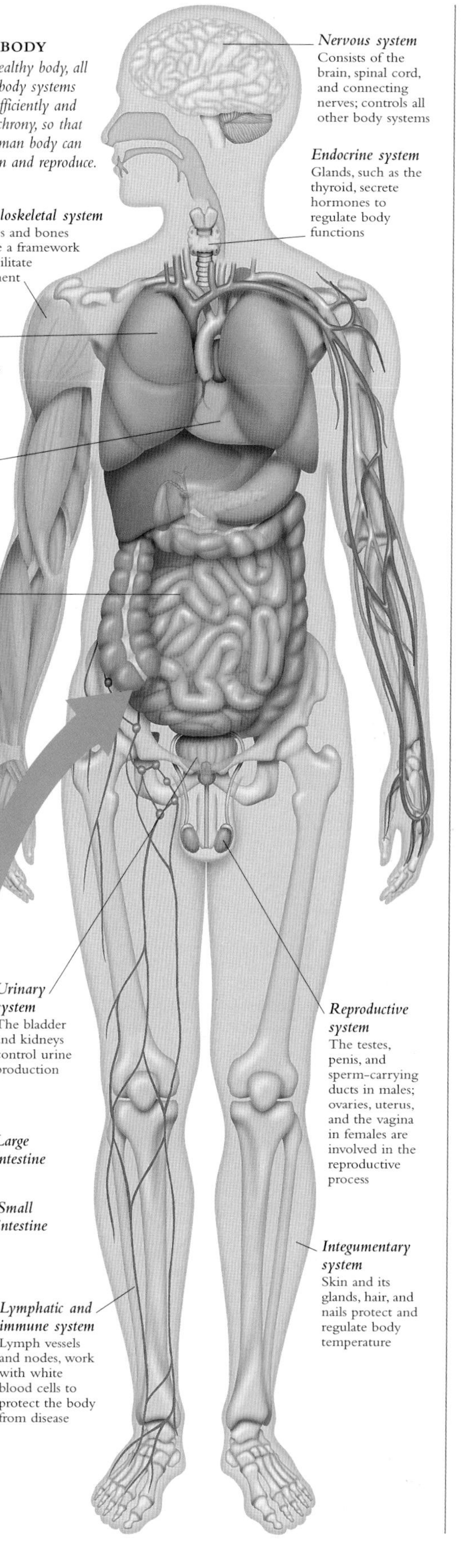

*Nervous system*
Consists of the brain, spinal cord, and connecting nerves; controls all other body systems

*Endocrine system*
Glands, such as the thyroid, secrete hormones to regulate body functions

*Musculoskeletal system*
Muscles and bones provide a framework and facilitate movement

*Respiratory system*
Lungs and airways provide the body with oxygen

*Cardiovascular system*
Heart and blood vessels transport blood around the body

*Digestive system*
Mouth, oesophagus, stomach, and intestines process food to provide fuel for the body

*Urinary system*
The bladder and kidneys control urine production

*Reproductive system*
The testes, penis, and sperm-carrying ducts in males; ovaries, uterus, and the vagina in females are involved in the reproductive process

*Integumentary system*
Skin and its glands, hair, and nails protect and regulate body temperature

*Lymphatic and immune system*
Lymph vessels and nodes, work with white blood cells to protect the body from disease

**MOVING THE BODY**
*Muscles provide the pulling power that bends the joints, lifts the legs, and propels the body forwards.*

# SUPPORT AND MOVEMENT

Humans are vertebrates, creatures that have a bony inner scaffolding with a central spine. Layer upon layer of muscles clothe these bones, forming solid flesh beneath the skin. Where bones meet, joints of various types allow this remarkable construction to move with great versatility. Although we are flexible, our sturdy core structure and limbs keep us balanced in our unique two-legged stance. Highly developed brain–muscle coordination enables us to fine-tune movement to a degree no other animal achieves.

Bones, joints, and muscles are all interdependent. If the skeleton were not provided with muscles it would collapse in a loose tangle of bones, like a puppet without its strings. Without joints, the body would for the most part be rigid and immobile. However, a few joints, such as those between the bones of the lower leg, provide stability, rather than flexibility.

## A BONY FRAMEWORK

Our long column of linked spinal bones (vertebrae) curves gently backwards and forwards over the centre of gravity. This arrangement provides the spine with greater resilience to shocks and gives stability to our upright posture. All other bones in the body link directly or indirectly to the spine, and each bone is designed for a particular purpose. They are the anchor points for muscles and the core around which soft tissues are wrapped. Some bones, such as those of the pelvis, are weight bearing, others have a protective function, girdling vital organs such as the heart and lungs and encasing the vulnerable brain.

Bones are long, short, flat, round, and various other shapes in between. Adult skeletons are readily distinguishable as either male or female by their size and small differences in their shape. In general, the bones of a male are slightly larger and heavier than those of a female because the muscles they have to support are heavier.

Bones usually stay strong and resilient well into old age, despite the stresses we often impose upon them. This is because they are able to renew themselves. Each bone contains living cells that are in a state of slow but continuous action, breaking down and renewing tissue. This process continues throughout life, keeping bones in good repair. If a bone breaks it is able to mend itself so successfully that within a few months it regains full strength, with few traces of the damage.

COMPACT BONE

SPONGY BONE

**BONE STRUCTURE**
*A long bone, such as in the arm, has a dense outer layer called compact bone, arranged in concentric rings. The inner layer, known as spongy bone, is an open network of rigid struts (above).*

## FLEXIBILITY

We can perform an amazing variety of physical feats because of the many ways in which our joints move. A joint is a highly elaborate piece of engineering. It involves bones that swivel, tilt, and slide; muscles that pull; and tough tendons and ligaments that hold everything securely together. From babyhood, one stage at a time, we find out how these joints make our bodies move. We learn to walk, run, and kick a ball. Later, we progress to movements requiring more sophisticated control and coordination, such as playing the piano or dancing. A few exceptional humans, such as athletes or gymnasts, train their bodies to perform movements that go beyond normal limits of flexibility and endurance.

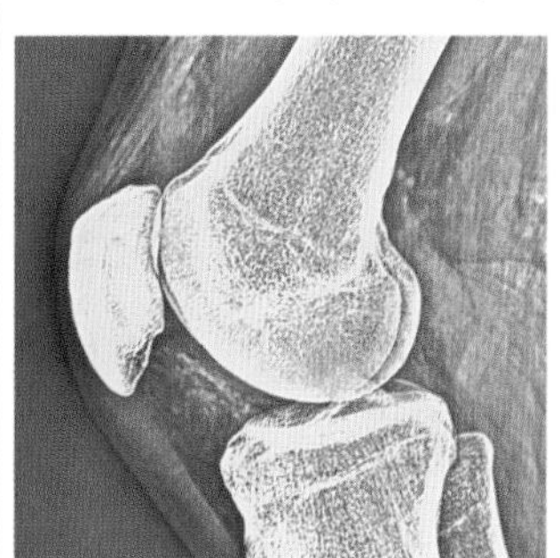

**MOVABLE JOINT**
*Flexible joints, such as the knee shown in this scan, allow us to move limbs and body in many different directions.*

## MUSCLE ACTION

The muscles that make us move are known as skeletal muscles. They vary greatly in size, ranging from the tiny muscles that move our eyes around in their sockets to the large muscles of the back and upper legs. Muscle tissue is dense and heavy. Together, the skeletal muscles make up about 40 per cent of our total body weight.

Muscles are attached to bones by cord-like tendons and make bones move by exerting a pull. Typically, they work in pairs, producing opposing movements. Control of every muscle movement that we make comes from the brain. Whether it is subconscious or voluntary, each movement is the result of a nerve signal reaching a particular muscle This signal stimulates the muscle fibres, which contract rapidly. However, much of the time muscle action is barely noticeable. For example, when we stand at a bus stop we may not be aware of any physical effort, but our muscles are at work. They automatically make constant tiny adjustments, correcting their tension to hold the spine erect, the head poised without wobbling on the neck, and body weight balanced centrally over the feet.

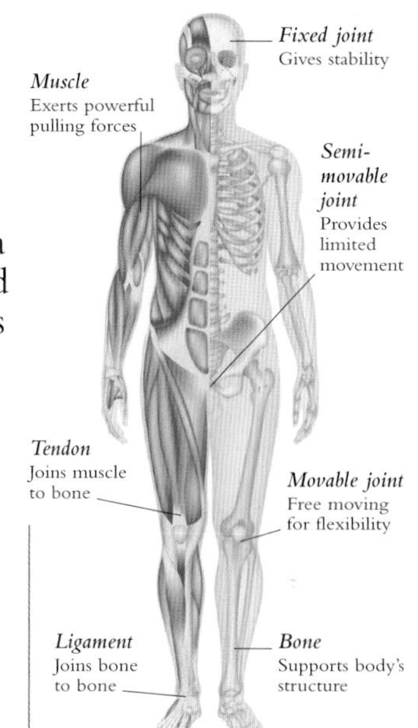


**MOVING FRAMEWORK**
*Body movement and stability depend on precisely controlled interaction between bones, muscles and joints.*

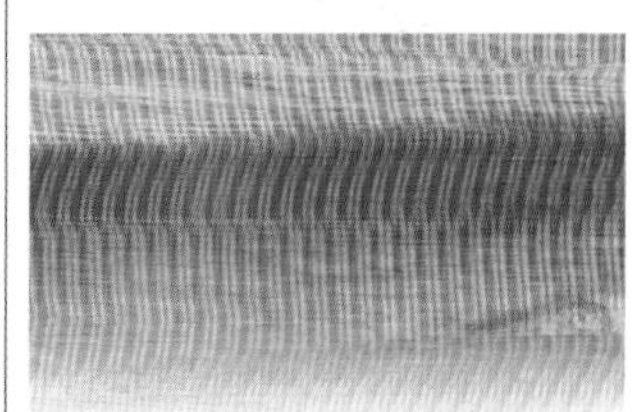

**MUSCLE FIBRES**
*When seen highly magnified, the long fibres of skeletal muscles have a striped appearance. These muscles are attached to bones and enable us to move.*

History

### Focus on movement

Human movement was revealed in detail for the first time through the work of a pioneering English photographer called Eadweard Muybridge (1830–1904). Following his revolutionary studies of animal locomotion, which excited worldwide attention, Muybridge turned to human subjects. In the 1880s, he took numerous sequence photographs, such as the somersaulting man shown right, that captured nuances of movement that are imperceptible to the onlooker.

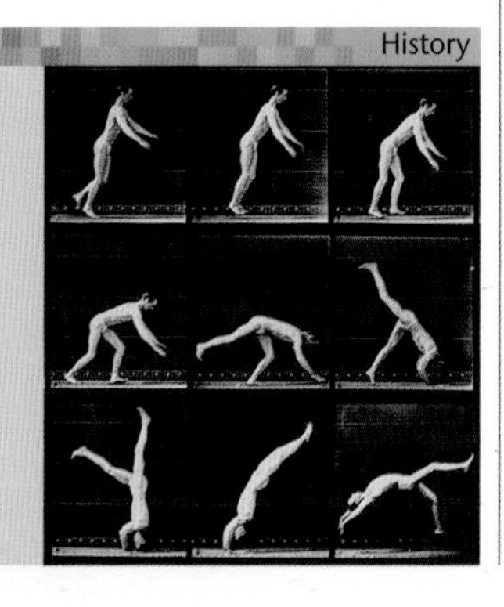

# THE SKELETON

The adult skeleton is made up of 206 individual bones. These come in all shapes and sizes from the tiny bones of the inner ear, measuring just a few millimetres, to the massive bones of the pelvis. Each bone links with others to form a sturdy, flexible framework. This structure is designed to perform many functions. The whole skeleton supports the soft tissues of the body and gives us shape. Specific groups of bones, powered by their attached muscles, provide the leverage that allows us to make a wide range of coordinated movements. As well as giving us shape and mobility, the skeleton provides protection for vital internal organs.

## SKELETAL ROLES

Our bones are arranged symmetrically on either side of the body and are organized in two main divisions. These divisions, the axial and appendicular skeletons, have different roles. The axial skeleton comprises the bones of the central part of the body. These bones include the skull, the spinal column, the ribs, and the sternum (breastbone) – 80 bones in all. Such bones have a predominantly protective role, surrounding some of the body's most vulnerable and vital parts. The bony helmet of the skull encases the brain, the bones of the spinal column contain the spinal cord, and the ribs form a cage around the heart and the lungs. Attached to the axial skeleton are the 126 bones of the appendicular skeleton, the main role of which is to provide the body with mobility. Bones in this division, which include those of the arms and legs, have many joints. Also part of the appendicular skeleton are the shoulder blades and the pelvis, which are the linking structures between the limbs and the core of the body.

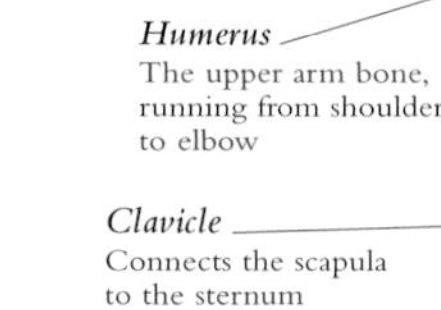

**COSTAL CARTILAGE**
*Pliable costal cartilage, a type of connective tissue, joins the ten upper pairs of ribs to the breastbone. The dark areas here are developing cartilage cells, surrounded by fibrous protein.*

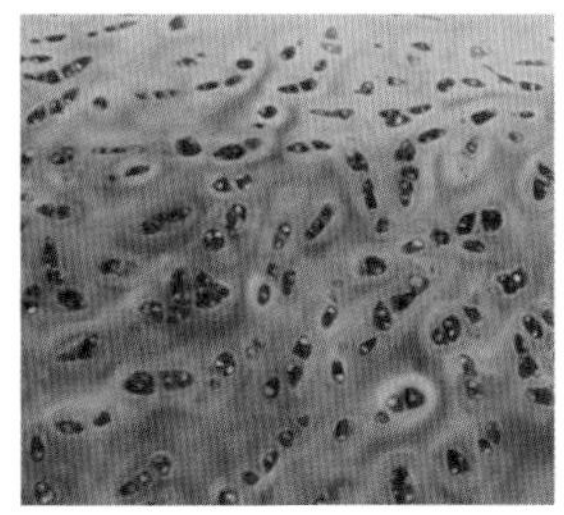
Axial skeleton
Appendicular skeleton

**SKELETAL DIVISIONS**
*The appendicular skeleton (blue) comprises the arms and legs and the bony girdles that link the limbs to the axial skeleton, the central frame of the body.*

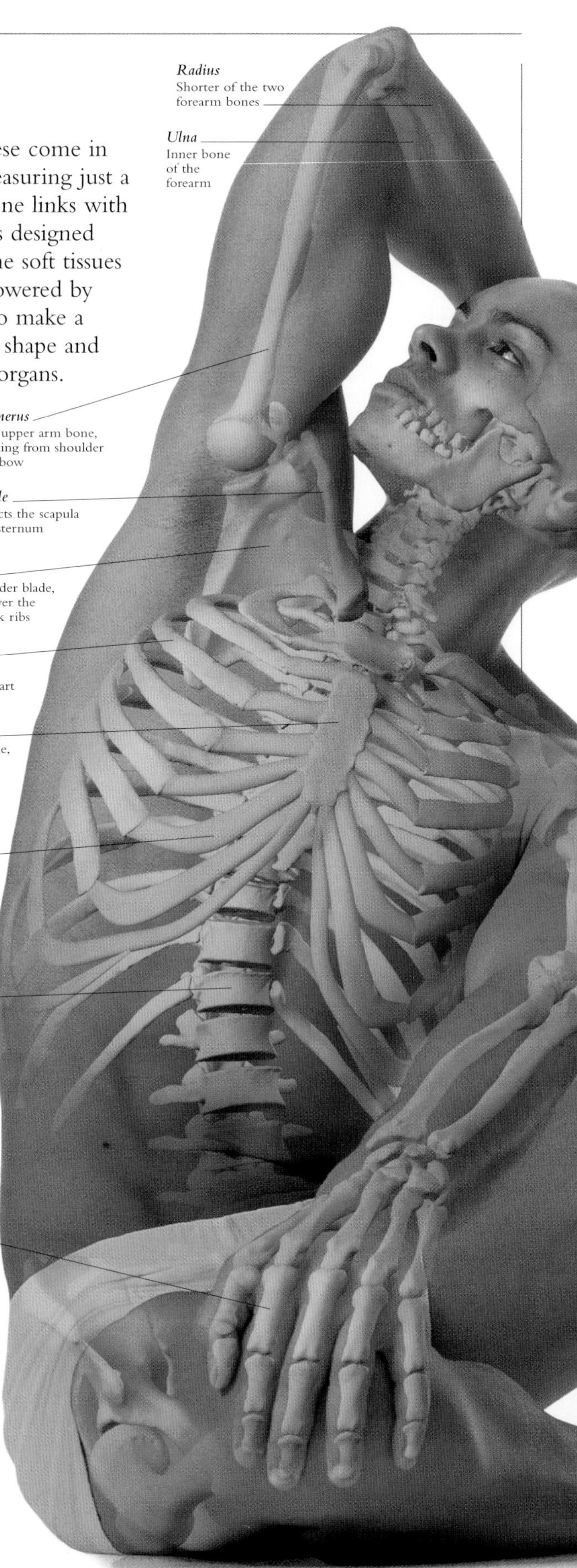
*Radius*
Shorter of the two forearm bones
*Ulna*
Inner bone of the forearm
*Humerus*
The upper arm bone, running from shoulder to elbow
*Clavicle*
Connects the scapula to the sternum
*Scapula*
The shoulder blade, situated over the upper back ribs
*Rib*
Twelve pairs shield the heart and lungs
*Sternum*
The breastbone, connected to the ribs by bands of cartilage
*Costal cartilage*
Springy connective tissue
*Spinal column*
The body's core structure, comprising a stack of linked bones
*Phalanx*
Finger bone, one of three that make up each finger (the thumb has two)

History

### Early X-rays

The first-ever X-rays of the human skeleton were taken in 1895. The technique was accidentally discovered by Wilhelm Roentgen, a German physicist. He observed that certain electromagnetic rays passed through soft body tissues but were blocked by dense bone. When Roentgen placed his wife's hand in front of the rays, the bones of her fingers, and her ring (shown here), appeared as shadow images on a photographic plate.

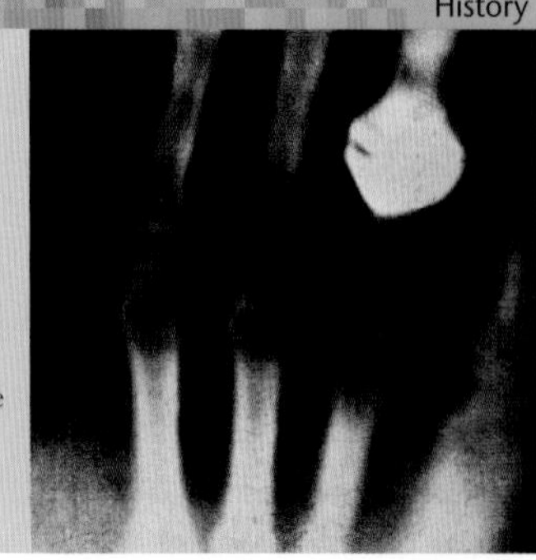

# TYPES OF BONE

Bones are of four main types: long, short, flat, and irregular. Long bones, such as those of the limbs, consist of a central shaft between two epiphyses (end sections). Short bones are roughly spherical or cube-shaped. Flat bones are thin sheets of bone, such as those that form the dome of the skull. Irregular bones fit into none of the above categories and have highly varied shapes. A further type, sesamoid bones, which are usually small and round, occur within tendons.

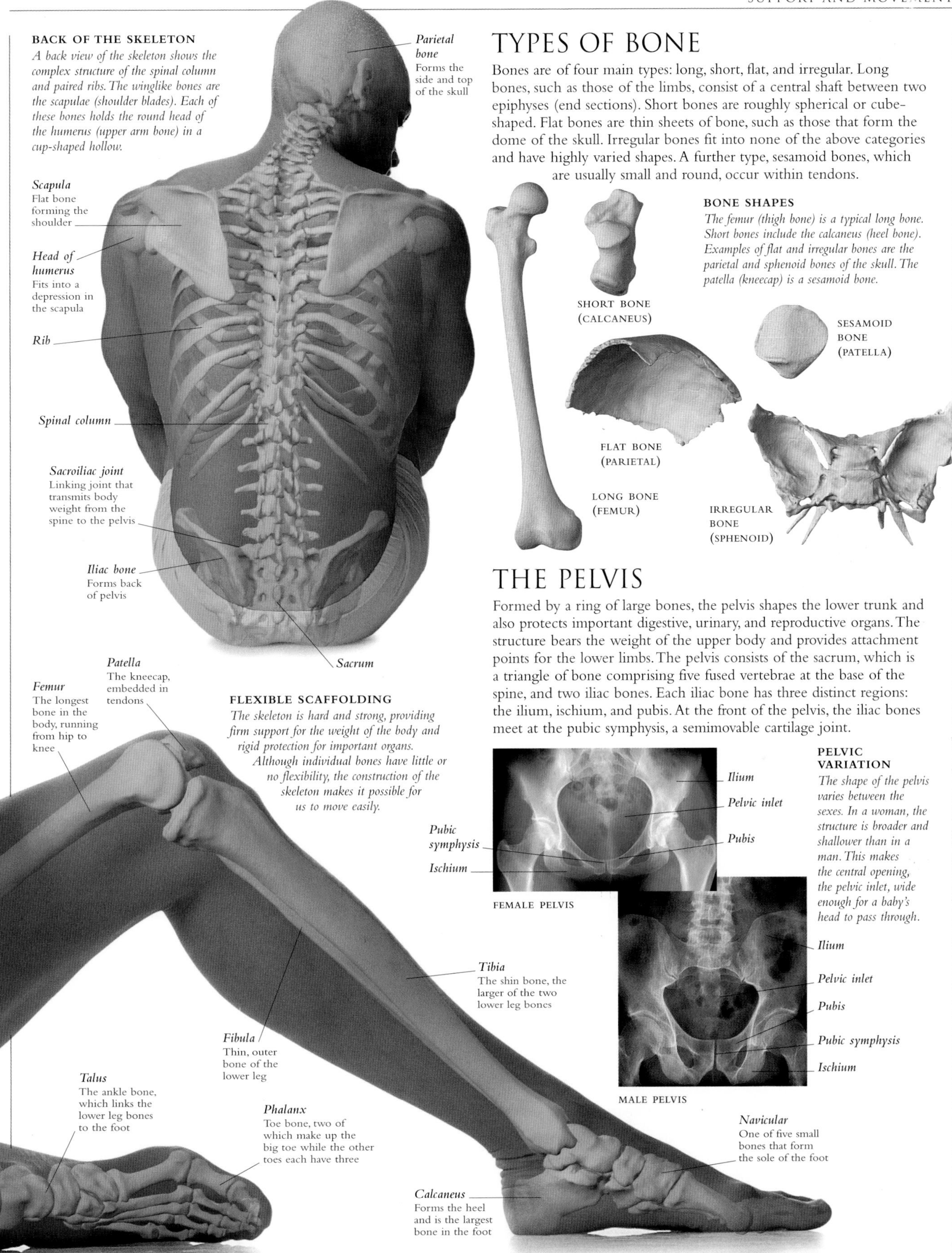


**BACK OF THE SKELETON**
*A back view of the skeleton shows the complex structure of the spinal column and paired ribs. The winglike bones are the scapulae (shoulder blades). Each of these bones holds the round head of the humerus (upper arm bone) in a cup-shaped hollow.*

**BONE SHAPES**
*The femur (thigh bone) is a typical long bone. Short bones include the calcaneus (heel bone). Examples of flat and irregular bones are the parietal and sphenoid bones of the skull. The patella (kneecap) is a sesamoid bone.*

**FLEXIBLE SCAFFOLDING**
*The skeleton is hard and strong, providing firm support for the weight of the body and rigid protection for important organs. Although individual bones have little or no flexibility, the construction of the skeleton makes it possible for us to move easily.*

# THE PELVIS

Formed by a ring of large bones, the pelvis shapes the lower trunk and also protects important digestive, urinary, and reproductive organs. The structure bears the weight of the upper body and provides attachment points for the lower limbs. The pelvis consists of the sacrum, which is a triangle of bone comprising five fused vertebrae at the base of the spine, and two iliac bones. Each iliac bone has three distinct regions: the ilium, ischium, and pubis. At the front of the pelvis, the iliac bones meet at the pubic symphysis, a semimovable cartilage joint.

FEMALE PELVIS

MALE PELVIS

**PELVIC VARIATION**
*The shape of the pelvis varies between the sexes. In a woman, the structure is broader and shallower than in a man. This makes the central opening, the pelvic inlet, wide enough for a baby's head to pass through.*

# THE SKULL

The skull's most important role is to enclose and protect the brain, but it also gives shape to the head and face, and houses the sense organs, forming sockets for the eyes and cavities for the sinuses and nasal passages. Two separate sets of bones form the skull. The eight bones surrounding the brain, known as the cranial vault, are mostly large curved plates of bone. A further 14 bones, of various shapes and sizes, make up the skeleton of the face. During childhood, the bones of the skull yield to facilitate growth, but by adulthood, all of the bones, except the lower jaw (*see below*), are fused together.



**SKULL MOVEMENT**
*The U-shaped mandible (lower jaw) is the only movable bone in the skull. It slots into the cranial vault, forming two hinges at ear level, which allow the jaw bone to move up and down, opening and closing the mouth. It also houses the bottom set of teeth.*

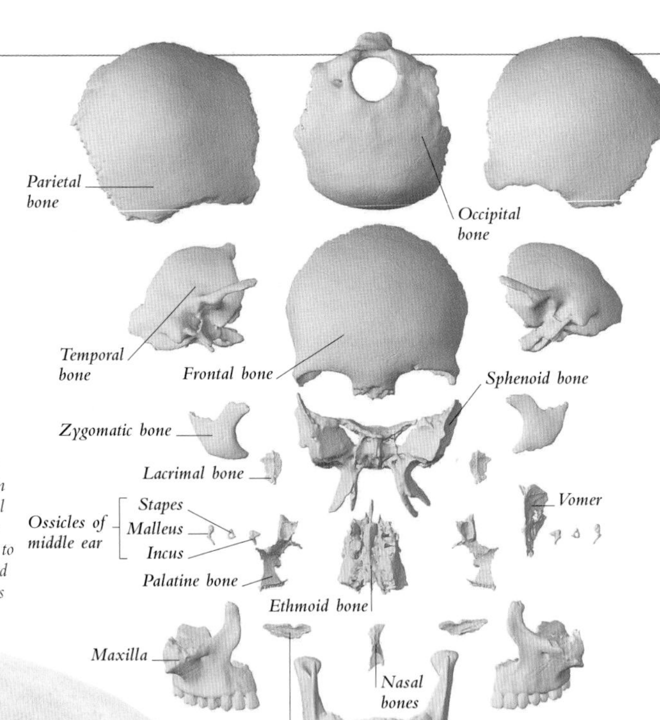


**BONES OF THE SKULL**
*The bones of the skull vary widely in size and shape – from the large, smooth, curved parietal and occipital bones to the tiny, intricate, jagged ossicle bones of the middle ear. The ossicles of the middle ear are the smallest bones in the body.*

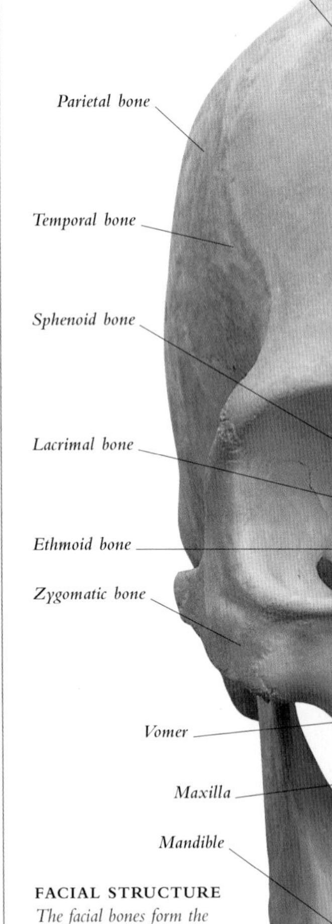


**FACIAL STRUCTURE**
*The facial bones form the skeleton of the face. The sphenoid and lacrimal bones form the eye sockets; the zygomatic bones are the cheek bones, the ethmoid bone and vomer give structure to the nasal cavity; and the maxillae and mandible contain sockets for the teeth.*

History

## Trepanning

The practice of trepanning consists of boring a hole in the skull of a living patient and removing a piece of bone, leaving the membrane surrounding the brain exposed. In Neolithic times, the practice was widespread, although the reasons for it are not known. It may have been performed to allow spirits to enter or exit the brain; to relieve headaches, infections, and convulsions, or even provide a cure for insanity; or to acquire rondelles (discs of bone) for charms or amulets. Incredibly, the practice of trepanning is still performed today in parts of Africa and South America.

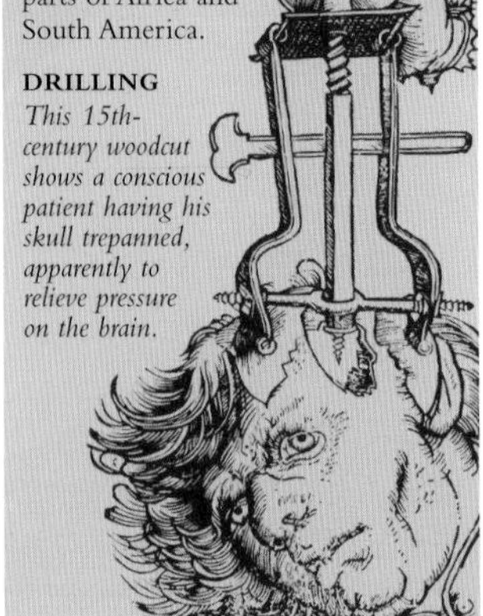

**DRILLING**
*This 15th-century woodcut shows a conscious patient having his skull trepanned, apparently to relieve pressure on the brain.*

# THE SPINE

The spine is made up of 33 bones called vertebrae, which are linked by a series of flexible facet joints. There are three main types of vertebra: cervical, which support the head and neck; thoracic, which secure the ribs; and lumbar, which support a large proportion of the weight of the upper body. The triangular-shaped sacrum and the tail-like coccyx consist of a number of vertebrae fused together. These bones sit at the base of the spine, providing a solid foundation for the spinal column.

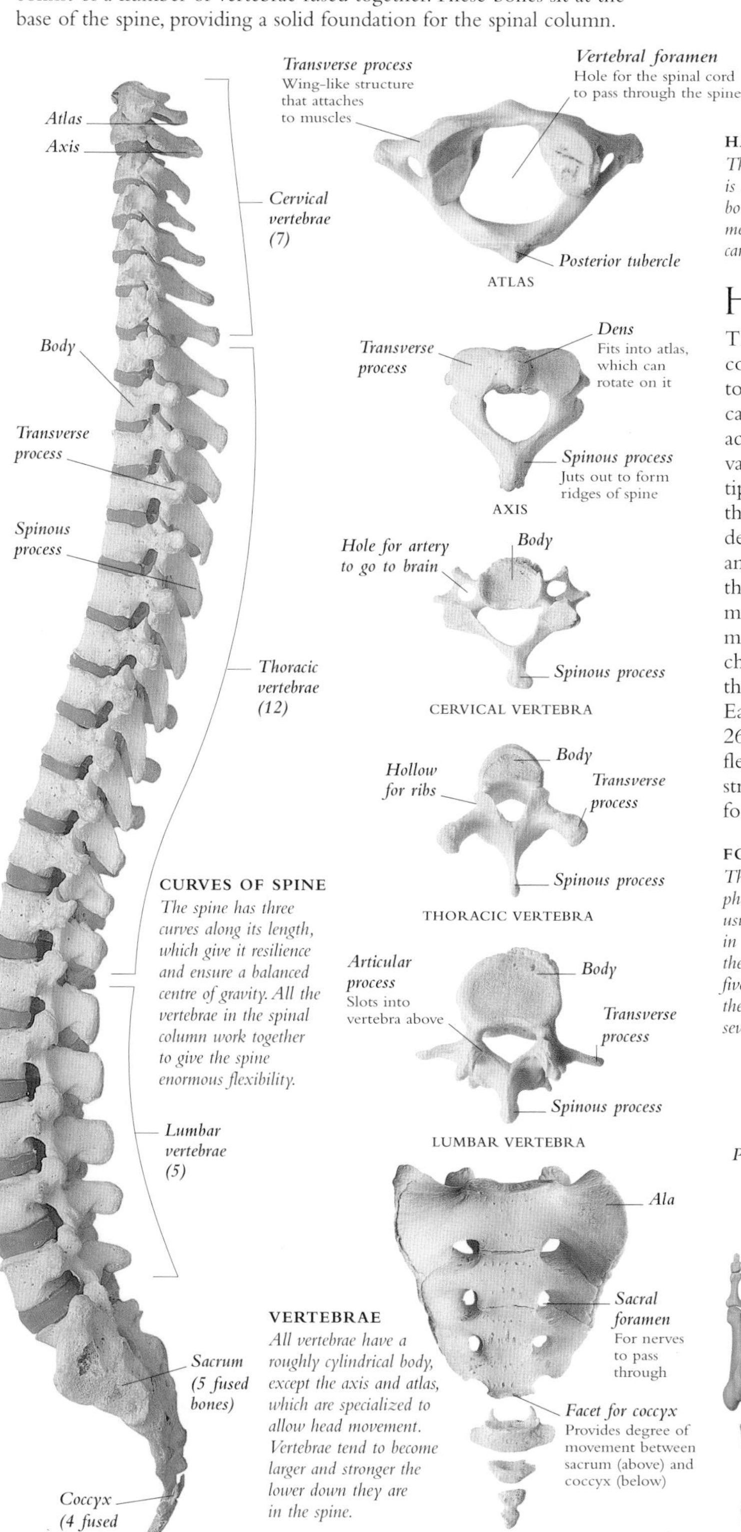


**CURVES OF SPINE**

*The spine has three curves along its length, which give it resilience and ensure a balanced centre of gravity. All the vertebrae in the spinal column work together to give the spine enormous flexibility.*

**VERTEBRAE**

*All vertebrae have a roughly cylindrical body, except the axis and atlas, which are specialized to allow head movement. Vertebrae tend to become larger and stronger the lower down they are in the spine.*

Hamate
Pisiform
Lunate
Scaphoid
Triquetrum
Trapezoid
Trapezium
Capitate
Metacarpals
Phalanges
Carpals

**HAND BONES**

*The hand (the back of which is seen here) is made up of three different types of bone: 14 phalanges (finger bones), five metacarpals (palm bones), and eight carpals (wrist bones).*

# HAND AND FOOT BONES

The hands and feet are similar in their skeletal make-up; both comprise an interlinking arrangement of small bones, which together form a fan-like structure. The hand is a versatile tool, capable of delicate manipulation as well as powerful gripping actions. The arrangement of its 27 small bones allows for a wide variety of movements. In particular, it is the ability to bring the tip of the thumb and the fingers together (having opposable thumbs) that gives human hands their unique dexterity. The feet and toes support and propel the entire weight of the body while it is on the move and also help to maintain balance during changes of position while the body is stationary. Each foot has a total of 26 bones, forming a flexible platform of strength and support for the entire body.

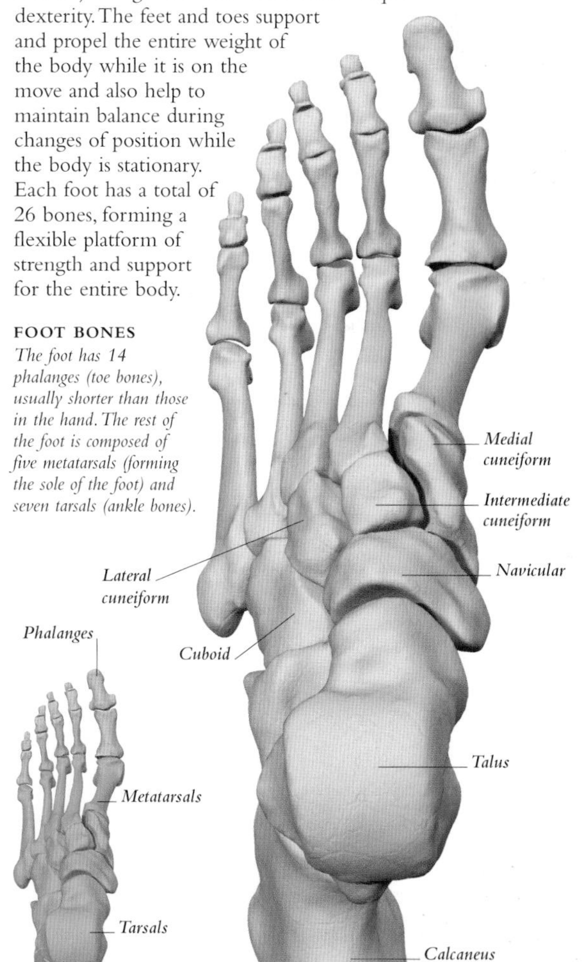


**FOOT BONES**

*The foot has 14 phalanges (toe bones), usually shorter than those in the hand. The rest of the foot is composed of five metatarsals (forming the sole of the foot) and seven tarsals (ankle bones).*

BODY

# BONE STRUCTURE AND GROWTH

Although bone looks solid and is extremely sturdy, it is surprisingly light; it also has a slight flexibility that gives some protection against shocks and jarring. These qualities are due to bone's remarkable structure and its fabric of elastic protein fibres. Bone is not inert but is constantly being remodelled, even after we reach full adult growth. One of bone's vital activities is the manufacture of most of our blood cells. Bone also serves as a reservoir for various minerals, such as calcium and phosphorus.

## A LIVING TISSUE

Bone is made up of living cells, as well as protein, mineral salts, and water. The cells of bones are of two main types: osteoblasts, which build new bone, and osteoclasts, which break it down. This process, which is lifelong, keeps our bones in a constant state of renewal and helps to minimize wear and tear until well into old age. Regardless of shape or size, each bone has an outer layer of compact bone, which is a dense heavy tissue, and an inner layer of light, spongy bone, which comprises an open network of interconnecting struts. Compact bone is made up of rod-shaped units called osteons. An osteon consists of concentric layers of hard tissue around a central channel containing blood vessels and nerves. Between the layers are tiny spaces housing bone cells and fluid that provides the cells with nutrients. In spongy bone, the gaps between the struts are filled with bone marrow, which in some bones is the site of blood cell manufacture (*see* p83).

**OSTEOBLAST**
*In this image, an osteoblast has become trapped inside a cavity in compact bone. These bone-producing cells maintain bone strength.*

**OSTEOCLAST**
*Large bone cells of this type constantly break down bone tissue so that it can be replaced with newer tissue.*

*Blood vessel*

*Periosteum*
Thin membrane covering the outer surface of bone

**COMPACT BONE**
*The units of compact bone, osteons, are arranged in rings. The white areas are channels for blood vessels and nerves. The dark spots are spaces containing bone cells.*

*Compact bone*
Dense, heavy outer layer of bone that is one of the toughest materials in the body

## BONE DEVELOPMENT AND GROWTH

When the skeleton first develops in a young fetus, it is made entirely of tough, springy cartilage. By the time a baby is born, however, much of the cartilage has hardened into bone tissue. The conversion of cartilage into bone, ossification, begins at sites called primary ossification centres in the shafts of long bones. In a newborn baby, the bone shafts are completely hardened, but the bone ends, known as the epiphyses, are still cartilage. Within these cartilage ends, hard bone gradually develops from secondary ossification sites. Between the shaft and the ends of a bone is a zone, the growth plate, which produces more cartilage to elongate bones. The process of ossification continues until the age of about 18 years. By adulthood, the processes of growth and ossification are complete and both the shaft and the ends of the bone have become continuous bone.

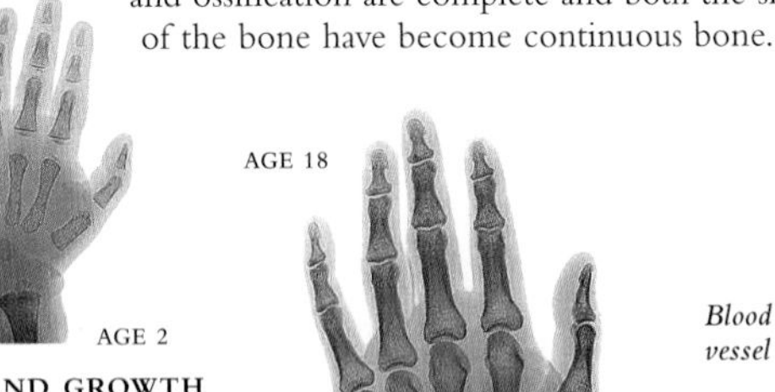


**HAND GROWTH**
*In a child aged 2, ossified hand bone shafts are opaque on an X-ray. The apparent gaps between the joints are cartilage, which looks transparent. By age 18, all bone is ossified.*

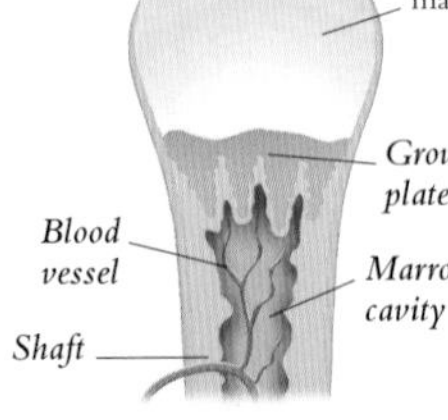


**LONG BONE OF A NEWBORN BABY**
*The shaft is mostly bone, while the ends are made of cartilage that will gradually harden.*

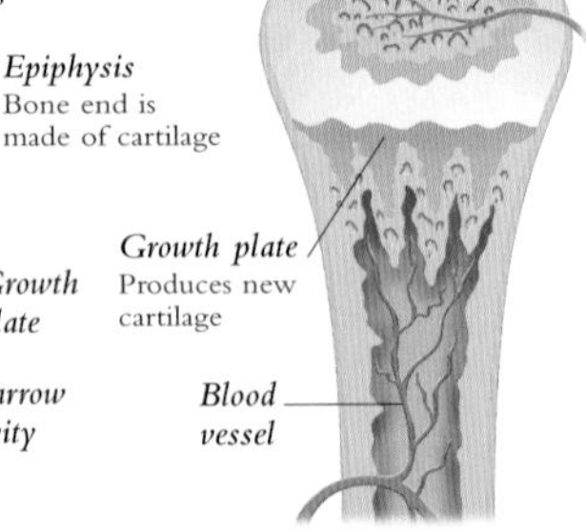


**LONG BONE OF A CHILD**
*Bone is forming in secondary ossification centres in the ends. A growth plate near each end of the bone produces new cartilage.*

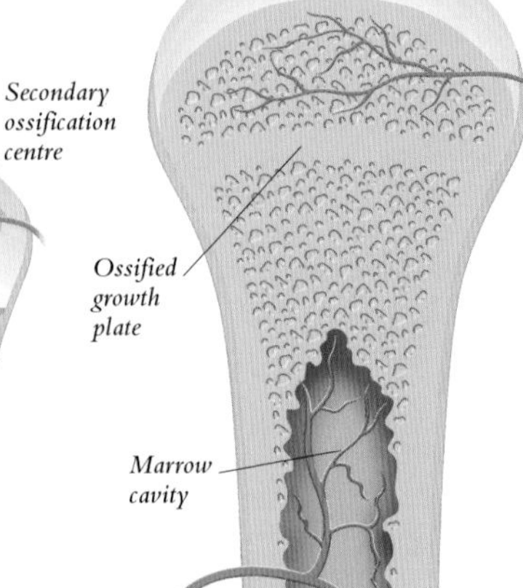


**LONG BONE OF AN ADULT**
*Growth is complete by age 18. The shafts, growth plates, and bone ends have all ossified and fused into continuous bone.*

# BONE REPAIR

Bone has an astonishing ability to mend itself after a fracture. The healing process begins within minutes of the break, when the blood clotting process is activated (*see* Repairing injuries, p105). Bone cells rapidly start building up a mass of new spongy tissue, called the callus, around the site of the damage. This tissue gradually becomes dense compact bone. A fracture in a long bone, such as in the leg or arm, normally takes about 6 weeks to heal in an adult. In children, the process is usually quicker. For some months after healing, a swelling remains over the site of the fracture. The thickened area is gradually whittled by the action of the osteoclasts, the cells that break down bone tissue, and the bone eventually regains its normal shape. Broken bones need to be returned to, and maintained in, their correct position in order for the ends to rejoin properly. For this reason, healing bones may need to be immobilized in a plaster or resin cast until the healing process is complete. If the fracture is severe, the broken ends may be pinned together with metal nails or plates.

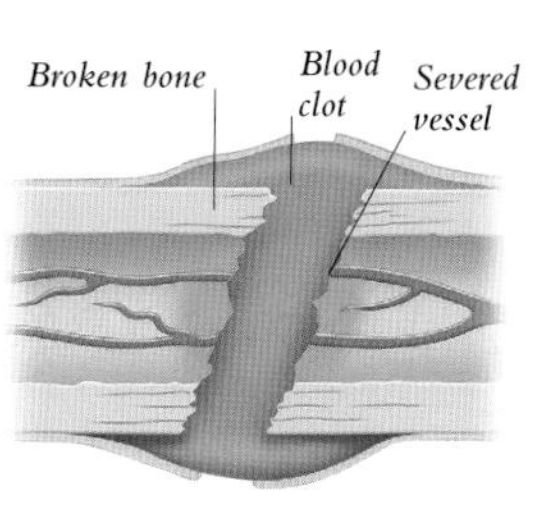


**IMMEDIATE RESPONSE**
*The first stage of repair begins almost at once with the formation of a blood clot. This seals off leaking blood vessels within the bone.*

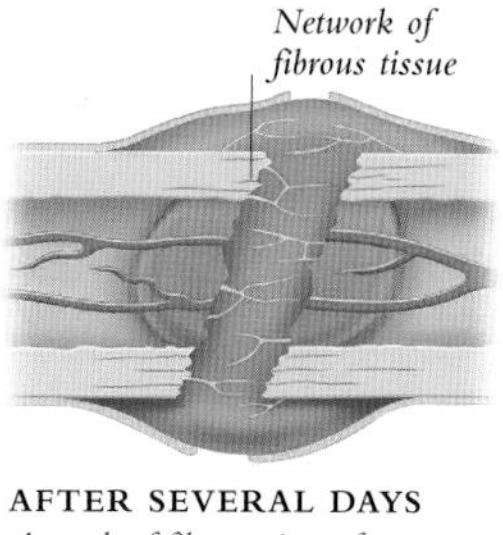


**AFTER SEVERAL DAYS**
*A mesh of fibrous tissue forms, reaching across the gap between the broken bone ends and gradually replacing the blood clot.*

*New spongy bone (callus)*

**AFTER 1–2 WEEKS**
*New, soft spongy bone, called callus, develops on the framework of the fibrous tissue, filling in the gap and eventually joining the bone ends.*

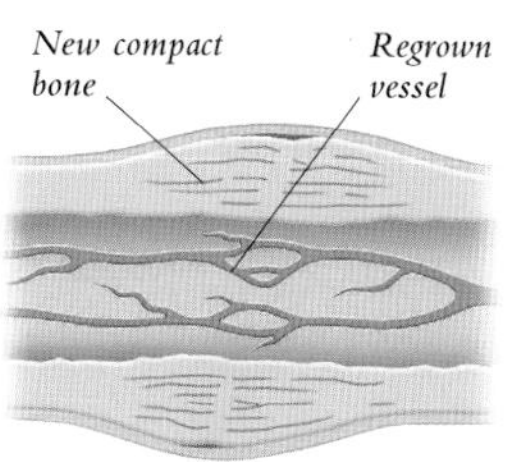


**AFTER 2–3 MONTHS**
*Repair is almost complete. Dense compact bone replaces the callus and blood vessels have regrown. The bulge will slowly shrink away.*

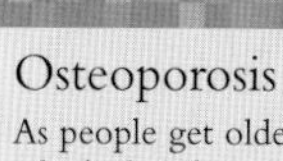

Health

## Osteoporosis

As people get older the rate at which their bone tissue is renewed slows down. By the age of 70 most people's skeletons have become about a third thinner and lighter than they were at age 40. This loss of bone density, called osteoporosis, makes bones more fragile and likely to break. The condition is linked to declining levels of sex hormones, and post-menopausal women are usually the most severely affected.

**FRAGILE BONE**
*The spongy bone tissue shown below is affected by osteoporosis. Its network of struts has become porous and brittle.*

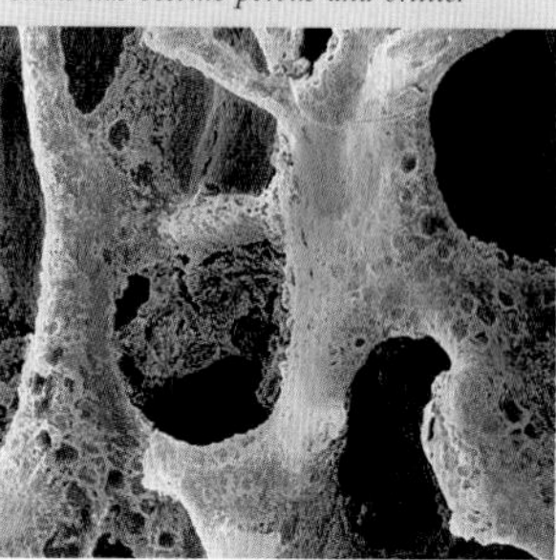

*Osteon*
Unit of compact bone comprising concentric layers of bone tissue

*Bone marrow*
The soft, fatty substance that fills the central cavities in bones and produces blood cells.

*Vein*

*Bone shaft*
Contains bone marrow and a network of blood vessels

*Artery*

*Spongy bone*
Open network of bony struts forming the bone's inner layer

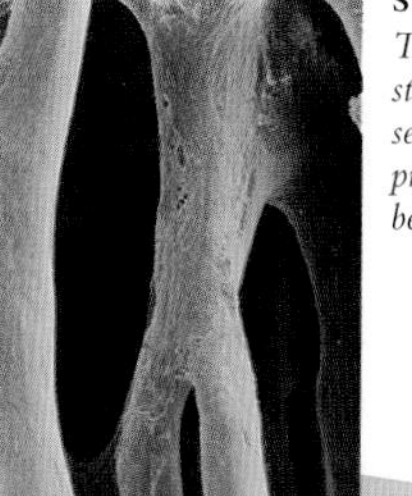

**SPONGY BONE**
*The lightweight honeycomb structure of spongy bone, as seen in this image (left), prevents the skeleton from being excessively heavy.*

*Epiphysis*
Forms each end of a long bone

**BONE MARROW**
*The spaces between the struts of spongy bone are seen here packed with bone marrow.*

**STRUCTURE OF A LONG BONE**
*A long bone such as the femur has a central canal filled with soft bone marrow and blood vessels. The canal is surrounded by a layer of spongy bone, which is wrapped around with a layer of tough compact bone. Covering the bone's outer surface is a thin membrane, the periosteum.*

Fact

## Ancient bones

Because bones are so hard, they can remain undecayed for hundreds of years after death. Over immense periods of time, bones can fossilize, their tissues being replaced with even harder minerals. Fossil bones often retain their shape so well that they are immediately recognizable. This 12,000-year-old skeleton of a Cro-Magnon human enables palaeontologists to reconstruct the appearance of one of our early ancestors.

# JOINTS

The point at which two bones meet is called a joint, also known as an articulation. Joints are classified by their structure or by the way in which they move. In the freely movable synovial joints, the surfaces that are in contact, called the articular surfaces, slide over each other easily. Semimovable joints, such as those in the spine, are more firmly linked and provide greater stability but less flexibility. Some joints, such as those of the skull, do not have any mobility at all.

## FREELY MOVABLE JOINTS

Most of the joints in the body are freely movable synovial joints. These joints are lubricated by synovial fluid secreted by the joint lining, thus enabling articular surfaces to move with minimal friction. Pivot and hinge joints move in only one plane (from side to side, or up and down, for example), and ellipsoidal joints are able to move in two planes at right angles to each other. Most joints can move in more than two planes, which allows for a wide range of movement. The shoulder, a ball-and-socket joint, is one of the most mobile and most complex joints in the body. It moves up and down, forwards and backwards, and can even rotate, allowing the arm to move in a complete circle.

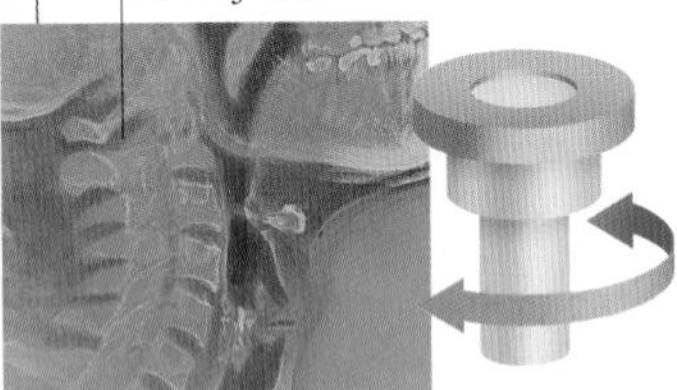


**PIVOT JOINT**
*One bone rotates within a collar formed by another. The pivot joint between the upper bones of the neck allows the head to turn.*

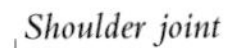
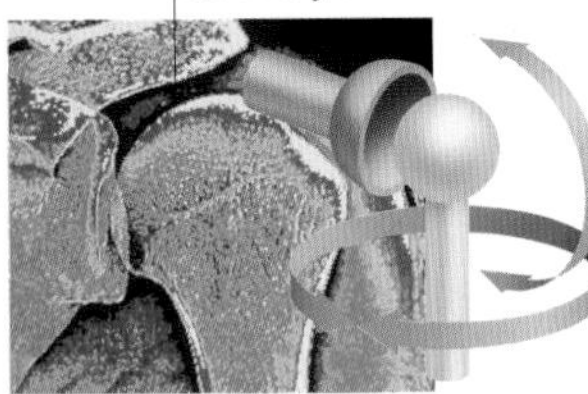


**BALL-AND-SOCKET JOINT**
*The ball-shaped end of one bone fits into a cup-shaped cavity in another, allowing a range of movement, such as in the shoulder.*

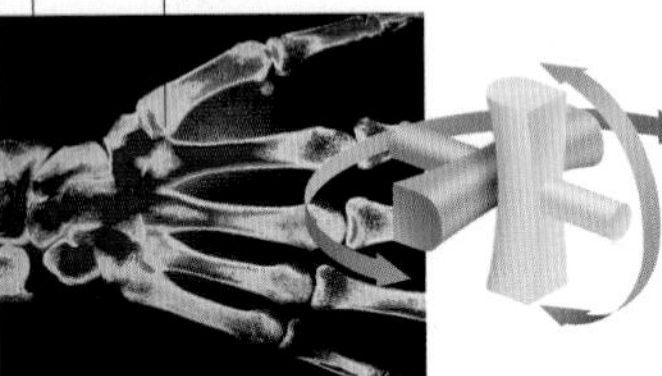


**SADDLE JOINT**
*Saddle-shaped bone ends that meet at right angles can rotate and move back and forth. Saddle joints are at the base of the thumbs.*

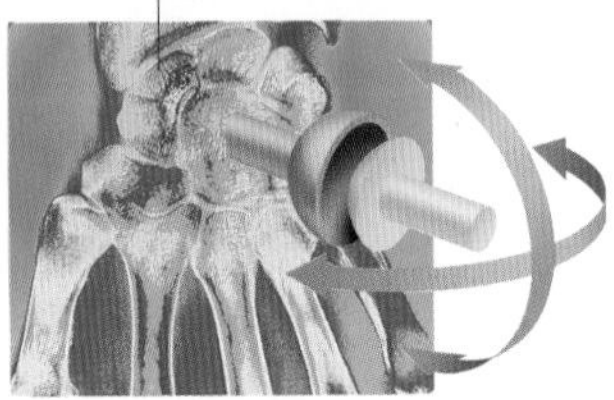


**ELLIPSOIDAL JOINT**
*The oval end of one bone fits into the cup of another, allowing varied movement, but little rotation, such as at the wrist.*

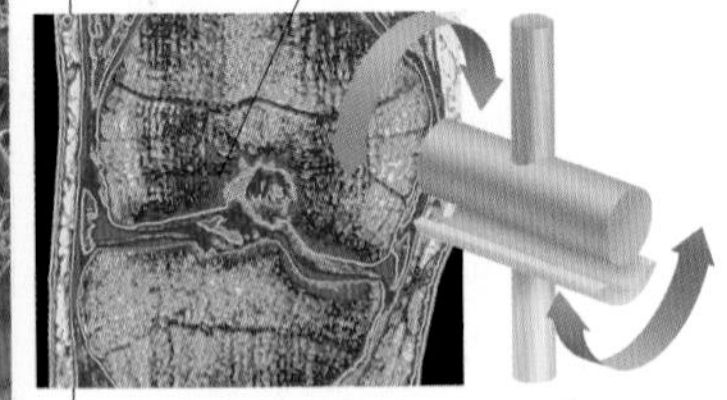


**HINGE JOINT**
*The cylindrical surface of one bone fits into the groove of another, allowing bending and straightening, for example at the knee.*

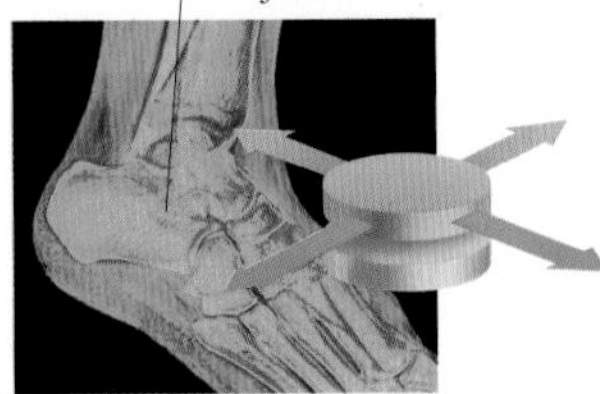


**PLANE JOINT**
*Surfaces that are almost flat slide over each other, back and forth, and sideways. Some joints in the foot and wrist are plane joints.*

## SEMIMOVABLE AND FIXED JOINTS

Not all joints in the body are as freely movable as synovial joints. In semimovable joints, the articular surfaces are fused to a tough pad of cartilage that allows for only a little movement. The joints in the spine are semimovable, as is the joint at the base of the pelvis. Other joints allow for no movement at all and are fixed in place. The separate plates of bone in the skull allow for growth during childhood, but in adulthood, once growth is complete, they fuse together. The sacrum, in the lower spine, is another example of joints that are fused together. Here, individual vertebrae form a solid triangular unit, providing stability and support.

*Pubic symphysis*

**SEMIMOVABLE JOINT**
*The pubic symphysis joint links the front halves of the pelvis. It is semimovable, allowing for only a limited scope for movement at the front of the pelvis.*

*Suture*

**FIXED JOINTS**
*The suture joints of the skull hold or connect the bones of the skull firmly together.*

## KEEPING JOINTS STABLE

The body's synovial (freely movable) joints need to be kept stable while allowing maximum flexibility. Most joints are stabilized by ligaments, which are tough bands of fibrous, elastic connective tissue. External ligaments attach to the bones on either side of the joint, forming a fibrous capsule that completely encases the joint. The fibrous capsule protects the joint from damage or injury, while keeping it stable, yet flexible. As well as external ligaments, the knee joint has internal ligaments. These so-called cruciate ligaments are thick, fibrous bands that cross over inside the joint, linking the ends of the bones that form the joint. Some of the joints in the body, such as the ankle and wrist, also have thickened sheets of connective tissue called retinacula that wrap, like cuffs, around the tendons. The role of the retinacula is to hold the tendons in and stop them from bowing when the muscles contract and shorten.

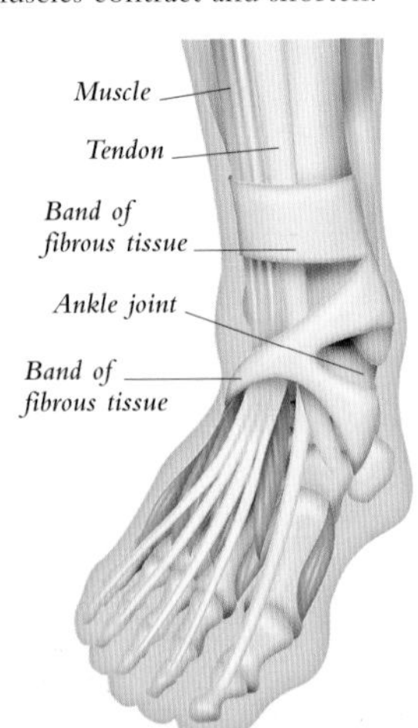


**STABILIZING BANDS**
*Bands of fibrous connective tissue, called retinacula, wrap around the ankle to provide the joint with maximum stability.*

Fact

### Flexible joints

The vertebrae (bones in the back) are amazingly versatile, keeping us stable and upright when needed, and also allowing us to bend and arch backwards (see below) and forwards. Individually, the vertebrae are rigid and immovable, but as an interlocking group, they join together and work as a team to give the spine its flexibility. A network of ligaments connects the bones and the surrounding muscle, providing support for the joints and controlling movement.

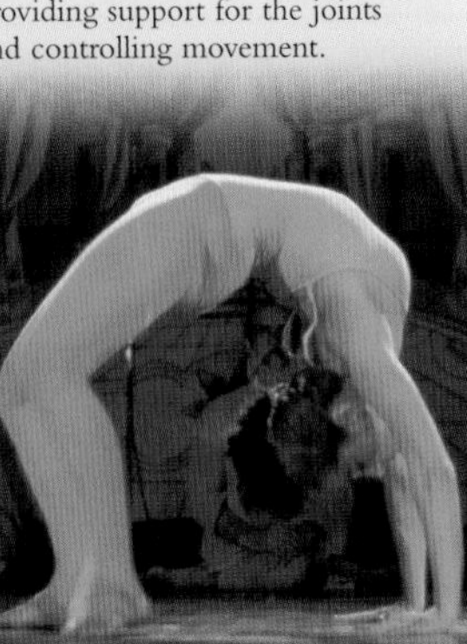
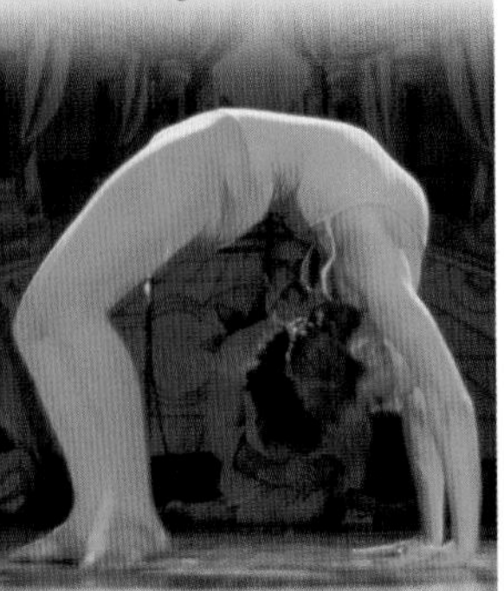

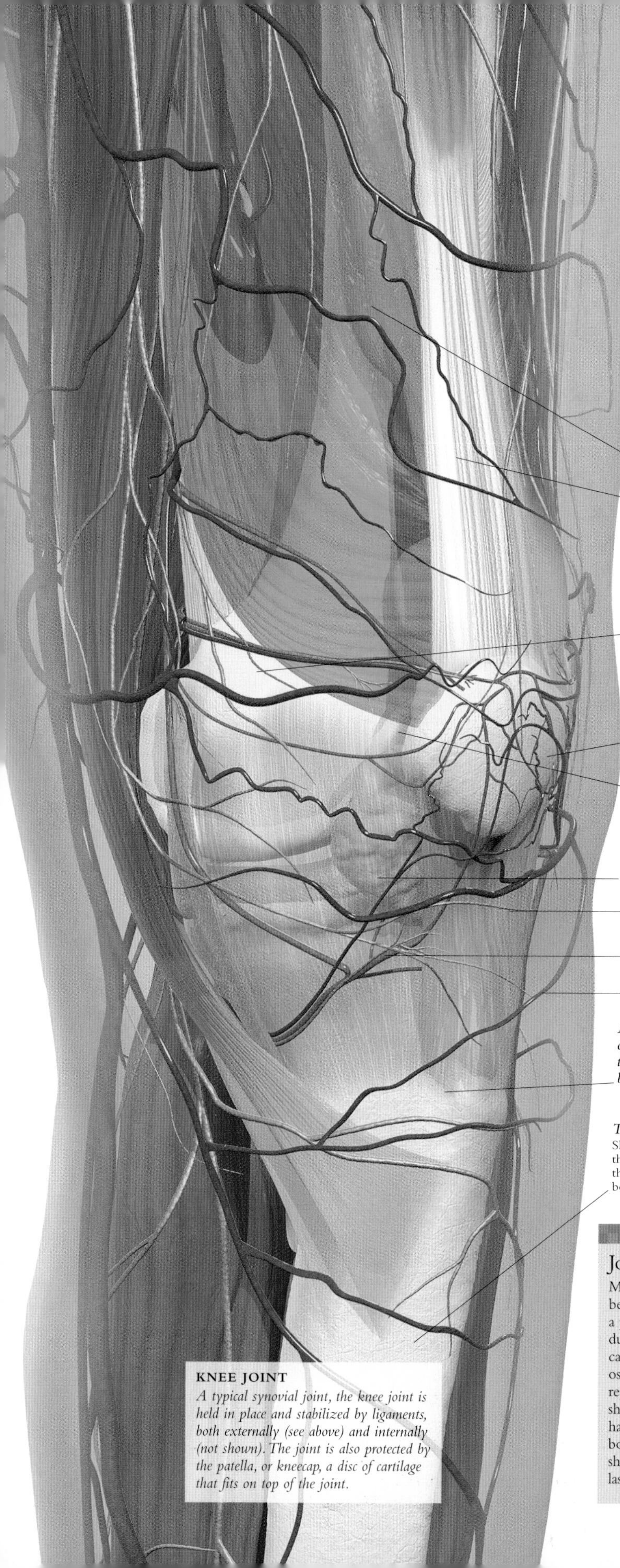

Muscle

*Patellar tendon*
Crosses over the patella, which is embedded in it

*Femur*
Thighbone, the bone of the upper leg

*Patella*
Protective disc of bone, also called the kneecap

*Synovial membrane*
Produces synovial fluid

*Pad of fat*

*Nerve*

*Artery*

*Vein*

*Attachment of patellar tendon to bone*

*Tibia*
Shinbone, the larger of the two leg bones

**KNEE JOINT**
*A typical synovial joint, the knee joint is held in place and stabilized by ligaments, both externally (see above) and internally (not shown). The joint is also protected by the patella, or kneecap, a disc of cartilage that fits on top of the joint.*

# INSIDE A JOINT

Synovial joints are structured so that they are movable, yet stable. The articular surfaces, where bone ends meet, are covered with smooth cartilage so that they can slide easily over each other. To ensure mobility, the ends of the bones are bathed in synovial fluid. This fluid also has a protective function and contains fats and proteins that nourish the cartilage covering the ends of the bones. The synovial membrane secretes and holds the fluid, reabsorbing and replenishing it continuously. The synovial membrane gradually thickens to form the outer fibrous capsule encasing the joint.

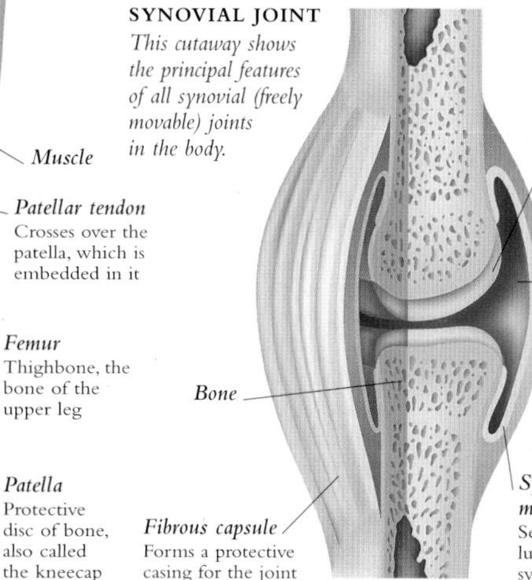
*Articular cartilage*
Covers bone ends, providing a smooth surface for bones to move over each other without causing damage
*Synovial fluid*
Lubricating fluid that bathes the bone ends, minimizing friction
*Bone*
*Synovial membrane*
Secretes lubricating synovial fluid
*Fibrous capsule*
Forms a protective casing for the joint

**SYNOVIAL JOINT**
*This cutaway shows the principal features of all synovial (freely movable) joints in the body.*

# SHOCK ABSORBERS

In the knee, the wrist, and the spine, shock-absorbing fibrous discs positioned between the bones of the joints protect the bones from damage. These discs are called meniscuses in the knee and the wrist, and intervertebral discs in the spine. The structure of meniscuses can be described as being like a jam doughnut. The inner core consists of a jelly-like material comprising mainly water and collagen (a material composed of protein fibres); the outer part is made of tough, fibrous cartilage.

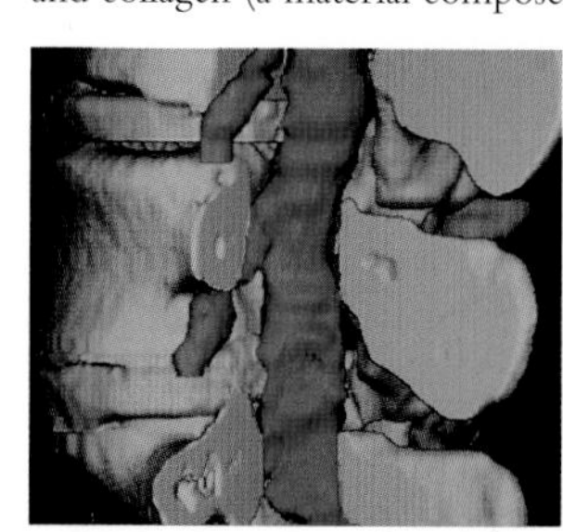

**SPINAL DISCS**
*In this CT scan, parts of the vertebrae have been cut away to reveal the intervertebral discs (yellow), which absorb shock and prevent damage to the spine.*

Health

## Joint replacement

Many joints in the body that have been damaged can be replaced with a prosthesis. Damage to joints is often due to conditions that damage the cartilage around the bone, such as osteoarthritis. Joints that can be replaced include the hip, knee, and shoulder joints. Here, a prosthetic end has been inserted into the humerus bone of the upper arm, creating a new shoulder joint. Such replacements can last for many years.

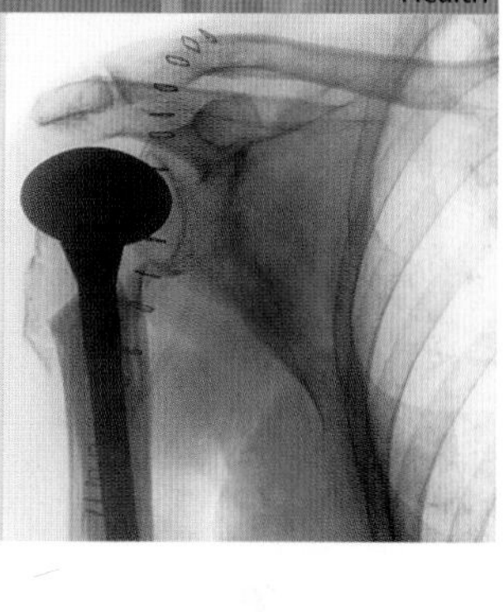

# MUSCLES

Muscles make up the bulk of the body – accounting for half of its weight. They consist of tissue that contracts powerfully to move the body, maintain its posture, and work the various internal organs, including the heart and blood vessels. These functions are performed by three types of muscle – skeletal muscle, cardiac (heart) muscle, and smooth muscle (see below). Most of the muscle in the body is skeletal muscle. Usually, each end of a skeletal muscle is attached to a bone by a tendon, a flexible cord of fibrous tissue.

## TYPES OF MUSCLE

There are three types of muscle in the body: skeletal, cardiac, and smooth. They each perform different roles and have differing structures. Skeletal muscle, which covers and moves the skeleton, consists of long cells, or fibres, that are able to contract quickly and powerfully. Cardiac (heart) muscle, is made up of short interlinked fibres capable of sustained rhythmic movement. Smooth muscle performs the unconscious actions of the body, such as propelling food along the digestive tract. Its fibres contract relatively slowly but they can continue in a state of contraction for long periods.

**SKELETAL MUSCLE**
*The strong parallel fibres that form this type of muscle can contract quickly and powerfully, but only for a short time.*

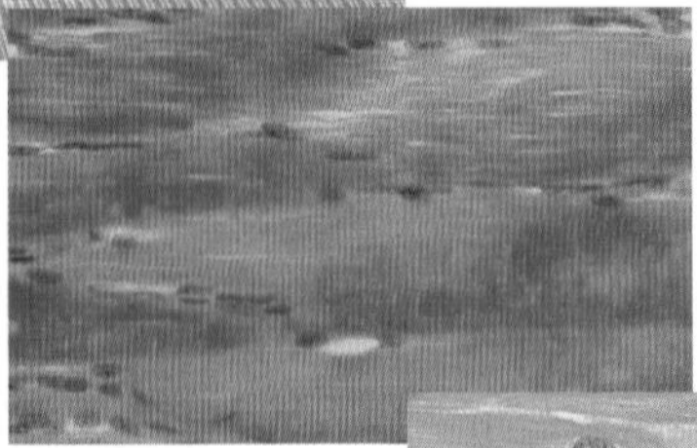

**CARDIAC MUSCLE**
*Short, branching, interlinked fibres form a network within the wall of the heart. Healthy cardiac muscle contracts rhythmically and continuously in order to pump blood around the the whole of the body.*

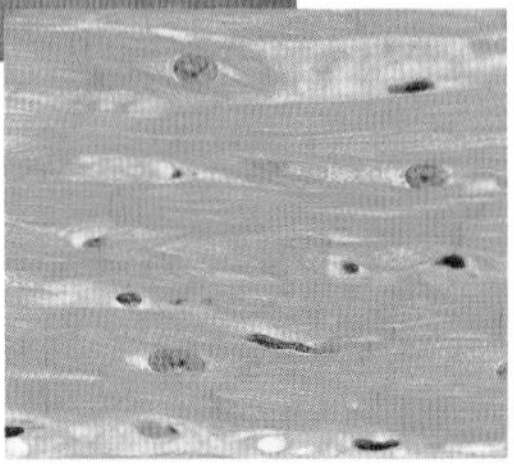

**SMOOTH MUSCLE**
*These short, spindle-shaped fibres are thinner than skeletal muscle fibres, and form sheets of muscle. Smooth muscle can contract for long periods.*

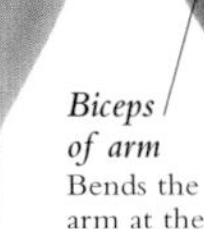


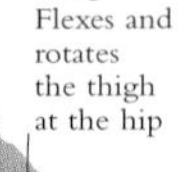

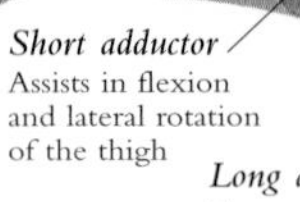

History

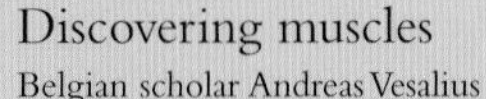

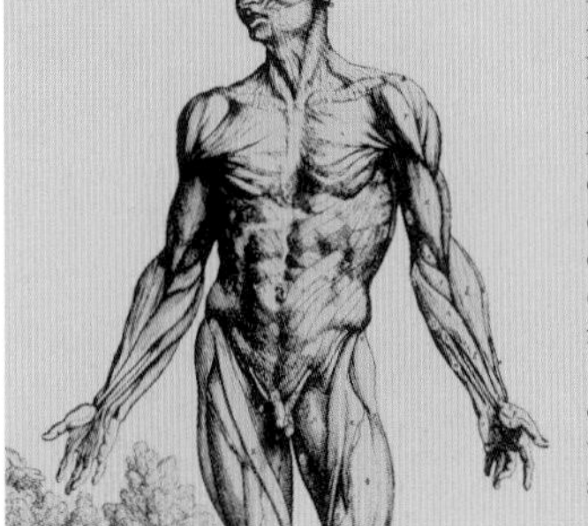

### Discovering muscles

Belgian scholar Andreas Vesalius (1514–1564) uncovered the intricate layers of the body's skeletal muscles by careful and accurate dissection. He established the science of anatomy, based on observation and realism. The famous engraving shown here was first published in 1543 in Vesalius' book entitled *On the Structure of the Human Body*, in which he describes cutting away "the skin together with the fat, and all the sinews, veins and arteries existing on the surface".

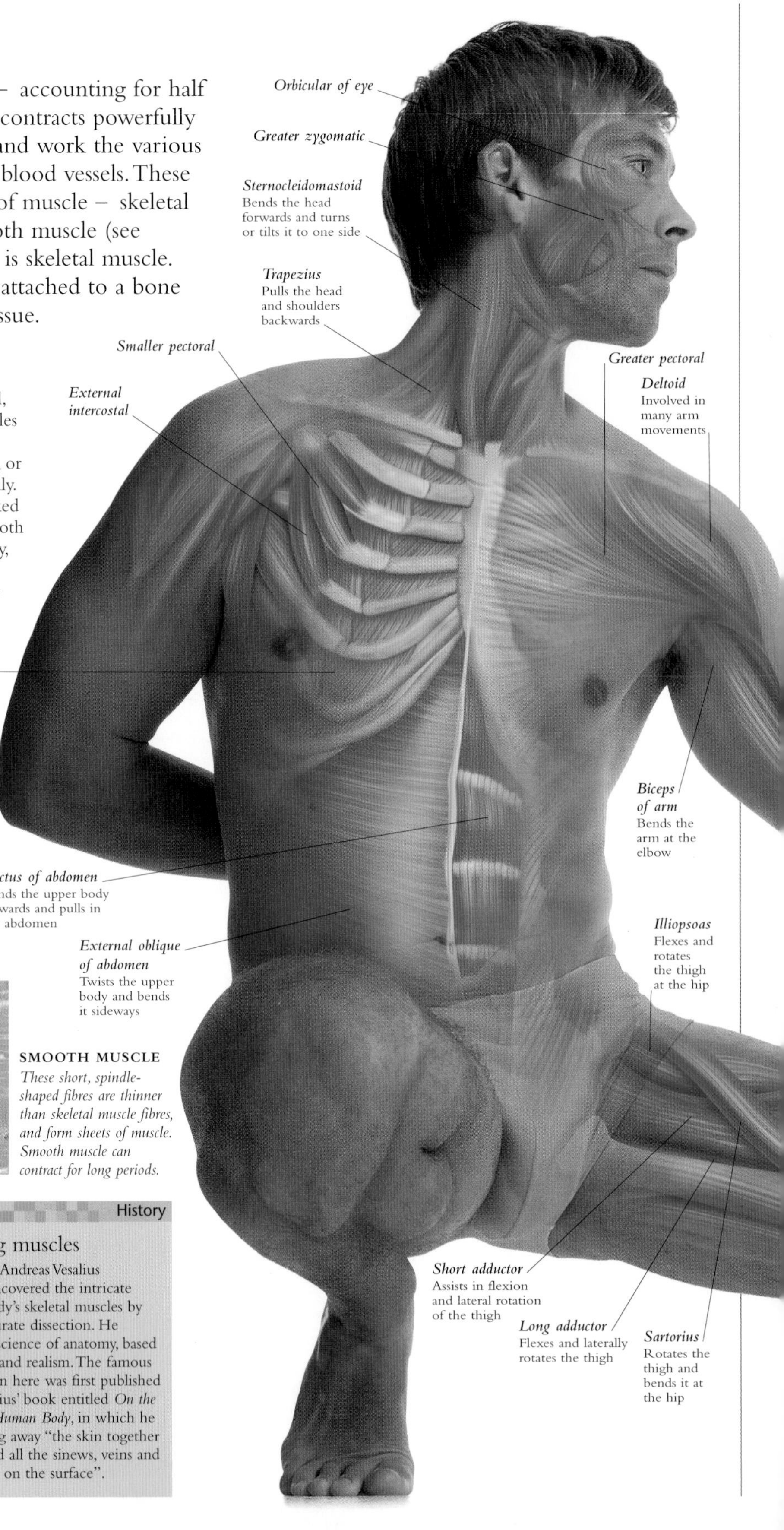

BODY

# SKELETAL MUSCLE

The body's skeletal muscles move the body and joints by contracting. In addition, they maintain a steady tension, or tone, that gives the body the support it needs to sustain posture. Our vast range of facial expressions, which is a significant means of communication, is also facilitated by our skeletal muscles. There are over 600 skeletal muscles in the body, in a variety of shapes and sizes – from large triangular slabs of muscle, such as the deltoid in the shoulder, to long, thin strips, such as the sartorius, which curves around from the hip to the inside knee. Skeletal muscle can contract quickly and powerfully. The shape, size, and length of a muscle all have a bearing on the strength with which it contracts, and this influences the amount of force it can generate. Muscle contraction is stimulated by nerve impulses, sent along pathways linking the brain to the muscle tissue.

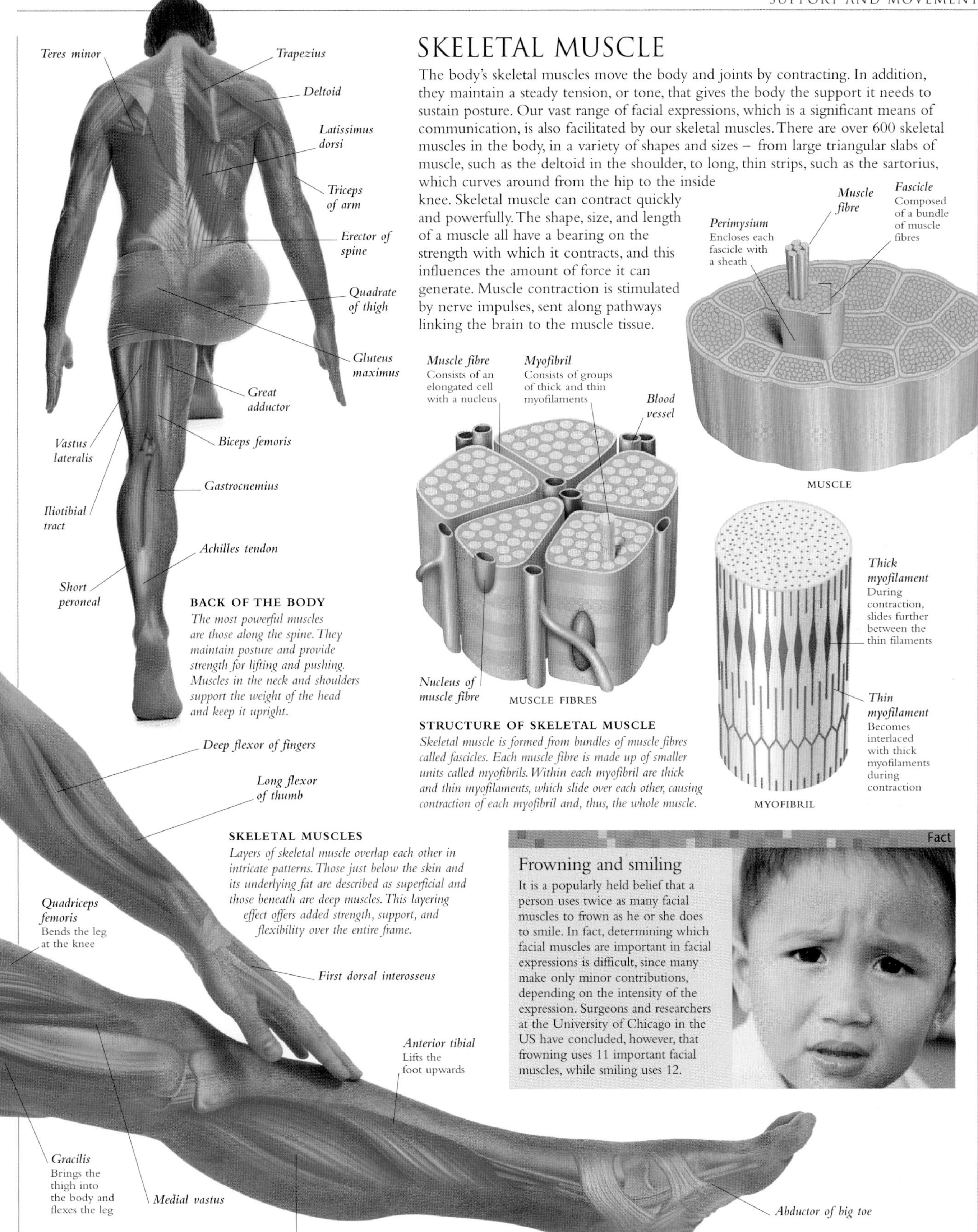


**BACK OF THE BODY**
*The most powerful muscles are those along the spine. They maintain posture and provide strength for lifting and pushing. Muscles in the neck and shoulders support the weight of the head and keep it upright.*

**STRUCTURE OF SKELETAL MUSCLE**
*Skeletal muscle is formed from bundles of muscle fibres called fascicles. Each muscle fibre is made up of smaller units called myofibrils. Within each myofibril are thick and thin myofilaments, which slide over each other, causing contraction of each myofibril and, thus, the whole muscle.*

**SKELETAL MUSCLES**
*Layers of skeletal muscle overlap each other in intricate patterns. Those just below the skin and its underlying fat are described as superficial and those beneath are deep muscles. This layering effect offers added strength, support, and flexibility over the entire frame.*

## Frowning and smiling

It is a popularly held belief that a person uses twice as many facial muscles to frown as he or she does to smile. In fact, determining which facial muscles are important in facial expressions is difficult, since many make only minor contributions, depending on the intensity of the expression. Surgeons and researchers at the University of Chicago in the US have concluded, however, that frowning uses 11 important facial muscles, while smiling uses 12.

## TENDONS

Muscles are linked to bones by tendons, which are cord-like structures made of collagen, a tough, fibrous protein. Tendons, unlike muscles, do not stretch but they have some flexibility. Where they join to a bone, tendon fibres pass through the bone's outer membrane, the periosteum, and embed themselves in the bone tissue. The link is extremely strong, and tendons resist great tension without snapping. Some tendons, including those in the hands and feet, are encased in fibrous capsules called synovial sheaths. The sheaths secrete a lubricating fluid that protects the tendons from friction where they move against the bone.

**TENDON FIBRES**
*Tendons are largely made up of collagen, a fibrous connective tissue. The fibres of a tendon are arranged in parallel bundles (as shown in the photograph), forming a cord-like structure of great strength.*

## HOW MUSCLES WORK

Skeletal muscles typically connect two bones, stretching from one to the other across a joint. An individual muscle produces movement when it contracts (shortens) and pulls on the bone to which it is attached. Muscles can only pull bones towards or away from each other; they cannot push. For this reason, many muscles are arranged in pairs, on either side of a joint, working in opposition to move body parts. While one muscle or muscle group contracts, pulling a bone in one direction, the muscle or muscle group on the opposite side of the joint relaxes. In order to reverse the movement, the first muscle relaxes while its counterpart contracts. Examples of paired muscles are the biceps and triceps in the upper arm. When the body is not moving, all muscles are held in a state of partial contraction. This natural tension, muscle tone, maintains posture.

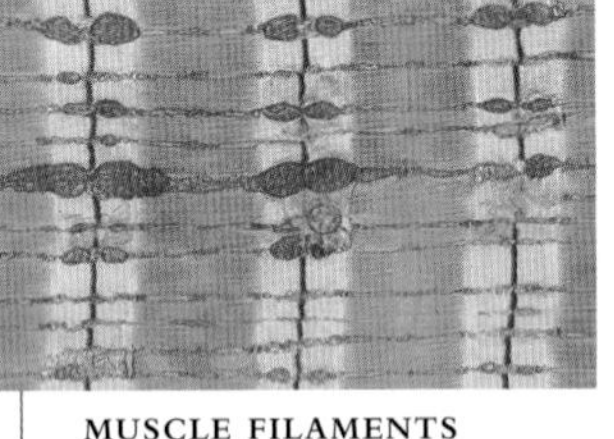

**MUSCLE FILAMENTS**
*The filaments that form skeletal muscle are arranged in a repeating pattern of stripes. In order to move the body, the muscle fibres need to contract and relax.*

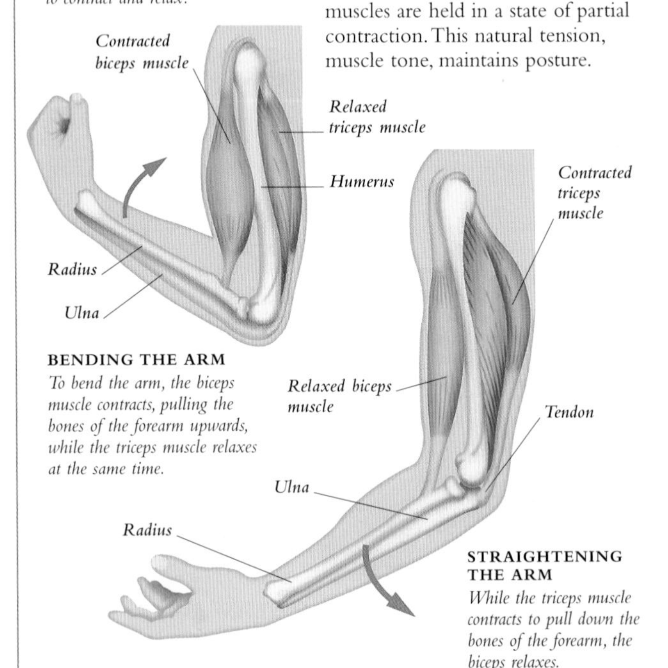


**BENDING THE ARM**
*To bend the arm, the biceps muscle contracts, pulling the bones of the forearm upwards, while the triceps muscle relaxes at the same time.*

**STRAIGHTENING THE ARM**
*While the triceps muscle contracts to pull down the bones of the forearm, the biceps relaxes.*

## LEVERS

Muscles pull on bones to make the body move according to the same principles that operate mechanical lever systems. A lever is a rigid bar that has one pivot point, the fulcrum; force applied to one part of the lever is transferred through the fulcrum to a weight (resistance point) on another part of the lever. Translated into bodily terms, this means that the bones serve as levers; the muscles apply force; the part of the body to be moved, such as a limb, provides resistance; and the joints at which bones meet function as fulcrums. There are three types of lever system in the body: first, second, and third class. These are classified according to the relative positions of the fulcrum, the force applied, and the weight being resisted.

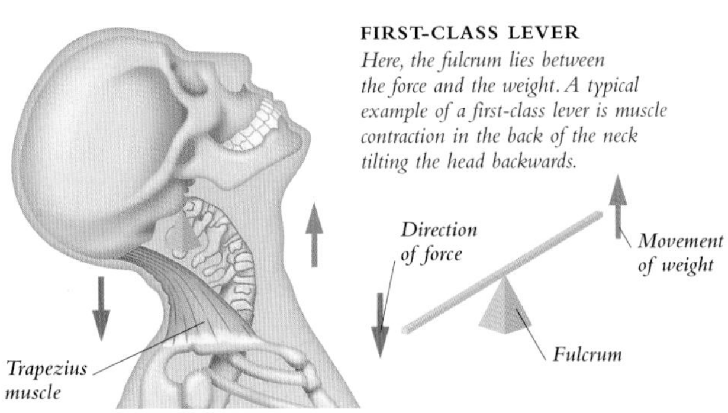


**FIRST-CLASS LEVER**
*Here, the fulcrum lies between the force and the weight. A typical example of a first-class lever is muscle contraction in the back of the neck tilting the head backwards.*

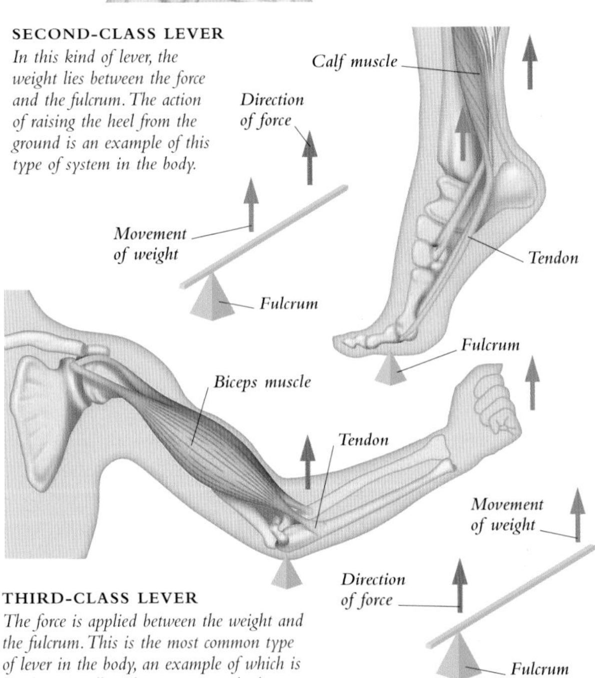


**SECOND-CLASS LEVER**
*In this kind of lever, the weight lies between the force and the fulcrum. The action of raising the heel from the ground is an example of this type of system in the body.*

**THIRD-CLASS LEVER**
*The force is applied between the weight and the fulcrum. This is the most common type of lever in the body, an example of which is bending the elbow by contracting the biceps.*

Fact

### Bodybuilding

The size and strength of the individual fibres in skeletal muscles can be gradually increased by physical exercise. Regularly performing multi-repetition exercises, especially using weights, is an effective way of developing well-defined muscles. A dangerous and illegal short-cut to bodybuilding is the abuse of certain drugs, anabolic steroids, that promote the growth of muscle tissue. Long-term use of these drugs can produce side-effects, including liver damage and reduced fertility. Anabolic steroids are prohibited by sporting and athletic authorities worldwide.

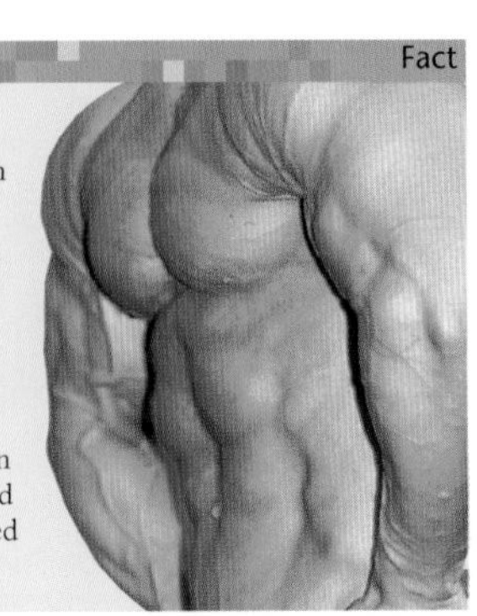

Health

## Muscle cramps

During vigorous physical exercise, sudden muscle cramps may develop. A common reason for these painful spasms is the accumulation of a waste product, lactic acid, in muscle cells. This waste builds up when the body is short of oxygen. Carried to muscle fibres in fine, branching capillaries (*see below*), oxygen is essential to muscle activity. If muscles run out of oxygen, for instance in a race, they make energy without it, which produces lactic acid.

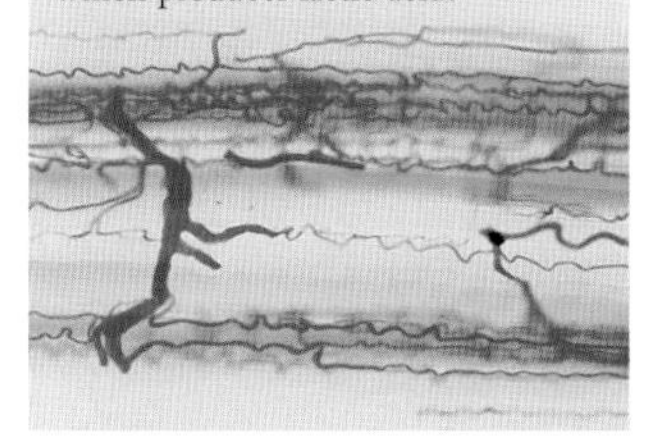

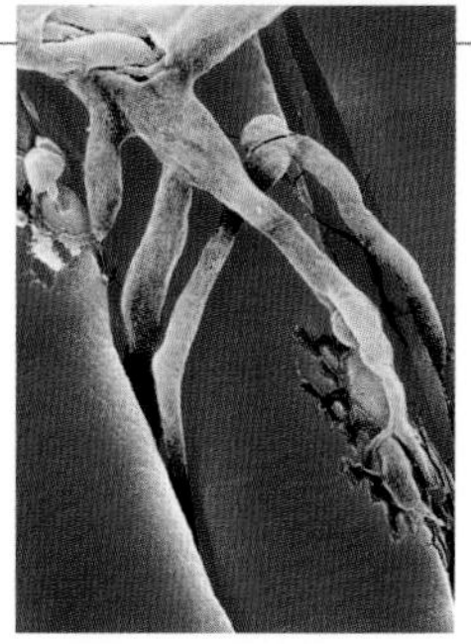

**NERVE JUNCTION**
*Impulses from the brain, triggering muscle action, reach muscle fibres through the endings of nerves (green).*

# NERVOUS SYSTEM CONTROL

Movement of the body depends not just on mechanical interaction of muscles and bones, but on signals from the brain and nerves. Contraction of skeletal muscles is often involuntary, but it is also likely to be the result of conscious thought. Once our brain has made the decision to move something – for example, to take a step forwards or bend the arm – it sends out electrical signals to muscles along nerve pathways. When the signals arrive at the appropriate muscle, the filaments within the muscle fibres respond by contracting. If signals from the brain cease, the muscle fibres are no longer stimulated and the muscle relaxes. Another link between the brain and body movement is an internal monitoring system, proprioception. Proprioceptors are types of sensory receptors, located in muscles and tendons, that collect information about the degree of stretch of muscles and tendons around the body. This information, which is passed to the central nervous system (the brain and spinal cord) helps to give us our sense of balance and our awareness of the position of various parts of the body in relation to each other.

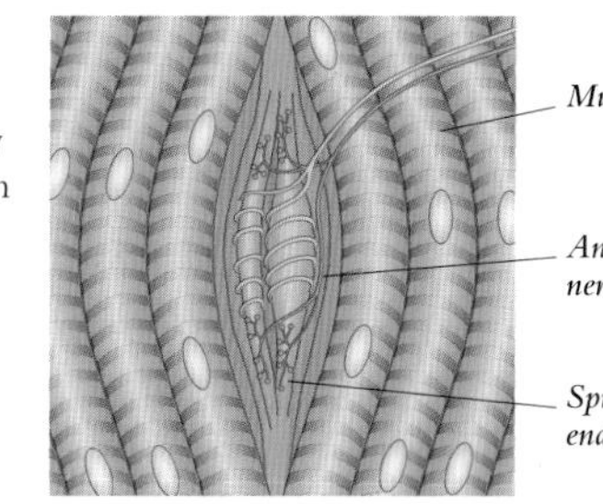


**PROPRIOCEPTORS**
*Two types of muscle proprioceptor are annulospiral sensory nerve endings, which wind around the fibres, and spray nerve endings, which lie on top of the fibres.*

# FAST AND SLOW MUSCLE FIBRES

Skeletal muscles are made up of two types of fibre: fast and slow. Fast muscle fibres contract rapidly, enabling someone to produce a burst of intense activity, such when making a sudden sprint or lifting a heavy weight. However, these fibres also tire quickly. Slow muscle fibres contract more slowly but can keep going for a long time. For example, a marathon runner uses slow fibres to sustain a steady, untiring pace over a long distance. The fibre types differ in the way they produce energy for muscle contraction. To be able to function, slow fibres need oxygen, obtained from circulating blood. The cells that make up slow fibres contain many mitochondria; these are structures that use nutrients and oxygen taken in by the body to create fuel for activity. Fast fibres have fewer and smaller mitochondria than slow fibres, and are able to produce energy without oxygen, although only in relatively small amounts. In most people, the proportion of fast and slow fibres in the skeletal muscles is about equal. However, some top athletes do seem to have a greater percentage of one type of fibre over the other, a genetic predisposition which possibly contributes to their particular talents. A sprinter or a basketball player may have more fast fibres, while a long-distance runner has more slow fibres.

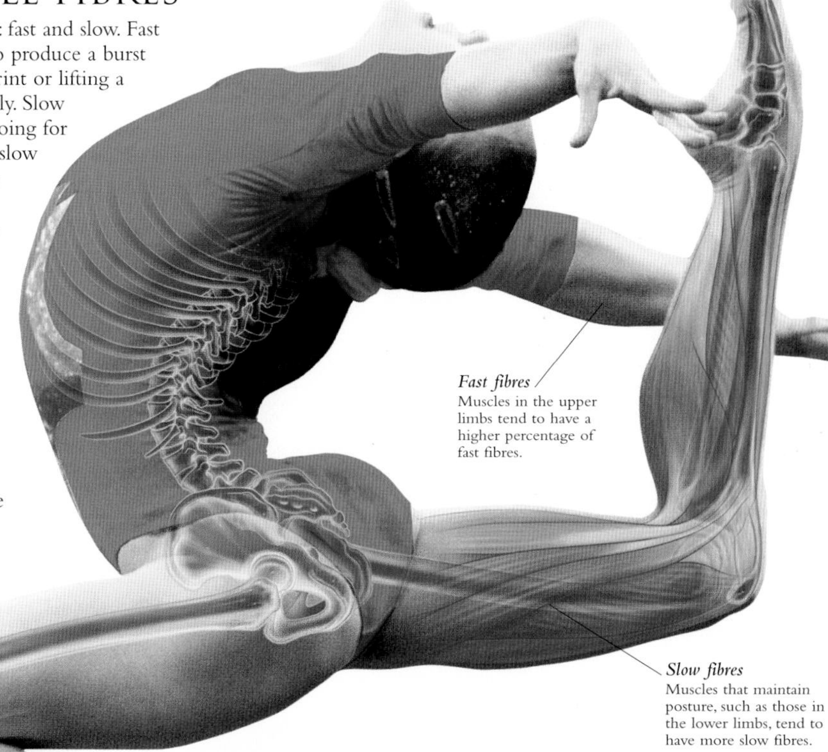

*Fast fibres*
Muscles in the upper limbs tend to have a higher percentage of fast fibres.
*Slow fibres*
Muscles that maintain posture, such as those in the lower limbs, tend to have more slow fibres.


**TYPES OF FIBRE**
*Like everyone, this gymnast has a mix of fast and slow fibres in her muscles. She needs the fast type to perform rapid movements and the slow fibres for more sustained exercises.*

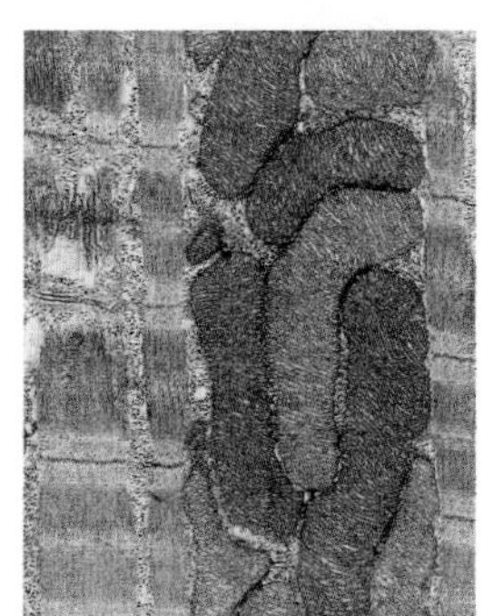

**ENERGY PRODUCERS**
*Large numbers of energy-producing mitochondria (centre of photograph) occur in slow muscle fibres.*

Fact

## Speed limits

All mammals, including humans, move by using the same mechanical principles. However, specialized adaptations between one species and another make a vast difference to physical performance. No amount of training could make a human win a race against a cheetah, the fastest animal on earth. This animal's long legs and spine, and the rapidity with which its muscles contract help it to reach speeds of 100km/h (62mph).

BODY

**USING OXYGEN**
*A swimmer breaks the surface of the water for a gulp of air. Humans cannot survive for more than a few minutes without replenishing their oxygen supply.*

# BREATHING AND CIRCULATION

Animals need a constant supply of oxygen; without this vital gas, they would die quickly. Oxygen keeps body cells working by releasing energy from the nutrients we obtain from food. Breathing draws in plenty of oxygen, but an efficient transport system is needed to carry the oxygen around the body. This role is performed by the heart and the circulating blood. Another important function of breathing and blood circulation is to remove the waste gas, carbon dioxide, produced by body cells.

The cycling of oxygen and carbon dioxide between the body and the atmosphere – respiration – is not just a matter of breathing in and out but involves every cell in the body. This complex process, which for the most part takes place without any conscious effort, needs the coordination of many body parts and functions. The brain, airways and lungs, heart, blood, and blood vessels are all essential components.

## AIR INTAKE

No one can survive for more than a few minutes without taking in air. Most of the time we are not aware of the action of breathing, although we draw breath over 20,000 times a day. The rhythmic, reflex action, as the lungs expand and shrink, is controlled by the brain. Breathing automatically adjusts in depth and rate in order to meet different levels of exertion. Breathing becomes noticeable only if it is difficult or we try to take deliberate control.

Every breath of air we inhale is filtered, warmed, and moistened as it flows through the cavities of the nose. After travelling down the pharynx (throat) and trachea (windpipe), inhaled air passes along ever-branching pathways until it is deep within the spongy tissues of the lungs. The moist interior of the lungs provides a damp surface through which oxygen can diffuse easily to reach the bloodstream. There are clusters of tiny air sacs, alveoli, in the lungs that together provide a huge surface area. This arrangement maximizes oxygen intake. Waste carbon dioxide leaves the body by the reverse route, entering the lungs through the alveoli. This gas is removed through the airways as we exhale.

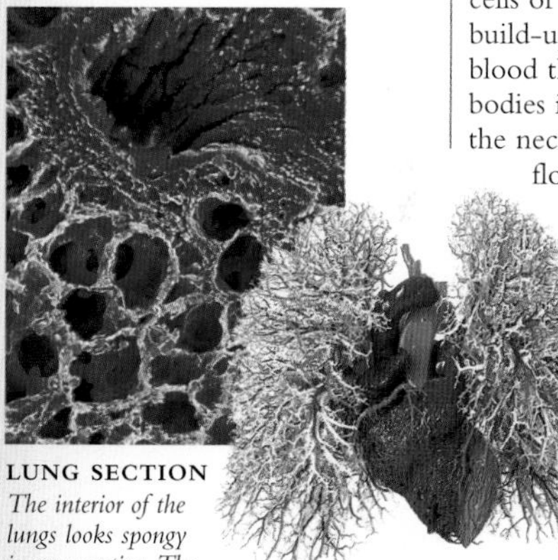

**LUNG SECTION**
*The interior of the lungs looks spongy in cross-section. The tiny airways seen here are bronchioles (top of photograph) and alveoli (bottom).*

**BRONCHIAL TREE**
*This resin cast shows how the air passages, bronchi, in the lungs branch to form an intricate network.*

## VITAL EXCHANGE

Human bodies are so large that simple gas diffusion through the lungs is not fast enough either to get oxygen to all our billions of cells or to prevent a dangerous build-up of carbon dioxide. The blood that travels around our bodies in a double circuit provides the necessary service. As blood flows through the lungs an exchange of gases takes place through the thin walls of the alveoli. The cells in blood collect oxygen and deposit carbon dioxide. From the lungs, oxygenated blood goes to the heart, which pumps it around the entire body. The oxygenated blood is carried through arteries, the largest of our blood vessels, then through smaller and smaller channels until it has reached every tissue. Once in the tissues, blood releases its supply of oxygen to cells that need it and picks up cell waste, carbon dioxide. To complete the circuit, the blood returns to the heart through a network of veins. The cycle then repeats as blood is pumped back to the lungs to rid itself of carbon dioxide and pick up more oxygen.

## BLOOD CIRCULATION

Under a microscope, it can be seen that blood is made from various types of cells floating in a watery fluid and that its colour comes from millions of round, red discs. These discs are red blood cells; they have special properties that enable them to collect oxygen in the lungs and release it in body tissues. The other components of blood take no part in oxygen transport. The fluid in which the cells float (plasma) carries nutrients; white blood cells help to protect the body from disease; and platelets are involved in blood clotting.

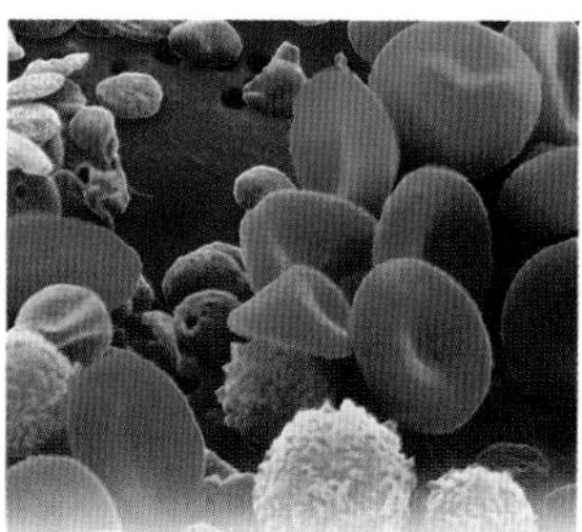

**BLOOD CELLS**
*A sample of blood shows a variety of cells: red cells carry oxygen, white cells fight disease, and tiny platelets aid clotting.*

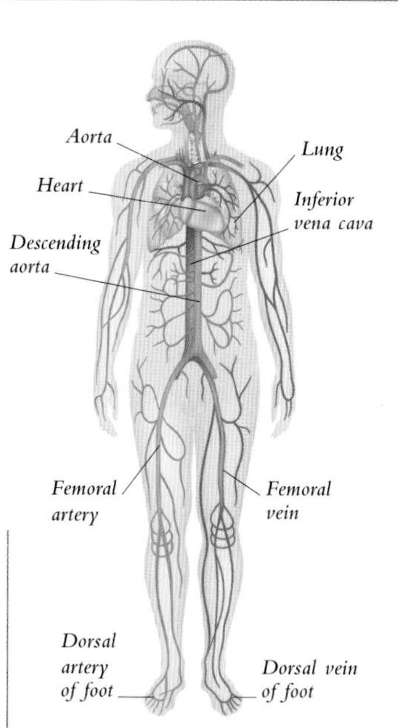


**OXYGEN CYCLE**
*Blood containing oxygen collected in the lungs is pumped into the arteries (red) by the heart and around the body. Oxygen-poor blood drains back to the heart through the veins (blue) and is pumped to the lungs.*

To get enough oxygen to all of the body's organs, all of our blood must pass through the lungs and around the body at least once a minute. Keeping this nonstop circulation going means that blood has to travel fast and therefore must be actively forced through the blood vessels. The heart is the pump that keeps blood flowing. Situated between the lungs, tilted a little towards the left side of the body, this hollow, muscular sac works tirelessly. Expanding and contracting with a steady rhythm, the heart fills with blood and then pumps it out again to the lungs and the body's labyrinth of blood vessels. The special heart muscle beats of its own accord, producing the characteristic sound heard in a stethoscope. Although heart action is automatic, the rate at which the heart beats is regulated by signals from the brain.

Fact

### Powerful lungs

Exercise increases the volume of the alveoli (air sacs) in the lungs, which means that more oxygen gets into the body with every breath. Because of their rigorous training, professional cyclists, such as Lance Armstrong (right), are known by sports physicians for their increased lung capacity. Some professional sports people are recorded as taking in up to 8 litres (14pt) of air a minute – people of ordinary fitness levels take in 5–6 litres (9–10pt).

# TAKING A BREATH

Breathing has two purposes: it gets oxygen into the body and removes carbon dioxide waste produced by body cells. Every minute, about 5–6 litres (9–10pt) of air pass into and out of the lungs. Oxygen from air entering the lungs diffuses into the bloodstream and is pumped through the circulation to all of the body's cells. In the body's tissues, the oxygen is exchanged for carbon dioxide waste, which is then returned, through the circulation, to the lungs to be exhaled.

## THE LUNGS

Healthy lungs are pink, soft, and spongy. The right lung is divided into three lobes, but the left has only two lobes because it shares chest space with the heart. Air moves into and out of the lungs through the nose and mouth, the trachea (windpipe), and a series of other air passages called bronchi and bronchioles. These airways branch out inside the lungs into an extensive, tree-like network of small tubes that end in alveoli (tiny air sacs). The lungs themselves are easily damaged, but three structures – the ribcage, spine, and a double-layered membrane, called the pleura – protect them from harm.

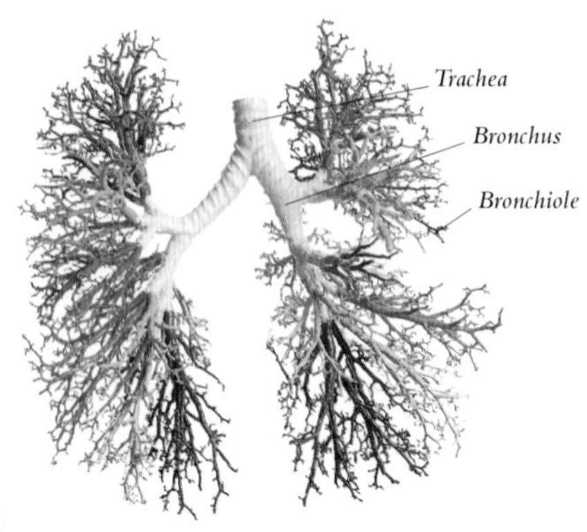


**AIR PASSAGES IN THE LUNGS**
*The largest airway in the respiratory system is the trachea. This branches into the two smaller bronchi which in turn divide to become bronchioles.*

Health

### Coughs and sneezes

The noisy, explosive force of coughing or sneezing is an effective way of clearing the airways and preventing harmful and irritating substances, such as dust, pollen, or mucus, entering the lungs and causing damage. Sneezing and coughing are often symptoms of viral infections such as the common cold and flu. Coughing can also indicate more serious disorders, such as pneumonia or damage to airways and lung tissues caused by smoking.

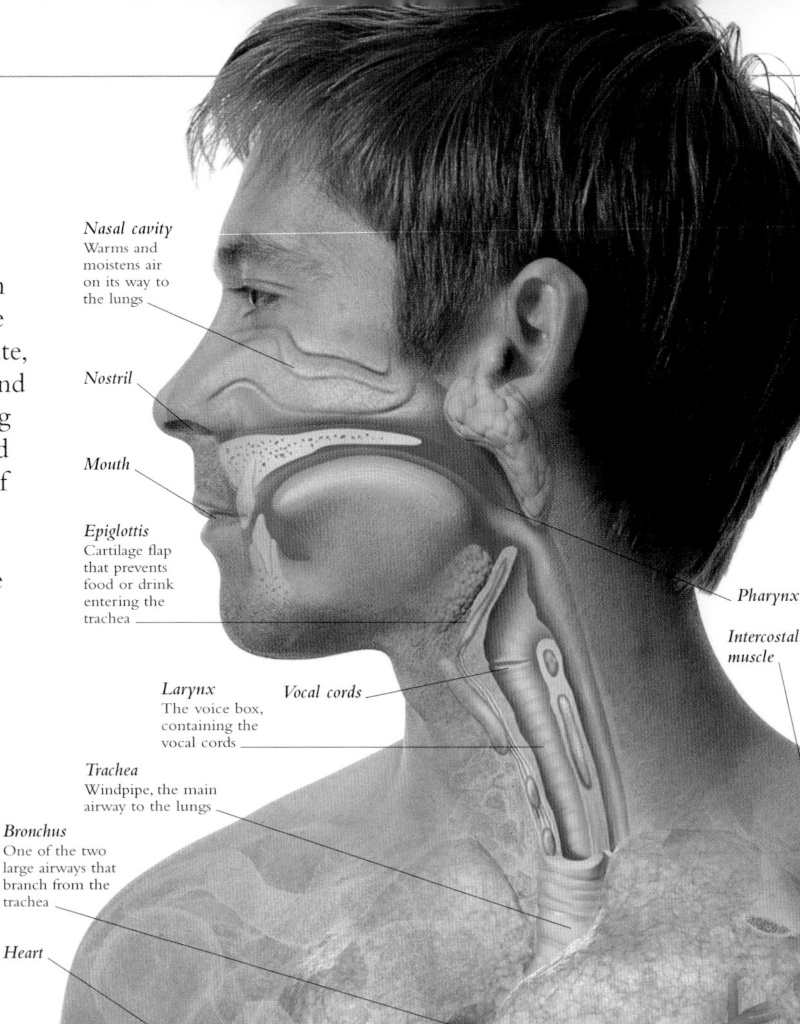


**THE RESPIRATORY TRACT**
*The nasal cavity and the pharynx (throat) form the upper respiratory tract. The larynx (voice box), trachea (windpipe), bronchi (air passages), and lungs make up the lower part.*

# GAS EXCHANGE

The tiny air passages in the lungs end in millions of alveoli – thin-walled air sacs – where gases are exchanged between the air and the blood. As oxygen enters an alveolus it dissolves in the moist lining and diffuses across the thin wall into a neighbouring capillary (tiny blood vessel). This happens rapidly: at a restful breathing rate, the blood in the capillaries comes into contact with air in an alveolus for only about 0.75 seconds but is fully oxygenated after about a third of this time. Once oxygen is in the blood, it binds with haemoglobin in red blood cells (a small amount of oxygen also travels freely in the blood) and is transported to the heart, where it is pumped to the body tissues. Carbon dioxide follows the opposite path to oxygen, and travels about 20 times as fast; it diffuses out of the capillaries and into the alveoli, where it is exhaled from the lungs.

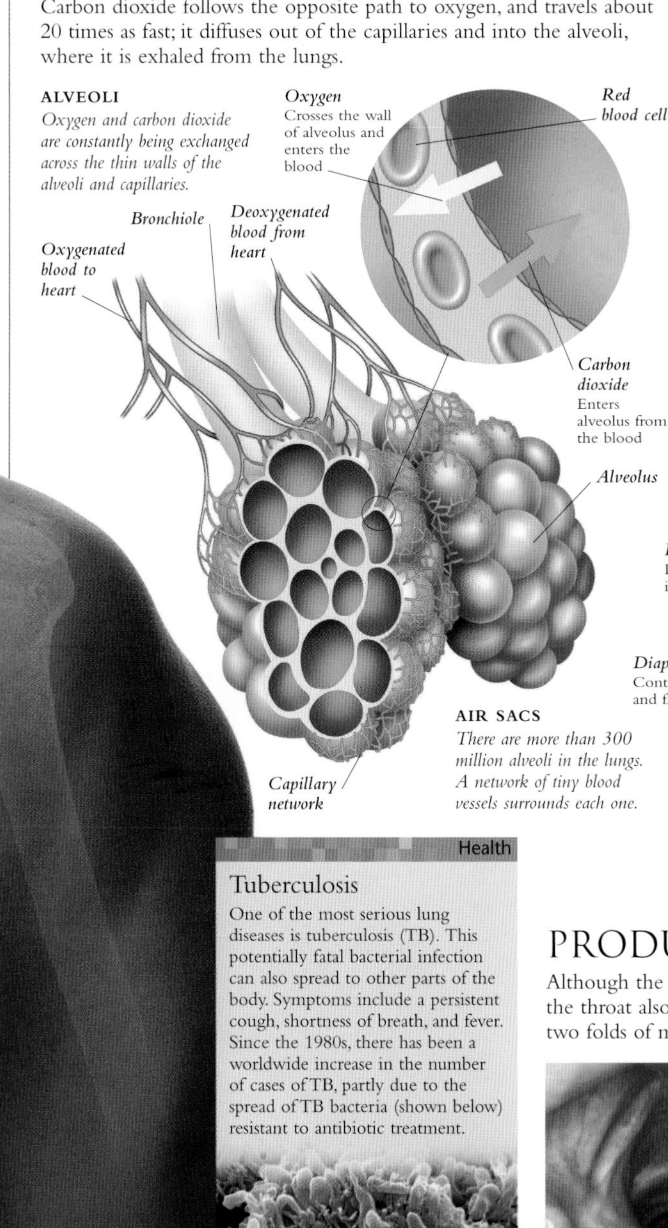


**ALVEOLI**
*Oxygen and carbon dioxide are constantly being exchanged across the thin walls of the alveoli and capillaries.*

**AIR SACS**
*There are more than 300 million alveoli in the lungs. A network of tiny blood vessels surrounds each one.*

Fact

## Free diving

In the sport of free diving, people train themselves mentally and physically to survive underwater on a single lungful of breath. The record for static free diving (staying still underwater for as long as possible) is more than 6 minutes for women and more than 8 minutes for men. Pearl divers in the South Pacific sometimes swim vigorously underwater without breathing equipment for up to 2 minutes at depths of 40m (130ft).

# GETTING AIR INTO THE LUNGS

When we breathe in, muscles work to expand the chest. The largest chest-expanding muscle is the diaphragm, a sheet of tissue that lies under the ribcage. At rest, the diaphragm arches upwards like a dome, separating the chest from the abdomen. When the diaphragm contracts, it flattens, pushing down the abdominal organs and increasing the space within the chest cavity. The intercostal muscles between the ribs also pull the chest up and out. The lungs increase in size, their internal pressure drops and air rushes in. Exhaling does not usually involve any work by the body. The diaphragm and intercostal muscles simply relax and return to their resting positions. This causes the lungs and chest wall to recoil like elastic. As they do so, the pressure in the lungs increases and air is forced back into the atmosphere.

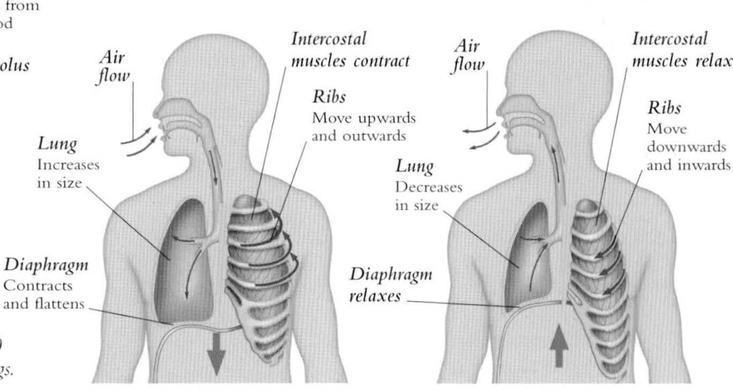


**INHALING**
*The diaphragm contracts and flattens and the intercostal muscles also contract. As the ribcage expands, pressure in the lungs drops and air rushes in.*

**EXHALING**
*The diaphragm returns to its dome-like shape and the intercostal muscles relax. Pressure in the lungs rises and air is forced out.*

Health

## Tuberculosis

One of the most serious lung diseases is tuberculosis (TB). This potentially fatal bacterial infection can also spread to other parts of the body. Symptoms include a persistent cough, shortness of breath, and fever. Since the 1980s, there has been a worldwide increase in the number of cases of TB, partly due to the spread of TB bacteria (shown below) resistant to antibiotic treatment.

# PRODUCING SOUNDS

Although the lungs evolved to deliver oxygen to the body, the passage of air through the throat also has the effect of producing sound. This effect is due to the vocal cords, two folds of mucous membrane within the larynx (voice box). When we speak, muscles in the larynx contract, pulling the vocal cords close together, and air is expelled from the lungs. The air forces its way through the small gap left between the vocal cords, causing the membranes to vibrate and creating the sounds of human speech. When the muscles relax, the vocal cords move apart and no sound is made. The length and tightness of the vocal cords affect the pitch of the voice – because men's vocal cords are longer and vibrate more slowly than women's, the male voice sounds deeper. The loudness of the voice depends on the force with which air pushes between the vocal cords.

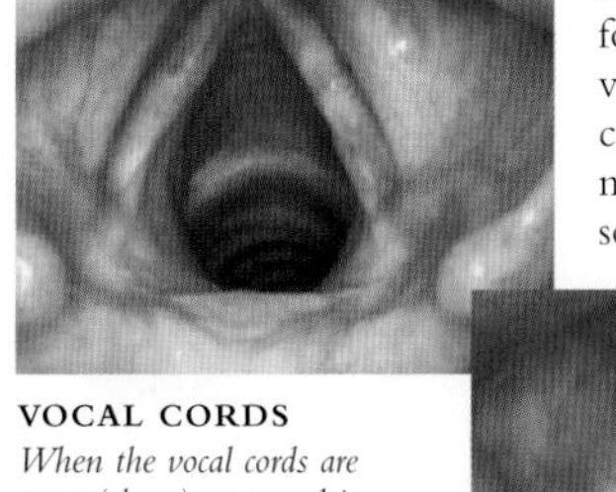

**VOCAL CORDS**
*When the vocal cords are open (above) no sound is made, but when they are closed (right, air vibrates across them to produce the sound of the human voice.*

# CIRCULATION

Blood vessels carry nutrient- and oxygen-rich blood from the heart to the rest of the body and return used blood for replenishment. The three main types of blood vessel are arteries, veins, and capillaries. The largest artery, the aorta, emerges from the heart and branches into a network of progressively smaller arteries that take blood all over the body. The smallest arteries form capillaries, where exchange of oxygen and nutrients for carbon dioxide and cell waste products occurs. The capillaries then join a network of tiny veins that merge into larger veins as they return blood to the heart.

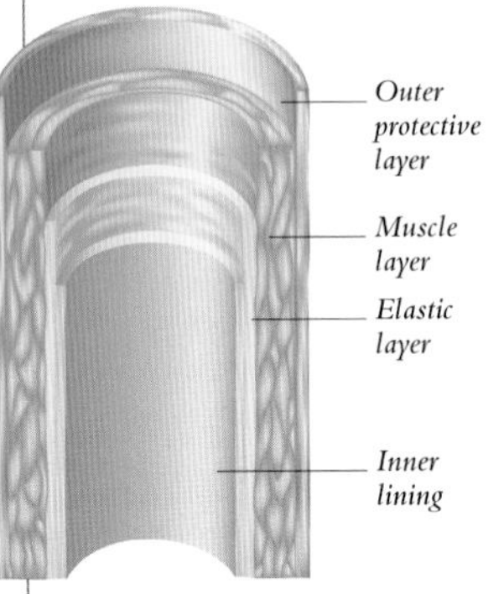


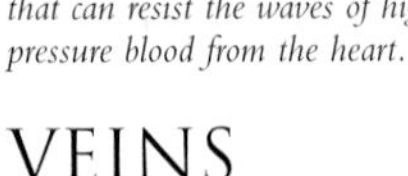

**STRUCTURE OF ARTERY**
*Arteries have thick, muscular walls that can resist the waves of high-pressure blood from the heart.*

## ARTERIES

Arteries are the primary supplier of blood to the body – taking blood away from the heart, picking up oxygen from the lungs, and then transporting the oxygen-rich blood to different areas of the body. They are large vessels with thick, elastic muscular walls. Some of the main arteries, such as the carotid (in the neck), aid the pumping of blood from the heart by contracting rhythmically, pulsing blood through the body. The largest artery in the body, the aorta, which carries blood out of the left side of the heart, is particularly elastic and is almost as wide as a garden hose.

## CAPILLARIES

Oxygen-carrying arteries divide into smaller blood vessels called arterioles, which themselves divide into tiny vessels known as capillaries. The capillaries join up to form venules, which in turn join to form veins. The veins carry deoxygenated blood back to the heart. Capillaries have a very important role in circulation, because it is in their web-like beds that the exchange of oxygen and nutrients for waste occurs. The capillaries are so small and delicate that their walls are only one cell thick – it takes 10 capillaries to equal the thickness of a human hair. It is the pores and gaps in the capillary walls that allow nutrient and waste exchange to take place.

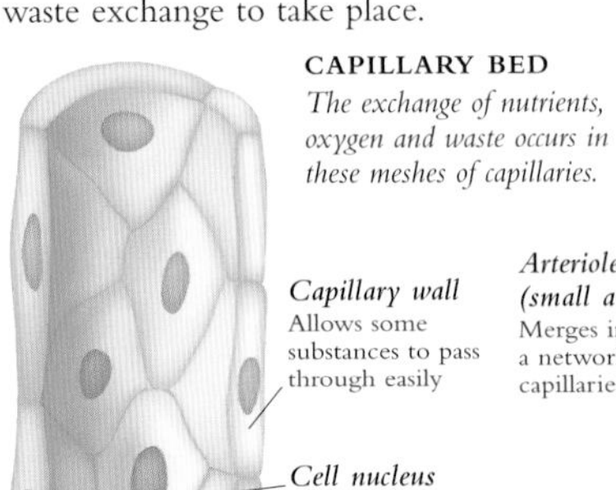


**CAPILLARY BED**
*The exchange of nutrients, oxygen and waste occurs in these meshes of capillaries.*

**STRUCTURE OF CAPILLARY**
*The walls of capillaries are only one-cell thick (here each cell and its nucleus are clearly visible), allowing nutrients, oxygen, and waste to pass through easily.*

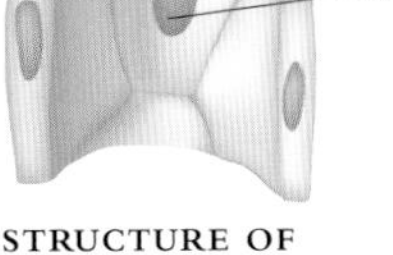

**IN A CAPILLARY**
*In this section through a capillary, individual red blood cells can be seen clearly.*

## VEINS

Veins are responsible for taking deoxygenated blood from the rest of the body back to the heart. They do not have to deal with such high blood pressures as arteries because they are carrying blood towards the heart rather than away from it. Consequently, they have much thinner walls, with less elasticity and fewer muscle fibres. This structure means that veins are often flatter than muscular arteries, which allows surrounding muscles to act on them, helping to squeeze deoxygenated blood along. The main veins in the body (such as the jugular vein in the neck and the main veins in the legs) contain one-way valves that keep blood flowing towards the heart but prevent it from going back the other way.

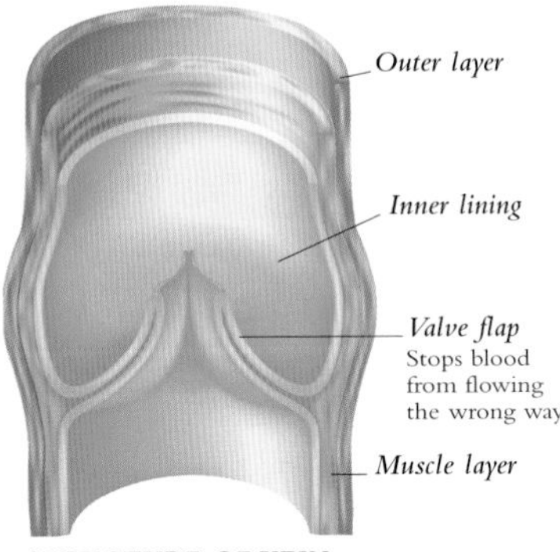


**STRUCTURE OF VEIN**
*Veins have thin walls that enable them to hold large volumes of blood. Large veins contain valves.*

Profile

### William Harvey

English physician William Harvey (1578–1657) was the first to prove that blood is pumped around the body in a closed circuit – without leaking or being consumed by organs (as was previously believed). Harvey's discovery led to a new way of treating patients whose lives were at risk as a result of blood loss or illness, by transferring blood into their veins from a healthy donor. His work led the practice of blood transfusions.

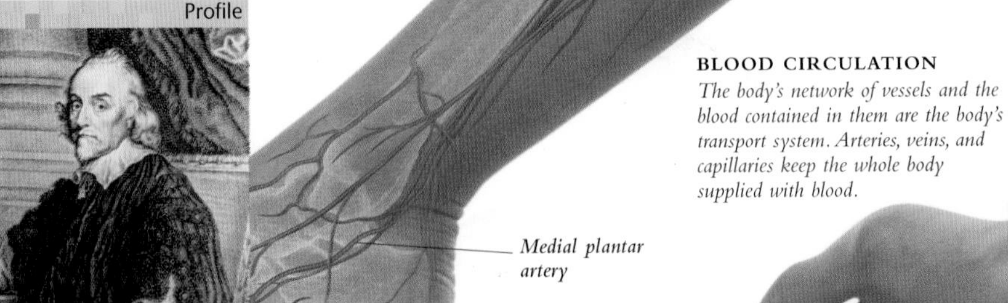


**BLOOD CIRCULATION**
*The body's network of vessels and the blood contained in them are the body's transport system. Arteries, veins, and capillaries keep the whole body supplied with blood.*

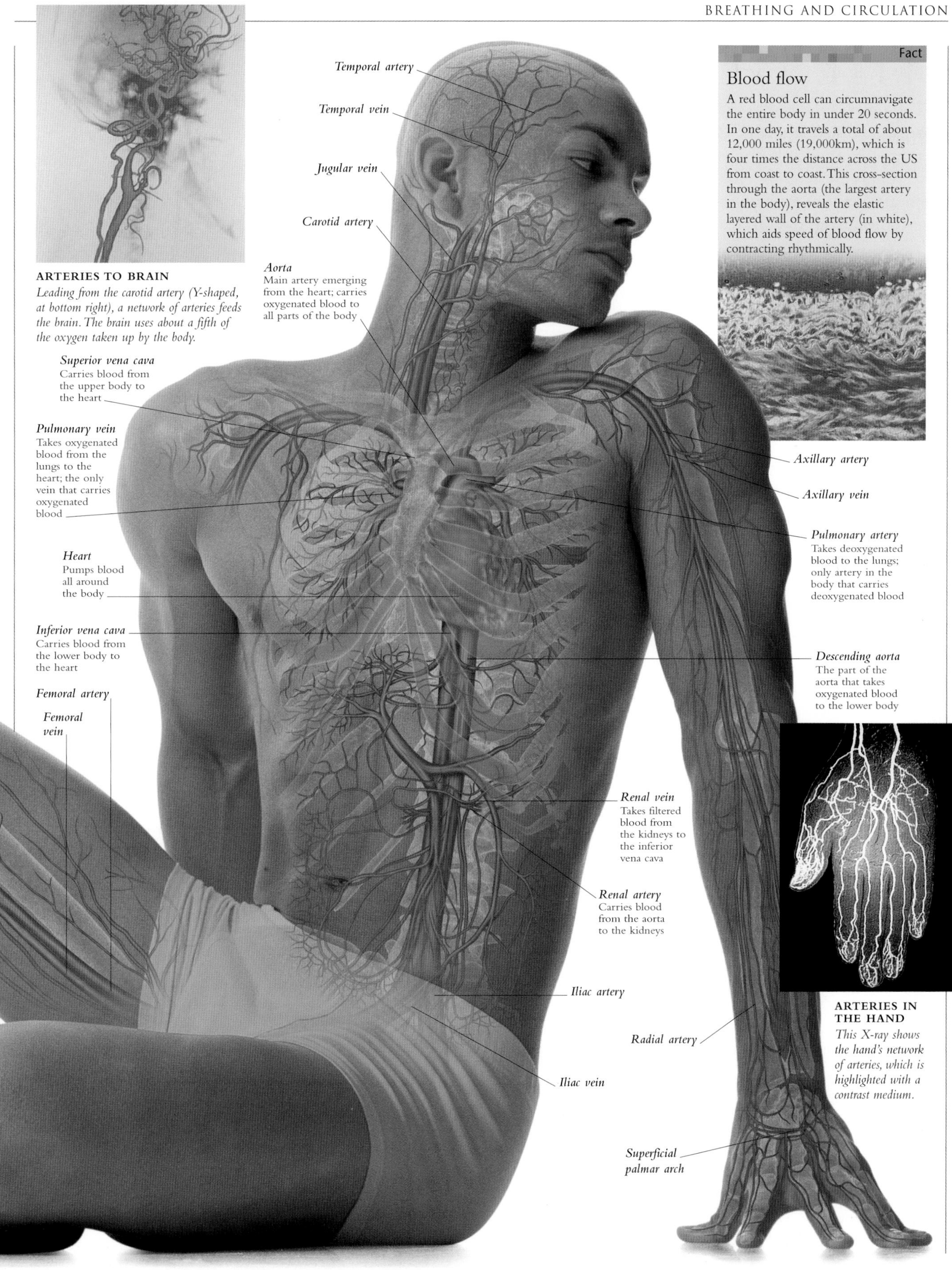
Temporal artery
Temporal vein
Jugular vein
Carotid artery
Aorta
Main artery emerging from the heart; carries oxygenated blood to all parts of the body
Superior vena cava
Carries blood from the upper body to the heart
Pulmonary vein
Takes oxygenated blood from the lungs to the heart; the only vein that carries oxygenated blood
Heart
Pumps blood all around the body
Inferior vena cava
Carries blood from the lower body to the heart
Femoral artery
Femoral vein
Axillary artery
Axillary vein
Pulmonary artery
Takes deoxygenated blood to the lungs; only artery in the body that carries deoxygenated blood
Descending aorta
The part of the aorta that takes oxygenated blood to the lower body
Renal vein
Takes filtered blood from the kidneys to the inferior vena cava
Renal artery
Carries blood from the aorta to the kidneys
Iliac artery
Iliac vein
Radial artery
Superficial palmar arch

**ARTERIES TO BRAIN**

*Leading from the carotid artery (Y-shaped, at bottom right), a network of arteries feeds the brain. The brain uses about a fifth of the oxygen taken up by the body.*

Fact

## Blood flow

A red blood cell can circumnavigate the entire body in under 20 seconds. In one day, it travels a total of about 12,000 miles (19,000km), which is four times the distance across the US from coast to coast. This cross-section through the aorta (the largest artery in the body), reveals the elastic layered wall of the artery (in white), which aids speed of blood flow by contracting rhythmically.

**ARTERIES IN THE HAND**

*This X-ray shows the hand's network of arteries, which is highlighted with a contrast medium.*

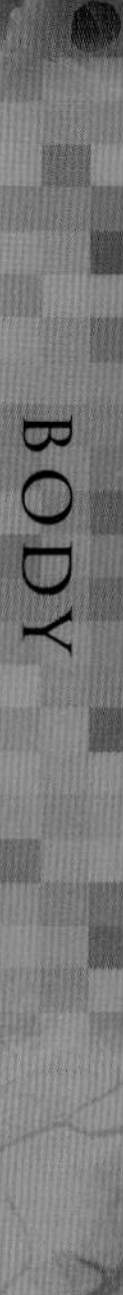

# THE HEART

The heart is the body's driving force. This powerful organ, the size of a large clenched fist, sits between the lungs, tilted towards the left side of the body. It works continuously, sending blood – about 5 litres (9pt) a minute – through the lungs and around the body, so that life-giving oxygen reaches every cell. In an average human lifetime, the heart beats over 3 billion times. The heart's special muscle contracts automatically, with no need for instructions from the brain.

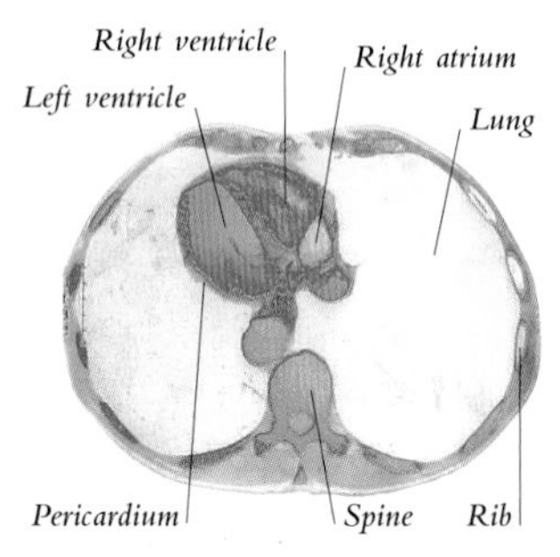


**LOCATION OF THE HEART**
*The heart (seen here from above) sits tucked between the lungs and is enveloped in a thick, double-layered membrane, the pericardium.*

## THE BODY'S PUMP

When circulating blood flows into the heart, it rushes through a series of inner chambers, pumped from one to another by the contractions of the heart muscle. There are two upper chambers, called atria (a single one is an atrium), and two lower chambers, known as ventricles. The right atrium fills with used blood that has deposited its supply of oxygen around the body. The left atrium collects blood enriched with oxygen picked up in the lungs. When full, the atria contract, squeezing blood into the ventricles below. The ventricles have a harder task than the atria, and for this reason their walls are thicker, especially on the left side of the heart. These chambers contract forcefully enough to push blood out of the heart and back into the blood vessels. Blood from the right ventricle flows through the pulmonary arteries to the lungs. The left ventricle sends oxygen-rich blood around the entire body.

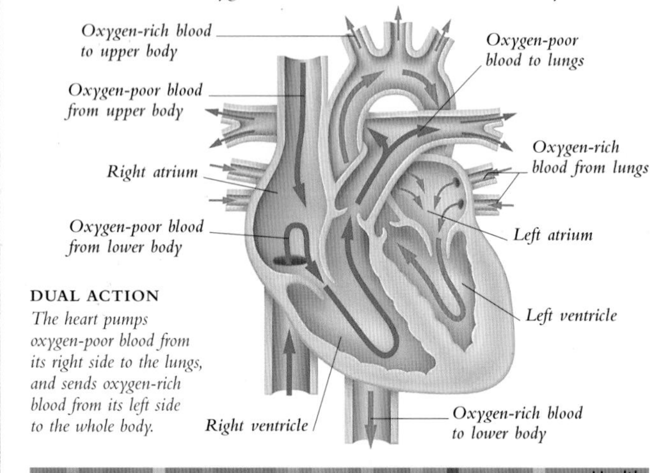


**DUAL ACTION**
*The heart pumps oxygen-poor blood from its right side to the lungs, and sends oxygen-rich blood from its left side to the whole body.*

Health

### Coronary artery disease

One of the most common causes of heart attacks is narrowing of the coronary arteries that supply the heart with blood (see ringed area in photograph). Coronary artery disease (CAD) is usually due to fatty deposits in the artery wall. The disease is often linked to obesity, a high-fat diet, lack of exercise, smoking, and a family (genetic) disposition. Coronary artery disease occurs more frequently among Western societies.

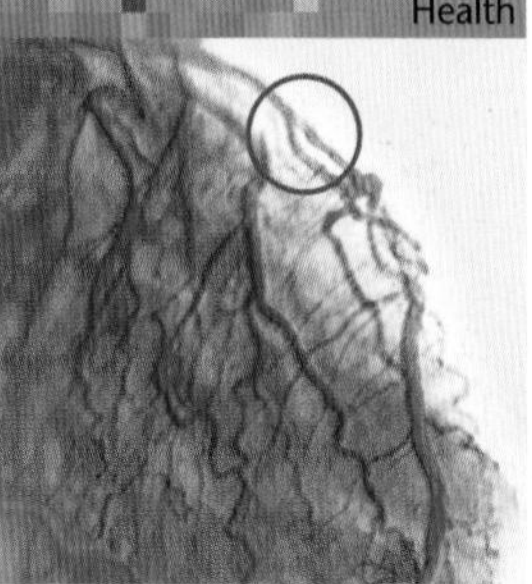

## CONTROLLING BLOOD FLOW

Blood flow between the atria and the ventricles is controlled by one-way valves. There are also valves at the openings to the blood vessels leading out of the heart. Each valve consists of two or three cup-shaped flaps of fibrous tissue called cusps, which prevent the blood from flowing backwards. When the atria contract, they push their load of blood against the valves leading into the ventricles, forcing the cusps open. As the ventricles fill and begin to contract, the pressure of blood rises on the other side of the valves and slams the cusps tightly shut. In the same way, blood leaving the ventricles builds up a high pressure that closes the valves at the ventricle exits. The closing of heart valves causes the familiar "lub-dub" sound of the heartbeat that is heard through a stethoscope.

**PULMONARY VALVE**
*The three cup-shaped cusps of the pulmonary valve close tightly together to prevent blood flowing backwards as it leaves the right ventricle.*

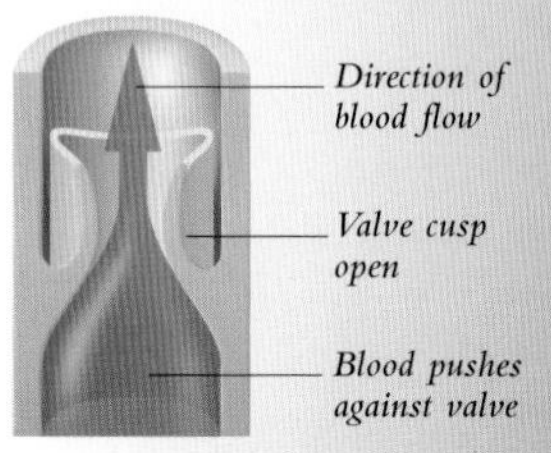


**HEART VALVE OPEN**
*As a heart chamber contracts it pushes blood up against a valve, forcing the cusps to open.*

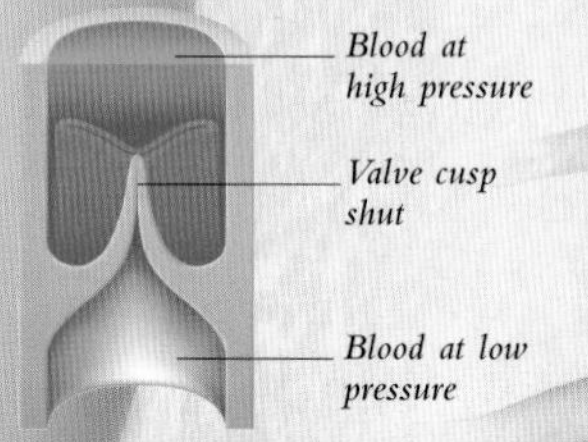


**HEART VALVE CLOSED**
*On the far side of the valve, rising blood pressure slams the cusps shut, preventing backflow.*

## BLOOD SUPPLY TO THE HEART

Like any other organ in the body, the heart is constantly hungry for a supply of oxygen-rich blood to ensure that it is always functioning efficiently. However, the heart cannot directly absorb any of the blood that it pumps ceaselessly through its chambers. Heart tissue needs its own separate blood supply. To serve this purpose, a network of blood vessels, known as the coronary system, spreads over the surface of the heart. The main blood supply lines serving the heart are the two coronary arteries. These arteries branch from the aorta, which is the largest blood vessel in the body, and subdivide into smaller blood vessels that penetrate the heart muscle. Once oxygen has been delivered to the heart, the coronary veins carry away the used blood and take it back to the heart's right atrium.

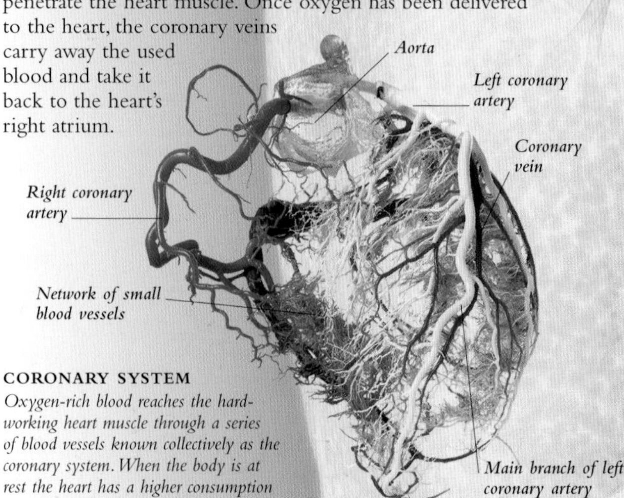


**CORONARY SYSTEM**
*Oxygen-rich blood reaches the hard-working heart muscle through a series of blood vessels known collectively as the coronary system. When the body is at rest the heart has a higher consumption of oxygen than almost any other organ.*

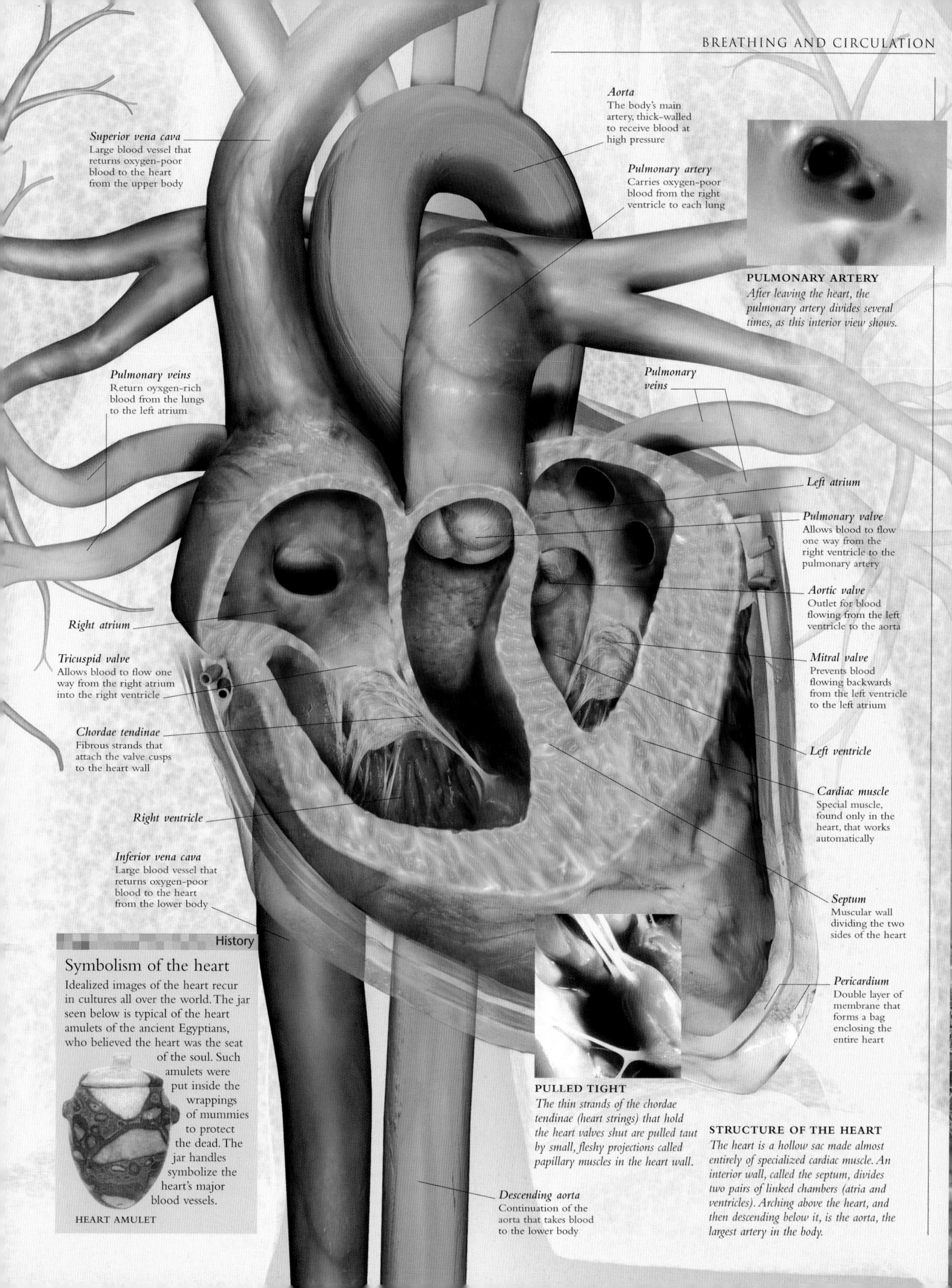

*Superior vena cava*
Large blood vessel that returns oxygen-poor blood to the heart from the upper body

*Aorta*
The body's main artery, thick-walled to receive blood at high pressure

*Pulmonary artery*
Carries oxygen-poor blood from the right ventricle to each lung

**PULMONARY ARTERY**
*After leaving the heart, the pulmonary artery divides several times, as this interior view shows.*

*Pulmonary veins*
Return oyxgen-rich blood from the lungs to the left atrium

*Pulmonary veins*

*Left atrium*

*Pulmonary valve*
Allows blood to flow one way from the right ventricle to the pulmonary artery

*Aortic valve*
Outlet for blood flowing from the left ventricle to the aorta

*Right atrium*

*Tricuspid valve*
Allows blood to flow one way from the right atrium into the right ventricle

*Mitral valve*
Prevents blood flowing backwards from the left ventricle to the left atrium

*Chordae tendinae*
Fibrous strands that attach the valve cusps to the heart wall

*Left ventricle*

*Cardiac muscle*
Special muscle, found only in the heart, that works automatically

*Right ventricle*

*Inferior vena cava*
Large blood vessel that returns oxygen-poor blood to the heart from the lower body

*Septum*
Muscular wall dividing the two sides of the heart

*Pericardium*
Double layer of membrane that forms a bag enclosing the entire heart

History

## Symbolism of the heart

Idealized images of the heart recur in cultures all over the world. The jar seen below is typical of the heart amulets of the ancient Egyptians, who believed the heart was the seat of the soul. Such amulets were put inside the wrappings of mummies to protect the dead. The jar handles symbolize the heart's major blood vessels.

HEART AMULET

**PULLED TIGHT**
*The thin strands of the chordae tendinae (heart strings) that hold the heart valves shut are pulled taut by small, fleshy projections called papillary muscles in the heart wall.*

*Descending aorta*
Continuation of the aorta that takes blood to the lower body

**STRUCTURE OF THE HEART**
*The heart is a hollow sac made almost entirely of specialized cardiac muscle. An interior wall, called the septum, divides two pairs of linked chambers (atria and ventricles). Arching above the heart, and then descending below it, is the aorta, the largest artery in the body.*

# THE HEART CYCLE

A single pumping action of the heart is called a heartbeat. When a person is at rest, his or her heart beats at a rate of 60–80 beats per minute, but during strenuous exercise this can rise to up to 200 beats per minute. Inside the heart, one-way valves prevent the blood from being pumped in the wrong direction. The characteristic rhythmic "lub-dub" sound of the heart is caused by these heart valves shutting tightly. A heartbeat has three phases. In diastole, the heart relaxes; during atrial systole, the atria (upper chambers) contract; and in ventricular systole, the ventricles (lower chambers) contract. The sinoatrial node (the heart's natural pacemaker) regulates the timing of these phases by sending electrical impulses to the atria and ventricles. Below, electrical activity is shown on an electrocardiogram (ECG) tracing.

**CONDUCTING FIBRES**
*Special muscle fibres in the walls of the heart conduct electrical impulses that regulate the heartbeat.*

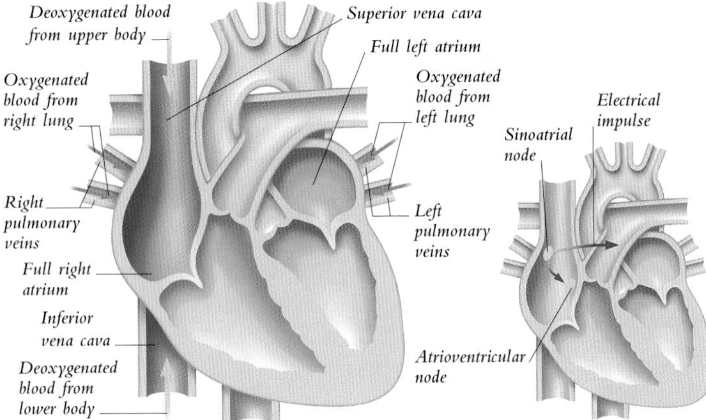


PASSAGE OF BLOOD

ELECTRICAL ACTIVITY

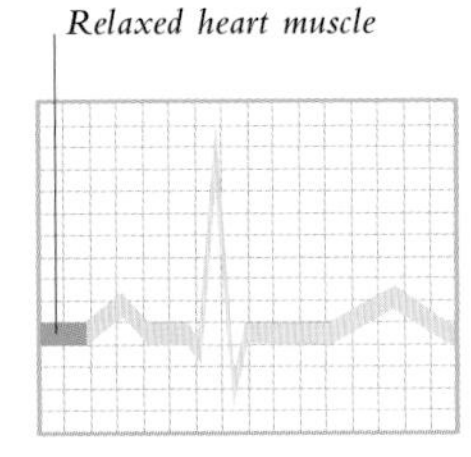


ECG TRACING

**DIASTOLE**
*The heart muscle relaxes and blood flows into the atria from the pulmonary veins and vena cava. Near the end of this phase of the heart cycle, the sinoatrial node emits an electrical impulse.*

Contracted right atrium

Contracted left atrium

Open tricuspid valve

Open mitral valve

Full left ventricle

Full right ventricle

Atrioventricular node

Electrical impulse

PASSAGE OF BLOOD

ELECTRICAL ACTIVITY

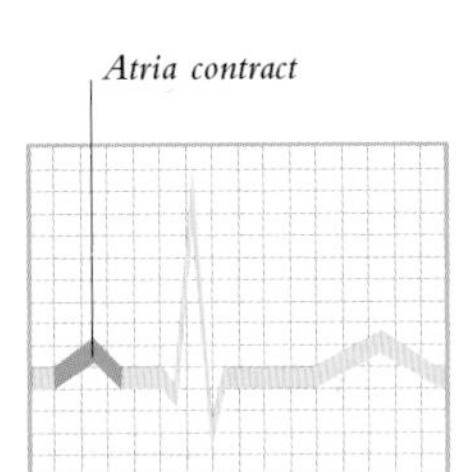


ECG TRACING

**ATRIAL SYSTOLE**
*The electrical impulse from the sinoatrial node spreads through both atria, causing their walls to contract and push blood into the ventricles. The impulse then reaches the atrioventricular node.*

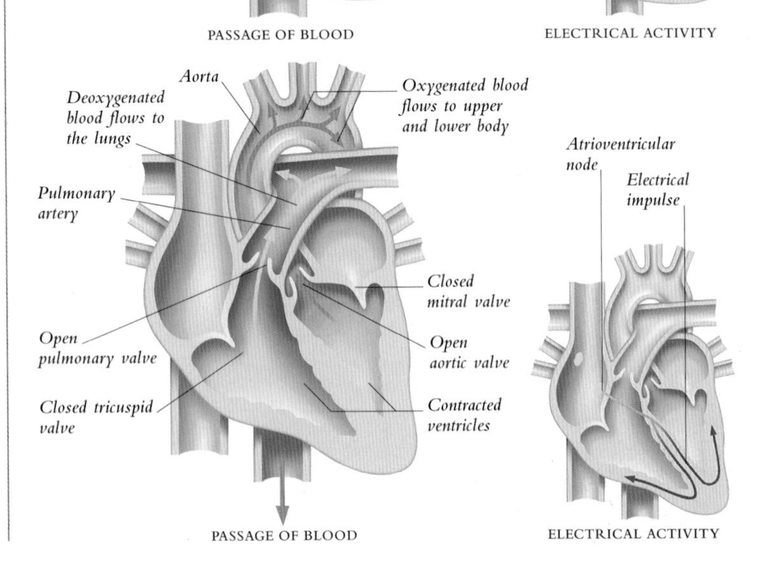


PASSAGE OF BLOOD

ELECTRICAL ACTIVITY

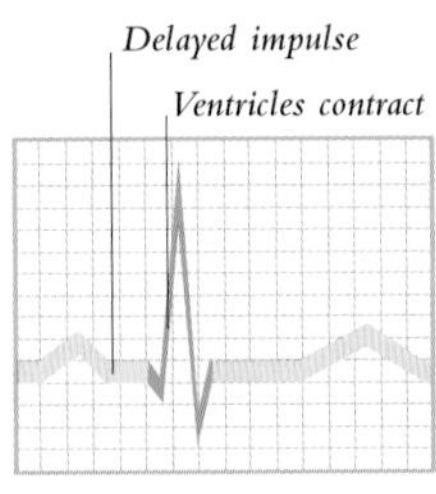


ECG TRACING

**VENTRICULAR SYSTOLE**
*The electrical impulse is delayed at the atrioventricular node. It then spreads through the walls of the ventricles, so that the ventricles contract at the same time, pushing blood into the aorta and pulmonary artery.*

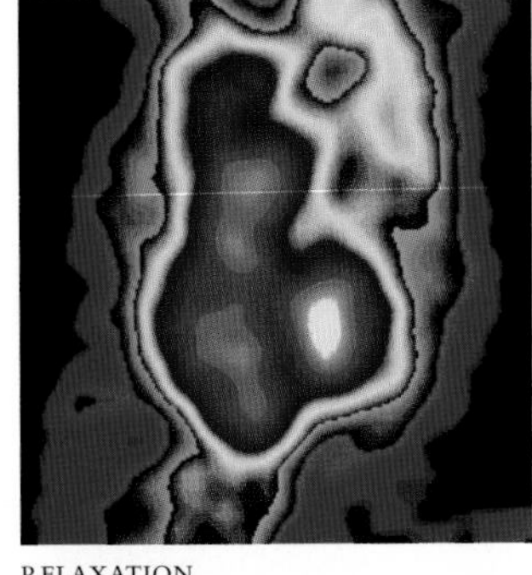

RELAXATION

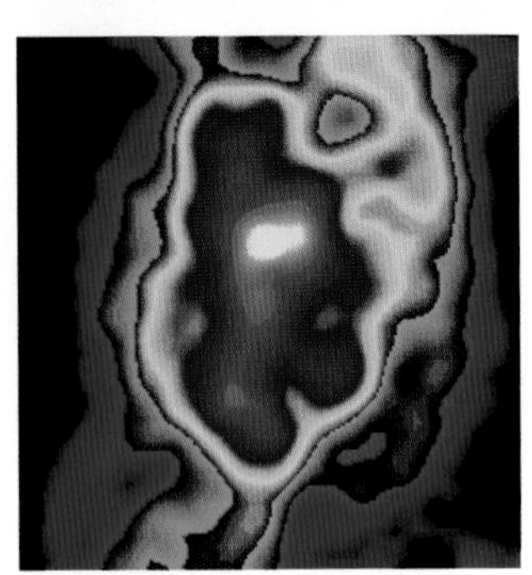

CONTRACTION

**DISTRIBUTION OF BLOOD**
*These two scans of the heart show how the distribution of blood (red) in the heart varies at different stages of the pumping cycle. At top, the heart is relaxed and in the process of filling with blood. Above, the heart is contracted and in the process of squeezing blood out.*

Health

## Pacemaker

An irregular heartbeat can be treated with a pacemaker that is surgically implanted in the chest. The device is inserted just under the skin in the chest (it is usually visible as a small bulge of skin) and supplies rhythmic electrical impulses along wires to the heart, to keep it beating regularly. An irregular heartbeat may be caused when the sinoatrial node (the area of the heart that initiates heartbeat) malfunctions, or when its impulses are blocked, for example by damage to surrounding tissue.

**LOCATION OF A PACEMAKER**
*This coloured X-ray shows a pacemaker (bottom right), attached by two wires (blue and red) to an enlarged heart.*

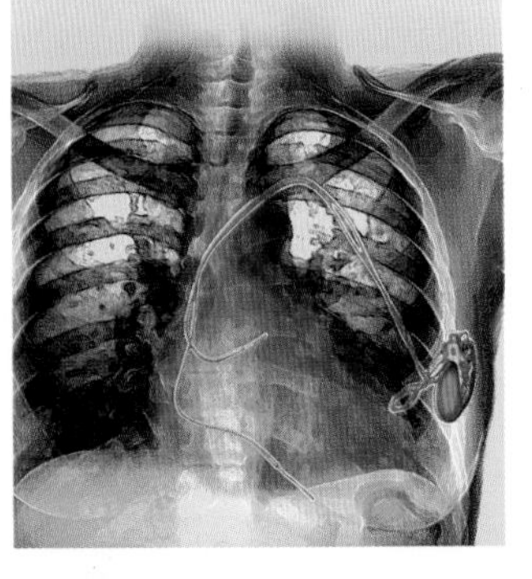

BODY

# BLOOD CIRCULATION

Blood circulates in two linked circuits: the pulmonary circulation, which carries blood to the lungs to be oxygenated, and the systemic circulation, which supplies oxygenated blood to the body. Arteries carrying blood from the heart divide into smaller vessels called arterioles and then into capillaries, where oxygen, nutrient, and waste exchange occurs. Capillaries join up to form venules, which in turn join up to form veins that carry blood back to the heart. The portal vein does not return blood to the heart but carries it to the liver (*see* p92). The heart powers both the pulmonary and the systemic circulations, shown in the illustration at right. In the pulmonary circulation, deoxygenated blood (shown in blue) travels to the lungs, where it absorbs oxygen before returning to the heart. This oxygenated blood (shown in red) is pumped around the body in the systemic circulation. Body tissues absorb oxygen, and deoxygenated blood returns to the heart in order to be pumped to the lungs again.

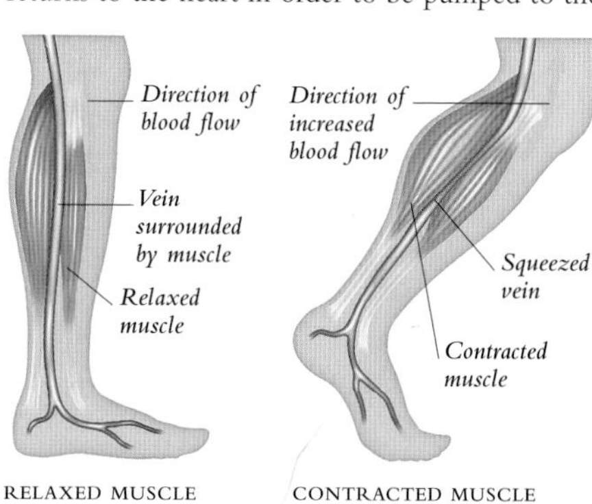


**VENOUS RETURN**

*The blood pressure in the veins is about a tenth of that in the arteries. A variety of mechanisms ensure that there is adequate venous return (blood flow back to the heart). Muscles contract and relax as we move, squeezing the veins that pass through them and pushing blood back to the heart.*

**A DOUBLE CIRCUIT**

*Here the pulmonary circulation, carrying blood to the lungs, is shown by green arrows and the systemic circulation, supplying oxygen-rich blood to the body, is shown by the yellow arrows.*

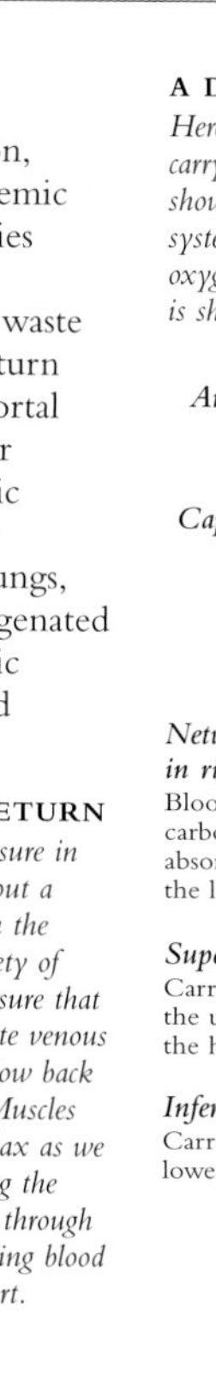


**HORMONE ACTION**

*Various hormones raise or lower blood pressure over a period of several hours and may remain effective for days.*

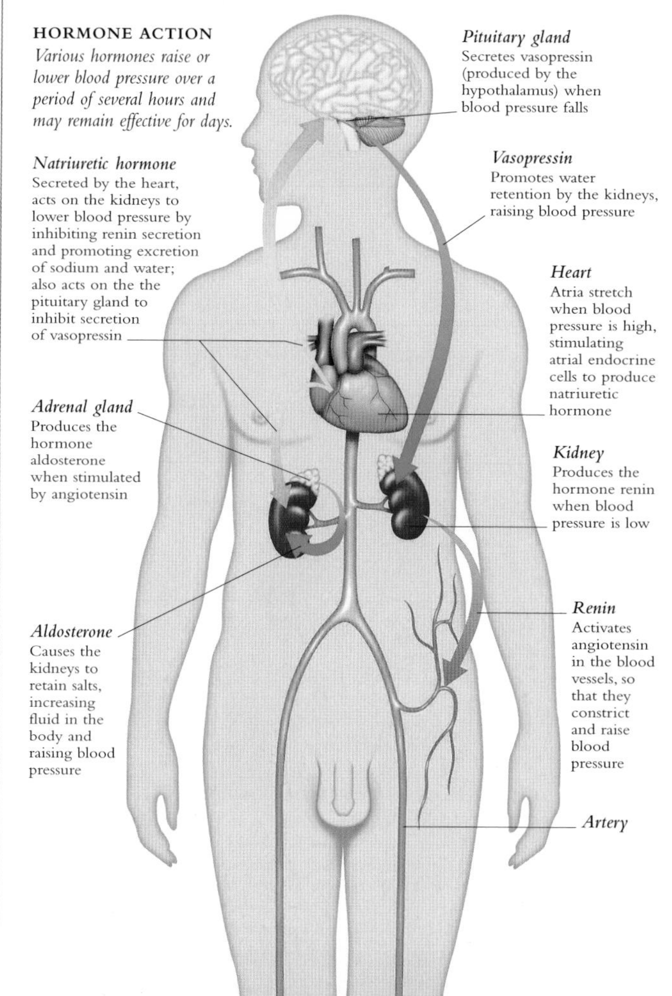


# BLOOD PRESSURE CONTROL

Blood pressure in the arteries must be regulated to ensure that there is always an adequate supply of blood, and therefore oxygen, to the organs. If arterial pressure is too low, not enough blood reaches body tissues. If, on the other hand, the pressure is too high, it may damage blood vessels and organs. Rapid changes in blood pressure (caused by heavy bleeding or a change in posture, for example) trigger responses from the nervous system within seconds. These autonomic nervous responses do not involve the conscious parts of the brain. Longer-term changes in blood pressure (caused by stress, for example) are largely regulated by hormones that affect the volume of fluid excreted by the kidneys. Hormonal responses work over a period of several hours.

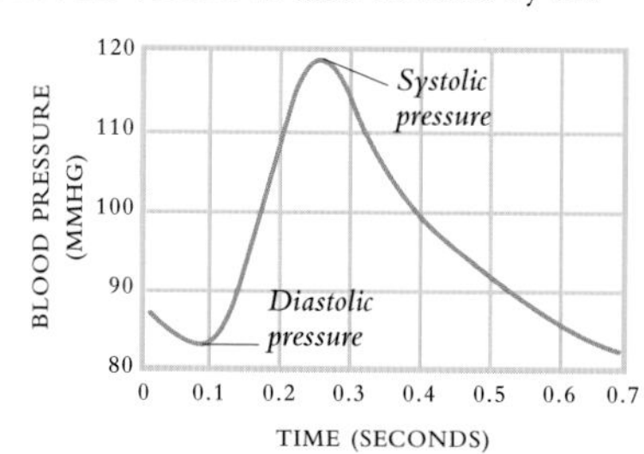


**BLOOD PRESSURE CYCLE**

*Arterial pressure is low while the heart fills with blood (diastolic pressure), but rises as the heart pumps blood out (systolic pressure). Pressure is measured in millimetres of mercury (mmHg).*

Fact

## Hypertension

Persistent high blood pressure, called hypertension, may damage the arteries and the heart. The condition is most common in middle-aged and elderly people. Genetic factors may contribute, as well as lifestyle factors such as being overweight and drinking excessive amounts of alcohol.

**NARROWED ARTERY**

*A build up of fatty deposits (brown) on the wall of this artery (red) has occurred as a result of hypertension.*

# BLOOD

The red liquid flowing in our network of arteries and veins has many roles. Blood is a supply line, transporting everything that the body needs to function efficiently. This essential fluid circulates constantly, through the lungs and around the entire body, taking oxygen and nutrients such as sugar, fats, and proteins to all body tissues. Blood also removes toxic wastes produced by cells, and helps to keep the body at the right temperature. Vitally, as part of our natural defence system, blood rapidly delivers disease-fighting cells to areas threatened by dangerous organisms.

**MAKE-UP OF BLOOD**
*This blood sample shows the separate ingredients.*

*Plasma*

*White blood cells and platelets*

*Red blood cells*

## BLOOD INGREDIENTS

Blood consists of a pale, straw-coloured liquid called plasma, in which float billions of red and white blood cells and platelets. Plasma, which makes up about half the volume of blood, consists mostly of water but contains various dissolved substances, including proteins, salts, and hormones. Red blood cells, by far the most numerous blood cells, are transporters of oxygen and carry away carbon dioxide, the waste product of body cells. The colourless white blood cells are part of the body's inbuilt defence mechanisms. The platelets, irregular-shaped cell fragments, are involved in blood clotting.

*Red blood cell*
Dimple-shape gives a large surface area for maximum oxygen absorption

## OXYGEN TRANSPORTERS

A single drop of blood contains about 5 million red blood cells. These cells carry the red pigment haemoglobin, from which they take their colour. Every haemoglobin molecule carries atoms of iron that attract and pick up oxygen in the lungs. As the red blood cells travel around the body they release their load of oxygen in the tissues. Highly flexible, the cells can squeeze through the tiniest blood vessels to reach every part of the body. Unlike most cells in the body, red blood cells do not have a nucleus.

*Neutrophil*
The most common type of white blood cell, which targets mainly bacteria

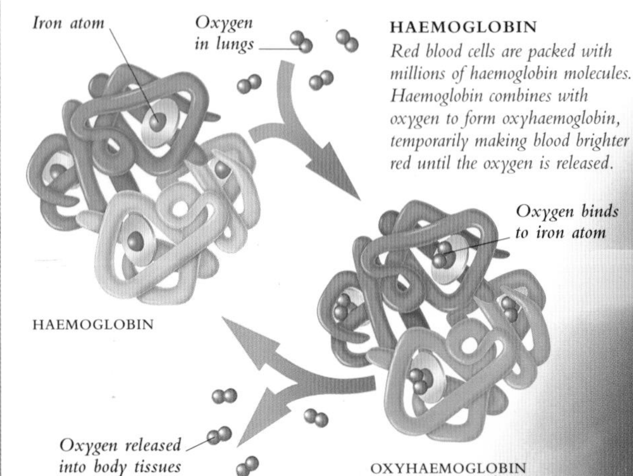


**HAEMOGLOBIN**
*Red blood cells are packed with millions of haemoglobin molecules. Haemoglobin combines with oxygen to form oxyhaemoglobin, temporarily making blood brighter red until the oxygen is released.*

Health

### Sickle-cell disease

The misshapen red blood cells seen here are caused by the inherited blood disorder sickle cell disease. A defect in the production of haemoglobin, the oxygen-gathering pigment, results in fragile cells that distort into a sickle shape when oxygen levels in the blood are low. These sickle cells may block narrow blood vessels, causing pain and preventing oxygen reaching body tissues. The disease is most often found in African–American people.

**IN THE BLOODSTREAM**
*Various types of cells tumble in plasma, as blood rushes around the body on an endless circuit. Red blood cells, which give blood its colour, predominate. White blood cells and tiny platelets are sparsely scattered in comparison, but they have greater versatility and more active roles.*

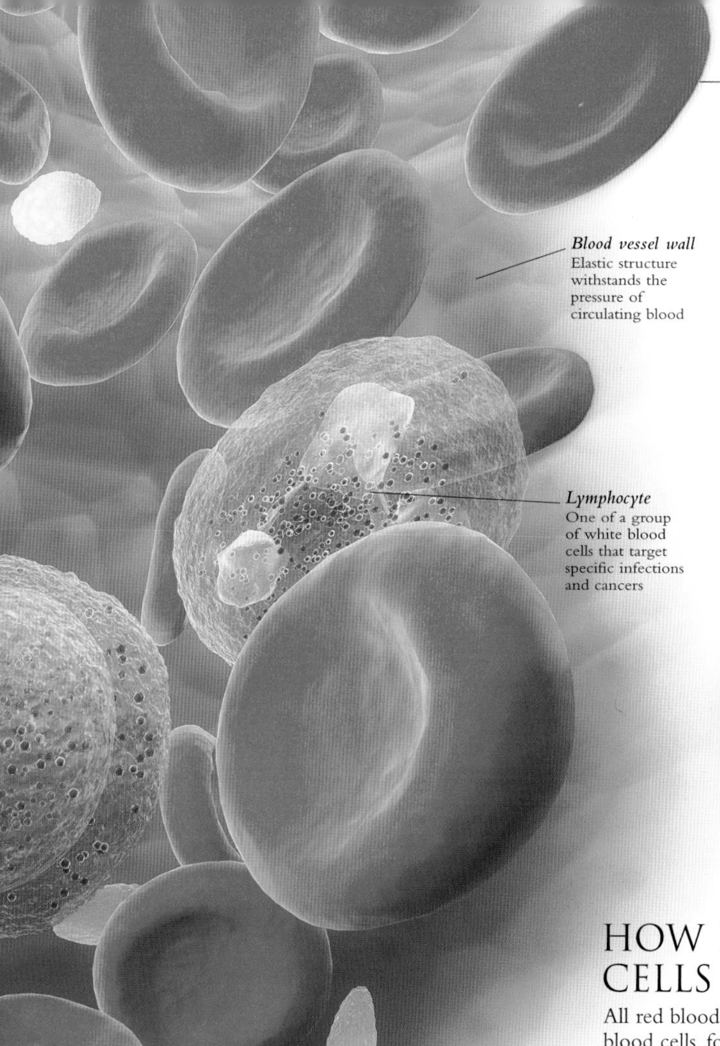
*Blood vessel wall*
Elastic structure withstands the pressure of circulating blood
*Lymphocyte*
One of a group of white blood cells that target specific infections and cancers
*Platelet*
Clumps with other platelets to seal damaged blood vessels

# PROTECTIVE CELLS

Billions of specialized cells circulate in the blood as part of the body's protective mechanism (*see also* Defence and repair, pp96–105). White blood cells of a variety of types have the task of tracking down and destroying harmful organisms. These blood cells migrate through the bloodstream, some engulfing and digesting bacteria and foreign particles, while others target cancer cells and specific infections. White blood cells are larger than red blood cells and have a nucleus. They are classified according to their role and appearance; the main types are neutrophils, eosinophils, lymphocytes, basophils, and monocytes. The other members of the blood's defence forces are platelets. These are not true cells but minute, disc-shaped cell fragments; like red blood cells they have no nucleus. Platelets move into action when a blood vessel is damaged, clumping together to plug the gap and stem bleeding.

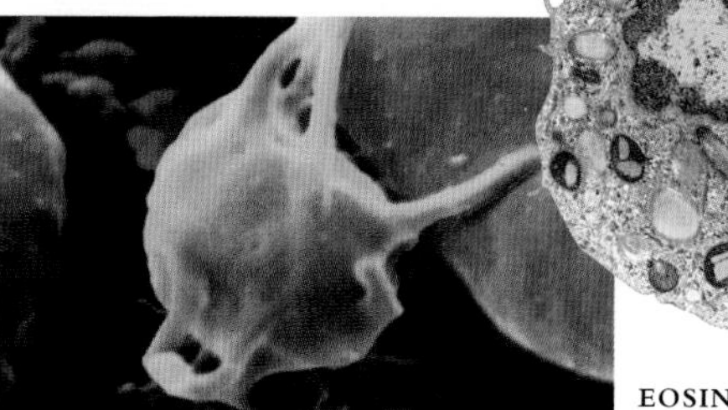

**PLATELET**
*A platelet prepares to take part in the blood-clotting process, forming spiky, adhesive extensions as it latches on and sticks to a red blood cell.*

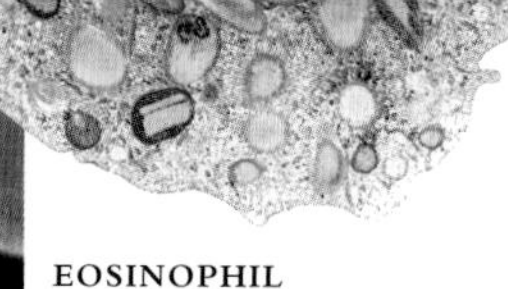

**EOSINOPHIL**
*The type of white blood cell shown in the photograph above contains numerous enzyme granules (in green), that react against foreign organisms such as bacteria.*

# HOW BLOOD CELLS ARE MADE

All red blood cells and platelets, and most white blood cells, form in the marrow of bones before passing into the bloodstream. The main blood cell production sites are in flat bones such as the breastbone, ribs, shoulder blades, and pelvis. Blood cells have short lives – some white blood cells last only hours and red blood cells are worn out after about 120 days – so constant fresh supplies are needed. Millions of new cells enter the bloodstream every minute. Red blood cells take a few days to mature in the blood before they become fully functioning.

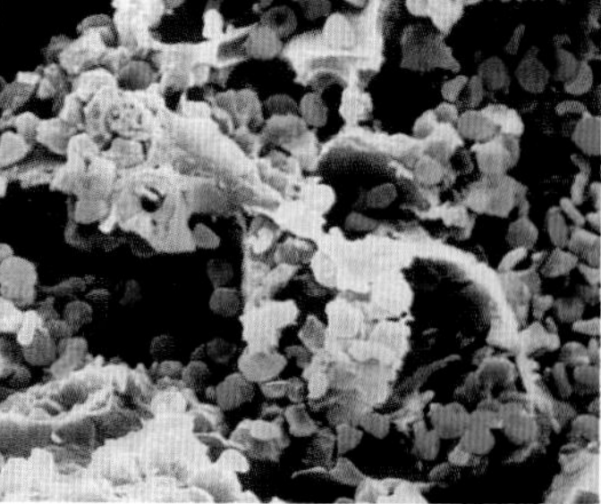

**BONE MARROW**
*Marrow, the soft, fatty substance filling the central cavity of bones, is a factory for the production of new blood cells.*

History

## Early blood transfusions

The first blood transfusions given to humans were carried out in the 17th century, long before the different blood groups were understood. Using a sheep as a blood donor, as shown in the illustration below, was attempted by a pioneering doctor in 1667. This particular patient survived, but such experiments were more likely to prove fatal.

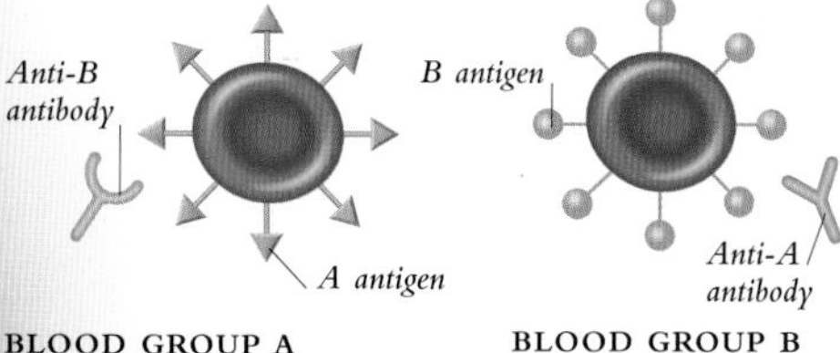
Anti-B antibody
A antigen
B antigen
Anti-A antibody

**BLOOD GROUP A**
*This blood type has A antigens on the red blood cells and anti-B antibodies in the plasma.*

**BLOOD GROUP B**
*Group B has B antigens on the cells and anti-A antibodies in the plasma.*

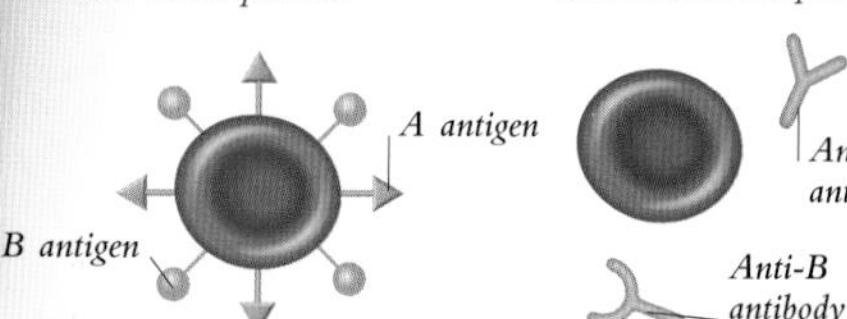
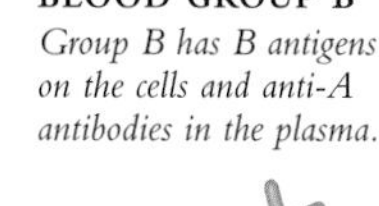
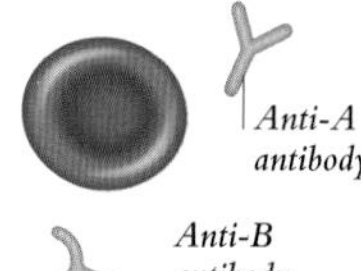
A antigen
B antigen
Anti-A antibody
Anti-B antibody

**BLOOD GROUP AB**
*The rare AB group has A and B antigens on the cells and no antibodies in the plasma.*

**BLOOD GROUP O**
*O blood has no antigens and contains anti-A and anti-B antibodies.*

# BLOOD GROUPS

Each person belongs to a blood group. There are many types of blood groupings, of which the best known is the ABO system. This system identifies four groups – A, B, AB, and O – by markers, called antigens, found on the surface of red blood cells. The body's immune system uses antigens to recognize the difference between its own cells and foreign cells. During a blood transfusion, a person must be given blood with the correct antigens. Otherwise, the immune system sees the new red blood cells as invaders and attacks them. In each blood group, proteins in the plasma, antibodies, stick to foreign blood cells, marking them for attack. Another method of blood typing is to identify the rhesus (Rh) antigen, found on the red blood cells of 85 per cent of people.

**TAKING IN FUEL**
*The sight and smell of food triggers the start of the digestive process by stimulating the flow of saliva in the mouth and the secretion of gastric juices in the stomach.*

# FUELLING THE BODY

Every process that takes place in the body is fuelled by energy derived from food. What we eat and drink also builds the structures needed to keep the body working efficiently and to make children grow. Food cannot be used by the body unless it is broken down into its simpler components, which is the task of the digestive tract. The extracted nutrients are used by cells or stored for later. Digestion and the cell reactions that make energy from food create waste products. Solid wastes are expelled from the digestive tract; watery wastes are dealt with by the urinary organs.

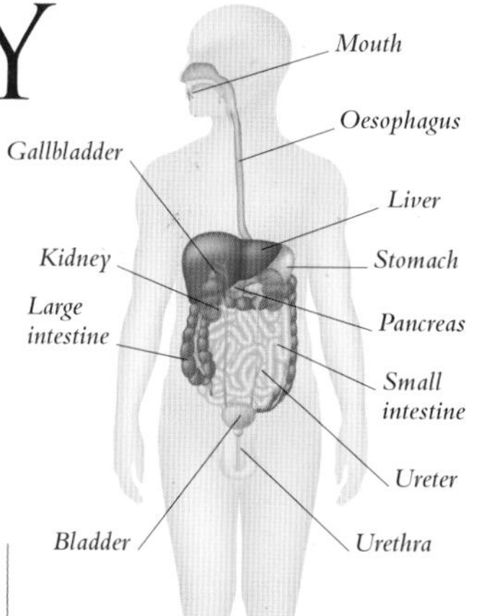


**DIGESTION AND EXCRETION**
*The digestive tract and its associated organs break down food and excrete solid wastes. The urinary organs filter the blood and excrete wastes and excess water as urine.*

Once we have taken in food, the body automatically acts to make use of it, and to get rid of waste products. The digestive and urinary organs are controlled by the brain. This control ensures that digestive juices to break down food are secreted at the right time and in the right place; that the muscles of the digestive tract are prompted into action; and that we are aware when waste products are ready to be eliminated from the body.

## DIETARY NEEDS

In order that our bodies function normally, our regular diet needs to include various components. Those needed in the largest quantities are carbohydrates, fats, and proteins. Carbohydrates, broken down into glucose during digestion, are our main source of energy. Fats are used to make the new cells needed for growth and repair, and are also stored as reserve fuel. Proteins, when broken down into their constituent acids, produce new proteins that are used to build cell structures and tissues; they are also used for fuel when necessary. Although needed in relatively small quantities, vitamins and minerals are important to our general health. These substances have varied uses, from playing a role in the functioning of the nervous system to being an essential component of bones and teeth. Another vital part of our diet is water, needed for chemical reactions and for transporting other substances around the body.

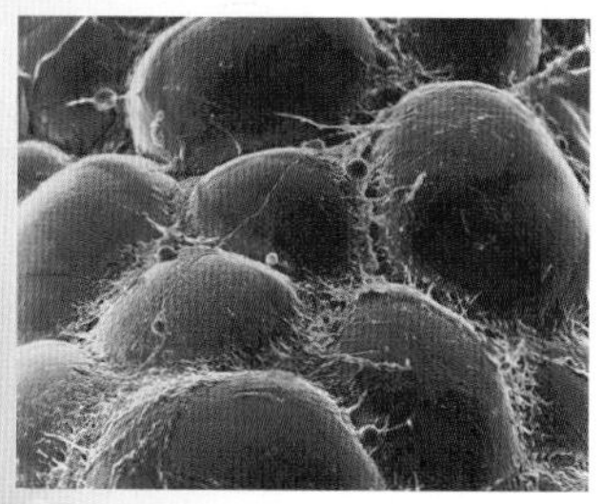

**FAT CELLS**
*Excess nutrients from food are stored as body fat in adipose (fat) cells. These cells, which form a thick, insulating layer beneath the skin, act as fuel reserves.*

## THE DIGESTIVE TRACT

The body's main site for digestion is a tube, about 7m (24ft) long, that extends all the way from the mouth to the anus. Various other organs, including the salivary glands, liver, gallbladder, and pancreas, also make essential contributions to the slow breakdown of food as it makes its way from one end of the body to the other.

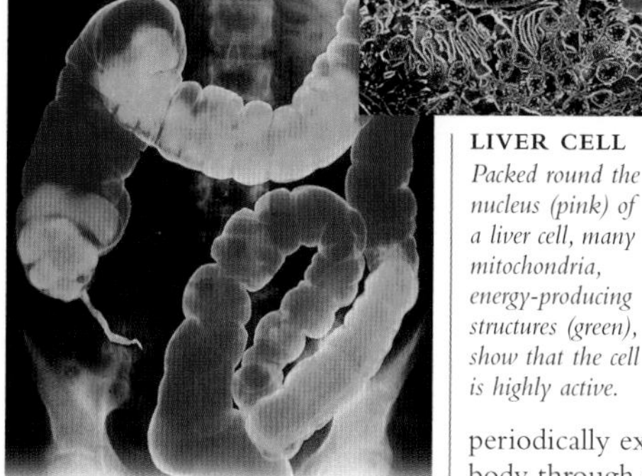

**LARGE INTESTINE**
*In the large intestine, undigested food is formed into faeces. Bands of muscle around the intestinal wall contract and relax to move waste material along the tract.*

The teeth and the chemicals in saliva start the digestive process, so that food slides easily down the narrow passage of the oesophagus to the stomach. In the stomach, a few hours of churning and mixing with acids so powerful that they can dissolve metal, reduces food to a thick liquid. The liquid is then released into the small intestine, the most important section of the digestive tract. In the loops and coils of this long tube, nearly all the useful components of food are absorbed. Waves of contractions in the intestinal wall keep the food moving along, while digestive chemicals complete the process of breakdown. Some of these chemicals are produced in the walls of the intestine itself, while others are added by the pancreas and the liver. By the time the food has moved on to the next part of the digestive tract, the large intestine, it is largely made up of unwanted materials and water. The large intestine, which is a much wider tube than the small intestine, has a relatively simple role. In this organ, the water is absorbed from the digested food, while the remaining waste is formed into faeces that are periodically expelled from the body through the anus.

**LIVER CELL**
*Packed round the nucleus (pink) of a liver cell, many mitochondria, energy-producing structures (green), show that the cell is highly active.*

## WASTE FLUIDS

Besides the elimination of faeces, the body has another system of waste disposal – urination. The urinary tract removes excess water, but its main function is to get rid of waste products formed by chemical processes in cells. The urinary organs are the kidneys, the bladder and their connecting tubes. The kidneys receive about 25 per cent of the blood pumped round the body. These organs filter out unwanted substances from the blood and excrete them as urine. The kidneys regulate the amount and composition of urine they make to keep body fluids and chemicals at the right levels. Urine trickles from each kidney through a tube, the ureter, and is stored in the expandable bag of the bladder. Except in very young children, emptying the bladder is under our conscious control.

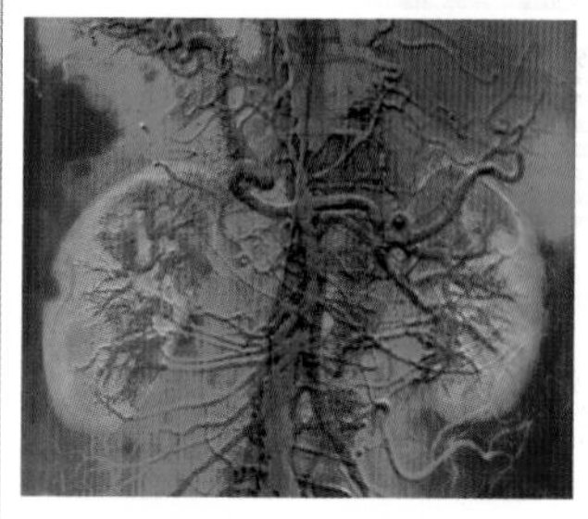

**BLOOD SUPPLY TO KIDNEYS**
*This image of the kidneys (yellow) seen on a coloured angiogram shows the major arteries that supply the kidneys with large amounts of circulating blood.*

Fact

### Diverse diets

Humans are adapted to eating a hugely diverse diet. There are many different types and combinations of food that can provide the body with the nutrients it needs for health, growth, and energy. Diets around the world vary according to tradition and what ingredients are available. These plump grubs, skewered ready for roasting in a Bangkok restaurant, are easily digested and highly nutritious, being rich in proteins and fats.

# BREAKING DOWN FOOD

Food consists mainly of water and the nutrients protein, carbohydrate, and fat. Before these nutrients can be used by the body, their large molecules must be broken down in the digestive system into units small enough for the body to absorb. Left over solid matter must also be packaged up, ready to be expelled as faeces. The breakdown of food starts in the mouth and continues through the stomach and intestines on the way to the anus, where it is excreted as faeces.

## THE DIGESTIVE PROCESS

The breakdown of food is a physical and chemical process. It begins in the mouth, where the teeth play an important role, and continues through the oesophagus to the stomach, where food is churned into a semiliquid called chyme. Most nutrient absorption takes place in the small intestine, where the breakdown is aided by bile and pancreatic juice. The large intestine continues to draw off nutrients and begins to package waste matter, ready for excretion at the anus.

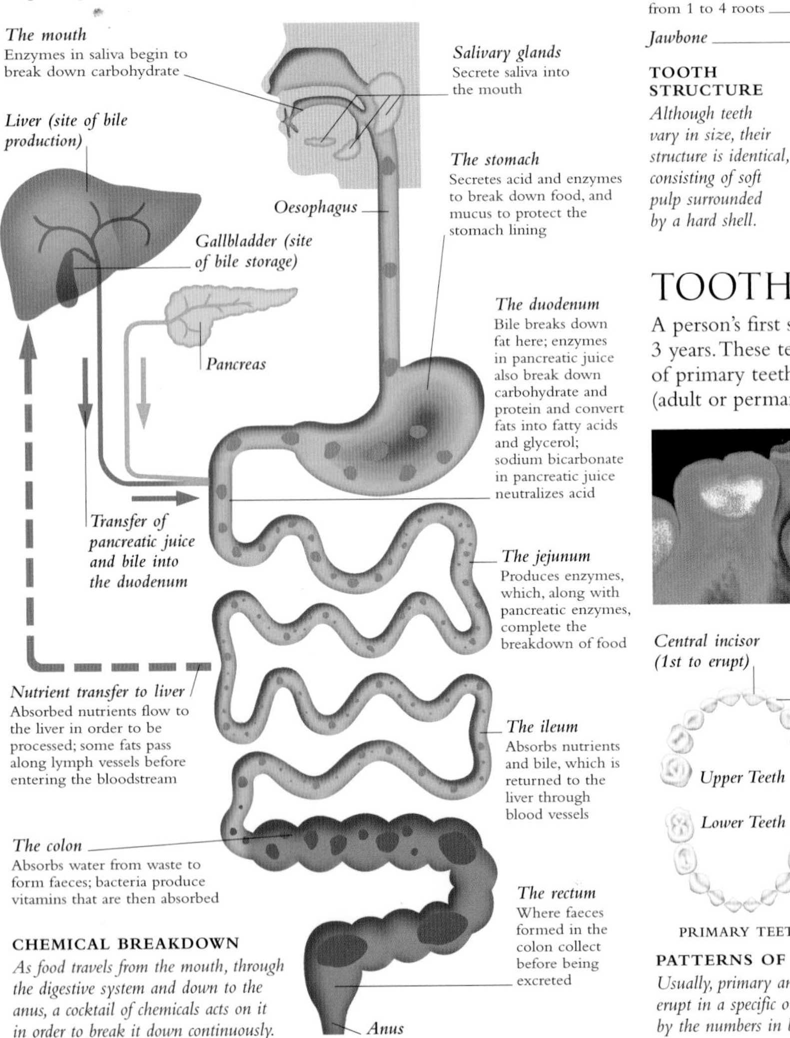
*The mouth*
Enzymes in saliva begin to break down carbohydrate
*Salivary glands*
Secrete saliva into the mouth
*Liver (site of bile production)*
*The stomach*
Secretes acid and enzymes to break down food, and mucus to protect the stomach lining
*Oesophagus*
*Gallbladder (site of bile storage)*
*The duodenum*
Bile breaks down fat here; enzymes in pancreatic juice also break down carbohydrate and protein and convert fats into fatty acids and glycerol; sodium bicarbonate in pancreatic juice neutralizes acid
*Pancreas*
*Transfer of pancreatic juice and bile into the duodenum*
*The jejunum*
Produces enzymes, which, along with pancreatic enzymes, complete the breakdown of food
*Nutrient transfer to liver*
Absorbed nutrients flow to the liver in order to be processed; some fats pass along lymph vessels before entering the bloodstream
*The ileum*
Absorbs nutrients and bile, which is returned to the liver through blood vessels
*The colon*
Absorbs water from waste to form faeces; bacteria produce vitamins that are then absorbed
*The rectum*
Where faeces formed in the colon collect before being excreted
*Anus*

**CHEMICAL BREAKDOWN**
*As food travels from the mouth, through the digestive system and down to the anus, a cocktail of chemicals acts on it in order to break it down continuously.*

## THE ROLE OF TEETH

The primary function of the teeth is to break down food ready for digestion. They are shaped differently according to their roles. The incisors at the front of the mouth are sharp and chisel-like to cut and hold food. On either side of the incisors lie the canines, which are longer and more pointed and are used for tearing food. Next, the premolars, with their two ridges, have a textured surface for grinding food. At the back of the mouth are the largest teeth, the molars, which have four or five cusps on their chewing surfaces, also used for grinding food.

**TOOTH ENAMEL**
*Calcium deposits on tooth enamel appear here in regular rows.*

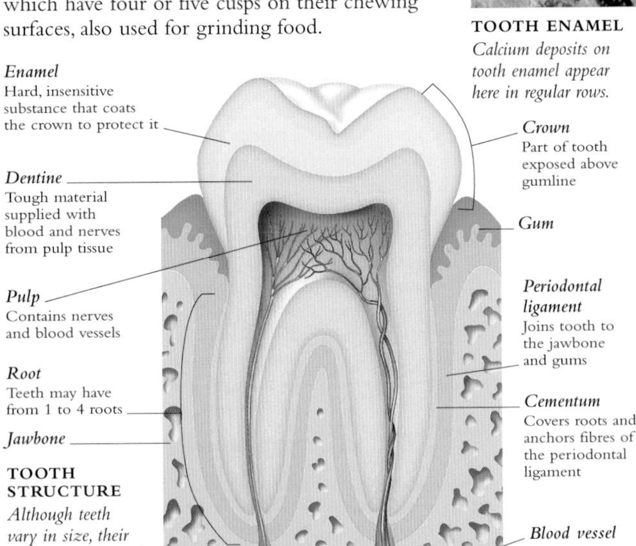
*Enamel*
Hard, insensitive substance that coats the crown to protect it
*Crown*
Part of tooth exposed above gumline
*Dentine*
Tough material supplied with blood and nerves from pulp tissue
*Gum*
*Pulp*
Contains nerves and blood vessels
*Periodontal ligament*
Joins tooth to the jawbone and gums
*Root*
Teeth may have from 1 to 4 roots
*Cementum*
Covers roots and anchors fibres of the periodontal ligament
*Jawbone*
*Blood vessel*
*Nerve*

**TOOTH STRUCTURE**
*Although teeth vary in size, their structure is identical, consisting of soft pulp surrounded by a hard shell.*

## TOOTH DEVELOPMENT

A person's first set of teeth appears between the ages of 6 months and 3 years. These teeth are known as primary, or milk, teeth. The enamel of primary teeth is relatively soft. As the jaw grows, a set of secondary (adult or permanent) teeth appears between the ages of 6 and 21. As the secondary teeth erupt, they displace the primary teeth, which usually fall out by the age of 13. In some cases, the third molars, or wisdom teeth, never appear.

**TOOTH ERUPTION**
*Here, a secondary tooth (purple) that has not yet erupted can be seen ready to emerge below a primary tooth (blue).*

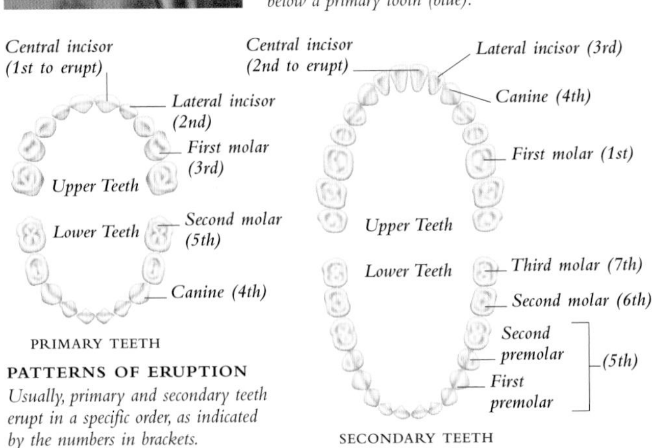
*Central incisor (1st to erupt)*
*Lateral incisor (2nd)*
*First molar (3rd)*
*Upper Teeth*
*Lower Teeth*
*Second molar (5th)*
*Canine (4th)*
PRIMARY TEETH
*Central incisor (2nd to erupt)*
*Lateral incisor (3rd)*
*Canine (4th)*
*First molar (1st)*
*Upper Teeth*
*Lower Teeth*
*Third molar (7th)*
*Second molar (6th)*
*Second premolar*
*First premolar*
*(5th)*
SECONDARY TEETH

**PATTERNS OF ERUPTION**
*Usually, primary and secondary teeth erupt in a specific order, as indicated by the numbers in brackets.*

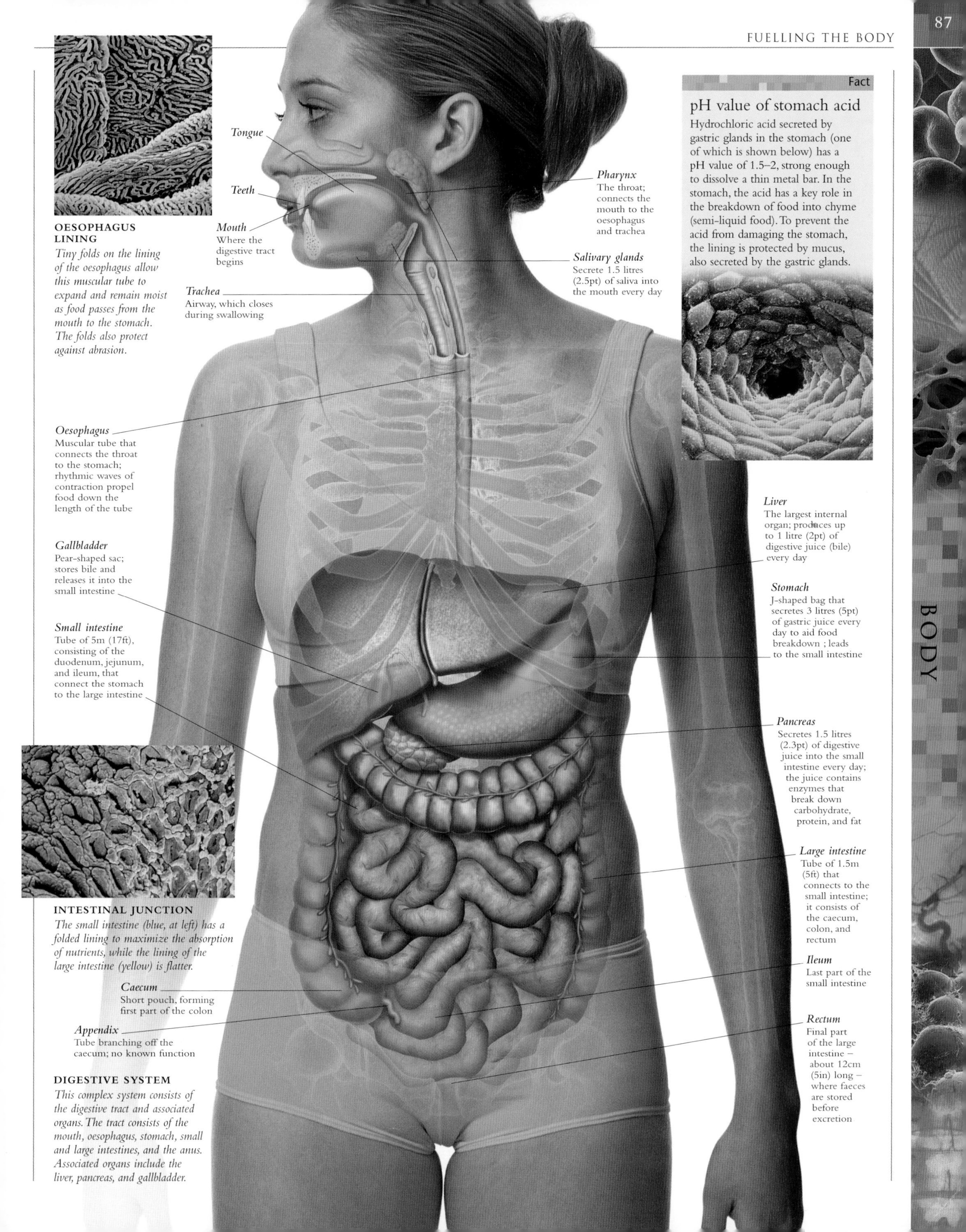
Tongue
Teeth
Mouth
Where the digestive tract begins
Trachea
Airway, which closes during swallowing
Pharynx
The throat; connects the mouth to the oesophagus and trachea
Salivary glands
Secrete 1.5 litres (2.5pt) of saliva into the mouth every day
Oesophagus
Muscular tube that connects the throat to the stomach; rhythmic waves of contraction propel food down the length of the tube
Gallbladder
Pear-shaped sac; stores bile and releases it into the small intestine
Small intestine
Tube of 5m (17ft), consisting of the duodenum, jejunum, and ileum, that connect the stomach to the large intestine
Liver
The largest internal organ; produces up to 1 litre (2pt) of digestive juice (bile) every day
Stomach
J-shaped bag that secretes 3 litres (5pt) of gastric juice every day to aid food breakdown ; leads to the small intestine
Pancreas
Secretes 1.5 litres (2.3pt) of digestive juice into the small intestine every day; the juice contains enzymes that break down carbohydrate, protein, and fat
Large intestine
Tube of 1.5m (5ft) that connects to the small intestine; it consists of the caecum, colon, and rectum
Ileum
Last part of the small intestine
Caecum
Short pouch, forming first part of the colon
Appendix
Tube branching off the caecum; no known function
Rectum
Final part of the large intestine – about 12cm (5in) long – where faeces are stored before excretion
BODY

**OESOPHAGUS LINING**

*Tiny folds on the lining of the oesophagus allow this muscular tube to expand and remain moist as food passes from the mouth to the stomach. The folds also protect against abrasion.*

Fact

## pH value of stomach acid

Hydrochloric acid secreted by gastric glands in the stomach (one of which is shown below) has a pH value of 1.5–2, strong enough to dissolve a thin metal bar. In the stomach, the acid has a key role in the breakdown of food into chyme (semi-liquid food). To prevent the acid from damaging the stomach, the lining is protected by mucus, also secreted by the gastric glands.

**INTESTINAL JUNCTION**

*The small intestine (blue, at left) has a folded lining to maximize the absorption of nutrients, while the lining of the large intestine (yellow) is flatter.*

**DIGESTIVE SYSTEM**

*This complex system consists of the digestive tract and associated organs. The tract consists of the mouth, oesophagus, stomach, small and large intestines, and the anus. Associated organs include the liver, pancreas, and gallbladder.*

# FROM MOUTH TO STOMACH

The upper parts of the digestive tract, comprising the mouth, oesophagus, and stomach, enable us to take in large quantities of food quickly. As soon as we start to chew, digestion begins, with the salivary glands releasing saliva to moisten and dissolve food. The tongue moves the moist food to the back of the mouth, triggering a series of involuntary processes. As we swallow, making room for another mouthful, the throat automatically accommodates the food and lets it pass into the oesophagus. A reflex action of the muscular wall of the oesophagus then propels food down to the capacious bag of the stomach.

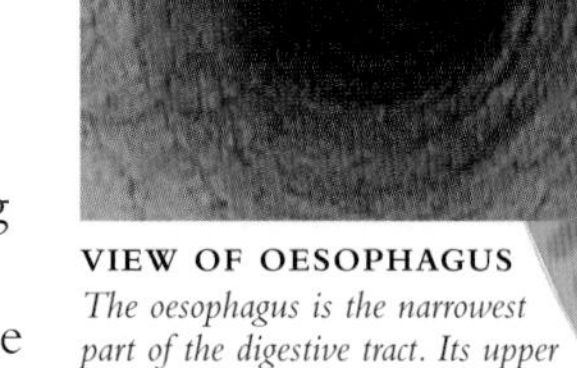

**VIEW OF OESOPHAGUS**
*The oesophagus is the narrowest part of the digestive tract. Its upper end lies flat until food enters.*

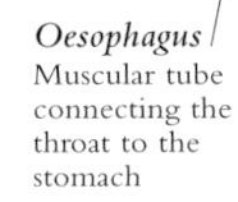

**Oesophagus**
Muscular tube connecting the throat to the stomach

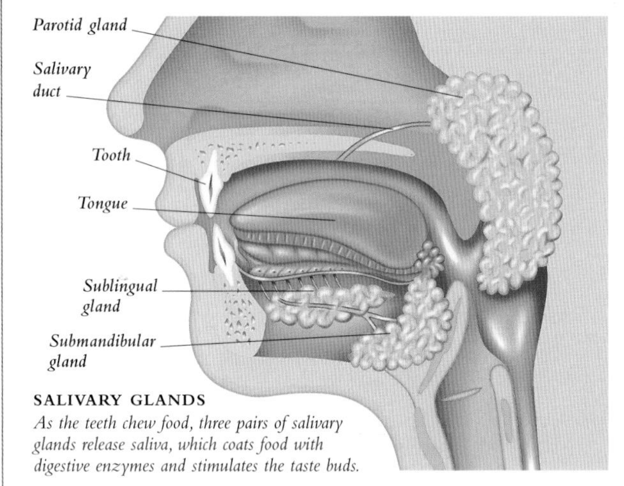


**SALIVARY GLANDS**
*As the teeth chew food, three pairs of salivary glands release saliva, which coats food with digestive enzymes and stimulates the taste buds.*

## INSIDE THE MOUTH

We use our incisors and canines (front teeth), to cut food and our molars (back teeth) to chew it. Chewing increases the surface area of food, making it easier to mix with saliva. Three pairs of salivary glands in the mouth release up to 1.5 litres (approx 2½ pt) of saliva a day through a series of ducts. This fluid moistens and softens food, and starts the process of digestion. Saliva is 99 per cent water, but also contains mucus, which lubricates food; antibodies, which form part of the body's defences against infection; and enzymes, which help to break down some of the elements of food before it is swallowed. The tongue is a muscular structure that moulds food into a soft bolus (ball). When the bolus is ready to be swallowed, the tongue pushes it towards the back of the mouth.

## SWALLOWING

Swallowing begins as a voluntary action as food is moved to the back of the mouth, but from this point on the process is automatic. The presence of the bolus of food triggers several involuntary reactions. The epiglottis, a flap of tissue at the back of the tongue, tilts downwards to seal the trachea (windpipe). This action prevents food from entering the airways and causing choking. At the same time, the soft palate (the back of the roof of the mouth) lifts upwards and closes the connection between the throat and the nasal cavity. Breathing halts momentarily while muscles in the wall of the throat contract to push the food into the oesophagus, a tube about 25cm (10in) long leading to the stomach. The food continues to move downwards, driven by peristalsis, waves of contractions produced by two sets of muscles in the wall of the oesophagus. The bolus, helped on its way by gravity and mucus from the lining of the oesophagus, takes around 8 seconds to complete its journey to the stomach.

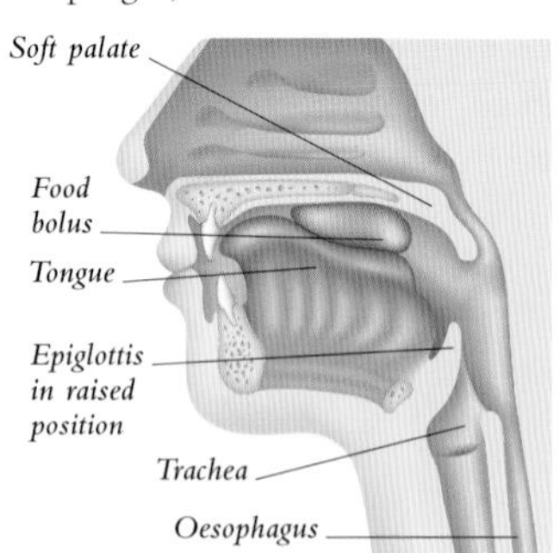


**READY TO SWALLOW**
*Before the food bolus reaches the back of the mouth, the epiglottis is raised in its normal position, allowing free flow of air from the nasal cavity to the trachea. The oesophagus is relaxed.*

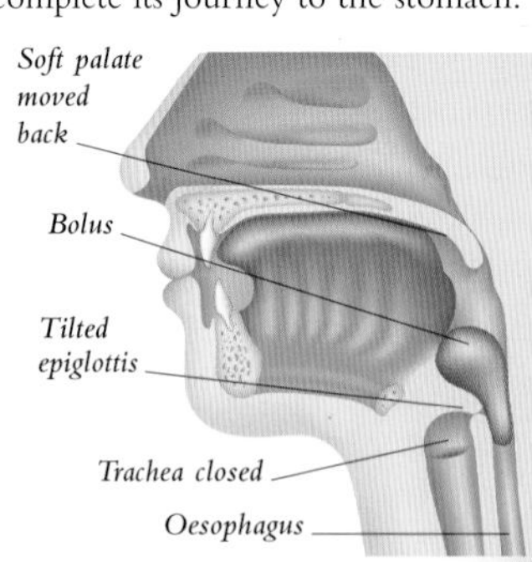


**SWALLOWING**
*As the swallowing reflex takes over, the epiglottis tilts, closing the trachea, and the soft palate lifts, closing the nasal cavity. Food enters the oesophagus and is pushed downwards.*

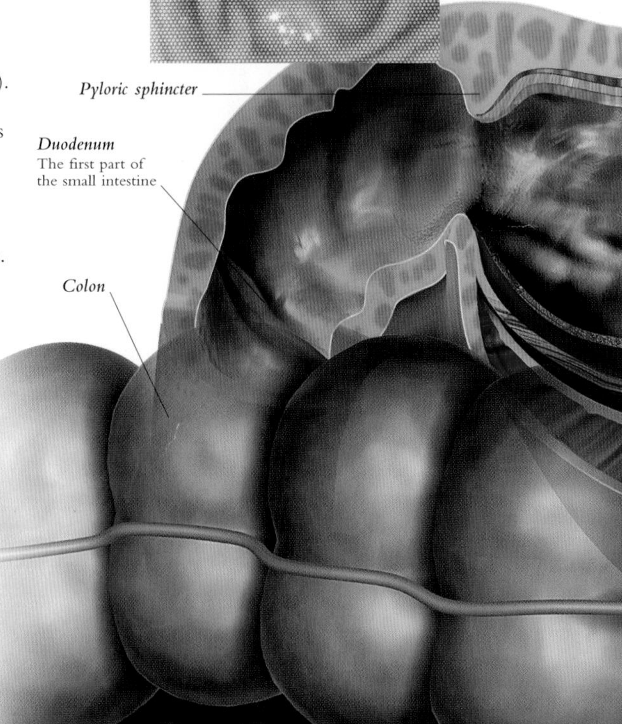


**PYLORIC SPHINCTER**
*This ring of muscle is a one-way valve between the lower end of the stomach and the start of the small intestine.*

# MIXING FOOD IN THE STOMACH

The stomach is primed for action by the taste, sight, smell, or even thought of food. The anticipation triggers increased secretion of the stomach's powerful gastric juices. When food enters the stomach, muscle contractions in the stomach wall begin. These muscles churn food, mixing it with the gastric juices and breaking it down into chyme, a semi-digested liquid. When chyme is digested enough, a muscular ring encircling the stomach exit, the pyloric sphincter, opens to release the liquid in spurts into the duodenum, the first part of the small intestine. Food usually remains in the stomach for about 4 hours, fatty meals staying the longest.

**TRANSITION AREA**
*The junction between the linings of the oesophagus (purple) and the stomach (yellow) is clearly marked by their differences in texture.*

***Muscle layers***
Three rings of muscle, running at different angles, contract in turn to churn food

***Mucosa***
Layer of mucus-producing cells that prevent the stomach being damaged by its own acids

***Lower oesophageal sphincter***
Ring of muscle that controls entry of food into the stomach

**MUCUS LAYER**
*The surface of the mucosa, the mucus-producing layer of the stomach, is seen here. The dark holes are the gastric pits through which acid juices enter the stomach.*

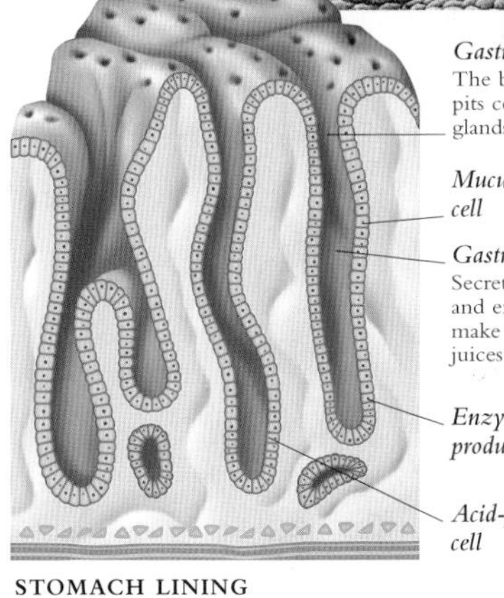


**STOMACH LINING**
*The inner lining of the stomach, the mucosa, is structured in thick folds. This lining contains cells that secrete protective mucus to prevent the stomach from digesting itself. There are also many cells and glands that secrete digestive acid and enzymes.*

***Submucosa***
Layer of loose tissue underneath the mucosa

Health

## Helicobacter pylori infection

The powerful acids produced by the stomach kill most bacteria, but one type, *Helicobacter pylori,* not only survives but is thought to infect half the world's population. The rod-shaped bacteria (shown below) flourish in the mucus layer of the stomach. In most cases, *H. pylori* causes no symptoms, but the bacteria may damage the stomach lining, leading to painful inflammation (gastritis) or peptic ulcers, when the lining is eroded by digestive juices.

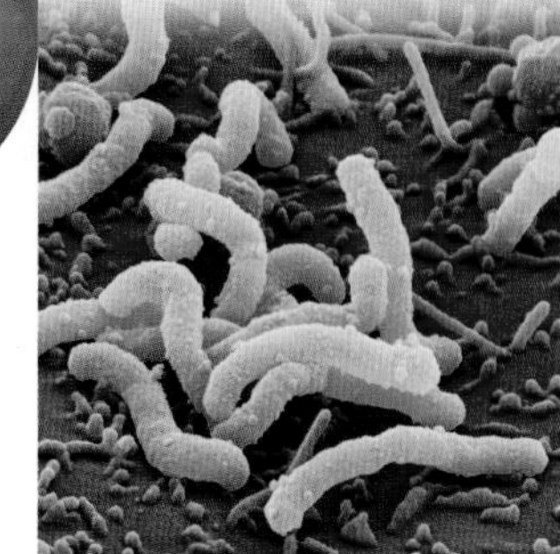

**THE ELASTIC STOMACH**
*Stretchy, muscular walls give the stomach the capacity to take in a lot of food at any one time. The inner surface of an empty stomach is folded into deep creases, rugae; as the stomach fills the rugae flatten out.*

# ABSORBING NUTRIENTS

After food leaves the stomach it has to pass through a series of processes that draw off beneficial nutrients and package up waste for excretion. Enzymes in the small intestine continue digestion and complete the chemical breakdown of food, aided by bile funnelled in by the biliary system. Nutrients are absorbed through the lining of the small intestine and enter the bloodstream. The material left, mainly waste, passes to the large intestine, where faeces are formed.

**DIGESTIVE COILS**
*A barium dye highlights the coiled small intestine, which links the stomach to the upper part of the large intestine (the colon). The small intestine is about 5m (17ft) long.*

## THE SMALL INTESTINE

The surface area of the small intestine is about the size of a tennis court. The organ is made up of three main parts: the duodenum, jejunum, and ileum. Short and C-shaped, the duodenum receives secretions from the liver and pancreas that aid digestion. The jejunum and ileum are both long and coiled, but the jejunum is thicker and slightly shorter than the ileum. In the small intestine, food, in the semi-liquid form of chyme from the stomach, is broken down by pancreatic juice, bile from the liver, and intestinal secretions so that nutrients can be absorbed and utilized. The inner lining of the small intestine has millions of projections called villi. Every villus contains a lacteal (lymph vessel) and a network of tiny blood vessels, which provide nourishment. Each villus is also covered by epithelium, or cell layer, that absorbs nutrients. The epithelial cells themselves have numerous projections called microvilli. Together, the villi and microvilli increase the total surface area of the small intestine for efficient and optimum absorption of nutrients.

**VILLI**
*Each of the thousands of finger-like villi in the small intestine is covered with epithelial cells (some are visible in white above), which absorb nutrients.*

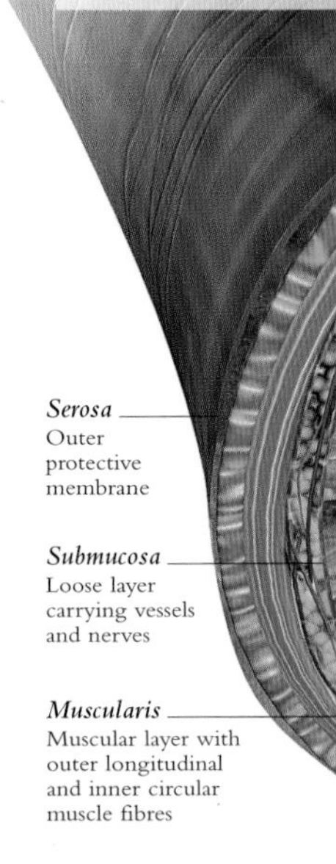

**STRUCTURE OF SMALL INTESTINE**
*The wall of the small intestine has four layers: the outer serosa, muscularis, submucosa, and mucosa.*

***Serosa*** Outer protective membrane

***Submucosa*** Loose layer carrying vessels and nerves

***Muscularis*** Muscular layer with outer longitudinal and inner circular muscle fibres

***Villus*** Maximizes surface area of mucosa

***Mucosa*** Inner layer that absorbs nutrients through projections called villi

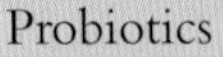

Issue

### Probiotics

So-called friendly bacteria, such as the one shown here (Lactobacillus acidophilus), live naturally in the small intestine. Their job is to reduce levels of harmful bacteria and create a healthy environment. Some people believe that taking supplements of friendly bacteria (known as probiotics) helps to boost the function and health of the small intestine. There is, however, no scientific evidence to support these claims.

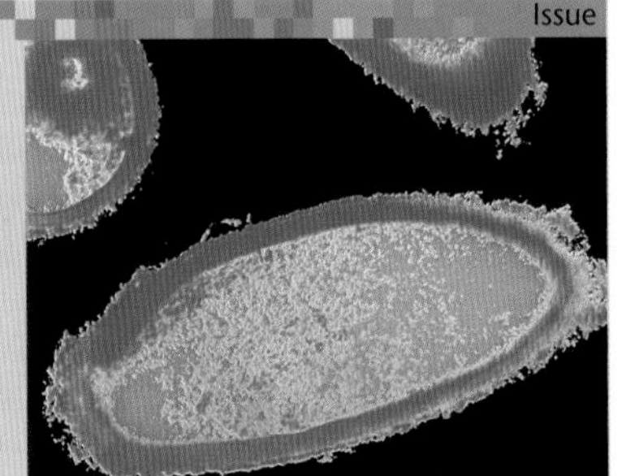

**DUODENAL FOLDS**
*The duodenum is the first section of the small intestine, after the stomach. These circular ridges, known as Kerckring's folds, help with the propulsion of food.*

## PERISTALSIS

A ball of food (bolus) is propelled through the small intestine by a sequence of rhythmic muscular contractions known as peristalsis. The muscular wall of the small intestine squeezes in behind the food in order to push it forwards into the next part of the intestine, where the muscle is relaxed. The flexibility of the muscular layer (the muscularis) of the small intestine facilitates this action. The contractions are strong and forceful in order to negotiate the many curves and folds of the small intestine. Peristalsis also occurs in the oesophagus. Other types of muscular action churn food in the stomach and make faeces in the colon, the main part of the large intestine.

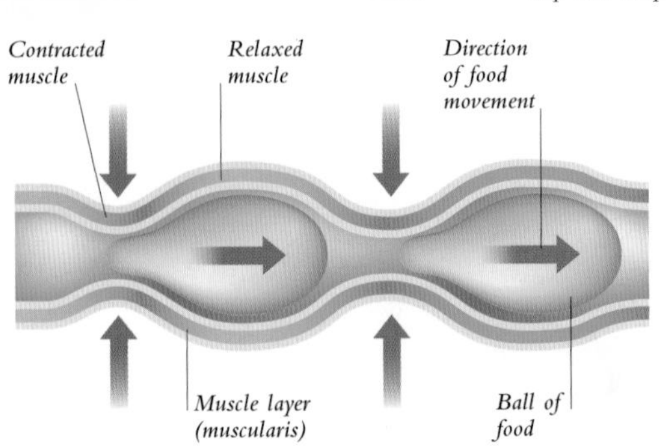


**WAVES OF CONTRACTION**
*Intermittent relaxation and contraction of sections of the muscular small intestine squeezes food along continuously. The rhythmic action is known as a "peristaltic wave". This action also occurs farther up the digestive tract, in the oesophagus, as well as farther down the tract, in the colon.*

# THE BILIARY SYSTEM

The role of the biliary system is to store bile produced by the liver (*see* p92) and deliver it into the small intestine during the digestive process. The biliary system consists of the bile ducts, the hepatic ducts leading from the liver, and the gallbladder (a pouch lying beneath the liver). Bile is used in the digestive process to break down fats. Hepatic ducts carry bile from the liver to the gallbladder, where it is concentrated and stored. When food is eaten, the gallbladder squeezes bile into the common bile duct, which empties into the duodenum.



**STRUCTURE OF THE BILIARY SYSTEM**
*When needed for digestion, bile leaves the gallbladder through the cystic duct, then travels through the common bile duct. This duct joins the pancreatic duct at the ampulla of Vater, the entrance to the duodenum.*

# THE PANCREAS

The pancreas is a large L-shaped gland situated beneath the liver and behind the stomach. At its broadest end (the head), it nestles in the curve of the duodenum. Its main part (the body) extends to the left and tapers off (the tail). The pancreas has two important functions. It releases a powerful juice that aids the digestive process and also secretes hormones that regulate the level of glucose in the body (*see* p118). Pancreatic juice is released through pancreatic ducts into the duodenum in response to the arrival of food in the digestive tract. The juice contains a number of enzymes that work together to break down fat, proteins, and carbohydrates. It also contains sodium bicarbonate, which neutralizes stomach acid.

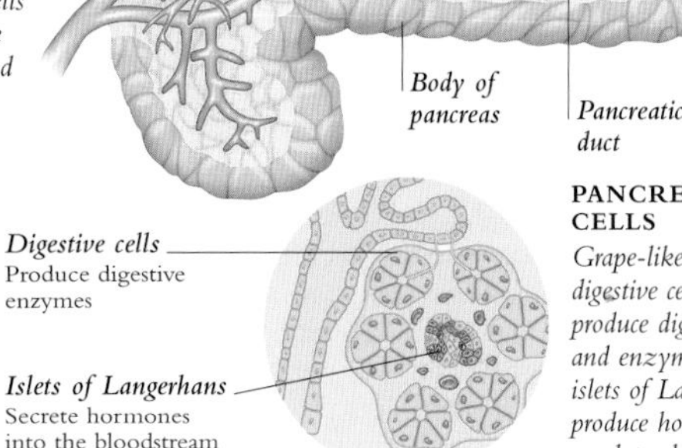


**STRUCTURE OF THE PANCREAS**
*The pancreas is an elongated gland that contains clusters of cells that produce enzymes and hormones.*



**PANCREATIC CELLS**
*Grape-like clusters of digestive cells called acini produce digestive juices and enzymes, while the islets of Langerhans produce hormones that regulate glucose levels.*

**CURVES OF LARGE INTESTINE**
*The caecum and the first part of the colon can be seen clearly here in bright yellow (left); the curve of the large intestine then travels clockwise to the rectum (at bottom).*

# THE LARGE INTESTINE

In the large intestine, which is made up of the caecum, the colon, and the rectum, faeces are formed before passing out of the anus. The caecum is a short pouch that connects the small intestine to the colon. The colon is the major part of the large intestine. Its main function is to convert chyme, liquid from the small intestine, into faeces. It draws off water from the chyme, reabsorbing it into the body. It also absorbs vitamins that are produced by billions of bacteria in the colon. The long, tubelike colon has bands of muscles that propel faeces along its length. Its inner lining secretes mucus to lubricate the inside of the intestine and ease the passage of faeces. Faeces are collected in the rectum before being passed out through the anus.

**STRUCTURE OF LARGE INTESTINE**
*Wider in diameter than the small intestine, the large intestine also has a less developed muscle layer.*



**COLONIC RIDGES**
*The triangular-shaped ridges of the colon (the main part of the large intestine) contract to force digested food along its length.*

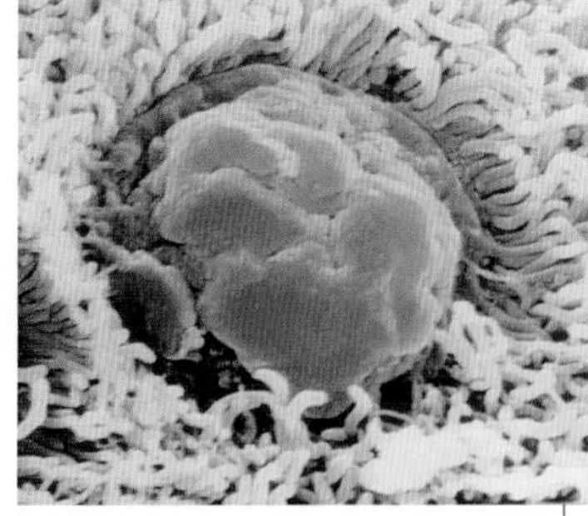

**LUBRICATING GOBLET CELLS**
*Embedded in the wall of the large intestine are goblet cells (in orange, above), which continuously secrete lubricating mucus in order to ease the passage of faeces.*

# THE LIVER

The versatile liver is one of the most chemically active parts of the body. This large wedge-shaped organ, which is situated in the upper abdomen, is estimated to have over 500 different functions that help to maintain the chemical balance of the body. One of the liver's key roles is its part in the processing of nutrients, although it is not directly connected to the digestive tract. The liver has remarkable powers of regeneration and may be able to repair itself following damage or disease.

Spleen
Spine
Liver

**SITE OF LIVER**
*In this scan of the abdomen viewed from above, the liver is the large, curved organ on the right of the image.*

## LINKS WITH THE DIGESTIVE TRACT

The liver's role in the digestive process is to process nutrients from food, which reach the liver in the bloodstream. During digestion, nutrients are absorbed in the digestive tract and pass into the blood through a dense network of capillaries (small blood vessels). The blood then drains from the capillaries into a much larger vessel, the hepatic portal vein, and is taken to the liver for processing. The portal vein divides into many tiny branches that deliver blood to the individual processing units – lobules – within the liver (*see also* Blood flow in the liver, opposite page).

Liver
Oesophagus
Gallbladder
Transverse branch of portal vein
Stomach
Spleen
Portal vein
Small intestine
Large intestine
Appendix
Rectum

**HEPATIC PORTAL SYSTEM**
*Nutrient-rich blood flows from the stomach and the intestines to the liver through a system of capillaries (small blood vessels) that merge into larger vessels, eventually becoming the portal vein.*

*Hepatic vein*
Drains blood from liver

*Bile duct*

*Gallbladder*
Sac where bile, a digestive fluid made by the liver, is stored

## A CHEMICAL FACTORY

Inside, the liver is made up of thousands of lobules, tiny processing units that together keep the liver functioning as a chemical factory. The lobules, which are just visible to the naked eye, are roughly hexagonal in shape; each one is formed from billions of cube-shaped cells, hepatocytes, that are arranged in columns radiating from a central vein. Hepatocytes are hardworking. One of their main tasks is to chemically modify nutrients into forms that can be stored in the liver or distributed around the body as fuel. Hepatocytes also secrete bile, a greenish-yellow fluid that drains into the digestive tract to break down fats. Other liver functions include cleansing blood to remove bacteria and worn-out cells; the breakdown of toxic substances such as alcohol into less harmful forms; and storage of vitamins and minerals.

**LIVER TISSUE**
*Rows of cells, organized in sheets, make up the tissue of the liver.*

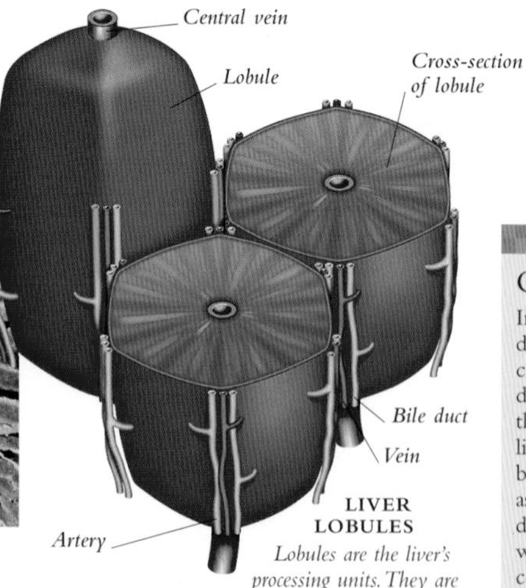


**LIVER LOBULES**
*Lobules are the liver's processing units. They are surrounded by blood vessels and bile ducts.*

**INSIDE A LOBULE**
*A vein at the centre of each lobule (seen here as a black circle) carries blood back to the heart.*

Health

### Cirrhosis

In the developed world, the liver disease cirrhosis is the third most common cause of death (after heart disease and cancer) in people over the age of 45. In cirrhosis, normal liver tissue is destroyed and replaced by bands of fibrous scar tissue (shown as a dark area in the photograph). The damage may be caused by infection with a hepatitis virus but in developed countries it is most often the result of excessive alcohol consumption.

# INSIDE LIVER LOBULES

Running between the layers of hepatocytes (liver cells) in the lobules are blood-filled channels called sinusoids. These small channels bring blood into direct contact with the hepatocytes, enabling the cells to obtain oxygen and to carry out their blood-processing roles. Once processed, blood drains into an exit vein running through the centre of the lobule. Bile secreted by the hepatocytes drains into small bile ducts that surround each lobule. The ducts lead to the common bile duct, the main route to the gallbladder where bile is stored until needed in the digestive tract.

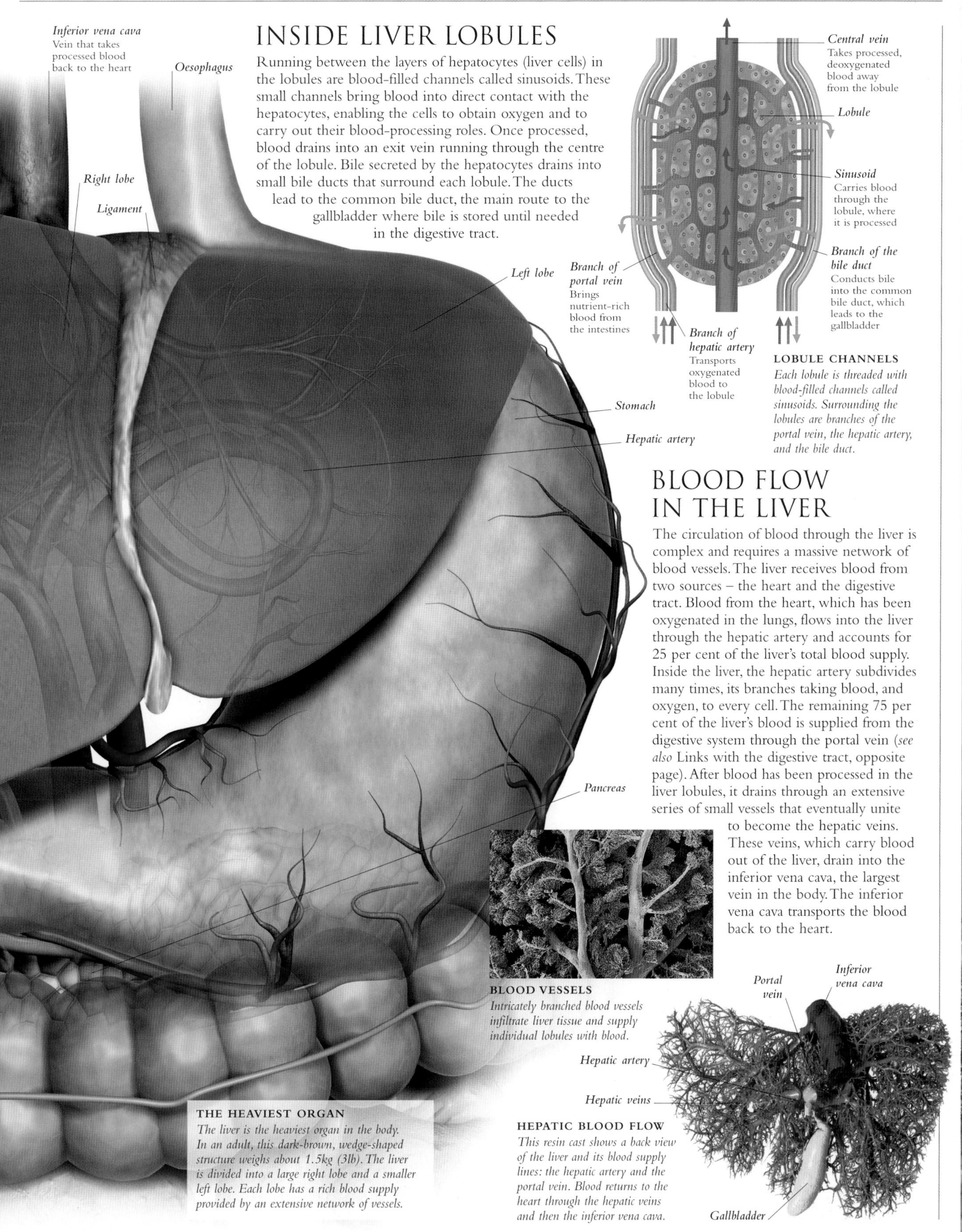


**LOBULE CHANNELS**
*Each lobule is threaded with blood-filled channels called sinusoids. Surrounding the lobules are branches of the portal vein, the hepatic artery, and the bile duct.*

# BLOOD FLOW IN THE LIVER

The circulation of blood through the liver is complex and requires a massive network of blood vessels. The liver receives blood from two sources – the heart and the digestive tract. Blood from the heart, which has been oxygenated in the lungs, flows into the liver through the hepatic artery and accounts for 25 per cent of the liver's total blood supply. Inside the liver, the hepatic artery subdivides many times, its branches taking blood, and oxygen, to every cell. The remaining 75 per cent of the liver's blood is supplied from the digestive system through the portal vein (*see also* Links with the digestive tract, opposite page). After blood has been processed in the liver lobules, it drains through an extensive series of small vessels that eventually unite to become the hepatic veins. These veins, which carry blood out of the liver, drain into the inferior vena cava, the largest vein in the body. The inferior vena cava transports the blood back to the heart.

**BLOOD VESSELS**
*Intricately branched blood vessels infiltrate liver tissue and supply individual lobules with blood.*

**THE HEAVIEST ORGAN**
*The liver is the heaviest organ in the body. In an adult, this dark-brown, wedge-shaped structure weighs about 1.5kg (3lb). The liver is divided into a large right lobe and a smaller left lobe. Each lobe has a rich blood supply provided by an extensive network of vessels.*

**HEPATIC BLOOD FLOW**
*This resin cast shows a back view of the liver and its blood supply lines: the hepatic artery and the portal vein. Blood returns to the heart through the hepatic veins and then the inferior vena cava.*

# FILTERING WASTE

Wastes produced as a result of chemical reactions in our bodies collect in the blood. In order for us to remain healthy, these wastes must be filtered out and excreted. The urinary system (or tract) performs this task, filtering blood and excreting waste products and excess water as urine. The system consists of two kidneys; the bladder; the ureters, which connect each kidney to the bladder; and the urethra, through which urine leaves the body. As the urinary system filters blood, it regulates the level of water in the body, and also maintains the balance of body fluids and substances, such as salts, within those fluids.

Health

## Bladder stones

Stones can form in the bladder if waste products in urine crystallize. Bladder stones are three times more common in men than in women, and much more common in people over the age of 43. Large stones, such as the five seen in this coloured X-ray of a bladder, usually need to be removed surgically; smaller stones may be broken down into fragments and passed in the urine.

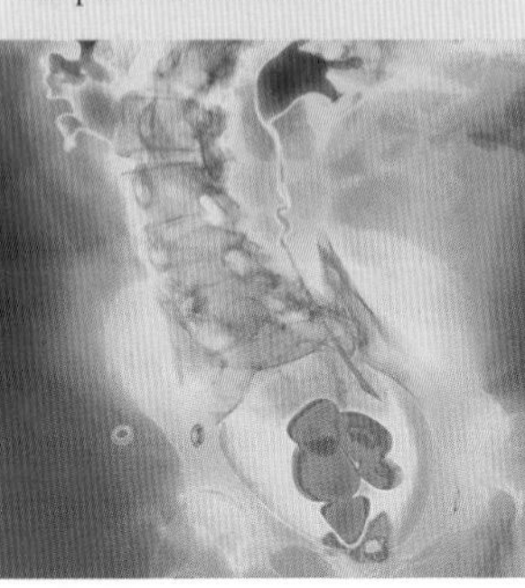

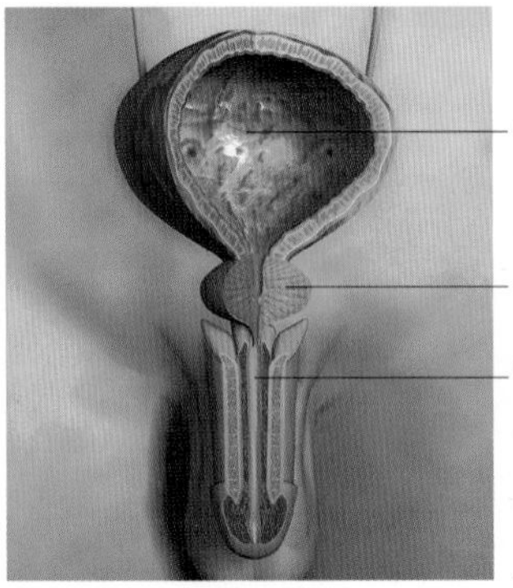

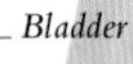

*Bladder*

*Prostate gland*

*Urethra*

**MALE BLADDER AND URETHRA**

*The male urethra runs from the bladder, through the prostate gland, to the tip of the penis.*

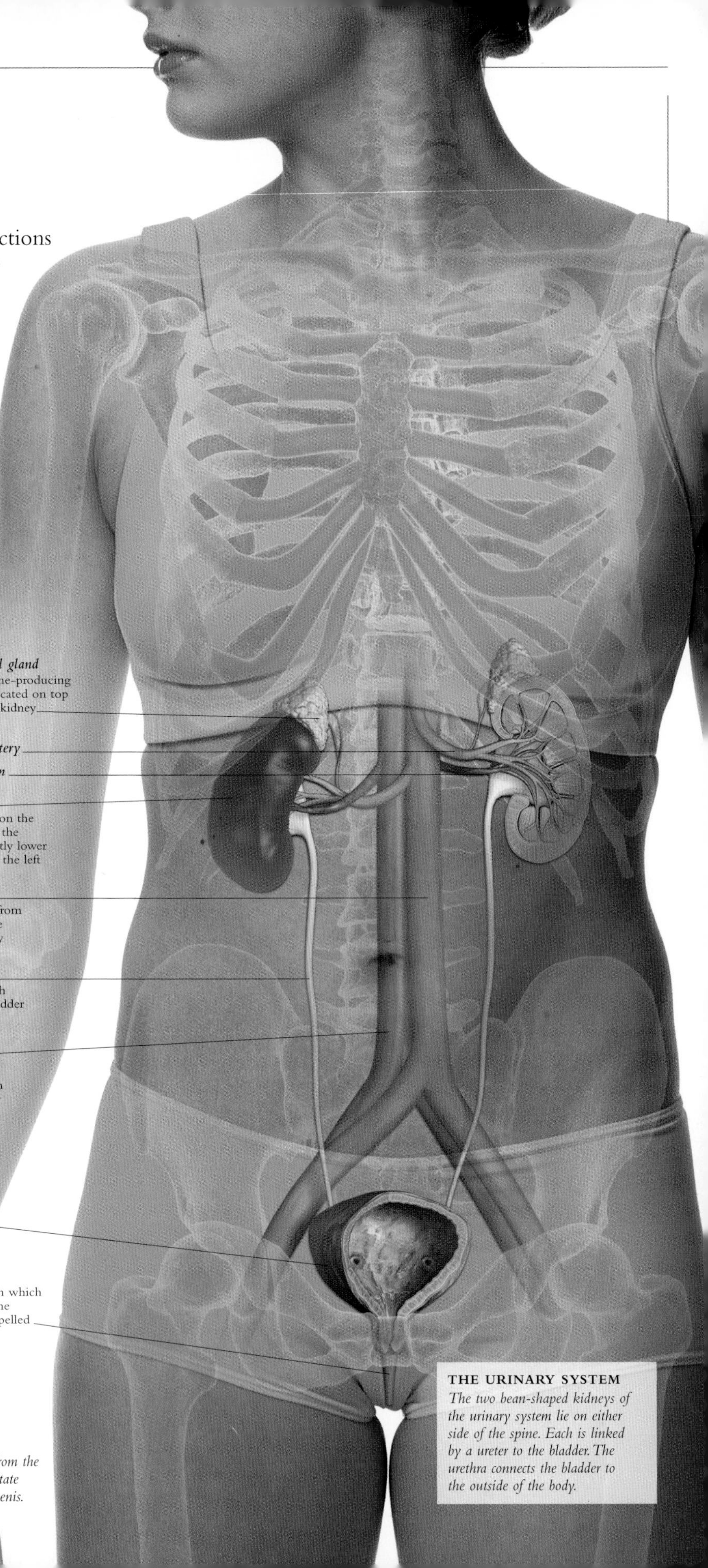

*Adrenal gland*
Hormone-producing gland located on top of each kidney

*Renal artery*

*Renal vein*

*Kidney*
The kidney on the right side of the body is slightly lower than that on the left

*Aorta*
Carries blood from the heart to the rest of the body

*Ureter*
Tube linking each kidney to the bladder

*Inferior vena cava*
Returns blood from the rest of the body to the heart

*Bladder*
A hollow organ with a thick, muscular wall in which urine is temporarily stored

*Urethra*
Tube through which urine from the bladder is expelled

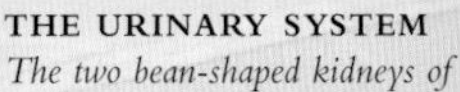

**THE URINARY SYSTEM**

*The two bean-shaped kidneys of the urinary system lie on either side of the spine. Each is linked by a ureter to the bladder. The urethra connects the bladder to the outside of the body.*

# THE KIDNEYS

The main role of the two kidneys is the production of urine, a fluid that contains many of the waste substances produced by the body. By altering the concentration and composition of urine, the kidneys can maintain a stable, balanced environment in the body, ensuring that the balance of body fluids is always correct. The kidney has three regions: the cortex, the medulla, and the renal pelvis. The outer layer, the cortex, contains many nephrons, the functioning units of the kidneys. Each nephron is made up of a glomerulus and a renal tubule. If the tubules of all nephrons were uncoiled and joined end to end, they would stretch for 80km (50 miles). The middle layer, the medulla, consists of cone-shaped groups of urine-collecting ducts. The inner region, the renal pelvis, branches out into cavities called major and minor calyces. Each minor calyx gathers urine from the medulla; the urine is then collected in major calyces and funnelled into the ureter, ready to be transported to the bladder.

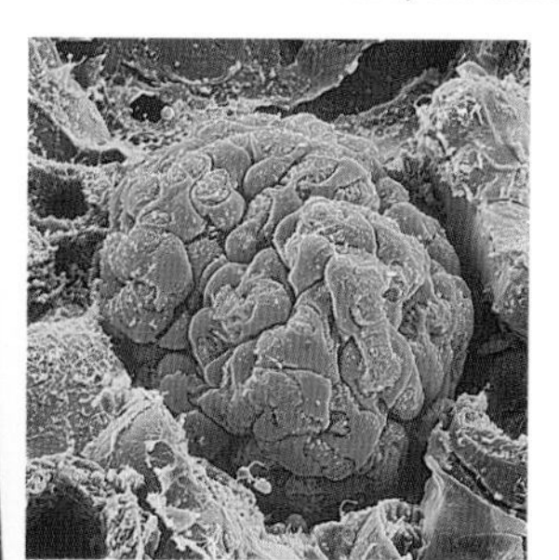

**GLOMERULUS**
*This tight cluster of capillaries is the part of the nephron called the glomerulus. It has a key role in filtering blood in the kidneys.*

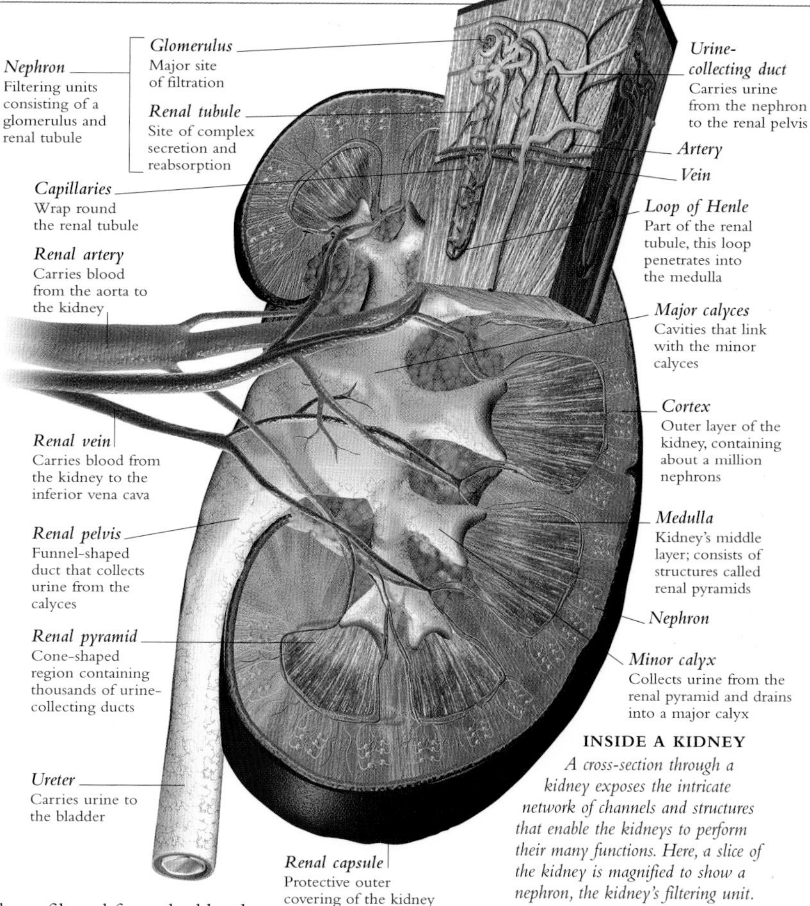

Nephron
Filtering units consisting of a glomerulus and renal tubule
Glomerulus
Major site of filtration
Renal tubule
Site of complex secretion and reabsorption
Capillaries
Wrap round the renal tubule
Renal artery
Carries blood from the aorta to the kidney
Renal vein
Carries blood from the kidney to the inferior vena cava
Renal pelvis
Funnel-shaped duct that collects urine from the calyces
Renal pyramid
Cone-shaped region containing thousands of urine-collecting ducts
Ureter
Carries urine to the bladder
Renal capsule
Protective outer covering of the kidney
Urine-collecting duct
Carries urine from the nephron to the renal pelvis
Artery
Vein
Loop of Henle
Part of the renal tubule, this loop penetrates into the medulla
Major calyces
Cavities that link with the minor calyces
Cortex
Outer layer of the kidney, containing about a million nephrons
Medulla
Kidney's middle layer; consists of structures called renal pyramids
Nephron
Minor calyx
Collects urine from the renal pyramid and drains into a major calyx

**INSIDE A KIDNEY**
*A cross-section through a kidney exposes the intricate network of channels and structures that enable the kidneys to perform their many functions. Here, a slice of the kidney is magnified to show a nephron, the kidney's filtering unit.*

# MAKING URINE

Urine is made up of unwanted substances that have been filtered from the blood by the kidneys' nephrons. Blood enters the nephron, through arterioles (small arteries) that branch from the renal artery, and is filtered through a dense cluster of capillaries called the glomerulus. The filtrate then enters the renal tubule, along which a complex process of secretion and reabsorption occurs. The beneficial substances, such as glucose, are reabsorbed for use in the body; the acidity of the blood is regulated; and water levels are adjusted. The waste product resulting from this process is a fluid called urine.

**THE PATH THROUGH A NEPHRON**
*Filtrate from the glomerulus flows through the renal tubule, which has three sections: the proximal convoluted tubule, the loop of Henle, and the distal convoluted tubule.*

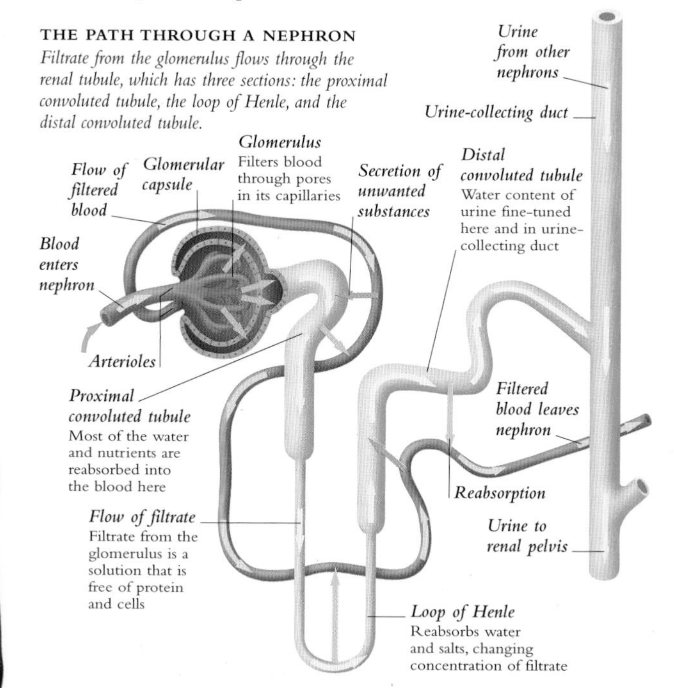

Urine from other nephrons
Urine-collecting duct
Glomerulus
Filters blood through pores in its capillaries
Flow of filtered blood
Glomerular capsule
Secretion of unwanted substances
Distal convoluted tubule
Water content of urine fine-tuned here and in urine-collecting duct
Blood enters nephron
Arterioles
Proximal convoluted tubule
Most of the water and nutrients are reabsorbed into the blood here
Filtered blood leaves nephron
Reabsorption
Flow of filtrate
Filtrate from the glomerulus is a solution that is free of protein and cells
Urine to renal pelvis
Loop of Henle
Reabsorbs water and salts, changing concentration of filtrate

# BLADDER CONTROL

When the bladder is full, nerves in the bladder wall send signals to the spinal cord. In older children and adults, the timing of urination can be regulated because this process is controlled by the brain. A sphincter (valve) of skeletal muscle in the bladder wall, close to the opening of the urethra, can be tensed to prevent urine from being expelled. (Infants lack this control and the bladder empties spontaneously when full.) When a person is ready to urinate, signals are sent from the brain to the bladder. A few seconds before urination, the pelvic floor relaxes, along with other muscles in the base of the abdomen, and the sphincters in the bladder wall also relax. The bladder wall then contracts and urine is expelled.

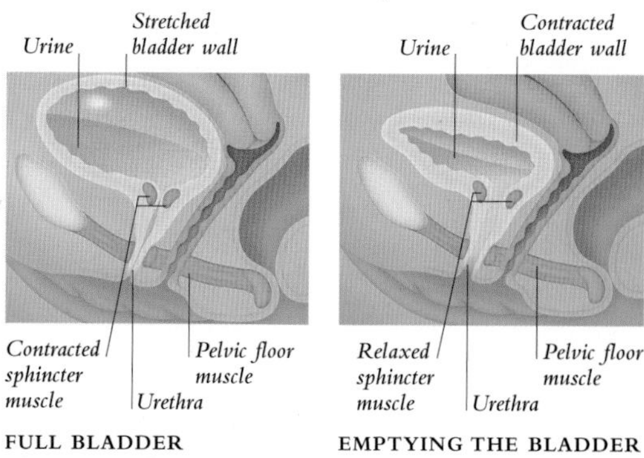

Urine
Stretched bladder wall
Contracted sphincter muscle
Pelvic floor muscle
Urethra
Urine
Contracted bladder wall
Relaxed sphincter muscle
Pelvic floor muscle
Urethra

**FULL BLADDER**
*The stretching of the bladder wall as it fills with urine triggers impulses that travel to the spinal cord, sending the message that the bladder needs to be emptied.*

**EMPTYING THE BLADDER**
*To empty the bladder, muscles in the bladder wall contract and the sphincter muscles also relax, forcing urine out of the bladder and down the urethra.*

**ON THE ATTACK**

*A macrophage (purple), a protective cell that has a role in the body's defences, attempts to destroy a cancer cell (yellow) that it recognizes as a foreign intruder.*

# DEFENCE AND REPAIR

We are exposed to many potential sources of harm every minute of the day. The air we breathe, everything we touch, and all we eat and drink are teeming with microscopic organisms that could cause diseases. The body can be damaged by accidental injury. Even our own cells can malfunction and cause illness. Besides these dangers, the body also has to cope with the effects of aging. Fortunately, we have inbuilt mechanisms for defence and repair that help to provide protection against diseases and degeneration.

The body automatically tries to avoid any obvious threats. The brain monitors information it receives through the senses and triggers a reflex action in response to certain situations. For example, we flinch from a source of pain and instinctively shun food that smells or tastes bad. Many harmful agents cannot be sensed and the body combats these with physical and chemical barriers.

## SOURCES OF HARM

Infections are our greatest health threat. A vast array of what we often refer to as "germs" – bacteria and viruses – can gain access to the body and cause trouble. Many types of parasite, such as fungi and intestinal worms, can colonize tissues and organs, damaging cells and absorbing vital nutrients. We are also surrounded by many environmental hazards, which are often difficult to avoid. Invisible, harmful radiation comes from sunlight and the rocks beneath us. Hazardous chemicals can enter the food chain or pollute the atmosphere. There is always the possibility of an accident, such as a burn or a fall. Sometimes, the body can be a danger to itself, attacking its own tissues or reacting to non-harmful substances such as pollen. Occasionally, the normal cell process of replicating in order to keep body tissues in a state of repair overaccelerates, causing cancer.

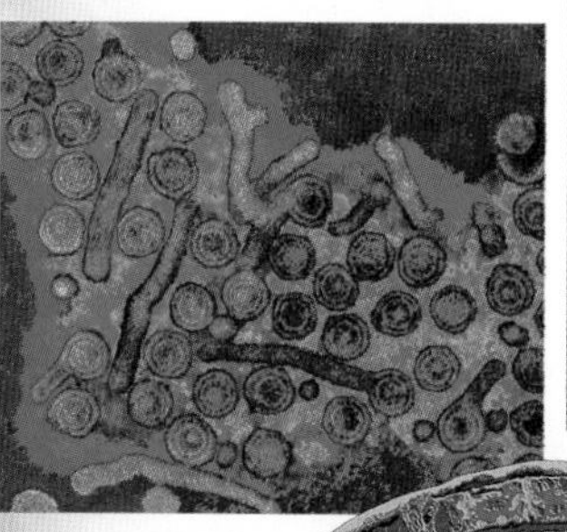

HEPATITIS B VIRUSES

LEGIONELLA BACTERIA

**INFECTIONS**
*The body's defences are primed to attack viruses and bacteria, but may not easily overcome the rapidly transmitted hepatitis B viruses (shown as blue circles, above) and the legionella bacteria that cause Legionnaires' disease.*

## BARRIER DEFENCES

Without our covering of skin we we would lose our main barrier against external sources of harm. Skin shields the internal organs, and its sensitivity allows us to be alert to dangerous levels of pain, heat, and cold. Other effective barriers are the membranes that line body openings such as the nose and mouth. The body also puts obstacles such as hair and protective chemicals in the way of infective organisms. Both on the outside and the inside, the body is colonized by communities of helpful bacteria and fungi that do not cause disease but instead act as further deterrents to invaders.

**SHEDDING SKIN**
*These layers of cells on the outer surface of the skin are worn out and are gradually being shed. New cells forming in the layers below will move up to take their place.*

## IMMUNE RESPONSES

The body is able to fight disease through complicated mechanisms known collectively as the immune system. Immune defences involve special white blood cells as well as the action of the lymphatic system. This system comprises a network of vessels and nodes (small nodules of tissue) that extends around the entire body and through which lymph (watery tissue fluid) drains. The immune system recognizes infective organisms as "foreigners". If such organisms bypass external barriers the body responds in a number of ways to disable or kill them. One part of the immune system works by releasing special proteins, antibodies, that destroy bacteria. Another type of immune response involves direct action by white blood cells, which seek out and destroy cancer cells or cells infected by viruses.

The immune system has the astonishing facility to remember specific infections that have entered the body. If the organism is encountered again, the body produces an extremely rapid response to deal with it.

Usually, certain symptoms let us know when the immune system is fighting infection. One indicator is a raised temperature; if the outer parts of the body are affected there may be redness and swelling. The immune system sometimes overreacts, producing antibodies against innocuous substances. This response can cause a variety of symptoms, such sneezing or rashes.

## GROWTH AND REPAIR

Some body cells have limited lifespans. They may be destroyed, shed naturally from the body, or simply be programmed to die within a certain time, sometimes after only a few hours. Every day, millions of new cells develop in the body to replace those that are old, damaged, or lost. Most types of cell are capable of making duplicate copies of themselves. This process, called mitosis, occurs fastest in childhood, ensuring that children attain their full growth. Mitosis continues in adulthood, although at a slower rate; the body would be eroded by wear and tear if new cells were not produced to keep skin, bones, and organs in good repair. However, not all replaceable body parts are made of living tissue. Hair and nails are made from dead, flattened cells hardened by the protein keratin.

Cell replication makes wound healing possible, as new cells gradually fill in the injured area. Before damaged tissues start to mend, the body activates its blood-clotting mechanism to plug the broken blood vessels.

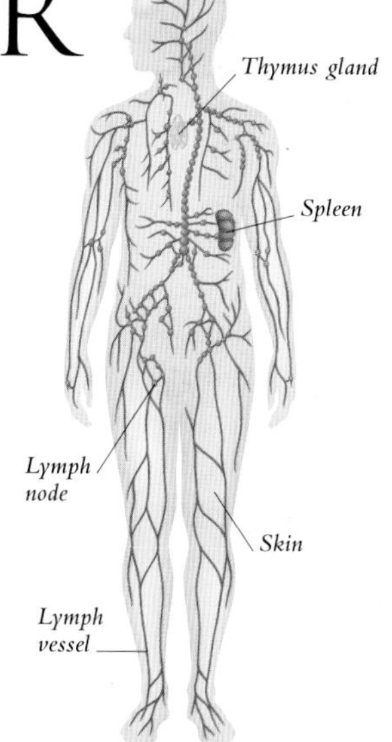


**THE BODY'S DEFENCES**
*The body has both outer and inner defences against diseases. Skin is the first barrier against infection. Internally, protection is provided by special blood cells and by the glands and vessels of the lymphatic system.*

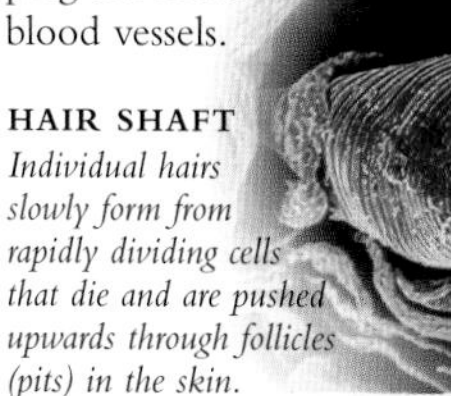

**HAIR SHAFT**
*Individual hairs slowly form from rapidly dividing cells that die and are pushed upwards through follicles (pits) in the skin.*

History

### Discovering germs

The discovery that certain types of microorganism cause diseases was made by a French chemist, Louis Pasteur (1822-1895), pictured below. Once this so-called "germ theory" was generally accepted by doctors and scientists, the causes of major diseases such as diphtheria and cholera were able to be identified.

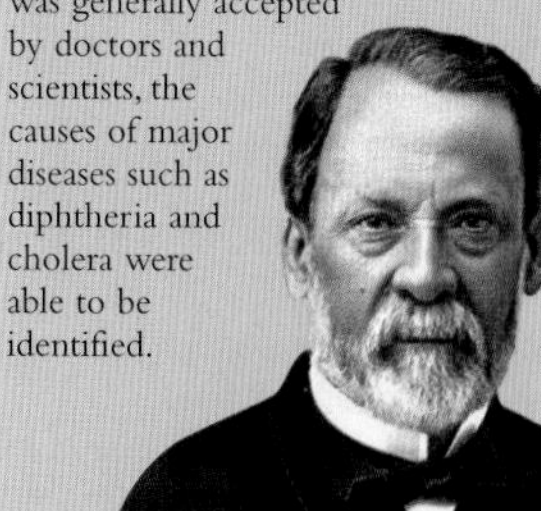

BODY

# FIRST-LINE DEFENCES

The body's first line of defence against possible damage is the skin, which provides a protective barrier between us and our environment. At body openings, where the skin does not provide cover, there are membrane linings to keep out harmful organisms. The body produces chemicals to lubricate and cleanse these barriers and further increase their effectiveness. On the surface of the skin and in internal organs, we have allies in the form of harmless microbes, which compete for body space with disease-causing organisms.

## THE SKIN

The body's outer covering has two main layers: the epidermis and the dermis. The top layer, the epidermis, is a tough, protective tissue with a surface made of dead cells. The cells are constantly sloughed off, making it difficult for small organisms to remain on the skin. This layer includes melanocytes, cells that produce melanin, a dark pigment. The pigment, which causes skin to darken in sunlight, helps to prevent damage from the ultraviolet radiation in sunlight that causes skin cancer. People with naturally dark skins have a lot of melanin and high protection against the sun; light-skinned people have the least protection. Beneath the epidermis is the dermis. This layer contains sebaceous glands, which secrete an oily substance, sebum, onto the surface of the skin. Sebum lubricates the skin and inhibits the growth of microorganisms. Another important skin function is to provide protection against extremes of temperature. A thermoregulation system, which uses special blood vessels and sweat glands, cools us down if we are too hot and conserves heat when body temperature starts to drop. The skin also contains billions of sensory cells that warn us about harmful levels of heat or cold as well as pressure and pain (*see* Touch, p128).

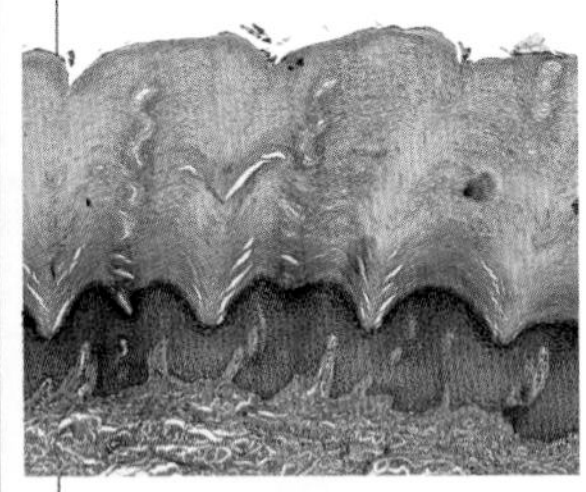

**SKIN SECTION**
*This skin from a finger clearly shows the tissue layers, with the epidermis (purple) at the top.*

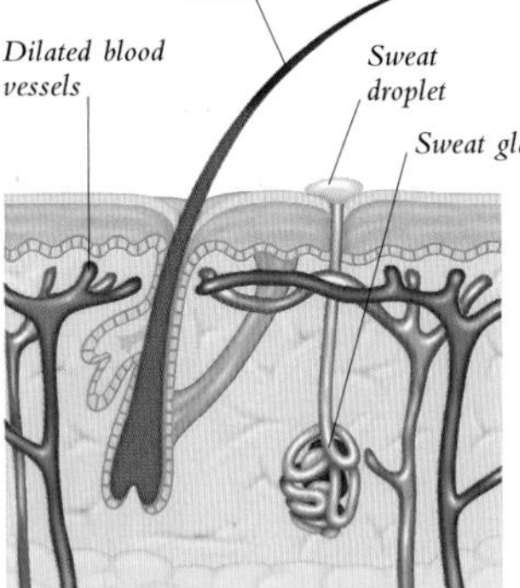
Hair
Dilated blood vessels
Sweat droplet
Sweat gland

HIGH BODY TEMPERATURE

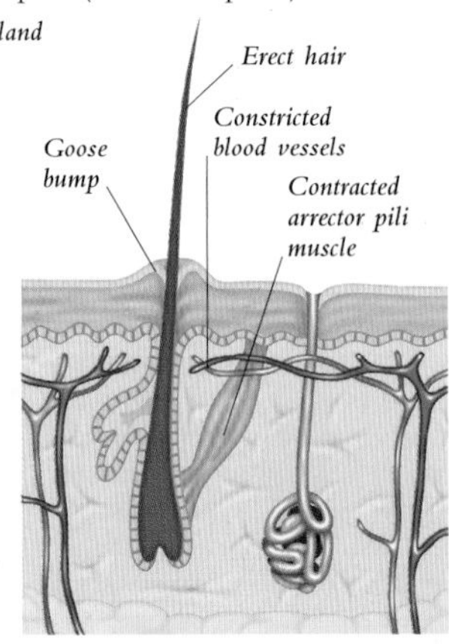
Erect hair
Goose bump
Constricted blood vessels
Contracted arrector pili muscle

LOW BODY TEMPERATURE

**THERMOREGULATION**
*To cool the body, blood vessels dilate to let more blood reach the skin and sweat glands produce sweat. To save heat, blood vessels constrict and arrector pili muscles pull hairs erect to trap warmth, creating goose bumps.*

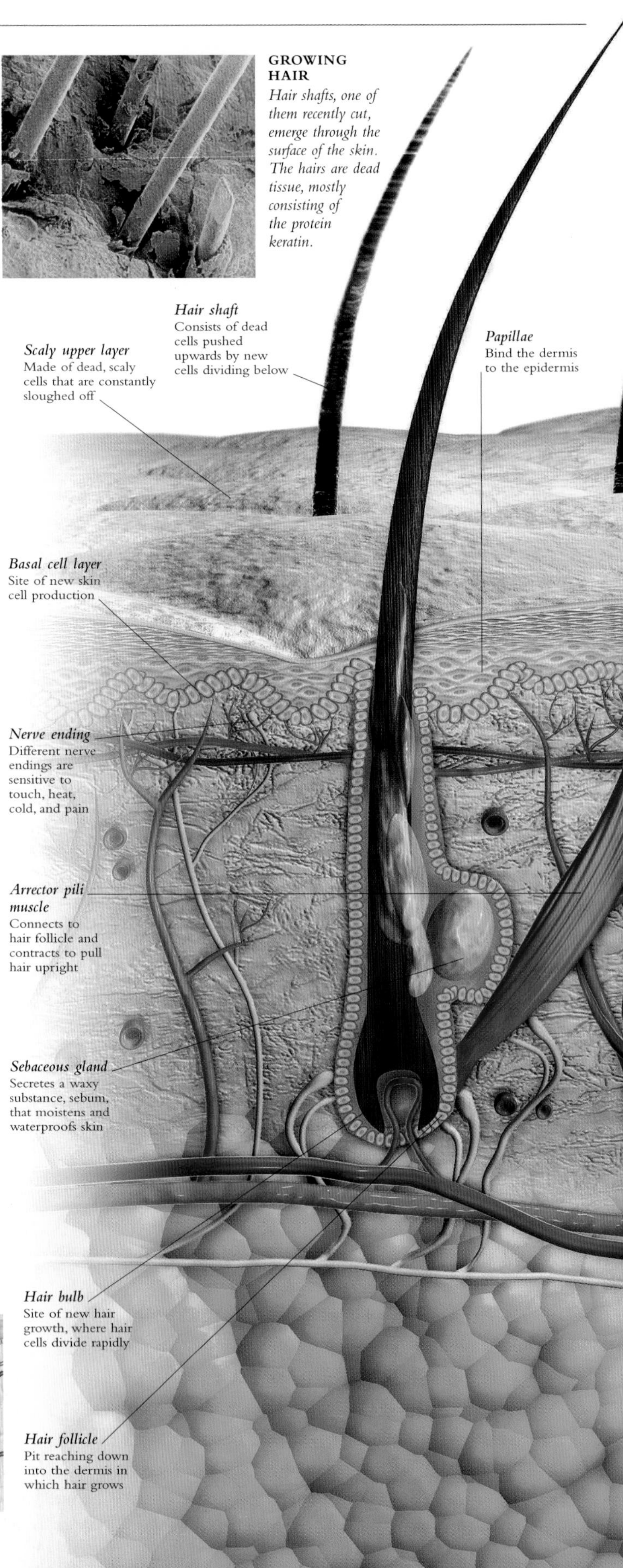
Hair shaft
Consists of dead cells pushed upwards by new cells dividing below
Scaly upper layer
Made of dead, scaly cells that are constantly sloughed off
Papillae
Bind the dermis to the epidermis
Basal cell layer
Site of new skin cell production
Nerve ending
Different nerve endings are sensitive to touch, heat, cold, and pain
Arrector pili muscle
Connects to hair follicle and contracts to pull hair upright
Sebaceous gland
Secretes a waxy substance, sebum, that moistens and waterproofs skin
Hair bulb
Site of new hair growth, where hair cells divide rapidly
Hair follicle
Pit reaching down into the dermis in which hair grows

**GROWING HAIR**
*Hair shafts, one of them recently cut, emerge through the surface of the skin. The hairs are dead tissue, mostly consisting of the protein keratin.*

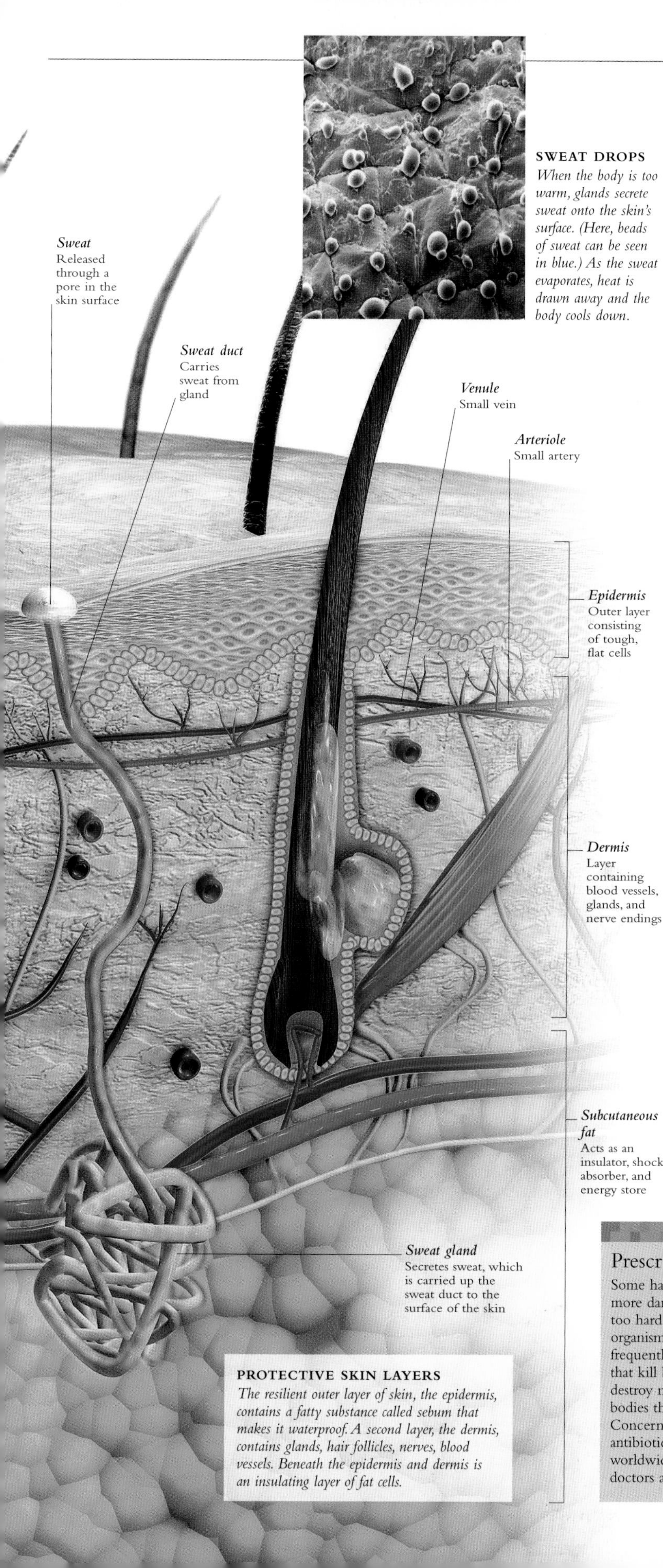
Sweat
Released through a pore in the skin surface
Sweat duct
Carries sweat from gland
Venule
Small vein
Arteriole
Small artery
Epidermis
Outer layer consisting of tough, flat cells
Dermis
Layer containing blood vessels, glands, and nerve endings
Subcutaneous fat
Acts as an insulator, shock absorber, and energy store
Sweat gland
Secretes sweat, which is carried up the sweat duct to the surface of the skin

**SWEAT DROPS**
*When the body is too warm, glands secrete sweat onto the skin's surface. (Here, beads of sweat can be seen in blue.) As the sweat evaporates, heat is drawn away and the body cools down.*

**PROTECTIVE SKIN LAYERS**
*The resilient outer layer of skin, the epidermis, contains a fatty substance called sebum that makes it waterproof. A second layer, the dermis, contains glands, hair follicles, nerves, blood vessels. Beneath the epidermis and dermis is an insulating layer of fat cells.*

# INTERNAL BARRIERS

In several areas of the body, such as the mouth, nose, eyes, anus, urinary tract, and vagina, inner tissues are exposed to the environment. These sites are protected by their own chemical and physical barriers. Glands in the mouth make saliva, a mix of mucus and antibacterial substances; tears wash and protect the eyes. Bacteria that get as far as the stomach are destroyed by powerful acid. Mucus in the genital tract traps organisms and tiny particles. Hairs and mucus in the nose do the same, and also in the airways, tiny hair-like projections – cilia – sweep dust and microorganisms trapped in mucus to the throat, where they are swallowed or coughed out. Even the ears have protective barriers: earwax and hairs in the canal of the outer ear trap dust and organisms.

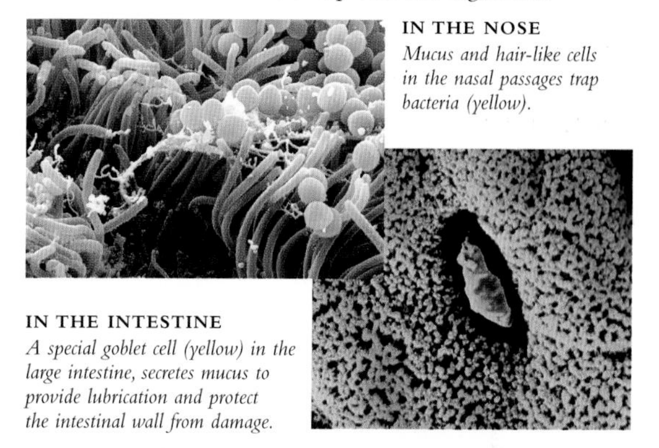

**IN THE NOSE**
*Mucus and hair-like cells in the nasal passages trap bacteria (yellow).*

**IN THE INTESTINE**
*A special goblet cell (yellow) in the large intestine, secretes mucus to provide lubrication and protect the intestinal wall from damage.*

# HELPFUL ORGANISMS

Not all organisms that live on and inside the human body are harmful – a significant proportion are positively beneficial. For example, two types of bacteria, lactobacilli and bifidobacteria, which live in the human intestine, manufacture essential vitamins. These "good" bacteria also maintain a healthy acid balance in the intestine, keep the number of disease-causing bacteria in check, and improve the body's immune function.

There are various other organisms living naturally in the body, usually without causing harm, that provide competition for invading organisms. Among such deterrents is the bacterium *Escherichia coli*, which inhabits the intestine; however, some types of *E. coli* can cause illness. The fungus *Candida albicans*, which mostly lives in the mouth and vagina, is another protective organism that itself sometimes causes disease.

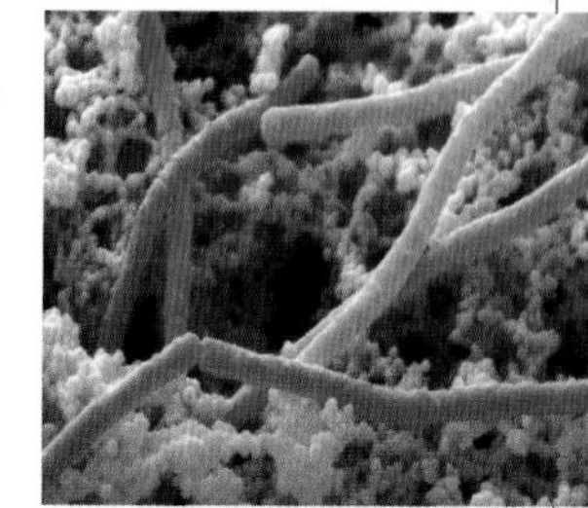

**GOOD BACTERIA**
*These lactobacillus bacteria (pink) help to stop harmful bacteria from overmultiplying in the intestine.*

Issue

## Prescribing antibiotics

Some harmful bacteria are getting more dangerous because we are trying too hard to wipe them out. These organisms are becoming resistant to frequently used antibiotics, the drugs that kill bacteria. Antibiotics also destroy many "good" bacteria in our bodies that help to combat disease. Concern about overprescribing of antibiotics has led health authorities worldwide to issue guidelines to both doctors and patients on the drugs' use.

# FIGHTING DISEASE

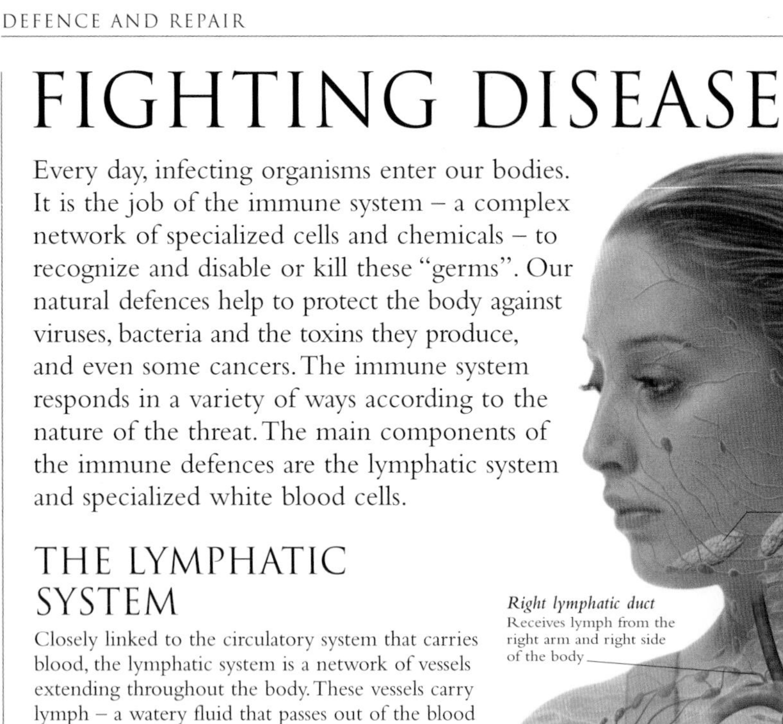

Every day, infecting organisms enter our bodies. It is the job of the immune system – a complex network of specialized cells and chemicals – to recognize and disable or kill these "germs". Our natural defences help to protect the body against viruses, bacteria and the toxins they produce, and even some cancers. The immune system responds in a variety of ways according to the nature of the threat. The main components of the immune defences are the lymphatic system and specialized white blood cells.

## THE LYMPHATIC SYSTEM

Closely linked to the circulatory system that carries blood, the lymphatic system is a network of vessels extending throughout the body. These vessels carry lymph – a watery fluid that passes out of the blood into body tissues. Most lymph goes back into the blood through the walls of small blood vessels, but the rest enters the lymph vessels. Lymph is kept in circulation by muscle contractions. Valves in the vessels prevent the fluid from flowing backwards. Nodes along the lymph vessels filter harmful microorganisms from the lymph. Filtered lymph returns through ducts into the blood to maintain the body's fluid balance.

**TISSUE IN LYMPH NODE**
*Circulating lymph filters through a fibrous mesh of tissue in the lymph nodes.*

Fact

### Immunity in babies

As a matter of survival, babies must quickly become immune to infection. They are helped by the thymus gland, which lies beneath the breastbone (white area, under the collarbone, below). The thymus makes white blood cells that play a key role in the immune system. At birth, the thymus is the biggest gland in the body; as a child matures the gland shrinks.

X-RAY OF BABY'S CHEST

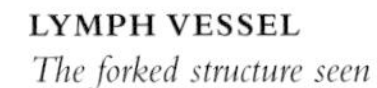

**LYMPH VESSEL**
*The forked structure seen in this lymph vessel (red) is a one-way valve that prevents lymph from flowing backwards.*

*Salivary glands*
Secrete substances that help to resist infection

*Right lymphatic duct*
Receives lymph from the right arm and right side of the body

*Axillary (armpit) nodes*
Filter lymph from the arms and breasts

*Subclavian vein*
Leads towards the heart and receives lymph from the thoracic and right lymphatic ducts

*Thymus gland*
Produces vital white blood cells

*Spleen*
Largest lymph organ in the body, which produces certain types of white blood cells

*Thoracic duct*
Largest lymph vessel, which receives lymph from most other lymph vessels in the body

*Deep inguinal (groin) nodes*
Filter lymph from the lower body

*Cisterna chyli*
Lymph vessel that receives lymph from the lower body

**LYMPHATIC NETWORK**
*The body's defence system includes lymph vessels, lymphatic tissue, and lymph nodes. Organs associated with the lymphatic network include the thymus gland and the spleen.*

BODY

# HOW LYMPH IS FILTERED

The bean-shaped nodes that act as filtering units for lymph occur in clusters in the groin, neck, and armpits, and behind the knee. Each node is filled with a network of meshlike tissue containing sinuses (cavities) packed with specialized white blood cells (lymphocytes). As lymph filters slowly through the sinuses, cell debris and disease-causing microorganisms and tumour cells are trapped, engulfed, and destroyed by the white blood cells. When part of the body is infected, nearby lymph nodes become swollen and tender as they act to limit the spread of infection. Sometimes the nodes may be visibly enlarged.

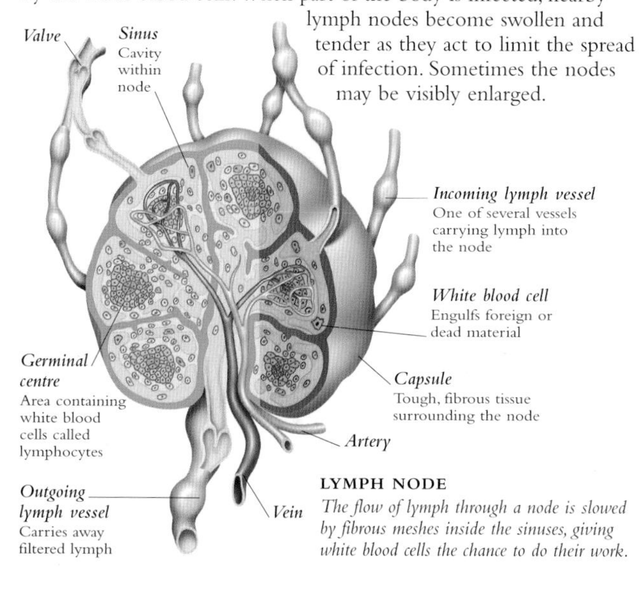


**LYMPH NODE**
*The flow of lymph through a node is slowed by fibrous meshes inside the sinuses, giving white blood cells the chance to do their work.*

# CELLS OF THE IMMUNE SYSTEM

Blood contains five main types of white blood cell: neutrophils, monocytes, basophils, eosinophils, and lymphocytes. Neutrophils, which belong to a class of cells referred to as phagocytes, patrol the bloodstream, engulfing foreign organisms. Monocytes, which are also phagocytes, perform a similar scavenging function in body tissues. Basophils release chemicals that trigger an inflammatory response; eosinophils play a role in allergic responses. Lymphocytes are more numerous in the tissues of the lymphatic system than they are in the blood. These cells target specific infective organisms, such as viruses, and are also capable of locking on to and destroying cancer cells.

MONOCYTE

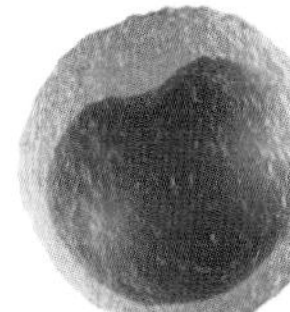

LYMPHOCYTE

**WHITE BLOOD CELLS**
*These five types of white blood cell each play a critical role in the body's immunity. They help to protect us against bacteria and bacterial toxins, as well as viruses and cancer.*

NEUTROPHIL

BASOPHIL

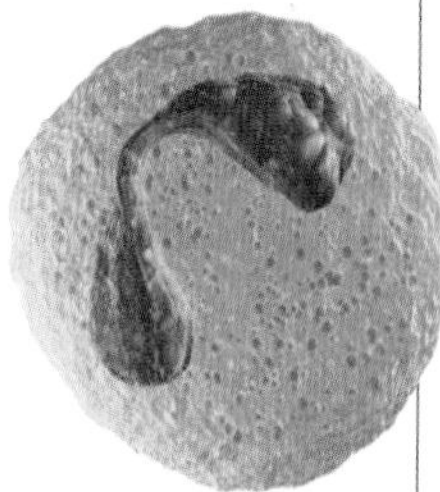

EOSINOPHIL

# THE INFLAMMATORY RESPONSE

When infectious organisms enter the body through broken skin or other damaged tissues, the immune system goes into action with the inflammatory response. The damaged tissue releases chemicals that attract phagocytes, infection-fighting white blood cells, to the scene. The chemicals also cause blood vessels near the site to widen, increasing blood flow, and to become more porous. These reactions bring more phagocytes into the area and allow them to pass easily through the blood vessel walls. The phagocytes surround, engulf, and destroy the invading microorganisms. The inflammatory response causes immediately obvious effects, including redness, pain, heat, and swelling at the site of infection.

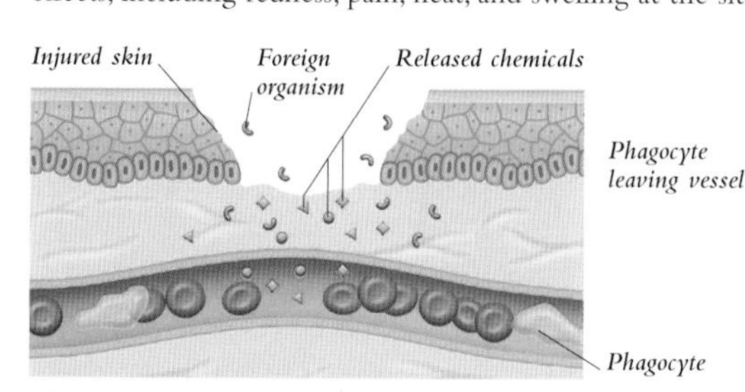


**FOREIGN ORGANISMS ENTER BODY**
*The skin is broken and foreign organisms enter. The damaged tissue immediately releases chemicals that attract phagocytes to the site of the injury.*

*Phagocyte engulfing organism*
*Inflamed tissue*
*Phagocyte leaving vessel*

**INFLAMMATORY RESPONSE TRIGGERED**
*Blood flow increases and blood vessels dilate (widen). This enables phagocytes to pass from the blood into the damaged tissue, where they destroy foreign organisms.*

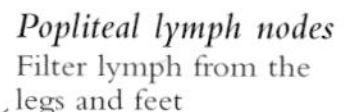

*Popliteal lymph nodes*
Filter lymph from the legs and feet

*Lymph vessel*
One of many carrying lymph from tissues to the main lymphatic ducts

Health

## Lymphoma

Sometimes, the white blood cells (lymphocytes) that help to protect us against cancer themselves become cancerous and grow uncontrollably. This cancer, lymphoma, has two main categories: Hodgkin's and non-Hodgkin's lymphoma. Lymphoma starts in a single lymph node but can spread to other nodes and body tissues. Symptoms include swollen lymph nodes, fever, and weight loss. Early treatment is often successful.

SECTION THROUGH LYMPHOMA

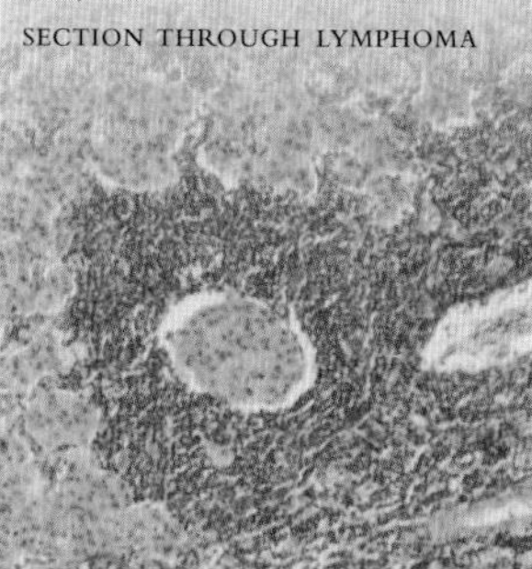

# ANTIBODY RESPONSE

The antibody response is a type of immune response that targets bacteria and toxins invading the body. When bacteria enter the body, cells called phagocytes wrap around them and take them towards white blood cells called B-lymphocytes (B-cells). Some of the B-cells are shaped specifically to lock on to bacterial antigens (proteins on the surface of the bacteria). These matching B-cells then multiply to produce two types of cell – plasma cells and memory B-cells. Plasma cells release antibodies that lock on to the antigens of the bacteria, coating the surface of the invading organisms so they cannot function properly. The antibodies also attract phagocytes that engulf the weakened bacteria, producing enzymes to break them down. Memory B-cells are programmed to memorize the bacteria's antigens, so the bacteria is recognized if it returns in the future.

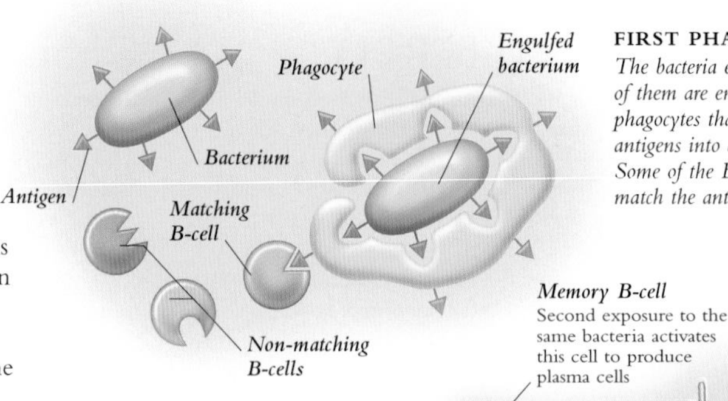


**FIRST PHASE**
*The bacteria enter the body. Some of them are engulfed by cells called phagocytes that bring the bacterial antigens into contact with B-cells. Some of the B-cells that they meet match the antigens.*

*Memory B-cell*
Second exposure to the same bacteria activates this cell to produce plasma cells

*Plasma cell*

*Antibody*

**SECOND PHASE**
*The matching B-cell multiplies to produce plasma cells, which produce antibodies to destroy the bacteria and memory B-cells, which are stored.*

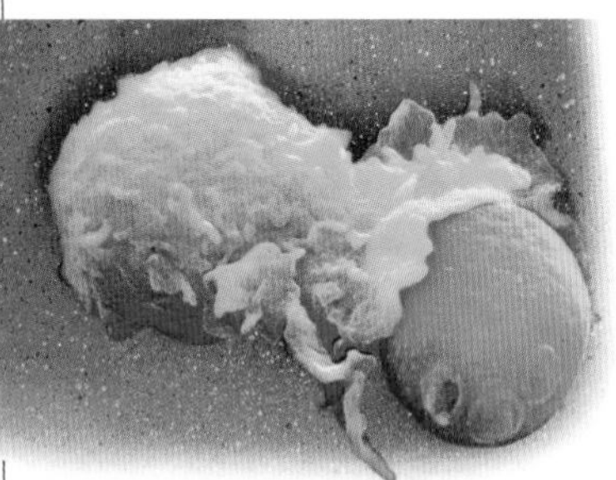

**DESTROYING AN INVADER**
*A white blood cell called a phagocyte (blue) starts to engulf a foreign organism. To destroy the organism, the phagocyte releases enzymes to break it down.*

**E. COLI BACTERIUM**
*If harmful bacteria, such as* E. coli *enter the body, the immune antibody response is quickly activated.*

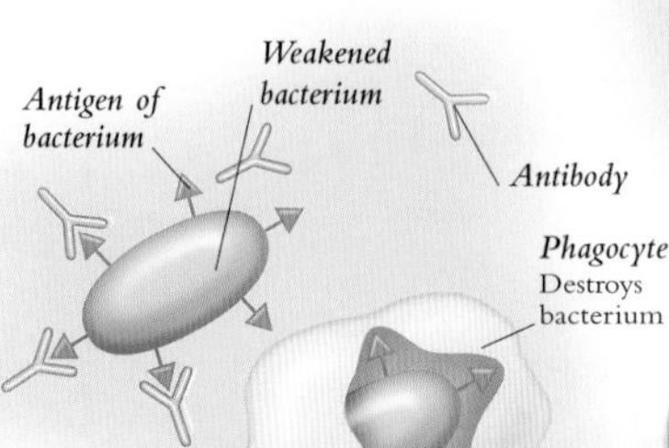


**THIRD PHASE**
*The antibodies released by the plasma cells lock on to bacterial antigens and weaken the bacteria; they also attract more phagocytes, to destroy the bacteria.*

# CELLULAR RESPONSE

The cellular immune response targets viruses, parasites and cancer cells. It depends on white blood cells called T-lymphocytes (T-cells). After recognizing a foreign antigen (protein), T-cells multiply and battle against infected cells or cancer cells. Infected or cancerous cells are engulfed by cells called phagocytes. These bring the invading antigens into contact with T-cells, some of which match up with these antigens. The matching T-cell multiplies to produce different types of cell, including killer T-cells, which contain toxic proteins, and memory T-cells, stored to protect the body against the same invader in the future. Killer T-cells lock on to the infected cell carrying the recognized antigen and release their toxic proteins. These proteins then destroy the cell.

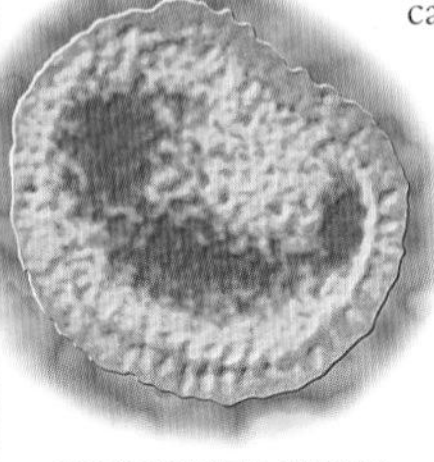

**INFLUENZA VIRUS**
*Proteins on the virus enable it to attach to and infect cells lining the respiratory tract.*

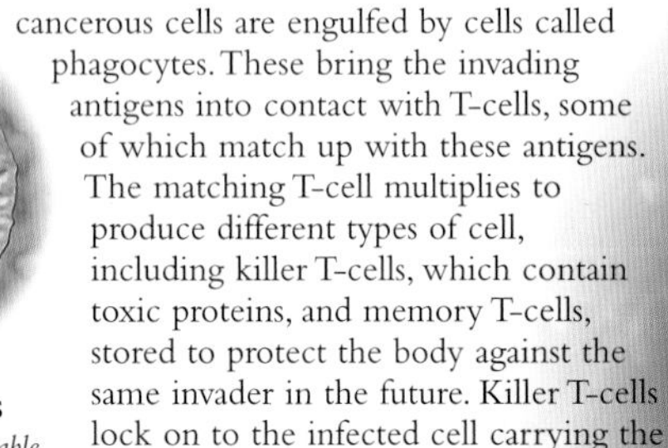

Health

## HIV and AIDS

When the human immunodeficiency virus (HIV) enters the bloodstream, it infects certain T-cells and reproduces rapidly within them, destroying the cells. If the infection is not treated, it may eventually lead to increased susceptibility to infections and cancers, called acquired immunodeficiency syndrome (AIDS).

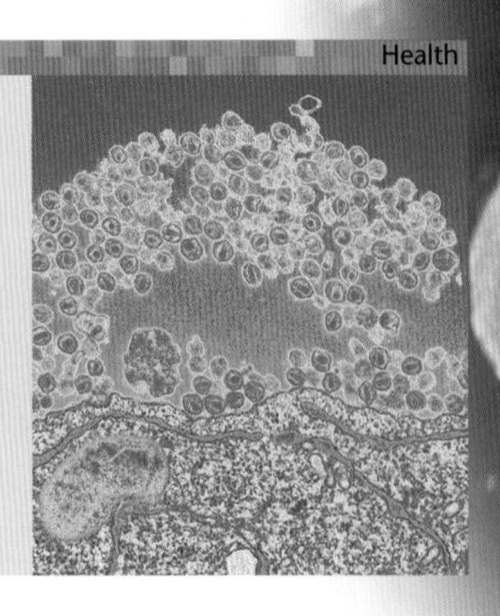

**BUDDING ACTION**
*Clusters of HIV particles (purple) bud away from the surface of an infected T-cell, leading to the destruction of the cell.*

**CANCER KILLER**
*A killer T-lymphocyte cell (orange) has attached itself to a cancer cell (purple). The T-cell releases enzymes that destroy the cancer cell; balls emerging from the cancer cell indicate it is in its death throes.*

BODY

# ALLERGIES

An allergy is an inappropriate immune response to a normally harmless substance, called an allergen. Common allergens include plant pollens, house-dust mites, small particles of hair or skin from animals (dander), and nut proteins. On first exposure to the allergen, the immune system becomes sensitized to it. During subsequent exposures, an allergic reaction occurs. Mast cells, located in the skin, nasal lining, and other tissues, are activated, releasing a substance called histamine that causes an inflammatory response, irritating body tissues and producing symptoms, such as a skin rash or respiratory problems. The severity of an allergic reaction varies among individuals.

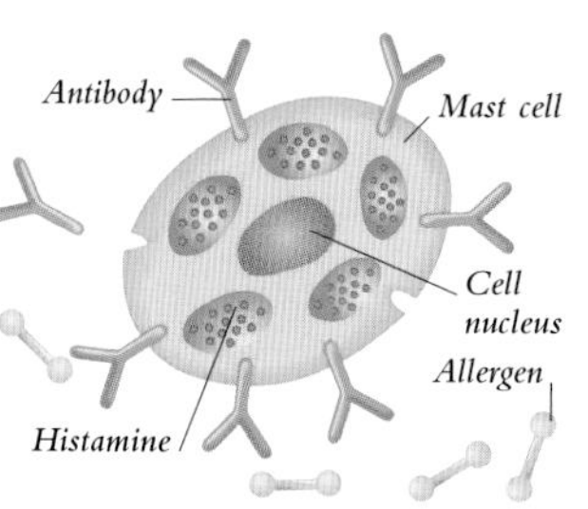


**EXPOSURE**
*When the body has a repeat exposure to an allergen, antibodies previously produced in response to it bind to the surface of mast cells, which contain histamine.*

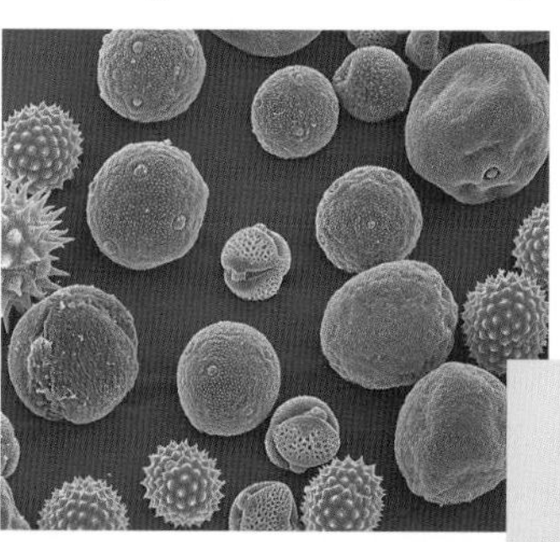

**POLLEN**
*A growing number of people are allergic to pollen grains (left), and suffer from respiratory problems (hayfever) as a result. The amount of pollen in the air (the pollen count) varies seasonally, so those people that are affected suffer more at certain times of the year.*

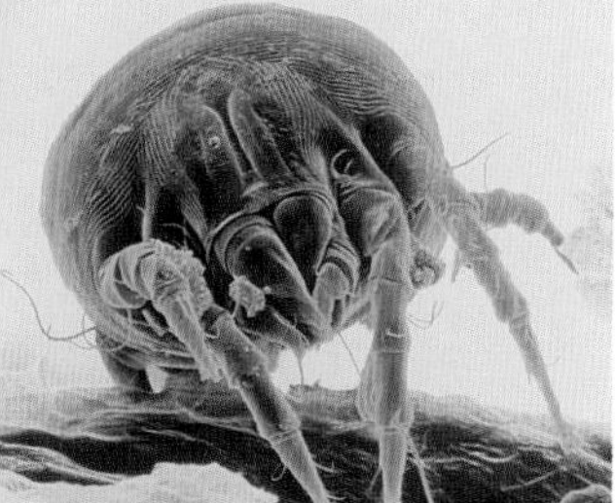

**HOUSE-DUST MITE**
*These mites feed on the dead scales of human skin in household dust; many people are allergic to their dead body parts and excrement in dust.*

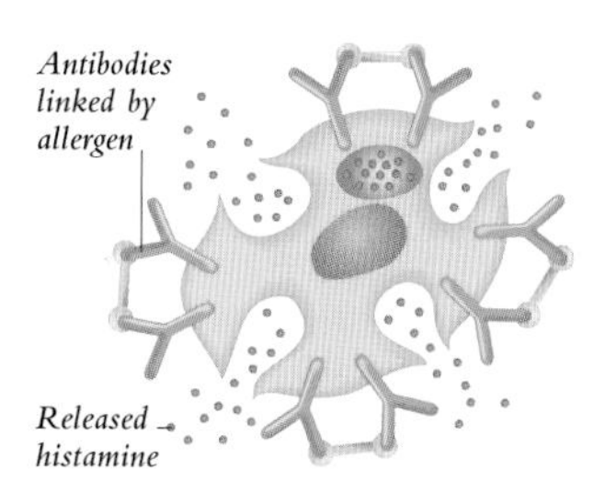


**RESPONSE**
*The allergens bind to and link two or more antibodies, causing the cell to release the histamine within. Histamine causes the symptoms of allergy.*

Issue

## Allergy is on the increase

A worldwide increase in the incidence of allergies such as asthma and hayfever has been blamed on environmental factors such as air pollution from road traffic. A lack of exposure to infections during childhood, which might provide protection in later life, has also been blamed for the rise in cases of allergies. Research has shown that the number of people that are being admitted to hospital with allergic reactions has increased dramatically over the last decade.

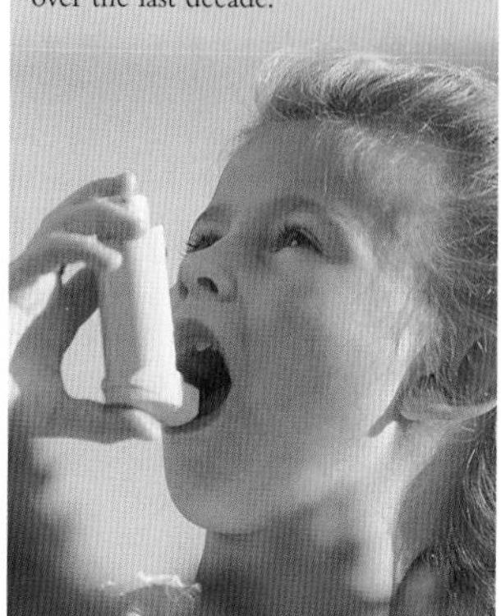

# IMMUNIZATION

To prevent an outbreak of a serious disease, such as measles, or to protect people who may be especially vulnerable to certain infections, such as influenza, immunization can be used. Immunization may be active or passive. Active immunization artificially creates a "memory" of the disease or infection before it has been acquired. It involves injecting a person with a vaccine containing harmless antigens that closely resemble those of the invading organism (the pathogen). The vaccine is usually a dead or weakened version of the pathogen. It stimulates the immune system to make antibodies and memory B-cells against the antigens of the pathogen. Immunity given by vaccine can last for many years. Passive immunization offers immediate, short-lived immunity. This may be required if no active immunization is available, or if rapid protection against an organism is vital. An extract of blood (antiserum) from a person or animal that has previously been exposed to the infection, and has therefore developed antibodies against it, is injected into the person. The antibodies in the antiserum act immediately against the organism if it is in the body, and can survive for several weeks.

History

## Vaccination

The caricature below shows Edward Jenner (1749–1823), an English doctor, administering a vaccine. Jenner developed the first effective vaccine after noting that milkmaids who caught cowpox (a mild disease) were immune to smallpox (a fatal disease). He scraped pus from a milkmaid's cowpox sore and inserted it into cuts on a boy's arm. Six weeks later he inserted smallpox pus into the boy's arm. The boy developed cowpox – but not smallpox. Jenner called his new protective method "vaccination", meaning "from cows".

**ACTIVE IMMUNIZATION**
*A vaccine is introduced into the body (usually by injection), stimulating the formation of antibodies. A memory is retained by the immune system, so that it will be able to act if the organism invades again.*

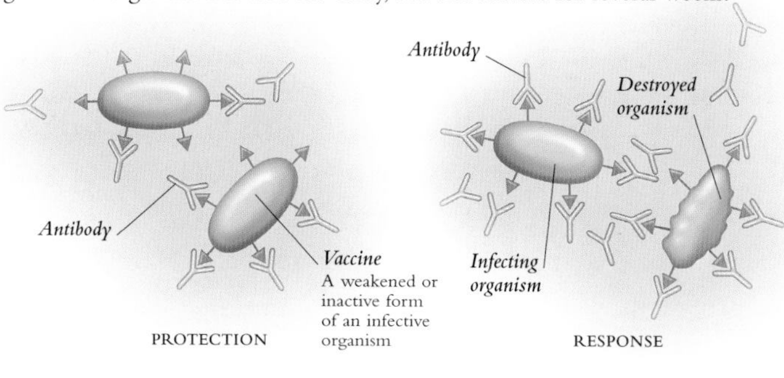


**PASSIVE IMMUNIZATION**
*Antibodies from an immune human or animals are introduced into the body. When exposed to the infectious organism, the antibodies destroy it and provide short-term protection.*

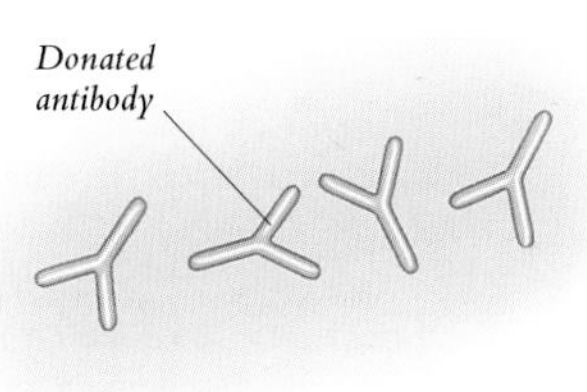


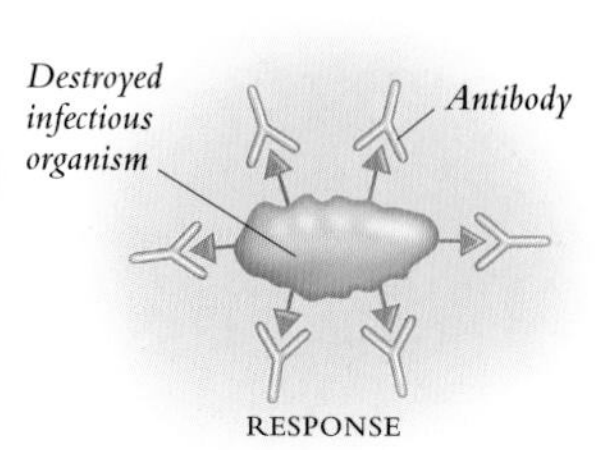


BODY

# REPAIR AND REPLACEMENT

The ability of the human body to regenerate itself is astounding. Our cells are in a constant state of flux, facilitating the growth of hair and nails, for example, and replacing cells that have been damaged or lost through wear or injury. Every type of cell has a different lifespan: some function for years; others degenerate within weeks. Each needs to be replaced, and this happens through a process of division and multiplication called mitosis.

## STAGES OF MITOSIS

The process of growth requires body cells to divide and multiply constantly. Cells also divide to replace those that have become worn out. Mitosis is a type of cell division that results in cells identical to the original cell. The process happens in stages (*see below*). Before the cell divides, its chromosomes coil up and are copied to form double identical strands of DNA (genetic material). The membrane that surrounds the nucleus of the cell (within which the chromosomes are found) breaks down and fine, thread-like fibres are spun across the cell. The double chromosomes line up on these fibres, which shorten to pull one of each chromosome to opposite ends of the cell. A nucleus membrane re-forms around each separated cell, and the cell constricts in the middle. It then splits in two, forming two identical cells.

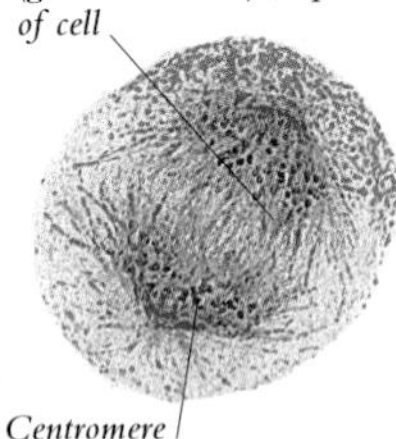
*Threads draw chromosomes (genetic material) to poles of cell*
*Centromere*

**EARLY ANAPHASE**
*In the first stage of mitosis, the cell sprouts several thread-like fibres (blue), which begin to draw the chromosomes, containing the genetic material, apart. Only the central part of the chromosomes, the centromeres (red dots), are visible.*

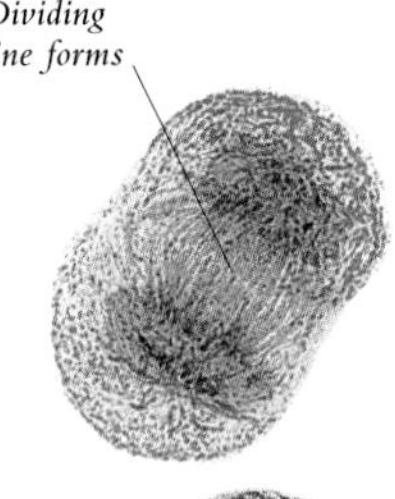
*Dividing line forms*

**LATE ANAPHASE**
*At this later stage of anaphase, the genetic material has separated as the thread-like fibres continue to pull the chromosomes to opposite poles of the cell. The protein helping this movement is dynein (purple).*

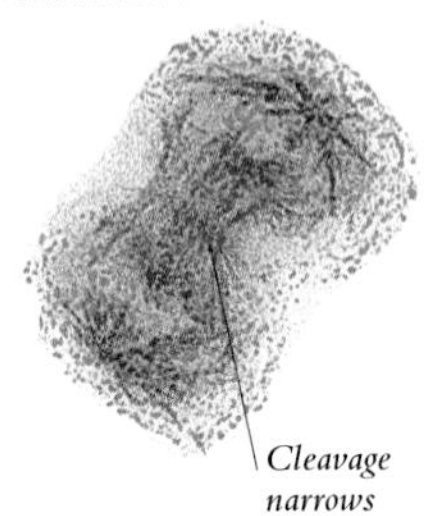
*Cleavage narrows*

**EARLY TELOPHASE**
*A distinct cleavage appears as the pulling action continues and the genetic material separates. A membrane forms around each of the two groups of chromosomes as two halves start to become two wholes.*

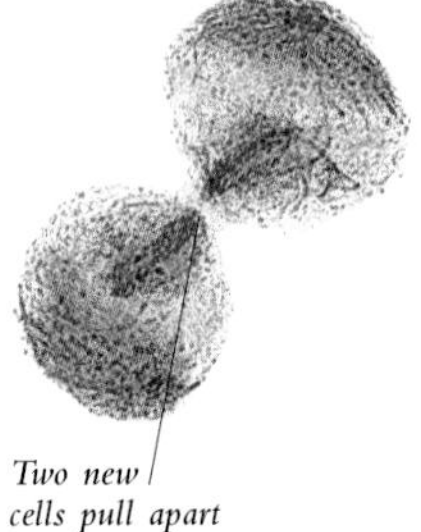
*Two new cells pull apart*

**LATE TELOPHASE**
*The two separate cells now draw away from one another, the active thread-like fibres now relaxing into each cell as the outer membranes close. New nuclei form and the chromosomes in each new cell unravel.*

*Cell membrane*
Breaks down as the cell begins the process of division

*Threads of spindle*
Connect the centre of each chromosome

*Centromere*
Point at which the chromosome pair splits to form single chromosomes

*Centriole*
Made of hollow tubules; duplicates prior to cell division

*Organelles*
Specialized structures, located within the cell's cytoplasm; organelles pull apart during cell division

**NEW CELLS FORMING**
*Shown here is the early telophase stage of mitosis, when a central cleavage forms in a cell, before it divides to form two new identical cells. Over 50,000 cells in your body will have died and been renewed in such a way by the time you have finished reading this sentence.*

## SKIN CELL REPLACEMENT

The skin constantly renews itself by shedding dead cells from the surface and generating new cells below. As a result, surface cells that are lost through wear, damage, or disease are quickly replaced. New cells are formed in the epidermis, the skin's upper layer, which serves as a tough, protective covering. In most areas of the body, the epidermis has four layers. In the lowest of these, the basal layer, cells are constantly dividing and new cells are being produced. As new cells move to the surface, their structure alters as they form intermediate layers of prickle cells (with spiky projections) and granular cells (whose cellular structure has begun to break down). It takes 1–2 months for cells to reach the surface. The topmost layer, at the surface, consists of dead, flat cells, which are continually being shed. On average, a person sheds about 30,000 of these cells every minute.

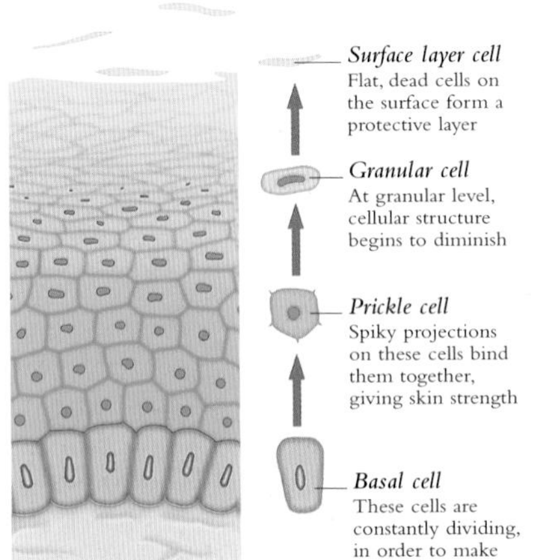
*Surface layer cell*
Flat, dead cells on the surface form a protective layer
*Granular cell*
At granular level, cellular structure begins to diminish
*Prickle cell*
Spiky projections on these cells bind them together, giving skin strength
*Basal cell*
These cells are constantly dividing, in order to make new cells

**SKIN GROWTH**
*As skin grows and renews itself, the cells move up the four different layers of the epidermis, their structure altering as they move towards the surface.*

*Cleavage*
Point at which the cell begins to divide

*Chromosome*
Contains most of the cell's genetic material

# HAIR GROWTH

Every single hair grows from a pit in the skin called the follicle, a specialized area of the epidermis that grows down into the dermis. A hair is generated from rapidly dividing cells in the bulb at the base of each follicle. The area under the bulb known as the papilla contains nutrient-rich blood vessels. Each follicle has growth periods followed by rest periods. However, the phases of different follicles are not synchronized. This means that at any one time some hairs are growing while others are being shed, which is why we do not generally notice hair being lost. A single hair grows between 6mm (¼in) and 8mm (⅓in) in a month.

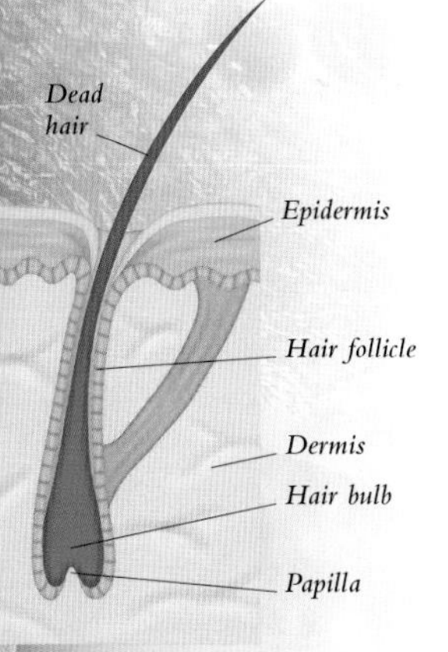


**REST PHASE**
*Cell activity in the papilla and hair bulb slows down and eventually stops. The hair then dies.*

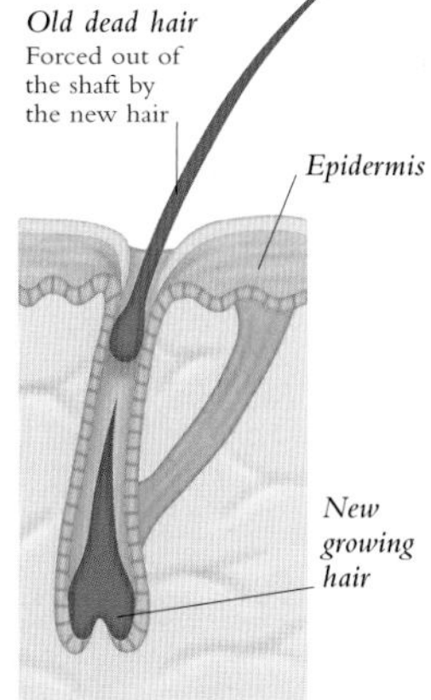


**GROWTH PHASE**
*Cells divide rapidly in the bulb, forming a new hair, which pushes out the old hair.*

Health

## Stem cell therapy

All cells in the body develop from stem cells. Because stem cells are able to renew themselves and form any one of a range of tissue types, research is ongoing into using them for tissue repair. Stem cells such as the ones shown here (from umbilical cord blood) could be used in the treatment of conditions in which tissues and organs have been damaged and the body is unable to regenerate them.

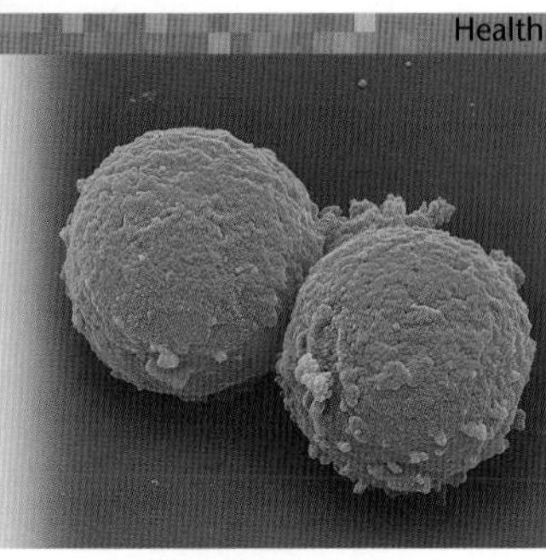

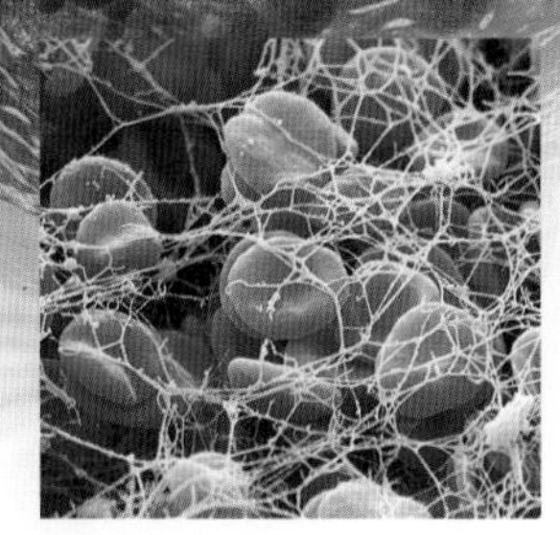

**BLOOD CLOT**
*When a blood vessel is injured, sticky strands of the protein fibrin form a web, enmeshing red blood cells and forming a clot.*

# REPAIRING INJURIES

When a part of the body is injured, the damaged area usually repairs itself, replacing the lost tissue in the process. Healing of injured skin takes place over a series of stages in which dead or damaged tissue is initially replaced by scar tissue and eventually by a batch of entirely new, healthy cells (*see right*). If an injury has penetrated the layers of skin very deeply, a visible scar may remain after healing has taken place. Bones also have the ability to repair themselves effectively after injury (*see* Bone repair, p65). Some nerves can regenerate if they are crushed or partially cut (*see* How nerves regenerate, p117). Certain tissues, such as heart muscle, cannot easily form new cells to replace scar tissue. This means that scar tissue often remains, which can affect heart function.

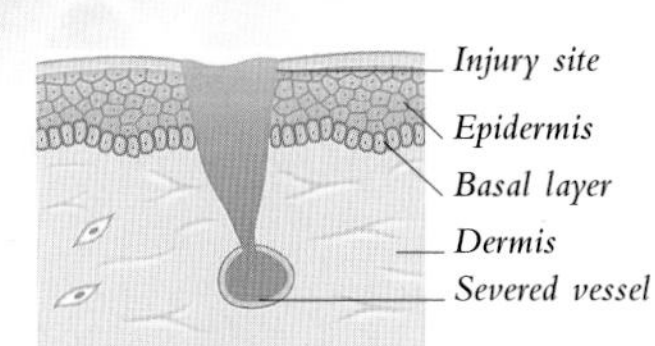


**SKIN INJURY**
*When the skin is broken, blood vessels in the dermis may be damaged, which results in bleeding at the injury site.*

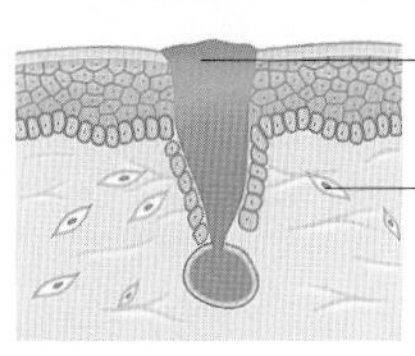


**CLOTTING**
*The blood forms a clot, and many repair cells, including fibroblasts, travel towards the injury site.*

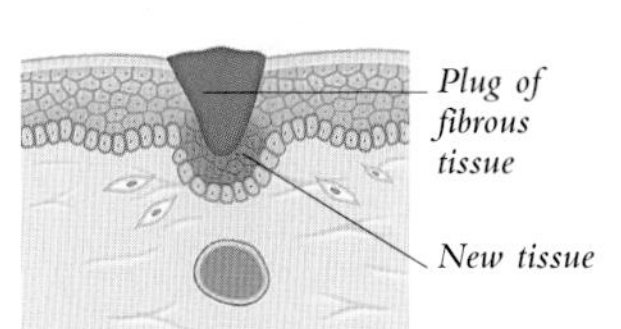


**PLUGGING**
*Fibroblasts produce a plug of fibrous tissue within the clot. Under the protection of the plug, new tissue begins to form.*

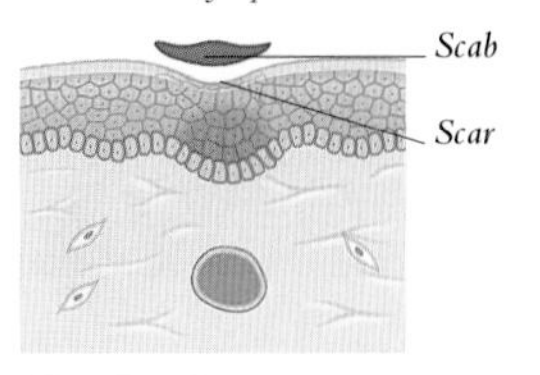


**SCABBING**
*The fibrous plug hardens to form a scab. The scab will eventually fall off, but may leave a scar at the site.*

**FINE CONTROL**
*Delicate movements, such as plucking guitar strings, are made possible by a highly complex system of nerves linking the brain with the rest of the body.*

# CONTROLLING THE BODY

The human body is constantly active, even when we are asleep. Just to keep us alive, numerous unconscious (involuntary) activities take place, such as the constant beating of the heart and the contractions of the intestinal wall during digestion. We also sense and move around in the external environment, which requires both voluntary and involuntary movements – all of these processes need to be coordinated and regulated.

Controlling the body is all about internal communication – areas of the body need to "talk" to one another in order to bring about a function or a movement, whether voluntary or involuntary. The regulation and control of processes and movements of the body is the responsibility of two systems – the nervous system, which consists of the brain, the spinal cord, and the nerves; and the endocrine system, which comprises specialized hormone-producing glands.

## NERVE LINKS

The nervous system facilitates communication by means of nerve links between the brain (the control centre of the body) and the rest of the body. Nerve fibres emanate from the brain, and many branch out to supply the upper body. Others enter the spinal cord to form a thick bundle of nerves. From the spinal cord, nerves branch out to serve all parts of the torso and limbs. The nervous system can be divided into two parts – the brain and spinal cord constitute the central nervous system, and the nerves that extend from the central nervous system comprise the peripheral nervous system. The entire nervous system is brought alive by the activity of many billions of nerve cells, or neurons. The brain alone has as many as 100 billion nerve cells. Nerve cells link together to form a complex network, and they communicate with each other by means of electrical impulses. A stimulus triggers a nerve cell to "fire" and the signal passes along nerve fibres to other nerve cells. Impulses are transmitted between nerve cells by chemicals called neurotransmitters. When your hand touches a hot surface, for example, nerve cells fire in response to the stimulus and send the impulse along the nerve network to the brain. The brain then processes the message and sends back a message to withdraw the hand. This complex process of messaging happens at top speed in a short space of time – impulses can travel along nerve fibres at up to 400km/h (250mph).

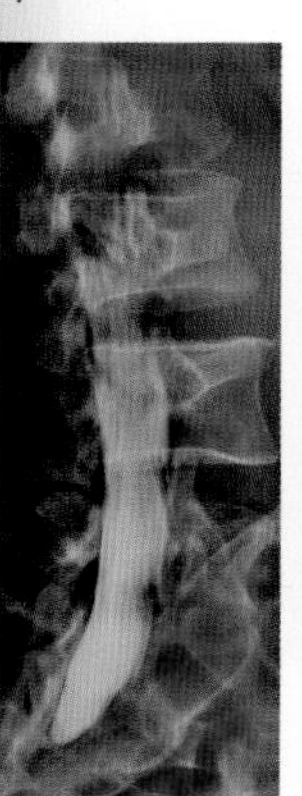

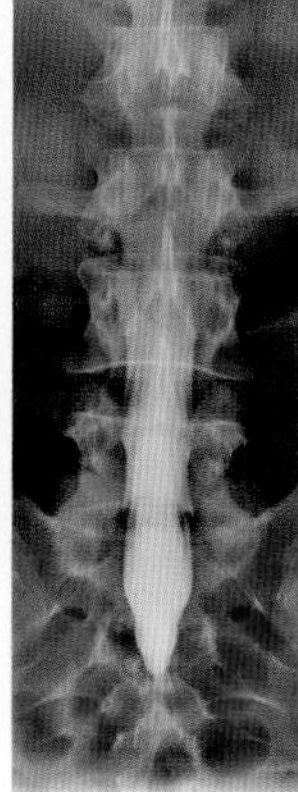

**SPINAL CORD**
*A thick cable of nerves, the spinal cord can be seen in white from the side (left) and from the back (right). It travels down the body through the spine's protective bones.*

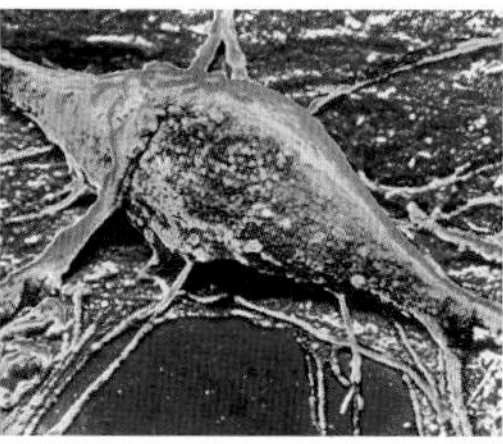

**NERVE CELL**
*Communication in the nervous system is facilitated by the action of nerve cells, which branch out to form an electrical network.*

## BRAIN POWER

The brain is the hub of all of the function and activity in the body. Humans have much more complex brains than any other animal, facilitating complicated internal functions and allowing us to make refined, precise movements, as well as enabling us to think, remember, reason, and plan, the so-called "higher" functions (*see* Mind, pp144–181). Different areas and structures in the brain have particular roles. The brain's highly folded surface, or cortex, is where the "higher" functions take place, and where refined movements are planned and controlled. Scientists have even been able to "map out" the cortex of the brain to define areas in which specific higher functions occur, such as language interpretation and movement sequencing. Inside the brain, "lower" functions take place in specific structures. The brain stem, for example, controls functions such as breathing and heart rate.

## HORMONES

As well as communicating by means of the nerves, the body also uses chemical messengers. This type of messaging is brought about by hormones, chemicals produced by the specialized glands and tissues in the body that make up the endocrine system. These hormones are released into the bloodstream and carried to certain areas of the body as and when they are needed, controlling a variety of the body's processes. For example, the thyroid gland in the neck produces the hormone thyroxine, which regulates the body's use of oxygen to create energy; and the pancreas in the abdomen is a large gland that produces the hormones glucagon and insulin, which are vital for regulating the level of glucose in the bloodstream. The most important of the glands is the pituitary gland, at the base of the brain. It regulates hormonal activity – stimulating hormone production from other glands and storing hormones when necessary.

*Thyroid gland* Regulates metabolism and calcium levels

*Brain* The body's control centre; site of the pituitary gland

*Spinal cord* Bundle of nerves running down the spine

*Adrenal gland* Influences energy use and stress response

*Pancreas* Regulates blood glucose level

*Testes (ovaries in women)* Secrete sex hormones

*Peripheral nerves* Link the brain and the spinal cord with the rest of the body

**CONTROL SYSTEMS**
*The nervous system consists of the brain, spinal cord, and peripheral nerves and the endocrine system consists of hormone-producing tissues and glands.*

**FAST-ACTING HORMONE**
*Here, the hormone epinephrine (adrenaline) is shown in crystallized form. This hormone is produced by the adrenal gland and triggers the "fight-or-flight" response to stress.*

Profile

### Charles Bell

Scottish anatomist and surgeon Charles Bell (1774–1842) was the first person to prove that nerves are not single units but consist of many separate fibres, bundled up and enclosed in a sheath. He also showed that a nerve fibre can transmit only motor or sensory stimuli, but not both, and that a muscle must be supplied with both types of fibre in order to achieve both voluntary and involuntary movement.

# NERVOUS SYSTEM

The brain and spinal cord, and the nerves emerging from them, form the nervous system, the body's most complex system. The nerves branch out to reach every part of the body. It regulates hundreds of activities simultaneously. The nervous system is the source of our consciousness, intelligence, and creativity and allows us to communicate and experience emotions. It also monitors and controls almost all bodily processes, ranging from automatic functions of which we are largely unconscious, such as breathing, to complex activities that involve thought and learning, such as playing a musical instrument.

## ORGANIZATION

The nervous system can be divided into two parts – the central nervous system, which is composed of the brain and the spinal cord, and the peripheral nervous system, made up of all the nerves that emanate from the central nervous system and branch out throughout the body. The role of the central nervous system is to regulate bodily activity, processing and coordinating nerve signals, while the job of the peripheral nervous system is to transmit these signals between the central nervous system and the rest of the body.

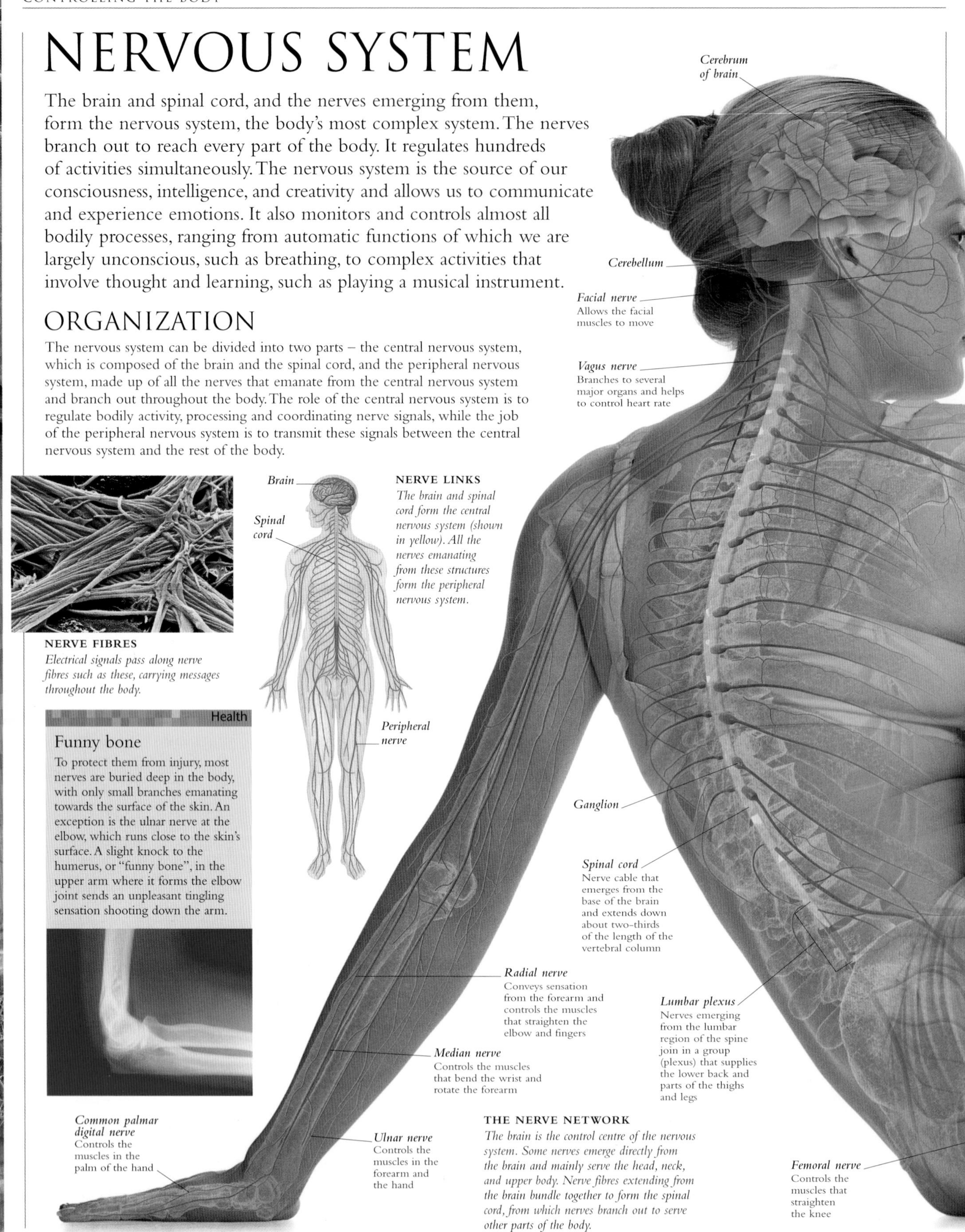
Cerebrum of brain
Cerebellum
Facial nerve
Allows the facial muscles to move
Vagus nerve
Branches to several major organs and helps to control heart rate
Brain
Spinal cord
Peripheral nerve
Ganglion
Spinal cord
Nerve cable that emerges from the base of the brain and extends down about two-thirds of the length of the vertebral column
Radial nerve
Conveys sensation from the forearm and controls the muscles that straighten the elbow and fingers
Lumbar plexus
Nerves emerging from the lumbar region of the spine join in a group (plexus) that supplies the lower back and parts of the thighs and legs
Median nerve
Controls the muscles that bend the wrist and rotate the forearm
Common palmar digital nerve
Controls the muscles in the palm of the hand
Ulnar nerve
Controls the muscles in the forearm and the hand
Femoral nerve
Controls the muscles that straighten the knee

**NERVE FIBRES**
*Electrical signals pass along nerve fibres such as these, carrying messages throughout the body.*

**NERVE LINKS**
*The brain and spinal cord form the central nervous system (shown in yellow). All the nerves emanating from these structures form the peripheral nervous system.*

Health

### Funny bone

To protect them from injury, most nerves are buried deep in the body, with only small branches emanating towards the surface of the skin. An exception is the ulnar nerve at the elbow, which runs close to the skin's surface. A slight knock to the humerus, or "funny bone", in the upper arm where it forms the elbow joint sends an unpleasant tingling sensation shooting down the arm.

**THE NERVE NETWORK**
*The brain is the control centre of the nervous system. Some nerves emerge directly from the brain and mainly serve the head, neck, and upper body. Nerve fibres extending from the brain bundle together to form the spinal cord, from which nerves branch out to serve other parts of the body.*

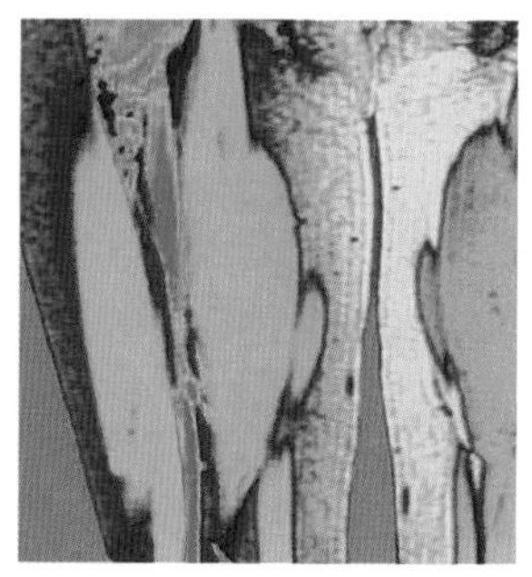

**SCIATIC NERVE**
*This MRI scan of the thigh shows the sciatic nerve (in green). These nerves have the largest diameter of all nerves in the body.*

# NERVES

Each of the nerves that runs through the body is made up of hundreds of nerve fibres that are grouped together tightly in bundles called fascicles. Most of these nerve fibres are insulated by sheaths made of a fatty substance called myelin (*see* p116). The fascicles are themselves wrapped in tough but elastic tissue called the perineurium, which serves to bind the groups together. An outer layer, called the epineurium, encloses several fascicles, together with blood vessels and fat-containing cells. The epineurium provides protection, shielding the nerve fibres from damage, and facilitates flexing and bending of the nerve, so that it can negotiate twists through the body.

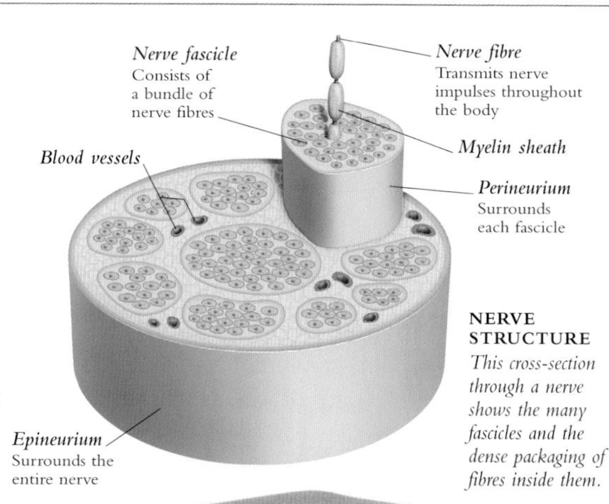


**NERVE STRUCTURE**
*This cross-section through a nerve shows the many fascicles and the dense packaging of fibres inside them.*

*Lateral plantar nerve* Allows toes to flex and curve

*Common peroneal nerve* Controls the muscles that lift the foot

*Tibial nerve* Supplies the calf muscles

*Sciatic nerve* Serves the hip joint and hamstring muscles; divides to form the tibial and common peroneal nerves

# THE BODY ELECTRIC

The sophisticated network of nerve fibres that branch out from the brain and spinal cord enable the human body to function as a result of the activity of nerve cells (or neurons). Nerve cells have a nerve cell body, which contains a nucleus and other structures such as mitochondria (energy-producing units). Most nerve cells also have one or more long projections (axons) that branch out from the cell bodies. The nerve cells carry electrical impulses, or signals, between the brain and other areas of the body, transporting these impulses along the axons to meet other nerve cells or to make contact with target organs (such as muscles) or sensory receptors. This electric network allows vital information to be relayed all over the body. (For more information on nerve cells, *see* p116).

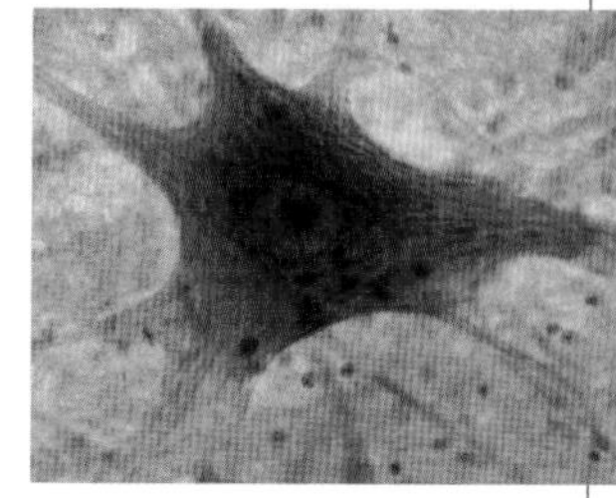

**NERVE CELL**
*Spidery projections branching out from the nerve cell body (the dark area in the centre) connect to form an electrical network.*

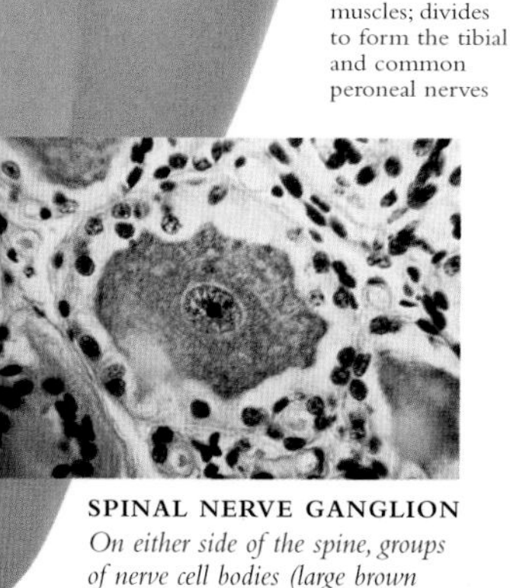

**SPINAL NERVE GANGLION**
*On either side of the spine, groups of nerve cell bodies (large brown areas above) make up ganglia.*

History

## Animal electricity

Italian anatomist Luigi Galvani (1737–1798) performed experiments on frogs' legs, aimed at investigating how nerves work. One experiment (right) involved bringing two rods of metal together (one silver and one bronze) and resulted in the frog's legs twitching. Galvani thought that this was generated by electrical activity in the frog's nerves and muscle tissue, but in 1800 Alessandro Volta proved that the two metals created the charge.

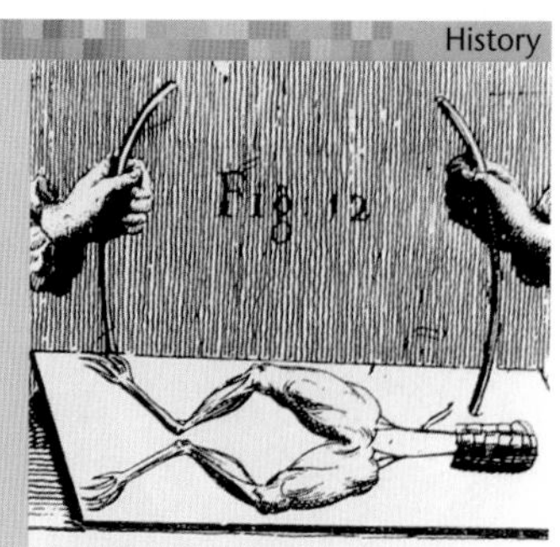


BODY

# BRAIN AND NERVE LINKS

The brain is the source of consciousness, thought, and reasoning. This incredibly complex organ coordinates information it receives from every part of the body and makes the appropriate responses. The brain has direct nerve connections with the upper body, but the main link between the brain and the rest of the body is the spinal cord. Like an electrical cable, the spinal cord runs down inside the spine from the base of the brain, transmitting and receiving messages. Along the cord's length, nerves branch out in pairs; these carry signals from the brain to activate body movement and take information to the brain about sensation.

**CEREBELLUM**
*This cross-section shows the convoluted layers of tissue in the cerebellum, the control area at the base of the brain.*

***Cerebrum***
The site of most of the brain's conscious functions

## THE OUTER BRAIN

The largest part of the brain is its massive outer structure, the cerebrum. Heavily folded into ridges and grooves, the cerebrum is divided into two halves, the cerebral hemispheres, by a deep central fissure. Each hemisphere is subdivided into four lobes: the frontal, parietal, occipital, and temporal lobes. The outer layer of the cerebrum is the cerebral cortex, also referred to as grey matter, which controls higher brain functions, including thought. The inner layer consists of the nerve tissue known as white matter. Tucked beneath the cerebrum are two smaller structures. One is the brain stem, which controls basic life processes such as heart rate and breathing. The other structure is the cerebellum, which is responsible for muscle coordination.

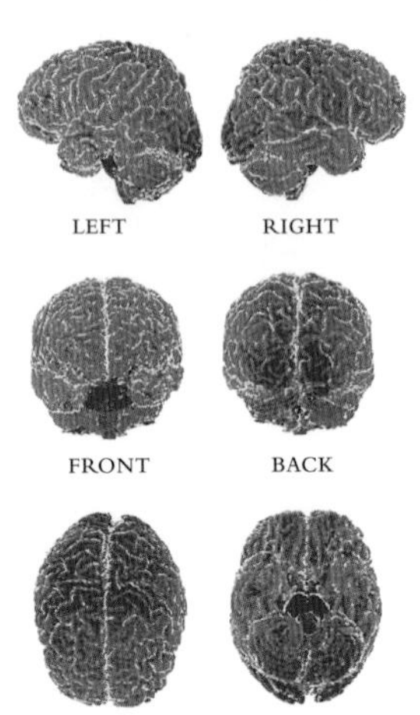


**VIEWING THE BRAIN**
*This series of scans shows the outer surface of the brain from all angles. Viewed from above, the hemispheres of the cerebrum are seen distinctly.*

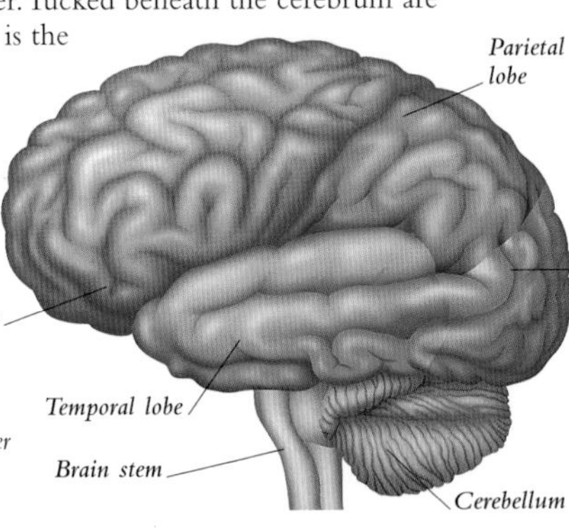

**LOBES OF THE BRAIN**
*The lobes of the cerebrum are named for the skull bones that cover them. Each lobe is associated with various aspects of brain function.*

Health

### Stroke

If brain tissues lose their blood supply they cannot function normally. This condition, called stroke, is a major cause of death, especially in the developed world. Causes of stroke include blockage of brain arteries by blood clots and bleeding within the brain. The symptoms of stroke develop rapidly and can include severe disabilities such as numbness, paralysis, and slurred speech.

**DAMAGED BRAIN TISSUE**
*This scan shows the cause of a stroke – damaged brain tissue (red area) due to an interrupted blood supply.*

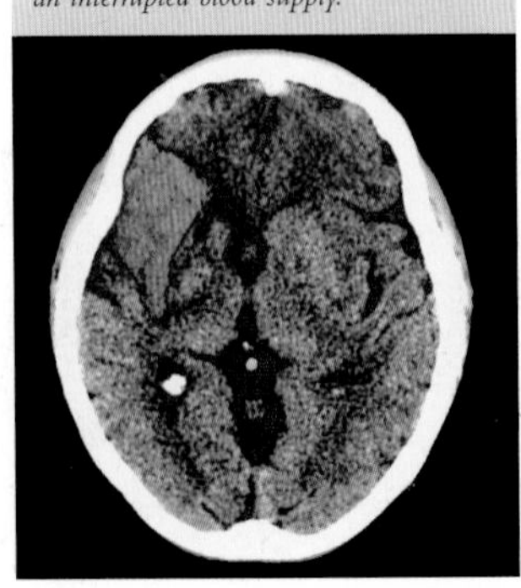

## THE SPINAL CORD

Running from the brain stem to the lower back, the spinal cord is a cable of nerve tissues about 45cm (18in) long in an adult, and roughly the thickness of a little finger. Like the cerebrum, the spinal cord comprises grey and white matter. At the core is grey matter, which is tissue made mostly of nerve cell bodies. Surrounding the grey matter is white matter containing nerve fibres that carry signals to and from the brain. Nerves emerge in pairs from the spinal cord and extend to the body and limbs (*see* Spinal nerves, p113). The cord is protected by the spinal column. Damage to the cord can cause numbness or paralysis in parts of the body.

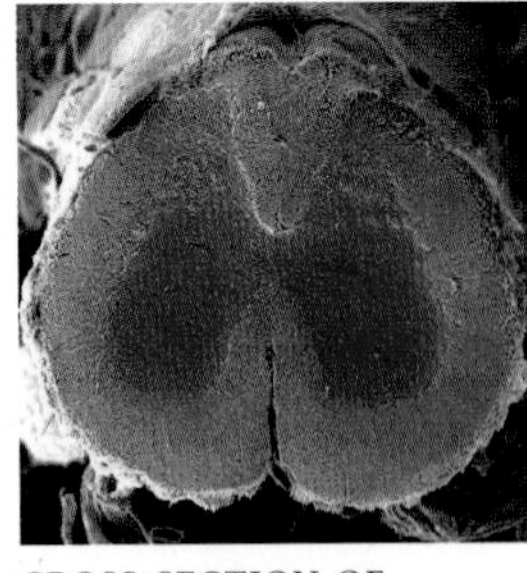

**CROSS-SECTION OF SPINAL CORD**
*The spinal cord consists of two types of tissue: grey and white matter. Grey matter (stained orange in this cross-section) comprises mainly nerve cell bodies. White matter (yellow) is made up of nerve fibres.*

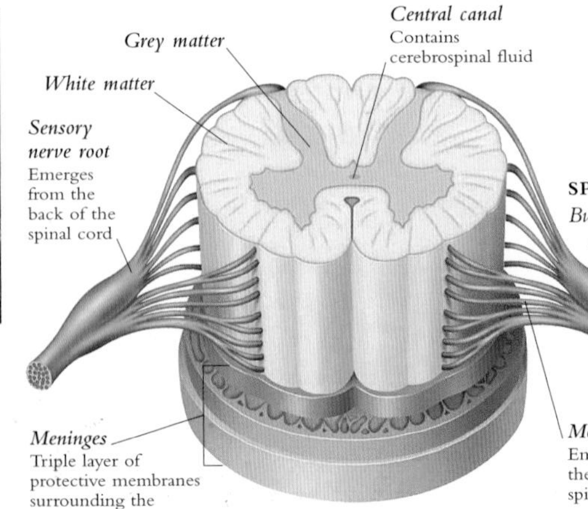


**SPINAL CORD**
*Bundles of nerve roots leave the spinal cord. These carry nerve signals to and from the central nervous system and the rest of the body.*

*Thalamus*
Area that relays nerve signals to the cerebral cortex

*Corpus callosum*
Bundle of nerve fibres connecting the cerebrum's two hemispheres

*Meninges*
Protective membranes covering the brain

*Skull*

*Cerebellum*
Involved in balance and body control

*Brain stem*
Involved in control of involuntary functions such as breathing and heart rate

*Hypothalamus*
Controls the body's endocrine, or hormone-producing, system

**BRAIN STRUCTURES**
*The largest part of the brain is the heavily folded cerebrum, which is divided into two halves connected by the corpus callosum. Other structures include the brain stem, cerebellum, thalamus, and hypothalamus.*

# PROTECTION

The brain and spinal cord are well protected from shocks and blows. Apart from being encased in bone, they are both covered by a triple-layered membrane, known as the meninges. The tissues of the brain and spinal cord are also cushioned by a circulating liquid, cerebrospinal fluid (CSF). Cerebrospinal fluid is produced by specialized cells within the brain. This liquid is mostly water but also contains some nutrients, such as glucose, to feed brain cells. CSF fills the inner chambers of the brain and circulates around the outer brain area, flowing into the space between the two inner layers of the meninges. CSF is reabsorbed into the blood and renewed 3–4 times a day.

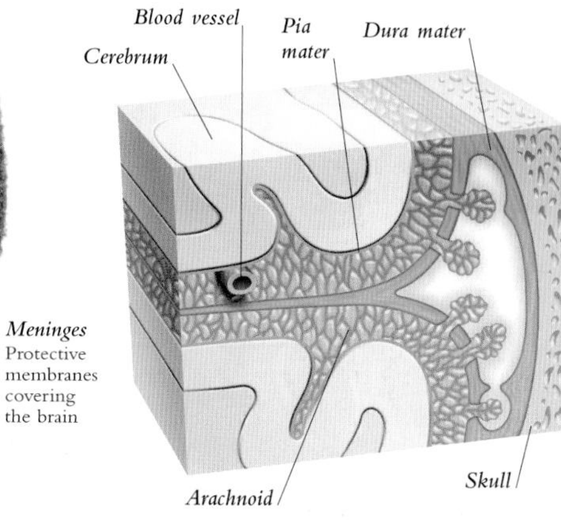


**MENINGES**
*Three membranes surround the brain and spinal cord: the dura mater, which forms the outer layer; the arachnoid layer in the middle, with weblike extensions; and the pia mater on the inside.*



# THE HUNGRY BRAIN

To fuel its complex functions, the brain is ever-hungry for energy-producing oxygen, which is carried in blood. To meet this need, about one-fifth of the body's circulating supply of oxygenated blood goes to brain tissues. The blood is delivered by four arteries, which come together in a network called the circle of Willis on the underside of the brain. The smallest blood vessels in the brain, capillaries, have a tightly constructed lining, referred to as the blood–brain barrier. This barrier controls the flow of blood into the brain, blocking the entry of harmful agents, such as viruses, that could damage the delicate tissues of the brain.

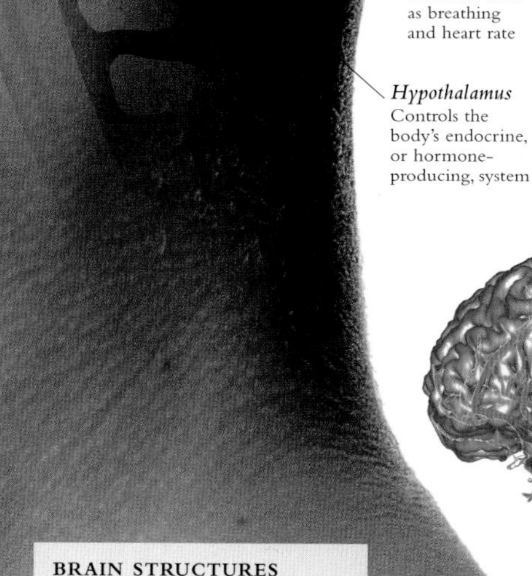

BLOOD VESSELS

**BLOOD SUPPLY TO BRAIN**
*Blood vessels extend throughout the brain (above). These vessels enter the brain at the circle of Willis (right).*

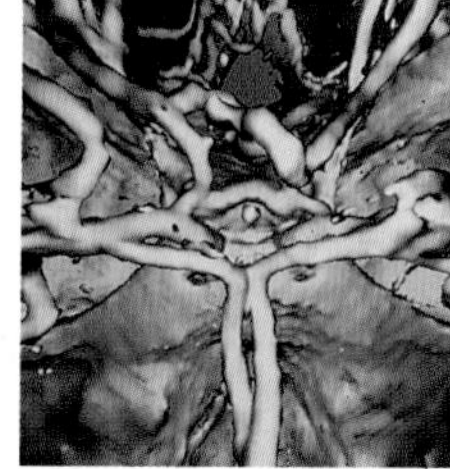

THE CIRCLE OF WILLIS

# INSIDE THE BRAIN

From the outside, the cerebrum looks like a lump of crinkled putty. However, a slice through the cerebrum reveals that beneath its outer layer of grey matter (the cerebral cortex) is white matter in which are embedded small islands of grey matter called basal ganglia. These structures are important in helping to control movement. Deep beneath the cerebrum is a group of structures known as the limbic system. They play an important role in emotions (*see* p164) and survival behaviour, such as the fight-or-flight response. Certain structures of the limbic system are involved in memory (*see* p156).

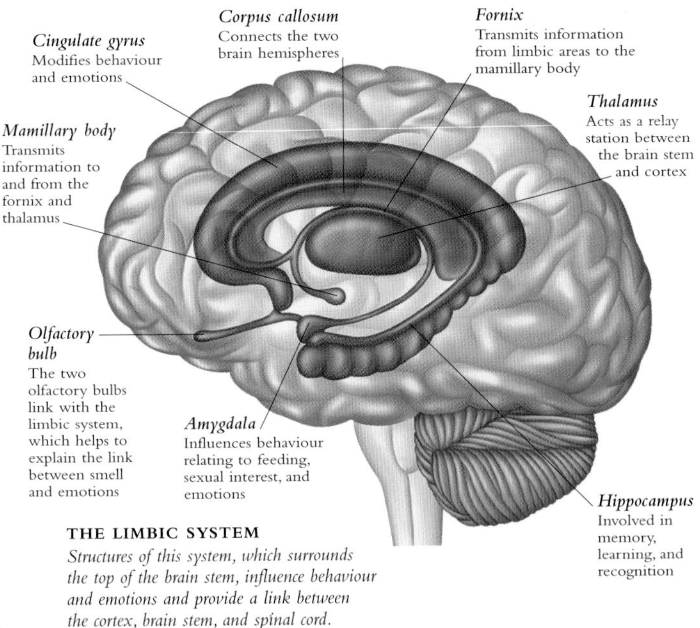

*Cingulate gyrus*
Modifies behaviour and emotions
*Corpus callosum*
Connects the two brain hemispheres
*Fornix*
Transmits information from limbic areas to the mamillary body
*Thalamus*
Acts as a relay station between the brain stem and cortex
*Mamillary body*
Transmits information to and from the fornix and thalamus
*Olfactory bulb*
The two olfactory bulbs link with the limbic system, which helps to explain the link between smell and emotions
*Amygdala*
Influences behaviour relating to feeding, sexual interest, and emotions
*Hippocampus*
Involved in memory, learning, and recognition

**THE LIMBIC SYSTEM**

*Structures of this system, which surrounds the top of the brain stem, influence behaviour and emotions and provide a link between the cortex, brain stem, and spinal cord.*

*White matter*
Consists mainly of nerve fibres, which transmit impulses

*Grey matter of cortex*
Consists mainly of nerve cell bodies; higher functions take place here

*Corpus callosum*
A large bundle of nerve fibres that connects the two hemispheres of the brain

*Basal ganglia*
Islands of grey matter that help to coordinate movement

*Brain stem*

*Cerebellum*

*Motor nerve tracts*
Carry impulses from the cerebrum to the spinal cord, crossing over in the brain stem

**BENEATH THE CORTEX**

*Under the cortex lies a network of nerve fibres and tracts that transmit impulses around the brain and to the rest of the body. The basal ganglia, which help control movement, are found deep inside the cerebrum.*

Fact

## Grey matter

Both the brain and the spinal cord contain areas of grey matter, as well as areas of white matter. The areas of grey matter are composed of a dense network of tightly packed nerve cells and support cells (neuroglia, see p116). The grey colour is due to the high proportion of nerve cell bodies compared to nerve fibres (white matter is composed of nerve fibres). In grey matter, nerve impulses are processed.

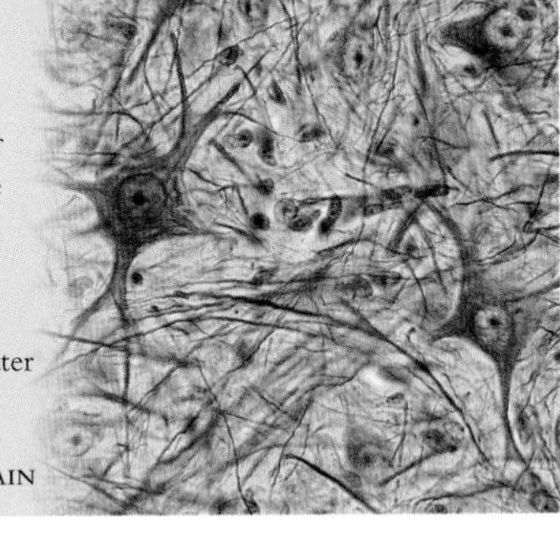

GREY MATTER OF BRAIN

# MAPPING THE BRAIN

Neuroscientists are able roughly to pinpoint those parts of the cerebral cortex (the outer layer of the cerebrum) that process nerve impulses concerned with particular "higher" human functions, such as intellect and memory. Areas of the cortex that are mainly concerned with detecting nerve impulses are known as primary areas, and the parts that are concerned with analysing impulses are known as association areas. For example, the primary visual cortex receives impulses from the eye in the form of a series of vertical, horizontal, and curved lines; and the visual association cortex forms images from these data, so that, for example, the vertical, horizontal, and curved lines become recognizable as an apple.

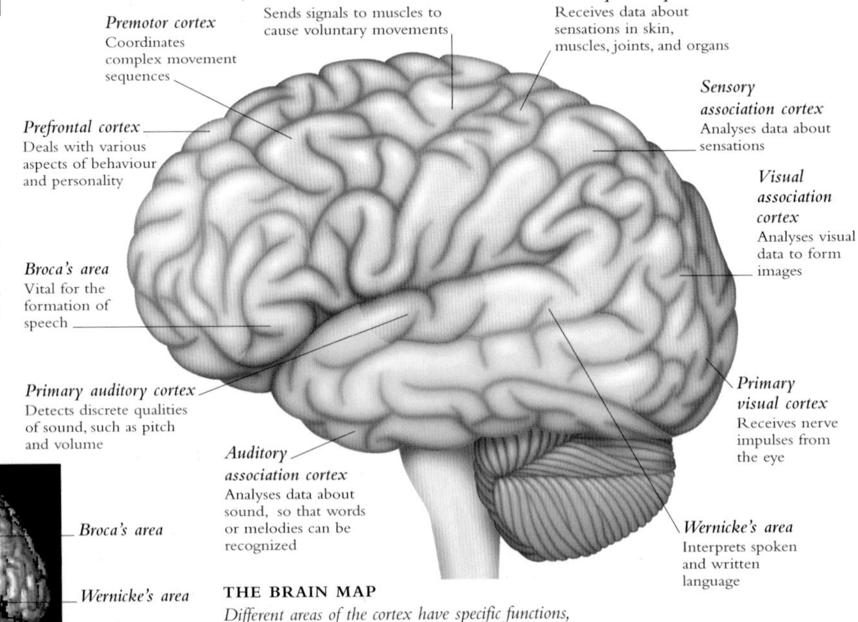

*Premotor cortex*
Coordinates complex movement sequences
*Motor cortex*
Sends signals to muscles to cause voluntary movements
*Primary sensory cortex*
Receives data about sensations in skin, muscles, joints, and organs
*Sensory association cortex*
Analyses data about sensations
*Prefrontal cortex*
Deals with various aspects of behaviour and personality
*Visual association cortex*
Analyses visual data to form images
*Broca's area*
Vital for the formation of speech
*Primary auditory cortex*
Detects discrete qualities of sound, such as pitch and volume
*Primary visual cortex*
Receives nerve impulses from the eye
*Auditory association cortex*
Analyses data about sound, so that words or melodies can be recognized
*Wernicke's area*
Interprets spoken and written language

**THE BRAIN MAP**

*Different areas of the cortex have specific functions, and the surface of the cortex can be mapped out accordingly. No specific areas have yet been identified as the exact sites of consciousness.*

**INTERPRETING LANGUAGE**

*In this scan, the orange "lit" areas are the parts of the brain that are most active while interpreting language and speaking. Known as Wernicke's and Broca's areas, they are in the left hemisphere of the brain.*

*Broca's area*

*Wernicke's area*

*Left hemisphere*

# CRANIAL NERVES

Branching directly from the underside of the brain are 12 pairs of cranial nerves. Most of these nerves supply the head and neck area and contain sensory and/or motor nerve fibres. Sensory fibres receive information from the outside world or internal organs, while motor fibres initiate contraction of voluntary muscles. The exception is the vagus nerve, which supplies certain organs in the chest and abdomen; it contains autonomic nerve fibres, which regulate involuntary functions.

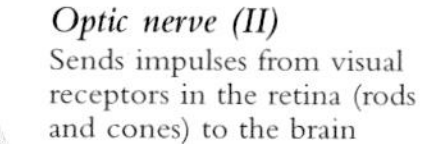

*Optic nerve (II)*
Sends impulses from visual receptors in the retina (rods and cones) to the brain

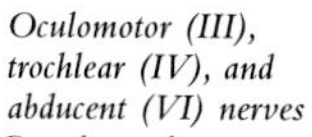

*Oculomotor (III), trochlear (IV), and abducent (VI) nerves*
Regulate voluntary movements of the eye muscles and eyelids, control pupil dilation and changes in the lens during focusing

*Olfactory nerve (I)*
Relays information about smells from the inside of the nose to the olfactory centres in the brain

*Vestibulocochlear nerve (VIII)*
Transmits data about sound and balance; also conveys information about the orientation of the head in space

*Trigeminal nerve (V)*
Contains sensory fibres that relay signals from the eye, face, and teeth; motor fibres innervate the muscles involved in chewing

*Glossopharyngeal (IX) and hypoglossal (XII) nerves*
Contain motor fibres that are involved in swallowing; and sensory fibres, which relay information about taste, touch, and heat from the tongue and the pharynx

*Facial nerve (VII)*
Relays information from the taste buds and the skin of the external ear; supplies the lacrimal glands of the eye and certain salivary glands; also controls muscles used in facial expressions

*Spinal accessory nerve (XI)*
Brings about movement in the head and shoulder; innervates muscles in the pharynx and larynx; involved in the production of voice sounds

*Vagus nerve (X)*
Involved in the control of many vital body functions, including heartbeat and formation of stomach acid

**CRANIAL NERVE FUNCTIONS**
*The 12 pairs of cranial nerves relay sensations from the outside world or send messages to initiate muscle contractions, mainly in the head and neck region. Some cranial nerves also convey information about the tension of muscles.*

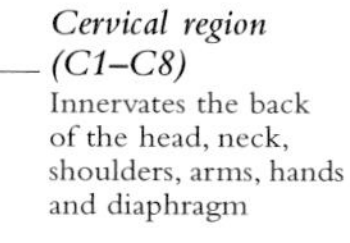

*Cervical region (C1–C8)*
Innervates the back of the head, neck, shoulders, arms, hands and diaphragm

*Thoracic region (T1–T12)*
Connected to the muscles between the ribs, deep back muscles, and regions of abdomen

*Lumbar region (L1–L5)*
Supplies the lower back, thighs, and parts of the legs

*Sacral region (S1–S5)*
Innervates the thighs, buttocks, legs, feet and the genital and anal areas

**SPINAL REGIONS**
*Spinal nerves that emerge from different regions of the spine are responsible for supplying specific areas of the body.*

# SPINAL NERVES

There are 31 pairs of spinal nerves, which emerge from the spinal cord and branch out through spaces between the vertebrae. Each nerve divides and subdivides into a number of branches. Two main subdivisions serve the front and the back of the body in the region innervated by that particular nerve. Spinal nerves are named according to the part of the spine from which they emerge – the cervical (neck region), thoracic (chest), lumbar (lower back), and sacral (sacrum and coccyx). The cervical nerves interconnect, forming two networks, or plexuses, that innervate the back of the head, the neck, the shoulders, and the diaphragm. The thoracic spinal nerves are connected to the muscles between the ribs, the deep back muscles, and parts of the abdomen. The lumbar spinal nerves supply the lower back, and parts of the thighs and legs. The sacral nerves innervate the thighs, the buttocks, the muscles and skin of the legs and feet, as well as the anal and genital areas.

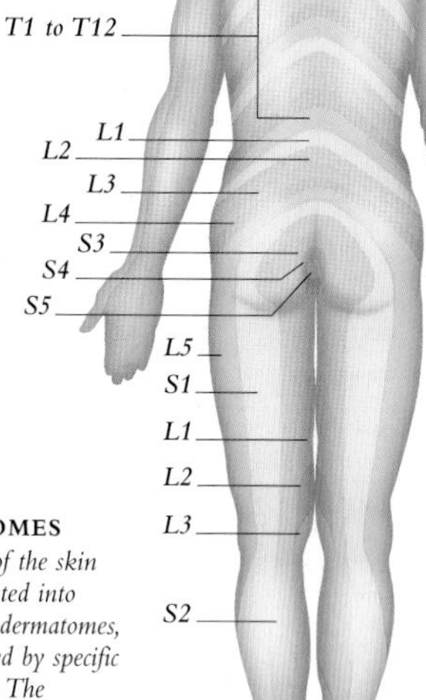


**DERMATOMES**
*The surface of the skin can be separated into zones, called dermatomes, that are served by specific spinal nerves. The dermatomes in the trunk are horizontal, while those in the limbs are vertical. Areas of sensation may overlap slightly.*

SYMPATHETIC

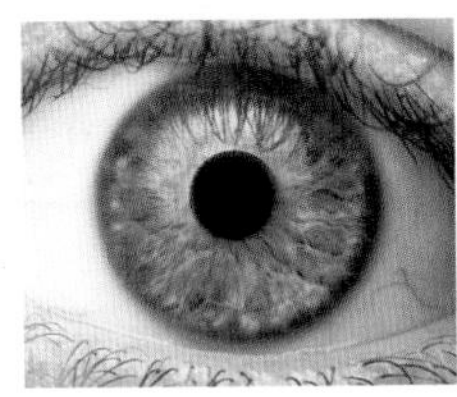

PARASYMPATHETIC

**AUTONOMIC NERVES**
*Sympathetic autonomic nerves stimulate the body for action; parasympathetic nerves have a calming effect. The pupils dilate or constrict under their control.*

# FUNCTIONS OF THE PERIPHERAL NERVES

The cranial and spinal nerves, which are part of the peripheral nervous system, carry information to and from the brain and spinal cord (the central nervous system). The peripheral nerves consist of bundles of nerve fibres, and these fibres may perform one of three different functions: sensory, motor, or autonomic. Sensory nerves carry information about inner bodily sensations and about events occurring in the outside world. Motor nerves send information from the central nervous system to skeletal muscles, facilitating voluntary movements. Autonomic nerve fibres carry instructions to organs and glands, and are not under conscious control. They are of two types: sympathetic, which stimulate organs and prepare the body for stress; and parasympathetic, which influence organs to maintain energy or to relax and restore energy.

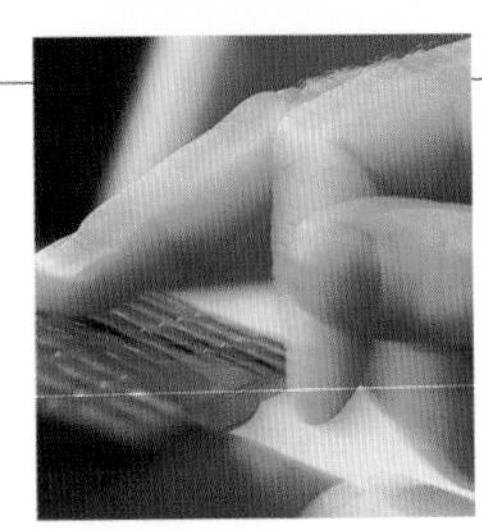

**MOTOR NERVES**
*Voluntary movements, such as typing, are facilitated by motor nerves, which transmit messages to skeletal muscles from the central nervous system.*

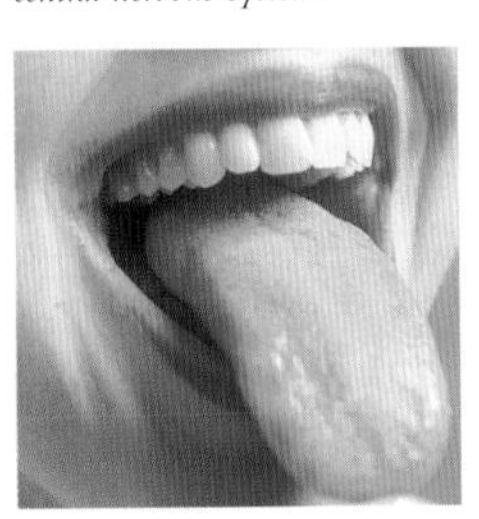

**SENSORY NERVES**
*Information about outside stimuli, such as through taste, touch, smell, vision, and hearing is relayed by sensory nerves to the central nervous system.*

Health

## Peripheral neuropathy

Damage to a cranial or spinal nerve commonly occurs as a result of the condition diabetes mellitus, but may also be associated with an infection. Damaged nerves may affect sensation (particularly at the extremities of the body, such as hands and feet), movement, or automatic functions.

DAMAGED NERVE FIBRE

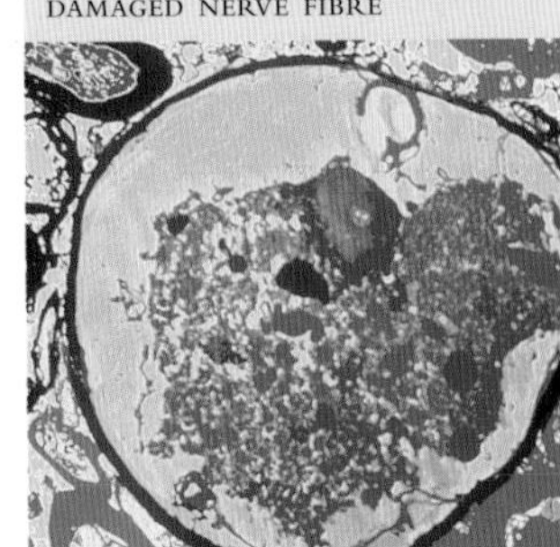

# VOLUNTARY AND INVOLUNTARY RESPONSES

The response of the nervous system to stimuli may be involuntary or voluntary. Autonomic nervous system responses are automatic: they control the body's internal environment and help to regulate vital functions, such as blood pressure. The two types of autonomic nerves, sympathetic and parasympathetic, have opposing effects (see right). Voluntary activities mainly involve the brain, which sends out the motor impulses that control movement. These signals are initiated by thought, and they may also involve a response to sensory stimuli. For example, sight and sense of position help to coordinate the action of walking. Reflexes mainly affect muscles normally under voluntary control, but are involuntary responses to stimuli, such as flinching a hand from a flame.

| AFFECTED ORGAN | SYMPATHETIC RESPONSE | PARASYMPATHETIC RESPONSE |
|---|---|---|
| EYES | Pupils dilate | Pupils constrict |
| LUNGS | Bronchial tubes dilate | Bronchial tubes constrict |
| HEART | Rate and strength of heartbeat increase | Rate and strength of heartbeat decrease |
| STOMACH | Enzymes decrease | Enzymes increase |
| LIVER | Releases glucose | Stores glucose |

**TWO TYPES OF RESPONSE**
*Sympathetic and parasympathetic nerves produce different responses in a particular organ. The sympathetic responses prepare the body to cope at times of stress; the parasympathetic responses help conserve or restore energy.*

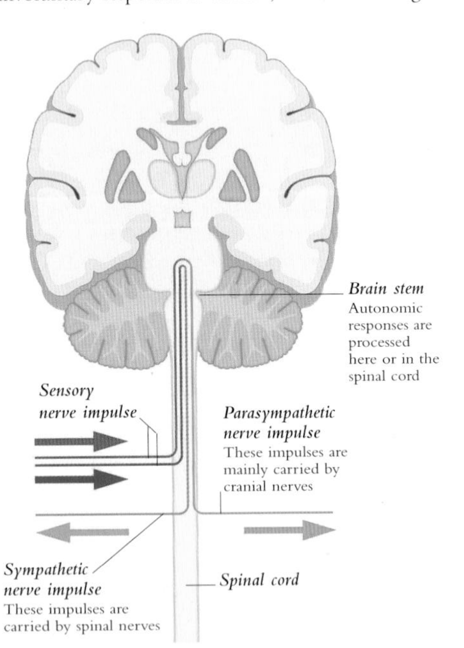
Brain stem
Autonomic responses are processed here or in the spinal cord
Sensory nerve impulse
Parasympathetic nerve impulse
These impulses are mainly carried by cranial nerves
Sympathetic nerve impulse
These impulses are carried by spinal nerves
Spinal cord

**AUTONOMIC RESPONSES**
*Information collected by internal receptors travels along sensory nerves to the spinal cord and the brain stem. Sympathetic and parasympathetic response signals have separate pathways.*

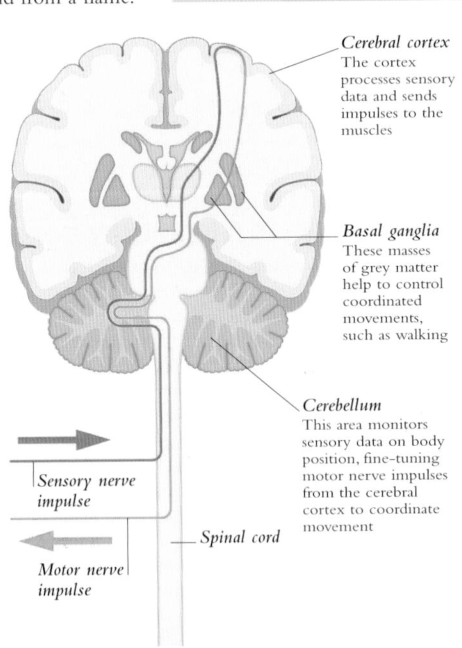
Cerebral cortex
The cortex processes sensory data and sends impulses to the muscles
Basal ganglia
These masses of grey matter help to control coordinated movements, such as walking
Cerebellum
This area monitors sensory data on body position, fine-tuning motor nerve impulses from the cerebral cortex to coordinate movement
Sensory nerve impulse
Motor nerve impulse
Spinal cord

**VOLUNTARY RESPONSES**
*The sensory impulses that are responsible for triggering voluntary responses are dealt with in many areas of the brain, and they travel through a complex nerve pathway.*

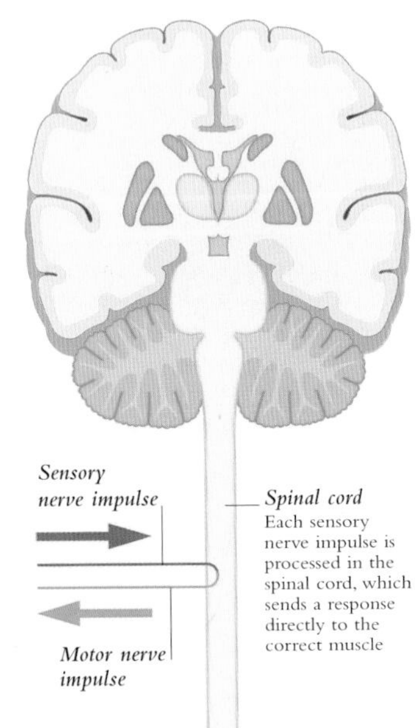
Sensory nerve impulse
Spinal cord
Each sensory nerve impulse is processed in the spinal cord, which sends a response directly to the correct muscle
Motor nerve impulse

**REFLEXES**
*Although some reflexes are processed in the brain, most are processed in the spine. This is the simplest nerve pathway, as the sensory and motor neurons are linked in the spinal cord.*

# MOVEMENT AND TOUCH

Areas of the brain's cortex that control movement and sense touch can be mapped as two continuous stripes running across the brain from side to side. The "stripe" where movement signals are processed, called the motor cortex, runs across the top of the cerebrum. The "stripe" running directly behind the motor cortex, known as the sensory cortex, processes touch signals. Areas on each side of the brain link with the opposite sides of the body, crossing over at the brain stem. Areas that are involved in complex movements, such as the hands and fingers, or body parts that are extremely sensitive to touch, such as the genitals, are allocated a larger proportion of the motor or sensory cortex (in relation to their size). On the whole, those parts of the body that are capable of complex movement are also highly sensitive to touch.

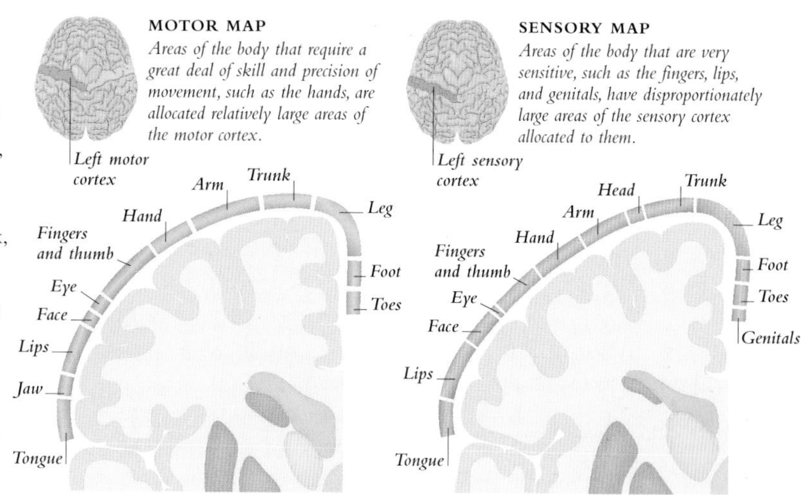


**MOTOR MAP**
*Areas of the body that require a great deal of skill and precision of movement, such as the hands, are allocated relatively large areas of the motor cortex.*

**SENSORY MAP**
*Areas of the body that are very sensitive, such as the fingers, lips, and genitals, have disproportionately large areas of the sensory cortex allocated to them.*

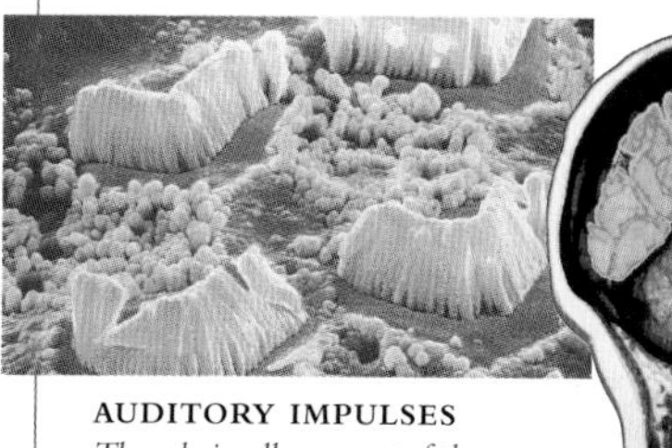

**AUDITORY IMPULSES**
*These hair cells are part of the organ of Corti in the inner ear, which converts sound waves into impulses to pass to the reticular activating system.*

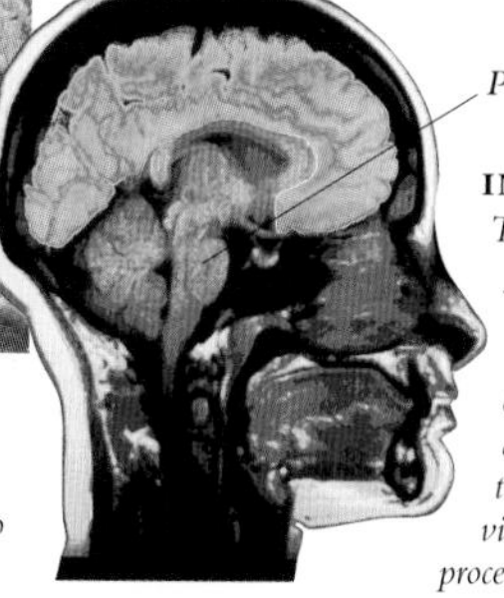


**IN THE PONS**
*The pons bulges from the front of the brain stem. It is in this part of the reticular activating system that auditory and visual impulses are processed and relayed.*

# STAYING ALERT

We are able to stay awake and alert as a result of the activities of the reticular formation, a crisscrossing, net-like core of nerve tissue that runs the length of the brain stem. (The term reticular means "net-like".) The reticular formation operates an arousal system – the reticular activating system – that is responsible for keeping the brain awake and alert. Pathways of nerve fibres in the reticular activating system detect incoming sensory information, for example from the ears or the eyes. These nerve fibres then send activating signals through the midbrain to the cerebral cortex in order to retain the state of consciousness. Damage to the reticular activating system may result in irreversible coma.

# SLEEP CYCLES

Nerve cells in the brain are never at rest. In a typical period of 8 hours sleep, they carry out different activities than those that occur during waking hours. The stages of sleep can be divided into rapid eye movement (REM) sleep, when dreams occur, and fluctuating phases of non-rapid eye movement (NREM) sleep, which is dreamless. Patterns of non-rapid eye movement (NREM) can be split into four stages. At stage one, we drift in and out of sleep and can be wakened easily, producing active brain waves (measurable electrical activity in the brain); at stage two, the brain waves slow down slightly; in stage three, slow waves and fast waves are interspersed as sleep begins to deepen; at stage four, the brain produces slow waves only and there is no eye movement or muscle activity. As sleep deepens, body temperature falls, breathing rate slows, and blood pressure is reduced.

NREM SLEEP: STAGE 1

NREM SLEEP: STAGE 2

NREM SLEEP: STAGE 3

NREM SLEEP: STAGE 4

REM SLEEP

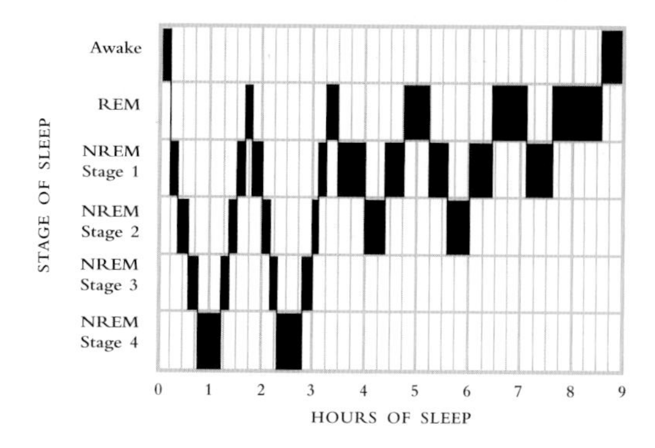


**A TYPICAL NIGHT'S SLEEP**
*The cycles in a night's sleep are made up of lengthening phases of REM sleep, when dreaming occurs, and four stages of NREM sleep. In stages 1 and 2 of the cycle, sleep is light and people tend to wake easily; in stages 3 and 4, sleep is deep and people are difficult to wake.*

**WELL-RESTED**
*The brain of a well-rested person has many active areas (red), indicating high brain wave activity and a state of alertness.*

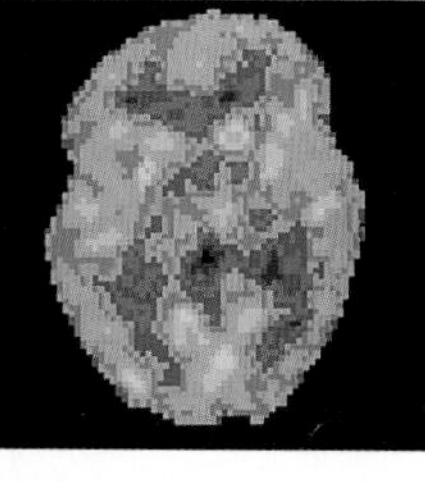

**SLEEP-DEPRIVED**
*In a sleep-deprived person, many brain areas are inactive (blue), resulting in sluggishness and forgetfulness.*

Health

## Sleep laboratories

Problems affecting sleep are very common and can be extremely debilitating. Studies in a sleep laboratory, or sleep research centre, can be beneficial in the diagnosis and treatment of sleeping disorders. Sleep studies can take place at night or during the day. The patient is put to bed and, as he or she sleeps, bodily processes and activities are monitored by a sleep technician. The brain's electrical activity is monitored, as well as breathing, heart rate, eye movement, limb movement, and snoring sounds.

# NERVE CELLS

The activities of the nervous system are initiated and carried out by neurons, nerve cells that carry impulses in the form of electrical signals. Each individual neuron forms connections with hundreds of other neurons, creating a dense network of communication. Some neurons react directly to messages received by bodily senses such as sight, hearing, and touch. These cells send off a rapid fire of nerve impulses that are picked up by other neurons and passed on in turn. The signals are directed to their final targets, the cells in muscles, organs, or glands, making the body respond. In addition to neurons, the nervous system contains vast numbers of other types of cell that help to nourish and protect neurons.

*Axon*
Nerve fibre projecting from the cell body that carries nerve impulses

**REACHING THE TARGET**
*Nerve fibres are seen here terminating in a muscle. The small end bulbs on the fibres fire impulses at the target tissues.*

## CONDUCTING NERVE IMPULSES

Nerve impulses travel along neurons in the form of electrical signals, racing down the axons (nerve fibres) at about 400 km/h (nearly 250mph). The transmission of an impulse reaches maximum speed if the neuron has axons that are coated in myelin, a fatty substance that provides insulation. In order to pass from one neuron to another, the signals must cross over synapses, tiny gaps between the cells. At the synapse, the impulse stimulates the release of neurotransmitters (chemical messengers), which transmit the impulse to the next neuron. Short projections called dendrites branching out from the cell body are ready to receive the impulse. The impulse then continues to travel from neuron to neuron until the target, such as a muscle cell, is reached.

**TRANSMISSION**
*A nerve impulse moves down the length of an axon and into the fine, branching terminal fibres. The impulse causes neurotransmitters (chemical messengers) to be released from nerve endings, which travel across the gap between one neuron and the next and excite the next cell to fire an impulse.*

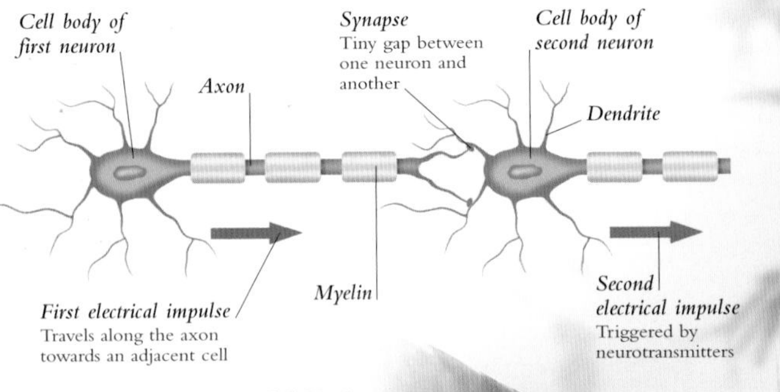


## CHEMICAL TRANSMITTERS

Neurotransmitters are stored in vesicles, tiny sacs in the nerve endings. When an electrical signal arrives at the nerve ending it causes the vesicles to dump their contents into the synapse. The transmitter moves across the gap to the next neuron and locks onto receptors in the cell membrane. Some receptors work like tiny gates, opening to let in ions (electrically charged particles). This triggers the opening of other ion channels and sets up a new electrical impulse to be sent to the next neuron.

**CROSSING THE GAP**
*When neurotransmitters cross a synapse they open channels in the next cell to let ions through.*

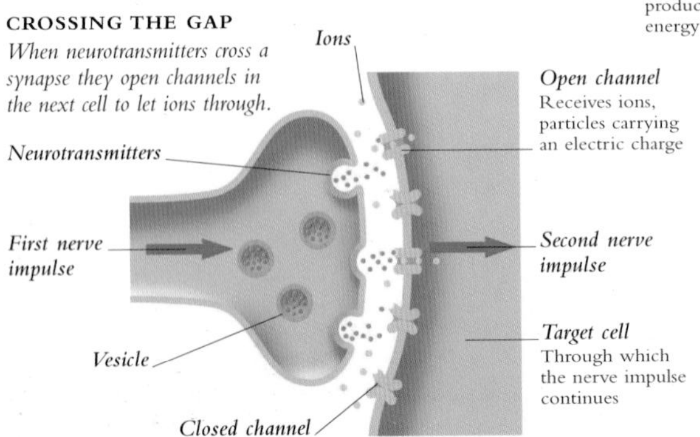


*Mitochondrion*
Small structure involved in the production of energy

*Nucleus*
Control centre of the neuron

*Cell body*
Materials needed by the neuron are made here

BODY

## SUPPORT CELLS

A large part of the central nervous system is made up of cells that do not conduct nerve impulses. These are glial cells, also referred to as neuroglia, which give protection and support to neurons. There are several types of neuroglia. Star-shaped astrocytes are found in brain tissue, where they outnumber neurons. Astrocytes have long, delicate projections that connect with blood vessels and regulate the flow of nutrients and wastes between neurons and the blood. Oligodendrocytes form the insulating myelin sheaths around nerve fibres in the brain and spinal cord, while Schwann cells do the same job in the peripheral nervous system. The smallest glial cells, microglia, destroy harmful organisms.

**ASTROCYTES**
*These cells, which make up over 50 per cent of brain tissue, give structural support to neurons and provide connections with blood capillaries.*

**OLIGODENDROCYTE**
*The plasma extensions of these cells wrap around neurons to form myelin sheaths.*

*Nucleus of Schwann cell*

***Node of Ranvier***
Gap in the myelin sheath of an axon, which helps the conduction of nerve impulses

***Myelin sheath***
Fatty coat that insulates the axons of some nerve cells, speeding transmission of impulses; in the peripheral nervous system, Schwann cells form this sheath

***Synaptic knob***
Terminal point of axon branch, which contains chemicals that transmit nerve impulses (neurotransmitters)

***Dendrite***
A neuron may have 200 of these small projections, which pick up impulses from other neurons

**NEURONS**
*Like most cells, neurons have a cell body containing a nucleus and various internal structures such as energy-producing mitochondria. Typically, neurons have one axon, which is a long, branching projection that transmits nerve impulses, and many shorter branches, dendrites, that receive impulses.*

## HOW NERVES REGENERATE

When a nerve is cut or crushed, the part below the damage dies away as nerve fibres lose contact with the cell body. Affected areas of the body may lose sensation and be unable to function. However, a nerve can sometimes repair itself by growing back along its original pathway. Regeneration is more likely to be successful if the outer myelin sheath remains intact and guides the regrowing fibre in the right direction. If the sheath is damaged, the fibre may form a clump, causing pain. Nerve fibres can regrow at a rate of up to 3mm a day. In the nerves of the brain and spinal cord, scar tissue forms, preventing regeneration.

*Myelin sheath* *Cut nerve fibre*
*Cell body*
INJURED NERVE
*Degenerating fibre*

*Nerve fibre sprouts*
*Cell body*
ATTEMPTED REPAIR

*New nerve fibre*
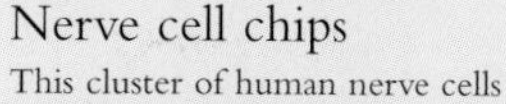
NERVE FUNCTION RESTORED

**REGROWTH**
*After a nerve injury, both the fibre and the myelin sheath below the damage degenerate. In an attempt at repair, the cell body stimulates the growth of new nerve sprouts from the remaining fibre. If one of these sprouts reconnects with the severed end, a new fibre may grow and restore nerve function.*

Fact

### Nerve cell chips

This cluster of human nerve cells (yellow) has been grown on a silicon chip. The experiment is an attempt to create an electronic circuit by combining living nerve tissue and computer technology. Such research projects may eventually lead to the development of chips that can be implanted in the brain to replace damaged nerve networks.

# HORMONES

As well as using nerve signals, the body has another method of control known as the endocrine system, which uses chemicals to activate cells. These chemicals, hormones, are produced in special glands and tissues throughout the body; they are released directly into the bloodstream and carried to specific targets. Hormone action stabilizes the body's internal environment and also regulates growth and reproduction.

## THE MASTER GLAND

A pea-sized structure at the base of the brain, often described as the master gland, controls most of the body's hormonal activity. This is the pituitary gland, itself controlled by hormones produced in another brain area, the hypothalamus. The pituitary hangs from the hypothalamus by a short stalk containing blood vessels and nerve fibres. It is divided into two areas: the anterior (front) and posterior (rear) lobes. The anterior lobe secretes growth hormone and stimulates hormone production in other glands. The posterior lobe receives hormones, such as antidiuretic hormone and oxytocin, from the hypothalamus and stores them for later release.

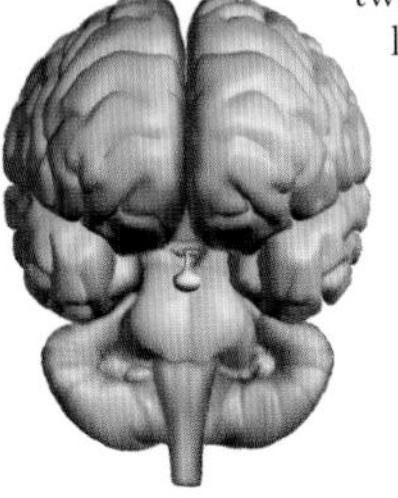

**SITE OF THE PITUITARY GLAND**
*The pituitary gland (small green structure in picture) is suspended by a stalk from the underside of the brain.*

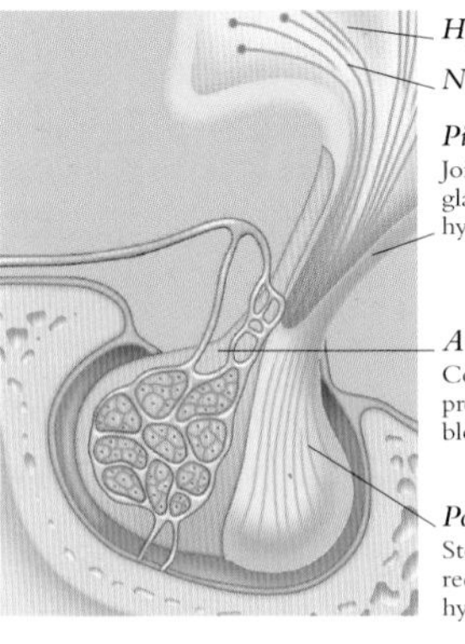


**INSIDE THE PITUITARY**
*The two inner lobes of the pituitary gland release a range of hormones that affect many parts of the body.*

## THYROID AND PARATHYROID GLANDS

The two-lobed thyroid gland is situated in the neck, resembling a bow tie in the way it wraps around the trachea (windpipe). Just behind the thyroid are the four closely associated parathyroid glands. The thyroid gland produces the hormone thyroxine, which regulates the rate at which body cells use oxygen to create energy. Another thyroid hormone is calcitonin, which controls the levels of the mineral calcium in the blood by slowing the loss of calcium from bones. The parathyroid glands secrete parathyroid hormone, which works together with calcitonin to keep calcium levels in the blood within normal limits. When calcium levels fall, parathyroid hormone is released to restore calcium levels.

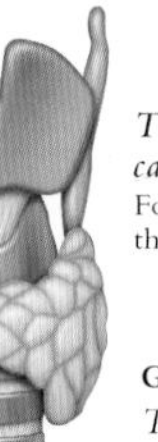


**GLAND STRUCTURE**
*The thyroid gland consists of two lobes of tissue that join together in front of the trachea. The four parathyroid glands are embedded in the back of the thyroid gland.*

## ADRENAL GLANDS

The paired adrenal glands lie on fatty pads above the kidneys. The hormones that these glands produce play an important role in the body's production and use of energy, and in our responses to stress. Each adrenal gland has two main parts: the cortex (outer layer) and the medulla (core). Among the hormones produced by the cortex are corticosteroids, which affect energy production, and aldosterone, which regulates the balance of salt. The adrenal medulla releases the hormone epinephrine (adrenaline) at times of stress or excitement. This chemical, which is extremely fast acting, prepares the body for action by increasing the heart rate and blood flow to the muscles.

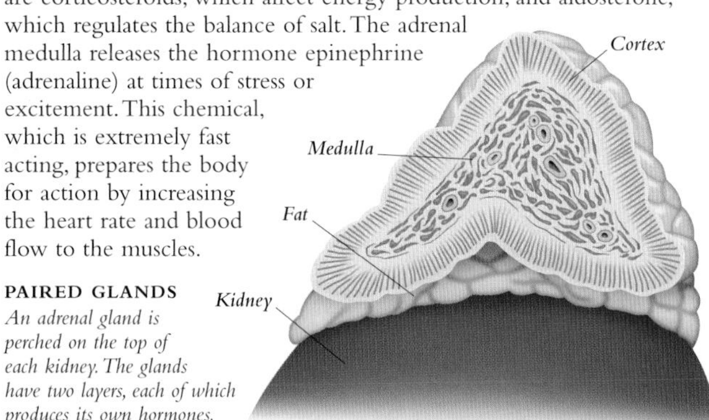


**PAIRED GLANDS**
*An adrenal gland is perched on the top of each kidney. The glands have two layers, each of which produces its own hormones.*

## THE PANCREAS

Situated deep within the abdomen, behind the stomach, the pancreas is an elongated gland with two functions. One function is the production of digestive juices, the other is the secretion of hormones. The hormone-making cells of the pancreas cluster in groups known as the islets of Langerhans. These cells produce glucagon and insulin, hormones that work together to keep blood glucose levels within a normal range. When blood glucose levels are low – perhaps after exercise or due to lack of food – glucagon raises them by stimulating the release of glucose stored in the liver. When blood glucose levels are high – perhaps after a meal – insulin lowers them by stimulating body cells to absorb glucose for fuel.

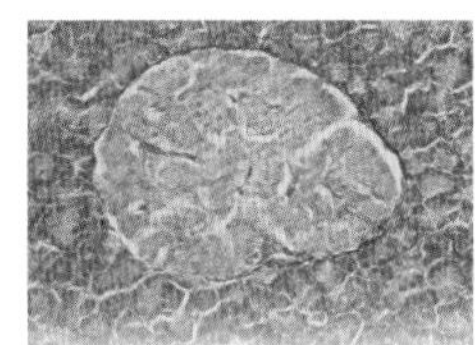

**ISLETS OF LANGERHANS**
*These clusters of pancreatic cells (mauve) produce hormones that control blood glucose levels.*

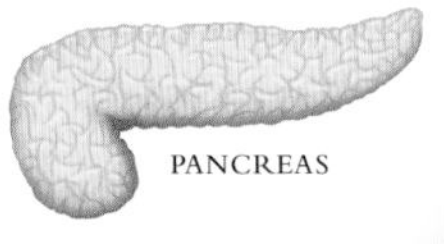


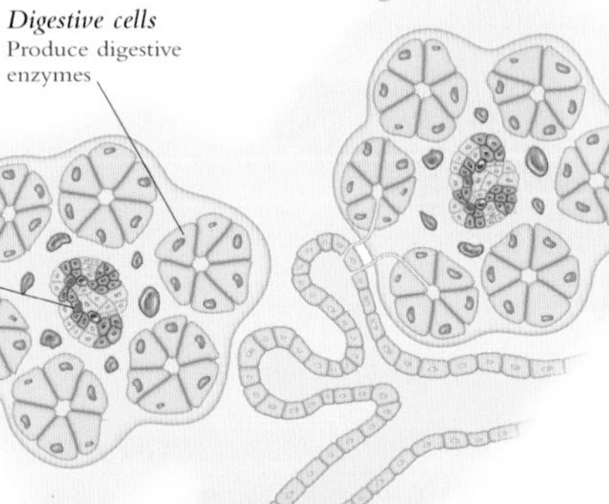


**PANCREATIC CELLS**
*The pancreas contains two specialized types of cell: those that produce digestive enzymes and those that secrete hormones.*

Health

### Insulin and diabetes

If the pancreas does not produce enough of the hormone insulin, or if the body is unable to use insulin properly, glucose levels in the body rise uncontrollably. This causes the disorder diabetes mellitus, symptoms of which include excessive urination and thirst. Diabetes may be treated, but not cured, by diet and regular insulin injections. Pen-like devices such as that illustrated (right, in use) meter out the correct dose of the hormone.

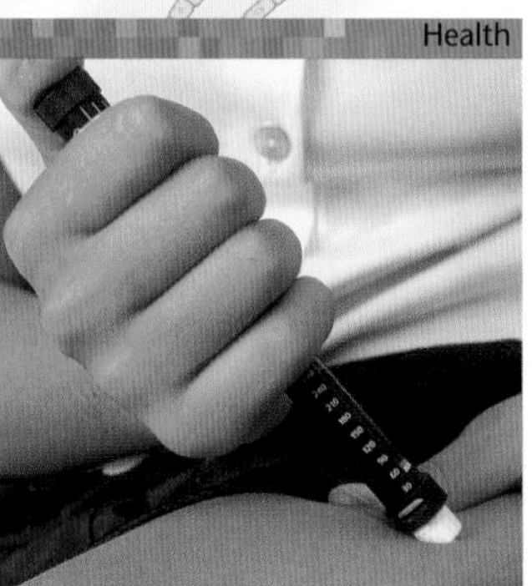

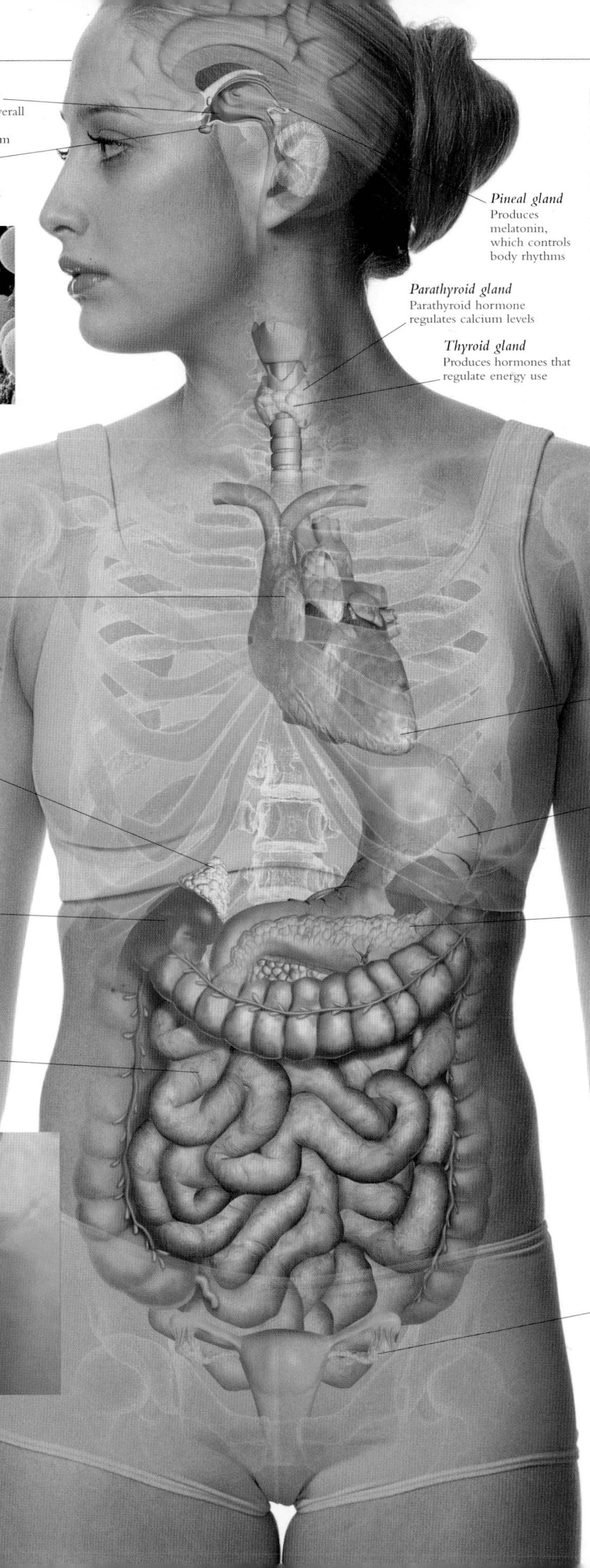

*Hypothalamus*
Brain area in overall control of the endocrine system

*Pituitary gland*
Regulates other glands in the body and secretes growth hormone and vasopressin

*Pineal gland*
Produces melatonin, which controls body rhythms

*Parathyroid gland*
Parathyroid hormone regulates calcium levels

*Thyroid gland*
Produces hormones that regulate energy use

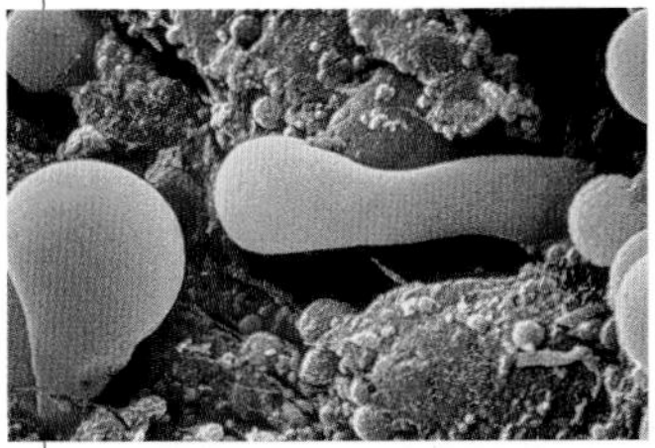

**THYROID HORMONES**
*The orange globules are hormones in the process of secretion in the tissues of the thyroid gland.*

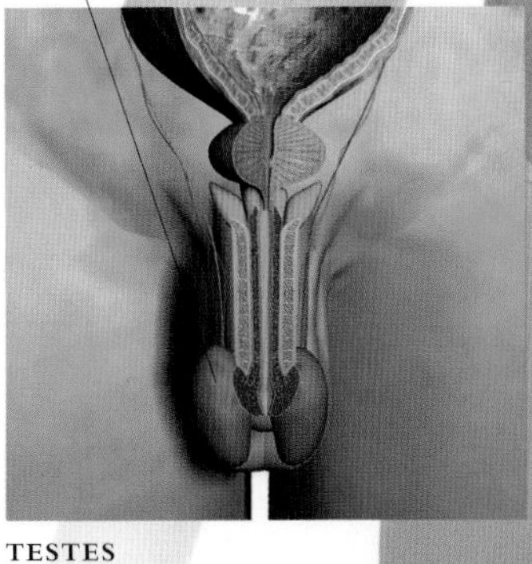

Health

## Melatonin levels

The hormone melatonin, produced mostly at night by the pineal gland, is related to our cycle of sleeping and waking. In dark winter days, the pineal gland secretes greater amounts of melatonin, which is thought to contribute to a form of depression known as seasonal affective disorder (SAD). Exposure to fluorescent light (below) may relieve SAD by decreasing melatonin secretion.

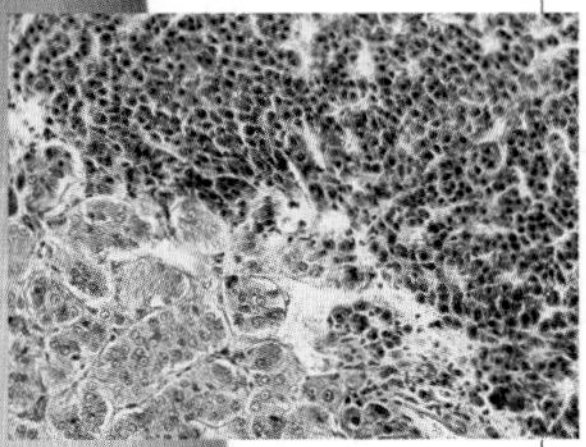

*Thymus*
Releases hormones involved in the production of white blood cells

*Adrenal gland*
Produces a number of hormones involved in energy use and the body's response to stress

*Kidney*
Produces the hormone erythropoietin, which stimulates production of red blood cells

*Intestines*
Lining contains cells that release hormones involved in digestion

*Testis*

*Heart*
Produces natriuretic hormone that controls blood pressure during exertion

*Stomach*
Glands in the stomach lining produce gastrin hormone, which stimulates production of digestive acids

*Pancreas*
Produces insulin and glucagon, which regulate blood glucose levels

**ADRENAL TISSUE**
*Clumps of hormone-secreting cells can be seen in the tissues of the adrenal gland's cortex (dark blue) and medulla (light blue).*

*Ovary*
Secretes the female sex hormones oestrogen and progesterone

**TESTES**
*The testes produce testosterone and other male sex hormones that control sperm production and the development of male sexual characteristics.*

**ENDOCRINE GLANDS AND TISSUES**
*Hormone-producing cells occur in tissues throughout the body – for example, the heart, stomach, and intestines – as well as in specialized endocrine glands, such as the pituitary gland in the brain and the adrenal glands above the kidneys.*

**SENSING THE WORLD**
*Our five main senses let us know that we are alive. Everything we see, hear, taste, smell, and feel keeps us informed about the world that surrounds us.*

# THE SENSES

All living things, including such diverse organisms as bacteria and plants, have a range of senses that allow them to respond to their environment. Humans, in common with other mammals, have five main senses: vision, hearing, touch, taste, and smell. Although these senses generally serve us very well in terms of survival, in some ways our sensory abilities seem to be limited. We cannot hear or smell as acutely as many animals, or see tiny objects from miles away like birds of prey. However, we do have the advantage of a large brain; when we receive sensory input, this brain lets us analyse the information to an incomparable degree.

**COMBINING SENSES**
*This professional tea taster needs to be able to smell his samples. He would find it hard to taste even distinct differences in flavour if he was deprived of his sense of smell.*

The brain is constantly bombarded with sensory information from outside the body. This information arrives through a number of different channels: the eyes, ears, nose, mouth, and skin. Specialized receptors at these sites respond to various forms of stimulus and relay messages down nerve pathways to the brain. No two people sense the world in the same way. Each brain is different and puts an individual interpretation on everything that is brought to its attention.

## SEEING

For most humans, vision is the dominant sense. What we see often overshadows much of the information we receive from our other senses.

The visible part of the eye does not have a great deal to do with the mechanics of seeing, although it lets in the light that we must have for vision. Eyes are complicated sensory organs that collect the light reflected off objects around us, focus it, and send signals to the brain about the intensity and colour of light they receive. Inside the eye, a clear, elastic lens adjusts to focus light rays on to a highly sensitive membrane, the retina, at the back of the eye. There, two types of light-sensitive cells react when light hits them, initiating a nerve signal to the brain. Every second, the brain analyses thousands of signals like this from the eyes, interpreting them as recognizable images and colours from the information it receives. To build up the complete picture, the brain uses clues such as perspective and draws on previous experience. Sometimes, the brain simply has to rely on guesswork to interpret what we are looking at; a wrong guess can give rise to strange visual illusions.

**THE EYE**
*Light entering the front of the eye is focused first by the domed cornea and then by the lens (yellow).*

## HEARING AND BALANCE

The sounds we hear consist of a continuous stream of vibrations, every one of which passes along thousands of nerve fibres to the brain for interpretation. By making use of this information, the brain helps us to do many things. For example, we can identify a sound and locate the object or animal that is making it; understand speech; and tune in to the rhythms and melodies of music. Our organs of hearing, which have a dual role as organs of balance as well, are hidden inside our heads. Here, a delicate network of bones and membranes, including the smallest bones in the body, carries sound vibrations deep into the coiled, snail-like chambers of the inner ear. Tiny hairs attached to sensory cells in the inner ear are moved by these vibrations, triggering nerve signals to the brain. What our brain registers and what we hear can be two different things. We have the ability to pick out what we want to listen to while screening out unwanted background noise.

**CELLS OF THE INNER EAR**
*Hair cells (yellow) in the inner ear move in reaction to sound waves, triggering nerve signals that are transmitted to the brain.*

## TOUCH

Touch is an essential sense that allows us to find out the size, shape, texture, and temperature of things around us. This sense also lets us feel pain. Without touch, normal day-to-day living would become nearly impossible. We would be unable to hold a pen correctly, or shake someone's hand, or find our way around in the dark. We would also be far more vulnerable to harm. Our sense of touch lies in the skin and comes from a variety of nerve structures that send messages to the brain whenever we come into physical contact with anything.

## TASTE AND SMELL

There are four basic taste flavours: salty, sour, sweet, and bitter. The sensory cells that detect them are dotted around various areas of the mouth and tongue. While these sensors tell us something about the flavour of the food in our mouths, taste also relies to a large extent on other senses. What we see and touch affects what we expect to taste; but the most important influence on taste is our sense of smell. Very few people can recognize even the familiar tastes of chocolate, coffee, and garlic without the aid of their noses. Our sense of smell is many times more sensitive than our sense of taste. We can detect thousands of different smells. The smell receptors are sited at the top of the nasal cavity. There are hundreds of different types, each sensitive to a different chemical substance. All the odours and tastes that we recognize are compiled in the brain from the combination of information received from the mouth and nose.

Usually regarded as the minor senses, taste and smell are necessary for survival, directing us to food that is good to eat and helping us to avoid harm. These senses can also affect our mood and emotional well-being.

**SENSITIVE FINGERS**
*The fingers have a high concentration of touch receptors and are among the most sensitive areas of the body.*

Fact

### Synaesthesia

Some people experience a sensory crossover, when messages about what they see, hear, smell, and taste become mixed. The condition, synaesthesia, means that a person may, for example, hear music and also see the sounds as certain colours. He or she may smell colours or taste words. Synaesthesia is usually present from an early age and the reasons for it are not fully understood. The condition cannot be treated but rarely causes distress.

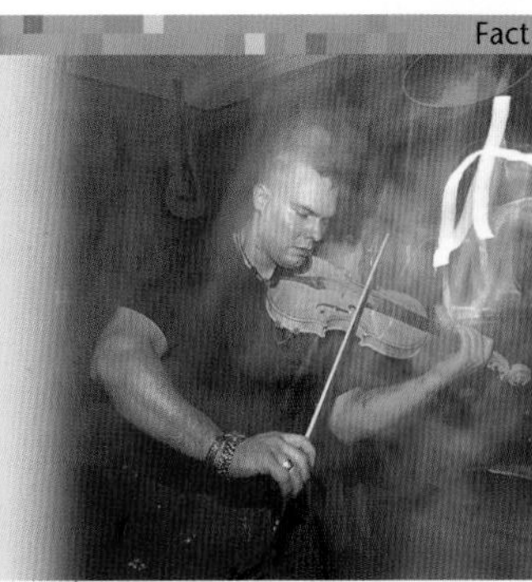

# VISION

We rely on vision for most of our information about the world. Simply by looking at something, we can judge its size and texture, what it would feel like to touch, whether it is harmful or not, and how far away it is. About 70 per cent of the body's sense receptors are grouped in the eyes. These receptors pass information to the brain, where it is analysed. What we see depends on our personal perspective. No two people see an object in quite the same way.

**FOVEA**

*This area of the retina, at the back of the eye, is the point where visual perception is at its sharpest.*

## HOW WE SEE

The eye works in a similar way to a camera. It has a transparent window at the front, the cornea, which gathers and refracts (bends) light rays reflected from everything in our line of vision. Behind the cornea is a clear, elastic lens, which automatically adjusts its shape to fine-focus the light rays. Inside the eye, the rays cross over and pass to the back of the eye, where the light creates an upside-down image on a layer of membrane called the retina. There are more than 126 million nerve cells in the retina; all are sensitive to light and some are specialized to distinguish colour. These cells respond instantaneously to light, converting the image on the retina into nerve impulses. The impulses are transmitted to the brain, flashing down nerve pathways that run from the back of each eye. Visual interpretation centres in the brain make sense of the light and colour messages and let us see the upside-down image the right way up.

**INCOMING LIGHT**

*Light rays entering the eye are bent by the elastic lens, which is held suspended at the front of the eye by delicate fibres.*

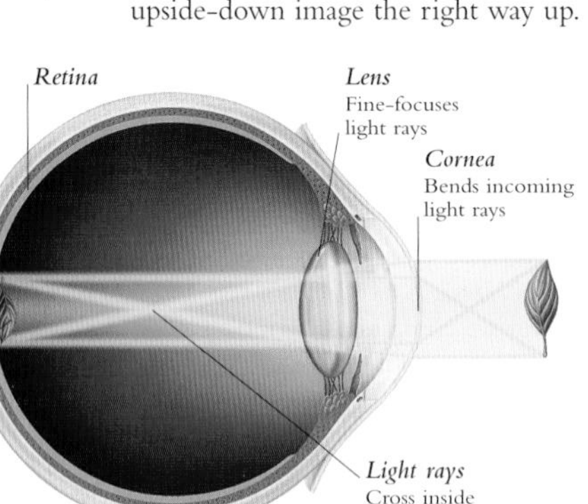


**EYE MECHANISM**

*Light rays entering the eye are focused by the cornea and lens to produce images on the retina.*

## PROTECTING THE EYES

The delicate eyeball is protected by the eyelids, which close like shutters to keep out harmful materials. Additional protection is provided by tears, which not only wash away dirt from the eye but also contain a natural antiseptic to help stop infections developing. Tears are produced by the lacrimal (tear) glands, which are situated above each eye, just beneath the eyebrow. These glands constantly secrete small amounts of salty, watery fluid that seeps out through small ducts and is washed across the surface of the eye several times a minute by blinking.

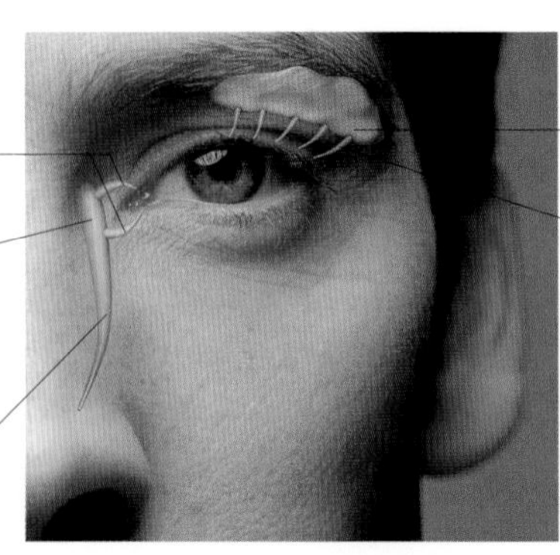


**TEAR APPARATUS**

*Tears seep into the eye from the lacrimal gland and drain away into the lacrimal sac. This sac empties into the nose through the nasolacrimal duct.*

*Fovea*
The part of the retina where light rays are most sharply focused

*Optic nerve*
Takes impulses from the retina to the brain

*Optic disc*
Point where nerve fibres from the retina meet to form the optic nerve

*Retina*
Innermost layer of the eye, where incoming light is converted into nerve impulses

*Eye muscle*
One of six surrounding muscles that rotate the eye in all directions

*Vitreous humour*
Clear, jelly-like substance filling the back of the eye and maintaining its shape

*Sclera*
Tough outer layer that forms the white of the eye

**EYE STRUCTURE**

*The eyeball has three layers. Outermost is the white sclera, modified at the front to form the transparent cornea. The middle layer, the choroid, holds the lens and most of the eye's blood vessels. The inner layer is the retina. Inside the eye are two cavities. The one in front of the lens, is filled with watery fluid called aqueous humour. The larger, back cavity contains jelly-like vitreous humour.*

**CHOROID**
*A dense network of blood vessels spreads through the choroid, the middle layer of the eyeball. These vessels supply the eye with essential nutrients.*

*Choroid*
Middle layer of the eyeball, the front parts of which comprise the ciliary body and the iris

**LENS CELLS**
*This magnified view shows the cell layers of the lens. The cells are transparent and have no nuclei to interfere with the entry of light rays.*

*Lens*
Transparent, fibrous disc of tissue that fine-focuses light rays for near or distant vision

*Iris*
Coloured area containing muscles that control the pupil

*Pupil*
Opening in the centre of the iris that lets in light

*Cornea*
Curved, transparent dome that bends incoming light

*Conjunctiva*
Thin, protective membrane covering the white of the eye and lining the eyelids

*Ciliary body*
Contains muscles that alter lens shape; its fine suspensory ligaments hold the lens in place

**CILIARY BODY**
*The pink folded structures and fine ligaments in this magnified image belong to the ciliary body. The muscle fibres of the iris are also shown (right).*

# THE AUTOMATIC EYE

The eye automatically adjusts to produce the clearest images on the retina under any conditions. To let in the maximum amount of light in a dim environment, the pupil widens. In bright light, the pupil becomes smaller to prevent the eye being dazzled. The pupil size is controlled by two sets of muscles – circular (running around) and radial (running crossways) – in the iris. The circular muscles contract to make the pupil narrow; the radial muscles contract to widen the pupil.

The eye also adjusts to give us the clearest view of both distant and near objects by changing the shape of the lens. This process, known as accommodation, fine-focuses light rays to produce sharp images. It is controlled by ciliary muscles surrounding the lens. To bring distant objects into focus, the muscles relax, while the ciliary ligaments that hold the lens suspended in front of the eye tighten, pulling the lens flat. To focus on near objects, the muscles tighten and the ligaments relax; the lens becomes slack and rounded, bending the incoming light rays.

**PUPIL ACTION**
*The pupil widens or narrows to let the right amount of light enter the eye. It becomes smaller in bright light and larger in dim light.*

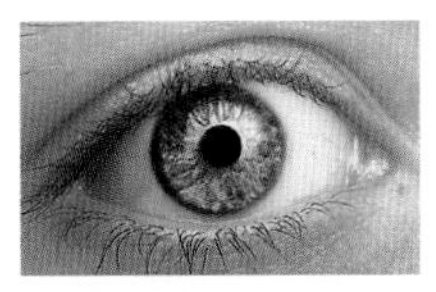
BRIGHT LIGHT

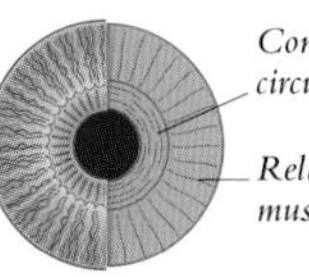


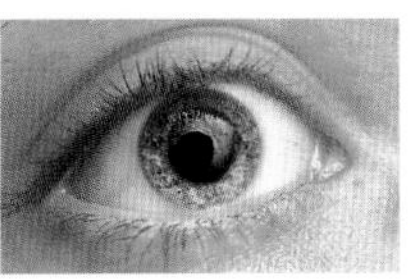
DIM LIGHT

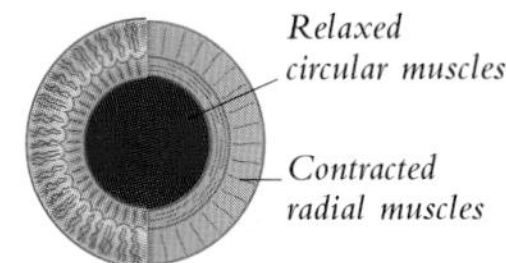


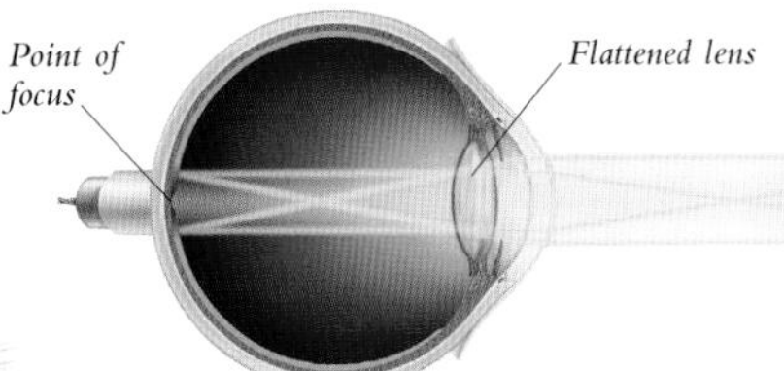

DISTANT VISION

DISTANT OBJECTS IN FOCUS

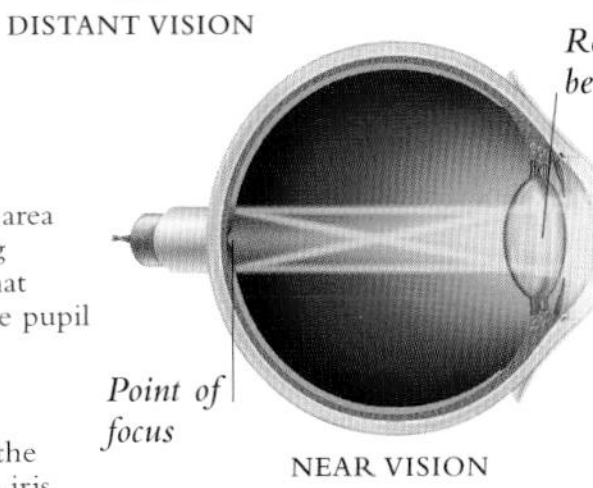

NEAR VISION

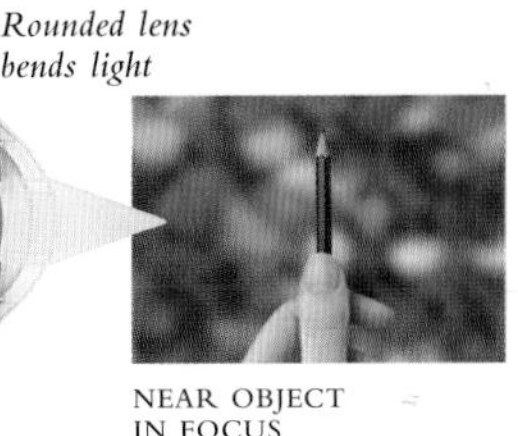
NEAR OBJECT IN FOCUS

**ACCOMMODATION**
*To keep distant objects in focus, the lens maintains a flattened shape. To focus on near objects, light rays must be bent more, so the lens is pulled into a more rounded shape by the ciliary muscles. This process is called accommodation.*

Fact

## Iris recognition

The patterns of each iris are unique. No two people have the same details; even in an individual the right and left eyes differ. This variation can be used for personal identification, in the same way as fingerprints. The iris is scanned by video camera and mapped by computer (see right) to pinpoint its features precisely. This information is encoded for storage on a database. Iris scanning is undergoing trials in such applications as airport security.

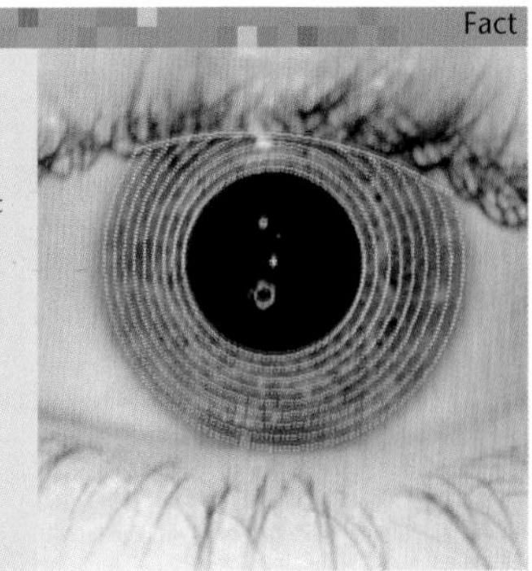

BODY

# MOVING THE EYES

The eyes are constantly on the move; even during sleep they flicker restlessly. During our waking hours, we visually track moving objects, let our eyes dart rapidly from one part of a scene to another, or follow words on a page. Each eyeball is moved in its bony socket by a set of six straplike muscles that are attached to the sclera, the tough, white outer layer of the eye. All these muscles have anchoring points at the back of the socket. Four of them, the rectus muscles of the eye, run in an almost straight line, while the other two, the oblique muscles, are set at an angle. The upper of the oblique muscles passes through a small cartilage pulley suspended from a corner of the socket. Between them, the muscles roll the eyeball up and down and from one side to the other. Because eye movements are small, the action of the muscles is precisely tuned. Three nerves that run from the base of the brain – the oculomotor, trochlear, and abducent nerves (*see* p113) – carry electrical impulses to activate the eye muscles.

Cartilage pulley

Eye muscle

Muscle attachment to sclera

Bony socket

**EYE MUSCLES**
*The eye muscles work in pairs to roll the eyes in various directions; the movements are tiny but precisely controlled.*

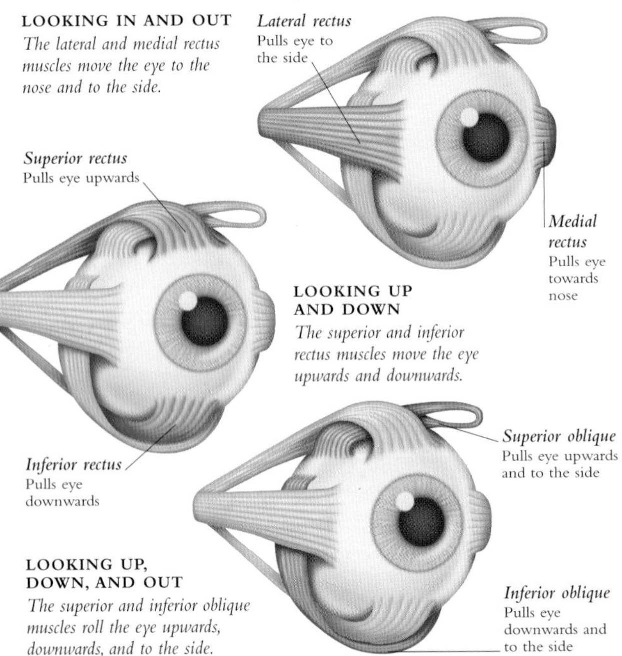


**LOOKING IN AND OUT**
*The lateral and medial rectus muscles move the eye to the nose and to the side.*

**LOOKING UP AND DOWN**
*The superior and inferior rectus muscles move the eye upwards and downwards.*

**LOOKING UP, DOWN, AND OUT**
*The superior and inferior oblique muscles roll the eye upwards, downwards, and to the side.*

Fact

## Seeing red

Only apes and some monkeys of the Old World (such as the orangutans shown right) are known to share the same form of colour vision as humans. Like us, the eyes of these primate relatives are sensitive to blue, green, and – a rarity in the animal world – red light. It is possible that this type of vision is an evolutionary adaptation designed to aid detection of foods, such as red fruit and green leaves, at the right time of year.

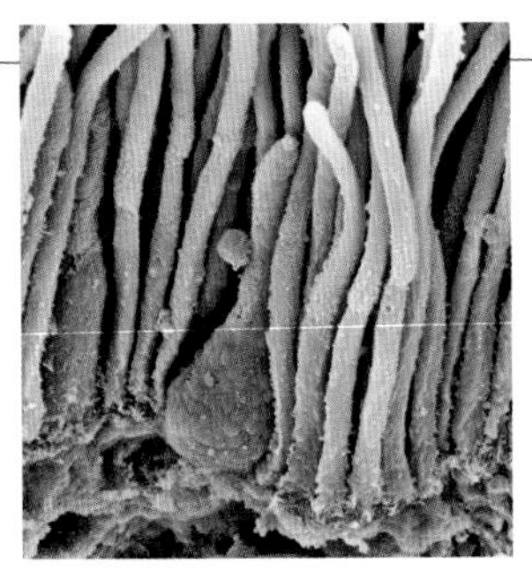

**RODS AND CONES**
*When light falls on the retina it activates rod cells and cone cells, which transmit impulses to the optic nerve.*

# RODS AND CONES

The retina contains two different types of light-receptive nerve cell known as rods and cones because of their shapes. Rods are by far the most numerous; there are some 120 million of them, distributed throughout the retina. Although they are extremely sensitive to all light, rods cannot distinguish colours and are mostly used for night vision. Cone cells, of which there are about 6.5 million, are responsible for colour vision; they are less light-sensitive than rods and work when stimulated by bright light. Cones are densest in an area of the retina called the fovea, the site of sharpest vision, where images are focused. There are three types of cone cell: red, blue, and green. Each type contains a different photopigment and is sensitive to a different wavelength of light. In response to light, both rods and cones produce signals that trigger nerve impulses.

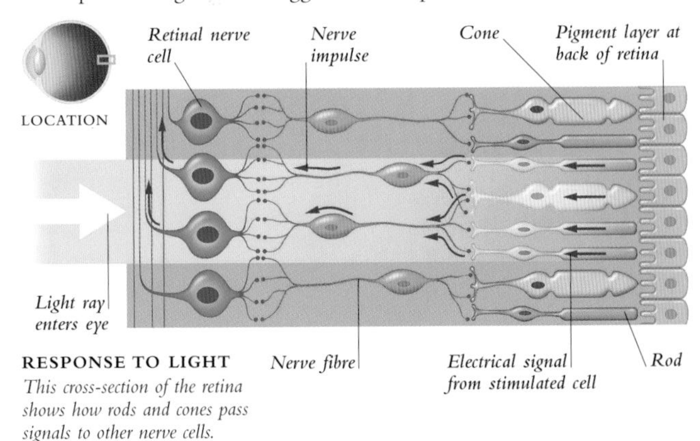


**RESPONSE TO LIGHT**
*This cross-section of the retina shows how rods and cones pass signals to other nerve cells.*

# DARK AND LIGHT ADAPTATION

When we move from a brightly lit place into the dark, our vision takes a few minutes to adjust to the change. Our eyes have to undergo the process referred to as dark adaptation before we can see clearly again. When light hits the rod and cone cells in the retina, the photopigments contained in the cells become bleached, which reduces their sensitivity to light. This reduced sensitivity is why colours can appear less intense in bright sunlight. It takes approximately 30 minutes of darkness for the pigments to regenerate and regain their maximum sensitivity. The eyes also need time to adapt when we move from darkness into bright light. Light adaptation is simply an improvement in perceiving areas of contrast, such as shadow, and is much quicker than dark adaptation.

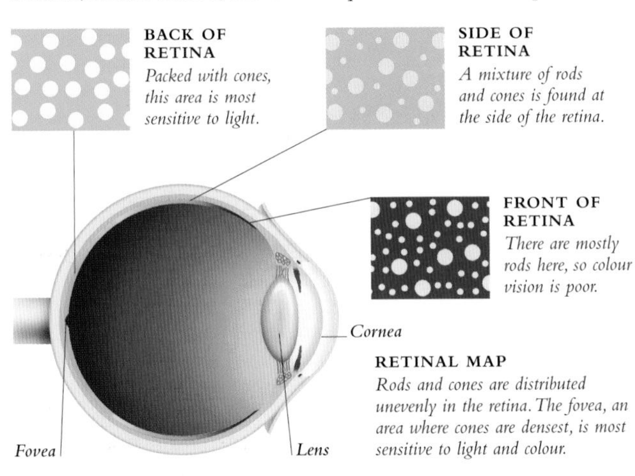


**RETINAL MAP**
*Rods and cones are distributed unevenly in the retina. The fovea, an area where cones are densest, is most sensitive to light and colour.*

BODY

History

## The colours of light

The mixture of colours that make up white light was discovered by English physicist Sir Isaac Newton (1642–1727). In the engraving seen below, Newton is carrying out an experiment in which rays of light pass through a prism, throwing the colours of the spectrum (the same as those of the rainbow) onto a screen.

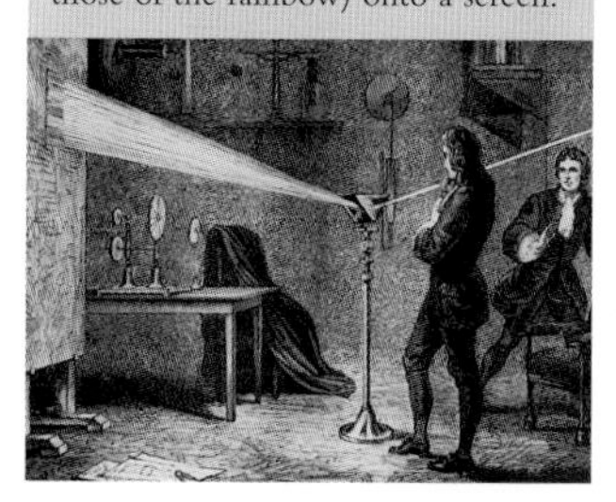

# VISUAL PATHWAYS

We can see because the visual processing areas in the brain, the right and left visual cortex, receive information from each retina. This information, in the form of electrical signals, passes down the two optic nerves, which meet at a junction called the optic chiasm. At this point, half of the nerve fibres from the left eye cross to the right side and half of the fibres from the right eye cross to the left side. The fibres continue along the optic tracts, which are the visual pathways to the brain. Information from the right half of each retina passes to the right visual cortex; information from the left half of each retina goes to the left visual cortex. The brain then combines the two messages to produce a complete, detailed colour picture.

**EYES AND BRAIN**
*This scan shows the optic nerves (yellow) that run from the retina of each eye to the visual cortex at the back of the brain.*

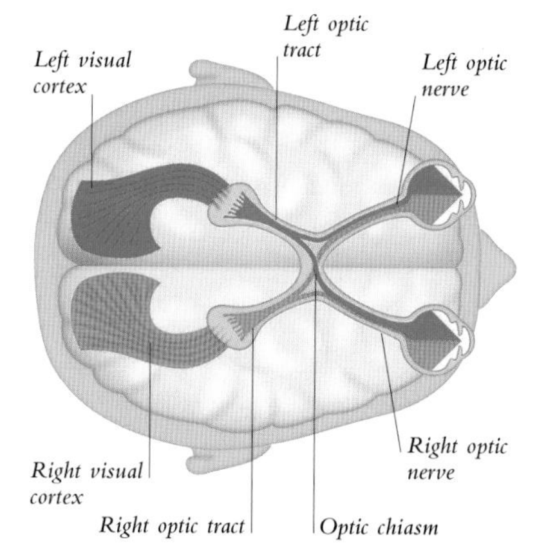


**BINOCULAR VISION**
*The right and left eyes send separate signals to the right and left visual cortex. The brain integrates the signals into a composite image.*

# SEEING IN 3-D

Because we have forward-facing eyes, the centre of our field of view is binocular, that is, it is seen by both eyes at the same time. Each eye looks at an object or a scene from a slightly different viewpoint and the two pictures overlap one another by about 120 degrees. This crossover gives us stereoscopic vision – the ability to see things, such as the die in the illustration, in three dimensions. The brain's visual processing centre receives separate electrical signals from the right and left eyes about the two different pictures and automatically amalgamates the information. As a result, we perceive only one picture. By comparing the variations in the signals, the brain can make accurate assessments about the depth of the object and how far away it is. This complex processing allows us to judge distances with great precision (an ability that greatly diminishes if one eye is closed).

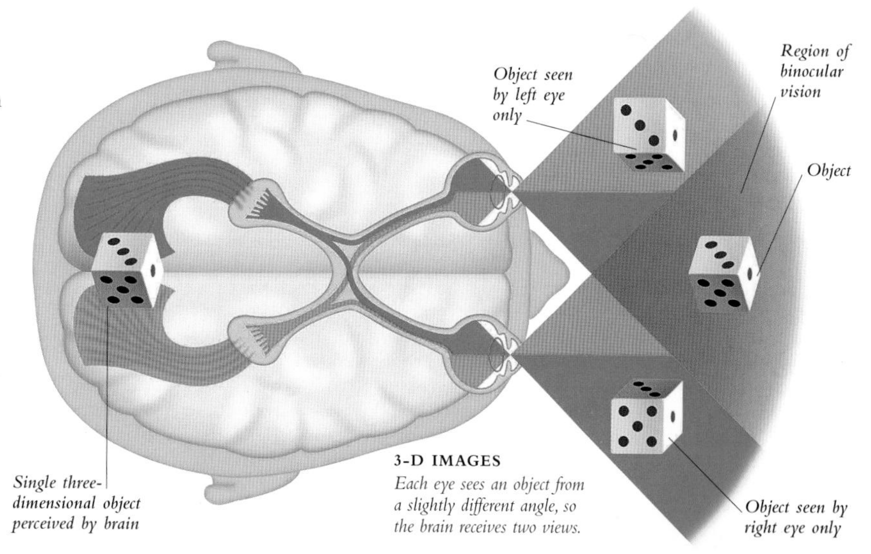


**3-D IMAGES**
*Each eye sees an object from a slightly different angle, so the brain receives two views.*

# TRICKING THE BRAIN

The brain attempts to make sense of everything that we see around us, putting visual information into some sort of context and using previous experience to make comparisons. If the information that we receive is ambiguous, the brain is forced to make guesses. For example, the picture below can be seen as either a white vase on a black background or as the silhouettes of two people facing one another. The brain cannot recognize both the vase and the faces at the same time and may switch back and forth from one to the other, trying to decide which is foreground and which is background. A particular pattern may give the impression that it is moving (*see top left*). This is because our eyes are scanning the image in rapid jerks, generating the transmission of staccato nerve impulses that the brain interprets as movement. Another brain-fooling optical illusion is the Ames room with its distorted dimensions (*see bottom left*). For centuries, artists and architects have made use of perspective, which allows the brain to interpret flat images as three-dimensional.

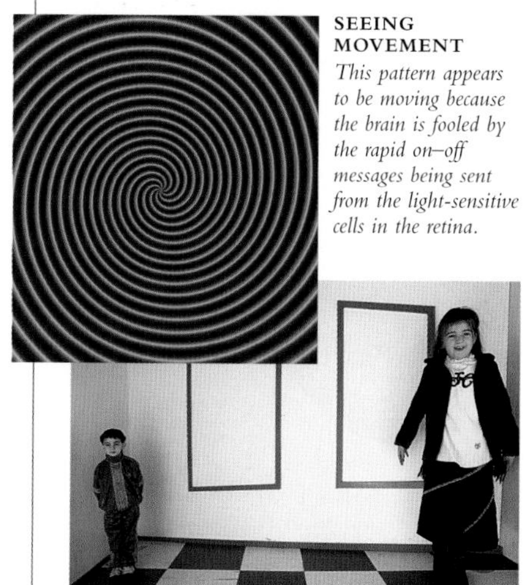

**SEEING MOVEMENT**
*This pattern appears to be moving because the brain is fooled by the rapid on–off messages being sent from the light-sensitive cells in the retina.*

**AMES ROOM ILLUSION**
*This room looks rectangular, but its floor, ceiling, and back wall are sloping. The boy is not smaller than the girl but farther away.*

**VASE OR FACES?**
*The eyes see two different images, and the brain jumps from one to the other; it cannot combine them.*

Fact

## Seeing television images

The pictures on a television screen are made up of strips of tiny dots, which glow red, blue, and green when stimulated by electrons. These colours match the three types of photopigments found in the colour-receptive cone cells of the retina. The electrons stimulate the dots by moving extremely quickly over the screen, and the brain integrates the colours to produce a clear image.

**TELEVISION SCREEN**
*This close-up view of part of a television screen shows the strips of red, green, and blue dots that make up the colour images.*

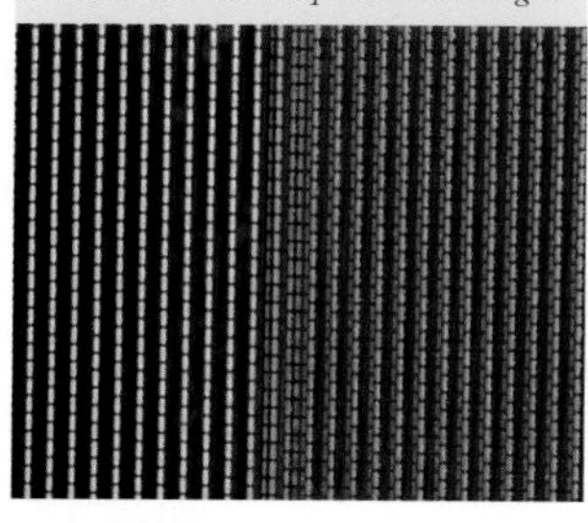

# HEARING AND BALANCE

In our communications with others, hearing is our most important sense. Humans spend much time talking and listening, and the ability to hear provides a fast and effective way of receiving information. Sounds can be protective, giving us immediate warning of danger. What we hear can also trigger strong emotional responses: music or laughter may lift our spirits; a baby crying can arouse caring instincts. Our ears also provide a constant flow of information about the body's orientation in space. This input gives us our sense of balance, allowing us to stand and move without falling.

## HOW WE HEAR

When an object moves, it disturbs molecules in the surrounding air and makes them hit adjacent molecules. The resulting ripple of air is a sound wave. A small movement, such as turning a page in a book, causes a slight air disturbance, whereas a large movement, such as dropping a boulder from a cliff, causes a big one. Sound waves make structures in the ear vibrate. The vibrations are sent as nerve impulses to the brain and interpreted as sound. The size of a sound wave is measured in decibels (dB). The quietest sounds that we can detect are around 10dB (for example, breathing). Sound levels of around 120dB and over (such as nightclub music) can cause permanent damage to the inner ear.

*Left auditory cortex*

**INTERPRETING SOUND IN THE BRAIN**
*There are auditory areas that interpret sound on both sides of the brain. Speech is mainly interpreted on the left side and music mainly on the right side.*

## HOW THE EAR PROCESSES SOUNDS

The part of the ear that we can see, the pinna, channels sound waves into the ear, but it is not an essential part of our hearing apparatus. People who lack a pinna can still hear perfectly well. Inside the ear is a tube, the ear canal, at the end of which is a delicate, fan-like membrane, the eardrum. Sound waves pass down the ear canal and make the eardrum vibrate. Beyond the eardrum is the middle ear, an air-filled space containing three tiny bones: the malleus, incus, and stapes. These bones pass on vibrations from the eardrum to a fluid-filled structure, the cochlea, in the inner ear. The spiral-shaped cochlea contains the organ of Corti, the receptor for hearing, which is filled with millions of hair cells that have protruding sensory hairs. When the fluid in the cochlea vibrates, the hairs move, stimulating nerves that carry sound messages to the brain in the form of electrical impulses.

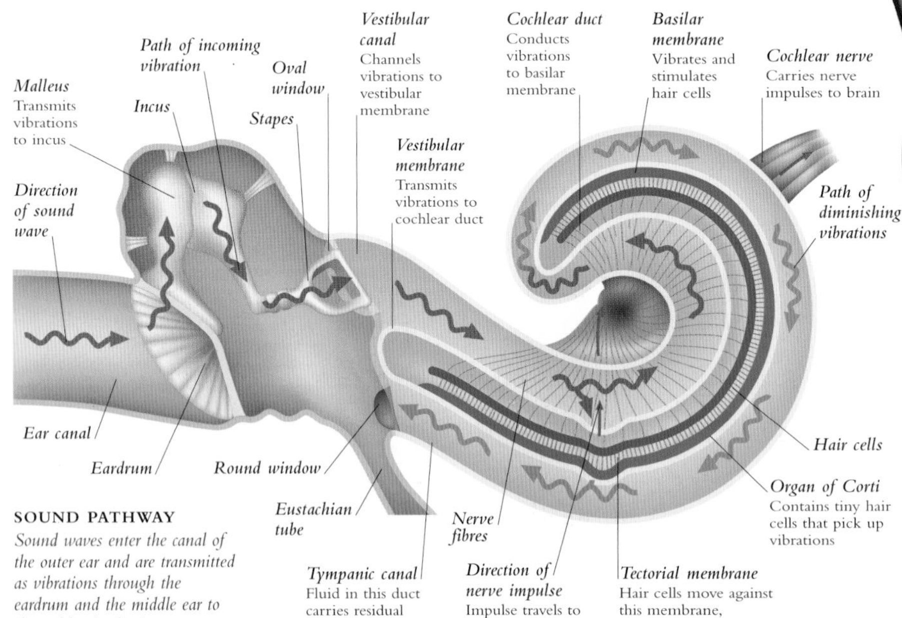
*Path of incoming vibration*
*Vestibular canal* Channels vibrations to vestibular membrane
*Cochlear duct* Conducts vibrations to basilar membrane
*Basilar membrane* Vibrates and stimulates hair cells
*Cochlear nerve* Carries nerve impulses to brain
*Oval window*
*Malleus* Transmits vibrations to incus
*Incus*
*Stapes*
*Vestibular membrane* Transmits vibrations to cochlear duct
*Direction of sound wave*
*Path of diminishing vibrations*
*Ear canal*
*Eardrum*
*Round window*
*Hair cells*
*Organ of Corti* Contains tiny hair cells that pick up vibrations
*Eustachian tube*
*Nerve fibres*
*Tympanic canal* Fluid in this duct carries residual vibrations
*Direction of nerve impulse* Impulse travels to the cochlear nerve
*Tectorial membrane* Hair cells move against this membrane, generating impulses

**SOUND PATHWAY**
*Sound waves enter the canal of the outer ear and are transmitted as vibrations through the eardrum and the middle ear to the cochlea in the inner ear.*

*Muscle*

*Cartilage* Gives the pinna its shape and resilience

*Pinna* Flexible outer flap of the ear, which channels sound waves into the ear canal

**INSIDE THE EAR**
*The ear contains sensory structures that detect sounds and information about our posture and movement. This information is converted into nerve impulses and sent to the brain.*

BODY

Health

### Vertigo

If the delicate apparatus of the inner ear is disturbed, the sense of balance can be affected. Being off balance causes vertigo – a false sensation of movement or an unpleasant feeling that everything around one is spinning. Causes of vertigo include an infection of the vestibular apparatus, often following a viral infection such as a common cold or flu. Vertigo is also a symptom of Ménière's disease, a disorder in which, for unknown reasons, excess fluid builds up in the inner ear.

# THE SENSE OF BALANCE

Several structures in the inner ear, referred to collectively as the vestibular apparatus, help us to stand upright and move without losing balance. This apparatus includes three fluid-filled loops set at right angles to each other – the semicircular canals – at the base of which sensory hair cells lie in small structures called cristae. The other structures that contribute to balance are two chambers (the utricle and saccule) within a fluid-filled area called the vestibule. These chambers both contain a macula, a sensory structure that is also full of hair cells. From the semicircular canals and the maculae, information about the speed and direction of the head's movement is transmitted by the hair cells to the brain in the form of nerve impulses.

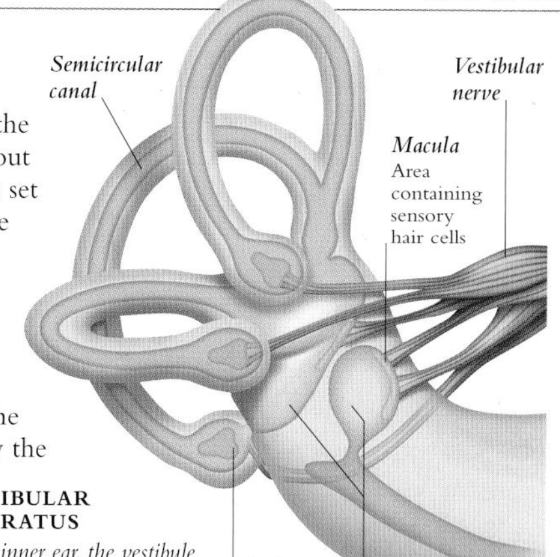

*Semicircular canal*

*Vestibular nerve*

*Macula* Area containing sensory hair cells

*Crista* Hair-filled sensory receptor

*Vestibule* Structure with two chambers, each containing a sensory area called a macula

**VESTIBULAR APPARATUS**

*In the inner ear, the vestibule and the semicircular canals contain sensory areas (maculae and cristae) that contribute to our sense of balance.*

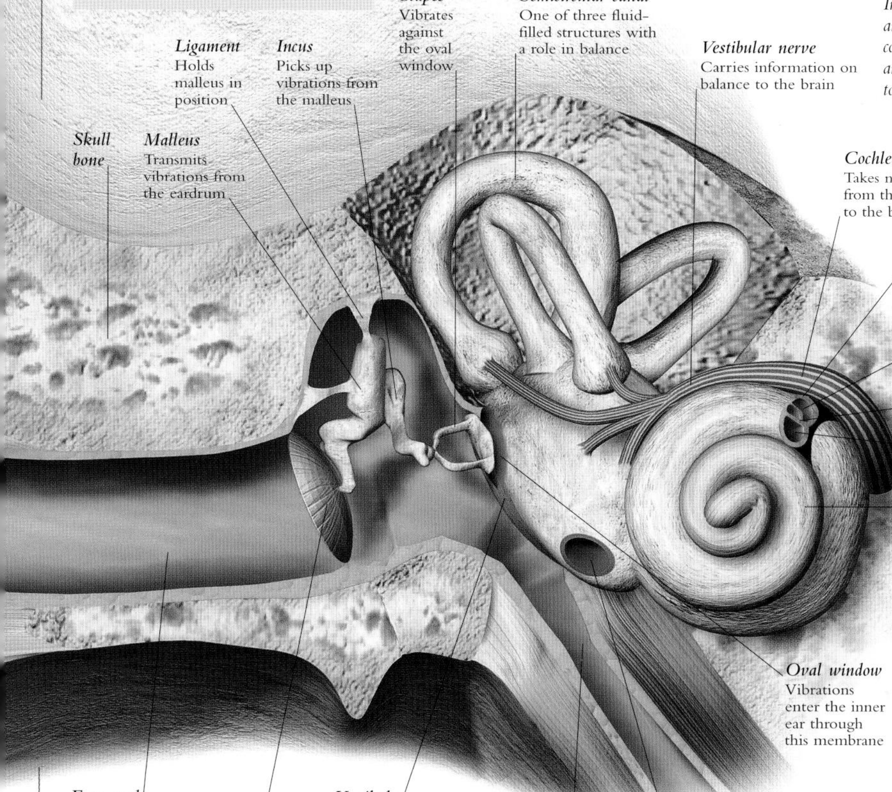

*Stapes* Vibrates against the oval window

*Semicircular canal* One of three fluid-filled structures with a role in balance

*Vestibular nerve* Carries information on balance to the brain

*Ligament* Holds malleus in position

*Incus* Picks up vibrations from the malleus

*Skull bone*

*Malleus* Transmits vibrations from the eardrum

*Cochlear nerve* Takes nerve impulses from the inner ear to the brain

*Cut edge of cochlea*

*Vestibular canal*

*Cochlear duct*

*Tympanic canal*

*Cochlea*

*Oval window* Vibrations enter the inner ear through this membrane

*Ear canal* Conducts sound waves to the eardrum

*Eardrum* Membrane that vibrates in response to sound waves

*Vestibule* Fluid-filled cavity that detects head position

*Eustachian tube* Connects the middle ear to the back of the nose and controls air pressure in the ear

*Round window* Vibrations leave the cochlea through this membrane

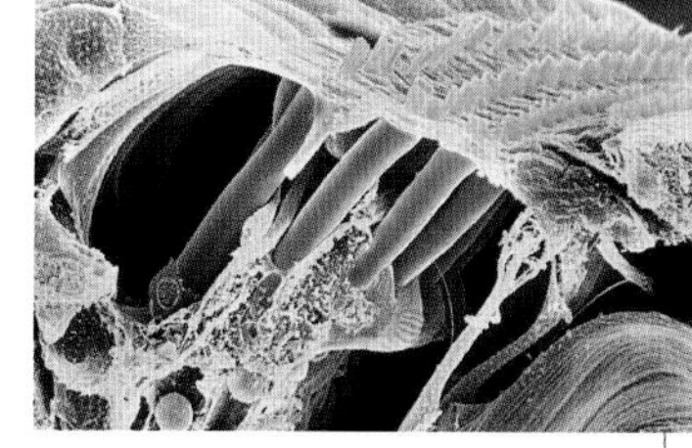

**ORGAN OF CORTI**

*This receptor for hearing contains hair cells (pink) with sensory hairs (yellow) that convert vibrations into nerve impulses.*

# DETECTING HEAD MOVEMENTS

The structures of the inner ear are able to detect several different types of head movement. Linear acceleration (for example, when we travel by car or use a lift) and changes in the position of the head in relation to gravity are detected by hair cells in the two maculae within the vestibule. Rotational movement (when the head is turned in any direction) is detected by the sensory receptors, cristae, in at least one of the fluid-filled semicircular canals. When the head turns, the fluid in the ear structures bends the sensory hair cells, stimulating them to produce nerve impulses.

**LINEAR MOVEMENT**

*When the head is upright, the sensory hairs on the hair cells, which are embedded in a gelatinous membrane, remain in an upright position. If the head is tilted, the pull of gravity causes the membrane to move. The movement pulls on the hair cells, so that they send out nerve signals.*

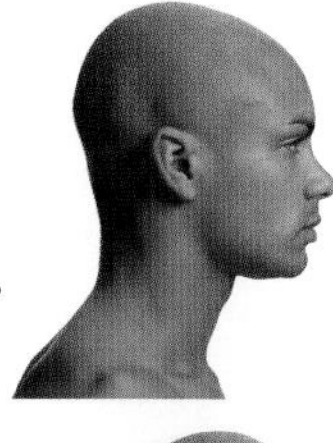

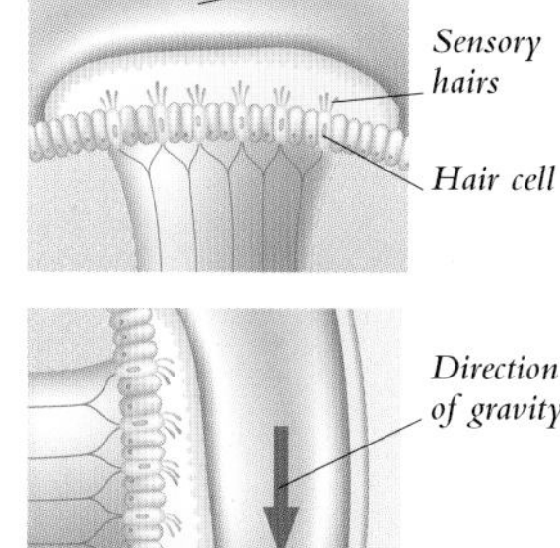


**ROTATIONAL MOVEMENT**

*The hair cells in each crista are embedded in a gelatinous cone-shaped mass, the cupula. When the head is still, the fluid in the canals does not move and the cupula is upright. If the head turns, the fluid moves and displaces the cupula, bending the sensory hairs, which produce nerve impulses.*

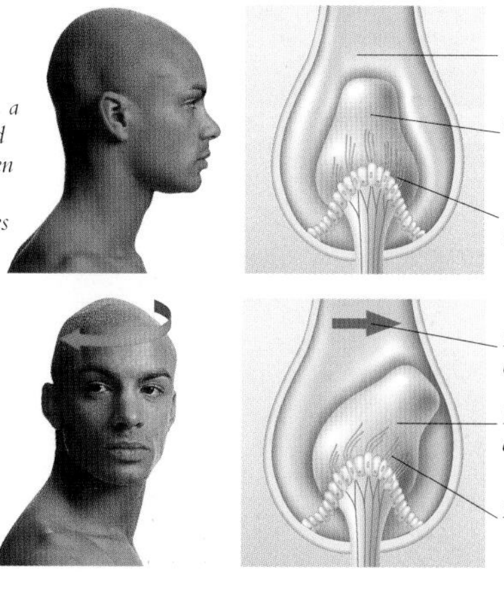

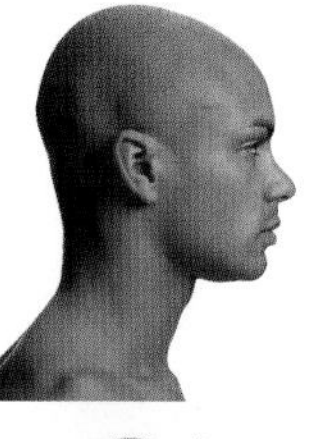

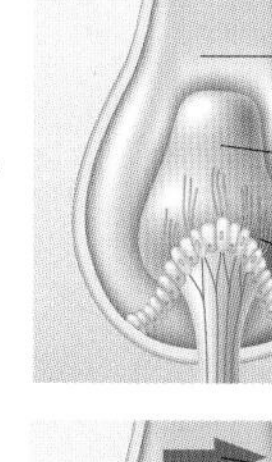

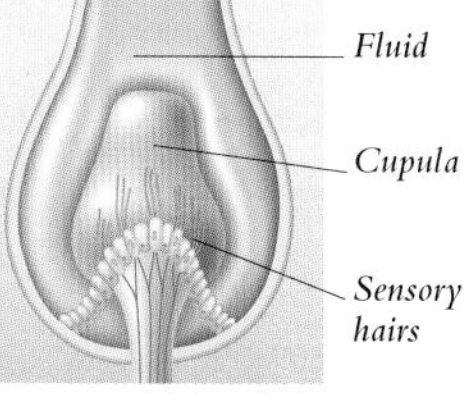


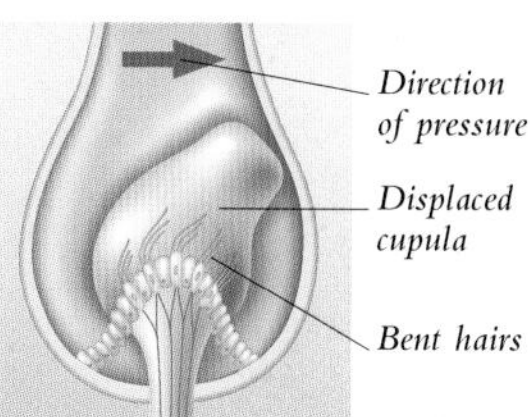

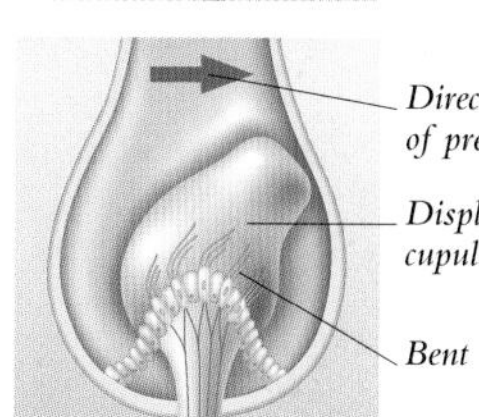

# TOUCH, SMELL, AND TASTE

Unlike any of the other senses, the sense of touch works all over the body. Touch tells us about our environment and whether we are in contact with something harmful. We need touch, and the associated ability to feel pain, as a survival mechanism. Equally important to our safety and enjoyment of life are the closely related senses of smell and taste. Both these senses function as chemical detection systems, picking up and reacting to molecules of odour and flavour. Smell works well on its own, but taste is relatively inefficient without the partnership of smell.

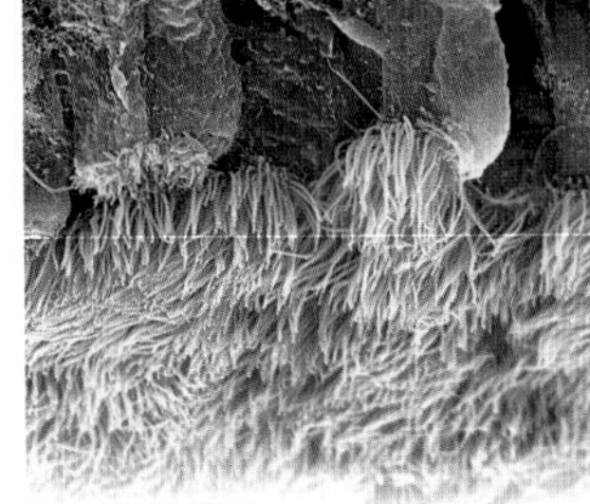

**CLEAR NOSE**

*These tiny hairs (cilia) in the nasal lining sweep the nose clear of mucus or particles. If the nose is blocked, odour molecules cannot reach the smell receptors in the nasal cavity.*

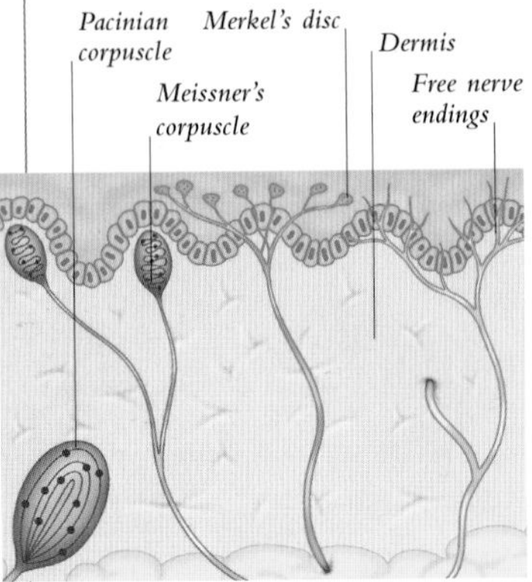


**TOUCH RECEPTORS**

*The skin's layers are full of touch receptors. Some receptors are contained in a capsule while others are free nerve endings.*

## TOUCH

We sense touch through our skin, which contains millions of sensory receptors throughout and below the dermis (second skin layer). These receptors can detect pressure, heat, and cold, and pass on nerve messages about pain. There are four main types of receptor. The most common are free nerve endings that branch out through the dermis like twigs and detect varying degrees of touch. The other types of receptor all have a structure at the nerve end that is formed from coiled nerve fibres or groups of cells. These receptors are all named after the people who first described them: Merkel's discs, Meissner's corpuscles, and Pacinian corpuscles. Merkel's discs lie just under the skin's surface and respond to continuous light touch and pressure. Meissner's corpuscles sit between the epidermis (top skin layer) and the dermis, mainly in hairless areas such as the fingertips, and respond to light touch. Pacinian corpuscles lie over the fatty layer deep under the skin, and detect vibration and deep pressure.

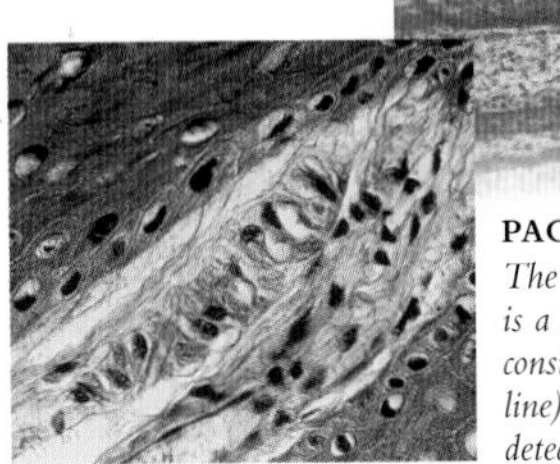

**MEISSNER'S CORPUSCLE**

*This type of touch receptor (shown here as a green structure) is common in the skin of the fingertips and the soles of the feet. Meissner's corpuscles respond to light touch.*

**PACINIAN CORPUSCLE**

*The oval structure seen above is a Pacinian corpuscle, which consists of a nerve ending (dark line) enclosed in membranes; it detects pressure and vibration.*

Fact

### The cause of pain

Pain is a complicated sense that occurs when specialized sensory nerve endings in the skin (nociceptors) are activated. If tissues are damaged, they release chemicals (such as histamine and prostaglandins) that stimulate the nociceptors to send pain signals to the brain. Some nociceptors react only to a severe stimulus, such as a cut; others are sensitive to firm pressure, such as pinching. The extent to which we feel pain is highly subjective.

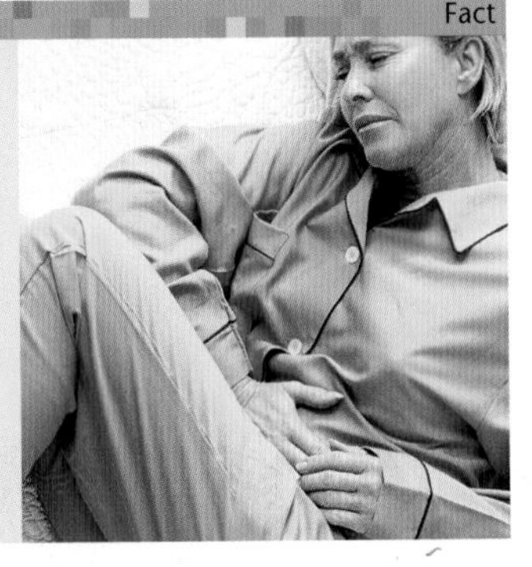

## SMELL

Substances that smell release odour molecules into the air. When we breathe in odour molecules they come into contact with two regions of fatty tissue at the top of each nostril, behind the bridge of the nose. These areas, called the olfactory regions, contain about 5 million nerve cells. The cells carry cilia (tiny hairs) that project into the mucous layer of the nose and are stimulated by specific odour molecules. When triggered by a smell, the cells send electrical signals to a site at the base of the brain, the olfactory bulb. The bulb in turn sends signals to areas deeper in the brain. One of these areas, the limbic system, is responsible for emotion. This association is why even fleeting smells can trigger powerful memories, and emotions.

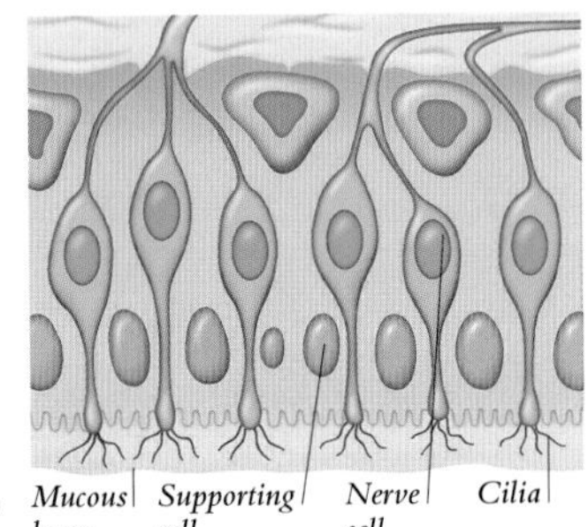


**OLFACTORY RECEPTORS**

*When odour molecules enter the nose they stimulate the cilia (tiny hairs) of nerve cells, causing signals to be sent to the brain.*

## TASTE

The organs of taste are the taste buds, of which there are around 10,000 in the mouth. Taste buds are located mainly on the surface of the tongue, but a few are scattered on the palate, throat, and tonsils. Each bud is a cluster of about 50 taste-detecting nerve cells contained within a pore. When food or drink is put in the mouth, it is dissolved by saliva, and chemicals are released that stimulate minute hairs on the taste cells. The cells send impulses along nerve fibres to the brain. Taste buds in the tongue are grouped in terms of taste types. We taste sweet things with the tip of the tongue, which is why we lick foods such as ice cream. Salty foods are tasted mainly at the front of the tongue. Sour taste buds sit at the edges, and taste buds that detect bitter tastes are at the back.

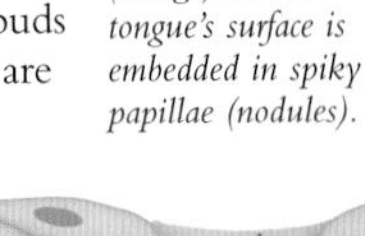

**TASTE BUD**

*A taste bud (orange) on the tongue's surface is embedded in spiky papillae (nodules).*

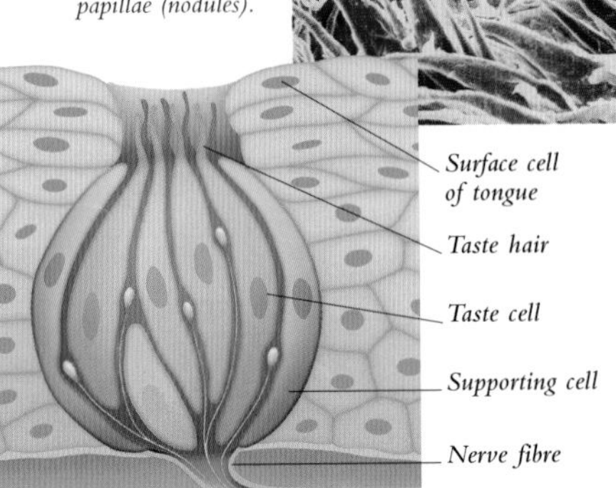


**SENSING FLAVOURS**

*Chemicals released during the breakdown of food and drink in the mouth enter pores in the tongue and reach tiny hairs on the taste buds. These hairs respond to flavours by generating nerve impulses, which send messages to the taste area of the brain.*

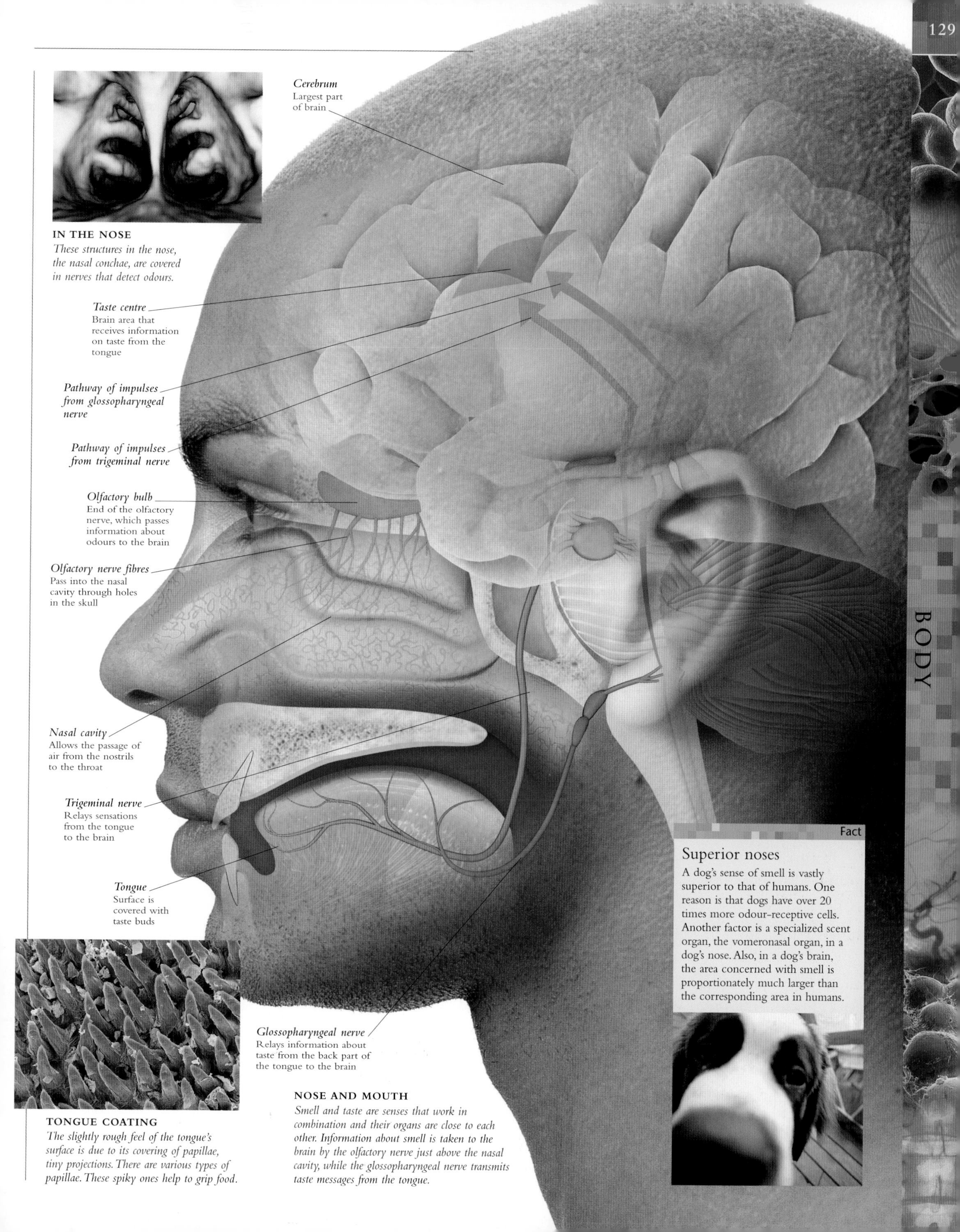

*Cerebrum*
Largest part of brain
*Taste centre*
Brain area that receives information on taste from the tongue
*Pathway of impulses from glossopharyngeal nerve*
*Pathway of impulses from trigeminal nerve*
*Olfactory bulb*
End of the olfactory nerve, which passes information about odours to the brain
*Olfactory nerve fibres*
Pass into the nasal cavity through holes in the skull
*Nasal cavity*
Allows the passage of air from the nostrils to the throat
*Trigeminal nerve*
Relays sensations from the tongue to the brain
*Tongue*
Surface is covered with taste buds
*Glossopharyngeal nerve*
Relays information about taste from the back part of the tongue to the brain

**IN THE NOSE**

*These structures in the nose, the nasal conchae, are covered in nerves that detect odours.*

**TONGUE COATING**

*The slightly rough feel of the tongue's surface is due to its covering of papillae, tiny projections. There are various types of papillae. These spiky ones help to grip food.*

**NOSE AND MOUTH**

*Smell and taste are senses that work in combination and their organs are close to each other. Information about smell is taken to the brain by the olfactory nerve just above the nasal cavity, while the glossopharyngeal nerve transmits taste messages from the tongue.*

Fact

## Superior noses

A dog's sense of smell is vastly superior to that of humans. One reason is that dogs have over 20 times more odour-receptive cells. Another factor is a specialized scent organ, the vomeronasal organ, in a dog's nose. Also, in a dog's brain, the area concerned with smell is proportionately much larger than the corresponding area in humans.

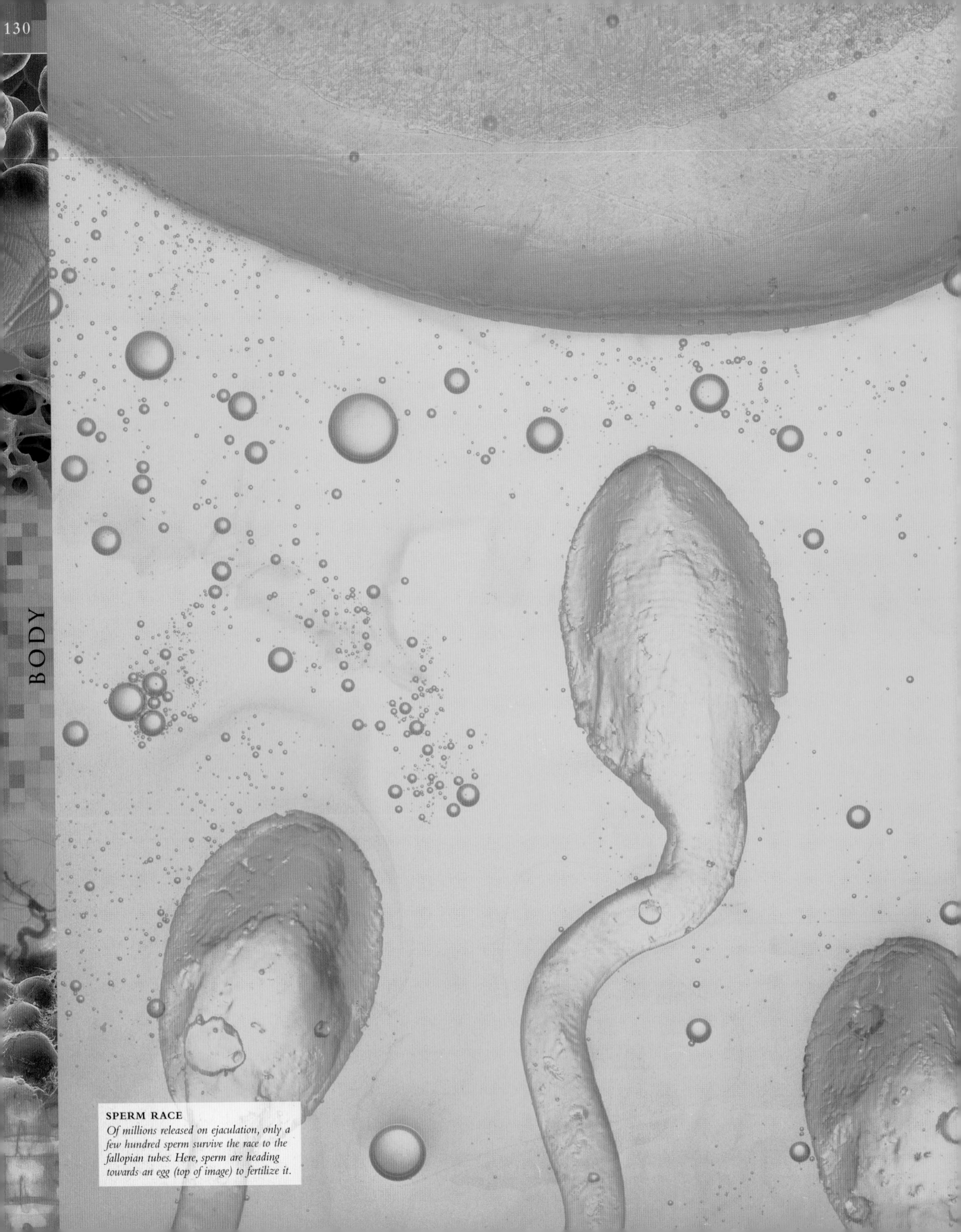

**SPERM RACE**
*Of millions released on ejaculation, only a few hundred sperm survive the race to the fallopian tubes. Here, sperm are heading towards an egg (top of image) to fertilize it.*

# REPRODUCING

Like all animals, humans need to reproduce. By combining their genetic material, males and females from each generation ensure their replacement. Although humans have longer reproductive lives and far more frequent periods of fertility than most species, we reproduce relatively slowly. Pregnancy lasts months, and it is rare for a woman to give birth to more than two children at a time. Humans appear to be among the very few animals that use sex for emotional as well as reproductive reasons.

The male role in reproduction is to make sperm and ensure that they come in contact with a female egg. The female stores the eggs and has the potential to perform a far more complex function. Once a woman has entered her fertile years, there is a possibility every month that one of her eggs could be fertilized, leading to pregnancy. If this occurs, her body adapts itself to provide a home for and nurture the developing baby, and then delivers it to the outside world.

## CREATING NEW LIFE

Fertilization, when a sperm fuses with an egg, is the beginning of life. Before this can happen, a series of crucial events must take place. The woman has to be ovulating, so that a mature egg is ready and in the right place. Sperm must be introduced into the woman's body, so sexual intercourse is necessary; and the sperm need to be healthy because they have a hard race ahead. The timing must be right, because neither egg nor sperm survive for long if fertilization does not occur. Although usually only one egg is released at a time for fertilization, millions of sperm rush to meet it. Most die en route but fertilization needs just one sperm to reach the egg and break through its outer protective layers.

In the last few decades women in the developed world have had the option of controlling their fertility so that they can choose when and whether to become pregnant. Both men and women can also be helped to procreate.

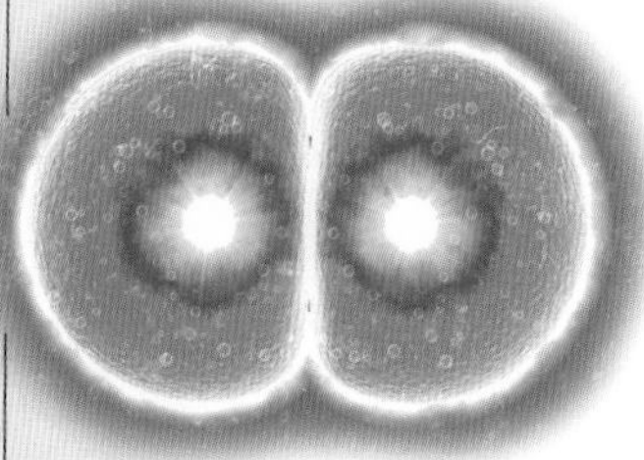

**EARLY LIFE**
*This egg cell has divided into two just hours after fertilization. Continuing to divide, the cells journey towards the womb where they implant and start a pregnancy.*

## INHERITING GENES

Each person has a unique appearance, although he or she is likely to have inherited certain obvious features, such as height, eye colour, or chin shape, from one or both parents. We all have physical traits that are shared with no one else. Our unique set of characteristics, as well as our sex, is acquired at fertilization, when sperm and egg cells join together. These sex cells each carry exactly half the number of chromosomes (structures containing genetic material) needed to make a new human. Once the cells have fused, a new cell is created containing a full set of chromosomes, half of which carry genes from the father and half of which carry genes from the mother. Almost immediately, the new cell starts to copy itself and replicate the genetic information. Whether a person develops into a male or a female depends on the particular type of sex chromosomes carried by the fertilizing sperm.

**SHARED INHERITANCE**
*Identical twins develop when a fertilized egg cell forms into two embryos. Such children share the same genes and are strikingly similar in appearance.*

## PREGNANCY AND BIRTH

Within an incredibly short time after penetration by a sperm, a fertilized egg cell, securely planted in the womb, has turned into an embryo with its own supply of nourishment. This tiny lifeform soon has the beginnings of all the vital organs. Before the end of the first month, the partly formed heart has started to beat. Just weeks later, the embryo, now a fetus, is unmistakably a human being. The changing shape of the mother's body is evidence of the events taking place within her womb. As the 9 months of pregnancy pass, the woman's breasts swell and her womb expands, nudging aside other internal organs to accommodate the growing body and lengthening limbs of the fetus. When the baby is ready to enter the outside world, an unstoppable series of events is set in motion. Labour starts as the womb contracts powerfully to push the baby down the birth canal and out of the mother's body.

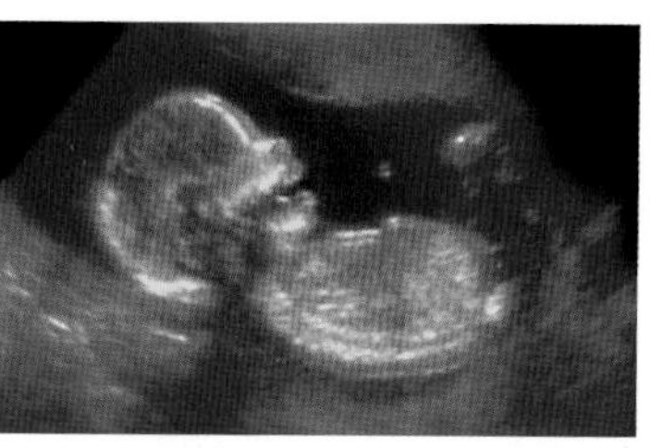

**IN THE WOMB**
*The 20-week-old fetus shown in this scan has all its internal organs in place, and its arms and legs have developed. The spine is seen as a curved yellow line.*

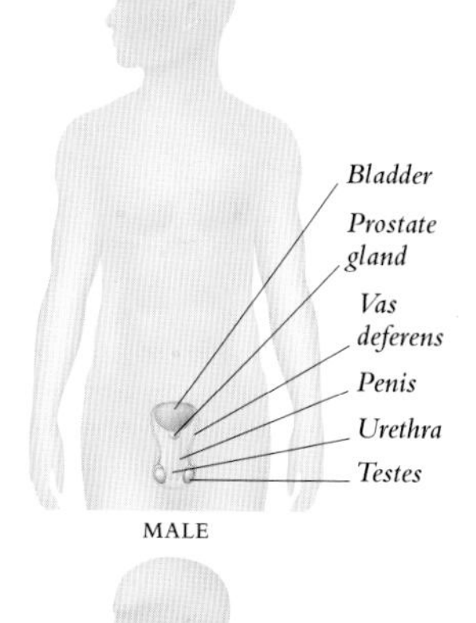


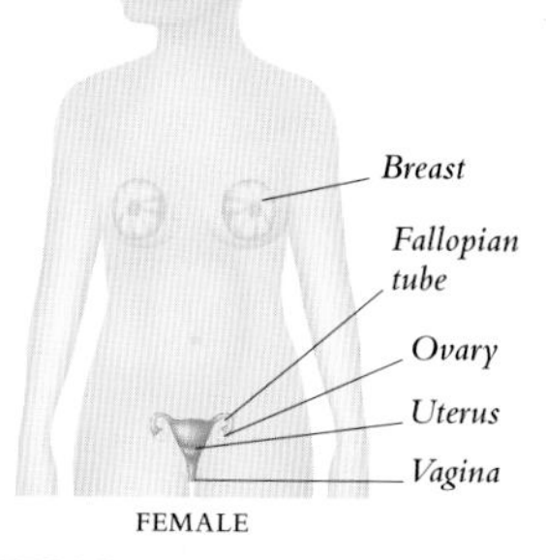


**REPRODUCTIVE SYSTEMS**
*The male sex organs make and transport sperm. The female organs store eggs and provide an environment for a fetus. Breasts produce milk and also have a sexual role.*

Fact

### Genetic fingerprint

The banded pattern shown below, known as a genetic fingerprint, is a computer analysis of a person's genes. It is produced by fragmenting a sample of genetic material, DNA, taken from body tissues and sorting the fragments by using an electrical charge. Because a person's fingerprint pattern is unique, the technique can be used for identification purposes.

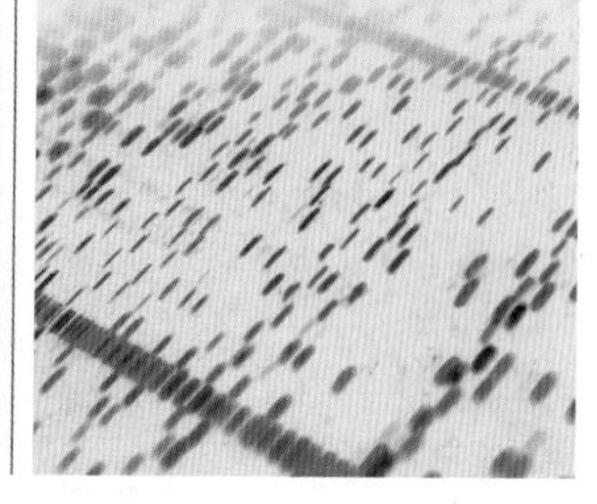

# FEMALE REPRODUCTION

The female reproductive organs enable a woman to have sexual intercourse, to become pregnant, and to give birth. Nearly all the organs are inside the body, rather than being external. Most important are the ovaries, two glands containing eggs (ova) that lie on either side of the uterus (womb). The hollow uterus has thick muscular walls that protect a fetus and contract powerfully to push the baby out during childbirth. The vagina, also muscular and highly elastic, connects the uterus to the external female sex organs: the sensitive clitoris and the folded tissues of the labia.

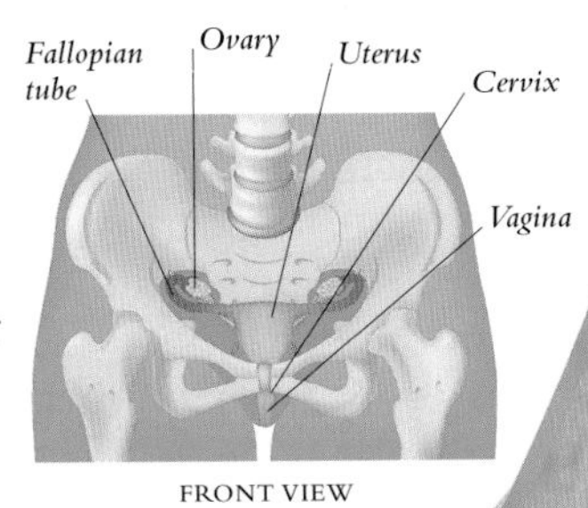

FRONT VIEW

## MAKING EGGS

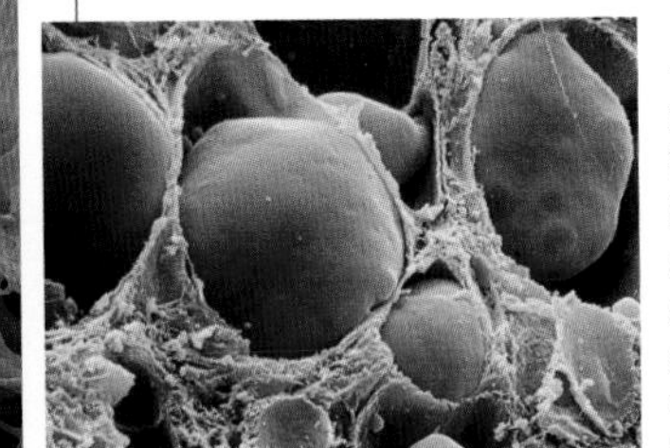
**INSIDE AN OVARY**
*Housed within their follicles in the ovary, several eggs are seen here at various stages of maturity.*

Human eggs, the female sex cells, develop in the ovaries before birth. When a baby girl is born she has around 2 million eggs in her ovaries. Each of the eggs is enclosed in a follicle, a sac containing nutrient cells. During the earlier years of childhood the eggs remain immature, and huge numbers of them degenerate and die. By the time a girl reaches puberty – usually between the ages of 10 and 14 – about 300–400 thousand eggs are left. Of these survivors, only about 400 become fully mature. A rise in the levels of sex hormones at puberty triggers the ripening of eggs and the start of monthly menstruation, a cycle that continues throughout a woman's reproductive life. Every month, several eggs start to develop. Usually, one egg reaches maturity and leaves the ovary while the rest die off. If the mature egg encounters a sperm in a fallopian tube, it may be fertilized there and have the potential to develop into an embryo.

## THE MENSTRUAL CYCLE

In a cycle repeating roughly every 28 days, an egg leaves one of the ovaries while the uterus prepares for possible pregnancy. This monthly cycle is regulated by the complex interaction of several hormones. At the start of the cycle, follicle-stimulating hormone (FSH), released by the pituitary gland in the brain, triggers development of egg follicles in the ovary. The follicles produce oestrogen, which in turn causes a surge of luteinizing hormone (LH). LH stimulates an egg that has reached maturity to burst out of the ovary and into the fallopian tube. The empty follicle, called the corpus luteum, secretes progesterone, which causes the lining of the uterus to thicken, ready to receive a fertilized egg. If fertilization does not happen, the lining is shed as menstrual blood.

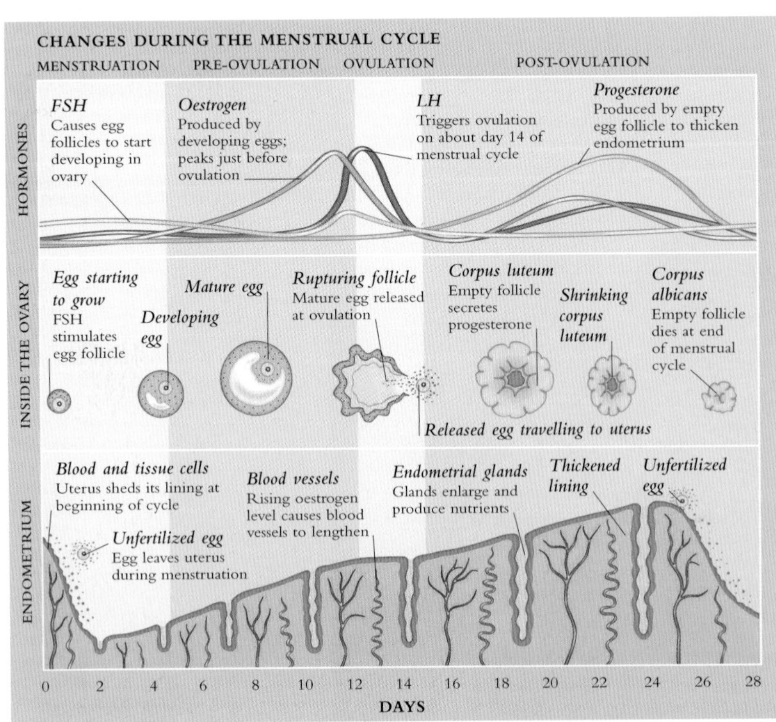


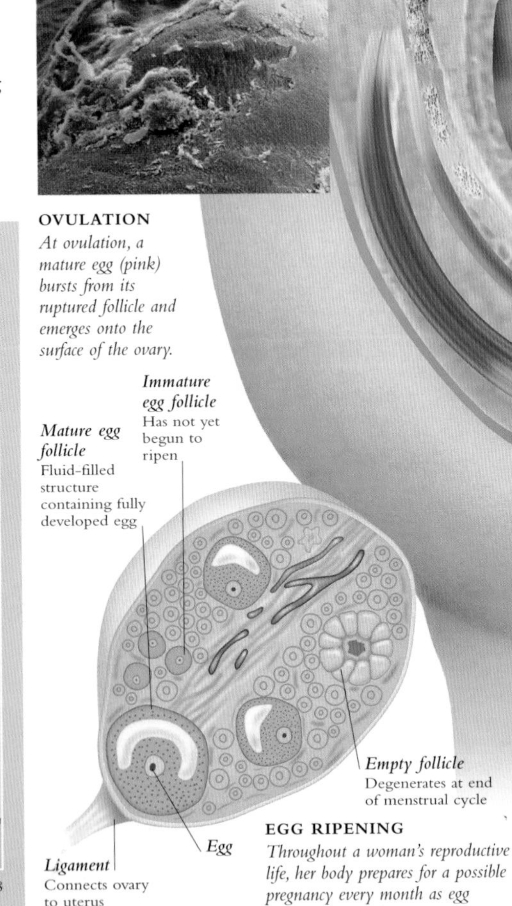

**OVULATION**
*At ovulation, a mature egg (pink) bursts from its ruptured follicle and emerges onto the surface of the ovary.*

**EGG RIPENING**
*Throughout a woman's reproductive life, her body prepares for a possible pregnancy every month as egg follicles ripen in her ovaries.*

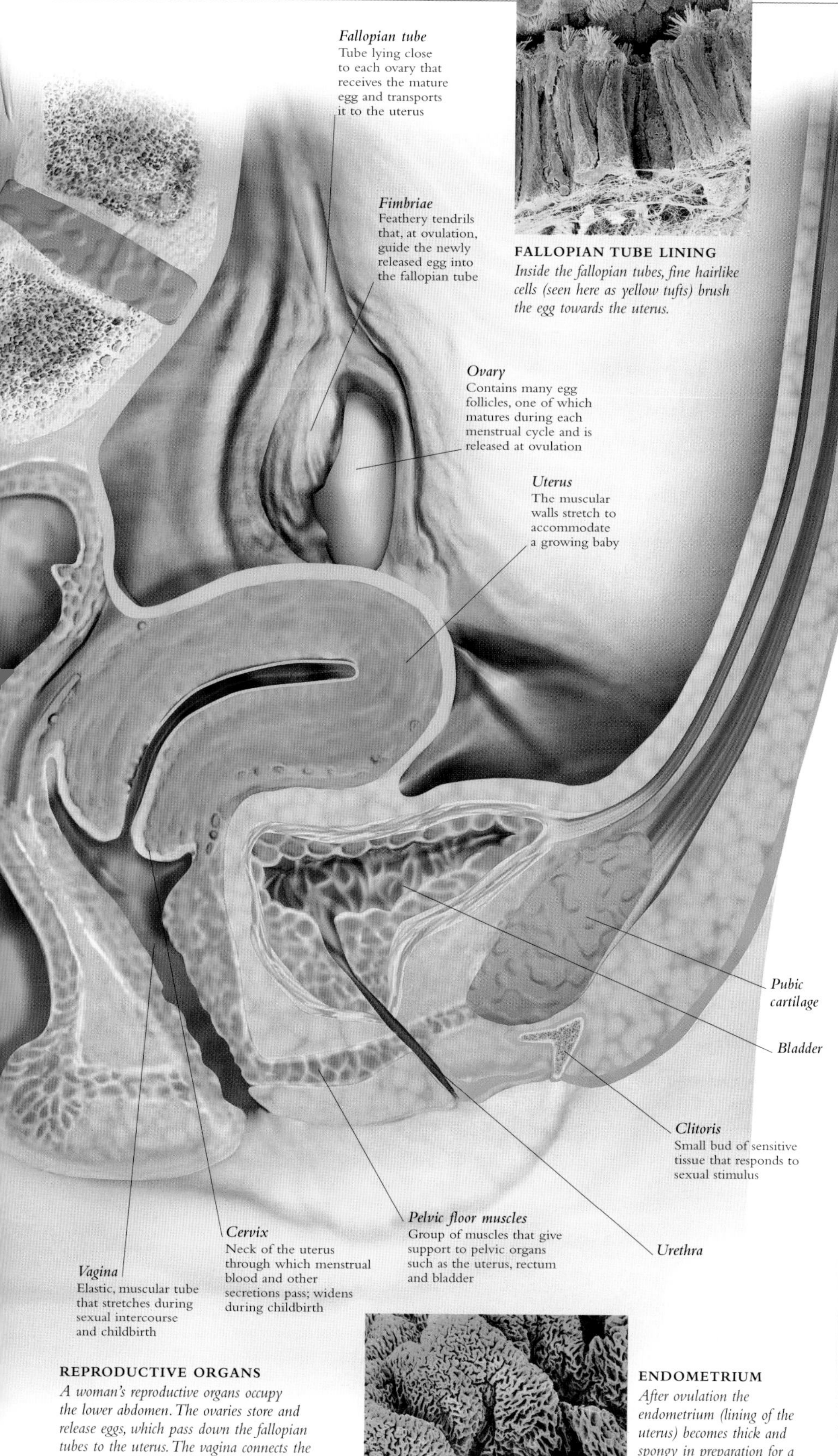
Fallopian tube
Tube lying close to each ovary that receives the mature egg and transports it to the uterus
Fimbriae
Feathery tendrils that, at ovulation, guide the newly released egg into the fallopian tube
Ovary
Contains many egg follicles, one of which matures during each menstrual cycle and is released at ovulation
Uterus
The muscular walls stretch to accommodate a growing baby
Pubic cartilage
Bladder
Clitoris
Small bud of sensitive tissue that responds to sexual stimulus
Urethra
Pelvic floor muscles
Group of muscles that give support to pelvic organs such as the uterus, rectum and bladder
Cervix
Neck of the uterus through which menstrual blood and other secretions pass; widens during childbirth
Vagina
Elastic, muscular tube that stretches during sexual intercourse and childbirth

**FALLOPIAN TUBE LINING**
*Inside the fallopian tubes, fine hairlike cells (seen here as yellow tufts) brush the egg towards the uterus.*

**REPRODUCTIVE ORGANS**
*A woman's reproductive organs occupy the lower abdomen. The ovaries store and release eggs, which pass down the fallopian tubes to the uterus. The vagina connects the uterus to the outside of the body. Folds of skin, the labia, protect the opening of the vagina and the sensitive clitoris.*

**ENDOMETRIUM**
*After ovulation the endometrium (lining of the uterus) becomes thick and spongy in preparation for a fertilized egg. If fertilization does not take place, the lining is shed.*

Issue

## Hormone replacement therapy (HRT)

Use of HRT to reduce menopausal symptoms and prevent osteoporosis (loss of bone tissue) may be linked with a slight increased risk of certain serious diseases. These risks, mostly associated with long-term use, are thought to include breast cancer and deep-vein thrombosis. However, in the opinion of many experts, use of HRT is justified by its benefits, provided treatment takes account of a woman's individual health history.

HRT TABLETS

# MENOPAUSE

With age, the ovaries become less responsive to the hormones released by the pituitary gland. Fewer egg follicles are able to develop in each menstrual cycle and dwindling amounts of oestrogen are produced by the ovaries. The result is that ovulation and menstruation become increasingly unpredictable until the point at which the ovaries stop functioning completely and a woman menstruates for the last time. This is the menopause, which usually occurs between the ages of 45 and 55. The hormonal fluctuations that happen before, during, and after the menopause cause a range of symptoms, including hot flushes, night sweats, vaginal dryness, and mood swings. Women are at greater risk of osteoporosis (loss of bone tissue) and heart disease after the menopause.

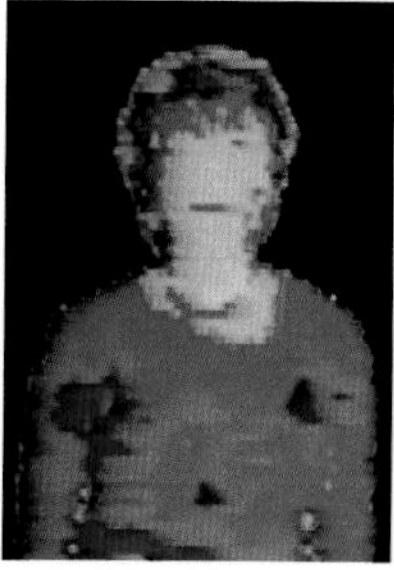
NORMAL

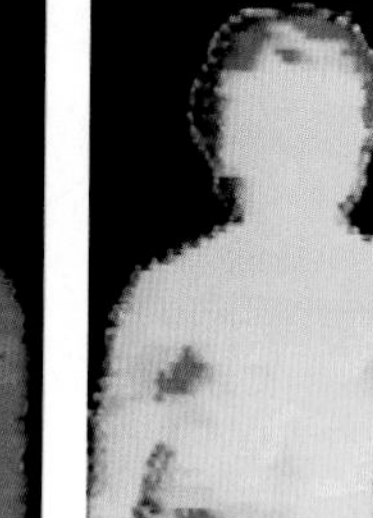
HOT FLUSH

**HOT FLUSHES**
*A surge of heat, felt most strongly in the upper body and head, is a common menopausal symptom. In these thermal images, increased skin temperature is represented by the white and yellow areas.*

# MALE REPRODUCTION

The main male reproductive organs are sited outside the body. A pair of sperm-producing glands, the testes, hangs suspended in a sac, the scrotum, underneath the pelvis. Through a series of tubes, the sperm – male sex cells – are conveyed to the other visible organ, the penis. When a man ejaculates during sexual intercourse, sperm in their millions are forced down the urethra, the tube that passes from the bladder to the end of the penis. The sperm enter the female's body and may combine with an egg. Sperm production begins at puberty and usually continues into old age.

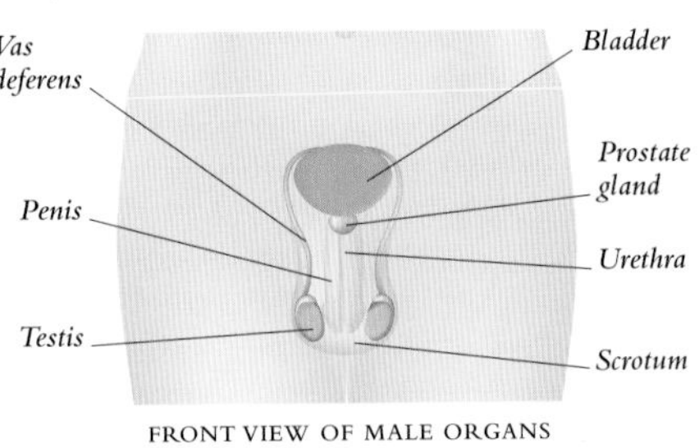


FRONT VIEW OF MALE ORGANS

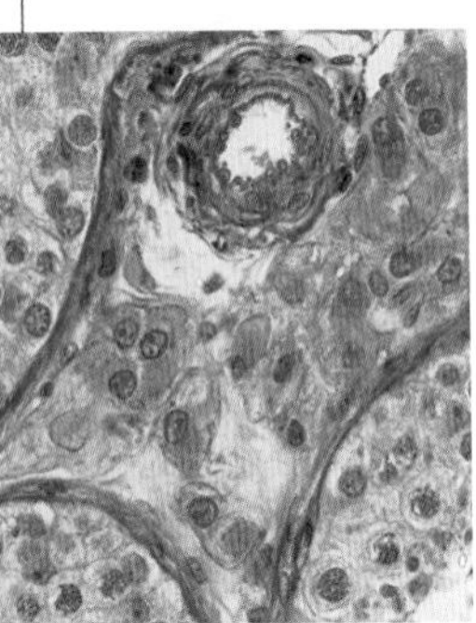

**TESTIS CELLS**
*In this microscopic view of tissue from a testis, the blue outline encloses large hormone-producing cells.*

## MALE GROWTH

As a male fetus develops in the womb, the immature testes form within the lower part of the abdomen and gradually move downwards. Most baby boys are born with their testes outside the body, although in some children one testis (or rarely both) fails to move down. No sperm are made until a boy reaches puberty between the ages of 12 and 15. The trigger for sperm production is a sharp rise in levels of the male sex hormone testosterone, which is secreted in the testes. The increase is stimulated by the production of other hormones in the brain. Testosterone causes physical changes, such as enlargement of the genitals, growth of body and facial hair, and deepening of the voice.

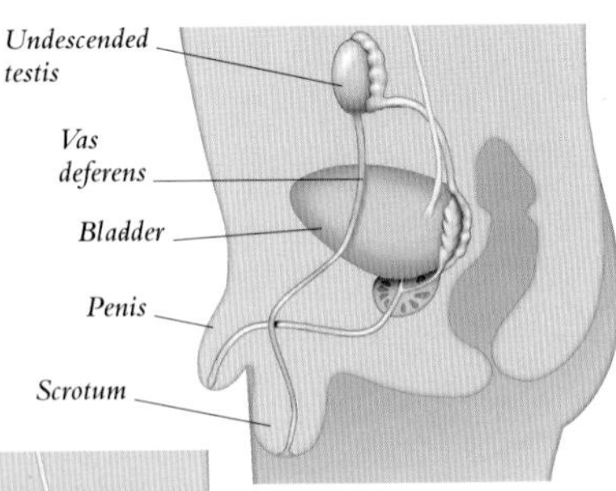


**DEVELOPING TESTES**
*Until about 2 months before birth, the testes are still within the abdomen. They then descend and move through a passage in the groin to the outside of the body. If they do not descend, the testes may require surgery in order to correct their position.*

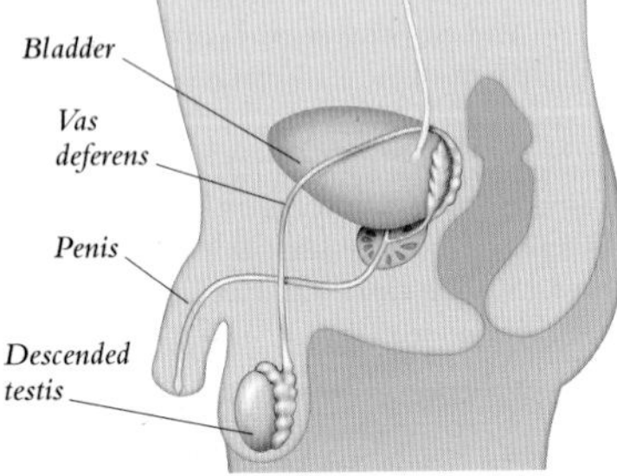


**DESCENDED TESTES**
*When the testes descend, each carries down blood vessels, nerves, and a vas deferens (sperm-carrying duct) to form the spermatic cord. The testes' position outside the body keeps them cool enough for sperm production.*

## TESTES AND SCROTUM

If the testes remained in the abdomen, they would be too warm to produce sperm, which require cooler conditions. Being suspended outside the body in the scrotum keeps the temperature of the testes at about 34°C (93.2°F), slightly below body temperature, which is optimum for sperm production. Sweat glands in the scrotal wall, together with the thin skin of the scrotum, help to maintain this temperature. In warm weather, muscles in the scrotal wall relax, allowing the scrotum to hang away from the body, keeping the testes at the right temperature. In cold weather, the scrotum draws closer to the body. Within each testis are about 1,000 tiny coiled tubes, the seminiferous tubules, where sperm are made. Behind each testis, a coiled duct, the epididymis, stores sperm and takes them to the vas deferens, a tube leading to the ejaculatory duct.

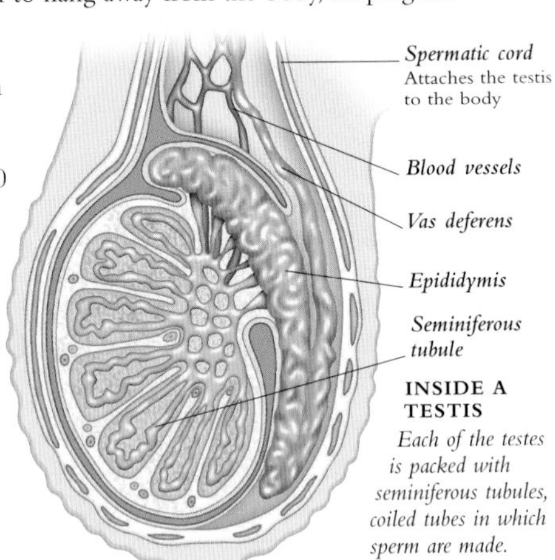


**INSIDE A TESTIS**
*Each of the testes is packed with seminiferous tubules, coiled tubes in which sperm are made.*

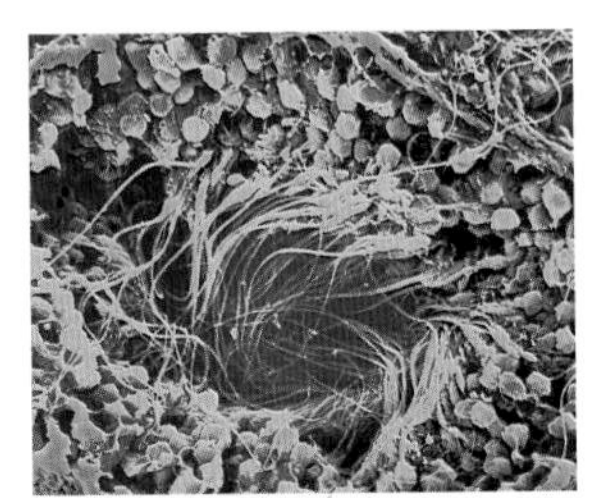

**DEVELOPING SPERM**
*These sperm are maturing inside a seminiferous tubule. The tails of the sperm are seen towards the centre.*

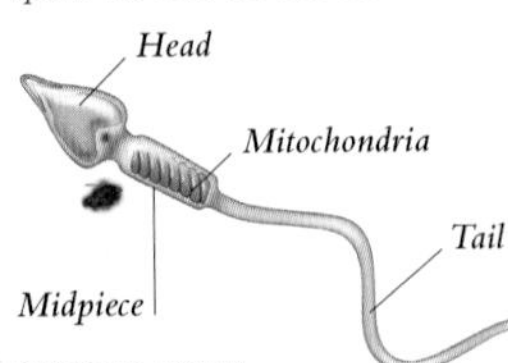


**SPERM CELL**
*The sperm's head contains DNA (genetic material), the midpiece contains energy-producing structures (mitochondria), and the tail propels the sperm in a corkscrew motion.*

## MAKING SPERM

From puberty, a man can produce as many as 120 million sperm a day. Sperm begin their lives as immature cells, spermatogonia, which replicate repeatedly. About half of these reach the next stage of development, to become spermatocytes, which in turn divide. Mature sperm cells produced by the spermatocytes each contain 23 chromosomes, half the genetic material needed to create a human being. Producing a sperm cell takes around 75 days. New sperm move from the seminiferous tubules through a series of tubes to the epididymis, where they are stored. From here, the sperm move to the vas deferens, where they may remain for several months. To form semen, sperm are mixed with fluids from the prostate gland and a pair of sacs called the seminal vesicles. Sperm not ejaculated will eventually die.

History

### Homunculi

Sperm cells were discovered by scientists in the late 17th century. The general belief in scientific circles was that sperm contained miniscule but fully-formed human beings, which were referred to as homunculi. After being ejaculated into a woman's body, a homunculus was thought to be incubated and to grow into a baby. It was not until the 1840s that scientists understood how a sperm cell penetrates the female egg and that after fertilization an embryo develops.

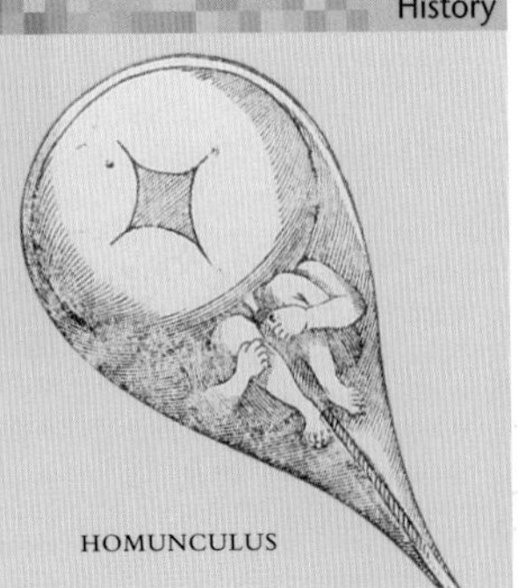

HOMUNCULUS

BODY

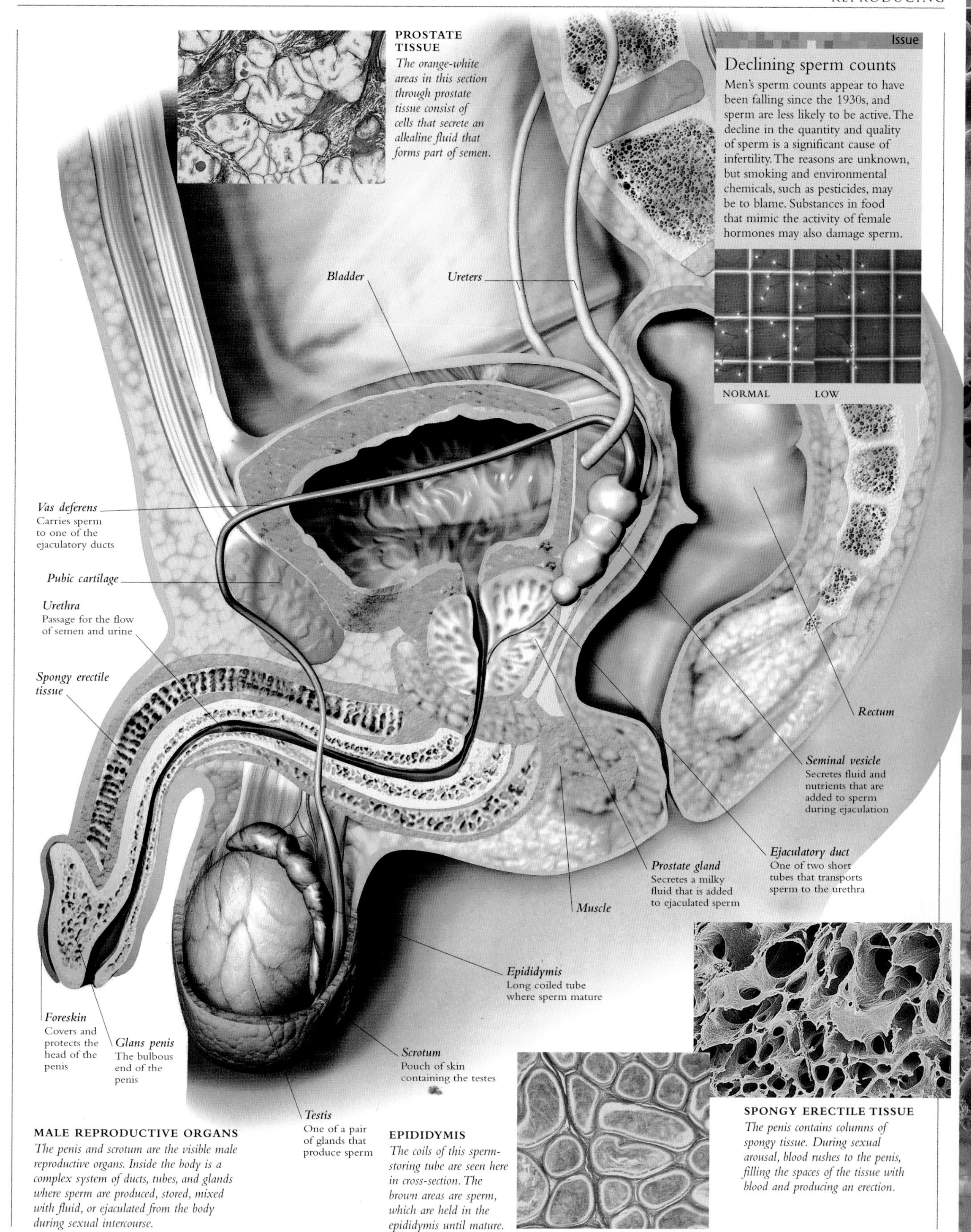
Bladder
Ureters
Vas deferens
Carries sperm to one of the ejaculatory ducts
Pubic cartilage
Urethra
Passage for the flow of semen and urine
Spongy erectile tissue
Rectum
Seminal vesicle
Secretes fluid and nutrients that are added to sperm during ejaculation
Ejaculatory duct
One of two short tubes that transports sperm to the urethra
Prostate gland
Secretes a milky fluid that is added to ejaculated sperm
Muscle
Epididymis
Long coiled tube where sperm mature
Foreskin
Covers and protects the head of the penis
Glans penis
The bulbous end of the penis
Scrotum
Pouch of skin containing the testes
Testis
One of a pair of glands that produce sperm
NORMAL
LOW

**PROSTATE TISSUE**
*The orange-white areas in this section through prostate tissue consist of cells that secrete an alkaline fluid that forms part of semen.*

Issue

## Declining sperm counts

Men's sperm counts appear to have been falling since the 1930s, and sperm are less likely to be active. The decline in the quantity and quality of sperm is a significant cause of infertility. The reasons are unknown, but smoking and environmental chemicals, such as pesticides, may be to blame. Substances in food that mimic the activity of female hormones may also damage sperm.

**MALE REPRODUCTIVE ORGANS**
*The penis and scrotum are the visible male reproductive organs. Inside the body is a complex system of ducts, tubes, and glands where sperm are produced, stored, mixed with fluid, or ejaculated from the body during sexual intercourse.*

**EPIDIDYMIS**
*The coils of this sperm-storing tube are seen here in cross-section. The brown areas are sperm, which are held in the epididymis until mature.*

**SPONGY ERECTILE TISSUE**
*The penis contains columns of spongy tissue. During sexual arousal, blood rushes to the penis, filling the spaces of the tissue with blood and producing an erection.*

# SEX AND CONCEPTION

The biological urge to reproduce is one of the strongest basic drives. Producing a new generation to ensure the continuation of the species fulfils one of the primary functions of the human body. However, sexual intercourse is not only the means by which males and females reproduce; it is also an outlet for emotional expression and is usually a profoundly pleasurable activity. A sexual relationship can therefore play an important role in maintaining the bond between partners and instilling a sense of personal well-being.

*Sperm tail* Propels the sperm with whip-like action

*Mitochondria* Provide energy needed for locomotion

*Sperm head* Contains the sperm's genetic material

*Granulosa cells* Cluster on the outer surface to nourish and protect the egg

*Outer membrane* Thickens once it has been penetrated

*Successful sperm* Can be seen losing its tail as its head enters the egg

## SEXUAL INTERCOURSE

Sexual intercourse involves a combination of physical and emotional responses. When sexual arousal occurs, blood flow to the genital area increases, resulting in swelling of the male penis and the female clitoris. The engorged penis becomes erect, enabling it to enter the vagina. The vagina moistens and widens in readiness for penetration. Once penetration has occurred, the vagina tightens around the penis, which moves backwards and forwards within the vagina, creating friction. At orgasm, which may occur in both partners, simultaneously or at different times, the male ejaculates, releasing sperm, and the female's vaginal walls contract rhythmically.

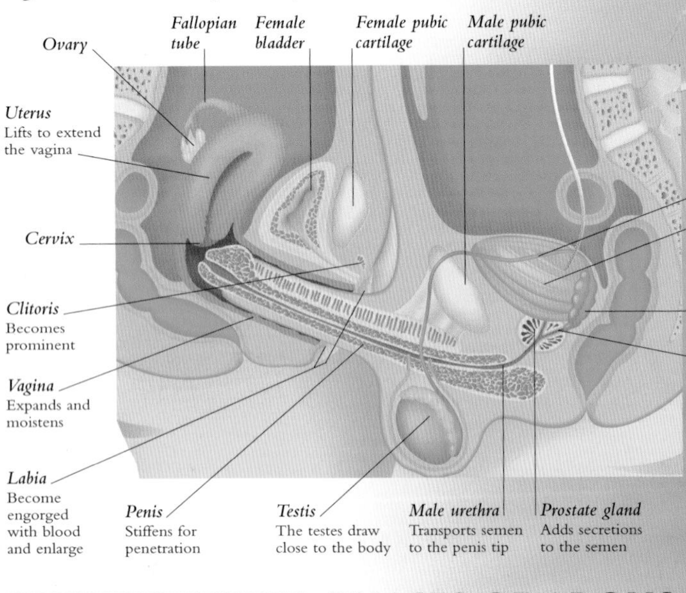


**PENETRATION**
*In order for penetration to take place, the penis stiffens and becomes erect to enter the vagina, which enlarges and moistens to accept it.*



**Fact**

### Testosterone levels

Men have 10 times as much of the hormone testosterone in their bodies as women. The hormone (shown in crystallized form below) contributes to the development of skin, organs, muscles, and bones, but it is most closely linked to sex drive. It is partly for this reason that men may have a higher sex drive than women.

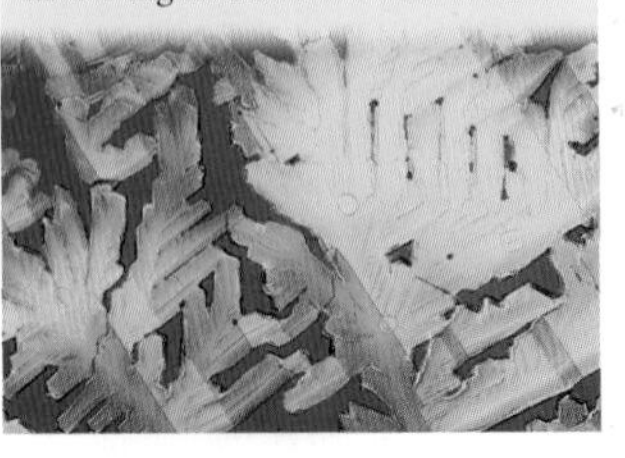

## PHASES OF AROUSAL

Sexual thoughts, the sight of a partner's body, and foreplay all contribute to what is called sexual arousal. Excitement causes breathing to quicken, heart rate to increase, and blood pressure to rise. In males, the penis becomes hard and erect. In females, the labia and clitoris swell, the vagina lengthens and becomes lubricated, and the breasts enlarge slightly and become highly sensitive to touch. The time it takes for arousal to occur varies for men and women, and also varies between individuals. In general, however, there are certain phases of arousal, during which sexual excitement mounts, on the way to orgasm, at which point excitement reaches its physical and psychological peak.

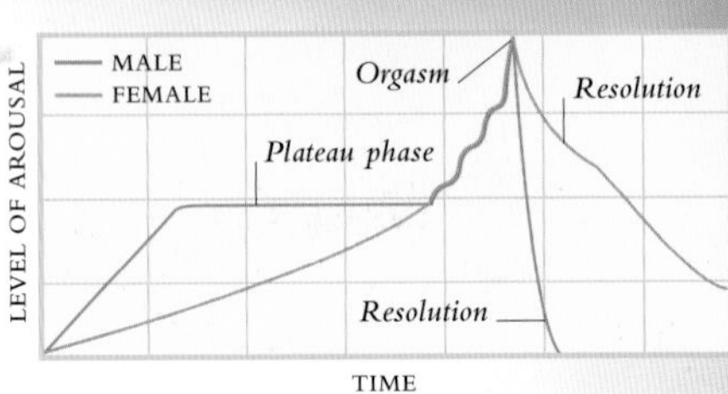


**MALE AND FEMALE AROUSAL**
*In men, sexual excitement rises rapidly to reach a plateau; in women, arousal is a more gradual process. In both sexes, arousal peaks at the point of orgasm, which may or may not occur simultaneously, and tails off rapidly into the resolution phase.*

# THE SPERM'S JOURNEY

After ejaculation, about 250 million sperm begin the journey up the female genital tract. Because many die or lose their way, however, only about 200 sperm succeed in reaching the egg. The journey to the fallopian tubes can take up to an hour, during which the sperm propel themselves along by whipping their tails back and forth, assisted by contractions in the walls of the vagina and the uterus. Once a sperm has reached the egg, it attempts to penetrate the egg's outer membrane, releasing enzymes that help to break down the membrane, to aid its entry.

**TRAVELLING SPERM**
*A human sperm must swim through thick, viscous fluid on its way to the egg; the equivalent of a human swimming through a pool of extra-thick syrup.*

*Nucleus of egg*

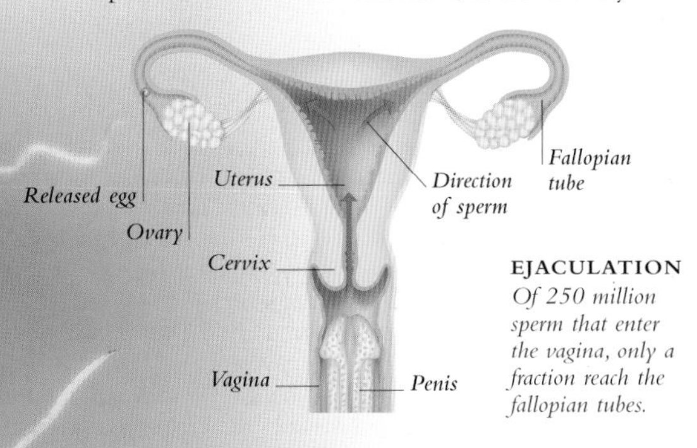


**EJACULATION**
*Of 250 million sperm that enter the vagina, only a fraction reach the fallopian tubes.*

# THE EGG'S JOURNEY

Pregnancy begins when an egg is fertilized by a sperm. Fertilization takes place in the outer part of a fallopian tube, when a sperm meets and fuses with an egg. Once it has fused with a sperm, the egg is called a zygote. Within 2 days of fertilization, the zygote starts its journey towards the uterus, propelled by the muscular action of the fallopian tube. As it travels, the cells of the zygote gradually divide to form a cluster of cells called a morula. After 4–5 days, the cluster develops a cavity and is called a blastocyst. This arrives in the uterus and embeds in the uterine lining. From this moment, the pregnancy is established, and the blastocyst is called an embryo.

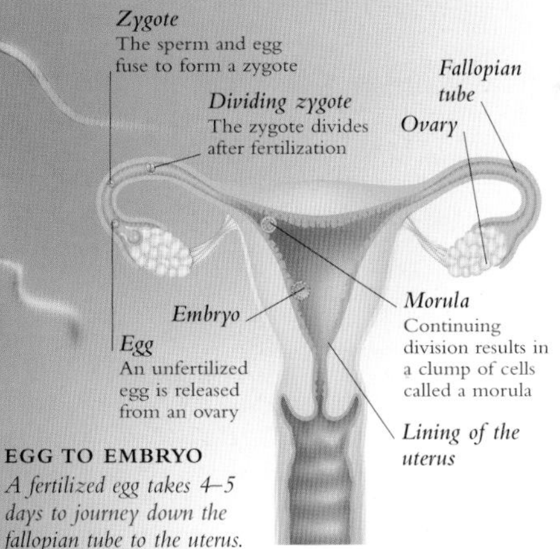


**EGG TO EMBRYO**
*A fertilized egg takes 4–5 days to journey down the fallopian tube to the uterus.*

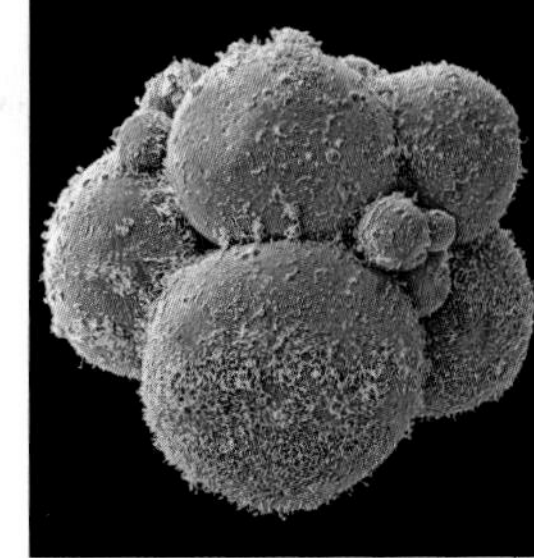

**CELL DIVISION**
*The zygote begins to divide soon after fertilization. As it journeys to the uterus, it continues to divide, forming a cluster of cells called a morula.*

**FERTILIZATION**
*The sperm burrow into the surface of the egg, in an attempt to penetrate the protective layer and enter. Only one sperm can be successful, however, losing its tail as it gains access. Once the egg has been penetrated, the outer surface thickens so that no more sperm can enter.*

Health

## Infertility

Many couples suffer from infertility. In about half of these couples, the cause is female infertility; in about 1 in 3, the cause is male infertility; and in the remainder, the cause is unknown. Here, a contrast dye shows damaged fallopian tubes, one cause of female infertility. One of the tubes is blocked and invisible; the other is narrow and distorted.

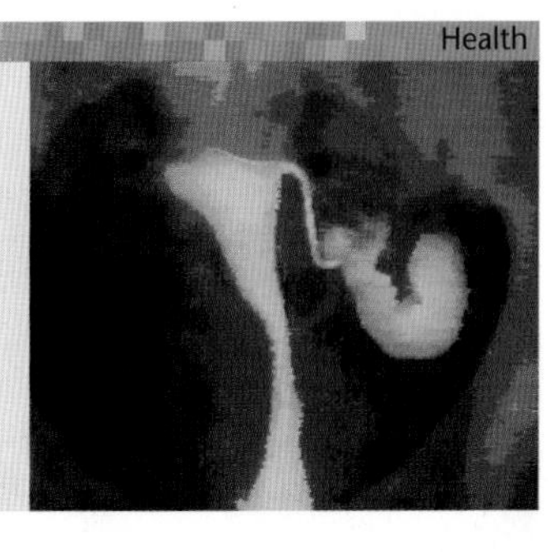

# GENES AND INHERITANCE

Genes control the growth, repair, and functions of each cell in the human body. They are sections of DNA (deoxyribonucleic acid), a long, coiled chemical molecule that is found in the nucleus of cells in the form of structures called chromosomes. Genes are arranged on 22 pairs of matching chromosomes, plus two sex chromosomes. Genes are the means by which characteristics are passed on from parents to children. DNA also provides cells with the instructions needed for building proteins required by the body.

*Y chromosome*
Presence of this chromosome indicates that the person is male. Males also have one X chromosome, which is much larger

## HUMAN GENOME

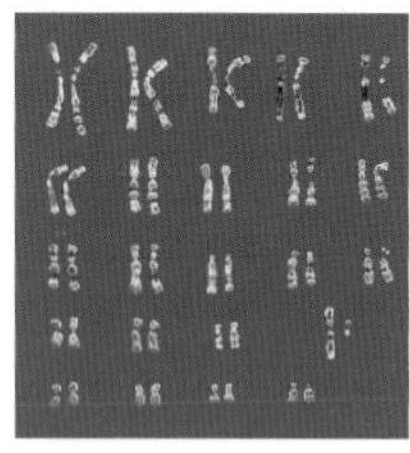

**HUMAN KARYOTYPE**
*Body cells contain 23 pairs of chromosomes. Together they are called a karyotype.*

The human genome is a human being's individual genetic make-up – the sum total of his or her entire complement of genes. An individual's genome is made up of half of his or her mother's genes and half of his or her father's genes. In total, the human genome contains about 30,000 pairs of genes, all of which are arranged on chromosomes, and can be represented in the form of bands of colour, forming rings around the "legs" of the chromosomes (*see right*). Genes in the human genome determine individual characteristics, such as hair colour, height, and size of nose, some aspects of behaviour, and also many disorders.

## MEIOSIS

Every cell in the human body is made by a process called mitosis (*see* p104), apart from sperm cells in males and egg cells in females. These cells are produced by a form of cell division called meiosis. In this process, the amount of DNA (genetic material) in the new cells is halved during two stages of cell division. The process ensures that a complete set of genes is obtained when an egg and sperm fuse. During meiosis, matching pairs of chromosomes exchange genetic material randomly. Each of the resulting sperm or egg cells then has a slightly different mixture of genes from each other. Here, only four chromosomes are shown to illustrate the stages of meiosis.

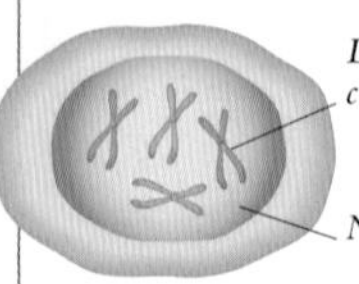


**STAGE ONE**
*In the nucleus of the cell, DNA in the chromosomes is duplicated to form X-shaped double chromosomes.*

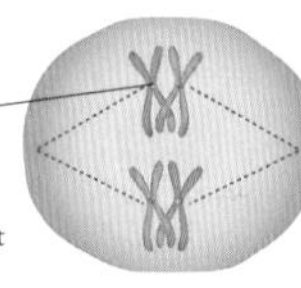


**STAGE TWO**
*The membrane around the cell nucleus disappears. Matching chromosomes touch in random places and may exchange genetic material.*

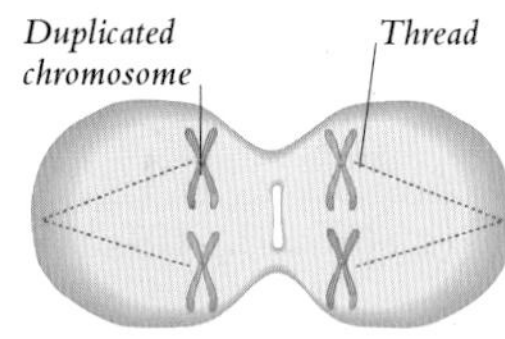


**STAGE THREE**
*Each duplicated chromosome has a mixture of genetic material. Threads pull the pairs of chromosomes apart.*

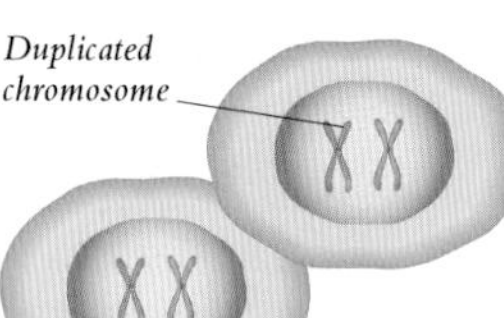


**STAGE FOUR**
*The cell divides to produce two new cells. Each new cell has a full set of 23 duplicated chromosomes from the original cell.*

Thread

Single chromosome
Pulled apart by threads

Chromosome

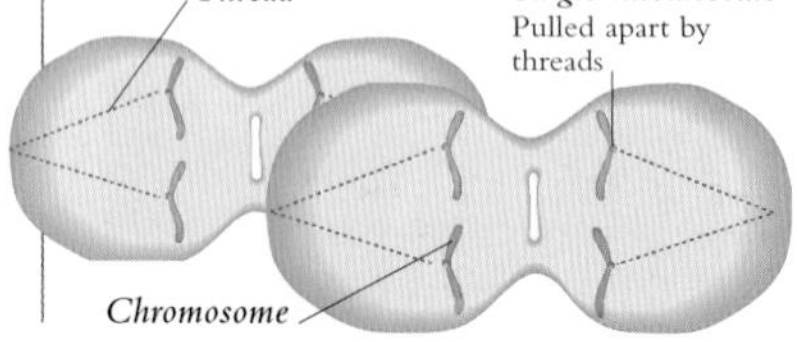

**STAGE FIVE**
*The duplicated chromosomes line up. More threads pull each chromosome apart to form two single chromosomes.*

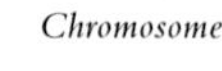


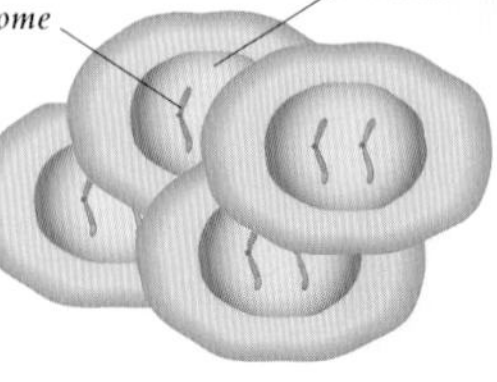


**STAGE SIX**
*The two new cells divide to produce four cells from the original cell, each with half the amount of genetic material.*

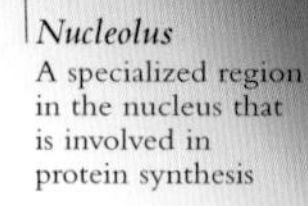

*Nucleolus*
A specialized region in the nucleus that is involved in protein synthesis

**CHROMOSOMES IN CELL**
*The chromosomes in this cell nucleus carry the DNA (genetic material). Among the chromosomes, an X chromosome and a Y chromosome (much smaller) are visible, indicating that this person is male.*

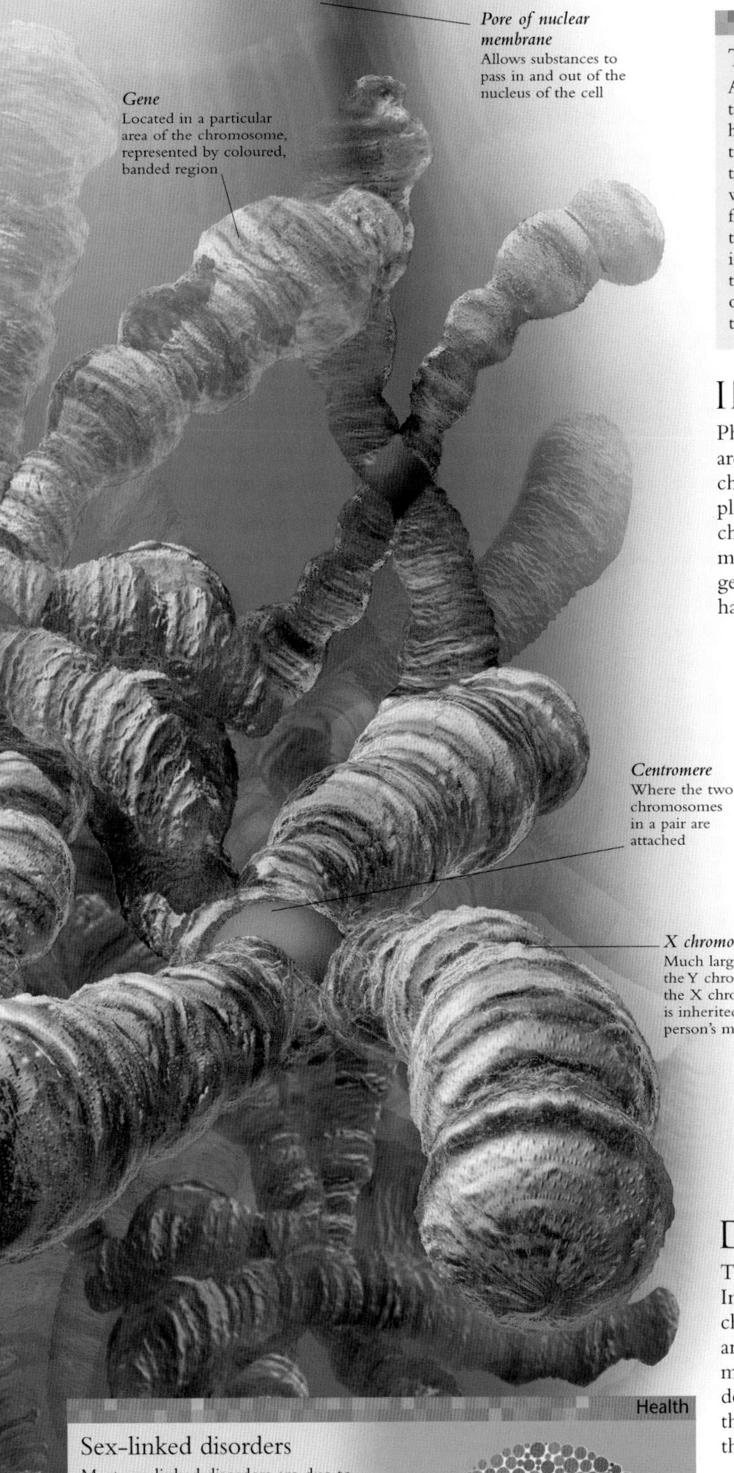


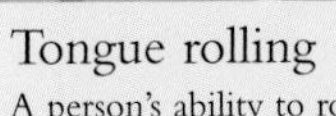

Fact

## Tongue rolling

A person's ability to roll his or her tongue into a "U" shape, as shown here, is considered to be an inherited trait. The gene for tongue rolling is thought to be dominant. A person who inherits the gene (from either father or mother), will be able to roll the tongue; and a person who has not inherited it, will not. Studies show that about 60 per cent of people of European descent have the ability to roll their tongue.

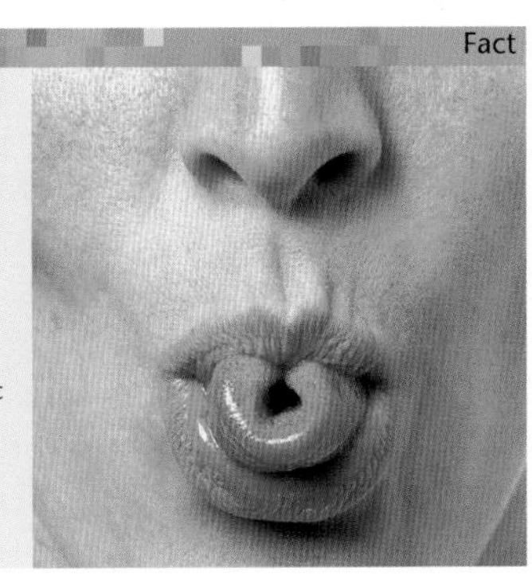

# INHERITED TRAITS

Physical characteristics, many disorders, and some aspects of behaviour are at least partly determined by genes that are passed from parents to children. Genes for each characteristic are always found at the same place on the same chromosome. At fertilization, each of the 23 single chromosomes in an egg cell and a sperm cell matches up into pairs, making a full set of 46 chromosomes that contains two copies of each gene. Half of a child's genes are inherited from his or her mother and half from his or her father.

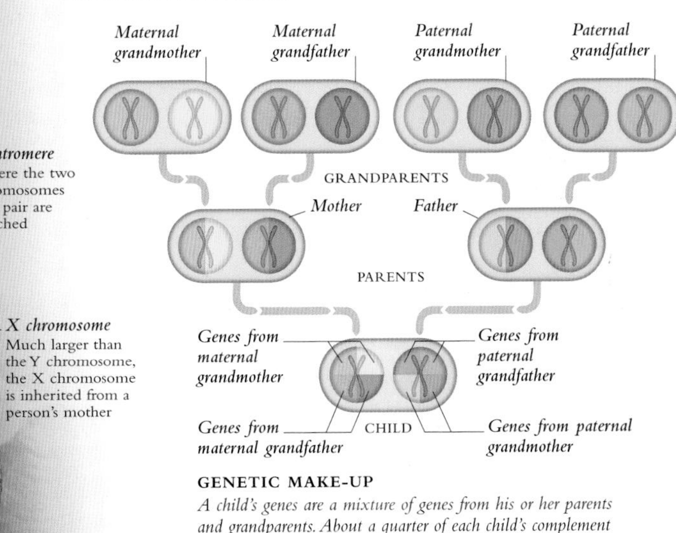


**GENETIC MAKE-UP**

*A child's genes are a mixture of genes from his or her parents and grandparents. About a quarter of each child's complement of genes are inherited from each grandparent.*

# DETERMINING GENDER

There are two sex chromosomes, X and Y, that determine gender. In addition to the other 22 chromosomes, females have two X chromosomes and males have an X and a Y chromosome. All eggs have an X chromosome (since they are produced by females), while sperm may contain either an X or a Y chromosome. The gender of the child depends on whether the sex chromosome in the sperm that fertilizes the egg is X (for a girl) or Y (for a boy). It is a male's sperm, therefore, that determines the gender of his child.

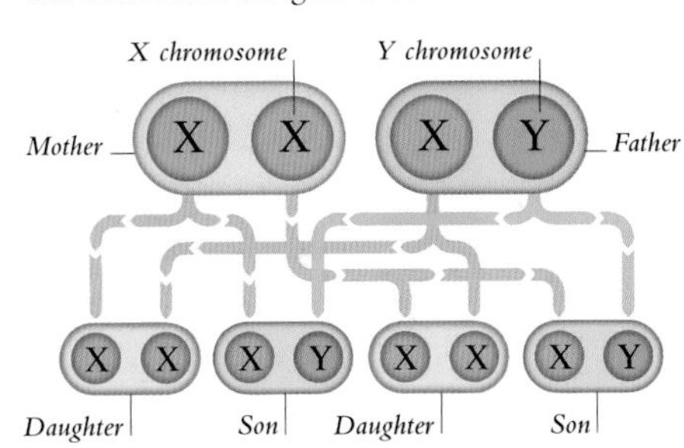


**BOY OR GIRL?**

*Boys have one X chromosome and one Y chromosome, while girls have two X chromosomes in all their cells (in addition to the other 22 pairs of chromosomes). The presence of a Y chromosome therefore indicates masculinity.*

Health

## Sex-linked disorders

Most sex-linked disorders are due to recessive genes on the X chromosome. Because males have only one X chromosome, they will be affected if they inherit the gene. Because females have two X chromosomes, however, they will not have the disorder (but will be carriers). An example of a sex-linked disorder is colour blindness, in which colours are confused. In this test (right), a person unable to distinguish red from green will see a number.

# PREGNANCY

Floating in fluid and cushioned by its mother's body, a new human grows within the seclusion of the uterus (womb). Over a period of about 270 days, a single cell, an egg barely visible to the naked eye, develops into a baby human being, ready to emerge into the world. The most critical period of pregnancy is the embryonic stage, when all of a baby's major organs and body systems are formed. This takes place in the 8 weeks after fertilization, a remarkably short time for such complex events. After this, the fetus begins to look recognizably human and the remaining time in the uterus is spent perfecting the finished product.

**IMPLANTING IN THE UTERUS**
*Fingerlike projections on the surface of the fertilized egg burrow into the lining of the uterus, and pregnancy begins.*

## A HUMAN TAKES SHAPE

In the first 8 weeks of pregnancy, a structure called the embryonic disc forms from the fertilized egg. The disc develops a series of folds that yield head and limb buds, and a stalk that eventually becomes the umbilical cord. A thin sheet of cells in the chest starts to flutter irregularly – this will later become the heart. At 8 weeks, the internal organs are formed – although immature, they are developing rapidly – and the embryo becomes a fetus. By 12 weeks, the fetus measures around 6cm (2.4in) from the head to the buttocks and is large enough to start expanding the uterus. At this stage, the fetus has begun to move its arms and legs, and from now on it becomes increasingly energetic. A woman is usually aware of her pregnancy long before she feels the baby moving, however. Just 3 weeks after fertilization has taken place, her breasts may be heavier and more tender than usual. Then her menstrual period fails to appear and she may start experiencing symptoms such as tiredness, nausea, vomiting, and an aversion to certain foods and smells.

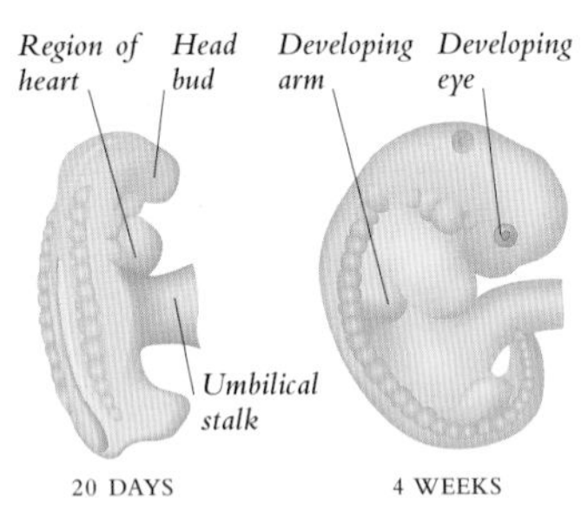


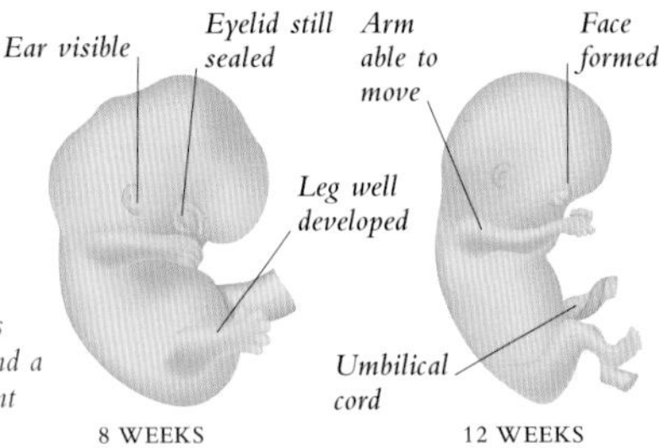


**THE FIRST 3 MONTHS**
*In the first 3 months, a tiny cluster of cells becomes a well-developed fetus with facial features, limbs, and a beating heart. At 12 weeks, the skin is a translucent red and the head is a third of the body length.*

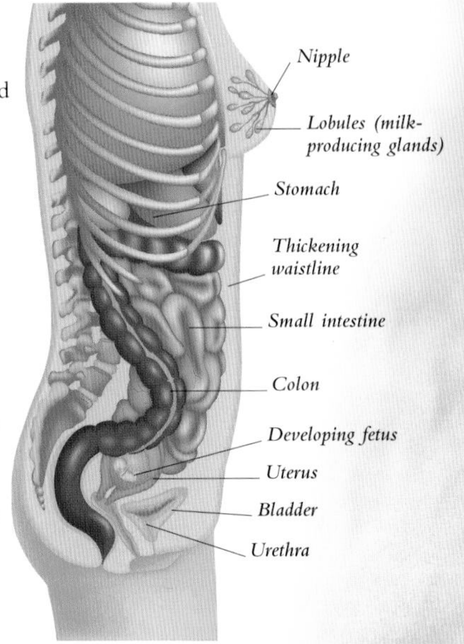


**THE 12TH WEEK**
*The waistline is thickening slightly. The breasts are heavier and feel tender; their surface veins are prominent and the skin surrounding the nipples has darkened.*

Health

### Pregnancy testing

Pregnancy is confirmed by a blood or urine test that detects a hormone, human chorionic gonadotropin (HCG), normally produced only by a developing placenta. Home pregnancy test kits are available, all of which use a sample of urine, although their methods of use vary.

**HOME TESTING**
*The result on the left is negative, the pink line showing only that the test has worked. A second line (right) appears if HCG is detected in a urine sample.*

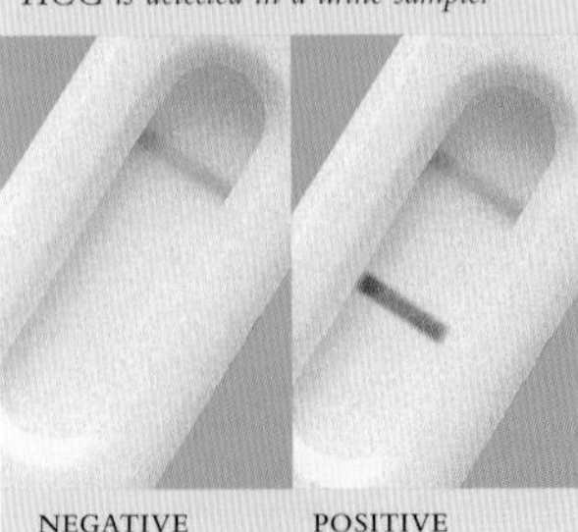

NEGATIVE POSITIVE

## NOURISHING THE FETUS

The growing fetus is nourished by an amazing organ, the placenta. This is a disc of tissue that develops from the cell cluster formed from the fertilized egg (*see* The egg's journey, p137). When the cell cluster attaches itself to the uterus, small fingerlike projections, villi, grow into the uterine lining. The villi branch, and an intricate network of tiny blood vessels grows within them. Meanwhile, the mother's tissues surround each invading villus with a separate system of blood vessels. The fetus is joined to the placenta by the umbilical cord. The maternal and fetal blood never mix in the placenta. They are separated by a thin membrane, called the chorion, which allows oxygen, antibodies, and nutrients to pass from the mother to the baby.

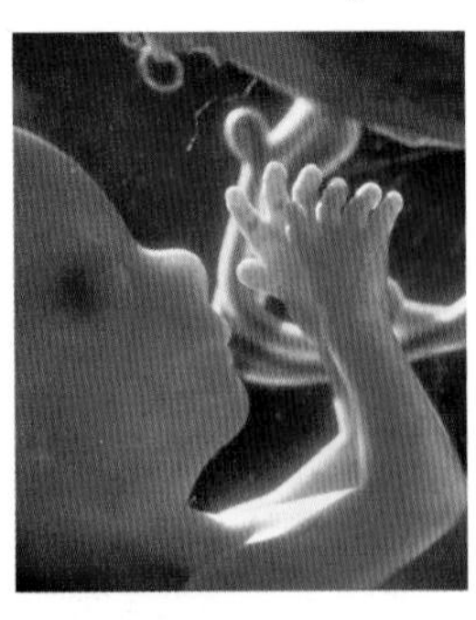

**LIFE SUPPORT**
*All of a fetus's nutritional needs are met by the placenta. The umbilical cord that connects the fetus to the placenta is about 60cm (24in) long.*

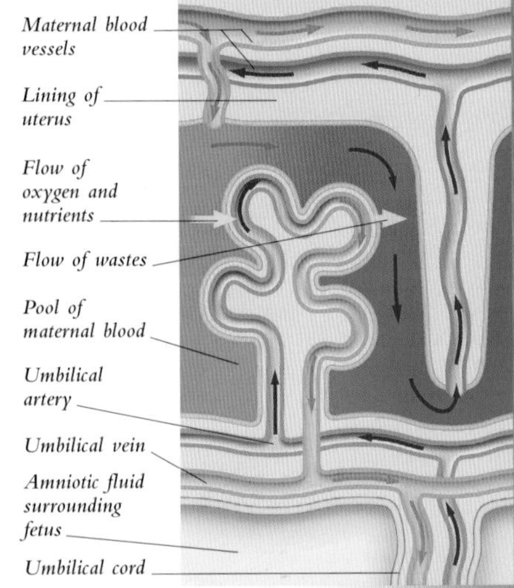


**BLOOD EXCHANGE**
*Inside the placenta, oxygen and nutrients pass from the mother's blood to that of the fetus. Waste products from the fetus's blood pass in the opposite direction.*

BODY

# THE EXPECTANT BODY

Pregnancy places many strains on the body. The uterus, which weighs 50–100g (1.8–3.5oz) before a woman conceives, increases in size to weigh about 1,100g (39oz). The growing fetus puts a huge load on the mother's heart and circulation. To sustain blood flow to the placenta, the mother's blood volume increases by 50 per cent and her pulse speeds up by about 15 beats a minute. To meet the oxygen needs of the fetus, breathing becomes deeper and the rate increases. In late pregnancy, the fetus enlarges during its final growth spurt to fill the pelvis almost completely. This compresses the mother's heart, stomach, bladder, and other organs, and causes her some discomfort. The movements of the fetus can be felt growing stronger and more insistent.

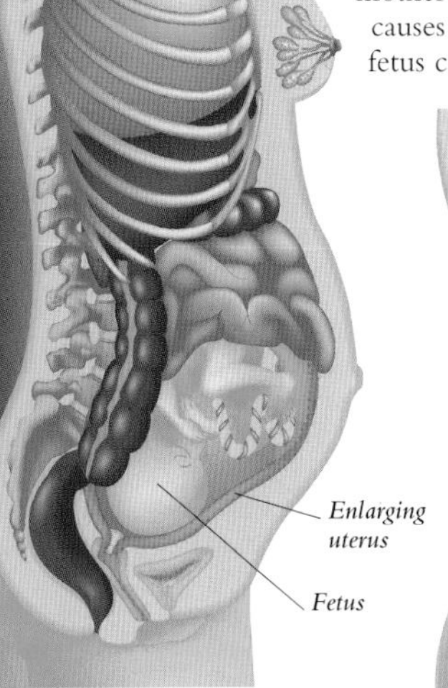


**THE 24TH WEEK**
*The mother's belly steadily swells. There may be a yellowish secretion, colostrum, from the breasts.*

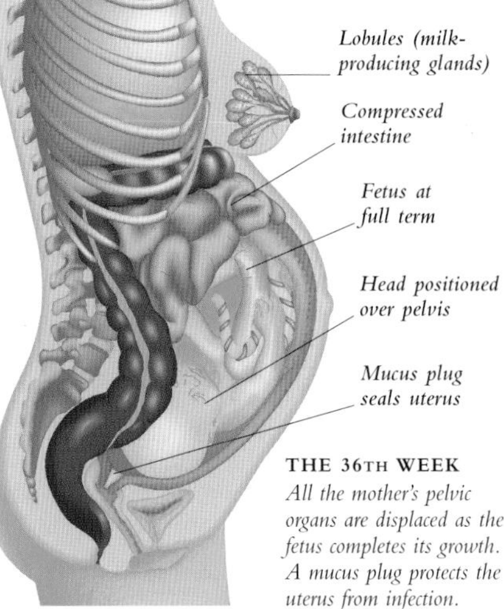


**THE 36TH WEEK**
*All the mother's pelvic organs are displaced as the fetus completes its growth. A mucus plug protects the uterus from infection.*

## BREAST CHANGES

Female human breasts are already relatively large compared to those of other mammals, but they grow and develop even more during pregnancy. Lobules (milk-producing glands), which are inactive in non-pregnant women, increase in number. Alveoli (sacs) inside each lobule expand as milk-producing cells increase. The breasts may start to secrete colostrum, a clear, yellowish fluid that is rich in protein and antibodies. Colostrum continues to be produced after birth and provides a highly nutritious first food for newborn babies. Milk secretion starts some 3 days after a baby is delivered.

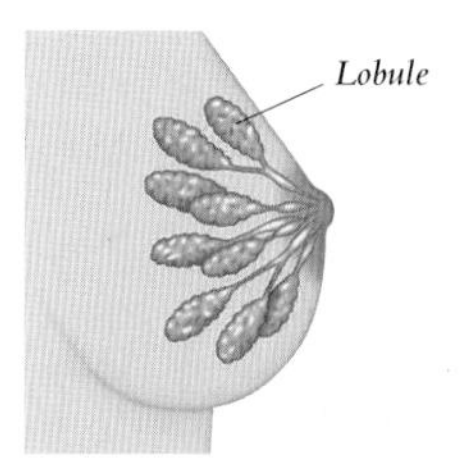


BEFORE PREGNANCY

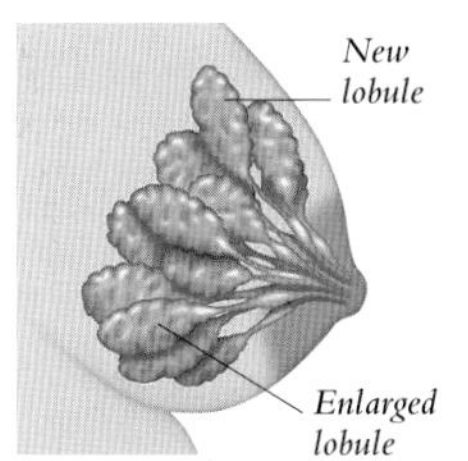


DURING PREGNANCY

**GETTING READY TO PRODUCE MILK**
*Before pregnancy, breast lobules are small and undeveloped. During pregnancy they enlarge and increase in number to form large, grape-like clusters.*

Fact

### Multiple pregnancy

Births of three or more babies are unusual in humans, but one birth in 80 or so is a twin birth. About a third of these are identical twins – when one fertilized egg divides into two. The rest are nonidentical (fraternal). These result from fertilization of two eggs with two sperm. Fraternal twins become more likely as a woman ages because she tends to release more than one egg each month.

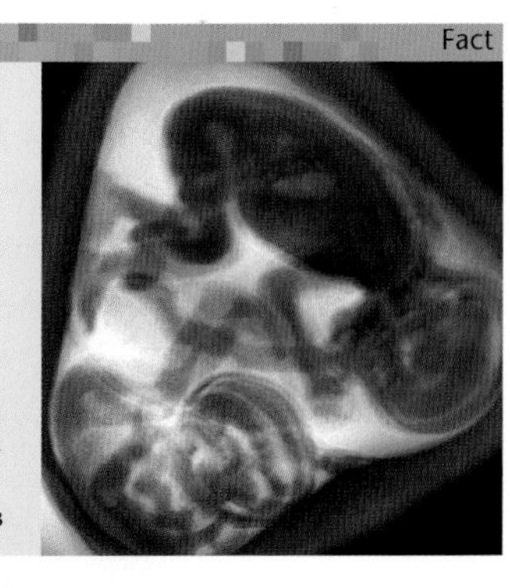

MRI SCAN OF TWINS IN THE WOMB

**FETAL SENSATIONS**
*A baby senses its surroundings while still in the womb. By 16 weeks, the baby can hear, and the sound of the mother's voice, heart, and stomach are never far away. Taste and smell are also well developed before birth.*

# CHILDBIRTH

Very few babies are born on the exact date calculated for their arrival. No-one can predict just when labour will start, but once the process begins, there is no going back. Compared with the birth process in other animals, a human birth is prolonged and difficult because of the large size of the baby's head and the narrow opening of the mother's pelvis. Labour is divided into three stages. During the first stage, the uterus contracts powerfully to pull the cervix upwards and to dilate (widen) it until the opening is large enough for the baby's head to pass through. The second stage is the delivery of the baby; and the third stage lasts from the birth until the delivery of the placenta.

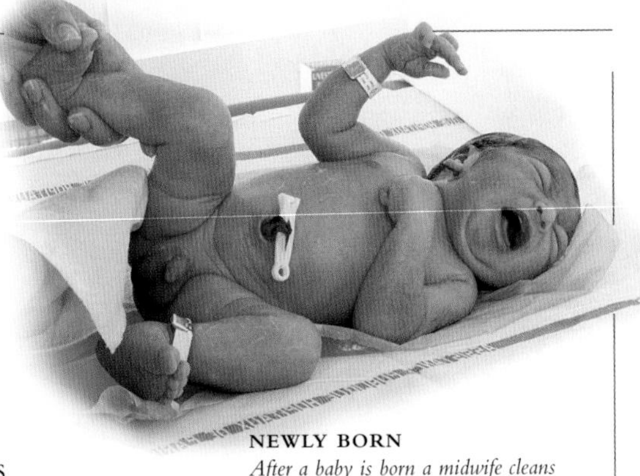

**NEWLY BORN**
*After a baby is born a midwife cleans away blood and mucus, clamps the umbilical cord, attaches identity tags, and evaluates the baby's condition.*

## GETTING READY TO BE BORN

In the weeks before birth, the baby's head descends farther into the pelvis, taking up virtually all the available space. The woman's bump is now lower in her abdomen, and there is greater pressure on her bladder, which means that she needs to urinate frequently. The muscles of the uterus may already have been contracting for a few weeks. These "practice" contractions become stronger and more frequent as labour approaches. The cervix (neck of the uterus) becomes thinner and, as it starts to open, the plug of mucus that seals the uterus is shed. This is known as the "show". At the beginning of labour, 96 per cent of babies are head downwards in the uterus. The rest are mostly in the breech position (buttocks first). A tiny proportion of babies at term lie obliquely or transversely (crosswise) in the uterus.

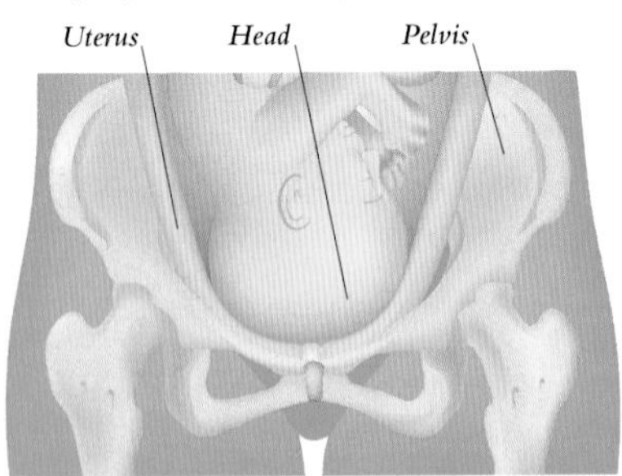
Uterus
Head
Pelvis

**BEFORE ENGAGEMENT**
*Engagement refers to the descent of the baby's head into the upper portion of the woman's bony pelvis prior to birth. In a first pregnancy, a baby's head may engage 2–4 weeks before delivery. In subsequent pregnancies, engagement may not happen until the onset of labour.*

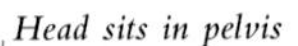
Head sits in pelvis

**AFTER ENGAGEMENT**
*When the baby's head drops into the pelvic cavity, pressure is reduced on the woman's stomach and diaphragm, making eating and breathing more comfortable. However, pressure on her bladder increases, causing more frequent urination. Engagement can be detected either by an internal examination or by feeling for the baby's head externally.*

Health

### Fetal monitoring

A baby's heart is monitored throughout labour; changes in heart rate may indicate that the baby is in distress. The heart may be monitored intermittently through the mother's abdomen with a hand-held ultrasound monitor or continuously with an abdominal belt attached to a machine (cardiotocographic monitoring). If a baby is thought to be in distress, an electrode is attached to its scalp to monitor heart rhythm precisely.

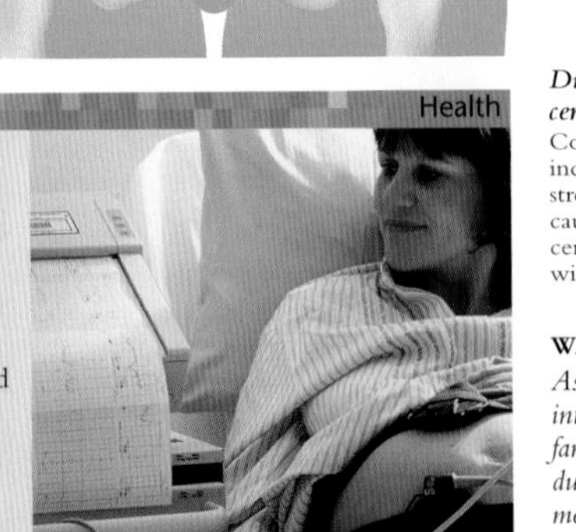

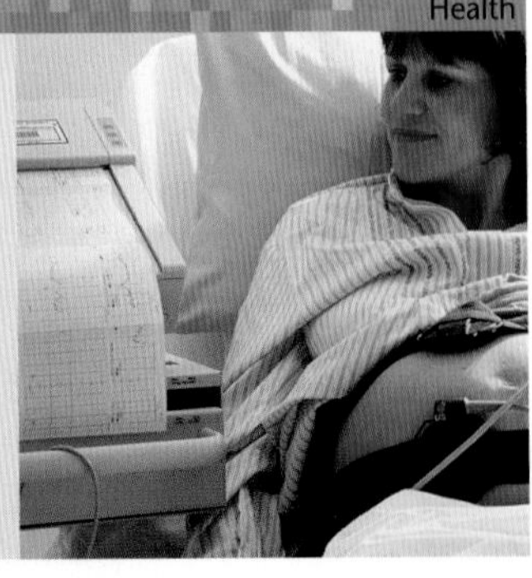

## GOING INTO LABOUR

Labour begins in response to the secretion of hormones that cause the uterus to contract. At first, contractions may be mild and infrequent but, as labour progresses, they become increasingly strong and painful, longer in duration (lasting up to a minute), and occur at regular intervals. Each contraction exerts a downward push on the baby and at the same time pulls on the cervix, gradually dilating (widening) the opening until it is about 10cm (4in) in diameter – wide enough to accommodate the baby's head. The membranes of the amniotic sac, the fluid-filled bag that has surrounded the baby during its time in the uterus, often rupture during this stage of labour. This event is commonly referred to as the waters breaking. There is a gush of fluid and contractions tend to become more forceful. The first stage of labour may be as short as 1–2 hours or it may be prolonged, lasting for more than 24 hours. Duration often depends on the number of children a woman has had.

**LABOUR IMMINENT**
*The baby is head downwards in the uterus, and the cervix is beginning to soften and open in preparation for birth. One of the early signs that labour will soon start is the "show", the passing of blood-stained mucus from the woman's cervix.*

Fetus

Uterus

Amniotic fluid

Mucus plug
Leaves the cervix up to 10 days before labour begins

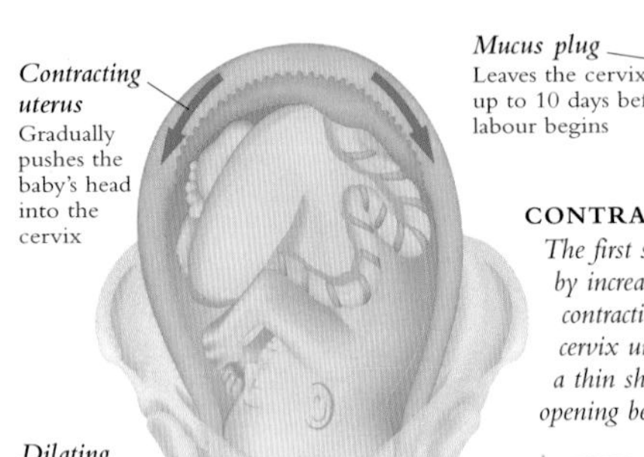
Contracting uterus
Gradually pushes the baby's head into the cervix
Dilating cervix
Contractions increase in strength, causing the cervix to widen
Bulging amniotic sac

**CONTRACTIONS BEGIN**
*The first stage of labour is marked by increasingly strong uterine contractions. These pull up the cervix until it has shortened into a thin sheet of tissue. The cervical opening becomes wider (dilates).*

**WATERS BREAK**
*As labour progresses, contractions intensify and the baby is pushed farther down. At some point during the first stage of labour, the membrane surrounding the baby usually ruptures to release a gush of fluid. A midwife can rupture the membrane artificially if necessary.*

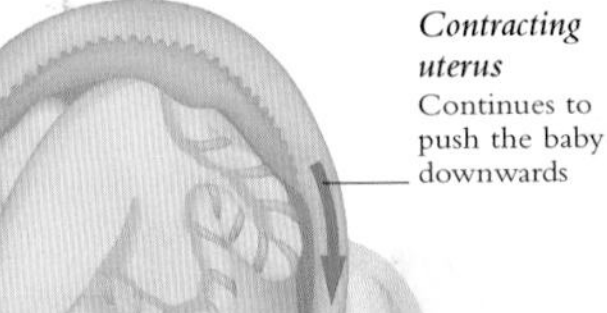
Contracting uterus
Continues to push the baby downwards
Dilated cervix
Diameter of about 10cm (4in) when fully dilated

BODY

# DELIVERY

The second stage of labour is the time from complete dilation of the cervix until the baby emerges into the world. The pressure of the baby's head on the floor of the pelvis gives the woman an overwhelming urge to bear down or push; she must synchronize her efforts with each contraction to help the baby's head along the birth canal. When the cervix is fully dilated, the baby's head usually drops sideways into the pelvis, aligning the skull's widest part with the widest part of the pelvis. The baby later turns so that it faces the mother's spine, allowing the shoulders to pass into the pelvis. When the baby's head appears at the vaginal entrance, an event referred to as crowning, the birth is imminent. Once the head is out, the rest of the baby's body is usually delivered quickly and easily. The placenta separates from the wall of the uterus and is expelled through the vagina.

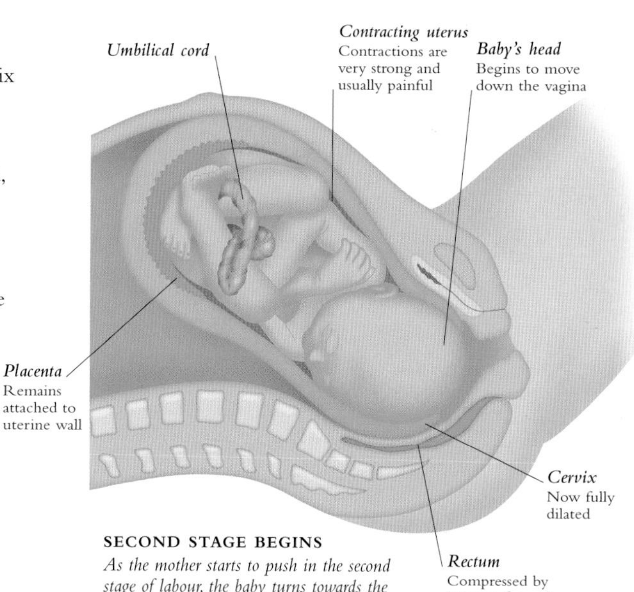


**SECOND STAGE BEGINS**
*As the mother starts to push in the second stage of labour, the baby turns towards the mother's spine with its chin pulled in against its chest. The baby begins to move out of the uterus into the vagina, which stretches in order to accommodate the head.*

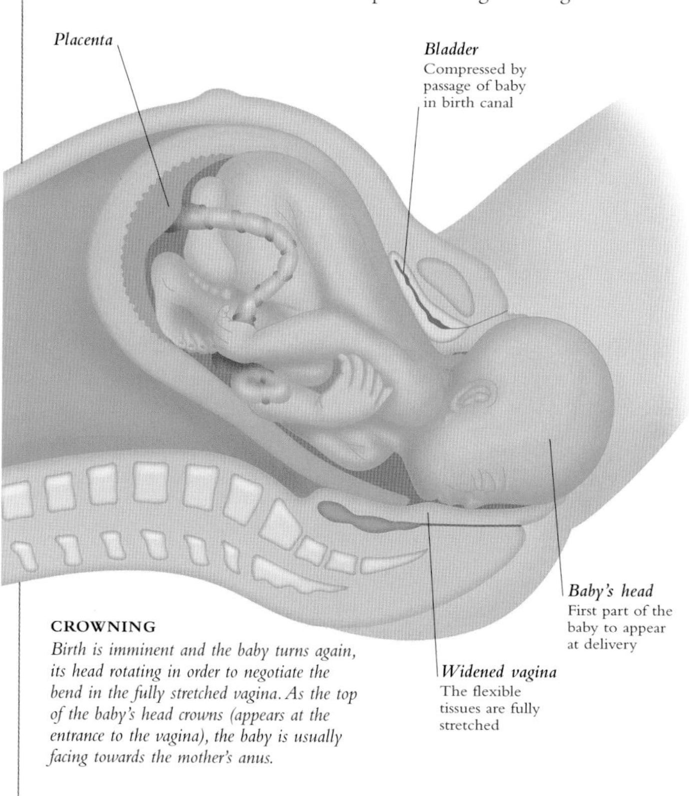


**CROWNING**
*Birth is imminent and the baby turns again, its head rotating in order to negotiate the bend in the fully stretched vagina. As the top of the baby's head crowns (appears at the entrance to the vagina), the baby is usually facing towards the mother's anus.*

*Placenta*
Will be delivered a few minutes after the baby's birth

*Baby's shoulders*
Follow smoothly after the delivery of the head

*Baby's head*
Guided and supported by a doctor or midwife as it emerges from the vagina

**THE HEAD EMERGES**
*A doctor or midwife gently guides the baby's head out of the mother's body, checking that the umbilical cord is not wrapped around the neck. The baby's body usually slides out easily.*

# UTERUS RETURNS TO NORMAL

After delivery, the uterus starts to shrink back to its normal size. During the process, which lasts several weeks, women may experience mild cramping pains. These cramps are often most pronounced during breastfeeding because the sucking action of the baby causes the release of the contraction-stimulating hormone, oxytocin. For around 6 weeks after delivery, there is a blood-stained discharge from the uterus, known as lochia, which consists of fragments of uterine lining from the site where the placenta was attached.

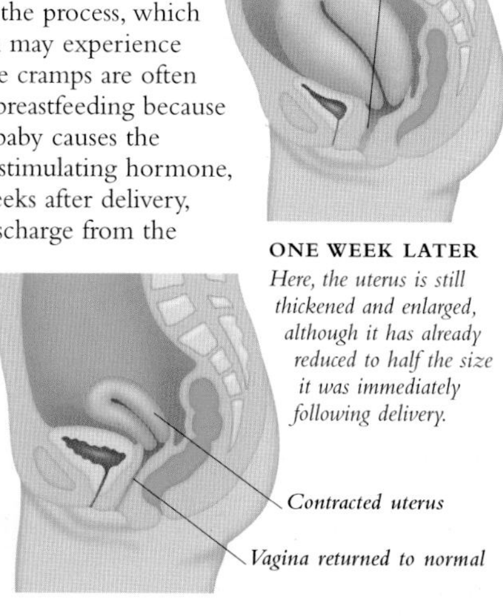


**ONE WEEK LATER**
*Here, the uterus is still thickened and enlarged, although it has already reduced to half the size it was immediately following delivery.*

**SIX WEEKS LATER**
*The uterus is back in its normal position although it has not yet returned to its pre-pregnancy size.*

History

## Caesarean section

Caesarean section, when a baby is delivered through an incision in its mother's abdomen, is supposedly named after the way in which the Roman emperor Julius Caesar (right) was born. However, the legend is doubtful, as few (if any) women could have survived such major surgery at that time. The first record of a successful caesarean section was made in Germany in 1500, when the wife of a pig gelder was operated on by her husband.

# MIND

# MIND

The brain is humanity's greatest resource. Most scientists now believe that what we know as the mind and the brain are a single entity, but the term "mind" is often used to refer to the brain's higher functions: information collection, storage, and processing. The mind is viewed as incorporating brain functions such as emotion, consciousness, language, and intellect, but not its purely physical ones, such as breathing. The human mind has been long misunderstood; even today, most of us seriously underestimate its power.

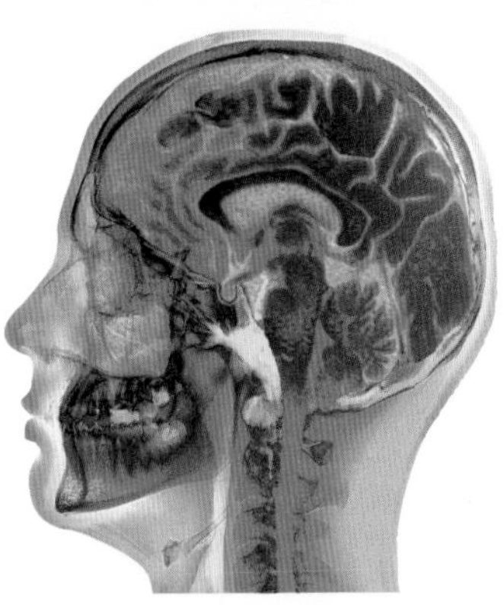

**A COMPLEX MACHINE**
*Many of the brain's mysteries have yet to be solved, despite centuries of research. However, imaging techniques such as MRI (above) are helping to increase our understanding.*

Fish learn to go to a certain place for food; rats learn the way through a maze; birds remember where they have stored hundreds of different food morsels. Even the humble sea slug learns to stop responding to a stimulus if it is repeated a number of times.

## UNIQUELY HUMAN

What sets humans apart from all other animals is their ability to think about thinking, to put together events that occurred at different times and places, to form abstract concepts, and to ponder on things they have not actually seen or done. Humans are not restricted to thinking about things directly cued by their current situation; they can work out solutions to problems without going through an active process of trial and error. They can categorize information in more than one way, set up pathways of thought, ponder on things, and reminisce. The transmission of cultural information across generations, from one individual to another, and from group to group allows each person to amass and pass on a vast array of facts, skills, and knowledge. We do not each have to reinvent the wheel, discover how to bake a chocolate cake, or calculate Pythagoras' theorem anew. Most other species must acquire the facts and skills to be stored in memory's nerve networks from direct experience.

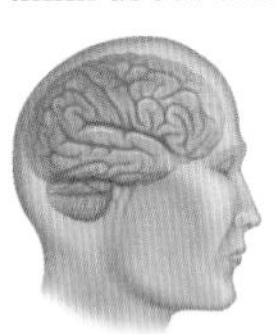

HUMAN

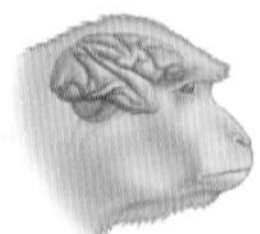

MONKEY

PIGEON

**BIG BRAINS**
*The human brain is proportionally much larger within the head than are the brains of all other animals.*

**KNOWLEDGE BANKS**
*Humanity has a vast base of shared knowledge upon which to build. We do not each need to reinvent the periodic table before we can learn about chemistry.*

## EARLY IDEAS

We have always had an instinctive desire to understand the brain's workings because, by so doing, we can better understand ourselves. Thousands of years ago, the mind was thought to exist separately from the body in the form of a vapour, gas, or intangible "spirit". Later, the Greek philosopher Empedocles (5th century BC) believed blood was the medium of thought and that intelligence depended on its composition. Democritus, a contemporary, thought that the human psyche was composed of light, fast-moving atoms dispersed through the body but found principally in the brain. The prevailing view in the East at this time was that the heart was the intellectual centre of the body.

Hippocrates and his followers (4th and 5th centuries BC) gave the brain a more significant role in human mental qualities. They believed that the brain was the source of phlegm, one of the four humours they considered vital for human emotion, cognition, and sensation. However, they understood little about brain function. Galen (2nd century BC) was one of the first to create a detailed picture of the brain's anatomy, drawing on his experience treating gladiators. He found three ventricles (large fluid-filled spaces) within the brain and believed that higher functions were based in the solid parts around them. By the 17th century AD, many anatomists thought mental faculties were in fact located in the ventricles. As study progressed, there was an increased interest in the role of the brain's cortex (outer layer).

**ARISTOTLE'S HYPOTHESIS**
*Greek philosopher Aristotle (4th century BC) shared the Eastern view that the heart was the source of thought and conscious sensations.*

## INCREASED UNDERSTANDING

Eventually, investigations into the mind moved firmly into the realms of science, the invention of the microscope and early research into nerve cells marking major turning points. From the early 19th century, there was a move to divide the brain into lobes with different functions, just as the body's organs have distinct roles. Damage to a specific lobe tended to cause the loss of a particular function. For example, damage to Broca's area in the left frontal cortex caused speech deficits, so it was deduced that speech was located in this area. The modern view of the brain is more complex. While functions are partitioned to some extent, they are also distributed across the brain through a network of complex connections. Speech, for example, involves both of the brain's hemispheres, not just the left cortex.

For years, scientists relied on evidence from brain damage to understand what the various parts of the brain did. However, newly developed imaging techniques now allow us to watch the brain in action. Computed tomography (CT) scans and magnetic resonance imaging (MRI) provide anatomical images of the brain, while positron emission tomography (PET) scans show the chemical and metabolic

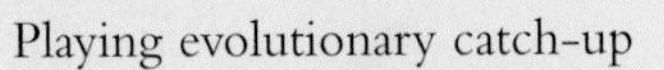

Fact

### Playing evolutionary catch-up

The human mind evolved in the African savanna, where our ancestors spent millions of years living in small groups of hunter-gatherers. Mental abilities helped them to survive the challenges of history. Modern life is vastly different, but genes take time to catch up and the human brain remains fundamentally that of our ancestors and their cousins, such as the Neanderthals (right). In some ways natural selection has been relaxed, so the human brain may have to work less hard nowadays.

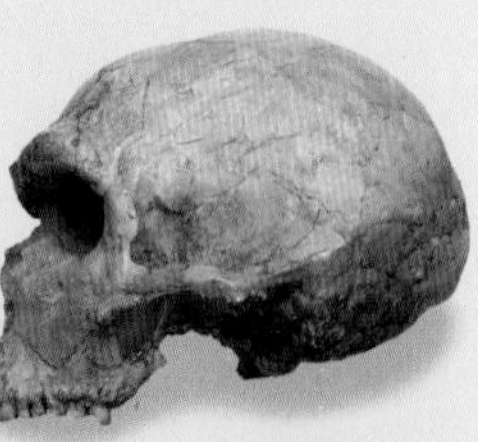

NEANDERTHAL SKULL

**THE SCIENCE OF PHRENOLOGY**
*In the 19th century, physicians began to try to locate areas of the brain responsible for specific functions.*

activity of the brain's tissues. Functional MRI is a recent innovation that enables us to see the mind at work. It shows what happens in the brain when a person is listening to music, reading, speaking, or thinking.

Brain research now involves a variety of specialists, including radiologists, psychologists, pharmacologists, neurologists, and even computer scientists. We have answers to some of the questions being asked around a century ago, but many areas of study from the late 19th and early 20th centuries remain the focus of investigation.

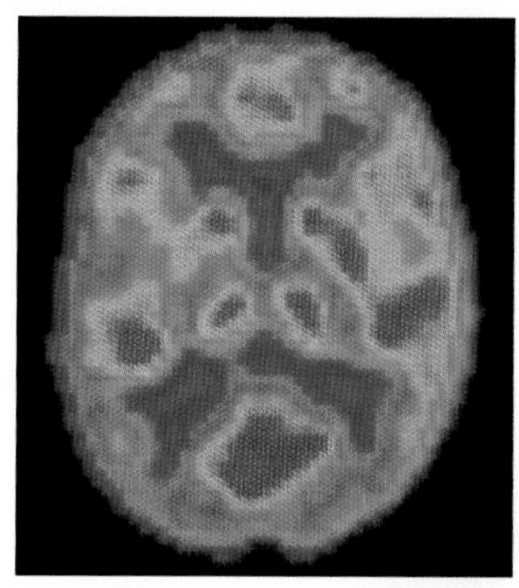

**TECHNOLOGICAL ADVANCES**
*Techniques such as PET scanning help us to understand the brain by showing the areas that are active when we carry out a specific task. In this scan of someone thinking, there is high activity in the red and yellow areas.*

## CONSCIOUS MINDS

Renowned psychoanalyst Sigmund Freud (1856–1939) had a profound influence on our understanding of the human mind. Although much of what he said has since been proved unfounded, he did get one thing more or less right: much of what happens in the brain takes place below the level of consciousness. Scientists have since taken Freud's idea and expanded on it. Our conscious sense of self, which seems to be at the steering wheel in command of our "free will", may in fact be little more than an illusion generated by the nuts and bolts of the brain's nerve activity, which makes all the real decisions. One of the special facets of the human mind is its ability to work out what's going on in other people's minds in order to manipulate or predict their behaviour. This ability may be an extension of human self-awareness: the unique higher form of consciousness that allows us not only to ponder on our own inner thoughts but also on those of others.

## TALKING MINDS

Humans have an instinct for language development and use, and the ability to use grammar is unique to the human species. The emergence of language was probably linked, historically, to our development of a sophisticated culture that incorporated highly complex tools, techniques, and spiritual beliefs. Language, therefore, may have proved a key component in our evolutionary success as a species. Whether or not the human mind could even exist without language is a matter of extensive debate because language seems to be a key part of the way we think. When we concentrate, we seem to speak to ourselves with an inner voice. Certainly, without our grasp of language, we would never have been able to get to the bottom of some of the mind's mysteries that we have now managed to unravel.

## ALL THE SAME?

While our brains' internal workings are essentially the same, how we use our minds and how we feel "in them" may be completely different for two separate people.

Health

### The power of the mind

Health can be improved by things we mentally link to healing, such as drugs. In drug trials, many people who receive a sugar pill still get healthier and this "placebo effect" is part of all cures. Expecting to get better has a powerful effect, and drugs often begin to "work" even before entering the bloodstream. The opposite is also true: told of a drug's side-effects, patients often have them; if told incorrectly we have a serious illness, we become ill. What we are told affects not only our physical health, but also our talents, personalities, and ability to cope.

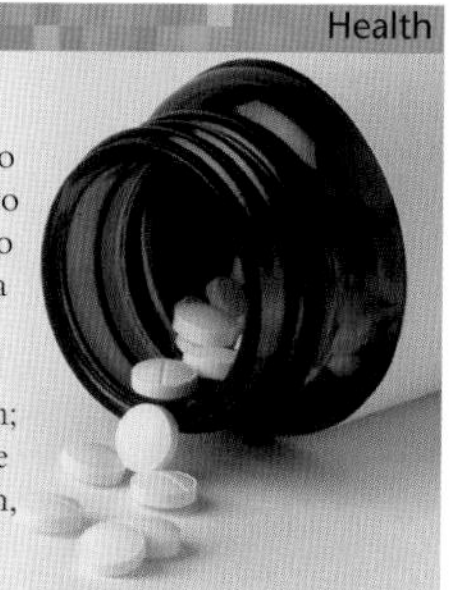

This uniqueness is the very essence of individuality. Every person feels both the result of and the owner of his or her mind. We feel that no-one else could understand what it is like to be us, even though we know that everyone else has the same workings and shares this feeling of uniqueness. Paradoxically, perhaps, one of the biggest things we have in common is the fact that we are different. It is clear that personality, intellect, and identity differ enormously from one person to the next. Yet much can still be learned about the generalities of the human mind and how it works by studying such differences.

## MUCH TO DISCOVER

The human brain is probably the most complicated mechanism in the entire universe. It has been studied scientifically for just 150 years and, in this short space of time, remarkable progress has been made. Those of us alive today have insights into the workings of the human mind that would not even have been dreamed of in our grandparents' youth.

Nevertheless, we are still a very long way off getting to the bottom of many of the mind's deepest and darkest mysteries. The issue of consciousness, in particular, remains as intractable as it ever has been. Will the answers to the remaining questions seem simple – or even obvious – to our grandchildren? Some scientists believe that we are still missing the "big idea"; in other words, the $e=mc^2$ of the mind that, once discovered, will completely change the way we see our world, not to mention ourselves, forever.

**EXERCISING THE MIND**
*All that we do, from breathing to feeling emotions, relies on brain power. Tasks such as learning to write or problem-solving most obviously tax the conscious mind.*

OPENING THE BLACK BOX

*Science may finally be on the verge of shedding light on some of the hitherto unanswerable questions on the workings of the human mind.*

# HOW THE MIND WORKS

The brain is more than just the most important organ in the human body; it is arguably the most complex structure in the universe. Our best tool for understanding the mind is the mind itself, but some philosophers view this idea as a paradox, making a true understanding of the mind's workings impossible. Nevertheless, science may finally be on the verge of opening the brain's "black box". Newly developed avenues of research are beginning to unravel more of the brain's secrets and show how mental functions are controlled by infinitely complicated wiring networks.

We share much of our brain's anatomy with other animals: principally the so-called "lower" structures, such as the brainstem, limbic system, and cerebellum, which are located deep within the brain. These areas are vital to everyday life; in fact, if the body stopped receiving commands from the brainstem, it would stop functioning and die.

## SOPHISTICATED BRAIN

Yet there must be something special about the human brain. We can do things no other animals can do: plan for the future, empathize, deceive, talk, invent complex tools, and draw on a shared store of information and skills. The clue to our increased capabilities appears to lie in the human brain's vastly enlarged cortex (outer layer). The prefrontal cortex is of particular significance because it is the site of rational thought.

**SENSORY INFORMATION**
*The processing of all the information that flows in through the sense organs goes further in the human mind than it does in the brains of other animals.*

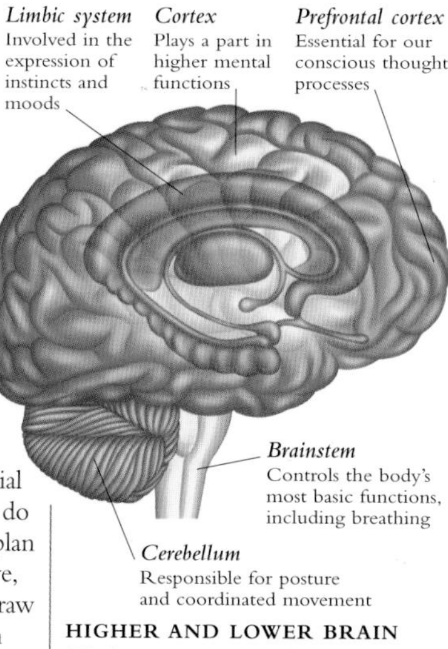
*Limbic system*
Involved in the expression of instincts and moods
*Cortex*
Plays a part in higher mental functions
*Prefrontal cortex*
Essential for our conscious thought processes
*Brainstem*
Controls the body's most basic functions, including breathing
*Cerebellum*
Responsible for posture and coordinated movement

**HIGHER AND LOWER BRAIN**
*The brain's cortex (outer layer) is the site of our conscious faculties, such as reasoning. The other, lower, areas deal with our emotions and unconscious body processes.*

The brain develops from the base upwards. The "lowest" areas, which control the body's housekeeping (feeding, breathing, digesting, urinating, and so on), are found mainly in the brainstem and they mature first. The areas involved in impulses, instincts, and primitive emotions – located in the limbic system – and those that control voluntary movement – in the cerebellum – develop next. The "highest" areas, associated with our cognitive processes, are the last areas to mature. This is why children can seem so impulsive and emotional; the higher areas of their brains, especially the prefrontal cortex, are not advanced enough to keep their basic urges and instincts in check.

## INFORMATION PROCESSING

Much of the power of the human brain is dedicated to processing the information that continually flows in through the sense organs and to deciding how the body should react to that information: which chemicals to release, which limbs to move, which facial muscles to pull, and so on. Information processing goes further in the human mind than it does in the brains of other animals. Apart from the role it plays in controlling the physiological aspects of our bodies, the brain is also involved in laying down and retrieving memories, processing thoughts and language, generating emotions, and providing us with our unique sense of self and innate conscious awareness. Even tasks the human brain performs on a second-by-second basis are remarkable. For example, we can instantly recognize sound waves received by our ears as meaningful speech.

We are totally oblivious to much of the work the brain carries out, but very occasionally we are aware of actively using our brains to process information, such as when we are searching to recall a word or a name or tackling a complex mathematical problem.

## MAKING CONNECTIONS

There are two ways of trying to understand the mind. The top-down approach asks what the mind does and why: what are memory, consciousness, and thought? Why did they evolve? Where are they found in the human brain? The bottom-up approach takes the brain apart and explores its inner workings: the nerve cells, brain chemicals, and circuits that magically make everything happen. The gulf between these two approaches is narrowing, but still exists.

The average human cortex comprises around ten billion tiny nerve cells; some experts think the missing piece of the scientific puzzle lies in the connections that form between them. As we live, experience, and learn, an almost infinite number of links develop between our brain cells. In fact, there are more possible connections than there are atoms in the whole universe. These complex wiring circuits are able to store and process information with more speed and subtlety than the most powerful computers in existence.

Such nerve connections enable the brain to undertake both the most rudimentary and the most baffling of tasks, such as reasoning about the workings of the world after interpreting only basic sensory input. Some scientists believe that this so-called "associationism" is the basis of all the brain's workings.

**INTERNAL CIRCUITRY**
*Experts believe that the key to the human mind lies in the connected nature of the brain's many cells.*

Fact

### The mind time-lag

The mind helps us react to events, but often it simply is not quick enough. A tennis player, for example, does not have time to wait for the ball to land before moving to return it. The secret lies in anticipation: the brain constantly predicts what is about to happen and plans a response on the basis of this prediction. If the tennis ball bounces off a lump in the ground, the prediction will be wrong and the racket will miss. Our minds tend to live in the future, with our conscious awareness always coming slightly later.

# CONSCIOUSNESS

Quite different from everything else that we know and understand, consciousness is a particularly difficult concept to pin down. Some have even referred to it as "the last surviving mystery". Stars and planets, fish and tigers, and rocks and atoms are all objective things that we can measure, study, and agree about. Consciousness, on the other hand, is private and subjective. There has been, and will continue to be, much debate on what consciousness actually is, how it functions, and exactly where it might be located.

## WHAT IS CONSCIOUSNESS?

The word consciousness is used in many different ways. We are conscious of our identities and our surroundings; we make a conscious effort to do well in our studies; we become politically conscious; and we talk of doing many things unconsciously. But consciousness is best described as the faculty that enables us to be aware of our own existence and think about our own thought processes. There seems to be a huge gulf between the world "out there" and each of our worlds "in here", and no physical measurement can capture the inner feelings of a personal experience. Most of us tend to see the mind and the brain as separate things; it is hard to accept that our conscious experiences may be no more than the firing of nerve cells within the brain.

**IS THIS ALL WE ARE?**
*Most people believe that their conscious feelings and experiences are more than simple brain activity.*

History

### Our mind as a theatre

Consciousness can feel like a theatre in which we each watch our own private show. René Descartes, a 17th-century philosopher, famously proposed that the brain processed sensory input so that it could then be viewed separately by the mind. This theory is known as the "Cartesian theatre". Most modern scientists reject this idea of a mind–brain split, but many people still believe in a central place in the brain in which everything "comes together".

## LOCATING CONSCIOUSNESS

Neuroscientists have never been able to locate a single brain area that, when damaged, removes consciousness yet leaves all other functions intact. The more scientists learn about the brain, the more they see it as a vast, integrated system in which information flows in and out along complex and parallel pathways. In vision, for example, information about colour, shape, position, movement, and size are all processed separately; there is no area in which they all come together in the brain. The existence of a specific area where consciousness "happens" therefore seems unlikely. Even if scientists accepted that a special consciousness area existed, it would still be hard to explain why its brain cells could cause conscious experience while similar brain cells elsewhere do not. The search for brain activity corresponding to what we consciously experience goes on, but critics say this type of research is misguided. They believe that, even if relevant brain activity is found, we will never find an area responsible for consciousness alone or, indeed, any special types of nerve cell.

**COMING INTO CONSCIOUSNESS**
*When our brains register something, for example a dog bounding towards us, it seems as if all aspects of the experience are pulled together by the mind.*

**NO CENTRAL "HEADQUARTERS"**
*In fact, each piece of information received by the brain (such as movement, colour, and shape) is processed in a separate area. There is no place where it all comes together; instead there is widespread brain activity, shown by the red and yellow areas in this scan.*

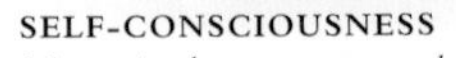

**SELF-CONSCIOUSNESS**
*Most animals see a stranger when they look in a mirror. Humans are one of the rare exceptions – we see ourselves.*

Issue

## Free will

We each feel as if we make conscious choices to act in particular ways. In reality, however, no experience – including that of free will – is essentially anything more than brain activity. We certainly think about an action before carrying it out, but we then tend to assume that it was the conscious thought that caused the action. In fact, both the intention to act and the action itself may result from earlier brain activity. Some studies controversially appear to confirm that conscious decisions are useless afterthoughts that simply give us a way of making sense of ourselves.

BRAIN ACTIVITY
Brain activity starts
Awareness of decision to act
Action occurs
-1000
-535
-200
-0
TIME (milliseconds)

**CONSCIOUS WILL**
*The results of one study suggest that a conscious decision to act comes well after the brain is already engaged in planning the action.*

# WHO AM I?

Unlike many other animals, adult humans have a powerful sense of self. When we look in the mirror, we see unique, conscious individuals, each with our own opinions, beliefs, and experiences. We feel as though there is an inner "self" that inhabits and controls the body. Yet there is no explanation for how this self receives information from the brain, nor how it exerts control over it. In fact, the more we learn about how the brain works, the less room there is for the idea of another controller inside it. Many scientists and philosophers believe that the concept of a conscious self is simply a useful illusion that gives our lives meaning: it is the "I" in all the stories the brain tells itself, but there is nothing that actually corresponds to the word "I". Most people find it very hard to accept that there is no such thing as a self, despite all the scientific evidence pointing to this. In fact, a majority of the world's peoples still find comfort in the idea of a soul or spirit that gives unity to a person's life and that may even continue to exist after his or her death.

*Enlarged liver*

**BELIEF IN THE SOUL**
*The Aztecs believed the liver was the seat of the human soul. This statue depicts the Aztec Lord of Death, his enlarged liver emphasizing his soul.*

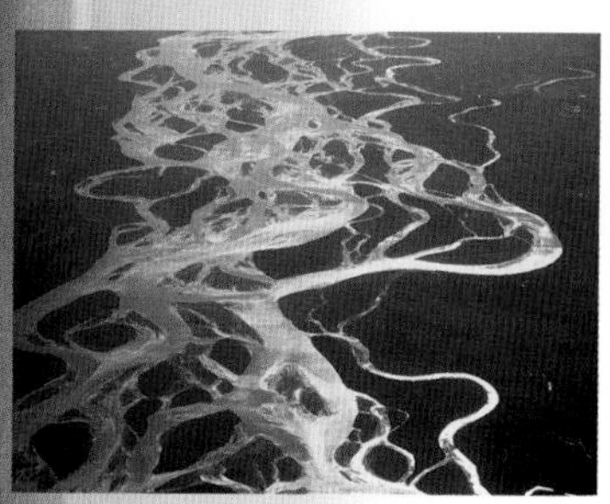

**NO SELF?**
*Buddhists deny the existence of an unchanging "self". They believe that our identities, like the course of this river, are constantly reshaped by events.*

# THE DEVELOPMENT OF SELF

We are not born with a sense of self. Babies show a degree of self-interest by crying if they are cold or hungry, but they do not recognize that other people have points of view, beliefs, or desires. In fact, it is not until their second year that babies start moving towards self-awareness. One of the first steps is acquiring the ability to follow another person's gaze: until about the age of 1, when an adult points at something, a baby will simply look at the adult's finger. Between 18 months and 2 years, children start to refer to "me" and "you", and by the age of 3 they can talk about their own preferences and those of others. However, 3-year-olds may thrust their heads under pillows and shout "I'm hiding" because they still do not understand that others cannot see exactly what they see. By about 4, children grasp that others may have different beliefs from their own and, by age 5, most have a fully developed sense of self. However, there are some people who do not fully develop a sense of self. Those affected by autism, for example, can recognize themselves and their own intentions, yet find it hard to apply the idea of "self" to others. As a result, they are unable to relate well to other people.

**ARE OTHER ANIMALS SELF-AWARE?**
*If, as some believe, self-consciousness only ever develops with language, then this dolphin cannot share our sense of self-awareness.*

Health

## More than one self

In dissociative identity disorder (DID), formerly called multiple personality disorder, several selves seem to inhabit the same body and may alternate unpredictably. According to recent research, the average person with DID has 13 identities; the possible number ranges from two to hundreds. Some psychiatrists feel the condition does not really exist. Others believe affected people may have been abused as children and that they created other selves to escape the pain.

**A DEVELOPING SELF**
*Young children believe that if they cover their eyes, nobody can see them. Learning that this is not the case is a major step towards self-awareness.*

# RECEIVING INFORMATION

The brain receives from the senses a vast quantity of information, which needs to be sorted before it can be used. A camera, like our eyes, can focus light – but it simply creates an image that it cannot interpret or experience. Similarly, no microphone knows that sound waves are the notes of a symphony or the cries of a child, no gyroscope can experience being upside down, and no chemical detector can smell flowers. But we do. The human brain is so good at turning sensory input into normal experience that we can fail to appreciate the enormous tasks it continually tackles without effort.

**MORE THAN ROBOTS**
*Human technology readily mimics our senses, but it grinds to a halt when faced with the unexpected.*

## THE ATTENTION CIRCUIT

Before we can use information, we need to focus our attention on it, which involves several brain areas locking in a circuit. The thalamus, a structure deep in the brain, relays sensory information to the appropriate part of the brain's cortex (outer layer) for processing – the visual cortex in the case of visual input. The information then enters short-term memory and is stored in the prefrontal cortex (*see* p156). The parietal lobe, a region of the cortex that gives us a physical and spatial sense of our own bodies, is also involved. Children have short attention spans because their brains are not yet fully developed. The ability to shift attention is as important as the ability to focus; without it, a person would be unreceptive to new sensory input and so unable to adapt quickly to new situations. Some people find it easy to monitor what is going on across their whole sensory field and are able to switch attention rapidly between different events; others are not so quick to notice new information. The reasons behind such differing abilities are unclear.

Health

### Inability to attend

In a rare condition called "neglect", a person fails to attend to one entire side of their world, often the left side. Affected people might shave only the right side of the face or eat just from the right side of a plate. While they are not actually blind to objects on the left, the full attention circuit is not activated on this side, making them effectively blind. This condition may be caused by damage to the brain's parietal lobes.

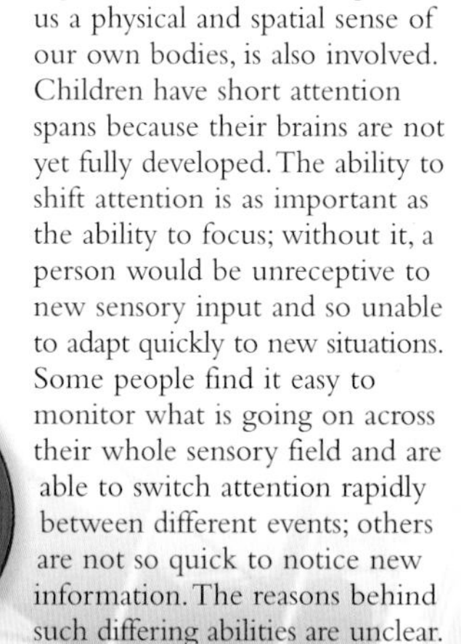

**THE FOCUS OF ATTENTION**
*Without the ability to focus, learning would be difficult and memory short.*

## FILTERING INPUT

The world offers the brain much more information than it needs; one of the first steps is to filter this information in a way that enables the brain to process it. As a result, we tend to perceive a whole thing before its intricate details. For example, we see a face as a unit instead of looking at its individual features. Because our brains are adept at focusing on only a small amount of the input from our sense organs, we can attend to faint input over clearer information when required. If, for example, we are involved in a conversation at a party and a more interesting discussion across the room catches our ear, we can filter out the detailed information of the close conversation and focus more on the one taking place farther away. Our ability to disregard irrelevant input is essential; without it, we would constantly be on full alert. This is why a person might startle on hearing the unexpected crash of material on a building site, but only the first time.

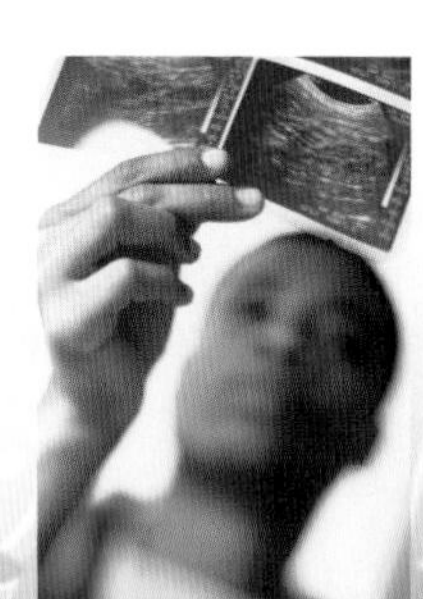

**EXPERTISE**
*An experienced doctor is able to ignore much of the detail on a scan, homing in only on what is relevant.*

**A PERCEPTUAL BLUR**
*Every second that we are awake and alert, the brain is bombarded with huge amounts of sensory information. It needs to work out which of this information is important and which is irrelevant.*

Fact

## Sensory deprivation

The brain finds it hard to stop perceiving, so when deprived of sensory input it tends to create its own. In the early stages of blindness, for example, people can suffer from hallucinations of varying intensities. Sensory deprivation can be deliberately induced in relaxation devices known as "floatation tanks", in which people lie in warm saltwater in complete silence and darkness. This pleasant experience may cause mild hallucinations. However, sensory deprivation can be used as a form of torture.

# FILLING IN THE GAPS

Although the world frequently presents us with too much information, it can also provide us with too little. Experience enables us to fill in information gaps in much the same way as it helps us to focus our attention when we are bombarded with vast quantities of irrelevant sensory input. There are many things out there that we need to be aware of, yet only tiny hints of their presence may exist; our brains must pull together fragments of information from all of our senses in order to detect these things. In a jungle, for example, we might hear twigs breaking underfoot and see a flutter of movement and a blur of yellow in the leaves. From these clues, all of a sudden we may perceive a tiger hidden amongst the greenery. In fact, humans are programmed to find links between information in the world even when they do not actually exist, the theory being that it is better to detect something in error than to fail to identify a hidden danger.

**LIFE OR DEATH**
*In battle, survival can frequently depend on detecting danger from the tiniest of clues.*

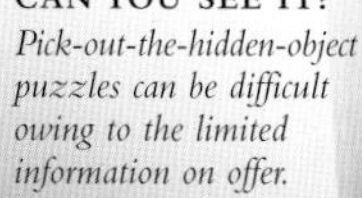

**CAN YOU SEE IT?**
*Pick-out-the-hidden-object puzzles can be difficult owing to the limited information on offer.*

# CATEGORIZATION AND FAMILIARITY

Whenever we look at a person, an object, or a scene, our brains instantly distil all the detail into a small number of categories. We see a house, for example, which is composed of windows and doors, and if we look closer we may see door handles, keyholes, and letterboxes. While we can recognize that each object is a different colour or shape, this unnecessary level of detail is quickly stripped away. In other words, if all we want to do is open a door, we do not need to register its colour – we simply need to recognize that it is a door. This process of categorization is also essential to speech perception (*see* p169). People perceive clear consonants even though they are often blurred by surrounding sounds; without this ability, we would never make sense of spoken language. In a similar way, we tag some of our life experiences as "familiar", perceiving certain people, things, and places as emotionally closer than others. In Capgras syndrome, this sense is lost. An affected person may see a relative as a duplicate rather than someone with whom he or she shares emotional ties. Such problems can also affect the familiarity of a pet or a person's own home.

**THE BRAIN'S PIGEON HOLES**
*To enable our brain to use information more than once or store anything in memory, we need to categorize everything we perceive.*

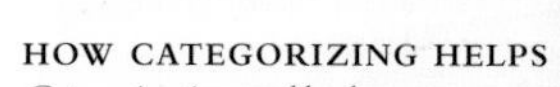

**HOW CATEGORIZING HELPS**
*Categorization enables humans to apply the experience of riding one horse to riding any other horse, regardless of its colour, age, or size.*

MIND

# LEARNING

The learning process occurs as a result of experience and it pervades our lives: we learn to ignore unimportant things, such as the drone of the traffic or the colour of the office walls, and to predict predictable things – like the fact that a car driving through a puddle will splash us if we stand too close to the curb. In general, we learn how to make good things happen to us and how to avoid things that could harm us. Memory is absolutely essential to learning; without this capacity, we would have to learn the same things over and over again every day of our lives. However, we only retain a small proportion of the vast quantity of information processed by the brain.

**CONSCIOUS LEARNING**
*At school, we are aware that we are trying to retain information and can even improve with practice. However, most learning is like the heartbeat: automatic and generally unconscious.*

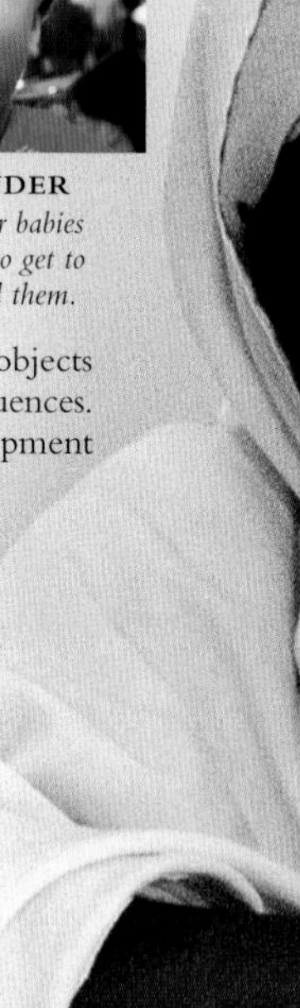

## LEARNING THROUGH LIFE

**A SENSE OF WONDER**
*The major challenge for babies and young children is to get to know the world around them.*

The brain goes through extremely rapid development in the womb, at times increasing by up to 250,000 nerve cells a minute. At birth, the brain has some 100 billion nerve cells. The first stage of learning involves the undoing of existing links between these cells so that only the relevant ones remain. Brain growth occurs rapidly: new connections are formed as facts, experiences, and associations are learned, and the weight of the brain trebles in the first 3 years. Children have to learn where their bodies stop and the world begins, that objects exist even when they cannot see them, and that actions have specific consequences. The fact that most children go through the same sequence of motor development and eventually reach a similar level suggests the unfolding of an inborn programme. But practice clearly accelerates progress and lack of normal experience delays it. Of course, for knowledge that must be consciously acquired (science, music, athletics, and so on), experience plays a far greater role. Learning never ends – we continue to acquire new information, skills, and experience every day of our lives.

**LIFELONG LEARNING**
*There are always new things to learn, and older people can actually pick up a lot from the young.*

Issue

### Birth of new brain cells?

Until recently, scientists thought all the nerve cells a person would ever have were present at birth, and that if any were damaged they would not be replaced. Recent studies, however, suggest that the brain can produce new nerve cells. When rats were given a learning task that required use of the hippocampus (a structure in the brain involved in learning and memory), new nerve cells appeared there. This suggests that learning may be associated with the birth of new brain cells, although similar studies in humans have yet to be completed.

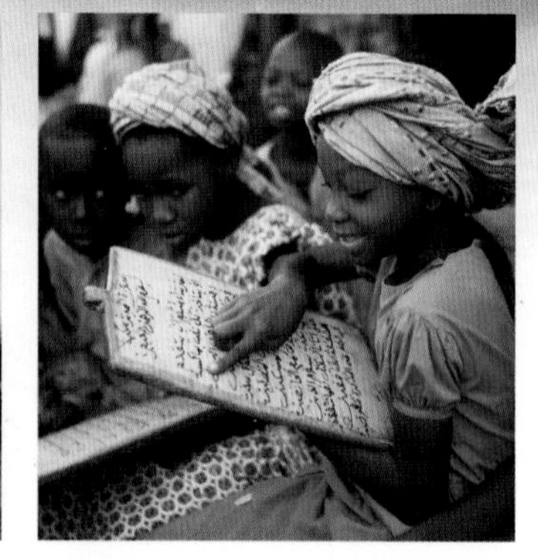

**RELATED EXPERIENCE**
*When learning to read, a child must associate written script with the spoken language he or she has begun to master.*

## MAKING ASSOCIATIONS

Humans have an amazing ability to associate experiences, sensations, objects, and events; this ability is the basis of most learning. Associated nerve cells are weaved into webs of experience and, when one item in the web is accessed, it brings with it a host of connected ideas. We relate new experiences to what we already know – a young boy may associate all houses with kittens and toy cars because he associates his own with them. As we grow up, we learn to adapt our own frameworks to better fit the experiences of others. We all tend to learn new skills by relating them to things previously mastered. Adults are particularly adept at this, having had time to grasp how experiences are related; they can be less effective at learning a skill unrelated to one that has come before.

# CONDITIONING

Conditioning is one way we learn direct relationships between two events – that one event always follows another or what we need to do to get a desired result. Learning through conditioning enables us to profit from experience. Both humans and other animals can be conditioned unconsciously and we can also condition others without being aware of what we are doing. For example, a parent who gives lots of attention to a child when he or she is naughty but ignores him or her when good unwittingly teaches the child to be naughty to gain attention. A recent study suggests that even an unborn child can be conditioned: the theme tunes of mothers' favourite soap operas were found to elicit a relaxation response in newborns.

**ONCE BITTEN, TWICE SHY**
*Humans learn from experience. If we are stung by a wasp as a child, we learn to avoid wasps; similarly, if an action has a favourable consequence, we tend to repeat it.*

History

## Pavlov's dogs

In the 1920s, Russian physiologist Ivan Pavlov was studying digestion in dogs. He noticed that the dogs started to salivate on hearing a buzzer that was sounded each time they were fed, even if no food was actually produced. Many responses are conditioned like this. If children with cancer receive ice cream as a treat before chemotherapy, it gradually becomes linked with the nausea that follows and they develop a dislike for it.

# LEARNING BY IMITATION

Humans can imitate an action with precision, honing a skill by repeatedly comparing it with the actions of others. This starts at an early age with simple gestures, such as holding up a toy phone as if to listen. The imitation becomes gradually more complex – a child gently rocks a doll to sleep or even, by the age of 4, stands at a toy cooker and pretends to make dinner. Mirror neurons, a type of nerve cell found in the brain, are one of the most exciting discoveries of recent years. A particular mirror neuron is activated when we carry out an action and exactly the same nerve cell responds when we see someone else carrying out that same action. Mirror neurons are programmed from birth to map other people's actions directly onto our own bodies so that even a minute-old baby can do it. Learning by imitation is vastly facilitated by this innate programming – we do not first have to learn how to imitate.

**MIMICRY**
*Very young children spend hours miming the actions of adults in pretend play.*

**MIRROR NEURONS**
*Humans have the invaluable ability to copy one another's actions while standing face-to-face.*

# LEARNING TO LEARN

Humans pick up conscious techniques for learning both through the teaching we receive and through practice. One of the first things we learn is that experiences and facts on which we focus attention are more likely to be retained than those we merely register in passing. For example, a passage from a book has a greater chance of being memorized if a reader blocks out everything else and concentrates solely on the meanings of the words. We find that it is also easier to focus on something intrinsically interesting than on something we find boring, so we can force ourselves to attend to more tedious items by creating mental links between them and subjects of greater personal interest. A motorbike fanatic, for instance, might learn the kings and queens of England by imagining them on different models of bike. Our learning strategies improve as our knowledge of where to look for information and what to ignore increases.

**PRACTICE MAKES PERFECT**
*Repetition is the most effective way of learning skills "by heart".*

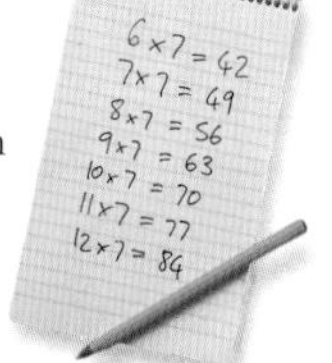


**LEARNING BY ROTE**
*The best way to tackle facts such as times tables is to go over them time and again so as to embed them firmly into long-term memory.*

# MEMORY

A critical part of our mental capacities, memory is often likened to a library with mile upon mile of shelves. When a book arrives, we must first decide if we want to keep it indefinitely. If so, it is logged, filed, cross-referenced, and placed on a shelf for long-term storage. Later, the book may be found and retrieved. There are so many shelves in the library that storage space is never a problem. People living today see more faces, visit more places, know more facts, and recount more stories than ever before, but the human memory goes on storing day after day. If problems do arise, they are in finding a memory when we need it. In other words, the book is there, but we are unsure which shelf it is on.

Input from senses → SENSORY MEMORY → Attention not paid → Information lost

SENSORY MEMORY → Attention paid → SHORT-TERM MEMORY → Memories not consolidated → Information lost

SHORT-TERM MEMORY → Memories consolidated → LONG-TERM MEMORY

**MEMORY FLOW**
*Information moves through up to three levels of memory store depending on how long we need to retain it.*

## THREE MEMORY STORES

Memory processes do not work in the same way in all situations. How they work depends on how long we need to store information – for a second, a matter of minutes, a few hours, several years, or even forever. The briefest and most transient memory store is the sensory store, which contains all information captured by the sense organs. Visual sensory memories last for about a tenth of a second; auditory ones for a few seconds at the most. Our attention processes capture just a small proportion of items from this sensory store and move them into the short-term memory, while all uncaptured items are lost. Short-term memory is more enduring and contains the items a person is thinking about at any one moment in time. Items in this store are easily accessible but most will be rapidly lost. Only memories that are consolidated through rehearsal or by being linked with other existing memories enter the final store – a long-term information repository of unlimited capacity.

## SHORT-TERM MEMORY

Experience is a stream of events, most of which are forgotten almost as soon as they occur. Yet, we need to hold events in our minds for long enough to act on them. To understand a sentence, for example, we have to hold the first few words in mind until we have heard it all. Without this ability to retain information, we would be unable to carry out even the simplest string of actions. The brain's short-term memory system allows us to put memories on hold before they are discarded or encoded for long-term storage. When we try to solve problems in our heads, we use our short-term "scratchpad" to hold partial solutions and to pull together different ideas and facts from both long-term memory and current experience. If the load gets too high, we may supplement this scratchpad with jottings on paper, but short-term memory often suffices. Unrehearsed items are lost from short-term memory after about 20 seconds. Long-term memories can be retrieved, further processed in short-term memory, and re-stored.

**WIPING THE SLATE**
*Short-term memory has very limited space; to store new information, its contents must be discarded or moved elsewhere.*

**HARDWORKING MEMORY**
*Everything that catches our attention as we go about our daily business – a distinctive number plate or dent on a car bumper, say – briefly enters our short-term memory.*

Fact

### Emotion enhances memory

Events experienced in a state of high emotion are likely to be remembered. A person involved in a fire may have a vivid memory of mundane details, such as the pattern on a mug he or she was given at the scene. Heightened memory is not caused by the emotion of the event itself but by the fact that it occurred while the brain was in an emotional state. Chemical changes that take place in the brain during intense emotion amplify all nerve activity, enabling the most inconsequential details to be recorded. The unusual amount of detail that is laid down when forming memories of emotionally intense experiences may account for the feeling that time passes slowly every time these experiences are relived.

# FORMING LONG-TERM MEMORIES

Incoming material must be processed through a set of distinct stages to become permanently encoded in the brain. A long-term memory – be it of a location, life event, skill, fact, or anything else – is formed when a particular pattern of nerve activity that has entered short-term memory is repeated. For a pattern of nerve activity to recur, the nerve cells involved must become linked through the memory consolidation process (below). Then, every time that specific pattern of nerve cells is fired, the network becomes more permanent. Once a "memory web" is consolidated fully it may last forever, but if it is not activated frequently enough it still tends to fade. The more nerve cell clusters and pathways there are in a particular memory, the more accessible that memory will be – each cluster of cells represents a different aspect of the memorized experience and acts as a "handle" to retrieve the entire memory.

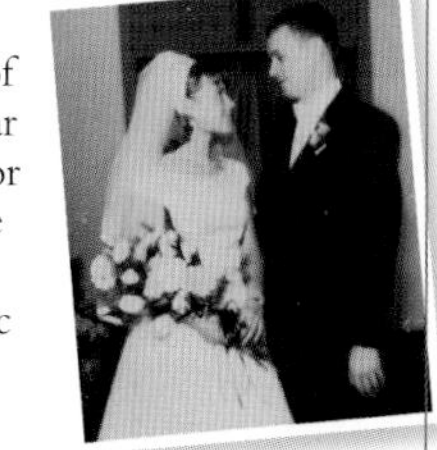

**MEMORABLE EVENTS**
*Many of our life experiences are held in the brain's network of long-term memory stores.*

**MASTERED SKILLS**
*Without a long-term memory, we would never be able to carry out previously learned skills.*

### HOW A MEMORY IS CONSOLIDATED

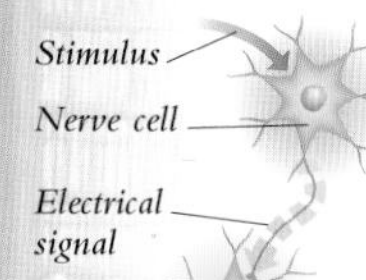


**CONNECTION**
*When an independent nerve cell receives a strong enough stimulus, it fires an electrical signal on to a neighbouring nerve cell.*

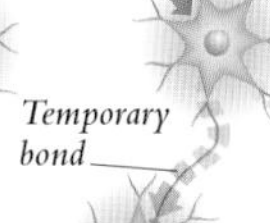


**LINKS FORM**
*A temporary bond forms between the two cells; they are likely to fire together in the future. Further firing draws in neighbouring cells.*

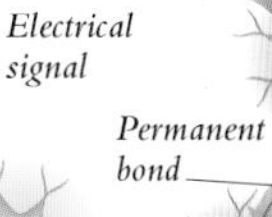


**STRONGER LINKS**
*With repeated firing, the nerve cells become firmly connected. They will always fire as one unit whichever cell is triggered first.*

**EXPANDING WEB**
*Through continued activation, other clusters of nerve cells are pulled into the network. The full web represents a single memory.*

MIND

# WHERE DO WE REMEMBER?

The critical brain structures involved in committing something to memory are the hippocampus (a structure in the centre of the brain) and the cortex (the brain's outer layer) that surrounds it. During learning, different areas of the cortex, such as the visual cortex and auditory cortex, process information about different dimensions of experience – what we see, hear, and smell, for example. Memories, especially those of personal life experiences, can comprise a great variety of such elements, so a single memory may be distributed over quite a wide area of the cortex. The general role of the hippocampus in memory consolidation appears to be to provide a cross-referencing system that links and, when necessary for memory retrieval, draws together all the different aspects of a memory from around the brain. Like the cortex, the hippocampus is also directly involved in storing certain types of memory, as are the putamen and the amygdala, located deep within the brain (see diagram, left).

*Prefrontal cortex*
Store for short-term memories

*Putamen*
Store for learned skills and procedures

*Cortex*
Store for memories of personal life events

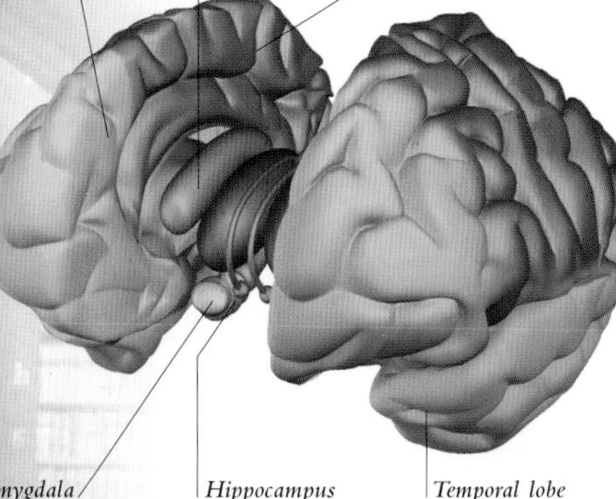

*Amygdala*
Store for unconscious emotional memories

*Hippocampus*
Store for spatial memories

*Temporal lobe*
Store for learned facts and details

**THE BRAIN'S MEMORY STORES**
*Different types of memory are known to be stored in different areas of the brain, although precise locations of individual memories have never been successfully identified.*

**LIBRARY OF THE MIND**
*Like books, memories are coded and stored for future reference, accessed if the cross-referencing system is effective enough, and finally replaced on the "shelves".*

## ACCESSING A MEMORY

Recognition occurs when we are faced with memory triggers; it is what makes a multiple-choice question easier than an open one. We are unlikely to be able to describe all road signs from memory, but we recognize them immediately. Recall involves more processing than recognition: we must generate various answers before judging which one "fits". People recall information most effectively when given reminder prompts, such as the initial letter of a required word. Prompts may also be unconscious – a song can remind us of another place or person and then set off a train of reminiscence. Even though the context of learning is eventually lost and what we know remains, contextual clues continue to account for the inexplicable surfacing of information. When confronted with a good memory trigger, the relevant "memory web" fires, drawing the memory out; the more links in the web, the better the chances of retrieval.

**RECOGNITION**
*Recognition is much simpler than recall and enables partial learning to suffice – this is the principle behind the idea of the police identity parade.*

**CONSTRUCTING A PHOTOFIT**
*Recalling a person's face from scratch is more complex than recognizing a face.*

**TRIGGERING A MEMORY**

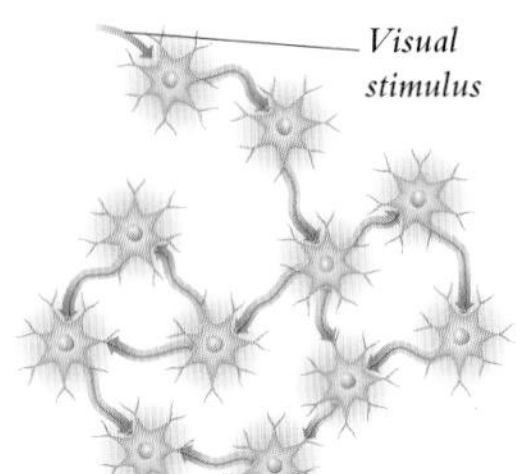


**ONE ROUTE TO RECALL**
*The memory of an old friend may be "hooked out" via any one part of its web. Here, the trigger is seeing a place once visited together.*

*Olfactory stimulus*

**ALTERNATIVE TRIGGERS**
*The same memory may also be accessed via other nerve clusters in the web – for example, through the smell of a familiar perfume.*

## WHY DO WE FORGET?

We can only forget something if we encoded the information in the first place, so sometimes we "don't remember" because information did not actually enter our long-term store. More often, we lose access to a memory; the book is still there but we have lost the filing card that tells us which shelf it is on. At times we can access part of the stored information, sensing that something is "on the tip of our tongue". We may even know that the word we want begins with a "b". In such cases, the information is likely to pop into our heads at a later time. Retrieval problems are more common when there is interference from another memory that uses the same cue. Most of us can remember our home phone number, and probably the one before it, but what about the one before that? We once knew it very well, but the current number interferes with the older memory. Difficulty in recall may also be due to connections in the brain becoming blocked or signals diminishing through lack of engagement with that memory or by damage. When they cannot be recalled memories start to fade, and eventually they disappear.

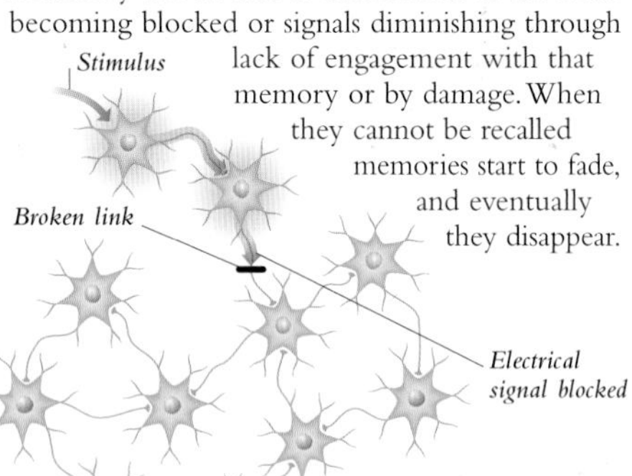


**ONLY ONE ROUTE TO A MEMORY?**
*When a link has faded or been damaged, and there are no alternative routes to complete the network, a memory is no longer accessible. If many links have formed between all the related nerve cells, the memory can usually still be activated.*

**WHERE IS IT?**
*Failure to remember is often a matter of failure to find where we "put" a memory rather than its complete disappearance.*

Health

### Anterograde amnesia

Damage to the hippocampus (a brain part involved in memory formation) causes anterograde amnesia (inability to form long-term memories). When affected people stop thinking about something, it vanishes. They rarely remember anything from after the brain damage: a man may be shocked to hear that his son is dead, no matter how often he is told. Earlier-formed memories can still be accessed.

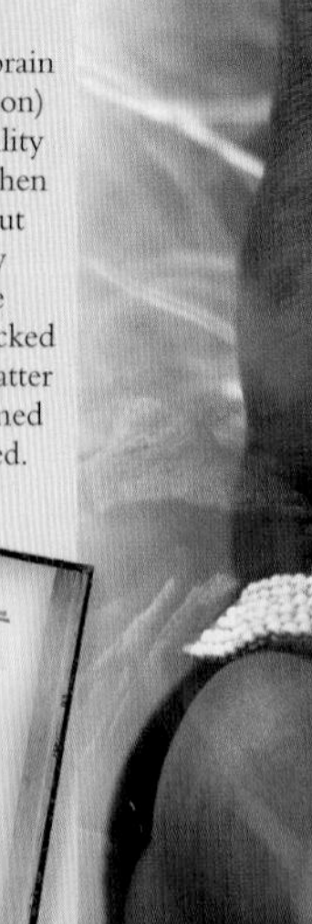

Fact

## Too young to remember

Few adults remember anything at all from the first 2 years of life and memories from before the age of 5 are difficult to access. Most people with younger siblings only have memories of their births if they occurred after the age of 3. As the hippocampus, the brain area involved in drawing together the parts of a long-term memory, does not reach maturity until about age 3, a younger child is not able to lay down explicit memories. Other explanations for childhood amnesia may be that small children do not have the language, organizational skills, or even sense of self that help to lay down structured memories. Nor do they rehearse items to be stored in memory.

# MEMORY AT DIFFERENT AGES

A small child's short-term memory is very limited. Toddlers use short sentences because they cannot plan long ones, and use actions to support memory, acting out stories rather than recounting them. Older people have another advantage over the young: they have many more memories and thus more links in the memory "web". As people age, though, they tend to find it more difficult to remember new things and to recall old ones. The difficulty in memorizing may be due to old brains being less "elastic" than young ones – the nerve cells are more firmly linked in familiar patterns so that greater stimuli are needed to form new links. Short-term memory peaks in a person's late teens and long-term memory peaks by the age of 30; both then show a gradual decline. Most people over 40 know that their memories are not quite what they used to be. They are more likely to forget where they put something and cannot always remember names. With age, it is also harder to learn new skills or to juggle tasks, yet we remember life experiences and self-performed tasks equally well at all ages. With use, memory can be maintained well. Old memories can be preserved by repetition and new ones formed by frequent exposure to exciting and novel experiences.

**"PLASTIC" MEMORY**
*Young people are much better than older people at learning and remembering totally new things.*

**JUGGLING TASKS**
*This chef is able to plan and juggle a multitude of tasks due to a highly effective memory; older people begin to find this type of job more difficult.*

**WITH AGE COMES WISDOM**
*This elderly Namibian storyteller is highly respected within her society for her long, strong memory and accumulated wealth of knowledge.*

# AMAZING MEMORY

Human memory is capable of incredible feats, and we all have the capacity to store massive quantities of information. Some people have an unusual ability to recall vivid details but, by using certain tricks, we could all use our memories better than we do. People make up rhymes, tie knots in hankies, set alarms, and write on their hands or scraps of paper. Memory aids are so effective that older people who use them often remember things better than the young. "Memory maestros" such as Dominic O'Brien, who, after one sighting, recalled a sequence of 2,808 playing cards with only eight errors, use more elaborate methods. Some visualize a route they walk regularly, "place" things to be remembered along it, and retrace the route to recall them. Others invent stories that connect items to be memorized or use mnemonics, which give the initial letters of things to be remembered; it is easier to recall one sentence than individual elements. The sentence "Richard Of York Gave Battle In Vain" is a prompt for a rainbow's colours (red, orange, yellow, green, blue, indigo, violet).

**MEMORY AIDS**
*The diagonal and curved sticks in this Micronesian navigation chart represent wave swells. Sailors used the chart to remind them of their own location and that of different islands.*

**VALUE OF MNEMONICS**
*Mnemonics are a good way of remembering details – for example, the different colours of the rainbow, in the correct order.*

MIND

# THINKING

When we think, we use existing information to generate new information. The way humans think sets us apart from other animals: we can classify information into more than one category, mull things over, and reminisce. But what sets us even further apart is our ability to think about thinking, connect distinct events, form abstract concepts, and hypothesize about the unseen. Because we are not restricted to thinking about things we are directly experiencing at any one moment, we can work out solutions to problems without going through a lengthy process of trial and error. The mental stages people go through when solving problems, being creative, deciding between alternatives, and reasoning are the focus of ongoing study.

## THINKING AND THE BRAIN

All high-level processing involves activity throughout the brain's cortex (outer layer), but the prefrontal cortex, at the front of the brain, is particularly important to thought. The brain is divided into two hemispheres, which, in humans – unlike in animals – have significantly different roles in thinking. Research has drawn attention to some functional differences: in most people, the brain's left side is specialized for language and its right for spatial information. However, the two halves do not function in isolation but communicate with each other through millions of nerve fibres known as the corpus callosum. Studies carried out on epilepsy patients, in whom the connections between the two hemispheres have been severed to stop severe seizures, confirm that the two sides have different capacities to handle verbal and spatial tasks, and that they can act somewhat independently when separated. Such studies have led to suggestions that everyone has two "minds": the right being intuitive, creative, spatial, and "exciting", and the left analytical, logical, verbal, and "dull". Most scientists now believe thinking involves coordinated activity across many brain regions. People with good nerve connections throughout the brain are the most effective thinkers.

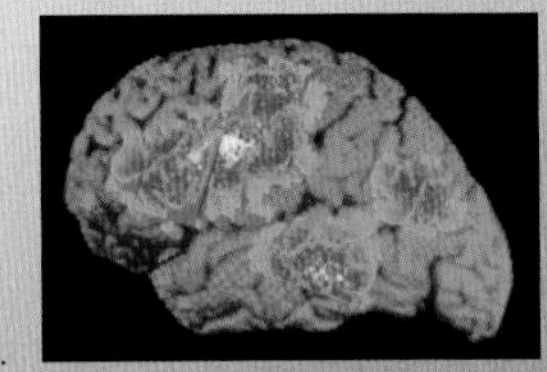

**AN ACTIVE BRAIN**
*The brain areas involved in thinking depend on the subject of thought, but the prefrontal cortex (far left) is always used. This scan of someone thinking about and saying verbs shows active areas, with most blood flow, in orange.*

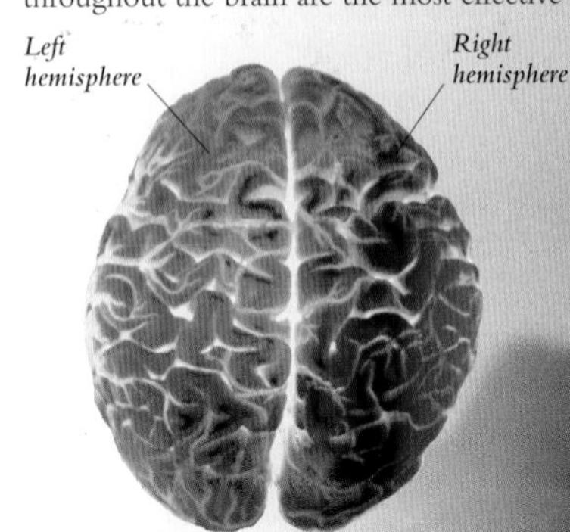


**THE TWO HEMISPHERES**
*A controversial and simplistic, yet popular theory suggests that each side of the brain has a different style of thinking, the left being more rational and the right creative.*

Fact

### Artificial intelligence

Attempts to build "intelligent" robots and machines are useful because they reveal the complexities of the human mind and provide insight into how we solve problems. One branch of the research has recently made a giant leap: computer networks modelled on real brain circuits have been programmed to break down the connections between their "cells," and create new ones. They can therefore "learn", like us, to carry out tasks from playing football to recognizing odours of vintage wines.

**ROBOCUP**
*Artificial Intelligence robots compete at this annual football event. By 2050, organizers hope to have assembled a robot team able to beat the best human side in the world.*

MIND

# CONSCIOUS AND UNCONSCIOUS THINKING

Thinking divides broadly into what is conscious and what is unconscious. Conscious thinking is only a fraction of what goes on in our brains – it is what we are aware of at any one moment and is therefore extremely limited. Unconscious thinking is constantly going on behind our conscious thoughts and determines much of our behaviour. Conscious thinking tends to be logical, sequential, and slowly processed; it is structured in terms of cause and effect and is directly related to the "here and now". Unconscious thinking, in contrast, is intuitive, nonsequential, fast, unexplained, and more easily makes associations between many different pieces of information. In essence, the unconscious mind "multitasks" more effectively than the conscious mind and is not limited to what is happening in the current context. Most rational thinking, including decision-making and reasoning, falls into the category of "conscious thinking". Problem-solving is often a mixture of conscious and unconscious thinking. Creative thinking can be at its most effective when taken out of the realms of conscious awareness. However, conscious processing remains an extremely important factor in creativity because it helps to exclude any unrealistic and unrealizable ideas that are generated by the brain.

**TIP OF THE ICEBERG**
*Conscious thought is just the tip of the iceberg. The great mass of unconscious thinking remains concealed below the surface.*

Issue

## What is genius?

Genius is not about speed learning or high IQs. A genius invents new ways to think, ignoring how past thinkers saw things. A genius looks at a problem in many ways. If asked for half of 11, he or she might give the answers "1 and 1" (11), "3" (ele+ven), and "X and I" (XI), as well as the usual "5.5". Most of us think rigidly, failing to see problems as unique and always comparing them with previous ones. A genius moves existing ideas into new patterns: Einstein did not invent the concepts of energy, mass, or speed of light, but he combined them in a novel way and saw something nobody else could.

# PROBLEM-SOLVING

All thinking involves some "problem-solving", but the term usually refers to working through a set series of stages to resolve a tricky yet defined situation. On a first attempt at a problem, people usually take many unnecessary steps. They mentally generate a small number of possible moves and only think ahead to a limited extent. This approach can be difficult: at some point, a step has to be taken that appears to move away from the goal. This type of "trial and error" is a popular, yet ineffective, technique. Instead, many experts believe that to solve a problem we must first work out the relationships between all its elements. We may have to change the way we view these elements and when we do this successfully we are said to have had "insight"; a fixation on one approach can lead to a lack of insight. As a child, the mathematician Gauss was told to add up all the numbers from 1 to 100 and quickly gave the answer "5050". He had noticed that the numbers formed pairs that added up to 101 (1 and 100, 2 and 99, 3 and 98). Because there were 50 such pairs, the answer was 50x101 (5050). A similar type of insight is required in real-life problems. How do you rescue a man who is hanging from an overpass over a busy road? If rescue attempts from below are unsuccessful, rescuers should mentally restructure the problem to try to reach him from above. Restructuring and insight are the basis of the "information-processing" theory of problem-solving. This approach involves defining the problem, generating a range of possible solutions, and then evaluating them. Surprisingly, we are more effective at this approach than computers; we may consider only a handful of possible solutions, but we tend to hone in on the right handful.

**MULLING IT OVER**
*A chess player can, by thinking through a few possible move sequences, choose a very good move that a computer might only arrive at by examining millions of possibilities.*

**TRANSFERABLE SKILLS**
*The skills required to tackle algebra seem specific but are the same as those we use to solve any problem.*

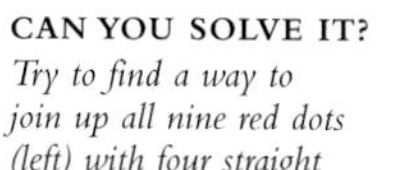

**CAN YOU SOLVE IT?**
*Try to find a way to join up all nine red dots (left) with four straight lines, without removing your pen from the surface of the paper. (For solution,* *see p163.)*

**EXPERT VERSUS NOVICE**
*In many fields, including mechanics, experts will spend time assessing a problem, then quickly solve it. Novices tend to plunge in, delaying a solution.*

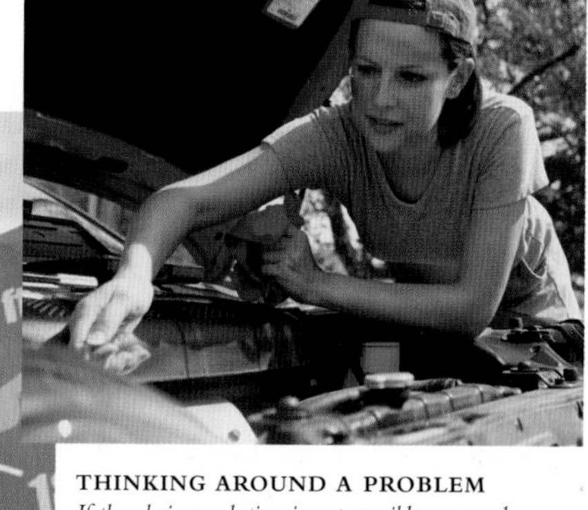

**THINKING AROUND A PROBLEM**
*If the obvious solution is not possible, we need to think again. Here, rescuers are trying to work out how to reach trapped miners when the main access route to the mine is blocked.*

MIND

# RATIONAL THINKING

The term "rational thinking" commonly refers to decision-making and reasoning. Coming to a rational conclusion is often more difficult than we imagine: to do so, we need both sufficient information about the subject or options under consideration and enough time to apply logic. People regularly depart from rational thinking. Some hope to win the lottery when the chances are 1 in 14 million; others will not fly in a plane following a recent crash because they are over-influenced by that rare event. Studies show that humans tend to avoid risk when trying to gain something yet seek it to avoid a loss. Asked if they would prefer £80 for certain or a lottery ticket with a high chance of winning £100 but a small chance of winning nothing, most people plump for the guaranteed £80. However, when asked to choose between losing £80 for sure or a good chance of losing £100 with a small chance of losing nothing, most gamble. We have a tendency to seek evidence to verify our assumptions, and in the process fail to consider things that challenge them. Oddly, if trying to discover if someone has broken a "rule" or cheated in some way, reason comes more easily. Many experts believe we have an innate "cheat-detecting" mechanism, but some think we acquire cheat-detection skills while growing up.

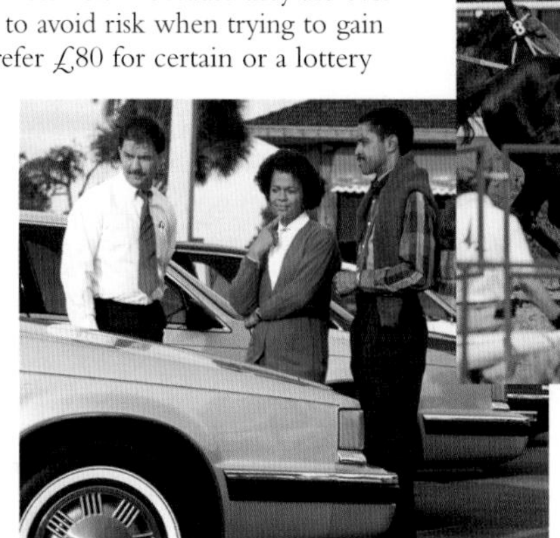

**MAKING CHOICES**
*We are constantly deciding between alternative courses of action. Even a task such as buying a car draws heavily on our powers of rational thought.*

**WEIGHING UP THE ODDS**
*Many people depart entirely from rational thought when placing a bet on a horse race, but seasoned gamblers reduce the risk of their wagers by applying reason to their calculations.*

**"EUREKA!"**
*Although Archimedes chanced upon the theory of water displacement on lowering himself into a bath, the inspiration did not appear from nowhere. He had already spent a great deal of time evaluating the problem.*

# CREATIVE THINKING

The process by which people arrive at new and useful ideas is seen as the pinnacle of human thinking. Valuable creative ideas are found in all areas of life: the invention of the wheel and the telephone, Picasso's masterpieces, and Gandhi's nonviolent methods for political change are examples. Creative thinking generally involves several stages: conscious thought; incubation (when the problem is set aside); sudden inspiration; and development and assessment of the idea. The unconscious incubation period seems to be critical, but its exact role is unclear. It may involve unconscious processing of the problem, shelving of misleading approaches, or gradual activation of the long-term knowledge store, eventually reaching relevant information. Accidental events have played a role in a number of scientific breakthroughs, including the discoveries of penicillin, X-rays, and the chemical treatment (vulcanization) of rubber. Some individuals also make deliberate use of chance: American writer William Burroughs cut up newspapers, randomly arranging the pieces to produce striking combinations of words. Some people believe that society's increasing reliance on technology and a few human experts means there is no longer any need for most of us to think creatively. Also, the more emphasis we place on rational thought, the less creative the world becomes. Humanity must be willing to use the creative power of the unconscious mind in order to respond to all the new challenges we face.

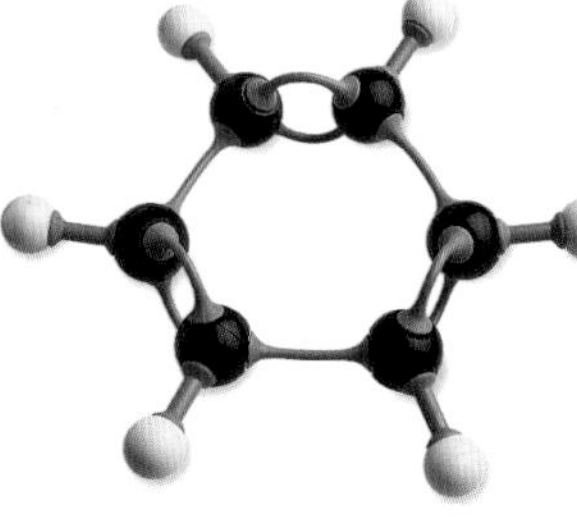

**SLEEPING ON IT**
*In 1865, while trying to work out the structure of the benzene molecule, German chemist August Kekule dreamed of a snake biting its own tail. This inspired vision told him benzene atoms must form a closed ring.*

**TIME: THE BIG IDEA**
*Gradual observation of the cyclical movements of the sun and moon led to the idea that time was measurable. This developed separately in societies all over the world.*

**GROUNDBREAKING DESIGN**
*The Eden project's "geodesic dome" was inspired by the two structures that contain the greatest volume per square foot and the greatest strength for the least weight respectively.*

MIND

Fact

## How Velcro was invented

In 1948, Swiss inventor George de Mestral returned from walking his dog to find that both he and the dog were covered in burrs (clinging plant seed-sacs). Mestral raced to his microscope to inspect a burr and found it was attached by way of little hooks to the tiny loops of his trouser fabric. He had a brain wave: why not develop a two-sided fastener, one side with firm small hooks like those of the burrs and the other with soft loops like those of the fabric? The idea initially met with derision, yet Mestral persevered and perfected the design. Velcro is a multimillion-dollar industry today.

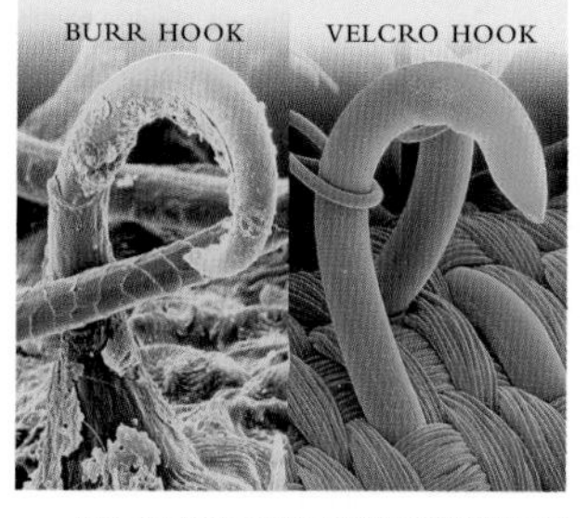

**VISUALIZATION**
*Much thinking involves visual images. Artists do not always paint what they see before them; some paint from images inside their own heads.*

**AN AUDITORY IMAGINATION**
*Mental representations do not have to be visual. A composer, for example, might experience vivid auditory "imagery" and from this write his or her music.*

# IMAGINATION

Thinking involves the manipulation of mental representations of situations. Imagination refers to an individual's ability to develop such representations, which vary greatly from one person to the next, and to manipulate them creatively. Some people are much better at this than others. However, even people who are adept at manipulating ideas in their heads tend to make notes or sketches when they are thinking because it is generally easier to work out relationships and restructure problems when considering something more concrete. Many people's imaginative thinking is based on visual images: they see things in their "mind's eye"; in fact, people who experience vivid visual imagery often believe that it would be impossible to think without using mental pictures – for example, to give somebody directions without picturing the route in one's head. However, the representations on which conscious thought depends do not have to be visual. Some people claim to have no visual imagination, yet have inner experiences that are verbal (inner speech), auditory (mental music, perhaps), or even olfactory (smells). A person who has permanently lost one of the senses may experience heightened mental representations relating to that sense. For example, in a situation strongly associated with a particular smell, such as when drinking coffee or walking in a park on a spring day, someone who has lost his or her sense of smell may be able to evoke olfactory "imagery" similar in quality and intensity to the actual smell previously experienced – a type of "smell memory".

# IMPROVING THINKING

Brainstorming is an active way of boosting creativity. Participants are urged to generate as many ideas as possible, no matter how outrageous, before evaluating them. Research indicates that this technique is most effective for solving problems in which many alternatives are required (for example, "think of as many uses for a brick as possible"). As a group procedure, participants must not criticize the ideas of other people until the evaluation phase, at which time some of the ideas are chosen for the group to develop. Studies show that brainstorming individually and then evaluating the ideas as a group usually produces better ideas than working in groups from the start because some people are to timid to put forward their ideas. "Lateral thinking" is another creativity-boosting technique, involving looking at a problem from as many perspectives as possible and being completely open to any and all ideas. Another useful tip is to focus on an ultimate goal and work back from it, which gives a better chance of achieving the goal than if we try working forwards. Occupying our bodies with physical exercise that requires no thought can free our minds to be creative – and the more we stretch our minds, the more new connecting pathways form in our brains, enabling us to process facts and manipulate ideas more efficiently in the future.

**BRAINSTORMING**
*Studies have shown that brainstorming is one of the most effective ways of coming up with innovative new ideas. It can be carried out by an individual or in a group.*

**ENERGETIC THINKING**
*Some people find that the mind is free to think more creatively when the body is engaged in repetitive physical exercise.*

Issue

## Do we think in our sleep?

The brain's visual areas are active during dreams, as are the areas responsible for our basic emotions, while areas that deal with rational thought and attention remain quiet. This may explain why dreams are so visual and emotional and lacking in coherence and logic. Dreaming may be the mind's way of processing information, helping us to make decisions and solve problems by testing our responses to scenarios without rational thought getting in the way. Other studies suggest dreaming may promote creative thinking by letting us link ideas that we would never connect while awake.

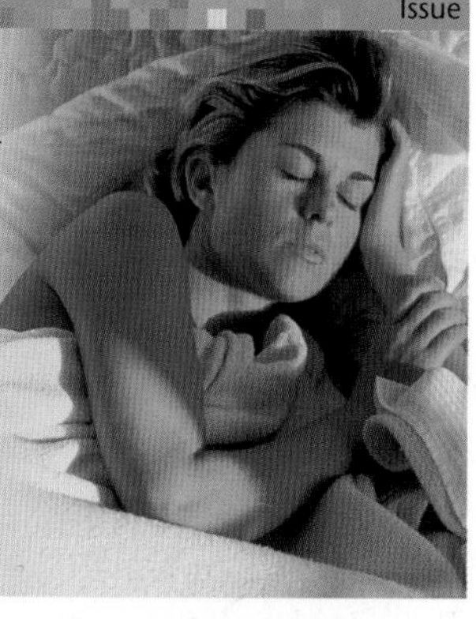

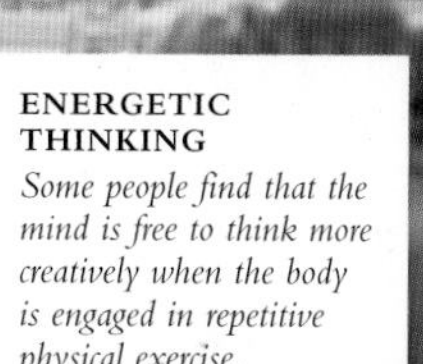

**CASTING THE NET WIDE**
*Lateral thinking involves embracing a wide range of ideas when problem-solving. This approach is more effective than honing in on one idea, just as using a net instead of a rod increases the chance of catching fish.*

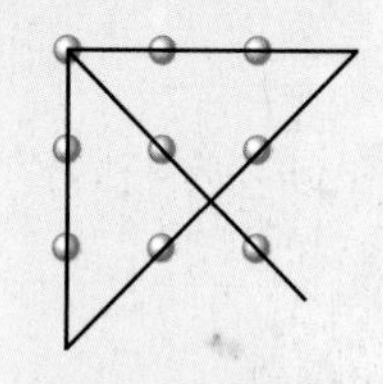

**SOLUTION FROM PREVIOUS PAGE**
*Most people do not "think out of the box", trying to stay within the square.*

# EMOTION

While humans are not always thinking, we are always alert to and immediately aware of any emotion-provoking stimuli around us. Over the course of a typical day, we experience a wide range of different emotions. An item in the news may provoke anger, a favourite piece of music may elicit joy, and a disagreement with a close friend may be distressing. Without emotion, our world would be numb, grey, and lacking in meaning or motivation. However, until recently, science had paid less attention to emotion than to thought, seeing it as somewhat irrational, primitive, and unnecessary. However, we now know that emotional experience is an essential part of how we function as human beings – and an attribute that we share with other animals.

Issue

### Happiness and the brain

Groundbreaking studies conducted by US psychologist Richard Davidson seem to show a correlation between a person's typical mood and the relative levels of electrical activity in the brain's two hemispheres. His research suggests that people with higher activity levels in the left hemisphere of the brain are generally happier, more enthusiastic, and more positive than those with higher levels in the right, who seem to be somewhat more anxious and withdrawn.

## WHAT IS EMOTION?

Emotion is a complex experience involving both the brain and the body. An emotion occurs when we are exposed to something that means something to us personally, such as the sight of a familiar postmark on a letter. We become aware of physical arousal, such as the heart beating faster. We also mentally evaluate the stimulus to decide if the body's initial reaction was "rational": does the letter contain a job offer, a tax demand, or news from an old friend? Emotions can be expressed in many ways, such as with a smile, a frown, a curse, or a rush to tear open the envelope, but our instinctive reaction may occasionally be suppressed. Our most intense emotional experiences tend to occur when the three important components of emotion (physical arousal, conscious awareness, and outward expression) are well coordinated. An emotion lasts as long as the stimulus that provoked it, so is often short-lived. Moods tend to be longer-lasting than emotions, essentially forming filters that colour our emotional experiences. They can last hours, days, or even longer. Joy and distress are emotions; happiness and sadness are moods. A sad mood casts a shadow over all emotions; a happy mood intensifies moments of joy and lessens moments of distress.

Fact

### Mood enhancement

Alterations in levels of certain brain chemicals, such as serotonin, can have a profound effect on a person's mood. Prescribed drugs such as fluoxetine (the principal constituent of Prozac) are based on this principle. By raising levels of serotonin in specific areas of the brain, the drugs lift depression and are even said to lead to "mood brightening" in people who are not depressed. Some herbal preparations, such as St John's Wort, are also known to have a mood-enhancing effect. Mood enhancement is not always a positive thing because it can make a person feel overconfident and not sufficiently aware of his or her real circumstances.

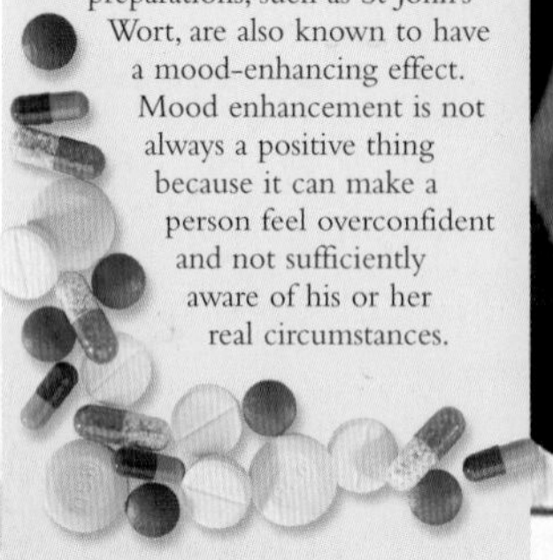

Automatically relayed signal
Thalamus
Registers fear signal
Amygdala
Sends signals to adrenal glands, which trigger release of hormones
Prefrontal cortex
Assembles slower, reasoned response that may override amygdala's response
Potentially overriding signal
MIND

**THE ANATOMY OF FEAR**
*Two circuits in the brain are involved in our reaction to fear. An immediate physical response is activated by the amygdala. However, the prefrontal cortex may override this response if it reasons that it was unnecessary or irrational.*

# TWO CIRCUITS OF EMOTION

Studies of fear have taught us a lot about the brain's role in emotion. When we encounter something frightening, the stimulus is registered in the thalamus in the centre of the brain, a relay station for sensory input. The thalamus sends a signal to the brain's amygdala, a structure that relays the message to the adrenal glands on the surface of the kidneys. These glands release hormones that immediately cause the body to start preparing for action. At the same time as the signal is conveyed to the amygdala, the thalamus also sends a signal via the sensory cortex to the prefrontal cortex, an area at the front of the brain that assembles a slower, more considered response. The body may already have begun its fear response through the amygdala, but signals from the prefrontal cortex can override this if the fear is seen to be irrational. For example, a loud rustling sound in a bush might cause the amygdala to initiate a fear response, but if the prefrontal cortex reasons that the rustle was simply caused by a cat, we can take control of our fear and relax.

**THE FEAR RESPONSE**
*If we are scared by a dog, the brain circuits involved in fear cause adrenaline to be pumped into the blood. The heart beats faster, breathing rate increases, and pattern of blood circulation alters, all of which sends oxygen-rich blood to the muscles to prepare us for flight.*

# CONTROLLING EMOTIONS

Involvement of the brain's cortex in emotional experience ensures there is an element of judgement and decision-making involved. The prefrontal cortex is not fully developed until a person is over the age of 20, which may explain why children are given to tantrums and teenagers can be more impulsive than adults. However, emotional control varies from person to person. Some always explode when they are angry – in such cases, the "calm down" message from the cortex may be too weak to modify the amygdala's immediate emotional response. A moderate degree of emotional arousal can sharpen awareness and improve performance, but excessively powerful emotions lead to a loss of mental focus. We can generally be taught to control anger and other excessive emotions. The links between the brain's cortex and amygdala can be strengthened through techniques such as breathing exercises, meditation, and yoga.

**UNRESTRAINED EMOTION**
*Children have less control over their emotions than adults do owing to the prefrontal cortex not yet being fully developed.*

**DEFUSING ANGER**
*An effective way of controlling anger is to separate oneself from the source of the anger and "count to ten".*

**INSTINCTIVE EMOTION**
*Humans have powerful emotional responses to a huge range of things, from personal life events to watching our favourite team.*

Health

## Phobias

A phobia is an excessive fear response to a generally harmless stimulus. Some objects of phobias could occasionally pose some sort of threat, such as spiders, dogs, and heights – even mice and enclosed spaces. However, people with phobias cannot control their fears even when they are at no risk whatsoever. Phobias occur when the brain is unable to send a signal to turn off an unnecessary fear response. Some phobias may be effectively treated by exposure to the feared object in safe conditions.

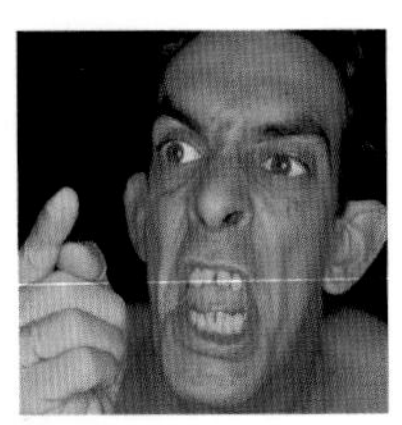
**ANGER**
*Anger is our response to a perceived threat, in which the body's resources are marshalled to confront it.*

**JOY**
*Joy is the most positive of our basic emotions and is probably the greatest motivator in life.*

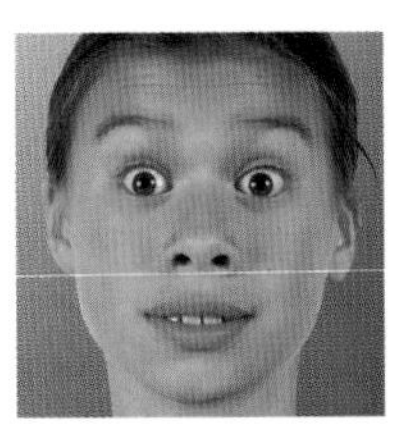
**SURPRISE**
*We raise our eyebrows in response to the unexpected, widening the eyes so as to take in as much as we can.*

**FEAR**
*Perhaps our most primitive emotion, fear is closely related to anger but causes us to withdraw from threats.*

**DISTRESS**
*An upsetting event readily produces a heavy feeling in the chest, tears in the eyes, and a lump in the throat.*

**DISGUST**
*Disgust is generated by a small area in the brain's cortex, leading to gagging and pursing of the lips.*

# THE BIG SIX

Psychologists have long debated how to classify emotions. Many now agree that there are six "basic", universal emotions (fear, anger, joy, distress, disgust, and surprise), produced by circuits embedded deep in the brain (*see* Two circuits of emotion, p165). Basic emotions are fleeting and occur when the brain areas involved in emotion are activated by an appropriate stimulus. They are out of our immediate conscious control; it is hard not to be surprised when someone creeps up behind us. The typical facial expressions of the basic emotions are universally recognized, even across very different cultures. They probably evolved early in human evolution and had considerable survival value. Basic emotions are also thought to be hardwired into the brains of mammals, but individual experience and cultural factors influence the development of emotions in humans.

History

## An early theory of emotions

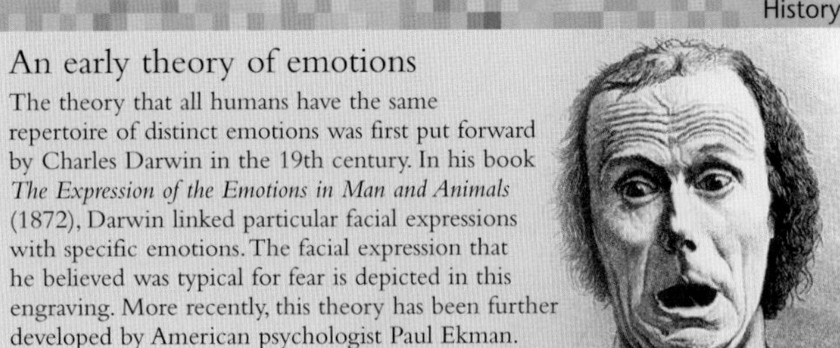

The theory that all humans have the same repertoire of distinct emotions was first put forward by Charles Darwin in the 19th century. In his book *The Expression of the Emotions in Man and Animals* (1872), Darwin linked particular facial expressions with specific emotions. The facial expression that he believed was typical for fear is depicted in this engraving. More recently, this theory has been further developed by American psychologist Paul Ekman.

# COMPLEX EMOTIONS

Human emotional experience is far more complex and diverse than the six basic emotions alone. We also have a repertoire of up to 30 other emotions. As well as anger, we experience irritation, guilt, shame, exasperation, and a whole range of unpleasant variants. Joy is related to satisfaction, bliss, ecstasy, pride, gratitude, and love. Some believe these so-called "complex emotions" are no more or less complex than the basic emotions, but that they evolved later on in the process of human evolution. Others believe that the complex emotions are blends of two or more of the six basic emotions. For example, guilt could possibly be seen as a combination of disgust (at oneself) and fear (of the consequences), just as a blend of blue and red makes purple. In a similar way, alarm might be a blend of fear and surprise. Complex emotions require more processing by the brain's cortex than basic emotions and are not thought to be shared by other animals. They probably evolved in order to help people survive as human society itself became increasingly complex.

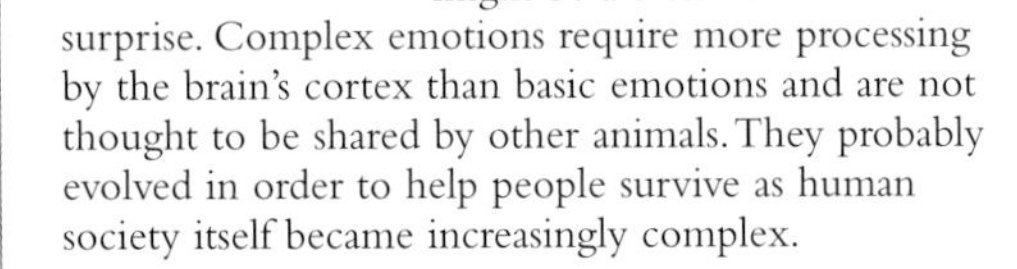
**BLENDING**
*Complex emotions may be a blend of the basic six – just as a range of hues can be created from a palette of six paints.*

**LIFE PARTNERS**
*We feel more comfortable making personal and business partnerships with people who freely express complex emotions (be they compassion, guilt, or even jealousy) because we feel they can be trusted.*

**A DEVELOPING MORALITY**
*Young children feel shame on being chastized by a parent; we all learn about relationships and about what is right and wrong through the experience of this and other complex emotions.*

Issue

## Why do we cry?

Crying is uniquely human. Other animals shed tears, but solely to wash the eyes. We cry with emotion, perhaps to elicit sympathy. There is some evidence that tears of emotion have a different composition from tears that keep the eyes healthy. So perhaps emotional tears rid the body of stress hormones – which may explain why we feel better after "a good cry". It seems strange that we also cry with joy, but there may be some overlap with the brain circuit involved in distress. Whether shed in distress or joy, tears must be a primitive emotional response because they are very hard to fake.

MIND

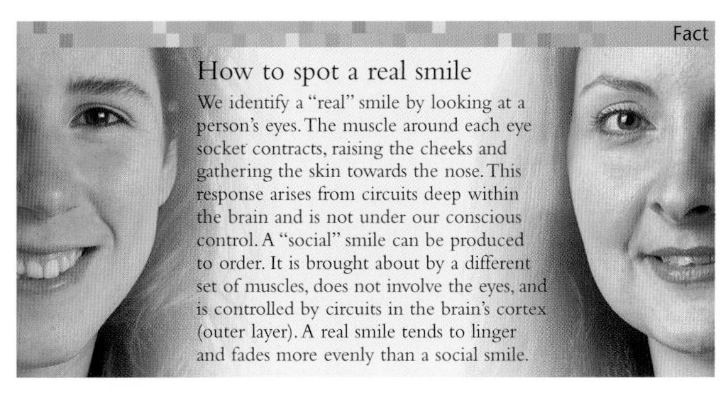

Fact

### How to spot a real smile

We identify a "real" smile by looking at a person's eyes. The muscle around each eye socket contracts, raising the cheeks and gathering the skin towards the nose. This response arises from circuits deep within the brain and is not under our conscious control. A "social" smile can be produced to order. It is brought about by a different set of muscles, does not involve the eyes, and is controlled by circuits in the brain's cortex (outer layer). A real smile tends to linger and fades more evenly than a social smile.

## THE EXPRESSION OF EMOTION

Gestures, posture, facial expressions, and eye contact tell us a lot about people and how they feel about us. Standing close may be a sign of romantic interest or an invasion of personal space; a shrug can express contempt, boredom, or disrespect. Maintaining eye contact during a conversation conveys sincerity, interest, or close attention, yet intense eye contact can be hostile. Emotional expression adds colour to conversations: we may slam a fist on a table to emphasize a point or raise our eyebrows in surprise, amusement, or disbelief. Controlling emotions is not easy, but we are usually able to control how far we express them, such as by manufacturing a smile or suppressing a look of disgust. The fact that humans display emotions outwardly can help us to convey our true feelings to one another, or to disguise them.

**IMPACT OF THE EYES**
*The eyes are used in the expression of all our emotions – useful when they are the only feature on view.*

**ADDING COLOUR THROUGH GESTURES**
*Close eye contact and emphatic body language show that someone is passionately involved in a debate.*

## READING EMOTIONS

Humans are adept at reading each other's facial expressions, which explains our tendency to mirror the emotions of others. Studies have shown that when we look at a smiling face, our facial muscles start to contract. But what about our ability to detect "false" emotions? Some people have a good sense of when a facial expression and other body language do not match; others find genuine emotions more difficult to detect. Yet the inability to read each other easily can be socially useful, allowing our relationships to run more smoothly. The lie detector (polygraph) test measures the physical components of emotion, such as increased heart rate, respiration rate, and sweating. The theory is that a nervous system response is easier to detect and harder to fake than any related change in facial expression, but the lie detector is unreliable in its current form. Some people show an emotional response simply because they are embarrassed or nervous. Moreover, habitual liars may not get emotional when they lie, so the lie detector cannot tell when they are expressing false emotions.

**IS HE LYING?**
*Even with complex lie-detection equipment, it can be difficult to tell if a person is trying to mislead.*

**MIRRORING**
*Watching someone cry, even in a film, may cause a lump to form in the throat or tears to well up.*

## GENDER AND EMOTION

It is generally believed that women express, identify, and understand emotions more easily than men. Recent studies suggest some differences between male and female brains, enabling women to distinguish more subtle changes in emotion. Women have linguistic and emotional functions in both the brain's hemispheres, but men are likely to have them only on one side. Another theory, now disputed, is that women have a wider corpus callosum, the band of tissue that transmits information from one side of the brain to the other. This could facilitate the incorporation of emotions (usually generated on the right) with language and thought (associated with the left). Yet in many cultures it is simply more acceptable for women to show emotion.

**TWO POLES OF EMOTION?**
*Differences in the ways that men and women express emotions are due both to biological and to cultural factors.*

# LANGUAGE

Understanding human behaviour is impossible without taking account of the huge influence of language. Language is the chief way in which humans share information across generations, with their contemporaries, and even with themselves. The ability to think and express ourselves verbally has immeasurably extended our control over the world: we are able to make sense not just of the present, but of the past and the future too. Through our grasp of words, we can communicate complex and novel ideas, and learn about things never seen or experienced. Humans also build relationships around language; it helps us to develop an awareness of our own minds and is a powerful tool to influence those around us.

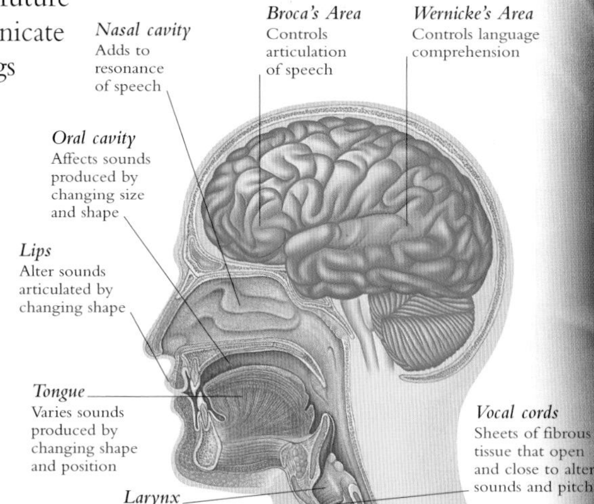
Nasal cavity
Adds to resonance of speech
Broca's Area
Controls articulation of speech
Wernicke's Area
Controls language comprehension
Oral cavity
Affects sounds produced by changing size and shape
Lips
Alter sounds articulated by changing shape
Tongue
Varies sounds produced by changing shape and position
Vocal cords
Sheets of fibrous tissue that open and close to alter sounds and pitch
Larynx
Contains the vocal cords

**THE ANATOMY OF LANGUAGE**
*Humans are born with specialized anatomy that enables us to produce and interpret speech sounds. The brain controls all of the language apparatus.*

## BUILT FOR LANGUAGE

The ability to communicate elaborately sets humans apart from all other species. We have highly specialized speech organs. In the throat is a voice box containing fibrous cords that move together or apart to vary the air passing between them. We also have very mobile facial muscles that enable us to shape precise sounds. The real key to language, though, is a complex brain that controls our speech organs, interprets language, and relates symbols to words. The human brain thinks of what to say and then translates this into words, and none of the noises made by other animals combine to form anything as complex as human language. Listening to, speaking, or reading just one word involves electrical activity in about a million nerve cells. The two language areas in the left side of the brain were discovered by Paul Broca and Carl Wernicke in the 19th century through studying the speech impairments of people with damage to each region. Wernicke's area makes sense of the speech we hear and the words we read. Broca's area processes our own speech and writing.

**CAN CHIMPS TALK?**
*Our nearest relatives can learn a limited vocabulary, but they do not have the innate human sense of how language works.*

Profile

### Noam Chomsky

Born in the US in 1928, Chomsky is one of today's most respected linguists. He is well-known for his theory of the Universal Grammar: that the human brain has an inbuilt programme enabling children to acquire language without tuition. He does not suggest that the words and rules of specific languages are inborn, but that all humans have the ability to identify different parts of speech (such as nouns and verbs) and understand the relationships between them. This view is widely accepted, although controversial.

## "INNATE" LANGUAGE

Children progress to intelligible speech in a very short time, suggesting that humans may be hardwired for learning language. The critical period for learning to discriminate between and reproduce the different sounds of a language with the accuracy of a native speaker occurs early in life. In a newborn baby, the brain contains millions of nerve cells that are sensitive to every sound a human could utter. Those frequently activated by the speech patterns and sounds of the baby's community form more complex connections. From about age 4, inactivated nerve cells start to die off. Almost every child becomes effortlessly fluent in his or her mother tongue merely by hearing it spoken and interacting with others. However, if deprived of such stimulation at an early age, children are likely to have problems learning to speak. This is illustrated by the well-documented case of Genie, a girl of 13 who was rescued from extreme deprivation in the early 1970s. Genie had spent her life shut in a room with little contact with others. She had hardly heard a spoken word when she was rescued and had almost no speech. With teaching, Genie built up a considerable vocabulary, but she was never able to grasp how to arrange words meaningfully. While this may in part be down to her emotional abuse, it seems likely that it was also too late for the unused language areas of Genie's brain to function normally.

**EARLY LEARNING**
*Speech starts to develop from a very early age as children absorb language and begin to reproduce the sounds they hear.*

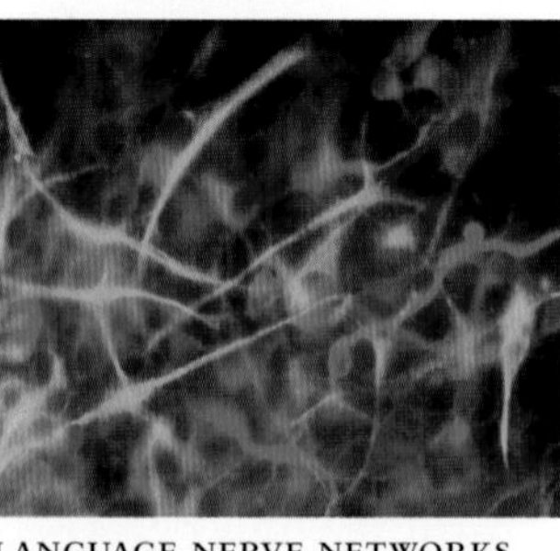

**LANGUAGE NERVE NETWORKS**
*At birth the brain has millions of nerve cells that are sensitive to all the sounds of speech. Only those frequently activated by a familiar language persist to develop into networks.*

**FREE-FLOWING SPEECH**
*Children instinctively begin to link words together, applying the rules of grammar in song and play. Rhythm and repetition reinforce language acquisition.*

**THE POWER OF WORDS**
*Humans are extremely social creatures. Language provides us with an excellent means of interacting through the conveying and sharing of information.*

## SPEECH PERCEPTION

Speech comprises sentences, which break down into words, syllables, and then consonants and vowels. One theory of how humans recognize speech is that, on hearing sounds, we unconsciously go through the motions of producing them and thereby identify what they are. It is certainly true that seeing people's lips move makes speech easier to "hear". In fact, what we see can even change what we hear. If the sound of a man saying "baba" is superimposed on a video of him saying "gaga", people hear "dada" (a sound halfway between the two). Basic speech sounds are always heard as "pure" sounds. For example, two consonants such as "b" and "d" can be electronically blended to create a series of intermediate sounds, but people only ever hear one or the other. Context is also important in speech recognition. Words are easily identified when part of a conversation but are harder to recognize if spliced out and presented alone. In one study, only 30 per cent of words could be recognized when isolated. Conversely, if some consonants in a conversation are replaced with random noise, the words are still recognizable.

**HEARING AIDS**
*It can be difficult to hear speech without seeing the speaker, and some people even say they cannot hear without their glasses.*

**FEELING SPEECH**
*Helen Keller, deaf and blind from an early age, learned to "hear" speech using tadoma, a method of feeling for a speaker's lip movements and the vibrations of his or her vocal cords.*

### High-speed understanding

Humans produce speech at around 12 speech sounds per second. However, we can understand speech in our mother tongue even when it is sped up to an astonishing 50 speech sounds per second. If a click is repeated 20 or more times per second, we are no longer able to hear a sequence of separate sounds, but simply a buzz. So how can we understand speech when it is spoken far too quickly for us to identify all of its component sounds? If we know the context of the speech, the brain is able to "fill in the gaps" between the sounds it can identify.

# HAVING A CONVERSATION

Whenever people engage in any type of conversation, their brains are in a continual flurry of activity. While we are listening to someone talking, our brains are having to take in and make sense of innumerable pieces of information: not only speech sounds, but also important visual clues such as lip movements, facial expressions, and gestures. In order to produce a spoken response, our thoughts must be clearly formulated and converted into intelligible strings of words in the correct grammatical order. The muscles that control the voice box and shape the lips and tongue to pronounce speech sounds then need to be given precise commands. This sequence of activities relies on many different areas of the brain functioning at lightning speed, and the coordination of these areas is absolutely vital. For the most part, however, we remain unaware of all the complex processes going on inside our heads.

*Motor area* Sends signals to muscles that produce speech

*Route of nerve signals*

*Visual area* Receives and analyses nerve impulses from the eye

*Auditory area* Detects and analyses speech sounds, pitch, and intonation

*Broca's area* Puts together spoken response

*Wernicke's area* Receives and interprets visual and auditory input

**ARTICULATE BRAIN**
*In language processing, sensory signals are received by the visual and auditory areas of the brain's cortex (outer layer) and passed to Wernicke's area for interpretation. Broca's area then produces a plan for speech, which is finally executed by the motor area.*

**PUTTING IT INTO WORDS**
*A casual chat may not feel like hard work, but it gives the brain's language centres as much work to do as an intense business discussion.*

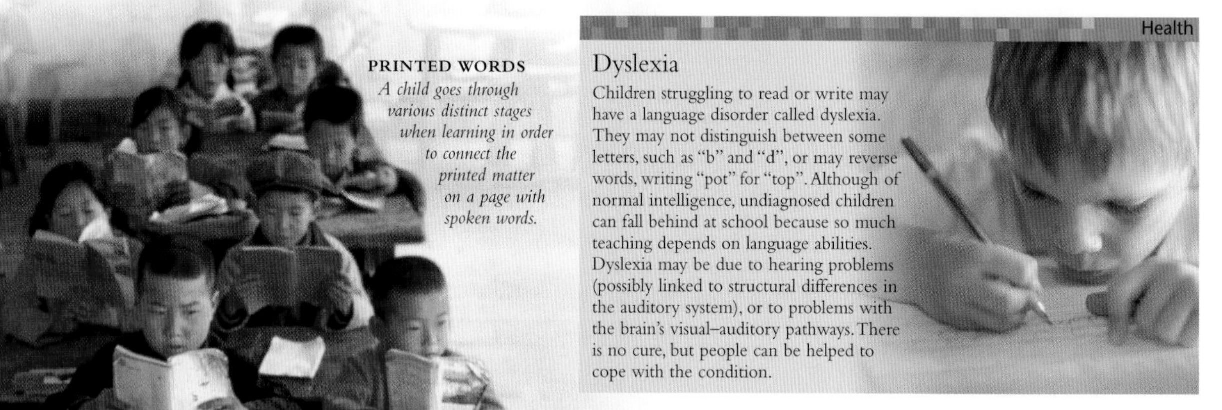

**PRINTED WORDS**
*A child goes through various distinct stages when learning in order to connect the printed matter on a page with spoken words.*

Health

## Dyslexia

Children struggling to read or write may have a language disorder called dyslexia. They may not distinguish between some letters, such as "b" and "d", or may reverse words, writing "pot" for "top". Although of normal intelligence, undiagnosed children can fall behind at school because so much teaching depends on language abilities. Dyslexia may be due to hearing problems (possibly linked to structural differences in the auditory system), or to problems with the brain's visual–auditory pathways. There is no cure, but people can be helped to cope with the condition.

# LEARNING TO READ

Most children are ready to learn to read by age 4 or 5, by which time they have instinctively acquired a large vocabulary and basic grasp of grammar. Connecting spoken and written words comes less naturally. Most children initially acquire a sight vocabulary, memorizing the "shapes" of whole words – a child may read his or her name simply by recognizing a pattern of letters or combination of brush strokes. However, this type of learning is very limited. Full comprehension of written words only comes with discovering how to decode the written symbols that make up words. Alphabetical languages in which there is an almost one-to-one correlation between letters and sounds, such as Spanish, are the easiest to learn to read. Others, such as English, are more complex because most sounds can be depicted by a number of letters and most letters can be pronounced in a variety of ways. The process is harder in nonalphabetical languages, such as Mandarin Chinese, in which characters or parts thereof may relate to units of meaning rather than to sounds. Yet all over the world, children learn to read in the same way: they break down written words into smaller parts (to groups of letters, then single letters; or to parts of characters, then single brush strokes) before building them back up again.

**FLASHCARDS**
*Many children initially learn to recognize certain words by connecting how they look with related images.*

## LEARNING ANOTHER LANGUAGE

Fluency in more than one language is easily achieved if a child is immersed in each from an early age. In the early years, the neurological network in the brain is fine-tuned to differentiate every speech sound that exists. At this stage it is no more difficult to learn a second language than a first, and a child raised by parents who speak different languages may learn two "mother tongues" simultaneously. The child may occasionally use words from one language when speaking the other, but such problems disappear over time as the languages' boundaries crystallize in the child's mind. However, this facility to learn languages is gradually lost; it is far harder to acquire new languages later on, such as when a child starts school or when an adult moves to a new country. Although many older people become at ease in a foreign language, they are unlikely to be able to reproduce the accent with total accuracy because the brain no longer makes such fine distinctions between sounds. Fluency in a second language depends on the age at which it is introduced and the extent of a person's exposure to it. Some adults seem to be better at learning new languages than others. The reasons for this are unclear but could be related to differences in the size of the brain's language-processing area. Some experts believe that this area is larger in certain people due to additional use during childhood.

**YOUTHFUL EASE**
*This little boy is likely to grasp the complexities of Chinese with relative ease as he is learning at such an early age.*

**ADULT LITERACY**
*It is generally much more difficult to pick up new language skills as an adult than it is as a child.*

**MULTILINGUALISM**
*Many cultures use a variety of languages in day-to-day life – road signs in Jerusalem are written in Hebrew, Arabic, and English.*

**GROWING UP BILINGUAL**
*Children of parents who speak different languages have an excellent chance of becoming bilingual at an early age with very little effort.*

## NONSPOKEN LANGUAGE

Speech is not the only way in which we communicate. Worldwide, many people use sign language, a formal method of communicating through hand movements. If a person uses signing as his or her main language, the language areas of the brain are activated just as in speech. Formal sign languages, such as American Sign Language (ASL), also have a specialized grammatical structure in the same way that spoken languages do; it is this underlying grammar that enables the expression of complex ideas. Hearing-impaired children learning sign language as a first language go through the same stages of language acquisition as hearing children. However, because they are unable to relate written symbols to sounds, they can have more problems learning to read. Recent studies suggest there is a "critical period" for acquiring sign language just as there is for acquiring any other. Although it may not appear this way to the untrained, signs do often relate visually to the things they are expressing – the signed word for "tree" in ASL, for example, depicts a tree trunk's vertical relationship with the ground; the Danish sign emphasizes the tree's canopy of leaves; and the Chinese sign shows the tree's upwards growth. Other types of nonspoken language have been devised for use in specialist fields such as computer sciences and mathematics.

**USING A SIGN LANGUAGE**
*These Native Americans are using a traditional sign language once used to enable people of different language groups to communicate freely with one another.*

**SENSITIVE FINGERS**
*Braille is a writing system based on patterns of raised dots, which enables blind people to read.*

**TALKING WITH ANIMALS**
*Animals, such as dolphins, can be trained to interpret and respond to commands given using hand signals, a form of sign language.*

Issue

### Do we need language to think?

We can have some thoughts without possessing language. Babies are able to grasp a simple concept like the softness of a toy, but they cannot discuss it. However, following a train of abstract thought is almost impossible without language. None of the ideas of science or religion could have been formulated without the words we use to express them. Whether the particular language we speak influences how and what we think is much debated. We clearly find words to express concepts already used in thought, but the ways in which we describe certain things, such as space and colour, are also likely to affect how we see the world.

**INDIVIDUALITY**
*The same life experiences and teachings affect each individual in a different way. This phenomenon is both a result of and a cause of our unique natures.*

# THE INDIVIDUAL MIND

A question frequently asked by psychologists is how, if we all share the same basic mental architecture and characteristic human traits, can we each be so unique in our personalities, motives, choices, abilities, and interests? Most people believe they are the way they are because of a combination of two factors: their individual life experiences and an inherent genetic bias. It is true that nature and nurture work together during childhood to mould the adults we become, but that does not mean that our behaviour is "determined" – we still have the ability to override our natural instincts.

A complex combination of human qualities makes somebody the individual he or she is. The most prominent of these features are probably a person's personality, intelligence, gender identity, and sexual orientation. Each one of these important facets of humanity originates within the human mind and, at the same time, exerts a profound influence upon it.

## ASPECTS OF INDIVIDUALITY

Personality refers to the very distinctive way in which every individual tends to react to his or her life experiences. Intelligence encompasses a wide range of factors, from artistic creativity to mathematical prowess, and from spatial awareness to good verbal skills; the most clearly intelligent people have high levels of ability across all of these areas. How we feel about our own gender and who we find ourselves attracted to, not to mention how these very personal characteristics fit into the "norms" of the societies we live in, are also vital factors in making us the individuals we are.

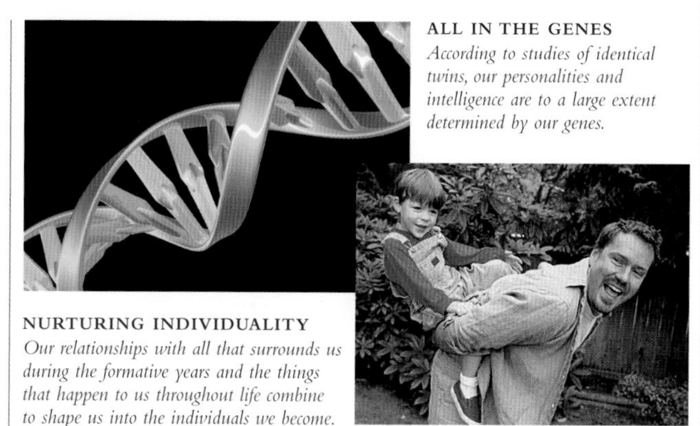

**ALL IN THE GENES**
*According to studies of identical twins, our personalities and intelligence are to a large extent determined by our genes.*

**NURTURING INDIVIDUALITY**
*Our relationships with all that surrounds us during the formative years and the things that happen to us throughout life combine to shape us into the individuals we become.*

## NATURE AND NURTURE

Individuality is based on a mixture of genes (nature), life experience (nurture), and probably random chance. Genetics builds the body, constructs the brain, and lays down circuits of nerves, giving us an instinct for language, a repertoire of emotions, and a large degree of personality. Experience then builds on all of these elements, allowing redundant nerve pathways in the brain to die off and new nerve connections to form. Random chance plays an unseen, but perhaps significant, role.

The nature versus nurture debate became an area of great controversy in the 20th century. However, the two things are inextricably linked, so cannot constitute a dichotomy. Research suggests that much of our personality, gender identity, intelligence, and sexual orientation is genetically influenced. In fact, about half of the behavioural and mental variation between people living in a similar environment is thought to be due to differences in their genes. So genes really can influence our IQs, how outgoing we are, and whether or not we are easily upset.

However, nurture also plays an critical role in personal development. The people who surround us during the formative years, added to our life experiences and the age at which they occur, profoundly shape the individuals we become. Research carried out into children raised in the wild is an extreme illustration of this. Some babies caught by wolves have been raised as members of the pack, growing up with no speech, no ability to walk on two legs, and a range of doglike instincts, such as a powerful sense of smell. It seems that nurture has, to some extent, moulded the children's brains into those of dogs.

## CAN WE CHANGE WHO WE ARE?

The brain has been described as "plastic" because it can remould itself. This plasticity does not mean we can change our personalities, intelligence, gender identities, or sexual orientation overnight, but we can control some aspects of our own minds. We can push ourselves to great achievements with adequate dedication, practice, and support, honing our skills and strengthening our nerve circuits to perfection.

We have a far better chance of moulding our minds if we truly know ourselves. Yet, for all our supposed self-awareness, we often have clearer views about others than ourselves. The Freudian view is that the main determinants of personality lurk in the unconscious, an area of the mind to which we do not have access, and that people have an infinite capacity for self-delusion. If this is true, asking people to "know" themselves, and therefore change, is doomed to failure. We are all ready to believe that other people are driven by unconscious desires and extremely deluded about their own motives, but are far less willing to accept that we ourselves may be driven and deluded in this same way.

Fact

### The value of individuality

Western culture tends to emphasize the individual as special and unique. In many Eastern cultures, individuals may be valued only as part of society and defined solely by their social status and relationships. In such cultures, displays of personal emotion are considered vulgar; a person should blend in rather than stand out from the crowd.

**SOCIETY OR INDIVIDUALITY?**
*While Western babies usually sleep alone from birth, Balinese children tend to sleep together or in parents' beds.*

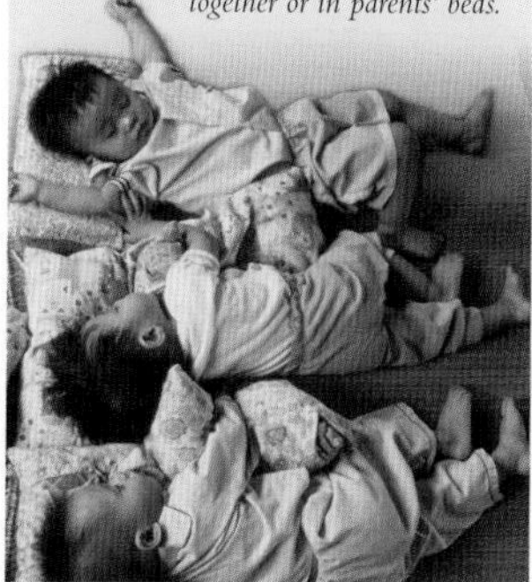

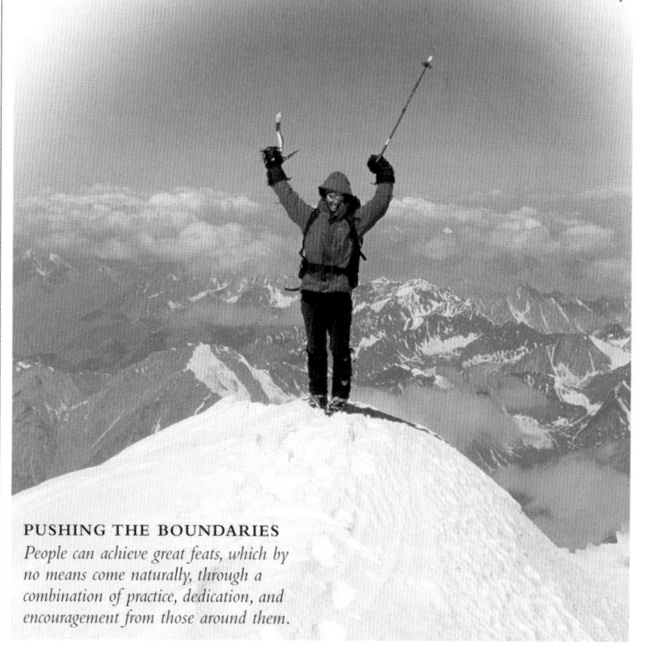

**PUSHING THE BOUNDARIES**
*People can achieve great feats, which by no means come naturally, through a combination of practice, dedication, and encouragement from those around them.*

# PERSONALITY

Although we can all feel happy or nervous, something that scares one person may exhilarate another. The fact that one person tends to react fairly consistently to his or her life experiences but different people react to things in different ways lies at the core of what personality is. Psychologists see personalities either as unique combinations of "traits" or as conforming to a limited number of "types". Temperament is closely related to personality; it is a person's characteristic disposition. Personality and temperament are influenced by many things and, in turn, colour all of our life experiences.

## PERSONALITY TRAITS

Just as we can create any hue from just three primary colours, some psychologists think all personalities are formed from a small number of personality "traits". These traits have been identified by asking people to describe themselves and others and standardizing the results using set questionnaires. Because characteristics do not fall into neat clusters, we cannot say exactly how many traits there are: one expert may claim that there are 16, while another opts for only two. Such variation is not a problem. At times it helps to consider fewer, wider traits; at others, more, precisely defined ones. Most experts now agree on five sliding scales, on which different "placings" combine to form limitless personalities. A person could fall at the sociable end of the "extroversion" scale, the conforming end of "openness", the middle of "neuroticism", and so on. Another could be a mirror image or differ in only tiny details; either way, every personality is unique.

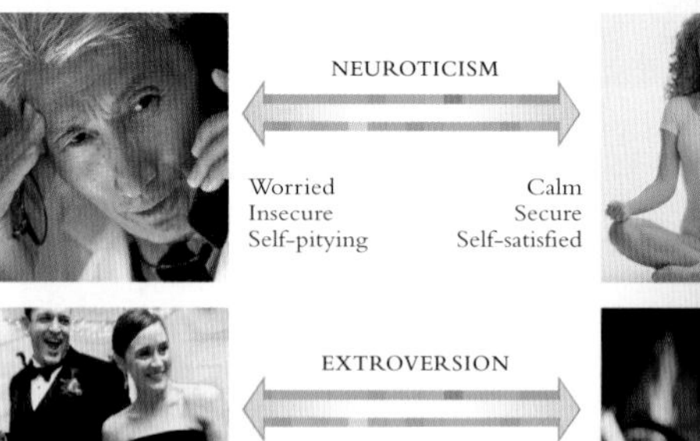


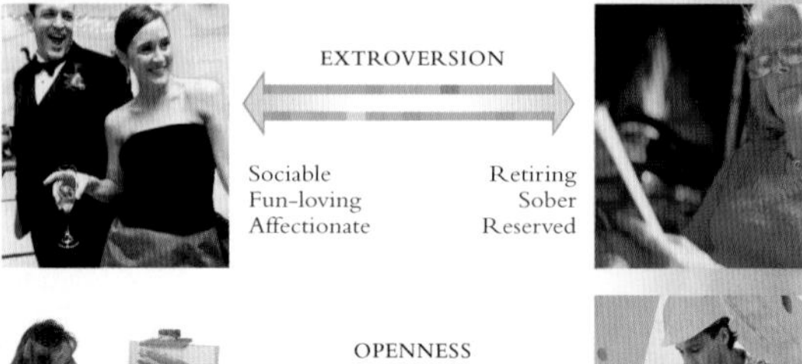

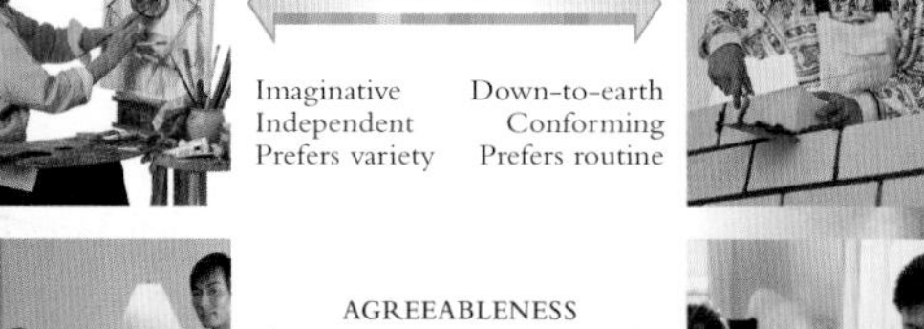


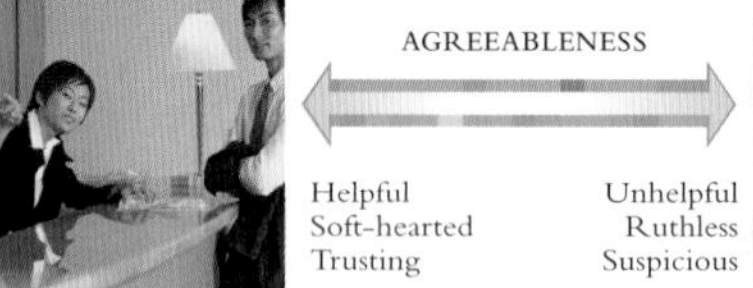


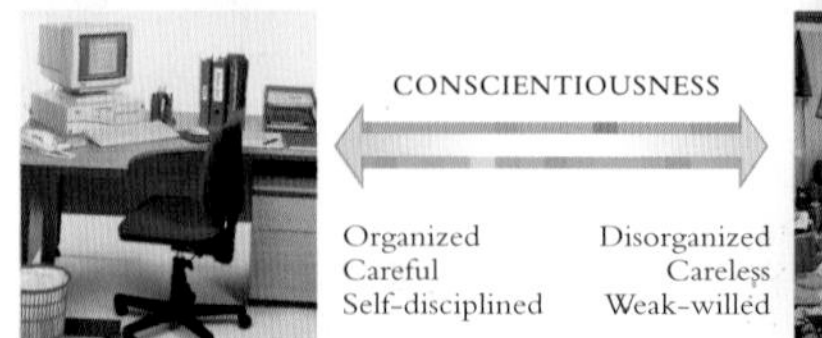


**THE "BIG FIVE" PERSONALITY TRAITS**
*Most experts now agree on five distinct traits, each of which works on a sliding scale. An individual is likely to sit at a different point on each scale, but very few people are found at either extreme.*

**STANDING OUT**
*The distinctive ways in which different people act when confronted with the same situation lie at the heart of the concept of personality.*

# PERSONALITY TYPES

Theories of personality types are more of a two-dimensional "all or nothing" approach than the "sliding scale" approach of trait theories. A well-publicized example of a type theory is the Type A–Type B pattern. "Type A people" are classified as achievement-oriented individuals who find it difficult to relax. They are highly driven to meet goals and tend to be vigorous, efficient, and easily frustrated. "Type B people" are the opposite: generally relaxed, unhurried, and down-to-earth. They may still work hard but are rarely as driven as Type A people. Links have been made between Type A people and increased risk of heart attack. The Type A–Type B pattern is now seen as too simplistic an approach to measuring personality, however, and other "type theories" have become more prominent. The "Meyers–Briggs Type Indicator" is a more recent type theory based on the work of the Swiss psychologist Carl Jung (1875–1961). This theory classifies people as falling at one extreme of each of four dimensions (extraversion–introversion; sensing–intuition; thinking–feeling; and judgment–perception), giving a total of 16 possible personality types, some of which are more common than others. Today, the Meyers–Briggs test is widely used by organizations to assess employees and their work-related behaviour.

**DISTINCT TYPES**
*A currently popular theory suggests that we can each be classified as one of 16 basic personality types.*

History

## The four humours

The idea of classifying people into "types" was developed by Greek thinkers around 400BC. They proposed that personality was determined by the dominant one of four bodily fluids (humours). A "sanguine" person was thought to have high levels of blood; a "phlegmatic" person high levels of phlegm; a "melancholic" person black bile; and a "choleric" person yellow bile. Hippocrates was the first person to apply this idea to medicine. As it turns out, however, black bile does not even exist.

**AN OLD THEORY**
*This 19th-century illustration depicts the four personality types proposed by the Greeks.*

# TEMPERAMENT

**PLACID OR PROBLEMATIC?**
*All children, even as babies, react very differently to situations. Some aspects of a baby's temperament are enduring.*

The term temperament refers to the usual vigour of a person's response. It is a useful way of assessing young children, whose personalities are not yet fully developed; research has shown that a baby's disposition in the early weeks is to some extent a sign of the adult he or she will become. A baby's reactions are fairly consistent: happy, easygoing babies tend to be easy to wean, put to bed, and bathe; irritable children are usually difficult in all these situations. Most experts agree that temperament results from inborn tendencies as well as from how a family reacts to a child. Children whose shyness exasperates their parents are much more likely to stay withdrawn than those whose parents help them to cope with new people and situations.

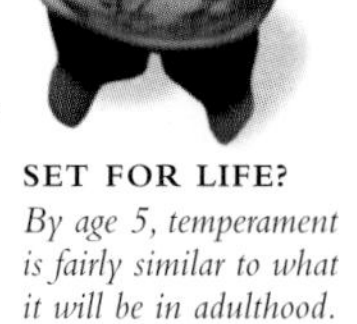

**SET FOR LIFE?**
*By age 5, temperament is fairly similar to what it will be in adulthood.*

# INTERACTION

Our highly individual personalities make each of us more compatible with certain people than with others. How different personalities interact adds colour to life. Understanding our own personalities and those of the people around us can help to improve our relationships, and for this reason, personality tests have become increasingly popular in the workplace. Research shows that groups made up of people with similar personalities may reach decisions faster but that mistakes can be made because not all viewpoints are adequately represented. On the other hand, in a team composed of people with wildly differing personalities, there may be problems reaching understanding, let alone consensus. Teams composed of varied yet open-minded people, able to appreciate the different styles and outlooks of others, generally have fewer conflicts in the long run and are far more likely to achieve satisfactory and balanced results.

**LEADERS AND FOLLOWERS**
*Some people act on impulse; others are more cautious. Teambuilding exercises aim to optimize these essential differences.*

Profile

## Sigmund Freud

Austrian psychoanalyst Sigmund Freud (1856–1939) provided great insight into the human mind, even though much of what he believed has since been disproved. He is renowned for his theory of three components of personality, although their existence is no longer accepted. The "id" required instant gratification of its impulses; the "ego" was realistic, arising from experience; and the "superego" was moral. Freud believed that these three parts were often in conflict, but that in a balanced character, the ego remained in firm control.

## GENES AND INHERITANCE

Studies have examined how far personality variation can be attributed to genetic factors by considering similarities in personality between twins; blood relatives; and adopted children and their birth and adopted families. Results suggest that almost half of personality variation may come down to our genes. However, nurture also plays a part, as shown by a US study from the early 1970s that suggested newborn boys cried more than newborn girls and were more demanding babies. The same results were not achieved in the UK. Circumcision was, at this time, routine in the US and not considered a traumatic procedure. Most doctors now agree that circumcision is painful and that circumcised babies cry more because they are in pain. Because mothers responded quickly to their crying sons early on, their sons learned this was an effective way of getting attention in the long-term.

**ALL IN THE FAMILY?**
*There is much debate over whether aspects of personality pass down through generations.*

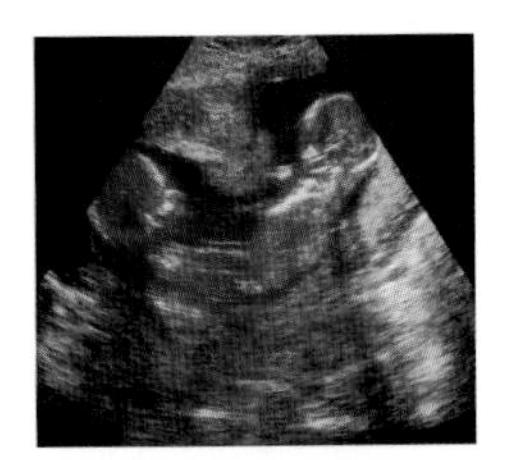

**TWINS BY NATURE**
*A number of studies have shown incredible personality similarities between identical twins who have been raised apart.*

## THE EFFECTS OF NURTURE

Our personalities are moulded by the circumstances in which we find ourselves. Studies showing that identical twins do not always react in the same way – even if reared together – clearly suggest that nurture is a dominant influence on personality. We are influenced by our family relationships from birth. As we grow up, our peers and the culture we were born into play important roles in making us who we are. The things that happen to us in life, such as illness, loss of loved ones, and changing schools, as well as the age at which these occur (losing a parent at the age of 5 has a different effect than losing a parent at 15) also have a huge impact. Even "events" that occur while we are still in the womb may affect how we develop as individuals. Because we are a social species, the views and behaviour of other people also influence how we develop, both directly (by training us, rewarding us, and setting "standards" for us to live by) and indirectly (by increasing or decreasing our self-esteem).

**LIFE CHANGES**
*Moving house is a momentous experience for a young child; without proper support, it may affect personality development.*

**OUR SURROUNDINGS**
*Where we grow up and who we mix with will influence who we become in both subtle and more obvious ways.*

**POSITION IN THE FAMILY**
*According to many experts, where a child sits in the family structure and hierarchy helps to define him or her as an individual.*

Health

### Personality disorders

There is a known connection between extremes of personality traits (*see* p174) and personality disorders. There are three broad classes of this type of disorder. The first is characterized by eccentric behaviour, for example paranoia; the second type features very dramatic behaviour, as seen in histrionic people who constantly crave stimulation; and people with the third type, such as those with obsessive–compulsive disorder, show anxiety. Personality disorders may become particularly obvious at times of stress.

**COMPULSIONS**
*People with obsessive–compulsive disorder (OCD) find it hard to control their behaviour; for example, some may clean obsessively.*

## MATERNAL ATTACHMENT

Many believe that the mother–child relationship forms a template for later relationships, and that any disruption of this early bond has long-term effects. Initially, there was a wide body of work supporting this hypothesis: children whose mothers had died or been hospitalized or imprisoned in the first two years of their lives shared a pattern of loss and depression. Few doubt that long-term maternal deprivation may have an adverse influence on a child's development, especially if there is no alternative adult with whom to bond. However, most now disregard the notion that short-term deprivation – caused, for example, by a mother going out to work – has similar effects. Recent studies show that short-term deprivation can cause unhappiness, but rarely has lasting effects on development if a child is given good daycare or another loving alternative to the mother. Even children who have been severely deprived may suffer few long-term effects if they are given sufficient attention and love in the end.

**A CLOSE BOND**
*The relationship between mother and child has traditionally been considered the most important in life.*

**SHORT-TERM SEPARATION**
*Contrary to popular belief, spending some time apart from the mother in the formative years is unlikely to have any long-term effects on a person's character.*

**A MIX OF INGREDIENTS**
*Our personalities are made up of many different ingredients, and it is virtually impossible to identify all the individual influences once we are fully developed.*

# HOW DIFFERENT INFLUENCES INTERACT

We are all like unique cakes. The ingredients that combine to make us who we are include our genes, environment, and life experiences, but we only ever see the end product. It is impossible to identify the individual factors that make people who they are – especially the relative importance of each. Our personalities go with us into every experience; in turn, how we react to a particular experience (such as by becoming fearful, stoical, or exhilarated) colours it, making its influence very personal. When knocked back by adversity, for example, one person will come out fighting while another loses self-confidence. There is reason to believe that coping teaches us to cope, in which case the fighter is better able to deal with adversity in the future, while the person who lost confidence is likely to crumble further. To study the relative influences on personality, many comparisons need to be made between different groups of individuals. The members of certain of these groups have some or all of their genes in common; others share much of their environment; some share neither or both. The results of such comparisons allow psychologists to discover the ways in which nature and nurture interact behind the scenes to create our unique personalities.

**IDENTICAL IN EVERY WAY?**
*Identical twins may have all their genes in common, yet they almost always grow up to be two totally individual people, even if they insist on dressing alike.*

**NURTURING CHARACTER**
*Life experiences combine with genetic factors in influencing who we are and how we behave. A happy childhood will make a person more secure and self-confident.*

Issue

## The roots of aggression

Some young children experience aggression in the home. They may shout and fight to get what they want or see their parents behave this way. There are clear correlations between a child's aggression with his or her peers and that shown by his or her parents. Pain and discomfort also influence aggression: frequently ill babies appear more likely to be aggressive adults. Yet there is also a genetic basis for differing levels of aggression, shown by the fact that adopted children often resemble their birth families. Weak messages from the brain's prefrontal cortex (*see* Controlling emotions, p165) and low levels of serotonin, a brain chemical, may also be associated with aggression.

# INTELLIGENCE

Our general capacity for high-level mental functioning, including learning, memory, and thinking, is known as "intelligence". The term is used, when referring to humans, in a similar way to how "performance" is used to describe cars. Both terms refer to overall effectiveness: the combination of a person's creative, verbal, and spatial skills; the combination of a car's acceleration, braking, and fuel efficiency. Someone may be very good at maths but have poor verbal skills, and measures of overall intelligence – such as IQ tests involving problem-solving items – take such variation into account.

## TYPES OF INTELLIGENCE

Intuitively, we judge people as "smart" or "stupid" depending on the extent to which they can cope with a range of mentally demanding situations. We also recognize that someone who is impressive in mental arithmetic may be totally unable to pack the car boot effectively. People with abilities across all intelligence areas are said to have good "general intelligence". Studies show that general intelligence affects everything a person does, although most individuals are still better at some things than others.

Where we have particular talents, there is a split between those with excellent verbal abilities but poor spatial skills and those who are the opposite. Some questions on IQ tests draw on prior knowledge (crystallized intelligence); others require people to handle totally novel problems (fluid intelligence).

In life, we also have to deal with emotionally laden situations. It has been suggested that the ability to empathize and work with others and to recognize and understand our own emotions constitutes "social intelligence". In fact, there is little concrete evidence that this type of intelligence exists.

**RECALLING FACTS**
*Quiz shows that draw on our general knowledge illustrate our levels of crystallized intelligence.*

**SOCIAL INTELLIGENCE**
*The ability to empathize and work with others is thought by some to be the most important form of intelligence and the basis of many of the other forms.*

Issue

### Spatial and verbal abilities

Most men are a little better at spatial tasks such as parking cars, visualizing how something might look from a different angle, and putting shapes together. Women tend to be better at verbal tasks. The differences are small and the overlap between the sexes is large. Mathematical ability has been associated with spatial skills, and there is evidence that the average man does better at maths than the average woman. Yet this gender difference has been disappearing over the last 40 years, suggesting that teaching methods or female attitudes towards maths may play a larger part than once thought.

## SPECIAL TALENTS

An essential part of being "us" is the things we can do. We may be academic, musical, sporty, and so on. Certain of these skills seem to come naturally to us; others we have to work at. Some experts argue that seven types of ability may be distinguished: linguistic, musical, logical and mathematical, spatial, physical, intrapersonal (awareness of our own feelings), and social. Literature on child prodigies abounds with stories of children who have remarkable talents. Very early language skills have been recorded: one child is said to have had a 50-word vocabulary at 6 months of age and a speaking knowledge of five languages by age 3. Child prodigies also exist in the worlds of music, the sciences, and the arts. Some, like Mozart, go on to later glory. The clearest example of exceptional skills in children is perfect pitch – the ability to name and sing a certain note without a reference tone. However, there is little evidence that most adults who are admired for their talents showed evidence of an exceptional natural ability during childhood. "Prodigies" certainly have special abilities, but they also, almost invariably, received vast amounts of education and training during early childhood. Certain areas, such as musical ability, are influenced to a great extent by genetic factors. Yet a genetic disposition is not sufficient in itself; it must be reinforced by parents and teachers.

**IMPACT OF TRAINING**
*Child "prodigies", including exceptional young gymnasts, do have special talents, but they have usually received a huge amount of personal attention and training as well.*

Issue

### Is intelligence increasing?

Studies indicate striking rises in raw IQ scores over the past 50 years – they appear to be increasing by around three points per decade. Yet such a rate of increase would imply that average people 100 years ago were functioning at the level of retarded people today, which cannot be the case. It is also strange that general knowledge and maths skills have been falling while IQ scores have been rising. Possible explanations for rising pass rates in IQ tests are better nutrition, greater use of computers and educational toys, and increased familiarity with IQ tests.

**THE BRAIN'S MAZE**
*The ability to find our way around is just one of the many facets of human intelligence. Some people are particularly good at orienting themselves as a result of a heightened spatial ability.*

# IQ TESTS

The first practical IQ (intelligence quotient) test was developed by French psychologists in 1904 to identify children who might not benefit from normal schooling. Simple tasks asked children to follow a finger with their eyes or shake hands with the tester. More difficult items included naming body parts, finding word rhymes, and repeating a string of numbers; some questions related to social skills. The score came out in terms of mental age: if a boy of 7 passed the same number of items as the average 7 year old, then his mental age was the same as his chronological age; a 7 year old who scored the same as the average 5 year old had a mental age of 5. An IQ derives from this mental age score and gives an indication of relative intelligence. The average IQ is 100; higher scores indicate better-than-average performance and lower scores worse. More recent IQ tests have built on this original model, but test a wider range of abilities. However, they barely touch on social skills. IQ tests are quite reliable: if retested, even after many years, people tend to do as well or as badly as they did on the first occasion. Those who perform well in the tests also tend to be successful in life.

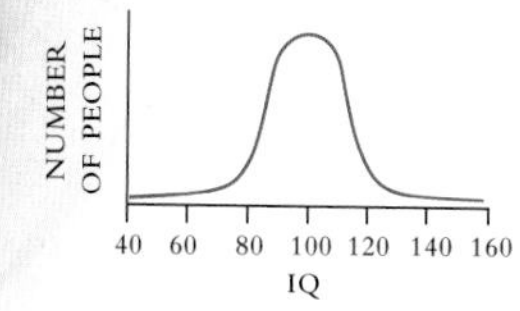


**INTELLIGENCE QUOTIENT**
*As is shown in this classic bell-shaped curve, about 80 per cent of people have an IQ falling somewhere between 80 and 120.*

**IQ TESTING**
*People all around the world sit formal tests so that relative intelligence levels can be studied. In many countries, children take these tests on a regular basis while at school.*

MIND

# INFLUENCES ON INTELLIGENCE

There is certainly a marked genetic component to intelligence, shown by the fact that identical twins tend to have similar IQs even when reared separately – often more similar, in fact, than fraternal twins reared together. However, environment and upbringing also play a crucial role: a child without access to a good education or a nutritious diet is at a significant disadvantage. One controversial claim is that genetic factors lead to differences between the IQs of different racial groups. However, there is no reliable evidence for this; any differences shown in tests could just as easily be the result of environmental factors. Aboriginal Australians, for example, have spatial abilities that surpass those of white Australians, but this could be as much the result of child-rearing practices as of genetic factors.

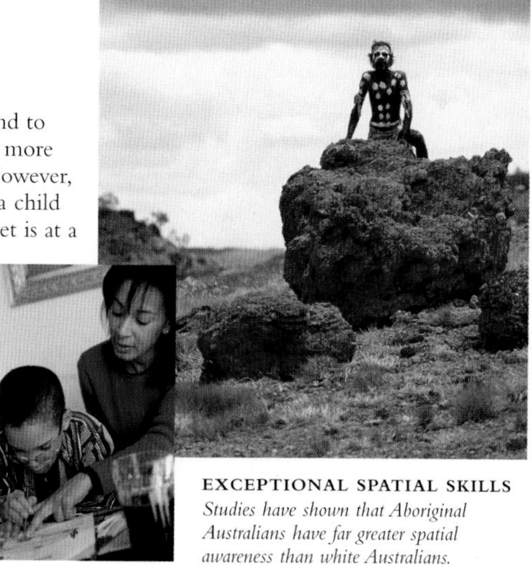

**NURTURING INTELLIGENCE**
*A child encouraged by an interested adult during the formative years is more likely to make the most of his or her intelligence.*

**EXCEPTIONAL SPATIAL SKILLS**
*Studies have shown that Aboriginal Australians have far greater spatial awareness than white Australians.*

Fact

### The Impact of self-esteem

Believing we can do something and knowing that others believe we can do it makes us more likely to succeed. For example, children who are told they are good at maths score higher in maths tests. Studies have shown that social groups considered inferior in their home country tend to do badly at school, but when they move to a culture that does not recognize their "inferiority", they perform better.

# GENDER AND SEX

Gender and sexuality are issues that most people tend not to consciously question. Yet who we feel we are, how we behave, and who we are attracted to have such a major impact on our lives that it is surprising how most of us are able to ignore these questions almost entirely. Gender role (how men and women should behave), gender identity ("I am a man/woman"), and sexual orientation (who we are attracted to) are deeply ingrained in our psyches before we even reach adolescence. If our own feelings about these personal issues coincide with the specific "norms" of the culture we live in, nothing is likely to bring them to our attention. Yet some people feel "who they are" does not fit neatly with society's views on gender or sexuality.

## GENDER ROLE

To a large extent, gender role is imposed on individuals by society's "rules" of male and female behaviour, which themselves result from fundamental differences imposed by our biology. Role division is very strict in some societies and less so in others, but even in Western society today – where there is a trend towards greater overlap – some separation remains. Gender role is unconsciously moulded from an early age. Children learn by seeing how men and women behave in the home, the media, and the community, and roles are gradually reinforced until they are second nature. Adults strongly influence the development of a child's gender role: they frequently use the words "boy" and "girl" to describe children, thereby highlighting gender differences; they encourage children to play with "gender-appropriate" toys (in the West, girls are often given dolls and boys toy cars); and boys are generally encouraged to take part in more vigorous play than girls. Chores are also frequently split along gender lines: in much of the world, girls are urged to help with the cooking and boys with hunting, fishing, or cultivating food. In society in general, more attention is paid to male assertiveness, girls are treated more protectively, and female displays of emotion are more tolerated. Some people rebel against the gender role given to them by society; others simply feel that it doesn't "fit". Increasingly, strict gender roles do not reflect the diversity and individualism inherent in societies.

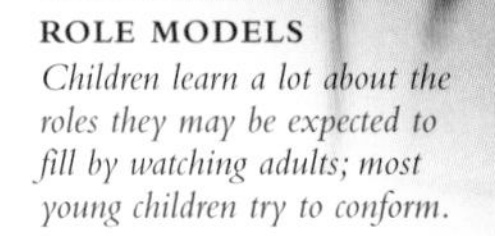

**ROLE MODELS**
*Children learn a lot about the roles they may be expected to fill by watching adults; most young children try to conform.*

**CULTURAL INFLUENCES ON GENDER**
*Miao women (see p452) are raised from an early age to embroider fine clothes. A girl who weaves and embroiders the most intricate patterns becomes the most sought-after bride in a community.*

**BREAKING STEREOTYPES**
*These female bodybuilders have chosen to build up their bodies in a manner traditionally thought of as more "male".*

Issue

### "Male" and "female" brains

British psychologist Simon Baron-Cohen argues that women's brains are programmed to be "empathetic" and men's to be "systematic". He does not suggest that all men are alike and total opposites of all women, who are also alike, but refers to a general tendency. He suggests that autism is an extreme version of the male brain. Baron-Cohen discusses the role genes may have played in determining the brain types, but critics argue that he plays down social and cultural factors that were inevitably also involved.

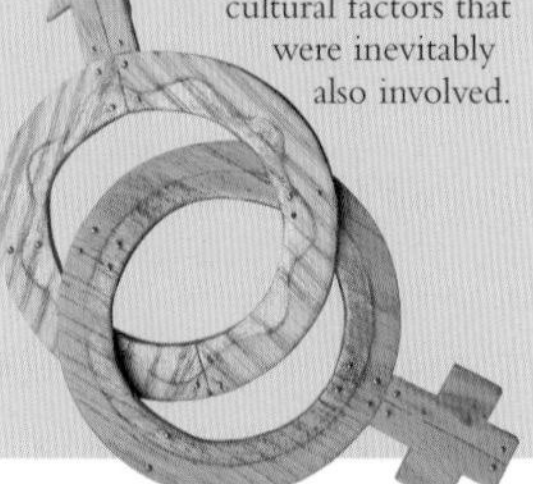

## GENDER IDENTITY

The feeling that we inhabit the correct body, or that we do not, is distinct from gender role (*see opposite*). Children classify themselves by gender long before they can talk. By age 2, they know their own gender and can pick out boys from girls in pictures; by age 3 or 4, they understand that when they grow up they will be the same gender as they are now; and, by 6 or 7, they understand that gender does not change just because appearances change – that if a boy puts on a dress he will not be a girl. Gender is "assigned" at birth according to a baby's genitals. But where does our feeling of belonging to one gender or other come from? Various studies suggest that the development of male gender identity is, like that of the male body, set by fetal hormones – particularly testosterone. If this hormone does its job properly, little boys seem to grow up feeling like little boys; when its effect is impaired, a genetic male grows up feeling female. There have been several cases in which healthy baby boys have been raised as girls – perhaps due to accidental removal of the penis during circumcision or undeveloped external genitals. Doctors have occasionally decided to perform sex-change operations, removing the testes and making the external genitals appear female. In such cases, the young children take on a female identity, but this rarely lasts to puberty. Most eventually switch back to being male because this is their overriding gender identity. Some people who are raised as their correct genetic gender can be confused about their gender identity, but the reasons for this remain unclear.

**EARLY CLASSIFICATION**
*It seems that, even before they speak, babies show more interest in children of their own gender than in those of the opposite.*

**SIMILARITY IN DIVERSITY**
*The concepts of gender and sexuality are frequently used to divide people into discrete groups, but our common humanity transcends our differences.*

**DIFFERENT IDENTITIES**
*These people were born biological males. They now choose to live as women, some having undergone gender reassignment operations.*

History

### Ancient attitudes

Attitudes towards sexuality change across time and cultures. In Ancient Greece, male bisexuality was simply part of the learning process, older men becoming "mentors" for teenage boys. In contrast, early Roman men were socialized to be dominant in all aspects of life, and sex was seen in terms of dominance and submission. Young men forced into "submission" would not go on to dominate the world. For this reason, relationships between older men and young boys were outlawed, though homosexual relationships with slaves or prostitutes were acceptable. Later, Roman culture fell into line with Greek culture.

## SEXUALITY

Most men have a male gender identity and most women a female gender identity. Sexuality is less clear-cut, involving a mixture of erotic attraction, romantic attraction, sexual behaviour, and self-identification, and a person can sit at a different point on each of these scales. There is a continuum of sexual attraction that runs from exclusive desire for people of the same sex to exclusive desire for those of the opposite, with various stages in between. In fact, most people probably do not fall at either extreme, even if their sexual behaviour does. There is a clear genetic basis for sexuality: over half of identical twins (who share all their genes) have the same sexual orientation, while just 22 per cent of nonidentical twins do. There is also some evidence of brain differences between gay and straight men, perhaps due to fetal hormones. However, biological factors may simply offer an individual an alternative route through life. In many cases, it seems children respond to their early environment by opting for "the most appropriate" of several biologically programmed patterns of development – in which case, nurture also has an influence on sexuality.

**LIFE PARTNERS**
*A majority of the world's societies are structured around monogamous heterosexual partnerships that people expect to remain in for life.*

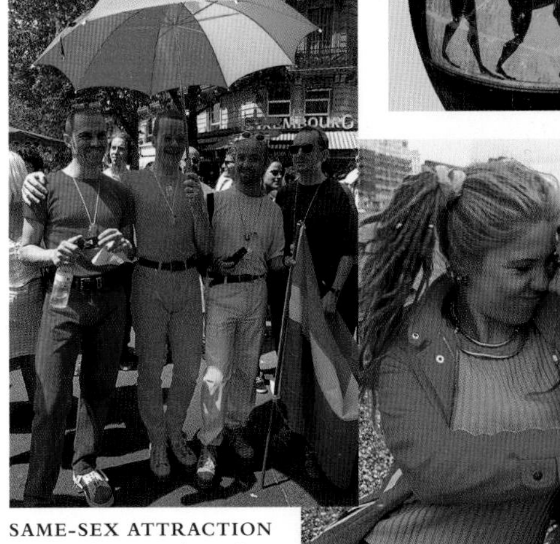

**SAME-SEX ATTRACTION**
*Most nonheterosexuals say they knew they were homosexual or bisexual by their teenage years, which seems to rule out a simple nurture view of sexuality.*

# LIFECYCLE

# LIFE CYCLE

The human life cycle starts with birth and ends with death, potentially more than 100 years later, and people in every society go through the same defining stages: childhood, puberty, and adulthood. Various events within each life stage, such as puberty and marriage, are marked and celebrated. How we do so, and the age at which major life events, such as marriage, are observed vary. Religion, tradition, social standing, and income, have traditionally defined the importance of each phase.

**OLD AND YOUNG**
*The two extremes of age come together in this grandmother and baby of the southern Africa !Kung tribe.*

In the first few years of life, humans grow and develop at a rapid rate. Throughout childhood, growth continues. By adulthood, the body is mature, but the process of aging starts almost immediately.

## THE BEGINNING

The first stage for everyone is birth, probably the most amazing part of the life cycle of any animal. Whether or not the new baby survives is, in part, a chance of geography. A baby born in the US in 2000 was 10–15 times more likely to survive than a baby born in central Africa. In the US, about 7 out of every 1000 babies born die in the first year. In contrast, 159 babies die before reaching 1 year in Niger, West Africa. Risks apart, birth is such an important occasion that most societies have traditions connected with it. It is usually a time of great rejoicing, although in many traditional cultures there is a period of waiting before the baby is named. This is designed to make sure that the baby is strong enough to survive.

Children go through incredible changes as they grow. From the helpless newborn, they become competent individuals within a few years. Childhood is a training ground for adulthood; in some parts of the world, children do household and agricultural tasks at an early age. Masai boys of East Africa, for example, are expected to herd cattle alone by the age of 8. In most developed societies, children may lead a longer, more sheltered upbringing, with several years of schooling before they start work.

**THE JOYS OF CHILDHOOD**
*For most children, like these three girls from northern Thailand, childhood is a time of happiness and freedom.*

## TRANSITION

Huge changes take place at puberty, the transition from child to adult. We develop physical features that make us sexually attractive, and in some societies, a person is regarded as an adult once he or she enters puberty. Girls may even be considered ready for marriage. To mark this important life stage, girls in many traditional societies undergo a period of seclusion during which their food, clothing, and behaviour may be subject to rules. Boys who live in societies still governed by tradition, often undergo a test of their manhood. Aboriginal Australians, for example, take part in their first "walkabout" and are introduced to the stories of their ancestors. In some cultures, maturity means assuming religious responsibilities, and this change may be marked with ceremonies, such as the Bar Mitzvah celebrated in Judaism. In most developed countries, puberty is not considered to be synonymous with the beginning of adulthood. Instead it is seen as part of the transitional period of adolescence. A person is usually considered to be adult at the age of 18 or 21.

**KEY TO THE DOOR**
*A decorated key, given at 18 or 21, represents entrance into adulthood in many Western societies.*

## MARRIAGE AND FAMILY

One of the major stages of the human life cycle is forming a bond with a partner. Having children is the biological motivation for pair-bonding, which many people formalize by marriage. Although most partnerships are between one man and one women, several cultures accept that a man or, more rarely, a woman may have more than one partner. Several countries also recognize the legality of same-sex partnerships. Denmark was the first country to legalize same-sex unions in 1989. In 2001, the Netherlands became the first country to give same-sex couples the same legal status in marriage as heterosexuals. In many parts of the industrialized world, more people are deciding to remain single, and couples may cohabit but not feel the pressure of religion or tradition to marry.

**TAKING PART IN FAMILY LIFE**
*In the West, the trend is for families to consist of one or two children and for only two generations to share a home.*

Since 1970, US censuses have shown a marked change in the composition of families and in living arrangements, with more and more children growing up in households with a single parent. In 1970, 3.4 million children lived in such households, but by 2003, the figure had risen to 12.2 million.

## CHANGING HEALTH

There are large discrepancies in the health and life span of people, according to wealth and geography. The main factors affecting national life expectancy figures are nutrition, water supplies, and health services. People in the richer, developed countries live much longer than those in developing nations. In western Europe, life expectancy is around 75 years or more. Life expectancy in many countries in Africa is substantially less, barely reaching 45 years in some of the poorest. In some parts of sub-Saharan Africa there has been a decline in life expectancy fuelled mainly by the HIV/AIDS epidemic (*see* p102). In the poorer countries, shorter life spans are often the result of diseases caused by infection or malnutrition. In developed countries, heart disease and cancer are the major killers, with lifestyle factors playing a large role.

## WORK

All societies expect their members to work during their adult lives and nearly all cultures regard with reduced respect those who cannot work.

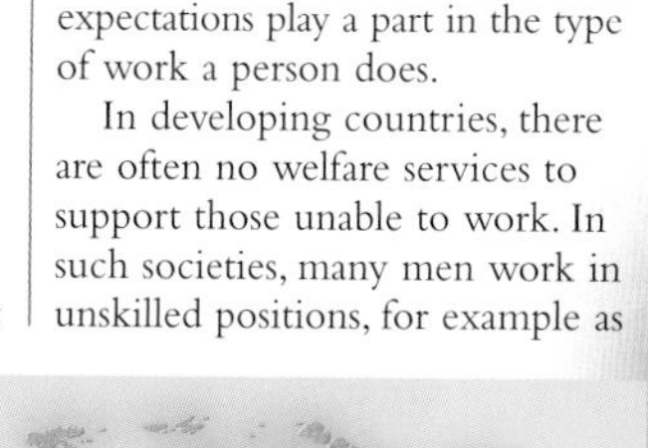

**WOMEN AND WORK**
*In many countries, women have physically demanding jobs – these Indian women work carrying bricks.*

The type and amount of work available depends on a country's economic status and technological development. In Japan, for example, the car and information technology industries are large employers of people. However, education and cultural expectations play a part in the type of work a person does.

In developing countries, there are often no welfare services to support those unable to work. In such societies, many men work in unskilled positions, for example as labourers. The hours may be long and the work hard. Holidays are unheard of, breaks limited, and protective clothing in short supply. Accidents are common and the work can cause early death. For example, in Bangladesh, the strain of operating a traditional rickshaw takes its toll on the bodies of the drivers, and injuries are common.

In countries such as the UK, the percentage of the population working in manufacturing and agricultural work has declined. Ever more people are working in offices and, as a result, often lead more sedentary lives.

## OLD AGE

In Europe and North America, people are expected to retire from their work at the age of 65 or earlier and live on a pension and their savings. Because extended families often do not live together, many elderly people eventually live out their lives in residential care homes or hospitals. For people who either meet or exceed their life expectancy, the final years are not always happy ones. In countries where there is no welfare system, such as India and China, elderly people continue to work as long as they are able. In some industrialized countries, the older generation is increasingly considered to be unproductive. In many African and Asian societies, however, the experience of older people is respected, and they are endowed with authority. The extended family is the basis of Chinese family life, and elderly relatives are likely to be cared for by their families in their last days. But this too is changing. In urban areas, senior citizens with savings and pensions may move into supported housing.

> Issue
> ### Can diet affect life span?
> Research in animals has suggested that it may be possible to extend life span by changing the amount we eat. The controversial evidence so far has come from experiments in mice. When fed a very low-calorie diet, they were healthier and lived 42 per cent longer than mice fed a normal diet. Whether or not the research can be extended to humans is uncertain, but some people are already adopting extreme calorie restriction – eating less than 500 calories a day, carefully selected to keep up levels of nutrients needed.

## DEATH

Death comes to us all at some time. Most people choose to die at home, surrounded by their family. However, in developed countries, more than half of all deaths happen in hospitals. All cultures give their dead a respectful cremation or burial and have a variable period of grieving. The life of the deceased may be commemorated with a memorial service or with something permanent (such as a tombstone) or with a living entity, such as a tree. In some cultures, death is seen as the start of a new life.

**MEMORIALS**
*These memorial tablets on Mount Haguro in Japan are just one example of how people commemorate the dead. Remembering is an important part of the grieving process.*

**LIFE SPANS AROUND THE WORLD**
*Average life expectancy varies considerably depending on place of birth. People in central Africa have the shortest life expectancy; Japanese people have the longest.*

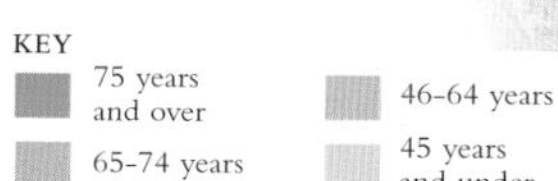

**CAREFREE**
*Childhood is a period of learning and developing new physical skills, without the restrictions of adult cares.*

# CHILDHOOD

The years up to adulthood are a time of the most fundamental changes in the mind and body. From being completely helpless as a newborn, the child passes through a series of stages to become physically mature and independent. Although the changes are dramatic, they occur over a very long period of time. Of all animals, humans have the longest period of dependency on their parents. The end of the childhood stage is marked by puberty and adolescence, a time of massive change.

At the end of 9 months in the womb, the newborn baby enters a very different environment – the outside world. In the mother's womb, the baby has been nourished and kept warm. Entering the world at birth, the baby has to face many new challenges, the first being to make the parents bond and care for him or her. Fortunately, humans are "programmed" to do this, and for the first few years of life the relationship with parents is the most important of all.

## EARLY DEPENDENCY

Unlike many animals, human babies cannot survive alone; their bodies are still very immature. Newborn babies cannot regulate their own body temperature nor can they feed themselves, and they have little control over their movements. Vison is poor, and on the first day the newborn can see only fuzzily: each eye's lens is not yet properly controlled. While foals can walk within 45 minutes of birth, for a human baby, these first

**BEING FED**
*Human babies are completely dependent on their parents to feed them and care for them. More than half the food taken in during the first year of life goes into providing energy for the developing brain.*

steps won't happen for about another year. As well as the immaturity of some of the body systems, the brain – the control centre of the rest of the body – can initially control only the basic functions needed to live: it has not yet developed the capacity to make sense of the world and control the body completely. The brain begins a rapid phase of development after the baby is born, driven by each new stimulus – such as a sight or sound – that the baby receives.

## GROWING AND DEVELOPING

From the earliest stage of infancy through to adulthood at around age 18, the body undergoes many changes. In the early months and

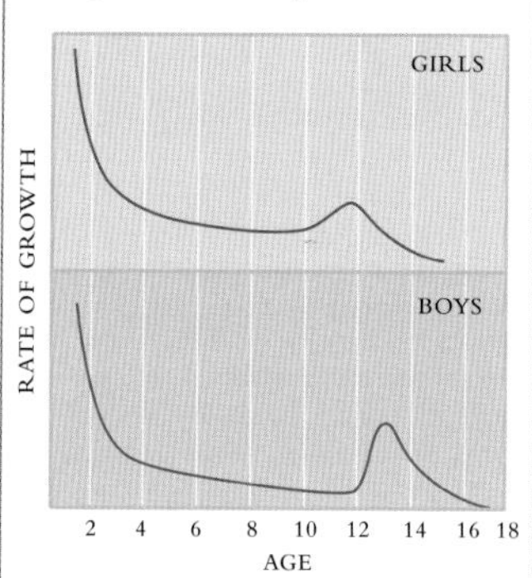


**GROWTH RATE**
*The rate at which children grow is greatest in the first year. After this, the rate of growth declines until the growth spurt of puberty, which happens around age 11–13 in girls and in boys a couple of years later.*

years, mental and physical progress can be measured almost by the day. Physical growth progresses rapidly in the first year, then continues more steadily until another growth spurt occurs that marks the start of puberty. Growth is stimulated by hormones, but achieving maximum potential height depends on adequate nutrition and good health. At various times, certain parts of the body grow more than others, which is why body proportions change during childhood. Until the age of 10, there is little difference in growth rate between girls and boys.

The growth spurt of puberty is triggered by sex hormones. Girls start their growth spurt at about age 10 or 11 and grow around 25cm (10in) during this time. Boys grow around 28cm (11in), but do not start their growth spurt until some 2 years later, when they are already taller. This accounts for the difference in average height of some 13cm (5in) between adult men and women.

Most of the increase in height from puberty onwards is due to extension of the long bones – such as the thighbones and shinbones. These bones grow from special areas of cartilage at their ends called epiphyses. Once bone growth is complete, the epiphyses fuse (*see* Bone development and growth, p64).

Childhood is also a time of learning and developing. All healthy children go through the same developmental processes: learning to walk and talk, to play with others, form friendships, and to think for themselves. Development depends on the explosive rate of change going on in the brain.

At birth, a baby's brain has billions of neurons (nerve cells), which form connections with others. By the time the child reaches age 2, each neuron may have up to 15,000 connection points, called synapses (*see* p116). Each new experience stimulates the formation of another new connection. As the child's experiences grow, so more neural connections are made. The more the experience (such as talking) is repeated, the more permanent the connections become. Some neural connections disappear because they are no longer needed. The process of learning and rewiring of neural connections is lifelong, but it is by far the most rapid in infancy.

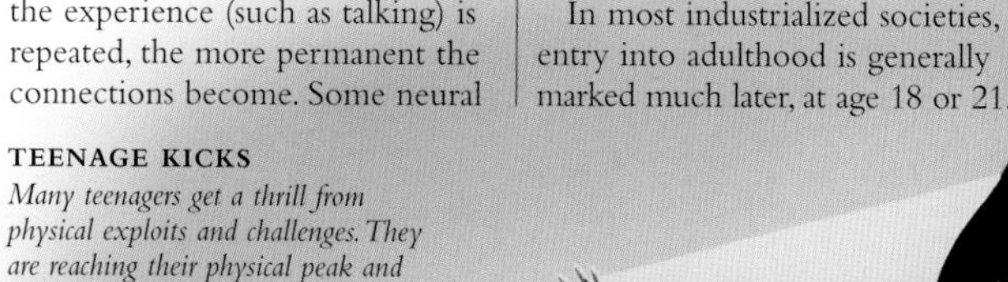

**TEENAGE KICKS**
*Many teenagers get a thrill from physical exploits and challenges. They are reaching their physical peak and may not mind taking risks.*

History

### Ancient Egyptian games

Several games have been found in explorations of Egyptian tombs. From these, it is obvious Egyptian children played some similar games to today's. Toys include spinning tops and board games. Snake is probably one of the earliest board games found. The stone board represents a serpent coiled with it's head in the centre. Players moved stones from the outside to the centre. The winner was the first to move his or her pieces around the snake and reach the centre.

SNAKE GAME

## TOWARDS ADULTHOOD

The years leading up to the full responsibilities of adulthood include several years of adolescence, although many traditional societies do not recognize such a stage. Rather, the person is seen one day as a child, the next day – or week – an adult. The years around puberty, when boys and girls become able to reproduce, are often marked in traditional societies by celebrations and rituals. At the end of the ritual, the child becomes an adult and has to take on adult responsibilities, which may include work, care of land, and even marriage.

In most industrialized societies, entry into adulthood is generally marked much later, at age 18 or 21.

# BIRTH AND EARLY CHILDHOOD

All newborn babies are helplessly immature and dependent on their parents. Unlike some animals, which can stand within hours of birth and become independent within months, the human baby needs constant care. Over the following months and years, this new individual acquires the ability to walk, talk, and play, to make friends, and to start to understand the world. The time at which children acquire these skills may vary, but all healthy children go through the same stages.

## BEING BORN

During a normal birth, the baby starts in a head-down position in the mother's womb and, as contractions intensify, gets pushed and contorted through the birth canal (*see* p142). This journey is a stressful experience for babies – levels of adrenalin (a chemical released at times of stress) are extremely high in newborns. The journey ends dramatically as the baby is born and confronted with sudden changes in light and temperature and takes its first breath. Not all births are the same. The place of birth – home or hospital – depends on geographical location, tradition, and the healthcare offered in the country where the baby is born.

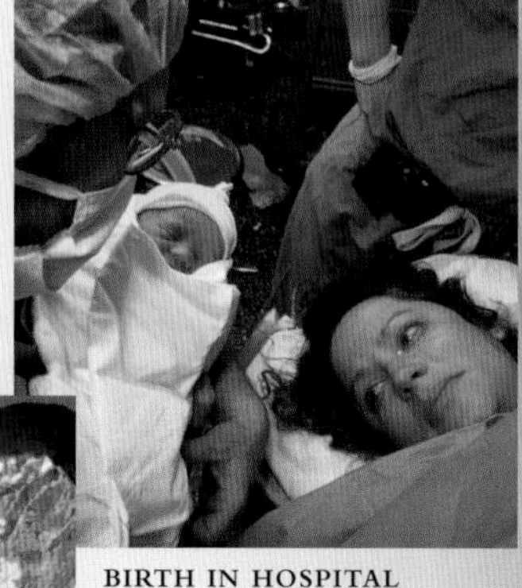

**BIRTH IN HOSPITAL**
*In many countries, women routinely give birth in hospital, where medical support is readily available in case of any problems or emergency. This baby has been born by caesarean section, in which a baby can be delivered quickly in an emergency.*

**BIRTH AT HOME**
*In rural Indonesia, experienced midwives support women giving birth at home.*

**BABY FEATURES**
*Adults have an innate attraction to baby features, such as round, protruding cheeks, big eyes, and a large head in relation to the body.*

## THE NEWBORN BABY

Despite 9 months in the womb, a newborn's brain and body systems are not all fully developed and ready to function. The reason is that the baby needs to be born before the head becomes too large to pass through the birth canal. The main bones of the skull are not yet fused, enabling them to overlap during birth (this explains the sometimes misshapen appearance of the newborn's head). There are two easily visible soft areas, called fontanelles, at the top of the head where the skull bones have not yet grown together. A newborn baby's movements are uncontrolled and shaky at first. As the brain develops, these movements become more deliberate. The skeleton is also immature: much of it is made of cartilage rather than bone and this ossifies (is converted to bone) over a period of years. Newborns' eyes are usually greyish blue, but they may change over a period of weeks as pigment develops. Newborns can see, but their colour vision is poor and most things seem out of focus for the first few days. However, they soon learn to recognize their parents' faces. Babies are most attracted by faces and movement. Hearing is better developed than vision.

Fact

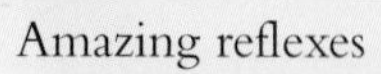

### Amazing reflexes

The newborn is capable of several reflex actions (automatic movements that need no conscious message from the brain). The gripping reflex is an echo of our primitive past (baby apes grip their mothers' backs when being carried). Some reflexes have an obvious survival value. Rooting, in which the head turns when the cheek is stroked, and sucking are essential in helping the baby to find the mother's nipple and feed.

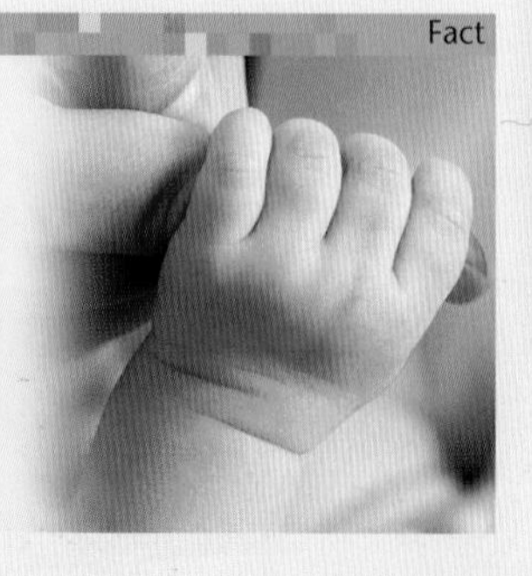

**NEWBORN VISION**
*Newborns focus best at a range of about 20–30cm (8–12in). They naturally prefer to look at faces which helps mother and baby to bond.*

# FEEDING AND NUTRITION

In the first few weeks of life, babies need a concentrated supply of nutrients for the tremendous rate of growth that takes place. Large amounts of protein and fat are particularly important for the brain and nervous system to develop and work properly. Breast milk, which consists of about 90 per cent water and 10 per cent protein, fat, sugar, vitamins, and minerals (such as calcium, iron, and zinc) meets all the nutritional needs of a newborn. In addition, it contains enzymes that are needed to digest milk and antibodies that bolster the baby's resistance to illness. Formula milk contains the same proportions of nutrients as breast milk but provides no protection against disease. As a baby grows, milk alone cannot meet all its energy needs, and by 6 months most babies are starting to eat solid food. Types of first solid food vary from one culture to another, but bland, easily digestible foods such as porridge and rice, and pureed fruits and vegetables are common. Different food types are introduced gradually to give the baby's still immature digestive system a chance to adjust. Certain foods cause particular problems: eggs and shellfish can cause allergies if introduced too soon, and babies' kidneys cannot process large amounts of salt. By age 3, most children are eating a diet similar to their parents, no matter what their culture.

**BREASTFEEDING**
*In addition to providing the baby's nutritional needs, breastfeeding can help promote bonding between mother and baby.*

**STARTING SOLIDS**
*Most children are beginning to eat solid food by the age of 6 months. The first foods are usually bland, soft, and easily digestible.*

# GROWING

Children grow rapidly in the first year of life – faster than they ever will again. A baby can be expected to grow by about 25–30cm (10–12in) in the first year, a massive one-third increase from birth. Along with a general increase in length and weight, body proportions also change. At birth a baby's head accounts for a quarter of the total length of its body, but by the age of 2 the head is only about a sixth of body length. Babies' arms and legs are relatively short compared with their trunks – again, this proportion changes after the age of 2 years. Initially, a baby's bones are composed mainly of cartilage, which is softer than bone. As childhood progresses, cartilage begins to harden through a process called ossification (*see* p64) although the ends of bones remain as cartilage for several more years to allow growth to continue. Although a child's growth depends on adequate nutrition, by the age of 2½ years, most children have attained approximately half their adult height.

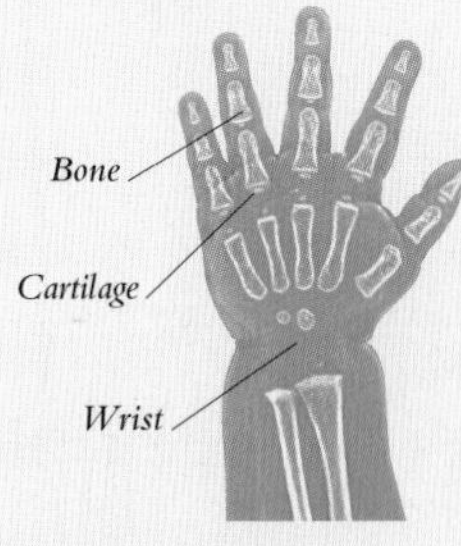


**BONE GROWTH**
*During the first few years, cartilage forms hard bone. The wrist of this 1-year-old has only two bones. By adulthood there will be nine bones in the wrist.*

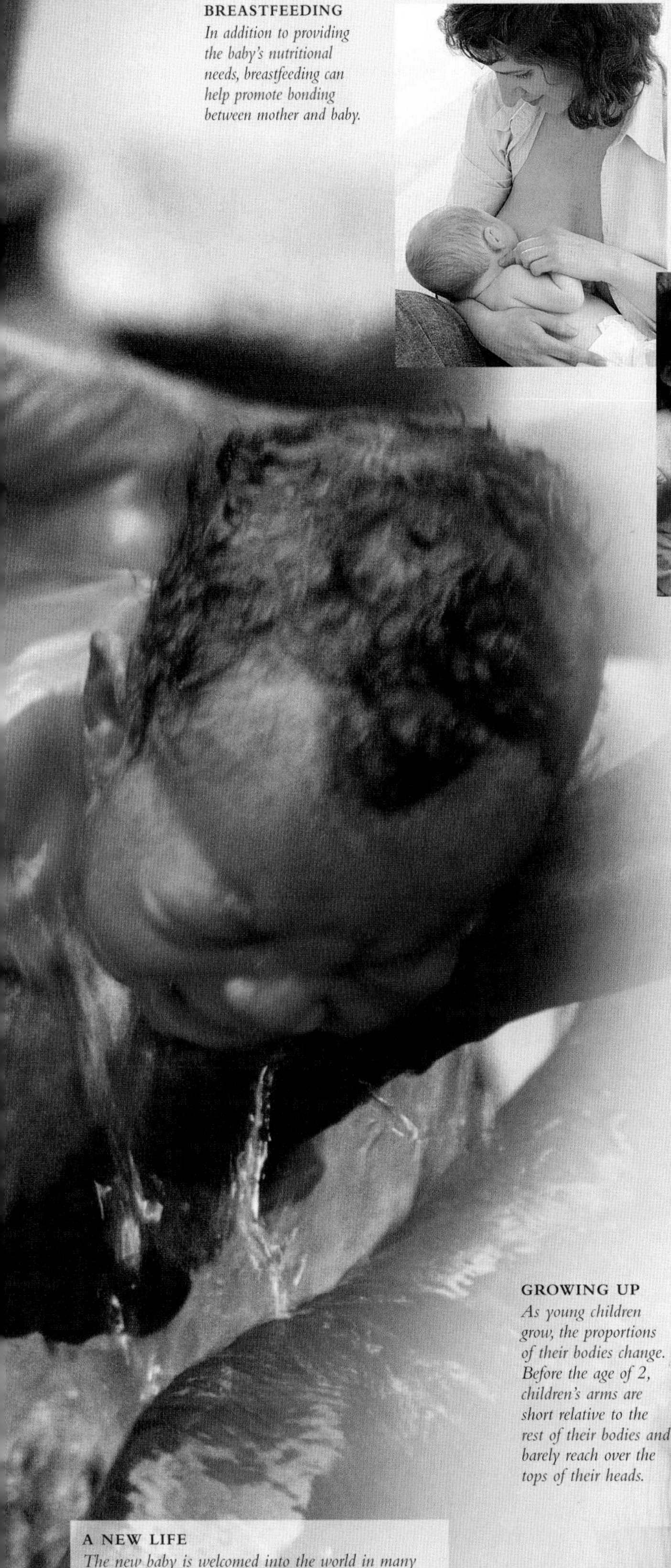

**A NEW LIFE**
*The new baby is welcomed into the world in many different ways. This newborn African baby is being washed by a midwife in a ritual process to cleanse and purify him before being returned to his mother.*

**GROWING UP**
*As young children grow, the proportions of their bodies change. Before the age of 2, children's arms are short relative to the rest of their bodies and barely reach over the tops of their heads.*

**FIRST STEPS**
*Around their first birthday, children are usually ready to take their first few steps unaided. Walking is a major leap in development, allowing greater freedom to explore.*

# EARLY DEVELOPMENT

During the first year, the brain almost triples in weight, reaching almost 1kg (2lb 2oz), and the nerves develop insulating sheaths around them (*see* p116), which dramatically speeds up the transmission of signals from the brain to the rest of the body. A dense network of neural pathways rapidly develops as nerve cells all over the brain become "wired up" to one another. As specific pathways are established, a baby becomes capable of new developmental feats. On average, a baby is able to hold his or her head up at 3 months, sit unsupported at 6 months, grab a toy with one hand at 7–8 months, pull up to standing at 10 months, and walk and speak a few words at 12 months. Although few children obey such a strict timetable, these milestones usually occur in roughly the same order and are independent of adult intervention. Even if parents give intensive training in a particular area, a child will not respond until the appropriate neural pathways are in place. Once they are in place, however, children benefit from parental encouragement and stimulation. This is particularly true of speech, which relies on mimicry. Children have a huge appetite for words – it is thought that 3-year-olds can learn 10 new words a day (*see* p168).

**BECOMING MOBILE**
*Crawling is often the first stage of becoming mobile and usually happens at around 8–9 months.*

**DEXTEROUS HANDS**
*By the age of 3, most children have gained sufficient dexterity and control of fine movements to be able to perform delicate operations such as doing up buckles or laces.*

Health

## Protection from disease

In the past, infectious diseases, such as diphtheria and measles, were among the main causes of death in children under 5 years. Today, the widespread immunization against these diseases means that infant deaths have declined sharply in developed countries. Children are commonly vaccinated against diphtheria, tetanus, whooping cough, and polio in their first year. Where immunization is not routine, death rates are much greater.

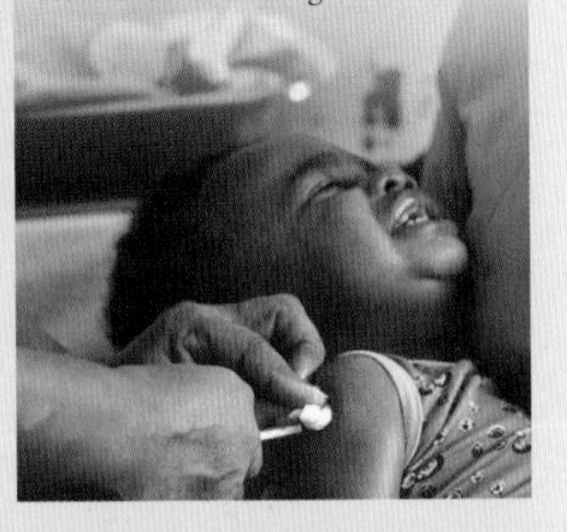

# EMERGING PERSONALITY

From around the age of 1 year, a child's personality becomes apparent. Some children are confident and bold, some are more timid. Some children are restless and some are placid. A young child's personality is partly genetic in origin and partly moulded by the environment. The emotional attachment between a child and the primary caregiver is of vital importance. Children who experience a loving, secure attachment are more likely to be happy and self-confident than are children who grow up in an unloving, critical, or tense environment. A child's relationships with siblings can also shape the developing personality. Some research indicates that personality may be influenced by the baby's exposure to specific hormones in the womb. For example, if the mother has a difficult or stressful pregnancy, the growing fetus may be exposed to high levels of the stress hormone cortisol, which has been linked to an anxious, irritable temperament in the child after birth. Testosterone (the male sex hormone) is also important: female babies exposed to higher than average levels of this hormone in the womb may be more tomboyish than other girls as they grow up.

**CONFIDENCE**
*Some children are naturally more outgoing than others and impose their personalities on all around them – even older brothers!*

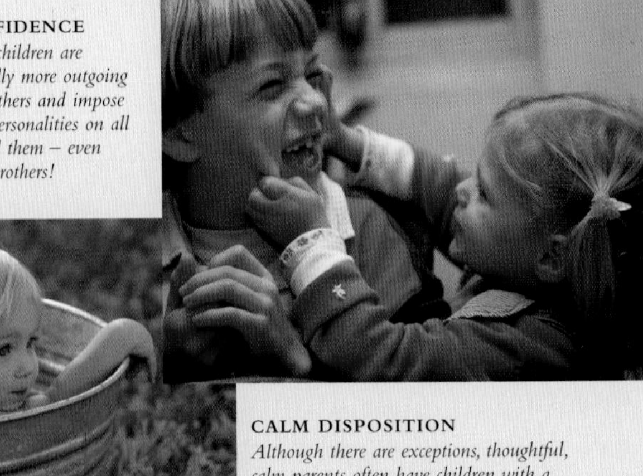

**CALM DISPOSITION**
*Although there are exceptions, thoughtful, calm parents often have children with a similar temperament. Genes and parental handling both play a role.*

# PLAY AND EXPLORATION

Once young children become mobile, they spend many of their waking hours exploring the world. It is normal for children between 1 and 3 years to be active all day. In the course of their explorations, babies and toddlers are constantly creating hypotheses about the world and testing them out. What may appear to a be random play is actually quite systematic experimentation. By pushing a food bowl off a table, reaching out to grasp a toy, or banging the side of a cot, for example, a baby develops a concept of him or herself as separate from the external world and discovers that each of his or her actions have specific consequences. Children have a natural drive to find out about objects. Even simple games such as hiding a toy under a cushion provide learning experiences. Whereas an 8-month-old baby believes that if a toy is hidden it no longer exists, a 10-month-old knows that if he or she lifts the cushion the toy will appear – this is known as the concept of object permanence – and the child will enjoy putting this to the test over and over again. Play also aids language development. Between 1 and 2, children become able to think in symbolic terms; they understand that one object can represent another, a shoe can be a boat, and the carpet the sea, for example. They also understand that words can represent objects. Exploring the world through play offers children more than just a greater understanding of their surroundings; it also allows them to develop socially and emotionally. Children who lack this freedom may have developmental problems later in childhood.

**PRETEND PLAY**
*Dressing up and role-playing games are great ways for young children to develop a sense of personal and gender identity.*

**EXPLORING THE WORLD**
*For children to develop to their full potential they need a certain amount of freedom to explore the world and discover that their actions have consequences.*

**PLAYING TOGETHER**
*Children learn social skills such as negotiation and cooperation through playing together. These are valuable skills later in life.*

# RITUALS AND CELEBRATIONS

Birth is traditionally a time for rituals and celebrations in which a child is named and welcomed into the community. The nature of these rituals vary from culture to culture. In Jewish communities, baby boys are ritually circumcised; in Hindu culture it is traditional to shave babies' heads as part of the birth ceremony. Sometimes there is a period of waiting before a birth is celebrated. For example, the Akamba people of West Africa wait 3 days before slaughtering a goat and naming a child. This delay is to ensure that the child is healthy and will survive before a name is given. In other cultures, the birth is celebrated immediately. Naming a child is a way of conferring both identity and paternity. Names may be chosen long before a child is born – as is common in much of Europe and the US – or they may be determined by the circumstances of birth. For example, in parts of Africa a child's name may reflect the fact that it rained on the day of birth.

**BIRTH GIFTS**
*Silver is a traditional gift to mark the birth of a baby.*

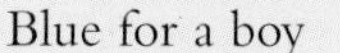

Fact

## Blue for a boy

Blue is a traditional colour for gifts or clothes for a boy in many places. The tradition stems back to ancient times when a boy was considered to be a greater asset than a girl. Blue was seen as the colour of heaven and therefore protective. Dressing a baby boy in blue conferred maximum protection against any evil spirits. In some cultures, this extended to hanging blue ornaments.

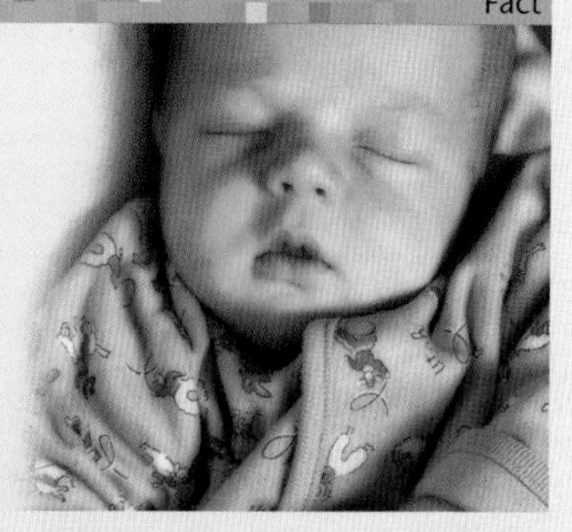

**BIRTHDAY CELEBRATIONS**
*Around the world people celebrate a child's birthday in different ways. Special foods, birthday cakes, and candles mark another year in a child's life.*

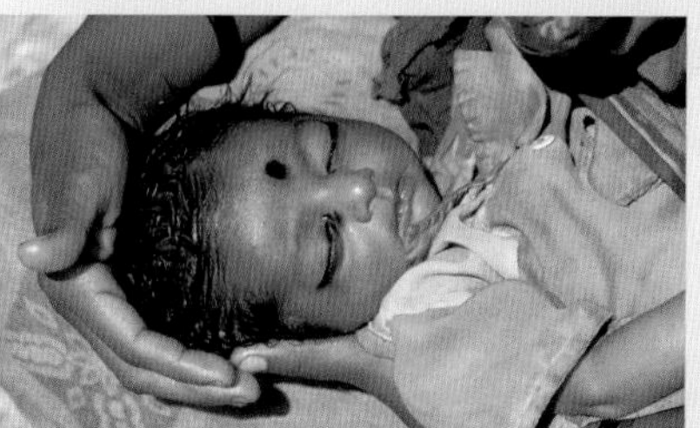

**THIRD EYE**
*Marking is a common ritual around birth. In Nepal, a baby is marked with black soot in the centre of the forehead as a protective sign to ward off evil.*

BIRTH RITUALS

## Islamic dedication

**In brief:** A Muslim baby is introduced to the faith soon after birth by a whispered dedication in each ear

A Muslim baby is often introduced to the religion within a few hours of being born. An elder of the Muslim community whispers the *azaan*, the call to prayer, into the child's right ear and the Islamic creed into the left. This is done to ensure that the very first thing a newborn child hears is the name of God. Often, the elder places a small piece of chewed date into the child's mouth. This action symbolizes the hope that the wisdom of the elder will pass to the child.

**WHISPERING ISLAMIC CREED**
*A newborn baby hears words from the Koran within hours of birth, either from its father or an elder of the community.*

BIRTH RITUALS

## Indonesian rituals

**In brief:** Rituals to protect the baby and wish him or her a good life are performed within 72 hours of birth

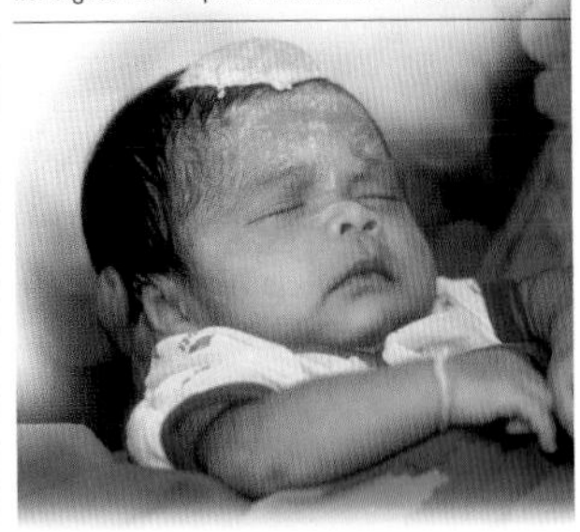

**PROTECTIVE RICE CAKE**
*A cake of rice and herbs is placed on the baby's head soon after birth. The cake, called a* buhung, *is left to fall off of its own accord.*

In parts of Indonesia, a birth ritual is performed using a cake of rice, herbs, and flour called a *buhung*. This is placed on the baby's head soon after birth. Once the rice cake has dropped off, if the baby is a girl, the cake is placed above the family's front door in the hope that she will marry early and be fertile. In another ritual, the placenta (afterbirth) is washed in tamarind-scented water and wrapped in a cloth. Symbolic objects are also placed in the cloth: a pencil and paper to encourage the child to study; a sewing needle for health; rice for luck; and a coin for wealth.

BIRTH RITUALS

## Beer ceremony

**In brief:** Northern Ugandan ritual symbolizing the importance of the baby's bond to the community

In northern Uganda, soon after birth and before being put to the mother's breast, a drop of home-brewed millet or banana beer is placed in the baby's mouth. Beer drinking is a communal event across East Africa and is seen as a symbol of community life. The act of giving a child beer as its first drink reinforces the importance of the community to the child and the fact that the child is part of a wider group of people than its family alone.

Whenever people are thought to be acting selfishly later in life, elders will remind them of the beer given to them at birth saying "beer before milk", which means "remember others before yourself".

BIRTH RITUALS

## Hindu cleansing

**In brief:** Performed in Hindu communities, washing and massage are believed to purify the newborn

In Hinduism, the newborn baby is considered impure, tainted by its time in the womb. The mother or birth attendant performs a series of massages and ritual washings with oil. Massage is believed to enhance the circulation, expel toxins, and aid the development of the digestive system. Often, very young babies are massaged with a soft wheat-dough ball dipped in almond oil and turmeric. Massage is performed daily from 3 weeks to at least 3 months.

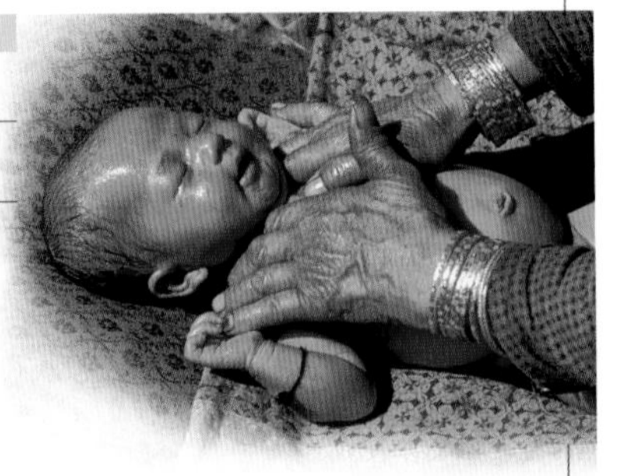

**MASSAGE RITUAL**
*A traditional midwife in India massages the newborn baby with a mixture of almond oil and turmeric. This process is believed to help expel toxins from the body.*

NAMING RITUALS

## Wodaabe shaving

**In brief:** Wodaabe nomads of West Africa carry out ritual head shaving before the child is named

For the Wodaabe nomads of Niger and Nigeria in West Africa, the baby naming ceremony is the most important ritual of childhood. First a goat is sacrificed in honour of the new life. Then the heads of the baby and mother are shaved, an act that is believed to strengthen the bond between them. Shaving the forehead is also believed to enhance the beauty of the face. In the evening of the ceremony, the baby is named by a village elder. Once the baby is named, the parents are not allowed to speak his or her name in case evil spirits hear it.

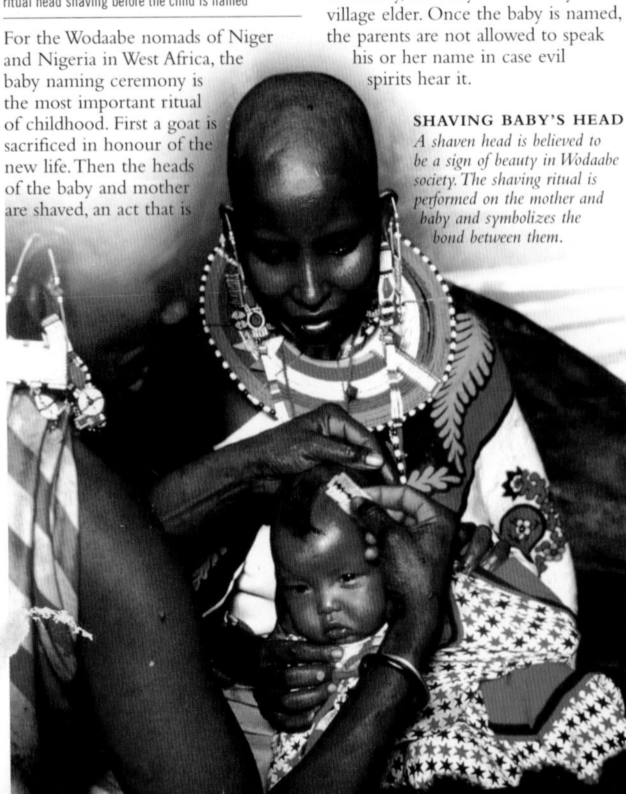

**SHAVING BABY'S HEAD**
*A shaven head is believed to be a sign of beauty in Wodaabe society. The shaving ritual is performed on the mother and baby and symbolizes the bond between them.*

NAMING RITUALS

## Christian baptism

**In brief:** Marking the baby with the sign of the cross admits him or her to the Christian faith

Baptism is the symbol of entry into the Christian faith. It represents the washing away of an old life and a commitment to a new life based on the teachings of Jesus. The earliest known baptisms recorded in the Bible are those of adults, who were baptised by full immersion under water and who subsequently took on a new Christian name. Early in the history of Christianity, the practice of baptising infants developed.

The ceremony varies in different branches of the Christian faith. In the Orthodox church, the priest immerses the baby three times in the font before anointing him or her with oil called chrism. This oil is also used in baptisms in the Catholic church. In the Anglican church the priest pours water, which has been blessed, over the baby's forehead. The priest then makes the sign of the cross and gives the baby his or her Christian name.

**BAPTISMAL FONT**
*The font, placed at the entrance to the church, contains water used in Christian baptisms.*

**HOLY WATER**
*A priest pours water over the baby's head, symbolizing the washing away of an old life and entrance into the Christian faith.*

LIFE CYCLE

NAMING RITUALS

## Brit Milah

**In brief:** In this Jewish ceremony, a baby boy is circumcised and given his formal name

In Judaism, a male baby is circumcised and given his Hebrew name at 8 days of age. The practice dates back 3,700 years and originates from a command given by God to Abraham. Today, male circumcision, in which the foreskin is cut from the penis, continues to be taken as a sign of Jewish cultural and religious identity.

The 8th day is chosen in the belief that a baby will have gained the physical and spiritual strength needed to undergo the rite by then. Brit Milah always takes place during daylight hours and is carried out by the *mohel*, a trained circumciser, either in the synagogue, or at the family home. The baby is laid on a cushion and held by a sponsor who has been chosen by his parents while the *mohel* cuts the foreskin with a double-edged knife. An empty seat is left next to the sponsor to allow room for the prophet Elijah, who is believed to visit every Jewish circumcision ceremony.

After the circumcision, the parents and *mohel* welcome the boy into the faith and bless him by praying for health, marriage, and good deeds. The baby is given a drop of wine and the *mohel* announces his formal name. Afterwards there is a celebratory meal.

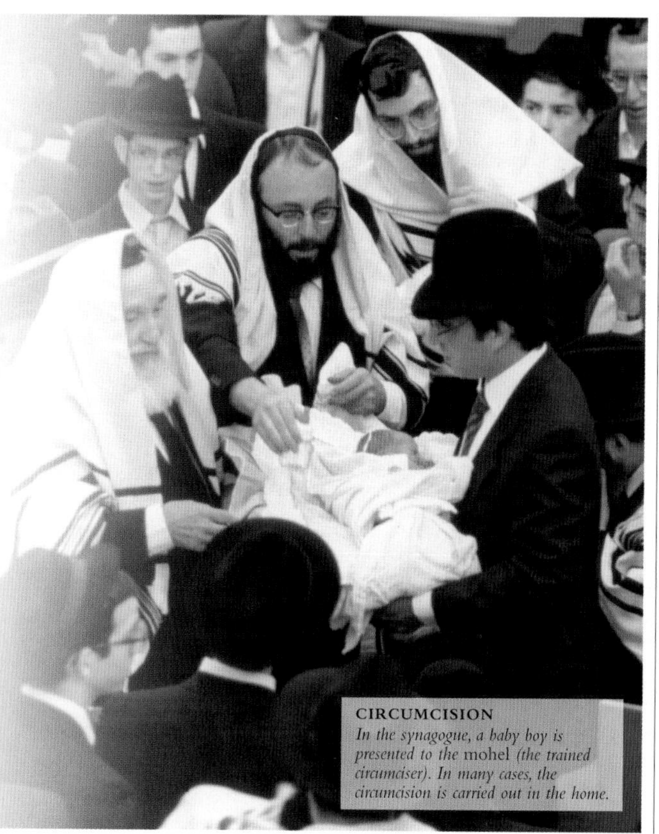

**CIRCUMCISION**
*In the synagogue, a baby boy is presented to the* mohel *(the trained circumciser). In many cases, the circumcision is carried out in the home.*

MARKING RITUALS

## Muslim circumcision

**In brief:** Circumcision (removal of the foreskin) may be performed soon after birth or later in childhood

Circumcision of Muslim boys is carried out to follow the example of the prophet Mohammed. It is held to be a sign of true belonging to the Muslim faith. The age at which boys are circumcised varies, but it is mostly conducted before puberty, commonly by 8 years. A party may be held. After circumcision, a boy can participate fully in all religious rituals and fasting.

**GUEST OF HONOUR**
*In Malaysia, a newly circumcised boy attends a celebration party.*

NAMING RITUALS

## Balinese naming

**In brief:** Formal naming of a baby in Bali takes place at a ceremony held when the baby is 3 months old

The largely Hindu population of Bali, in Indonesia, performs a series of rituals at various intervals in the first year of a baby's life. Formal naming takes place when the baby is about 3 months of age. Up to this point the baby is not seen as an individual and is often referred to by both family and friends as "mouse".

The most important part of the naming ceremony involves holding the child over a bowl of water containing a leaf for wisdom, coins for wealth, grains of rice and kernels of corn for diligence, and jewellery representing desire.

Whatever the baby takes from the water with his or her right hand is said to characterize the baby's future life. After the ceremony the baby is allowed to be put on the ground for the first time since birth.

Fact

### Teknonyms

In Bali and many other parts of the world, parents assume the name of their first-born child and are called by this name thereafter. This new name is called a teknonym. In Fiji, spouses can call each other by their original names but everyone else must call them by their teknonym. In East Africa the system is reversed. Spouses must call each other "mother of" and "father of" the name of their first child, while everyone else can refer to the parent's original name.

**BALINESE GOLD**
*For the official naming, children in Bali are often decorated with gold, representing a wish for future wealth. The gold disc on this baby girl's head also confers protection.*

MARKING RITUALS

## Chudakarana ceremony

**In brief:** This Hindu ritual is believed to cleanse the child; it symbolizes a loosening of ties with the mother

In the Hindu Chudakarana ceremony, the child's head is shaved in a family ritual. The ritual takes place when a child is an odd-numbered age, most commonly 3 years. The shaving is believed to cleanse the body and soul of the child – baby hair is associated with the child's time in the womb and with being unclean. Hair-shaving also symbolizes the loosening of the tie between mother and child and the beginning of the father's role in the child's life and his or her education.

For the ritual, the mother, father, and child sit in front of a fire. The child is bathed and dressed in new clothes. The father shaves off the first locks of hair, quoting sacred mantras – phrases with symbolic meaning – and a verse in which he asks the razor to be kind and gentle to the child. A barber completes the hair-shaving but leaves a few tufts behind, the number of which is determined by family tradition. After the hair is shaved off, it is mixed with cow dung and buried or thrown into a river. This disposal is believed necessary to prevent anyone being able to use the hair in "sorcery" against the child.

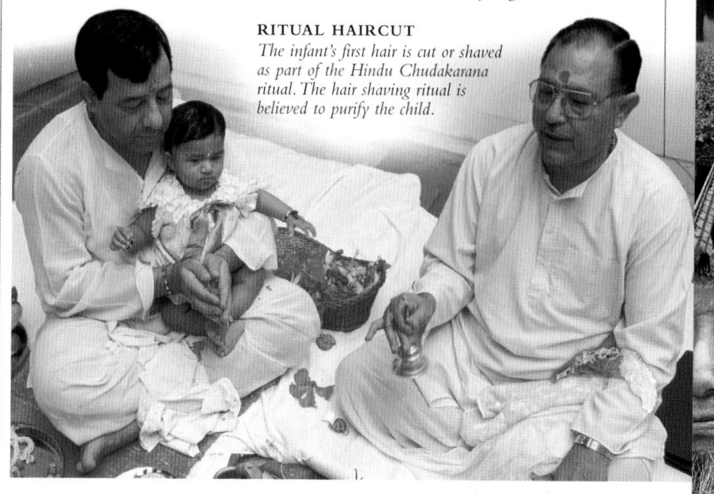

**RITUAL HAIRCUT**
*The infant's first hair is cut or shaved as part of the Hindu Chudakarana ritual. The hair shaving ritual is believed to purify the child.*

PROTECTION RITUALS

## Aboriginal smoke ceremony

**In brief:** Aboriginal people use aromatic smoke to cleanse and purify their babies and young children

Australian aboriginal babies are regularly passed through smoke in the belief that this will protect them and promote resistance to disease. The ritual stems from the belief that the earth is their mother and that humans therefore come from the earth and return to the earth when they die.

Aboriginal tradition says that during life people may become polluted and this contamination would be harmful for the land if people took it back into the earth at death. For this reason, purification and cleansing are important rituals for people as well as for their land.

The "smoking" is normally carried out by a baby's grandmother. Aromatic herbs and leaves are used to make a fire. The baby is held in the clouds of smoke above it for a few seconds before being massaged. The smoke removes impurities from the youngster and from the area. Sick children, too can be cleansed. They are smoked daily before being covered in charcoal paste to aid healing.

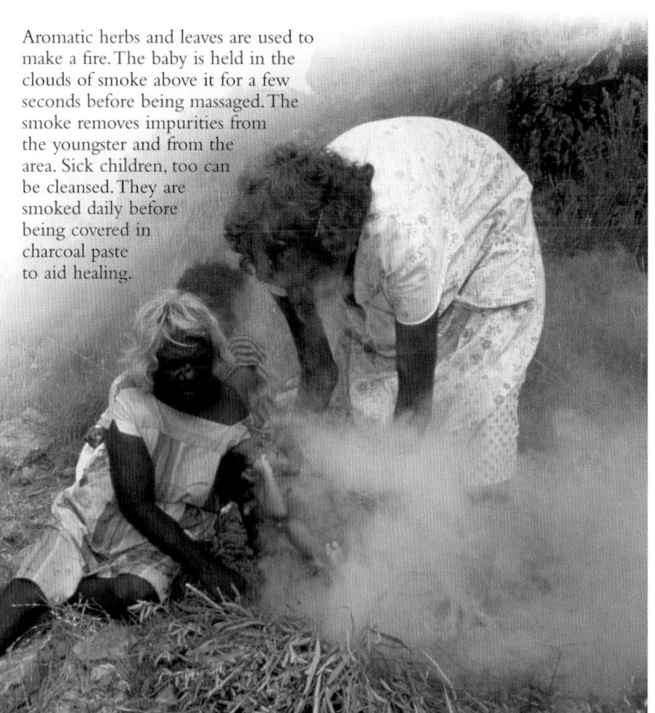

**ABORIGINAL SMOKING**
*A baby is held in the smoke of a fire of aromatic leaves for a few seconds at a time to purify and to protect him.*

PROTECTION RITUALS

## Tempering

**In brief:** In Russia, regularly exposing the body to cold water is believed to help people withstand disease

It is a traditional Russian belief that the body should be immersed regularly in very cold water to help people to withstand diseases in the cold climate. This process of cold-water immersion is called tempering or *zakalivanie*. Babies as young as a few months old are dipped into icy waters by their parents for a few seconds at a time.

TEMPERING A 9-MONTH-OLD BABY

PROTECTION CHARMS

## Birth amulets

**In brief:** Worn in many parts of the world, amulets are believed to protect the child from harm

Amulets are trinkets believed to ward off evil. In Africa, birth amulets are made up by a traditional healer using sacred herbs. The amulets are tied around the child's waist or wrist, or both, as protection from evil and the jealousy of others. In the Philippines, such charms are called *bigkis,* which means "strong bond". They are tied around the child by its mother to strengthen the bond between them. In Islamic cultures, quotations from the Koran are tied into small parcels, which are attached to necklaces.

**BANGLADESHI AMULET**
*A protective charm is tied around an infant's waist by its mother. Amulets have been worn as protection against evil since ancient times.*

**TIGHTLY WRAPPED**
*In a newborn baby is tightly wrapped by a nurse. This practice is common in Asian hospitals such as this one in Mongolia.*

EARLY CARE

## Swaddling

**In brief**: Tightly wrapping a baby is believed to comfort him or her and is practised in many societies

Swaddling – wrapping a baby tightly in cloth or a blanket – is common across eastern Europe, the Middle East, and Asia. In some places, a child may be bound for up to a year.

Swaddling also helps infants to remain on their backs while they sleep. In this position, they are believed to be at less risk of sudden infant death syndrome. Recent research shows that swaddled babies are more likely to sleep for longer through the night, because they feel secure and the startle reflex, which can wake them, is limited. US experts have also found that swaddled infants have twice as much deep sleep than unswaddled ones.

**KEEPING WARM**
*Where temperatures regularly drop below freezing, swaddling helps keep babies warm. These Mongolian infants are well protected.*

EARLY CARE

## Sleeping arrangements

**In brief:** A baby may share a room or a bed with its parents, mother, or siblings, or it may sleep on its own

A baby's sleeping place may be defined by custom, by income, family size, or style of housing. Often, there is no space for children to have a room of their own. In many parts of the world, babies and young children share a bed with their parents or their brothers and sisters. The most common sleeping arrangement is for mother and child to be in one bed and the father in another.

Babies in the US and most of Europe are often put to sleep in a cot in a room on their own. In some societies, sleeping arrangements are strictly defined by custom. For Mongolian nomads, family tents (also called *yurts*) are divided into female and male sides. All the children sleep in the female side of the tent in their early years, but boys move over to the male side of the tent at the age of 5.

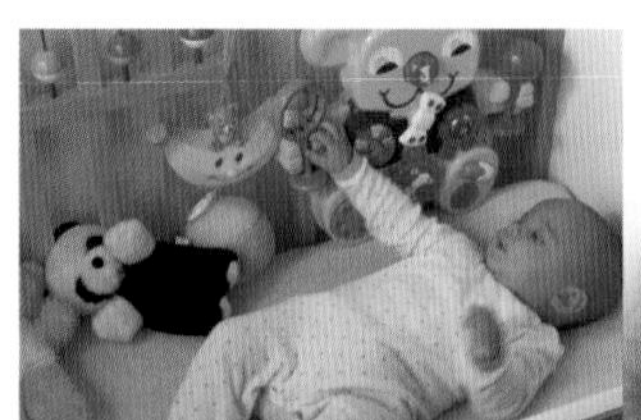

**SURROUNDED BY TOYS**
*In most Western societies, parents put children to sleep on their own, with cuddly toys to keep them company.*

**COSY ALASKAN CRIB**
*An Alaskan baby may have its own crib, such as this one covered in reindeer skin, but it will remain close to where its parents sleep.*

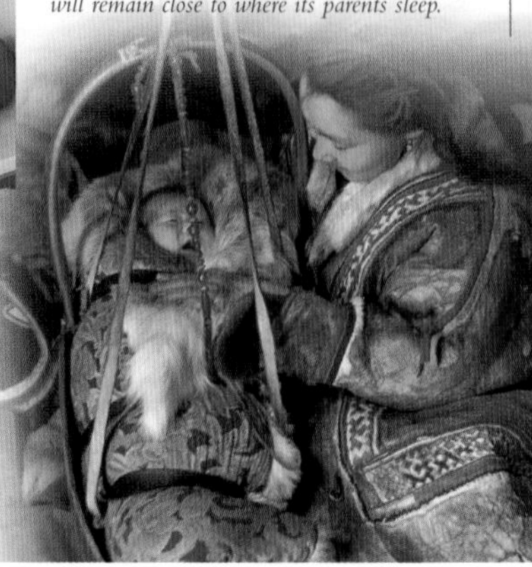

EARLY CARE

## Breastfeeding

**In brief:** Attitudes to breastfeeding vary widely; there is variation in the mother's diet, in frequency of feeding, and in how the colostrum is viewed

Breastfeeding is not an instinct for a mother or child. Where babies are born in hospitals, a nurse often oversees the process of teaching the mother to breastfeed. In many traditional societies, this role is often played by the woman's mother.

There is a variety of attitudes towards all aspects of breastfeeding, including attitudes to the colostrum (the fluid that is produced in the first 3 days after birth), to the frequency of feeding, and the diet of the mother while she is breastfeeding. In several countries, the colostrum is considered bad for infants. In parts of China, it is thrown away, and the baby is given to another woman to feed for the first few days; in Somalia, the baby is given camel's milk; in Bangladesh, babies are given honeyed water until their mothers' produce milk. In Japan, mothers massage their breasts to speed the conversion of colostrum to milk.

Mothers may choose to feed on demand (putting the baby to the breast to feed whenever he or she cries) or using timed feeds. Mothers in the !Kung tribe of southern Africa constantly carry their children and put them to the breast on demand, on average every 13 minutes. In rural Nepal, where mothers often do hard agricultural work, the baby is left in the care of other women, and the mother returns every couple of hours to breastfeed him or her.

**FEEDING IN COMFORT**
*In the early days of breastfeeding, a mother may find it easier to feed her baby without other distractions.*

**GETTING ON WITH LIFE**
*Breastfeeding can be done anywhere. This young mother in Eritrea breastfeeds while she studies English.*

Issue

### Breast or bottle?

The World Health Organization (WHO) recommends that babies are exclusively breastfed for the first 6 months of their lives. In developing countries, breastfeeding is safer for babies than alternatives such as formula or animal milk. This is because breast milk does not need to be mixed with water, which, in areas where sanitation is poor, may contain harmful bacteria which can cause diarrhoeal diseases. In countries where there is clean drinking water, bottle-feeding is safe and practical, and enables others to help.

**SHARING FEEDING**
*In countries with clean drinking water, fathers can help feed their baby using a bottle.*

EARLY CARE

## Transportation

**In brief:** Many methods are used to transport young children, from simple slings to elaborate pushchairs

In many cultures, babies are carried by someone – mother, grandmother, or sister – for most of the first year of life. Being kept close to the body in this way has a calming effect. It also enables mothers to continue with domestic tasks as this arrangement keeps both hands free. In Guatemala, mothers weave slings from brightly coloured cotton and carry their babies in front of them. In Vietnam, mothers tie their children to their backs with a square of fabric that is supported by shoulder straps. In the West, parents have come to rely on increasingly sophisticated pushchair designs.

**BACK SLING**
*In Kenya, mothers tie their children to their backs in a brightly coloured cloth called a* kikoy.

**PUSHCHAIRS**
*Prams and pushchairs can be expensive, but many people find they provide a practical solution to carrying older babies and toddlers.*

EARLY CARE

## Weaning

**In brief:** Babies are weaned from milk to solid food over a period of months or, in some places, even years

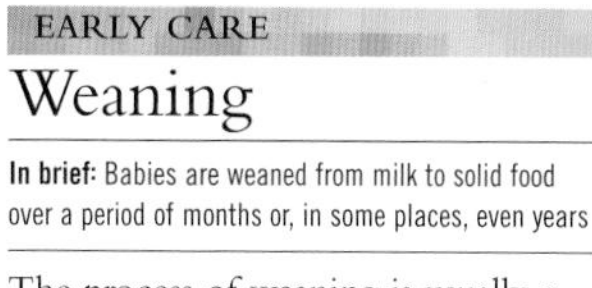

The process of weaning is usually a gradual one of adding in solid food alongside breast milk and then substituting food for a session of nursing. This process can last up to 4 years. In Bangladesh, which is said to have the highest proportion of breastfed babies and the longest weaning periods in the world, the average age of weaning is 36 months. In some societies, weaning is defined by custom and takes place at the same age for all children. In Mozambique a weaning ceremony called *madzawde* takes place when a child reaches 2 years old. The midwife is invited to the ceremony and gives sacred leaves to the parents, who then discard anything associated with the breastfeeding period. From this time on, the child does not drink the mother's milk because this now belongs to the next baby. It is believed that drinking milk will make the weaned child ill.

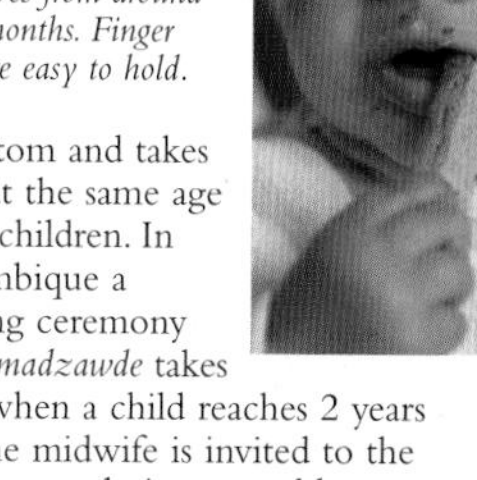

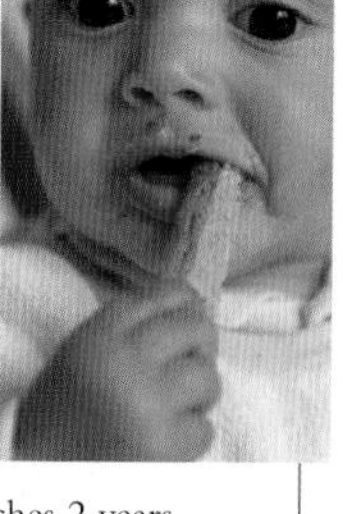

**FINGER FOOD**
*Young babies can begin the process of feeding themselves from around age 6 months. Finger foods are easy to hold.*

**FEEDING BY MOTHER**
*In China, watery rice with added fish protein is the most common weaning food. This mixture, called* congee, *is bland but nutritious.*

# LATE CHILDHOOD

Between the ages 5 and 11, a child continues to grow and develop, creating stronger links with the world outside the family. For most children, learning becomes formalized with the start of school, and they continue to learn practical skills that will serve them well in adulthood. Children also begin to understand the emotions and needs of others over this period of time. They develop their own moral values, too, and they learn about the values of their own society. In some societies, children of this age are already looking after other children or working.

## PHYSICAL SKILLS AND DEVELOPMENT

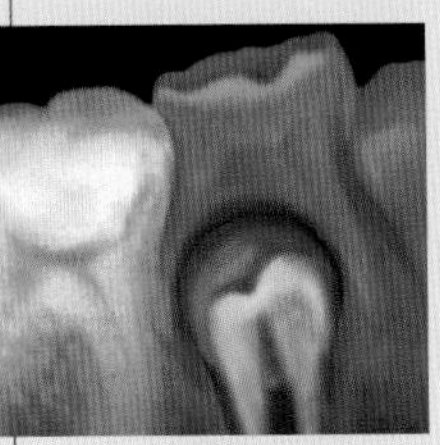

**ADULT TEETH**
*Permanent teeth usually start to appear at age 6. This X-ray shows a permanent tooth ready to replace a first tooth.*

From ages 5 to 11, before the adolescent growth spurt, boys and girls grow at the same rate. Their facial features start to look more mature as the facial bones change shape, with the forehead and chin becoming more pronounced. The first set of teeth are gradually replaced by a second, permanent set. By the age of 6, the growth of the head and brain is about 95 per cent complete. Children enter their 6th year of life already able to walk, run, jump, and skip. As the years progress, their physical skills become more refined. Children develop a gradually increasing and refined sense of balance, coordination, and spatial skills. Powers of concentration also improve as the brain continues to develop. From the age of 7, children are more able to persevere with tasks. They also begin to develop logic skills and can work out problems for themselves. By 10 or 11, children are able to analyse problems, and their levels of understanding improve so that they can solve hypothetical problems. By the end of this stage of childhood, surges of hormonal activity have started to initiate the adolescent growth spurt and sexual development.

Health

### Childhood obesity

Growing social prosperity in many parts of the world has contributed to epidemic rates of child obesity. In the 21st century, children in many countries eat more than ever before and exercise less. Being overweight in childhood carries the risk of diabetes and heart disease in adult life. In the US, nearly a quarter of children are defined as being overweight – a figure that is twice what it was in the 1970s. In China, obesity in children has increased by one-fifth over the same period, while it has tripled in Brazil.

**BALANCE AND CONTROL**
*By the age of 7 or 8 most children have a well-developed sense of balance as well as good control of their movements. Being able to ride a bicycle, without falling off even when distracted, is one sign of this improved balance and agility.*

## SCHOOL AND LEARNING

Many children start school at around age 6, but school is not the only place that children learn. Parents and family are also important teachers. In the early years of childhood, children will have been learning by the example of others. By the age of 5 or 6, learning outside the home starts to take on more significance. Studies have shown that children who start school early achieve more later in life and are less likely to drop out of education. One of the most important achievements of this age is learning to read and write. Unlike speech, these skills must be taught and consciously learned. Some children learn to read with little difficulty; others may find it a slow, frustrating process. The process may be enormously complicated in one language, yet relatively straightforward in another. As well as going to school, children learn practical skills, and in some societies these skills may be at least as important as the skills learned at school. In traditional nomadic societies children as young as 5 or 6 are learning to herd cattle and by 7 or 8 are expected to look after the animals on their own. Educational opportunity may differ for girls and boys depending on where they live. Girls do not always have the equal access to learning that most have in the West.

**SKILLS FOR LIFE**
*Practical skills may be more important than reading and writing in some societies. In Mongolia, children learn to ride a horse from a very young age.*

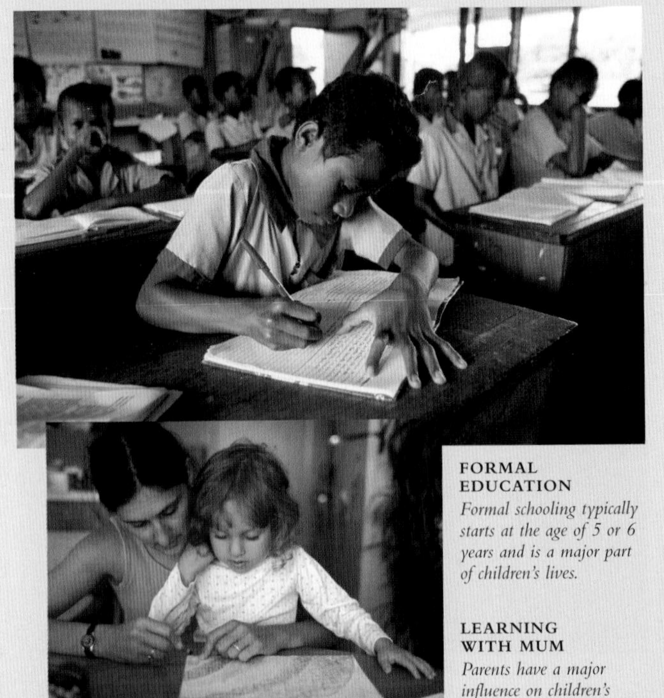

**FORMAL EDUCATION**
*Formal schooling typically starts at the age of 5 or 6 years and is a major part of children's lives.*

**LEARNING WITH MUM**
*Parents have a major influence on children's learning at this early stage of their lives.*

**FREE TO PLAY**
*Most children enjoy physical activities. These youngsters' strength and stamina are at such a level that they can play for several hours without appearing to get tired.*



# FRIENDSHIPS

Children's friendships are key building blocks for development into adult life. Attachments that are formed during the years 5–11 help to teach social skills, solve problems, and develop self-confidence. From 7 years onwards, youngsters start to develop relationships which are deeper and more longer lasting as they begin to be able to see another person's point of view. Children's friendships are training grounds for adult relationships. Playing with friends, for example helps them to develop a sense of humour and such skills as the ability to negotiate. Early in life, friendships are made and broken on an almost daily basis. However, by 9 and 10 comes a growing loyalty and commitment. Some children have many friends while others prefer the company of just one or two. Children with large families may not feel they need as many friends. Some youngsters find friendships difficult and may be victims of bullying.

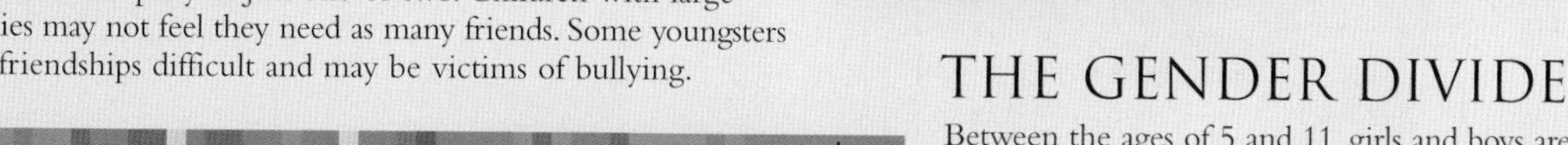

**FRIENDS**
*Children enjoy trying new activities with friends. These American boys take huge pleasure from fishing and being together.*

Issue

## Child labour

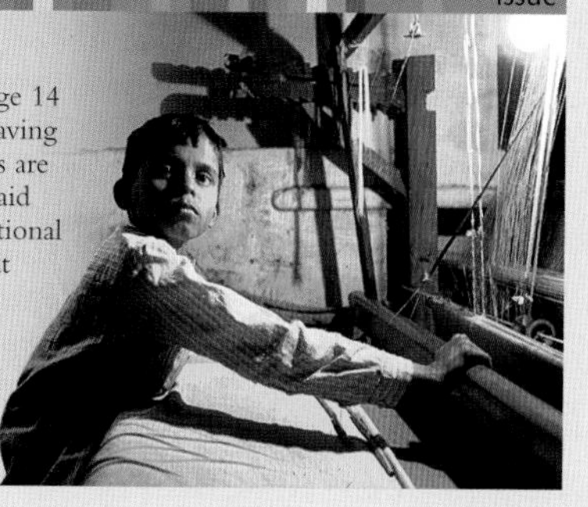

Millions of boys and girls under age 14 are employed in India's carpet-weaving industry. Sometimes, these workers are as young as 8 or 9. Children are paid only a minimal wage. The International Labour Organization estimates that worldwide at least 100 million youngsters between 8 and 15 are leading working lives; many of them have been forced into work to pay off family debts.

# THE GENDER DIVIDE

Between the ages of 5 and 11, girls and boys are often treated very differently. It is during this time that children chose to move into different groups according to their gender. Although in the US and most of the developed world girls have just as much opportunity as boys, in many traditional societies girls' tasks are often very different from those of boys. While boys are still allowed the freedom to play outside and take part in physical activities, girls are often made to do household chores such as looking after younger siblings, cleaning, cooking, and collecting water.

**GIRL'S WORK**
*Girls in India are often expected to carry out chores, such as collecting water. For these girls there is little time for play.*

PLAYING

## Language games

**In brief:** Rhyming and other language games help to promote language development in children

Children all over the world like to repeat nonsense rhymes and make up nonsense languages. Ludlings is the name for language games that follow rules to make new words out of existing ones. In Brazil, for example, a ludling called *jerigonzo* inserts a "p" in front of every vowel. Sweden and Russia have the most known varieties of ludlings. Children play ludlings for enjoyment, and to be able to talk to each other in secret, but the games naturally help to develop language skills. Rhymes with actions also help to promote language development: children can learn the rhymes without having to understand them. The actions then help children to fit meanings to the words.

CHILDREN PLAY A RHYMING GAME

PLAYING

## Logic games

**In brief:** Games such as chess help to develop logic and critical thinking skills in children

Games that develop logic skills have been around for centuries. They include chess, backgammon, mahjong, and mancala. Chess first developed in northern India in the 5th century. Today the game of chess played in Europe is just one version of many similar board games found in Japan, China, and Myanmar (Burma). Chinese children seem to have a particular flair for chess. Russia also has a strong history of chess excellence. After the Russian revolution, the government set up special chess schools to foster talented young players. Mancala is a logic game that is played in every country in Africa. It is thought to be the oldest board game in the world; early examples of mancala were found in excavations of ancient Egyptian temples. In the game, players move pieces across hollows and, in the process, try to capture an opponent's pieces. Chess and mancala both demand skills of critical thinking and problem-solving. Developing a strategy is crucial: players must think ahead and have an overall plan of how to defend their pieces and capture those of their opponent.

**MANCALA**
*This game of logic and strategy is played all over Africa. Instead of using a board and pieces, children often make holes in the ground and use pebbles as pieces.*

**PLAYING CHESS**
*Chess encourages mathematical skill, concentration, and logical thought. Some children become highly proficient at a young age.*

PLAYING

## Pretend games

**In brief:** Imitation games help children to work out what adults do and learn about different roles in life

Most children use role-play and pretend play in their games. By pretending to be a doctor, mother, shopkeeper, or soldier, for example, children imitate the things that adults do and say. In this way, they both exercise their memories to re-enact situations and start to learn about different roles in life. There is also evidence that role-play is an effective way for children to express their fears and anxieties and learn to control them. Pretend games may therefore help to overcome problems. They can have a healing effect on children who have suffered violence or abuse. Children show great inventiveness in pretend play and often improve on reality by inventing idealized fantasy worlds.

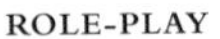

**ROLE-PLAY**
*Children observe adult behaviour and then incorporate what they see into their games.*



PLAYING

## Team games

**In brief:** Games that involve being part of a team help children's social development

Children all over the world play ball games such as soccer, cricket, and hockey. Playing team games is good for children's fitness and social development. Child psychologists suggest that 8 years is the ideal time for children to start playing team games. By this age, children have the conceptual skills to understand the rules of the game as well as the physical maturity for the exertion involved. Being part of a team teaches children about cooperation and achievement as well as sportsmanship and how to lose gracefully. The negative physical and psychological impact of team games can also be significant. Each year 75,000 children in the US visit hospital with soccer-related injuries. Over-competitive games can have detrimental effects on children's self-esteem. In the UK research shows that children start playing soccer for fun but drop out when it becomes too competitive.

Competition is not an element of all team games. In the Congo, Pygmy children play team games where there is no concept of winning, only the aim of having fun. Children don't always have to participate in team games to enjoy them. Supporting a professional team can give children a sense of identity and, in some cases, this can become a way of life.

**AMERICAN FOOTBALL**
*Children derive huge pleasure and self-confidence from playing games such as football – a popular game with many American children.*

**CRICKET**
*Team games such as cricket help children to understand rules and to learn to cooperate. These children play near the Taj Mahal in India.*

PLAYING

## Traditional games

**In brief:** Many games, such as hopscotch and skipping, have been passed down through generations

Some games have lasting and worldwide appeal. The early Romans played marbles, skittles, dominoes, and hopscotch. The original Roman hopscotch grids were more than 30m (100ft) long and were used during military exercises to improve soldiers' footwork. It is believed that Roman children copied these training manoeuvres, drawing their own smaller grids and adding a points system.

Hopscotch is called *marelles* in France, *Tempelhüpfen* in Germany, *hinkelen* in the Netherlands, *ekaria dukaria* in India, *pico* in Vietnam, and *rayuela* in Argentina.

The board games ludo and snakes and ladders both originated in India in the 16th century (ludo was known as *paschi*) and then spread to Europe when explorers brought the games home in the 18th century.

Jacks is common worldwide. In Korea it is traditional for the whole family to play a version of jacks with pointed sticks on New Year's Day; all ages take part and players who lose are given forfeits.

Every culture has its own games that have been passed down from one generation to the next. In Jamaica a game called "Bull Inna Pen" has a long folk history. It is a version of "chase" played around the theme of a hen protecting her chicks from a bull. Some of the oldest games are the most enduring: 2,000 years after its invention, marbles is still popular. Players even use special words, such as "shooting" or "knuckling down" to describe rolling a marble.

Health

### A sedentary childhood

In the West the time that children spend watching television, surfing the internet, and playing video games is increasing. The average American child spends 1,023 hours a year watching television, and 6½ hours a day on all media-related activities. The amount of time spent on physical activity is falling. This is partly responsible for the rise in childhood obesity.

**SKIPPING GAMES**
*Skipping over a rope in time to a chant, rhyme, or song has always been a favourite pastime among children.*

**HOPSCOTCH**
*This traditional hopping and jumping game is played by children in many different countries, from France to Argentina and Vietnam.*

LEARNING

## Going to school

**In brief:** Formal education is considered important in every society and usually starts at about 5 years

Starting primary school brings great changes in children's lives. It means that they spend most of their time away from their parents with peers of a similar age. For most children, school is their first experience of learning in a structured way and being assessed on how well they perform. What children learn at school depends largely on the culture in which they live. Literacy and numeracy skills form the basis of most educational systems. Children are typically taught about the culture and history of their country, and many school curricula incorporate religious education. Physical education is a regular part of most school timetables. In India children receive tuition in yoga from an early age.

**REFUGEE SCHOOL**
*In situations where formal education is impossible, adults set up makeshift outdoor schools.*

Getting to school is not always easy. In Zimbabwe, children from rural areas frequently have to walk for 1½ hours to get to class. In some developing countries teachers run schools with minimum equipment under trees, in their own homes, or in refugee camps. If teachers are limited in number, shortened classes may be run in shifts to allow more children to attend them. In Hawaii, the need to wear a uniform (including shoes) was found to prevent children from going to school so a decision was taken to allow children to attend classes barefoot. There are currently 113 million children worldwide who do not attend primary school. There are several reasons: some families cannot afford books and pens; some need their children to go to work to earn money. In other cases, large class sizes and harsh forms of discipline deter families from sending their children to school. Two-thirds of primary-age children not in school are girls whose families may withhold schooling for cultural reasons or so that girls can help out at home. These children miss out on the good start to life that school provides. Research has found a direct link between low levels of education and literacy and low levels of health. People who are better educated tend to live longer and suffer less from preventable disease.

**LEARNING IN A CLASSROOM**
*The classroom is the traditional setting for learning. Classrooms vary across cultures from sparsely equipped rooms to high-tech environments full of computer equipment.*

**BOARDING SCHOOL**
*Some schools provide meals and accommodation and children return home to visit their parents only during school holidays.*

Issue

### Single-sex schools

Evidence suggests that single-sex education has benefits for both girls and boys. Girls educated in single-sex schools tend to be better at expressing themselves, more competitive, less prone to peer pressure, and more likely to have a high self-esteem. For boys the benefits appear to be academic. One piece of research found that boys' pass rates in languages rose from one-third to 100 per cent after a change to single-sex classes. In societies in which girls and boys are traditionally segregated at a certain age, single-sex schools give girls the opportunity to study for longer.

LEARNING

## Home education

**In brief:** Some parents choose to teach their children at home, even where state education is freely available

Home schooling is common in the US and among expatriates living abroad who want to keep their children in touch with the curriculum of their home country. In the US about 1 million children are educated at home, most of them from families who have a strong Christian faith and who want to maintain a Christian ethos in their children's education. Some parents decide to home school for social reasons and believe they can create a better environment for learning than is found in local schools.

Children who are educated at home do well academically, but they may lose out on developing social skills, such as interacting with other children from different social backgrounds.

**PERSONAL ATTENTION**
*Home-schooled children may benefit from the more personal attention given by tutors or parents who teach them.*

LEARNING

## Religious education

**In brief:** Children may be educated in their faith in mainstream school or out of school

All religions make provisions for children to understand and learn the faith, whether as part of an out-of-school activity or in a religious school. Children often start to take part in religious celebrations from an early age and may begin formal learning at any age, but usually around age 5 or 6. In the Christian faith, parents may chose to send children to "Sunday school" or bible classes outside of normal school. In Islam, children may attend extra lessons at the mosque, learning Arabic and quotations from the Koran. In Japan, children often attend lessons at the temple, where they learn Buddhist sayings and exercises. Many Jewish children have lessons at the synagogue in which they learn some Hebrew.

**STUDYING AT THE MOSQUE**
*Muslim children often attend the mosque to learn Arabic and readings from the Koran.*

**SUNDAY SCHOOL**
*In the Christian faith, religious education may continue at Sunday school. Children may also become involved in the religion by joining church choirs.*

Religious schools that provide children with the bulk of their education exist in many parts of the world. These schools range from the missionary schools that exist in parts of Asia and Africa to Islamic religious schools called *madrassahs*, which aim to give children an in-depth knowledge of the religious books, prayers and language of Islam.

Fact

### Buddhist monks

In the Buddhist faith, boys may spend time living as monks in a monastery. In Thailand, for example, when a relative dies, the family may send their sons to the monastery for a time as part of mourning and to help the spirit gain entry to paradise. Novice monks learn how to meditate and practise the standards of Buddhism. Some stay for the rest of their lives, becoming ordained and adopting all the precepts of Buddhism including celibacy and purity.

LEARNING

## Practical skills

**In brief:** Most children learn practical skills through imitation; these skills are important for adult life

Much of children's everyday learning does not take place in the classroom. Instead it involves picking up practical skills from their parents or other adults, skills they will use for the rest of their lives. Just as the age of 5 is the stage when children usually start school, it is also the age when parents begin to give children small tasks to do. Skills can range from helping clear plates from a dinner table in a typical American family to herding the family's goats in parts of Africa.

Children learn skills by watching and imitating. Among the Gisu people of Kenya, for example, 5-year-old boys follow behind their older brothers as they herd cattle. The young boys wave sticks and imitate their brothers, who beat the cattle to move along. By the age of 7, they use the same technique in reality to herd sheep and goats, and by 9 years they too are in charge of cattle. Skills such as building fires, laying traps, shooting arrows, fishing, digging the fields, peeling vegetables and fetching water are

**LEARNING TO FARM**
*In traditional rural societies, children's help is vital to the family in earning money or growing crops. These children in the Philippines help their mother to plant rice.*

often self-taught. Children as young as 5 years watch their brothers and sisters, and, starting with miniature versions of adult tools soon build up on their experience.

In most traditional rural societies, children's domestic tasks are a serious contribution to the household economy. In rural China, 80 per cent of children who drop out of school are girls whose families say they need the extra help to run the home and farm. Sometimes, children learn a skill with the aim of earning money for their families. In Pakistan, children learn how to weave carpets, which are then sold for family income. In more prosperous households, however, children may use practical skills to earn pocket money for themselves.

There are strong gender values attached to the practical tasks taken up by boys and girls. Across the Middle East girls look after siblings, do embroidery, and help with the cooking. Boys fetch water and firewood. Even in the US, boys are more likely to do outdoor work such as taking out the rubbish and clearing leaves, while girls focus on indoor domestic skills such as cooking.

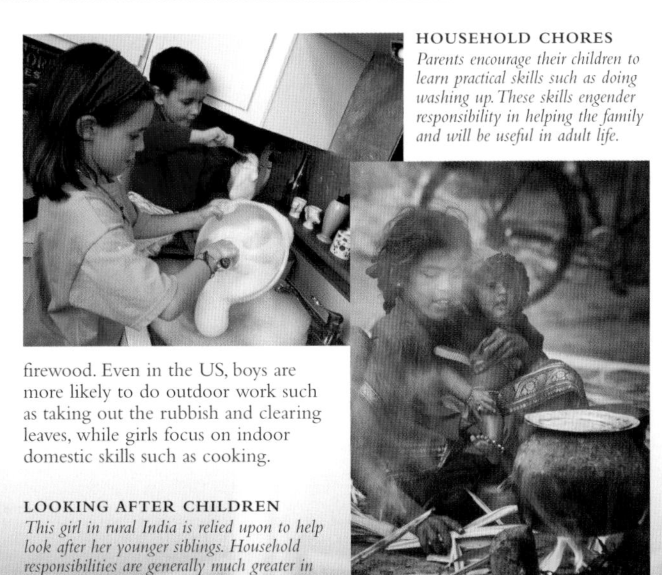

**HOUSEHOLD CHORES**
*Parents encourage their children to learn practical skills such as doing washing up. These skills engender responsibility in helping the family and will be useful in adult life.*

**LOOKING AFTER CHILDREN**
*This girl in rural India is relied upon to help look after her younger siblings. Household responsibilities are generally much greater in traditional societies than elsewhere.*

History

## Scout movement

Englishman Robert Baden-Powell founded the scout movement in 1907. It has since become the largest voluntary membership organization in the world. There are now more than 28 million scouts – boys and girls – worldwide. Baden-Powell intended the scout movement to complement formal education, helping children to develop their full potential in a range of practical skills and challenges otherwise unavailable.

**ROUNDING UP ANIMALS**
*Children learn skills by imitation. This boy in Colorado, US learns to round up sheep by imitating his father, using a rubber snake instead of a stick.*

# ADOLESCENCE

**INDIVIDUALITY**
*Adolescence is a time when young people may feel a need to express their individuality. Like this Japanese girl, it is often through what they wear.*

Adolescence is a long stage, which may start as early as 8 years and can last until a person reaches 18 to 20 years – the age at which a person is considered an adult in the West. The word adolescence comes from the Latin word *adolesco*, meaning "I grow", and signifies a time of many changes. The most important of these is puberty – the time when reproductive maturity happens. The start of puberty is celebrated in many cultures with rituals and ceremonies. In some societies, children are regarded as adults once they reach a certain age, which may be even before they reach puberty. During adolescence a person also enters into the final stage of developing physical and emotional maturity.

## PHYSICAL CHANGES

Puberty is marked by a series of physical changes in both boys and girls. There is an obvious growth spurt that starts at the age of 10 or 11 in girls and 12 or 13 in boys. The limbs lengthen more than the torso and more fat is laid down, especially in girls. Height increases by about 25cm (10in) in girls and 28cm (11in) in boys (by the time boys start their growth spurt they are already taller than girls). Once bone growth is complete, the epiphyses at the ends of the long bones fuse (*see* p64). Most of the physical changes of puberty are driven by the sex hormones oestrogen (*see* p132) and testosterone (*see* p136). These hormones drive the start of reproductive life in girls with the monthly release of eggs and the onset of menstruation; this is known as menarche. The equivalent stage in boys is called spermarche; this refers to the beginning of sperm production in the testes. Other obvious physical changes are the growth of breasts in girls and body hair in both sexes. Vocal changes in boys usually occur around age 14 to 15. The vocal cords enlarge from 8mm (3/8in) to 16mm (5/8in) in a single year, lowering the voice pitch by several octaves.

**HEIGHT CHANGES**
*During adolescence, all young people develop the physical characteristics of adulthood.*

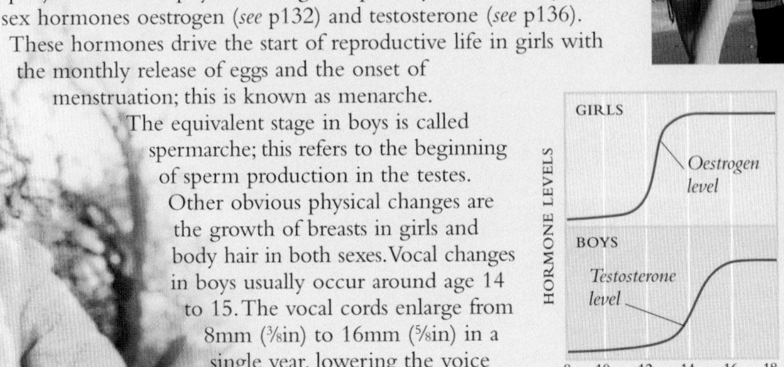


**HORMONE LEVELS**
*Oestrogen levels in girls surge at around age 12, reaching a peak 1–2 years later. In boys, the equivalent surge – in testosterone – begins later.*

## MOOD AND BEHAVIOUR

Sex hormones exert a profound influence on emotions as a child progresses towards adulthood, influences that are just as great as the physical changes. Moods may fluctuate wildly from hour to hour. Adolescents often pursue a relentless programme of testing their world to the limit – sometimes creatively and sometimes less usefully – with parents, teachers, and their own bodies. This desire to experiment, coupled with the need to break with the parental nest, leads to adolescents taking risks in many areas. Experimentation, sometimes in response to peer pressure, is a natural part of development. It may include trying out sex, driving while not qualified, drinking alcohol, smoking, and taking drugs.

**RISK-TAKING BEHAVIOUR**
*Adolescents are often oblivious to potential dangers. As a result, they are much more likely than adults to take part in hazardous activities.*

Fact

### A chaotic brain

Evidence suggests that during puberty the brain undergoes some "rewiring". The part of the brain that enables people to read facial expressions is one of the areas affected. In one study, volunteers were shown happy, sad, and angry faces. Younger children were better at recognizing the emotions than teenagers. Brain scans showed a shrinkage of one area of the brain, the prefrontal cortex, in teenagers. It is this area that is responsible for controlling emotions (*see* p165).

**BECOMING A WOMAN**
*How a girl experiences adolescence depends on the society in which she lives. During this stage in her life, she develops the capacity to bear children.*

# SEX AND RELATIONSHIPS

As children go through adolescence, so dawns an increasing sexual awareness and curiosity. Powerful sexual urges drive both girls and boys to test their ability to attract the opposite sex, to initiate new relationships, and to experiment sexually. Sexual awareness can generate much anxiety, coming as it does with fluctuating moods and generalized concern about physical attractiveness. Both girls and boys struggle with their sense of identity, moods, a tendency to return to childish behaviour, and worries about whether they are normal. Adolescent boys and girls often have widely divergent interests. While young teenaged boys frequently act in a "macho" manner, girls may be concerned with feminine pursuits. Such behaviour helps adolescents to define their sexual identity, but makes it difficult for them to establish common ground in relationships. At this stage, compatibility, friendship, and shared interests, the mainstays of relationships, take a back seat to the sexual drive. In some cultures, boys and girls are kept separate at this time; a girl may have to dress in clothes that cover her completely, and she may be forbidden to spend time alone with a boy who is not her brother.

**FIRST KISS**
*Adolescents experience powerful sexual urges, and their first real relationships start now.*

**CLOSE FRIENDSHIPS**
*During the adolescent years, many friendships are cemented and new ones begun. Teenage girls in particular often spend time in groups before they start pairing up with boys.*

**MALE IDENTITY**
*Style of dress is often used to define a person's identity and to express his or her membership of a particular group. This behaviour is typical of that of young adolescents, boys and girls.*

Issue

## Teenage parenthood

The rate of teenage pregnancies varies around the world. In most developed countries, the rate is relatively low; girls are expected to finish their education and not have children until they are in their twenties or older, by which age they are considered to be emotionally mature. In some countries, teenage parenting is the norm. For example, in parts of Africa and Asia, girls as young as 16 may be married and have more than one child. In these societies, extended families and close communities can offer practical support.

PREGNANCY TESTING

# NEW EXPERIENCES AND RESPONSIBILITIES

With their dawning maturity, adolescents are encouraged to accept the responsibilities of adulthood, in their family life, their educational and/or work life, their spiritual life, and in their personal relationships. Adolescents may explore a number of training and career options before settling for their life path. In some Western societies, adolescents take time between school and further education to explore the world around them. For the first time, these young adults-to-be will be legally able to learn to drive, to have sex, to drink alcohol, to vote, to marry, and to have children. The significance of the threshold of adult life, with its freedoms and increasing financial independence, may be marked by reaffirmation of religious beliefs. In some cultures it is marked by a series of rituals that often include separation, trials (for boys), and for girls a period of seclusion during which time they learn about their responsibilities as an adult.

**LEARNING TO DRIVE**
*For many young people in the West, learning to drive is a rite of passage. However, the age at which it is legal to do so varies.*

**RELIGIOUS RITES**
*Puberty may be a time when a young person is expected to take a more active role in religious life. Many religions have rituals, such as this Bar Mitzvah.*

**BECOMING A MAN**
*For boys, adolescence can be a period of highs and lows. Although the physical changes of adolescence happen later than girls, they are just as pronounced.*

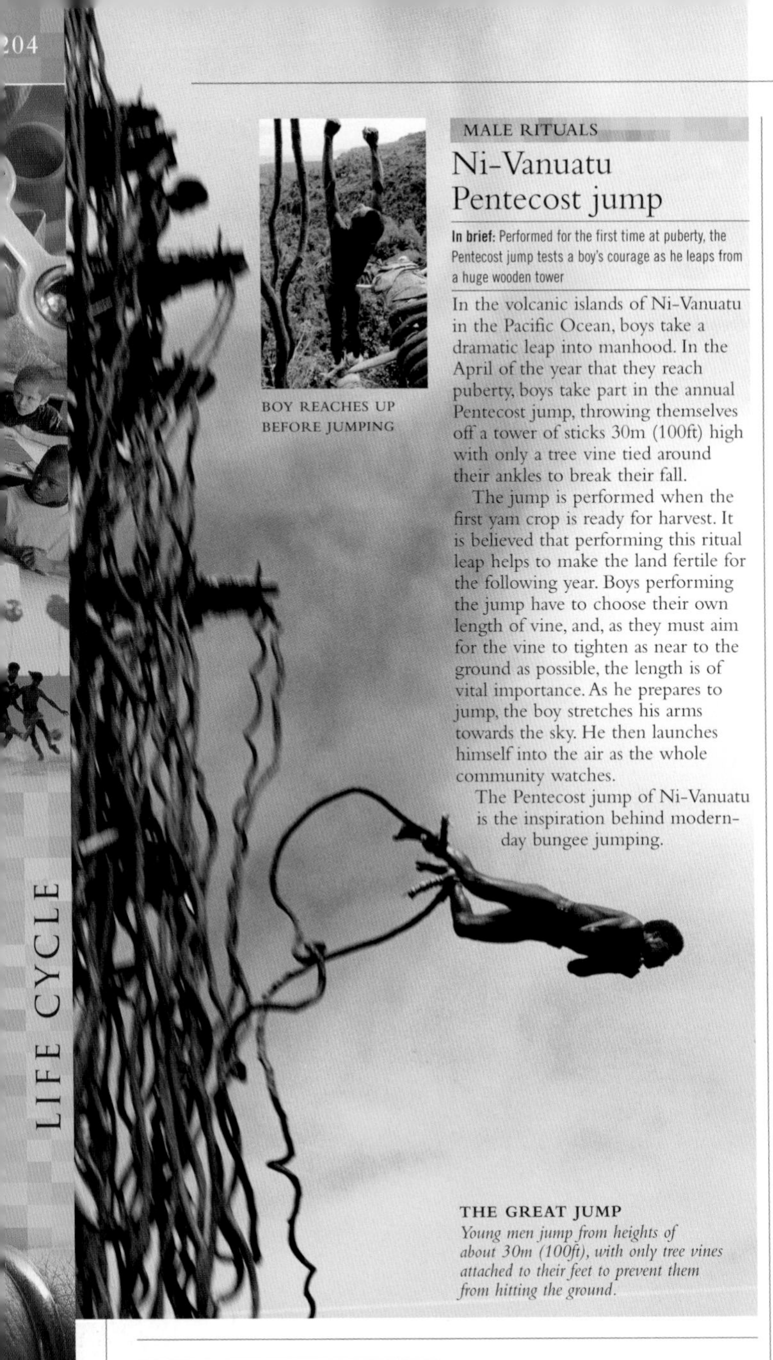

BOY REACHES UP BEFORE JUMPING

MALE RITUALS

## Ni-Vanuatu Pentecost jump

**In brief:** Performed for the first time at puberty, the Pentecost jump tests a boy's courage as he leaps from a huge wooden tower

In the volcanic islands of Ni-Vanuatu in the Pacific Ocean, boys take a dramatic leap into manhood. In the April of the year that they reach puberty, boys take part in the annual Pentecost jump, throwing themselves off a tower of sticks 30m (100ft) high with only a tree vine tied around their ankles to break their fall.

The jump is performed when the first yam crop is ready for harvest. It is believed that performing this ritual leap helps to make the land fertile for the following year. Boys performing the jump have to choose their own length of vine, and, as they must aim for the vine to tighten as near to the ground as possible, the length is of vital importance. As he prepares to jump, the boy stretches his arms towards the sky. He then launches himself into the air as the whole community watches.

The Pentecost jump of Ni-Vanuatu is the inspiration behind modern-day bungee jumping.

**THE GREAT JUMP**
*Young men jump from heights of about 30m (100ft), with only tree vines attached to their feet to prevent them from hitting the ground.*

MALE RITUALS

## Australian aboriginal initiation

**In brief:** Rituals include marking ceremonies and initiation into secret knowledge by older men

Australian Aboriginal initiation ceremonies may last for many months or even years. They are seen as a way of mourning the death of a child while also celebrating the birth of an adult.

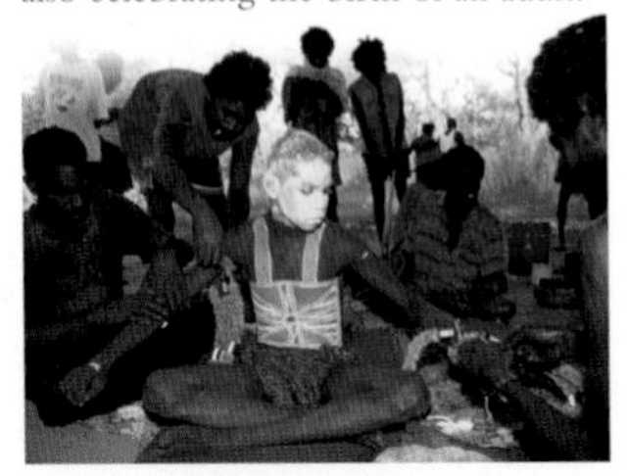

As part of the initiation, secret ritual knowledge is disclosed by older men to boys. Such knowledge includes the dreamtime, the spiritual beliefs that Aboriginal people hold about the creation of their ancestors. Women and outsiders are not allowed to witness the dances and songs through which the knowledge is passed down the generations. Other ceremonies involve marking the boy's body – all initiated men must have a permanent sign of adulthood, and they may be circumcised, tattooed, ritually scarred on the chest, or have a tooth removed. Older men also tutor boys in hunting, marking, and making wooden tools.

**INITIATION PAINTING**
*The father of a young boy about to enter adulthood paints his son's chest with pictures illustrating stories that have been passed from one generation to another.*

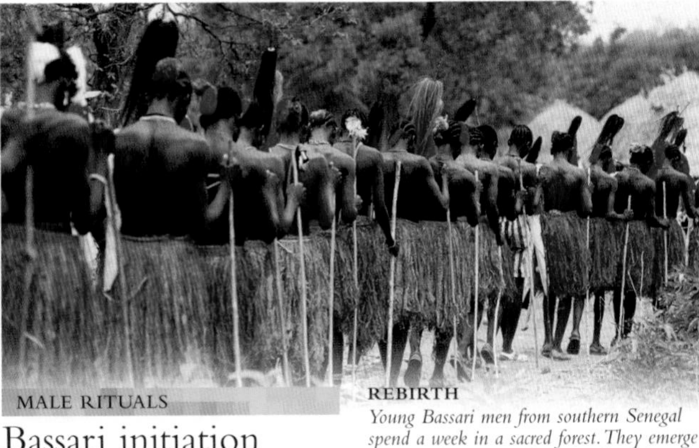

**REBIRTH**
*Young Bassari men from southern Senegal spend a week in a sacred forest. They emerge from the forest in a silent, transformed state.*

MALE RITUALS

## Bassari initiation

**In brief:** Groups of boys are subjected to a period of seclusion and ritual away from their village; after several months they return as men

Among the Bassari people of southern Senegal, the transition from boyhood to manhood is marked by a ritual called Koré. During the ritual, which may last several months, boys are taken from their families to live in a communal house away from their village. They are looked after by older initiates and are taken into a forest that is believed to be sacred. Here, it is believed, the boys "die". After a week of a series of rituals they emerge from the forest in a trancelike state and are fed and bathed as if they were babies. They are not allowed to look at anyone or to show any signs of emotion during this period.

The culmination of the Koré ritual is a masked dance in which the new initiate fights a masked warrior. Whether he wins or loses, the initiate is then considered a man. He is given a new name and assumes a new status among the people of the village.

**CONFRONTATION**
*The climax of the Bassari male initiation is a duel between each new initiate and a masked man.*

MALE RITUALS

## Masai circumcision

**In brief:** The Masai of Kenya believe circumcision (removal of the foreskin) is symbolic of rebirth

For the Masai people of East Africa, circumcision is the final part of a ritual that marks the passage from boyhood to manhood and novice to warrior. It is usually carried out between the ages of 15 and 18. The boy must first spend a week looking after cattle alone and taking purifying cold showers. At dawn on the next day, he is circumcised. The boy's father, whose status changes from warrior to elder at the same time, gives him cattle of his own. The pain of circumcision is said to symbolize the depth of the break between childhood and adulthood. After the ceremony, the new warrior wears black clothes for 8 months. Other boys who are circumcised at the same time become members of one age-set and are bound by ties of loyalty and obligation for the rest of their lives.

**BIRD HEADDRESS**
*Markings painted on these boys' faces, and the headdress of bird's feathers, indicate their intermediate status between boy and man.*

MALE RITUALS

## Bar Mitzvah

**In brief:** In the Jewish faith, boys take on religious responsibilities at age 13, and parental responsibility lessens; it is marked by a religious ceremony

Bar Mitzvah marks the beginning of adulthood for Jewish boys, the phrase meaning "son of the commandment". The term is also used to describe the ceremony held at this time. According to Jewish law, children do not have to obey the various commands relating to moral life believed to have been given by God. However, at the age of 13, boys are expected to become responsible for their own moral behaviour. Parental responsibility for their sons' actions lessens at this time, something that is reflected in the Bar Mitzvah blessing said by parents: "Blessed is God who has now freed me from bearing full responsibility for this person". The Jewish community marks boys' coming of age with a ceremony, which is held at the synagogue on the first Saturday after the boy's 13th birthday. During the ceremony, the boy reads a passage from the Torah and chants blessings. In some cases, he is asked to give a speech, in which he is expected to begin with the words "Today I am a man".

**BAR MITZVAH CLASS**
*Long before their 13th birthdays, Jewish boys attend classes at the synagogue to learn Hebrew in preparation for their Bar Mitzvah.*

**READING THE TORAH**
*An Israeli boy celebrates his Bar Mitzvah at the Wailing Wall in Jerusalem. During the ceremony he reads from the Torah.*

Fact

### Bat Mitzvah

A coming-of-age ritual for girls, to mirror the Bar Mitzvah for boys, was first introduced in the 1920s, a time of reform in the Jewish faith. When a girl reaches the age of 12, Bat Mitzvah is celebrated with a party and the giving of gifts at the girl's home. Depending on the views of the congregation, a girl may be invited to read from the Torah in the synagogue. The centrepiece of Bat Mitzvah is a candle-lighting ceremony that is held in the evening. The girl usually lights 14 candles, each one honouring a member of her family, starting with her grandparents and ending with herself. Each dedication is accompanied by a poem or song.

MALE RITUALS

## Native American vision quest

**In brief:** Some Native American peoples hold a ritual in which boys spend a period in the wilderness

Traditionally, certain Native American peoples, including the Navajo and the Sioux, embarked on a trial called a vision quest for the first time at around puberty. The quest involves a period of physical hardship and solitude in the wilderness with the aim of attracting a vision from a spirit.

In the Navajo vision quest, boys go in to the wilderness for 4 days and stay there without food, drink, or shelter; they attempt to stay awake the whole time. The spirit the boy sees is believed to give him songs and rituals that he can use to protect himself from danger during adulthood. The same spirit is believed to remain as the boy's guardian for the rest of his life.

Traditionally, Sioux vision quests were embarked on at various stages of life, often during adulthood.

**SIOUX INDIAN VISION QUEST**
*As part of the vision quest ceremony, a buffalo skull is raised to the sky. Buffalo were once central to the lives of the Sioux.*

MALE AND FEMALE RITUALS

## Balinese tooth-filing

**In brief:** In Bali, tooth-filing is believed to eradicate wild spirits and is done at various stages of life

Pubescent Balinese girls and boys participate in tooth-filing ceremonies to mark their entry into adulthood. Tooth-filing symbolizes a move away from the "animal" qualities of childhood. Adults are expected to exert control over lust, greed, jealousy, drunkenness, anger, and confusion. Tooth-filing is carried out by a Hindu priest in July and August. Girls and boys wear white and yellow to represent purity; their teeth-filings are collected in a coconut shell, which is buried behind the family shrine.

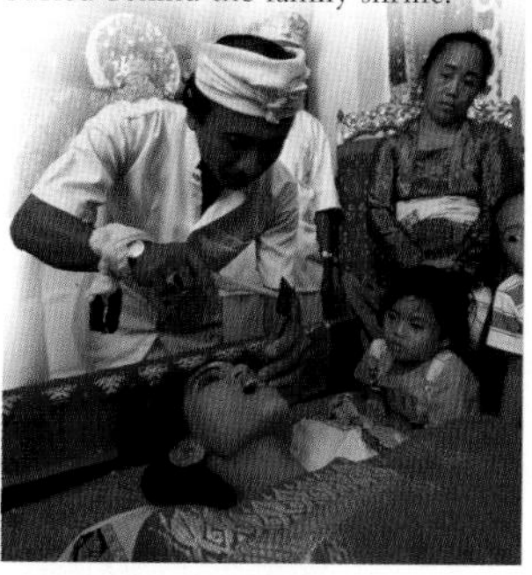

TEETH BEING FILED BY A BRAHMAN

FEMALE RITUALS

## Apache sunrise ceremony

**In brief:** Apache girls celebrate coming-of-age with a ceremony symbolizing health, purity, and strength

After her first menstrual period, an Apache Native American girl and her family hold a sunrise ceremony, which takes place over a whole weekend. On the Friday night, the girl's godmother dresses her in a yellow dress, an ostrich feather (to represent a long, healthy life), and a shell on her forehead (to represent purity). Starting at sunrise the next day, the girl enacts the story of the "White Painted Woman", which is a central figure in Apache folklore.

In between acting the story, the girl grinds flour and bakes corn cake. That night she stays awake, praying; in the morning, the girl offers a piece of the corn cake to the sun, and the rest to her family and godparents. This ritual is a sign of taking on womanly duties. The girl's family then bless her with pollen, which is believed to be sacred and to have healing qualities. By the end of the sunrise ceremony, the girl is believed to have shown the endurance and resolve needed for womanhood.

**PAINTED FACE**
*The tale of the "White Painted Woman", a symbol of fertility and longevity, is enacted by the Apache girl during the ceremony.*

SACRED POLLEN

**SKILLS OF WOMANHOOD**
*As part of the puberty ritual, an Apache girl bakes corn cake under the guidance of her grandmother.*

FEMALE RITUALS

## Balinese purification ceremony

**In brief:** A Balinese girl's first menstrual period is marked by rituals symbolizing purity and fertility

In Bali, when a girl first starts to menstruate, she enters a period of seclusion in her home. On the 5th day of seclusion, she undergoes a purification ceremony. The girl, dressed in ornate clothes and a gold headdress, is carried to a temple where she receives blessings and takes part in a set of rituals to purify her.

Palm leaves and broom, which represent fertility, are used in the ceremony. A Hindu priest pours water through a basket used to steam rice (rice is a symbol of marriage and fertility) on to the girl's hands. She then wipes her hands on her head to purify herself. The girl and her family then return home for a feast.

**TEMPLE BLESSING**
*A young girl stands in the Balinese Hindu temple as part of her purification ceremony. She wears a traditional ornate headdress.*

FEMALE RITUALS

## Scarification

**In brief:** Scarring of parts of the body, especially the abdomen, is common in some African societies

Ceremonial scarring of the body and face is associated with reaching puberty in many African societies. Among the Karo people of Ethiopia, it is traditional to honour a girl's first menstruation with a scarification ritual. The scars are created by cutting the abdominal skin with a knife and then rubbing the cuts with ash to leave a raised and darkened weal. The scars are believed to accentuate the curves of a girl's body, give her additional beauty, draw attention to her abdomen, and signify that she is eligible for marriage. Going through the pain of scarification also shows that a girl has the womanly qualities of bravery and endurance. Among the Yoruba people of Nigeria, girls may be scarred on their cheeks at puberty. The pattern of scars not only denotes their new status as women but also signifies which clan and area they are from. In Botswana, the !Kung hold scarification rituals for boys when they kill their first animal.

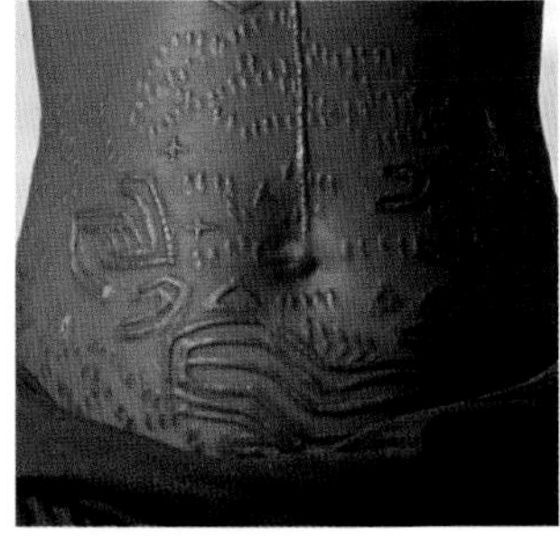

**SCARRED ABDOMEN**
*Scarring is a sign of beauty in several African societies. This pattern of scarring has been produced by cutting the skin with a knife.*

FEMALE RITUALS

## Krobo seclusion ritual

**In brief:** After a 3-week seclusion period, Krobo girls are welcomed as women at an "Outdooring" ceremony

The tribal people of the Shai and Krobo, from eastern Ghana, mark the passage of a girl from childhood to womanhood by a 3-week period of seclusion. A group of girls of roughly the same age go through the ritual, known as Dipo, at the same time.

First, the girls discard their old clothes and have to wear red cloth to represent menstruation. Their heads are shaved in a ritual that symbolizes cleansing and purification, and the leaving behind of their childhood. Over the following 3 weeks, specially appointed Dipo guardians ("mothers") teach the girls many aspects of becoming a woman. These include skills from cooking to dancing.

During one part of the ceremony a priestess holds the girls over a sacred stone called *Tekpete*. From this ritual, it is believed that the priestess can tell whether or not the girl is a virgin. Afterwards, the girls are taken to a nearby river for ritual bathing in order to purify them.

The seclusion ritual ends with a ceremony known as "Outdooring". For this ceremony, the girls, who are now considered to be women, perform dances to show off their new skills to prospective husbands. Dipo rituals have been performed by the Krobo people since the 11th century.

**SACRED STONE**
*The girl is held by a priestess over a sacred stone. From this, the priestess knows whether or not the girl is a virgin.*

**DANCING**
*At the end of the seclusion ritual, the girls perform a dance to show their gracefulness.*

Issue

### Female circumcision

Removing parts of the female genitals is still practised in some African countries and other parts of the world, despite worldwide campaigns against this ritual. Circumcision of young girls, also known as female genital mutilation (FGM), is performed at puberty. The immediate health consequences may include severe bleeding and infection. There are other long-term consequences, including urinary incontinence, infertility, and problems with childbirth.

**EMERGING AS WOMEN**
*At the end of the seclusion period, girls walk in readiness to be welcomed as women. White calico cloths and chalky white paste are symbolic of purity.*

FEMALE RITUALS

## Southern Indian menstruation ritual

**In brief:** Girls in Tamil Nadu, southern India, undergo a series of rituals at first menstruation; they include washing in turmeric, seclusion, and feasting

When they first start to menstruate, girls from Tamil Nadu, in southern India, go into a period of seclusion for 9, 11, or 13 days. They rest inside a small hut built out of fresh leaves by women in the community.

During her seclusion, a girl is shown washing rituals using turmeric water and is instructed by other women how to behave during menstruation. Because menstruation is considered to be a time of impurity, girls are taught that, for the duration of their menstrual period, they must avoid touching food, plants, and flowers. During this time, girls must not look at birds or see men before bathing, and they must carry leaves from the neem tree and an object made of iron to ward off any evil spirits.

After the period of seclusion and instruction, the women prepare a large feast and give the girl presents. They paint her feet with a mixture of turmeric, red ochre, and limestone. This is the same mixture that will be used to decorate the girl's feet on her wedding day, and therefore symbolizes her entry into womanhood and her readiness for marriage.

CELEBRATIONS

## Balls and proms

**In brief:** Some US and European adolescents mark their coming of age with sophisticated parties

The debutante ball started in Europe in the 17th century as an occasion where girls from wealthy families were introduced into public society. This "coming-out" event signalled the fact that a girl had reached adulthood and was available for marriage. The debutante ball opened a season of balls and parties where young girls would meet eligible men, agree to dance with those they liked by signing them on to a dance card, and often end the season by announcing their engagement. Today, debutante balls continue among the rich elite in countries across Europe and in the US and as more general forms of celebration in Australia and the Philippines. The high school prom was created in America in the early 20th century as a way for local children to learn the manners and customs that are associated with the rich. Today, the high school prom is a universal celebration of the end of school life. Teenagers dress up and go with a date of their choice to a dance held at their school.

**GOING TO THE PROM**
*High-school proms are part of the coming-of-age ritual for American teenagers. Young women and their partners dress for the occasion.*

**VIENNESE BALL**
*Debutantes in Austria attend a ball with selected boys to mark their entry into womanhood. The girls wear white.*

**DEBS READY FOR THE BALL**
*In parts of English society, girls continue the old tradition of attending debutante balls to celebrate womanhood and meet young men.*

CELEBRATIONS

## Quinceanera

**In brief:** Girls throughout Latin America celebrate their 15th birthday as their entrance into adulthood

Across Latin America, girls mark their 15th birthday with a celebration called *quinceanera*. The girl dresses in white or pastel colours and is accompanied to church by her godparents and 15 maids of honour. She sits by the altar for a thanksgiving mass service and is presented with a gold medallion bearing a religious Catholic image and an inscription of her name and the date. At the end of the service, she places flowers by a statue of the Virgin Mary. In some Hispanic countries it is traditional for the girl to change her shoes during the service, from flat heels to high heels. A young boy brings in the new shoes on a pillow, and her father helps the girl to change her footwear. After the church service, the family hosts a large party or fiesta in honour of their daughter. The party is opened by father and daughter dancing a waltz. The music, dancing, and eating continue all night. Throughout the celebrations, the girl is feted by her family and friends for having passed from girlhood to womanhood.

**QUINCEANERA ADMIRERS**
*At 15, a girl celebrates her coming of age in Hispanic communities with a special mass followed by a party. Age 15 is seen as the time of transition from child to adult.*

**KIMONO-CLAD FRIENDS**
*Japanese girls celebrate womanhood dressed in traditional kimonos. A special Coming-of-Age Day is celebrated every year.*

CELEBRATIONS

## Coming-of-Age Day in Japan

**In brief:** Held annually in Japan, Coming-of-Age Day allows young men and women to mark entry to adulthood

Every year on the second Monday of January there is a national holiday in Japan called *Seijin no hi* (the "Coming-of-Age Day"). All those who have turned 20 in the previous year celebrate reaching the age of official adulthood at which they can vote, drive, smoke, and drink alcohol.

After going to a Shinto shrine to ask for blessing, the 20-year-olds attend celebrations with songs and speeches in their local towns. They then go on to separate parties. Boys dress in suits and girls usually wear traditional kimonos.

Coming-of-age events have a long history in Japan. In the 16th century girls would have a Mogee ceremony in which they marked their adult status by plucking their eyebrows and dying their teeth black – techniques that are thought to enhance their beauty. Girls would also take part in an all-day archery competition, said to test their endurance. In some Japanese towns a short archery competition is still a part of *Seijin no hi*.

**ADULT RESPONSIBILITIES**

*In adulthood, people take on new roles and responsibilities. Becoming a parent is probably one of life's most challenging but rewarding tasks.*

# ADULTHOOD

The adult body represents the peak of human development. Most people are physically strongest in the years of early adulthood, with their organs working at their most efficient. From the mid-20s onwards there is a physical decline. However, at a personal and social level, humans continue to grow in stature for many years to come. During adulthood, people usually form lifelong relationships, and may marry and have children. They also take on responsibilities of work, among others. These experiences bring a maturity and wisdom that is rarely seen in young adults.

Reaching adulthood means taking on new responsibilities, becoming independent of parents, and making one's own lifestyle choices. In many societies, but not all, it means leaving the family home. The stage at which an individual is deemed to have reached adulthood varies from culture to culture. In some parts of the world, people as young as 14 are treated as adults. In most Western societies, however, people are not considered adult until they reach 18–21 years of age.

## PEAKS AND TROUGHS

The early adult years represent the pinnacle of many of our physical capabilities. Internal organs are generally at their most efficient. Lung capacity is greatest around age 21, declining by a third by 60. Young adult hearts are fittest, but as heart muscle alters with age, the strength of its contractions diminishes. As a result, the work that the heart can do drops by around 20 per cent over the next 30–40 years. From middle age onwards, the body's immune system (natural defences) becomes less efficient, one reason for the greater frequency of cancers and other disorders after the age of 50. Bone density peaks in the 20s, dropping slowly thereafter. People who take regular exercise preserve their bone mass better than people who are inactive. The most dramatic decline in bone density occurs in women after the menopause (*see* p133), which normally occurs anywhere between the ages of 42 and 60. At this time, sex hormones, which normally help to maintain bone mass, wane.

For women, menopause marks the end of the fertile years. Women are physically able to have children from a very young age. Physically, the best time for childbearing is in the early twenties, and in some societies, this is usual. In their twenties, women usually produce healthy eggs and fertility is at its peak. However, in Western societies, many women now delay childbirth until they have settled careers and financial independence – often when they are well into their thirties. Getting pregnant becomes more difficult at this age however; infertility rates rise and the chances of conception fall dramatically from the age of 35 onwards. While women have a limited time in which they can have children naturally, men can continue to father children for many years, even though there is some decline in the quantity and quality of sperm as men get older.

**PEAK STRENGTH**
*Muscle size and strength peak in early adulthood. With age, the number of muscle fibres (which make up muscles) declines.*

Fact

### Mental capacity

Contrary to popular belief, there is no massive decline in nerve cell (neuron) numbers in most areas of the brain as we age. In fact, neurons continue to make fresh connections, important for learning and memory, throughout adulthood. Any decline in mental capacity in older age is usually caused by disease. In tests, people in their 70s and 80s show only slight reductions in memory.

NERVES CELLS IN THE BRAIN

## ROLES AND RELATIONSHIPS

The roles that we have as adults, and the relationships made, are more complex and enduring than those of childhood. Adults take on many new roles: they may become a husband or wife, a parent and, eventually, a grandparent. The starting point for many of these roles is finding a partner for life. In the formation of a partnership, humans are often considered to be monogamous. But within this broad description lies a wide spectrum of "mate" patterns, some cultures allowing more than one husband or wife. In traditional societies, people are encouraged to marry early. Family life varies from culture to culture, as do patterns of having children. In the UK, the average woman is 29 years old when she first becomes a mother. In many traditional societies, girls may marry and start having children by age 16.

**MARRIAGE**
*For many, marriage is an important milestone in adulthood.*

The role of parent is probably one of the most rewarding and challenging of adulthood. In most traditional, non-industrialized societies, extended family networks and stable communities help adults to meet these challenges. However, in much of the industrialized world, where families are often smaller and general community support weaker, raising a family is often more difficult.

In industrialized societies, much of people's daily lives is taken up with work outside the home. As well as providing income, many people identify themselves by what they do, and work confers status and identity.

**CONTROL CENTRE**
*Most adults spend at least half their waking hours at work, often in stressful environments. This steel-mill operator spends his day in front of a control panel.*

## GETTING OLDER

By middle age many people reassess their roles in life. They may no longer feel needed as parents; for example, mothers may have to reassess their roles when children leave home. People may think about changing jobs or retiring. In most industrialized societies, people retire and expect several years of leisure. Some societies revere elderly people, who are seen as wise and full of knowledge. In the rapidly-moving industrialized world, elderly people may be seen more as a burden, especially as life expectancy increases and their numbers grow.

**AN ACTIVE OLD AGE**
*An elderly man in Karimabad, Pakistan continues to work at the loom. Retirement is not an option for everyone.*

# FINDING A PARTNER

Humans, like other animals, have an inbuilt drive to form sexual partnerships. The biological basis for this drive is the need to procreate – to have children and pass our genes on to the next generation. The ways in which we find and choose our partners are influenced by where we live and our beliefs and culture. However, the biological urge is the same the world over. From adolescence onwards, chemicals in our brain drive us to fall in love. By adulthood, most people are looking for a partner. In some parts of the world, a partner is chosen by the individual's family – sometimes during childhood. In these cases, the importance of finding the perfect partner is seen as too important to leave to chance.

## HOW PEOPLE MEET

The most common way that people in Western societies meet is through casual encounters, such as at work or through friends and family. However, in many countries a partner may be initially selected by the family or by a paid matchmaker. Until around 50 years ago, these methods were common worldwide. Studies in the UK have found that the settings best-suited for meeting a prospective partner are those that combine at least two of three key features: shared interest, alcohol, and sociability. In many Mediterranean countries, the old tradition of promenading at dusk helps to provide a chance for young men and women to meet. However, in today's increasingly mobile industrialized societies, finding a life partner can be difficult. More and more people are turning to professional introduction agencies, advertising, and the internet to find a suitable partner. Internet dating is popular in the US and Europe and increasingly in India and Pakistan.

**SPEED-DATING**
*Many groups now organize events in which people have up to 30 "dates" lasting just a few minutes in a bar or club. Speed-dating is becoming an increasingly popular way to meet people.*

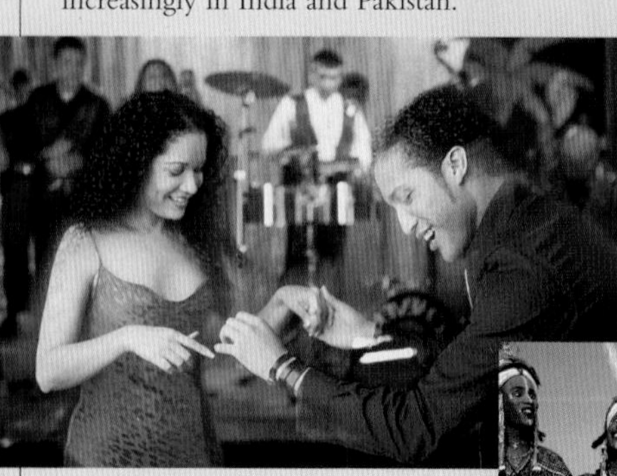

**CHANCE MEETING**
*Many young couples meet through friends or at bars and dances. In urban areas especially, chance meetings at social occasions often provide the starting point for a new romance.*

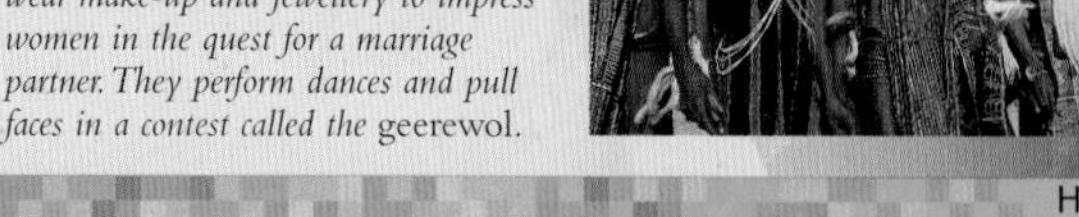

**ANNUAL FESTIVALS**
*Wodaabe men from Nigeria and Niger wear make-up and jewellery to impress women in the quest for a marriage partner. They perform dances and pull faces in a contest called the* geerewol.

History

### Japanese matchmakers

Under the feudal system in Japan, from the 12th to 14th centuries, marriages were often used as a political and diplomatic means to maintain peace and unity among feudal lords. As a result, the role of the matchmaker, known as a *nakodo*, was established. He would be employed by both families to arrange the best match and would be present at the wedding (*right*). *Nakodos* are still used by some families in Japan today.

BEAUTY

*Like women the world over, this woman from the Tuareg tribe in North Africa, uses jewellery and make-up to make herself attractive to a prospective partner. Tuareg women are free to choose their future husbands.*

# RULES OF ATTRACTION

Whether we are aware of it or not, from adolescence onwards we send out a host of signals that help attract a prospective "mate". The simplest of these signals is body shape. A narrow waist and wide hips in a woman are highly attractive to men because they are a sign of fertility. For women, physical characteristics may be less important. They tend to be attracted to men who are likely to be able to provide for their children – so strength, maturity, and resources (in other words, wealth) rank higher than good looks. More subtle factors may also play a role in attraction. Pheromones are natural chemicals produced by the body; they have been well researched in animals, where they indicate whether or not a female is sexually receptive. Scientists have now discovered that humans also respond to these chemicals, although the effects are much more subtle. When men and women were exposed to tiny amounts of pheromones, they found the wearer to be more sexually attractive. In women, there was also a subtle positive effect on emotional state and mood.

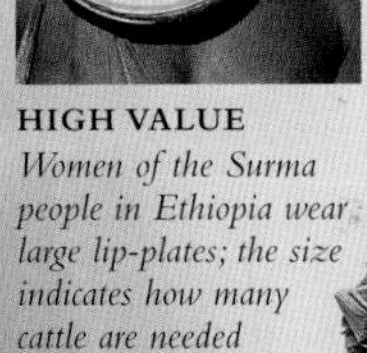

HIGH VALUE

*Women of the Surma people in Ethiopia wear large lip-plates; the size indicates how many cattle are needed for marriage.*

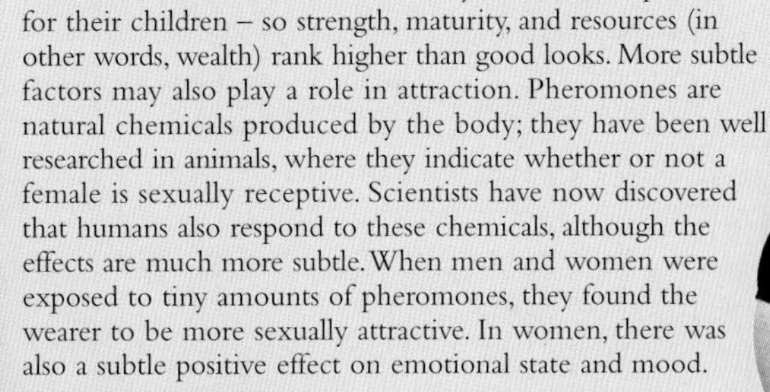

BIG HIPS

*Marilyn Monroe was known for her hour-glass figure. In evolutionary terms, this is a sign of fertility.*

LOVE OF MONEY

*Wealth is attractive – in evolutionary terms it is the modern-day sign of a man's ability to provide for a family.*

Issue

## Scent and attraction

Scientists have found that people have very different individual preferences for body odours. In one study, men nearly always preferred the smell of women who had an immune system very different from their own. There may be good reason for this: children born to parents with different immune systems have the best chance of fighting off illness themselves.

# FALLING IN LOVE

There are many scientific studies and theories about falling in love. According to one of them, there are three stages of love: lust, attraction, and attachment. Lust drives us to search for a mate, attraction helps to focus our attention on a suitable partner, and attachment ensures that the partners stay together to rear young.

Studies of what happens in the brain have shown that the sight of an attractive face stimulates the pleasure centre of the brain. As people fall in love, the brain releases a chemical similar to amphetamine, explaining the rush of excitement, increased heart rate, loss of appetite, and poor sleep. The rush of emotion in the first few weeks and months of meeting a partner do not usually last. Researchers suggest the attraction stage lasts 18 months to 3 years. After that time, staying together means forming a binding attachment, based on shared interests, friendship, and commitment. The suggestion is that after 3 years any child is sufficiently biologically secure for a couple to part and start again.

BIKERS IN LOVE

*People are often drawn to partners who look similar to themselves and have similar interests.*

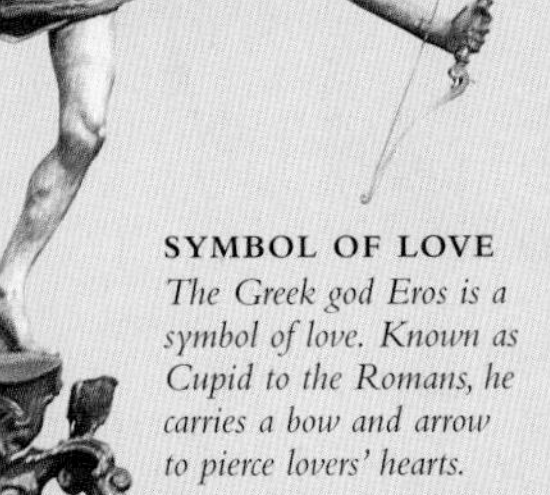

SYMBOL OF LOVE

*The Greek god Eros is a symbol of love. Known as Cupid to the Romans, he carries a bow and arrow to pierce lovers' hearts.*

# MARRIAGE

Nearly all societies formally recognize partnerships by marriage. This ritual is celebrated with a ceremony that may last a few minutes or several days. Although traditions vary from one country to another, marriages have several features in common, in particular the commitment made by the couple to each other in front of their family and friends. Marriage may be preceded by a period of engagement. At the marriage ceremony, contracts are made and vows exchanged that acknowledge the permanence of marriage. Ceremonies are often full of symbolism, for example of love, fidelity, and prosperity.

**GOLD BANDS**
*The plain gold band of the wedding ring symbolizes eternity. Diamond engagement rings symbolize faithfulness.*

## VOWS AND CONTRACTS

In many cultures, a new couple must make certain vows and sign contracts when they get married. Wedding vows often focus on what the couple will do for each other while they are together and on their mutual love for each other. Wedding contracts focus on the legal and economic aspects of marriage, and on what rights each party has if the marriage disintegrates. Jewish marriage contracts, or *ketubah*, have been used for over 2,000 years. Orthodox Jews still use the original Aramaic text, which outlines the responsibilities of a husband towards his wife in Jewish law. Some other modern Jewish communities have revised and amended the original text to make it more relevant to today's couples. Muslim marriage contracts also have a long history. They detail the rights and obligations of the husband and wife, and they protect the interests of the bride by stating the amount of money that she is entitled to if the couple divorce.

**JEWISH MARRIAGE CONTRACT**
*The marriage contract signed at a Jewish wedding is known as a* ketubah. *It is beautifully decorated and often framed and displayed in the couple's new home.*

**SIGNING THE CONTRACT**
*In most marriages, the couple sign a contract. This Malaysian bride signs the marriage contract in front of a witness.*

## TYPES OF MARRIAGE

Most marriages are monogamous, a union of one man and one woman. However, there are certain cultures where a man or woman takes more than one spouse, a practice known as polygamy. It takes two forms: polygyny describes one man having more than one wife, while polyandry is when a woman has more than one husband. The latter is uncommon but does happen in Nepal and Tibet. In some African polygamous families, everyone lives as one household, while in others the different wives live separately and the husband shares his time between them. Although it is not widespread in the Islamic world, up to four wives are permitted under Islamic sharia law.

**MANY WIVES**
*These identically dressed women in Nigeria are all married to the same man. Polygamy is widely practised in Africa, helping to ensure population growth.*

History

### The wedding dress

White has long been accepted as the traditional colour of the wedding dress in the West, but wedding gowns were not always white. England's Queen Victoria was one of the first to wear white when she married her cousin Albert in 1840. A white dress was a symbol of wealth because it could only be worn once. Later, the white wedding dress also became an indication of a woman's virtue, the colour being a symbol of purity.

**19TH-CENTURY WEDDING DRESS**

## MARRIAGE RITUALS

Marriage ceremonies, in all their variety, involve many rituals with many layers of meaning. In some cases, the original meaning of a ritual has been lost, but the ritual itself remains. Symbols of fertility and eternity are common themes. In India, the bride's traditional red sari represents fertility. In the West, the practice of guests showering the bride and groom with rice or confetti at the end of the ceremony symbolizes the hope that the couple may be blessed with many children. In ancient times, the bride carried herbs – it was believed that strong-smelling herbs and spices would ward off and drive away evil spirits, bad luck, and ill-health. Later, flowers became a common feature in weddings across many cultures and took on meanings all of their own. Orange blossoms, for example, symbolize happiness and fertility; ivy is a symbol of fidelity; and lilies represent purity. In Hinduism, the exchange of garlands in front of the guests indicates a lifelong commitment. A ritual that is central to many marriage ceremonies is an exchange of rings or the tying of thread around the couple's wrists. Both signify a long-lasting, binding relationship. In the West, the wedding feast, which follows many wedding ceremonies, often includes a symbolic wedding cake.

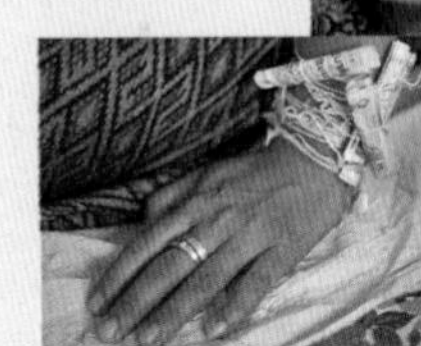

**GIFTS OF MONEY**
*In Cambodia, it is traditional for wedding guests to tie string around the wrist of the bride. This bride at a Buddhist ceremony has also received money.*

**WEDDING BOUQUET**
*Although the bridal bouquet is a Western tradition, flowers have been used since ancient Roman times. White flowers signify purity.*

**WEDDING CAKE**
*Exotic cakes are a common feature of the wedding feast. This traditional French wedding cake, called a croquembouche, is a tower of cream-filled pastry balls.*

**WEDDING DAY**
*The practice of showering newlyweds with confetti dates back to an older Eastern tradition of throwing rice, a symbol of fertility.*

# DOWRY AND BRIDEWEALTH

In most cultures, an economic exchange takes place when a man and a woman get married. This exchange is either bridewealth or a dowry. Bridewealth refers to a payment made by the groom, or his family, to the family of the bride. It is common in many parts of the world, particularly Africa, where it is usually paid in cattle. The groom "buys" the rights to his wife's labour and, crucially, to her children. Dowry refers to the wealth that a bride must bring with her when she gets married and sets up a new home with her husband. It is most common in Europe and India and may be composed of cash or items for the family home.

**CASH FOR THE BRIDE**
*In the Middle East, a cash dowry may be paid by the bride's family. It is intended as an early inheritance to help the couple set up home.*

# THE MARRIAGE HOME

When a couple get married it is common for them to set up a new family home of their own. New couples have three options: they can live near the groom's parents (virilocal), near the bride's parents (uxorilocal), or in a totally new location (neolocal). In Africa, virilocal residence is the most common choice and often the new couple build their new house within the compound of the groom's father; they will live together as an extended family. In most modern societies and in urban settings throughout the world, neolocal residence has become the norm; this means a new couple set up their own home independently of their parents.

**MOVING HOME**
*After marriage, a Gamo bride – from the highlands of Ethiopia – is transported to her new home to live with her husband's parents.*

Fact

## Separation and divorce

Rates of divorce have more than doubled in the West since the 1950s. In the US, about half of all marriages eventually fail. Divorce has become easier in many societies, but in some places and religions, it is very difficult for couples to divorce. In Catholicism, divorce is not generally recognized, although the church can issue an annulment if the couple can prove the marriage was invalid. In Italy, where the vast majority of people are Catholic, the divorce rate is just 12 per cent. In Cuba, in contrast, 75 per cent of marriages end in divorce. In the Islamic faith, although divorce is not encouraged, there are guidelines on how it should be done.

MARRIAGE CEREMONIES

## Secular weddings

**In brief:** Couples with no religious convictions may marry in civil ceremonies, creating their own vows

**MARRIAGE WITHOUT RELIGION**
*An increasing number of couples marry in ceremonies without religion, still exchanging vows and sharing the moment with friends.*

Couples wishing to have a legally binding marriage ceremony with no religious content can have a civil wedding that takes place in a registry office or another licensed venue.

In many parts of the world it is not legal to get married out-of-doors because licensed venues must be covered and permanent.

Humanist or "do-it-yourself" weddings, in which couples write their own vows, can be performed in any location – from a mountain top to beneath the sea – however they need to be backed by a civil ceremony to make them legally valid.

MARRIAGE CEREMONIES

## Hindu weddings

**In brief:** The traditional Hindu marriage ceremony includes many symbolic acts and a number of stages

During a Hindu marriage ceremony, the bride and groom are said to become the embodiment of the god Vishnu and the goddess of wealth, Lakshmi. As part of the ritual, the bride puts her foot on a stone in the corner of the wedding canopy, and the groom instructs her to be as strong as a stone or mountain in their future life. The couple are tied together, and they walk around a sacred fire four times to represent the four goals of life: duty, wealth, pleasure, and spiritual freedom from the cycle of life. The groom leads on the first three circles, signifying that he is expected to guide his wife towards these goals. The bride leads the groom on the final circle, signifying that the spiritual journey through life is an individual one. Then the couple take seven steps and make a vow at each one. On the last step they look up and ask the moon, sun, and gods to witness their marriage and make it as long as the life of the stars. They place offerings on the sacred fire and are blessed by a priest. Finally, the bride and groom promise their hearts to one another by placing a hand on each other's chest. They also promise to feed each other. This promise marks the end of a day-long fast and symbolizes the way in which the couple will provide for each other.

**TRADITIONAL DANCE**
*In some parts of India, Hindu women perform a traditional dance with sticks in the street after the wedding celebration.*

**RED AND GOLD**
*Red is the traditional colour for a Hindu bride. As part of the day-long ceremony, the bride and groom sit together to be blessed.*

MARRIAGE CEREMONIES

## Christian weddings

**In brief:** All Christian weddings share in common the exchange of rings; vows may be spoken

Western and Orthodox Christian weddings take place in a church. The couple exchange rings, and it is traditional for the bride to wear white as a symbol of purity. In the Anglican tradition, the couple make vows to each other, promising love, faithfulness, and solidarity through all of life's circumstances, including health, wealth, sickness, and poverty. After the vows a minister announces that the couple are "man and wife". In Greek Orthodox weddings, no vows are spoken; instead, a number of symbolic acts are performed. First, at the door of the church, the priest blesses both of the wedding rings. The couple's sponsors then exchange the rings three times between the couple's fingers. The number 3 represents the Holy Trinity of Father, Son, and Holy Spirit, which is central to the Christian faith. At the altar, the couple are given floral crowns joined by a ribbon, and they then share wine from a cup, and walk around the altar three times as an illustration of the journey of marriage. Finally, the priest blesses the couple and separates their hands as a sign that only God can separate them from each other through death.

Fact

### The meaning of the veil

The origins of the wedding veil may be traced back to times when a groom would throw a blanket over the head of the woman of his choice when he captured her and carried her off. The veil was also important for arranged marriages – it covered the bride's face until the man was committed to her at the ceremony.

**WHITE WEDDING**
*In many Christian wedding ceremonies, a bride traditionally wears a white dress. In this wedding in Manila, Philippines, the bride's parents and friends straighten her train as she arrives at the church.*

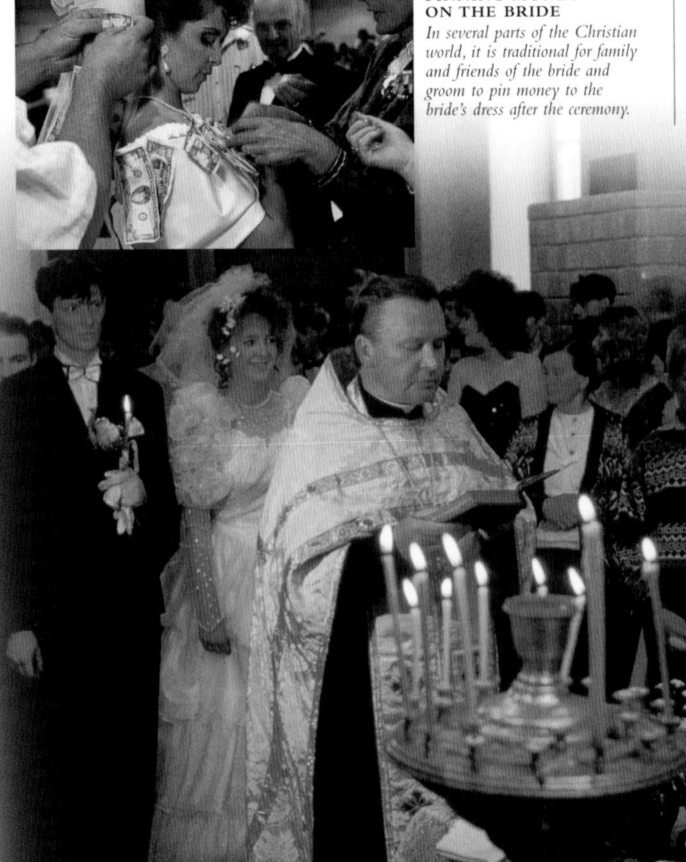

**PINNING MONEY ON THE BRIDE**
*In several parts of the Christian world, it is traditional for family and friends of the bride and groom to pin money to the bride's dress after the ceremony.*

**ORTHODOX WEDDING**
*The bride and groom stand behind the priest in this orthodox ceremony in Poland. The ceremony is carried out by candlelight.*

TRADITIONAL BOAT WEDDING
*The bride is transported to her wedding on a canal boat in Zhouzhuang. The wedding ceremony may be performed on the boat.*

MARRIAGE CEREMONIES

## Chinese weddings

**In brief:** Chinese weddings are usually held on an auspicious day; customs include serving tea

Chinese wedding customs vary, depending on factors such as geography, wealth, and social status. For most Chinese people, however, the dates of a marriage proposal and wedding are considered important; a fortune-teller is often hired to choose times that are astrologically favourable. On the morning of a wedding, it is traditional for the groom and his companions to arrive at the bride's house in decorated cars to take the bride to the wedding. Before he can enter the house, the groom must answer questions posed by the bridesmaids; he may even be asked to perform by singing a love song or demonstrating his strength with push-ups. The groom then offers gifts to the bridesmaids and is permitted to enter. Inside, the bride serves tea to her family before leaving for the ceremony. During the ceremony, the couple worship heaven and earth, and they pay respects to the groom's ancestors. Tea is served to the groom's family after the wedding ceremony, and the newly-weds are given gifts wrapped in red paper.

A young boy gives the bride a tangerine for good fortune, and an older woman leads the bride to a bridal room, where she waits for the groom. Inside the bridal room, the groom lifts the bride's veil, and the couple drink wine from fragile wedding cups linked with red string. The couple exchange cups, cross arms, and sip the wine a second time.

BRIDE SERVES TEA
*As part of the wedding ceremony a bride serves tea to the wedding guests. The red bridal headdress is traditional for the Yao people.*

DECORATED CAR
*On the morning of the wedding, the groom's friends cover the car with flowers. The car is used to collect the bride and take her to the ceremony.*

MARRIAGE CEREMONIES

## Korean weddings

**In brief:** Traditional Korean weddings include symbols of fertility; the wedding is usually at the bride's home

EXCHANGING GIFTS
*The Korean bride and groom traditionally exchange gifts as they marry. The gifts may be rings, jewellery, or watches.*

A traditional Korean wedding takes place at the bride's house, where the couple stand at opposite sides of a table as they take vows. The table is covered with symbolic items: bamboo and pine needles for constancy, chestnuts for longevity, a male and female goose for faithfulness (geese stay with one partner for life). Red and blue candles symbolize the man and the woman. Under the table are a rooster, whose morning call represents the start of a new marriage, and a chicken, whose egg-laying symbolizes fertility. After the ceremony the couple spend 3 days in a bedroom with a paper door: friends make holes in the paper to peep into the room.

MARRIAGE CEREMONIES

## Shinto weddings

**In brief:** The traditional Japanese Shinto wedding uses rituals to symbolize the longevity of the marriage

Drinking sake (rice wine) is an important part of a Japanese Shinto wedding. The couple enter the shrine where the marriage is to take place and stand on either side of the altar. The priest purifies members of the bride and groom's families by waving a paper staff. He then asks them if they approve of the marriage. Girls wearing red and white dresses bring forward three cups of sake. The bride and groom pass the cups between each other, taking three sips from each one – a ritual that is repeated three times. The number 3 is believed to be significant because it cannot be divided. The couple then repeat an oath of faithfulness and obedience, and sake is shared among the guests. After this part of the ceremony, the bride and groom enter a sanctuary where they make an offering of twigs from an evergreen tree, an act that symbolizes a long-lasting marriage.

SYMBOLIC DRESS
*The bride wears a hat that is believed to conceal the horns of jealousy and carries a small knife for ritual suicide should she dishonour her husband.*

MARRIAGE CEREMONIES

## Jewish weddings

**In brief:** Jewish weddings usually take place under a canopy; other customs vary according to place

Jewish weddings take place under a canopy called the *chuppah*, which represents the marital home. The bride, who wears a veil, comes to meet her husband under the canopy where she circles him seven times. After the marriage commandments, a legal marriage contract is agreed, and the bride accepts a ring from the groom. Then the couple say seven blessings, praising God and their marriage. They drink from a wine glass, which the groom then crushes under his foot to symbolize the destruction of the Temple and the fragility of marriage. Finally, the couple eat together, having fasted all day.

ORTHODOX WEDDING
*The rabbi reads from the* ketubah, *(marriage contract). Jewish weddings can be held anywhere; a canopy is erected to represent the couple's new home.*

History

### Wedding rings

The exchange of wedding rings was a Roman practice adopted by various faiths, including Judaism. In the past, some Jewish communities would loan a magnificent ring to the bride, decorated with a miniature building and inscribed with the words *Mazel Tov* or "Good Luck".

ANCIENT WEDDING RING

MARRIAGE CEREMONIES

# Muslim weddings

**In brief:** Muslim ceremonies include readings from the Koran, but other aspects vary around the world

Muslim weddings commonly include decoration of the bride, a formal marriage agreement, and introduction of the bride and groom. These rituals are followed by family celebrations, which may happen on a single day or be spread over many.

In Pakistan and India, women gather in a celebration known as *mehndi* to sing songs and decorate the bride with turmeric and henna. Among the nomadic Tuareg people of Mali in North Africa, blacksmiths, who are believed to have mystical powers, prepare the bride by rubbing black sand into her hair.

In South Asia, including India, the formal marriage agreements are known as *nikaah*; the groom's father proposes marriage and the bride's father accepts. An imam (priest) reads from the Koran and mediates a discussion between the male members of the two families over the amount of *mehar*, the compulsory marital gift from groom to bride. When the terms of the marriage are agreed, the wedding contract is written and presented to the couple for their consent.

Although the bride and groom have usually met before their wedding day, there is a moment when they are formally introduced to each other, as if for the first time. In parts of Southeast Asia, for example the Philippines, the groom touches the bride's head when they meet. The couple sit together during a feast but are expected to look unhappy. The bride, in particular, must express sadness at leaving her family.

In Afghanistan, when a couple are introduced, they sit next to each other with their heads covered by a cloth and a copy of the Koran between them. The couple are given a mirror to see each other's faces. If a woman is marrying a man who lives outside her country and who is not present, she can use a photograph for this ceremony. In Indonesia, following the formal introduction, a couple are led in an elaborate procession to a hall where their families are waiting to celebrate with a sumptuous feast.

Men and women usually sit, eat, and dance separately at Muslim weddings and it is traditional for two parties to be hosted: one by the bride's family to mark her send-off and, later, one by the groom's family to celebrate the bride's arrival.

In North Africa, marriage parties are marked by camel races, poetry reading, and singing. Among the Tuareg people, the guests put up and take down a tent for the couple every day, rebuilding it ever larger as a symbol of a growing family.

Only after a party at the groom's home may a new bride visit her own family. In the Sudan, when a woman visits her family for the first time, she takes gifts of a sheep, rice, butter, and sugar to show that she is living in a prosperous household.

**UNDER THE VEIL**
*In Pakistan, both bride and groom wear elaborate headwear and veils to cover their faces before the wedding.*

**WEDDING PROCESSION**
*A procession is a common part of a Muslim wedding. This bride and groom in Indonesia walk through rice fields to the wedding hall before the ceremony.*

MARITAL GIFT PURSE

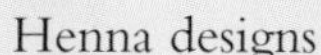

Fact

## Henna designs

Islamic brides-to-be have their hands and feet decorated with dark brown henna paste that fades to a deep red colour. The elaborate henna designs, which are applied by female friends and family members, include the name of the groom. It is believed that the darker the stain of henna on the skin, the deeper the love between future husband and wife.

**PREPARING TO MARRY**
*In North Africa, the hair of a Tuareg bride is rubbed with fine black sand and then braided. Finally, a yellow paste mixed with a red pigment is applied to her face.*

**ARRIVAL OF THE GUESTS**
*Tuareg nomads from North Africa arrive at a wedding on camels and donkeys – camels for the men, donkeys for the women.*

MARRIAGE CEREMONIES

## African tribal weddings

**In brief:** Tribal weddings in Africa take several forms, but all usually involve a ritual to symbolize the separation from the parental home

The main features of African tribal weddings are rituals associated with the bride leaving her family to live with the family of the groom and the accompanying celebrations. In Namibia, the groom's family "kidnap" the bride (usually by arrangement), dressing her in leather and covering her with butter. Among the Masai people of East Africa, a groom goes to the bride's home to collect her, at which time the bride's father spits on her as a farewell blessing. At the home of her new in-laws, the bride is greeted with a feast and dancing.

Part of a Zulu wedding ceremony involves the bride taking a knife and symbolically cutting herself away from her family; she then walks towards the groom. Shortly after this part of the wedding, the groom undergoes rituals that mark his manhood. He fights his peers with sticks but tends to any injuries that he inflicts on them, representing his responsibilities of defending and protecting his people. In Somalia, marriage ceremonies only finally come to an end with a wedding party at the birth of a couple's second child.

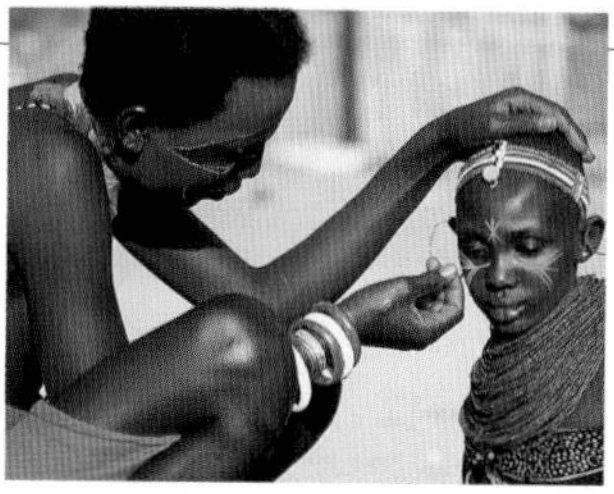

**ADORNING THE BRIDE**
*A Masai warrior helps his wife-to-be to prepare for the marriage ceremony by decorating her face with natural pigments.*

**ZULU BRIDE**
*A veil is used to cover the bride's face in many societies. This Zulu bride wears a veil made of knotted red and white thread.*

**ZULU WEDDING PROCESSION**
*The couple's family and other members of the community form the procession in a Zulu wedding.*

MARRIAGE CEREMONIES

## Gay weddings

**In brief:** Marriage of same-sex couples is legal in a small number of countries

In 2001, the Netherlands became the first country to give full civil marriage rights to gay couples. Gay marriages have been allowed in Belgium since 2003. Although Norway, Sweden, and Denmark allow gay couples to register their partnership and to qualify as a legal partnership for mortgage, taxation, and divorce purposes, they are not permitted to adopt children, share a surname, or seek fertility treatment. In Denmark, gay couples may celebrate their legal relationship in church but without a designated religious ceremony.

A GAY COUPLE MARRIES IN DENMARK

MARRIAGE CEREMONIES

## Group weddings

**In brief:** Several societies encourage mass weddings for economic or symbolic reasons

The Unification Church of South Korea, which also has followers in many countries, including the US, is a religious cult known for conducting mass weddings. In 1995, the leader, the Reverend Moon, married 35,000 couples in matches he had made personally. In China, group marriages allow large numbers of people to marry on the few especially auspicious days in October. Mass weddings may also be held to mark political events. For example, in the year 2000, the former leader of Iraq, Saddam Hussein, oversaw a wedding to commemorate the invasion of Kuwait. Group weddings may be held for economic reasons; the government of Syria, for example, hosts them to encourage couples to marry without fear of debt.

**MUSLIM GROUP WEDDING**
*Women in bridal gowns wait for their grooms at a mass ceremony in Dhaka, Bangladesh. This type of wedding is also popular in other countries such as Sudan and Pakistan. The government may organize and pay for it.*

**MASS WHITE WEDDING**
*Over a hundred couples married in this ceremony in Taiwan. Mass weddings reduce the cost to families.*

# FAMILY

When a couple have children, they bring a new generation into a web of people related by biology, marriage, cohabitation, and adoption – a family. The way families are organized varies a great deal across the world. Different values are placed on certain relationships, and different rules dictate how people respond to, and are obligated to each other. There are also varied religious interpretations of the meaning of family life. In the West, "family" may mean a small unit of mother, father, brother, and sister living in the same house. In Africa, "family" may be a large group of people who venerate the same dead ancestor. What is common, however, is that people count on their families as their source of greatest emotional and economic support. When this support breaks down, they find it harder to negotiate the world at large.

Fact

### Fertility rates

Some of the highest rates of fertility (number of children per woman) are in the poorest countries in the world. In parts of Africa, such as Mali, Rwanda, and Somalia, women have more than seven children each. But these countries also have some of the highest rates of death among babies and children under 5 years. One of the aims of international campaigns to educate women and improve health in these countries is to reduce infant death rates.

**EARLY PARENTHOOD**
*Women in much of Asia, such as this woman in Bangladesh, start having children while they are still young, and they often have many children. Large families are a bonus in rural communities, where children can help look after the land and animals.*

## HAVING CHILDREN

The age at which women start having children varies widely. So too does the average number of children they have. In many parts of Africa and Asia, having large numbers of children is linked to rural lifestyles – where children provide much-needed economic input – and to early marriage. In Bangladesh, half of women are married and bearing children by the age of 18. The lowest numbers of children are born to Russian, Estonian, Czech, and Latvian women – an average of 1.2 per woman. In developed countries, the number of children born to a couple is falling, and women are leaving childbirth until later. This is usually due to later marriage, the growing expense of having children, and women placing greater importance on career choices. The optimal fertility period for women is between the ages of 18 and 24: by the age of 30, a woman's fertility has dropped by 30 per cent and at the age of 40, by 60 per cent. Menopause brings an end to the natural childbearing years (*see* p133).

**OLDER PARENTS**
*In much of the developed world, the trend in recent years has been for couples to start having children later in life. Many people now wait until their thirties or even early forties before starting a family.*

**FAMILY FOCUS**
*Families are central to most people's daily lives, even though globally there are many different definitions and types of family. This Mongolian family expresses the closeness of family life.*

History

### Kibbutz life

In the early days of kibbutzim – rural communities unique to Israel – idealists believed that the nuclear family unit was obsolete and that children should be reared together, away from their parents. It was believed the entire kibbutz should exist as one large family unit. Up until the 1970s, children slept in special children's houses with someone appointed to tend to their night needs. However, parents and children alike found the regime distressing. Today, children on every kibbutz live and sleep with their parents, at least into their teen years, and the children's houses have become daycare and activity centres.

## TYPES OF FAMILY

In the West, a couple move away from their parents to establish their own nuclear family. Sometimes several related nuclear families share the same home; this is known as a compound family. In many societies, a spouse moves to join his or her in-laws, forming an extended family consisting of a group of siblings, plus their spouses, children, and the grandparents. Extended families also occur in societies that allow polygamous marriages, where a man may live with several wives and their children. Many other family types exist. In the Caribbean, matrifocal families – a mother and her children from a series of partners – are common. Many people in the West do not live in a family at all: in 2001, the number of people living alone in the UK exceeded the number living in nuclear families.

**EXTENDED FAMILY**
*In Rajasthan, India, extended families are common. These provide support to mothers who often have several children.*

# FAMILY RELATIONSHIPS

Cultural values influence the bonds between siblings, parents, grandparents, and cousins and give priority to some relationships above others. In West African Yoruba culture, for example, all children are expected to leave their parents and live with their grandparents for a time to ensure their moral development. In Peruvian Quechuan communities, the relationship between grandchildren and their mother's father is considered much more important than that with their father's father because grandchildren inherit only from the mother's side. Marking a difference between uncles exists in many societies: a father's brother is often called "father" in recognition of the superior status of the father's side. Many cultures recognize the in-law relationship as especially problematic, and some put restrictions on the interaction between a man or woman and his or her in-laws so as to avoid conflict. Among Aboriginal Australian communities, for example, a husband can neither approach his mother-in-law nor say her name. In Papua New Guinea, a husband must not say the name of any of his wife's relations, and must even avoid looking at them.

**TWINS**
*Being a twin is probably the strongest of all bonds. Twins sometime feel they understand each other so well, they do not need others. In some cultures, twins are seen as special and are revered.*

**GRANDPARENTING**
*The relationship between grandparent and grandchild is cherished in many societies, such as the United Arab Emirates.*

**SIBLING RELATIONSHIPS**
*Brothers and sisters can provide instant playmates. The relationship between them is the longest-lasting of all family ties.*

# INHERITANCE ISSUES

Humans like to keep their property – land, animals, houses, furniture, jewellery, money, knowledge, and power – in the family. Most societies around the world have laws to regulate inheritance. In societies that have a patrilineal system of inheritance, property passes from father to son. In matrilineal societies, such as the Ashanti people of Ghana, sons inherit from the mother's side of the family – in fact, from their mother's brother. In some cultures, all children receive an equal inheritance; in others, the eldest son inherits all (a practice known as primogeniture). Sometimes, as with the Tlaxcala people of Central Mexico, the youngest son is the prime inheritor (a custom called ultimogeniture).

**INHERITED LAND**
*In some parts of the world, such as the UK, land has been passed down through the generations, traditionally to the eldest son. Some societies have markedly different inheritance rules.*

Issue

## Foreign adoptions

Adopting a child from abroad often involves a protracted legal process. Sweden has one of the highest rates for adopting children from abroad (as does the US). Of the 35,000 or so registered foreign adopted children in Sweden, 70 per cent are girls and the majority from Asia. Research shows that one-quarter of foreign adoptees go on to have problems with language development, learning difficulties, and acute crises over their sense of identity.

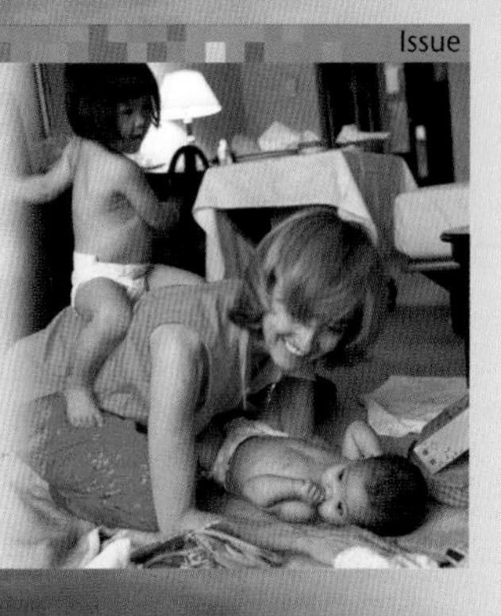

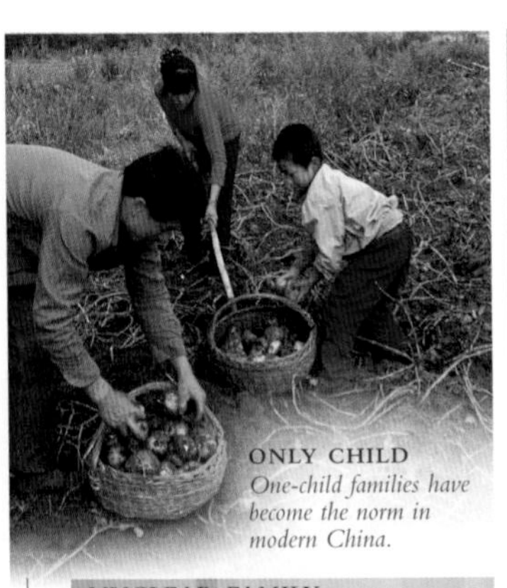
**ONLY CHILD** *One-child families have become the norm in modern China.*

NUCLEAR FAMILY

## Chinese families

**In brief:** Chinese families are restricted by law to having only one or two children

In 1980 China introduced strict laws to curb population growth. Urban families can have only one child and rural families can have only two. Financial incentives, family planning, and compulsory abortions are used to enforce the law, which has prevented an estimated 250 million births in the last decade. Opponents say the policy has led to widespread infanticide, as families prefer their one child to be a boy. Certainly the male to female ratio seems to be rising in China – some regions now have about 131 males to every 100 females.

NUCLEAR FAMILY

## Inuit families

**In brief:** About a quarter of Alaskan Inuit children are adopted, forging strong links between families

Among the Inuit of Alaska, around 25 per cent of children are adopted by couples in the same community as the biological parents. The Inuit say that adoption ensures that all couples have children to raise, including couples who are either infertile or too old to have more children. Adoption is also seen as a way of sharing the work of childcare, ensuring that all children get the love and attention they need. It is particularly common for a first-born child to be adopted, with the belief that the mother will be more ready for motherhood with her second child. Often a first child is adopted by the mother's parents, meaning that the child becomes the mother's "brother" or "sister". There are strong ties between couples who adopt each other's children, and the adoptive parents give gifts of money and support to the biological parents. Adoption therefore ties together an Inuit community with bonds that extend beyond the family.

**ADOPTIVE PARENTS** *Many Inuit children are adopted by their grandparents and brought up as siblings of their biological parents.*

INUIT MOTHER AND CHILDREN

NUCLEAR FAMILY

## Iban families

**In brief:** Iban families live in family units within a communal longhouse

The Iban people of Malaysia and Borneo live in communal longhouses, often next to a river. Each longhouse is made up of apartments occupied by separate families, with the connecting verandah serving as a shared space. The word for family and a longhouse apartment is the same: *bilek*.

New longhouses are founded by groups of brothers and sisters with their spouses. When their children marry and start families, new *bileks* are added to the side of the longhouse. As a result, families in the centre of a big longhouse tend to be more closely related than those at the ends.

Newly married couples stay in either the groom's or the bride's family apartment until they have children of their own. In order to become a true *bilek* and qualify for an apartment, the couple must first own a loom. Once a wife has a loom, she weaves patterned bedclothes for her husband. The Iban believe that their weaving patterns come from heaven, and that the husband's handmade bedclothes allow him to reach heaven in his dreams. Weaving thus seals the couple's relationship and future family life in the spiritual realm.

**IBAN LONGHOUSE** *The traditional longhouse in Sarawak, Malaysia, is usually built on the river's edge and is home to an entire community.*

**IBAN APARTMENT** *Many different nuclear family groups live in separate units within a traditional longhouse.*

EXTENDED FAMILY

## Mormon families

**In brief:** A small proportion of Mormons live as polygamists, with husbands marrying several wives

**MULTIPLE WIVES** *Although the practice of polygamy, which is also known as plural marriage, is illegal in many US states, the law is seldom enforced.*

The Mormon practice of marrying more than one wife began in the mid-19th century, when Joseph Smith, founder of the Mormon church, preached that polygamy was a form of "celestial marriage". Since then, the Mormon church has split several times over the issue. Today most Mormons adhere to US law and live as monogamists. About 5 per cent of Mormons in the largely Mormon state of Utah live in polygamous families – an informal arrangement that is not recognized by the law. Such Mormons lead discreet lives and disapprove of flaunting their lifestyle.

EXTENDED FAMILY

## Tibetan families

**In brief:** Some Tibetan families are made up of one wife, several husbands, and all their children

In Tibet there exists one of the rarest forms of family structure in the world: fraternal polyandry, in which a group of brothers marry the same wife. The eldest brother normally has authority over his siblings but all participate as sexual partners once they reach adult age. Tibetans see several benefits in this unusual family arrangement. It allows some of the brothers to travel from home for long periods while herding livestock, secure in the knowledge that their wife is being protected by another brother. Also, farmland tends to stay in the hands of the family unit, rather than being divided between male heirs. Fraternal polyandry remains popular in Tibet, despite a ban imposed by the ruling Chinese government.

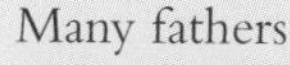

Fact

### Many fathers

Children of a Tibetan polyandrous parents tend not to know who their true father is (although their mother might know). All the children in a family call their mother's oldest husband "father" and the other husbands "father's brothers". When boys grow up, they may repeat the pattern and marry the same wife, meaning that their children have many fathers and many grandfathers.

**MAN AND WIFE**
*A Tibetan woman at home with one of her six husbands (left). A few Himalayan families include not only multiple husbands (polyandry) but multiple wives (polygyny) as well. The large families are better able to make use of the land's meagre resources.*

EXTENDED FAMILY

## Somali families

**In brief:** Somalis greet each other with a summary of their family and clan background

The importance of family and descent in Somalia is revealed by the everyday greeting, "Whom are you from?" In reply, Somalis refer to membership of both an extended family and a clan of families with a common ancestry. Even small children are expected to answer with full knowledge of their descent and their place in the wider clan structure. One in five marriages in Somalia is polygamous, a man having more than one wife. Each wife has her own house in a family compound, and the husband moves regularly between the houses to sleep.

**SOMALI WIFE**
*The head of a Somali household is usually a man, but women take charge of child-rearing and domestic duties.*

EXTENDED FAMILY

## Mediterranean families

**In brief:** In Mediterranean countries such as Greece and Italy, families are often large and in close contact

In Mediterranean countries such as Greece and Italy, relatives tend to live close to each other and frequently meet up to share meals and discuss family matters. These cultures place great value on marriage; of all European countries, Italy and Greece have the lowest number of cohabiting rather than married couples and the lowest numbers of single-parent families. Family size is also important, and Mediterranean couples tend to have more children than elsewhere in Europe. Nevertheless, trends are changing – increasing numbers of young people are moving away from home, choosing not to marry, or having fewer children.

**FAMILY EATING**
*Family meals are regular occurrences in the Mediterranean. Lunch can last a whole afternoon as relatives catch up on gossip.*

**FAMILY GATHERING**
*This large Burmese family has come together to celebrate a boy's* shin-pyu *– his initiation into the order of Buddhist monks.*

EXTENDED FAMILY

## Burmese families

**In brief:** Burmese extended families share a house and celebrate Buddhist occasions together

Extended families are common in all ethnic groups in Burma. A single household might include a husband, wife, their children, uncles, aunts, in-laws, cousins, and grandparents. Religion is very important, and Buddhist customs and rituals help to bind the family together. According to Buddhist teaching, children regard parents as sacred, and one of the most important acts of worship for children is to show obedience to parents. During the month of *Thadingyut* (the end of Buddhist Lent), children are expected to show additional respect by bowing to their parents when they approach them; the family also worship daily at a family shrine in their home.

Burmese extended families draw together for religious celebrations. The most important family occasion is when a boy joins a temple to study Buddhism. All his relatives come to bless him at the start of his time in the temple.

In recent years, political turmoil in Burma has put pressure on many aspects of family life. Large numbers of Burmese people have become refugees, causing families to fragment. The number of households headed by a woman has risen because many husbands and fathers have died in political conflict. Poverty has driven some families to sell their children for domestic service and even prostitution.

# DAILY LIFE

Sleeping, washing, preparing food, eating, working, and relaxing are activities most adults engage in daily. Yet, how much time is spent on each varies enormously. The country a person lives in, his or her social standing, gender, income, interests, and family commitments affect what proportion of each day is taken up by providing and preparing food, and whether there is time or money left for recreation. For members of religious orders, for people with a strong faith, or for those who live in fundamentalist countries, prayer may dominate their lives. In rural India, women's days are spent finding firewood, cooking, and tending a plot of land. In Japan, by contrast, so-called "salarymen" spend little time at home: getting to and from work on public transport takes many hours, and they are expected to spend their evenings socializing with colleagues.

## HOME LIFE

Home may be a building in which a person spends much of his or her time or it may just be where he or she returns to sleep. It may contain an area for worship, or it might double as an office or as a workshop. An entire social group may share one building, several generations of the same family may live together, or a nonrelated female may share the family home to look after the children and cook and clean. For men in the West, home is a place to practise hobbies or maintain the car or the fabric of the home. Traditionally, a woman's place has been regarded as being that of homemaker, and even for those who wish it were otherwise, in Western or developing countries, the cost and scarcity of childcare, and perhaps the existence of a partner with a greater earning capacity, may mean it is the mother who takes care of the family and runs the household. In Sudan, the divisions between the sexes go further. For example, the laws or customs of the Beja people, in the east of the country, prevent women and men mixing freely in society and extend to the home, where there are women-only and men-only areas. In some cultures, it may also be the custom for women to eat after the men have finished.

**WASHING IN THE RIVER**
*In homes without running water, a woman may spend much of her day collecting drinking water. In India, women use rivers for washing clothes and bathing.*

**POUNDING GRAIN**
*Women in parts of Africa spend many hours a day collecting and preparing food for their families. These Mende women from Mali, Africa, are using a large pestle and mortar to grind grain.*

**HOME REPAIRS**
*Since the 1980s, do-it-yourself has become an accepted weekend pastime in much of Europe and North America, for men and women alike.*

Issue

### Working motherhood

Economics, personal preference, and social pressures influence whether, and how soon, a mother returns to work after giving birth. In developed societies, a woman may be made to feel guilty if she returns to work while her children are young and also if she does not. The working mother may employ someone to take care of her family or she may place the child in daycare. In this way, a woman can continue to develop her career.

**GOING TO MARKET**
*At this market in Damnoen Saduak, Thailand, the buying and selling of essentials is done from floating stalls.*

## WORKING LIFE

Between the ages of 16 and 65 years, most people spend at least 8 hours a day at the workplace. This commitment affects how they organize their lives, influences how others see them, and determines how much leisure time they have, and how much money they earn. However, work means different things to different people. A woman in rural India may spend the daylight hours breaking rocks at a roadside, whereas had she been born and educated in the West, she would be likely to work in an office using computer equipment. There may be family or social pressure on a person to do, or not to do, a type of work (*see* Caste system, p262); and a country's economic situation or the gender of the worker may further limit the choice. People in very low-paid positions may do more than one job to survive financially. In the West, the traditional "position for life" is becoming ever rarer – increasingly people have short-term contracts or work for several employers. Flexitime is popular with employees, although businesses have yet to be convinced of the benefits; and the predicted mass move to home-working, using the internet, has not yet occurred.

**TEA PICKERS**
*Most pickers on tea plantations are women. The hours may be long and the work can be hard, as well as labour intensive.*

**TOWARDS RETIREMENT**
*In many countries, a person works for as long as he or she is able and then is looked after by the family. In Western countries, the age of retirement may be determined by law.*

Health

### Impact of shift work

People who work shifts – especially night shifts – are at risk of health problems. Fatigue, stress, and sleep disturbances are common. There may be an impact on relationships and friendships, which may result in emotional or social problems. Research has also shown that, compared with other employees, shift workers are 40 per cent more likely to suffer from heart disease; are at greater risk of stomach disorders; and are more likely to die younger.

**AMBULANCE WORKERS**
*Emergency staff, such as ambulance crews, provide a round-the-clock service. However, shift work may affect their health and relationships.*

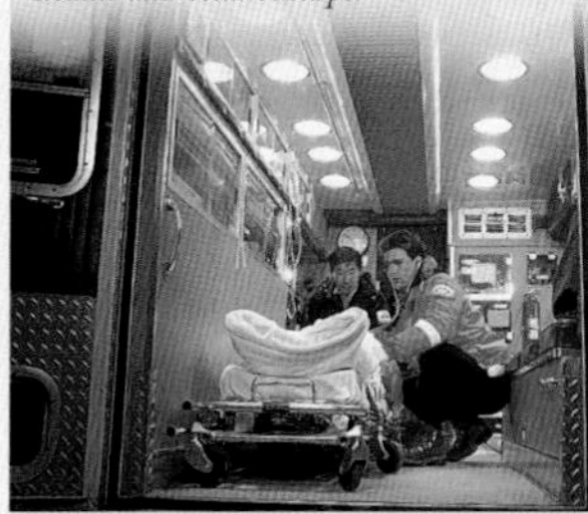

## LEISURE

In developed countries, leisure time is a precious commodity. It is a chance to relax, spend time with friends, and to recuperate from the stresses and pressures of working life. In the West, it is time that is being used to reverse the effects of having a sedentary job and taking the train or car to work, by being a member of a gym or a sports club. However, the distinctions between work and leisure time may be becoming blurred. Employees with high-pressure jobs and responsibility may have a contract that says he or she has to be contactable, even outside of the workplace, and while on holiday. In some cultures, there is not a great distinction between the place of work and the home. In these societies, as in many others, socializing revolves around the extended family. In countries from Greece to China, coffee shops and teahouses are where men spend their leisure time. Sport is also enjoyed worldwide. The popularity of football is such that it has become a sort of lingua franca that crosses language barriers.

**PLAYING GAMES**
*In the Mediterranean and Middle East, men while away their leisure hours in the company of other men, often in cafés. Backgammon, shown here, is a popular game.*

**AT THE CINEMA**
*Going to the theatre or cinema are popular leisure activities in many parts of the world. Cinema attracts individuals as well couples, friends, or entire families.*

HOME LIFE

## Preparing meals

**In brief:** In some societies preparing meals is still considered to be women's work and may take all day

FOOD PREPARATION
*Women in rural India spend a large part of their day preparing meals for their families, a task that involves cooking on open fires.*

Preparing a meal for the family in rural India or Africa can take a large part of the day. In contrast, a Western urban household may spend only minutes preparing and cooking meals. Using a microwave, a ready-prepared meal may take a few minutes from refrigerator to table. However, to cook the same meal in rural India, without electricity and running water, can take 5 hours. In many developing counties, all food is prepared by women. They often have to collect firewood and water and light a fire before they can start cooking. To cook rice, it first needs to be threshed, to remove the husk, and cleaned.

HOME LIFE

## Eating arrangements

**In brief:** Many societies have strict codes that determine who sits where at meals and who eats when

The way people arrange themselves to eat often reflects hierarchy in the group. It is common in Islamic societies for men and women to eat separately. In the Maldives, women wash the men's hands, serve them, and watch them eat before having their own food. Across South Asia, the seating plan reflects status, with the most distinguished person present seated farthest from the door and facing it. Chinese tradition places the more important guests on the right of the host, whereas at very formal European dinners, revered visitors were traditionally seated "above" the salt on the table. By contrast, status is represented by height in much of Africa: men sit on chairs and eat at a table, with women and children on mats on the floor. In Bolivia, dining is a social activity – eating and drinking alone is considered bad manners.

FAMILY EATING TOGETHER
*Chinese families usually eat together, helping themselves from a communal pot. This meal is a typical, spicy dish from Sichuan Province.*

MEN EATING TOGETHER
*In Mali, Africa, a middle-class family eats lunch with friends. Each uses only the right hand.*

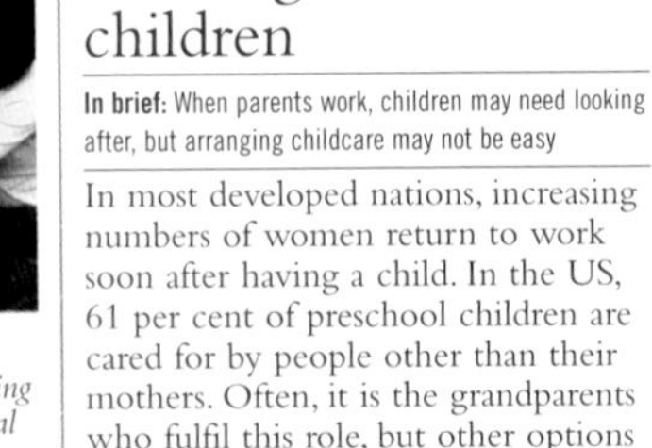

HOME LIFE

## Looking after children

**In brief:** When parents work, children may need looking after, but arranging childcare may not be easy

In most developed nations, increasing numbers of women return to work soon after having a child. In the US, 61 per cent of preschool children are cared for by people other than their mothers. Often, it is the grandparents who fulfil this role, but other options include daycare centres and nannies.

In African societies, however, it is common for mothers to take their children with them as they go about their daily chores. As children grow up, they are entrusted to the care of a "babysitter", a female relative from a poorer family who may forgo going to school to look after children.

A RWANDAN MOTHER AND HER BABY

HOME LIFE

## Cleaning the home

**In brief:** House-cleaning not only has a practical function, it may also have a symbolic purpose

As dawn breaks in the Andes, the first task of the day is to sweep the house. It is the women and children who take the brooms and start to clear the ever-present dust from the compound. This work ensures both physical and spiritual cleanliness – sweeping is believed to get rid of any harmful spirits that may have entered the house during the night.

Cleaning the home can also be symbolic of sweeping away the old, ready for the new. In the parts of the world where Chinese New Year is celebrated, every family is expected to thoroughly spring-clean their home in preparation for this special date.

CLEANING CHORES
*In Guinea Bissau, West Africa, women sweep and clean without the help of labour-saving devices.*

HOME LIFE

## Keeping clean

**In brief:** In some cultures, bathing has a religious and social function as well as to being hygienic

Even in the poorest societies soap is an essential household item, but out of necessity the same soap is used for washing the body, hair, clothes, and utensils. In most developed nations there is a range of products for these tasks. Soap was first invented in Babylon in 2800BC by mixing oil and ash, the same basic ingredients that are used today. The Romans brought bathing to Europe and built luxurious bath houses, similar to the single-sex *hammams* (steam baths) still used in Islamic countries. Most religions have laws on keeping the body clean. Muslims, for example, must wash before prayer, and if water is not available they can use sand. Likewise, cleansing without soap and water can be achieved through sweating and steaming, a technique used in North American sweat lodges, Russian banias, and Finnish saunas. In Finland, the 5 million Finns have 2 million saunas between them.

STEAM BATH
*Bathhouses, such as this one in Istanbul, are still popular in many societies. They are not only a place to get clean, but also offer a chance to relax and socialize with friends.*

WASHING CLOTHES IN THE RIVER
*Washday for the Himba of Namibia means going to the river. The women use stones to pound the clothes. They leave the clean items to dry the sun.*

Health

### Access to clean water

For some people, drinking water may be a source of sickness and death. More than 1 billion people do not have access to clean and safe drinking water. Instead, they have to use stagnant and polluted sources of water. This water often contains microorganisms that cause diarrhoea or parasites, such as river flukes, that can enter the bloodstream and cause schistosomiasis (bilharzia).

LIFE CYCLE

WORK

## Growing and harvesting

**In brief:** Technology can make a difference to farming output but it is not a guarantee of a good harvest

Billions of people around the world spend their daily lives growing and harvesting food, either for themselves and their families, or to provide an income. The job of growing food crops, raising animals, and catching fish is often a physical one, and farmers and fishermen often work long hours. Types of farming vary, from the small-scale, subsistence farming found in large parts of Africa and Asia to the vast commercial farms that rely on machinery in parts of Europe and the US.

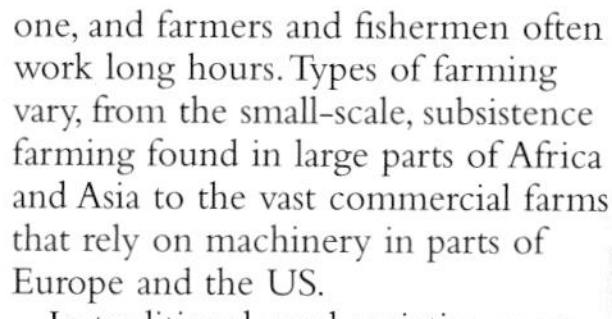

In traditional rural societies, most of the family is involved in farming, the girls and women tending the plants and boys and men looking after the animals. The day may start at 3 or 4am and finish at sunset. Fishermen around the world often fish at night, bringing their catches ashore ready for a morning market.

Developments in technology have changed many peoples' day-to-day lives. In the Sudan, for example, the introduction of ox-ploughs to people who were traditionally nomadic means that farmers can cultivate larger plots of land. Traditionally, they worked the land with hoes. Using animals makes farming physically easier and the people now produce enough to be able to sell their crops. In contrast, in Cuba, the end of communism and the collapse of large-scale mechanized farms has meant that farm workers have had to return to traditional farming methods. For example, in the 1980s there were 1,000 pairs of oxen in Cuba. By the beginning of the 21st century there were 300,000.

**ANIMAL LABOUR**
*In Vietnam, oxen are widely used to plough the fields, allowing farmers to work larger fields than was possible in the past.*

**FISHERMEN'S CATCH**
*After spending many hours at sea, fishermen sort their catch on the quayside. These men are on Penghu Island in the Taiwan Straits.*

**GRAPE HARVESTING**
*Mechanical harvesters speed up the grape harvest at large vineyards; they reduce the need for lots of pickers in the short harvesting season.*

WORK

## Manual labour

**In brief:** Some people may have no choice but to do manual labour, for others it may be an opportunity

In India, the caste system (*see* p262) prescribes that the higher castes will not perform manual labour. Because it is lower castes that take on manual work – to the extent that their names are linked to specific jobs such as "tanners", "blacksmiths", or "cleaners" – manual labour is seen as demeaning and polluting. In developed countries, manual labour is also considered a lowly occupation and is associated with hazardous conditions. In Canada, manual industry – including mining, forestry, and construction – accounts for 14 per cent of jobs but 33 per cent of work-related injuries. The number of manual labourers on farms in the UK has fallen, leaving fewer people to do the same amount of tasks, resulting in longer working hours. The worker may become exhausted, and being tired increases the likelihood of having an accident. Farming has the worst safety record of all UK industries: on average, there is one death every week from a farming accident. Manual labour can, however, represent a great opportunity to improve a person's circumstances. Every year, many thousands of Nepali migrant labourers take up jobs on construction sites in the Gulf States, where they earn more money in a season of work than they could hope to make in a lifetime of doing the same labour at home. All work has risks associated with it but the world's oldest couple (100 years and 101 years) attribute their good health and longevity to a lifetime of labouring in the paddy fields of Thailand.

**PHYSICAL LABOUR**
*In India, men and women are employed as road-builders. Where machinery is limited it is a physically demanding occupation that can also be dangerous.*

**PRODUCTION LINE**
*Women are the main employees in jobs that require dextrous fingers. These women work in an electronics assembly line in Hong Kong.*

Issue

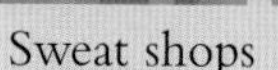

### Sweat shops

Many garments intended for sale to people in industrialized countries are made in developing countries, such as India and Bangladesh, in unregulated textile factories . The working conditions in these centres of production would fail to meet even the most basic of health and safety regulations in industrialized counties. In Bangladesh around 1.5 million people work in the clothing industry and in this type of plant. These employees may put in as many as 70 hours in one week and yet still receive only minimal pay.

**INDUSTRIAL ENVIRONMENTS**
*Massive pipelines carry chemicals, water, and waste products. In countries with health and safety regulations, workers require protective clothing and headgear.*

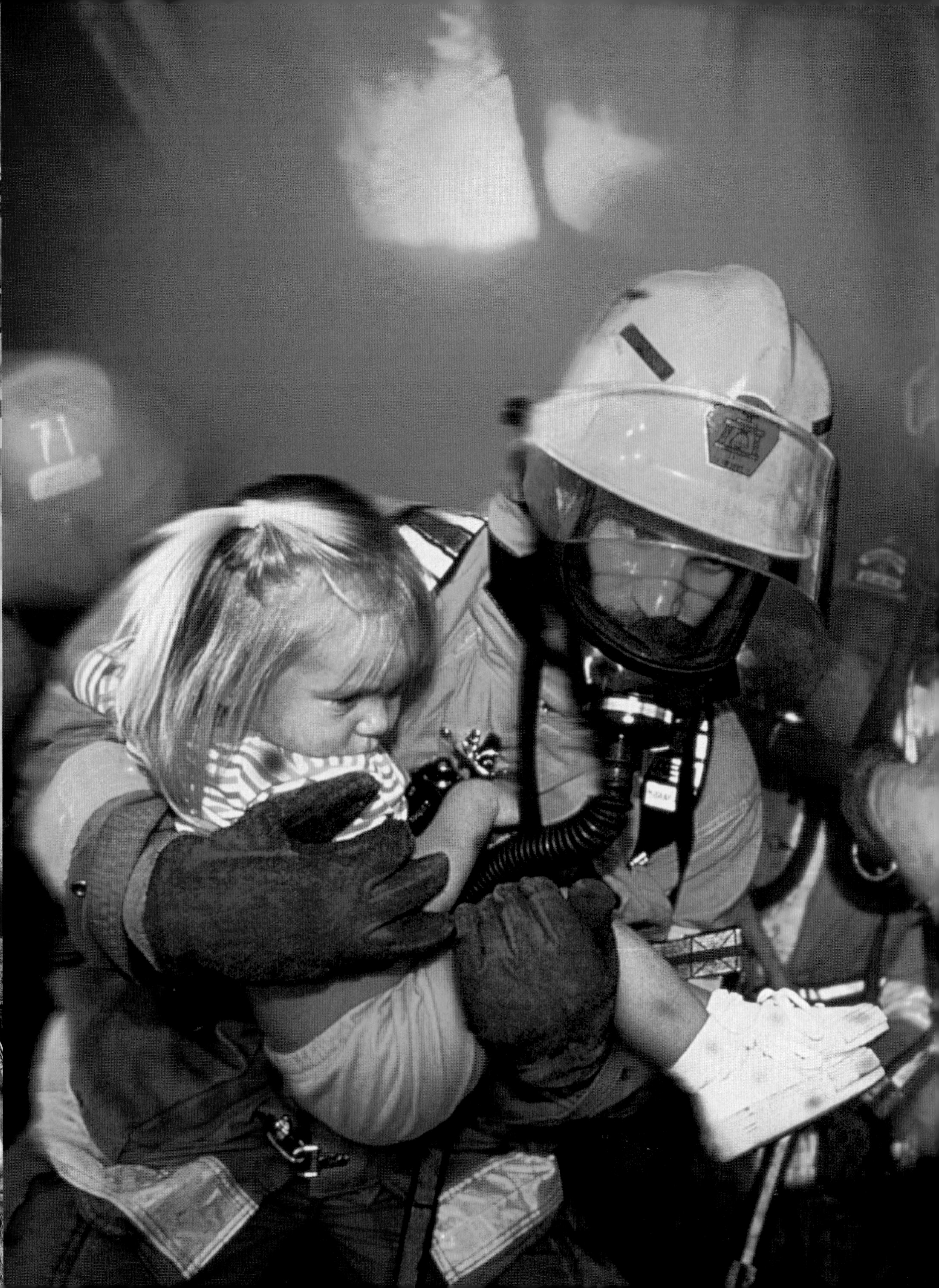

WORK

# Emergency work

**In brief:** Workers in the emergency services, such as police, firefighters, and paramedics, may have to work in dangerous situations; shift work is usual

People who work in the emergency sector include police, firefighters, paramedics, lifeboat crews, and mountain rescue teams. Such workers are frequently called out to unexpected and sudden events that have to be dealt with urgently.

Working in the emergency sector is often very rewarding, rarely dull, and is recognized and respected in the community. However, there are health risks with this type of work. The unpredictable nature of the work can cause stress, and although emergency workers are trained not to put their lives at risk, the work can be hazardous. According to the International Labour Organization, emergency workers often work in the most hazardous environment of all workers, next to that of military personnel in combat.

Emergency workers must also often contend with long working hours, a shrinking workforce and, in many countries, a lack of workplace rights including the right to strike. The role of workers in the emergency sector is constantly changing under the impact of changes in technology and in the nature of industrial activities. The working lives of certain parts of the emergency sector may also be influenced by increasing levels of crime and violence, especially in large cities and other urban areas.

Shift work is common in all types of emergency work. However, unpredictable working hours frequently disrupt the sleep–wake cycle, which can lead to a higher-than-average accident rate (*see* Shift work, p223). Numbers employed in the emergency sector vary from country to country. In the UK, for example, there is around one police officer per 400 population. In some parts of Africa, one firefighter serves more than 33,000 people.

Emergency workers usually have training in first aid – they are often the first on the scene of an incident where there are casualties. Increasingly, workers are being trained to deal with a potential terrorist attack. Specialized emergency workers work as rescuers in mountainous areas or as lifeboat crews – they need specific skills for the environment in which they work. Some specialist teams exist to deal with earthquakes and avalanches, where specialist techniques and equipment may be needed to save people who are trapped.

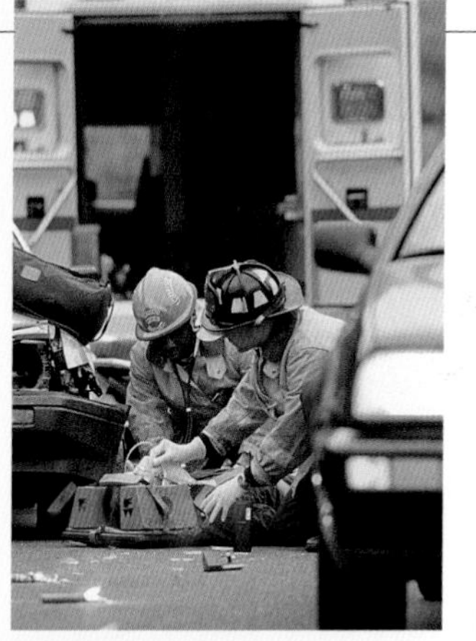

**EMERGENCY TREATMENT**
*Emergency workers such as paramedics are often the first at the scene of a road accident and may be able to give life-saving treatment before a victim is taken to hospital. Many emergency workers have first-aid training.*

Fact

## Death in the line of duty

As a result of the terrorist attacks on the World Trade Center in New York on 11 September 2001, 343 firefighters lost their lives. As many as 600 emergency workers had to take long or permanent leave, or were able to return to only limited duty, as a result of health problems and stress. Before this event, some fire stations had not suffered a fatality.

**RISKING LIVES**
*New York firefighters search for survivors in the aftermath of the attacks on the World Trade Center, September 2001.*

**EARTHQUAKE RESCUERS**
*Emergency workers search the rubble for survivors following an earthquake in Taiwan. Rescuers face potential dangers, such as falling masonry and potential aftershocks.*

**RIOT POLICE**
*Police often face confrontation as part of their work, and body armour is increasingly necessary. These fully-protected police officers with riot shields prepare to face demonstrators at an antiglobalization march in London in 2003.*

**RESCUE FROM THE WAVES**
*In the UK, the men who operate lifeboats are volunteers. When alerted to an emergency, they leave their normal paid jobs to attend. Here, the crew of the lifeboat rescue a swimmer.*

**SAVING LIVES**
*A firefighter in California, US carries a child to safety from a burning building. Fire crews often work in hazardous situations but are usually highly-trained.*

WORK

## Office work

**In brief**: Office workers usually have sedentary desk-based jobs; stress is common

Sitting in an office presents fewer hazards than doing factory-based or manual work, but it still subjects the body to stress. Using a mouse and a keyboard, answering the telephone, reading a computer screen and paper documents, involve different muscle groups. Yet people often use the same bad posture for all these tasks, straining the body and sometimes developing an increasingly common condition known as repetitive strain injury (RSI). The problems are compounded by working in stressful environments and for long hours: 28 per cent of US office workers regularly work more than 50 hours per week.

**DESKBOUND OCCUPATIONS**
*The very technology that enables people to do a greater variety of tasks more efficiently may mean staff work long hours and subject their bodies to great stress.*

WORK

## Service work

**In brief**: Workers in the service industry are often undervalued; abuse is common in some areas

The service industry provides services, such as transport, restaurants, and hotels, rather than goods. In general, service work is undervalued and, although things are changing, people typically work long hours for little reward. They may also find their health suffers, if, like hairdressers, they are on their feet all day, or if they work shifts that interrupt their sleep patterns, as is the case for many cleaning staff. Hotel and restaurant employees may also have temporary contracts and no grievance procedures. Service workers may earn only a basic salary and have to rely on tips, or they may have more than one job to make a living wage. People who work in pressurized situations, such as fast food outlets, may suffer stress and a higher-than-average incidence of injuries. Part-time workers in the US may not receive health insurance. The conditions of domestic service workers, for example nannies, cleaners, and cooks are no better. Their sector of the industry remains unregulated, and abuses can result. A thriving trade of people-trafficking preys on this lack of accountability. People pay vast sums to be smuggled into another country and, being there illegally, they are open to exploitation.

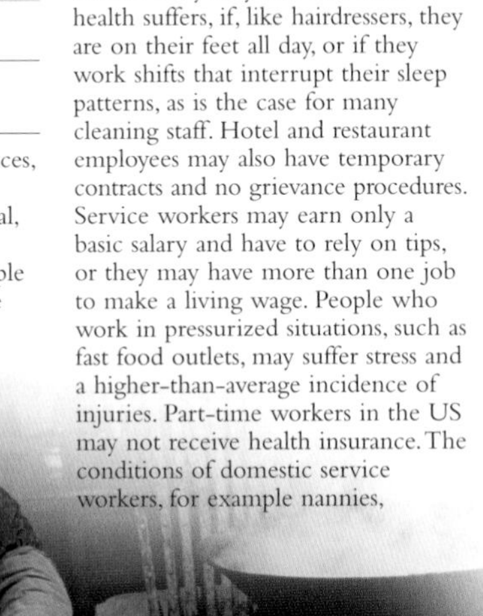

**WAITING AT TABLE**
*Restaurant workers may work split shifts and have no job security. In France, waiting staff have campaigned for better protection. In prestigious restaurants, they may be well paid.*

**PREPARING NOODLES**
*In many parts of the world, conditions for people working in sectors such as food preparation like this Chinese noodle stall worker are unregulated.*

WORK

## Professional work

**In brief**: Providers of professional services may have high levels of job satisfaction but work under pressure

Professional workers, such as lawyers and doctors, have usually spent many years training. They often work long hours and under pressure. However, their qualifications give them status in the community, mean they have more employment opportunities than non-professionals, and may mean they can practise abroad. In South Africa, 15 per cent of doctors have left to work in Australia, the UK, or the US, where they earn more and fill the gaps left by nationals who drop out of the profession. This "brain drain" occurs across Africa – in the Democratic Republic of Congo, for example, there are just 400 doctors, a particular cause for concern because of the need for doctors to deal with large numbers of people with HIV and AIDS.

**PROFESSIONAL AT WORK**
*A doctor examines a child at a clinic in Kenya. Trained medical professionals are in short supply across the African continent.*

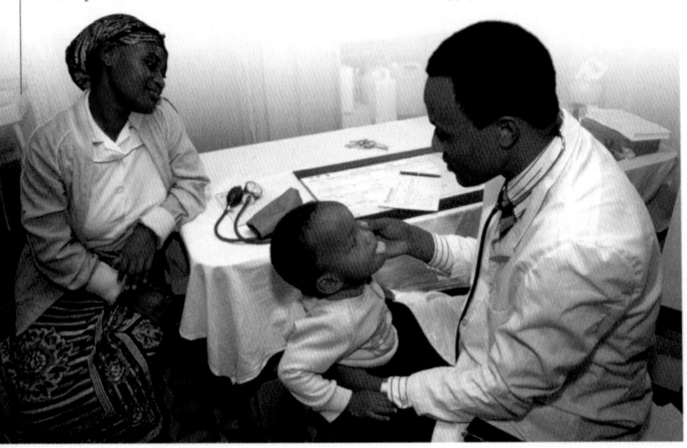

WORK

## Creative work

**In brief**: To excel, artists, artisans, and performers need talent but also commitment and self-discipline

Behind a seemingly effortless performance of professional music and theatre are years of hard work and training by those taking part. To become an opera singer, for example, requires training in singing, acting, and languages, and only a tiny fraction who start this career end up being well-known, high-earning soloists. In India, for example, girls who aim to be classical dancers are advised to start training no later than the age of 6. An early start is necessary because it takes many years to develop the muscle tone and control required to excel. Dedication, self-discipline, and patience are essential to all successful artists, whether they write, paint, or are involved in creating other forms of art. It took Tolstoy 6 years to write his great novel *War and Peace*, while the painter Cézanne once said it had taken him 40 years to get to the stage where he could draw one picture outline in an hour.

**CRAFTSWOMAN AT WORK**
*A young potter in Germany puts the finishing touches to a vase she has fashioned. Patience is a prerequisite for someone who wishes to become a successful artisan or artist.*

**CANTONESE OPERA SINGER**
*Performers in a Chinese opera will have begun their training at a very early age. They will also have spent many years learning such skills as singing and dancing.*

LEISURE

## Socializing

**In brief**: Spending time with friends and family on an informal basis is an important part of most societies

Most cultures have a designated place for socialising. Across the Middle East, men gather in coffee houses, where smoking a shared *hookah* pipe is the accompaniment to conversation and relaxation. In Oman, the *hookah* has been outlawed; the government declared it was preventing people from working. In China, people meet at tea houses, often bringing their own cup and tea leaves to top up with the hot water provided. In Australia, people devote an average of 2 hours a day to seeing friends over a meal or drink, or talking on the phone. However, they also spend around 2½ hours each day watching television, either alone or with their family.

**CAFÉ CULTURE**
*Two men enjoy the sunshine on the Greek island of Naxos. In Mediterranean countries, cafés are popular places to meet and socialize.*

LEISURE

## Self-improvement

**In brief**: Each year, people engage in leisure activities leading to physical and mental improvement

Every year, people in the UK sign up for a range of non-work-related courses and classes, from flower arranging to yoga. Many of these activities are physical; 14 per cent of Britons work out in some way at least once a week and in 2003, they spent a total of £125,000 million on keep-fit activities. But others invest in reading as a path for emotional, spiritual, and social self-improvement. In 2003, people in Britain spent £38 million on self-help books, 66 per cent of which were bought by women. In the US, top-selling self-improvement titles include those on how to get rich and how to break the habit of underachievement.

**ART CLASS**
*The desire to learn a new skill or improve an existing hobby continues throughout life. In urban centres, the opportunities and subjects on offer can seem unlimited.*

**WORKING OUT IN LEISURE TIME**
*In the West, there is ever-increasing interest in both getting and keeping fit. People may join a gym to help them reach their goal.*

LEISURE

## Playing sport and games

**In brief**: Games and sports are a way to relax. They range from chess to occupations such as baseball

For a few professionals, playing sport and games is a full-time way of making a living, but for most people, games are a competitive but relaxing hobby to watch or to take part in. Although some sports, such as football and cricket, have international appeal, the popularity of others is distinct to certain countries. Kite-flying, for example, is a sport that is especially revered in Pakistan. An annual competition sees experts trying to cut each other's kites out of the sky by covering the string of their own in crushed glass. The flyer whose kite is left at the end of the competition is hailed as a hero.

Buzkashi is a team sport that is unique to central Asia. During the game, horsemen vie for possession of the carcass of a calf. They score points when they gallop the length of the field and succeed in planting the carcass in the goal area.

In Mongolia, the skills of archery have been extolled in literature and poetry through the centuries, and all Mongolians acquire at least the basics. Capoeira is a 400-year-old art form that was developed as a method of expression and self-defence by the African slaves in Brazil. Today it is Brazil's second largest sport after soccer. Spectators form a ring, and clap and play music to accompany the performers while they put themselves through intricate sequences of stylized acrobatics and martial arts moves.

Bull fighting, where the matadors (fighters) try to outwit, daze, injure, and kill bulls, is the Spanish national game. However, this sport is viewed as brutal by people in other countries.

For some people, the excitement and appeal of sports and games is enhanced by betting on the outcome. In Bali, betting on cock-fighting, for example, is a multi-million rupiah business. The people placing the bets have to know a complicated vocabulary of bets and odds, which changes during the match. Following the betting procedure during the match requires both concentration and many years of practice.

**BASEBALL MATCH**
*A player gets ready to pitch during a game. The sport is often seen as the US national pastime because it is so popular.*

**CHESS**
*Chess is a game of skill and strategy popular across all age groups. Most people play it to relax, but there are competitions for professionals.*

**CAPOEIRA PERFORMERS**
*The Brazilian sport of capoeira is a rich blend of martial arts, dance, and ritual. It puts emphasis on core strength, flexibility, and balance, making it an ideal workout.*

Health

### Sports injuries

People who play a sport, whether professionally or as a leisure activity, are at risk of damaging their body. There are two basic types of injury. The first type is acute (sudden) and traumatic, involving a single blow from a single application of force, such as breaking a leg during skiing. A second type – overuse or chronic injury – is usually the result of repetitive training. Runners may develop inflammation (swelling) of the tendons for example. Despite these risks, health experts agree that people who lead active lives are likely to live longer and be healthier.

**DRAWING A CROWD**
*Board games may be played in the home or, in Germany and Austria, outside with giant pieces. From Greece to China, groups of older men playing in cafés is a typical sight.*

# OLD AGE

Many scientific attempts have been made to combat our inescapable mortality, inspiring documentary film makers and science fiction writers. However, even the most advanced technology is unlikely to keep the human body alive for eternity. The human body ages as a consequence of cumulative damage to the body's cells and tissues. Eventually, as the body's repair systems falter, we succumb to age-related diseases, such as cancers, stroke, heart disease, and arthritis. The effects of aging can be slowed by a healthy lifestyle, but not halted. Although humans are the longest-lived mammals, like all other species we have a finite life span.

## PHYSICAL CHANGES

As the body ages, physical changes take place. For many people, one of the first signs of aging is reduced hearing and changing vision. As a result of noise-induced damage to the cochlea (the inner part of the ear), hearing becomes less acute from middle age. Vision also deteriorates as the lens of the eye becomes more rigid and the retina (the light-receptive disc at the back of the eye) degenerates. The skin starts to lose its elasticity, developing wrinkles and lines. Blood pressure rises as the walls of the blood vessels harden and become less elastic. The lungs also stiffen with age. As the skeleton ages, joints suffer from wear and tear. Women in particular are at risk of osteoporosis (loss of bone density). The immune system, which helps defend the body against infection, also starts to become less efficient. Although memory often declines, the brain is still capable of learning and new nerve connections continue to be built throughout life.

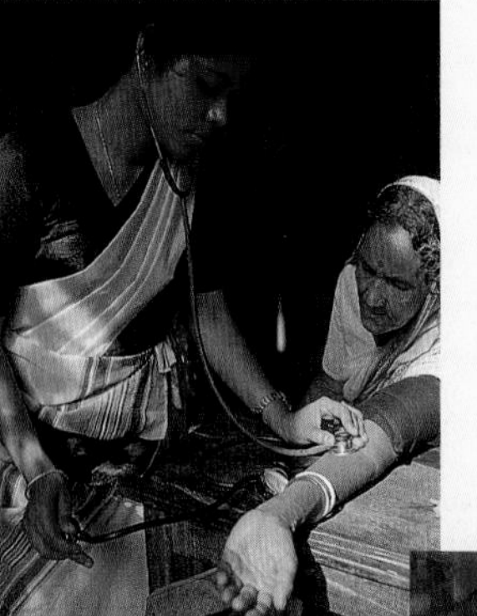

**HEALTH CHECK**
*Regular health checks, for example for blood pressure, are vital for older people.*

**OSTEOPOROSIS**
*Reduced bone density, or osteoporosis, is shown as an orange area in this false-coloured X-ray of a hip.*

## THE SCIENCE OF AGING

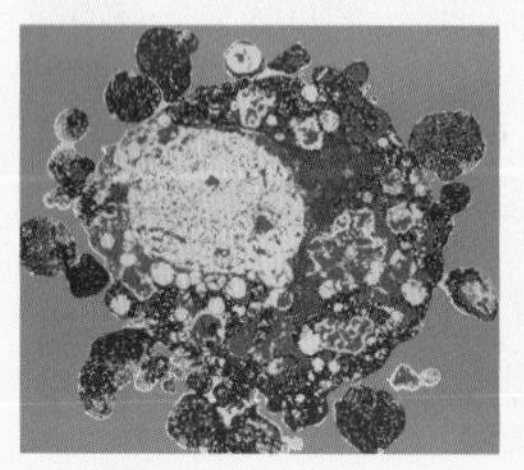

**DYING CELL**
*During our lives, cells undergo programmed death and are replaced. Eventually errors occur, which may contribute to aging.*

Throughout our lives, the various types of cell that make up the tissues and organs of our bodies have to be renewed. For example, the cells lining the digestive tract get replaced every 72 hours, while blood cells are replaced every 10 days. However, this renewal process is not foolproof, and as the body ages, new cells begin to contain errors within the DNA (*see* p52). In addition, substances called free radicals – highly reactive byproducts of ordinary metabolism – damage cell components, including DNA. Damage to DNA within the cell nucleus affects the reproductive core of the cell, while damage to DNA in mitochondria (the cell's powerhouse) interferes with cell function. Some enzymes (chemicals that speed up reactions in the body) have been found to counteract free radicals. It is believed that people who live to the age of 100 or more may naturally have more of these enzymes.

Health

### Changing colour vision

Cataracts, a clouding of the normally transparent lens of the eye, is a common disorder of old age. One effect is a distortion in colour vision. The French painter Claude Monet is believed to have developed cataracts in later life, and the effect may be seen in his paintings. Subjects he painted several times changed colour. Tones became muddy and whites and greens became yellow.

MONET'S WATER LILIES

LIFE CYCLE

# DEFYING AGE

The physical effects of aging can be reduced by specific lifestyle factors such as exercise and diet. People who remain physically active and mobile can sometimes delay the effects of aging and there are various examples of old people who continue to be active even into very old age. In some societies, people lead more physically active lives than in the West and avoid some age-related disorders that Westerners suffer. Similarly, a diet that is balanced and reduces the risks of heart disease and obesity is likely to help people retain good health into old age. However, there is also a link between genetics and aging – longevity tends to run in families. One of the oldest people whose age was reliably documented was a French woman, Jeanne Calment. She died in 1997 at the age of 122, famously outliving her lawyer who had hoped to make money from a real estate deal with her. She had taken up fencing at age 85, cycled at 100, and recorded a rap CD at 121. Evidence suggests that around 120 years is the maximum human lifespan.

**100 YEAR-OLD SKIER**
*Kaizo Muira from Japan skies every year, and has done since the age of 14. He celebrated his 100th birthday on the slopes.*

**TAI CHI IN THE PARK**
*In Vietnam, many people practice Tai Chi. This is a traditional form of exercise that is good for maintaining strength and flexibility at any age.*

Fact

### Longest lives

Some of the oldest people in the world live in the islands of Okinawa, Japan. Many Okinawans live active, independent lives well into their 90s and 100s. Their longevity is partly credited to their diet, which is low in calories and fat and high in fruit and vegetables. Other aspects of lifestyle thought to be important are exercise, a strong spiritual belief, and firm social and family ties (which helps to reduce stress). Compared with North Americans, Okinawans have 80 per cent less breast cancer and prostate cancer, and less than half the ovarian and colon cancers. Studies also found Okinawans have healthy arteries and a low heart-disease risk.

# CHANGING ROLES AND CARE

Old people are sometimes viewed as being a burden on family and society, but it was not always so. In pre-industrialized times, elderly people probably enjoyed a higher status than they do now. In many parts of the world, such as much of Asia, elderly people are revered as the repository of accumulated learning and wisdom. In tribal Africa, old people have very defined roles as elders in the community. In many parts of the Western world, people are encouraged or forced to retire from work and may then find their role and status in the community disappears. Combined with declining finances and poor health, old people may eventually become depressed and feel unneeded, however this is not always the case. Old people in Japan, for example, are both respected and indulged. It is assumed without question that they will be cared for, usually by the daughter-in-law. Age is a byword for experience, authority, and the right to behave as they choose. When people do eventually need care, it may be provided by the family or community. Increasing numbers of elderly people in Western societies need residential care as they become unable to look after themselves.

**HELP WITH BATHING**
*For elderly Japanese people bathing is very important. Nursing homes provide help for people who are no longer able to look after themselves.*

**GRANDPARENTING**
*Many older people become grandparents or even great grandparents. They may find a role helping to care for their grandchildren.*

**ENJOYING LATER LIFE**
*Life expectancy in India is around 61 years. Only 4 per cent of people are older than 65. In many countries, women live longer than men do.*

# DEATH

In most Western settings, people grow up protected from death, and many do not see a dead person until a parent or grandparent dies, and often not even then. In the past most people used to die at home, and death was seen as a normal part of life. This is still the case in traditional societies, but in Europe and the US at least 70 per cent of people die in hospital. Death may be expected, after a terminal illness, or happen suddenly and come as a great shock to loved ones. Our attitudes to death and our beliefs about what happens at death and afterwards depend, to a large extent, on our religious faith and our culture.

**LAST WILL**

*People often draw up a will at some stage in their lives to express their wishes for how their property and belongings should be disposed of after death.*

## PREPARING TO DIE

In the West, most adults avoid discussing death or thinking about it too deeply as it is easier to deny if it remains a remote concept. Death is nonetheless inevitable, and when it occurs cultural rituals become particularly important. Being prepared for death means accepting the finality of death and ensuring that one's wishes will be acted on. These wishes are often expressed in a will, which is a legal document that sets out what should happen to an individual's possessions after death. Most people in the West draw up a will at some stage in their lives. In some countries, such as the US, people with long-standing or terminal illnesses may draw up a "living" will, to state under what conditions they would not want to be resuscitated. It is hard to know what a dying person thinks and feels. Different people face death in different ways, and attitudes depend in part on spiritual and religious beliefs and resources. In most societies, families gather together to comfort and support the dying. In some cultures, children are included; in others, death is considered to be an adult affair.

**GATHERING OF FAMILY**

*Having close family nearby is a comfort to people who are very ill and close to death. Some people die suddenly, leaving no chance for the family to gather.*

Issue

### Near-death experience

Many people have claimed to have had a near-death experience (NDE), in which they thought they had died. Such people have reported feeling intense joy or peace, and many have said they felt they were travelling down a tunnel with a light at the end. For some, NDEs confirm a belief in life after death. However, scientists explain these experiences as a "trick" of the brain. The centre of the eye's visual field is more active than the periphery, which may explain the tunnel effect. The brain may also be flooded with natural morphine-like chemicals, explaining the sensation of joy.

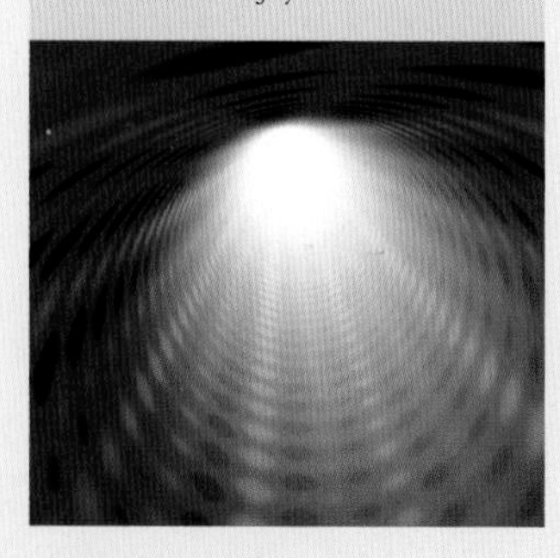

## WHAT HAPPENS AT THE MOMENT OF DEATH?

The moment of death happens when the heart stops beating and the brain shuts down. Leading up to this moment, breathing becomes irregular with very quick breaths being followed by very long ones. High blood pressure can force fluid into the lungs. When the heart stops pumping, air no longer passes in and out of the lungs. Starved of oxygen, the tissues cannot function. Within 10 seconds without oxygen, the brain's electrical activity reduces. Within 4 minutes, the brain is damaged irreversibly. The pupils of the eyes promptly widen and no longer react to light. The body cools from its normal temperature of 37°C (98.4°F), and muscles stiffen within 4-6 hours. Diseases kill because the heart or lungs fail, or they cease to function when another vital organ, such as liver or kidney, is critically impaired. It may not be clear which organ has collapsed. Sometimes, several organs fail simultaneously.

| | |
|---|---|
| Heart disease | 38.7% |
| Infections | 19.0% |
| Cancer | 12.2% |
| Injury | 8.9% |
| Respiratory disease | 6.3% |
| Death at or around birth | 4.4% |
| Digestive disease | 3.5% |
| Other | 7.0% |

**CAUSES OF DEATH**

*Heart disease is the major cause of death worldwide. Causes of death vary from country to country.*

Fact

### The wake

A wake was originally held to "watch over the corpse" – literally staying awake – and was an Irish tradition. Because someone was watching, family and friends could arrive at any time during the night for the funeral and, as they had often travelled long distances, they were given refreshments. The wake was a pre-funeral celebration, and became the forerunner of the modern post-funeral party. The word comes from Old English waeccan, to wake. The idea of "watching over" was literally to ensure that the person was dead and did not waken during the vigil.

# BELIEFS ABOUT DEATH

The ways in which funeral rites are observed in a society or among a group of people reflects the beliefs held about death, including beliefs about an afterlife. Christians believe that if they follow the teachings of Jesus, they will go to heaven when they die. Buddhists believe that life is a cycle and that the dead are reborn at different "levels" of life depending on how they have lived their life. In Judaism, the entire body must be buried soon after death so that the soul can go to heaven and face God on Judgement day. Hindus believe that they are reincarnated after death. Because the body is not needed, Hindus elect for cremation, with the ashes being committed to water to purify them. For some people, death signifies finality: atheists attend funerals only to show respect for the deceased.

**BURIAL BLESSING**
*In some faiths burial is favoured. In this Christian burial in Vietnam, mourners dressed in traditional white place a blessing on the coffin before it is buried.*

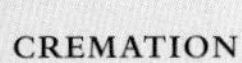

**CREMATION**
*In the Hindu faith, the deceased is normally cremated. This method of disposal reflects a belief that the body is not needed in the afterlife.*

# GRIEVING AND MOURNING

Many cultures observe a structured mourning period, in which there is a prescribed number of days during which people conduct their grief. For example, in Islam men do not cut their hair for 40 days after a death. In Hinduism, members of a household do not cook for 7 days after a death in the family, instead relying on friends to provide them with food. In some traditional societies, a special mourning ceremony is held some time after the funeral. Public mourning has become more widespread in recent years. For example, people express their grief after the death of a national leader or monarch with flowers laid at special sites. There are various recognized stages of grief that psychologists believe people go through (*see right*) after a bereavement. The final stage of acceptance may not be reached for months or even years. However, most people can be helped through their grief by friends and family and, if necessary, special counselling.

ANGER AND INTENSE EMOTION
↓
DENIAL
↓
GUILT
↓
DEPRESSION AND SADNESS
↓
ACCEPTANCE

**STAGES OF GRIEF**
*Up to 5 stages of grief are recognized, but not everyone goes through all these stages after a loss.*

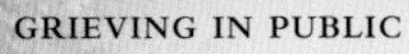

**GRIEVING IN PUBLIC**
*An outpouring of grief is a common reaction at the time of a death. In some societies, public grieving is normal and encouraged. These women share their grief at a funeral in Harare, Zimbabwe.*

**LAYING FLOWERS**
*A woman in the Veneto region of Italy lays flowers on a gravestone. Such acts of remembrance help people to come to terms with the death of a loved one.*

**REMEMBERING THE DEAD**
*Tombstones are a way of honouring the dead and keeping their memory alive as here in the Heiliger Sand Jewish cemetery in Worms, Germany.*

FUNERAL RITUALS

## Buddhist funerals

**In brief:** Buddhists believe in rebirth and the importance of the soul; cremation is usual

Most Buddhists cremate their dead, a ritual that stems from the belief that Buddha was cremated. Rituals vary from place to place. In Thailand, people gather at the home of the deceased for 3 days where they eat, play games, and hold remembrance ceremonies. On the 3rd day the body is carried to the crematorium. The procession is often led by a man with a white banner, followed by elderly men with flowers in silver bowls and then by 8 to 10 monks. At the crematorium, the monks chant holy verses; then the coffin is placed on a pyre. Mourners throw candles, incense, and fragrant wood beneath the coffin. In the Mahayana tradition of Buddhism, practised in China, the period directly after death is important: it is known as the "intermediate period". During this time, before the body is "reborn", actions can influence the form of the rebirth. For this reason, mourners pray and perform ceremonies to help the soul find a good path.

**CREMATION IN CHINA**
*Most Buddhists are cremated: the soul of the deceased is seen as more important than their body. A picture of the person is placed in front of the coffin.*

**VIETNAMESE PROCESSION**
*In this elaborate funeral procession in Vietnam, the body of the deceased is carried in a lace-covered coffin by more than 20 pallbearers.*

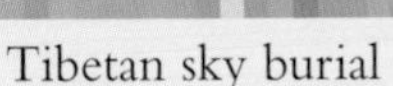

Fact

### Tibetan sky burial

In Tibet, one of the most common ways of disposing of the body of the deceased is by allowing it to be devoured by birds, in a ceremony known as "sky burial". Men dressed in white robes take the body to a high peak in a wilderness area, where they slowly proceed to chop the body up, throwing the flesh to one side. Often, several bodies are disposed of at the same time. Very soon, vultures arrive and swarm down to eat the flesh or carry it away. When there are only bones left, the men grind them up and mix them with barley flour for the crows and hawks that have been waiting their turn. Eventually, nothing at all is left of the body.

FUNERAL RITUALS

## Muslim funerals

**In brief:** The body is buried within 24 hours of death; the grave is dug to face Mecca

Muslims bury their dead as soon as possible after the death, preferably within 24 hours. First, the body is washed by relatives (a male relative washes and male body, and female relatives a female body). Washing is done at home or at the mosque. While the bathing is taking place, a funeral prayer is recited. The body is then wrapped in white sheets and taken to the burial site. The grave must be dug so that it points to Mecca, and the head of the deceased is placed at the end closest to the Holy City. Muslim funerals are simple affairs and most Muslims will place a simple headstone on the grave.

**PREPARING THE BODY**
*Muslim women express their grief over the body of a deceased relative. The burial is usually performed within 24 hours of death.*

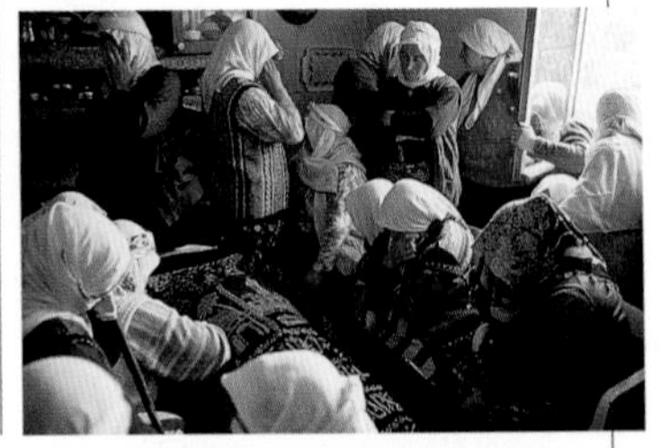

FUNERAL RITUALS

## Confucian funerals

**In brief:** Elaborate rituals are held and the body is washed and bound, ready for life in the next world

**SOUTH KOREAN FUNERAL**
*The funeral procession is led by someone who sings mournful funeral songs, followed by people bearing colourful funeral banners.*

Confucian funerals are very elaborate. In Korea, where most people are Confucians, a funeral can last for several days. The body is first bound, hands and feet tied, and covered with a coat. The next day, it is washed and prepared for burial. Three spoonfuls of rice and a metal coin are placed in the deceased's mouth. The body is then wrapped in a traditional hemp or silk death dress, known as *suui*, and placed in a coffin. Three days of mourning follow. On the 3rd day, there is a huge procession to the burial site. The grave site is exorcised of evil spirits and the coffin is lowered into the ground. When the family return to their home they put up a picture of the deceased and embark on 3 more days of mourning.

FUNERAL RITUALS

## Christian funerals

**In brief:** Although rituals vary, Christian funerals involve a church service followed by burial or cremation

The precise nature of Christian funerals varies around the world and according to the branch of the faith. They all share certain features in common: a period of mourning at home, a church ceremony, and a brief ceremony at the grave side. Since Christians believe in the resurrection, many speeches or sermons given at funerals refer to it.

Many Christians bury their dead, but some prefer cremation. Catholic funerals usually contain a mass. First, the body of the deceased is washed and often embalmed. The body may then lie in an open coffin for visitors to come and pay their last respects. This is called a "wake". In many countries, the wake is also a time of feasting and drinking. The body is carried or driven to the church in a big procession and a mass is recited. The coffin is taken to the grave site and lowered into the ground. At this point, the priest often makes a speech about the deceased, recollecting the events of his or her life. There is often a blessing as soil is thrown over the coffin. Protestant funerals tend to vary more than Catholic ones, and they are often designed by the family of the deceased. There are bible readings, hymns, a sermon, and then a speech. In many parts of the world, Christian funeral customs have been blended with local funeral customs.

**PRIEST LEADS PROCESSION**
*The body of the deceased is carried by pallbearers in a coffin. After the church service, the coffin is taken to the graveyard for burial.*

**MEXICAN FUNERAL**
*Huge numbers of people attend Mexican funerals, all participating in the procession to the burial site.*

FUNERAL RITUALS

## Dogon funerals

**In brief:** The dead are placed in caves; masked dancers chase the spirit away from the village

The Dogon people of southwest Mali have large and elaborate funerals. The people believe that when a person dies, his or her spirit is released and it is important to lead the spirit away from the village to prevent it from causing trouble. An important role of Dogon funerals is, therefore, to lead the spirit away and help it on its route to the realm of the spirits.

The body of the deceased is traditionally buried in caves high up in cliffs, above the Dogon villages. The body is hauled up with ropes and then left to decay in the stone chambers. Most Dogon funerals include a ceremony in which the men perform mock battles, designed to chase the deceased's spirit out of the village. The men chase around the village, armed with spears and guns, and fire bullets into the sky. All Dogon funerals include ceremonies with masked dancers. These ceremonies typically last for 3–5 days, and the masked dancers dance on the rooftop of the deceased's house, in the village square, and in the surrounding fields. Women are not allowed to participate in these dances and must stay away at a safe distance. Many different masks are used (*see right*), some representing particular spirits and ancestors, others representing men, women, and animals. The dance is symbolic of the order of the world and an attempt to put it back into balance after the turmoil caused by death.

**CAVE BURIAL**
*Wrapped in a burial shroud, a village elder's body is hauled up the escarpment to its final resting place in a burial cave high above the Dogon village in southwest Mali.*

Fact

### Dogon death mask

Masks are an important part of Dogon culture and are thought to be one of the main links between the living and the realm of spirits. Masked dancers perform at Dogon funerals to entice the spirit away from the earth and towards the spirit world. Hundreds of masks may be worn at any one funeral. One common mask that always appears is Satimbe, which represents Yasigi, the first ancestress of the Dogon people.

SATIMBE MASK

**DOGON CAVE**
*Burial caves are built in the cliffs. They are thought to have been built by people who lived in the area before the Dogon.*

**BURIAL CHAMBER FLOOR**
*Dogon burial chambers were built centuries ago in the Badiagra cliffs in Mali. The caves are littered with the bones of the dead, which have been left here for many years.*

FUNERAL RITUALS

## Ghana funerals

**In brief:** People in Ghana celebrate the lives of the deceased; coffins are often elaborately made

Funerals in Ghana, particularly among the Ashanti people, are as much about celebrating the life of the deceased as mourning his or her death. It is thought that if people have a good funeral then they will be remembered by the living. Most funerals take place on a Saturday, when friends and family can travel. In contrast to the meagre amounts of money spent on daily living, families compete with each other to spend ever greater sums of money on food, drink, and music. The body of the deceased is washed and dressed in his or her best clothes and put on display to visitors. As most Ashanti people are now Christians, the body is then placed in a coffin and paraded to the church, where a mass is said and stories told. The body is usually buried. After the funeral, family and friends party late into the night.

**FANTASTIC COFFINS**
*As the ostentation of Ghanaian funerals has increased, people have begun to design elaborate coffins in the shape of fish, animals, cars, and so on.*

CAR COFFIN

FUNERAL RITUALS

## Bobo funerals

**In brief:** Masked dancers perform at the funeral to drive the spirit of the deceased to the afterworld

The funerals of the Bobo people, who live in the Upper Volta region of Burkino Faso, are large affairs dominated by masked dancers. The people of this region believe that, after death, the spirit needs to be driven into the afterworld. First, the body is taken to a burial tomb by masked dancers. In some Bobo areas, dancers in wooden masks appear at the burial site and must be given millet beer. There is then a period of weeks or months when the village is in mourning, and certain ritual restrictions are in place. The restrictions come to an end at the funeral ceremony, in which masked dancers exhort the deceased's spirit to leave the village and move to the realm of the ancestors. Once this ritual is completed, Bobo people believe that the village has been purified and can return to normal.

**FUNERAL DANCE**
*A masked dancer performs at a funeral ceremony, exhorting the spirit of the deceased to leave the earthly realm and find its way to the realm of the spirits.*

FUNERAL RITUALS

# Hindu funerals

**In brief:** The body is usually cremated and ideally the ashes of the deceased are scattered at a holy site

The way in which Hindu funerals are conducted varies between the types of communities. In India, Indonesia, Malaysia, and Nepal they do, however, share a number of things in common.

Hindus believe in reincarnation: that at death the spirit leaves its former body and returns to earth in another body on its journey towards Nirvana. Most Hindu funeral practices seek to aid the departed spirit in this process of rebirth. To this end, the vast majority of Hindus are cremated. There is a belief that the burning fire releases the spirit from the physical body and that cremation is therefore an important first step in its journey.

In India, the cremation is preceded by a short period of mourning around the corpse immediately after the death. Women, in particular, gather to weep and wail, while the deceased's sons take gifts to the priests, or brahmans, in order to remove the sins of their father. Because it is believed that the person who receives these gifts will also receive the sins of the deceased, the sons often have to make a large payment to persuade the priest to take the gifts. Soon afterwards, the body is washed and then is dressed in traditional white clothes. Later the same day, the funeral pyre is built or the body may be taken to a crematorium. If the family lives near the holy city of Varanasi, the body may be taken to the burning ghats, the terraces which lead down to the river Ganges. It is believed that this immense watercourse cleanses all sins and thus ensures the spirit a successful rebirth in its next life.

The eldest son is expected to light the fire and to pray for the wellbeing of the deceased's spirit. Priests recite scriptures from the Vedas or from the *Bhagavad-Gita* (sacred Hindi texts). Once the corpse has burned, the ashes are scattered in the Ganges or other holy place, and the family returns home to continue mourning. All mourners change their clothes before re-entering the house, to avoid bringing in the pollution of death. Over the next 20 to 30 days, the fathers-in-law of the sons of the deceased supply the mourners with food. Their actions ensure that the mourners eat well even though they are distracted by grief.

**MALAYSIAN PROCESSION**
*Two oxen, animals that are believed to be sacred in the Hindu faith, pull a cart bearing an altar through the streets in Malaysia. The procession takes the body to the crematorium for cremation.*

**FUNERAL CEREMONY, INDIA**
*Flowers cover the body of the deceased in a funeral procession in New Delhi. The body will be carried from the family home to the funeral site where cremation will take place.*

**PREPARING THE BODY, NEPAL**
*The eldest son prepares the body, which is set on a wooden raft. The body is cremated and set afloat down the river, a process that is believed to both free and purify the spirit.*

Fact

## Feeding the spirits

Hindus believe that, during the first 10 days after death, the spirit of the deceased is in a transition period. Devotees consider that it is looking for, or in some cases that it is creating, a new body with which it can enter a new life. During this time the children of the deceased perform the Shraddha ceremony. Every day they make offerings of rice balls, known as *pinda*. This type of food is thought to help in the transition process by providing the nourishment the spirit will need during this challenging period.

**DISPERSING THE ASHES**
*Ashes are usually scattered at a holy site such as the river Ganges. If the relatives are unable to do this themselves, they can pay a holy man to perform the ritual.*

**CREMATION, BALI**

*In Hindu Bali, preventing the spirit from returning home is all-important. Corpses are carried to cremation grounds on enormous towers. On arrival, the body is transferred into an animal-shaped coffin and cremated.*

GRIEVING AND MOURNING

## Gamo customs

**In brief:** Large-scale gatherings are held to mourn the deceased, and may last a whole day

When someone dies in the Gamo Highlands of Ethiopia a huge mourning ceremony, known as *yeho* takes place. After the body has been buried by the family, the entire village descends on the communal mourning ground. Men and women from several different communities parade together, the men carrying spears and chanting war chants, the women weeping and wailing. The men charge around the mourning ground, brandishing their spears and chanting. As many as 5,000 men may be running around the mourning ground at one time. Close family members of the deceased are expected to mourn more strongly than others. They show their grief by beating their chests, scratching their bodies, or throwing themselves to the ground. A *yeho* lasts a full day.

**GRIEVING BROTHER**
*Gamo men charge around the community mourning ground in the* yeho *ceremony. The deceased's brother is restrained by a kinsman.*

GRIEVING AND MOURNING

## Jewish rituals

**In brief:** Grieving follows a 7-day ritual; every year thereafter a candle is lit in memory of the dead

After the death of a Jewish person, the body is buried within 24 hours. A 7-day mourning period, known as *shiva*, then begins. Prayers are said every day at the parent's or child's home. Close family members sit on special low chairs and rip their clothes to show their anguish. Family and friends visit and take part in the prayers. For the first 7 days, the house of the mourners is usually full of people, which eases the immediate shock and pain of the family's loss. After the *shiva*, mourning becomes more private. Orthodox Jews will not go out to parties or listen to music for a year after the death of a family member. Every year, the date of a person's death is marked by the lighting of a special candle known as a *Yahrzeit* candle.

YAHRZEIT CANDLE

REMEMBERING THE DEAD

## Japanese customs

**In brief:** Memorial customs include special services on certain anniversary days after the death

Death has great significance in Japanese culture and is traditionally marked by memorial services. The particular customs and rituals vary around the country.

Memorial services for people take place at frequent intervals, usually on the 7th, 49th, and 100th days after a death. In most cases, the family also holds services on the first anniversary of the death and then on subsequent anniversaries that have the numbers 3 or 7 in them (3, 7, 13, 17, and so on) until the 33rd anniversary of the person's death. Particularly devout people also observe monthly services for the dead. Relatives pay a priest to carry out a special ritual.

**JIZO-SON**
*These small stone figures are worshipped in the Buddhist Shugendo cult in memory of children who have died.*

LIFE CYCLE

MEMORIAL CUSTOMS

## Madagascan bone-turning rituals

**In brief:** Bone-turning is performed to maintain contact with ancestors; bones are removed and returned to tombs

The Merina and Betsileo people of Central Madagascar perform regular rituals to maintain contact with deceased relatives. It is believed that, after death, the deceased become *razana*, or ancestors, from which all life ultimately flows. It is therefore important to maintain a good relationship between the living and the dead. One way this is done is through the bone-turning ceremony, known as *famadihana*. During this ceremony, the body is removed from the burial tomb and either wrapped in a new shroud or moved to a new tomb. The ceremony is an occasion for singing, dancing, and feasting.

**PARADING THE BONES**
*The remains of the deceased are paraded around the village to help maintain the relationship between the living and the dead.*

MEMORIAL CUSTOMS

## Mass memorials

**In brief:** Most societies have mass memorials to mark occasions when large numbers of people have died

Many countries have mass memorials to remember terrible events when large numbers of people were killed. These memorials may be material edifices, such as statues or museums, or they may be annual remembrance days, often accompanied by particular symbols of remembrance.

There are many war memorials at sites around the world. Some, such as the Vietnam memorial in Washington, DC list the names of every person known to have died. In Hiroshima, Japan, the Memorial Peace Park has been built to remember those who died during the bombing of the city. As well as the large memorials erected by nations or communities, people often make their own offerings, such as flowers, at significant sites.

**REMEMBERING HIROSHIMA**
*Strings of brightly coloured origami paper cranes symbolize peace and hope at this site in Hiroshima, Japan.*

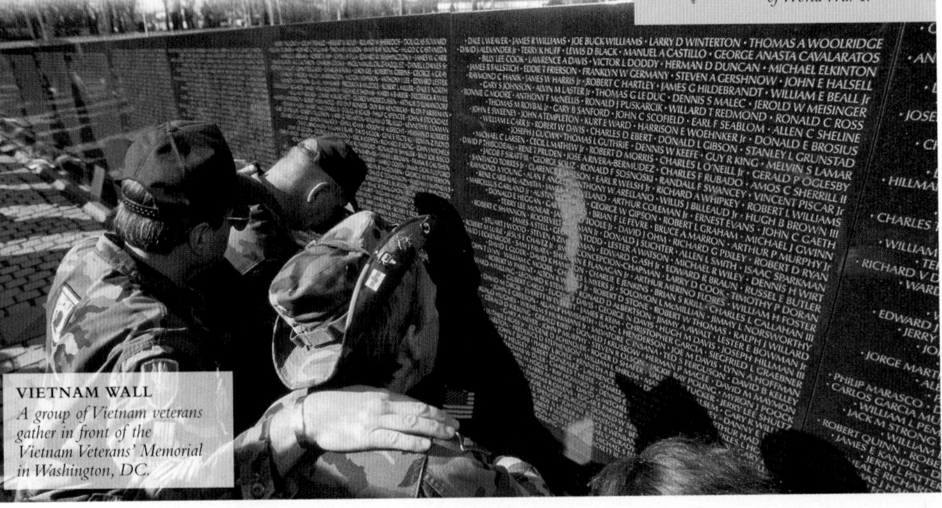

**VIETNAM WALL**
*A group of Vietnam veterans gather in front of the Vietnam Veterans' Memorial in Washington, DC.*

History

### Remembrance poppies

Poppies have come to symbolize lives lost during the two World Wars in several countries, including the US, Canada, Australia, and the UK. The poppy's association with war stems from a poem written by Canadian John McCrae during World War I: "In Flanders fields where poppies blow/Between the crosses, row on row...."

**DRIED POPPY**
*This poppy was preserved by a soldier from the battlefields of World War I.*

ANNUAL FESTIVALS

## Chinese Ching Ming festival

**In brief:** A week-long festival is held every Spring to show respect for deceased family members; food and other offerings are placed on tombs

Every year on the 106th day after the winter solstice, Chinese people celebrate the Ching Ming (literally "clear and bright") festival. It is traditional to seek the blessings of the ancestors before a new undertaking, and the date of the Ching Ming festival is in April, at the official start of Spring. This is seen as the season of new beginnings, and a time when ancestral blessing is needed most.

Families show their respect to the ancestors by visiting the graves of deceased family members and tidying them up, putting on flowers, lighting candles, and making offerings of food. Often, the day turns into a family picnic and is usually a happy occasion. The Ching Ming festival is also known as "Grave-Sweeping Day" and "Spring Remembrance".

**PLACING GIFTS ON TOMBS**
*Family members leave offerings on tombs during the week-long festival of Ching Ming.*

ANNUAL FESTIVALS

## Obon festival

**In brief:** In this 3-day festival, ancestors are believed to revisit their families

**FAMILY AT OBON FESTIVAL**
*On the last day of the Obon festival, the family leads the spirit of their deceased relative back to the grave.*

The Obon festival, which is of Buddhist origin, is an important tradition for Japanese people. It is a time when the living pray for the peaceful resting of the souls of their ancestors and when ancestors are believed to visit their families. On the first day, people clean their homes and make offerings on small family altars. They then light paper lanterns inside their houses and go to their relatives' graves to call the spirits back home. The next 3 days are spent "welcoming" the spirits with food, drink, and dancing. Finally, a procession leads the spirit back to the grave. In some places, people make little boats with candles or floating lanterns to guide them.

ANNUAL FESTIVALS

## Fete Gede

**In brief:** Thousands of people walk through the streets of Haiti in this annual Vodou festival

The focal point of Fete Gede, an annual Vodou festival in Haiti, is a huge procession, during which thousands of white-clothed people parade through the streets. Haitians believe the spirits of the dead join in this procession by "possessing" them. Possessed individuals go into a trance-like state and start talking in nasal voices. As the procession continues and the music and drumming gets louder, more people become possessed. It is believed that the ancestral spirits, through possession, can heal illnesses and resolve problems.

**VODOU POSSESSION**
*During the annual Fete Gede festival, thousands of people parade through the streets. Many of them go into a trance-like state, believing themselves to be possessed.*



ANNUAL FESTIVALS

## Mexican Day of the Dead

**In brief:** Celebrated in November each year, the Day of the Dead in Mexico is marked by festivals and candle-lit graveyard vigils

The *Dia de Muertos*, or Day of the Dead, is one of Mexico's most important festivals. In fact, the festival takes place over 2 days, November 1 (All Saints Day) and November 2 (All Souls Day). It is a time for celebration, when the living remember their deceased relatives with joy. At this time, people believe that the spirits of dead family members come back to visit the living. Traditionally, the first day is for people who died as children and the second day for those who died as adults. Families welcome the spirits by clearing an area in their house and setting up a special altar on which they arrange a wide variety of offerings. These may include selections of the deceased's favourite foods, sweets, candles, flowers, small figurines, and photographs. Some people even set aside a towel and soap for the spirit to wash before it eats and tequila and cigarettes to have afterwards.

Mexicans use this time to tidy up graves and to decorate them. In some places, all-night vigils take place in graveyards. During the day, there is music and dancing and parades with people in death-oriented fancy dress.

**PHONE CARD**
*The Day of the Dead is so enshrined in Mexican culture that everyday items like phone cards may be decorated with symbols of death.*

**SKELETON PARADE**
*Music, dancing, and parades are central features of the Day of the Dead in Mexico's towns. The skeleton is seen as a symbol of fun.*

**NIGHT VIGIL**
*In some parts of Mexico, people spend all night at the graveyard, bringing food and candles, to keep watch over their dead.*

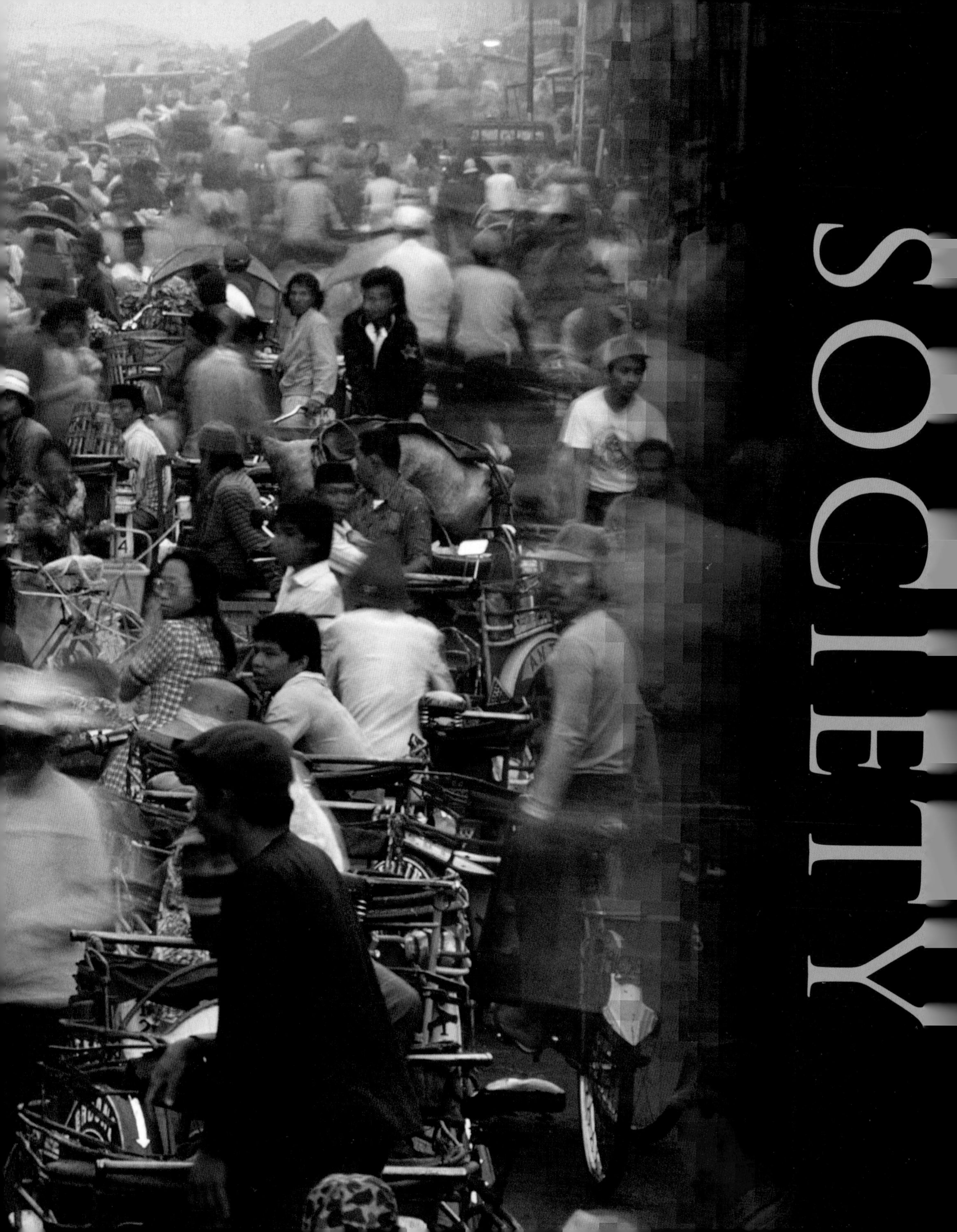
SOCIETY

# SOCIETY

Human beings are by nature social animals and have always formed themselves into groups. Affiliations are organized on the basis of common rules and networks of relationships – be they family, power, religion, or trade. Such an arrangement makes up the structured community known as a society. Although certain social norms, for example the encouragement of partnerships such as marriage, taboos on incest, and the division of labour by gender or age, are common to all societies, the customs and habits practised by groups vary widely.

Since humankind's earliest days, we have lived and worked in social groups. The family, with its infinite variety of extended relationships, is the most basic social organization. Our predecessors, however, learned that living in a larger group was a more effective way of meeting their material and spiritual needs, and a better defence against threats. The resulting villages, made up of a dozen or so families with a shared, self-sufficient lifestyle of agriculture or herding, proved to be such a successful form of social structure that some still exist today.

## BECOMING MORE COMPLEX

Social change can sometimes be described as "opportunistic" – in other words, societies adapt in response to their environments, both physical and social. Such natural adaptation can make societies more complex. This type of social change began to happen about 12,000 years ago with the gradual domestication of plants and animals – the catalyst for the advent of agriculture (*see* Early agriculture, p34). The ability to grow food meant larger populations could settle and be supported and there was a 20-fold increase in the world's population. Such growth created the need for centralized controls to regulate the conflicting needs of such large numbers of people and to organize the food production necessary to feed them.

The food production methods that arose led to food surpluses, role specialization, and "nonproductive" groups, such as priests, artisans (who produced goods), bureaucrats (who collected taxes and administrated), and soldiers (who protected and gained territory from rival groups). These new roles carried varying levels of authority and resulted in a social hierarchy, which replaced previous, more egalitarian, relations.

**SELF-SUFFICIENT SOCIETY**
*Small villages comprise family groups. Such societies achieve self-sufficiency through a basic division of labour.*

A more recent example of this opportunistic type of social change occurred in northern Europe in the 18th century. The effects of the Industrial Revolution (*see* Economic revolution, p43) were mass rural-to-urban migration, capitalism, and rapid technological advances, which, in turn, led to the development of the modern state.

## SOCIALIZATION

Learning to fit in with society, or socialization, starts in the family. It is later shaped by peer groups and, more formally, by educators. As a person grows up, he or she learns about the social norms that exist within the society: the rules governing conduct that help to create common behaviour, values, and goals. These norms of society, and those of its subgroups, influence every aspect of its members' lives, ranging from their religious and political beliefs to their moral values, the type of house in which they live, what they wear, and even the food they like and how they choose to eat it.

**REBEL OR CONFORM**
*In refusing to fit in with the norms of society, members of street gangs are still socialized – albeit into "antisocial" forms of social grouping.*

There are always social divisions and conflicts of interest whenever people coexist, and not everyone will obey the rules all of the time. Young people, for example, may rebel and join "antisocial" groups, such as street gangs, which exist uneasily within wider society.

When a norm is "broken" – if someone commits theft or murder, for example – sanctions, ranging from social disapproval to prison or even capital punishment, are applied to restore social harmony. However, social norms are not always just. As societies develop, structures that are based on inequality, such as racial segregation, can be recognized as unacceptable and challenged.

## MIXING CULTURES

Few societies today are completely isolated. From the villager in the most traditional of societies to the most sophisticated city-dweller, everyone on the planet is connected both to people within their groups and to others outside them by a variety of technological and economic systems. Yet, despite the fact that migration is on the increase, the idea of a global society means little to most people. We are generally concerned only with what affects us directly, namely the customs, religions, values, and lifestyles of our smaller subcultures.

In today's globalized world, though, people who are brought up in one society may choose to live part of their life in another – or even in more than one. This type

Issue

### Social isolation

In developed countries, there is a trend towards the breakdown of the family unit. Why this is happening is debatable, but some people, especially the elderly, find themselves becoming increasingly socially isolated as a result. Mental breakdown or economic misfortune may cause people to drop completely out of society. Sometimes, feelings of loneliness or isolation can aggravate existing mental conditions, leading in extreme cases to suicide.

**CHANGING ALLEGIANCE**
*New American citizens pledge allegiance to the flag to show their loyalty to their adopted society.*

of migration brings with it challenges for the migrants, their children, and the societies they join. Migration raises questions of identity and loyalty, and how much loyalty and conformity is demanded differs between societies. Even societies that seem to welcome and encourage multiculturalism can have problems accommodating many different ethnic groups.

## SOCIAL ROLES

Society can be thought of as a giant multifunctional organization. It produces the goods and services its members require, it keeps order, and it provides for people's health, education, and protection. To achieve the smooth functioning of a society, the individual members are expected to play different roles. We cannot all be landowners or leaders, but each of us – whether we clean the streets or work on a factory production line – is an essential component in society.

As with social behaviour, social role is to some extent determined by a person's upbringing, social position, and education. Race and gender may also affect the choice of roles that are available.

Providing that a society is not rigidly caste-based (*see* p262), there is usually scope for a person to change roles. However, the more prestigious and well-paid careers, such as law or medicine, tend to require long periods of formal study. Other professions may require preparation in the form of training, perhaps in technical colleges or as an "on the job" (vocational) apprenticeship.

## POWER IN SOCIETY

Modern societies are highly organized; however, attaining and maintaining this structure requires authority and the power that goes with it. This capacity is generally shared by various public offices, including judges, politicians, civil servants, industrialists, and army generals. While social order is most obviously shaped by politicians and the state, others are able to exercise their might less formally, by influencing government and public opinion behind the scenes. Business people, television executives, and newspaper editors, for example, can exercise power through their wealth or access to mass communication. Some individuals have authority over a much smaller domain, such as a headmaster in a school. While the use of force is rare, the threat or potential to use force is usually an unspoken part of the power equation in any society.

**SPECIALIZED ROLES**
*The larger and more complex the society, the greater need there is for its members to possess different skills to fill a variety of positions – from shoe-shiner to high-powered business person.*

Fact

### Posthumous success

Some people achieve a far greater degree of fame and success, and therefore status, following their death – especially if they happen to die young. The Dutch painter Vincent Van Gogh (1853–1890) is an example of this phenomenon. His life was characterized by poverty and a lack of artistic recognition, but the quality of his work and his early death ensured his position in the history books as one of the world's most highly valued painters.

**ENFORCING AUTHORITY**
*The authority of a police officer is seen as legitimate because it is vested in the role, not in the individual.*

## SOCIAL STATUS

Most people care how they are perceived by others, and few can escape the instinctive habit of placing themselves and others in a ranked order according to a range of characteristics. In developed societies, the principal symbols are usually material wealth and social status, but the standards by which people judge each other differ from one society to another. Wealth and power are irrelevant in a simple, foraging society, where charisma, skill, and generosity are the most highly prized social virtues. In some countries, youth and physical appearance are most important; in others, experience and age are venerated. Being very well-known can also confer social status, unless the person is famous for antisocial reasons – mass murder, for example.

In more rigid societies, social status is fixed and people tend to accept their social positions more easily. In a more mobile society, people often feel the need to display any emblems of status they may have, such as expensive cars and homes and designer clothes.

**WORKING TOGETHER**
*Although some social relationships are self-interested, a society cannot function if it is not founded on a base of mutual cooperation and community support.*

**OIL PRODUCTION**
*Oil is a valuable raw material in any economy, being the starting point for the production of many different goods. Oil-producing countries are usually wealthy.*

# ECONOMIES

The economy is the name given to the vast and complex web of human activities concerned with the production, distribution, and consumption of goods and services, involving everything from the basic necessities of life to the most expensive luxuries. Economies can be structured in many different ways. They range from the small-scale economies of people living in nomadic herding communities, where the economic system remains small and local, to the global economy of the world based on international trade in raw materials, goods, and services between countries.

**MONEY SUPPLY**
*National governments regulate the supply of money to support its economic policies. The US dollar is one of the most widely recognized currencies.*

In producing any product, many individuals and companies, often from several countries, combine to make a finished item. No single company makes a computer, for example. For this reason, the economic system has to carry out the complex task of coordinating the activity of many inputs in the production process. In the majority of societies, the market (the system in which goods are bought and sold) plays a large role in organizing what is produced because it allows people to respond to incentives. Prices are an important influence on how consumers spend their money and on the decisions of producers as to what to produce.

## LOCAL AND GLOBAL

Many people in the developing world, especially parts of Asia and Africa, have little more than the basic necessities and frequently not even these. In such societies, the economy is largely organized along family lines. It is also highly dependent on self-sufficiency. The family grows much of its own food and provides its own shelter. Very limited use is made of the outside market. This system is in stark contrast to the economic life of most people living in countries where average incomes are higher.

**SELF-SUFFICIENT**
*In many parts of Africa a large proportion of the population aims to be self-sufficient, selling surplus produce at local markets, such as this one in Cameroon.*

Profile

### Adam Smith

In 1776 the Scottish economist Adam Smith (1723–1790) published *An Inquiry into the Nature and Causes of the Wealth of Nations,* one of the most important books on economics ever written. In it, he stated that markets guide the economy like an "invisible hand", and that even though everyone looks after his or her own interests, the market ensures that society as a whole benefits. He argued that governments should not intervene in the economy and should adopt a policy of *laissez-faire,* or "hands-off".

Here, most production is organized by firms employing thousands of workers. Events from all around the world – be they economic, political, or natural causes like the weather – impinge on economic life and may cause prices to rise or fall, generating profits or forcing companies to contract, or to go out of business altogether.

## GOVERNMENTS

The market is never simply left to its own devices. Governments also play a crucial role in regulating economic activity and providing a legal framework in which it can operate. In many countries, governments tend to be heavily – and often directly – involved in the production of goods and services. This is a non-market mode of production. Typical areas of economic activity where the government plays a large role are the provision of healthcare and education services, policing, much of the transport system, such as roads and railways, and infrastructure, such as ports and airports. In some cases, the government itself undertakes production; in others it combines with the external market and uses it to obtain what it deems is necessary.

There is a great difference of opinion among experts as to how much can and should be done by governments in order to improve the performance of a nation's economy. Economic booms and slumps appear to be wasteful as well as causing great hardship to individuals. Some efforts at stabilization to avoid large swings in the economy are common in most countries.

**GOVERNMENT POWER**
*Governments play an important role in regulating the economy and framing the laws in which it operates.*

Economic policy, or how an economy is managed and run, operates primarily at the national level. There are some important exceptions. The European Union has a common tariff policy and a common currency for many of its members. International organizations, such as the World Bank, the International Monetary Fund (IMF), and the World Trade Organization (WTO), attempt to develop common approaches to international trade, flow of money, and aid to developing countries. There are many doubts as to who benefits and who loses from these commonly agreed initiatives.

## INVESTMENT

A basic feature of any economy is the way it devotes resources to meeting current needs compared to maintaining or adding to future capacity. In traditional societies, these decisions come down to something like setting aside seeds for future planting. In advanced industrial societies, investment in future production involves savings institutions, financial markets, and the provision of money to both existing and new companies. These provisions include bank loans, bonds, and encouraging shared private and public ownership in a project. The ability to improve future output can, for example, depend on having new equipment and machinery. Investment in research, development, education, and training are vital considerations when increasing productivity.

**INVESTING IN TRANSPORT**
*Government investment is usually required to fund expensive projects, such as trains like this new Japanese high-speed "Maglev" train.*

# HOW ECONOMIES WORK

Economic systems involve groups of people (from individuals through communities to nations) producing or buying the goods they need. Goods may be produced from raw materials and then used directly by the producer, exchanged, or sold. The collective decisions of a community on what they want to buy influence prices, which in turn influence the supply (by producers) and demand (from consumers). Economies, therefore, are often led by the supply of raw materials that can be manufactured into finished items, people to produce things, and industries, such as banking, to service them. There are many different types and sizes of economy.

Fact

### Finite resources

According to the United Nation's Food and Agriculture Organization, more than 70 per cent of the world's fisheries are depleted or almost depleted. At the same time, more and more people depend on these fish stocks for their food and their livelihoods. Competition over access to fishing grounds has already produced confrontation, even violence in the case of Iceland and the UK in the 1970s, as countries that are dependant on fishing try to keep out larger nations anxious to exploit all available stocks.

## RAW MATERIALS

Timber, coal, oil, and other raw materials are harvested or mined and then refined and transformed into useful end-products. Recently, the emphasis on digital technology in information and other services and the transfer of much manufacturing, from shoes to steel, away from Western industrial economies to emerging economies, seem to be reducing the role of raw materials in economic life. Yet the ever-rising demand for reliable energy sources and lack of decline in consumer interest in material goods suggest that the role of raw materials is as important as ever. Some raw materials are reproducible and can be replenished. Others, like iron, occur in fixed amounts, with some uncertainty as to the size of remaining stocks. In spite of a growing world population and higher income levels, many raw materials have fallen in relative price, partly as a result of the invention of substitutes such as plastics. Forecasts of economic decline due to world exhaustion of some raw materials have proved to be too pessimistic, so far. However, projections on the likely availability of fuel for cooking in low-income countries and above all, on potential water shortages, are a serious source of concern.

**TEA CROP**
*The world demand for tea can easily be met by increasing production in East Africa and the Indian subcontinent, although there are environmental implications in such an increase.*

**COPPER MINING**
*Copper is one of many minerals with finite resources, although new deposits are regularly discovered.*

**DRILLING FOR OIL**
*As the world's demand for oil continues to rise, new reserves are eagerly exploited.*

SOCIETY

# PRODUCING GOODS

The first half of the 20th century was marked by great changes in transport and communications, with many new inventions and developments. In contrast, the last 50 years have seen more change in the way things are produced rather than what is produced. Computer control has made possible industrial robots able to manufacture goods, while many new processes, from quality control to automated fabrication, have reduced labour costs in many industries. The ability to switch specifications more easily has resulted in a much greater variety of goods that can be purchased at individual outlets. With the notable exception of the digital economy, exemplified by personal computers and mobile phones, product improvements are now more common than new products.

**SARI PRODUCTION**
*Even the production of the traditional Indian sari has been transformed by new technology.*

**MORE CARS**
*New computer technology has transformed car production in recent years, delivering far higher specifications at a cheaper price.*

**CALL CENTRES**
*Consumer helplines are sometimes based in locations far from the consumer, such as this call centre in Delhi, India.*

# SUPPLYING SERVICES

Although manufacturing has been the basis of many economies in the industrialized world for many years, there has been a recent shift towards the supply services. Such growth is most noticeable in high-income economies such as the US and is the result of a combination of labour-saving technical changes in manufacturing, along with the migration of much of manufacturing to emerging and developing countries like Malaysia. There are several main growth areas in the expanding service sector: healthcare and educational services, maintenance (including cleaning), financial and legal services, security, and administration. These newer service industries are joining some well-established ones, such as tourism, which is itself one of the largest economic sectors. Some services must be supplied locally while others can be supplied over great distances, even internationally. Satellite communication opens up the possibility of providing global services, especially those that are related to information and entertainment.

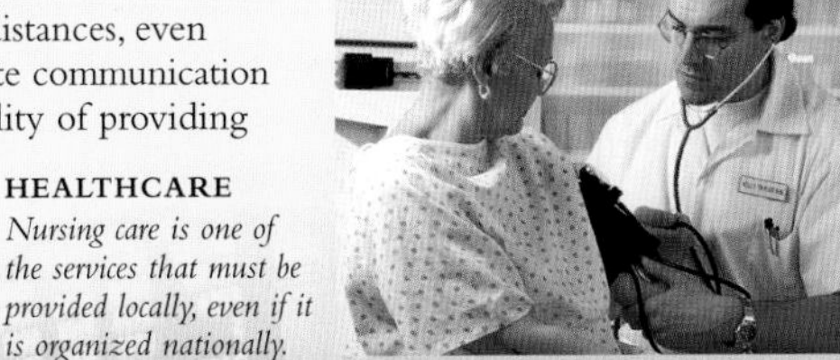

**HEALTHCARE**
*Nursing care is one of the services that must be provided locally, even if it is organized nationally.*

**STOCK EXCHANGE**
*Financial services such as banking, stock dealing, and insurance now operate on an international basis, with dealers regularly trading over many time zones.*

# BANKS AND MONEY

The invention of money was one of the great discoveries of humankind. Complex social organization would not be possible without it. Once a society reaches a certain level of complexity, people begin to use mechanisms of exchange, be they shells or animal teeth, as money. In the past, gold and silver played an important role, but these have now been largely replaced by paper money with no intrinsic value. The institution of money facilitates saving and plays a crucial role in borrowing and lending, making possible giant investment projects as well as small personal loans. In developed industrial societies, central banks control the interest rate at which commercial banks can borrow and the amount of money in circulation. Such control is a major factor in dealing with inflation and in regulating levels of economic activity. In Europe, many countries have adopted a common currency, the Euro. A shared currency encourages international trade and removes pressure to manipulate money supply for political ends.

**BANKING**
*Without the banking system and other services, our modern economy would be unable to function effectively.*

**WOMEN'S CREDIT BANK**
*In developing countries, various local, small-scale systems for encouraging savings and making loans are in place, such as this credit bank in Bangladesh.*

# HOW MARKETS WORK

The formidable task of coordinating thousands of decisions and activities needed to produce goods and services can be done through the market mechanism. In markets, individuals and companies buy and sell a product or a service at prices that are largely driven by the interaction of supply (how much is available) and demand (how many people want it). Prices respond to changes in these two factors and have the effect of making demand equal supply. Markets always operate in a legal and institutional framework. Some are competitive, with many buyers and sellers, while others are under the influence of giant firms with few competitors. Markets are an efficient and a reasonably fair way of organizing certain economic activities. However, they are not effective when it comes to activities such as public services. Many products, such as wheat, coffee, oil, and other raw materials, are subject to various international agreements to control supply with the goal of keeping prices higher or more stable than they would otherwise be. In times of war or other emergencies, many countries abandon markets.

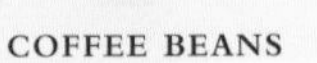

**COFFEE BEANS**
*Coffee prices follow market demand but can fluctuate widely, harming the economies of coffee-producing countries.*

**THE NASDAQ INDEX**
*NASDAQ (the National Association of Security Dealers' Active Quotations) provides a computerized system of stock prices to US brokers.*

**PETROL CRISIS IN ZIMBABWE**
*Shortage of petrol in Zimbabwe in recent years has led to huge price rises and queues of motorists at the petrol pumps.*

History

## Stock markets

The development during the 17th century of joint stock companies, in which individuals invested in return for a share in ownership and profits, led brokers to set themselves up in business trading shares. In London, brokers met in coffee shops until, in 1773, they bought their own building, the predecessor of today's Stock Exchange.

THE ORIGINAL LONDON STOCK EXCHANGE, 18TH CENTURY

# WORLD TRADE

International trade in goods and services is of huge importance to most small national economies. Larger economies, such as the US, have also seen international trade grow in importance in recent years. Falling transport costs, sophisticated consumer preferences (a result of higher incomes), global advertising, and economies of scale have all contributed to this growth. Free-trade groupings of countries such as the European Union (EU) have stimulated trade between certain nations. There are now several major trading blocs and trade agreements, all of which serve to boost trade. Trade in services, such as helplines and call centres, have been aided by satellite communication. By cutting costs and increasing choice, world trade benefits consumers but can potentially hit companies and workers if their markets are lost to producers abroad.

**CONTAINER SHIPS**
*Relatively low transport costs have helped stimulate trade between nations.*

**TRADE BLOCS**
*There are six trading blocs with agreements to encourage trade within them. The countries above the line control 70 per cent of world trade.*

KEY

- European Union
- Mercosur
- Association of South-East Asian Nations
- Southern African Development Community
- Economic Community of West African States
- North American Free Trade Agreement

# WEALTH DISTRIBUTION

Even with the existence of a large and prosperous middle class, great inequality exists between the richest and poorest members of society in most industrial economies, as shown by indicators such as infant death rates and life expectancy. Equally great are the differences in wealth between the richest and poorest countries. For example, the average person in Mozambique earns one six-hundredth of the average person in Switzerland. How wealthy a country is depends on natural resources, industrial strength, population size, and political stability. About a quarter of the world's people live in "absolute poverty" – they have little or no food to eat, are often homeless, and are poorly clothed. Most of these people live in southern and eastern Asia and sub-Saharan Africa. Often, there are also great disparities in wealth between individuals within a nation. Most societies make some effort to mitigate the worst effects of poverty through income taxation and welfare provision (*see* p278). These measures often have only limited success, however. The reality is that we live in a world of staggering levels of economic inequality, with luck playing a central role in who gets what.

**BANANA CROP**
*Countries relying on a single crop such as bananas or coffee are often very poor.*

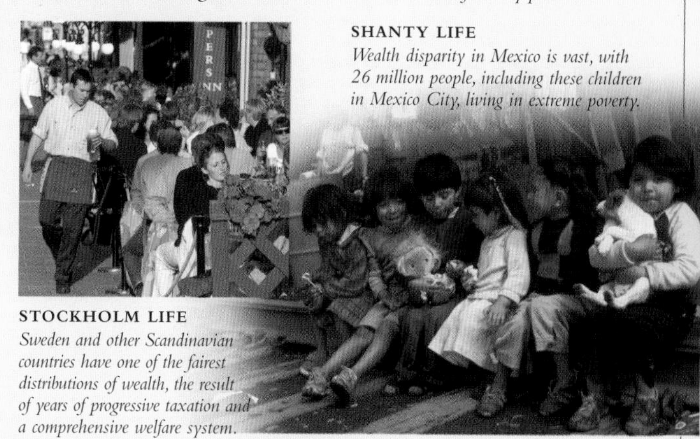

**SHANTY LIFE**
*Wealth disparity in Mexico is vast, with 26 million people, including these children in Mexico City, living in extreme poverty.*

**STOCKHOLM LIFE**
*Sweden and other Scandinavian countries have one of the fairest distributions of wealth, the result of years of progressive taxation and a comprehensive welfare system.*

**NOMAD**
*Operating one of the simplest forms of economy, nomadic herders provide only for themselves, with little need to trade with their neighbours.*

**HONG KONG**
*Economic growth in Hong Kong has relied heavily on its strong textile industry. Shoppers in this "special administrative region" of China have plenty of choice in the large shopping malls.*

# TYPES OF ECONOMY

Economies range from the simplest subsistence-farming and nomadic herding communities through the communal kibbutz and cooperative to the modern capitalist and socialist states. All national economies have some private market activity alongside some government-controlled activity. Political ideology tends to exaggerate the differences between the various economies: the claim of some Asian economies to be socialist or communist has more to do with politics than with economic planning. The major contrasts today therefore are not between market and socialist economies but between a largely self-sufficient economic life in developing countries and a more specialized economic activity that depends on extensive use of the market in the more developed economies.

Fact

## Aid and debt

Many countries in sub-Saharan Africa rely on aid from developed nations to survive. Mozambique, for example, one of the world's most aid-dependent nations, has been ravaged by decades of civil war, floods, and famine. Aid accounts for 60 per cent of national earnings and pays for food to feed 7 million of its 18 million population. But such levels of aid create huge debts, which have to be repaid at some point in the future.

WAITING FOR FOOD, MOZAMBIQUE

TRADITIONAL ECONOMIES

# Agricultural subsistence

**In brief:** Self-contained agricultural communities live off their own land, practising various types of farming

Subsistence agricultural communities live off the land. They prepare the land, plant the seeds, and work to see their investment of seed and labour come to fruition with each crop. In these communities, farmers usually produce only enough to support themselves and their families; any surplus is sold or exchanged in local markets or with neighbours, or stored for later. The markets are usually well developed in agricultural societies, although such markets suffer when harvests are poor.

Farming techniques and practices vary greatly among these communities. In India, for example, most farmers prepare the land with an ox-drawn plough, whereas in Africa it is more common to use hand-held hoes. Some communities plant just one or two crops, while others prefer a variety of grains, fruit, and vegetables. Where the land is poor, people use manure or fertilizer to improve productivity. Likewise, where water is scarce, the farmers devise irrigation systems.

Despite the variation in techniques, most agricultural societies share similar social and economic systems. Agricultural communities tend to be fairly large because food supply is generally reliable and sustained, and therefore is conducive to promote population growth.

Any surplus produced may be passed up the social hierarchy to finance a non-productive elite of chiefs, kings, or priests. Their role is to take care of the spiritual or magical elements of production, by praying for rain or appeasing the gods. Today, subsistence agriculture is practised in much of sub-Saharan Africa and southern and eastern Asia, with pockets in Europe and Central and South America.

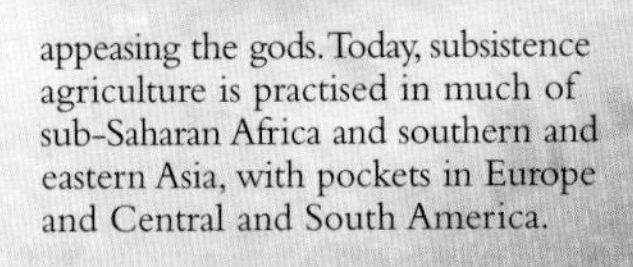

**BRINGING IN THE HARVEST**
*After harvesting, workers in the Yunnan province of China have to carry the crop back to the farm and prepare it for consumption, storage, or sale. There are no machines available to lessen the workload.*

**WOMEN'S WORK**
*In Rwanda, women work on the land in order to provide enough food for their families to eat. Children who are too young to be left alone are carried on their mothers' backs.*

**OXEN POWER**
*Few subsistence communities have access to tractors or mechanical aids. As a result, farmers such as these in Tamil Nadu, India, rely on oxen to carry out the heavy chores.*

Issue

## Genetically modified?

The success of genetically modified (GM) cotton in the US, China, and Argentina has led some farmers in sub-Saharan Africa to investigate the use of GM crops in their fields to increase yields and profits. Although some experts believe GM crops could help alleviate poverty, others are sceptical. The critics of this theory believe that by switching to GM crops, farmers will become trapped in a cycle of dependence on the companies producing GM seeds.

**BALINESE RICE TERRACES**
*People living in hilly areas often make terraces so that they can farm the slopes. These rice terraces in Bali are kept moist year round by rainfall and irrigation.*

TRADITIONAL ECONOMIES

## Hunting and gathering

**In brief:** In an economy based on hunting and gathering, also called foraging, food is acquired through hunting animals and gathering edible products

Hunting and gathering is the oldest economic system on the planet – as old as the history of humankind on Earth. Our earliest ancestors were hunters and gatherers (*see* p28) and remained so until farming began with the domestication of cereals and animals about 10,000 years ago. Today, hunting and gathering is carried out by a small number of peoples living in some of the most marginalized and inhospitable parts of the world.

Hunting takes many different forms, and techniques also vary widely: the San Bushmen of the Kalahari desert in southern Africa hunt with bows and arrows, the Mbuti Pygmies of the Ituri rainforest of the Democratic Republic of Congo hunt with nets, and the Inuit of Alaska hunt whales from boats with harpoons. In most cases, it is the men who hunt, while the women gather fruits, vegetables, nuts, and other edible products. The exception to this division is net hunting, in which women have the important role of beating the bush and chasing animals into the net.

Most of these societies share a number of common traits. They are almost all nomadic, moving from one place to another. They are also almost all very simple societies with little hierarchy or social organization. People who live by hunting and gathering only produce enough food for themselves. There is little surplus, and their economic system does not offer opportunities to use the surplus, such as to exchange it for status or more wives. Instead, for many, the economic system is built around sharing. There is little concept of personal property; most items must be shared with other members of their group. Hunters and gatherers thus form some of the most egalitarian societies on Earth.

**GATHERING THE FOOD**
*The forest is a source of food for Aboriginal Australians. It is women who do the gathering while the men hunt for game.*

**SEA HUNTING**
*Aboriginal Australian hunters use pointed spears to hunt the dugong, a large mammal that lives in shallow, tropical waters.*

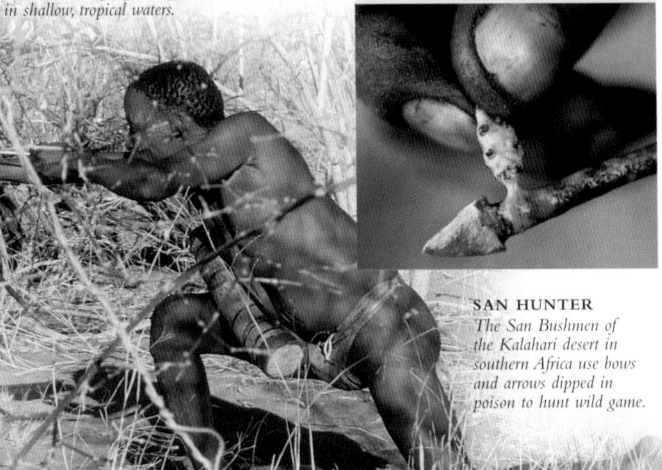

Fact

### Hunters' poison arrows

The San Bushmen of the Kalahari desert use poison extracted from beetle larvae to help kill their prey. The hunter collects the larvae and squeezes the liquid from them, smearing it on the shaft of the arrow, just below its tip. One arrow may require the poison of 10 larvae. Once the hunter has fired the arrow and injured the animal, it may take several hours or even days for the poison to take effect and kill it.

**SAN HUNTER**
*The San Bushmen of the Kalahari desert in southern Africa use bows and arrows dipped in poison to hunt wild game.*

TRADITIONAL ECONOMIES

## Nomadic herding

**In brief:** Herders use their animals for personal consumption or exchange them for other goods

Nomadic herders live off their animals. Their herds of beasts – generally cattle in sub-Saharan Africa; camels in North Africa and western Asia; or reindeer (caribou) in northern Europe, northern Asia, and North America – provide them with most of their food and drink in the form of meat and milk.

Herders use the animal hides to make clothes, shoes, and other items. Sometimes the hides are stretched over a simple wooden frame to make the herders' homes. The animals' bones and horns are used as tools or implements, and their urine may also be used medicinally, usually as an anaesthetic (painkiller).

Livestock, whether it is sheep, goats, camels, or cattle, need water and fresh vegetation. The need for water and grass necessitates a semi-nomadic existence for herders. Consequently, societies whose economies are based on herding tend to cover a large geographic area. Depending on their environment, many nomadic herders also grow small amounts of grain, fruit, and vegetables, or they may gain such products through trade or barter with their agricultural neighbours.

In herding societies, the principal measure of wealth is the animal that is herded, whether it is a camel, cattle, or reindeer. East African cattle-herders, such as the Masai, often say that the cow is like their bank account and that any calves are the interest on their investment. Their economic system is based on exchange. Cattle can be exchanged for virtually anything in a herding society, from grazing rights and grain to bullets and brides. For example, a Masai man may pay at least 10 cattle for a wife.

Herding communities use knowledge and skills that have been developed over thousands of years. But many herders now also make use of modern veterinary knowledge and medicines and use this to increase the productivity of herds.

Most herding societies produce only small amounts of surplus produce. For this reason, such societies have a very low degree of social hierarchy. They are usually egalitarian or have a simple age-grade system (*see* p260), in which role and status is defined by age.

**BEDOUIN NOMADS**
*For the Bedouin of the Sahara and Arabian deserts, camels are used for transport, their hides for shelter, and their meat for food.*

**REINDEER HERDING**
*The nomadic herders of North America and northern Europe and Asia herd reindeer (caribou), using them as transport to pull their sledges through the snowy landscape.*

TRADITIONAL ECONOMIES

## Horticultural subsistence

**In brief:** Small-scale subsistence farming communities may use "slash and burn" techniques

Horticultural subsistence production involves growing crops in small plots of land. In most places, the location of this land changes every few years as the village moves to a new area or expands in a new direction. The communities also hunt or rear a few animals and gather wild plants.

Horticultural communities are common in Papua New Guinea, the Amazon, and central Africa. Many communities live in forests and practise "slash and burn" methods, in which they cut and burn down small patches of forest to clear land to grow crops. When the land loses its fertility, they move on to another area.

People usually live in villages made up of close blood relations. Their economic systems are based on gift exchange. In Papua New Guinea, people create ties of indebtedness by giving gifts, as it is necessary to return another gift to the giver, often with interest. Men who can manage social relations and judge the right amount of gifts to give to the right people can become "big men" and assert leadership in the community.

**WEEDING CORN**
*Some subsistence farmers, such as the Txukahamai of Brazil, grow sweet potatoes and other crops in small forest clearings.*

**CLEARING LAND**
*Some horticulturists practise "slash and burn" techniques to clear shrubs and trees in order to grow crops.*

ECONOMIC COMMUNITIES

## Kibbutzim

**In brief:** A kibbutz is a type of Israeli community based on mutual aid, equality, and cooperation

A kibbutz (literally a "communal settlement") is a rural community unique to Israel. The first kibbutzim were founded at the start of the 20th century by Jewish pioneers from Europe. Today, there are some 270 kibbutzim with a total population of over 130,000 people. Most kibbutzim focus on agriculture, but increasingly there is a trend towards small industry, including processed foods, metalwork, and plastics. They account for some 33 per cent of Israel's agricultural production and nearly 7 per cent of manufactured goods production.

The kibbutzim were founded on the basic principles of mutual aid, equality, and cooperation in production, consumption, and education.

Members are allocated work in one of a number of areas, with no work carrying greater status than any other. Many people work in helping with the running of the kibbutz itself, doing the cooking, cleaning, and administration that is necessary to run such a community. Whatever their work, all kibbutz members are provided with free accommodation and food, as well as childcare, schooling, and social activities.

**EQUAL DRIVERS**
*Because kibbutzim were founded on the idea of equality, very few jobs are considered to be exclusively for men or women.*

ECONOMIC COMMUNITIES

## Craftwork communities

**In brief:** Certain communities have particular craftwork skills that can be traded

Some communities do craftwork to make the clothing and utensils they need for daily life. Some of the most common crafts are basketry, metalworking, pottery, tanning, weaving, and woodwork.

In many communities, individuals carry out craftwork alongside their other productive activities. In some societies, craftwork has become a specialized type of work carried out by a distinct group of people. Such groups are organized into castes, according to occupation. The most obvious example is India, where certain castes carry out craftwork while higher castes engage in agriculture (*see* p262).

Similar arrangements are found in other parts of the world. For example, among the Dogon people of West Africa, metalwork is carried out by a separate caste of people that mainly keeps itself apart from farmers.

However separate craftworking groups might appear, they are always linked, economically and socially, to an agricultural or herding society with whom they exchange their products for food. Today, most of these relationships are based purely on buying and selling products.

**DOGON CARVING**
*Dogon blacksmiths from West Africa produce craft items, which they trade within their community.*

**SILK WEAVER**
*In small communities in Uzbekistan, silk weavers produce exotic fabrics, which they trade with the wider community.*

ECONOMIC COMMUNITIES

## Communes

**In brief:** Modern communes exist on the basis of shared principles, values, and ideals

Modern communes have been set up by groups of like-minded people throughout Europe, North America, and Australasia. All of them have in common the principles of communal living and various degrees of shared ownership of production and consumption. Many also share an idealism for democracy, social justice, and personal self-expression and self-exploration, and promote a permissive lifestyle and that is socially and ecologically sustainable.

Beyond such similarities, there is a wide degree of variation among individual communes. Some are organized like small villages in rural areas, while others are based in converted houses in the middle of cities. Some have a particular religious orientation, such as the Christian Bruderhof communities or the many Buddhist communes, while others are more influenced by New Age ideas or are firmly secular. Some insist that all resources are pooled equally, while others allow a degree of private ownership and differentiation. In some communes, the majority of the people work on and for the commune, for example in farming, manufacturing, education, or tourism. In other communes, most people have regular jobs outside the commune but focus on the shared living experience they create within it.

Most of the people who live in these modern communes have made an intentional choice to do so. They see these communities as an antidote to what they perceive to be the overly materialistic and individualistic way of life in Western society.

**BRUDERHOF SHOP ASSEMBLY**
*Some communes have a religious orientation, such as the Christian Bruderhof communities found in several parts of the world, including the US, Germany, and Australia.*

**FINDHORN**
*The Findhorn community in northern Scotland is home to about 200 people who are committed to an ecologically sustainable way of living.*

ECONOMIC GROUPS

## Cooperatives

**In brief:** These collectively-owned organizations benefit people by pooling resources and buying power

A cooperative is an organization that is owned and controlled by the people who use its products, services, or supplies. Cooperatives have been set up all over the world – in both developed and developing countries – and have focused on a wide range of activities, from farming and craftwork to housing and insurance. Many socialist economies have encouraged cooperatives in the past.

Most cooperatives are set up by people living close to each other. Cooperatives often have a strong community element to them, providing a social as well as an economic focus. The key principles behind cooperatives are self-reliance and working together to achieve increased buying power and economies of scale. A craftworkers' marketing cooperative may consist of a number of individual craftworkers who pack, process, and sell their product together, reducing costs and improving their power in the market. A farmers' supply cooperative may consist of a number of farmers who buy fertilizers, seeds, and other materials and share the savings among them.

**WOMEN'S BREAD-MAKING COOPERATIVE**
*By sharing the costs of flour and other ingredients, and using a communal oven, this women's cooperative in Gambia, West Africa has reduced the price of bread substantially.*

**FARM WOMEN'S COOPERATIVE**
*This fruit and vegetable stall sells the proceeds of a women's farming cooperative in Maryland, US. The cooperative provides a marketplace for produce.*

**COFFEE COOPERATIVE**
*These coffee workers in Costa Rica work as a cooperative, sharing both the costs and labour involved in producing the beans in order to reduce prices.*

ECONOMIC SYSTEMS

## Socialist economies

**In brief:** A socialist economic system is based on state ownership and control of the economy; few socialist economies remain worldwide

Socialist (also called communist) economies are based on a belief in the common ownership of industry and property in order to achieve a fairer and more equal society. To that end, a strong, centralized state must be created to "command and control" every level and every aspect of the economy, from supply through production to distribution and sale. Such a state, however, requires tight political control over its people, reducing liberty in favour of the common good. The world's first communist state was set up in Russia in 1917. Rapid industrialization and the collectivization (state control) of agriculture transformed its economy.

**KIM II SUNG**
*The dictatorial leader of North Korea, Kim Il Sung, kept his country isolated in order to preserve its socialist purity.*

The bureaucracy and planning required were both hugely wasteful, however, leading to the social and economic collapse of the communist system in 1991. Today, only China, Cuba, and North Korea are socialist economies. Since 1978, China has introduced substantial market-economic reforms while retaining state control over society in order to reduce poverty and produce a higher standard of living for its 1.3 billion people. The economy is now open to foreign trade and investment, and private enterprise is encouraged.

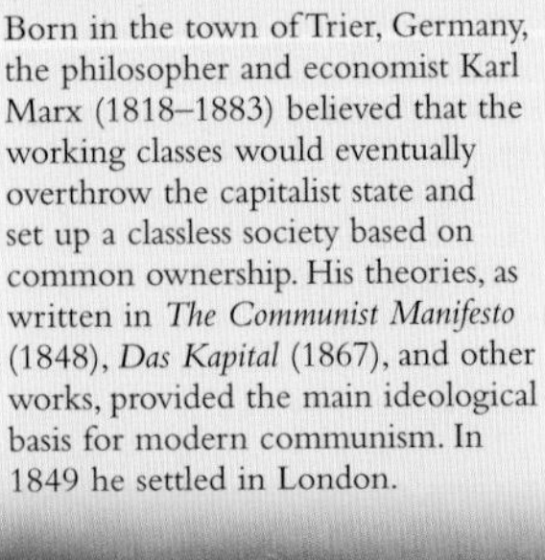

**CUBAN CARS**
*Imports of US cars, such as these Fords and Oldsmobiles, stopped when Cuba became a communist state in 1959.*

Profile

### Karl Marx

Born in the town of Trier, Germany, the philosopher and economist Karl Marx (1818–1883) believed that the working classes would eventually overthrow the capitalist state and set up a classless society based on common ownership. His theories, as written in *The Communist Manifesto* (1848), *Das Kapital* (1867), and other works, provided the main ideological basis for modern communism. In 1849 he settled in London.

ECONOMIC SYSTEMS

## Developing economies

**In brief:** Some poor countries have had limited economic development; factors such as low productivity hamper their effectiveness

Over the past century, only a few countries have managed to make themselves rich enough to challenge the economic dominance of the established industrial economies of western Europe and North America. The rest have suffered from low productivity caused by a lack of investment, limited technology, and primitive infrastructure. These factors prevent them from being competitive in either domestic or international markets. It is a huge challenge for developing countries to generate the surpluses that are needed to purchase the improved technology and machinery needed to raise production. Among the few that have managed it are the "Tiger economies" of Southeast Asia, as well as a few states in South America, where strong state direction of the economy and targeted investment have achieved rapid industrialization and economic growth in the past 30 years.

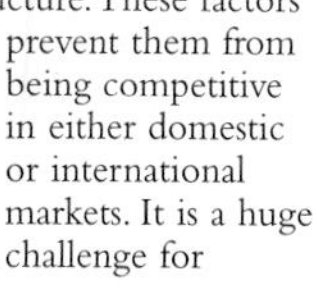

**THE MARKET IN ACTION**
*In recent years, India has opened up its markets to the world in order to reduce its massive poverty.*

**VIETNAM GARMENT FACTORY**
*Vietnam has reformed its socialist economy by encouraging private enterprise, following the same path as China.*

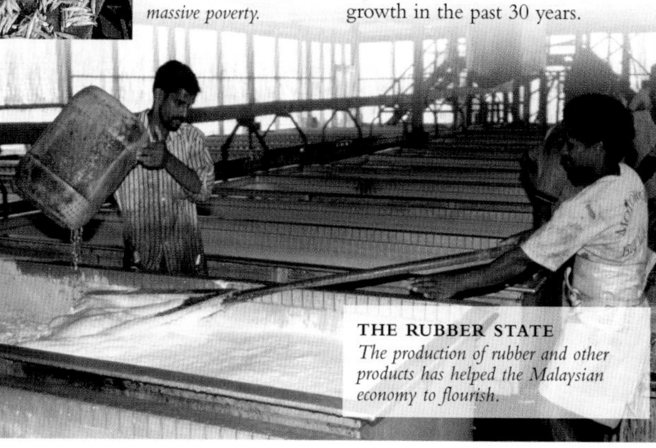

**THE RUBBER STATE**
*The production of rubber and other products has helped the Malaysian economy to flourish.*

ECONOMIC SYSTEMS

## Government-regulated capitalism

**In brief:** Countries such as South Korea operate a market economy but have close state involvement

Most capitalist economies depend on some state activity to make them work, but in a few, the state has played a major role in creating and sustaining market growth. South Korea is a notable example, growing from a poor rural economy 50 years ago to a world-beating, hi-tech economy today. This growth was achieved through a system of close relations between government and business at the expense of working people's rights. Workers were denied the right to strike or to form trade unions, their wages were kept low, and they were forced to work long hours. The state controlled foreign trade, prioritizing the import of raw materials and machinery necessary to promote rapid industrial growth and encouraging exports to raise the money to pay for them.

**SOUTH KOREAN SHOPPERS**
*Although South Korea has a vibrant home market, most of its goods, such as clothing and electronic items, are produced for export.*

**SOUTH KOREAN SHIPYARD**
*South Korea is the world's biggest shipbuilder, launching 45 per cent of all new ships. The country also produces cars and electronic goods.*

ECONOMIC SYSTEMS

## Free-market capitalism

**In brief:** A market-driven economy functions with minimal state regulation and interference

Free-market or "laissez-faire" economies are driven by signals, such as price changes, given by the market. According to one of its main founders, the 18th-century Scottish economist Adam Smith (*see* p245), the only role of the state is to provide the social and legal framework in which the market can function and it should interfere as little as possible. Historically, most free-market economies have in fact relied on the state to support their operations in the face of economic and social problems. Today, the term free-market capitalism is mainly used in order to distinguish the US from the other major economies, such as France, where the state has a major role. As with all capitalist economies, demand drives supply, which determines price.

**CONSUMER CHOICE**
*Free market economies encourage the production of a massive variety of goods, often with many different brands of the same product, such as these trainers.*

**SHOPPING MALL**
*Large shopping centres, such as this mall provide the impetus for consumer demand, vital to any free-market economy.*

Fact

### Advertising

Producers rely on advertising to sell their goods, particularly when consumers are faced with many different brands or labels. As a result, advertising has become a substantial industry in its own right, employing the most creative designers and copywriters to sell their client's goods in a crowded marketplace.

**SOCIETY'S HIERARCHIES**
*Like this statue, society has several layers. A leader has ultimate authority but delegates it to state structures, such as the army. Individuals also play a vital role.*

# SOCIAL STRUCTURES

A basic function of society is to coordinate the activities of its members so that their individual and shared objectives can be achieved. Ever since the first humans walked the earth millions of years ago, we have needed to organize ourselves. At first, society's structures were concerned only with the basic necessities of life: eating, sleeping, and surviving in a hostile world. Yet, as the population grew and humanity began to form communities, more complex social structures were needed in order for society to function efficiently.

Social structures have principally been concerned with maintaining order, minimizing conflict, and giving people an acceptable level of security and predictability in life. They also provide a formal way of passing techniques for the efficient functioning of society down the generations. Certain structures are also put in place to look after members of society who cannot help themselves, such as children, the sick, and elderly people.

## ROLES AND STATUS

In order to avoid conflict, every society requires a broad agreement about who is entitled to exercise authority in a range of situations. In addition, because the benefits of society tend to be unequally distributed, some general acceptance of who is entitled to what is an integral part of any society.

In Europe, ranking has historically been based on "class", which is usually defined by inherited wealth and status. Yet the rigid line between "upper" and "working" class has blurred in recent years. Successful or aspirational people are theoretically able to change class or even ignore it altogether, with effort and ability counting for as much as – if not more than – status and wealth.

This blurring of class boundaries is less true in other societies. In India, for example, the caste system remains important, with social status closely tied to family background.

**PRIVATE ADVANTAGE**
*Education at a top fee-paying school, such as Eton in the UK, traditionally guaranteed a high social status in adult life. In meritocratic societies, this advantage has eroded somewhat.*

In most modern societies, status is to a great extent based on an individual's occupation, although precisely which occupations are valued varies according to the specific society. High status does not always equate with wealth.

## POLITICAL ORGANIZATION

Although human political and power structures can be categorized in various ways, the main, and by far most complex, unit of political organization in the world today is the nation state. Some states consist of people who share a common history, language, religion, and ethnicity. Many others have brought together more diverse groups of people who differ in one or more of these important features. When such a cultural mix occurs, conflict is a likely outcome, with one group often being subjugated by another.

**INDIAN PARLIAMENT**
*Parliamentary democracy is a common type of political structure in the world today. The Indian parliament comprises two houses and all new bills must be passed in both.*

Nation states vary enormously. Most are democracies or aspiring democracies of one form or another, although many countries in the world still run in a variety of other ways – including as monarchies, theocracies, and military dictatorships. Without exception, for any type of power structure to be effective, it must have at least some degree of popular legitimacy. Within a state, there is usually a division between three main branches of government, each of which has a different remit: the legislative (lawmaking) branch, the executive (administrative) branch, and the judicial branch.

A few nation states have enough power to operate on a global scale, dictating economic policy to other states or acting as global policemen. International organizations, such as the UN, also exercise power globally.

## CONTROL

A principal role of any government is to prevent and resolve conflict, a task only achievable if there is an understanding among the people about what is socially acceptable. As well as these informal sets of conventions, governments manage societies by embodying popular consensus in formal laws.

Conflict management is of great importance to modern societies, as is managing dissent. Throughout our history, there have always been individuals who do not agree with some of the "rules" that are imposed by the societies in which they live. Such people generally make their views known through peaceful means of protest, although some may break the law or act in a way that other members of society consider harmful. In most societies, the police and the judiciary are at the forefront of maintaining order.

Agreement about the law and how it is enforced is especially hard to achieve in newly formed or socially divided nations where popular consensus is either weak or nonexistent. Where consensus and coexistence are not possible, conflict can occur. This may be at a local level between noisy neighbours, at a national level between different ethnic groups, or at an international level between warring states.

**IN DEFENCE**
*The main task of a society today is to defend its citizens from attack without having an army that acts oppressively at home.*

History

### Japanese feudalism

From 1192 to 1868, real power in Japan lay not with the emperor and his court but with the military administration of the shogun. The shogunate was a feudal, dynastic government drawn from the leading families of the nation. It held absolute power. Although conservative in outlook, the shoguns ensured peace and economic growth for long periods of time, but at the expense of Japan's isolation from the outside world. During the feudal period, only the noblemen and those who held samurai status (granted as a reward) were able to own land. These ruling classes governed over peasants, artisans, agricultural workers, and merchants.

**PAYING HOMAGE TO THE SHOGUN**

# ROLES AND STATUS

Division of labour and well-defined roles help a community to function as a team and so make the best use of their resources. As early societies became more complex, they became more hierarchical. Today, the concept of rank is common to all modern societies, from the caste system of India to the European class system. Societies have different ways of structuring themsleves. In certain societies, some individuals have considerable choice as to how they will function in the society. Others are more tightly constrained. Differences in role and status are a fact of gender in many societies. Although in many industrialized nations, women aim to match status with men, there is still an underlying discrimination in many women's experience.

Fact

### Nobel Prize

Founded in 1901 by Alfred Nobel, the Swedish inventor of dynamite, the Nobel Prize is the highest accolade a person can win for achievements in science, literature, or politics. The winner receives more than a million US dollars, instant fame, and – perhaps most important of all – a dramatic elevation in social status. Awards such as the Nobel Prize, Oscars, and Olympic medals serve to confirm a person's position at the top of a professional hierarchy.

## SOCIAL HIERARCHIES

All societies have formal and informal ways of ranking people according to power, esteem, or ability. The main divide in European societies until well into the 20th century was between a ruling class, defined by wealth and ownership of land, and a labouring class, whose members lived by earning a wage. Between these was an intermediate and less easily defined middle class. These class divisions were sharply defined, and there was very little chance to move from one class to another. Since the second half of the 20th century, however, the boundaries between classes have blurred, and most industrialized societies have become more meritocratic, meaning that people can climb the social ladder by being successful. In other societies, however, such as Hindu India, where the caste system operates, social status remains closely tied to family background. Small societies tend to have few hierarchies, but large societies can include multiple overlapping systems, and a person's status in one hierarchy may have little bearing on his or her status outside it. For example, a professional football player may be highly regarded in sporting circles but unheard of elsewhere. Hierarchies, such as those based on wealth or political power, cross national boundaries. Others are purely local. Among the Masai of East Africa, for example, the ability to jump high from a standing position increases status.

**UNTOUCHABLE**
*In India's rigid social structure, people of different castes seldom socialize and almost never intermarry. The lowest caste (Untouchables) are consigned to menial occupations, such as cleaning.*

**OLYMPIC GOLD**
*A gold medal is an instant badge of status in sporting circles, but it does not guarantee higher status in other hierarchies based on class or wealth.*

## OPPORTUNITY AND ASPIRATION

In a meritocratic society, the social hierarchy is fluid, and a person's status and wealth should, in theory, depend on effort and ability rather than social background. In practice, no society is purely meritocratic because inequalities of wealth become entrenched and lead to inequalities in the education system. Even so, the concept of equality of opportunity remains a cornerstone of modern industrial societies such as the UK and US, and it is theoretically possible for a person of any gender, class, or ethnic background to rise through the social hierarchy. As well as making society more fluid, meritocratic principles lead to an increase in social aspiration. People who may once have resigned themselves to their fate as housewives or working-class labourers now aspire to climb the ladder and become rich and successful. Social aspirations are an important motivator. This fact is exploited by marketing professionals and advertisers who seek to link their products with lifestyles to which their target audiences aspire.

**EDUCATION**
*Equal educational opportunities are essential to meritocratic ideals. In some countries, up to half the population may attend university.*

**WOMEN'S OPPORTUNITIES**
*Gender divisions have become less important in industrial societies, and many women now work for a living. Even so, the burden of childcare still falls on women, with the result that many have interrupted careers or a poorly paid, part-time job.*

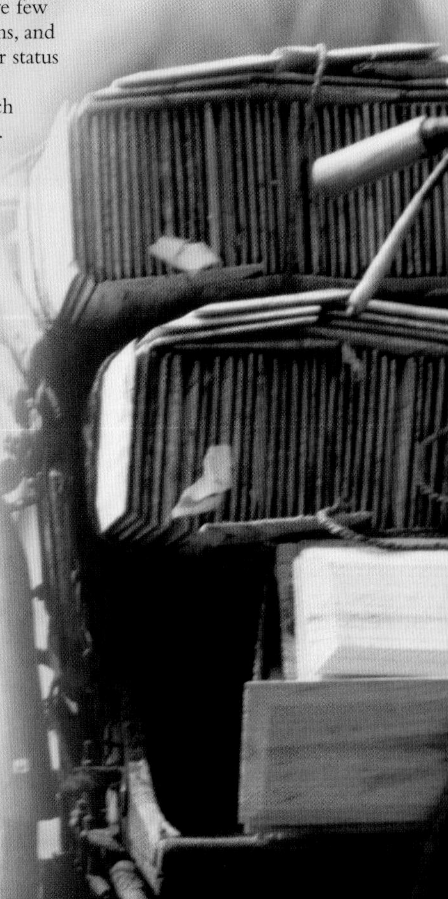

**HERO FOR THE MOMENT**
*High status can come and go in a day. This man's moment of glory came after winning the annual Palio horserace, which is held in Sienna, Italy.*

**CELEBRITY STATUS**
*Thanks to the modern media, film stars and international celebrities have all the status of royalty or political leaders but none of the responsibilities.*

# STATUS

In most societies, the main determinant of status is occupation. High earnings and a position of power in an organization almost always raise status, but the nature of the occupation also matters. In some societies, for instance, teachers and religious leaders are highly respected but have a low income. High status does not necessarily mean a person is liked: many political leaders are at the top of the social hierarchy but privately despised. Likewise, drug barons, Mafia godfathers, and warlords earn their status through fear rather than popularity. Status is not a purely individual attribute; it also attaches to groups. In nearly all societies, membership of a social group – from a teenage clique to an African tribe or the English aristocracy – is a badge of status, often advertised by a code of dress, a way of speaking, and shared group rituals. Status can also attach to places. As every city estate agent knows, house prices in certain neighbourhoods reflect the social cachet of the area as much as the convenience of the location or the specification of the property.

**VILLAGE CHIEF**
*In Yakel village, Vanuatu, the village chief shows his status by wearing a modest feather headdress and sitting centre stage at village meetings. In nonindustrial societies such as this one, elders command a great deal of respect for their wealth of knowledge.*

**TOWN MAYOR**
*Costume and ritual play exactly the same role in displaying status in all societies, from hunter-gatherer communities to modern cities. This Afrikaner mayor in South Africa flaunts his title by taking the best seat at a public ceremony and wearing a flamboyant chain.*

Issue

## Serotonin and status

Some scientists think the feel-good factor that comes from success and high status is due to elevated levels of the brain chemical serotonin, which plays a key role in controlling mood. Primates with a high social rank tend to have a higher serotonin level than their inferiors and are less inclined to impulsive or depressive behaviour. Scientists have even found that raising a monkey's serotonin level with drugs makes the monkey rise in status. Whether the same is true of humans is far from certain.

**APE STATUS**
*The open mouth of this chimpanzee is a gesture of submission towards a higher status chimpanzee.*

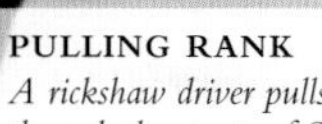

**PULLING RANK**
*A rickshaw driver pulls his passenger through the streets of Shanghai, China. Although China is communist, disparities of wealth and status exist in the cities.*

SOCIAL HIERARCHIES

# Age grades

**In brief:** Some small-scale societies are organized by age; people move through different grades throughout life, and each age grade carries a particular a role

Age-graded social structures are most common in regions with a history of intertribal warfare. Every few years, boys around the age of puberty are initiated into a group, or set, usually through circumcision. As they grow up, they pass through a series of distinct age grades such as youth, warrior, and elder. Each age grade has clearly defined social roles and is taught tribal values and traditions by older members of the community. Members of an age grade form powerful bonds of loyalty, cutting across kin and making the group highly effective as a warrior unit.

The best-known examples of age-graded societies come from cattle-herding communities in East Africa. In the past age grades also existed in the native communities that lived on the North American plains. This system became established after the introduction of horses and rifles by the Spanish, which fuelled a pattern of horse raiding and warfare among different tribes. Age grades known as sodalities formed, which strengthened the tribes' fighting skills, especially during the summer bison hunts. Age grades are still found among the Masai, where cattle raiding is a source of conflict. Masai men pass through a celibate warrior stage, during which future political leaders are elected. At the end of the warrior stage, men become elders and can then marry.

Issue

## Pensions and protests

In age-graded societies, social status tends to rise with age. In industrial societies, by contrast, people may lose status when they retire, causing them to feel undervalued by society, especially if they live apart from their family and depend on state support. The feeling of dissociation from society has led to vocal protests against benefit cuts among older people in some countries.

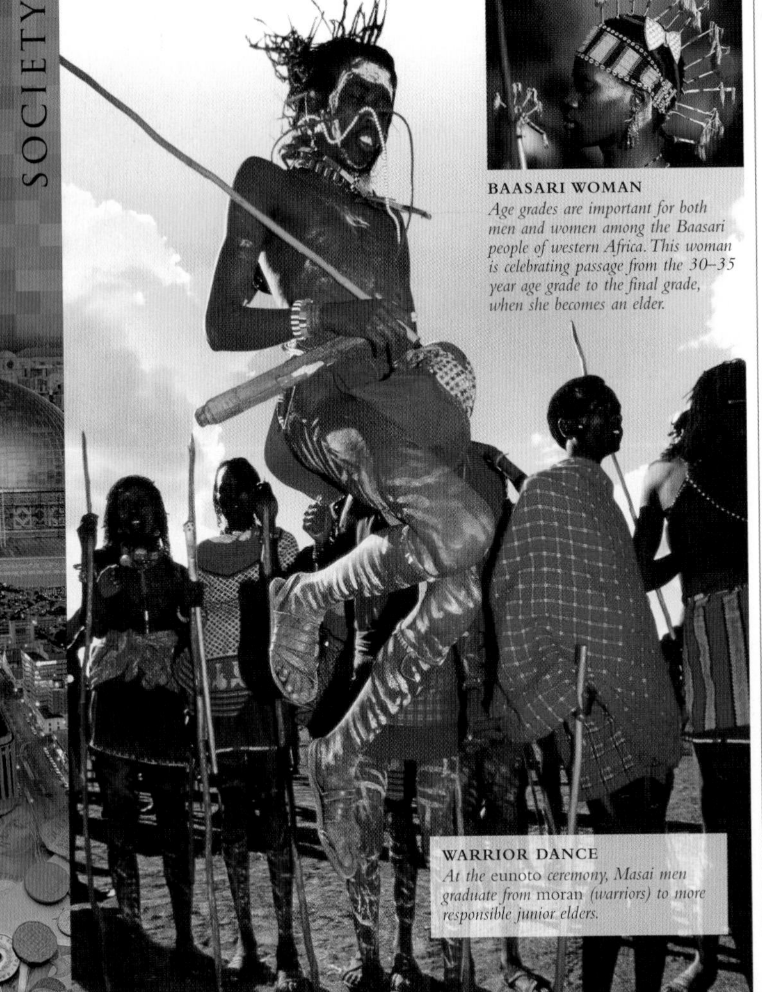

**WARRIOR DANCE**
*At the eunoto ceremony, Masai men graduate from* moran *(warriors) to more responsible junior elders.*

**BAASARI WOMAN**
*Age grades are important for both men and women among the Baasari people of western Africa. This woman is celebrating passage from the 30–35 year age grade to the final grade, when she becomes an elder.*

SOCIAL HIERARCHIES

# Egalitarian

**In brief:** Societies that are egalitarian have little or no hierarchy; people who live in these societies are unlikely to accumulate wealth and power

In egalitarian societies, no person or group has appreciably more wealth, power, or prestige than any other. Differences in certain skills are acknowledged, but while certain individuals may be highly esteemed, they cannot transform their special skills into wealth or power.

Egalitarian societies are mostly found among small, mobile bands of hunter-gatherers (*see* p252), such as the Inuit, the !Kung people of the Kalahari desert in southern Africa, and the Ainu of Japan. Such societies have little or no concept of private property and believe that natural resources are free or common goods.

There are very practical reasons why the member of an egalitarian society do not accumulate wealth or power. First, the nomadic way of life naturally prevents the accumulation of material possessions. Second, foraging for food requires a large area of land that members of the community use collectively, so land ownership by individuals is not possible. And third, survival depends on resources being shared rather than privately owned.

**INUIT HUNTER**
*Hunter-gatherers such as the Inuit see land as a common resource, freely open to all members of the community.*

**!KUNG THATCHERS**
*Women and men have similar roles among the egalitarian !Kung people of southern Africa. These women share the job of thatching a hut.*

SOCIAL HIERARCHIES

# Ethnic groups

**In brief:** People are often defined by ethnic group; in many societies ethnicity also defines status, and there may be discrimination and differences in opportunity

Ethnic group is a broader concept than tribe and a more accurate term than race. It refers to cultural identity that marks out a particular group within a nation or region, such as Kurds in the Middle East or Basques in Spain. The group is distinguished by a distinctive language, religion, history, geography, or culture.

Ethnic diversity is a feature of most modern nations – the legacy of conquest, colonialism, and economic migration. Many states deal with the potential divisions that can occur in a multiethnic nation by encouraging toleration and assimilation. Some states or ethnic groups, however, have attempted to prevent assimilation by enforcing discriminatory laws (such as the apartheid system); by the use of "ethnic cleansing" (as in the expulsion of Muslims from Bosnia); or even by genocide (as in Nazi Germany in the 1940s and Rwanda in the 1990s).

Colonialism exacerbated ethnic tensions as colonial rulers sometimes played ethnic groups against each other, using the old principle of "divide and rule". Following independence, new national boundaries created further problems by paying little regard to the historical distribution of ethnic groups.

**RWANDAN VICTIMS**
*Almost a million Tutsis and moderate Hutus were slaughtered in Rwanda by Hutu extremists in 1994. These Hutus returned to burnt out villages.*

SOCIAL HIERARCHIES

# Gender stratification

**In brief:** Division of labour between the genders and some degree of inequality occur in nearly all societies

Nearly every society is divided to some extent along gender lines, with men and women playing different and often unequal roles. The extent to which gender roles differ varies enormously, from societies where men and women have theoretically equal opportunities to societies where they lead separate lives and take pains to avoid social contact.

The degree of gender stratification is often determined by the political organization, economic arrangement, and type of kinship organization (descent through the male or female line) of a society. Gender status tends to be less stratified among hunter-gatherers (*see* p252) than among agricultural economies. More intensive farming practices reinforced the division between women's private (domestic) work and men's public labour. Because a greater value is placed on the male role, these societies see women as having inferior social and political status. In societies that are matrilineal, a woman's status tends to be higher. This is because membership of a descent-group, land allocation, inheritance, and overall social identity all pass through the female line.

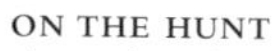

**ON THE HUNT**
*Among the Baka pygmies of Africa, hunting big game is considered a male activity.*

**HOLDING THE BABY**
*Childcare is a largely female task in all societies. The extent to which men contribute varies from one culture to another.*

History

## Votes for women

In the early 20th century, the English feminist Sylvia Pankhurst led a campaign calling for women to be given the vote (suffrage). Pankhurst's followers – the suffragettes – took increasingly desperate measures to get noticed, including chaining themselves to the railings outside 10 Downing Street. Full suffrage was finally granted in 1928, and since then, women's rights have improved profoundly in many countries.

SUFFRAGETTES, JUNE 1911

**AFGHAN WOMEN**
*In Muslim societies, women refrain from contact with men outside the family and dress modestly in public. Devout women may cover the whole body.*

SOCIAL HIERARCHIES

# Class system

**In brief:** Industrial societies were traditionally divided into upper, middle, and working classes

The term class first became widely used in Europe during the 19th century. Society had formerly been divided into feudal landowners and peasants, but the industrial revolution of the 18th century led to a massive growth in towns and the emergence of an urban working class whose members laboured in the booming new manufacturing industries. Much of the land and property remained in the hands of a wealthy elite, whose members led a life of leisure and derived an income from their property without having to work. Between this upper class and the working class was a middle class made up of educated professionals such as doctors, teachers, and lawyers, as well as businessmen and self-employed tradesmen. Class divisions were very sharp and there was little social mobility. A person's class was largely fixed at birth and was obvious in their style of dress, their accent, their occupation, and their lifestyle. The sharp divisions and conspicuous inequalities led to antagonism and partly inspired the Marxist revolution in eastern Europe.

Since the early 20th century, class distinctions have faded in Europe and the US; most of the antagonism between the classes has declined. At the same time, standards of living have risen and social mobility has improved, leading to the evolution of a more meritocratic society.

**HARD LABOUR**
*Manual jobs in industry are still perceived as working class, though salaries of oil-rig labourers now rival those of teachers and doctors.*

**A LIFE OF LEISURE**
*Polo was invented in Asia but spread to upper-class England during colonial rule of India in the 19th century.*

SOCIAL STRUCTURES

# Meritocracies

**In brief:** Social rank and wealth depend on achievement in meritocratic societies

In a meritocracy, wealth and status depend in theory on effort and ability rather than on class. A meritocracy is different from an egalitarian society in that significant inequalities and a social hierarchy do exist. However, the hierarchy is more fluid than in a class-based society and rank is more closely tied to wealth. Meritocratic principles underlie many modern democracies, but a true meritocracy is difficult to achieve without eliminating inherited wealth and educational inequalities.

GRADUATION DAY

SOCIAL HIERARCHIES

# Caste system

**In brief:** In Hindu society, the caste system provides a fixed hierarchy; historically it has determined every stage of an individual's life, from work to marriage

Caste is the main social system that governs much of Hindu India.

Hindus are born into four main castes (also called varnas), which followers believe were established at the time of creation. These varnas include the high-caste Brahman (priests) followed in descending order by Kshatriya (warriors), Vaisya (merchants), and Sudra (peasants).

Hindus believe Brahmans emerged from the mouth of Lord Brahma, who, in their faith, is the supreme being overseeing the universe. The other castes are considered to come from other parts of Lord Brahma's body: Kshatriyas from his arms; Vaisyas from his thighs; and the Sudras from his feet. Below these varnas are Dalits, also called untouchables or Scheduled castes. Each varna has thousands of subcastes (or jatis).

Historically, a person could not convert out of the caste or marry outside of the group into which he or she was born. People who defied these rules were usually ostracized by their community. Caste also determined a person's work. Brahmans were priests and teachers, Kshatriyas served as soldiers, Vaisyas were shopkeepers, and Sudras were farmers. Dalits did degrading and polluting jobs such as street-cleaning or disposed of the carcasses of dead animals.

Food taboos and even conversation were linked to and governed by caste. Brahmans were not permitted to eat food that had been touched or prepared by a person of a lower status. A low-caste person could not speak to someone of high-caste status.

Although caste remains a strong social system in India, restrictions are relaxing, especially in urban areas. The changes are the result of external influences such as television and other media the expansion of the education system. Discrimination was outlawed in 1989. An example of how things are changing is in employment: there are now teachers who are Dalits and farmers who are Brahmans.

**WARRIOR CASTE**
*Kshatriya (warrior caste) men in India historically served as soldiers, or (as right) as a palace guard.*

**BRAHMAN**
*The vermillion mark on the man's forehead (left) denotes a high-caste Brahman. He may be a teacher or a priest.*

**ARRANGED MARRIAGE**
*Hindus have been expected to marry within their caste. Love matches were rare: marriage was usually arranged by the couple's families.*

Fact

## Victims of crime

Dalits ("untouchables") are often the victims of crime: in 2000 more than 25,000 crimes against Dalits were reported. Many more incidents are probably not reported out of fear of reprisals. Tensions between Dalits and people from higher castes usually result from a perception that Dalits have violated caste system rules, such as marrying or working outside the caste boundaries. Dalits have been paraded naked through villages; lynched, hung, or beaten to death; Dalit women have been raped.

**CONTRASTING STATUS**
*High-caste Hindus are likely to have wealth and be generally respected. Dalits, however, are often reduced to begging on the streets.*

**DALIT LABOURERS**
*In India, there are about 160 million Dalits ("untouchables"). They often do the most menial jobs. These Dalit women unload bricks from a kiln for very low pay.*

# POWER

In politics, a government is a group that controls a nation or another political entity. Historically, the members of such a ruling elite relied on armies to keep themselves in a position of authority and in turn used this authority to gain the wealth they needed to fund those armies. Today, however, a greater variety of power structures exists. Some societies are led by tribal councils, others by multiparty democracies (governments elected by the people). There are also regimes such as military dictatorships, in which the population has no say over who is in control, and opportunities for influencing or changing the power structure are extremely limited.

## SCALE OF POWER

Human political organization can be divided into bands (the least complex), tribes, chiefdoms, and states. Bands are small, nomadic, kin-based groups that are found among foraging (hunter-gatherer) communities. Tribes are generally associated with small-scale food production based around a fixed area. The members of this type of group live in semipermanent villages and share a common descent and cultural characteristics. Both bands and tribes rely on collective agreement to make decisions. Chiefdoms, although also kin-based, are characterized by a powerful central leadership, a permanent political structure, and an emerging social stratification.

**FAMILY TIES**
*The San people of the Kalahari desert live in family bands, the smallest type of unit recognized in society.*

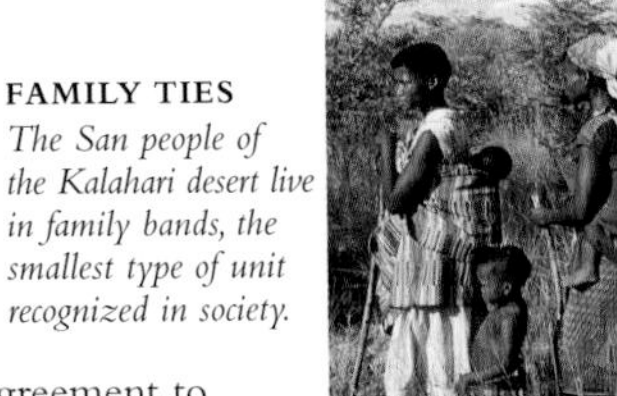

The largest political unit is the nation state. This type of organization does not rely on common descent or kin relations for the allegiance of its members; instead, a state encompasses many communities under a centralized government. The government rules and organizes all the people over whom it has authority, some of whom may belong to social classes that themselves have power. A state's leaders have the authority to make and enforce laws, draft people for war, and collect taxes with which to pay for their actions.

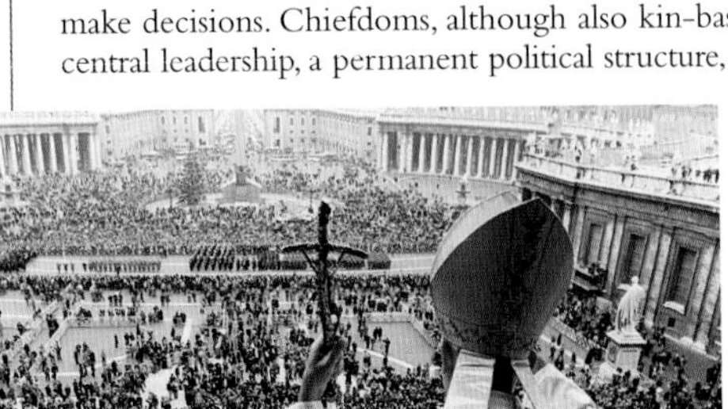

**RELIGIOUS LEADER**
*The Pope leads prayers in the Vatican City. Throughout history, many societies have been organized around a religion or a religious leader.*

History

### Spartacus's revolt

Republican Rome depended on slaves to provide labour; in fact, there were more slaves than there were Roman citizens. In 73BC, a man named Spartacus, who had been sold into slavery, escaped with 80 others. More slaves joined him, attracted by his goal of gaining freedom. They eventually numbered more than 70,000. After 3 years of conflict and a huge loss of life, the revolt was crushed and the authority of the Roman elite was restored. Although it had not been his primary aim, Spartacus's actions forced the Romans to change how military power was allocated, which had far-reaching political effects.

## THE NATION STATE

Political scientists generally describe a "nation" as a group of people who share aspects of language, culture and/or ethnicity, while "nation state" refers to a political structure in which a national group exercises complete authority within its borders. A defining feature of nation states, as opposed to feudal systems or empires, is the separation of economic and political life. Government systems within nation states may differ widely in their institutional arrangements, but most make a distinction between the legislative (lawmaking), executive (administrative), and judicial branches. Although most states today claim to be nation states, many are made up of lots of nations, of which no single one is dominant. In some, including Canada, the US, India, Eritrea, and China, efforts have been made to create a unique national identity that encompasses the different groups within the state's boundaries. In some states, such as the former Yugoslavia and many countries within Africa, divergent nations have been forced under the umbrella of a single state, sometimes with disastrous consequences. In almost all cases, the different nations within such a state compete for dominance.

**REFUGEES ON THE MOVE**
*In 1947, the nation states of India and Pakistan were formed and the millions who found they were on the "wrong" side of the new border had to move.*

**JUDICIAL BRANCH**
*In France, as in most nation states, the legal system and the judges sit separate from economic and political structures.*

**LAW ENFORCEMENT**
*The police and military have the power to maintain order in a nation state. Here, Libyan police academy graduates prepare to enforce the law.*

Issue

### Triumph of democracy?

When the Cold War ended, some said that liberal market democracy had triumphed and that the ideological struggle between left and right was over. Many Westerners predicted that there was no alternative but for liberal democracy to sweep the globe. Others felt this view ignored the possible expansion of non-Judaeo-Christian systems, such as those based on Islam. As a system, liberal democracy is less contested today than it has been for over 100 years, but the alternative systems will not simply disappear.

SOCIETY

# GLOBAL POWER

The ways in which nations handle relations with other nations is not rigid, but reflects how they balance their own interests with those of others. Some states are able, by virtue of their economic or military influence, to exercise power on a global level. States with great economic or military power can often force other nations to adopt certain national or international policies as the price of doing business. Global power is also exercised through international institutions. The largest international organization is the United Nations, which all countries may join. Other bodies, such as the African Union and the Organization of American States, are open to a limited regional membership. Some groups base membership on other criteria. For example, the British Commonwealth is open only to states that were once part of the British empire. The G8 ("Group of Eight", comprising the US, Germany, France, Russia, the UK, Japan, Italy, and Canada) and OPEC (the Organization of Petroleum Exporting Countries) are groups based on economic cooperation. Their member countries meet to establish treaties and set trading standards and practices. Structures such as NATO (North Atlantic Treaty Organization) are founded on shared military power. The end of the Cold War (*see* Alternatives to capitalism?, p44) led many international agencies to refocus, establishing a new relevance to the new era.

**CHANGING RELATIONSHIPS**
*Presidents George W Bush and Vladimir Putin would once have been on opposite sides of the Cold War, but global politics are constantly changing.*

**MILITARY POWER**
*The economic strength and military might of the US government enables it to influence the actions of nations and organizations worldwide.*

**KEEPING THE PEACE**
*The deployment of UN forces provides a neutral presence that can keep warring groups apart for as long as is necessary to negotiate a peace agreement.*

**POWER RESTS IN FEW HANDS**
*Leaders of the eight "G8" nations, plus some other invitees, meet annually. In an informal forum, they work out a common approach to economic, trade, and other global issues.*

**UNITED NATIONS**
*The flags of member nations fly at the UN's New York headquarters. Most independent states belong to the UN and participate in its decision-making.*

Fact

## Role of NGOs

In November 1999, Médecins Sans Frontières, the international medical aid agency, became the first ever nongovernmental organization (NGO) to win the Nobel Peace Prize. The award signalled a period of prominence for NGOs, whose structure and organization enables them to take actions independent of state sovereignty. Although this means that NGOs are free to criticize human rights violations and to provide independent aid relief, it can also make it difficult to hold them accountable for their actions.

COMMUNITY STRUCTURES

## Bands

**In brief:** The simplest political structure is the band, a small, mobile, kin-based group of hunter-gatherers

Bands are the oldest and least complex form of political organization. They contain just 30–50 members and are based around a subsistence economy of foraging (hunting and gathering). The men usually fish and hunt; the women gather roots and berries. Social relations are characterized by cooperation, and food is shared among members.

Band-based groups live in a variety of environments and include Australian Aborigines, Kalahari San, Arctic Inuit, and Central African Mbuti "Pygmies". In foraging societies, only two social structures are significant: the nuclear family and the band. Band membership is flexible (members can join the band of either parent) and seasonal (band size changes according to food availability).

**COMMUNITY LIFE**
*An Inuit band comes together and pools its resources in order to feed everyone at a community gathering.*

Band leadership is informal and is usually provided by older men, who are respected for their wisdom, skills, and experience. The headman of a band has little official power and can only try to persuade band members to follow his advice. Political, social, and religious decisions are often indistinguishable within bands.

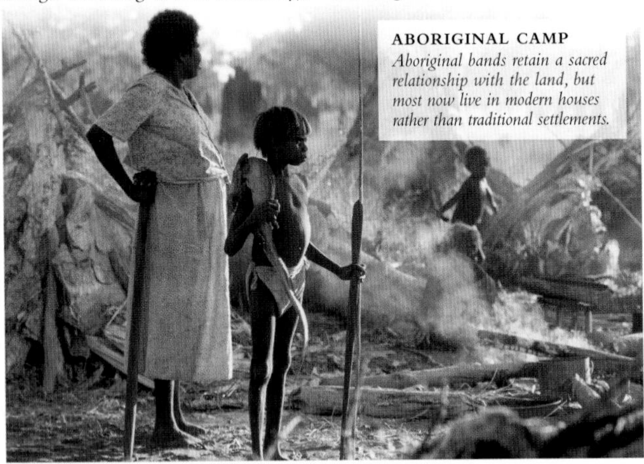

**ABORIGINAL CAMP**
*Aboriginal bands retain a sacred relationship with the land, but most now live in modern houses rather than traditional settlements.*

COMMUNITY STRUCTURES

## Tribes

**In brief:** A tribe is a crop-growing or animal-herding group that is village-based and part of a network

Tribes are food producers, rather than food collectors like bands. They live in village-based, self-ruling groups, each of which is descended from a common ancestor. Tribal societies range in size from the Amazonian Yanomani, with 20,000 members, to the North American Navajo, with 160,000 members, and the Dinka of the Sudan, with 500,000 members.

Tribes are most commonly bound by a branching mechanism known as "segmentary lineage organization" (SLO). This acts as a temporary form of regional political organization, uniting tribal people in a series of larger networks. These networks can be village-based or may be spread over a far wider geographical area, with only a distant ancestor in common.

The size of the SLO depends on economic, religious, or military needs. The closer the hereditary relationship between the people in a single SLO, the greater the mutual support; the more distant the common ancestor, the greater the chance of hostility. The pastoral, or animal-rearing, Nuer of southern Sudan are an example of this type of tribal structure. When necessary, they can swiftly organize themselves to protect or expand their territory, often at the expense of their neighbours, the Dinka. Certain other tribes, including the Masai in East Africa, organize tribal life through other structures, such as age grades (*see* p260) and secret societies.

**TANZANIAN MASAI**
*Masai tribal elders meet in public assemblies where community matters are discussed in order for a consensus to be reached.*

Fact

### Village headmen

The village headman is a feature of the Yanomani tribe in the Amazon region of South America. Village heads, of whom there are over 200, are responsible for mediating internal conflicts and representing their villages to outsiders. Their authority is dependent on general respect for their bravery, generosity, and charisma, and they cannot enforce their decisions.

COMMUNITY STRUCTURES

## Chiefdoms

**In brief:** A transitional form of political structure between tribe and state is known as a chiefdom

The first chiefdoms developed around 6,500 years ago, but few still exist today. Historically, many marked themselves apart from tribes by their ability to build monuments (Stonehenge or the Easter Island statues, for example). The best-known chiefdoms are those recorded in Polynesia.

Chiefdoms are larger and more complex than tribes and so require a more centralized political structure to unite their various local communities (each ruled by its own chief) and to regulate large-scale food production.

Chiefs, who frequently inherit their position, have the authority to enforce decisions, settle disputes, and draft manpower, helped by an elite of close kin. Religious cults and priesthoods developed within chiefdoms to buttress the authority of chiefs. Polynesian chiefs rely on *mana* (mystical powers) for their authority.

A typical example of a chiefdom was recorded in precontact Hawaii, whose society was divided into three levels of people: the Ali'i (chiefs who were descended from gods), the Konohiki (land administrators), and the Maka'ainana (commoners). Social rank in this chiefdom was based on closeness of relation to the chief, and access to resources depended on one's position in the social hierarchy.

Chiefdoms are characterized by an economy that produces more than is necessary for subsistence. In theory, everyone hands in a share of what he or she produces, and this is stored and returned at communal feasts or during times of famine.

**COLLECTIVE AGREEMENT**
*Samoan chiefs meet to agree on the appointment of a new chief. Outside of Polynesia, however, there are few remaining chiefdoms.*

**WHERE TRADITION SURVIVES**
*This traditional Zulu chief from Swaziland in southern Africa still has nominal power over his tribal "kingdom".*

SOCIETY

NATIONAL STRUCTURES

## Monarchies

**In brief:** A government that is ruled by a single hereditary leader is known as a monarchy

In an absolute monarchy, power is concentrated in the hands of a single person who rules alone or with the help of unelected advisors. An absolute monarch is both the head of state and the head of government, and there are no checks on his or her power. Up to the 17th century, monarchies were a common form of government around the world. They were especially strong in the 14th and 15th centuries, when hereditary kings and queens ruled many parts of Europe.

Today, constitutional monarchies, in which the hereditary monarch remains head of state but the government is run by an elected body (such as a parliament), are more common than absolute monarchies. In certain types of constitutional monarchy, such as those of Jordan and Saudi Arabia, the monarch retains a lot of political power. In others, such as England and Sweden, the role of the monarch is primarily ceremonial and all the political power rests with the elected parliament.

**ROYAL WEDDING**
*Inhabitants of the Polynesian kingdom of Tonga celebrate the marriage of their king.*

**THRONE DAY**
*Representatives from the Moroccan provinces swear allegiance to their king, Mohammed IV, in 1999.*

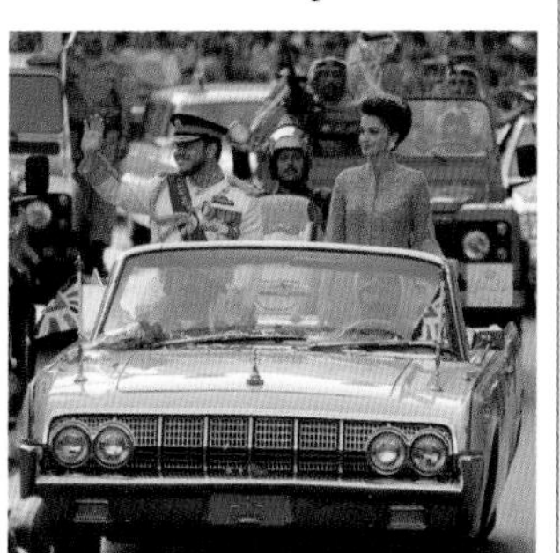

**A KING IS CROWNED**
*King Abdullah and Queen Rania of Jordan meet the crowds after the official crowning ceremony, known as an investiture, in 1999.*

NATIONAL STRUCTURES

## Theocracies

**In brief:** A theocracy is a form of government in which the roles of religious leader and ruler are combined

In a theocracy, religious leaders have ultimate political authority and the church is the state. Theocracies have encompassed diverse religious beliefs, from the sacrificial practices of the Aztecs to the Buddhism of Bhutan and Tibet. Today, only Iran has true theocratic rule, although the Vatican City is also classed as a theocracy. Iran's president and parliament are elected by popular vote, but a council of Islamic leaders has the power to veto policies.

IRANIANS SUPPORTING THEOCRACY

NATIONAL STRUCTURES

## Military states

**In brief:** When power rests solely with an unelected military elite, this structure is a military state

Most modern military dictatorships are formed after a coup d'état, when military leaders often justify their takeover as a way of bringing political stability. These military regimes tend to portray themselves as nonpartisan. They claim to be neutral parties that will provide interim leadership in times of economic or political turmoil or civil war, and an alternative to corrupt and ineffective politicians. In practice, however, such regimes often use force in order to remain in power.

Military dictatorships are not accountable to an electorate, and in the past have been brutal. In addition, they do not have a reputation for providing long-term stability or a healthy economy.

Since the 1990s, military dictatorships have become less common because all such regimes around the world have come under increasing international pressure to reform. As a result, some armies that previously ruled states now prefer to remain in the background or to legitimize their leadership by holding general elections, although the independence and validity of these is questionable.

**CONGOLESE COUPS**
*In 1996, Laurent Kabila became President of the Democratic Republic of Congo after a military coup. However, on 21 January 2001, Kabila himself was killed in a coup attempt.*

**MYANMAR PROTESTS**
*Despite much opposition, such as from Nobel Prize winner Aung San Suu Kyi, Myanmar (once Burma) has been ruled by a military council since 1958.*

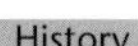

History

### A military coup in France

In 1958, tension surrounding the Algerian war of independence led to a military coup in France. The army rebelled when the government tried to withdraw from Algeria, which was then a colony. The revolt led to the collapse of the fourth French republican constitution and to the establishment of the Fifth Republic by the World War II hero, General Charles de Gaulle. The new leader restored stability, although military unrest continued into the 1960s.

**RULE BY FORCE**
*The military in Pakistan has stepped in and taken control of the government several times, most recently in October 1999.*

SOCIETY

NATIONAL STRUCTURES

## Single-party states

**In brief:** One political party dominates and no opposition parties are allowed in a single-party state

Single-party states often pay lip service to democracy by holding regular elections, but without a choice of ruling party, these elections are largely symbolic. There are many reasons why states adopt a one-party system. A dictator may want to retain a façade of legitimacy and democracy, or a national emergency might lead to the downfall of democracy.

In some situations, a single-party system may be considered the best way to unite ethnically, linguistically, and religiously diverse societies. The fear is that a large number of parties would make government impossible because the people would vote along ethnic lines, and therefore no single party would gain a majority.

A sudden and rapid transition to democracy can also sometimes result in a one-party system because new voters are overwhelmed by the multiplicity of choices.

Although a leader of a single-party system may come to power with the best of democratic intentions, he or she frequently becomes at the very least authoritarian, and sometimes unquestionably dictatorial. Without official opposition, a leader can argue that he or she is the only person who knows what is best for the people and is therefore the only person who is sufficiently qualified to run the country.

**THE POWER OF MONEY**
*The images of China's leaders on the country's banknotes are a daily reminder of the power and authority of the Chinese leadership.*

**SINGLE-PARTY CUBA**
*Fidel Castro came to power in 1959 in a revolution opposing dictatorship in Cuba, but he went on to establish yet another dictatorship.*

**CULT OF PERSONALITY**
*Thousands of North Koreans show their support for leader Kim il-Sung at a planned rally, under a poster bearing his image.*

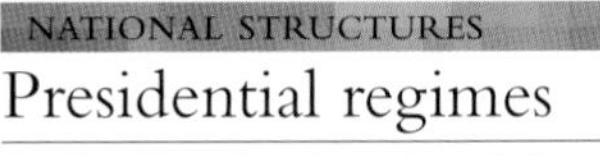

NATIONAL STRUCTURES

## Presidential regimes

**In brief:** In a presidential regime, the democratically elected president of a state has become a dictator

In some nations, political leaders who were initially elected through democratic means to their position as head of government subsequently become dictators, using force or illegal means to remain in power. Elections may continue to be held, in order to give the government a veneer of legitimacy, but these elections are unlikely to be either free or fair. In fact, such leaders often rig elections so that they continue to win despite an overwhelming opposition.

Presidential regimes tend to arise in countries in which there is a high degree of corruption and the military can be easily swayed to support a dictator. Leaders often argue that they are retaining control for the good of the country in order to prevent the government from being taken over by unstable or reactionary elements.

**INFLUENCING THE VOTE?**
*The military stands by to protect voters in Zimbabwe's 2002 elections. The military presence also reminded voters of what would happen if they did not vote the "right" way.*

NATIONAL STRUCTURES

## Transitional states

**In brief:** A nation state that is restructuring and has a temporary government is said to be transitional

It is almost impossible to change any system of government quickly. Any time that a country and its people undergo a major change of political system, there is necessarily a certain period of readjustment as new policies and administrative systems are devised and then put into place.

A transitional period may follow a war in which the old government has been overthrown, or it may occur after the emergence of a new country. The collapse of the Soviet Union, for example, gave rise to many new states. Many of these new nations lacked independent government and, with no history of democracy, they initially had transitional governments. In some cases, an international body will take over the temporary running of a country while a new system is put into place. When East Timor voted for independence from Indonesia in 1999, the United Nations Transitional Administration in East Timor was established. This structure exercised legislative and executive authority for 3 years, until the government could take over.

Transition is not always positive, however. Some "transitional" states never complete a transition from war to peace and remain riven by violent conflict and invasion for decades.

**AFGHAN ELECTION**
*After many years of war, in 2003 the transitional Afghan government began the move towards democracy by holding provincial elections in the capital, Kabul.*

**TURNING POINT**
*By toppling the statue of Saddam Hussein, American troops in Iraq marked the end of his dictatorship and the start of a transitional state.*

**TRANSITION IN RWANDA**
*After the civil war, Rwanda had 3 years of transitional government. Local cooperatives were also set up and international help provided.*

Fact

### Somalia – a failed state

Located in the Horn of Africa, the country of Somalia was created in 1960 by the arbitrary merger of two former colonies – British Somaliland and Italian Somalia. The postcolonial state was never stable, and by 1991 it had completely collapsed into civil war and warlordism. Despite numerous peace initiatives and international intervention, Somalia remains without an effective government.

NATIONAL STRUCTURES

## Parliamentary democracies

**In brief:** When the head of government is chosen indirectly, this is called a parliamentary democracy

The distinguishing characteristic of a parliamentary democracy is the fact that the government is chosen from representatives who have been elected to a parliamentary assembly. Typically, the party that has the majority in parliament chooses the chief executive, who is known as the prime minister. However, in some parliaments there are so many parties that none holds a majority. In this case, parliamentary members decide who to elect.

Unlike in presidential democracies (*see* p270), in parliamentary systems the administrative and lawmaking branches are combined. This structure usually leads to more discipline among party members, who almost always vote along party lines. The government in a parliamentary democracy therefore tends to retain firm control over its parliamentary majorities, which reduces the ability of parliamentary assemblies to challenge the government's wishes.

In a parliamentary system, the head of government and the head of state are separate roles. The former has the power to enact laws, while the latter is often a purely ceremonial role.

**GERMAN PARLIAMENT**
*Germany has a parliamentary form of government. Its parliament is based in the Reichstag building in the capital Berlin.*

**FREE AND FAIR**
*In 1994, after the fall of apartheid, South Africa held its first democratic elections in which each citizen could participate.*

NATIONAL STRUCTURES

## Multiparty democracies

**In brief:** A multiparty democracy is a system in which political parties vie for votes to form a government

A multiparty system is composed of several recognized political parties, each of which competes for votes. This system prevents the leadership of a single party from setting policy without challenge. If the government includes an elected congress or parliament, the different parties may share power according to a system of either proportional representation (PR) or "first-past-the-post".

In PR, each party wins a number of seats proportional to the amount of votes it receives. In first-past-the-post, the electorate is divided into a number of districts, each of which elects by majority vote one person to fill one seat. The first-past-the-post system does not tend to generate many parties, but instead moves naturally towards a two-party system. PR, by contrast, enables a number of major parties to form.

A two-party system requires voters to align themselves in large blocs. These blocs can often be so large that the members cannot agree on any overarching principles, a state of affairs that allows centrists to gain control. On the other hand, if there are multiple parties, each with substantially less than a majority of the vote, they are forced to compete for the support of smaller parties. In this way, these smaller parties can acquire inordinate political influence.

**BALANCE OF POWER**
*France has a multiparty democracy in which the president (right) is elected by popular vote and then appoints the prime minister (centre).*

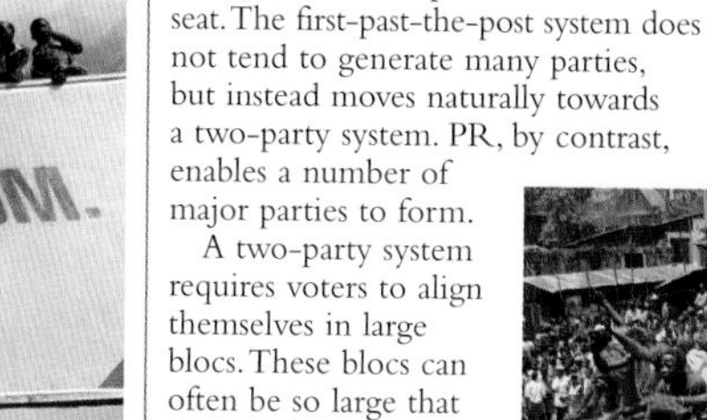

**ELECTION FEVER**
*Sierra Leone successfully held elections just 6 months after the end of the civil war in 2002. The country now has a multiparty democracy.*

GLOBAL STRUCTURES

## Global institutions

**In brief:** The power and the scope of international organizations are increasing ever more rapidly

Although nation states remain the key sources of power, international bodies are taking on greater roles in national and international governance. Such institutions include the United Nations (UN), World Trade Organization (WTO), International Monetary Fund (IMF), and European Union (EU), all of which were founded after World War II. The World Bank was set up in 1944.

Their precise goals differ, but they have a common remit – to foster international cooperation and establish the rule of law for international affairs. They were also set up to help prevent a recurrence of the devastating economic, trade, and military conflicts of previous decades.

The UN was founded directly on the principle that world order is based on the equality of states regardless of economic and political weight. More recently, the UN has placed a greater emphasis on protecting human rights and has become involved in peacekeeping.

With the end of the Cold War, international institutions have had to develop a new relevance. The EU, in particular, has metamorphosed from a trade body into a federal government. Similar groups of nations may follow suit, despite doubts about how viable supranational structures really are. It is unclear if such bodies could ever actually enjoy genuine popular legitimacy.

**SECURITY COUNCIL**
*The United Nations' security council comprises 15 states: 5 permanent members and 10 rotating members.*

**UNDER FIRE**
*The World Bank, and its president, have come under criticism for its heavy-handed intervention in the economies of developing countries.*

Issue

### Is the trend inevitable?

Over the past few decades, international organizations have become increasingly involved in the domestic affairs of many of the world's states. The trend towards global economic and political integration, globalization, was thought by many to be the inevitable evolution of international affairs. Recently, however, this notion has been challenged by antiglobalization movements (see right) who seek to give nations and their populations greater independence from international institutions and commercial enterprises.

XALAPA
cada vez somos
ALTO LUCERO
PAN
VICENTE
PRESIDENTE
SALVADOR GARCIA
DIPUTADO FEDERAL
Cambio que a Ti te Conviene
ICENTE FOX
FOX

NATIONAL STRUCTURES

# Presidential democracies

**In brief:** In a presidential democracy, the head of state and the head of government are the same person

The central principle in a presidential democracy is that the administrative (executive) and lawmaking (legislative) arms of government are separate. The executive branch is led by a directly elected president, while the members of the legislative branch are elected separately. The benefit of this form of government is that the two branches act as a check on each other's power, therefore neither has the complete authority to enact laws. In addition, both branches are directly accountable to the people because they may be thrown out of office at the next election.

**SWEARING IN**
*The Greek president and prime minister are elected and sworn in by the Greek parliament.*

Presidential systems do not distinguish between the roles of head of state and head of government; both are performed by a president who is directly elected by the people. In presidential democracies, therefore, the president is always an active participant in the political process and not merely a symbolic figurehead.

In some presidential systems, such as those in South Korea or Taiwan, there is an office of prime minister or premier. However, unlike in parliamentary democracies (*see* p269), the premier is responsible to the president, not to the legislature. In the US, in the late 19th century, there was much speculation that the position of Speaker of the House of Representatives would eventually become a quasi-prime minister, and that Congress would evolve into a type of parliament. These changes clearly did not happen. More recently, it has been suggested that the office of White House Chief of Staff, the President's principal aide, has become a de facto US prime minister of sorts, with his or her dominance or weakness in the US governmental system depending on whether the current president has a hands-on or hands-off approach to governance.

Other countries with presidential systems include Mexico, Sri Lanka, Greece, Egypt, Kenya, and many South American nations. Despite having a president with considerable powers, France is here classed as a multiparty democracy (*see* p269) rather than a presidential one owing to its slightly more cooperative form of government. In this country, the president's domestic policy must be approved and co-signed by the prime minister. In turn, the prime minister requires the support of the majority in the National Assembly, which is not always of the same political persuasion as the president.

Profile

## Boris Yeltsin

On 12 June 1991, Boris Yeltsin (1931–) became the first elected president of Russia after being made unelected president a year before. To prevent the communists from returning to power, Yeltsin established the Commonwealth of Independent States and ended the existence of the Soviet Union. Yeltsin and the new Russian congress were divided over the balance of powers in the new constitution. Yet, after a failed coup in 1993, Yeltsin's reforms – which gave considerable powers to the president – were approved.

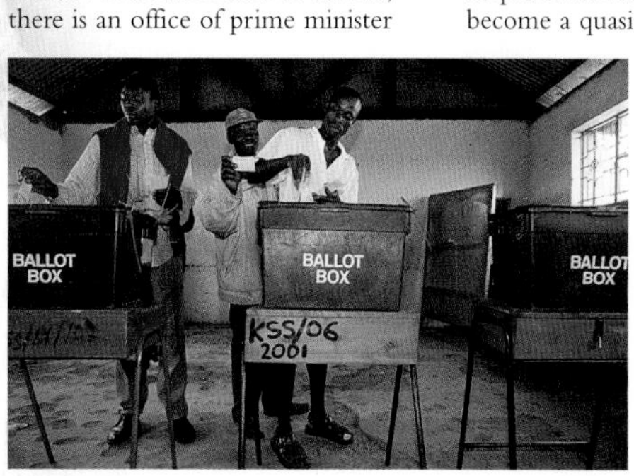


**KENYA ELECTIONS**
*In Kenya, voters elect the president, the parliament, and the civic leaders at the same election. Kenya has a president and a parliament but no prime minister.*

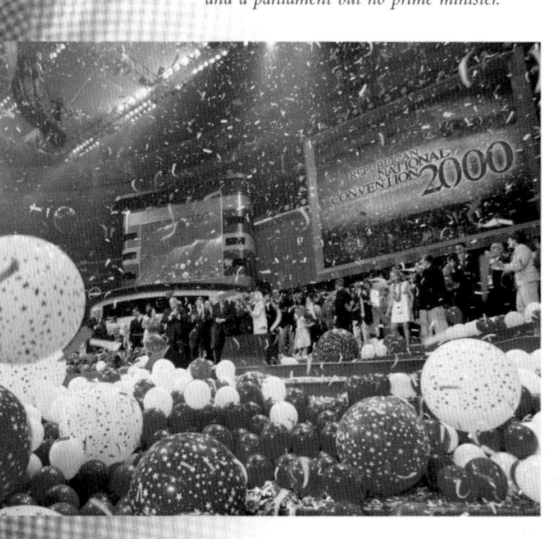


**PRESIDENTIAL POWER**
*The 2000 election campaign of Vicente Fox was successful. Laws enacted by the Mexican president cannot be changed, but he can only be elected for one 6-year term.*

**PARTY POLITICS**
*In the US, the main political parties present all candidates wishing to run for president at national conventions. During primary elections, each party chooses a candidate to represent it.*

# CONTROL AND CONFLICT

Wherever people establish themselves, they develop codes of behaviour that allow them to co-exist. However, not everyone will agree with all the rules. This means there is always the potential for some people to break them and act in a way others consider socially harmful. What is considered socially acceptable behaviour differs between societies and is subject to change. A serious transgression may be defined as a crime. All countries and societies have laws that define types of crime and that set out the punishment for committing them. Today, the most important categories include violence against people, and crimes against property and public order.

Profile

### Rosa Parks

In 1955, the US citizen Rosa Parks refused to give up her seat on a colour-segregated bus for a white man in Montgomery, Alabama. She was convicted of "disorderly conduct" but her case went to appeal in 1956 and the Supreme Court ruled that segregating bus services was unconstitutional. This was a major step in overturning the US laws on racial segregation.

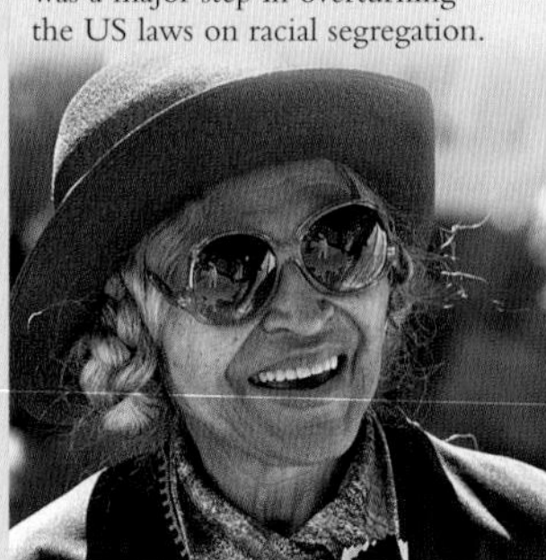

## SOCIAL CONTROL

In most countries, a combination of social controls and a formal criminal-justice system limits behaviour that is illegal or antisocial. Levelling mechanisms are a form of social control – the person who stands out is subject to ostracism or ridicule and is made to conform. Morals and values are fundamental controls on the behaviour of an individual and are therefore important in maintaining order and stability. Morality – the ability to judge right from wrong in relation to actions that may affect others – and values, such as respecting other people's property, are often influenced by the family and by religion. Most religions have codes of behaviour they expect members to follow, but also certain activities that are discouraged or forbidden. Religion also helps to keep the status quo – people will accept hardship or inequality if they are promised rewards for doing so in the next life.

**SPIRITUAL HELP**
*A congregation at a church in Bukavu, Democratic Republic of the Congo, asks God to guide their actions.*

## CONFLICT

Conflict covers a range of issues and behaviours, from neighbours who cannot agree on acceptable noise level, to friction at an international level, for example the building of a wall by Israel to separate it from the West Bank. There are also many activities that are legal in some countries and illegal in others. Such types of behaviour include taking drugs, drinking alcohol, having or performing an abortion, having sex outside marriage, and helping someone to commit suicide. As new threats to societies emerge, or existing ones proliferate, such as terrorism, alternative ways of controlling society are developed. Such checks on behaviour may take the form of international cooperation or, in the case of cross-border conflict, special bodies that arbitrate at an international level (*see* International law, p274 and Resolving conflict, p276).

**AN INDIVIDUAL SPEAKS OUT**
*During a politically motivated protest march in Israel, a female participant makes her position clear to a passerby who stopped to challenge her opinons.*

# LAW AND LAW ENFORCEMENT

Custom and individual morality are not sufficient to ensure an orderly and reasonably secure society. While important on a day-to-day basis, these informal social forces must be backed up by a more formal legal system. There is a need to protect people from others and from the arbitrary acts of social institutions (which include the government). Such protection requires a system for enforcing contracts, ensuring rights, resolving conflicts, and defining responsibility. The courts interpret and apply the law and have an influence in making it. Ideally, everyone is treated equally by the legal system of their country, but in practice, the better off a person is, the more skilled the lawyer he or she can pay to represent him or her. The ability to impose penalties and carry them out through the use of force, including prison, is the ultimate means of translating the decisions of the courts into action. However, the police, and other law-enforcement officials, have limited resources and have to prioritize their work. Usually the police are constrained in how they fight crime and disorder by a combination of general political pressures, public opinion, and specific government directives.

**IMPRISONMENT**
*Imprisonment can be a deterrent to people who may be tempted to commit a crime and a chance to rehabilitate offenders. This guard works in a Texas prison.*

**CROSS-BORDER RULINGS**
*After the 1994 Rwandan civil war, the UN set up several courts, among them the International Criminal Tribunal, which judged military personnel.*

Issue

## Capital punishment

More than 80 countries punish some categories of crime with death. In the US, 37 states can impose the death penalty for certain crimes as can US federal and military courts. Supporters of the death penalty claim it acts as a deterrent and prevents violent criminals reoffending. Opponents deny there is evidence to support the deterrent claim; there is also a risk of innocent people being executed.

# DISSENT AND PROTEST

A person or a group that disagrees with another or with, for example, the way a country is run, has several options. The choices include keeping quiet or taking action clandestinely, especially if there is a fear of reprisals. If the circumstances merit it, regardless of possible backlash, or if the society is an open one, the person or group may choose to express dissatisfaction openly. Individuals may choose to speak out by opting not to buy products of a country or organization whose policies they disagree with. A survey in 2003 found that 39 per cent of Britons were far more likely than they had been 5 years previously to use their purchasing power to make a point about an issue that concerned them – from child labour to an environmental question. People can also write to a representative of the organization they disagree with, or to a regulating body, or they can join in a letter-writing campaign. Sometimes people take to the streets. One example of such action occurs in Buenos Aires, Argentina. Every Thursday since 1977, the now elderly Mothers of the Plaza de Mayo have marched in the square in front of the presidential palace, demanding to know the fates of family members and friends who disappeared during the military dictatorship of the mid-1970s to early 1980s. Occasionally, peaceful demonstrations turn into riots, for example the anti-globalization protests in Geneva, Switzerland in 2003 and Genoa, Italy in 2001, where there were violent clashes between demonstrators and the police.

**FACING THE TANKS**
*In Tiananmen Square, in Beijing, China in 1989, a lone protester defied the military might of the government, while around the world people watched the dramatic events on television.*

**RIOTING IN BELFAST**
*In Northern Ireland, ahead of the annual Orange Order parade by Protestants, police officers from the Royal Ulster Constabulary face congregating nationalists.*

**CONFLICT ON THE STREETS**
*People can take to the streets to show dissent. This demonstrator being arrested in Washington, DC, US was protesting against a march by the Ku Klux Klan.*

SOCIETY

LAW AND ENFORCEMENT

## Informal court systems

**In brief:** Resolving disputes or meting out justice can be done through community-based systems

Justice has been meted out for many centuries through informal systems. In Islamic social life, for example, men must hold an informal consultation – a *shura* – before making a decision. For some people, traditional courts represent old-fashioned hierarchies and power relations; however, evidence suggests that they can be adapted to the modern world. In Gujarat, India, women lead sessions to resolve disputes relating to the community, family, and marriage, and award compensation. In Uganda, a system based on local "resistance councils" was introduced in 1986. Under this arrangement, villagers can take action at a local level in issues of justice and development, and can also feed their decisions up to judges and MPs.

Fact

### Religious law

Islamic countries follow a religious code for living called sharia, part of which derives from the Koran. Some countries enforce sharia as well as their own state law. In Pakistan, the formal legal system is based on English common law but because the country is an Islamic state there is provision to abide by the rules of sharia law where appropriate. Iran, however, practises sharia law only.

**VILLAGERS AS JURY**
*A return to the traditional Gacaca courts enabled Rwandans to judge those accused of killing their families during the genocide, which took place in 1994.*

LAW AND ENFORCEMENT

## Formal court systems

**In brief:** A formal framework of laws regulates the activities of a society and the people living within it

Many societies have a formalized legal system, which may exist alongside informal systems. Most formal systems recognize criminal law and civil law.

In criminal law, the main aim is to deter and to punish. An act such as murder is a criminal matter, and it may be government officials who bring the person to trial. Civil law usually deals with disputes between individuals or organizations. Either side can bring a case, and both employ professionals to represent them. Civil law cases are usually brought to a formal court to claim money. The main difference between civil and criminal law is that a criminal case must be proved "beyond all reasonable doubt" while a civil case has to be proved on a "balance of probabilities".

One example of these differences was the 1995 criminal case against O J Simpson in the US. A criminal court failed to prove "beyond reasonable doubt" that the former football star had killed his wife. At a later civil trial, a jury found that on the "balance of probabilities" he had murdered her and awarded compensation to the wife's family.

**BAGHDAD COURTROOM**
*Under Saddam Hussein, Iraq suffered years of a corrupt judicial system. At the end of the war, the courts reconvened on a small scale.*

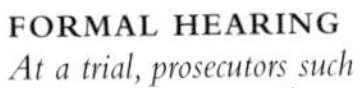

**FORMAL HEARING**
*At a trial, prosecutors such as the two here, put the case against the accused.*

LAW AND ENFORCEMENT

## International law

**In brief:** International courts have power to make rulings to settle disputes that arise between countries

Since the end of World War II, the United Nations (UN), supported by the International Court of Justice in the Hague, Netherlands has taken the lead in settling disputes between countries and establishing conventions, from the Law of the Sea to territorial rights in air space.

More recently, international law has adopted a new role in protecting individual rights. In extreme cases, it has been used to justify intervention by outside forces, as has been the case in parts of former Yugoslavia, Liberia, and in the Congo in the 1990s. In 2003, the UN established the International Criminal Court at the Hague. This permanent international court has jurisdiction over serious human rights infringements, including war crimes, acts of genocide, and aggression against sovereign states.

International trade has its own body to establish and police the rules, the World Trade Organization.

**HARMONY AT THE COURT**
*The figure of a woman, on the insignia of the International Court of Justice, represents agreement.*

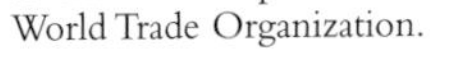

**INTERNATIONAL JUSTICE**
*The International Court of Justice, in the Hague, Netherlands, can settle disputes between nations about borders, for example.*

LAW AND ENFORCEMENT

## Punishment

**In brief:** Punishment is a way of deterring someone from transgressing, or penalizing a person who does

When a society sets out a code of conduct, it recognizes that not everyone will obey it. For most people, knowing that they will be punished if they break the rules deters them from doing so.

Speeding drivers who are caught once may be warned, but the second time they may be fined or lose their licence for several months. For more serious crimes, people are likely to be jailed. Alternatives to incarceration exist, however, such as electronic tagging, where an offender wears a electronic device that tracks his or her movements. Another alternative is work in the local community, such as a residential home, on an unpaid basis. Some punishments can be brutal. In Iran, people accused of drinking alcohol may be flogged; in Sudan, a thief may have a hand amputated. The death penalty exists in some countries.

**PROBLEMS IN PRISON**
*Numbers of prisoners continue to rise in all countries, causing overcrowding in detention facilities, including in this Chinese women's prison.*

History

### Banishment

In the past, expulsion from the homeland – either temporarily or permanently – existed as an alternative to capital punishment. By the 18th century, European states were routinely banishing criminals to penal colonies. Until 1946, Devil's Island, (*right*), off the coast of French Guiana, was a French penal colony from which escape was the only way out; but very few of those who tried, survived.

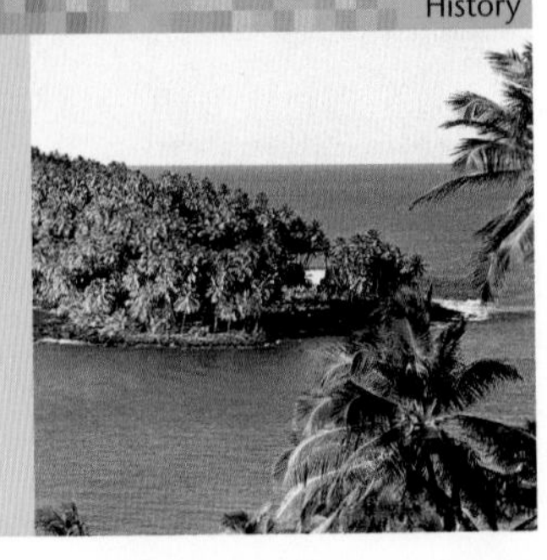

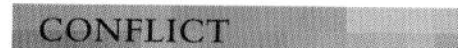

CONFLICT

## International war

**In brief:** Traditionally, soldiers were the casualties of war; in modern conflicts it is civilians who are at risk

The ultimate expression of conflict between nations is war. The effects are always devastating, but as technology has become more sophisticated so the risks have moved – from the soldiers to the population of the country in which the war is being waged. At the start of the 20th century, between 10 per cent and 15 per cent of those who were killed in wars were civilians. In World War II, more than 50 per cent of those who died were civilians. By the end of that century, however, civilians comprised over 75 per cent of the 7 million casualties.

The change is the result of fighting moving from battlefields, with soldiers in the frontline, to centres of population. When the politicians who direct the war target army bases, for example, it is almost impossible to prevent civilians being killed or hurt.

**TECHNOLOGY AND WARFARE**
*Equipment used in war continues to increase in sophistication. Above, a crew monitors radar screens onboard a guided missile cruiser.*

The effects of chemical and nuclear technology, however, are even more indiscriminate. The bombs dropped in Japan in 1945, for example, caused thousands of deaths (Nagasaki, 70,000 victims; Hiroshima, 140,000 victims) and those who were in these areas and survived continue to suffer the effects of exposure to radiation, such as cancers including leukaemia.

**TRADITIONAL WEAPONS**
*Ninety years after the first tanks were built, they are still in use as a mobile, armoured weapon that can operate in all conditions.*

CONFLICT

## Civil war and revolt

**In brief:** Civil war is war between the peoples of a nation; revolution is the overthrowing of a regime

Civil war occurs when different communities within a nation fight, often as a result of ethnic or religious differences. A revolt, or revolution, applies to a situation in which a regime (usually a repressive one) is overthrown by the people it governs.

Civil war may involve all parts of society, and it can be inhumane and bloody. In the war in Sierra Leone from 1991 to1999, at least 50,000 people were killed, and a further 100,000 people were mutilated. Many victims were children.

A protracted civil war can ruin the economy of the country; hospitals and schools may be destroyed and medical supplies might not get through. Another disruptive effect of civil war is that persecution of the inhabitants, or their wish to escape the fighting, causes movements of people, both within the country and out of it. Civil war now accounts for more than 90 per cent of conflicts worldwide.

A revolt or uprising may be less bloody than civil war. Historically, revolts have been led by a minority of politically active people. Revolts challenge the legitimacy of a regime and seek to provide an alternative. For example, in 2004, rebellion broke out in Haiti and caused the overthrow of the president, Jean-Bertrand Aristide.

**CIVIL WAR IN KOSOVO**
*A man returns home to find his town was destroyed during the Yugoslavian civil war which raged from 1991 to 2001.*

History

### Revolutions in Russia

In 1917, uprisings by soldiers, workers, and peasants wanting social equality and economic democracy in Russia resulted in two revolutions. The first, in February, ousted the monarchy. The second, in November, created the world's first communist state. Vladimir Ilich Lenin (*right*) headed the new government and became leader of the Union of Soviet Socialist Republics when it was founded in 1922.

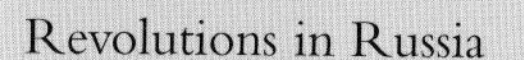

CONFLICT

## Terrorism

**In brief:** Use of systematic acts of violence and terror to achieve a political goal is called terrorism

Terrorism – the use of violence and intimidation in the pursuit of political aims, is emerging as the biggest danger to world stability.

One of the aims of terrorism is to bring about political change. In 2004, for example, terrorists planted a series of bombs on the railway system in Madrid, Spain on the eve of a national election. As a result of the attacks, the Spaniards voted in a leader who said he would withdraw the country's troops from Iraq, where they were part of peacekeeping coalition forces.

Terrorism also aims to overwhelm the normal functioning of society by causing panic and uncertainty among inhabitants. This was the effect of the attacks on the World Trade Center in New York in September 2001. Americans around the world were incredulous and shocked. The raids had taken place in their country seemingly without warning and by an unknown enemy; most US citizens suddenly felt vulnerable. The attacks have had far-reaching effects. Around the world, governments have introduced or increased general and specific security strategies and measures to counter the threat of a possible attack by terrorists. These range from low-level initiatives, such as checking the bags of visitors to more stringent controls on people wishing to enter a foreign country.

**SARIN GAS ATTACK**
*In 1995, 12 people were killed and 5,500 were injured in an attack on the Tokyo subway by members of a terrorist group.*

**SEARCHING THE RUBBLE**
*Terrorism reached a new level with the attack by al Qaeda on the World Trade Center in New York on 11 September 2001; more than 2,700 people died.*

SOCIETY

INTERVENTION

## Resolving conflict

**In brief:** Mediators may help resolve conflict between individuals, organizations, or nations

The potential for conflict between individuals and groups of individuals is ever present, and societies have developed mechanisms and agencies to resolve differences. For example, a couple who argue continually, and whose relationship is at risk, can seek help from a dedicated counsellor. Among the Abkhazian people of Georgia, conflict is resolved through elders who act as mediators between the parties. When a dispute has ended, the mediator brings the two sides together and slaughters a bull. The blood symbolizes the new brotherhood between the two factions.

In industrialized societies, trade unions may represent employees in a given sector. If the union wins the approval of its members to strike (*see* p277), mediation from an independent body may be needed to resolve it. In the UK, for example, this function is provided by the Advisory, Conciliation and Arbitration Service (ACAS).

When conflict is between nations, a key individual, such as a president or monarch, or a body such as the United Nations, may defuse or resolve crises and protracted conflicts.

Profile

### Jimmy Carter

In 2002, Jimmy Carter received the Nobel Peace Prize "for his decades of untiring effort to find peaceful solutions to international conflicts, to advance democracy and human rights, and to promote economic and social development". During his time as the 39th president of the US (1977–1981), Carter mediated in the Camp David Accords between Israel and Egypt.

**PREVENTING VIOLENCE**
*Mediators can bring together opposing forces in negotiations. In this instance in the Ivory Coast, West Africa, French soldiers stand guard over rebels and mediators in case of possible violence.*

**TROOPS AS PEACEKEEPERS**
*During the war in the Balkans in the 1990s, the United Nations brought in troops to act as peacekeepers. Here, UN tanks patrol Sarajevo.*

PROTESTS

## Sanctions

**In brief:** Imposing sanctions, often by stopping free trade, is used to put pressure on a government to change its behaviour

Governments and organizations have a variety of ways open to them to put pressure on leaders of countries whose policies they do not agree with. In 1970, South Africa was banned from international cricket because its apartheid laws did not allow black players. The country only returned to international cricket in 1991, after the release of Nelson Mandela (*see* p44) and the introduction of laws that banned discrimination in sports. In 1962, the US imposed full economic sanctions against Cuba, measures that were still in force 40 years later.

Economic sanctions can be effective; however, they can have a negative impact on human rights (*see below*). Individuals can also take similar action to make their disapproval known (*see* Consumer choice, opposite page.)

**SIERRA LEONE**
*Sanctions were imposed on Sierra Leone diamond exports to prevent profits being used to buy weapons.*

Issue

### Sanctions and human rights

Imposing economic sanctions can affect the quality of food and the availability of drinking water in the targeted country. Sanctions can cause disruption in the distribution of food, sanitation, and medical supplies; inhibit the working of health and education systems; and undermine the right to work. They can also restrict the opportunities for seeking asylum.

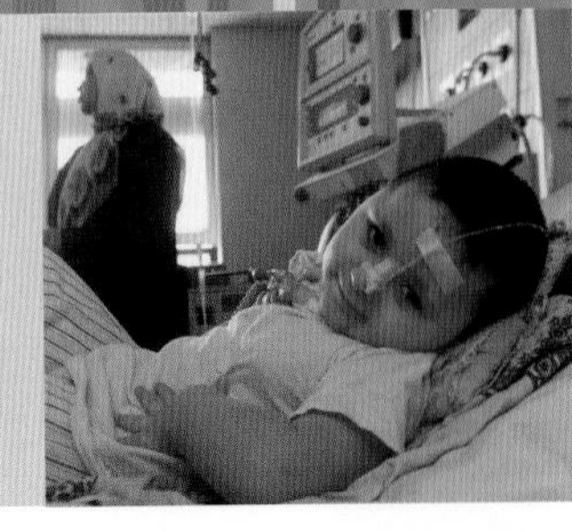

**A SCARCITY OF ESSENTIALS**
*In 2000, the United Nations imposed sanctions on Iraq, permitting it to sell oil only to buy food and medicines. Here, workers close the pipeline in Baghdad.*

INDIVIDUAL DISSENT

## Demonstrations

**In brief:** When many people are dissatisfied with a government, they can take to the streets to demonstrate

Taking to the streets to demonstrate is an internationally recognized way for a large number of people to show dissatisfaction with a government; with changes a regime is proposing; or when a ruling party is not taking action fast enough. The success of such measures depends on how much support there is for the protesters in the society and on the power and determination of the government to resist the demonstrators' demands.

Demonstrations are not a new idea. At the end of the 19th and beginning of the 20th centuries, the suffragettes used this method to secure votes for women in Britain (*see* p261). In 1893, after studying law in the UK, Gandhi (*see below*) went to South Africa. Here, appalled by the treatment of Indian workers, he organized mass demonstrations of protest. In the 1960s, the efforts of US civil rights protesters won the right for African-American citizens to vote. Twenty years later, protests in the Philippines in 1986 ousted the President, Ferdinand Marcos.

Demonstrators are not always peaceful; they may riot, as the antiglobalization protesters did in Geneva, Switzerland, in 2003.

**THE RISE OF ECO-ACTIVISM**
*In the last quarter of the 20th century, people began to take direct action to prevent organizations and governments damaging the environment. These protesters set up home in areas of threatened woodland in the UK.*

Profile

### "Mahatma" Gandhi

Mohandas (later called "Mahatma", Great Soul) Gandhi was born in India in 1869. In 1915 he joined the Indian National Congress. In order to achieve his goal of self-rule for India, Gandhi practised passive resistance and noncooperation; he also encouraged cottage industries to revitalize the domestic economy. In 1947, India achieved independence. Gandhi was murdered a year later.

**ANTIWAR PROTESTERS**
*The decision by US President George W Bush to invade Iraq in 2003 led to antiwar protests worldwide, before, during, and after the end of hostilities.*

INDIVIDUAL DISSENT

## Strikes

**In brief:** Workers may take strike action (refusing to work) in disputes with employers over pay or conditions

If a conflict arises between workers and employers, attempts are usually made to come to an agreement through negotiation. However, workers may decide to strike if employers are unwilling to negotiate or if negotiations fail. In many countries, workers join trade unions, which aid collective bargaining. In some companies, union membership is compulsory, but it may be illegal or discouraged elsewhere.

Strike action can be successful, but strikes may become protracted and costly both for the workers and their families, who lose out on pay, and for the employer who may experience serious business losses while workers are on strike. In some places, especially parts of Africa and Latin America, strikers may be victims of violence.

**ACTION BY SHIPYARD WORKERS**
*French employees have a history of taking action to complain about conditions. Here, workers stop a train bringing parts to a port.*

**PICKETING IN THE US**
*In Michigan, US, General Motors union members parade outside the factory carrying placards. Pickets such as these may attempt to stop other employees from working.*

INDIVIDUAL DISSENT

## Consumer choice

**In brief:** Individuals can put pressure on a regime by not buying products originating in that country

FRUIT MAY BE SUBJECT TO BOYCOTT

Consumers may feel they can take action as individuals. For example, they can choose to buy, or not to buy, a product or products from a country whose policies they disagree with, or items or animals involving a method of production or capture that they oppose. At various times, British shoppers have boycotted apples and wine from France, as a protest against actions by the French government. Western consumers have also chosen not to buy tins containing tuna caught by net because of the number of dolphins killed by this method of fishing. There have also been protests worldwide against companies such as McDonald's and Nestlé and against oil producers, such as Shell.

INDIVIDUAL DISSENT

## Petitions

**In brief:** People who oppose or support something can organize or sign a petition to show their opinions

Petitions involve the collection of signatures from supporters of a cause and presenting them to an official representative of the organization that has been targeted. Such petitions can be an effective method for people to take action or to get involved with a campaign. They can range from local initiatives, such as students protesting against the introduction of a uniform at their school, to international action, for example, the vigorous efforts to persuade the big banks to cancel debt in developing countries.

**INDIVIDUALS IN ACTION**
*In 2003, 3 million Venezuelans signed a petition calling for a referendum to vote President Chavez out of office.*

# WELFARE

Many people, at some time in their lives, find they are unable to provide adequately for themselves. They may need help with food, clothing, shelter, healthcare, or other personal necessities. Most children too young to look after themselves count on their families for support and in many parts of the world, the family support network exists for life. In some countries, however, the family is not in a position to help – perhaps for financial or geographical reasons – in which case people, if they do not have adequate resources themselves, may have to rely on private, charitable, or state welfare. Such structured provision, however, is unavailable in many parts of the world.

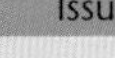

Issue

### Caring for children

When a child loses one or even both parents and is unable to be supported by an extended family, he or she may be taken into care. Some are placed in orphanages or homes run by the state or by charitable or religious groups. In areas of conflict, children's homes (such as the Afghan one shown below) play an especially vital role. Others are placed with new families and are fostered in short-term placements or adopted. In some countries, the state has the right to remove a child from his or her parents, temporarily or permanently, if it considers the child to be at risk from abuse, cruelty, or neglect, or if the child is considered a risk to him- or herself.

**NURSING CARE**
*In developing countries, nurses – including these Rwandan Red Cross workers and nurses from other aid agencies – play a vital role in educating people about health issues.*

## GLOBAL VARIATION IN WELFARE

Throughout history, most people have traditionally relied upon informal family, tribal, or community support. This remains the case today for the majority of the world's population, most notably the impoverished millions of Africa, Asia, and the rest of the developing world. In such regions, state or private provision is often rudimentary or nonexistent. Outside agencies, such as international aid agencies and nongovernmental organizations, can offer limited help, although rarely on a permanent basis. In wealthier countries, a combination of public provision that is paid for out of taxes and distributed according to need, and private provision that is paid for by fees or personal insurance has ensured generally high levels of welfare available to everyone from cradle to grave. The scale of public welfare provision varies considerably, from about 40 per cent of national output in some European countries to barely half that in Japan and the US, where private provision is far more widespread. However, private provision has its limitations, because those most in need are usually the least able to afford it. In these cases, the state takes a leading role in welfare provision, directing funds and services to those most in need.

**FAMILY SUPPORT**
*In most of the world, welfare starts in the home, with family members caring for elderly or disabled dependants.*

**WELFARE STATE**
*A refugee camp is a microcosm of a welfare state, but shelter, food, healthcare, and sometimes even education are in this case provided by aid agencies.*

# THE VALUE OF EDUCATION

Traditionally, education has been divided into the purely academic and the vocational. Academic education develops the intellectual capacity of a student, passing on a body of knowledge to open up and expand the student's mind. That knowledge can range from the most basic skills of literacy and numeracy through to high-level abstract thought. By contrast, vocational training uses a different process; it gives students specific practical abilities that are then developed through work experience. In early societies, people learned solely from the adults around them and from the experiences of life itself. Later, religious leaders took on a major role in providing the first formal systems of education, partly because clerics were often the only educated people in society and partly in recognition of the formative influence schooling has on a person's life. Recently, vocational education has grown in importance as a way of preparing people for work. However, if too much stress is placed on the economic advantages of education to increase a nation's "human capital" and thus its potential wealth, the crucial role of education in helping people to think for themselves can be obscured. Without the basic ability to think as a foundation, all forms of training will be far less effective.

**STARTING EARLY**
*Children learn social skills as well as academic knowledge when they start formal schooling.*

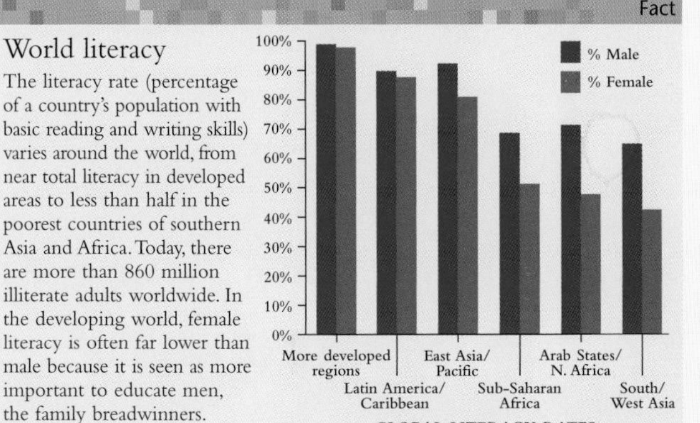


### World literacy

The literacy rate (percentage of a country's population with basic reading and writing skills) varies around the world, from near total literacy in developed areas to less than half in the poorest countries of southern Asia and Africa. Today, there are more than 860 million illiterate adults worldwide. In the developing world, female literacy is often far lower than male because it is seen as more important to educate men, the family breadwinners.

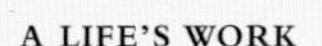

GLOBAL LITERACY RATES

**A LIFE'S WORK**
*Vocational training is considered to play a useful role in preparing people for starting work.*

# HOW EDUCATION IS PROVIDED

A person's education invariably begins in the family and in the local community. Formal education is then usually provided at three levels. Primary or elementary schooling is aimed at children from the age of 4 or 5 up to about 11 and is compulsory in almost every country. At this level, most emphasis is given to the role education can play in preparing children for adult life through acquiring the essential skills of communication and comprehension and a basic knowledge of the workings of the world. Secondary education takes children up to the age of 15–18 and helps them to develop analytical skills through more focused teaching in a wider range of subjects. Tertiary or postschool education is usually split into two types. Further education is a term that refers to technical and vocational training that is explicitly geared towards gaining skills and expertise for a particular job or career. Higher education is more academic in content and taught to degree level at a university or college. In many places, tertiary education also encompasses adult education, providing adults with new skills, often as a way of helping them return to work after a period of absence through illness, unemployment, or child-rearing.

**PRIVATE EDUCATION**
*Parents who pay to send their children to private schools, such as Roedean in the UK, believe the education provided will bring social and economic advantages later.*

**ROLE OF RELIGION**
*Religious schools of all faiths, including this Catholic school in India, provide a high level of teaching throughout the world.*

SOCIETY

# HOUSING AND SHELTER

A house is not just a shelter from the elements but also a home from which people organize and live their lives. In all societies, therefore, people consider obtaining a secure home one of life's priorities. All but the most basic of houses, however, are expensive to build or buy. As a result, most housing around the world is provided by the state, by private landlords, or by charities and philanthropic organizations and is rented rather than owned outright. The gap between an individual's assets and the costs of initial construction and maintenance is often filled by state housing provision or rent subsidy. About 1.6 billion people (one-quarter of the world's population) have no home at all or live in cramped conditions in substandard housing with inadequate power, water, or sanitation. One billion live in rural areas, the other 5 billion in overcrowded cities, drawn there by the promise of work and of a better future for themselves and their families.

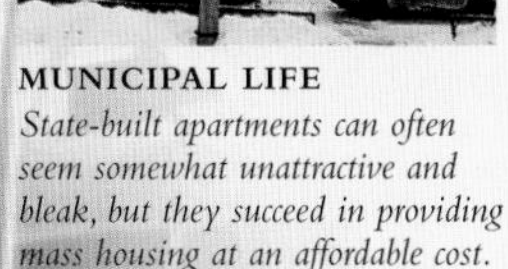

**MUNICIPAL LIFE**
*State-built apartments can often seem somewhat unattractive and bleak, but they succeed in providing mass housing at an affordable cost.*

**COMMUNITY LIFE**
*Many people live in housing built by the local community. Brunei's government has helped villagers to construct longhouses, in which they live communally.*

# BASIC HEALTH MEASURES

In many parts of the world, medicine is a luxury few can afford or have proper access to. For such people, the most effective form of healthcare and the main contributor to improved levels of health is the rise in living standards that is made possible by sustained economic development. Access to clean water and effective sewage systems has played the biggest part in reducing killer diseases. Better diets, sanitation, and hygiene can also dramatically reduce illness and premature death. Even where the medical science exists to control or eliminate diseases, inadequate levels of development can prevent its effective use. In poorer countries, conditions that are easily preventable, such as cholera, diarrhoea, and malaria, are widespread. So, from a world perspective, healthcare needs to focus on promoting economic development and raising living standards to enable the effective control of diseases. Immunization programmes, coordinated internationally by the World Health Organization (WHO) and other health agencies, make a huge impact on health levels. Smallpox was completely eradicated by 1980 and current global initiatives aim to eradicate polio, measles, and tetanus. Even in the more developed countries, basic health problems still exist and tend to be tackled through government health drives.

**HEALTHY WATER**
*Having access to clean water is probably still the most effective form of healthcare in the world.*

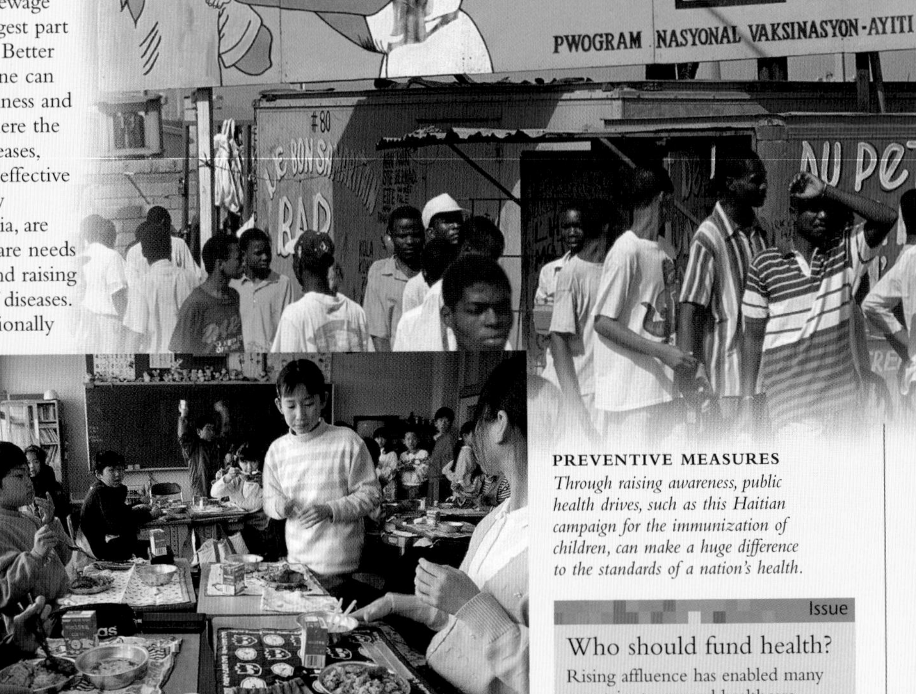
VAKSINEN TIBEBE
POU BA YO LAV
PWOGRAM NASYONAL VAKSINASYON-AYITI

**PREVENTIVE MEASURES**
*Through raising awareness, public health drives, such as this Haitian campaign for the immunization of children, can make a huge difference to the standards of a nation's health.*

**FOOD AND HEALTH**
*Schools that provide nutritious meals at lunchtime, for example this one in Tokyo, do much to improve children's health.*

# FUNDING HEALTHCARE

Social development brings a shift in healthcare, as with all other welfare services, away from a reliance on the family and local community to more formal structures operated by professionals. Rudimentary care and the alleviation of symptoms are replaced by scientific diagnosis, treatment, and cure. Such a shift brings with it a huge increase in public expenditure. Today, most countries with formal healthcare systems spend the equivalent of between 5 and 15 per cent of national output on health, with the government generally paying between one half and three-quarters of this amount. As countries become wealthier, they tend to spend proportionately more on healthcare. Funding varies greatly around the world. In the UK, for example, the National Health Service is free at the point of access and is funded through taxation, although many now use private provision to avoid long waits for hospital treatment. In France, whose healthcare system was rated the most efficient in the world by the WHO in 2000, patients pay for their treatment but the majority of the cost is reimbursed by the state. In the US, there is an enormous disparity in service offered between private provision funded through insurance, typically paid for by employers, and the statefunded provision available to the poor and the elderly. Worldwide, for people with access to health services, the first port of call is the local primary care service provided by doctors, nurses, and midwives. If these professionals are unable to solve the problem, or in the case of an emergency, the next port of call is the hospital. Hospitals provide secondary care, including specialist diagnostic and therapeutic services.

Issue

## Who should fund health?

Rising affluence has enabled many countries to expand healthcare provision, but it has also brought with it the issue of personal responsibility. For example, smoking causes many illnesses, and nearly one in five people in the US is clinically obese, mostly as a result of an unhealthy lifestyle. Do people who do not take responsibility for their own health have a right to state-funded healthcare? In addition, the Western trend towards private healthcare and medical insurance, enabling richer people to receive more efficient treatment, can actually make it more difficult for those with less money to receive the care they are entitled to.

**SECONDARY CARE**
*Hospitals provide an essential range of specialist medical services. Those that are state-funded are the largest and most costly part of any state healthcare system.*

**LOCAL KNOWLEDGE**
*Community resources, such as this local Indian midwife, continue to play an important healthcare role in many countries.*

SOCIETY

## HEALTH TRENDS

Although modern medical science has an international impact, many countries still rely on traditional medicine: health practices incorporating plant and other natural medicines, techniques such as acupuncture, and spiritual therapies. This dependence on traditional medicine is partly out of necessity – according to the WHO, more than a third of people in developing countries lack access to essential modern medicines, such as antibiotics. Traditional medicine places much emphasis on maintaining general wellbeing and this holistic approach has attracted interest in more developed countries, where it is usually known as complementary medicine. Healthcare is not a fixed human need but adapts in parallel with other changes. Socioeconomic developments (such as urbanization and industrialization), medical improvements (such as new vaccines), and population shifts all affect healthcare. The combined impact of these and other changes is that life expectancy is on the up around the world. In the developed countries, this has been the case for the past 150 years, as both infant mortality and premature death rates have declined. This progress has now spread to most other regions of the world, apart from those poorer countries now battling with HIV/AIDS. However, even in the richest countries, new health problems, such as obesity and heart disease, are beginning to emerge.

**ANCIENT REMEDIES**
*Traditional herbal medicine, as practised by this Chinese herbal doctor, remains popular and is often used alongside Western medicine.*

**A GROWING TREND**
*One obvious health trend in the developed world is the rise of obesity in young people. Carrying excess weight leads to ill-health and heart disease in later life.*

## LOOKING AFTER THE POOR

The world's first social security system was introduced in Germany in the 1880s by Otto von Bismarck. The cost of the state-run scheme was met by employers through contributions deducted from the wage packets of their employees, with any shortfall being made up by general taxation. This mix of compulsory contribution and general taxation has remained the model for most modern social security systems, although there is considerable variation in practice and structure around the world. What all systems attempt to do, however, is to offer a safety net for those people who are unable to provide for themselves. Initially seen as a form of insurance for workers unable to work for reasons outside their control, social security – alongside free healthcare and compulsory education – became a key plank of the welfare state systems that developed in many industrial countries during the 20th century. Almost all of these systems have grown in scope and complexity over the years and require a considerable amount of bureaucracy to administer them. As costs have risen, there has been increasing debate over whether benefits should be paid at the same rate to everyone who is eligible or whether they should be means-tested and paid according to the wealth of the recipient, enabling them to be targeted at those most in need.

**NO SECURITY**
*Even in relatively advanced countries such as Russia, some people fall through the gaps in the social security system and are forced to beg for basic sustenance.*

Profile

### Otto von Bismarck

In 1871, Otto von Bismarck (1815–1898) became the first chancellor of a united Germany. He introduced the first social security system in 1881, partly to quell popular support for socialism and partly to win the allegiance of workers. The scheme provided sickness and accident cover, unemployment benefits, and pensions for the elderly and disabled. Similar schemes were soon introduced in the rest of Europe, Australia, and New Zealand.

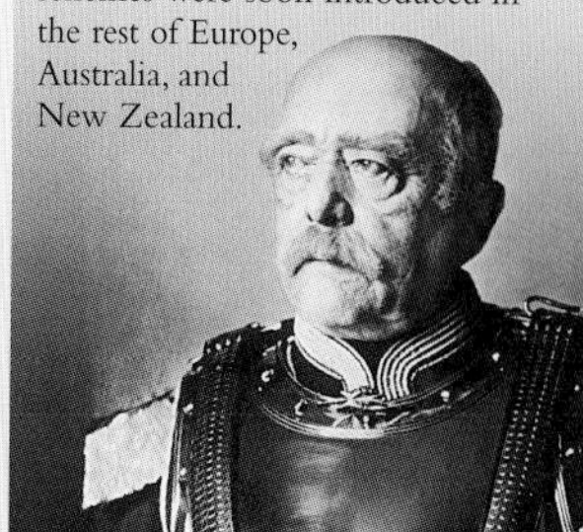

## AGING POPULATION

The rise in life expectancy across the developed world is putting great strain on pension provision, as the proportion of pensioners increases relative to people of working age. As a result, both workers' contributions and general taxation have to rise to pay for the shortfall. Since the late 1970s, therefore, the trend has been away from publicly funded pension provision towards increased reliance on private schemes. Most governments are also trying to limit their long-term liabilities, usually by raising the age of pension entitlement and/or reducing the level of state benefits. Such moves to curtail state provision are accompanied by official encouragement for private schemes organized through employers and individual pension arrangements. However, many of these private schemes have themselves encountered difficulties, not least because a good proportion of their funds are invested in stock markets subject to bouts of extreme volatility. As pensions are paid from the wealth that exists at the time, the solution to the long-term pension crisis lies in increasing a nation's wealth through better economic performance and higher output. Rising life expectancy in the developing world will have an even greater financial impact on communities because there is usually no state pension provision whatsoever.

**NEW PURSUITS**
*Increasing social provision for the elderly enables many, such as this bingo player, to live independent and rewarding lives well into old age.*

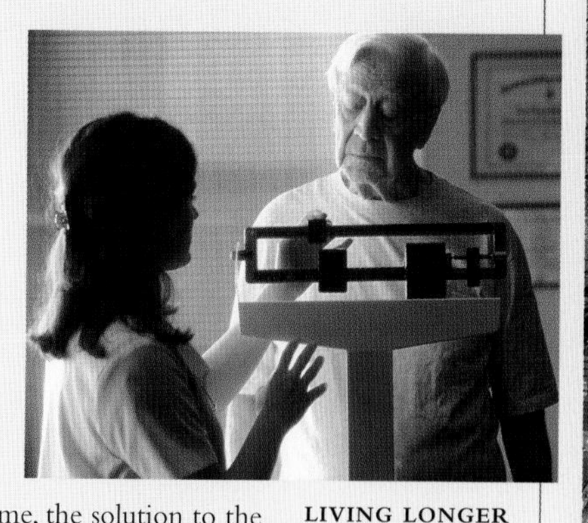

**LIVING LONGER**
*The trend towards an increase in life expectancy puts a strain on health and social services, as well as on pensions and other benefits.*

# CULTURE

# CULTURE

Culture includes what people believe, how they behave, how they shape their environment, and what they understand about the world. Most elements of culture are passed on by the family or community. People absorb this knowledge unconsciously as they grow up; the values and habits of their social group seem normal, while those of other groups may seem strange or even threatening. Some people, however, choose to adopt or reject certain elements.

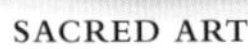

**SACRED ART**
*In many cultures, making and viewing art has been used as an aid to worship or prayer. These Buddhist monks are making a mandala to transmit spiritual energy.*

A shared culture strengthens social bonds. For individuals, it provides general rules for behaviour that they could not develop by themselves, even over a lifetime.

## BASICS OF CULTURE

Every culture is based on three elements: ideas, customs, and objects. These elements reinforce each other to provide a worldview and define the place of individuals within it.

Common ideas are the cornerstone of a culture. People find it easier to relate to one another if they share an understanding of how the world works, how to distinguish right from wrong, and even aesthetic preferences, such as a common ideal of beauty.

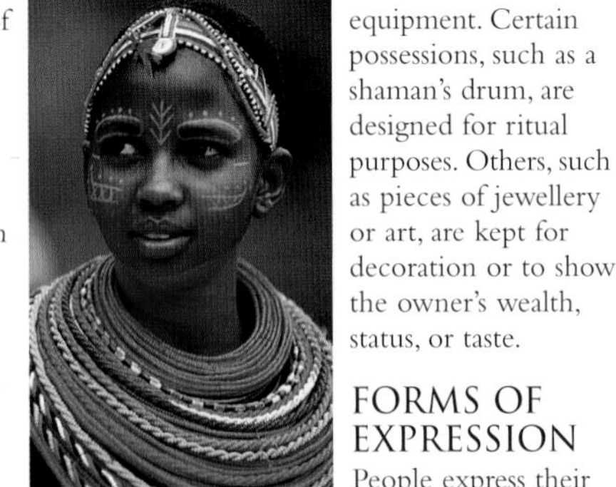

**SOCIAL SIGNAL**
*For many people, like this Masai girl, clothing and body decorations convey information about status and wealth.*

People's worldviews are revealed in their customs. Some customs develop for practical reasons to ensure health and safety. They include rules about what is safe to eat, how to dress for protection in harsh climates, and hygiene habits. Certain customs are designed to ensure that people treat each other fairly. Others may provide a way for individuals and groups to assert their identity, especially when meeting strangers.

A culture is also defined by the objects with which people surround themselves. Many such objects have symbolic as well as practical value; for example, the tools of a person's trade, such as a doctor's stethoscope or white coat, may be status symbols as well as essential equipment. Certain possessions, such as a shaman's drum, are designed for ritual purposes. Others, such as pieces of jewellery or art, are kept for decoration or to show the owner's wealth, status, or taste.

## FORMS OF EXPRESSION

People express their culture in their beliefs, communication styles, clothing and body adornments, and art and science.

Religions and other belief systems form a framework for existence by giving explanations of how the universe was created, the origin of human beings, and what happens after death. They also allow people to mark life stages such as entering the world, maturity, marriage, and death. In addition, belief systems define the difference between right and wrong and set out rules for guiding or controlling people's behaviour.

Language and nonverbal forms of communication enable people to interact and express themselves. A shared language is one of the strongest elements that bind a group together; in some cases, as with Bengali, people may even fight for the right to speak it. Gestures, "body language", and symbols such as religious artefacts or national flags can convey messages even more strongly than words.

Clothing and body adornment may show instantly who a person is and his or her social status. It may also be a potent symbol of group identity; for example, people may preserve their traditional forms of dress to maintain their culture or reconnect with their heritage. On the other hand, some people adopt styles that conflict with accepted dress codes as a sign of rebellion against their society's values. In addition, many people have their own personal style, which is a form of self-expression.

Art and science are the most sophisticated forms of cultural expression. Many types of art, such as portraits and popular novels, are based on simple responses to the everyday world; on the other hand, art may involve highly intellectual exploration of forms and ideas. Science is a formal method for collecting, analysing, and processing data. It includes many disciplines, with both practical and theoretical aspects. Both art and science may be seen as bringing prestige to their societies; for this reason, they are often controlled by elite groups.

## GROUP IDENTITY

Agreement on a wide range of matters reduces conflict and aids cooperation. Such shared values may exist in groups of any size – from the "microculture" of specific families, workplaces, communities, or gangs, to larger groups such as social classes, up to national and even global culture.

Cultural norms, such as codes of morality or dress, may be enforced by group leaders, such as political officials, ministers of religion, or senior family members. They may also be upheld by ordinary people to ensure group cohesion. Anyone who does not conform to these values may seem upsetting

**REBELLION**
*These young Japanese people have adopted an extreme form of Western dress as a form of rebellion against their society's traditions.*

Fact

### Eating customs

Many cultures have rituals to do with eating. Some of these customs evolved for practical purposes, such as eating with only one hand, which is kept clean for this purpose. Mealtimes also have the social functions of bringing the group together and reinforcing relationships and hierarchy. Special meals may be an important part of festivals and ceremonies.

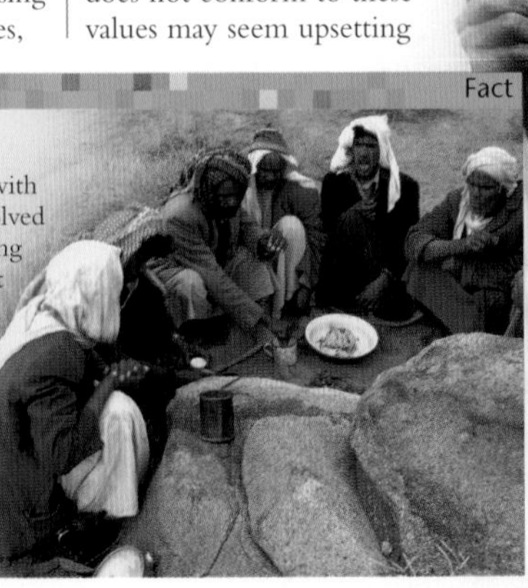

BEDOUIN SHARING A MEAL

or threatening even if they pose no physical danger. For example, fans of rap or heavy metal may be seen as intimidating just because of their loud music, and minority religious groups with a distinctive lifestyle, such as the Amish or Orthodox Jews, may be viewed with suspicion even though they do not force their beliefs on anyone else in society.

## HOW KNOWLEDGE IS TRANSMITTED

There are many ways by which culture can be received and passed on. Much of what we know about the world is taught by our elders as we grow up. Such learning may be part of daily life or may be transmitted formally in school. Another source, especially of new information, is people of our own age. Children can also transmit knowledge to older people – for example, by showing their parents how to use the internet.

For most people, the effects of their early cultural shaping last for life. It is possible, though, for anyone to make his or her own choices in beliefs and actions.

**MOCHICA EARRING**
*The Mochica civilization of Peru (100BC–AD700) left artefacts such as this earring; they also left lingering traces in modern culture, such as place names.*

Education, the influence of mass media, and the loosening of social bonds have allowed many people to adapt or even shake off their initial cultural experience.

## WHEN DIFFERENT CULTURES MEET

Many societies across the world have absorbed influences from other cultures. This process has gone on throughout history. In some cases, it has been an unintentional side-effect of trade or learning. In certain instances, however, one people has actively tried to impose its way of life on another.

The most severe cultural clashes have occurred when one people has invaded or made war on another. For example, when the Spanish conquered South American territories during the 15th–17th centuries, they wiped out the Inca empire. In less extreme cases, the invaders' culture was simply overlaid on that of the conquered people. This process occurred in the early Islamic empire of the 7th and 8th centuries AD, where the Muslims tolerated the customs and even the beliefs of the conquered Byzantines and Persians.

The diffusion of influence can also be a two-way process, as in the relationship between the Europeans and the Japanese during the 19th century. Japanese society adopted Western-style government and legal systems; in turn, their prints and clothing were admired by European artists such as Claude Monet and Vincent Van Gogh.

**CULTURAL CROSSOVER**
*Indian film-makers sometimes rework Hollywood films. One of the most successful of these adaptations is* Judaai, *inspired by the blockbuster* Indecent Proposal.

Today, the effects of mass travel and the influence of the media have left almost no society in the world untouched. The most dominant culture is that of the US. Its goods, technology, and entertainment have spread to even the remotest regions and have had a profound effect; for example, people all over the world may be exposed to American films.

Mass emigration of workers to other countries, on a temporary or permanent basis, may also produce extensive contact between cultures. Two countries where such contact occurs are the Persian Gulf states of Qatar and the United Arab Emirates. Expatriate workers from other Arab countries, the Indian subcontinent, Southeast Asia, Europe, and the US, all add their customs as well as their skills to the indigenous culture.

**PASSING ON SKILL**
*Indigenous peoples across the world feel the need to preserve their ancestral traditions. This Aboriginal Australian man is teaching two boys how to hunt.*

A further influence on many cultures is humanitarian activity. Medical organizations such as the Red Cross and Medecins Sans Frontières may operate in poor or war-torn areas where nobody else will go. A huge array of charities supply technological and agricultural help to the poorest areas. Some support "fair trade" arrangements so that small-scale farmers can enjoy better trading conditions when selling their goods to rich countries.

**MOBILE SNAPSHOT**
*Mobile phones have evolved rapidly in the last 20 years. Users have also adapted, learning new skills such as text messaging and taking photographs with their phones.*

## ADAPTATION AND INNOVATION

Cultural changes can have both beneficial and negative effects. Advances in medicine may enable many people to enjoy longer, healthier, and more productive lives than ever before. However, other changes, such as fast food and huge numbers of television channels, could lead to health problems such as obesity or social problems such as an increase in violence.

People can use aspects of culture such as scientific skills, literature, or philosophical or spiritual ideas to understand their existing situation and to deal with new problems or opportunities. If their innovations are sufficiently useful, or appeal to enough other people, they may spread through society. Affluent, industrialized societies may readily accept both technological and social advances, such as sexual equality. Other groups, such as strict political or religious communities, may see changes as harmful and insist on keeping to traditional customs.

**PRAYER AT THE GANGES**
*Pilgrimage is a powerful experience for believers. For Hindus, the holiest pilgrimage site is the River Ganges.*

# BELIEF

Throughout history, people have held religious or other beliefs. Such beliefs offer explanations for mysteries in human life, the natural world, and the universe. For many people, religion is the most important aspect of life; for some, it is little more than a social convention. Still others have rejected formal religion, opting for a human-centred morality. Beliefs shape much of human thought, action, and experience. They are central to culture, and for many people are the element that defines who they are.

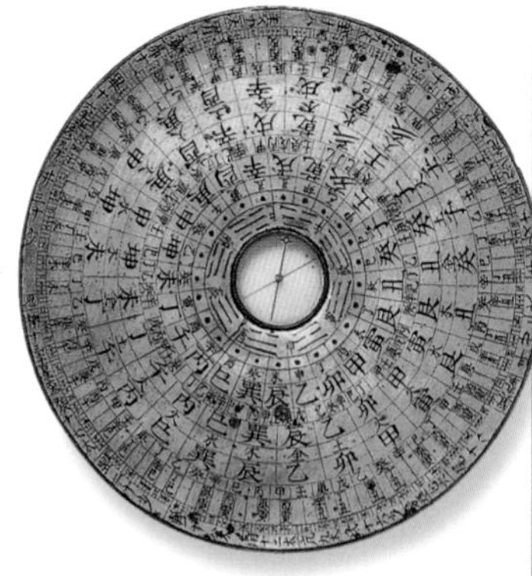

**FENG SHUI COMPASS**
*Feng shui is a traditional Chinese belief still practised today. It is based on the idea that people, objects, and landscapes are infused with chi (natural energy). This compass is used to align objects in harmony with chi.*

There is evidence that human beings have had spiritual beliefs for many thousands of years. There are also clues to suggest that other hominids, such as Neanderthal man, may also have held such beliefs; for example, artefacts from Neanderthal graves seem to show that these people carried out rituals when burying their dead.

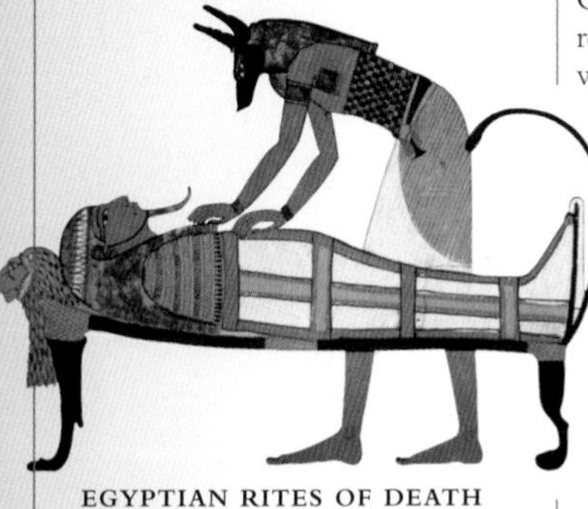

**EGYPTIAN RITES OF DEATH**
*Coping with death is important in almost all religions. The ancient Egyptians believed that the jackal-headed god Anubis guided people's souls after death.*

## ORIGINS OF BELIEF

The first religions helped people to understand the world. Some beliefs, after being explored and redefined, gave rise to modern sciences; for example, beliefs in gods who existed in the stars led to astronomy. Many religions had explanations for the origin of human life, what happens when we die, and why misfortunes occur. Rules for daily life – in particular, what you could eat and whom you could marry – were designed to keep people healthy and create stable family groups. Stories about the ancestors, handed down from one generation to the next, helped to bind a community together.

## WORLD SYSTEMS

Most belief systems arose within specific communities or among small groups who came together in faith. A few, however, were taken up by the most powerful people of their time and spread much further. For example, in the 3rd century BC the Indian emperor Ashoka became a Buddhist. He renounced war and set out rules for his society based on respect for all forms of life. He also sent out missionaries to other countries including Sri Lanka, Myanmar (Burma), and China and as far as Greece. In the 4th century AD Christianity became the official religion of the Roman empire; it was then carried across the empire by soldiers and traders. From the 7th century, the Muslims spread from the Arabian peninsula into southern Europe, northern Africa, and further into Asia.

As trade routes grew and new lands (such as the Americas) were discovered, many beliefs spread across the world. Some groups, such as the Zoroastrians, kept the faith within their community, while others, such as Christians and Muslims, converted more people to their religion.

## BELIEFS TODAY

Millions of people still follow religions. In Western societies many have abandoned these faiths, but others have returned to them as a guide and a source of comfort in a complex world. The most extreme believers, known as fundamentalists, create communities governed by strict religious laws. A few strive to impose their views on the whole world by any means, including the use of force or even terror.

Many members of traditional communities still hold the ancient belief that humans have a direct relationship with nature and with the spirit world. They consider all parts of the environment, even stones and water, to have their own spirit. This belief is called animism. It may involve specialists known as shamans, who contact the spirits or travel to the spirit world for purposes such as healing sick people.

The need for belief is still strong even in people who do not follow organized religions. Some belong to sects – offshoots of a religion, such as the Mormons and the Jehovah's Witnesses, whose beliefs sprang from Christianity. Others follow entirely new beliefs such as Scientology or Rastafarianism.

Religion can sometimes give rise to problems. Certain community members, particularly women, may be oppressed. Intolerance of other faiths may lead to conflict. A few believers reject scientific learning; for example, Creationists deny the theory of evolution. For these reasons, millions of people follow nonspiritual belief systems. Such people may choose to be atheists, who reject any belief in gods; agnostics, who hold that we cannot know whether or not gods exist; or humanists, who believe that morality is a human responsibility rather than a spiritual matter.

Issue

### God and the brain

Neurologists have been discovering how spiritual belief affects the brain. In one experiment, Buddhists were asked to meditate, then scans were taken of their brains. It was found that the activity of the parietal lobes, which control our sense of ourselves in time and space, was reduced during meditation. This effect could cause blurring of the boundary between "self" and the outer world, and a sense of being at one with God and the universe. Believers of many faiths have described such feelings during meditation, trances, or prayer.

**JOURNEY OF A GOD**
*The worship of Mithras (left), the ancient Persian god who mediated between heaven and earth, is an example of a belief that spread far from its place of origin. It was adopted by the Romans and carried through their empire, reaching as far as the British Isles.*

**PEACE SIGN**
*Many people hold certain ethical beliefs as strongly as religious people hold their faith. The peace symbol, created for the movement to ban nuclear bombs, represents a belief that war, poverty, and abuses of power should be abolished.*

# WORLD BELIEFS

**EXPRESSION OF FAITH**
*Some people show their faith openly, for example by their dress. This ring, for a Muslim, bears text from the Koran.*

All belief systems provide a structure for human life and thought. Two of their main functions are to define human beings' worth and place in the world and teach people how to treat others; the so-called "Golden Rule", found in a wide variety of religions and other moral systems, is the clearest example of such guidance. Other elements common to many systems are a form of discipline contained in prayer and ritual, and myths and legends to entertain people as well as to explain mysteries. Elements of a belief may persist when the belief itself has largely died away; for example, in Western societies even people with no outward signs of belief may observe festivals based on religious holy days, such as Christmas.

## MORAL GUIDANCE

An essential element of both religious and nonreligious moral systems is the need to treat others properly. This idea is defined in the Golden Rule. It is expressed as "Treat others how you would wish to be treated yourself" or "Do not do to others what you would not wish them to do to you". The rule prompts people to imagine themselves in someone else's place. It is meant to promote kindness, honesty, and trust, and to limit abuses of power. The Golden Rule first appears in ancient teachings ranging from Confucianism and Hindu faith to Native American beliefs, and is a cornerstone of Christianity. It has led to the view that all humans are equal in importance and every person has the right to be treated with fairness and respect. Today, this idea is enshrined in the United Nations' Universal Declaration of Human Rights.

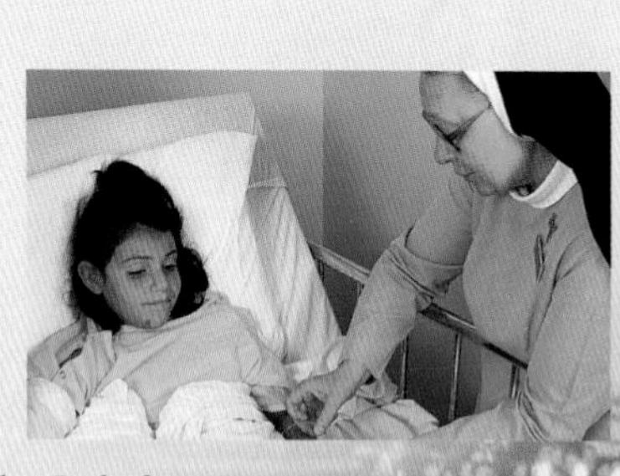

**PROVIDING CARE**
*Many religions stress the importance of caring for people in need. This Christian nun is working as a nurse in a hospital.*

## WAYS OF WORSHIPPING

In many belief systems, group activities such as public rituals, festivals, and pilgrimages are important both to reaffirm faith and strengthen the bonds of a community. Rituals are repeated, ceremonial sequences of actions that take place during prayer or holy days. Even some non-religious people recognize the need to mark special occasions such as marriage or naming a baby. Festivals take place on holy days or mark times of year such as New Year. Pilgrimages are journeys to sacred sites; the effort of travelling to the site is intended to test the strength of one's faith. Acts of physical endurance, such as fasting, may also be carried out as tests of faith.

Prayer and meditation are more private forms of spiritual discipline. Prayer is an appeal to a deity. People may pray to ask forgiveness for their sins, give thanks for blessings, or deal with spiritual problems. Meditation involves calming and focusing the mind to free it from daily concerns.

**MUSLIM PRAYER**
*This man, although praying publicly, is focusing on his own contact with Allah.*

**PILGRIMAGE**
*These pilgrims are climbing Mount Fuji in Japan, home of the goddess Sengen-sama. They wear special white clothes and straw hats. The climb involves visiting the 88 temples on the mountainside.*

**SUN RITUAL**
*A southern African shaman carries out a form of divination based on the setting sun.*

## MAGIC AND DIVINATION

In ancient societies, people used magic to contact spirits or influence situations and divination to find out the future. Magic has two forms: contagious magic, in which a spell is performed on an object that has been in contact with the person or thing to be targeted, and sympathetic magic, involving the use of objects that resemble the target. Divination involves predicting the future from phenomena such as the movement of the stars and planets (astrology), or by methods such as using Tarot cards. Magic and divination are still used in traditional communities in places such as Siberia, and by people with neo-pagan beliefs, but are forbidden in religions such as Judaism and Christianity, in which believers are required to submit to God's will. People with rational belief systems, such as atheists, may also disapprove of them. However, many people with little interest in formal religion find a spiritual element, or at least entertainment, in activities such as astrology.

**TAROT CARDS**
*The mystical symbols on Tarot cards date back to 16th-century Italy. The cards are laid in specific patterns to give predictions.*

## MYTHS AND LEGENDS

**HEROIC STRENGTH**
*People no longer believe in the ancient Greek gods, but legends about them are still familiar in Western culture. This vase shows the half-divine hero Herakles (whom the Romans called Hercules).*

Stories about the supernatural are called myths or legends. They are a way for societies to pass on their rules and beliefs. Some explain how the universe was created and where humans came from. Others are based on types of character. One type is the Hero, who performs brave deeds, such as Herakles (Hercules). Another is the Wise Man, such as Väinämöinen in Finnish myth. A third is the Trickster, who fools his enemies and even the gods by his cunning, such as Anansi Spider in West Africa. In the oldest belief systems, legends may be the main form of spiritual and social teaching. Myths may also persist in religions with scriptures. For example, in Celtic legend King Arthur was associated with the Holy Grail, the cup that Jesus reputedly used at the Last Supper. In medieval Christianity, the quest for the Grail symbolized the soul's striving for purity. People still enjoy tales with mythic themes, from the Buddhist story of the Monkey King to *The Lord of the Rings*.

**KING ARTHUR**
*Legends about King Arthur were popular in medieval times. This picture shows the Round Table, at which he and his knights gathered.*

Issue

### Cults

There is no agreed definition of a cult, but the word is usually used of a faith group that has extreme beliefs and exerts intense control over its members' lives. Many cults engage in harmful or even fatal acts. In 1995, for example, the Aum Shinri Kyo cult released the nerve gas sarin in the Tokyo underground, killing 12 people and injuring thousands more. This act has led to protests against the cult leader (below).

**CATHOLIC INITIATION**
*Some beliefs, such as Roman Catholic Christianity, have highly elaborate rituals and symbolism. These priests are being elevated to the rank of cardinal; they prostrate themselves to express their total submission to God and the Church.*

WORLD RELIGIONS

## Christianity

**Founded:** 1st century AD

**Estimated number of followers:** 2 billion

**Distribution:** Worldwide, particularly in Europe, North and South America, southern Africa, Australia

Christians believe that Jesus, a Jew born in ancient Palestine around 4BC, was both the Son of God and fully human. Jesus is part of the Trinity: God as the Father, Son, and Holy Spirit. He taught people to love God and to care for others, including poor and sick people and social outcasts. Jesus' teachings led to conflict with the Jewish and Roman authorities, and he was crucified. Christians believe that he returned from the dead and that through his death he atoned for humanity's sins.

Christianity is the largest religion in the world. There are thousands of forms (denominations), but the main branches are the Roman Catholics (about 1,200 million), Protestants (360 million), and Orthodox Christians (170 million). The Orthodox branch is the oldest; it originated in Constantinople (now Istanbul) and is based in Greece, Russia, and eastern Europe. Roman Catholic Christianity dates from 1054AD, when the western part of the Church (based in Rome) separated from the eastern Orthodox part. Protestant Christianity developed in the 16th century as a reaction to the Catholic tradition. It held that people had a direct link to God, in contrast to the Catholics' priestly hierarchy.

Many people, especially Roman Catholics, venerate Jesus' mother, the Virgin Mary. Saints are also revered. They include the apostles, whom Jesus chose to help him spread his message; martyrs, who died for their beliefs; and people who dedicated their lives to Christianity.

The Christian Bible includes the Old Testament (the Jewish Bible) and the New Testament. This part is the main Christian scripture. It centres on the Gospels of Matthew, Mark, Luke, and John, which provide four accounts of Jesus' life and work.

Festivals commemorate events in Jesus' life; the primary ones mark his birth (Christmas), death (Good Friday), and resurrection (Easter Sunday). In addition, Sunday is a day of worship. Followers of the main traditions, particularly Catholics, attend church and take communion, in which they consume bread and wine symbolizing Jesus' body and blood. This ritual originated at the Last Supper, the meal Jesus shared with his apostles just prior to his death, when he asked them to take bread and wine in memory of him.

The holiest places are Bethlehem, the site of Jesus' birth, and Jerusalem, where he died. Other centres are Istanbul for Orthodox Christians and the Vatican for Roman Catholics.

Christianity has had a deep effect on world culture. In the 4th century AD it became the religion of the Roman empire. Over the next 1,500 years, it spread all over the world. In medieval Europe, centres of religious study also preserved Greek and Roman learning. State patronage combined with faith inspired some of the world's finest art, music, literature, and architecture. In addition, Christianity still forms a basis for laws and ethics today.

**GUADALUPE PILGRIM**
*A pilgrim carries a picture of the Virgin Mary to the shrine of Our Lady of Guadalupe, in Mexico. This shrine, dating from the 16th century AD, is the oldest and most important site of pilgrimage in the Americas.*

**BAPTISM**
*Some Christians, like these Pentecostal worshippers, baptize people by immersing them in a lake, river, or pool, as an echo of Jesus' baptism in the River Jordan.*

**ORTHODOX PRACTICES**
*The traditions of Orthodox Christianity date from the first few centuries of the Christian era. Services involve rich imagery, ceremony, and music, but are held in the ordinary language of the congregation.*

**HOLY CROSS**
*The Cross symbol has various forms, such as the Orthodox cross, with one large and two small bars.*

**MODERN FAITH**
*This cross, in Groom, Texas, was built in 1995 but is already a focus for pilgrims. At 58m (190ft), it is said to be the largest cross in the northern hemisphere.*

Fact

### The Black Madonnas

Hundreds of statues and paintings in Roman Catholic Europe show Mary and baby Jesus with black-painted faces and hands. Many of these images date from medieval times. Several, such as Our Lady of Czestochowa in Poland (below) and Our Lady of Montserrat in Spain, are still venerated today. Some historians consider that black figures were originally created as images of the Egyptian goddess Isis and her son Horus, and that this tradition was perhaps adapted by the Church to attract pagans to Christianity.

**COLLECTIVE PRAYER**
*Men at prayer fill this mosque in Saudi Arabia. Collective worship, called al-jum'a, takes place at noon every Friday. It consists of prayers and a sermon.*

WORLD RELIGIONS

# Islam

**Founded:** 6th century AD
**Estimated number of followers:** Around 1.2 billion
**Distribution:** Worldwide, especially the Middle East, North Africa, Central Asia, and Southeast Asia

Islam originated in the 6th century AD in the Arabian peninsula. Muslims believe that Allah sent various prophets, including Abraham, Moses, and Jesus, to teach people to live a righteous life, but they were largely ignored. Allah then selected the last and greatest prophet, Muhammad, and made to him a series of revelations. These teachings were written down and became the Koran.

Because the Koran is taken to be the actual word of God, Muslims read it in the original Arabic even if Arabic is not their native language. Muslims also refer to writings called the *hadith* (the sayings of the prophet).

Muhammad was born in Mecca around AD570. He received the first revelations when he was about 40 years old. He started to preach in Mecca but was persecuted, so fled to Medina; the Muslim calendar starts from the date of this flight, or *hijra*. Muhammad eventually returned to Mecca, and the city remains at the heart of Islam.

The name "Islam" is Arabic for "submission", and reflects the fact that the Koran teaches people to live in submission to Allah. The faith governs every aspect of life. Its main elements are known as the Five Pillars. The first is the *shahada*: the affirmation "There is no God but Allah; Muhammad is his prophet". The second is *salat* (prayer). Muslims pray five times a day. They can say prayers anywhere, but always face Mecca. Many also attend prayers in a mosque on Fridays. The third pillar is *sawm* (fasting) through Ramadan, the ninth month of the Muslim year. During the day people abstain from eating, drinking, smoking, and sexual intercourse. The daily fast ends at sunset with a meal called *iftar*, and the month ends with the festival of *Eid al-fitr*. The fourth pillar is *zakat*, in which people give a proportion of their annual income to charity. The fifth is the *Hajj* (the pilgrimage to Mecca), which Muslims try to undertake at least once in their lives.

**CALL TO PRAYER**
*Although people may pray anywhere, a man called a* muezzin *gives the call to prayer from the top of a minaret, in a mosque.*

There are two main forms of Islam. The larger is Sunni Islam, followed in most of the Muslim world. The other is Shia Islam, centred in Iran and Iraq.

Islam contributed greatly to world culture, especially from the 8th to 13th centuries AD. Arts included calligraphy, literature, and architecture. Muslim scholars believed that scientific study deepened understanding of Allah. They built on Greek and Latin learning to make huge advances in mathematics, science, and philosophy.

**ISLAMIC STUDY**
*Learning the Koran and studying texts on it are important parts of Muslims' education.*

**FINDING MECCA**
*Muslims use a special compass called a* qibla *to help them find the direction of Mecca for their daily prayers. This medieval qibla has the days of the month inscribed on it.*

**THE KA'BA**
*This structure, at the centre of the Great Mosque in Mecca, has been a Muslim shrine since the time of the prophet Muhammad. Walking around the* Ka'ba *seven times is one of the main acts of the* Hajj.

WORLD RELIGIONS

# Buddhism

**Founded:** 6th–5th centuries BC

**Estimated number of followers:** 360 million

**Distribution:** Worldwide, but mostly in Tibet, Nepal, Mongolia, China, Japan, Sri Lanka, and Southeast Asia (state religion of Thailand and Bhutan)

Buddhists follow the teachings of Prince Siddharta Gautama, or Buddha ("enlightened one"). Prince Siddharta was born in northeast India around 560BC. While still a young man, he attained enlightenment. In this process, he learned the ultimate truth that only by overcoming selfish desire can the cycle of birth, life, and death be broken. He also learned what are known as the Four Noble Truths: the universality of suffering, its origin, its cure, and how to find that cure by following a "middle way" of good thoughts and moral actions. Buddha had the chance to attain Nirvana, the state of freedom from suffering and desire, but instead turned back and dedicated himself to teaching others what he had learnt.

After Buddha's death, two traditions emerged. Theravada, the main form in Southeast Asia, is the older; it closely adheres to the Tripitaka (three baskets) of Buddha's teachings. Mahayana, followed mainly in north and east Asia, emerged around 250BC and includes a wider range of teachings. Mahayana also stresses the importance of the *bodhisattva*: someone who attains the threshold of Nirvana but turns back from it in order to help others. Chief of the bodhisattvas is Avalokiteshvara, who is seen as the embodiment of compassion.

Buddhism involves training the mind to reach enlightenment rather than showing faith in gods. Believers pay homage to Buddha's image but revere him as a teacher rather than a deity. People also give offerings of flowers as a reminder that life is impermanent; light to dispel darkness from the mind; and incense, which signifies the lasting fragrance of Buddha's insights. Another important practice is meditation – stilling the mind in order to gain insight. Buddhists also chant prayers or mantras (single words or phrases, which are repeated many times) to focus their concentration.

**OFFERING OF LIGHT**
*Monks light candles as offerings in front of the great stupa at Bodhnath, Nepal. Stupas are burial mounds containing relics of Buddha. The eyes on each side of the stupa represent Buddha's "all-seeing eyes".*

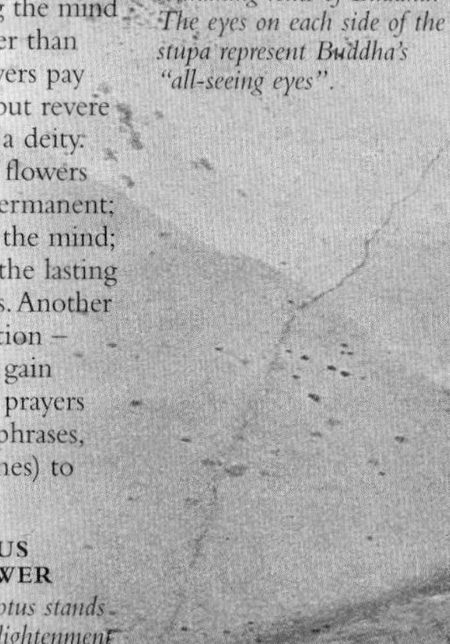

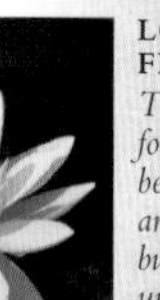

**BUDDHA'S HANDS**
*Images of Buddha include symbolic gestures. The right hand pointing to the ground signifies Buddha asking the earth to bear witness to his enlightenment.*

**LOTUS FLOWER**
*The lotus stands for enlightenment because its roots are fixed in mud but its petals grow up into the air.*

**PRAYER FLAGS**
*In Tibet, people fly poles or strings of prayer flags to ensure good fortune for all sentient beings. Each flag has a mantra on it; each flutter is a recitation of the mantra. The flags are usually flown on important days such as Losar (New Year) or as protection from harm.*

Profile

## The Dalai Lama

For 6 centuries the Dalai Lamas have been the principal spiritual leaders for Tibetan Buddhists. His Holiness the present (14th) Dalai Lama had to leave Tibet in 1959 after the Chinese invaded; since that time he has lived in Dharamsala, northern India, with around 100,000 of his followers. The Dalai Lama has campaigned constantly to preserve the Tibetan Buddhist way of life and worked to promote peace all over the world. He was awarded the Nobel Peace Prize in 1989 and has become one of the world's most widely respected spiritual authorities.

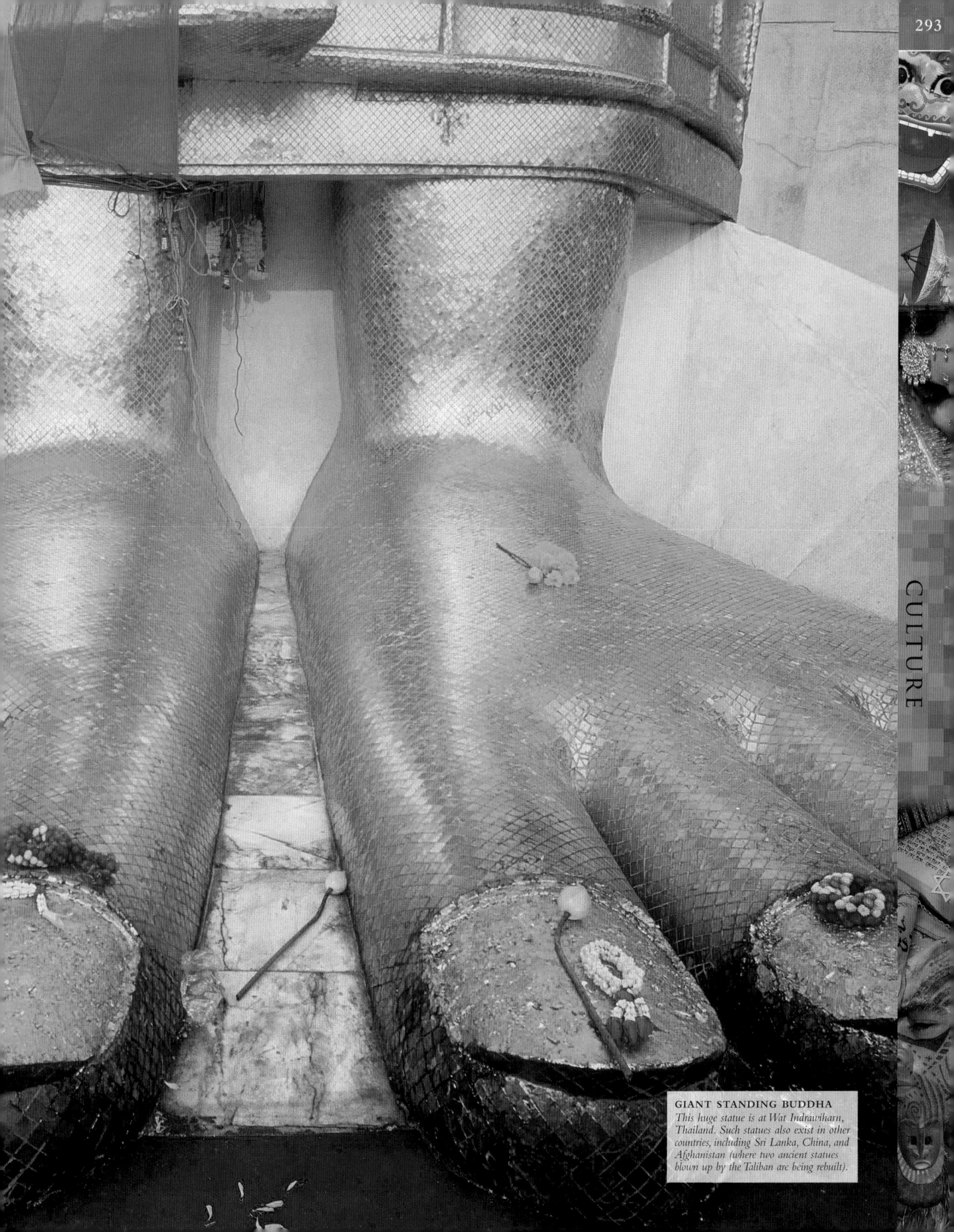

**GIANT STANDING BUDDHA**
*This huge statue is at Wat Indrawiharn, Thailand. Such statues also exist in other countries, including Sri Lanka, China, and Afghanistan (where two ancient statues blown up by the Taliban are being rebuilt).*

WORLD RELIGIONS

# Hinduism

**Founded:** About 1,800BC

**Estimated number of followers:** About 800 million

**Distribution:** Worldwide, but concentrated in India

Hinduism is the oldest world belief system still widely practised today. The earliest known believers were people of the Indus Valley civilization, in what is now Pakistan, around 1,800BC. The name "Hindu" comes from the Sanskrit name for the River Indus, *Siddhu*. The word "Hinduism" was first used by Western scholars in the 19th century AD, to refer to the collection of religious traditions existing in the Indian subcontinent.

Hinduism has absorbed beliefs from many diverse systems. Followers do not all believe the same things but most subscribe to certain fundamental ideas.

At the centre of the faith is the belief in one god or "absolute spirit", whose name is Brahman. He is seen as the ultimate reality, existing within everything in the universe and beyond time and space. In addition, Brahman exists in the human soul as *atman* (self). Heaven is believed to be the union of Brahman with atman.

**OM SYMBOL**
*The syllable OM is said to contain all of time and everything beyond time. It is repeated daily and used in meditation.*

**RIVER GANGES**
*The River Ganges is sacred for Hindus; it is considered to be a goddess (Ganga) and a consort of the god Shiva. Pilgrims bathe in the river to wash away their sins.*

Brahman governs the universe through the *trimurti*, or three principal qualities: Brahma, the creator; Vishnu, the preserver; and Shiva, the destroyer. The trimurti have millions of forms as gods and goddesses, which illuminate different aspects of Brahman. Many people worship one deity above all others. Most follow Vishnu or an *avatar* (incarnation) of the god – usually Krishna or Rama. Some believers follow Shiva. Others worship the Mother Goddess, or one of her various forms, such as Parvati or Kali. Hindus usually worship their particular deity in their homes, using a family shrine, but may attend a *mandir* (temple) as well.

Every soul is believed to undergo *samsara*, a constant cycle of death and rebirth. Each life is ruled by *karma*: the principle that the balance of good and bad acts in one life determines the state of the soul in the next life.

**GANESHA**
*One of the best-loved deities, Ganesha is the son of Shiva and the goddess Parvati. He is the god of learning and the "remover of obstacles".*

The end of the process is *moksha*, in which the soul is freed from samsara and united with Brahman.

Hindus can earn merit for their next life by *dharma* ("appropriate living"). Dharma governs the whole lifetime. Families teach it to their children by rules, example, and religious stories and myths. In fact, Hindus usually refer to their belief system as *Sanatana Dharma* ("eternal dharma").

Dharma also involves following the rules for one's *jati* (social caste). The castes form a hierarchy with Brahmins at the top and Dalits ("untouchables") at the bottom. The caste system was abolished by law in India in 1949 but is still significant for many people.

History

## Sacred scriptures

Hindus have several sacred texts. The Vedas, dating from about 1500BC, contain hymns, prayers, and rites for various gods. The Upanishads, written between 800 and 400BC, show how *atman* (the human self) can unite with Brahman. They also discuss *samsara* (the cycle of rebirth) and *karma* (the law by which acts in one life are rewarded or punished in the next). At the heart of Hindu faith is the *Bhagavad-gita* (Song of the Lord), a dialogue between the god Krishna and a warrior called Arjuna. It shows the three paths to salvation: action, such as religious observances; insight; and devotion to a god.

**BHAGAVAD-GITA**
*A scene from the* Bhagavad-gita *shows the god Krishna driving Arjuna's chariot while counselling him.*

**CELEBRATING HOLI**
*The spring festival of Holi is a time of pleasure for Hindus. A major feature is throwing brightly coloured powder and water balloons at people, as these revellers have done.*

WORLD RELIGIONS

## Judaism

**Founded:** c2000BC

**Estimated number of followers:** 14 million

**Distribution:** Worldwide, especially the US, Israel, and Russia

Judaism is the oldest monotheistic religion (one having a single god) and is the parent of Christianity and Islam. Jews trace their ancestry to Abraham, the leader of a tribe from Mesopotamia (now Iraq); his son Isaac; and Jacob, father of the 12 ancient tribes of Israel.

Jews believe that they are the chosen people of God (Yahweh). They believe that God made a covenant with them, promising them a land of their own (Israel) in return for obeying his laws. These laws were given to the prophet Moses and set down in the *Torah*, which comprises the first five books of the *Tanakh* (Jewish Bible). The 613 laws lie at the heart of the Jewish faith, covering many aspects of daily life, from ritual and morality to hygiene. The most important are the Ten Commandments, which constitute the fundamental rules for society.

The centre of religious life is the home. Jewish families take care to teach children traditions such as eating *kosher* food (food prepared according to Jewish law) and observing the Sabbath (day of rest).

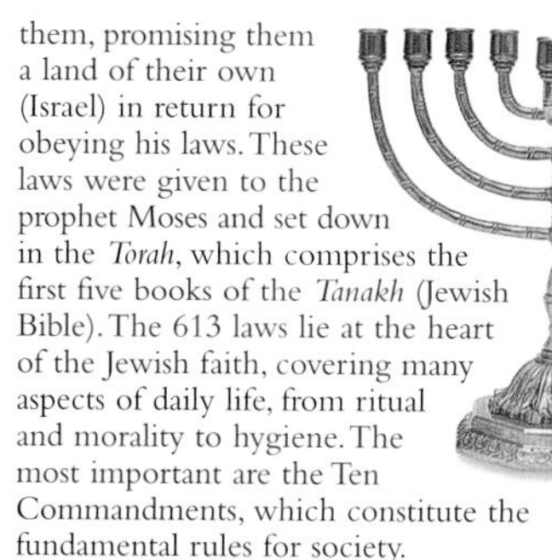

**MENORAH**
*A special* menorah *(ritual candlestick) is used at Hanukkah. This 8-day festival marks the Jewish victory over the Seleucid kings, who had defiled Solomon's temple.*

Jewish people may feel a communal bond even though they vary widely in beliefs. There are several main forms of belief, from that of the *haredim* (very orthodox Jews) to liberal and secular Judaism.

Throughout history, Jewish people have been persecuted. Many await the coming of a Messiah (anointed king), who will end war and establish God's kingdom on earth. Unlike Christians, Jews believe he is yet to appear.

Despite their small numbers, Jews have had a huge influence on world culture. Eminent Jewish people include the scholar Maimonides (1135–1204), the physicist Albert Einstein (1879–1955), and many lawyers, politicians, artists, and humanitarians.

**WESTERN WALL**
*This wall in Jerusalem was part of the Temple of Solomon. It is the centre and holiest site of the Jewish world.*

**READING THE TORAH**
*During worship, which takes place in a synagogue, several people may read from the Torah. The text is handwritten on a long scroll.*

History

### Pesach (Passover)

This 7-day festival celebrates the Jews' release from slavery in Egypt. On the first two nights, the tale is told of how God killed the first-born children and animals in Egypt but "passed over" those of the Jews. A ritual meal, *Seder*, is eaten. It includes *matzah* (unleavened bread); salt water, to signify the tears of slavery; and foods that symbolize suffering and then hope.

**FAMILY RITUAL**
*It is traditional for children to read from the* Haggadah, *which explains the order of the* Seder *ritual.*

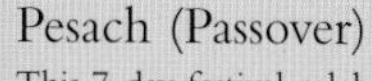

WORLD RELIGIONS

## Chinese religions

**Founded:** Confucianism and Taoism date from the 6th century BC

**Estimated number of followers:** 225 million

**Distribution:** Worldwide; mainly China and eastern Asia

The religious tradition in China is a mixture of four ancient belief systems. It comprises the "Three Ways" of Confucianism, Taoism (or Daoism), and Buddhism (which arrived in China in the 1st century AD), together with the indigenous folk religion.

Confucianism is a system of ethics created by Kong Fuzi, or Confucius (551–479BC). The system is based on five principles: respect for all people; love of family; mutual generosity between friends; welcome to strangers; and loyalty to the state. Confucius taught that the proper conduct of relationships, including respect for one's parents and ancestors, would create order and harmony within the family, and by extension in society.

Taoism is based on the writings of Lao-Tzu (Laozi), who lived in the 6th century BC. These writings are collected in the *Tao Te Ching*.

**T'AI CHI CH'UAN**
*In Taoist belief, the body is an energy system through which chi (vital force) flows. The movements of T'ai Chi Ch'uan are thought to focus and channel chi.*

Lao-Tzu taught that the Tao ("Way") is the hidden force that flows through and controls the cosmos. Believers try to become one with the Tao. They practise the art of *wu-wei*, or "going with the flow" and not trying to control or resist events. This practice leads to correct behaviour, creating peace and harmony within people and the world. Taoism also has a popular form that involves healing and exorcism rituals, festivals, and the quest for immortality.

Folk religion is still observed in the form of festivals (particularly the Chinese New Year), use of magic charms, and ancestor worship. One traditional art associated with folk religion is feng shui. This practice is a way of aligning buildings and objects in harmony with natural energies.

**YIN AND YANG**
*This symbol unites yang (light, active, male) and yin (dark, passive, female) energy. The two forces oppose but complement each other.*

Chinese people may blend the Confucian moral system with Taoist respect for the natural world and the Buddhist beliefs about the soul. The Three Ways were official religions of the Chinese state for over 2,000 years, until the early 20th century.

All of the religions, except for Buddhism, include the idea that there are vital forces permeating the cosmos. The main force is chi (or qi), the life energy from which everything is made. Chi comprises the opposing forces of yin (passive energy) and yang (active energy). It is considered important to keep yin and yang in harmony.

**DANCING DRAGON**
*The dragon is a symbol of water, yang (active energy), and heavenly power. It is a major feature in Chinese religion and festivals.*

**LAI-SEE**
*At the Chinese New Year, children are given* Lai-see*: red envelopes containing money.* Lai-see *are thought to bring luck, health, and happiness.*

CULTURE

WORLD RELIGIONS

## Sikhism

**Founded:** 15th century
**Estimated number of followers:** 23 million
**Distribution:** Worldwide, but mainly northern India

Sikhs view their religion as a revelation from God (known as *Sat Guru*, or "true teacher") to his disciple, Nanak. The word *Sikh* means "disciple".

Guru Nanak taught that all religions share the same truth. He set out a doctrine of salvation based on a direct relationship with God rather than ritual or priestly hierarchies. He accepted the Hindu doctrine of *karma* (*see* p294) but rejected the caste system, believing that all humans are equal before God. Guru Nanak also taught that a holy life centred on serving the community.

There were 10 gurus. The last one, Gobind Singh, set up the *Khalsa* (pure community) to protect Sikhs against persecution from the Mughals (Muslim rulers). He asked all Sikhs to wear five symbols of allegiance: *kirpan* (sword), *kangha* (comb), *kes* (uncut hair), *kara* (metal bracelet), and *kachch* (breeches). Sikhs still wear these items today.

At his death, Guru Gobind Singh passed his authority to the Sikh holy book, the *Guru Granth Sahib*. The book comprises the words of the gurus and of some Hindu and Muslim teachers. It is revered as a guru and is the focus of worship. A copy is kept in each *gurdwara* (religious and social centre).

**HOLY PROCESSION**
*Sikh guardians, carefully carrying the* Guru Granth Sahib *(Sikh holy book), process towards the Golden Temple.*

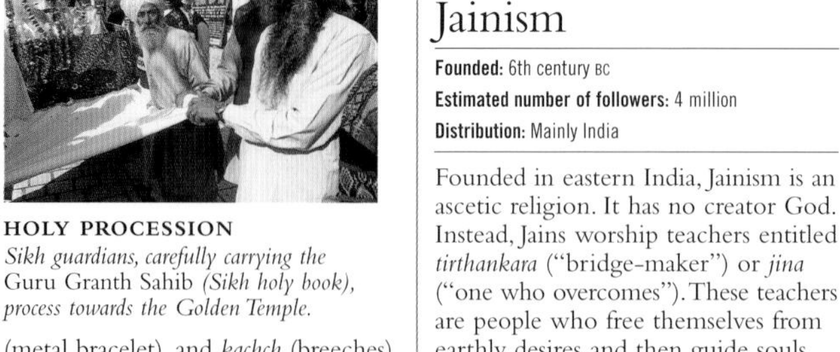

**GOLDEN TEMPLE**
*The Golden Temple at Amritsar, in the Punjab, is the most revered pilgrimage site. It holds the original Sikh holy book.*

WORLD RELIGIONS

## Jainism

**Founded:** 6th century BC
**Estimated number of followers:** 4 million
**Distribution:** Mainly India

Founded in eastern India, Jainism is an ascetic religion. It has no creator God. Instead, Jains worship teachers entitled *tirthankara* ("bridge-maker") or *jina* ("one who overcomes"). These teachers are people who free themselves from earthly desires and then guide souls across the "river of transmigration" from one life to the next. The 24th *jina* was Mahavira, or "the great hero" (540–468BC), the founder of Jainism.

The principal belief, and the way by which Jains are said to find release from the cycle of rebirth, is *ahimsa* (non-violence towards all living things). This doctrine has deeply influenced many non-Jains, including Mahatma Gandhi.

**RESPECT FOR LIFE**
*Jain monks wear masks and carry brooms to sweep the ground as they walk, to avoid inhaling or stepping on any living creatures.*

WORLD RELIGIONS

## Baha'i World Faith

**Founded:** 19th century
**Estimated number of followers:** 6 million
**Distribution:** Worldwide, but mainly India, Africa, and South America

The Baha'i faith arose out of Islam, in 19th-century Persia, but became a separate faith. The founders were Mirza Ali Muhammad, known as the Bab ("Gate"), who announced the arrival of a great spiritual leader, and the leader himself, Baha'u'llah ("Glory of God").

Baha'ism contains elements from all the world's earlier religions. Believers promote equal rights for both sexes, compulsory education, and social harmony. They regard their beliefs as a divinely revealed plan that will bring world unification and peace through a global legislative body and language.

**BAHA'I CENTRE**
*The Shrine of the Bab, in Haifa, Israel, holds the Bab's remains. It is one of the most holy sites for Baha'is.*

WORLD RELIGIONS

## Zoroastrianism

**Founded:** Around 12th century BC
**Estimated number of followers:** 150,000
**Distribution:** Worldwide, but mainly Iran and India

Zoroastrianism is the oldest and most influential of the world religions. Some features, notably the opposition of good and evil, were adopted by later religions such as Judaism.

The faith was founded by the prophet Zarathustra (Zoroaster in Greek), who lived in Persia (now Iran). He taught that the world is essentially good but is locked in a perpetual battle between the force of good, represented by the supreme god Ahura Mazda, and the force of evil, represented by the bad spirit Angra Mainyu. His words were set down in the *Avesta*, the Zoroastrian scripture.

Early believers were persecuted, but Zoroastrianism became the official religion of the Persian empire for 1,000 years. When the empire fell to the Muslims in the 7th century AD, further religious persecution forced some to flee to India, where they were known as Parsis (people from Persia).

**NAVJOTE CEREMONY**
*In* Navjote, *7-year-old children are initiated into the faith. Like this girl, they have a sacred thread* (kusti) *tied around their waist.*

Zoroastrians pray five times a day in the presence of fire, which symbolizes righteousness. They also worship in temples before a sacred fire. They see it as their duty to support good through thoughts, words, and deeds. For this reason, many serve their communities in charity, education, and medicine.

WORLD RELIGIONS

## Shintoism

**Founded:** 4th century BC
**Estimated number of followers:** 110 million
**Distribution:** Japan

Shinto (the way of the gods) is Japan's oldest religion. Unlike most religions, it has no founder or holy scriptures, although the main elements appeared from the 4th century BC onwards.

Shinto faith centres on the worship of *kami* (spirits). Kami are believed to inhabit most natural features, such as water or mountains, and exceptional people, such as the Buddha or Japan's Emperor. While most Shinto worship relates to earthly kami, heavenly kami responsible for creating the world are mentioned in later texts; the most important of these spirits is the sun goddess Amaterasu.

Shinto shrines are built to house kami. They are marked by a *torii*, a gateway that symbolizes the border between the human and kami worlds.

Japanese people practise Shinto and Buddhism alongside one another; for example, they may have a Shinto wedding and a Buddhist funeral.

**NATURE WORSHIP**
*Respect for nature is a central part of Shinto faith. This monk is worshipping on a mountainside.*

**SYMBOLIC GATE**
*The torii is the gateway to a Shinto shrine. It marks the border between the sacred area and the ordinary world. This torii, on Miyajima Island, is one of the best-known examples in Japan.*

TRADITIONAL BELIEFS

## Shamanism

**Founded:** Possibly about 100,000BC

**Estimated number of followers:** Not known

**Distribution:** Worldwide, but mainly northern Europe, northern Asia, Southeast Asia, Africa, the Americas

Shamanism exists in many traditional societies around the world and involves a wide variety of beliefs and practices. In most cultures, however, shamans perform similar tasks, such as curing sick people, predicting the future, and handling disputes.

The shamans act as a link between the human and spirit worlds. They travel to the spirit world by entering a trance; to reach this state, they take hallucinogenic plants or perform acts such as fasting, drumming, or dancing. In the spirit world, they may consult spirits or dead people, rescue the wandering spirits of sick people, or collect items to cure illness. Shamans often have spirit helpers, usually animals, who provide them with extra power.

Shamanism may also exist in modern, urbanized societies, where some people have adopted it as a tool for personal discovery.

SHAMAN'S RATTLE *Some shamans use the sound of a rattle to shift their consciousness into the spirit world.*

CHUKCHI SHAMAN *A shaman from the Chukchi people of Siberia uses a drum made of animal skin. Drums may be seen as "steeds" that carry the shaman to the spirit world.*

Health

### Healers fighting AIDS

In some parts of Africa, health authorities are turning to shamans and other traditional healers to help them combat AIDS. Some of the healers' herbal medicines, when given to people who already have AIDS, may help the patients' bodies to fight off associated infections. In addition, because the healers work so closely with their communities, they can be very helpful in teaching people about sexual health.

TRADITIONAL BELIEFS

## Animism

**Founded:** Possibly 100,000BC

**Estimated number of followers:** Not known

**Distribution:** Worldwide in traditional indigenous communities

Animism is probably the oldest belief system in the world. Today, it survives among ancient communities such as the Ainu in Japan and the tribal peoples of northeastern India.

In its simplest form, animism is the belief that the world is inhabited by spirits that are invisible to the senses but accessible by other means, such as rituals. The term "animism" also covers animatism: a belief in a universal force that inhabits objects and gives them powers, such as giving herbs their healing properties.

Animists may revere useful or beautiful objects, such as trees and mountains, but fear dangerous forces such as predatory animals. People may perform rituals or carry charms to please helpful spirits such as fertility gods, or to guard themselves against harmful forces such as disease spirits.

Animism prompts people to respect the natural environment and consider the consequences of their actions. For example, when killing an animal or felling a tree, an animist may first ask permission of its spirit.

CHARM *The Nte'va people of West Africa make charms like this one to protect them from illness.*

HMONG HAT *The Hmong people of Vietnam make special hats decorated with silver and lucky symbols to protect babies from evil spirits.*

TOTEM POLE *These highly carved poles, made by Native Americans of the northwest Pacific coast, include images of totem animals, which have a legendary relationship with specific clans.*

TRADITIONAL BELIEFS

## Vodou and related beliefs

**Founded:** Vodou founded in West Africa around 4,000BC; related beliefs founded about 18th century AD

**Estimated number of followers:** 60 million

**Distribution:** Mainly Haiti; also parts of West Africa, the Caribbean, and the Americas

Vodou is based on the ancient beliefs of West African peoples; the name comes from the Yoruba word *vodu* (spirit). It is still an official religion in the West African country of Benin. The faith was carried to the Americas by slaves. It survived on Haiti in the 18th century, and spread to the West Indies and the southern states of the US. In South America and Cuba it has given rise to Umbanda, Candomblé, and Santeria, beliefs that blend West African tradition with Roman Catholic rituals.

Vodou is based on God and groups of *lwa* (spirits). Some lwa are identified with Christian saints; for example, Erzulie (the spirit of love) is identified with the Virgin Mary. Each lwa is recognized by distinctive behaviour.

People contact lwa to harness their power for health, good fortune, and protection from evil. In ceremonies, a *houngan* (priest) or *mambo* (priestess) invokes a lwa; it then "mounts" a worshipper, who enters a trance. An animal may be sacrificed for the lwa.

The popular idea of "voodoo", with features such as zombies and voodoo dolls, comes from films and has almost nothing to do with the actual beliefs.

VODOU IN HAITI *A vodou believer carries a human bone in his mouth during a ritual for Guede, the lwa (spirit) of the Dead. This ritual takes place in a cemetery, where people bring food and other gifts for the lwa.*

COMBINED BELIEFS *Santeria mixes Roman Catholic and Vodou beliefs. This shrine has a statue of the Virgin Mary together with Vodou elements, such as glasses of water for ancestors.*

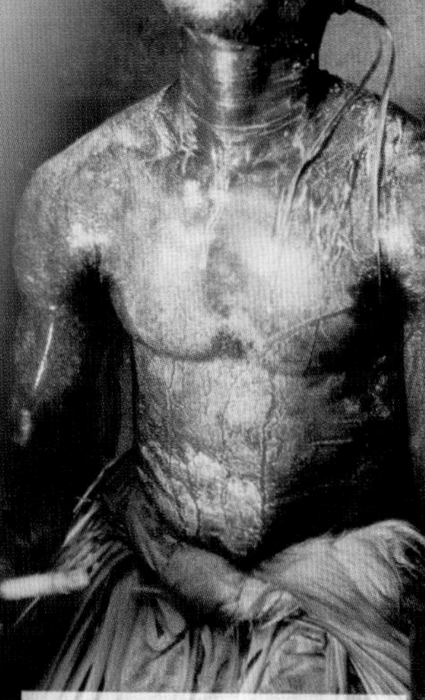

JOINING WITH A SPIRIT *Some rituals involve people dancing until they enter a trance. A person in a trance is believed to be possessed by a lwa (spirit) and behave as that spirit.*

TRADITIONAL BELIEFS

## Aboriginal Australian beliefs

**Founded:** About 40,000BC
**Estimated number of followers:** Not known
**Distribution:** Australia

The Aboriginal peoples of Australia and the surrounding islands are thought to have one of the most ancient cultures in the world – at least 40,000 years old. Over the millennia, they have developed a complex, subtle spirituality centred on their environment. There are several hundred groups, each with their own myths and rituals, but all share certain core beliefs.

The main feature of Aboriginal belief is the state known in English as the Dreaming. This state is eternal. It included the dawn of time, when the spirit Ancestors walked the earth; it also continues now, as an ever-present spiritual plane of existence.

When the Ancestors appeared, the earth was barren. They travelled across it, shaping it by acts such as hunting, fighting, camping, and dancing. They also made the stars, elements, people, plants, and animals. Once they had finished, some of them sank into the earth and became sacred landforms.

Each Aboriginal group preserves the songs telling how their own patch of territory was created and relating the deeds of the Ancestors who are associated with the area. The songs of successive areas may be sung in sequence, forming a track called a songline. The songlines, in turn, make a network across Australia. People can follow trails across the landscape, even in unfamiliar areas, by referring to these songs.

**CREATION LEGEND**
*Dot paintings are traditionally made to convey information about the Dreaming. Many paintings show stories about a particular landscape.*

**CEREMONIAL PAINTING**
*To prepare for ceremonies and rituals, Aboriginal people paint themselves with earth-based pigments such as clay or ochre.*

**SPIRITUAL CENTRE**
*Uluru (Ayers Rock) has been sacred to Aboriginal peoples for thousands of years. It is a point where many songlines (spiritual paths) connect.*



**DRUIDS**
*A group of modern druids, gathered at Stonehenge in the UK, observe the summer solstice – the shortest night of the year.*

TRADITIONAL BELIEFS

## Neopagan beliefs

**Founded:** 18th–20th centuries
**Estimated number of followers:** 1–4 million
**Distribution:** Mainly Europe and North America

Neopagan systems are revivals of European pagan (pre-Christian) beliefs. Interest in these beliefs has grown since the 1960s, especially among Europeans trying to reclaim their cultural heritage. The main forms are Wicca, Druidism, and Ásatrú (Norse paganism).

Wicca gains its name from the Old English for "witch" or "wizard". It is thought to have originated among the ancient Celts and date from the 7th century BC, and was revived in the UK in the 1940s. Wiccans worship a supreme God and Goddess. Other practices include benign (white) magic and rites focused on the Earth's natural cycles.

Druidism also originated with the Celts; it was revived during the 18th century AD. Druids revere trees, above all the oak, and mistletoe. They practise divination based on ogham script, in which each letter represents wisdom linked to a certain type of tree. Druids also mark the solstices and other important times of year, with rituals at ancient monuments such as Stonehenge.

Ásatrú has been recognized as a religion since the early 1970s. It has its own texts, such as the poetic and prose Eddas (which tell the old Norse myths) and deities, including Thor, Odin, and Freya. Followers gather in groups called Kindreds to honour their ancestors and celebrate the seasons. They also practise divination using runes (letters of the old Norse alphabet).

**WICCA RITE**
*In Wiccan rituals, participants form a circle and call upon the spirit guardians of the north, south, east, and west to form a sacred space.*

**RUNIC CHARM**
*Followers of Ásatrú use Norse runes written on leaves or stones as protective charms, or for divination.*

MODERN BELIEFS

## New faiths

**Founded:** 20th century
**Estimated number of followers:** Not known
**Distribution:** Some, such as Scientology, worldwide; others, such as Cao Dai, centred in specific countries

During the 20th century, various new belief systems arose around the world. Some mix ideas from existing faiths, while others are based on original forms of spirituality or philosophy.

One example of a combined system is Cao Dai, followed in Vietnam. It mixes elements from many religions, together with spiritual revelations received by the founder in the 1920s. Another is Falun Dafa (Falun Gong), founded in China in 1992. Although it is not a full belief system, Falun Dafa is a spiritual discipline based on Taoist and Buddhist concepts.

Some systems have spiritual beliefs mixed with philosophical or political ideas. For example, Scientology includes both Eastern spiritual tradition and elements of Freudian theory. Rastafarianism mixes Christianity with Jamaican nationalism.

A few faith systems are totally original. One such system is Eckankar, founded in 1965. It is based on a belief in God (called Sugmad) and includes the idea that souls can travel through different planes of existence.

**ROOTS OF STRENGTH**
*Many people of the Rastafarian faith grow their hair in long dreadlocks as a sign of purity and strength.*

MODERN BELIEFS

## Sects of major faiths

**Founded:** Various dates through history

**Estimated number of followers:** From a few thousand to several million

**Distribution:** Worldwide

Most major religions have offshoots called sects. These faith groups share most of the tenets of the main religion but also hold certain unique beliefs.

Many sects are subsections of denominations (major, well-established forms of a religion). Christianity, for example, consists of the Orthodox, Roman Catholic, and Protestant denominations, which in turn include thousands of sects. One example of a Christian sect is The Religious Society of Friends (Quakers), who focus on a direct relationship with God, based on praying in silence, rather than ritual and religious hierarchy.

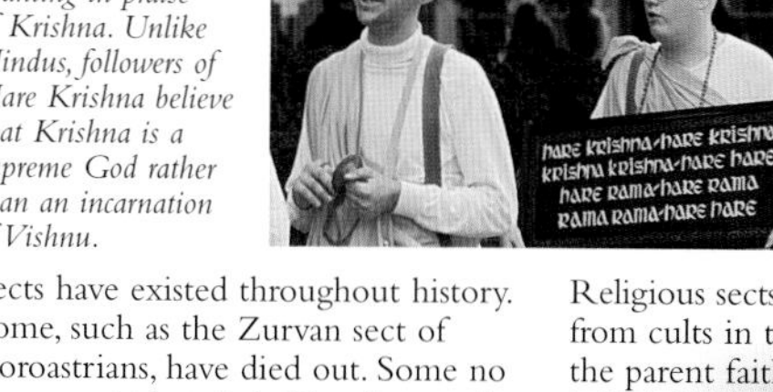


**HARE KRISHNA**
*These believers are chanting in praise of Krishna. Unlike Hindus, followers of Hare Krishna believe that Krishna is a supreme God rather than an incarnation of Vishnu.*

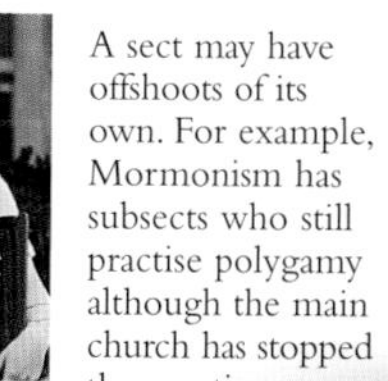

Sects have existed throughout history. Some, such as the Zurvan sect of Zoroastrians, have died out. Some no longer exist but their ideas live on; for example, the Pharisees, a Jewish sect from 2,000 years ago, have long since disappeared, but their ideas underlie later Rabbinical Judaism. Others, such as the Ismaili sect of Shia Islam, are still active. Certain sects developed into accepted denominations or even new faiths; for example, the Baha'i faith grew from Islam but is now a distinct religion. A few sects are now popular worldwide, notably the International Society for Krishna Consciousness (ISKCON), or Hare Krishna, which is derived from Hinduism.

A sect may have offshoots of its own. For example, Mormonism has subsects who still practise polygamy although the main church has stopped the practice.

Religious sects may be distinguished from cults in that they still resemble the parent faith and hold its central beliefs. No Christian sect, for example, would deny the importance of Jesus. Certain groups, however, have practices more usually associated with cults, such as close control of their followers' lives and very coercive methods of recruitment. One sect that has attracted criticism for such practices is the Unification Church (the Moonies).

**MORMON FAMILY**
*The old Mormon practice of polygamy (marriage to many wives) still occurs in more fundamentalist subsects, but is no longer recognized by the main branch of the faith.*

**GROUP WEDDING**
*Unlike mainstream Christians, people in the Unification Church of the Rev. Sun Myung Moon (Moonies) may take part in group weddings involving thousands of couples.*

NONRELIGIOUS BELIEFS

## Humanism

**Founded:** 18th–19th centuries

**Estimated number of followers:** 3 million

**Distribution:** Worldwide

Humanism is a belief that "man is the measure of all things". This system stresses reason rather than faith, and places humanity, rather than any gods, at the centre of life. Most humanists are secular (nonreligious), but some Jewish and Christian people also embrace a humanist system of ethics.

Some of the ideas underpinning humanism have existed for centuries. Ancient Greek philosophers questioned many aspects of life, including the role of the gods. In China, Confucius (*see* p.295) developed a moral system based on respect and duty to others. Much later, a major influence was the logical, rational mode of thinking devised by European intellectuals such as the English philosopher Francis Bacon (1561–1626), which was based on reasoning rather than faith.

Humanism in its present form arose in the 18th and 19th centuries. At this time, advances in scientific thinking, such as Darwin's theory of evolution (*see* p.31), challenged religious doctrine. In addition, many people questioned why problems such as poverty and oppression still existed in societies whose ethical values were founded on religious teachings.

During the 20th century, religion became separate from politics in many countries, and the public domain (especially education) grew increasingly secular. Mass travel, and media such as radio and television, made people more aware of different belief systems from around the world.

Now, increasing numbers of people feel that faith need not have anything to do with morality, and deplore what they consider to be the negative aspects of religion, such as dogmatism and intolerance of other people's lifestyles. Modern humanists, who count many distinguished thinkers among their members, are often outspoken supporters of social reform, pacifism, human rights, and privacy laws.

**A SECULAR CELEBRATION**
*Various humanist organizations offer non-religious, personally tailored ceremonies for life events such as births and weddings.*

NONRELIGIOUS BELIEFS

## Atheism and agnosticism

**Founded:** Western varieties date from 19th century

**Estimated number of followers:** Not known

**Distribution:** Worldwide

Neither atheism nor agnosticism is a full belief system, because they have no fundamental philosophy or lifestyle requirements. These forms of thought are simply the absence of belief in, or denial of, the existence of deities.

Atheism is commonly divided into "weak" and "strong". Weak atheists have no faith, simply because the feeling is not there. Strong atheists conclude, from existing evidence and argument, that gods do not exist.

Nonbelief has existed for centuries. For example, Buddhism and Jainism have been called atheistic religions because they do not advocate belief in gods. Western atheism, as a movement opposing religion, developed in the 19th century, and became widely accepted by intellectuals. In the 20th century, atheism became the official belief system in communist states such as the USSR, China, and Albania.

Agnostics hold that the existence of gods can be neither confirmed nor denied and is a matter beyond human understanding. They also believe one should not make claims for ideas that cannot be upheld by rational enquiry.

Profile

### Bertrand Russell

The British philosopher, mathematician, and activist Bertrand Russell (1872–1970) was a lifelong radical. During World War I Russell was fired from his teaching post at Cambridge University, England, and sent to prison for his pacifist views. After the war he was a journalist and lecturer. Russell declared that Christianity was "the principal enemy of moral progress in the world". He described himself as an atheist and a humanist, and held that science and social reform provided the only hope for human improvement.

**INFORMATION SOCIETY**
*In many modern societies, people are subjected to a huge variety of verbal and visual signs, from advertisements to the brake lights on vehicles.*

# COMMUNICATION

The most obvious factor that distinguishes humans from other animals is the variety of ways in which we communicate. Human interactions are a rich blend of language, facial and body movements, and symbols; we also assess the context of any message to obtain the maximum amount of information. Our style of communication is one of the clearest markers of identity. The language and gestures that we use show the community or nation to which we belong; in addition, each of us has a personal vocabulary of commonly used words and gestures reflecting our individual concerns.

**CALLING LONG DISTANCE**
*Electronic forms of communication such as satellite telephones can now be used in even the most remote parts of the Earth.*

Many animals have a set of sounds and physical signals that they use in particular contexts, such as attracting a mate or warning others of danger. Animals such as dogs and primates, which interact extensively, may use sets of signals to express emotions such as pleasure or fear, or to show dominance or subordination. Even these animals, however, tend to use the same signal every time they encounter a specific situation. Human beings, on the other hand, can talk about the past and the future, discuss abstract ideas, and put entirely new concepts into words.

## LANGUAGE

Speech is the most versatile form of communication. It is a system with infinite possibilities, but every language has a basic three-level structure. Each comprises about 20 to 60 phonemes (sounds), many thousands of words, and a grammar (the set of rules used to form words and build sentences).

In early childhood, every human has the potential to learn one or more languages: we develop ranges of sounds, repertoires of words, and sets of grammatical rules for each. In the process, we create a language afresh from what we hear around us. Everyone's language remains individual; even siblings can be distinguished easily by their choice of words and pronunciation. If there are 5,000 languages, there are 6 billion voices in the world today.

Nobody knows when or why language evolved, although several theories have been put forward (*see* Origins of language, *above right*). Over the centuries, many languages changed and absorbed new words from others as people migrated and came into contact with different cultures; one example is Maltese, which is based on Arabic but includes many words from Italian. Certain languages spoken in remote areas changed relatively little; for example, modern speakers of Icelandic can still read medieval sagas without too much difficulty.

Some languages were a medium through which beliefs and learning spread internationally. In Southeast Asia, Sanskrit has been a language of science and religion for over 2,000 years. In eastern Asia, Chinese is a major cultural influence. At various times in European history, Greek, Latin, Arabic, German, and English were successively used for science. Until the mid-20th century, French was the language of diplomacy. No language in history, however, has had the global reach that English has today.

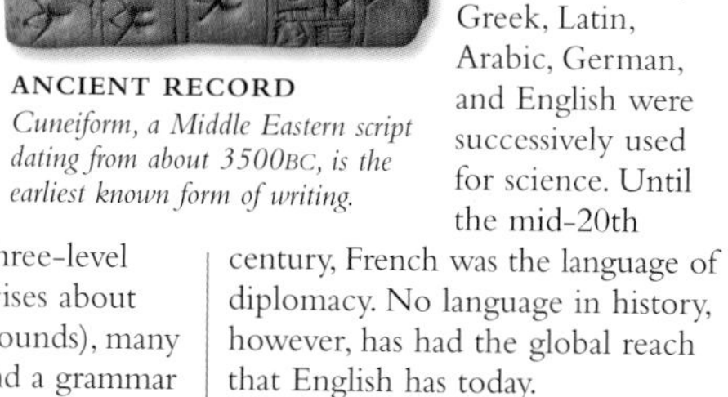

**ANCIENT RECORD**
*Cuneiform, a Middle Eastern script dating from about 3500BC, is the earliest known form of writing.*

## NONVERBAL FORMS

In addition to language, humans have a large repertoire of facial expressions, gestures, and other nonverbal signs with which they can communicate.

Issue

### Origins of language

Various theories exist to explain why language developed. Many scientists have assumed that it was first used as an aid in activities such as hunting. A more recent theory suggests that language evolved as a substitute for the common primate activity of mutual grooming, which reinforces social bonds, and was used for the same purpose. An animal can only groom one other at any time, but talking allows humans to "groom" several people at once.

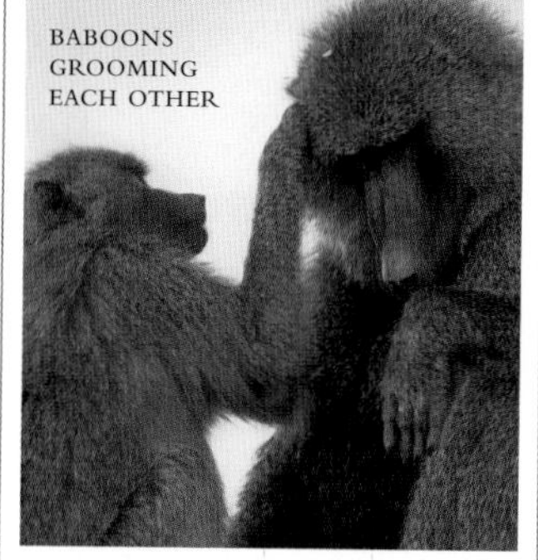

**BABOONS GROOMING EACH OTHER**

Some signals are involuntary and common to all cultures. Examples include the facial expressions for basic emotions such as joy and fear (*see* p166), or the instantaneous "eyebrow flash" that occurs when we see someone we know and like. Other signals, called gestures, are learned and vary from one culture to another. For example, when beckoning someone, an English or French person will turn the palm up and curl the fingers, while a Chinese or Japanese person will turn the palm down and make a scratching movement. In addition, everyone uses some gestures that are unique to them but recognized by family, friends, and other regular contacts. Understanding nonverbal signals can give us great insight into other people's thoughts and feelings and enable us to respond appropriately. For this reason, many people today seek to learn about this form of communication for purposes such as business, travelling abroad, and finding sexual partners.

**SHOWING THE WAY**
*Many road signs are designed to be understood without words. The symbol, and the colour and shape of the sign itself, all indicate the type of hazard or conditions ahead.*

Symbols, such as flags and logos, are another form of nonverbal communication. They are often used to represent abstract entities such as companies and nations.

## NEW TECHNOLOGY

Recording information enables people to communicate across time and space. Writing was the first invention for this purpose. The ability to read and write is a form of power, and for centuries these skills were restricted to certain high-status people, such as priests and officials. With the invention of large-scale printing, knowledge could be spread much more widely. The advent of radio and television allowed messages to reach whole populations at once. These mass media are usually controlled by governments and international corporations, although ordinary people are allowed limited input (for example, in interviews). The internet is a much more democratic medium; it enables individuals to share information with others across the world.

# WORLD LANGUAGES

Nobody knows when humans first began to use speech. Ancient writings form a precious record of human history, but they represent only a fraction of the languages that have existed. Languages have changed continually through the ages; many have been carried across the world by trade and conquest and are now spread through mass media such as television. Language also adapts to reflect people's changing identity, social structures, and knowledge about their environment. At least 5,000 languages are spoken in the world today. A few, such as Mandarin Chinese, English, and Spanish, are spoken by thousands of millions; however, at the other end of the scale, some languages are on the verge of extinction.

## EARLY LANGUAGES

The earliest evidence of language comes from ancient manuscripts and inscriptions. These records begin with the cultures that first used writing, more than 5,000 years ago: the Sumerians (in what is now southern Iraq) and the Egyptians. Many more languages were soon being written down. They included Chinese, Sanskrit, Pali (used by the Buddhists of Southeast Asia), classical Hebrew, and ancient Greek. Others have disappeared, including Akkadian (which replaced Sumerian), Hittite (used in Turkey), and Pyu (used in Myanmar). Later came Latin, Ethiopic, Aztec, and Mayan. These written languages were most often used for religious scriptures, inscriptions on public monuments, and commercial records. They can, therefore, give information about the most important individuals and activities of those ancient peoples. Some may also provide clues about the relationship between different modern languages; for example, Sanskrit is the ancestor of Hindi and Bengali. A few of these languages are still familiar, if only in specialized contexts, such as religious worship, government, law, or academic study.

**EGYPTIAN SCRIPTS**
*The Rosetta Stone, dating from 196BC, was the key to translating hieroglyphs. It gives the same text in hieroglyphs, common script, and Greek. Scholars read it by finding familiar names and comparing with Greek.*

**REBIRTH OF HEBREW**
*For almost 2,000 years Hebrew had existed almost solely as a religious and literary language. When the state of Israel was founded in 1948, it was adopted as the national language. Hebrew is now used in daily life, even for such prosaic activities as personal banking.*

**MAYAN INSCRIPTION**
*This carving is an example of the Mayan script of Central America. The script is more than 2,000 years old and was used until the 16th century.*

## THE SPREAD OF LANGUAGE

Language first spread as people migrated to new lands. One example is the Austronesian language family, carried from southeastern China across the Pacific. In some cases, a people seized land from existing inhabitants; in this way, Bantu languages spread across southern Africa and English and Spanish across the Americas. Languages may also be spread by trade. If speakers of different languages often talk together, they may develop a lingua franca (a common language), such as Swahili in East Africa, or a pidgin (a simple language including elements from two or more others). A pidgin may be expanded, used by settled communities, and learned as a mother tongue; it then becomes a creole. One example is Tok Pisin, the former pidgin English of Pacific trade, now the national language of Papua New Guinea.

**PORTUGUESE SPEAKERS**
*Demonstrators in Porto Alegre, Brazil, carry placards in Portuguese, the national language. Portuguese was carried to Brazil as a result of trade and conquest.*

### A linguistic coincidence

The linguist R M W Dixon, while researching Aboriginal Australian languages in the 1980s, found that the word for "dog" in the Mbabaram language (formerly spoken in northeastern Australia) was identical to the English word "dog". It is rare for words to develop completely independently in different languages.

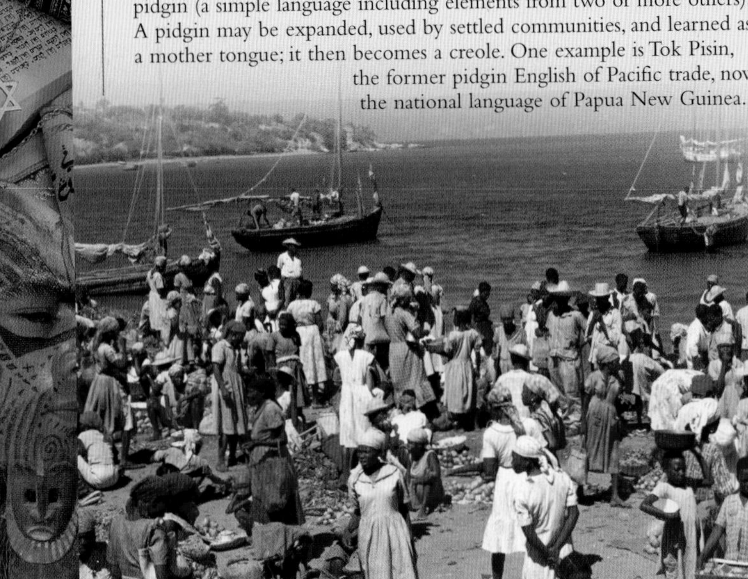

**MEETING OF MINDS**
*The pidgin languages that were once used in the Atlantic slave trade have a flourishing legacy in the Caribbean. One French-based pidgin developed into Haitian Creole, which is now the mother tongue of millions in Haiti and neighbouring islands.*

**LEARNING LETTER FORMS**
*A Tibetan child learns to write with a bamboo pen and a wooden tablet. In places where pens and paper are scarce, tablets are useful because they can be re-used.*

# WRITING SYSTEMS

Writing developed in three parts of the world independently: Egypt and the Middle East (5,000 years ago), China (3,500 years ago), and Central America (3,000 years ago). Early scripts (known as ideographic scripts) comprised a series of pictures representing ideas. Two developments followed: the pictures were simplified, and they were used to represent objects that looked like (or whose names sounded like) the original meanings of the pictures. Modern Chinese and Japanese use similar systems, known as logographic scripts. They have thousands of characters and take a long time to learn. In later writing systems, each letter denotes a specific sound (phoneme). One group of scripts, the alphabets, has letters for both consonants and vowels. Another group, the abjads, is based on consonants, with vowels sometimes added as extra marks. A third group has letters made up into syllabic blocks. All have a limited range of characters and all are relatively quick to learn.

**REVIVED TRADITION**
*The Mongolian script was banned while Mongolia was under Soviet rule but has since been revived in schools.*

Fact

## Scripts for blind readers

To enable blind people to read, scripts consisting of raised shapes are used. The letters are read by touch. The best-known script is Braille, formed from patterns of dots. It was invented in 1829 by Louis Braille, a Frenchman who became blind as a child. Another is Moon, a system of lines. This script was invented by William Moon in Brighton, England, in 1847. Both Braille and Moon are used in many countries.

## Examples of writing systems

**ALPHABETS**
*An alphabet has separate letters for each consonant and vowel in a language. Alphabets are used in a wide variety of modern languages.*

human — ROMAN
челове́к — CYRILLIC (USED IN RUSSIAN)
άνθρωπος — GREEK
인간 — KOREAN
ადამიანი — GEORGIAN
MONGOLIAN

**ABJADS**
*These systems are like alphabets but almost all of the letters represent consonants. Vowels may be indicated by marks above and below the letters.*

إنْسان — ARABIC
אדם — HEBREW

**SYLLABIC SYSTEMS**
*In syllabic writing systems, each character represents a syllable (a combination of consonant and vowel sounds). Many syllabic systems also include separate marks to indicate vowel sounds and linking elements between the characters.*

मानव — DEVANAGARI (USED IN HINDI)
มนุษย์ — THAI

**LOGOGRAPHIC SYSTEMS**
*In these systems, each basic character represents one word or part of a word. Some characters also indicate pronunciation.*

人 — CHINESE; ALSO KANJI (USED IN JAPANESE)

# DESCRIBING THE WORLD

At its most basic, language is used to define and describe the world. Some words for enduring features, such as names for plants, animals, and rivers, may stay the same over centuries, or be preserved when the original language has been forgotten. Others, such as words for fashionable clothing or music, may change rapidly. Words also show what is important to a speaker or community. Each language defines the world in its own way; for example, speakers of different languages may identify different colours in the spectrum. All speakers of a language share thousands of common words, but many also use a specialized or "focal" vocabulary for subjects of particular interest. Some focal vocabularies develop because of where the speakers live, such as those for camel-breeding in Arabic, eucalyptus forests in Australian languages, Arctic weather in Norwegian, and local medicinal plants in many languages. In urbanized societies, medicine, machine maintenance, and software are examples of subjects for which a focal vocabulary is used.

**NAMING THE STARS**
*In the Middle Ages, the Arabs were among the foremost astronomers in the world. Arabic influence can be seen in some star names, such as Betelgeuse (the giant's shoulder), Sayf (sword), and Rigel (foot) in the constellation of Orion.*

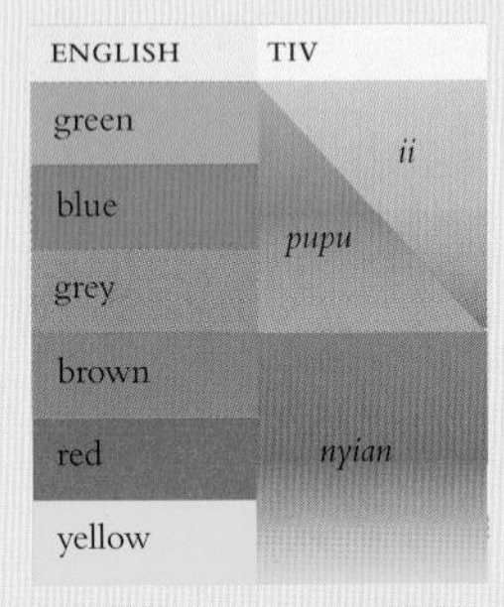

| ENGLISH | TIV |
|---|---|
| green | ii |
| blue | ii / pupu |
| grey | pupu |
| brown | nyian |
| red | nyian |
| yellow | nyian |

**DEFINING COLOURS**
*All humans with normal vision see the same range of colours, but different peoples may have different names and definitions for colours. For example, English has more names for colours than Tiv (a language of southern Nigeria).*

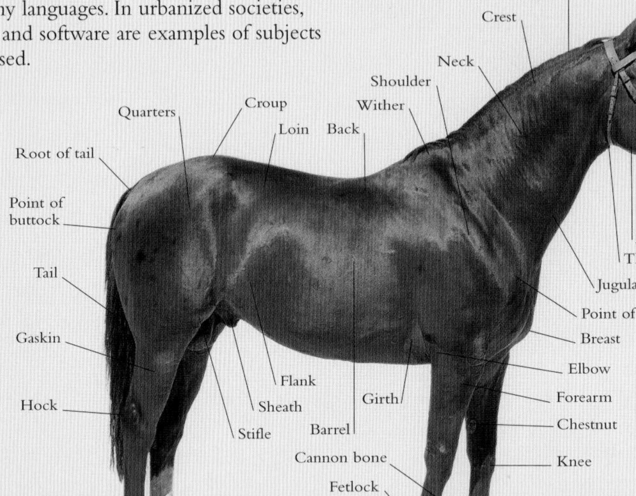
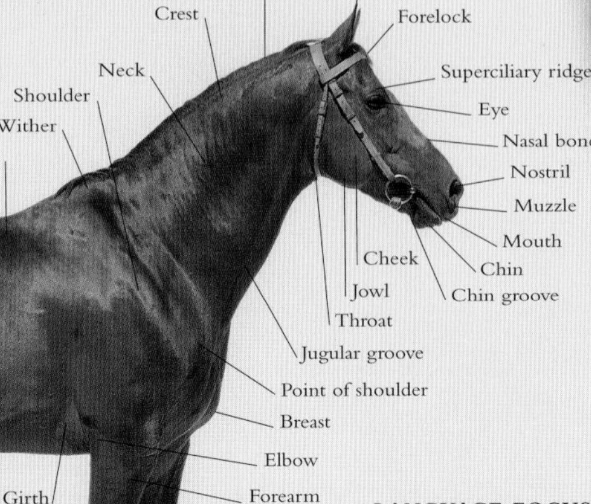


**LANGUAGE FOCUS**
*Everyone has a vocabulary that is especially rich in certain areas of interest. These sets of specialized words are called a focal vocabulary. Horse breeders, for example, have an extensive focal vocabulary to describe the breeds and physical characteristics of horses.*

# LANGUAGE AND IDENTITY

Language is one of the most important markers of identity. We often use the same name for a language and for its speakers, such as "French" or "German". Nationalists may fight to have their language community recognized as a distinct political entity; in contrast, some states forbid the use of minority languages in schools or in public. Within nations or communities, some subgroups, such as African–Americans in US inner cities, may use their own form of speech to set themselves apart from the rest of society. Surnames show a person's family relationship and, in some cases, area of origin. Personal names may be more private; in some cultures, knowing someone's first name gives you power over them. Some people even choose an alternative name, such as a nom de guerre (battle name) or nom de plume (author's name), as part of a special public persona.

**POWERFUL SECRET**
*Many Roma people have a secret name, which is given by their mother at birth. The mother never speaks this name, even to the child, in order to conceal the child's identity from evil spirits.*

**BASQUE BABY**
*The Basque people of the French/Spanish border are fiercely proud of their language (Euskara), a symbol of their distinctive, self-contained community. Basque is not known to be related to any other language.*

**MAKING A MARK**
*Graffiti artists advertise their presence by spray-painting a special signature, called a "tag", on places that are hard to reach, such as the sides of trains and the tops of tall buildings.*

Fact

## Ritual language

Languages may be preserved in religious scriptures and rituals for centuries after they have disappeared from everyday use. Examples include the Hebrew of the Jewish sacred books; Latin, used by the Roman Catholic church; and Sanskrit, the language of the Hindu scriptures. One unusual case is Avestan, from prehistoric central Asia. Zoroastrians in Iran and India still recite Avestan hymns 2,500 years later, although few of them understand the words.

CULTURE

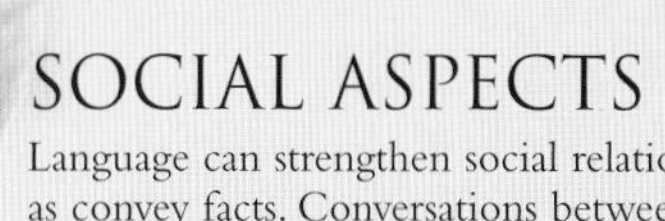

# SOCIAL ASPECTS

Language can strengthen social relationships as well as convey facts. Conversations between parents and children, or "small talk" between acquaintances, may not provide much new information but are vital for reinforcing the bonds between individuals. Familiar legends and stories strengthen a community's sense of their history and values. Language can also highlight social distinctions. People speak in different ways to friends, to those younger or older than themselves, and to employers. In some societies, rulers and elders are addressed with special words or forms of speech that show respect. In certain places, such as Japan, women and men use different forms of a language. Speakers may also use words or languages that are considered "prestigious" (such as Western people using Latin or French words) to make themselves appear more cultured or important. In addition, language may be linked to power or prosperity: learning a global or national language may give people access to political power or improve their job prospects.

**PASSING THE TIME**
*Everyday conversations may serve to reinforce bonds between people as much as to pass on information.*

**STORYTELLING**
*Telling stories is not just entertainment. It is a way to pass on beliefs and values; it also makes us think about problems and how to solve them.*

# WORLD LANGUAGES TODAY

It is possible to compare languages in several ways. How many speak them as a mother tongue? Mandarin Chinese heads this list. How many speak them as a first or second or additional language? By this criterion English takes first place, because wherever it is not the mother tongue it is likely to be a second or third language. Some languages are widely used but have relatively few native speakers; for example, Malay has only 18 million and Swahili about 5 million, but many more people in several countries speak them as common or official languages. On the map below, the colours represent the world languages with the most native speakers (*see also* pp306–310), showing countries with particularly large numbers of speakers. This simple map, however, obscures the facts that these languages may also be used in other areas, many communities usually speak more than one language, and millions of people are bilingual or multilingual.

**WORLD LANGUAGES**
*This map shows the extent of the languages with the most native speakers. Some of these languages have spread across the world. Others are used by large numbers of people but in relatively confined areas.*

Issue

## Endangered languages

As more and more people benefit from using English and other major world languages, fewer people are teaching traditional local languages to their children. Governments are still enforcing the use of national languages, and speakers of minority local languages still face disadvantage and discrimination. Many dialects and minority languages are being forgotten, year by year, all over the world. Usually, when the last speakers die, a language is lost forever. It is estimated that 2,500 languages are likely to disappear in this way during the next 100 years.

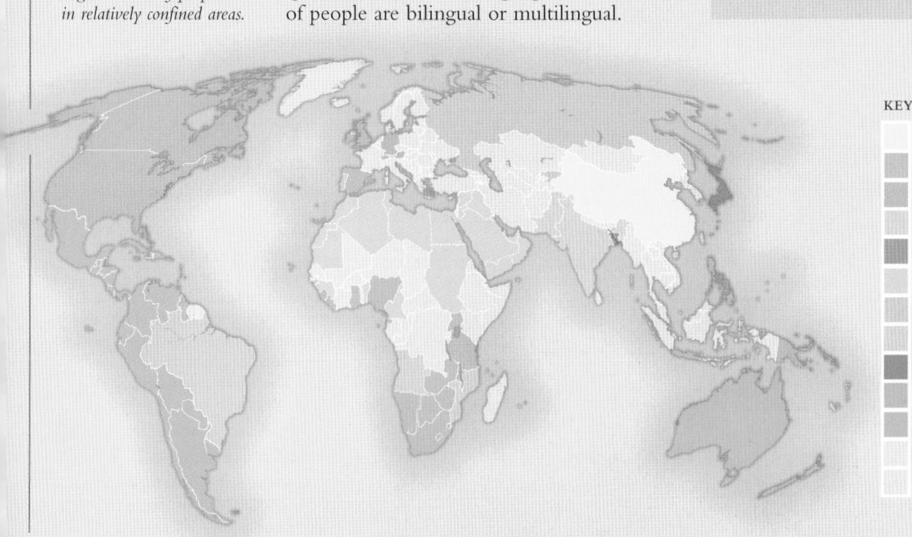

**INCREASED EXPOSURE**
*Owing to travel and business demands, people around the world are becoming more and more familiar with English and other widely used languages.*

## WORLD LANGUAGES

# Mandarin Chinese

**Estimated number of native speakers:** 800 million

**Countries with most speakers:** China, Taiwan, Indonesia, Malaysia, Singapore

Mandarin, or *putonghua* (common speech), has more native speakers than any other language in the world. It is the national language of China and is also spoken in other parts of eastern Asia.

One distinctive feature of Mandarin Chinese is that words may be pronounced with four tones of voice to show differences in meaning. The language has a logographic writing system with thousands of characters, each of which represents a meaning and corresponds to a syllable.

The oldest writings, found on oracle bones used for divination, date from about 1400BC. Literature appeared 1,000 years later with a collection of poetry called the *Book of Songs*. Chinese has the world's oldest printed text: a Buddhist sutra printed in AD868. Some works, such as the 12th-century epic *Water Margin*, are world famous.

In the time of the Chinese empire, the language was spread by officials and scholars. It influenced Korean, Japanese, Vietnamese, and Thai. Today, it is spread by travel and mass media. Several other languages are also known as Chinese, including Wu of Shanghai and Yue or Cantonese.

**BEAUTIFUL SCRIPT**
*Some non-Chinese people like to use Chinese script as decoration in tattoos.*

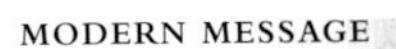

**MODERN MESSAGE**
*Although many Chinese symbols are ancient, they can be combined to make new words for the modern age.*

Fact

### Chinese computers

Typing methods for Chinese are very different from those for other scripts. Several keys are used to form each character. In some systems, the keys represent specific shapes (as shown on the left); in others, the keys represent sounds. Some computers enable users to write by hand using keypads and handwriting-recognition software.

**PRACTISING LETTERS**
*Calligraphy is still a highly valued skill among writers of Chinese. Artists use traditional brushes made from bamboo and animal hair to produce the strokes of each character.*



## WORLD LANGUAGES

# English

**Estimated number of native speakers:** 350 million

**Countries with most speakers:** US, UK, Canada, Australia, South Africa, New Zealand, Ireland

English is a true global language. As well as being the mother tongue in countries such as the UK and US, it is an official language in many parts of the world, such as India, South Africa, the Caribbean, and Nigeria. Linguists estimate that up to 1,800 million people can speak English to some extent.

The English language developed among Germanic invaders who settled in Britain around AD400. The oldest form, Old English or Anglo-Saxon, was similar to German. Some examples of the literature survive, such as the epic poem *Beowulf* (8th–10th century AD). After the Normans conquered England in 1066, the language was influenced by Norman French. During the Middle Ages, English absorbed further words from Latin and Greek, especially in technical subjects such as medicine and law.

One of the greatest influences on modern English was the King James Version of the Bible, which was first published in 1611 and was widely used until the 20th century. The playwright William Shakespeare (1564–1616) also had a profound influence; he is thought to have coined up to 3,000 new words. From the 17th century, the English language spread through the British empire, especially to the Caribbean, India, parts of Africa, North America, Australia, and New Zealand. After the end of the empire, in the mid-20th century, US power prolonged its influence. The language now spreads on the internet and in science, the media, trade, and advertising. English words are found in many other languages, such as *wapuro* (word processor) in Japanese; in turn, a variety of foreign words have entered English, such as *xocoatl* (chocolate) from the Nahuatl language of Mexico.

There are English dialects around the world, the oldest being those of England and Scotland. Other varieties include Black English, used in the US. There is also a standard form based on UK English, which is used worldwide.

**A GLOBAL LANGUAGE**
*Children all over the world learn English, often to help improve their job prospects.*

**FOREIGN NEWS**
*English is used globally in the media, as in these Czech newspapers.*

**ENGLISH IN AVIATION**
*As international aviation developed, the authorities realized that one language was needed for universal use; they chose English.*

WORLD LANGUAGES

## Spanish

**Estimated number of native speakers:** 225 million

**Countries with most speakers:** Mexico, Colombia, Argentina, Spain, US, Venezuela, Peru, Chile, Cuba, Ecuador, El Salvador, Honduras

Spanish is spoken all over the world. Approximately 450 million people use it, including 225 million for whom it is their mother tongue. It is the national language of Spain, Mexico, and most of Central and South America. It is also the second language of the US, where the historic Spanish-speaking minority has been swollen by millions of recent migrants.

The language originated in the Iberian peninsula as regional forms of Latin during the time of the Roman empire. From AD711–718 Spain was conquered by the Arabs and for the next 500 years it was part of the Islamic empire. Many Spanish words have Arabic origins, especially those to do with government or high culture. Various place and river names are also derived from Arabic, such as the Alhambra (*qalat al-hamra*, red fortress) and the Guadalquivir (*wadi'l kabir*, great river) in Andalucia.

Spanish as we know it began as the dialect of the kingdom of Castile in the northern Iberian mountains. The first literary landmark is the epic *Poema del Cid* (c1205). As Castile annexed the Muslim territories of central and southern Spain, the Castilian dialect spread through the whole peninsula.

Christopher Columbus made his first voyages to the Americas in 1492; from then on, Spain led the conquest of Central and South America. Soldiers, missionaries, and traders spread the language. The Spanish also reached the south and west of what is now the US. Their influence can still be seen in place names such as Florida (land of flowers) and Nevada (snow-covered).

The Spanish language has also been spread by world-famous literature such as *Don Quixote* by Miguel de Cervantes (1547–1616); the plays of Federico García Lorca (1898–1936); and the films of Pedro Almodóvar (1949–). Regional dialects in Spain include Aragonese and Leonese. In Central and South America, each country has its own form of Spanish and local dialects.

**ARGENTINIAN SHOPPERS**

*The Argentinians are descended from many different peoples but all use Spanish as their national language.*

**DON QUIXOTE**

*This statue shows Don Quixote and Sancho Panza, two of the most famous characters in Spanish literature.*

**SPANISH IN THE US**

*A group of demonstrators marching in Washington State, US, display their message in Spanish as well as English.*

WORLD LANGUAGES

## Hindi/Urdu

**Estimated number of native speakers:** 220 million

**Countries with most speakers:** India, Pakistan

Hindi and Urdu are spoken by more than 400 million people around the world, about half of whom use them as a mother tongue. The two languages are structurally the same. However, because Hindi is used principally by Hindus and Urdu by Muslims, they have different writing systems.

The two languages are descendants of Sanskrit, the classical language of India, which dates from the second millennium BC. The first identifiable ancestor of Hindi, the speech of Delhi and its region, was recorded in the 12th century. Urdu was a special form of this language, which developed in medieval courts and armies. A variant, Hindustani, was later adopted for local administrative use in British India.

Hindi is written in the indigenous Devanagari alphabet. It is one of the 11 regional administrative languages of India, and, together with English, is also a national language. In addition, Hindi and its eastern variant, Bhojpuri, are spoken by millions of people in expatriate Indian communities around the world.

**ALTERNATIVE SCRIPTS**

*A post office sign in Kashmir gives the same information in Hindi (top), English (centre), and Urdu (bottom).*

Urdu is written in a Persian form of the Arabic alphabet. Its vocabulary is influenced by Persian and Arabic. It is the official language of Pakistan but only a minority, centred in Karachi, speak it as their mother tongue. Urdu is also spoken by several million people in the larger cities of India, with a special concentration at Hyderabad. One form is used as a lingua franca in India's biggest cities, Calcutta and Bombay.

**LANGUAGES OF INDIA**

*This advertisement is written in the national languages of India: Hindi and English.*

WORLD LANGUAGES

## Bengali

**Estimated number of native speakers:** 180 million

**Countries with most speakers:** Bangladesh, India (West Bengal), UK

Bengali is spoken in and around the Ganges delta, in the northeast of the Indian subcontinent. It is the official language of Bangladesh, where most speakers are Muslims, and the Indian state of West Bengal, where most are Hindus.

The language has two main forms: *sadhu bhasa*, the literary language; and *colit bhasa*, the colloquial form. The Bengali alphabet is related to Devanagari, used for Hindi.

In Bangladesh, Bengali is a focus of nationalist pride. From 1947 to 1971, when the country was part of Pakistan, people fought to have it accepted as a national language. There were clashes with the police, in which some protesters died for the sake of their language. Bengali is also spoken in expatriate communities all over the world, especially in the UK.

**BRICK LANE**

*One of the largest concentrations of expatriate Bengali speakers is in Brick Lane, London, UK.*

**READING BENGALI**

*Women in West Bengal, India, attend a reading class. Many adults are illiterate; however, levels of literacy in the state are higher than for India as a whole.*

Profile

### Rabindranath Tagore

The poet and philosopher Rabindranath Tagore (1861–1941) was the first world-famous writer in Bengali. He won the Nobel prize for literature in 1913 and was admired by eminent poets such as W B Yeats. His work focused on love, spirituality, and humanitarian ideals. He also wrote songs, including the national anthems of India and Bangladesh.

CULTURE

WORLD LANGUAGES

## Arabic

**Estimated number of native speakers:** 165 million

**Countries with the most speakers:** Saudi Arabia, Gulf states, Yemen, Lebanon, Palestine, Israel, Syria, Jordan, Iraq, Egypt, Sudan, Libya, Algeria, Morocco

Arabic is the main language of the Middle East and North Africa, and is used across the Muslim world. It owes its importance to being the language of the Koran, the sacred text of Islam.

The language originated in northern Arabia. The highest form, Classical Arabic, emerged in the 7th century AD, when the Koran was compiled. As the Muslims conquered new territories, Arabic spread through North Africa, into Spain, and as far as Southeast Asia.

During this time, a standard form arose among soldiers and bureaucrats. Standard Arabic, which is based on classical texts, is still the language of science and literature, and is also the medium of government, broadcasting, publishing, cinema, and international communication.

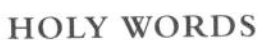

Arabic has influenced languages wherever Islam has spread. Swahili is full of Arabic loan words, as are Persian, Urdu, Turkish, and Malay. Cultural terms such as *madrasa* (school) and *adab* (literature) are widely borrowed. It also left its mark on European languages because the Arabs took up the scholarly tradition of ancient Greece and Rome and passed it on to medieval Europe. Some words relate to science (algebra, chemistry, zero); others, to trade (sugar, cotton, tariff) and rulership (admiral). A few are now used in new contexts; for example, *mujahidin* actually means "holy warriors" but is now also the name for the Afghan resistance force who fought the former Soviet Union in the 1980s.

Fact

### Arabic script

Arabic is based on 28 letters, each representing one sound. The short vowels a, i, u, and other signs, may be added as small marks to guide pronunciation. With the rise of Islam, Arabic calligraphy became an art form, with the letters shaped into elegant graphic patterns. The script was used to create beautiful copies of the Koran and to adorn the walls of mosques and other buildings.

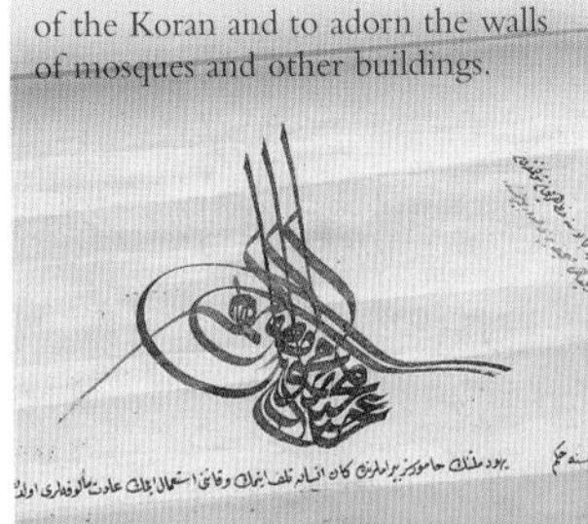

**SAUDI EMBLEM**
*The national flag of Saudi Arabia bears the Islamic script "There is no god but Allah; Muhammed is the prophet of Allah".*

**HOLY WORDS**
*A Nigerian boy learns verses from the Koran. Muslims all over the world revere the Koran, and most study it in the original Arabic.*

**MODERN MEDIA**
*The Muslim world uses a standard form of Arabic in international media such as broadcasting, as is the case at Al-Jazeera news station, Qatar.*

WORLD LANGUAGES

## Russian

**Estimated number of native speakers:** 175 million

**Countries with most speakers:** Russia, other states of former Soviet Union, states of eastern Europe, Israel

Russian is spoken across large parts of eastern Europe and northern Asia, and also in Israel. It has about 277 million speakers, including 175 million for whom it is their mother tongue.

The language is based on the 13th-century dialect of Moscow. It was first written in Old Slavonic, the classical language of Orthodox Christianity. Tsar Peter the Great (1672–1725) reformed the script for secular use in 1708, creating the Cyrillic writing system.

Russian power spread the language both westwards and eastwards from the 15th century onwards. In the 19th and 20th centuries, large numbers of Russian speakers migrated to parts of central Asia and Siberia, reaching as far as the Pacific coast. They included political exiles, criminals, and members of religious sects; some of the migrants settled in these areas. The Soviet Union took control of states on the Baltic coast and in eastern Europe and central Asia. Many people in these countries still speak Russian as an official language or as a lingua franca.

Many works of Russian literature are world famous. They include the novels of Fyodor Dostoyevsky (1821–1881), the poetry of Aleksandr Pushkin (1799–1837), and the plays of Anton Chekhov (1860–1904).

PETER THE GREAT

**TRANS-SIBERIAN RAILWAY**
*This railway, running from Moscow to Vladivostok, links Russian-speaking areas across nearly 10,000km (6,000 miles) and seven time zones.*

WORLD LANGUAGES

## Portuguese

**Estimated number of native speakers:** 176 million

**Countries with most speakers:** Brazil, Mozambique, Angola, Portugal, Guinea-Bissau, Macau, East Timor

Portugal is a small country with only 10 million people, but its language is spoken by much larger numbers in its former territories across the world. The biggest group of Portuguese speakers, 158 million people, exists in Brazil.

Portuguese originated in the 12th century as a dialect of Galician, the language spoken in the northwest of the Iberian peninsula, and spread southwards. During the 15th and 16th centuries it spread around the world with traders and colonists. Loan words provide evidence of this adventurous history. For example, the words *arroz* (rice), *alcachofra* (artichoke), and *algodão* (cotton) come from Arabic; *inhame* (yam) is taken from a Bantu language of southern Africa; and *maíz* (maize) comes from Carib. In addition, several Portuguese-based creole languages exist in Africa and Asia.

Portuguese looks like Spanish but sounds very different, with its nasal vowels and *sh* and *zh* sounds. The two major varieties are those of Portugal and Brazil, which differ from one another only in minor ways.

History

### The age of exploration

Portuguese was the first European language to spread beyond Europe. Sailors opened up seaways across the world. Vasco da Gama (c1469–1524), for example, was the first European explorer to sail from Africa to India; his journey was immortalized in the national epic poem *Os Lusíadas*, by Luís de Camões (1524–1580).

**PORTUGUESE LEGEND**
*Dancers in the Brazilian city of Parintins act out a Portuguese Catholic legend dating from the 19th century.*

WORLD LANGUAGES

## Japanese

**Estimated number of native speakers:** 120 million
**Countries with most speakers:** Japan

Japanese is spoken only by Japanese people and has no known linguistic relatives. However, it has been deeply influenced by Chinese. The language has many loan words from Chinese. The earliest form of Japanese script, *kanji*, consisted of Chinese characters; it dates from the 4th century AD and was used in early and medieval literature. Modern Japanese uses three scripts: *kanji*, *hiragana*, and *katakana*. Most words are written with one or two *kanji* characters, to which *hiragana* characters are added to show prefixes and suffixes. *Katakana* is used to spell loan words from foreign languages.

Japanese has distinct forms to express status and courtesy. Plain speech is used by socially superior people. Lower-ranking people use honorific speech to address others and humble speech to describe themselves. There is also a distinct difference between men's and women's speech. However, Japanese is unusual in that some of its greatest literature was written by women. The supreme work is the 11th-century *Tale of Genji* by Lady Murasaki, a servant of the Imperial court.

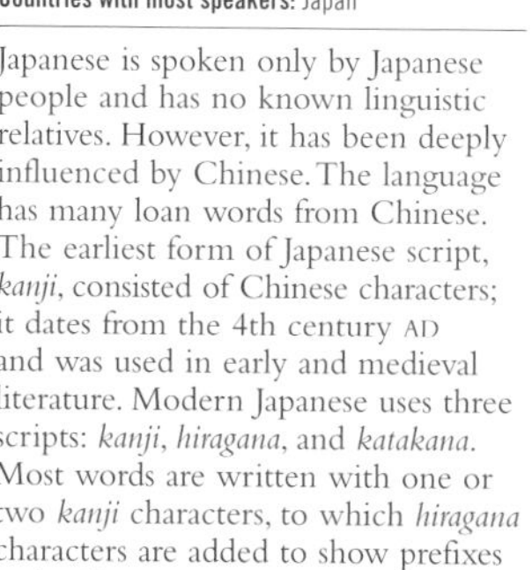

**FEMININE ACCENT**
*Women's speech is traditionally more polite and refined than that of men. Certain words are considered "rough" and avoided.*

**DIFFERENT DIRECTIONS**
*A cluster of advertisements illustrates different writing styles. Japanese was traditionally written vertically; today it may also be written horizontally.*

WORLD LANGUAGES

## Javanese

**Estimated number of native speakers:** 75 million
**Countries with most speakers:** Java (Indonesia)

Javanese is spoken mainly on the island of Java in Indonesia. It is counted as a world language despite being based in such a small area because it has so many speakers. The national language is actually Indonesian (officially called *Bahasa Indonesia*), a form of Malay.

Java was once a Hindu kingdom and its oldest historical records are in Sanskrit, the classical language of India. The first inscription in Javanese is dated to AD806. Javanese has been written in three scripts: first, one of the Indian type; then Arabic script, after the arrival of Islam; and today the Latin alphabet.

Javanese is unusual in that it has politeness built into its structure. Both tone of voice and different words are used to indicate the relative status of speakers and other persons. As an example, the following verbs, shown in bold type, all mean "gave": *Aku* ***ngekeki*** *kancaku buku*, "I gave my friend a book" (indicating that speaker and friend are equal); *Aku* ***njaosi*** *bapakku buku*, "I gave my father a book" (indicating higher status for the speaker's father); *Bapak* ***maringi*** *aku buku*, "My father gave me a book" (showing lower status for the speaker). Young people in Java sometimes prefer Malay to Javanese, because when speaking Malay they do not need to indicate status.

**RESPECT FOR ELDERS**
*One form of Javanese speech is used to show respect for elders and other high-status people.*

**HISTORIC SCRIPT**
*The old Javanese form of writing resembles Indian scripts but with simpler letterforms.*

WORLD LANGUAGES

## German

**Estimated number of native speakers:** 120 million
**Countries with most speakers:** Germany, Austria, Switzerland, Kazakhstan, Poland, Hungary, Italy

About a quarter of all Europeans speak German as their mother tongue. The language was a lingua franca in much of eastern Europe until the middle of the 20th century. It is still important internationally as a major language of science and scholarship.

German was actually two languages: Low German, spoken in the northern plains, and High German, used in the southern hills. Low German is still spoken, but High German has become the standard language.

One distinctive feature of German is that any speaker or writer may make new compound words freely from existing ones. For this reason, perhaps, German has fewer loan words than other European languages. Another unique characteristic is the traditional Fraktur script, which was commonly used in German printing until the 1930s.

The standard forms of the language in Germany and Austria are distinctly different. Regional and minority dialects also differ widely. Swiss Germans speak a unique form called *Schwyzerdütsch*; it is a focus of local pride, though it has no official status and is seldom used in writing.

History

### Martin Luther's Bible

The first text of modern standard German is Martin Luther's translation of the Bible, completed in 1534. Luther (1483–1546) wished to reform the Church and direct Christians back to Biblical teachings. The Bible sold in great numbers throughout Germany. Its language became the accepted model for German prose.

**FRAKTUR**
*The Fraktur script was used from the 16th to the early 20th centuries. It is now used mainly as decoration.*

WORLD LANGUAGES

## French

**Estimated number of native speakers:** 70 million
**Countries with most speakers:** France, Canada, Belgium, Switzerland

French has a relatively small number of native speakers compared to other world languages, but it is global in its reach. For centuries it was the main lingua franca for European travellers. Until recently, French was the world language of diplomacy. It is also an official language for international bodies including the European Union.

The first record of French is from a treaty of AD842. Its position as the only national language of France was decreed in 1539. The language spread to many other parts of the world with French conquests and is still a second language in much of Africa. French-based creoles are spoken by several million people in Haiti and other Caribbean islands.

French has had a major influence on world culture, as can be seen from words such as *couture* (fashion) and *cuisine* (cooking). However, the language is protected from external influences by the Académie Française. This official body monitors its development and tries to limit the absorption of words from English and other languages.

**IMPECCABLE TASTE**
*The influence of French cuisine has led to French words becoming widely used in restaurants and cooking.*

**FRENCH IN AFRICA**
*A boy in Cameroon studies French. Like many African nations, Cameroon belongs to the international community of French-speaking countries.*

# NONVERBAL COMMUNICATION

A great deal of human communication takes place without the need for words. On a personal level, this type of interaction includes body language, posture, and gestures, which all convey powerful messages about a person's feelings and intentions. Some of these signals are universal, while others are specific to certain cultures or situations. The use of symbols, such as flags and company logos, is another form of non-verbal communication. Understanding these messages can give people insight into the thoughts or culture of others and improve relationships. People or groups can also learn how to utilize nonverbal signals consciously, to give their messages maximum impact.

Fact

### Signs of thought

According to practitioners of neuro-linguistic processing (NLP) skills, eye movements can give clues to people's thoughts. A person glancing upwards is thinking visually; looking up to the right (upper picture) shows that the person is visualizing something, while looking to the left (lower picture) shows that the person is remembering an image. Someone looking directly right or left is formulating or remembering words. A person glancing down and to the left is focusing on an inner dialogue; someone looking down and right is checking his or her feelings.

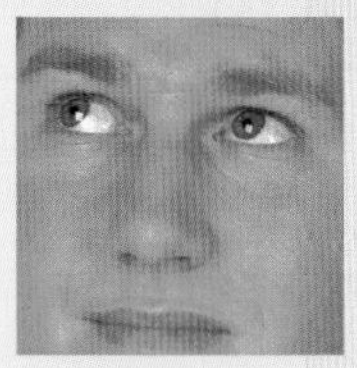

IMAGINING

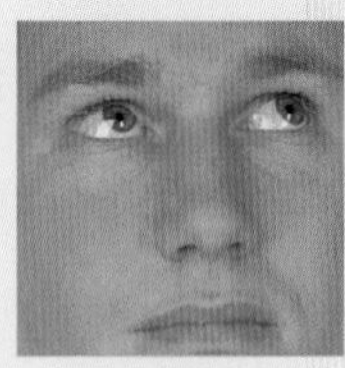

REMEMBERING

## BODY LANGUAGE

In contacts between individuals, body language – movements, posture, facial expression, and use of personal space – is the most powerful form of communication. Even when a person is speaking, it is estimated that only 7 per cent of his or her message is conveyed through the words; 38 per cent comes from the tone of voice, and a full 55 per cent from the body language. This form of communication usually shows someone's repressed or subconscious feelings about other people or situations. It includes signs of social dominance or submission, aggression, anxiety, and sexual attraction. These basic signals are recognized by most people in the world and even by domestic animals, such as dogs. We can learn to "read" body language by noticing clusters of signals (rather than isolated actions). For example, the signs showing that a person likes us might include direct, friendly eye contact together with leaning closer and pointing the hands and feet towards us.

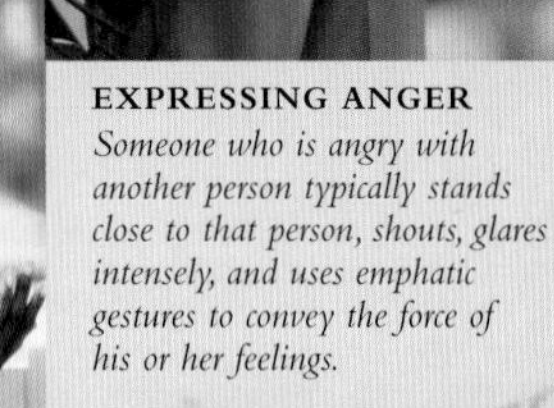

**EXPRESSING ANGER**
*Someone who is angry with another person typically stands close to that person, shouts, glares intensely, and uses emphatic gestures to convey the force of his or her feelings.*

**PAYING ATTENTION**
*Attentive listeners convey their interest and sympathy by leaning towards, moving close to, or touching the speaker.*

## GESTURES

A gesture is a deliberate nonverbal signal. In everyday life, gestures are most often used to convey particular types of message, such as greetings, expressions of respect, or insults. Many of these actions are specific to one or a few cultures, such as the *hongi*, the Maori greeting in which people touch their noses together. Other gestures have different meanings depending on the context; for example, making an "O" shape with the thumb and forefinger could mean "OK" or it could signify "no money". Certain gestures act like punctuation marks in speech, emphasizing the words. Formalized systems of gestures have been devised for situations in which people cannot usually be heard, such as hand signals for scuba divers or the "tic-tac" system used by bookmakers on racecourses. The most complex gesture systems are the sign languages created for deaf and deaf–blind people, which are used in the same way as spoken language but involve forming words with hand movements and facial expressions.

**SILENCE**
*This gesture, used to tell someone to be quiet, is usually made with a "shh" sound but can easily be understood even when the person receiving it is out of earshot.*

**SIGNALLING**
*This police officer is using a clear, simple system of arm signals to stop, move, and direct traffic.*

CULTURE

History

### Mathematical symbols

The basic symbols +, -, x, ÷, and = originated in medieval Europe. The plus symbol, +, may have developed from the abbreviation for the Latin word *et* ("and"). This sign, and the minus symbol, -, were used in trade to show excesses or shortfalls in the weight of goods. The multiplication symbol, x, originated in England in about 1600. The symbol ÷ was first used for division in the 17th century; it may represent a fraction with dots instead of numbers. The "equals" symbol, =, was introduced by the English mathematician Robert Recorde, in 1557. He used two parallel lines because, as he said, "noe .2. thynges can be moare equalle".

# SYMBOLS

A symbol is an image that conveys a message or stands for a person, object, or concept. The most basic symbols are designed to be understood instantly, such as the stylized male and female figures on toilet doors or the skull and crossbones meaning "danger". Others may need to be learnt at first but are then easy to understand, such as highway signs, symbols for chemicals, and icons on computers. Writing systems (*see* p303) are more complex systems of symbols, in which letters, characters, or other marks represent sounds or words. Other systems based on symbols include numbers and other mathematical signs and musical notation. Many symbols are recognized across cultural boundaries. The most common examples are religious symbols, such as the Star of David and the Taoist "yin-yang" sign, and national flags. Certain global organizations, such as the Red Cross, also have emblems that are instantly recognized. In addition, some of the world's most successful companies are represented by simple logos that have become widespread and are now familiar sights in many cultures.

**ROAD SIGNS**
*Signs for traffic need to be simple and graphic in order to convey information instantly.*

**SIGN OF SUCCESS**
*The symbols for certain commercial brands have become familiar all over the world. This logo is based on a letter in English but is recognizable even to people who use other languages and scripts.*

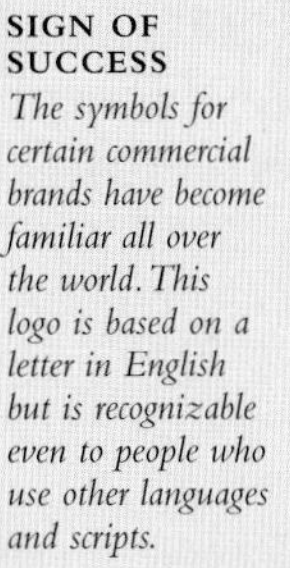

**NATIONAL FLAGS**
*Flags are potent symbols of nationhood. These flags represent various European countries as well as Russia.*

BODY LANGUAGE

## Affection

**In brief:** People express love and warmth towards others by closeness and comforting touch

Affection is expressed within close relationships such as parent and child, lovers, or friends. The form it takes in each case often involves comforting touch for both people. The most obvious example is the cuddling that parents, particularly mothers, give their children. This care finds an echo in the behaviour of loving couples; when one partner is touched or stroked by the other, he or she has a sense of affection and being nurtured. The action and response help to cement pair bonds. More general signals of affection include "tie signs" that publicly demonstrate a bond, such as physical closeness, hugging, or walking arm in arm. Girls may sometimes show affection by grooming each other's hair – another tie sign.

**CLOSE CONTACT**
*These two friends show their ease with one another by their casual physical contact.*

**STEADY PARTNERS**
*Settled couples may express their mutual affection openly but calmly – for example, with a gentle hug as they walk along.*

BODY LANGUAGE

## Sexual attraction

**In brief:** Sexual signals draw attention to attractive areas of the body and show a focus on the other person

People who feel sexual attraction for someone send out both unconscious and deliberate signals. One distinctive unconscious reaction is dilation, or widening, of the pupils (*see below*); another is pointing the hands and feet towards the other person. Touching the mouth, or glancing repeatedly at the other person's mouth, suggests an unconscious wish to kiss. Other signs of sexual attraction include "preening" actions that draw attention to physical assets, such as running the fingers through the hair. A more conscious, playful sign is exchanging glances and giving a person extra attention.

**FLIRTING**
*By touching her mouth, this woman is unconsciously signalling a desire to be kissed.*

**LOOK OF LOVE**
*People in love often gaze intently into each other's eyes, ignoring everything else around them.*

Fact

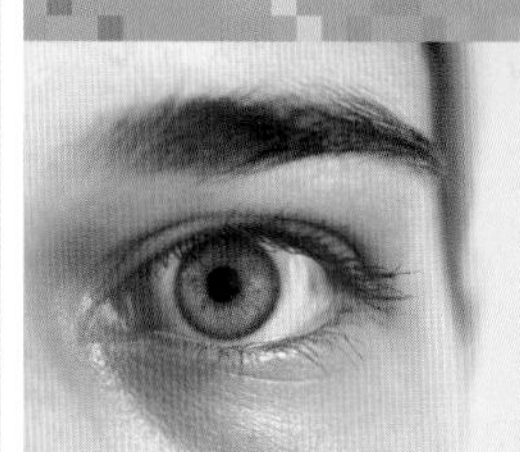

### Widening of the pupils

When a person finds someone else attractive, his or her pupils dilate, or widen. This response occurs as a reflex reaction. It is also unconsciously noticed by other people. For example, if men see two images of a woman's face that are identical except for pupil size, they instinctively prefer the one with the larger pupils.

BODY LANGUAGE

## Empathy

**In brief:** Empathy or agreement is shown by movements that unconsciously echo those of the other person

People tend to like others who are like themselves because they provide a feeling of security. Unconsciously, people create empathy with others by altering their body language to match. Breathing and facial expressions change; in particular, a person who is in sympathy with someone else tends to "mirror" the other person's posture and movements. It is possible to encourage a feeling of empathy with someone consciously by noticing these elements in his or her behaviour and subtly changing one's own actions to match them.

**MATCHING POSITIONS**
*Friends unwittingly show sympathy with each other by adopting similar body postures while talking.*

BODY LANGUAGE

## Anxiety

**In brief:** Signs of anxiety show suppressed "fight or flight" reactions or serve to release nervous tension

People feel anxious when facing threats or uncertainty, as in preparing to take a driving test. In such situations, the adrenal glands release epinephrine (adrenaline) to prepare the body for action; the resulting stress is called the "fight or flight" response. It may be shown by pale skin, tense muscles, and rapid breathing. If a person cannot take action, he or she may displace anxiety into activities such as biting the nails.

**SIGNS OF DISCOMFORT**
*Stress can often be seen by a tense, frowning face. Anxious people may also cover their mouths, perhaps to stifle sounds of alarm.*

BODY LANGUAGE

## Threat

**In brief:** Threatening actions are used to make a person appear bigger, stronger, or intimidating

Threatening body language is used by someone who is trying to exercise power and dominance over another. It is a deliberate attempt to unsettle or frighten an opponent. A raised fist is the most obvious, universal symbol of threat because it suggests a physical attack. More commonly, a person who wishes to intimidate someone else stands taller and pushes the chest out to appear larger. He or she also stares intensely at the opponent. Certain peoples, such as the Maoris, have even developed threat signals into a stylized display for combat or sporting events. The most subtle threat signal is tightening of the jaw and facial muscles, combined with a quiet but angry tone of voice.

**PUBLIC AGGRESSION**
*A member of an extreme political organization gives the fascist salute as he marches. His shouting and raised fist make him appear forceful and intimidating.*

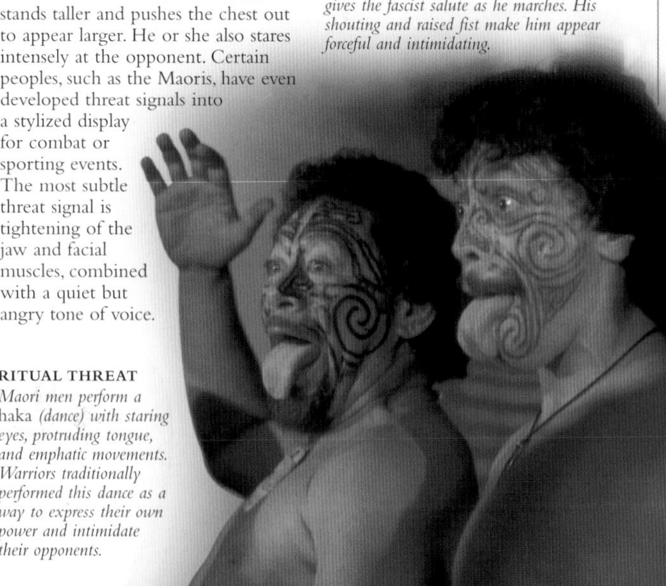

**RITUAL THREAT**
*Maori men perform a haka (dance) with staring eyes, protruding tongue, and emphatic movements. Warriors traditionally performed this dance as a way to express their own power and intimidate their opponents.*

BODY LANGUAGE

## Status

**In brief:** Posture and body movements can show a person's social dominance or subordination

Body language instantly shows people's social standing. Dominant people tend to stand tall, look others in the eye, and take up space; for example, when seated they may lean back and put their hands behind their heads. Subordinate people may have a lowered gaze and slumped posture, as if to shield themselves from attack. They may also smile more than others, to deflect possible aggression.

**NATURAL CONFIDENCE**
*Dominant people adopt poses that take up a lot of space and leave their bodies undefended, as they have no fear of attack.*

GESTURES

## Respect

**In brief:** Gestures of respect show recognition of someone else's higher social position

Showing respect to another person emphasizes that person's importance. Most gestures of respect are designed to make the giver temporarily appear smaller than the other person; they include removing one's hat, bowing, and kneeling. They are often used in traditional settings to reinforce the rules of a hierarchy. For example, Roman Catholics kneel to kiss the Pope's ring. Students of Eastern martial arts may kneel and bow their heads at the feet of their teacher. This gesture shows trust as well as respect; it is thought to imitate baring the back of the neck to be cut by a sword. The strongest gestures of respect are used by people worshipping God; for example, Muslims touch their foreheads to the ground, and pilgrims of various faiths may prostrate themselves at shrines. Certain gestures have evolved from older displays of respect. One example is the military salute, which developed from doffing the hat. Another is the Eastern practice of touching the hand to the forehead, which echoes the act of touching the forehead to the ground. A further way to show respect is to inconvenience oneself briefly for someone else's sake. Schoolchildren may be expected to stand when a teacher or other adult enters a room. People may stand in the presence of a president or other dignitary, or may show respect for a state or its head by standing when the national anthem is played.

**REVERENCE**
*A group of devout Buddhists show love and reverence for their religious leader by bowing to receive his blessing.*

**MODIFIED GESTURE**
*The military salute developed from the act of removing a hat to show respect to a superior. The lower-ranking person salutes first; the superior returns the gesture as a courtesy.*

GESTURES

## Greetings

**In brief:** Gestures of greeting show respect, trust, and goodwill towards others

Cultures across the world have different gestures for greeting people, but all such gestures have the same purpose: to show respect and goodwill. One of the most common gestures, principally used in Europe and North America, is the handshake. This gesture may have originated several centuries ago as a sign that someone was not carrying weapons. It was made popular by the Quakers in the 17th century as a way to show friendliness and equality. The strength of the grip may be significant; a firm handshake, as in the greeting that politicians give voters, is designed to inspire goodwill and trust (although it can sometimes seem over-confident to the person on the receiving end).

**SHOW OF SOLIDARITY**
*People engaged in political struggles often greet each other by raising their forearms and clasping the other person's hand, to express their solidarity and shared strength.*

**HUMILITY**
*In a Japanese greeting, each person bows deeply to show honour and humility to the other. The person of lower status usually starts the bow and bows lower and longer.*

Another widespread form of greeting is kissing the cheek. Many people kiss to show friendship or affection, but in some areas, such as Mediterranean and Arab countries, the cheek kiss is used even in formal situations, such as heads of state greeting a visiting dignitary. Formal kissing differs from friendly kissing in that both parties stand a short distance apart and lean towards each other, and the actual kiss is brief.

Hugging is a form of greeting most often used by family members or close friends. Hugs usually expose the front surface of the body to contact, so they are only used with people we trust. In cultures with a more reserved style of behaviour, such as among Chinese and Japanese people, hugging is uncommon and is never done in public.

Some modern cultures have created their own forms of greeting gestures. People involved in political movements may clasp each other's raised hands as a gesture of solidarity. American sportsmen have given the world the "high five", and some young African-Americans use a range of hand greetings including touching fists and various sequences of grips.

**MAORI GREETING**
*The* hongi *is a Maori greeting in which people press noses to exchange the "breath of life".*



**HIGH-FIVE**
*The "high-five" (slapping raised hands together) was first used by American sportsmen, but it has since spread widely in Western soceties.*

GESTURES

# Insults

**In brief:** Insult gestures vary around the world, but all convey contempt; many are sexually obscene

Humans around the world have a rich variety of ways to show dislike or contempt for one another. However, most insulting gestures have similar functions: to suggest that someone is inferior or stupid or to show defiance or contempt. Such gestures are likely to anger, upset, or disgust recipients who recognize them.

One common way to demean someone is to show them a part of your body that is considered "dirty". In Asian countries such as Thailand, it is rude to expose the soles of the feet; in the Arab world, it is offensive to show the soles of the shoes or touch someone with the shoe. In the US and UK, an increasingly common rude gesture is "mooning" – pulling down the trousers or even the underwear and then bending over to expose the buttocks to someone else.

**FOOT FAUX PAS**
*In countries such as Thailand, it is very rude to expose the soles of the feet in the direction of someone else because the feet are the lowest part of the body and are considered dirty.*

**DANGEROUS HORNS**
*The "horns" sign is often used as a salute by heavy metal music fans. However, in Italy it is very rude, meaning "your wife is cheating on you".*

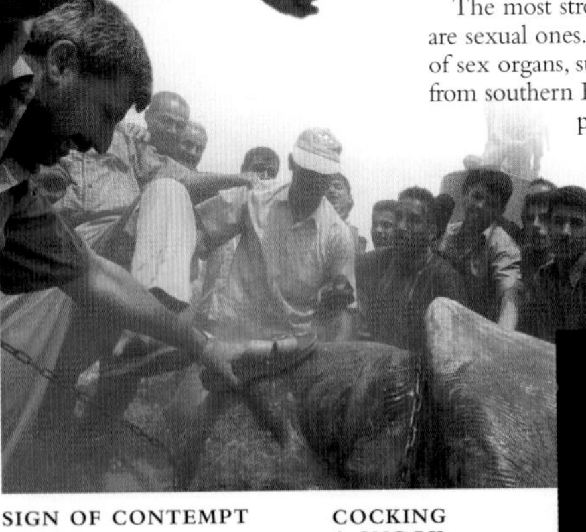

**SIGN OF CONTEMPT**
*Crowds attack a statue of the deposed Iraqi dictator, Saddam Hussein. They hit the statue's head with their shoes as a sign of deep contempt.*

Another way to insult someone is to use gestures more appropriate to summoning a servant or talking to a child, such as snapping the fingers to call someone or patting the head or face of an adult. A third way is to imply that a person is stupid or mad, such as the European "donkey ears" gesture of placing the open hands by the ears and waggling the fingers.

The most strongly offensive gestures are sexual ones. Some reflect the shape of sex organs, such as the "fig" sign from southern Europe, in which a person makes a fist and pushes the thumb through the fingers to echo the shape of female genitals. Other signs imitate sexual intercourse, such as "flipping the bird" in the US, in which the extended middle finger is flicked upwards. In France and Italy, a stronger version is used, with the fist and forearm jerked up. In Southeast Asia, hitting one fist against the open palm of the other hand has a similar meaning.

Certain hand gestures have different meanings in different cultures. In Europe, tapping one's temple can mean "you're mad", but in the US it means "you're smart". A thumbs-up sign in most countries can mean "good" or "good luck", but in Iran is an offensive sexual gesture. An "O" sign made with the thumb and first finger might mean "OK" in the US, but it means "money" in Japan, "sex" in Mexico, and "you're homosexual" in Ethiopia, and is a gross insult in Brazil. The "horns" sign, which is a greeting when used by young American males but an insult in Italy, is another gesture that may cause offence in the wrong context. Today, as more people travel the world for business or leisure and encounter different societies, it has become increasingly important for everyone to be aware of culture-specific gestures and thus avoid accidentally causing offence.

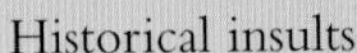

History

## Historical insults

Certain insults have ancient origins. One is the Greek *moutza*, in which an open hand is thrust in someone's face. This gesture dates from Byzantine times, when people used to push excrement into criminals' faces. Another is the British V-sign, thought to come from the Battle of Agincourt in 1415. French soldiers would cut off the first two fingers of English bowmen; archers who still had their fingers would wave them in defiance.

**COCKING A SNOOK**
*The "cocking a snook" gesture, found in many societies, is used for mocking someone else.*

**RUDE BOY**
*Sticking out the tongue is one of the earliest rude gestures that people learn. It is thought to originate from babies rejecting food or the mother's breast.*

**MOONING AROUND**
*"Mooning", in which people pull their trousers down or show their bare buttocks, is usually a gesture of disrespect or contempt.*

GESTURES

## Triumph and defeat

**In brief:** Certain gestures, and some unconscious signals, show triumph at an achievement or dejection at defeat

Expressions of triumph and defeat are most often seen after contests, such as sporting matches. Winners may show their dominant status (*see* p313) by making themselves look bigger – for example, by punching the air, jumping up and down, or running around with their arms wide. Even when alone, people may make triumphant gestures to celebrate an achievement. Losers, in contrast, show their dejection by a slumped posture, hanging head, and small movements. Defeated people also tend to shield the front of the body, which contains the genitals and internal organs. "Winners" feel no such need for self-protection, so they walk tall with their chests out and shoulders back.

**SUCCESS**
*An Aboriginal boy, having just caught a duck, raises his prize in the air for everyone to see.*

**WINNER AND LOSER**
*The winner of this wrestling match shows his suddenly heightened status by expansive, theatrical gestures. In contrast, the loser appears to be crushed by his defeat.*

GESTURES

## Emphasis

**In brief:** Emphatic gestures are used while a person is speaking, to stress a point or express strong feeling

In many cultures, when people wish to emphasize what they are saying they use gestures. The amount of emphasis that is naturally used varies from one culture to another, but most people use their hands more than usual if they wish to reach into a listener's personal space and increase the effect of their message. To draw a listener's attention, a speaker may hold up a forefinger, or may make a "clutching" gesture, as if encouraging the listeners to grasp the point. In order to stress certain words or phrases, he or she may beat the air with a forefinger or hand, like the conductor of an orchestra. To appeal for attention or understanding, a speaker may make a submissive gesture with the hands spread out and the palms vertical or turned upwards. The larger and more forceful the gesture, the stronger the message conveyed.

**STRESSING A POINT**
*A US politician giving a speech in a campaign uses his raised finger to draw attention to a specific statement.*

**PLEA FOR SYMPATHY**
*A speaker who is trying to gain the listener's sympathy or agreement will hold the hands out wide in a silent appeal.*

GESTURES

## Sign languages

**In brief:** Fully gestural languages enable deaf or deaf–blind people to communicate complex information

There are about 200 sign languages for use by deaf people, and their families and friends, around the world. These systems are fully fledged languages rather than just simple sets of gestures, and have their own rules for grammar, punctuation, and sentence structure. Some of them, such as American Sign Language, are officially recognized as distinct languages. All sign languages are based on hand gestures. Some words are shown by single gestures; others are spelled out using the fingers, with signs for each letter. Deaf–blind people may use a form of spelling in which letters are tapped on to another person's hand.

In many forms of sign language, people use facial expressions and body movements in the same way that hearing speakers use tone of voice. For example, people who wish to stress a specific point may widen their eyes and raise their eyebrows while signing.

**EMPHATIC EXPRESSION**
*A deaf woman signing to some friends uses facial expressions and body movements to reinforce her hand gestures.*

GESTURES

## Formalized systems

**In brief:** Special sets of gestures are used by people who are temporarily unable to hear others, such as divers

For situations in which people cannot hear each other but need to exchange simple messages, formalized systems of signals have evolved. In scuba-diving, sailing, and guiding aircraft, the signals convey vital messages such as warnings or calls for help. Diving signals are given with the hands; other types are conveyed with flags or batons. In some sports, signals are used if people are usually out of earshot. One example is the tic-tac system, which bookmakers use to pass on betting odds in horse-racing; another is the system by which baseball umpires talk to players.

**UNDERWATER SIGNALS**
*Divers use clear hand signals to exchange important information, such as where they are going, how much air they have, and whether or not they are all right.*

SYMBOLS

## Emblems

**In brief:** Emblems are graphic symbols representing particular countries, organizations, or concepts

An emblem is a graphic image used to represent a group or concept. Some emblems are used by nations, powerful families, and religious groups, such as the two-headed eagle of Imperial Russia, the totem animals of Native American clans, or the Jewish Star of David. Such emblems may have been created to advertise the group's identity both to friends and to enemies. They can be a focus of intense pride, as with the flag of the US. The modern equivalents are company logos; an effective logo will be strongly and instantly associated with the company, and will attract business away from competitors. Certain emblems stand for concepts that are recognized all over the world. One is the Red Cross flag, representing humanitarian aid. Another is the looped red ribbon to show support for people with AIDS.

**DANGER SIGN**
*The colour red and the image of a skull and crossbones are both widely recognized symbols that denote a lethal hazard.*

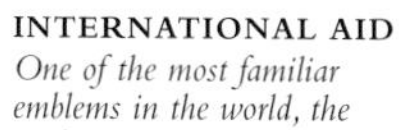

**INTERNATIONAL AID**
*One of the most familiar emblems in the world, the Red Cross is used internationally to show that medical help is available.*

# MASS MEDIA

The rapid expansion of mass media – communication systems that provide information quickly to huge numbers of people – has been one of the major features of the last century. Branches of the media, especially the press, radio, and television, have become powerful global industries with a pervasive influence in society, producing a wide range of news, advertising, opinion, and entertainment. A more recent development is that ordinary people can obtain or share information directly by using the internet.

**INFORMATION GENERATION**
*Many children today are already adept at accessing, sharing, and creating information using computers and internet technology.*

## THE GROWTH OF NEW MEDIA

For centuries, newspapers were the only mass medium. The invention of the telegraph in the late 19th century made instantaneous communication across long distances possible, but only for a few. It took a series of further inventions in the first half of the 20th century – first radio, then film and television – to make true mass communication possible. These technologies, although powerful, are also costly to run so, until recently, their use was largely controlled by governments and media corporations. Now, some people argue, the growth of the internet may have brought a new revolution. For the first time it is possible for individuals to reach a potentially global audience directly by using low-cost personal computers. The internet remains beyond the reach of a large proportion of the world's population, but companies and individuals are continuing to work on improvements that should make the technology cheaper and more widely available.

**HOME RADIO**
*By the 1930s radios had become part of the furniture of many homes, particularly in Europe and North America.*

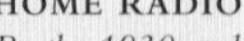

History

### The Gutenberg press

The printing press and movable type were invented by the Chinese, but the first person to achieve large-scale book printing was the German Johannes Gutenberg (1390s–1468). The first mass-produced book was the Bible, translated into German by Martin Luther (*see* p309). For the first time, books and documents could be made relatively quickly and cheaply for ordinary people.

## USES OF MASS COMMUNICATION

As the new mass media technologies have arrived, each has been greeted with hope and fear. Some hoped that they would bring us education, the best of the world's culture, and an understanding of the world's problems. Others feared they would be used to pander to our prejudices, corrupt our morals, and degrade our tastes. There is some evidence that both sides have been right. For the first time even illiterate people have been able to use radio and television to learn what is going on across the world almost as soon as it happens, and the best theatre and music are available to much wider audiences. In addition, the media have been used to spread vital public information, such as advice on preventing the spread of AIDS. On the other hand, through propaganda the media helped to sustain dictators such as Hitler and Stalin, and played a key role in the 1994 genocide in Rwanda. Many also suggest that television advertising has created desires for goods that people do not need and cannot afford.

**INFLUENTIAL WORDS**
*Newspapers are inexpensive to produce and can be circulated to huge numbers of readers. As a result, they and their owners may have a potent influence on public opinion.*

**MEDIA SATURATION**
*Today, many events involving famous or infamous people, such as film premieres or court cases, are reported exhaustively by media organizations from all over the world.*

CULTURE

GLOBAL MEDIA

## Newspapers

**In brief:** The first mass medium, newspapers are still produced locally, nationally, and internationally

From 59BC in ancient Rome a kind of official newspaper, the *Acta Diurna* (Daily Events), was posted in public places throughout the empire. National newspapers have existed since the 17th century. Today, despite the profusion of other media, papers still have a role in society – often by giving easy access to more specialized, local, or trivial news than radio or TV can provide.

**A RANGE OF NEWS**
*Some newspapers are widely circulated in many countries. Others are produced to serve specific communities or other interest groups.*

GLOBAL MEDIA

## Posters and billboards

**In brief:** Whether cheap posters or large billboards, these media convey powerful visual and verbal messages

Posters, and large images on billboards, can be extremely effective ways to convey simple messages by means of striking statements and bold images. They are often used by companies and governments for advertising or propaganda. Alternative groups may also use posters as a form of protest (as students did in Beijing during China's Cultural Revolution) or simply write over them in order to subvert their original message (as anti-advertising campaigners do).

**UBIQUITOUS IMAGE**
*The largest companies can advertise their products in many different societies, often far from their original target population.*

History

### Propaganda

Mass communication provided 20th-century dictators with a new means of indoctrination. In totalitarian societies such as Nazi Germany and Soviet Russia, posters on every street promoted the ideals of those societies and denigrated their enemies. In some countries the mass media are still primarily sources of propaganda, but satellite TV and the internet can now provide alternatives to this form of misinformation.

1917 SOVIET POSTER

GLOBAL MEDIA

## Digital media

**In brief:** The newest forms of mass communication, digital media can allow ordinary people as well as big corporations to share information

The internet and other digital media provide instant news, as well as information for purposes ranging from academic study to entertainment. Unlike most other media, which largely provide a one-way flow of information, digital media enable readers or listeners to make their own contributions. At present, few conventional media companies have online services that allow full interactivity (in part because it is not clear how providing such facilities can make them money), but "citizen media" sites are emerging across the world. They range from the Indymedia network serving the antiglobalization movement to OhMyNews, a South Korean online newspaper with 26,000 "citizen journalists". People usually need computers to gain access to the internet, but new mobile phones may enable many more people to explore this medium.

COMPUTER MOUSE

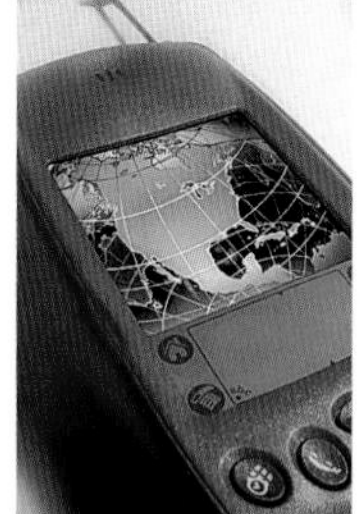
**AT YOUR FINGERTIPS**
*The latest mobile phones enable users to share speech, words, and pictures.*

GLOBAL MEDIA

## Radio

**In brief:** An inexpensive medium, radio can reach people even in remote parts of the world

Across most of the developed world, television has displaced radio as the main medium of entertainment and information, leaving radio mainly for news, music, and talk shows. (The latter allow listeners to contribute directly to programmes by giving opinions over the telephone.) However, certain international radio stations, such as Voice of America and the BBC World Service, are still influential sources of news and other information. Radio remains vital in the developing world because radio programmes are cheaper to produce than television programmes; in addition, radios are inexpensive, and some models can run on solar power or even by clockwork.

**FAR-REACHING SIGNALS**
*In remote communities, radio may be the only way to obtain news about the wider world.*

GLOBAL MEDIA

## Television

**In brief:** Television is perhaps the most heavily used medium, especially in Western countries

In the space of half a century, television has gone from being an expensive novelty to becoming the main source of news and entertainment for billions of people. Television-watching varies from country to country and person to person, but across much of the West it averages 2 hours a day, more than any other leisure activity, and it is still higher among children.

This heavy use has caused concerns. Some people have feared that the violence often seen on television might make impressionable children more violent, but despite thousands of studies the evidence that television is directly harmful in this way has been inconclusive. It was also feared that the rapid spread of inexpensive US programmes across the world might lead to the gradual destruction of traditional cultures, but local programmes remain more popular.

Thanks to cable, satellite, and digital technology, viewers around the world can receive larger numbers of channels than ever before, but this expansion has largely provided more of the same material. Advances in production, such as viewing on demand, have been slower to arrive.

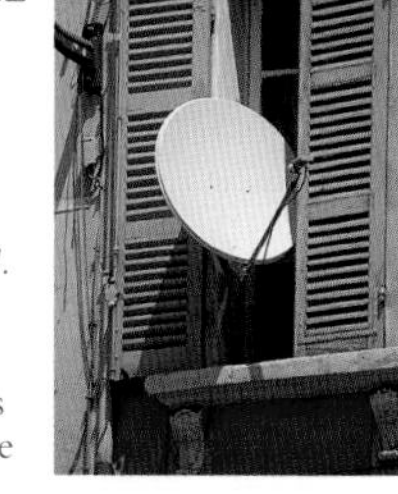
**EXTENSIVE INFLUENCE**
*Satellite TV has enabled the spread of programmes, and cultural influences, all over the world.*

**HYPNOTIC VISION**
*The flickering images on television screens are thought to have a mesmerizing effect. They can capture the attention of viewers wherever they are.*

Fact

### Wireless technology

One of the newest developments in internet communications is a broadband telecommunications network known as wi-fi (wireless fidelity). This service transmits data over short distances, using radio waves sent from points called "hot spots". Wi-fi networks are cheap to set up and transmit data very fast, so they can provide internet facilities in poor or inaccessible areas where it would be difficult and expensive to install telephone lines.

**CONNECTING KABUL**
*Afghanistan is a poor, mountainous country. For many people there, wi-fi is the only way to obtain internet access.*

**A WEALTH OF DECORATION**
*Some clothes and accessories are worn as symbols of status and affluence. This Yemeni Jewish bride is displaying her wealth of silver, coral, and amber.*

# CLOTHING AND ADORNMENT

Humans are the only species to make and wear clothes. Our early ancestors wore animal skins, and since then people around the world have created a huge variety of clothing and other forms of adornment. What we wear reveals a lot about who we are. It sends out messages about our way of life, community, and religious beliefs, as well as our occupation and status in society. We choose special outfits to attract other people and mark important occasions, or to show that we have reached a particular stage of life.

Clothes are usually combined to form an outfit. The effect is often enhanced with accessories, body decorations such as make-up and tattoos, and an appropriate hairstyle.

Unlike other mammals, humans do not have natural protection from the environment, such as fur or a layer of insulating fat. Humans have thick hair only on small areas of the body, and when naked we only feel comfortable at a temperature of more than 21°C (70°F). We need clothes in order to survive in very cold, hot, wet, or dry climates; otherwise, we could die from exposure in only a few hours.

**ADAPTING FROM ANIMALS**
*The oldest forms of clothing include plumage and animal skins, like the headdress worn by this Kikuyu man in Kenya.*

## EARLY CLOTHING

It is thought that people first wore clothes around 70,000BC. The garments probably consisted of animal skin or dried leaves or bark, draped or tied on the body. Any such clothes perished without trace a long time ago, but there is indirect evidence for their existence: genetic studies suggest that human body lice, which need the warmth from clothing, evolved at this time. Bone and ivory sewing needles dating from 30,000BC have been found in Russia. During this time, an ice age gripped large parts of the Earth, which suggests that early clothing was chiefly used for insulation.

**BASIC DRESS**
*Even people who normally wear few or no clothes, like this Amazonian man, decorate their bodies with accessories.*

## CHANGING STYLES

Originally, clothes and accessories were made by the wearers or by local craftspeople. In some places, such as parts of India and Pakistan, the act of making beautiful clothes was believed to bring good luck.

Today, although some clothes and other items are still made by hand, most are mass-produced. Such products are cheap, but their wide availability reduces the diversity of clothing styles around the world. For example, the popular informal attire of shirt, jeans, and trainers has been adopted by millions of people in many cultures.

## MODERN FUNCTIONS

Some outfits are designed primarily for comfort or protection. They may be made for a specific climate, such as the warm, insulating clothes worn in cold areas. Certain items are designed to ensure safety in hazardous occupations, such as hard hats and ear defenders for construction work, camouflage gear and bulletproof vests for military combat, and fire-retardant suits for rescue work.

Many styles of dress have social functions. One of the most basic of these functions is to protect a wearer's modesty. Clothing worn for this purpose is designed to conceal sensitive or "taboo" areas of the body such as the genitals or female breasts. Some religions or cultures require people (especially women) to wear garments that cover their entire bodies for the sake of humility and sexual chastity.

In contrast, some people in liberal societies flaunt their sexuality with revealing clothes. A person may wear such items to feel confident as well as to attract a potential mate. Many people adopt distinctive outfits for other personal reasons, such as to challenge social values or simply to express their individual taste.

Another important function of clothing and adornment is to give out signals about one's social rank. Animals can use their fur, feathers, or skin to show their sex or status, but humans have only a limited ability to do so (for example, by blushing), so we have developed an array of artificial body decoration. Luxurious clothes or expensive accessories identify the wearer as a wealthy or influential member of society. Conversely, some people wear very simple clothes or even go naked (as Jain and Hindu holy men do) to show that they reject materialistic and social values.

Styles of dress may also reinforce group identity. Many festivals and rituals include special dress codes as part of ancient traditions. The dress may be a visual representation of a people's history or traditions, or it may be used to show political or national allegiance. Standardized clothing – a uniform – enables people to demonstrate solidarity with others in the same company, organization, or country.

**TEAM PLAYERS**
*Participants in team sports or competitions usually wear special outfits showing their team or national identity.*

Fact

### The power of masks

Masks are used in rituals around the world. Their main function is to disguise the wearer's identity for social or spiritual purposes. Some masks are worn during religious rituals. Others are worn for plays or for re-enacting myths. The mask shown below is used in Balinese stories and represents a female demon called Rangda.

# TYPES OF DRESS

People use a wealth of materials and objects to modify their appearance. The most obvious are different kinds of clothes. In addition, people may wear accessories such as jewellery, transform their skin with body decoration, and style their hair in various ways. Some of these changes are simple and temporary, but others may take hours or be permanent. Throughout history, styles of dress have developed to suit human needs or fashions. Outfits may be commonplace or unique, incorporating hand-made or mass-produced items. Certain types of adornment, such as jewellery, have great value or may even be works of art.

## CLOTHING

Clothes are body coverings made from fabric or other flexible materials. The materials are chosen for particular qualities, such as durability or beauty; for example, clothes for manual work are often made from hard-wearing textiles such as denim, while luxury clothes are made from rich fabrics such as silk. Another important element is style. The most basic forms of clothing are all-in-one body coverings made from simple lengths of fabric, such as the Indian sari. More complex styles include tailored suits or haute couture. The most highly specialized clothing is that made for activities such as deep-sea diving, mountaineering, and space travel. Clothes may be decorated by methods such as needlework, batik, and printing.

**HARD-WEARING JEANS**
*Denim jeans were originally designed as cheap, durable clothes for American farm workers. In the 20th century, they became popular as classless leisure wear.*

Fact

### Intelligent textiles

Scientists are currently developing so-called "intelligent" textiles with special properties. For example, Phase Change Materials (PCMs) react differently in heat and cold to keep wearers comfortable. "Smart clothing" has built-in electronics such as a mobile phone. Future uses could include healthcare; for example, sensors in a shirt could monitor a person's heartbeat.

**EAGLE FEATHERS**
*In Navajo culture, eagle feathers, like the ones in this dancer's regalia, have great value. Navajo elders may award eagle feathers for achievements such as saving a life.*

## ACCESSORIES

Many people adorn their clothing and bodies with accessories. Jewellery is one of the oldest and most widespread types of accessory. It is usually made from objects or materials of particular beauty or rarity, such as beads, shells, precious stones, ivory, glass, and metals. People may wear it to show their affluence and social rank, or as a charm for healing or protection. Gold and silver may have both financial and spiritual value. For example, in China some Miao women wear more than 10kg (22lb) of silver jewellery both as a sign of wealth and prestige and to ward off evil spirits. Other kinds of accessory include natural articles such as feathers and flowers, which often have symbolic importance. Modern accessories include watches, handbags, spectacles, and mobile phones, which may have practical uses but also reveal the owner's status.

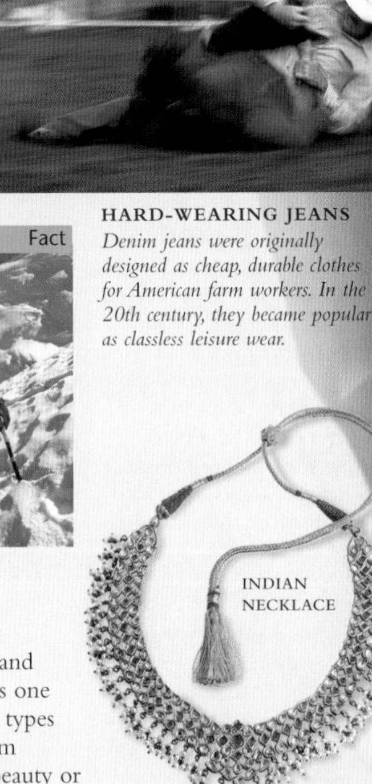

INDIAN NECKLACE

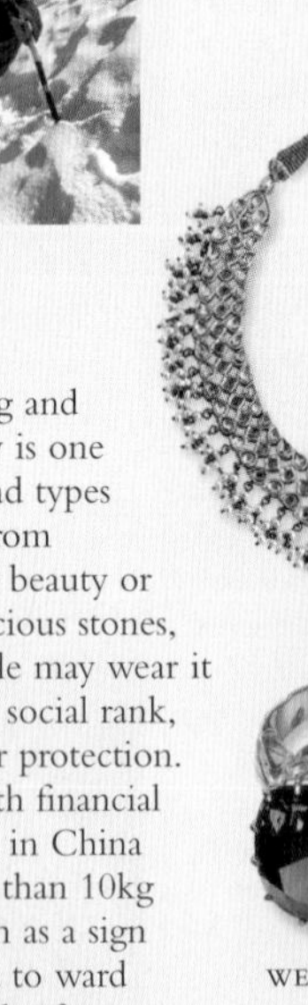

WESTERN RING

NATIVE AMERICAN EARRING

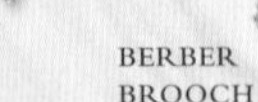

BERBER BROOCH

**JEWELLERY**
*In societies all over the world, people make finely crafted jewellery from gold and silver, gems, and stones such as turquoise. Jewellery is often worn to show the wearer's wealth and status.*

BEDOUIN WRIST AND ANKLE BRACELETS

ENHANCING SEXUAL APPEAL
*In both traditional and modern cultures, people often dress to attract potential partners. This Wodaabe man from Niger is dressing for the Yaake charm dance, in which the men vie to attract women.*

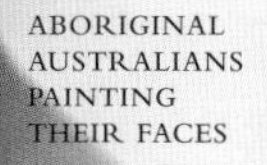

ABORIGINAL AUSTRALIANS PAINTING THEIR FACES

**FACE-PAINTING**
*Using cosmetics or face paint is the easiest way to decorate the skin. It may be done for rituals or to enhance attractiveness.*

WOMAN USING MAKE-UP

**HAIRSTYLES**
*Changes to the hair's style and colour may last for several weeks or months. Radical changes may, therefore, represent a strong expression of the wearer's identity.*

## FACE AND HAIR ADORNMENT

The head is the most individual and noticeable part of a person. For this reason, two of the most universal forms of adornment are the use of cosmetics and the adoption of different hairstyles. Cosmetics are used widely throughout the world but especially by women in North America and Europe. Make-up may be applied to enhance the eyes and lips or to hide "flaws" such as wrinkles. Married Indian women traditionally paint a small red dot (the *sindoor*) on the forehead each day to show that they have a husband. In many traditional societies, face-painting is common to both sexes. A modern equivalent is practised by sports fans who paint their faces in their team colours. Hair grows continually, so regular haircuts enable people to modify their appearance in striking ways. In particular, Afro hairstyles have become an art form; weaves, locks, braids, and extensions are both highly decorative and a proud reminder of African heritage.

## BODY DECORATION

In many cultures, people decorate their skin directly, through painting, tattooing, scarring, or piercing. Body painting is probably the most ancient technique. People often do it as a special procedure for rituals. Many Aboriginal Australian peoples, for example, decorate their bodies with spiritually significant designs including detailed clan motifs and references to their ancestors and land. People may create permanent decorations by scarification (making scars). Among Aboriginal Australians and some African peoples, it is traditional to cut the skin and rub in irritant substances to create swirling lines of raised, damaged flesh. This technique is used for decoration and to show tribal identity; in addition, the painful nature of the process may be an endurance test, as in some coming-of-age ceremonies. In many cultures, people pierce parts of the body, particularly the ears, to form a secure fixing for jewellery. Piercing of other areas, such as the nose, eyebrow, and navel, has become popular in Western subcultures for self-expression. Tattooing is another widely used form of decoration. For permanent tattoos, the skin is punctured to receive ink or dye. Semipermanent *mehndi* tattoos, made with red or black dye from henna plants, are a traditional decoration in the Middle East, the Indian subcontinent, and Central Asia, and are also fashionable in the West.

**SKIN-PIERCING**
*Piercing may provide a secure fixing for jewellery or other decorations. The act of piercing may also be performed as a test of courage or a sign of religious devotion.*

**FULL-BODY TATTOO**
*For yakuza (members of the Japanese underworld) it is the custom to have full-body tattoos as a symbol of their endurance and loyalty to the organization.*

PRACTICAL PURPOSES

## Cold climates

**In brief:** Cold-weather clothing is made from insulating fabrics and is often worn in layers to trap warmth

Nowhere is clothing more important than in very cold places, such as Arctic regions and the high Asian steppes. It must provide good insulation against extremely low temperatures and protection from snow and ice, while letting the wearer move comfortably.

Cold-weather clothing must allow perspiration to escape because damp clothing is a poor insulator and can chill the skin. It must also enable the wearer to adjust the level of warmth easily. For these reasons, many outfits are based on the "layer principle": several layers of clothing are worn, and air trapped between the layers provides insulation. As an example, in winter the Buryat people of Mongolia wear a snug vest or tunic covered with a long, fur-lined robe fastened over the shoulder to create a double thickness of cloth.

Many peoples living in cold regions traditionally made clothing from the fur and skins of animals such as deer, mink, seals, and reindeer, which provided natural insulation. Most modern clothing is made from synthetic materials. Some fabrics, such as Gore-tex®, keep out wind but let moisture escape from the skin. Others have a structure that traps air. Silk undergarments may also be worn as a lightweight insulating layer.

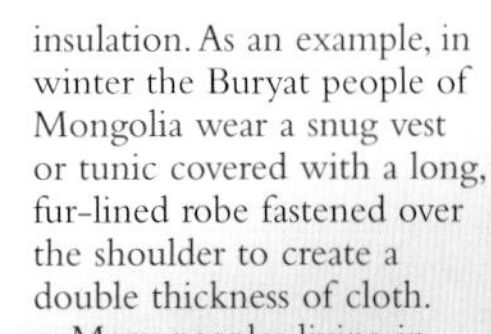

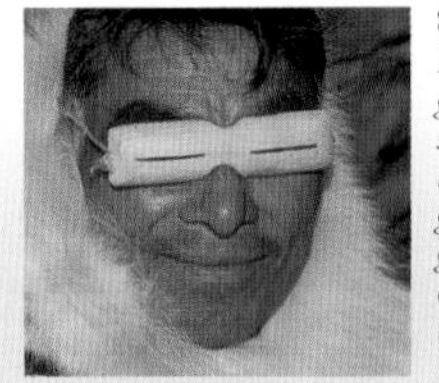

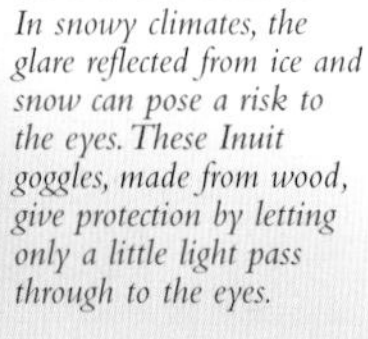

**SNOW GOGGLES**
*In snowy climates, the glare reflected from ice and snow can pose a risk to the eyes. These Inuit goggles, made from wood, give protection by letting only a little light pass through to the eyes.*

**WARM FURS**
*The Inuit people of northern Canada traditionally make clothes from caribou skin, sometimes with the fur turned inwards, or sealskin, which is water-resistant.*

**FELT CLOTHING**
*This Sámi woman is wearing traditional clothing made of felt fabric. Felt is used by many peoples because its dense structure traps warm air and provides insulation; in addition, it keeps out cold wind.*

PRACTICAL PURPOSES

## Rainy climates

**In brief:** Rainproof outfits include materials that repel water and accessories that shield users from rain

Peoples living in very wet climates have traditionally made waterproof outer garments using natural materials. In British Columbia, Canada, the Nuu-chah-nulth weave hats and capes from cedar bark fibres. In Northeast India, the Khasis weave huge hats from bamboo and sugar cane. Other fabrics include waxed or oiled materials that repel water. Modern synthetic fabrics let perspiration escape so that the skin does not become chilled.

**RAIN SHIELDS**
*Different peoples use a variety of accessories as shields against rain. Umbrellas (right) can be used by more than one person, while hats (below) keep the hands free.*



PRACTICAL PURPOSES

## Jungles

**In brief:** People living in jungles wear few or no clothes because fabric would trap moisture close to their skin

People living in jungles often go naked, except perhaps for a loincloth or other genital covering, because even loose-fitting clothes would trap humidity. In addition, their skin is naturally resistant to insect bites so they do not need clothes for protection. However, they do use body painting, stylized haircuts, and accessories to express their identity.

**SPARSE CLOTHING**
*Men of the Yagua people, in the Amazonian jungle of Peru, traditionally wear little except skirts and wigs made of long grass.*

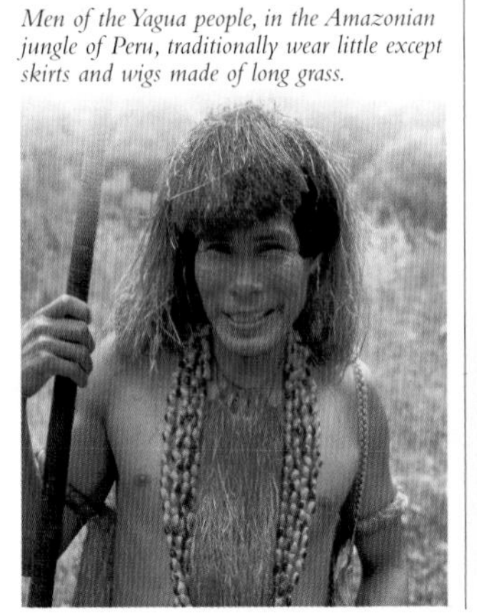

PRACTICAL PURPOSES

## Deserts

**In brief:** In deserts, people wear long, loose garments to shield their skin from the sun and let cool air circulate

Many of the peoples living in desert areas wear clothes made of natural fabrics such as cotton, which draws sweat away from the skin and thus cools the body. They may also wear long, flowing robes that let cool air circulate around the body.

Facial protection is also vital in deserts to keep sand out of the nose and mouth, and because wind-blown sand is a major hazard that can cause temporary blindness or eye injury. Nomads in the Sind region of Pakistan wrap long, embroidered scarves around their faces, while the Arabs of North Africa have a rich variety of traditional headdress styles. Similarly, Berber women from Morocco and Tunisia wear woollen veils with tie-dyed motifs.

The decoration of desert clothing, as with many forms of dress, owes much to the environment. For example, villagers in some parts of northern India, Pakistan, and Afghanistan stitch small mirrors into their clothes to reflect the sky and give the illusion of water. The mirror-work is also believed to ward off evil spirits, which are thought to flee as soon as they see their reflections.

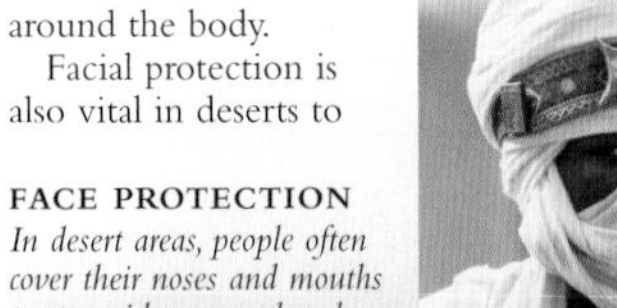

**FACE PROTECTION**
*In desert areas, people often cover their noses and mouths to stop airborne sand and dust from entering.*

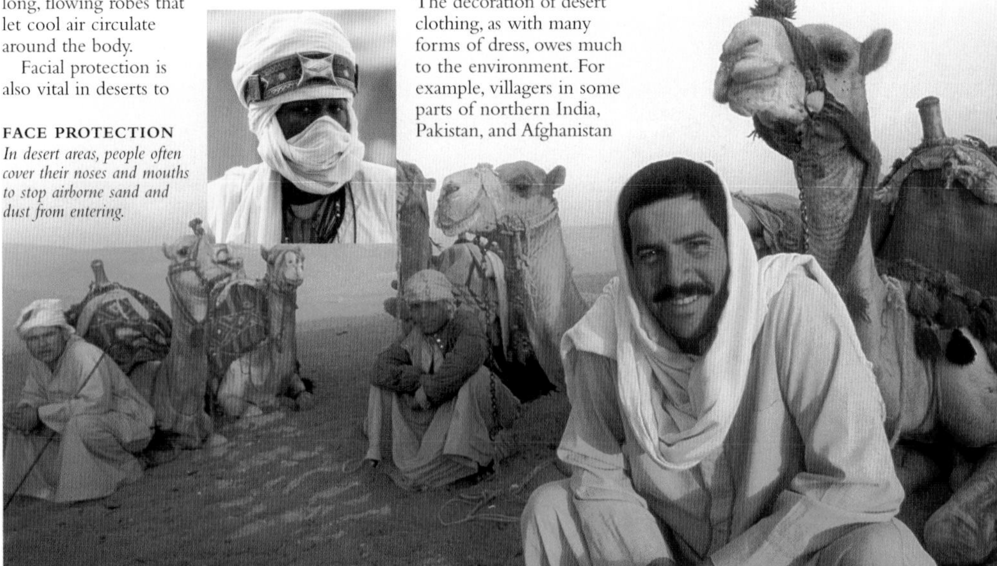

**COMFORT IN THE DESERT**
*These Egyptian men wear long, loose clothes to keep their skin cool. The pale colours reflect the sun, adding to the cooling effect.*

MODESTY

## Social decency

**In brief:** Garments are used to conceal the genital area and any other parts considered to be indecent

Various parts of the body, especially in women, are seen as taboo in different cultures. People in most societies cover their genitals, often with loincloths. In many cultures, it is considered indecent for women to show their breasts; in Muslim cultures, their hair must also be covered; and Roma women wear long skirts because menstruation is thought to "pollute" a woman's lower body.

**BARE MINIMUM**
*At the beach, people may feel freed from usual clothing rules, but most wear some sort of swimsuit, even if it is only minimal.*

SYMBOLISM

## Religious dress

**In brief:** People of many religions wear items that express particular values, such as faith and humility

Dress can be a potent symbol of religious faith. Christian and Buddhist monks prefer simple robes that reflect their focus on spiritual life. Sikh men, who consider hair a gift from God, grow a beard and protect their hair with a cotton or silk turban. The turban also symbolizes commitment to the Sikh values of humanity, trust, faith, and service to the commuity. Orthodox Jewish men wear a skull-cap (the *kippah* or *yarmulka*) as a reminder of God's constant presence. Special dress may also be worn as part of worship; for example, Roman Catholic women sometimes wear veils while they are in church as a sign of submission to holy authority.

Religious clothing rules sometimes cause controversy. In strict Islamic countries, such as Algeria, all Muslim women have to wear headscarves or wear the veil (*hijab*), and they may be punished severely for failing to do so. In contrast, Muslim women in more secular countries, such as Egypt, may choose to wear the veil in order to affirm their identity as Muslims, or to preserve their privacy.

**HIDING THE BODY**
*This Muslim woman's veil and long, loose robe are designed to hide her face and body shape at all times. Islam requires women to wear this sort of clothing for modesty.*

**CHRISTIAN PIETY**
*Nuns cover their hair, and may shave their heads, to show that they focus their attention on spirituality rather than on displaying physical beauty.*

DISPLAY

## Sexual attraction

**In brief:** People who wish to attract partners often wear clothing that shows off their physical assets

Clothing and other body adornments can emphasize sexuality in various ways. Often, particularly in Western societies, it is used to show off parts of the body that are considered to be sexually attractive or beautiful, such as women's faces, cleavage, and legs or men's muscles and body shape. This sort of display may be an overt sign of sexual availability. More often, however, it is simply an expression of confidence; people are most likely to show off their bodies if they feel comfortable with their appearance and powers of attraction. Women are most likely to enhance their sex appeal, often by wearing miniskirts, underwired bras, high-heeled shoes, make-up, and perfume.

**STYLIZED DISPLAY**
*In Western fashion, the models' clothing and even their bodies are used to create a very stylized version of sexual display.*

**SHAPE AND MOVEMENT**
*The fringes on this South African dancer's traditional outfit accentuate the motion of her hips.*

**FLAMBOYANT MALE**
*Some men choose clothing that makes them look elegant, strong, or slightly rebellious, to help them in attracting potential partners.*

Cosmetics, in particular, have been used in many cultures and by both sexes, to give a desirable impression of youth, health, and perfection.

Concepts of beauty vary between cultures, so attire that one group finds alluring may be strange or distasteful to another. For example, some African peoples and Inuit perforate the lower lip and cheek to insert a wide disc, or labret, made from wood, glass, ivory, or bone; the resulting facial distortion may be a sign of beauty among these peoples but is shocking to foreigners. Similarly, revealing female garments that are acceptable in the West may be taboo in other countries where it is still normal for women to cover up their bodies.

In various societies around the world, people may wear extravagant costumes in communal displays of sexual identity, such as beauty pageants and carnivals. One such event is the Yaake charm dance performed by Wodaabe men (*see* p416) to impress female referees, who judge the men on their colourful make-up, wide eyes, and broad smiles.

Some people may flout accepted dress codes for their sex or may even adopt the clothing of the opposite sex. This practice, known as cross-dressing, is sometimes performed by shamans (*see* p297) in traditional cultures, possibly to help them break out of their usual identity and pass more easily into the spirit world. In modern societies, it is popularly assumed that cross-dressers are homosexual. In fact, many simply want to try out what it feels like being someone of the opposite sex. Cross-dressing may also be a form of entertainment, as with the female *Takarazuka* performers of Japan, some of whom dress in stylized Western male clothing.

History

### The face of the geisha

Geisha are female Japanese artists who entertain men with their beauty, charm, and intelligence. Over several centuries, the elements of the geisha's make-up, such as the red lips and pale skin, have become symbols of femininity and erotic appeal.

**SUPREME STATUS**
*The clothing and gold adornments of the Ashanti monarch are more luxurious and elaborate than those of anyone else, thus clearly showing his supremacy.*

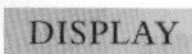

DISPLAY

# Wealth and status

**In brief:** Elaborate, luxurious dress and accessories are used to identify the wearer as a very high-status person

The most impressive outfits of all are often those worn by society's leaders: rulers, nobles, priests, and political officials. Very rich people, as well as those lower down the social scale, may also seek to show off their wealth or enhance their prestige by wearing fine clothes or costly accessories.

Prestige is often associated with rich fabrics such as velvet, silk, ivory, and fur. High-status people may also wear the skins, feathers, teeth, or claws of wild animals such as bears, leopards, or eagles. Regalia made from these parts may enhance status by symbolically transferring the creatures' perceived qualities, such as strength, to the wearer.

Colour may be regarded as a status symbol if the right to wear a certain shade is limited to an elite group or important person. Purple, for example, is widely associated with royalty, originally because of the rarity and value of purple dye. High-ranking Sioux warriors wore blue because the colour was associated with Skan, the supreme sky god.

**SPIRITUAL EMINENCE**
*The pointed yellow hat of this Tibetan lama is a symbol of his high spiritual office.*

In some societies, decoration conveys messages about social importance. In Ethiopia, the width and complexity of the motifs on a person's *shamma* (shawl) indicate his or her rank at a glance. In several Native American tribes, the chief's ceremonial shirt was traditionally painted with stories of his past exploits to communicate his success as a ruler.

Rare or valuable accessories, such as a monarch's crown or a chieftain's ceremonial dagger or shield, are classic markers of prestige. Other status symbols gain their potency from the fact that they are unique; examples include clothes designed for particular famous individuals (such as Marilyn Monroe). Less wealthy people may still be able to afford ostentatious articles such as gold jewellery and expensive watches. Such items are sometimes regarded as a store of value that can be inherited by relatives.

The body itself may be adapted to show a person's social standing. In some traditional societies, high rank is shown by body painting or tattooing. In affluent modern societies, expensive make-up, immaculate styling of the hair and nails, and even cosmetic surgery make people feel special. They also subtly suggest that users have enough time and money to spend on pampering themselves.

Fact

## Luxury within reach

People in many cultures often wear luxurious clothes and accessories to enhance their social status. By doing so, they may attract the admiration, respect, or envy of others. In Western culture, luxury items include haute couture clothing and expensive lingerie. Until recently only the richest individuals could afford these items, but now "designer" clothes are attainable for increasing numbers of people. The wide availability of cheap imitations has blurred the social boundaries even further.

**MARKS OF DISTINCTION**
*In Polynesian and Maori society, tattoos signify that a person has high social status, like this Western Samoan chieftain.*

**FINERY AND FUN**
*In England, horse-racing at Ascot is associated with the elite. It is traditional to wear formal, expensive clothing, but some women express a sense of fun with eye-catching hats.*

DISPLAY

## Festival costumes

**In brief:** At festivals, people wear costumes to express traditional values or enact stories, or just for show

Festivals are major events in the life of a community and call for especially fine or detailed dress. Such outfits are worn for short periods only but may take a huge amount of expertise, time, or money to prepare. Some are treated with reverence or even as sacred items.

Costume-making skills are often passed on through a family or between community members, so the costumes may be powerful symbols of ancestry and group identity. In Indonesia, the complex designs of batik cloth are traditionally produced from memory and may be specific to one village. Different parts of West Africa have jealously guarded techniques for dyeing and weaving ceremonial dress.

Most festival clothing is designed to produce maximum impact during public performances such as theatre, music, or dance displays, religious rituals, and sporting contests. For example, during Carnival, a festival with religious origins held in the Caribbean and South America, people wear eye-catching costumes adorned with feathers, sequins, and glitter.

In festivals around the world, people wear special clothing to act out stories or imitate mythical beings, as with the Biblical processions staged in Spain during Holy Week or the ghost and monster costumes that Americans wear for the pagan festival of Halloween. Masks feature in many occasions, including Bolivia's *La Diablada* (Dance of the Devils), in which dancers re-enact the rescue of trapped miners from an army of demons.

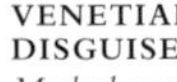

**VENETIAN DISGUISE**
*Masked revellers have always been a feature of the Venice Carnival in Italy. The disguise allows people to step out of normal social roles.*

**TIBETAN TRADITION**
*Monks of the pre-Buddhist Bön religion wear ornate costumes for the Full Moon Festival.*

**CREATIVE SKILL**
*Participants in the UK's Notting Hill Carnival, in London, wear bright, extravagant costumes, many of which are homemade.*

GROUP IDENTITY

## National dress

**In brief:** People may wear a traditional or ceremonial style of dress to show their national or ethnic identity

Most of the world's older nations have a form of traditional national dress. However, in large or more recently founded states such as India, Russia, China, and Nigeria, regional or ethnic styles are usually more important than a standard national costume.

Some people wear their national or regional dress simply because it is their custom. Others, however, choose to wear it to show their pride in their heritage. In either case, this sign of group identity can strengthen bonds between members of a community.

Certain patterns or materials evoke past events, religious beliefs, or legends. For example, the traditional dress worn by the Miao people of Yunnan Province, China (*see* p450) symbolizes their connection with their ancestors. The Miao decorate the fabric with painstaking embroidery that takes 1–2 years to complete, as a way to show respect for their dead relatives.

Sometimes, minorities in modern societies, such as Native Americans, preserve traditional dress to reaffirm an endangered cultural identity. A few "traditional" styles, however, are not as old as they seem. For example, tartan cloth, or plaid, is an ancient form of dress in Scotland, but most plaids were created in the 18th and 19th centuries, when there was a huge revival of interest in Scottish history.

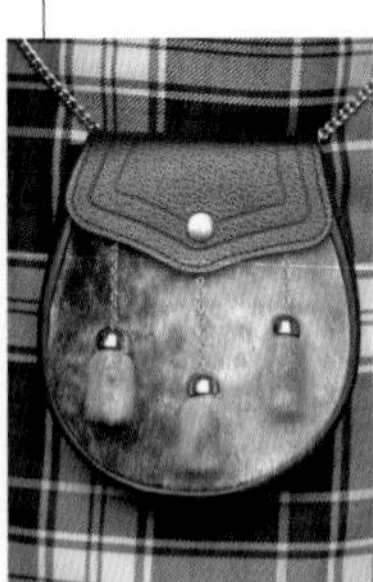

**SCOTTISH DRESS**
*The tartan kilt and the sporran (pouch for carrying small items) are items of traditional Scottish highland clothing for men. Scots still wear them, usually as formal dress.*

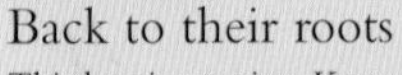

History

### Back to their roots

This boy is wearing *Kente* cloth, the traditional dress of the Ashanti people of West Africa. *Kente* cloth originated in about the 12th century AD. The style is simple but the colours and designs of the fabrics are complex, with each pattern having its own history and symbolic meaning. In recent decades, *Kente* has become popular with African–American people wishing to express pride in their African heritage.

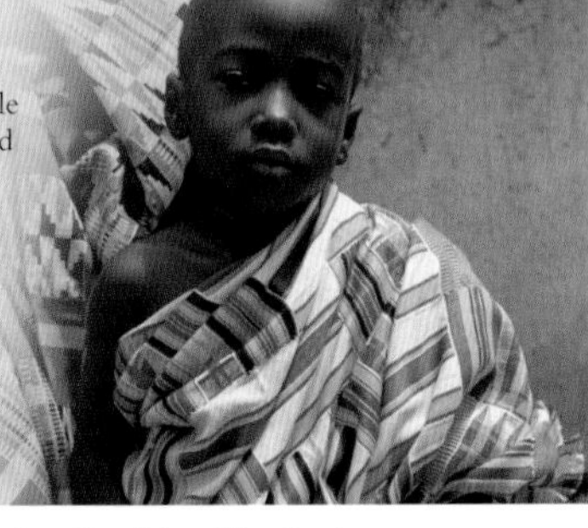

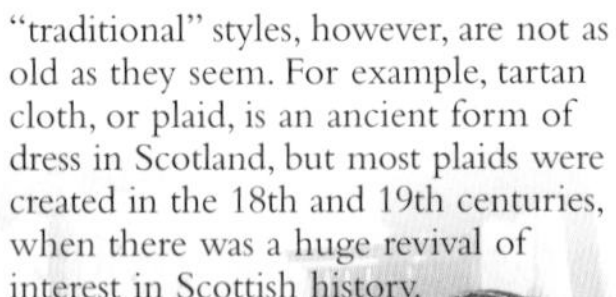

**EVERYDAY CUSTOM**
*These Rajasthani women still wear their traditional dress every day for comfort as well as adornment.*

GROUP IDENTITY

## Subcultures

**In brief:** Groups who rebel against the values of their society often wear outfits that flout normal dress codes

Minority groups often adopt a distinct style of dress as a rebellion or protest against social norms. In the West, such groups include leather-clad bikers, punks with spiky hair, and wearers of the scruffy "grunge" look in the 1990s. Nonconformist clothes may also give out a political message; for example, hippies opposed to the Vietnam War adopted the dress of pacifist traditions such as Buddhism. However, people who flout conventional dress codes risk persecution from mainstream society.

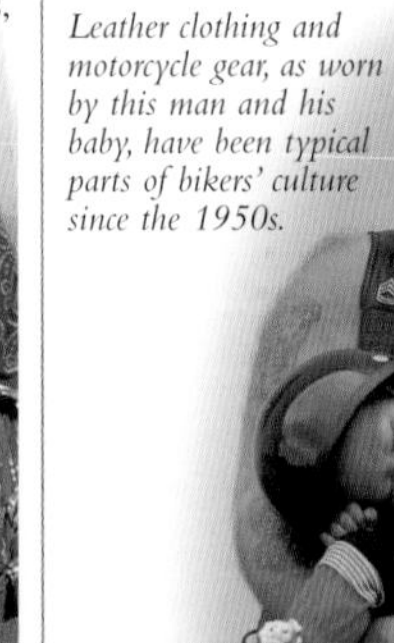

**BIKERS**
*Leather clothing and motorcycle gear, as worn by this man and his baby, have been typical parts of bikers' culture since the 1950s.*

GROUP IDENTITY

## Uniforms

**In brief:** Uniforms are identical outfits that emphasize group bonds and/or mask individuality

A uniform is identical clothing worn by the members of an organization, particularly the military, police, and emergency services. It partially hides the wearer's individual identity and so emphasizes group cohesion; there may be a discreet name badge, but usually a uniformed person is anonymous.

Historically, the main function of military uniform was to identify the soldiers of each side and to intimidate opponents during battles; for this reason, it could be impressive. Today, it serves more to encourage a sense of discipline in wearers. In some cases, uniform has a political purpose: massed ranks of armed forces feature prominently in state occasions around the world, serving to emphasize national pride.

Other people, as well as the armed forces, may wear uniforms for reasons to do with group cohesion or discipline. Schoolchildren often have to do so; the uniform may also help to prevent bullying of children who are too poor to afford the latest fashions.

After the Communist revolution in China, the leader, Mao Zedong, ordered all men and women to wear a simple, mass-produced suit as a symbol of social unity. The "Mao suit" fell from favour in the 1980s, although some people still wear it.

Sometimes, people force others to wear uniforms as a way to suppress their identity. The most extreme examples are the uniforms worn by inmates of prisons or, in the past, people held in concentration camps.

**CEREMONIAL GUARD**
*The Swiss Guards, who form the Pope's bodyguard, wear this distinctive traditional uniform for ceremonies.*

**SCHOOL WEAR**
*Many schoolchildren, like these girls in Nepal, have to wear uniforms. The clothing is designed to promote "school spirit". It may also remind children that they are subject to school rules.*

GROUP IDENTITY

## Professional dress

**In brief:** People in various professions wear smart or ceremonial dress to show their occupation and status

A smart suit and (for men) a tie has long been the internationally accepted dress for business. However, some people, including lawyers, medical workers, academics, and airline crews, have a special uniform that shows the wearer's office and qualifications. It may incorporate a standard system of symbols, such as the braid on a pilot's jacket or the fur-trimmed hoods of academic gowns. Some objects used in professional workplaces can also be associated with those professions, such as a doctor's stethoscope.

**LEGAL TRADITION**
*In the UK, legal figures such as the Lord Chancellor (below), judges, and barristers have traditionally worn horsehair wigs as a symbol of their office and authority.*

LIFE EVENTS

## Coming of age

**In brief:** Some societies have distinctive dress to show when people have left childhood or reached adulthood

Many societies have special forms of dress for young people on the brink of adulthood. Some forms represent a break with childhood. Others signify maturity, introduction to adult society, or readiness for sex or marriage. In some cases, young people undergo coming-of-age rituals together. They wear special, identical dress to show that they are a distinct peer group.

**MODERN DEBUTANTES**
*These young American women wear white gowns to mark their introduction into smart society as adults.*

LIFE EVENTS

## Weddings

**In brief:** Outfits for brides and grooms symbolize their commitment and express qualities such as wealth

Among the most elaborate forms of dress, wedding costumes highlight the roles of all participants, particularly the bride and groom.

The couple's clothes often express their suitability in terms of beauty, purity, or social status. Jewellery and fine cloth advertise wealth. The bride's virginity may be shown by a colour, such as white, or a symbol, such as garlands of flowers, a nose ring, or a veil. Various items are used to seal and bless the union. In Korea, for example, the groom offers the bride's family a box containing red and blue silk for her dress. Married people are usually identified by a particular accessory, such as a wedding ring in Western societies or the gold *mangalasutra* necklace worn by Indian women.

**FULANI WEALTH**
*This Fulani woman wears sumptuous gold earrings to show the wealth that she brings to the marriage.*

**HINDU WEDDING RING**
*Wedding ceremonies in various societies, including Hindu and Christian communities, focus on the exchange of rings by the bride and groom as a symbol of the couple's union.*

LIFE EVENTS

## Bereavement

**In brief:** Many bereaved people wear sombre clothing; others, however, wear outfits that symbolize life

When someone has died, mourners may wear special dress at the funeral and for some time afterwards to show respect for the person. In Western society, mourners usually wear black clothes in a sombre, formal style. Taoist Chinese people cover their faces with rough hessian veils. Jewish people make a tear called the *keriah* in their clothes to signify a broken heart. Some forms of dress, in contrast, assert vitality in the face of death. For example, Dogon men wear spectacular masks at funeral dances and some Christians wear bright clothes to signify resurrection.

**MOURNING WHITE**
*In devout Muslim society, the female relatives of a dead person will wear white clothes and veils for a year after the death.*

**JOINING ART AND SCIENCE**
*The City of Arts and Sciences in Valencia, Spain, designed by Santiago Calatrava, combines aesthetic beauty with technological innovation.*

# ART AND SCIENCE

The arts and sciences are the highest expressions of human creativity. Art is the use of specific media, such as paint, language, and music, to beautify the self or environment, to explore ideas, or to create something new for its own sake. Science is a systematic method of establishing facts, used to investigate aspects of the world around us. Both art and science can be used for practical purposes or involve more intellectual, abstract enquiry. Achievements in these fields are often a source of collective pride and can reaffirm traditional values. In certain cases, however, innovations may challenge prevalent views in society or may even result in punishment from authority figures.

**PERFECT PROPORTION**
*The shape of a Nautilus shell conforms roughly to the Golden Section. The shell, and each turn of its spiral, fit into rectangles whose sides are in the ratio 1.618:1.*

Both art and science may demand close observation of the world, problem-solving skills, schooling in an existing body of knowledge, and innovation. Both are used to shape our environment and thinking, and in turn are given form by the society in which they are produced.

## EARLY KNOWLEDGE

In prehistoric societies, art and science were rooted in everyday life. People made paintings on rocks, possibly for rituals such as contacting spirits, and made sculptures such as the Willendorf Venus, a small, stone female figure dating from about 24,000BC, which may have been a fertility symbol. The people would also have needed to gather practical knowledge about the world, for example by discovering good plants for food or medicine and learning the habits of animals, in order to eat well and stay healthy.

## EVOLVING THOUGHT

From 4000BC onwards, the first civilizations appeared, such as the Indus Valley people of the Indian subcontinent, the ancient Egyptians, the Mesopotamians, the Olmec of Central America, and the culture of predynastic China. As various societies developed, more efficient food production freed some people to focus on specific skills, such as art, astronomy, and mathematics. Information was collected and recorded in a systematic way. Units of measurement were devised to define space and time; for example, the Babylonians (c1800–1590BC) were the first to use a numeral system based on groups of 60 to measure time and count degrees in a circle. Various societies produced advances such as irrigation systems.

The greatest range of cultural achievement occurred in ancient Greece (8th–4th centuries BC). The Greeks raised observation to a high level in anatomy, the study of nature, and sculpture. They developed the skill of establishing facts by logic. They also discerned mathematical patterns such as the ratio called the Golden Section. Rectangles and spirals that conform to this ratio (roughly 1.618:1) are still regarded as aesthetically pleasing. The Golden Section was used in buildings such as the Parthenon in Athens. It also occurs in nature, as in the spirals of a Nautilus shell. The Romans carried on the Greek tradition in art and made innovations in literature, architecture, and engineering. Greek and Roman achievements would be unsurpassed for a thousand years.

**FROZEN IN MOTION**
The Discus Thrower *shows the Greeks' skill in blending accuracy with visual rhythm.*

## CHALLENGING IDEAS

In many societies, art and science have been practised in the service of religion or produced for an elite. For example, in medieval Christian Europe, the early Islamic empire (8th–13th centuries AD), and Tang-dynasty China (AD618–907), advances in knowledge reinforced and enlarged on accepted values. However, during the Renaissance, people began to challenge general understanding. The Copernican theory that the Earth orbited the sun, published in 1543, dislodged humanity from its assumed place at the centre of the universe. Andreas Vesalius (1514–1564) questioned ancient Greek learning and fell foul of Christian moral strictures in his attempts to provide insights into the anatomy and functions of the human body.

In the UK during the 19th century, Charles Darwin's theory on the origin of species and Sir Charles Lyell's evidence for the ages-long processes of geology challenged religious teachings on the creation of the world. These discoveries were fiercely contested by Church authorities and still provoke controversy today. In the art of this time, European painters such as Gustave Courbet had some of their work rejected by art experts because it portrayed plain, stark reality.

SPIRAL GALAXY

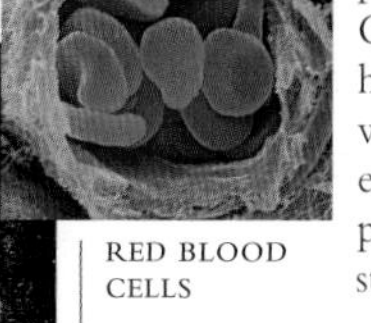

RED BLOOD CELLS

**POWERFUL VISION**
*Modern scientific instruments enable us to see objects invisible to the naked eye, from tiny blood cells to distant galaxies.*

## THE MODERN WORLD

Today, the range of achievement is huge. Some people have created innovative art such as the music installation *Longplayer,* based at sites in the UK, Egypt, and Australia, which is set to play for 1,000 years without repeating itself. We can see tiny structures inside cells and view the farthest parts of the universe. Scientific advances have produced improvements in healthcare, such as mass vaccination, that have doubled the lifespan in many countries.

Arts and sciences are still closely associated with elites. Some forms, such as film and medicine, are part of huge industries. However, many ordinary people also have the time, money, and freedom of expression to enjoy the arts, and have access to scientific information. In addition, many can share the achievements of other cultures – both the technical skills of the West and the knowledge and arts of indigenous societies.

Profile

### Leonardo da Vinci

A genius in art and science, Leonardo da Vinci (1452–1519) was one of the foremost figures of the Renaissance. His dazzling paintings and drawings include the most famous works of Renaissance art. He made extensive studies of human anatomy and was one of the first geologists. He also invented a variety of machines, including a helicopter, a hang-glider, a robot, and a submarine – concepts that were centuries ahead of their time.

# ART

Humans practise an enormous range of artistic activities involving the expression of ideas through a variety of media, from painting and sculpture to literature and music. Every culture produces some form of art, and in most societies art is held in high regard. The reasons for this esteem vary according to the type of community. In traditional societies, art is often closely integrated into other activities; for example, a dance may be part of an initiation ceremony. In Western societies, art is usually much more clearly separated from other aspects of life, and it is more often made and enjoyed for its own sake. Throughout the world, people can experience art on many levels and appreciate it for the comments it makes on important issues, for the intellectual challenge it poses, or for pleasure.

Fact

### An unbroken tradition

Rock painting is the oldest surviving form of art. In certain parts of the world there are still artists who work in this way. Using pigments made from local earth, rocks, and minerals, Aboriginal Australians, and the San of the Kalahari in southern Africa, make images of people and animals. The paintings often have powerful religious meanings for their makers. Many are also beautiful and are crafted with a high level of skill.

## EARLY ACHIEVEMENTS

Some of the earliest human artefacts are works of art, such as prehistoric rock paintings, possibly created during religious rituals. Later peoples, such as the Egyptians 5,000 years ago, early Chinese dynasties (from 3500BC), and Mayans (AD300–900), produced lasting monuments to glorify gods or rulers; they also created everyday items, such as pots and wall paintings, with beautiful images of the world as they saw it. Some civilizations developed styles that influenced artists for centuries afterwards. For example, the aesthetic principles of the ancient Greeks and Romans influenced the artists of the Renaissance, and many Europeans were familiar with ancient Greek and Latin literature.

**SPIRITUAL PROTECTION**
*The "terracotta army" found in Xi'an, China was made to protect the first Emperor, Qin Shihuang Ti (259–210BC), after his death. The figures have lifelike faces and detailed uniforms, and may have been based on real people.*

**INDIVIDUAL PORTRAIT**
*An ancient Egyptian wall painting depicts a foreman and his son. As well as painting sacred subjects, the Egyptians created pictures showing everyday life and individual people.*



## ART AND CULTURE

The relationship between art and its audience varies depending on the art form. Some forms, known as indigenous art, are rooted in the beliefs of both creator and community. This type of art includes works such as Navajo sand paintings or the masks carved by many African tribes. Outsiders can appreciate their beauty, but their true audience is the local group of believers. High art, by contrast, may be challenging for the audience. People spend years studying the subtleties in the paintings of Rembrandt, from the 17th century, or the poetry of pre-Islamic Arabia. However, this type of art deals with universal ideas and emotions, may hold its value over time, and may be appreciated by people from different cultures and eras. Popular or mass art is designed to have instant, if fleeting, appeal. Genres such as pop music or the Hollywood movie, produced with the backing of entire industries, are often dismissed by educated people as "mere entertainment" – but at their best they too can incorporate subtleties and complexities of word and image.

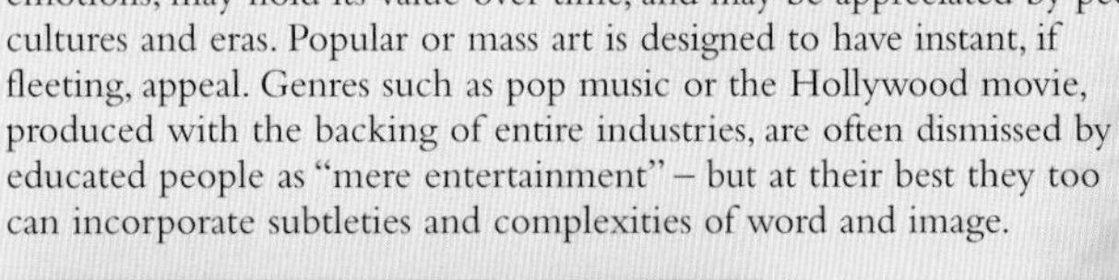

**INDIGENOUS ART**
*Dot painting is a traditional art practised by Aboriginal people in Australia. The painters use motifs that have been developed over many centuries.*

**HIGH ART**
*Work classified as high art represents the greatest aesthetic achievement of a whole society. For example, Shakespeare's plays are considered to be one of the highest examples of European theatre.*

**POPULAR ART**
*There is a huge variety of popular music. Some forms, such as pop, are designed for mass appeal. Other forms are more sophisticated; for example, some DJs in clubs mix and sample existing pieces of music to create complex new compositions.*

KEEPING TRADITION ALIVE
*Many people today, like this company of Mexican folk dancers, take care to preserve traditional arts so that their culture is not submerged by modern global civilization.*



# TYPES AND USES OF ART

The main artistic disciplines include visual arts, such as painting and sculpture; verbal arts, such as literature; and performance arts, such as opera and dance. Even cooking and perfume-making may be regarded as art if a high level of skill is involved. All of these forms may be put to various uses. Perhaps most simply, art may be a way to examine and interpret aspects of the world. For example, novels, photographs, and portraits may be valued for their fidelity to life and for the insights that the artist gives into his or her subjects. Some art has a deep spiritual significance. Sacred art has existed from the beginning of human history and in all societies. Examples include Christian church music and Buddhist *thangka* art (paintings of Buddha or deities), which are used to glorify gods or to provide a focus for worship or contemplation. Art may also have important social roles. Grand buildings, or public sculptures or murals, may commemorate notable events in history or symbolize civic pride. Some works are luxury items for important or rich people, like the Fabergé eggs of Imperial Russia or the sculptures for the medieval court in Benin, Africa. Sometimes art carries a political message. It may be produced to support a regime, like the "social realist" art of former communist states. It may also be used to attack oppression and injustice, like the plays of Chilean writer Ariel Dorfman (1942–). Finally, art may be produced for its own sake, to explore media and techniques.

LUXURY ITEM
*The Fabergé eggs, made with precious metals and gems, were created in the 19th century to amuse the Russian Imperial Family.*

SACRED MEANING
*The curving lines of Le Corbusier's church at Ronchamp, France, built in 1955, are designed to give the idea that the Roman Catholic Church is a mother "embracing" the worshippers.*

RADICAL ARTIST
*The American artist Andy Warhol (1928–1987) made paintings, prints, and films that explored the nature of modern society.*

POLITICAL MESSAGE
*In the former Soviet Union, the "social realist" style of art was used to convey political messages. A famous example is the sculpture* Worker and Collective-Farm Girl *(right), emphasizing the workers' nobility.*

VISUAL ARTS

## Painting and drawing

**In brief:** The disciplines of painting and drawing are among the oldest and most diverse arts

Our earliest ancestors started to draw on cave walls many thousands of years ago. At first, people made pictures using materials from their surroundings, such as earth pigments and charcoal. Nobody knows why the paintings were made; they may have had spiritual significance, like the ancestral art of the Aboriginal Australians. The earliest "high" art showed deities and mythical figures or reflected the social order. The ancient Egyptians, for instance, would depict the Pharaoh as the largest figure in a painting. For centuries, the most enduring and highly prized art was created for religious purposes; for example, Orthodox Christian icons are used to remind worshippers of Jesus' presence. Art was also made for aristocrats, such as medieval European illuminated manuscripts, or as a leisure pursuit for the upper classes, such as Chinese calligraphy and silk painting. During the Renaissance, Italian masters such as Giotto, Leonardo da Vinci, and Titian produced works that broke new ground in realism, developing the art of perspective and exploring the potential of oil paint. A major break from realistic tradition came with impressionism in the 19th century, in which paint was handled freely to create the impression of natural light and open-air views. In the 20th century, artists pushed the boundaries further, into cubism, expressionism, and abstraction. Today, painting and drawing can be a hobby or a life's work and can embrace any style.

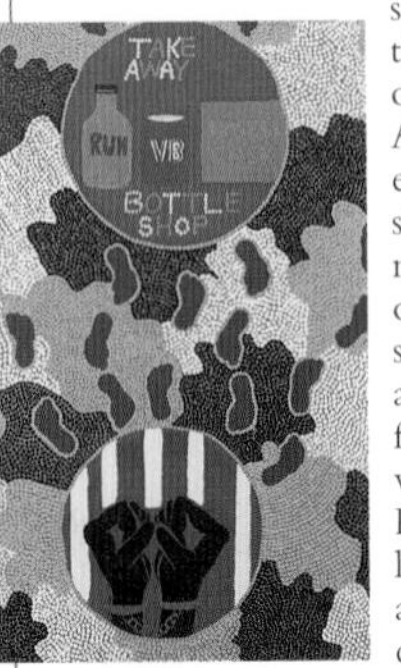


**SOCIAL MESSAGE**
*This modern example of Aboriginal dot painting illustrates the risks associated with alcoholism, such as imprisonment.*

**SACRED IMAGES**
*Orthodox Christian icons are images of Jesus or the saints. The stylized forms represent a spiritual rather than a visual reality.*

**A PAST MASTER**
*The great French painter Henri Matisse (1869–1954) used intense colours and sinuous lines to create expressive and visually harmonious forms. A dramatic example of his style is* Dance*, painted in 1909.*

VISUAL ARTS

## Sculpture

**In brief:** Sculpture is used to create lasting, powerful works, sometimes on a monumental scale

Sculptures can be made using many different methods – by carving hard materials such as wood or stone, by casting metals, or by modelling soft materials such as clay. Because they are three-dimensional, such works often have a lifelike quality and a powerful impact, and in many cultures they were used to depict gods or ancestors. African metalworkers developed bronze casting thousands of years ago, using it to make stunning religious sculptures, while the Native American people of the northwest US carved totem poles to represent their spiritual and human ancestors. Sculptures often represent the most perfect art of their time. Figures such as the Charioteer of Delphi are among the finest works produced by the ancient Greeks. In later times, people from many countries have admired Michelangelo's *David*, with its superbly realized face and body. From the 19th century, however, sculptors broke away from the ideal of naturalism, either to revisit ancient styles such as African art or to experiment with pure, abstract form. Today, sculpture is popular as public art because statues and other sculptures function as focal points in streets and parks. They may be created for every purpose, from commemorating a statesman or hero to simple decoration.

**SET IN STONE**
*The faces of four former American Presidents are carved into the granite of Mount Rushmore, South Dakota, in the US, to commemorate the first 150 years of US history. Each face is 18m (60ft) high.*

**MODEL OF PERFECTION**
*The greatest examples of Renaissance art, such as Michelangelo's* David*, blended classical Greek and Roman technique with contemporary expressiveness.*

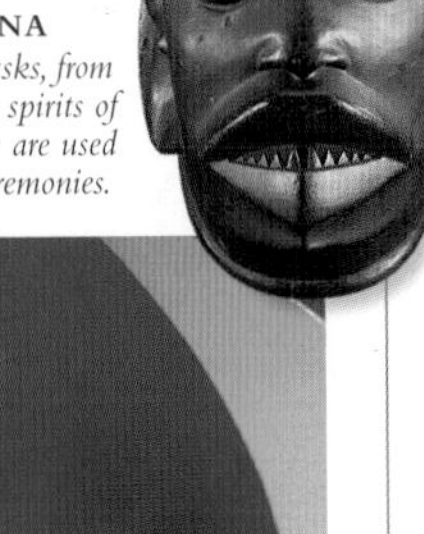

**RITUAL PERSONA**
*Makonde helmet masks, from Kenya, represent the spirits of tribal ancestors. They are used in male initiation ceremonies.*

**COMMUNITY ART**
*Abstract sculpture is a highly rarefied form of art; however, many people enjoy it and abstract works often adorn public spaces such as parks.*

VISUAL ARTS

## Photography

**In brief:** Available to professionals and amateurs, photography is used for reporting, advertising, and art

Since photography was invented in the 19th century, it has been used for both practical and aesthetic purposes. Early photographers, such as the Frenchman Louis Daguerre (1787–1851), found that variations in light and viewpoint could alter the impact of an image. In the 20th century, as the technology improved, photography became widely accessible. It is essential in journalism; some images of news events have attained iconic status. Its power is also important in advertising. As art, photography is most commonly used to portray the world, as in the striking images of André Kertész and Henri Cartier-Bresson. Other artists, such as Man Ray and Alfred Stieglitz, have employed experimental techniques and manipulated images to create abstract work.

**CAPTURING HISTORY**
*This photograph shows Jack Ruby killing Lee Harvey Oswald, the man who assassinated US President John F. Kennedy.*

VISUAL ARTS

## Architecture

**In brief:** Architects use both aesthetic and technical knowledge to produce a wide range of building forms

The design of buildings is one of the oldest arts, with a wide range of styles across the world. Architects have to combine aesthetic judgment with hard, practical knowledge about the capabilities of their materials and the nature of the sites. Such skills have developed over thousands of years. The Greeks and the Egyptians formulated the principle of the Golden Section, a mathematical ratio that was considered to produce pleasing proportions. The Romans developed the arch and pioneered the use of concrete, making possible huge structures such as the Colosseum in Rome. In the Middle Ages, master masons in Europe created the Gothic style, as seen in Chartres Cathedral, France, in which pointed arches and flying buttresses supported tall structures with vaulted ceilings and huge windows. Renaissance architecture was marked by a combination of Greek and Roman knowledge with innovation, as in the work of Italians Palladio (1508–1580) and Filippo Brunelleschi (1377–1446). During the 18th century, architects in the UK and the US again returned to classical principles in the Georgian style. By the end of the 19th century, more effective methods for producing iron and steel, and new ways to build metal frames for tall buildings, led to the first skyscrapers. Today, computers and materials such as titanium allow the design of uniquely shaped buildings, such as Frank O. Gehry's Guggenheim Museum in Bilbao, Spain. In contrast, there is also a continuing tradition of vernacular architecture (the use of local styles and materials), especially in rural areas, to create buildings such as houses and barns. Log cabins in the European Alps, reed houses in southern Iraq, timber-framed houses in Germany, and turf-roofed buildings in Scotland all belong to this tradition.

**REED HOUSE**
*The Marsh Arabs of southern Iraq traditionally use reeds to make buildings. Some of these structures are large and beautifully crafted.*

**CLASSICAL INSPIRATION**
*Filippo Brunelleschi's huge dome for the Cathedral of Santa Maria del Flore, in Florence, was inspired by the structure of the 2nd-century Pantheon in Rome.*

VISUAL ARTS

## Film-based arts

**In brief:** Film, and related arts such as animation, may take many forms, from mass entertainment to high art

In just over a century, film has evolved from a primitive technology that could only produce short, silent, black and white movies to today's wide-ranging art form and global business. There are many film studios around the world. The largest are based in Mumbai, India, and Hollywood; their products include epics such as *The Lord of the Rings*, and often involve sophisticated camera work and complex computer-generated special effects. On a smaller scale are "art-house" films, such as those of Ingmar Bergman, Krzysztof Kieslowski, and Ozu Yasujiro, which feature beautiful cinematography and intellectually absorbing storylines. Some films involve animation, in which characters are created from drawings or models, or by computer. Film is one of the most accessible art forms, available on TV, video, DVD, and the internet.

**TECHNICAL MAGIC**
*The 1999 film* Stuart Little *used state-of-the-art animation techniques to create lifelike, talking animal characters.*

**INDIAN EXTRAVAGANZA**
*The Indian film industry, based in Mumbai, is thought to be the largest in the world. Many of the films are musicals with lavish costumes and dancing displays.*

VISUAL ARTS

## Decorative arts

**In brief:** Practical forms of art are used in all societies to enhance the appearance of everyday objects

All societies find ways to beautify the objects in their living environment. Pottery, textiles, and wooden utensils were among the first decorated items. Different cultures around the world have developed their skills in different decorative arts. Turkey, Iran, and Afghanistan are well known for their carpets; China, for its ceramics; and many Native American peoples, for their textiles. There have also been international styles and fashions, often beginning in Europe and spreading around the globe. For example, the Europeans spread the ornate Baroque style of the 18th century to South America. In modern societies, design is an important element in almost every product, from furniture and utensils to computers, kitchen appliances, and vehicles. Today, mass travel and communication have allowed styles to spread rapidly between cultures. Now it is easy for Western tourists to buy indigenous people's crafts from other areas and, conversely, for Western ideas to influence other artists across the world.

BIRD VASE BY PICASSO

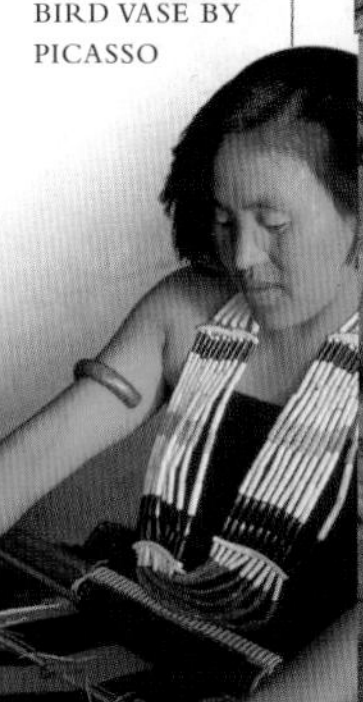

**SKILFUL WEAVING**
*In many ancient societies, such as the Naga people of northeast India, weaving is done by women. The work is often highly patterned and complex.*

Issue

### Afghan war rugs

Since the Soviet occupation of Afghanistan in the late 1970s, carpet weavers have been supplementing traditional motifs, such as flowers and animals, with new images of war, including helicopters, tanks, guns, and armoured cars. Similar motifs have appeared on carpets while the Americans have been occupying the country.

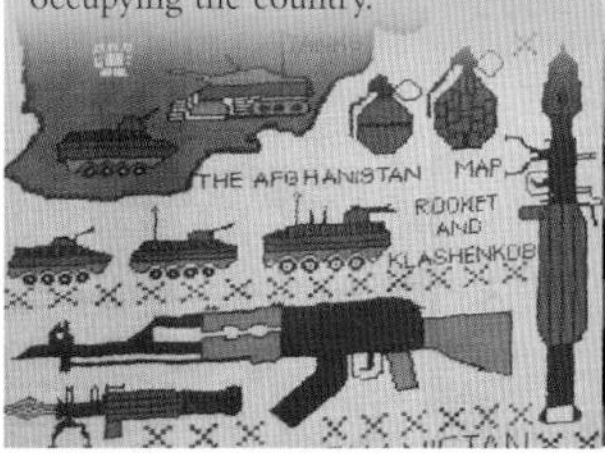

LITERATURE

## Poetry

**In brief:** The ancient art of poetry is used for many forms of expression, from personal feeling to epic tales

In poetry, verbal patterns and imagery are used to give aesthetic pleasure as well as to convey ideas. One of the most diverse arts, poetry ranges from simple lyrics and nursery rhymes to epics. It is also an ancient art; rhymes and rhythms may have evolved to help people remember ideas and stories long before writing developed. Poetry that includes intense imagery is perfect for expressing emotions, as in the Bible's "Song of Songs" , the love poems of William Shakespeare (1564–1616), or the nature poetry of the 8th-century Chinese masters Du Fu and Li Bo. On a far larger scale, epic poetry is used to tell great historical or mythical stories. Some of the world's foremost examples are the historical epics of the ancient Greek poet Homer, composed in the 8th century BC, and the *Divine Comedy* of the Italian poet Dante (1265–1321), a depiction of the medieval cosmos. Modern poetry is usually small-scale; two notable exceptions are the verse novel *The Golden Gate*, by the Indian poet Vikram Seth, and *Omeros*, by the Nobel laureate Derek Walcott.

Profile

### Li Bo

One of China's most famous poets, Li Bo (AD701–762) lived during the Tang dynasty. He had a highly romantic style, rich in natural and Taoist imagery. More than 1,000 of his poems survive today, and some of them are world famous. According to legend, Li Bo died by falling into a river while drunkenly trying to embrace the moon's reflection. Taoist believers count him among the immortals, as the god of poetry.

LITERATURE

## Prose fiction

**In brief:** Fiction includes novels and short stories; it is the world's most popular form of literary art

Two major forms of fiction are novels and short stories. The first novels were written in China and Japan in the 10th and 11th centuries; the first European novel was *Don Quixote*, by the Spaniard Miguel de Cervantes (1547–1616). By the 19th century, Russian writers were creating portraits of entire societies, such as Leo Tolstoy's *War and Peace*. Writers such as France's Emile Zola (1840–1902) used novels and shorter stories as a tool of social criticism, while authors such as Charlotte Brontë (1816–1857) explored psychological matters. During the 20th century, some great writers developed new techniques; Ireland's James Joyce (1882–1941) employed the "stream of consciousness" to suggest characters' thought processes, and the Argentinian short-story writer Jorge Luis Borges (1899–1986) played with ideas of reality and identity. More conventional short stories are still widely read, such as the fairy tales by the Danish writer Hans Christian Andersen (1805–1875). Today, prose fiction is still the most popular literary form, with millions of books sold every year.

**MERMAID'S TALE**
The Little Mermaid *by Hans Christian Andersen is a popular traditional tale and is now world famous as an animated film.*

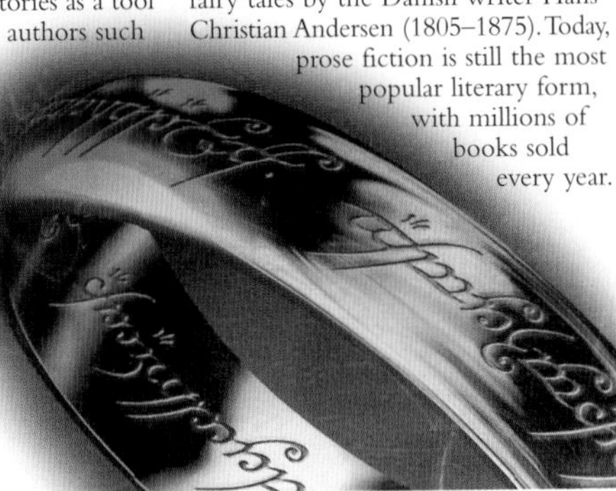

**RULING RING**
The Lord of the Rings, *written by J.R.R. Tolkien, is one of the bestselling and most widely read books of all time.*

PERFORMING ARTS

## Theatrical arts

**In brief:** Theatre is an enduring, diverse art form in which even ancient styles are still performed today

Drama, in which stories are told with words and gestures, has existed for thousands of years. It may have grown out of storytelling and dance, practised as part of religious rituals or to pass on tribal history. By the time of the ancient Greeks, plays were being staged for aesthetic enjoyment. The Greeks created the genre of tragedy, in which serious themes were explored through emotional crises in high-ranking characters, and comedy, featuring humour and low life. In other cultures, drama was used to tell myths and religious stories, as in Indonesian *wayang kulit* puppetry, the medieval Christian morality plays of Europe, and the Noh drama of Japan. These forms of theatre are often highly stylized, with stock characters and gestures. Later playwrights, such as the 17th-century French dramatists Pierre Corneille and Jean Racine, wrote tragedies that closely followed classical Greek models. In Spain, both comedy and tragedy were developed by writers such as Pedro Calderón (1600–1681) and Lope de Vega (1562–1635). The greatest of all dramatists is said to be William Shakespeare (1564–1616). His plays are typified by exquisite language and deep insights into human nature. The first modern dramatists, such as the Norwegian Henrik Ibsen (1828–1906), began to reflect more realistic views of life and relationships. Some, such as the Irish/French Samuel Beckett (1906–1989), broke the bounds of character and plot to examine the essence of human existence. Dramas from all periods and cultures are performed today, in a variety of styles. Other forms of theatre include musicals (some of which are hugely popular in many countries), mime, and puppetry.

**MASTER OF SILENCE**
*The silent drama of mime has been made popular by its most famous performer, the French artist Marcel Marceau (1923–).*

**DRAMATIC LOCATION**
*This amphitheatre in Syracuse, Sicily was built by the ancient Greeks in the 5th century BC. It is still used for festivals of classical Greek drama.*

**SHADOW PLAY**
*The* wayang kulit *(shadow puppets) of Indonesia have been employed for centuries to enact traditional stories. They are principally used in tales from the Hindu epics the* Ramayana *and the* Mahabharata.

**CLASSICAL INDIAN DRAMA**
*Kutiyattam is a 1,000-year-old art form from Kerala, India. It is based on Sanskrit epics, which are enacted using stylized gestures, music, and singing.*

PERFORMING ARTS

## Dance

**In brief:** Dances are beautiful, expressive sequences of movement used in rituals, as high art, or for pleasure

Every society has its traditional dance forms. Folk dances may be performed by everyone in the community, often at festivals or at celebrations such as weddings. Other forms are confined to specialists, such as the ritual dances performed by shamans. Such dances may be performed to tell stories or to communicate with spirits. Several cultures have complex classical traditions of theatrical dance. Ballet dates from the 18th century, and developed primarily in France and Russia; styles include the traditional romantic dances and the experimental forms developed in the US during the last century. *Bharta Natyam* is a classical Indian system with stylized movements of the eyes, hands, limbs, and body. Dance may also be a form of personal or erotic expression, as in flamenco and belly dancing. Certain dances have attracted disapproval for being "indecent", including the waltz in 19th-century Europe (because men and women danced so close). Other forms are associated with subcultures, such as break dancing, which began on New York streets in the 1970s. Today, people still watch and perform traditional dance or popular styles such as tango, salsa, and jive. Many also enjoy dancing as informal self-expression.

**BALLET**
*Traditional Western ballet combines graceful movement with a range of stylized gestures to convey stories.*

**FEMININE FORM**
Raqs sharqi *("belly dancing") developed in the Middle East. It was originally danced for women only, but now dancers perform for mixed audiences.*

**EXPRESSION OF THANKS**
*A Native American man from the Shoshone people performs a traditional dance at a festival. The Plains peoples perform such dances to give thanks to the sun, animals, and plants.*

PERFORMING ARTS

## Music

**In brief:** The emotional and aesthetic appeal of music is used for ritual, social, and even commercial purposes

Music has been a part of human life for thousands of years. Drumming is perhaps the most ancient form, and is often used in rituals; the rhythm and vibration are thought to induce an altered state of consciousness in both performers and listeners. Indigenous peoples around the world have their own musical traditions, often centuries old. Some use unique forms of singing, such as the throat-singing of Central Asia. Many use particular instruments, such as the *kora* (African harp) of Mali, West Africa; the Australian didgeridoo; or the *kheng* pipes of the Vietnamese Miao people, which are thought to "speak" to listeners. The most complex musical forms have particular scales and harmonic rules, which differ from one culture to another; for example, the traditional Western scale comprises seven basic notes, plus sharp and flat tones, whereas the scales used in Indian classical music have many more. Work by Western classical masters such as Mozart and Beethoven, or Indian classical music (*marga*), may inspire deep emotional responses; it may also provide intellectual pleasure as listeners follow the intricate tonal patterns. Another classical art is opera, which combines music and drama. Western opera involves the use of a highly demanding vocal technique (*bel canto*), while Chinese opera is based on complex sets of movements. Jazz also requires a high level of artistry. Having developed as the music of African-American people, it has now become a classical art. In contrast, popular music is usually produced for instant mass appeal, although some styles, such as electronic music, are highly innovative. Today, music of all kinds is often performed live but more commonly reaches listeners through radio, television, and recordings. Music production has grown into an enormous global industry; work by the biggest stars can be heard or bought anywhere in the world. People can also record songs from the radio or download them from the internet.

**CHINESE OPERA**
*The Chinese style of opera has been popular for centuries. Features include a falsetto singing style, a range of special gestures, and the use of make-up to define characters.*

**CLASSICAL INSTRUMENT**
*In classical Indian music (marga), musicians use a tambura to produce the characteristic drone sound underlying the melody.*

History

### The development of jazz

Jazz began in the 19th century when African-American musicians mixed the harmonies of Western folk music with African rhythms. During the 20th century, different strains evolved, both popular and avant-garde; some musicians, such as Louis Armstrong (1901–1971), combined mass appeal with technical virtuosity. Since then, jazz has influenced rock, classical, and other forms of music.

LOUIS ARMSTRONG

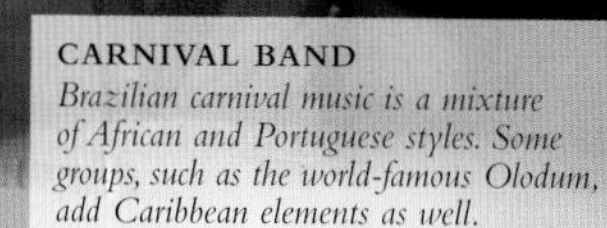

**CARNIVAL BAND**
*Brazilian carnival music is a mixture of African and Portuguese styles. Some groups, such as the world-famous Olodum, add Caribbean elements as well.*

# SCIENCE

Scientific endeavour is best understood as a specialized way of acquiring knowledge of certain kinds. It is based on direct observation of the universe, using measurement and classification, collecting data, and testing ideas by experiment. Data is analysed using the universal language of mathematics. The physical and life sciences (sometimes called natural sciences) follow this method fairly closely. The social sciences also use techniques such as analysis and experimentation, although in a looser way. Interdisciplinary studies, such as geography, draw on both natural and social sciences. Scientific knowledge may not provide the answer to every human problem, but it is a powerful tool for explaining and predicting elements of the world around us.

## THE FIRST SCIENTISTS

Various ancient civilizations, such as the Chinese, had a systematized understanding of nature, and scholars in India had advanced knowledge of mathematics. In Greece, during the 4th century BC, Aristotle first developed the basic tools of science: empiricism (finding out facts by observation and experiment) and induction (inferring a general principle from specific observations). For centuries, Greek learning was lost to Europe; however, it was preserved by Arab scholars such as Ibn Rushd, also known as Averroes (1126–1198), who combined it with their own studies. The first statement of the scientific method (*see opposite page*) is attributed to Roger Bacon (1214–1294), who drew on Arab translations of Aristotle. During the Renaissance, Galileo Galilei (1564–1642) and Leonardo da Vinci (1452–1519) refined the method, paving the way for modern science.

History

### Units of measurement

The first units of measurement were needed for tasks such as building, weighing goods, and measuring time. They were based on common items such as stones, ropes, seeds, and body parts; for example, the foot was one of the first units of length used in Mesopotamia and Egypt. As societies traded goods and knowledge, units became standardized. In the 1950s, an international system (SI units) was devised; this system includes standard definitions for the kilogram, metre, and second.

CAROB SEEDS

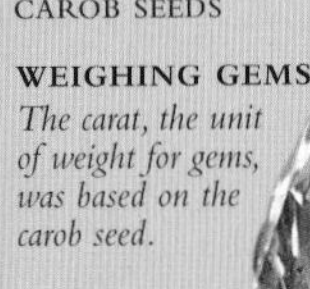

**WEIGHING GEMS**
*The carat, the unit of weight for gems, was based on the carob seed.*

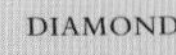

DIAMOND

**CALCULATING TOOL**
*In the past, the abacus was widely used for counting, addition, and subtraction. Different types existed in Europe, China, and Japan. The traditional Chinese abacus (above) is still used today.*

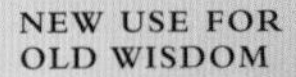

**NEW USE FOR OLD WISDOM**
*The San people of southern Africa have amassed detailed knowledge about the habits of animals and the uses of plants. Their information is now being applied in fields such as drug development and ecology.*

# THE SCIENTIFIC METHOD

The modern scientific method is based on four steps: observation, hypothesis, experimentation, and independent verification. First, a scientist makes observations. He or she then suggests an hypothesis (a working assumption) that explains the observations and is consistent with past observations, and predicts the results that should arise according to the hypothesis. The scientist tests the predictions by further observation or through experiments, which he or she repeats to ensure consistent results. Other scientists may also perform the experiments to verify the findings independently. If the hypothesis is proved, it is accepted as a theory; if found to be false, it is modified. A distinguishing feature of scientific theories is that they may be falsifiable: in other words, there may be some experiment or discovery that could eventually disprove them. If observations cannot be explained by existing theory, a further theory may be proposed to cover both new and existing evidence. In rare cases, a new theory completely revolutionizes a world view.

Make observation
FORMULATE HYPOTHESIS
Test hypothesis
FORMULATE NEW HYPOTHESIS
Acceptable
Wrong
VERIFY HYPOTHESIS
Acceptable
ACCEPT AS NEW THEORY

**FORMULATING A THEORY**
*This diagram shows, in a very basic form, the steps that scientists follow in order to arrive at a new theory.*

# TECHNIQUES AND INSTRUMENTS

The basic technique in science is the observation of particular phenomena under controlled conditions. For this task, scientists need accurate viewing, monitoring, and measuring instruments. An internationally standardized system of measurement, SI units, allows scientists anywhere in the world to make accurate, consistent measurements. During the 20th century, unprecedented technological progress produced increased accuracy in instruments that had been developed in previous centuries, such as telescopes and microscopes, but also resulted in the creation of completely new instruments, notably computers. To process and analyse data, scientists deploy mathematical tools such as formulae and probability theory. The development of powerful supercomputers has enabled scientists to process huge amounts of data and thus perform highly complex tasks, such as solving equations that could not be solved by mathematical analysis or generating models of complex phenomena such as protein reactions and climatic systems.

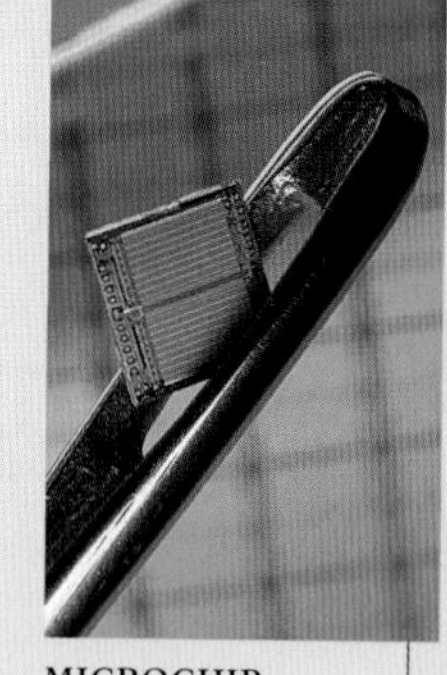

**MICROCHIP**
*In addition to being used in computers, microchips are now a vital part of many everyday machines and appliances.*

# SCIENTIFIC DISCIPLINES TODAY

Science today encompasses many diverse fields, which grow as more data is collected and become more complex as new theories are added. There are, however, three main groups and several widely recognized disciplines. One group comprises the physical sciences, such as physics, chemistry, and astronomy; these disciplines deal with the basic aspects of matter and energy. Another consists of the life sciences, which cover humans, animals, and plants. The most recently developed group comprises the social sciences, in which various facets of human experience are studied. Some scientists pursue abstract knowledge ("pure" science) for its own sake, although such work may also have practical benefits. Others use "applied" science for practical purposes, such as medicine, engineering, and manufacturing.

**MEDICAL IMAGING**
*Scanning techniques, such as magnetic resonance imaging (MRI), can reveal body structures in great detail without the need for invasive operations.*

**LIGHT SIGNALS**
*This scientist is exploring the new technology of optical computing, in which computers would work much faster by using laser beams rather than electronic signals.*

Issue

## Science and controversy

Scientific advances have often led to disputes with authorities or wider society. For example, Andreas Vesalius (1514–1564) transformed anatomical study but challenged existing beliefs by dissecting corpses rather than relying on ancient Greek teachings. Today, contentious issues include the genetic modification of plants and animals; the use of nuclear power; and the causes of global warming.

**GENETIC MODIFICATION**
*These mice carry a jellyfish gene making them fluorescent. The mixing of genes from different species has caused public concern.*

**SPACE SHUTTLE**
*The Space Shuttle orbiters are the world's only reusable space vehicles. They are used to launch satellites, as a base for experiments, and to ferry people and supplies to the International Space Station orbiting the Earth.*

PHYSICAL SCIENCES

## Mathematics

**In brief:** The symbols, formulae, and operations of mathematics underlie all other scientific disciplines

Mathematics may be defined as the study of number, quantity, shape, and space. Such concepts are expressed in symbols, which form a "language" used in all sciences. The major disciplines of arithmetic, geometry, and trigonometry developed thousands of years ago for practical use in trade, measurement, and astronomy. Calculus, which is used to measure rates of change (for example, in acceleration) was discovered in the 17th century. In the 19th century, statistics was developed as a tool for interpreting data. Mathematics is still evolving; one of the newest fields is chaos theory, which is used to analyse systems such as economies and the atmosphere.

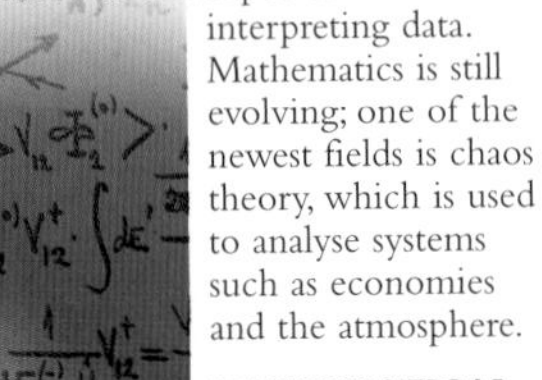

**MATHEMATICAL EXPRESSION**
*Mathematics is a form of expression based on symbols, formulae, and particular operations.*

PHYSICAL SCIENCES

## Physics

**In brief:** Knowledge from physics influences many other sciences and has enabled huge advances in technology

Physics concerns the properties and interactions of matter and energy: the basic elements of the universe. It is deeply rooted in mathematics; in turn, it comes into a range of other sciences, such as engineering, chemistry, geography, astronomy, and biology. Many of the disciplines included in physics have practical uses, including optics (the study of light), fluid dynamics (the study of liquids and gases), ballistics (the study of projectiles such as bullets and rockets), and electronics. Discoveries in physics have underlain some of the greatest advances in human civilization. In the 17th century, Sir Isaac Newton (1642–1727) formulated basic laws of motion and gravity. In the 19th century, Michael Faraday (1791–1867) and James Clerk Maxwell (1831–1879) investigated electricity and magnetism; their work made it possible to harness electricity. Further advances have enabled people to see inside the body without surgery, travel faster than sound, walk on the moon, and study the structure of atoms (the basic constituents of matter). Modern physics allows scientists to investigate matter and energy at very large or tiny scales. Some theories have led to radical new understanding; for example, the special theory of relativity, published by Albert Einstein (1879–1955), states that mass and energy are equivalent and the speed of light is constant, and quantum theory states that matter may behave as waves or as particles.

**LONG SPAN**
*The laws of physics help to explain the forces that act on structures such as bridges; such knowledge is essential in engineering.*

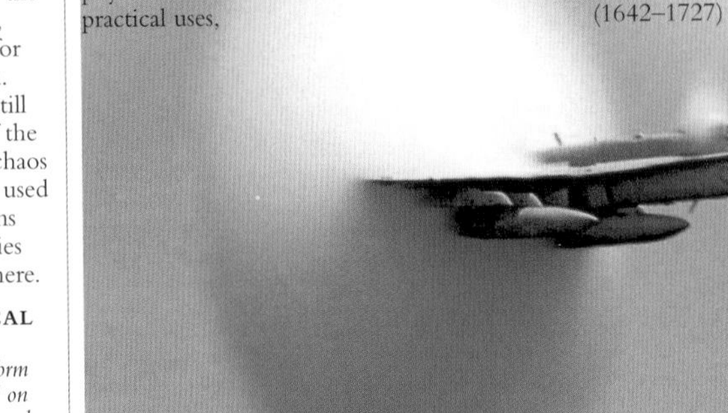

**THE SOUND BARRIER**
*A cloud of vapour condenses around a US fighter plane as it goes supersonic. Knowledge about sound waves and atmospheric effects has enabled people to build aircraft that fly much faster than sound.*

PHYSICAL SCIENCES

## Chemistry

**In brief:** The many uses of chemistry include analysing the nature of matter and creating new substances

Chemistry is the study of matter – its composition and properties, and the changes that occur when substances react with one another. Potters and metallurgists in ancient cultures made practical use of chemical reactions, but chemistry only emerged as a scientific discipline from the 17th century. John Dalton (1766–1844) put forward the theory that matter consists of atoms, and that atoms of different substances may be combined to form compounds (molecules). The Swedish chemist J.J. Berzelius (1779–1848) devised the system of letters and numbers used to define elements and compounds (such as $H_20$). Dmitri Mendeleev (1834–1907) devised the periodic table, in which the chemical elements are arranged according to their atomic weight. Chemistry has thousands of applications in today's world. Some involve identifying and analysing unknown substances, such as pollutants in a water supply. Many more involve synthesis: the manufacture of new substances such as plastics, drugs, fertilizers, and textiles.

**AGRICULTURAL SCIENCE**
*Chemistry is important in the development of artificial and organic crop treatments such as fertilizers and pesticides.*

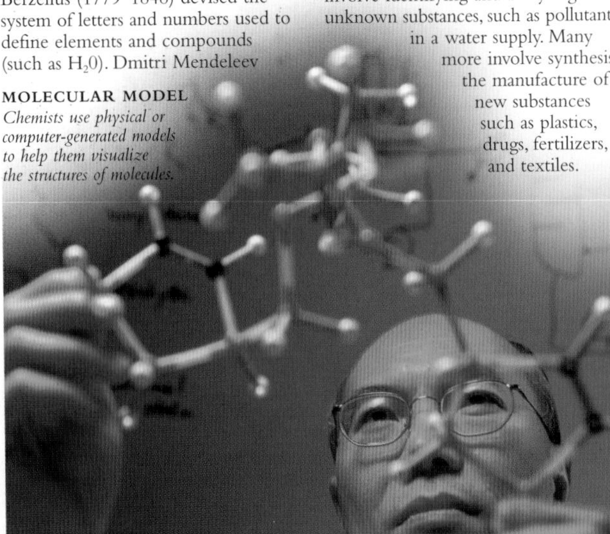

**MOLECULAR MODEL**
*Chemists use physical or computer-generated models to help them visualize the structures of molecules.*

PHYSICAL SCIENCES

## Geography

**In brief:** The various forms of geography concern both natural and man-made landscapes

The discipline of geography has both physical and social aspects. Physical geography is the study of natural elements in landscapes, and human geography is the study of the way in which humans interact with their surroundings. The ancient Greeks were the earliest geographers. Eratosthenes (c276–194BC) was the first person to calculate the circumference of the Earth, and the famous *Geography* by Ptolemy (cAD85–165) included maps of the known world. In the Middle Ages, Muslim scholars preserved and built on Greek knowledge, and travellers such as Ibn Battuta (1304–1377) gave accounts of various lands and peoples. From the 15th to the 17th centuries, European voyages to the Far East and the Americas led to advances in cartography (map-making), such as the projection devised by the Flemish cartographer Gerardus Mercator (1512–1594) to represent the globe. Today, physical geography includes areas such as geomorphology, the study of the processes by which landforms are created and shaped, and biogeography, concerning the distribution of plants and animals over the Earth. It is strongly linked to other physical sciences such as geology. Practical applications include civil engineering and navigation. Human geography has links with economics and sociology; uses include population studies, town planning, and planning healthcare provision. Both physical and human geographers now employ a range of technology; for example, aerial and satellite photography and satellite navigation systems are used to collect information, and computerized geographic information systems (GIS) are used to analyse data.

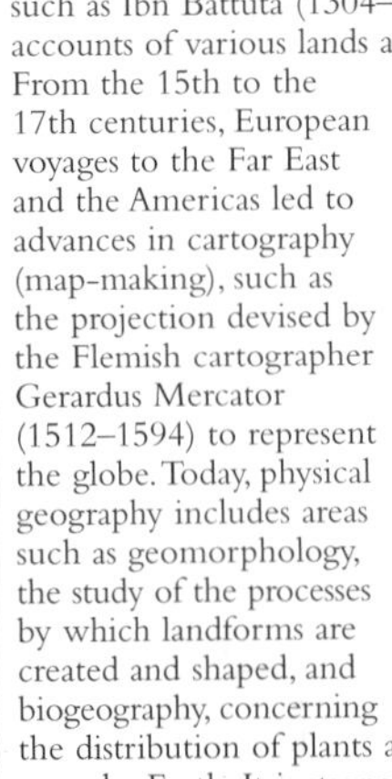

**STUDY OF LAND USE**
*Satellite photographs, like this image of a reservoir and farms, can reveal patterns of land use.*

**ROCK FORMS**
*Physical geography provides explanations for processes such as the erosion of rocks.*

PHYSICAL SCIENCES

## Geology

**In brief:** Geologists study the Earth's structure and history, together with human impact on landforms

Geologists study the Earth's surface and interior. Their areas of interest include the composition of rocks and minerals, how landforms are created and shaped, and why volcanoes and earthquakes occur. For centuries people have used practical knowledge of the Earth in areas such as mining, but the science of geology dates from the 18th century. The British geologists James Hutton and Sir Charles Lyell suggested that changes in the Earth occurred over millions of years and were still taking place. The German Alfred Wegener (1880–1930) put forward the theory of continental drift: that there was one landmass that broke into pieces, which drifted to form the continents. Today, geology is applied to find oil, gas, and minerals and predict natural disasters. It is also used to study interactions between landforms and constructions such as tunnels and dams.

**FINDING OIL**
*Oil companies use geologists to help them identify and exploit sources of oil and natural gas.*

**EARTHQUAKE STUDY**
*Seismology, a branch of geology, is the study of earthquakes and other earth movements. It enables scientists to analyse and predict earthquakes.*

PHYSICAL SCIENCES

## Environmental science

**In brief:** The different fields of this science concern the environment and humanity's relationship to it

This holistic science concerns the relationship between humans and their environments, both natural and man-made. Environmental science includes such fields as ecology (the ways in which organisms interact with their habitats), the physics of the atmosphere, energy production, the chemistry of manufacturing processes, and the psychology of consumerism. Scientists aim to control environmental damage resulting from human activities. Such damage first became evident during the industrial revolution in the 18th and 19th centuries; it has been made worse by industry, intensive agriculture and fishing, overuse of natural resources, and high population growth. Areas of study include using energy efficiently, finding clean sources of power, and developing technology for use in poor or remote areas of the world.

**CLEAN ENERGY**
*To reduce environmental problems, scientists are developing pollution-free forms of transport, such as this solar-powered car.*

Issue

### Wind farms

One nonpolluting source of power is the wind. Large groups, or "farms", of wind turbines are increasingly used to generate electricity. They are usually situated on high or open ground or in the sea. Wind farms produce no harmful gases and use a constantly renewable energy source. However, they have been criticized as noisy and unsightly. The turbines may also pose a hazard to wild birds, which can be injured or killed by the blades.

PHYSICAL SCIENCES

## Meteorology

**In brief:** The science of meteorology is vital in weather forecasting and preventing environmental disasters

Meteorology involves studying the composition of the atmosphere and the changes that produce weather. Its most important application is weather analysis and forecasting. For centuries, people relied on observing phenomena such as cloud types and even animal behaviour. Accurate measuring became possible when Galileo Galilei (1564–1642) invented the thermometer. Later developments included the invention of the barometer (to measure atmospheric pressure) and of temperature scales. Robert Boyle (1627–1691) carried out work on gases that advanced the understanding of atmospheric physics. Gaspard-Gustave Coriolis (1792–1843) identified the force produced by the rotation of the Earth, which causes the movement of air and sea currents. In the 19th century, the first weather stations were set up. Since the 1960s, satellites and computers have enabled huge amounts of data to be processed. Today, meteorology is vital for aviation and navigation, predicting hazards such as hurricanes, and assessing climatic threats such as global warming.

**WATCHING A HURRICANE**
*Satellites are an indispensable tool for providing images of weather systems and large-scale formations such as hurricanes.*

PHYSICAL SCIENCES

## Astronomy

**In brief:** One of the oldest sciences, astronomy covers the nature, processes, and history of the universe

Astronomy is the study of planets, stars, galaxies, and other forms of matter and energy in the universe. It was first developed by ancient peoples such as the Chinese, Egyptians, Babylonians, and Mayans, for practical purposes such as navigation, timekeeping, and determining dates for festivals. The Greeks devised a geometrical model of the universe, with the Earth at its centre. This theory lasted until the 16th century, when Nicholaus Copernicus (1473–1543) developed the concept that the Earth revolved around the sun. Isaac Newton (1642–1727) defined laws of motion and gravity that helped to show how planets orbit stars. Astronomy is still based on observation, with telescopes collecting data from even the farthest parts of the universe. The latest ideas include theories on the possible shape of the universe and on the way in which it came into being (such as the "big bang" theory).

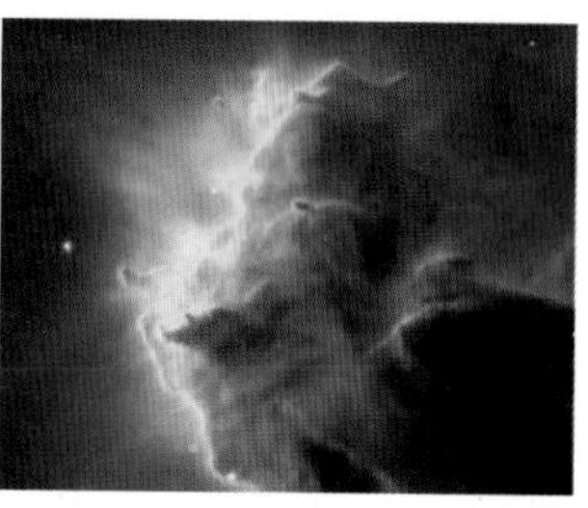

**BIRTH OF A GALAXY**
*Telescopic views of nebulae are providing astronomers with detailed information on the ways in which stars and galaxies form.*

**SPACE TELESCOPE**
*Astronauts service the Hubble Space Telescope. This telescope, orbiting the Earth at the edge of the atmosphere, has given unprecedented views of stars, galaxies, and nebulae.*

LIFE SCIENCES

## Biology

**In brief:** The science of life, biology encompasses the study of all organic structures, both animal and plant

Biology, the science of life, has existed in some form for thousands of years. Practical knowledge about animals and plants was vital in hunting, foraging for food, and agriculture. Various peoples, notably the ancient Greeks, made systematic observations of the natural world. For example, in the 4th century BC Aristotle wrote several works on the physiology of animals and another philosopher, Theophrastus, produced the first text on botany. During the Renaissance, artists' and scientists' curiosity led them to make detailed portraits of animals and plants and dissect corpses for anatomical study. In the 17th century, pioneers of microscopy, such as Robert Hooke (1635–1703) and Anton van Leeuwenhoek (1632–1723), made the first studies of objects invisible to the naked eye, such as capillaries and cells. In the 19th century, two huge advances in understanding resulted from Charles Darwin's theory of natural selection and Gregor Mendel's work on the way in which genetic characteristics are inherited. With the discovery of DNA structure in 1953 by Francis Crick and James Watson, and in 2003 the complete sequencing of the human genome (all of the DNA that makes up a human being), biologists now have a code for creating people. Such discoveries can, however, lead to controversy. In biology, learning has often been restrained by prevailing moral doctrines; for example, religious bans on human dissection hampered understanding of the body's structures and functions for centuries. In the 21st century, advances in genetics have raised new practical and moral issues – for example, how to regulate the genetic engineering of organisms including ourselves.

**PRACTICAL USE**
*A knowledge of biology and genetics is essential in many areas of agriculture, such as animal breeding.*

**ELEPHANT FAMILY**
*By studying behaviour and social interactions in animals, biologists can find out about the animals themselves and sometimes shed light on human behaviour as well.*



LIFE SCIENCES

## Medicine

**In brief:** Medical science has been practised in many cultures through history; it now includes a huge number of techniques to preserve health and prevent disease

Humans have practised various forms of medicine for thousands of years. At first, healing often involved magic, but gradually more rational methods evolved; the ancient Chinese used an extensive range of drugs, and Indian texts describe surgical instruments. In Greece, Hippocrates (460–377BC) taught that diet and exercise could prevent disease, and Galen (AD131–201) carried out the first experiments to determine the workings of the body. Their work, though flawed in places, went largely unchallenged until the Renaissance. In 1543, the Belgian anatomist Andreas Vesalius (1514–1564) made the first accurate study of the human body. Later pioneers included William Harvey (1578–1657), who discovered the way in which blood circulates; Louis Pasteur (1822–1895), who invented vaccination; and John Snow (1813–1858), who identified the way in which diseases spread. Today, medicine is highly specialized and complex. A huge pharmaceutical industry produces drugs for almost every condition. In affluent societies, screening enables doctors to identify and treat diseases, such as cancer, at an early stage. Imaging techniques, such as MRI and PET scanning, and "keyhole" operations have superseded invasive surgery in many cases. However, these new technologies are expensive, and coupled with the fact that people are living longer, put immense financial pressure on public medical bodies. Another major consideration for doctors and researchers today is the role of ethics in new technologies such as gene therapy and assisted conception.

**SURGERY**
*Operations involving new surgical techniques, such as microsurgery and laser surgery, have lessened the levels of discomfort and reduced the recovery times for patients.*

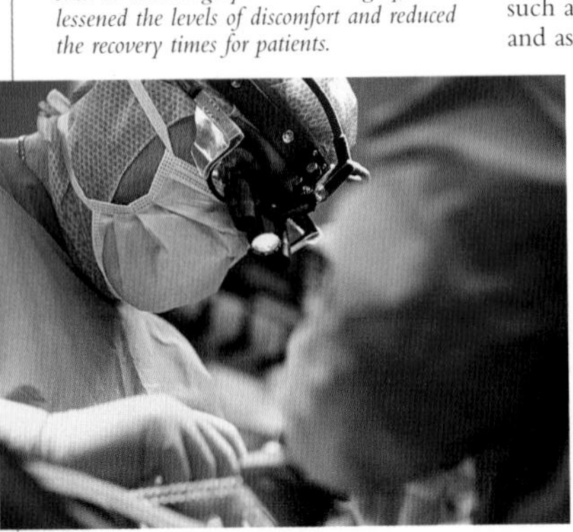

Issue

BOLDO HERB

### Medicinal plants

Plants have long been used to cure illnesses and treat wounds; for example, the Boldo herb was used in Chile to treat gallstones. Some drugs are also derived from plants, such as the anticancer drug vincristine, made from the rosy periwinkle of Madagascar. Scientists are now reassessing certain old remedies, such as echinacea (used by Native Americans), and looking at newly discovered plants that could be used in drugs.

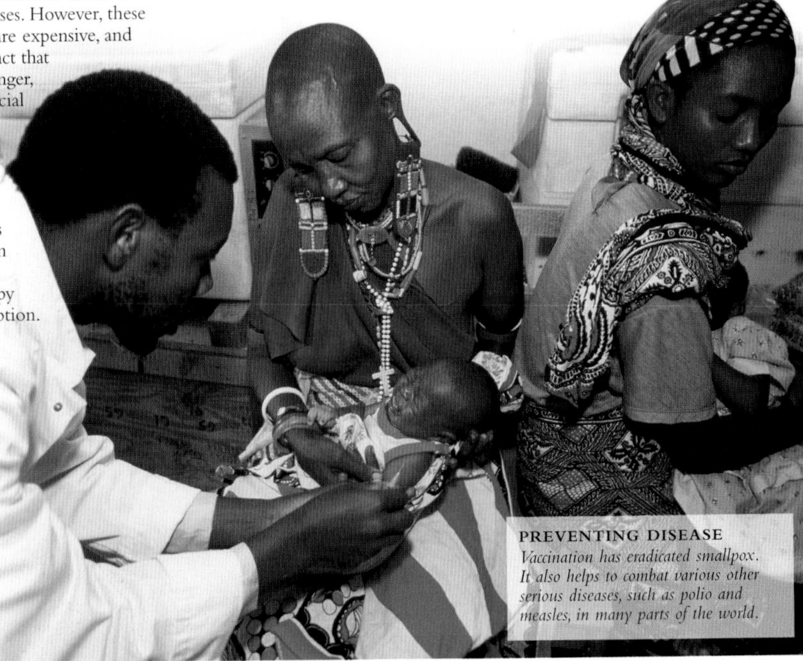

**PREVENTING DISEASE**
*Vaccination has eradicated smallpox. It also helps to combat various other serious diseases, such as polio and measles, in many parts of the world.*

SOCIAL SCIENCES

## Psychology

**In brief:** The study of human thinking and behaviour, psychology has uses in business as well as healthcare

Psychology is the study of individuals, either alone or in groups, with the aim of explaining and predicting their thoughts and actions. Mental processes that interest psychologists include perception, cognition, learning, and thinking. Early psychology concerned matters such the soul and freedom of the will, and until the late 19th century, the discipline was regarded as a branch of philosophy. The German scientist Wilhelm Wundt (1832–1920) was the first to use experiments in psychology. The science developed into schools such as psychoanalysis, founded by Sigmund Freud (*see* p175), which was used to explain people's unconscious drives; behaviourism, which involved observing behaviour; and the Gestalt and humanistic schools, which focused on understanding behaviour. Modern psychology draws on biology (based on new brain-imaging technologies) and cognitive models. It is used to assess and treat mental problems, and is also applied in business and marketing.

Profile

### Abraham Maslow

The humanistic psychologist Abraham Maslow (1908–1970) is most widely known for his theory of the hierarchy of human needs, first set out in 1943. Maslow represented needs as a pyramid. At the base were physiological needs, such as satisfying hunger; then came needs for safety, love and belonging, and esteem from other people. The topmost needs were for self-actualization (realizing personal potential) and transcendence of the ego. He believed that people had to meet lower needs before addressing the higher ones.

SOCIAL SCIENCES

## Sociology

**In brief:** Sociological studies illuminate the rules and processes that connect people in groups of all sizes

Sociology is the study of humans as members of groups, from families and religious communities to bureaucracies. Sociologists consider facets that unite or divide people, such as ethnicity, class, and gender. They employ both quantitative methods (such as statistical analyses) and qualitative methods (such as interviews) to understand social conditions and changes. The discipline first emerged in the late 19th and early 20th centuries; Émile Durkheim (1858–1917) pioneered the use of observation and statistical evidence, and Max Weber (1864–1920) was the first to theorize about social organisation. In modern sociology, there are three main theories: conflict theory (individuals struggle to maximize their own benefits), functionalism (individuals play specific roles in society), and interactionism (individual identity is created through interactions with others). Sociological studies are increasingly used by governments and policy-makers in attempts to tackle social problems such as crime, divorce, and substance addiction.

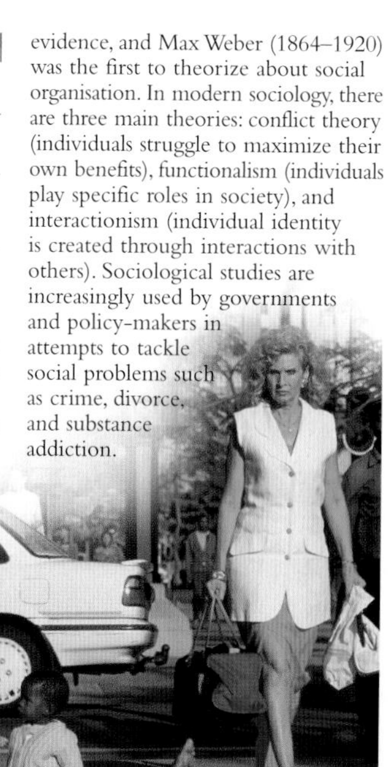

**SOCIAL DIFFERENCES**
*One major area of sociological study is the examination of differences in classes or levels within a society.*

SOCIAL SCIENCES

## Anthropology

**In brief:** In this study of humanity, techniques range from living with a community to forensic science

Anthropologists study biological, social, and cultural aspects of humanity all over the world and in all historical periods up to the present. The science of anthropology developed in the late 19th and early 20th centuries. One of the first prominent anthropologists was Bronislaw Malinowski (1884–1942). He started the practice of living with different communities in order to collect data about them; he also developed the theory of functionalism (in which social institutions are thought to form interdependent parts of a stable society). Another is Claude Lévi-Strauss (1908–), who suggested that kinship and other relationships in different societies had similar underlying structures, and regarded culture as a form of symbolic communication. In modern anthropology, one major branch is physical anthropology, in which primate behaviour, human evolution, and the physical variations in modern human groups are studied. This field includes forensic anthropology: the study of human bones from any period (including the present) in order to determine the cause of death. Another branch is sociocultural anthropology; this discipline focuses on kinship, social networks, religion, and similarities and differences in culture. A third branch, linguistic anthropology, is the study of the ways in which language varies across space and time, how it relates to culture, and how it is used socially. Some social scientists also regard archaeology, the study of material remains of human societies, as a branch of anthropology.

**READING THE BONES**
*Forensic anthropologists can reconstruct models of faces using human skulls, to find clues about life and death in various cultures.*

**AZTEC ARTEFACT**
*Anthropology has links to archaeology. Remains from ancient peoples, such as the Aztecs of Mexico, can provide information about the cultural history of their descendants.*

**EXCHANGE OF IDEAS**
*Economists sometimes make use of ideas from other disciplines; for example, they may use psychology to investigate trading decisions.*

SOCIAL SCIENCES

## Economics

**In brief:** This discipline deals with the way humans use resources and define the value of goods and services

Economics is based on studying the ways in which limited resources are allocated to produce commodities and services, and how they are distributed to meet human desires and needs. In microeconomics, the focus is on the behaviour of individuals and firms. In macroeconomics, the economy as a whole is studied to help people understand interactions between factors such as employment and inflation. Underlying economics is a theory of value. The classical labour theory of value states that commodities and services are worth the amount of labour that went into them, and that price reflects the availability of labour. Market theory holds that price and value are the same. Economics is inseparable from politics and ideology. For example, in the 1940s John Maynard Keynes (1883–1946) advocated that governments should use financial measures to protect their peoples from adverse economic conditions. To help them understand the complex patterns of today's societies, economists now borrow concepts from biology and information sciences.

**MONEY TALKS**
*Through most of history, money has been the most important consideration in economic decisions.*

# PEOPLES

# PEOPLES

We are not only members of the human species. We also belong to states, nations, and ethnic groups. Most of us have an ethnic identity and a feeling of belonging to a particular people. The members of a people might have a shared language, customs, or history – or all three. But the diversity of human culture is infinite. Experts estimate that we speak more than 6,800 languages and fall into a comparable number of peoples. This section on human diversity provides a snapshot of more than 300 of these groups.

An ethnic group can be regarded as a "people" regardless of whether it has enclosed its territories within political boundaries to form a nation state, or whether the group is just one of many peoples within a country. Some peoples consist of a handful of individuals, carrying the memory of a dying culture. Other peoples are established in nation states that still bear their name – the Portuguese of Portugal, for example. Each ethnic group has a unique culture, of which language is a key element. The number of languages spoken in an area can indicate ethnic diversity. Europe is the primary home of a mere 3 per cent of living languages (around 230), while Mexico alone has 288 languages, implying the same number of peoples.

## BODY VARIATION

The human species shows a wealth of physical diversity. Some of it is aligned with cultural differences, some of it is not. East Asians are renowned for their almond eyes, but unrelated groups also possess them, including the San (Bushmen) of Africa. Some clear physical differences occur between peoples whose isolation has increased their distinct features. Among these cases are the unusually tall Dinka of Sudan and the "Pygmy" peoples of Central Africa. So, there are some connections between physical appearance (determined by genetic kinship) and culture. However, belonging to an ethnic group is not generally about resemblance owing to genetic inheritance. The Magyar people of Central Europe, for example, are distinguished from their neighbours through the use of a Finno-Ugrian language, which is unlike the Indo-European tongues spoken by others in the region. However, studies show that only 10 per cent of Magyar genes can be traced to their Central Asian, Finno-Ugrian ancestors, pointing to centuries of intermarriage.

In reality, migration, interaction, cultural spread, and the evidence of our genetic makeup undermines the physical method of separating ethnic groups; it is impossible to tell someone's ethnic background simply by looking at them.

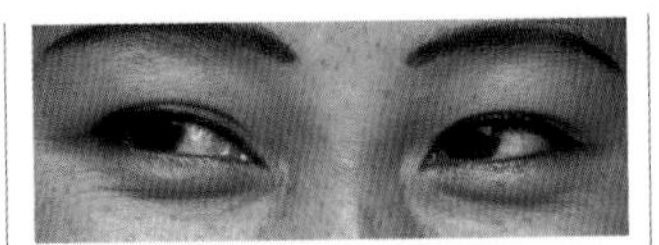

ALMOND EYES

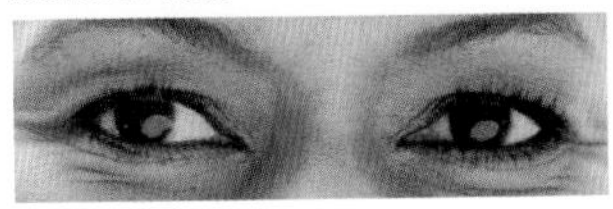

NON-ALMOND EYES

**DIFFERENCE IN EYE-SHAPE**

*Almond eyes occur in unrelated peoples such as Inuit, eastern Asians, and the San. Experts therefore suggest that all humans originally had this characteristic and that those with non-almond eyes, such as Europeans, have lost it.*

Issue

### The myth of race

A "race" is a category into which people can be placed according to inherited physical features, hence ancestry. The concept of race has been undermined by genetics. The genetic history of our species is dominated by intermixing of genes, so that there has never been such a thing as a genetically "pure" race.

## WHAT IS A PEOPLE?

If members of a people are not necessarily related genetically, then what exactly is it that unites them? Although people may feel bound by biological kinship, the true story of ethnicity has more to do with a shared language, and a set

**GENES AND MIGRATIONS**

*The genetic family tree below can be compared to what we know about early human movements across the globe (right). The colour coding of the migration lines (right) matches the colours of the different branches of the human family tree (below).*

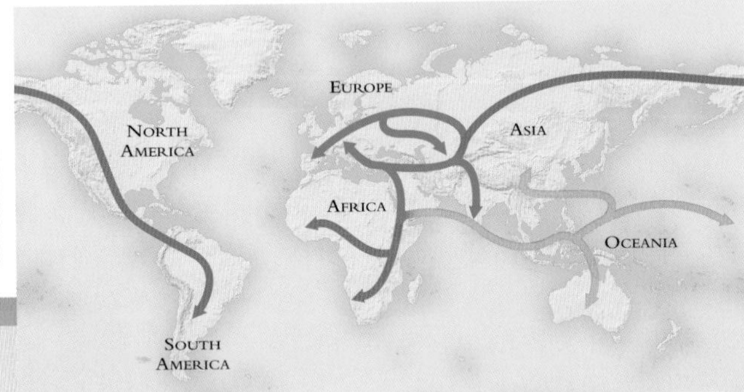


### Genetic history

Although there has been much intermixing of genes between peoples throughout history, geneticists can preceive an underlying pattern of kinship using genetic markers. These markers are distinctive portions of DNA possessed by some humans and not others, with which we can trace our genetic history. The result is this tentative family tree of the species. It shows, for example, that most Pacific Islanders and Aboriginal Australians have certain markers that most other people do not. According to the genetic data, all branches of the tree share at least one marker with a hypothetical woman called Eve, who lived in Africa 150,000 years ago.

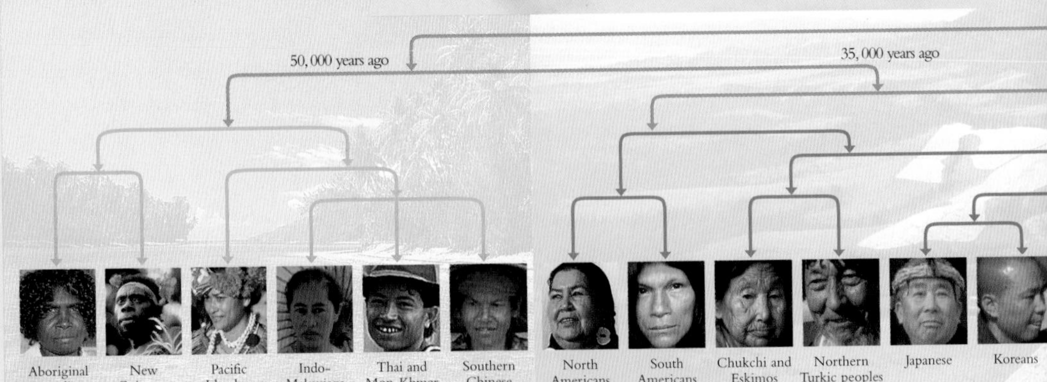

of traditions, including religion, a sense of homeland, and a common history. These elements add up to create what could be described as a communal ethnic identity.

## LANGUAGE AND ETHNICITY

Groups who speak different languages are, broadly speaking, divided, while people who speak the same language are, at least in some way, united. One language does not always equate to a single ethnic group, however. For example, Spanish-speaking Chileans are not Spanish by ethnicity. Likewise, the use of a separate language does not necessarily create a meaningful split in ethnicity. Many of the French-speaking Quebecois, for example, identify themselves as "Canadian," so language is obviously not enough on its own to capture the full essence of "Canadianness."

**GREETING CUSTOM**
*Hongi, or the gentle pressing of nose and forehead, is how Maori customarily greet each other.*

## TRADITIONS

Ethnic identity often finds its best expression through traditional customs. The ways in which related people mark the milestones of their lives, interact with strangers, and spend their free time can be truly uniting features. Dancing is a universal human pastime, and is often seen as a quintessential way to express ethnicity, whether it is the hypnotic pogo-ing of the Masai, the energetic legwork of Irish dancing, or the elegant fan dances of the Japanese geisha. Not all Irish people know how to perform a gigue, but they all recognize it as a part of Irish culture.

History also plays a significant role in determining a person's ethnic identity. Many cultures look back into their past in order to explain their present life situation. Sometimes this can take the form of a shared story of origin (oral history) or a creation myth. A lot of people view this as evidence of their ethnicity.

Religion can be a very powerful ethnic characteristic. In some cases, most tragically in Bosnia and Herzegovina, the adherence to a religion overrides other ethnic traits. Although Croats, Serbs, and Bosniaks all share a mutually intelligible language and comparable day-to-day lives, their history as Roman Catholics, Orthodox Christians, and Muslims was enough to prompt bloody ethnic war in the 1990s and the creation of ethnically based divisions of land. Conversely, religion can also breed a sense of super-ethnicity. Bonds of faith draw together Muslims across the world, for example, with Arabized Berbers of Morocco feeling brotherhood with Muslim Malaysians of Southeast Asia.

**DISTINCTIVE SPORTS**
*Only in Vietnamese communities is there a festival of traditional Vietnamese wrestling. The fact that no-one else shares this form of wrestling encourages a sense of community.*

Issue

### Disappearing peoples

Some cultures are in critical danger. The most badly affected are those trying to pursue a traditional lifestyle at odds with modern economics, and on the fringes of government control. Within two generations, a people can lose its way of life and its language. There are only around 100 speakers of Yukaghir, the language of a Siberian people (right), and the speakers are all older than 50.

## HOMELAND

Living near to other people over generations promotes a sense of neighbourhood. Even if genes and ancient migrations are ignored, the place of birth can provide a sense of belonging. This attachment to a territory may be fulfilled when an ethnic group creates a formal nation state for itself. For stateless peoples, however, territorial belonging can also turn into a

"EVE"

150,000 years ago, in Africa

35,000 years ago

50,000 years ago

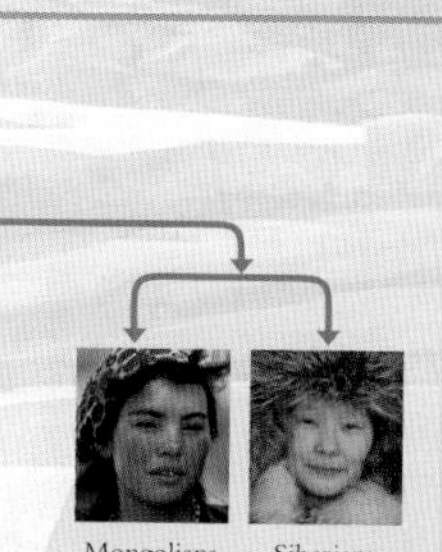


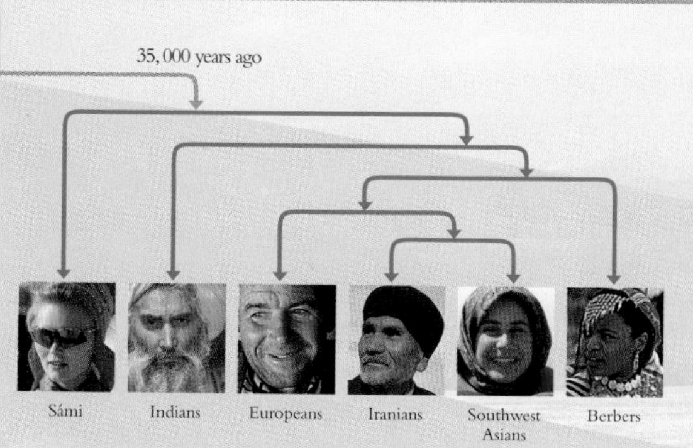


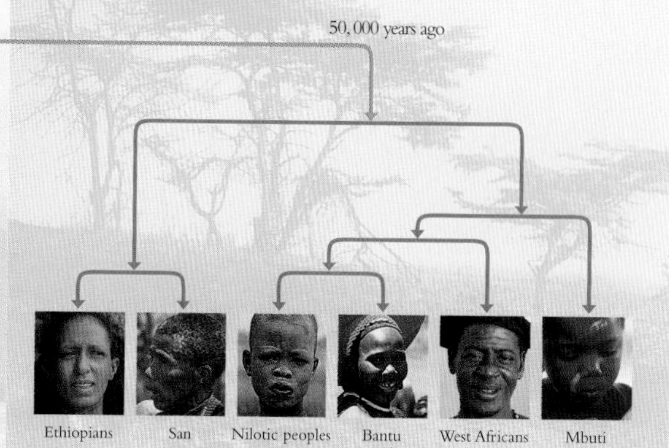


PEOPLES

claim for an ethnic homeland, and that is when ethnic feeling becomes political; fights over rights to land between ethnic groups are a common feature of history.

## ORIGIN OF TRIBES

A tribe is a community whose members are in some way related. In many societies, tribes still provide a modern network of kinship that influences social, economic, and political decisions. Early humans are thought to have evolved in tribal groups, within which tribal behaviour, such as bonding and group unity when faced with attack, evolved. Such behaviour may have been favoured by evolution because of the benefit of cooperative action. An individual who can rely on the help of fellow tribespeople in times of conflict, illness, or hunger stands a much greater chance of survival. The wealth of ethnically specific rituals, ceremonies, dances, and festivals that recur in anthropological studies, may have their origins in cementing fellow feeling within a group. Inclusion in the group also implies the exclusion of others, polarizing people into "them" and "us". Even societies that are no longer viewed as "tribal" usually retain elements of behaviour that reinforce this sense of belonging.

## FORCES OF CHANGE

Since ancient times, the migration of peoples has changed ethnicity. Migrants influence the people in their adopted land, but their own ethnicity also changes as they grow separated from their ancestral home. Immigrants can displace a native community, but a more common pattern is mingling of cultures and mixing of genes. The result can be the loss of one group and the domination of another. The less powerful group may become passively absorbed, or "assimilated" into the ethnic mainstream of the bigger group, or the migrants may become urbanized, losing rural traditions as they move to the city. A powerful group might even actively suppress minority groups in a process called ethnocide. However, sometimes, new groups are created, with blended traditions; or the result can be multiculturalism, with neighbours each preserving their distinct customs.

**SEASONAL RITES**
*A Bedik man of Senegal, West Africa, calls for a blessing on next season's crop. People with a strong connection to the land hold seasonal festivals to ensure a good harvest.*

SWEDISH FOOTBALL FANS

PAPUAN SING-SING GOER

**TRIBAL COLOURS**
*Peoples assert their ethnic identity in many, varying ways. Here, two Swedish girls attend a football match with the national flag on their faces, while a Papua New Guinea man wears face paint for a pantribal* sing-sing.

## ASSIMILATION

Assimilation can happen to an ethnic minority whether it is the indigenous culture or a migrant community. It can occur passively, owing to the action of individuals, such as marriage between different groups breaking down ethnic differences. Globalization and forms of modern communication are often seen to be encouraging the most complete cultural assimilation ever known; for example, the music television channel MTV is now well known among a majority of the world's youth. Western music, fashions, and morality appear frequently around the globe, but ethnic groups tend to adapt the new information into their own unique story.

## URBANIZATION

Today, migrants tend to move to cities, a trend termed urbanization. Bringing with them their own traditions, ways of cooking, speaking, praying, playing, and socializing, people moving to cities add to the cultural mix of their

**HARVEST FESTIVAL**
*Hungarians celebrate the wine harvest. Such farming festivities are lost when a people becomes urbanized.*

**ENDURING DISTINCTIONS**
*Never conquered by outsiders, the Amhara of Ethiopia have not undergone any assimilation, and continue to practise their Orthodox form of Christianity.*

adopted home. New ethnic groups can form from the different peoples added to the melting pot. Modern cities, such as New York, Paris, Sydney, and Rio de Janeiro, have a distinct character, contrasting with the land around them. However, although urbanization results in vibrant cities, many urbanizing people leave behind their distinctive lifestyles in the countryside. They lose ethnic events connected with the harvest, or social systems and codes of behaviour associated with hunting and gathering. Economic development and urbanization both force people to abandon old ways of living and encourage them to seek a new way of life in another society.

**ETHNIC BLEND**
*Brazilians blend their diverse backgrounds to create a composite culture. These musicians mix African and Latin rhythms.*

## ETHNOCIDE

Destruction of a minority culture is a phenomenon called "ethnocide". In such cases, minority groups are forced to drop their traditions and accept those of their more powerful neighbours or invaders. Language is often the first victim. Linguists predict that half of the tongues spoken today will have disappeared by the end of the 21st century – a rate of two a week. Some groups simply abandon their language as it is of little use in a majority society. Repressing the distinctiveness of ethnic minorities was considered an essential part of the historic European concept of the nation state. It claimed that an effective country was one inhabited by a single people. An ethnic group was then forcibly moulded from the diverse peoples in the country. When Italy was unified in 1861, fewer than 10 per cent of people in the new kingdom spoke Italian. Today, Italian regional character remains strong, but no one would deny the existence of a clear Italian ethnicity. More recent examples of forming ethnically cohesive nations however, are not seen as benevolent. In the 1980s, the government of Bulgaria sought to "Bulgarianize" the Bulgarian Turkish community; spoken Turkish was banned in public and Muslim names were forbidden.

Nation-building was carried out across the globe by empires, and attempts were made to adapt the native cultures of the colonies to those of the colonists. European languages became the lingua francas in the colonized countries, while religious missionaries left permanent marks across the world. New religions often form a strong part of a people's ethnic identity. The Pacific island states are renowned for their fervent Christianity, which has become an essential part of their Pacific ethnicity.

The most extreme form of ethnocide is called genocide. Recent decades have seen some of the worst examples. The Nazis of Germany attempted to eradicate the Jewish community of Europe, killing 6 million people. In Rwanda in 1994, genocide claimed the lives of 800,000 people: Tutsis, as well as Hutus who tried to protect them. Even in the 21st century, there have been ethnically motivated massacres: Arab militants targeted black communities in western Sudan in 2004. Both ethnocide and genocide are rarely totally successful. Persecution usually serves only to strengthen ethnic identity, cement resolve, and promote the survival of customs and ways of life.

History

### Emerging nations

The expansion of the Russian empire, followed by the further growth of the Soviet Union, swamped many smaller nations. Some 100 of these peoples survive to this day. Several of them newly enjoyed independence on the breakup of the Soviet Union in 1991. Peoples such as the Lithuanians, here demonstrating for independence earlier in 1991, were free to forge new states.

## MULTICULTURALISM

In the 20th century, the "melting pot" concept of ethnically blended countries was popular. However, the cultural loss through ethnocide and assimilation has been a high price to pay. An alternative process is multiculturalism, the maintenance of diversity within a nation state. Many societies are increasingly multicultural. Avoiding assimilation, some immigrants expand the concept of their original ethnic identity. Especially in the West, hybrid ethnic identities have appeared, such as Asian–American, Black–British, and Greek–Australian. Few would consider emigrating back to the land of their ancestors. Yet many, often in the face of discrimination and ignorance, choose to assert their ethnic distinctness from the people around them.

**LAND RIGHTS**
*Native Americans still rally for the right to practise their culture on ancestral lands.*

## ETHNIC AUTONOMY

In countries where ethnic groups are minorities due to the creation of states and empires, the trend today is towards greater autonomy and recognition of ethnic rights. The international community is keen to promote ethnic diversity: languages are protected, cultural practices revived, and political power is devolved. Although there are many peoples who do not have fair representation, there are many more who are being granted autonomy. Some recent efforts to end several long-running conflicts have generated increased reconciliation. In Australia, Aboriginal peoples have had their rights to ancestral lands at least partially recognized by the state. In Sri Lanka, the peace efforts are focused on establishing an autonomous homeland for the ethnic Tamils living there. The many peoples of East Timor reclaimed their land in 2002 as a completely independent state.

Issue

### Cultural travel

The appeal of apparently "exotic" culture is often a highlight for people holidaying abroad. Money paid by tourists for crafts, songs and dances, is often a crucial part of a people's economy, as it is for the Ndebele of Zimbabwe (below). Foreign interest can ensure that a culture survives and is a source of ethnic pride. However, some ethnic minorities are in the powerless position of being exhibited before tourists and forced to perform a caricature of their former culture.

## PEOPLES

The pages that follow present a selection of the thousands of peoples around the world. Taking cues from languages, customs, traditions, religions, kinship, and territorial belonging, the selection of peoples represents both the diversity of smaller ethnic groups and the majority of the most prominent peoples in each region. In this way, the 70,000-strong Huli tribe of Papua New Guinea are represented alongside the 280 million Americans to provide a broad picture of ethnicity, from well-defined, ancient ethnic minorities to the emerging ethnicities of the modern world.

**ARRIVAL OF THE HORSE**
*The introduction of the horse was a key stage in the development of North American culture, both of natives and of settlers, from the Prairie Provinces of Canada to Mexico.*

# NORTH AMERICA

The modern nations of North America are composite cultures formed by peoples who have flocked there from all quarters during the last five centuries. Their new identities – "Mexican", "American", "Canadian" – overlay far more ancient native cultures. Native peoples, the First Nations of Canada, the tribes of the US, and the indigenous peoples of Mexico were developing their civilizations in isolation for millennia beforehand. Many were extinguished entirely. Some are now reasserting their identity.

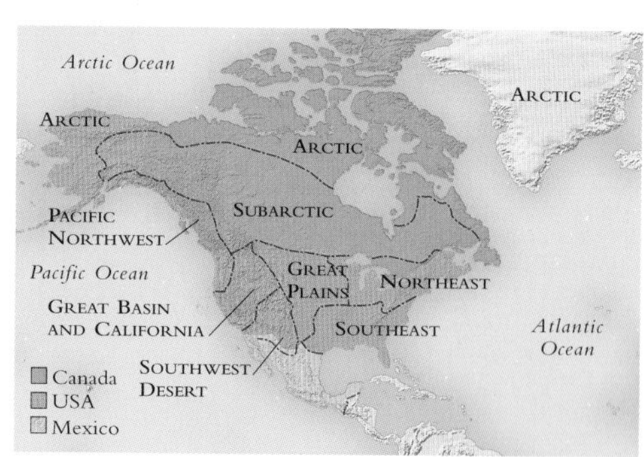


**THE REGION**
*Today's North Americans are citizens of three countries. Some traditional territories, however, cross political boundaries. Native peoples are here divided instead into geographical regions.*

When Europeans first set foot in the Americas in 1492, they found two continents full of culture and civilization that had been evolving in isolation from the rest of the world for at least 12,000 years.

## MIGRATION

The first peoples to reach North America arrived many thousands of years ago from northeast Siberia, having crossed the frozen Bering Sea. Many remained hunter-gatherers and fishermen, but where conditions permitted, people became farmers. A settled existence allowed some, such as the Anasazi of Arizona and Mound Builders of the Mississippi, to develop complex societies. A few, notably in Mexico, established states through warfare, trade, the organization of labour, and the spread of religion. These included the Olmecs, Toltecs, Aztecs, and Mayans. Europeans arrived in force in North America from the 16th century. Along the east coast, English-speaking settlers founded farming communities, and in the southeast established tobacco and cotton plantations, which in time were responsible for the importation of millions of African slaves. France established a chain of forts between Canada and Louisiana, while the southwest of the continent became Spanish dominions. In Mexico, explorers, conquistadors, settlers, and missionaries sought riches, souls, farmland, and Native slaves.

**MAYAN TEMPLE**
*The Mayan people built temples to the plumed serpent, Kulkulkan, in Chichen Itza, Mexico.*

**INUIT ART**
*Inuit used soapstone to create their art. From the Arctic to the tropical jungle, Pre-Columbian art flourished.*

## TRIBES AND NATIONS

When the Europeans arrived, Native peoples in North America had diversified into many peoples speaking hundreds of languages. The peoples ranged from small hunter-gatherer bands in the Arctic, through settled, complex agricultural societies in the southeast, to the large, militaristic state of the Aztecs, which dominated central Mexico.

## CHANGE

The last five centuries have seen massive changes to the continent's Native peoples, and to the emerging cultures of the settlers from Europe and Africa. In Mexico, many Native peoples, although not all, were absorbed into the expanding mestizo (mixed native and European) culture. In Canada and the US, famers originating in Europe pushed Native Americans aside as they turned the forests and prairies into farming and cattle country, and into cities. The bitter US Civil War of the 1860s resulted in the liberation of African slaves. Racial prejudice and segregation, though, were not eliminated. African-Americans only attained legal equality 100 years later after the Civil Rights Movement. In the more remote parts of Mexico, indigenous groups have retained their language and culture. In the south, Mayan peoples continue to demonstrate unrest. Native peoples in Canada and the US, pushed onto reservations, have confronted inevitable changes. Some have exploited resources on reservation lands or opened casinos. Many others, however, are marginalized. Both Canada and the US continue to experience immigration from all parts of the globe. Many groups form localized communities and maintain their Old World customs. Some parts of the US have begun to lose their Anglo identity as Latin Americans enter the country. Hispanic culture is now growing rapidly throughout North America.

**NASA WORKERS**
*North American culture now exports itself globally, leading global enterprises such as space exploration.*

History

### Coast to coast railroad

In 1869, the Union Pacific Railroad joined railways starting respectively on the Atlantic and Pacific coasts, thus forming a transcontinental railway. It opened up commerce and stimulated industrialization across the continent. It also spelled the end of countless Native cultures in the West, as industrialized European-style culture became dominant.

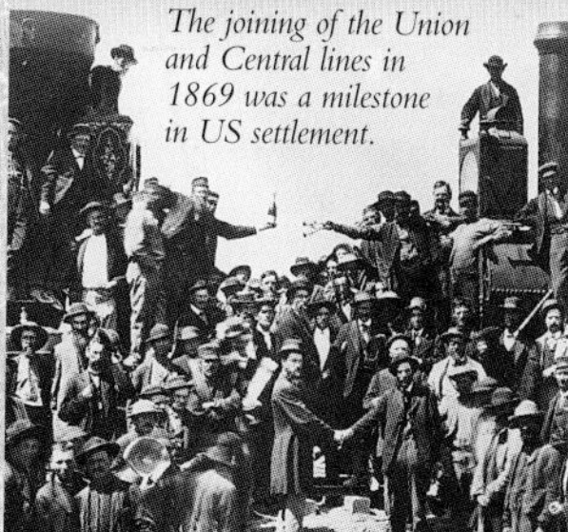
**RAIL UNION**
*The joining of the Union and Central lines in 1869 was a milestone in US settlement.*

**ADOBE HOUSE**
*Native Americans, as a rule, no longer live in the characteristic dwellings of their ancestors. Some people in the southwest, however, still live in adobe houses, made from sand, clay, and straw.*

NORTH AMERICA

## Americans

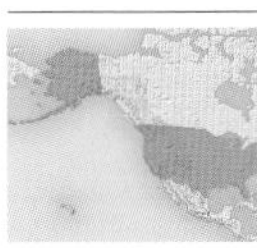

**Population** 289 million
**Language** English, Spanish, French, German, Italian, Chinese, Filipino
**Beliefs** Several, including Protestant and Catholic Christianity, Judaism, Islam, nonreligious beliefs
**Location** US; military bases/govt offices abroad

More than any other country, the United States of America sees itself as the land of opportunity. Today, in the larger US cities, within the space of a couple of streets one can pass from Chinatown to the Italian quarter and then from a Hispanic to an African–American or a Korean-dominated area. Spanish is spoken as a first language by nearly 20 million Americans. Despite this diversity, however, most Americans observe national rituals, such as Thanksgiving dinner at the end of November. Devotion to baseball and US football is also a unifying factor.

To generalize, Americans are individualistic (witness the gun lobby's insistence on the right to bear arms), with a culture of competitiveness, hard work, and respect for material success. They tend to be talkative, direct, and informal when making new acquaintances, although the right to privacy is deeply ingrained. There is widespread hostility to intrusiveness from officialdom.

The US is the most religiously diverse nation in the world. Christian faith is strongest in the "Bible Belt" of the Midwest and the South, yet there are more American Muslims than Episcopalians, Presbyterians, or Jews. Los Angeles has the greatest variety of Buddhists in the world. The US still cherishes the myth of the "melting pot", although there are two important exceptions. In the past, Native Americans were dispossessed and almost eradicated (although now there is growing interest in their traditions). For most African–Americans, their history is rooted in slavery. The practice was abolished in the 19th century, but it was not until the civil rights campaigns of the 1950s–1960s that the vexed issue of racial equality was properly addressed.

**ETHNIC MIX**
*Cities such as New York include immigrant communities from all over the world. Many groups live in specific areas, but otherwise people from every ethnic background travel and mix freely together.*

**OSCAR**
*The Oscar statuette symbolizes the global reach of the American film industry. The annual Oscar ceremonies promote the most successful films and film-makers from around the world.*

**BAND LEADERS**
*New Orleans brass bands typically play at funerals and parades. Their music was one of the forerunners of jazz and rock. It survived alongside these forms and is still popular today.*

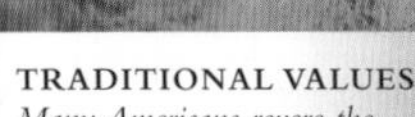

**RIDING THE FENCES**
*The cowboy is an iconic figure across the world, thanks to Western films. Real cowboys are less glamorous but still have a strong identity based on traditional horse and cattle ranching.*

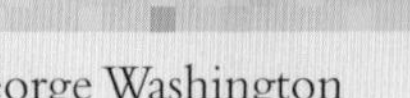

**TRADITIONAL VALUES**
*Many Americans revere the traditional community values still followed in small towns, such as respect for one's elders.*

Profile

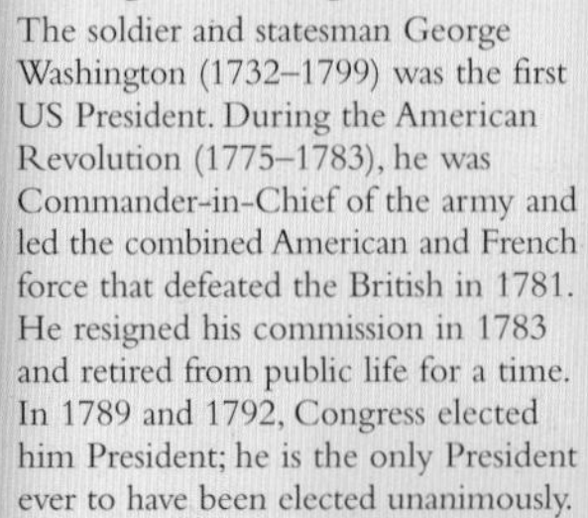

### George Washington

The soldier and statesman George Washington (1732–1799) was the first US President. During the American Revolution (1775–1783), he was Commander-in-Chief of the army and led the combined American and French force that defeated the British in 1781. He resigned his commission in 1783 and retired from public life for a time. In 1789 and 1792, Congress elected him President; he is the only President ever to have been elected unanimously.

**HONOURING THE FLAG**
*Respect for the US flag is an integral part of being American. Displaying the flag is an important element of celebrations such as this St Patrick's Day Parade in New York.*

NORTH AMERICA

## Canadians

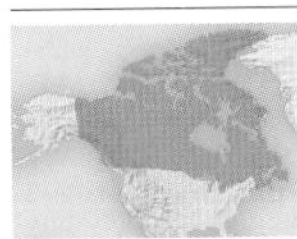

**Location** Canada

**Population** 31.3 million (people of British and French descent make up more than 50% of the population)
**Language** English, French
**Beliefs** Christianity: Roman Catholic, Protestant

Canada boasts a cosmopolitan population mix, but two ethnic subgroups are dominant: British Canadians and French Canadians.

From the 18th century onwards, Europeans flooded into the open expanses of Canada, while the Native Americans suffered disease, dispossession, and alienation. Most of the early migrants were from the UK, and continued political ties to the UK have given Canadian culture a distinctly European feel (although proximity to the US has inevitably exerted an influence).

Québec, an eastern province of Canada, formally became a French colony in 1608. The French ceded Québec to the British in 1763, but the French stayed. Today, the French-speaking inhabitants of Québec (the Québecois) are defiant about their distinct culture within Canada.

Recent immigration from Asia has helped the modern trend towards a new Canadian national identity. In the 2001 census, almost a third of respondents described their ethnicity simply as "Canadian".

Throughout the 20th century, Canadians adopted an increasingly suburban lifestyle. Around two-thirds of Canadians now live in just 5 per cent of the huge territory, clustered in cities along the St Lawrence Seaway and the nearby Great Lakes.

With a vast untamed hinterland that is mostly devoid of habitation and stretches to the North Pole, rural exploits, such as lumberjacking, are ingrained in the Canadian self-image. The Mounties (The Royal Mounted Police) provide an endearing and enduring symbol of the isolated frontier spirit.

Canadians are also well known for their devotion to ice hockey, a high-speed and often violent sport.

Issue

### Québec

Calls for a fully independent Québec have so far not been heeded, and in 2003 the separatist Parti Québecois was relegated to the opposition. However, social reforms in the 20th century favoured Québecois, and speaking French has been compulsory in Québec's government since 1976.

**TWO CULTURES**
*The French-influenced architecture of this hotel in Québec city contrasts with the more British uniforms of the guards.*

**NATIONAL PASSION**
*A group of Canadians on Alta Lake prove that ice hockey can be played indoors or outdoors.*

**ETHNIC MIX**
*Vancouver children show that Canadians originate from all over the world (especially eastern Asia and Europe).*

**TIMBER INDUSTRY**
*A worker fells a tree in one of the forests of British Columbia. Logging is an important primary industry in Canada.*

NORTH AMERICA

## Métis

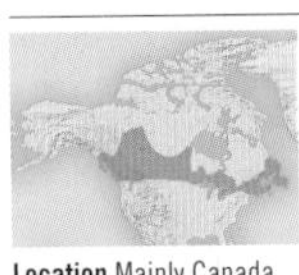

**Location** Mainly Canada
**Population** 100,000

**Language** English, French, Cree, Salteaux (a variety of Chippewa), Michif (a language with origins in English, French, and Cree), Athabaskan languages (some northern Métis)
**Beliefs** Roman Catholic

The Métis are the descendants of the offspring of French Canadian, Scottish, English, and eastern Native American fathers, and Cree, Chippewa, and Athabaskan mothers.

In the 18th century, many traders and trappers established alliances with the local Native Americans through marriage. Some went on to form communities in the Red River area of present-day Manitoba. Their offspring, the first Métis, became known as "Red River Half-Breeds".

Many Métis today are farmers, small businesspeople, and government employees, although in the past they were mainly hunters, trappers, and traders, as well as skilled craftsmen and cartwrights.

History

### Métis rights

Relations between the Métis and the Canadian government have not always been smooth. In 1972, a Métis leader and the (then) prime minister, Pierre Elliot Trudeau (below), met to discuss Métis land and treaty problems. In 1982, the Métis were acknowledged as an aboriginal group. This change enabled them to pursue their traditional rights legally.

ARCTIC AND SUBARCTIC

## Alaskan Athabaskans

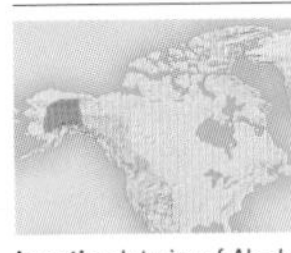

**Location** Interior of Alaska (US)

**Population** 5,600
**Language** Athabaskan languages
**Beliefs** Christianity: Episcopal, Russian Orthodox, Evangelical Protestant; traditional spirit beliefs

The Alaskan Athabaskans are one of the major groups of Native Americans in the Alaskan subarctic interior. These peoples were little disturbed by incursions from white people due to the harsh environment they inhabit. As a result, a large portion of Alaskan Athabaskans continues to fish and to hunt caribou and moose in the subarctic forest. Recently, some Alaskan Athabaskans have been forced to contest the rights to their lands with mining and logging companies.

Rifles and motorized transport have replaced bows, arrows, and snowshoes, but Alaskan Athabaskan hunters continue to follow a set of beliefs that includes having respect for the animals they hunt.

The Tanaina, one of the Alaskan Athabaskan peoples, have an unusual system of faith. They are members of the Russian Orthodox Church (the US purchased Alaska from Russia in 1867), but they believe in supernatural figures that are "shared" with other Athabaskans.

**DRYING CHAR**
*An Alaskan Athabaskan woman hangs fish out to dry. Fish is an important source of food for Athabaskans, and Arctic char is a common catch.*

ARCTIC

## Yupik

**Population** 21,000
**Language** Yupik, English, Russian (in Siberia)
**Beliefs** Christianity: Russian Orthodox (in both Alaska and Siberia), Catholic, Moravian; traditional spirit beliefs
**Location** Western Alaska, northeastern tip of Siberia

Although the coast-dwelling Yupik are linguistically closely related to the Inuit, they consider themselves to be separate. They do, however, feel they are the same as the Yuit, who live on Lawrence Island in the Bering Sea and on the northeastern tip of Siberia. Although the Cold War made passage across the Bering Sea difficult, the Alaskan Yupik have made "friendship visits" to the Siberian Yuit since 1950.

**TATTOOS**
*In Alaska, Christians forbade facial tattooing, but Siberian Yupik, such as this elder, continue the practice.*

Many Yupik are Russian Orthodox Christians, a legacy of the Russian discovery of Alaska in the 1700s and Yupik contact with Orthodox missionaries in the 1800s. Outsiders invaded Yupik country in greater numbers in the early 1900s with the discovery of gold in the Yukon and the advent of commercial fishing. The presence of white people caused influenza epidemics in 1900 and 1919, which decimated the Yupik people.

Since the 1960s, schools, health care, transportation, and social services, mostly subsidized by the government, have appeared in Yupik communities. Furthermore, cabins and wood-frame houses have replaced traditional sod houses. Nevertheless, the Yupik remain exceptionally isolated. Today, many Yupik fish, hunt, or work in government-funded employment, while a few make a living carving wood and ivory for the tourists.

**ICE FISHING**
*A shark is caught (for scientific purposes) through a hole in the ice, using ancestral Yupik expertise.*

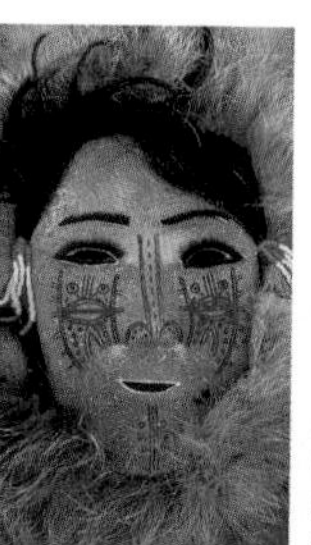

**DANCE MASK**
*Yupik masked dances honoured the entities that made Arctic life possible. This mask is made of caribou hide.*

ARCTIC AND SUBARCTIC

## Chipewyan

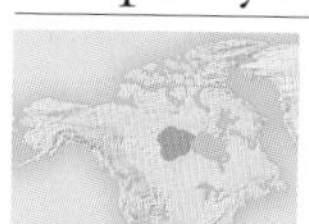

**Population** 5,000
**Language** Chipewyan (an Athabaskan language), English
**Beliefs** Roman Catholic Christianity, spirit beliefs (including faith in visions and dreams)
**Location** Canada (western subarctic forests)

Many Chipewyan remain committed to the fishing and caribou-hunting way of life of their ancestors, although they are sedentary or seminomadic, and use mechanized transport and rifles. They also observe a number of traditional practices, including the strict segregation of boys' and girls' playgroups from the age of 10 years.

The erosion of the Chipewyan's nomadic culture began in 1717 when they encountered English commercial entrepreneurs. Soon, the Chipewyan became important participants in the subarctic trade, exchanging furs and hides for guns, metal tools, and cloth.

In earlier times, the Chipewyan were organized into small, nomadic bands. They lived in tepees and wore one-piece trouser-and-moccasin outfits, but bought clothing, tents, and log cabins have replaced these.

ARCTIC

## Inuit

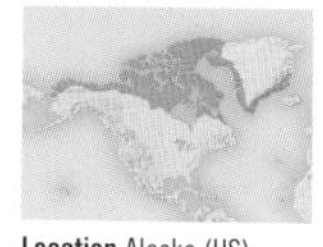

**Population** 100,000
**Language** Inuktitut, English, Danish (in Greenland)
**Beliefs** Anglican, Moravian, Greenlandic Evangelical Lutheran, traditional spirit beliefs
**Location** Alaska (US), Arctic Canada, Greenland

Inuit, along with related Arctic peoples, are also known as Eskimo, but some interpret this word to mean "eaters of raw flesh", and many reject the label. Arriving across the Bering land bridge in around 3000BC, the Inuit now inhabit a vast area of the Arctic, but their language and culture is fairly consistent. Inuit did once, however, live in different types of houses: rectangular, earth-covered homes (Alaska) or domed dwellings made of stone, earth, and whalebone (Greenland). In Canada, Inuit still sometimes build their temporary ice shelters called igloos, but they have always lived for part of the year in more permanent homes.

**SETTLED LIFE**
*These Inuit people live in a modern settlement on Victoria Island in the Canadian Arctic.*

Inuit subsisted mainly on fish, caribou, and moose and still eat these animals today. Whaling is also a vital part of the economy, and recently, Inuit have contested International Whaling Commission quotas.

The Inuit traditionally travelled over snow and ice in dogsleds and by sea in either the kayak or the larger *umiak*. Today, however, many Inuit use snowmobiles, rifles, and modern camping equipment. The Inuit also once utilized walrus and whalebone tools, often carved with exquisite care. However, they now make these mainly for the art market.

Traditional Inuit religion is organized around complex shamanic beliefs, many of which are described in a well-developed oral tradition that features epic tales and song duels.

**HUNTING EXPEDITION**
*An Inuit hunter, from the Qaanaaq (Thule) District of Greenland, traverses the ice using traditional means: a sturdy sled and a large team of dogs.*

**ICE HOUSE**
*Some Canadian Inuit still build ice igloos for shelter during hunting expeditions.*

**SUN HAT**
*This elaborate wooden helmet was once used to protect the wearer from the effects of the sun.*

PACIFIC NORTHWEST

## Tlingit

**Location** Alaskan panhandle (US)

**Population** 25,000
**Language** Tlingit (a language related to Haida), English
**Beliefs** Russian Orthodox, other Christian denominations, traditional spirit beliefs

Many modern Tlingit men are fishers, and women work in canneries and produce handicraft items – such as woven baskets and wooden carvings – for tourists. The Tlingit originally followed a complex religion revolving around a clan system traced through the maternal line. There were two "super-clans" called Raven-Crow and Eagle-Wolf, and marriage within a super-clan was forbidden. The Tlingit were also known for the 18m (60ft) log ships from which they mounted fishing expeditions and war parties. In 1971, Tlingit activism gave rise to the Alaska Native Claims Settlement Act, which returned 40 million hectares (100 million acres) of land to Alaskan native peoples.

**TOTEM POLE**
*Tlingit social organization is expressed in the crests represented in masks and carved totem poles, such as this "owl pole".*

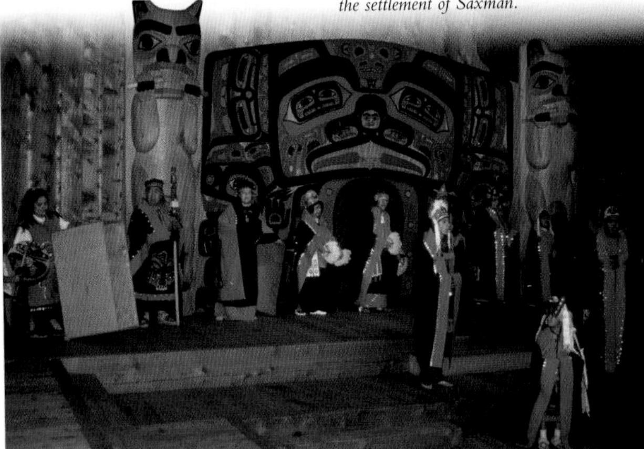

**REVIVING TRADITIONS**
*Tlingit dancers perform in the Beaver Clan House, part of a recreated Tlingit village in the settlement of Saxman.*

PACIFIC NORTHWEST

## Makah

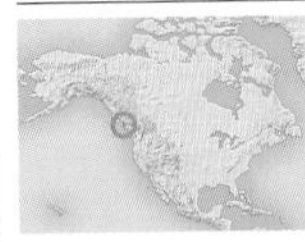

**Location** Neah Bay (Washington state, US)

**Population** 1,000
**Language** English, Makah (one of the 7 languages, of the Wakashan language family, which also includes Kwakiutl and Nootka)
**Beliefs** Christianity, traditional spirit beliefs

The Makah are a Wakashan-speaking people, one of a number of groups who constitute the Nootka. Originally, they had a class system incorporating nobles, commoners, and slaves, and Makah chiefs, who enjoyed ceremonial privileges and the ownership of salmon streams. They also engaged in a form of potlatch (*see below*). The Makah are traditionally fishermen and whale hunters, but recently there has been conflict with the International Whaling Commission over their right to hunt whales.

Following World War II, the Makah, who had ceded their land to the US government in 1855, bought an old army camp at Neah Bay in the Cape Flattery area of Washington and created a reservation there. For decades the bay was a commercial fishing port; now it is a sport-fishing centre.

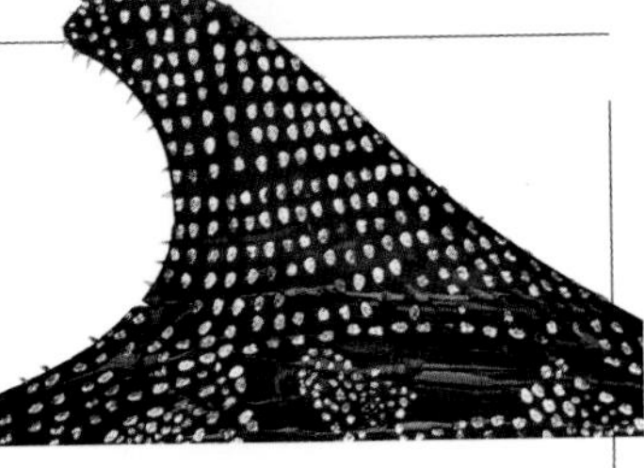

**WHALE-FIN ORNAMENT**
*A whale fin, inlaid with hundreds of sea-otter teeth, probably belonged to a person of high status – perhaps the whaling captain.*

The Provincial Museum in Victoria (British Columbia) has helped the Makah and other Nootka peoples to preserve their culture and language.

**WOLF DANCE**
*The wolf has sacred meaning for the Makah. Members of each family don carved wooden wolf masks and blankets to perform a unique wolf dance, specific to their clan.*

PACIFIC NORTHWEST

## Haida

**Location** Alaskan panhandle (US), Haida Gwaii islands (Canada)

**Population** 3,500
**Language** English, Haida (a language related to Athabaskan languages; almost extinct)
**Beliefs** Christianity, traditional spirit beliefs, (both with an emphasis on reincarnation)

The Haida are culturally related to the Tlingit and Tsimshian peoples. Their language, Haida, is thought to be related to Athabaskan tongues. Haida society was organized around a clan system with stratified ranks of noble, commoner, and slave. There were two major clans, determined through the maternal line. The clans were further subdivided into smaller local groups, or lineages, which each had their own chief and functioned independently from the other lineages.

The Haida traditionally relied on hunting and fishing (particularly for cod, halibut, and salmon). They were also skilled woodworkers and were widely known for their finely crafted canoes and elaborately carved totem poles (featuring crests representing significant family events). The Haida, in common with the Tlingit and other northwestern coastal peoples, held potlatch ceremonies, during which important social and political issues were decided (*see right*).

The arrival of Europeans in 1774 (and the later arrival of mass-produced goods), brought many changes for the Haida people. From the late 18th century up until the early 20th century, smallpox was a recurrent problem. Loss of traditional land was another difficulty, and the Haida of British Columbia and Alaska have, during the course of the last century, been involved in several land disputes. In Alaska, for example, a reservation established for the Haida peoples in the 1930s was dissolved in 1952. However, in 1971, the Alaska Native Claims Settlement Act gave three village corporations representing the Haida and neighbouring groups a total of 9,300 hectares (23,000 acres) of land.

In recent years, the Haida of two communities living on reserves at Masset and Skidegate (in the Queen Charlotte Islands, now known as Haida Gwaii) have formed the Council of the Haida Nation. They have revived many of the Haida arts, including the manufacture of totem poles and canoes. They have also reintroduced traditional dances.

**HAIDA DANCE**
*Holding a decorated paddle, a youngster from Haida Gwaii (Queen Charlotte Islands) dances to the drumbeat.*

**CREATION MYTH**
*This carving by noted Haida artist Bill Reid shows the raven delivering the first people into the world by releasing them from a mussel.*

Fact

### Potlatch

In common with the Tlingit and other northwestern coastal peoples, the Haida held potlatch ceremonies. Potlatches involved nobles wearing an elaborate headdress (below) competing for prestige by giving away or destroying valuable property. The Haida political economy, based largely on potlatch, was destabilized when mass-produced trade goods appeared in the late 19th century.

NORTHEAST

## Innu

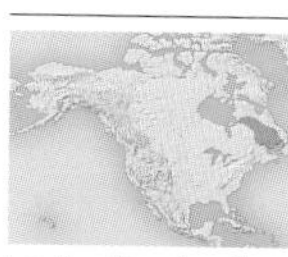

**Population** 12,000
**Language** Montagnais, Naskapi (both Native American languages of the Algonkian family), English
**Beliefs** Traditional spirit beliefs
**Location** The subarctic Labrador region (Canada)

The Innu, a group consisting of the Montagnais and Naskapi peoples, is the Algonkian-speaking people living in Canada's Labrador subarctic. The name Innu means "human being".

Prior to the 19th century, the Innu lived in skin tents and hunted caribou, which provided essential clothing and food. In the late 19th and early 20th centuries, however, there was competition for caribou from white fur trappers. In the 1930s, fur prices fell and caribou herd sizes dropped dramatically, causing terrific suffering for the Innu.

Since the 1950s, the Innu have been subject to several government-sponsored relocation schemes, which have produced spectacular social collapse. In a two-and-a-half year period between 1991 and 1994, there were 203 attempted suicides in the Mushuau Innu community of Utshimassits (population: less than 500). In 1993, the Canadian Commission of Human Rights declared the Innu "the victims of ethnocide or cultural genocide" by the Canadian and Newfoundland governments.

Today, many Innu retain at least part of their traditional relationship with the land. They fish and hunt caribou just like their ancestors did, but they have better technology to negotiate their inhospitable domain, such as guns, steel tools, traps, and motorized transport.

A number of Innu are now fighting for rights to the lands that were once theirs. In recent years the Innu also successfully resisted the foundation of a NATO Tactical Fighter Weapons Training Centre at Goose Bay, citing low-flying bombers as the reason for the decreased numbers of caribou.

**SNOWSHOE**
*Wide, sturdy snowshoes were essential to the Innu when travelling over snow. They consisted of an animal skin over a wooden frame.*

**SMOKED FOOD**
*At an Innu hunting camp in southern Labrador, a woman smokes beaver meat – a traditional Innu food – over an open fire.*

**HUNTING BY CANOE**
*A hunter travels by motorized canoe over Burnwood Lake, southern Labrador. Hunting is very important to the Innu culture.*

NORTHEAST

## Micmac

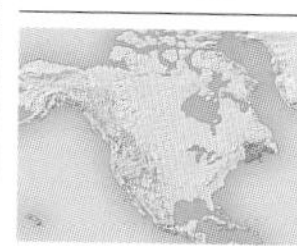

**Population** 11,000
**Language** Micmac (a language of the Algonkian family), English
**Beliefs** Traditional spirit beliefs (including faith in one supreme being and in several lesser gods), Catholic Christianity
**Location** Canadian Maritime provinces, Gaspé Peninsula (Québec), Maine

Formerly nomadic fishermen and hunters, the Micmac are now mostly wage labourers. Many of them work in construction or in factories, and some are employed in the commercial fishing industry or otherwise own small businesses. Attempts over the last hundred years to transform the Micmac into farmers have generally failed (although in earlier decades a few took up potato raising). When a Micmac dies, the deceased's home is occupied until after the burial. This prevents the soul of the departed returning to the abode. The Micmac were possibly the first Americans to encounter Europeans.

**MOCCASINS**

**HARVESTING WILD RICE**
*While one man pushes the canoe through the wild rice plants, the other uses sticks to knock the rice grains into the boat.*

NORTHEAST

## Chippewa

**Population** 160,000
**Language** Chippewa, Ojibwa (both Native American languages of the Algonkian family), English
**Beliefs** Roman Catholic and Anglican Christianity, traditional spirit beliefs
**Location** Minnesota, Wisconsin, Michigan, Manitoba, Quebec, Ontario

The Chippewa, historically also known as Ojibwas, are concentrated north and west of the Great Lakes, both in reservations and in major cities. Some Chippewa of Minnesota's White Earth Reservation harvest and market wild rice just as their forebears did. At the Walpole Island Reserve in Ontario, the population includes not only the Chippewa but also the Ottawa and Potawatomi – descendants of refugees from Michigan, Indiana, and Ohio. The three peoples have a very close relationship, and their community is called The Three Fires.

The Iroquois and Sioux were the enemies of the Chippewa people in early times. But, with their distinctive birch-bark canoes, the Chippewa were the masters of their vast land of lakes and rivers. Periodically, the Chippewa fought the white people, notably alongside Chief Pontiac's Ottawa in 1763 and with Chief Tecumseh's Shawnee in the years leading up to the British-US conflict of 1812. Their last battle with the US Army took place in 1898, at Leach Lake (Minnesota).

**RITUAL DRESS**
*Chippewa men sometimes wear a wolf pelt on special occasions. The wolf is thought of as a "helper spirit".*

NORTHEAST

## Iroquois

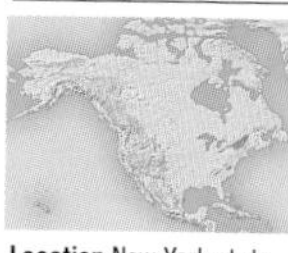

**Population** 20,000
**Language** Iroquoian languages, English
**Beliefs** Handsome Lake Religion (a mixture of traditional Iroquois and Quaker beliefs), other Christian denominations
**Location** New York state (US), Quebec (Canada)

The Iroquois League, also known as The Five Nations, comprised Mohawk, Seneca, Onondaga, Cayuga, and Oneida peoples, all linked by related languages and a common clan system. It was founded in the late 16th century. In 1710, the League admitted the Tuscarora people, and from 1722 onwards The Five Nations became known as The Six Nations.

The Iroquois traditionally lived in rectangular, barrel-roofed longhouses and, as well as hunting, planted beans, maize, and squash (the "three sisters"). Women played an important role in political processes.

Today, many Iroquois work in New York, where there are more jobs than on the reservations. Mohawk construction workers, because of their lack of vertigo, played a large part in constructing the New York skyline.

**MOHAWK DRUMMERS**
*On the streets of Montreal, Mohawk drummers take part in a display of traditional dancing and music making.*

NORTHEAST

## Mennonites

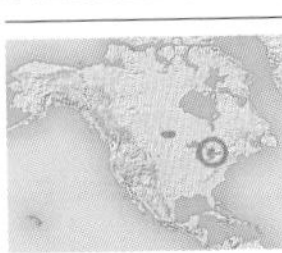

**Population** 400,000
**Language** English, German (spoken by some conservative Mennonites)
**Beliefs** Mennonite Anabaptist Christianity (a strict religion that maintains a central tenet of adult baptism)
**Location** Central Canada; Pennsylvania, Ohio (US); Belize; Paraguay

The Mennonites originated within the Anabaptist movement, a radical organization that started in 1525 in Switzerland. Some Mennonites came to the US as early as 1683, but most emigrated during the 19th century. They came from German-speaking communities in Russia, Poland, and Prussia, many settling as farmers in the Manitoba prairies (central Canada).

Traditional Mennonites reject infant baptism in favour of adult baptism, and live in communities organized around *Ordnung* (strict convention). Some forbid the use of instruments and vocal harmony in church and still sing hymns in German. Today, such conservatism is rare, and only around one-third of Mennonites live on farms. Some parents, however, still withhold their children from secondary school, insisting they be socialized at home.

Although the Mennonites reject conflict and discourage relations with other communities, they have often attracted hostility. During World War II, many registered as conscientious objectors. Over this period, some Mennonite schools were closed because teachers taught in German rather than in English, and vigilantes painted Mennonite houses yellow to represent supposed cowardice.

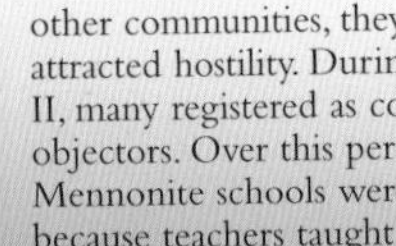

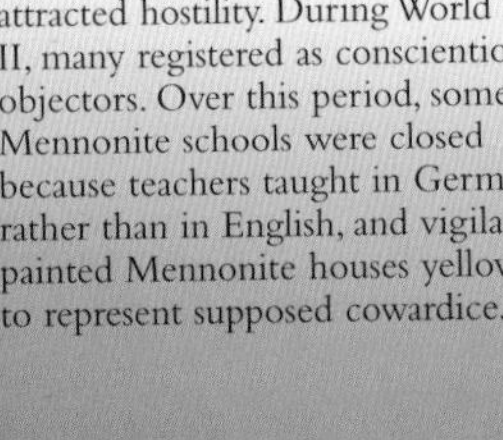

**FARMING FAMILY**
*Conservative Mennonites continue to live a simple rural life. They eat plain food and wear unsophisticated clothes.*

**RURAL TRANSPORT**
*A conservative Ontario group of Mennonites shuns motorized transport, preferring horse-drawn vehicles.*

NORTHEAST

## Amish

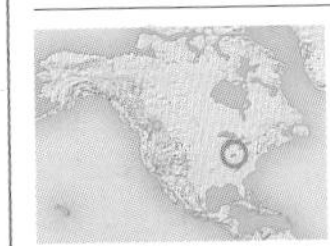

**Population** 130,000
**Language** English; also a dialect of German
**Beliefs** Amish Anabaptist Christianity (a strict religion that maintains a central tenet of adult baptism)
**Location** US (Pennsylvania, Ohio, Indiana, New York)

The Amish (also called Pennsylvania Dutch) were originally a Mennonite splinter group. They began to migrate to North America in the early 18th century. They are best known for their conservative clothing and lifestyle. They shun the use of cars, telephones, and electricity and rely on horses for farming. The Amish mostly live in strict, religious communities. They have little to do with the outside the world, and many speak a dialect of German.

A number of Amish forbid their children more than a primary-school education. In the past, some parents were sent to prison because of this. Since 1972, however, it has been recognized that the refusal of Amish parents to permit their children to have a secondary education is based on religious principle and therefore constitutes part of their human rights.

Within Amish society, adherence to *Ordnung* (convention) is important, although the most severe punishment for infringement is *Meidung* (shunning).

**LEARNING TO SEW**
*In traditional Amish communities the role of women has not changed: their place is in the home. Therefore skills, such as sewing, are passed down from mother to daughter.*

**AMISH QUILT**
*A maple-leaf design is a typical Amish pattern.*

**RAISING A BARN**
*Amish men build the frame of a new barn, without the benefit of modern technology.*

NORTHEAST

## Potawatomi

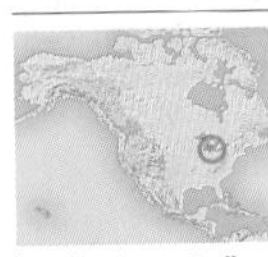

**Population** 28,000
**Language** Potawatomi (an Algonkian language closely related to Chippewa), English
**Beliefs** Several Christian denominations, some traditional beliefs
**Location (ancestral)** Michigan, Wisconsin (US)

In 1846, most Potawatomi were relocated from their ancestral home in Michigan and Wisconsin to Kansas. There are now different bands of Potawatomi. The Citizen band, who were moved from Kansas to present-day Oklahoma, largely assimilated into US culture. However, the Prairie band, those who remained in Kansas, retained many traditional values. In earlier times, the Potawatomi survived by hunting, fishing, and gathering wild rice.

**TOBACCO POUCH**
*Colourful, ornate pouches were originally used to carry tobacco, which had a ceremonial function.*

NORTHEAST

## Cree

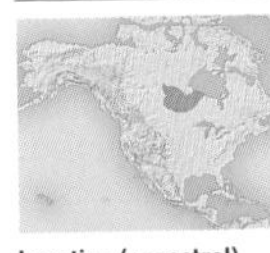

**Population** 150,000 (mainly scattered groups)
**Language** Cree (which is one of the Algonkian languages), English
**Beliefs** Roman Catholic, traditional spirit beliefs
**Location (ancestral)** NE subarctic Canada

The Cree once occupied a vast area in the northern tundra, the eastern forest, and the northern plains of Canada, fishing and hunting moose, caribou, or buffalo. Some Cree are still engaged in these pursuits, but now they do this on a commercial basis rather than for subsistence. Likewise, saddle horses, motorized toboggans, and pickup trucks have replaced the traditional pack animals and dogs.

During the mid-20th century, the Cree – who formerly lived in small, nomadic bands – were concentrated into villages. This policy resulted in high unemployment, alcohol abuse, and other social problems. In recent years, the discovery and exploitation of oil on Cree lands has been a source of tension between the Cree and the oil companies (and among the Cree themselves).

The Cree were enemies both of the Chippewa to the north and of the powerful Blackfeet to the south and west. For many years the Cree withheld the trade in firearms to both, eventually driving the Blackfeet from much of the Canadian plains country. The Cree also played an important role in the westward spread of the fur trade.

**CREE WINTER TENT**
*A kerosene lamp illuminates a trapper's winter tent. The trapper's equipment is silhouetted against the tent wall.*

**CREE FACE**
*Dark hair is a typical feature that testifies to Cree origins.*

**HUNT MEETING**
*In James Bay (north Québec) hunters meet to discuss tactics in advance of an autumn goose hunt.*

GREAT PLAINS

## Sioux

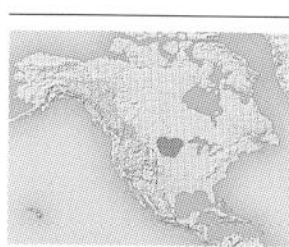

**Population** 153,000
**Language** Dakota, Nakota, and Lakota (varieties of a common Siouan language); English
**Beliefs** A number of Christian denominations alongside traditional spirit beliefs
**Location** Prairie provinces (Canada); northern Midwest states (US)

Many of the people known as the Sioux prefer to call themselves Lakota, Dakota, or Nakota, depending on the dialect of Sioux spoken. In around 1750, they were forced west onto the northern prairies in a domino effect of migrating peoples started by Europeans settling in New England. The Sioux then roamed with the buffalo across the grasslands. From 1750 to 1881, the popular image of Sioux as buffalo hunters and horseback warriors was a reality. Since then, settlers from Europe have decimated the buffalo herds and moved the Sioux to reservations with few jobs. Many Sioux moved to cities and left much of their culture behind. One trend today is towards the Sioux reclaiming their land in the Dakotas for buffalo ranching.

Sioux no longer live in tepees or hunt for a living, but they still smoke the ceremonial pipe. Sioux culture, along with the image of a mounted Great Plains warrior, is a focus of pan-Native-American identity.

**SIOUX WOMAN**
*A Lakota woman wears ethnic dress and a feather in her hair.*

History

### Wounded Knee

As the bison herds dwindled during the 1880s, many Sioux turned in desperation to the new Ghost Dance religion (*see* p359). Whites felt threatened by the ghost dancers and this tension led to the massacre at Wounded Knee, South Dakota, on 29 December 1890, an event that is now recognized by the Oglala Sioux Nation as a tribal holiday.

SIOUX GHOST DANCE

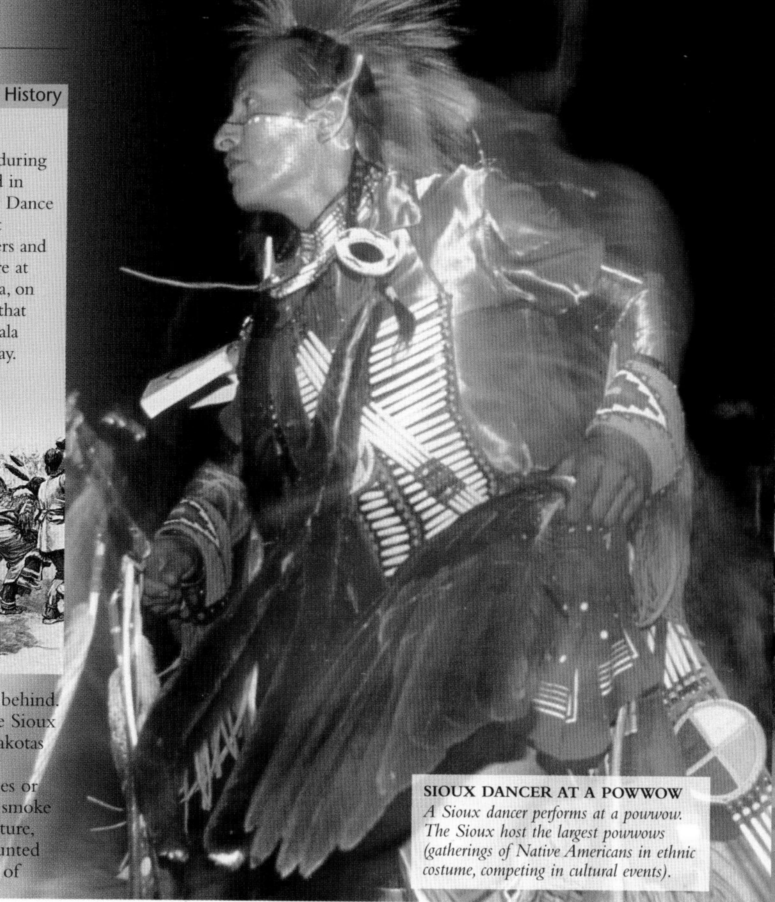

**SIOUX DANCER AT A POWWOW**
*A Sioux dancer performs at a powwow. The Sioux host the largest powwows (gatherings of Native Americans in ethnic costume, competing in cultural events).*

GREAT PLAINS

## Blackfeet

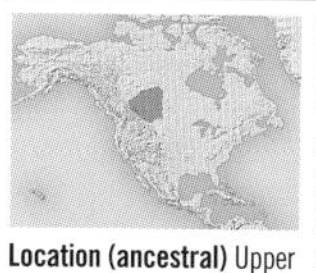

**Population** 25,000
**Language** Blackfeet (an Algonkian language); also English
**Beliefs** Roman Catholic Christianity, traditional spirit beliefs
**Location (ancestral)** Upper Missouri (US, Canada)

Most Blackfeet live in reservations in Alberta, Canada, and Montana, US, in areas that were once part of their traditional lands. Some raise cattle and sheep, but many exist in poverty and dependency. A large number also live and work away from the reservation, a trend that began in World War II when hundreds of Blackfeet migrated to Canadian cities to work in aircraft plants and shipyards.

In the 18th century, the Blackfeet, allied with both the Atsina and the Sarsi peoples, dominated the Upper Missouri region. They were swift to adapt to the arrival of firearms and to horse riding, and they were greatly feared by other local tribes – as well as traders and trappers. However, the Blackfeet were decimated by smallpox epidemics and by starvation (when the numbers of buffalo declined in later decades).

**REVERED BUFFALO**
*A decorated buffalo skull is an expression of the veneration of the buffalo by the Blackfeet.*

GREAT PLAINS

## Cheyenne

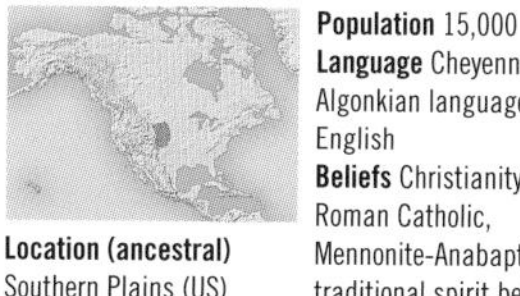

**Population** 15,000
**Language** Cheyenne (an Algonkian language), English
**Beliefs** Christianity: Roman Catholic, Mennonite-Anabaptist; traditional spirit beliefs
**Location (ancestral)** Southern Plains (US)

The Cheyenne live on two US reservations: one in Oklahoma (Southern Cheyenne), the other in Montana (Northern Cheyenne). In Montana, logging, ranching, and growing crops such as alfalfa are the most important economic activities, while in Oklahoma, raising wheat, ranching, and the exploitation of oil provide the greatest amount of employment. Many people, however, are entirely dependent on government assistance. Marginalization and social deprivation cause acute problems on both reservations.

The Cheyenne have a vibrant ritual life. They perform ceremonies such as the Renewal of the Sacred Arrows, which is a ritual whereby sacred arrows empower the men of the tribe. Even today, following a death, Cheyenne people distribute a person's property before he or she is buried, according to their tradition.

**BATTLE VICTORY**
*Warriors celebrate victory at a reenactment of the Battle of the Little Big Horn.*

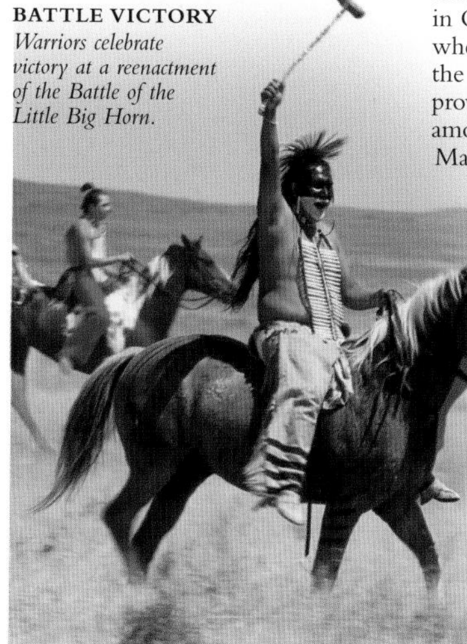

Profile

### Little Wolf

Little Wolf (*circa* 1820–1904) was a chief of the Northern Cheyenne. In 1877, the government moved the group to a reservation in Indian Territory (today's Oklahoma). In September 1878, Chief Little Wolf left with 355 people to walk the more than 650km (400 miles) home. En route they were overtaken by a cavalry force. They surrendered in March 1879.

**BUFFALO CLOAK PAINTING**
*A robe celebrates the 150th anniversary of a treaty made in 1851 between the US and the Cheyenne (and other Native Americans).*

The Cheyenne were once nomadic bison hunters and horse raiders. They participated in several conflicts with the US, which culminated in Custer's defeat by the Northern Cheyenne and Sioux at the Battle of the Little Big Horn (1876).

GREAT PLAINS

# Comanche

**Population** 12,000
**Language** English, Comanche (nearly extinct, spoken by only 1% of tribal members, all of whom are elders)
**Location (ancestral)** Southern Great Plains (US)
**Beliefs** Christianity, traditional spirit beliefs

The Comanche today are centred on the town of Lawton in Oklahoma. Many make a living by farming and the lease of mineral rights. In the past, however, they were buffalo-hunting nomads and the most feared tribe on the southern plains. The modern Comanche are proud of their illustrious past and, as well as organizing regular powwows (cultural celebrations) for tribal members, have recently devised a flag that expresses symbolically their pride in their heritage. They have also begun to issue Comanche car license plates that identify cars registered to tribal members living on tribal lands.

The Comanche have had to work hard to sustain their tribal integrity. In 1901, their reservation was broken up into allotments of 160 acres per person, the remainder being opened up to white settlement. Furthermore, after World War II many Comanches dispersed throughout the US in order to work in the defence industry. As a consequence, many of the tribe's traditional institutions no longer exist. Today, one of the major concerns of Comanches is the survival of their language, which at present is spoken by only 1 per cent of tribal members, all of whom are elders. From the late 19th century until the mid-20th century Comanche children, like many Native American children of the time, were removed from their homes, sent to boarding schools, and punished for speaking their language. The US government discovered the worth of the Comanche language during World War II when some Comanches were employed by the armed forces as wireless operators, their language being used as a secret code. Little, however, was done to rescue the language until in 1993 the Comanche Language and Cultural Preservation Committee was formed.

**TODAY'S CHIEF**
*Today's Comanche leaders such as Wallace Coffey (left) continue to promote Comanche culture and interests.*

History

## Ghost dance movement

In the summer of 1890, many Native American peoples of the West, including the initially sceptical Comanche, heard of the teachings of the Paiute prophet Wovoka. Some believed that Wovoka was the Christ of whom missionaries had spoken, and many adopted his Ghost Dance, believing it would bring back the buffalo, resurrect the dead, and get rid of the White people.

**VISIONARY OBJECT**
*A club of the Arapaho people is carved with the face of a spirit helper a ghost dancer had seen in a vision.*

**CALUMET**
*This stem and pipe, or calumet, was a universal symbol used for many different ceremonial purposes.*

**POWWOW**
*A Comanche dancer performs at a powwow as part of the Oklahoma Red Earth Festival. The Comanche were instrumental in developing the powwow in the southern plains in the 20th century.*

The Comanche language is closely related to Shoshone. The languages diverged in the late 17th and early 18th century when the ancestors of the Comanche, who had by now started to ride horses, swept out of the Rocky Mountains to follow the buffalo herds, driving the eastern Apaches off the southern plains. During the 18th and 19th centuries, the Comanche were considered by many to be the most skilled and renowned horsemen on the continent, and they were considered fearsome warriors. The 19th-century advance of the Texas frontier, and the migrations of settlers to California in the years following the discovery of gold in 1849, placed considerable pressure on Comanche hunting grounds. Many bands, along with their Kiowa allies, took to war – marking the beginning of a battle that continued to rage for over 25 years (*see right*).

Comanche continue to affirm their identity with rituals and powwows. The Comanche Homecoming Powwow has been held annually ever since 1952, when Korean War veterans were welcomed home with a traditional Victory Dance.

Profile

## Quanah Parker

Hostilities between the Comanche people and the US ceased only in 1875 with the surrender of Chief Quanah Parker. Parker, who was the son of a white captive woman and Chief Nokoni of the Quahadi Comanches, subsequently became the first chief of all the Comanche, a justice on the US Court of Indian Offenses, an advocate of "white civilization", and a friend of President "Teddy" Roosevelt.

PLAINS

## Mandan

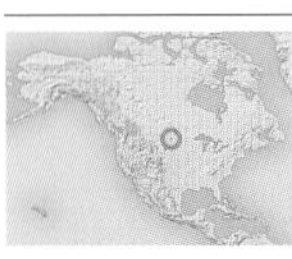

**Population** 1,000
**Language** English, Mandan
**Beliefs** Roman Catholic, Congregationalist, traditional spirit beliefs (with ceremonies such as the sun dance and the bear ceremony)
**Location (ancestral)** Upper Missouri (US)

The Mandan, originally one of three settled horticulturist groups along the upper Missouri River, traditionally planted maize, pumpkins, and squash and traded with the nomadic peoples of the plains. Today, the Mandan live on a reservation at Fort Berthold, North Dakota. Some Mandan still engage in agriculture but many have sold or leased their lands to white ranchers.

**MANDAN WOMAN**
*An elegant Mandan woman wears the traditional dress of her people.*

SOUTHEAST

## Cherokee

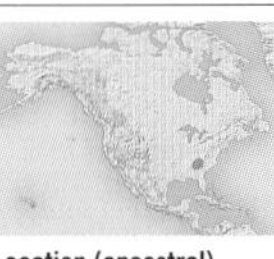

**Population** 130,000 (in the early 18th century there were 20,000, living in 64 communities)
**Language** Cherokee, English
**Beliefs** Mostly Baptist
**Location (ancestral)** Southeastern US

There are two Cherokee branches: the Eastern Cherokee (southeastern US), and the Western Cherokee (Oklahoma). The eastern division practise subsistence farming, as their ancestors did, although they also work on logging camps and seek seasonal wage labour; some receive government assistance. Many in the western division rent their land to ranchers. In both divisions, however, poverty is widespread, and three-generation households are common.

Profile

### Sequoyah

George Guess, or Sequoyah, a Cherokee man born in 1776, enlisted in the US army in the early 1800s. He saw that, unlike the white soldiers, he and the other Cherokees were unable to write letters home or read military orders. He therefore set about creating a Cherokee alphabet. By the 1820s, not only many people, but also many publications were using it, including the first Native American newspaper, the *Cherokee Phoenix*, in 1828.

In the 18th century, the Cherokee rapidly expanded, defeating the Tuscarora, Shawnee, Catawba, and Creek in a series of wars that gave them supremacy over much of the southeast. However, defeat by the Chickasaw in 1768, and later wars with white people, weakened their influence. In 1828, the Georgia Cherokee were forced to move to Indian Territory (in present-day Oklahoma). During the 1,300-km (800-mile) "Trail of Tears", over 4,000 people died.

**LACROSSE**
*The Cherokee enjoy lacrosse. The traditional equipment was a netted stick and deerskin ball.*

**ANCIENT CRAFT**
*In a recreated Cherokee village, a woman demonstrates the art of traditional belt weaving.*

**INTRICATE WEAVING**
*The Cherokee are skilled weavers, producing textiles such as this patterned blanket.*

SOUTHEAST

## Sea Islanders

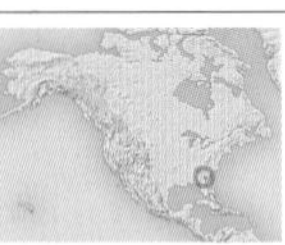

**Population** 125,000
**Language** English, Gullah (the only surviving form of a plantation Creole once spoken over much of southern US)
**Location** Islands off the southeastern US coast
**Beliefs** Predominantly Baptist and Methodist

The Sea Islanders, who live on the coastal islands from South Carolina to northern Florida, are better known as Gullah or Geechee, though the Islanders reject these names.

The sea islands are a repository of African culture from the time of slavery, as exhibited by the myriad myths and beliefs of their inhabitants. However, resorts and residential developments are now displacing the African-American population in many parts of the region, and the Sea Islanders are in danger of losing their culture and language. Traditionally, Sea Islanders regarded themselves as fishermen and "rivermen", but historically, many worked for logging companies and on mainland docks.

**WEST AFRICAN RHYTHMS**
*Dreadlocked Sea Island drummers perform at the 5-day Gullah festival, which is held every year in South Carolina.*

SOUTHEAST

## Louisiana Creoles

**Population** 100,000 (living mainly in downtown New Orleans and Bayou Têche)
**Language** Louisiana French, or Creole; English
**Beliefs** Roman Catholic Christianity, Vodou, spiritualism
**Location** Southern Louisiana (US)

The Creoles of southern Louisiana are the descendants of people from a number of groups, including African slaves, French and Spanish colonists, Native Americans, and Haitians. The original Creole language came from French, and many Creole people still speak a form of this language today, although it is different from that spoken by their neighbours, the Cajuns.

Creole culture in southern Louisiana is evident in the seafood element of its cuisine and in the local music, notably in some forms of New Orleans jazz and in the music known as zydeco. Zydeco bands are characterized by accordion-playing, the use of a *frottoir* (metal washboard), played with thimbles, spoons, and bottle openers, and the singing of soul and rhythm and blues in Creole French.

Religious practices, although mainly Catholic, also incorporate spiritualism and Vodou. Creoles traditionally advance themselves within society through membership of Mardi Gras carnival crews, burial societies, and a black Catholic society called The Knights of Peter Claver. The Mardi Gras Indian parades (when Creoles dress extravagantly in the attire of imaginary Native American tribes) and the festival known as the Zulu Parade (which parodies key differences between black and white people) are two distinctly Creole institutions.

**ORNATE COSTUME**
*A Creole wears a flamboyant costume as part of the "Creole Wild West" tribe, one of the many involved in the New Orleans Mardi Gras festival.*

SOUTHEAST

## Lumbee

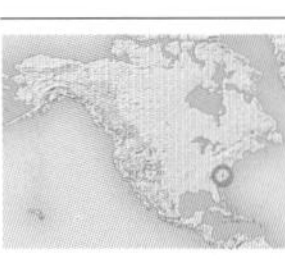

**Population** 30,000
**Language** English (Lumbee were probably originally familiar with Algonkian, Eastern Siouan, and Iroquoian languages)
**Location** Robeson County, (North Carolina, US)
**Beliefs** Baptism, other Christian denominations

Today, most Lumbee live in Robeson County in North Carolina, many as low-paid tenant farmers and factory workers. The ancestors of the Lumbee were probably remnant groups of a number of mainly Siouan-speaking peoples, including the Catawba, Quapaw, Tuscarora, and Yamasee. After the Civil War, the Lumbee insisted on their legal identity as Native Americans. Today, a minority reject the label Lumbee and insist on Tuscarora, a name that they feel reflects their Indian past.

**CEREMONIAL ATTIRE**
*Lumbee ceremonial dress includes an elaborate headband that partially covers the face.*

PEOPLES

SOUTHEAST

## Cajuns

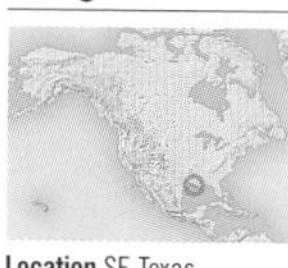

**Location** SE Texas, southern Louisiana (US)
**Population** 800,000 (descendants of around 18,000 Catholic French Canadians)
**Language** Cajun French (still widely spoken today), English
**Beliefs** Roman Catholic

The Cajuns are the descendants of the French colony of Acadie, which was founded in the present-day Canadian provinces of Nova Scotia and New Brunswick. Refusing to swear allegiance to the Crown after the fall of Canada to the British in 1755, the Acadians were expelled. Over the next 20 years, some of these people congregated in the sparsely populated bayou country (marshy land) around the Mississippi River. They formed small farming, fishing, and trapping communities and, rather remarkably, remained isolated for almost 200 years.

The Cajuns have developed a distinctive cuisine that includes jambalaya (rice with chicken and shrimps), gumbo (okra soup), and crawfish. They also have an emotionally charged music, sung in Cajun French. This includes waltzes and two-steps in which the fiddle, accordion, and triangle feature. Today, Cajun people are no longer entirely separated from mainstream North American culture, but they still usually live in small, self-contained communities. They are proud of their identity and many still speak French.

**MARDI GRAS**
*Chickens are captured during the Cajun Mardi Gras celebrations.*

**MAKING MUSIC**
*The fiddle is often a key feature of Cajun music, which includes waltzes and two-steps.*

SOUTHEAST

## Creek and Seminole

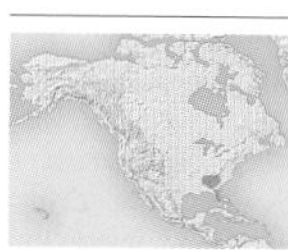

**Location** Oklahoma, Florida (Seminole only)
**Population** 61,000
**Language** Creek or Seminole (closely related dialects of the Muskogean language family), English
**Beliefs** Baptist and other Christian denominations

The Creek are Native Americans who in the 1830s were moved from Georgia and Alabama to Oklahoma. In the 1700s, a group of Creek migrated to Florida and became known as Seminole ("runaway"). But in the 1800s, most Seminole were moved back to Oklahoma and – along with the Creek, Choctaw, Chickasaw, and Cherokee – became part of "The Five Civilized Tribes". A few Seminole evaded relocation; their descendants now remain in Florida.

**SEMINOLE DOLL**
*Dolls made by the Seminole provide a record of what the people once wore. This doll is dressed in a cape and a full-length skirt.*

SOUTHWEST DESERT

## Navajo

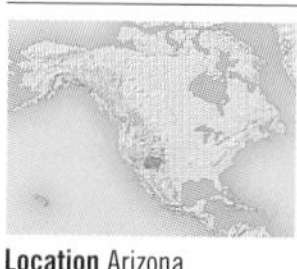

**Location** Arizona, New Mexico, Utah (US)
**Population** 200,000
**Language** Navajo (an Athabaskan tongue closely related to Apache), English
**Beliefs** Christian denominations, Native American Church, traditional spirit beliefs

The Navajo people are divided into 64 clans, and members trace their descent through the maternal line. Marriage is forbidden between fellow clan members. The Navajo reservation covers an enormous area of more than 64,000 sq km (24,500 sq miles). This, combined with the fact that the Navajo are the most populous of all US Native American groups, has allowed these people an autonomy and dynamism denied to most other, more beleaguered Native Americans. Today, many Navajo work as silversmiths, sheep farmers, and stockmen, as well as working in the mineral and tourist industries.

Navajo culture is an interesting mix. From their Athabaskan heritage comes the language, the sand paintings used in healing rituals, and the concept of *hozoji* (a state of harmony with supernatural powers). The influence of the Pueblo Native American people is evident in Navajo mythology and in the blankets they make, which often feature patterns of symbolic and religious significance.

Navajo served as radio operators for the US marines in WWII. Their language was used as a code, and the Japanese were unable to decipher it.

**JEWELLERY**
*This turquoise-and-silver pin is an example of Navajo jewellery design.*

**PASSING ON TRADITION**
*The Navajo are known for their expert handicrafts. The art of traditional skills, including weaving, are passed from generation to generation.*

History

### Defeat of the Navajo

In 1863, US Colonel Kit Carson raided the Navajo stronghold of Canyon de Chelly. The Navajo were taken to Bosque Redondo Reservation in New Mexico, home of their enemies, the Mescalero Apaches. However, in 1868, they were allowed to return to a reservation set aside for them. This area included Canyon de Chelly, today a national monument.

**CANYON DE CHELLY**
*The steep cliff faces and wall caves that make up Canyon de Chelly, Arizona, formed the backdrop of the 1863 defeat.*

**DESERT CAMP**
*Many Navajo men work as stockmen. These ones are setting up camp for the night in Monument Valley, a Navajo Nation tribal park in Utah and Arizona.*

SOUTHWEST

## Apache

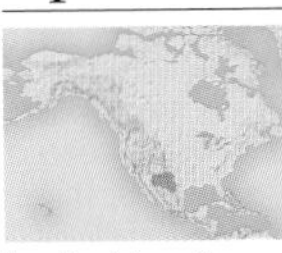

**Location** Arizona, New Mexico, Oklahoma (US)

**Population** 20,000
**Language** English, Apache (an Athabaskan language)
**Beliefs** Roman Catholic, Protestant (Assemblies of God, Baptist, Pentecostal Miracle Church), Mormon, traditional spirit beliefs

"Apache" is the name given to several related Athabaskan-speaking peoples who live in the desert and mountain country of the southwestern US (although in late prehistoric times they also occupied much of the high plains country). Most of these peoples are presently settled in reservations. The Apache include the Chiricahua, Jicarilla, Mescalero, White Mountain, San Carlos, Tonto, Lipan, and Kiowa Apache. They are closely related to the Navajo people (*see* p361).

When Spanish invaders first encountered the Apache in the late 16th century, they were mainly nomadic peoples, hunting and trading with the settled Pueblo peoples of the Rio Grande valley in New Mexico. By the 1660s the Apache had acquired the horse; soon afterwards, they were carrying out raids on their neighbours and going deep into Mexico. They were equally defiant when later confronted by US armed forces.

Although most Apache today regard themselves as Christians, many still give credence to the supernatural powers associated with various natural phenomena. These powers include Life Giver (the sun), Changing Woman (the source of life), and her twins: Slayer of Monsters and Child of Water. These spirits are sometimes equated with God, the Virgin Mary, and Jesus respectively. The *gaan* mountain spirits are also considered important in Apache tradition. In 1991, the Apache Survival Coalition filed a lawsuit against the construction of an observatory on Mount Graham in Arizona, a site considered sacred by many Apache.

History

### Tenacious resistance

The US found itself embroiled in guerrilla wars after it assumed political control of Apache lands in Texas, New Mexico, and Arizona in the mid-19th century. The Apache leaders continued their resistance; Geronimo (*see below*), the last one, was only finally subdued in 1886.

**WAR CLUB**
*The war club was widely used by Apache men in hand-to-hand fighting.*

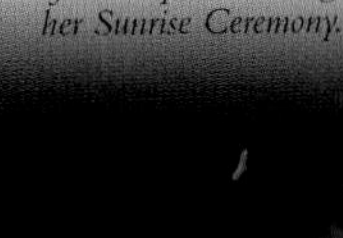

**PUBERTY RITUAL**
*A girl receives a blessing of sacred pollen during her Sunrise Ceremony.*

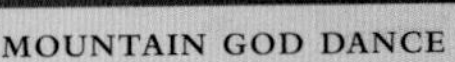

**MOUNTAIN GOD DANCE**
*The Mescalero Apache dance group perform the Dance of the Mountain Gods. They thank the gods for the four elements: land, sky, water, and fire.*

SOUTHWEST

## Hopi

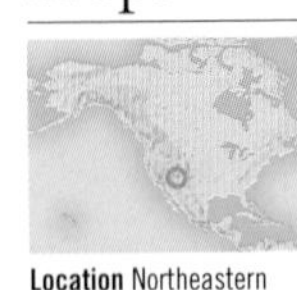

**Location** Northeastern Arizona (US)

**Population** 7,000
**Language** Hopi (an Uto-Aztecan language)
**Beliefs** Traditional beliefs (organized around more than 300 *Kachina* – nature spirits who bring rain, good harvests, and health)

The Hopi have occupied the southern edge of Arizona's Black Mesa for more than 800 years. They are one of several peoples in the Southwest known as *Pueblo* peoples (from the Spanish word for "town", which is used to refer to the indigenous communities). The pueblos consist of rectangular, often multistorey, houses built from adobe, a sun-dried brick. Those of the Hopi are said to be among the oldest in the Southwest. Many houses are grouped around plazas and are interspersed with *kivas*: large, sunken rooms used as ceremonial centres or men's houses.

Today some Hopi, like their ancestors, cultivate maize, rice, squash, and cotton, but most depend on wage labour and commercial ranching, or draw welfare payments. Traditionally, the Hopi were also potters, producing beautiful vessels with a yellow hue. Only shards of the original vessels now remain. However, in the late 19th century a Hopi woman named Nampeo revived this craft tradition.

The Hopi remain a deeply religious and secretive people. Their spiritual lives centre on *Kachina* ceremonies, in which Hopi men take on the identity of *Kachina* (nature spirits). The men also carve wooden dolls that represent the *Kachina*. Another important Hopi spiritual ceremony is the Snake Dance. This ritual spectacularly involves the handling of rattlesnakes, which are said to commune with deities responsible for bringing rain.

The Hopi are organized into clans. Men and women, by custom, have to marry outside their clan of birth. Clan membership is inherited from the mother's line. Formally, the clan is headed by an elder woman known as the "clan mother". In practice, though, it is mainly men who have political authority, inheriting office from their mother's brothers.

**HOPI BOY**
*At about eight, Hopi boys and girls begin initiation rites that guide them through adolescence to adulthood. In these rites they learn the principles of being a Hopi.*

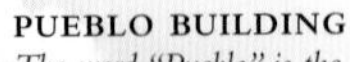

**PUEBLO BUILDING**
*The word "Pueblo" is the name of the towns built by Hopi (below) and related peoples, and the name of the culture area around the Rio Grande.*

**HOPI POTTERY**

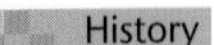

History

### Marriage customs

In Hopi culture, women traditionally showed their status by the way they dressed their hair. Married women simply wore braids, but unmarried women marked their status by wearing the "squash blossom" style, comprising a bun wrapped around a U-shaped bow on each side of the head. This style also served to advertise the woman's competence in household matters.

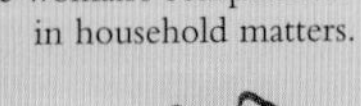

***KACHINA* DOLL**
*The Kachina dolls were traditionally made for women and children. Today, many are sold to tourists and collectors.*

SOUTHWEST

## Paiute

**Population** 7,500
**Language** Paiute (an Uto-Aztecan language related to Comanche), English
**Beliefs** Christianity: Mormon, Native American (or Peyote) Church; traditional spirit beliefs
**Location** Nevada, Utah, eastern California

The Paiute are the most widespread of the peoples in the desert area known as the Great Basin. In this region they saw few white people until the 1860s. In the 1880s, with their way of life coming under serious threat, they sought salvation in a millennial cult called the Ghost Dance (*see* Comanche, p359), which spread through the peoples of the western US.

The Paiute have a rich body of myths and legends, many of which are morality tales about animal ancestors. They are also well known for their Mourning Ceremony, known as "Cry", in which a deceased person's property is publicly burned and song cycles are performed.

Many Paiute earn a meagre living through wage labour on ranches or in hay and grain production, sometimes supplementing these activities with hunting and gathering.

**CRADLEBOARD**
*This type of baby carrier is common to several plains peoples. It features a rigid protective board above the baby's head. They are still made and sold today.*

**RANCHING LIFESTYLE**
*Paiute used to live by hunting rabbits and other small game. Much of their land is now pasture, so many work on ranches.*

SOUTHWEST

## Tohono O'odham

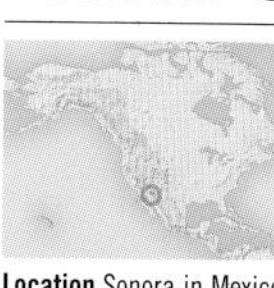

**Population** 20,000
**Language** Pima-Papago (an Uto-Aztecan language related to Huichol), Spanish, English
**Beliefs** Roman Catholic Christianity mixed with traditional spirit beliefs
**Location** Sonora in Mexico, SW Arizona reservations

Tohono O'odham, meaning "desert people", is the name preferred by many of this people, otherwise known as Pima and Papago. They live in Arizona and just across the border in Mexico. Today most of those in the US subsist on reservations in service-sector jobs or welfare. Many in Mexico, where they are called Pima Bajo, work on isolated ranches or in mines and sawmills.

Much traditional Tohono O'odham culture still survives. The Coyote and Buzzard totemic divisions still have ritual significance. The exchange of gifts between individuals is considered more appropriate than buying and selling. Headmen are chosen for their generosity and their softly-spoken and humorous natures.

MEXICO

## Mexicans

**Population** 104 million
**Language** Chiefly Spanish; various Mayan, Nahuatl, and other regional indigenous languages
**Beliefs** Mainly Roman Catholic Christianity, some Protestant
**Location** Mexico; also southern US

A succession of advanced cultures that thrived in Mexico were abruptly ended by Spain's conquest of the Aztecs in 1521. Colonialism did not, however, lead to complete subjugation. Native Americans and mestizos (people of mixed indigenous and European ancestry) adopted Roman Catholic Christianity but subverted it by giving Christian saints the characters and powers of native gods. In the annual cycle of saints' days, fiestas, and carnivals, traditional dress, music, and dance are blended with religious rituals to bring the nation's communities together.

The traditional ingredients of maize, chillies, tomatoes, and beans play a major role in cooking, with tortillas (thin pancakes of corn flour, or wheat in the north) retaining top spot as the flexible staple in many dishes.

The people have a distinctive modern culture. Despite the number of US programmes on television, millions at home and abroad watch Mexican *telenovelas* (soap operas). The songs of lost and unrequited love sung by *mariachis* – romantic figures in Mexican cowboy suits – also remain very popular.

**MEXICAN FACE**
*The Mexicans are a mix of more than 50 indigenous peoples, mestizos, and those of European descent.*

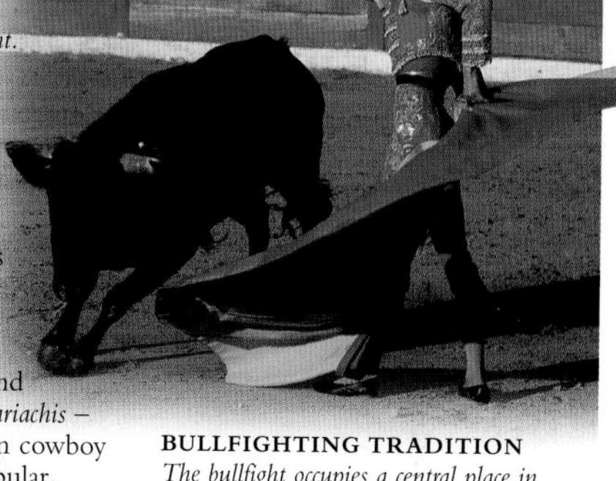

**BULLFIGHTING TRADITION**
*The bullfight occupies a central place in Mexico's history and culture. Although it now attracts some criticism, it is still seen as a display of skill and machismo. The largest bullring in the world is in Mexico City.*

**RELIGIOUS DEVOTION**
*The Virgin of Guadelupe, the dark-skinned Madonna and symbol of Mexico, is believed by millions of Mexicans to be an earth goddess.*

Fact

### Day of the Dead

The Day of the Dead is celebrated elsewhere in the world as All Saints' or All Souls' Day on 1 November. However, nobody celebrates it as spectacularly as the Mexicans. This Mexican national holiday reveals a sophisticated and irreverent acceptance of death as Mexicans spend time in graveyards picnicking and holding all-night vigils for their ancestors. They also build shrines in their homes in honour of deceased relatives.

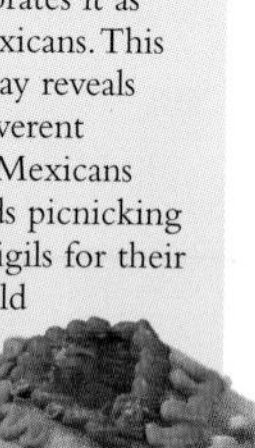

**SUGAR SKULLS**
*Sugar skeletons or skulls, and other items with a death motif, play a popular part in the celebrations.*

MEXICO

## Tarascans

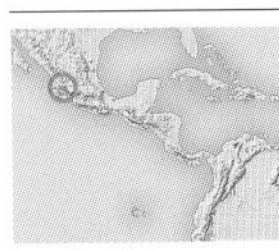

**Population** 175,000
**Language** Tarascan,
**Beliefs** Folk Catholic Christianity (which emphasizes fiestas and patron saints), beliefs barely influenced by pre-Columbian religions
**Location** Michoacan state (Mexico)

The Tarascans today predominantly engage in subsistence agriculture. Many live in wooden cabins within compounds surrounded by dry-stone walls, although they are replacing many with dwellings made of brick and concrete. Most still speak Tarascan, but the overwhelming majority also speak fluent Spanish.

Tarascans have distinctive beliefs and practices concerning birth and the raising of children. A baby is only baptized after a 40-day period of rest and isolation for mother and child, called *patsákuni*. The baby is then swaddled for 6 weeks and kept in physical contact with the mother or immediate female relatives.

**WONDERFUL WEAVING**
*Tarascans are known in Mexico for their expert weaving of cloth and straw, as well as for their distinctive pottery.*

Tarascan history before the colonial period is sparse. The Tarascans were skilled in metallurgy and the building of monumental structures. Spanish conscription, diseases, and forced labour imposed profound changes.

**TARASCAN FAMILY**
*Women in a Tarascan family often wear traditional, brightly coloured woven skirts as they go about their everyday tasks.*

MEXICO

## Zapotecs

**Population** 500,000
**Language** Zapoteco (related to language of the neighbouring Mixtecs)
**Beliefs** Folk Catholic Christianity (images of Christian saints resemble the Zapotecs' old gods)
**Location** Oaxaca state (southern Mexico)

Calling themselves the Ben'Zaa or Cloud People, the Zapotecs are, along with their neighbours the Mixtecs, the descendants of a people who occupied the city of Monte Albán (southern Mexico) from AD200 until its ruin in AD900.

Zapotec communities are mostly organized around communal lands, with villages built around a plaza in which ceremonies are held. Most people identify with their community of origin rather than a perceived ethnicity. There is also a large Zapotec population in Oaxaca, an important market town with a growing tourist sector. However, as Zapotecs in Oaxaca come into contact with mass culture, much that made them interesting to visitors is disappearing. Plastic and metal dishes are fast replacing local pottery, while factory-made attire is displacing *huipiles* (traditional back-strap woven blouses), shrouds, and sombreros, as Zapotecs attempt to counter claims by mainstream Mexicans that they are poor and "backward".

Today, the Zapotecs are Catholics, but images of Christian saints look like their old gods. San Pedro resembles the old rain god Cocijo.

**CLOTHING**
*Many Zapotecs favour mass-produced clothes instead of these handmade textiles.*

**ANCIENT URN**
*This Zapotec urn, found in the ruins of Monte Albán, Mexico, dates from AD200.*

Fact

### Making mescal

Mescal, made in Oaxaca, is related to tequila. Zapotecs traditionally use it in rituals and ceremonies, for example to bless crops. Producing mescal involves harvesting the heart of the agave plant. This is then baked, mashed, fermented, and distilled – usually in wooden casks (below) and copper stills.

**MESCAL AGING IN WOODEN CASKS**

MEXICO

## Nahua

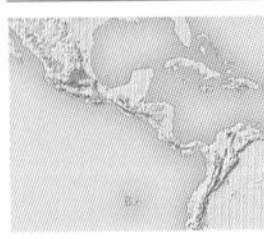

**Population** 1.75 million
**Language** Nahua (an Uto-Aztecan language)
**Beliefs** Folk Catholic Christianity (including rituals that originated in non-Christian religious practices)
**Location** Dispersed throughout central Mexico

The Nahua, or Mexicano as many prefer to call themselves, are the speakers of various dialects of Nahua, the language spoken by the Aztecs. Nahua-speaking communities are usually found living in the hilly or mountainous districts of central Mexico into which the Spanish language has only barely penetrated.

Most Nahua cultivate maize; some use horse- or mule-drawn ploughs; others employ slash-and-burn techniques, clearing vegetation then planting seeds in the bare earth with the aid of a dibble stick. Wheat, barley, beans, chilli peppers, squashes, onions, and tomatoes are also grown, and many Nahua raise cattle and sheep, and make sugar loaf.

The Nahua have a rich ceremonial life, much of it organized around *cargo*, a system of religious and political offices. Although most are pious Catholics, many Nahua also visit shamans of both sexes.

"Colonization" legislation in 1883 and 1894 meant the loss of lands to outsiders for many Nahua, although some was returned during the Mexican Revolution in the early 20th century. Since World War II, however, a number of Nahua farmers have been exploited by cattle ranchers and have again been dispossessed.

**WORSHIPPING THE VIRGIN**
*In a blend of indigenous and Catholic ritual typical of Latin America, Nahua people bring their own ethnic style to Guadelupe to worship at the Virgin's shrine.*

**NAHUAN GIRLS**
*Girls join a procession on a Catholic Saint's day in the town of Cuetzalan, Mexico.*

**APPEALING ARTWORKS**
*Tourists are drawn to the products of several Nahua crafts, notably pottery.*

History

### The Aztecs

In the 15th and 16th centuries, the Aztecs, who were a Nahua people, built an empire in present-day southern and central Mexico. The success of their empire was based on a high-productivity system of agriculture. The Aztecs were also known for their architecture and stone carvings (below).

MEXICO

## Huichol

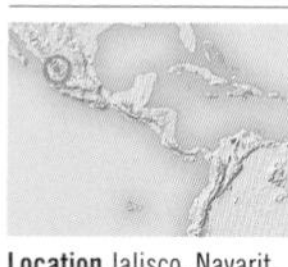

**Population** 20,000
**Language** Huichol (which is closely related to the Cora and Nahua languages)
**Beliefs** Spirit beliefs, influenced by Christianity (mainly Catholicism)
**Location** Jalisco, Nayarit, Zacatecas, Durango

The Huichol are relatively isolated. Most live in dispersed homesteads in the mountainous Sierra Madre Occidental, where they hunt, fish, and practice slash-and-burn horticulture. The majority of Huichol communities are situated on communal lands to which members have rights of use but not actual ownership. Most ranchos are centred on family shrines, and the influence of Christianity is limited in many households. Marriages between close cousins are encouraged.

In spite of measles and smallpox epidemics, the Huichol successfully resisted the Spanish until the 1720s. Thereafter, they were given relatively privileged status, being exempt from paying tribute and allowed to govern their own affairs. This special status was awarded because Huichol lands constituted a buffer zone between Mexican communities to the south and more dangerous Native American peoples to the north.

In the late 19th century there was an invasion of mestizo (people of mixed Spanish and Native American ancestry), cattlemen and colonists but the Mexican Revolution (1910–1920) gave the Huichol and their Cora neighbours the chance to expel these immigrants. Since then, programmes like "Plan Huicot" – designed to integrate the Huichol and Cora into Mexican culture through agriculture, education, and healthcare – have enjoyed only partial success.

**FESTIVE SOMBRERO**
*A Huichol man smiles for the camera as he sports his feathered sombrero and embroidered poncho, all part of the ethnic Huichol costume, to a local festival.*

**BEAUTIFUL BEADWORK**
*Although often produced for the tourist trade, Huichol beadwork is considered a high form of craftsmanship in Mexico.*

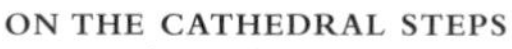

**BACKSTRAP LOOM**
*A craftswoman uses the customary backstrap loom to create a traditional motif.*

Fact

### Huichol textiles

Many Huichol are accomplished in handicraft, including weaving, embroidery, and the making of objects like baskets, cradles, and hats. Whatever form the work takes, it usually expresses the spiritual beliefs of the Huichol people. Imagery used in embroidered art, for example, is sometimes inspired by dreams or the natural world, and representations of deer are common.

HUICHOL EMBROIDERY

MEXICO

## Tzotzil (Mayans)

**Population** 135,000 (Tzotzil Mayans only)
**Language** Tzotzil (a Mayan language)
**Beliefs** Roman Catholic Christianity (influenced by spirit beliefs and more recently by Evangelical Protestant Christianity)
**Location (all Mayans)** Belize, Yucatán (Mexico), Guatemala

The Tzotzil language, one of 30 Mayan tongues, is spoken in approximately 20 municipalities in Chiapas state, each of which has a distinctive style of Tzotzil dress. Each municipality also traditionally had a unique ceremonial life organized around elaborate fiestas. Until recently, these costumes and ceremonies gave Tzotzil communities a conservative and introverted feel.

The fiestas of San Lorenzo, San Sebastián, and the Virgen del Rosario Menor are still important among the mainly Catholic Tzotzil, as are the mountain and cave shrines and the activities of shamans. However, over the past few decades, many Tzotzil have rejected the beliefs and practices of their parents, and the system of community authority based on *cargo* (temporary politico-religious offices) and *compadrazgo* (godparent relations), and have instead embraced evangelical forms of Protestantism.

Traditional Tzotzil communities are still predominantly agricultural, with members cultivating maize, beans, and squash, and rearing domestic animals. More recently, however, government-built roads and commercial development of agriculture and hydroelectric power have brought about considerable transformations to the region, and the Tzotzil have emerged as merchants, haulage workers, and administrators.

**ON THE CATHEDRAL STEPS**
*A group of Tzotzil Mayan women congregate on the steps of San Cristobal de las Casas cathedral in Chiapas state, Mexico.*

**TOWN OFFICIALS**
*Municipal officials within the traditional Tzotzil Mayan system of authority have to wear ceremonial costumes from time to time.*

**DISTINCTIVE DRESS**
*Tzotzil Mayans in the village of Zequentic wear blue ponchos with a geranium design.*

**DANCE OF THE CONCHEROS**
*Concheros (shell wearers) perform with musicians in front of the old Basilica in Guadalupe. Their dance predates the Spanish invasion.*

# CENTRAL AND SOUTH AMERICA

A long history of migration has given this part of the world many diverse peoples and cultures. From the 16th century, the Europeans added their genes and beliefs to the melting pot, as well as introducing slaves from Africa and Asia. Inhabitants today range from forest-dwelling tribes who still practise subsistence farming to city-bred stockbrokers.

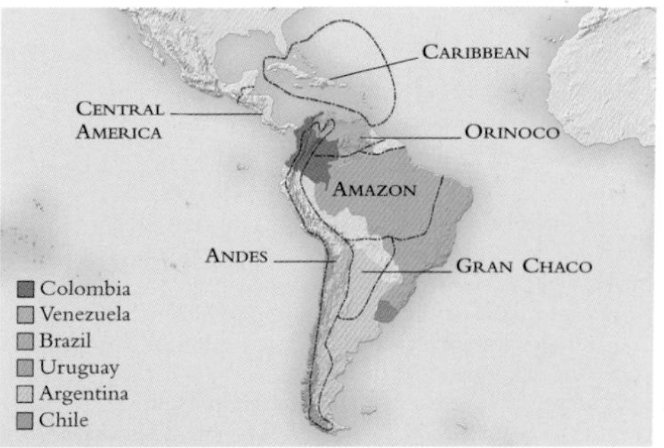


**THE REGION**
*The vast area encompasses Central America, the Caribbean islands, and the countries of South America.*

Farming communities developed in the Andes and the Amazon basin before 2000BC. These early dwellers almost certainly reached South America by way of Mexico and the narrow isthmus of Central America.

## MIGRATION AND SETTLEMENT

In this region, complex patterns of migration have produced peoples with varying cultures, speaking hundreds of languages, which are classified into many language families.

The Europeans, who first arrived in the 16th century, came from several nations and divided the continent, by conquest and treaty, into colonies. The Spanish claimed much of the Central and South American mainland, although the Portuguese established control of Brazil. The Dutch, French, and British took control of the Guianas in northern South America (now Guyana and French Guiana). The British came to govern the majority of the Caribbean islands during the 18th and 19th centuries.

The colonial powers brought African slaves or indentured labourers from Europe and Asia, the descendants of whom now constitute the majority of some countries – notably dark-skinned people in much of the Caribbean, and East Indians in Guyana. Free settlers also arrived: Italians went to Argentina, Japanese to Brazil, Lebanese to Central America and the Caribbean, and Mennonites (*see* p356) to Paraguay and Belize.

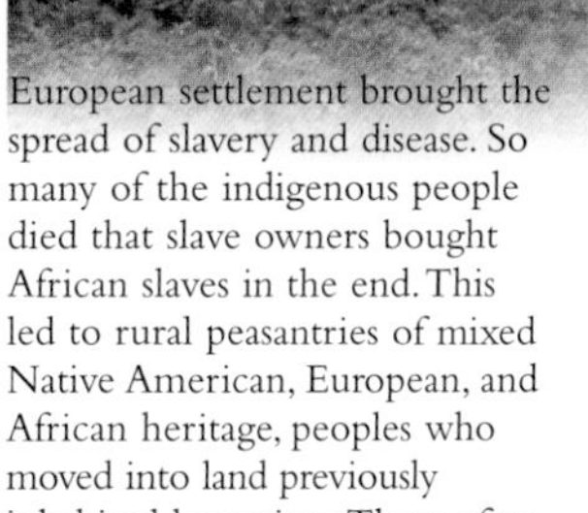

**HONDURAN MAN**
*Some Hondurans have very dark skin as they descend from the slaves introduced in the 16th century by Europeans.*

## TRIBES AND NATIONS

The first Europeans recorded the presence of a wide range of human culture. In the northern part of Central America, Mayans were among the corn-growing village communities organized around the worship of mountain spirits. By contrast, the rainforests of the Amazon, the Guianas, and lowland Central America were inhabited by peoples who planted cassava, hunted, and fished.

Some of the people of central Brazil's savannas also planted cassava, as well as establishing complex forms of social organization. Others of these groups spent much of the year "on trek", hunting and gathering in a nomadic existence. Tupían peoples with large hierarchical villages and complex religious lives also lived in Brazil, but they were wiped out during the early colonial period.

Pastoralist peoples living in communities in the Andes, such as the Incas, raised llamas and alpacas and grew root crops. They had long-distance trading partners and built large settlements.

To the south, the peoples of present day south and central Chilean lowlands, also pastoralists, as well as fishermen, hunters and farmers, resisted Inca domination.

AZTEC GOD MASK

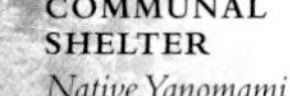

**COMMUNAL SHELTER**
*Native Yanomami communities still live in traditional circular constructions.*

European settlement brought the spread of slavery and disease. So many of the indigenous people died that slave owners bought African slaves in the end. This led to rural peasantries of mixed Native American, European, and African heritage, peoples who moved into land previously inhabited by natives. These often desperately poor mestizos (people of mixed Native American and European ancestry) now tend to occupy social positions above those of the indigenous peoples, while the most elite positions are most often reserved for inhabitants of purely European descent.

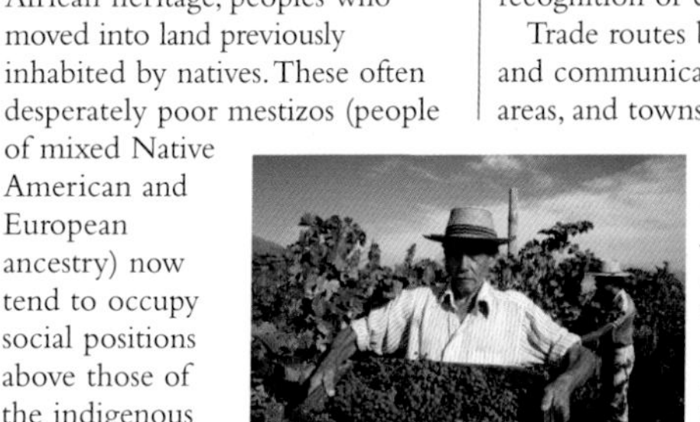

**HARVEST TIME**
*A man harvests grapes for Chilean wine, which is becoming increasingly popular worldwide.*

## CHANGE

Settlement by the colonial powers was motivated by the search for wealth and their practices began the destruction of the forests. In the Andes, the indigenous Quechua- and Aymará-speaking populations extracted silver and tin, initially for colonial overseers and later as workers for multinational companies.

The change to production for trade has caused reorganization of local economies, and ecosystems have been transformed beyond recognition or even disappeared.

Trade routes bring settlement and communications into remote areas, and towns grow into cities.

However, many indigenous peoples are economically marginalized. Many of them have adopted the ways of life promoted by the nation-state and now downplay their Native American heritage. Others, however, resist absorption, though often at the cost of discrimination, land disputes, and armed conflict.

Fact

### Shrines and shamans

This statue of the Virgin de Izamal rests on a shrine surrounded by flower arrangements. The town of Izamal was an ancient Mayan pilgrimage site, and the Catholic Church incorporated many of the religious traditions practised here before the invasion of the Europeans. The icon is a good example of this fusion of practices, which is characteristic of the region.

CARIBBEAN

## Cubans

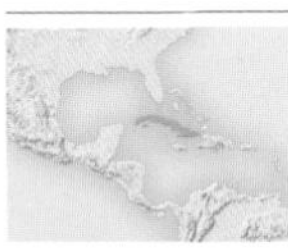

**Population** 11.7 million
**Language** Spanish
**Beliefs** Catholic and Protestant Christianity, Judaism, Santería (a blend of African and Christian rituals), atheism (official creed until 1992)
**Location** Cuba, southern US (particularly Florida)

Cuba's original inhabitants were the Arawak, who were virtually wiped out by the Spanish conquest in 1514. Today, people of mixed origins (of African and European ancestry) form 51 per cent of the population. The rest are later arrivals: white people of Spanish origin (37 per cent), black people (11 per cent), and Chinese (1 per cent).

**MIXING RHYTHMS**
*Cuban music blends Spanish forms with African beats. Artists continue to merge new sounds with traditional ones to create distinctive Cuban versions of rock, reggae, and rap music.*

The modern Cuban national identity, forged in the 19th century, is based on the independence struggle against Spain and drew on African and Hispanic influences to promote a "tropical" culture. The Communist revolution of 1959, headed by Fidel Castro, ended racial segregation and emphasized collective well-being and pride in the island's Afro-Cuban roots.

The nation's idea of its distinctiveness has been boosted by trade embargoes imposed by the US since 1961 in a bid to isolate Castro's regime. Cubans have turned the resulting economic scarcity into an art form, displaying imagination and vigour to preserve the colonial architecture and keep vintage US cars on the roads. Although short of consumer goods and civil liberties, Cubans are proud of their internationally regarded state health structure, education system, and national sports teams (especially those competing in baseball, boxing, and athletics).

Cubans have inherited rich cultural traditions. Song and dance styles, such as *son*, *mambo*, *chachacha*, and, *conga*, interweave Spanish forms with compelling African rhythms. Elements of these styles have been blended into salsa, an exported dance that captivates audiences worldwide. Writers, painters, musicians, and film-makers have long thrived in Cuba's heady cultural environment.

**WELL PRESERVED**
*As a result of the US restrictions on imports, the Cubans still preserve and run cars such as this 1950s US automobile.*

**SPANISH ORIGIN**
*The pale complexions of this couple indicate that they are, like many Cubans, descended from Spaniards.*

CARIBBEAN

## Jamaicans

**Population** 2.7 million
**Language** English, English patois
**Beliefs** Protestant Christianity (Baptist, Anglican), Rastafarianism, Judaism, Islam, Hinduism, spiritual cults
**Location** Jamaica; also the US and the UK

Jamaica's mostly black population consists of former slaves freed by the original Spanish colonialists and those shipped in from Africa by the subsequent British rulers to work on sugar plantations. A history of slave rebellions remains a source of national pride. Emancipation in 1834 created demand for cheap indentured labour. The majority of workers came from India between 1845 and 1917, while others entered from China, the Middle East, Portugal, and South America. Over 1,000 Germans arrived in the 1830s in search of land.

Jamaica is popularly associated with reggae, beach resorts in Montego Bay, rum, highly spiced "jerked" pork and chicken, and world-class cricketers such as Courtney Walsh. Less well appreciated is a deeply devout society that has produced some 250 religious denominations. Jamaican people are laid-back, welcoming, and outward-looking. They delight in using the constantly evolving local patois (based on a rich mixture of English, Spanish, and African languages) to outwit adversaries in debate. However, they face problems such as a narrowly based political system, glaring social inequalities and serious, usually drug-related crime.

Fact

### Reggae

Originating in Jamaica, reggae developed in the 1960s and 1970s and was popularized largely by Bob Marley (1945–1981). Its lyrics are based on Rastafarianism, a movement that promotes repatriation to Africa and the worship of the Ethiopian emperor Haile Selassie (1892–1975). Reggae is still widely enjoyed today. These Jamaicans sing at one of Kingston's many reggae clubs.

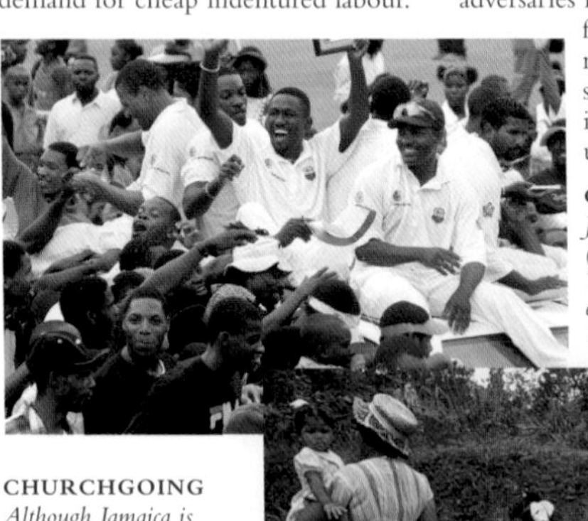

**CRICKET TRIUMPH**
*Jamaicans such as Wavell Hinds (seen here, in the centre, next to Trinidadian Brian Lara in the cap) play a major role in the West Indian cricket team.*

**CHURCHGOING**
*Although Jamaica is famous as the base of the Rastafarian faith, many people practise traditional or evangelical Christianity. Many Jamaicans, like this family, attend church every week. They also keep Sunday as a day of rest and prayer.*

**RAFT TRAVEL**
*Bamboo rafts were traditionally used for carrying bananas from plantations to the ports. The rafts are still used on rivers such as the Rio Grande.*

**CARNIVAL DISPLAY**
*The colourful pre-Lenten Trinidadian Carnival is the largest event of its kind in the Caribbean.*

CARIBBEAN

## Trinidadians and Tobagans

**Population** 1.3 million
**Language** English, Hindi, Creole, French, Spanish, Chinese
**Beliefs** Catholic Christianity; also Hinduism, Protestant Christianity, Islam, Rastafarianism, Santeria
**Location** Trinidad and Tobago; also UK

Like many Caribbean islands, Trinidad and Tobago were once populated by native Arawakan-speaking peoples. By the late 17th century, however, harsh Spanish rule and European diseases had devastated them. The islands were later ruled by the British, who brought African slaves, mainly to Tobago, to compensate for the population collapse and to work on sugar plantations.

Trinidad has the greater mix of peoples. After the abolition of slavery in 1845, thousands of workers arrived, mostly from India, some from China. Syrians, Europeans, and free West Africans came as well, but in smaller

**OIL WORKER**
*Trinidad and Tobago are rich in oil and gas. The oil industry is a major employer and contributor to the economy.*

numbers. Higher wages and later an oil boom attracted workers from around the Caribbean.

Today, ethnic tension is low, but most African and Asian Trinidadians support different political parties and worship separately. However, at the pre-Lenten Carnival, ethnic and social barriers are forgotten. Composers vie to write the smartest calypso song or the catchiest *soca* (contemporary Afro-dance-oriented number), and steel bands (invented in Trinidad) practise for months to outplay each other.

CARIBBEAN

## Puerto Ricans

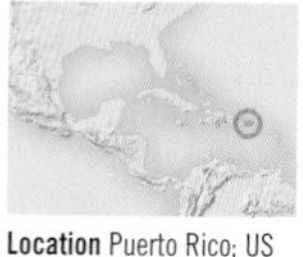

**Population** 3.9 million
**Language** Spanish, English
**Beliefs** "Latinized" Roman Catholic Christianity (a fusion of Spanish, Native American, and African traditions); also evangelical Protestant Christianity
**Location** Puerto Rico; US (especially New York City)

The original inhabitants of Puerto Rico, the Taíno Native Americans, whose descendants date back to at least 900BC, were decimated (and later wiped out) after the Spanish conquests of the late 15th century.

The 16th and 17th centuries saw African slaves shipped in to work in sugar

**STANDARD OF LIVING**
*A woman prepares a meal outside her kitchen window. Although she appears poor, most Puerto Ricans have a higher standard of living than other Caribbean people.*

mills and on plantations, and although few pure black people remain, most of the population has African ancestry. New waves of European immigration in the 18th century reinforced the Spanish character of the islands. However, Spain's military defeat to the US in 1898 led to all Puerto Ricans being granted US citizenship in 1917. Ever since that time, the population has been uneasy with this arrangement, resisting independence but also shunning full statehood. The island remains a US commonwealth, which has brought economic benefits. There has been heavy migration to the US, and US fast food, television, and baseball have been absorbed into Puerto Rican life.

Popular song and dance forms such as *la bomba* keep African rhythms alive, and *Jíbaros* (rural workers), still perform the *seis* – songs and dances of Spanish origin. *Jíbaros* are afforded mythic status in national song and literature.

**PRESERVING CULTURE**
*Despite the strong US influence, Puerto Ricans are keen to preserve their identity, particularly as expressed through song and dance.*

CARIBBEAN

## Créoles

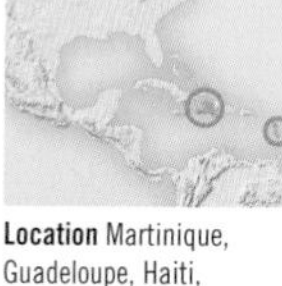

**Population** 7.5 million
**Language** Créole French, French, English
**Beliefs** Roman Catholic Christianity, Vodou (in Haiti); also obeah (a kind of sorcery) and other African-influenced belief systems
**Location** Martinique, Guadeloupe, Haiti, St Lucia, Dominica

The term "Créoles" has been widely applied to the peoples of Haiti, Martinique, and Guadeloupe, most of whom speak the Créole language (which has a French-based vocabulary

**DOMINICAN SCHOOLCHILDREN**
*Schoolchildren walk home through the crowded streets of downtown Roseau in Dominica. Schools here are underfunded and close shortly after midday.*

**AFRICAN INFLUENCE**
*A Haitian sells flowers at the roadside. Many Créole people, like this man, are of West African descent.*

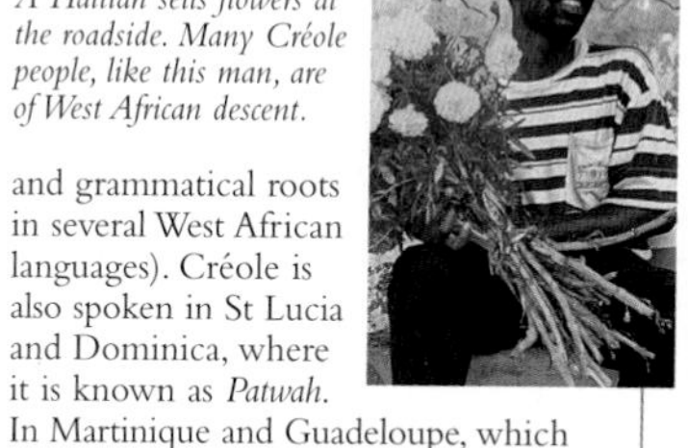

and grammatical roots in several West African languages). Créole is also spoken in St Lucia and Dominica, where it is known as *Patwah*. In Martinique and Guadeloupe, which are still under French control, French is the main language. Haiti has been independent of French rule since 1804, and the requirement of competence in French has lessened; Créole, now an official language of the country, has come to the fore. While Martinique and Guadeloupe are relatively wealthy, Haiti is the poorest country in the western hemisphere.

**VODOU RITUAL**
*Worshippers take a ritual bath during a Vodou pilgrimage in Haiti. About half of Haiti's population follow Vodou.*

CENTRAL AMERICA

## Tz'utujil (Mayans)

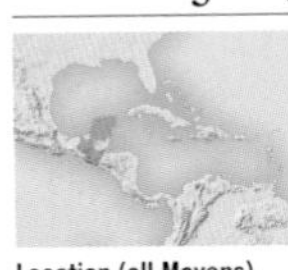

**Population** 82,000 (Tz'utujil Mayans only)
**Language** Tz'utujil (a Mayan language)
**Beliefs** Roman Catholic Christianity (influenced by spirit beliefs), Evangelical Protestant Christianity (in more recent years)
**Location (all Mayans)** Belize, Yucatán (Mexico), Guatemala

The Tz'utujil are one of more than 30 Mayan peoples of Central America. Many Tz'utujil live in the town of Santiago Atitlán in Guatemala. The inhabitants of Atitlán, who are known as Atitecos, are mainly agriculturalists, cultivating maize, chickpeas, avocados, and tomatoes. The Atitecos formerly conducted local politics according to the so-called *cofradía* system, in which religious leaders controlled civil government, presided over the town's communal lands, delegated labour, and held exclusive power to direct important rituals. A number of these rituals are concerned with sacred bundles – traditionally, wrapped human remains that are associated with both Catholic saints and pre-Christian Mayan deities.

For much of the 1980s, the Guatemalan army occupied the town of Atitláu, resulting in considerable violence. Some 1,700 Atitecos out of a total population of 20,000 were killed. In 1990, the town's inhabitants at last secured the removal of the army garrison after a massacre which left 13 people dead. The factionalism resulting from the occupation and the subsequent violence has weakened the *cofradía* system, as has the spread of Protestantism. Mayan evangelists of various denominations have been vocal in their opposition to the local belief in *Maximón*, a religious figure to whom the Atitecos pay homage with gifts and sometimes money. Effigies of this figure are dressed in human garb, and a large cigar placed between his lips.

**MAM MAYANS**
*The people of the town of Todos Santos belong to the Mam culture and are famous for their brilliantly coloured costumes.*

**FLOWER VENDORS**
*Mayans sell flowers in the market of Chichicastenango, a town in Guatemala inhabited mainly by the Quiché Maya.*

History

### Ancient cosmology

Ancient Mayan calendrical systems and computations of celestial bodies were used mainly for divination and regulating sacred rituals. They were arguably far ahead of their Old World contemporaries in terms of accuracy and precision.

**MAYAN CALENDAR**

CENTRAL AMERICA

## Garifuna

**Population** 300,000
**Language** Mostly Garifuna (an Arawakan language, related to Goajiro), Spanish, Creole English
**Beliefs** Ancestral spirititualism mixed wih Catholic Christianity
**Location** Honduras, Belize, Nicaragua, Guatemala

Of African descent but speaking a Native American language, Garifuna are the descendants of Carib Indians from St Vincent and the escaped black slaves who joined them in the 17th and 18th centuries. Also known as the Black Caribs, they were considered troublesome by the British, who in 1797 forcibly transported nearly 5,000 of them to the Bay Islands off Honduras. Further persecution and war have left the Garifuna in scattered communities over four countries.

Garifuna households are headed by older women, many of whom have had several spouses. The men typically work in logging camps, in ports as dock workers, and as sailors, most returning periodically to their villages.

**SETTLEMENT DAY**
*A parade celebrates the Garifuna's settlement in 1832, in the safe haven of Belize.*

CENTRAL AMERICA

## Kuna

**Population** 50,000, mainly in San Blas, with a small number in nearby Colombia
**Language** Kuna (a Chibchan language)
**Beliefs** Traditional spirit belief system mixed with Christianity
**Location** San Blas (Panama), Colombia

Most Kuna villages are on the tiny islands of the San Blas archipelago, although there are also a few in Panama City, the city of Colón (in Panama), and in neighbouring Colombia.

*Molas*, the famous reverse appliquéd cloth panels sewn by Kuna women, form a key part of traditional Kuna dress, which many women wear on a daily basis after reaching puberty. *Molas* are also the mainstay of Kuna income: they are sold not only to tourists visiting the region but also to keen art collectors abroad.

The Kuna are also renowned for their innovative grassroots approach to development in the San Blas region. They have received international funding to help them protect both the region's rainforest, in which they cultivate cassava, coconut trees, and other plants, and the waters that supply them with fish and lobster.

**KUNA BEAUTY**
*A Kuna girl wears elaborate jewellery and covers her traditionally short hair with an ornate red and yellow scarf.*

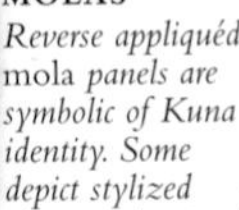

**MOLAS**
*Reverse appliquéd mola panels are symbolic of Kuna identity. Some depict stylized plants and animals.*

CENTRAL AMERICA

## Miskitu

**Population** 150,000
**Language** Miskitu (a Misumalpan language, possibly related to the Chibchan family)
**Beliefs** Traditional spirit beliefs mixed with Christianity
**Location** Eastern Nicaragua, NE Honduras

Miskitu people are principally coastal people who subsist on a diet of fish, turtle meat, cassava, plantain, and rice. They have given their name to a region of Nicaragua and Honduras called the Mosquito Coast. The Miskitu are an "assimilating" people: by choosing partners from a variety of groups, they have absorbed an exotic mix of peoples over the centuries, including Native American groups and Afro-Caribbean people. They regard any children of such mixed marriages as simply Miskitu. Moravian missionaries converted the Miskitu to Protestantism in the 19th century. Although almost all Miskitu are now Christians, most also still believe in the spiritual authority of *sukias,* or shamans, who interpret their dreams and the domains of spirits for them.

**HOUSES ON STILTS**
*Miskitu houses around islands off the coast of Nicaragua serve as bases for coconut-collecting, turtle-hunting, and storm shelter.*

The Miskitu came to prominence in the 17th century, when they would exchange local produce and slaves with English-speaking traders for iron tools and guns, with which they raided their neighbours. Soon, warlord leaders emerged among this once egalitarian people to coordinate these raids and to redistribute booty and trade goods. The most important of these warlords came to be known as kings, and some were even crowned by British authorities in Jamaica or Belize.

In the late 19th and 20th centuries, the Miskitu became workers in the region's logging camps, mines, and banana plantations, which were owned mainly by North Americans. Today, many coastal Miskitu are employed in extracting marine resources, notably shrimp and lobster.

**TURTLE PEOPLE**
*Trade in endangered turtles is forbidden. Nevertheless, the Miskitu, once known as the "turtle people", can continue their small-scale turtle-hunting for domestic consumption.*

Issue

### Contra conflict

The Miskitu became embroiled in international politics during the unrest in Nicaragua in the 1980s. When the socialist Sandanista government came to power in 1979, the Miskitu felt alienated by their development strategy, which seemed to favour mainstream, Spanish-speaking mestizo people (of European and Native American ancestry). Some Miskitu joined the US-backed "contra" rebels in fighting against the government. Peace was finally achieved when a coalition government was elected in 1990.

**DISPLACED MISKITU**
*Displaced Miskitu refugees pack to return home to Nicaragua after the end of fighting between contra rebels and the Sandanista government.*

ORINOCO REGION

## Goajiro

**Population** 110,000
**Language** Goajiro (an Arawakan language related to Mehinaku, Garifuna, and Kampa)
**Beliefs** Traditional spirit beliefs
**Location** Goajira peninsula (Colombia and Venezuela)

Living in inhospitable savannas, desert zones, and mountain ranges means that the Goajiro remain an isolated people. Even their own *miichipala* (hamlets) are at some distance from one another. They practise horticulture, hunting, fishing, raising livestock, and diving for pearls and lobster, but many now supplement this with periods of wage labour and smuggling.

Goajiro kinship is matrilineal, as is *eiruku* (clan) membership: it passes from mother to daughter, or from a mother's brother to her son. Shamans remain key figures in the Goajiro culture, presiding over the *yonna* dances, which celebrate the successes of clan members.

**BRIDE-TO-BE**
*A traditional Goajiro marriage entails a payment to the bride's mother's family by the groom.*

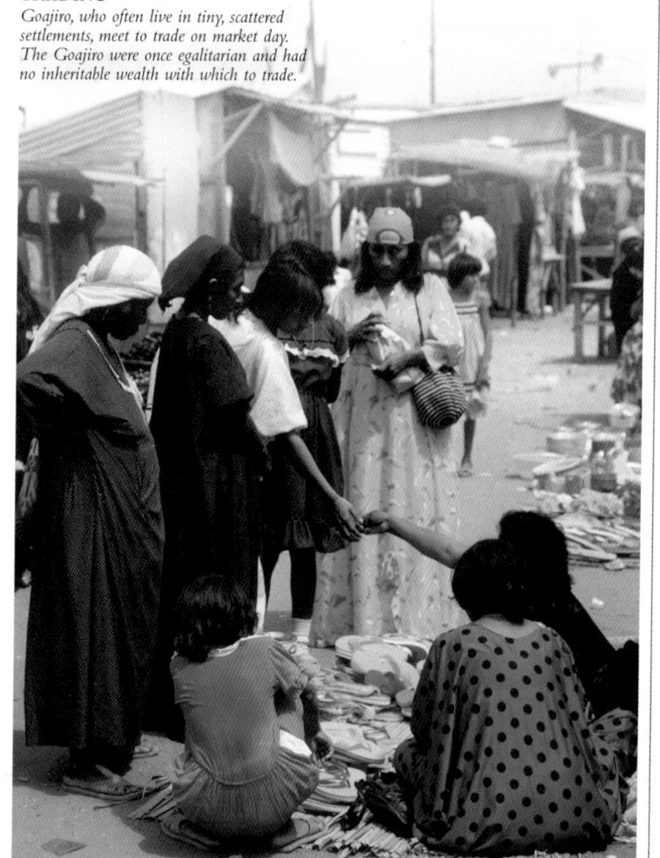

**TRADING**
*Goajiro, who often live in tiny, scattered settlements, meet to trade on market day. The Goajiro were once egalitarian and had no inheritable wealth with which to trade.*

ORINOCO REGION

## Piaroa (Caribs)

**Population** 10,000
**Language** Piaroa (a member of the Carib family)
**Beliefs** Spirit beliefs (centred around a creator called Wahari), Christianity (many have converted to New Tribes Mission)
**Location** Orinoco river basin (Venezuela)

The Piaroa are one of 30 or more peoples speaking related languages who are loosely called Caribs. Many Piaroa still live in dispersed seminomadic communities occupying *isode* (conical or rectangular communal houses). Traditionally, Piaroa people preferred marriages with relatively close cousins, although choices were restricted to the mother's brother's child or the father's sister's child (other cousin marriages were deemed incestuous). Such close marriages protected the community from potentially predatory strangers. There is now a trend towards larger, central settlements, which feature schools, health clinics, electric light, and running water.

The Piaroa still hunt, fish, and cultivate cassava and other crops, but many also grow cash crops, raise cattle, collect vines for rattan furniture, and work for wages in gold mines.

Many Piaroa believe in a creator called Wahari, who was incarnated as a tapir, but in recent decades many have been converted to Christianity. Some scholars claim that the Piaroa shamans are losing their influence to a new generation of teachers and nurses.

COLOMBIA

## Colombians

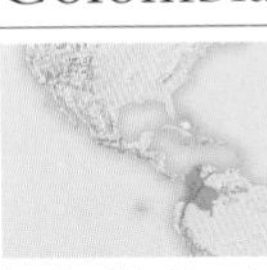

**Location** Colombia and neighbouring countries
**Population** 44.2 million
**Language** Spanish (a Romance language) in addition to more than 180 indigenous languages and dialects
**Beliefs** Roman Catholic; also Anglican, Mormon, Lutheran, various evangelical sects

The Colombian peoples bear the stamp of the period of Spanish colonial rule, which dates from 1499 when treasure-seekers poured into the country in search of the legendary city of El Dorado. Around 58 per cent of the population have mixed Spanish and Native American ancestry (called mestizos), a further 20 per cent are of largely Spanish origin; 14 per cent have both black and white ancestry; 7 per cent are descendants of African slaves; and 1 per cent are descended from a variety of indigenous peoples.

In contrast to other Latin American countries, Colombia experienced little immigration in the late 19th and early part of the 20th century; subsequently, small numbers of immigrants have arrived from Europe, the Middle East, and eastern Asia.

Colombia has a reputation for violence and criminal activity. In part, this reputation stems from a history of bloody feuds and revolts, such as the Liberal revolt against Conservative rule at the end of the 19th century and violence from ruthless drug cartels against the state. An ongoing war – involving the army, right-wing paramilitaries, and left-wing guerrillas – has resulted in the kidnapping and killing of civilians. The effect of the war has been a weakening of society and its institutions, little development, and no sense of national identity.

Despite this instability, Colombia has perhaps the richest variety of music in South America. The music incorporates the rhythms of the Caribbean, Cuban salsa, Jamaican calypso, and the Spanish-influenced sounds of the Andes.

**COFFEE FARMER**
*Colombian farmers wear broad-rimmed sombreros much like those worn in Mexico.*

**COUNTRY BUS**
*Buses are the main form of transport in Colombia. Old-fashioned wooden buses, or* chivas, *negotiate the perilous country roads.*

Profile

### Gabriel García Márquez

Colombian Gabriel García Márquez (1928–) received the Nobel Prize for literature in 1982. His novels and short stories weave together myths, dreams, and reality and reflect the ethnic mosaic of the country. His most famous work is *One Hundred Years of Solitude*, published in 1967. Like most important writers in Latin America, García Márquez sides with the poor and the weak.

**PICKING COFFEE**
*Coffee is grown on the fertile hillsides of Colombia. It remains one of the country's most important exports.*

**FESTIVAL OF FLOWERS**
*A man carries a floral display as part of the annual* Feria de Flores *(Festival of Flowers) in Medellín.*

PEOPLES

VENEZUELA

## Venezuelans

**Location** Venezuela; also North America
**Population** 25.7 million
**Language** Spanish, English, 27 distinct indigenous languages
**Beliefs** Christianity, local religions (syntheses of Christianity, African beliefs, and indigenous mythology)

Venezuela was colonized by Spain in the 16th century. For the native population, centuries of disease and exploitation followed, which reduced the people to the marginalized tribal groups they are today. Venezuelan black people are either descendants of plantation slaves or of more recent migrants who came to work in the oil fields from the 1920s onwards.

The oil boom in the 1970s attracted further workers from Guyana and Trinidad, which was followed by European migration in the 1940s and 1950s. Most of the 800,000 immigrants at this time were from Italy, Spain, and Portugal and were attracted by jobs and settlement grants. There are also around 1 million illegal migrants from neighbouring wartorn Colombia.

The long-standing US involvement in the country's oil industry has had an impact on the way of life. The most obvious legacies include baseball, beauty contests, and, among the young, cars. Television channels are dominated by American-made programmes, but Venezuelan *telenovelas* (soap operas) still attract national and Hispanic audiences worldwide, as does Caracas salsa music. Traditional folk music flourishes, reflecting the mixed identities of the inhabitants of the Caribbean, Andean, and Llanos (central plain) areas. People from each region have different versions of the *joropo*, the energetic national dance for couples, but it is the beat of drums that dominates Afro-Venezuelan fiestas on the coast.

**NATIVE FESTIVAL**
*Venezuela is a country of immigrants; however, there are shared traditions, as is shown by this festival near Caracas.*

**MAKING MUSIC**
*The Venezuelan government subsidizes the highly popular national orchestra.*

**HUNTING CAPYBARA**
*These giant, semiaquatic rodents are related to guinea pigs. People living in rural areas capture these animals for food.*

**CAPITAL CITY**
*Caracas is a city of contrasts, with towering glass skyscrapers in the centre and sprawling shanty towns on the surrounding hills.*

ORINOCO AND AMAZON

## Yanomami

**Population** 30,000
**Language** Yanomami
**Beliefs** Traditional spirit belief system: nature is regarded as sacred and the fates of humans and the environment are inextricably linked
**Location** Brazilian-Venezuelan border

The Yanomami are probably the largest population in the Americas that is still isolated from the influence of Western national culture. This is despite the increasing numbers of *garimpeiros* (gold miners) that have appeared in Yanomami territory in recent years.

In the 1960s, these communities came into contact with white people for the first time. Then, as now, they were shifting cultivators, subsisting on cassava, bananas, forest fruits, and prey captured by hunting and fishing. During the sixties, social scientists wrote books and made films about the Yanomami. They often emphasized the people's warlike nature – in particular the feuds between villages over women kidnapped and taken as wives.

Some experts said the Yanomami were exemplars of a "savage society" of a type that had ceased to exist elsewhere in the world. These accounts were later tempered by reports from other anthropologists that highlighted Yanomami humanity and, more recently, by books and articles that have sought to discredit some of these earlier statements. Even though in August 1993, a group of gold miners murdered 70 Yanomami, in the Brazilian part of Yanomami territory, the stereotype of the violent Yanomami continues to persist.

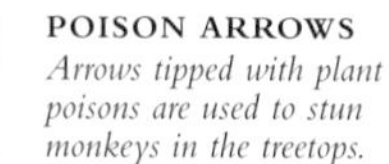

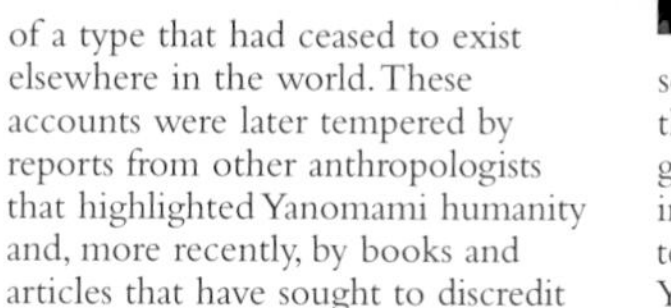

**POISON ARROWS**
*Arrows tipped with plant poisons are used to stun monkeys in the treetops.*

**FATHER, SON, AND PET**
*Yanomami children learn from their parents and other relatives. Knowledge is handed down between generations by demonstration and word of mouth.*

**HAMMOCKS**
*Yanomami sleep in hammocks hung from the ceiling of a hut. Several families share one of these structures.*

Issue

### The fight for the forest

The Yanomami live in the rainforest that straddles the Brazilian-Venezuelan border. From 1970 until the 1980s, gold miners illegally flocked to the area. They brought diseases and began to destroy the forest (right). Although the miners were expelled in 1992 after the creation of the "Yanomami Park", there is pressure for mining to resume.

**AFLOAT IN A FLOODED FOREST**
*Boats are essential in the wet season, when much of the Amazon floods. Outboard motors are increasingly used for long trips.*

ORINOCO

## Tukano

**Language** Barasana, Tukano, Bará, Desana, Tatuyo, Tuyuka, Siriano, Yurutí, Carapana, and Piratapuyo
**Beliefs** Traditional spirit beliefs; Roman Catholic and Evangelical Protestant Christianity
**Location** Vaupés river (Colombia)
**Population** 4,500

The Tukano live along the banks of the Vaupés River (and a number of its tributaries), in southeastern Colombia and northwestern Brazil. They plant manioc (a plant whose roots are used as a source of food), and the men fish (using traps, nets, poisons, lines, and bow and arrow) and hunt (using guns, bows and arrows, blowpipes, and poisons) in the surrounding rainforest.

Although the term "Tukano" can mean people who only speak Tukano, it is also used to refer to speakers of Barasana, Tatuyo, Siriano, Desana, Tuyuka, Yurutí, Piratapuyo, Bará, and Carapana, who all live in the same region and share a distinctive culture with the Tukano proper.

These peoples belong to separate communities, each associated with a different language. Members of a community are expected to marry outside of their group, as unions "within a language" are considered incestuous. As a result, people often speak as many as 10 languages with Tukano serving as a lingua franca.

The Tukano are well known for their *Yurupari* rituals. These are masked ceremonies in which sacred trumpets (said to sound like the voices of spirit beings) with apparently phallic significance are paraded before the longhouse (communal dwelling). The trumpets are made of twisted bark and a hardwood mouthpiece. Women and children risk execution if they see these trumpets and masks.

ORINOCO

## Saramaka

**Population** 22,000
**Language** Saramakan (a Creole influenced by Portuguese, English, and west African languages)
**Beliefs** African-influenced beliefs; also Christianity: Moravian, Roman Catholic
**Location** Surinam; also refugees in French Guiana

The Saramaka are "maroons" – a group of slaves who fled mainly Portuguese-owned plantations into the forests of Surinam during the late 17th and early 18th centuries.

They formed fiercely independent communities organized around the cultivation of cassava, taro, and plantain, and hunting and fishing.

In the 19th century, many Saramaka established contact with the Dutch colony, and several became boatmen, logging workers, and labourers.

The Saramaka have a matrilineal society, and most women have their own home. Many men have two or more wives, but the women live in separate houses. During the late 1980s, there was violent conflict between the Saramaka and the government of Surinam. Many Saramaka fled to French Guiana, where some still live as refugees.

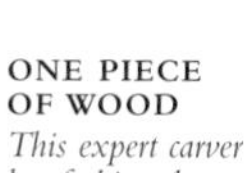

**ONE PIECE OF WOOD**
*This expert carver has fashioned a chair from a single piece of wood.*

**PALM-THATCH HUTS**
*Saramaka people live in huts made of palm leaves. This style of housing is reminiscent of West African huts.*

**CARNIVAL TIME**

*Brazil's annual Carnival is celebrated in many cities, but Rio's is the most noisy and spectacular. Thousands of people dress in flamboyant costumes and dance through the streets to samba music.*

BRAZIL

# Brazilians

**Location** Brazil
**Population** 179 million
**Language** Portuguese, Spanish, 215 surviving indigenous peoples speak about 180 languages
**Beliefs** Mostly Roman Catholic; also Candomblé (religion with African roots) and Evangelical Protestant

During Portuguese colonization of Brazil (from 1530), most of the indigenous peoples were decimated by disease and exploitation and then replaced by the millions of African slaves needed on the sugar plantations. From 1700 onwards, a population of displaced Native Americans, mestizos (people of mixed Spanish and Native American ancestry), people with one black and one white parent, and poor white people spread inland to become gold and diamond miners, cattle ranchers, forest workers, or squatters. This dynamic mix of cultures epitomizes Brazilian society.

In the late 19th and early 20th centuries, there were high levels of immigration from Europe – especially from Italy – as well as from Japan and Arab countries. Many European Jews arrived later, escaping from the Nazis. These newcomers settled mainly in the southeast and were crucial for Brazil's industrial expansion.

The samba and bossa nova music and dances capture the sensuousness of the Brazilian character, and it is popular local melodies and folk music that give a distinctive feel to the works of Brazil's most celebrated composer, Heitor Villa Lobos (1887–1959). The annual Rio Carnival is a global symbol of Brazilian exoticism, but it is victory in five football World Cups (1958, 1962, 1970, 1994, and 2002) that is the greatest source of national pride. Football was brought to Brazil by Scottish engineers at the end of the 19th century, and the nation now boasts some of the world's best players.

Nearly 80 per cent of Brazilians now live in cities, yet a third of the country's workforce is still employed on the land. The country has more than 200 mllion cattle and produces more beef than the whole of the US.

Brazil also has one of the world's most unequal societies. However, the poor now have a champion in former factory worker Luis Inácio "Lula" da Silva, elected president in 2002.

**CHILDREN OF THE SLUMS**
*The impoverished slums of Rio de Janeiro are a stark contrast to the city's prosperous neighbourhoods. Many Brazilians regard them as a breeding ground for criminals.*

**AFRICAN ANCESTRY**
*Brazil's character comes largely from a fusion of African and Portuguese influences. There are more people of African descent here than in any nation outside Africa.*

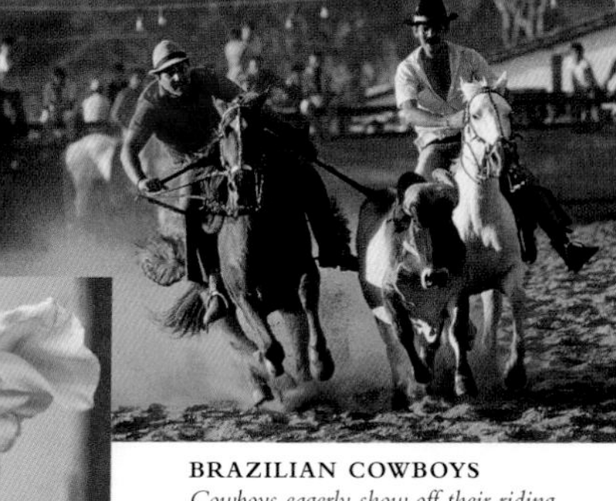

**BRAZILIAN COWBOYS**
*Cowboys eagerly show off their riding skills by rounding up cattle at a public performance. Rodeos are still a major way of life in rural Brazil.*

History

## Tupí resistance

Before the Portuguese arrived in the 16th century, the most prominent inhabitants of Brazil were the indigenous Tupí people. When colonization began in 1530, native land was seized, and the people were used mostly as labourers. However, there were some pockets of resistance against this domination. In Maranhão in 1538, for example, the Tupí attacked and managed to destroy the local Portuguese colony.

16TH-CENTURY TUPÍ CHIEF

AMAZON

## Mehinaku

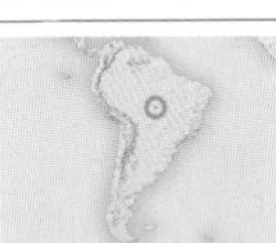

**Population** 170
**Language** Mehinaku (a member of the large Arawakan language family and related to Campa, Goajiro, and Garifuna)
**Location** Xingu, Mato Grosso region (Brazil)
**Beliefs** Traditional spirit belief system

The Mehinaku people live in the Xingu National Park, which is in the Mato Grosso region of Brazil. The land changes dramatically with the seasons, affecting the way the people travel. In the dry months, between May and August, they can cross the baked flood plains on bicycles. Between September and April, much of the land lies under water, and the bicycles give way to canoes. The people live much as their ancestors did, by cultivating manioc and fishing (often with poisons that stun the fish). They live in circular villages organized around plazas on which they conduct village rituals and wrestling (the most popular sport).

The Mehinaku believe that they do not have Western technology because the Sun Creator gave the *kajaiba* (outsiders) milk, machines, cars, and planes, while to the people of the Xingu – the Mehinaku and their neighbours – he gave manioc, bows and arrows, and clay bowls.

AMAZON

## Kayapó

**Population** 2,500
**Language** Kayapó (one of around 10 indigenous South American languages, together known as Gê languages)
**Location** Mato Grosso region (Brazil)
**Beliefs** Traditional spirit belief system

The Kayapó traditionally live in circular villages. Women and children occupy straw-roofed houses around a plaza in which a men's house stands. Boys generally join an age-set that acquires progressively more privileges as its members grow older, including permission to enter the men's house. The Kayapó men join one of two ritual societies, and each man generally joins the same group as his wife's father.

Although the Kayapó spend much of the year in horticulture, during the dry season men, women, and children all participate in an extended hunting and gathering trip. In order to sustain this way of life, the Kayapó need access to a considerable amount of land. Since the 1960s, the appearance of plantations and cattle ranches owned by Portuguese-speaking Brazilians (and in more recent decades the arrival of loggers and miners) have threatened this access. In 1989, the Kayapó persuaded the World Bank to abandon funding for the planned Altimira Dam, which threatened Kayapó lands. In 1991, they won control of 44,000 sq km (17,000 sq miles) of rainforest threatened by mining and logging interests.

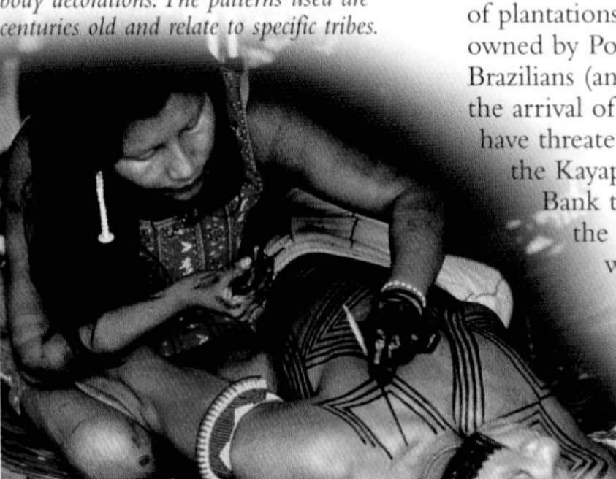

**BODY DECORATION**
*The Kayapó are known for their elaborate body decorations. The patterns used are centuries old and relate to specific tribes.*

**TRIBAL CHIEF**
*The chief of a Kayapó tribe wears a headdress made of bright macaw feathers.*

**ARMED HUNTER**
*During the dry season, the Kayapó traditionally engage in hunting trips. They are highly skilled with bow and arrow.*

AMAZON

## Bororo

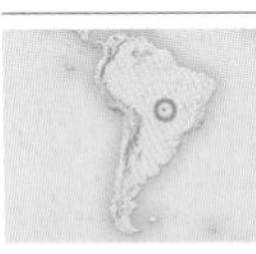

**Population** 700 (split into an eastern and a western division)
**Language** Bororo (a Gê language related to Kayapó)
**Location** Mato Grosso region (Brazil)
**Beliefs** Traditional spirit belief system

The Bororo live in the relatively dry southeast of the Amazon basin. Their society is composed of two halves, with the social organization reflected in the physical layout of the villages. Each community is arranged in a circle. Women and children from one social half occupy the north side, while women and children from the other half inhabit the south side. Membership of these groups is determined by the half to which the mother belongs. Spouses must be taken from the other half, and after marriage, the women remain with their maternal kin group. The men – rather than crossing the village circle to join their wives – spend most of their lives in the men's house (found in centre of the village plaza). This layout remains consistent although the Bororo move frequently.

Contacts with gold and diamond prospectors in the 18th century, and subsequently with Brazilian settlers, have decimated the Bororo population, reducing it perhaps from as many as 15,000 in the late 18th century. However, their traditional social system has remained resilient.

**PARROT FEATHERS**
*Amazon peoples, such as the Bora of Peru (left), adorn themselves with parrot feathers. Bororo men also use parrot feathers in ceremonial headdresses.*

**HIGHWAYS ON THE WATER**
*In the wet season, canoes can be a useful means of transport. In the dry season, however, the Bororo become nomadic and trek across the savanna.*

ARGENTINA

## Argentinians

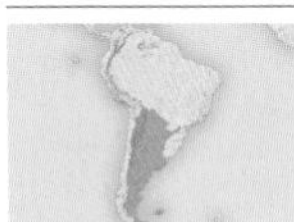

**Population** 37.5 million (over 80% in urban areas)
**Language** Spanish, Italian, Welsh (small communities in Patagonia)
**Location** Argentina
**Beliefs** Predominantly Roman Catholic, Protestant

Argentinians stand out in South America as the most European of the continent's peoples. The country's population at independence (in 1816) was under half a million – comprising the ancestors of Spanish colonialists, the remainder of the Native American population they had conquered, the mestizos (people of mixed Spanish and Native American ancestry), and a smaller group descended from slaves. Nation-building was conceived in political terms: settlers were needed to create a modern state, and northern (Protestant) Europeans were initially targeted as ideal recruits. Northern Europeans, Poles, and Russians are indeed now part of the Argentinian mix, as are Arabs, Jews, and Japanese. However, during the period of mass immigration between the 1850s and the 1940s (when over 3.5 million people arrived), about 45 per cent of those entering the country came from Italy and 32 per cent from Spain. This influx from southern Europe encouraged the view that national identity or *Argentinidad* was both ethnic and cultural

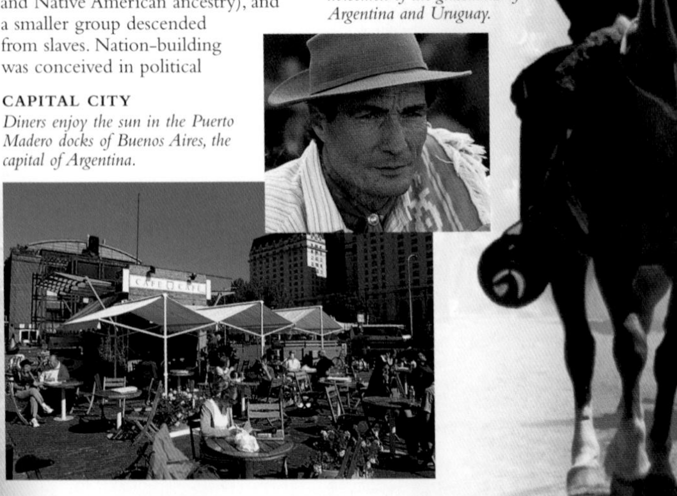

**CAPITAL CITY**
*Diners enjoy the sun in the Puerto Madero docks of Buenos Aires, the capital of Argentina.*

**GAUCHO**
*The gauchos are the nomadic horsemen of the grasslands of Argentina and Uruguay.*

GRAN CHACO

## Guaraní

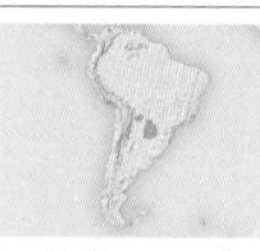

**Location** Paraguay; also Brazil, Bolivia

**Population** 4 million
**Language** Guaraní (a Tupían language), Spanish; Guaraní-Spanish bilingualism is common
**Beliefs** Predominantly folk form of Roman Catholic Christianity

Guaraní was the name given to the Tupí – Native Americans who once lived in present-day Paraguay (and also parts of Brazil and Argentina). It is also the language spoken by the peasants who live in much of the eastern part of Paraguay. The Tupí culture, remembered in historical literature for warfare and cannibalism, has gone. In contrast, the language, which has survived (and may be spoken by people with little or partial Native American ancestry), has become a symbol of Paraguayan identity. Guaraní speakers usually also speak Spanish. However, the two languages are used in different social situations: Spanish is generally saved for formal occasions; Guaraní is the language of intimacy.

Most Guaraní speakers today are self-sufficient farmers raising livestock or growing oranges, wheat, or sugar cane. Many live in families in which the mother is a central figure. Ritual kinship, such as relationships with godparents, is also important.

**CLIMBING TO THE CANOPY**
*A Guaraní child adeptly scales the slender branch of a tall tree near the village of Potrero Guassu, Brazil.*

**HEADDRESS**
*A Guaraní man in a feathered headdress shows one of the many faces of the Paraguayan Guaraní-speaking people.*

URUGUAY

## Uruguayans

**Location** Uruguay; also Brazil, Argentina

**Population** 3.4 million
**Language** Spanish, Portunol/Brazilero
**Beliefs** Christianity: Roman Catholic (although numbers of churchgoers are small), Protestant (evangelical); Judaism

Until the 16th century, Uruguay was inhabited mostly by the indigenous Charrúas tribe, but these people were largely exterminated during Spanish colonization. In 1828, the country gained independence from Brazil and Argentina (although Argentina's cultural and economic influence remains strong). Waves of immigrants (mostly Italians and Spanish, with smaller groups of Jews, Lebanese, and Armenians) arrived from the 1830s onwards.

Although most music and dance traditions (polkas, tangos, and waltzes) are European and Argentinian in origin, they are given a twist to make them distinctly Uruguayan. The same is true of the folk dances and songs sung in Spanish, Portuguese, or a hybrid of the two – *Portunol*, or *Brazilero*. The most important festival is Carnival Week, where the people of Montevideo dress up and deliver themselves to the rhythms of *la morenada* (Afro-Uruguayan) drummers. The holiday of *La Semana Criolla* is an opportunity for urbanites to play at being gauchos (fabled cowboys of the pampas) by indulging their appetite for meat at *asados* (barbeques).

**ETHNIC MIX**
*Uruguay's population includes various ethnic minorities, but most people are of European and Argentinian origin.*

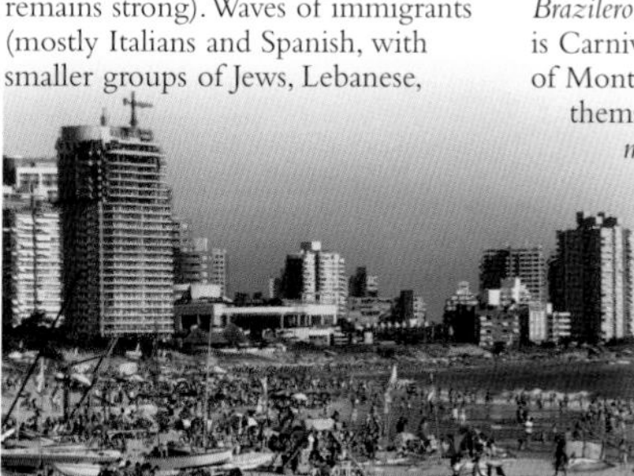

**CITY ON THE SEA**
*The city of Punta del Este in southeastern Uruguay is a popular beach resort.*

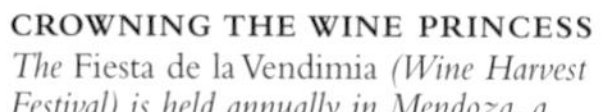

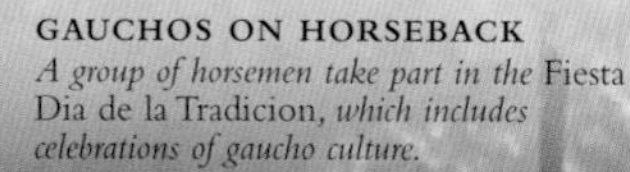

**COLOURFUL SASH**
*The gaucho traditionally wear a colourful sash called a* faja. *It is tied around the waist, with one end left hanging over the thigh.*

in character, with Catholicism as one key ingredient (although perhaps only 10 per cent of the population are now practising Catholics) and language another. *Lunfardo*, the street slang of Buenos Aires, is heavily influenced by Italian, which many Argentinians still speak, and by Portuguese and African languages. *Lunfardo* can now often be heard in the lyrics of tango, a national style of music and dance. Some rural communities, such as the Welsh-speakers found in Patagonia, have preserved their own language.

Cultural symbols (such as the tango) and passionate nationalism (focused above all on football) reinforce the stereotype of the Argentinian people as individualistic and ostentatious. A powerful mythical element of national identity (although far removed from Argentinians' mainly urban lifestyle) is the lone gaucho, riding the range on horseback, fiercely self-reliant yet given to *gauchadas* (acts of spontaneous generosity and helpfulness) in which Argentinians take particular pride.

Argentina has good social welfare and education systems, and practically all Argentinians are literate.

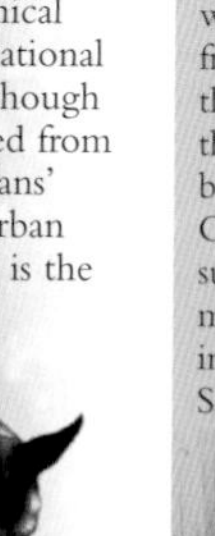

**GAUCHOS ON HORSEBACK**
*A group of horsemen take part in the* Fiesta Dia de la Tradicion, *which includes celebrations of gaucho culture.*

**CROWNING THE WINE PRINCESS**
The Fiesta de la Vendimia *(Wine Harvest Festival) is held annually in Mendoza, a wine-producing province. The celebrations include crowning a "wine princess".*

History

### Maté: a national obsession

Maté, a stimulant drink similar to tea, was first made by the Guaraní people, from the yerba-maté bush (which they called *caa*). It was adopted by the Spanish in the 16th century, and became so popular that the Catholic Church tried to ban it, but without success. Drinking maté is still a major element of traditional culture in Argentina, Uruguay, and other South American countries.

CHILE

## Chileans

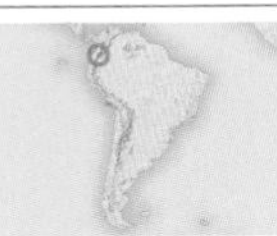

**Population** 15.8 million
**Language** Spanish and a number of indigenous languages
**Beliefs** Roman Catholic Christianity; also Evangelical Protestant Christianity, Judaism
**Location** Chile

Prior to the Spanish invasion (from 1535), the Inca empire conquered most indigenous cultures in the area of modern Chile. They did not overcome the fierce Araucanos (the ancestors of today's Mapuche), however. Over time, the Araucanos interacted with Spanish settlers, and centuries of interbreeding produced the mainly mestizo (mixed Spanish and Native American) population of today. Compared to that of other South American countries, 19th-century migration from Europe was modest. Yet the impact of incoming English, Italian, and French traders was great, because they intermarried with the local white elites and forged an affluent, cosmopolitan culture that today retains its European character. There was a larger-scale German and Swiss colonization of the southern lakes region, and their legacy still pervades many communities in the south. Despite this variation, Chileans take pride in their Spanish-influenced folk culture, expressed in dance and song forms such as the *cueca* and *tonada*. Chileans also have a passion for football, with emotions running high during matches with Argentina. The land, poverty, stoicism, and hospitality of the Chileans was immortalized by Pablo Neruda (1904–1973), a winner of the Nobel Prize for Literature.

**HONOURING THE VIRGIN**
*Men carry the Virgin Mary surrounded by flowers during one of the many colourful religious festivals held in Chile.*

**RODEO RIDER**
*A Chilean man wears the typical poncho and hat of a gaucho (a farmhand hired for his skilled horsemanship).*

Issue

### The cost of growth

Chile's economy – particularly the remarkable growth recently – has been based on extracting natural resources: first mining silver and copper from the deserts of the north (below) and then taking timber and pulp from the southern forests. Government policy has consistently supported this, with little regard for environmental issues such as pollution and lost biodiversity.

ANDES

## Paez

**Population** 70,000
**Language** Paez (part of the Chibchan language group, a set of Native South American tongues)
**Beliefs** Catholic Christianity influenced by traditional spirit beliefs
**Location** Mostly in the Cauca valley, Colombia

The Paez live mainly as farmers in the Cauca Valley, where they cultivate potatoes, coffee, hemp, plantains, cassava, and coca. Most live in sun-dried brick houses or in wattle-and-daub dwellings centred on a town plaza. This plaza is the site of communal labour projects known as *minga* and of rituals including puberty rites for girls and dances to mark the death of children.

**PAEZ MAN**
*Like many native South Americans, the Paez have adopted the dress of Spanish colonists, like this traditionally shaped sombrero.*

In 1971, the Paez – along with the neighbouring Guambianos and several other minority groups – founded the Indigenous Regional Council of the Cauca (CRIC), whose main aim was to recover lands that were lost during the previous century to wealthy landowners and non-indigenous sharecroppers. By 1993, CRIC had recovered large tracts of land, but had lost hundreds of activists to hired killers, police, military, and guerrillas.

ANDES

## Jívaro

**Population** 35,000
**Language** Jívaro (part of the Jebero-Jívaroan language group)
**Beliefs** Traditional spirit beliefs; also Roman Catholic Christianity
**Location** Forest areas of Ecuador

The Jívaro have a fearsome reputation as headhunters, and the shrunken heads of their victims occupy pride of place in many ethnological museums. Although the Jívaro people no longer collect heads, those who live in the remote, heavily forested mountain homelands are still well beyond the reach of Ecuador's state authorities. In the past, the Jívaro also resisted subjugation by the Incas and the Spanish.

The Jívaro subsist primarily on cassava, sweet potato, plantains, game, and fish. Most of them live in nuclear households, although some men have two or more wives. The Jívaro enjoy a rich spiritual life, focused on seeking *arutam* (souls), sometimes with the aid of hallucinogens (especially datura). However, they are very much engaged with worldly concerns as well. Protests by Jívaro people against environmental damage by multinational oil and gas companies, for example, have been particularly vocal. Also, a recent Jívaro community radio project has raised literacy among the Shuar (one of the larger Jívaro groups) by 90 per cent.

**JIVARO WOMAN**
*Gender roles are sharply defined in Jívaro life. Women cook, cultivate, tend animals and children, and make manioc beer.*

**HUNTING WITH A BLOWPIPE**
*The blowpipe is the chief weapon of the Jívaro. Darts are tipped with a paralyzing toxin (curare) derived from rainforest vines.*

ANDES

## Otavaleños

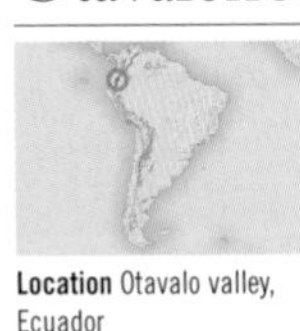

**Population** 60,000
**Language** Quechua (language of the Incas; some Ecuadorean varieties are known as Quichua)
**Beliefs** Christianity influenced by traditional spirit beliefs
**Location** Otavalo valley, Ecuador

Otavaleños are one of many peoples conquered by the Incas in the 15th century. Like the other vanquished peoples, they are broadly described as Quechua because they speak the Quechuan language. Previously, however, they spoke a Chibchan language related to Paez. Otavaleños are subsistence farmers, cultivating potatoes and raising livestock. Some work in the textiles industry (large numbers of Otavaleño textiles are exported to fashion buyers in North America), and today several are prosperous business people. Tourists now flock to Otavalo, attracted by the distinctive local costume, with its mixture of pre-Hispanic, Spanish colonial, and modern influences.

**SPINNING WOOL**
*The Otavaleños are known for their woven alpaca- and llama-wool textiles, notably ponchos, shawls, and rugs.*

ANDES

## Campa

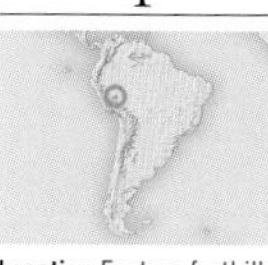

**Population** 37,000
**Language** Campa (an Arawakan language)
**Beliefs** Evangelical Christian denominations, Seventh Day Adventist Church, traditional spirit belief system
**Location** Eastern foothills of the Peruvian Andes

The seven groups that make up the Campa live in the Montaña country on the eastern flanks of the Andes, surviving by swidden (slash-and-burn) horticulture and hunting and fishing. Calling themselves "Ashaninka" ("our kind"), the Campa consider that they are different from their neighbours, the culturally and linguistically very similar Matsigenka (who also like to refer to themselves as Ashaninka).

The Campa have had a violent and troubled history. Disease, slave raids, and other social atrocities decimated many Campa groups during the early 20th-century rubber boom. It was only in the 1960s that the slave trade was finally curtailed in Campa country. In recent decades, thousands have died or disappeared during the confrontation between the Peruvian government and Shining Path (Communist Party of Peru) guerrillas. Possibly a quarter of all Campa have fled their homes.

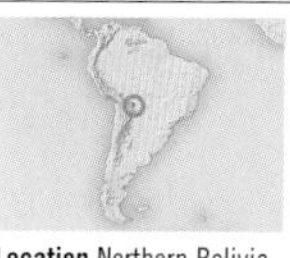

**IN A CUSHMA OF COTTON**
*Campa people wear coarsely woven cotton tunics, called* cushmas*, made from cotton plants grown near the village.*

**TRADING FOR SALT**
*Campas' neighbours, the Matsigenka, meet white settlers to trade salt on the River Manu, Peru. The Matsigenkas' simple canoes are hollowed out of tree trunks.*

ANDES

## Aymará

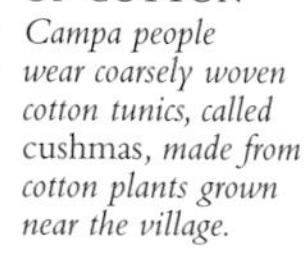

**Population** 2 million
**Language** Aymará (one of three languages in the Aymaran language family)
**Beliefs** Catholic Christianity influenced by traditional spirit beliefs
**Location** Northern Bolivia, southern Peru

The Aymará are a Native American group living on the Altiplano (the high plateau of the Andes), particularly in the area around Lake Titicaca. Many Aymará are pastoralists, tending flocks of sheep, llamas, and alpacas, although in the Potosí region of Bolivia, some have worked in silver and tin mines.

The most important unit of social organization in Aymará communities is usually the *ayllu*, which consists of several extended families who own shared land. The majority of marriages take place within this group. Long-distance trade is also important to the Aymará people, and many sustain important trading relationships with what are called *compadres* (ritual kin) and *ayni* (formal exchange partners).

One typical group of Aymará, the Laymí, live in the area around the tin-mining settlement of Llallagua. The patron saint of this town is the Virgin of the Assumption, whose annual feast occurs on 15 August. This celebration is remarkable for the procession of dancers who gather in the streets, drinking heavily and revelling with intense physical exertion – despite the altitude of 3,700m (12,000ft) above sea level.

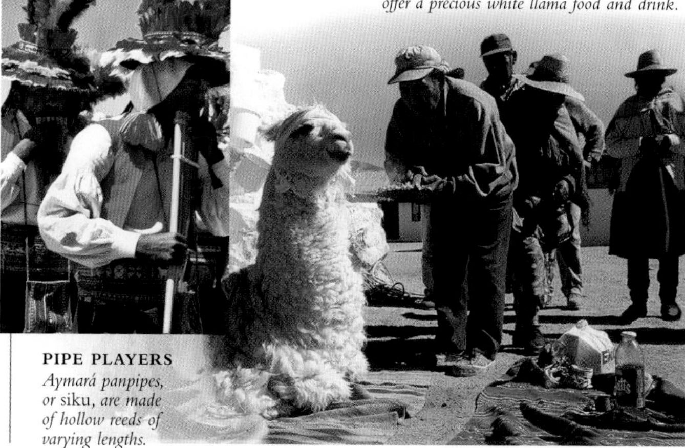

**OFFERINGS TO A WHITE LLAMA**
*Llamas were domesticated by native South Americans around 7,000 years ago, and are invaluable to their way of life. Here, men offer a precious white llama food and drink.*

**PIPE PLAYERS**
*Aymará panpipes, or siku, are made of hollow reeds of varying lengths.*

ANDES

## Taquileños

**Population** 1,200
**Language** Quechua (introduced when the Incas conquered Taquile in the 13th century)
**Beliefs** Catholic Christianity influenced by traditional spirit beliefs
**Location** Taquile island (Lake Titicaca, Peru)

Taquileño people live on the island of Taquile, in Lake Titicaca, which straddles the border between Peru and Bolivia. Taquileños grow potatoes, corn, broad beans, and quinoa, with tourism supplementing their income. They are famed for their costumes and knitted and sewn textiles. The colouring of their costumes reveals marital status: married men wear *chulos* (stocking caps) with red pom-poms, while unmarried men have white pom-poms. Married women wear multilayered, dark-coloured skirts, while unmarried ones wear brighter colours.

The original inhabitants of Taquile were from the Tiahuanuco civilization, which some believe dates as far back as the 2nd millennium BC. During the 1930s, the Peruvian government used the island as a prison, but in 1937 the Taquileño people bought it back.

**WARM IN WOOL**
*Wool caps give welcome protection against the cold Andean nights. Today many are sold to tourists on day-trips.*

**SPINNING WOOL**
*Taquileño costumes are worn for daily tasks as here but are shown to full effect during carnivals.*

**TRAVELLING TO MARKET**
*A Mapuche family travel to market in a traditional ox-drawn cart. Most Mapuche people are small-scale farmers.*

ANDES

## Mapuche

**Population** 1 million
**Language** Mapudungun
**Beliefs** Traditional spirit belief system (*machi, or shamans, are influential and able to ward off evil forces sent by kalku,* or witches)
**Location** Southern central Chile

The group of people known today as the Mapuche are really three distinct Araucanian peoples (a group of Native Americans now found in southern central Chile): the Picunche, the Huilliche, and the Mapuche proper. All three peoples speak the Mapudungun language. The lands that they occupy are exceptionally diverse and support a wide variety of subsistence farming, including the cultivation of wheat, maize, potatoes, beans, squash, and peppers. They also raise sheep, cattle, horses, guinea pigs, and llamas (wealth was once measured in numbers of llamas). Harvests of wheat and maize are often celebrated with rituals known as *ngillatun*.

Both the Incas and the Spanish peole found that the Mapuches were difficult to subdue and, out of the three Araucanian peoples, they only succeeded in conquering the Picunche. However, by the end of the 19th century, the Mapuche and Huilliche had been pacified and placed on thousands of small reservations.

**VENETIAN TRADITION**
*Venice has some of the oldest continuous cultural traditions in Europe. The annual* Regata Storica *(historical regatta) shown here first took place in the 13th century.*

# EUROPE

People have used the term "Europe" for thousands of years, yet what it actually means, both culturally and geographically, is still unclear. This uncertainty arises chiefly because Europe has a blurred boundary with Asia, and Russia and Turkey are counted as part of both Europe and Asia. However, people generally accept that Europe encompasses some 44 countries, and that its peoples can, in theory at least, lay claim to a European identity, as well as to the identity of their ethnic group or their nation state.

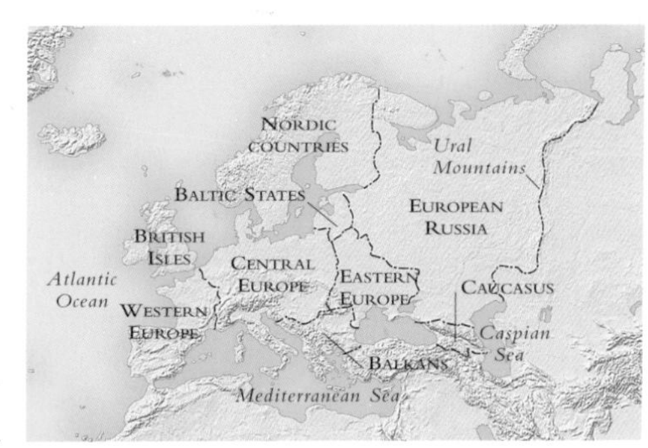


**THE REGION**
*Europe stretches from the Arctic in the north to the Mediterranean in the south, and from the Atlantic in the west to the Ural Mountains and Caspian Sea in the east.*

Europe's cultural landscape was formed by waves of migrants arriving first from southwestern Asia and the steppes of Central Asia, and more recently from southern and eastern Asia, Africa, and the Caribbean.

**CELTIC BROOCH**
*Celtic artefacts such as this Irish silver button are found throughout Europe – a sign of one-time Celtic dominance.*

## MIGRATION AND SETTLEMENT

Of the peoples arriving in Europe during the great ethnic migrations from the eastern steppes, the most influential were the waves of Indo-Europeans. They included, or developed into, the Celts, Greeks, Italic and Germanic peoples, and Slavs. Their migrations spanned several thousand years, and lasted until the Middle Ages.

The Celts spanned the continent from the British Isles to northern Italy, Spain, and Austria, but their dominance was long overshadowed by the Roman empire. The gradual collapse of this empire (from the 3rd to the 6th century AD) led to Germanic peoples from northern Europe establishing kingdoms from Spain, through France, and into Italy. During medieval times, the lure of land and riches drew Vikings and Normans as far south as Sicily, while Slavs founded powerful kingdoms in the east.

On the edges of the continent were other peoples, who were not part of the Indo-European grouping. Northern Russia, for example, became home to Finnic speakers such as the Karelians (*see* p383), Mari, and Udmurts (*see* pp396–397). In addition, Turkic migrants arrived in European Russia in the 13th century, and the Jewish people spread across mainland Europe.

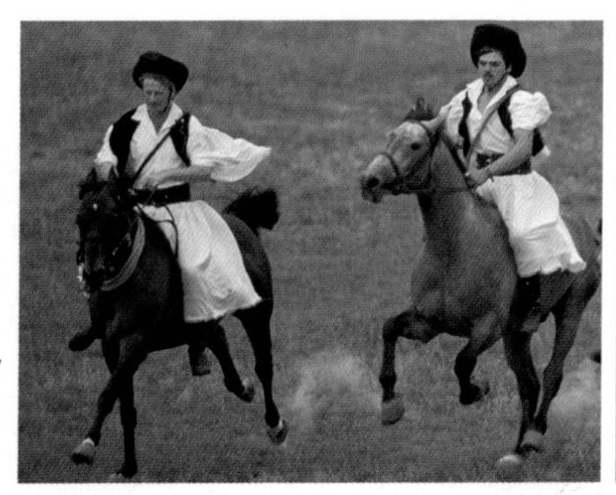

**MAGYAR RIDERS**
*The Magyars of Hungary are descended from nomadic horsemen who migrated west from the Asian steppes.*

## TRIBES AND NATIONS

Europe is unusual globally in that its political boundaries tend to align with its ethnic groups. This feature is largely due to the establishment in Europe of the world's first nation states, such as Portugal, France, and England. The states were all single administrations that aimed to bring together people of shared language and culture. An effort was made to "make" nations of France and Italy, for example, while the disparate city states and principalities of the Holy Roman Empire submerged under Prussian influence into a single Germany. The model worked less well in places such as the ethnically mixed Balkans, because Slavic peoples were deeply divided by differing religions and cultures. In fact, the Austro-Hungarian empire, Yugoslavia, and the Soviet bloc (the alliance of eastern European countries dominated by the old Soviet Union) disintegrated in the attempt to override such tribal divisions. Certain regions were left with boundaries of national hostility between areas of similar culture, such as Serbia and Croatia.

The effort to unify lands also left minority communities: the Celtic populations of Britain and France, the Basques of northern Spain, dispersed populations in the Balkans, and the many non-Russian peoples of the Russian empires. Most of these isolated groups have since sought to gain autonomy, some resorting to violence in the name of their cause. However, there are also minority communities in Europe, such as the Sorbs and Bavarians (*see* p391), who have managed peacefully to retain their ethnic identity within their larger nation state or states.

Issue

### A new nationhood

While Europe becomes more unified under the European Union, some of its nations are also reasserting their independent identities, such as the Celtic peoples in the UK and France, and the Basques and Catalans in Spain. The Scots have even restored their own Parliament, which held its first session in 1999.

**SCOTS CHEERING THE RETURN OF THEIR OWN PARLIAMENT**

**MULTICULTURAL CITY**
*The latest wave of migrants to Europe comes from far afield. Many European cities, such as Dublin (right), now have thriving Asian communities.*

## CHANGE

Centuries of division, climaxing with the horrors of World War II, have urged Europeans towards unity. A key step was the fall of the Soviet bloc, which allowed the merging of eastern and western cultures. The creation of the European Union (EU) not only promotes a "national" European identity, as symbolized by the euro currency, but also upholds the cultural and linguistic identities of its component peoples.

Adding to Europe's diverse mix of peoples is a recent influx of immigrants. Like the Celts, Slavs, and Turkic peoples before, they are likely to form new composite cultures within Europe's ever-changing landscape.

NORDIC COUNTRIES

## Icelanders

**Population** 286,000
**Language** Icelandic (related to Old Norse; the alphabet has 36 letters including two not used in any other language)
**Location** Iceland
**Beliefs** Mainly Evangelical Lutheran

Icelanders are descendants of the Vikings, who arrived in Iceland in the 9th century AD. There is also evidence of an early influx of Celtic settlers. Iceland has the oldest functioning legislative assembly in the world, the Althing (established AD930). In 1380, governance of Iceland was transferred to Denmark, and it was not until 1944 that all formal links between the countries were severed. The law says that Icelandic surnames must be constructed by adding the suffix "-son" or "-dóttir", for son or daughter, to the father's or mother's first name. In 1980, Icelanders elected the world's first female president.

**COD FISHERIES**
*Fishermen such as these still provide much of Iceland's income, despite dwindling Atlantic cod stocks.*

NORDIC COUNTRIES

## Norwegians

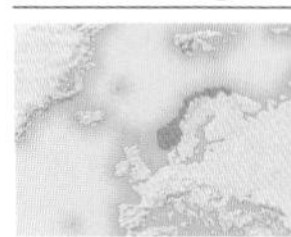

**Population** 4.5 million in Norway
**Language** Norwegian (a Germanic language with two forms, Nynorska and Bokmål)
**Location** Norway, Svalbard (Norwegian Arctic islands)
**Beliefs** Mainly Evangelical Lutheran

While the people of Norway can trace their descent back to the Germanic tribes, the roots of Norwegian statehood lie in the Viking era. The territory was unified under the rule of King Harald (AD860–940), and Christianity first arrived in Norway around this time. From the 14th century until 1814, Norway was unified with Denmark. More recently, in the 1960s, Norway's economy was boosted by the discovery of gas and oil deposits in territorial waters. The Norwegians are an independent people and have twice voted against European Union membership, in 1972 and 1994.

The Norwegian language is divided into two different forms: Nynorska (meaning "new Norsk", which is based on Norwegian dialects) and Bokmål (meaning "book language", which is originally derived from Danish and has been adapted a number of times since).

The Norwegians have a special fondness for skiing, camping, and enjoying the beauty of the fjords. They have also retained much of their cultural heritage: traditional folk singing, dancing, and storytelling are still popular today. Seafood is widely eaten, and a Christmas speciality is *lutefisk* (dried cod soaked in lye). Norwegian people also benefit from an extensive and advanced social welfare system, an excellent standard of living, and one of the highest life expectancies in the world.

Well-known Norwegians include artist Edvard Munch, playwright Henrik Ibsen, composer Edvard Grieg, and the first person to reach the South Pole, Roald Amundsen.

**MIDDAY TWILIGHT**
*Norwegians in the north of the country contend with extremes of day length. The town of Tromsø experiences midnight sun in summer, but in winter it is dark at midday.*

**RURAL LIFE**
*Norwegians have never been short of timber, and buildings in rural areas are wooden. Remote parts of Norway such as this are reached in winter by snowmobile.*

NORDIC COUNTRIES

## Swedes

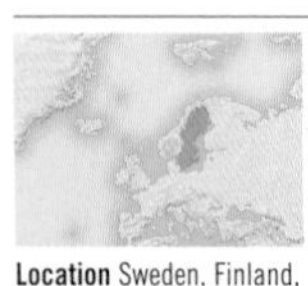

**Language** Swedish (a northern Germanic language; part of the Indo-European language family)
**Beliefs** Mainly Evangelical Lutheran (87% of the population); also Roman Catholic, Eastern Orthodox
**Location** Sweden, Finland,
**Population** 8.9 million

The Swedes originate from Germanic tribes that inhabited regions of Scandinavia after the Ice Age. In Viking times, Sweden consisted of several competing kingdoms, which were eventually unified to form one single state under one king. In 1939, the country assumed a stance of political neutrality, which it retains today. This has helped Sweden prosper, and the Swedes enjoy excellent social welfare provision. The Swedish people are great nature lovers, and they are also known for their contributions to scientific research. Furthermore, it was a Swede – the 19th century chemist and engineer Alfred Nobel – who founded the Nobel Prize.

**SMORGASBORD**
*As much a tradition in Denmark as it is in Sweden, the cold buffet meal known as smorgasbord includes crispbreads, pickled fish, salads, pies, and cured meat.*

**MIDSUMMER CELEBRATION**
*Two days of festivities mark the summer solstice. People enjoy pagan customs such as dancing around a maypole.*

NORDIC COUNTRIES

## Danes

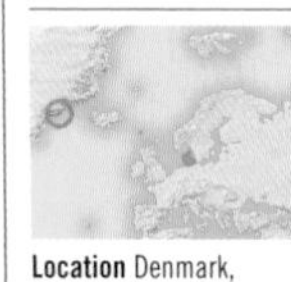

**Population** 5.4 million in Denmark
**Language** Danish (a northern Germanic language, part of the Indo-European family)
**Location** Denmark, Greenland, Germany
**Beliefs** Mainly Evangelical Lutheran

The Danes are closely related to their Scandinavian neighbours. The Danish people had extensive contact with other Europeans early in their history, which is reflected by the arrival of Christianity in Denmark in the 10th century.

The Danes live in a thoroughly modern society, and for this reason traditional ways are less prevalent than in many other European countries. They are also a tolerant people. For example, in 1989, Denmark became the first country in Europe to legalize same-sex marriages. The Danish outlook is perhaps best described by the concept of *hygge*, which means cosy and snug – striving to create an intimate mood. *Hygge* is reflected in many facets of life, including a preference for small pubs and cafés. Denmark boasts a modern welfare state in which few have too much and even fewer have too little. Living standards are generally very high.

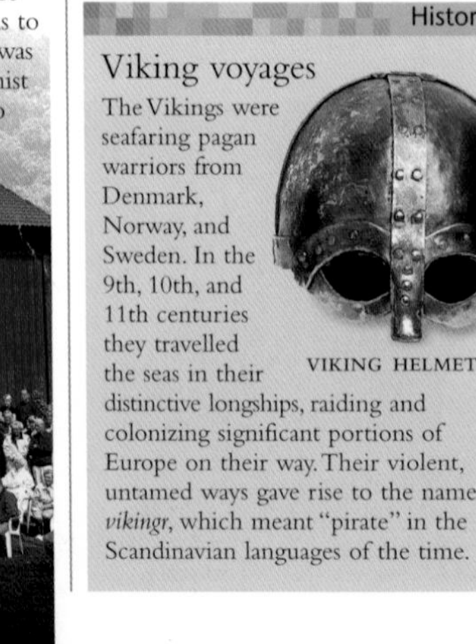

History

### Viking voyages

The Vikings were seafaring pagan warriors from Denmark, Norway, and Sweden. In the 9th, 10th, and 11th centuries they travelled the seas in their distinctive longships, raiding and colonizing significant portions of Europe on their way. Their violent, untamed ways gave rise to the name *vikingr*, which meant "pirate" in the Scandinavian languages of the time.

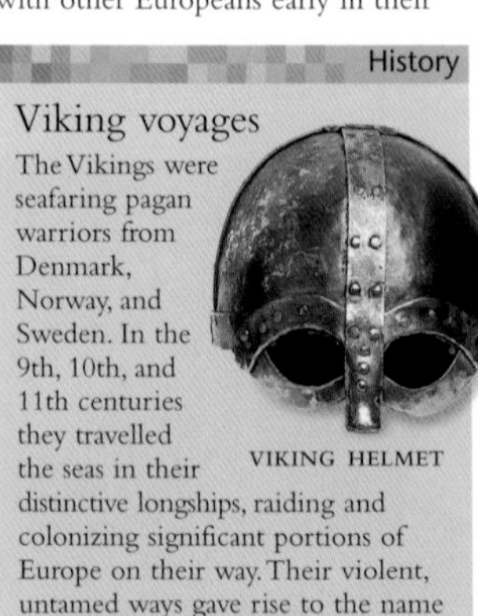

**VIKING HELMET**

**NORDIC SKIERS**
*A Danish couple enjoy watching a cross-country ski event being held in Sweden. There are few hills suitable for this in Denmark.*

NORDIC COUNTRIES

## Finns and Karelians

**Population** 5.2 million
**Language** Finnish and Karelian (languages of the Finno-Ugric group related to Estonian and Hungarian)
**Location** Finland, Sweden, Karelia (in Russia)
**Beliefs** Mainly Evangelical Lutheran

Finland lies between the East and the West geographically, culturally, and politically. To the east, in Russia, live the Karelian people, who are closely related to the Finns and also speak a Finnic language. To the west, Finland's strong association with Sweden from 1323 to 1809 has meant that Swedish religious, judicial, and parliamentary systems now form the basis of Finnish statehood. By reputation, Finns love nature and are not usually given to small talk. Sauna is a quintessential part of Finnish culture: Finland has two million privately owned saunas.

*Sisu* is a concept familiar to all Finns. It means inner strength – in particular the persistence and patience that promotes personal survival. *Sisu* is often used to describe Finnish sportspeople or when discussing the Finnish troops' resistance to the Soviets during the winter of 1939–1940. Although generally regarded as a positive trait, "bad" *sisu* describes either stubbornness and stupidity or poor behaviour in children.

**WINTER FUR**
*Thick furs keep out the winter cold of northern Finland. Finns are a nature-loving people, who escape to their summer cottages and ski-slopes at weekends.*

**REINDEER RACE**
*During winter, the lakes in some regions act as surfaces for races where reindeer pull along skiers.*

BALTIC STATES

## Estonians

**Population** 1.4 million
**Language** Estonian (a Finnic language related to Finnish and, more remotely, to Lapp), Russian
**Location** Estonia and northwestern Russia
**Beliefs** Evangelical Lutheran, Russian and Estonian Orthodox

**ESTONIAN FOLK SINGERS**
*These singers in traditional costume express national sentiment in song, as Estonians did in the "Singing Revolution" of 1988–1991.*

From the 13th century until 1918, the Estonians lived under the rule of various foreign powers. The Danes, Germans, Poles, Swedes, and Russians have all left their imprint on Estonia. Since independence in 1991, the country has undergone a transformation. A new constitution was announced in 1992, and the first elections to the single-chamber parliament were also held that year.

Estonians enjoy a rich tradition of folk poetry known as *runo-songs*. Choral singing is another cultural highlight, and some Estonian song festivals feature thousands of singers.

NORDIC COUNTRIES

## Sámi

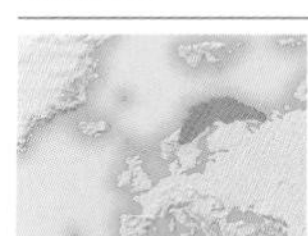

**Population** 75,000
**Language** 10 Sámi languages or dialects (all related to Finnish), Norwegian, Swedish, Finnish, Russian
**Location** Lapland (Arctic Scandinavia, northern Finland, northwest Russia)
**Beliefs** Mostly Evangelical Lutheran and Orthodox Christianity, shamanism

The Sámi, also known as Lapps, are the indigenous people of Lapland, an area covering the northern parts of Norway, Sweden, and Finland as well as the Kola Peninsula in Russia. The Sámi are genetically different from the rest of the peoples of Europe. They are distributed across a vast area, with very varied environments, living conditions, cultural traits, uses of the Sámi language, and levels of contact with other ethnic groups. Today the Sámi have cultural autonomy and their own flag; and each Sámi has a personal *joik* (folk song). When the Sámi sing, they do not sing about a subject. Instead, they say that they *joik* their subject.

Historically, they are a nomadic people, surviving by herding reindeer, hunting, and fishing. However, more recently, they have been encouraged to pursue a more settled lifestyle. The Sámi mostly live in small family groups and are a peace-loving people. Theirs is said to be the only known culture that has never been to war.

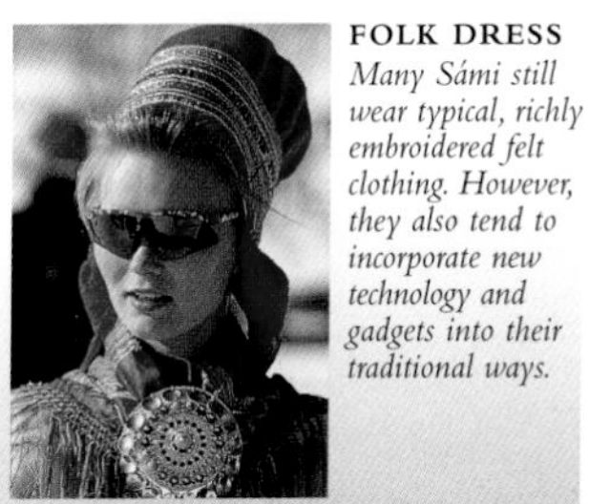

**FOLK DRESS**
*Many Sámi still wear typical, richly embroidered felt clothing. However, they also tend to incorporate new technology and gadgets into their traditional ways.*

**THE LAST NOMADS**
*In a continent of mainly urbanized people, some Sámi continue to herd reindeer, an activity that demands a nomadic lifestyle.*

BALTIC STATES

## Latvians

**Population** 2.4 million in Latvia
**Language** Latvian (a Baltic language related to Lithuanian), Russian
**Location** Latvia and Russia
**Beliefs** Evangelical Lutheran, Roman Catholic, Russian Orthodox

The Latvians (or Letts) originated from Baltic tribes that settled in the region in around 2000BC. Since the 9th century AD, Latvians have been ruled by Vikings, Germans, Poles, and – since the end of the 18th century until very recently – Russians. By the end of the 1980s, ethnic Latvians comprised just 50 per cent of Latvia's population. An independent Latvian government was established in 1991.

**TRADITIONAL DANCERS**
*National costume, song, and dance strengthens Latvian identity. Riga stages a Song Festival every 5 years.*

BALTIC STATES

## Lithuanians

**Population** 3.5 million
**Language** Lithuanian (a Baltic language, and the most archaic of today's Indo-European languages); also Russian, Polish
**Location** Lithuania, Poland
**Beliefs** Mainly Roman Catholicism

Lithuanians, by stereotype, are an often gregarious and emotional people. They are descended from the Baltic tribes that settled in the area in 2000BC. They resisted encroachments by neighbouring powers and formed their own state in 1253. In 1386, Lithuania entered into a union with Poland, and by the early 16th century the kingdom of Poland-Lithuania had become the largest state in Europe.

The Lithuanians adopted Catholicism as a result of their link with Poland. It became a key feature of national identity after the region succumbed to the rule of Orthodox Christian Russia in 1795. The regime imposed an extensive "Russianization" policy on Lithuanians. During World War II, Lithuania suffered 3 years of Nazi occupation, and all of the 220,000 Lithuanian Jews were put to death. The Soviet Union reoccupied Lithuania in 1944, and 250,000 people were deported to Siberian *gulags*. Lithuania reclaimed independence in 1991 and has since begun its "return to Europe", economically, politically, and socially.

**SAINT GEORGE**
*This carving depicts the patron saint of the Lithuanians.*

**AMBER SELLERS**
*Residents of Lithuania have traded locally available amber since the Bronze Age.*

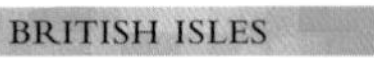

BRITISH ISLES

## Scots

**Population** 5.35 million
**Language** English; Scots, Orcadian (related to English), Scots Gaelic (Celtic language)
**Location** Mainly Scotland; also England, Canada, US
**Beliefs** Church of Scotland (Presbyterianism), Church of England (Anglicanism)

There are several strands making up modern Scots, including the pre-Roman Picts, the "Scots" invaders from Celtic Ireland, the Anglo-Saxons, and Nordic settlers on the far northern islands. The idea of Scottishness consequently lacks the "ethnic" coherence enjoyed by the Welsh, for example. Also, the Scots Gaelic language is not a crucial part of national identity since English is universal.

Scotland was incorporated into the UK by the 1707 Act of Union, and for almost 300 years it was subordinated to the English Parliament. This connection did not dampen Scots nationalism, however, which had been forged by over 1,000 years of antagonism with the English. In 1999, the nationalist movement was rewarded with the devolution of lawmaking powers to a revived Scottish Parliament.

Traditional Scottish dress is well-known worldwide. The kilt, a wraparound skirt worn by Highland men, is a familiar Scottish symbol, especially when worn with a sporran (purse), a *lèine* (shirt), a *còta* (coat), a *bonaid* (hat), and the bagpipes. Although the tartan patterns on the kilts are now ascribed to the various Scottish clans, the clan designations were lost during an 18th-century ban imposed by the British on Scottish dress (in an effort to stamp out Highland rebellions).

Scots made up a large portion of early migrants to the US, Canada, Australia, and New Zealand. All over the world, Scots observe tradition by celebrating Burns Night (25 January), in honour of Scotland's famous poet, Robert Burns (1759–1796). A Burns supper includes haggis (stuffed sheep stomach), which often featured in Burns's works.

**WATER OF LIFE**
*A Scot tends a copper still in the process of distilling whiskey, a spirit made from malt-flavoured barley.*

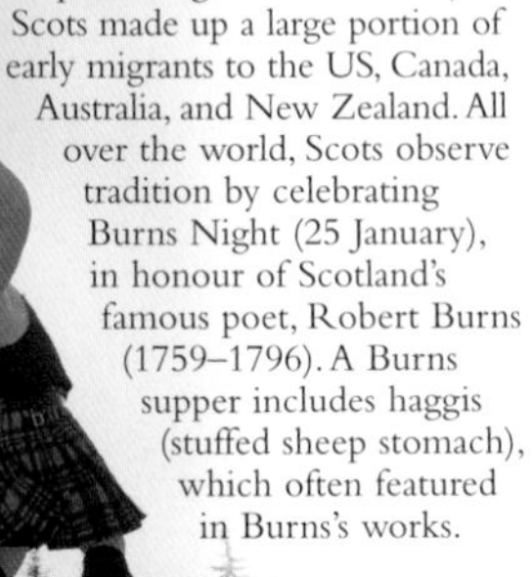

**HAMMER THROW**
*A kilted competitor at one of Scotland's many Highland Games, which feature traditional athletics events.*

BRITISH ISLES

## Irish

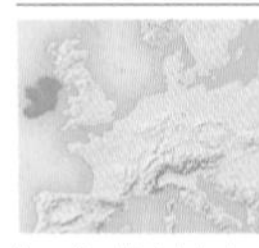

**Population** 5.3 million
**Language** English, Gaelic (an official language of the Irish Republic)
**Location** Mainly Ireland; also UK, US, Australia
**Beliefs** Roman Catholic (mainly Irish Republic), Protestant (principally Northern Ireland)

Issue

### "The Troubles"

Conflict in Northern Ireland is linked to the UK's historically Protestant governance of the region. Catholic Republicans wish for a united Ireland, while Protestant Unionists fight to maintain the link with the UK. One potential flashpoint occurs if the (Protestant) Orange Order parades are permitted to march through Republican areas.

**ORANGE DAY PARADE**

Before the arrival of the Celtic Gaels sometime between 600BC and 150BC, pre-Celtic tribes inhabited Ireland. Later invasions by Vikings and Anglo-Normans created the modern Irish people. Their ancestral Celtic language, *Gaeilge* (Gaelic), carries these historic influences and was revived strongly from the 19th century onwards.

Although rural communities still exist, Irish life is largely urbanized. Connections with European culture are strong in the cities, and the economy is increasingly service-based. Despite this, folk music, folk dancing (complete with costumes, exemplified by the *ionan* tunic), are still popular. Among Ireland's best-loved sports are home-grown games such as Gaelic football and hurling, a sport related to hockey. Poetry is also an important part of Irish tradition and history, first in Gaelic, then in English. Irish poets, as well as novelists and dramatists, remain influential worldwide.

In the 19th century, harsh conditions prompted hundreds of thousands of Irish to flee their homeland. As a consequence, a high proportion of emigrants to the New World and Australia were Irish. Many assert their Irishness on St Patrick's Day (17 March), usually with a glass or two of one of Ireland's most famous exports, Guinness stout.

**BOOK OF KELLS**
*Irish literature was already at least 400 years old when the illuminated gospel called the* Book of Kells *was completed in the 9th century. Although it is written in Latin, it is viewed as a monument of early Irish civilization.*

**TRADITION OF MUSIC**
*The banjo, guitar, and fiddle have joined traditional Irish instruments, such as the harp, in today's Irish folk music.*

BRITISH ISLES

## English

**Population** 47.4 million
**Language** English (a Germanic language with much French vocabulary)
**Beliefs** Anglican Protestant, other Christian denominations
**Location** Mainly England; also Wales, Scotland

The English evolved slowly as a people as influxes of northern German and Danish invaders intermarried with the native people of England from the 5th century AD onwards. The English language, once similar to those spoken by many Germanic and Nordic tribes at the time, has since absorbed many other strands, principally French.

Several key events in English and British history redefined the bond between the English people and their government. The signing of the agreement of the Magna Carta in 1215, the English Civil Wars of 1642–1651, and the Glorious Revolution of 1688–1689, each eroded the power of the monarchy and lent a powerful sense of institutionalized "fairness" to the people. Furthermore, notions of Englishness (a reserved nature, a sense of fair play) were spread worldwide by the British empire. In fact, "Englishness" is often dwarfed by "Britishness", a concept that incorporates the other nations of the UK and its various immigrant peoples.

The English have invented some of the world's most popular sports, including football, rugby, and cricket. The arts are also highly valued, and English literature, drama, and music regularly achieve global acclaim.

**VILLAGE CRICKET**
*Englishmen playing cricket on a village green beside an old pub is a dearly-held ideal for some.*

**ENGLISHMEN'S HOMES**
*It is said that an Englishman's home is his castle, meaning the English value the privacy of home life. The feeling is reflected in modern English housing developments.*

**SEASIDE**
*The English still enjoy sunny days on the local beach. The huts, although not as popular as they once were, maintain English people's much-valued privacy when changing into swimwear.*

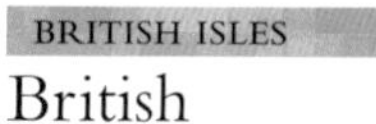

BRITISH ISLES

## British

**Population** 60 million
**Language** English (a Germanic language with some French words), Scots (regarded by some as a dialect of English), Welsh
**Beliefs** Christianity, Islam, Hinduism, Judaism
**Location** Great Britain, Northern Ireland

Today, Scots, Welsh, and particularly English, national sentiment is divided between that of their smaller nation and that of Britain. Most Protestant people in Northern Ireland also identify themselves as British, or at least citizens of the UK. Britons whose family arrived recently from outside Europe, from South Asia or the Caribbean, for example, see themselves increasingly as British, but not as English, Scottish, or Welsh.

**NATIONAL IDENTITY**
*Britons of a range of ethnic backgrounds greet the Queen with UK flags on her first visit to a British mosque. Multiculturalism may be a feature of an emerging British identity.*

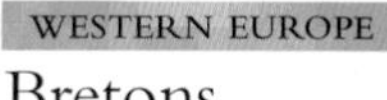

BRITISH ISLES

## Welsh

**Population** 1.85 million (plus small Welsh-speaking communities in North America and Argentina)
**Language** Welsh (Celtic, related to Breton), English
**Beliefs** Anglican, other Christian denominations
**Location** Mainly Wales; also England, Scotland

The Welsh are Celts who survived in the western portion of Britain when northern German tribes invaded from the fifth century AD onwards. Around half a million Welsh people speak their language, *Cymraeg*. Although all schoolchildren now learn Welsh, the main language is English. However, in recent years, Welsh has experienced a renaissance. While Welsh nationalism is strongly linked to the people's Celtic heritage, it is also grounded in more modern history. Welsh identity was revived in the 18th and 19th centuries, long after the country's absorption into the English crown in the 13th century. The Welsh have since reinvented many of their traditions, such as the annual *Eisteddfod*, a literary competition that today claims to be the largest folk festival in Europe.

After the Industrial Revolution, coal-mining became wedded to the national culture, but modern economic realities have brought closure of the mines and accelerated the shift to a service-based economy.

**ENGAGEMENT GIFT**
*The giving of love spoons is a Welsh custom. Traditionally, a young man carves one as a present for his intended bride.*

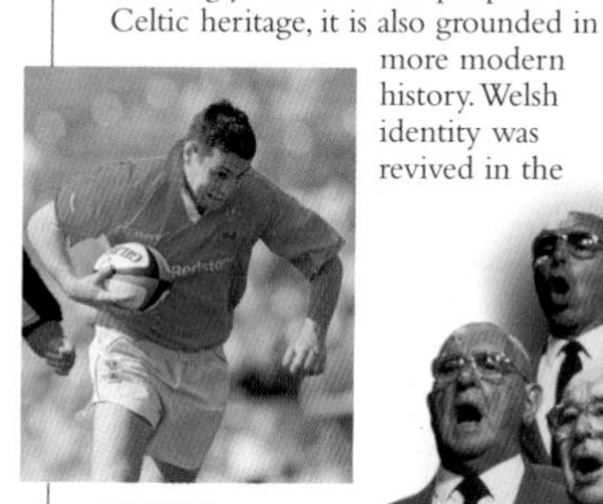

**RUGBY**
*In coal-mining days, each mining village had a rugby club. There is still immense pride in the national team.*

**MALE-VOICE CHOIR**
*The Welsh are renowned for their love of singing. Male-voice choirs sing hymns and national songs.*

WESTERN EUROPE

## Bretons

**Population** 600,000 (the total population of Brittany is 3 million)
**Language** Breton (a Celtic language related to Welsh but not to French), French
**Beliefs** Catholicism
**Location** Brittany (northwestern France)

The Bretons inhabit a peninsula of northwestern France that was once an independent duchy but was allied to the French crown in the 15th century. As French citizens, however, the Bretons are not recognized as a separate nationality.

For centuries after joining France, Bretons continued to speak their language and practise their cultural traditions. Yet, during the 19th century, the growth of industry, improved communications, rural exodus, and the advent of free, compulsory education meant that speaking French became a necessity, even in rural Brittany. Schools punished children severely for using Breton in class, and instruction was exclusively in French. By the time of World War II, Breton was rarely used as a mother tongue.

Today, Bretons are trying to revive their language and traditions. Symbols of Breton culture, such as traditional dress (especially the *coiffe*, a woman's decorative lace cap), cuisine (mainly crêpes), and Celtic festivals, are often promoted for tourism. Brittany has high unemployment and depends heavily on tourists for income.

Religious observance is generally higher in Brittany than in most other parts of France.

**OYSTER INDUSTRY**
*Brittany is an oyster-farming area. Many Bretons eat them all year round, but particularly during family celebrations.*

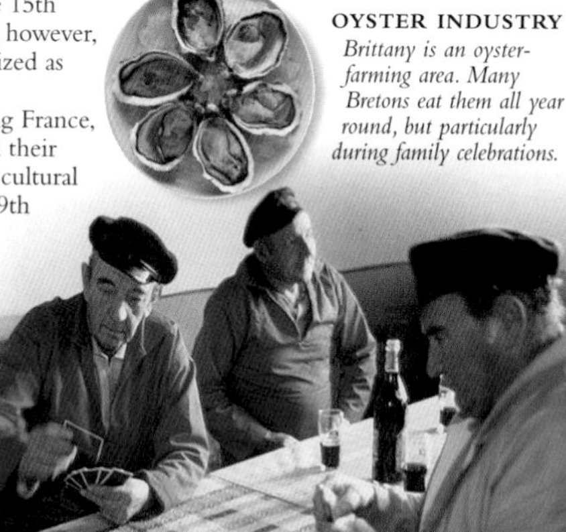

**BRETON SAILORS**
*A group of Breton men enjoy a glass of wine and a game of cards in a Brittany café.*

WESTERN EUROPE

## French

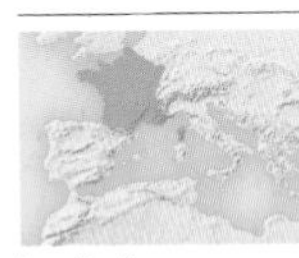

**Population** 60 million
**Language** French (a Romance language related to Spanish and Italian); also Alsatian, Corsican, Basque, Catalan, Breton, Gascon, Arabic, Berber
**Location** France, overseas departments and territories
**Beliefs** Roman Catholic

The ancestors of the French include the Celtic Gauls, present from the 6th century BC, and the Germanic Franks, who ruled from the 5th century AD. Modern France includes regions with their own cultures, from Alsace in the north to Corsica off the south coast.

Despite these mixed origins, or perhaps because of them, the French have worked hard to become united. France was one of the first nations to attain political unity, in the late Middle Ages. French (*see* p309), which was designated the national language in 1539, is an important symbol of that unity.

Another unifying factor was the Roman Catholic Church. Each King of France was anointed with holy oil and called "the most Christian King", while the country became known during the Middle Ages as "the eldest daughter of the Church". Today, Roman Catholic Christianity is still the main religion among the French. Belief, however, has been a personal affair since the Revolution of 1789. The revolutionary leaders tried to outlaw religion altogether. They failed, but agnosticism and atheism became acceptable forms of belief, and the concept of secularity as a principle of public life has endured.

The French see themselves as a highly cultured and civilized people, and French culture – particularly in the forms of literature, architecture, *cuisine* (cooking), and *couture* (fashion) – is admired internationally. The French government spends substantial sums of money on encouraging cultural production in France and promoting it elsewhere in the world.

**ALSACE**
*The region of Alsace, on the German border, is neither fully French nor German. It has its own culture, shown by its timber-framed houses (left).*

**EVOLVING NATION**
*These French people are of north and west African, as well as European, ancestry. They are enjoying a music concert in a Paris street.*

**FRENCH BREAD**
*French bakers make a number of distinctive breads, none more iconic than the baguette, or "French stick".*

**GOURMET NATION**
*The French are proud of producing the finest of food. In country markets, small producers sell cheese, fruit, and vegetables.*

History

### Bastille Day

On 14 July 1789, the people of Paris stormed the Bastille, the prison that stood for the tyranny of the Ancient Regime. Today, 14 July is a national holiday in France. It marks the beginning of the French Revolution and the national values of "liberty, equality, fraternity", which are still a source of pride for ordinary people.

WESTERN EUROPE

## Catalans

**Population** 10 million
**Language** Catalan (a Romance language closely related to that spoken in the Balearic Islands); also Spanish (Castilian form), French
**Location** Catalonia (Spain and France), Andorra
**Beliefs** Roman Catholic

The Catalans occupy the eastern Pyrenees, on the border between France and Spain. The main vehicle of Catalan culture is the language, which has survived better on the Spanish side than on the French. In the mid-20th century it was outlawed by Spanish dictator General Franco, but since his death in 1975 it has become a potent symbol of Catalan identity and is now widely used in daily life. In Roussillon, on the French side, Catalan was almost totally wiped out by monolingual schooling in the late 19th and early 20th centuries. It is now being taught again, even in the state system, but is no longer a mother tongue there.

Catalans of both nationalities are proud of their heritage. Catalan music and dance, particularly the *sardana*, are popular, and the seafood-based culinary traditions are also widely enjoyed.

**SARDANA DANCE**
*Catalan music and dance – particularly the* sardana, *in which participants dance together in a circle – is widely appreciated.*

WESTERN EUROPE

## Basques

**Population** 1 million
**Language** Basque, or Euskara (an isolated language with no known linguistic relatives); also Spanish, Catalan, French, English
**Location** Basque Country (NE Spain and SW France)
**Beliefs** Roman Catholic

Basque villages are said to be centred on two things: the Roman Catholic church and the *pelota* court. Based on tennis, and adopted from France, *pelota* (which means "ball") was born as a game played with bare hands and a ball on a tennis court. The game has since evolved into a number of forms with varying balls, catching implements, and court designs.

Because the Basque language was only given a written form in the 16th century, Basque culture has a strong oral tradition. Improvised poems are often sung with accompaniment from dancers and music played on traditional instruments, including the *txistu* (Basque one-handed flute).

**SPORTING TRADITION**
*In the most impressive form of* pelota, *players wear a special basket on the hand to catch and accelerate the ball.*

WESTERN EUROPE

## Spanish

**Population** 40 million
**Language** Castilian Spanish (a Romance language related to Portuguese); also Catalan, Aragonese, Galician
**Beliefs** Roman Catholic
**Location** Spain, including the Canary Islands

The Roman Catholic Church has long been central to Spanish identity, but some of the most striking architecture in southern Spain is Moorish. The Moors, or Muslims, ruled Spain for over 500 years (until the 15th century); the historic struggle between Muslim and Catholic powers is still played out every year in the *moros y cristianos* processions that take place in many Spanish towns and villages. Other popular festivals reflect the continued importance of the Catholic Church, particularly those linked with Holy Week (Easter).

Spanish daily life tends to follow a different timetable to most of western Europe; meal times are considerably later than would be normal elsewhere, with lunch between 2 and 4pm and dinner any time after 10pm. Spanish bars and cafés are open late into the night, while most shops and services still tend to be closed for two or three hours in the afternoon – the time of the traditional siesta. Although less well known than French cuisine, Spanish dishes such as paella and tortilla are nevertheless enjoyed by people across the world.

**ABRIL FERIA**
*Flamenco dancers enjoy the week-long* Abril Feria, *or April Fair, in Seville.*

**GOING OUT FOR TAPAS**
*The Spanish have distinctive eating habits. The evening meal, served at 10 or 11 o'clock, may come in the form of tapas – many separate light snacks.*

**THE RUNNING OF THE BULLS**
*One of Spain's most famous festivals takes place in Pamplona, when bulls are released from their corral. Daredevils run along with the bulls as they stampede through the streets to the bullring.*

WESTERN EUROPE

## Portuguese

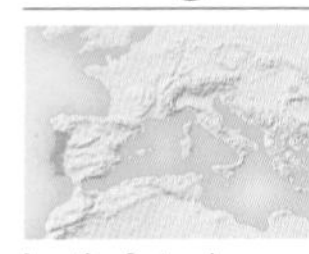

**Population** 10 million
**Language** Portuguese (a Romance language related to Spanish and Italian)
**Beliefs** Mainly Roman Catholic Christianity; also Protestant Christianity, Judaism
**Location** Portugal, the Azores, Madeira

The Portuguese people can trace their ancestry to Celtic, Phoenician, Roman, Visigoth, and also Moorish invaders. During the 15th and 16th centuries, it was the Portuguese who led the world in exploring previously uncharted lands and seas. They pioneered the concept of colonization by Europeans and can claim to have founded the first modern nation-state in European history. In the late 1960s, more than one million Portuguese people were employed elsewhere in western and northern Europe. Since then, the Portuguese have developed their economy, thus reducing the need for such extensive economic emigration.

Port wine originated in the Oporto region of northern Portugal in the mid-18th century, when producers began to add brandy to the local wines. This resulted in a fortified wine that could be kept in cellars for two to three years, thus allowing it to be sold as "mature" port wine. The success of Port wine has turned Oporto, the "granite city", into the centre of one of the world's most famous wine-producing regions.

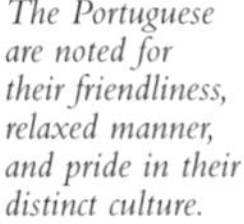

**PORTUGUESE SMILE**
*The Portuguese are noted for their friendliness, relaxed manner, and pride in their distinct culture.*

**FADO SINGING**
Fado *singers express the Portuguese concept that they call* saudades *– an emotion that combines joy, sorrow, tragedy, and loss.*

WESTERN EUROPE

## Maltese

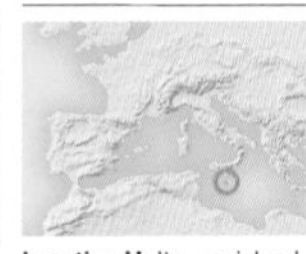

**Population** 400,000
**Language** Maltese (an Arabic-based language, the only one to be written using the Roman rather than Arabic script); also English
**Beliefs** Roman Catholic
**Location** Malta, an island between Italy and Libya

The Maltese are friendly but tough, as befits a people who have traditionally earned their livelihood from the land and the sea. They also cherish their unique culture, in which religion plays an important role. In 1530, the Holy Roman Emperor, Charles V, granted Malta to the Knights of the Hospitalers and the island became a Christian fortress. Malta now has a greater percentage of Roman Catholics in its population than any other country. Each year the citizens of Malta's towns and villages celebrate the anniversary of their local saint. These *festas* are times of celebration, processions, and fireworks, when the streets and local churches are decorated in bright colours.

The Maltese are very superstitious and it is common to see them using the sign of the cross to ward off the "evil eye", while depictions of thick-browed eyes adorn boats and buildings across the island for the same purpose.

**EYE OF HORUS**
*Maltese boats set sail with the traditional protective symbol of the thick-browed eye, or "Eye of Horus", painted on the prow.*

CENTRAL EUROPE

## Belgians

**Population** 9.5 million: 6.4 million Flemings, 3.1 million Walloons
**Language** Flemish (a Dutch dialect), Walloon (a dialect of French)
**Beliefs** Predominantly Roman Catholic
**Location** Belgium, made up of Flanders and Wallonia

Until the 19th century, Belgium was divided into Flanders, inhabited by Dutch-speaking Flemings, and Wallonia, home to French-speaking Walloons. Both have learned to live with the threat of assimilation by powerful neighbours. This legacy has given Belgians a mix of individuality, a wish for domestic stability, and the ability to make compromises. Flanders' tough, racing cyclists, the Flandriens, encapsulate the Belgian spirit. Up to a million spectators line the roads for major cycling events.

FLEMISH CYCLING EVENT

CENTRAL EUROPE

## Dutch

**Population** 16 million
**Language** Dutch, Frisian (both Germanic languages of the Indo-European language family)
**Beliefs** Nonreligious, Calvinist Protestant, Roman Catholic
**Location** Netherlands

Dutch identity is closely associated with tolerance, secularization, social welfare, and moral leadership in international relations (the Netherlands is the world's fourth largest foreign-aid donor). The Dutch have evolved systems of polite discussion, orderly consultation, pragmatic compromise, egalitarian decision-making, and government planning. These qualities arose from the need to work together to manage water and reclaim land; about 60 per cent of the population now lives in the 27 per cent of the country that is below sea level. The Dutch combine a firm European outlook with a strong cultural and economic orientation towards the English-speaking world. The Dutch import more English-language books than any other non-English-speaking nation. In addition, there are over 800 museums in the Netherlands, making it the country with the highest museum density per square kilometre. The Dutch are also rightly proud of being the first country ever to issue regular newspapers: the *Haarlems Dagblad*, published since 1656, claims to be the oldest newspaper in the world.

Another embodiment of Dutch character is the annual speed-skating competition called the Eleven-City Tour. It is both a folklore festival and a national experience cherished by all. The race follows canals and lakes for some 200km (125 miles) and places emphasis on cooperation and finishing the course, rather than on winning.

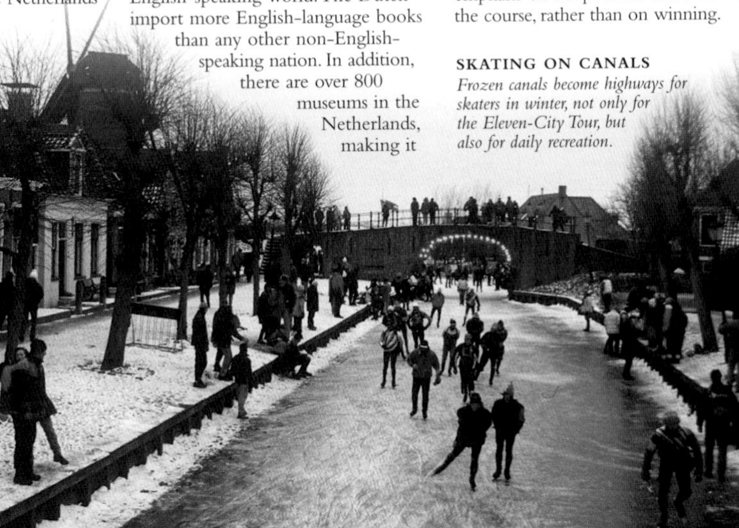

**SKATING ON CANALS**
*Frozen canals become highways for skaters in winter, not only for the Eleven-City Tour, but also for daily recreation.*

CENTRAL EUROPE

## Austrians

**Population** 8.1 million
**Language** German; also Croatian, Slovenian, Magyar
**Beliefs** Roman Catholic, Protestant, nonreligious beliefs
**Location** Austria, South Tyrol (Italy)

The nation state of Austria was created out of the ruins of the once-great Austro-Hungarian Empire in 1918, but it was not until after World War II that a distinct national identity was promoted. All Austrians speak German, although the Austrian – and even more so the Viennese – dialect can pose problems in communication with standard German. Today, the Austrian population is composed of people who embrace national identity, as well as those who would rather think of themselves as residents of their *Burgenland* (region), such as Tyrol or Vorarlberg. Austrians enjoy skiing, and their country was host to the first slalom competition in 1905. Many famous Alpine skiers were Austrian, including pioneer Matthias Zdarsky (author of the first ski instruction book), and 1970s Olympic champion Franz Klammer.

**FASCHING FESTIVAL**
*Austrians wear animal costumes and sing traditional songs during the annual festival of Fasching, which is the Austrian version of the pre-Lent festival, or Mardi Gras.*

Issue

### Austrian and Italian Tyrol

Residents of the mountainous Tyrol region are often seen as typical Austrians. They tend to have strong regional consciousness and some still dress up in ethnic costume (below). However, not all of the Tyrol is within Austria. The southern part became a part of Italy after 1918, although people there still speak German. After a long struggle, South Tyroleans now enjoy significant minority rights in Italy.

CENTRAL EUROPE

## Swiss

**Population** 7.2 million
**Language** Swiss German, French, Italian, Romansch (a Romance, Latin-based language with similarities to French and Italian)
**Beliefs** Roman Catholic, Protestant
**Location** Switzerland

The Swiss fall into four distinct groups: the Swiss-Germans (the majority), Swiss-French, Swiss-Italians, and a small Romansch-speaking community. Despite cultural connections with the three countries bordering the Swiss Confederation, the Swiss identity is strong, although political differences between the different language groups have increased in recent years.

Protestantism remains the dominant religion in most urban areas. Religion still plays an important part in Swiss daily life, and in the mainly Catholic rural areas it dictates the timing of festivals.

The emergence of financial service industries has made the Swiss among the most affluent people in the world. With this wealth comes a high standard of living and very efficient public services.

Swiss cuisine is renowned for its cheeses and chocolate. Emmenthal and Gruyère cheeses and the art of fondue are well-loved the world over. Owing to the Alpine location, skiing is a particularly popular pastime in Switzerland, attracting millions of tourists every year.

**SWISS CHEESE**
*A Swiss man makes Gruyère cheese in front of visitors at a demonstration dairy. Swiss cheese-makers now prepare their famously holey cheese both in modern dairies and as a cottage industry.*

**ALPHORNS**
*Swiss folk culture includes Alphorn festivals (right),* Schwingen *(wrestling) competitions, and also enjoyment of the singing style called yodelling.*

CENTRAL EUROPE

# Sorbs

**Population** 60,000 (including those in Poland and the Czech Republic)
**Language** Sorbian (a Slavic language with two main dialects), German
**Location** Mainly Lusatia (eastern Germany)
**Beliefs** Roman Catholic, Protestant

The Sorbs are the smallest group of Slavic people in the world. It is feared by many that the remaining Sorbs will soon become Germanified.

There are two distinct groups: the Upper Sorbs, whose language is most similar to Czech, and the Lower Sorbs (or Wends), who speak a language with more similarities to Polish.

SORB EASTER PARADE

CENTRAL EUROPE

# Poles

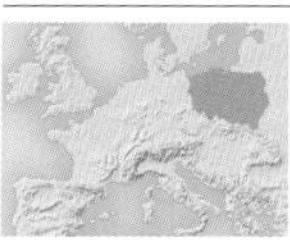

**Population** 38 million (in Poland; plus communities in western Europe and North America)
**Language** Polish (a Slavic language related to Czech), Russian, Belarussian
**Location** Poland; also Lithuania, Belarus
**Beliefs** Roman Catholic

Along with the Czechs and the Slovaks, the Polish people constitute the Western Slavs. The term Poles, derived from the Slav tribal name Polanie, means "people of the plain."

From the 16th to the 18th centuries, Poland was joined with Lithuania in a commonwealth, an official union based on the earlier marriage of a Polish queen to a Lithuanian prince. What was once a strong power eventually fell apart under pressure from enemies such as, Russia, Turkey, Sweden, the Cossacks, and the Tatars. The Poles did not fully achieve an independent state until the end of World War I and the demise of the Russian and Austro-Hungarian Empires.

In the years after World War II, when Poland was under the guardianship of the Soviet bloc, several anticommunist crises occurred. They culminated in 1980 with the creation of Solidarity, the trade-union turned national-opposition movement that steered Poland out of communist control. Solidarity eventually dropped out of the Polish political mainstream in 2001.

**POLISH FACE**
*Polish people are Slavs, descendants of peoples who migrated into Europe in the 6th and 7th centuries.*

**CHORAL TRADITION**
*Female choir members sing in traditional, brightly embroidered folk costumes, complete with colourful headscarves. There is a strong tradition of choral performance in Poland.*

Four meals a day is the norm in Poland: breakfast, a light snack, a hearty lunch after work, and a light supper. Polish cuisine features much use of cheese in stuffings and toppings. Typical foods are *pierogi* (dough balls filled with potato and cheese), pickled cabbage, meat dishes, and beetroot, particularly in *barszcz* (beetroot soup).

The Polish people are proud of their Roman Catholic heritage, and particularly of their most famous son, Karol Wojtyla – Pope John Paul II.

CENTRAL EUROPE

# Germans

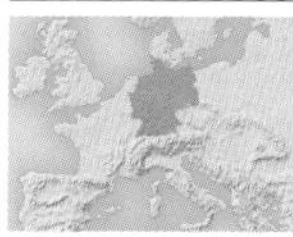

**Population** 100 million (including communities in Kazakhstan and elsewhere)
**Language** German, Friesian
**Beliefs** Protestant (especially Lutheran), Roman Catholic, other Christian denominations
**Location** Germany, Poland, Kazakhstan, Hungary, Italy

The German people were historically divided among myriad small states, which stretched from the Baltic Sea in the north, to the southern Alps, and deep into eastern Europe. The arrival of a single Germany in 1871 set apart the Germans living there from the German-speaking peoples of neighbouring countries (especially Switzerland and Austria).

Germany is still composed of culturally distinct regions called *Länder*. Examples are Bavaria, Friesland, and Westphalia.

**FRIESIAN BUILDING**
*Friesland is one of Germany's many distinct regions, called Länder. The Friesians build using characteristic architectural styles (left), and also have their own distinct language.*

Germans are evenly split between people who follow Protestantism, those who are Roman Catholic, and those who follow other creeds or none at all. Germany was the birthplace of Martin Luther (1483–1546), the man behind the Reformation, and Lutheranism is particularly strong in the north. In Bavaria, Catholicism is dominant, and Bavarian people maintain a particularly distinct regional character that is greatly influenced by the local geography of dense forests and high mountains.

German living standards are generally high and society is liberal. German education is well respected the world over, and many Germans have been leaders in their fields of study, such as the composers J S Bach (1685–1750) and Richard Wagner (1818–1883), the philosophers and theorists Friedrich Nietzsche (1844–1900) and Karl Marx (1818–1883), and the painter Franz Marc (1880–1916).

German food generally has an emphasis on meat and winter vegetables. German *Würste*

**THE MUNICH OKTOBERFEST**
*Bavarians toast each other at the annual* Oktoberfest *beer festival. They wear lederhosen (leather trousers), one of the most celebrated ethnic costumes in Europe.*

(sausages) are particularly famous, as is *Sauerkraut* (pickled cabbage). Germans are also famed for their great love of beer. The annual Munich Oktoberfest attracts thousands of thirsty beer-lovers from all around the world.

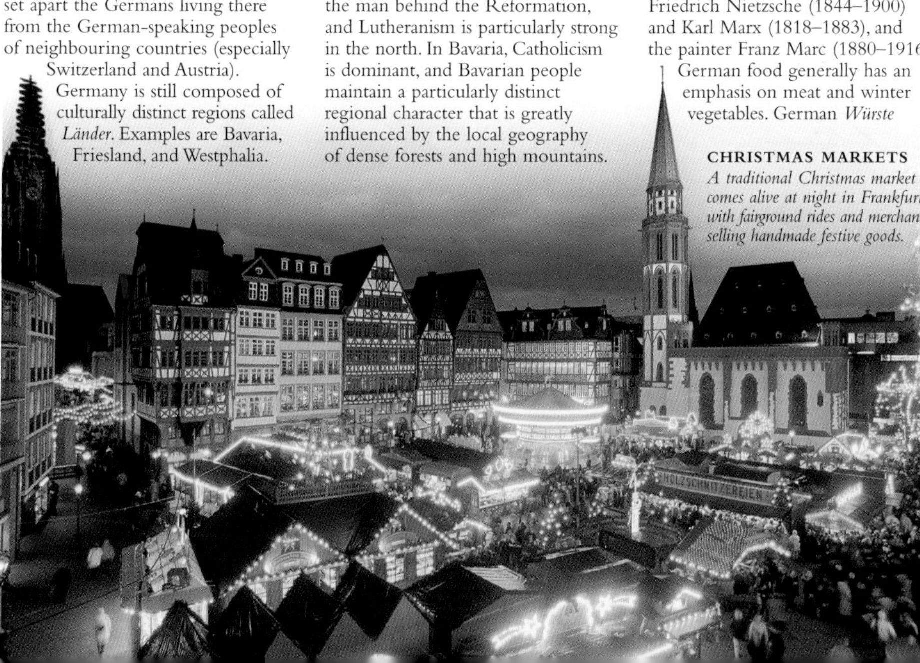

**CHRISTMAS MARKETS**
*A traditional Christmas market comes alive at night in Frankfurt, with fairground rides and merchants selling handmade festive goods.*

History

## Reunification

Germany was fragmented between East and West after World War II. People on each side, already separated by old regional differences, began to grow apart once more, influenced by contrasting economic and cultural backgrounds. The focus of reunification in 1990 was the uniting of Berliners from either side of the Berlin Wall (below). There is still, however, a distinction between the more affluent West Germans and their East German (*Ossi*) counterparts.

**ST PAUL'S FEAST**
*Sicily, like much of Italy, is strongly Catholic. Each town has its own patron saint. Here, the people of Palazzolo Acreide honour St Paul with a shower of confetti on the saint's annual feast day.*

CENTRAL EUROPE

# Italians

**Population** 58 million
**Language** Italian (a Romance language of the Indo-European family; dialects differ widely between regions of Italy)
**Location** Italy; also US and elsewhere in Europe
**Beliefs** Predominantly Roman Catholic

The Italians derive their name from the ancient Itali people, who lived in what is now Calabria (southern Italy). They have contributed much to the cultural, artistic, religious, and political evolution of Western civilization. In particular, the humanist ideals of the Italian Renaissance have greatly influenced the modern world.

In 1861, the great city-states, for example, Florence, Venice, and the Kingdom of Naples, as well as Sicily and Sardinia, were unified under King Victor Emmanuel. In the 1920s, Italians came under the rule of Benito Mussolini, who established a Fascist dictatorship. After defeat in World War II, Italy returned to democracy. Weak coalitions and frequent changes of government have characterized postwar politics in Italy.

Fittingly for a nation whose land surrounds the Vatican, or Holy See (the administrative body for the Roman Catholic Church worldwide), Italians are proud of their Catholic heritage. They are also renowned for their cuisine and their contemporary style. Italy is a world centre for opera, fashion, and interior design; and the Milanese, Florentines, and Venetians are widely held to be among Europe's most cultured citizens. The people of the less wealthy south of Italy, and of the islands of Sardinia and Sicily, can claim a distinct culture and their own traditions. Although well-known as the birthplace of the Mafia, the south of the country is also the spiritual home of some of the Italian diaspora's most famous sons – from baseball's Joe DiMaggio to entertainer Frank Sinatra – and to one of the Italians' most famous exports, pizza.

**LIFE ON SCOOTERS**
*Since the Vespa and Lambretta became design classics in the 1940s, scooters have been a mainstay of social life for many Italians.*

**FOOTBALL**
*The ups and downs of a football match fire the hearts of this Sicilian family.*

**PASTA**
*Some claim that Marco Polo introduced pasta to Italy, inspired by China's noodles. Historians, however, think pasta has been a part of Italian life since Etruscan times, in the 4th century BC. It now comes in many forms, such as tortellini, above.*

History

## Italian Renaissance

The Renaissance, which began in Italy in the 14th century and later spread throughout Europe, marked a "rebirth" of classical ideals in art, architecture, philosophy, and literature. The first architect to employ the new classical style was Brunelleschi, who designed the *duomo*, or cathedral, of Florence. Works by other Renaissance men, such as da Vinci and Michelangelo, remain artistic touchstones to this day. The Renaissance also marked a renewed interest in studying humankind and nature.

THE DUOMO, FLORENCE

CENTRAL EUROPE

# Czechs

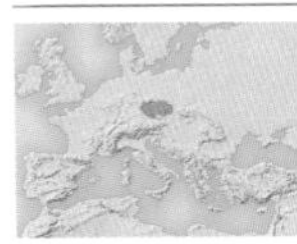

**Population** 9.7 million
**Language** Czech (mutually intelligible with Slovak)
**Beliefs** Roman Catholic, Czech National Hussite Church (after Protestant Bohemian reformer Jan Hus, 1370–1415)
**Location** Czech Republic, Slovakia

Geographically and culturally the most western of the Slavonic peoples, the Czechs are descended from one of many tribes that journeyed from Central Asia to Central Europe from the 3rd to the 6th centuries AD.

Czechs are proud of their musical heritage: they produced the renowned 19th century classical composers Bedrich Smetana and Antonin Dvorák and boast many museums devoted to folk music and dance. Pilsner beer is a popular drink in Bohemia, as is wine in southern Moravia. Goulash and knedliky (dumplings) are typical Czech dishes. Carp forms part of the traditional Christmas meal.

**FOLK FESTIVAL**
*A girl wears Czech costume at a festival in the region of southern Moravia.*

CENTRAL EUROPE

# Slovaks

**Population** 4.6 million
**Language** Slovak (a Slavic language mutually intelligible with Czech)
**Beliefs** Roman Catholic, Protestant
**Location** Slovakia, Czech Republic, Hungary

Slovaks were separated from their close kin, the Czechs, by Hungarian occupation (1025–1918). When Hungary was in turn occupied by the Ottoman Turks, the Slovak capital, Bratislava, became the Hungarian capital. Folk music was a vital way of ensuring the survival of the Slovak language, of which the Slovak people are particularly proud. Folk traditions continue to be an integral part of everyday life, especially in the east. Slovaks still celebrate pagan festivals, such as the burning of the winter goddess Morena on summer bonfires.

**PAINTED HOUSE**
*Old wooden houses in rural areas are a symbol of Slovak culture. Slovak houses are sometimes painted with geometrical designs.*

CENTRAL EUROPE

# Magyars

**Population** 11.6 million (mainly in Hungary)
**Language** Magyar, or Hungarian
**Beliefs** Roman Catholic; also large Calvinist Protestant minority
**Location** Hungary, Slovakia, NW Romania, N Serbia

Magyar (pronounced *Maw-djar*) are also known as Hungarians, a name taken from the Ungars, one of the principal tribes that migrated from the steppes of Central Asia, arriving in Central Europe in AD896. They speak a language most closely related to languages of Siberian nomads, such as the Khanty. Its relative strangeness has been a barrier to integration with their European neighbours. Water polo is the national sport – the Hungarian team won gold at the 2000 Sydney Olympics. Magyars, nonetheless, have the lowest life expectancy and some of the longest working hours in Europe, and smoking remains highly popular.

**TRADITIONAL DRESS**
*These days worn mainly at folk festivals, traditional dress preserves Magyar culture in countries where Magyars are a minority.*

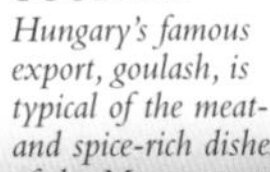

**GOULASH**
*Hungary's famous export, goulash, is typical of the meat- and spice-rich dishes of the Magyar.*

**FOLK ENTERTAINMENT**
*Folk traditions of dancing and music influenced the famous 19th century Magyar composer Franz Liszt.*

EUROPE

# Roma

**Population** 12 million
**Language** Romani (a group of several Indic languages related to Rajasthani and Gujarati of northwestern India)
**Beliefs** Traditional Roma beliefs, Christianity, Islam
**Location** Slovakia, Hungary, Romania, FYR Macedonia, Bulgaria, across Europe

Commonly known as Gypsies, the Roma people originally came from India, but they left the area in the 11th century in search of a better quality of life and arrived in Europe a couple of centuries later. They are now a minority in every country in Europe.

The Roma fiercely value their freedom. They like to live in mobile homes and are well-known for their beautifully decorated caravans called *vardos*. *Vardos* used to be pulled by horses, and horses are still extremely highly valued among Roma today. Roma traditionally have a nomadic lifestyle and in the past, they chose professions that they could practise on the move. Many of the men were musicians, while it was common for women to dance and to read fortunes for their non-Roma neighbours. Many Roma still practise these professions. Although a lot of Roma now live in fairly settled communities, they still travel for large fairs and festivals, such as the International Gypsy Festival, or *Khamoro*, held every May in the Czech Republic.

Many Roma have adopted the religion of their neighbours, but traditional beliefs still remain in most communities. Roma have strong opinions on the concept of *marimé* (impurity) and will go to great lengths to ensure that they do not come into contact with items that they consider as "impure", such as food cooked by non-Roma people, cooking implements that have been washed in an inappropriate container, and even women who are menstruating.

**PORTRAIT**
*A Russian Roma woman of the Volga region wears a typical Roma headscarf.*

**KHAMORO FESTIVAL**
*Roma people, both young and old, enjoy the Khamoro, or "Sun" festival in Prague. The event features concerts, workshops, and seminars.*

**MARRIAGE FESTIVITIES**
*Muslim Roma dance through the streets of FYR Macedonia in celebration of a wedding, which, according to Roma custom, is an arranged marriage. Traditional Roma society is closed to outside marriages.*

Issue

## Outsider treatment

Roma people, such as the family below, now live on the margins of a number of European societies. They have always been treated as outsiders by their European neighbours and have suffered much persecution and marginalization throughout their history. Almost everywhere, their basic civil rights are threatened, and because they do not have their own homeland, they have no government to provide protection for them or to speak on their behalf.

PEOPLES

CENTRAL EUROPE

## Slovenes

**Population** 2 million (88% of Slovenia's population)
**Language** Slovenian (a Slavic language related to Serbo-Croat, but with much Italian and German vocabulary)
**Beliefs** Roman Catholic
**Location** Slovenia, enclaves in Austria, Italy

Long domination by other powers held back the emergence of a strong Slovene national culture until the 19th century. Writers are held in high esteem. The poet France Preseren (1800–1849), for example, is seen as the father of the nation. In one of his poems, known by heart by many Slovenes, he expresses *hrepenenie*, a feeling of melancholic longing, which many claim as a national trait.

Slovenes tend to prefer living in a rural environment. Outdoor activities, such as cycling, hiking, and gardening, are the favourite family weekend pastimes, and skiing is the most popular sport in the country.

**COUNTRY WEDDING**
*An accordion player plays a Slovenian dance tune for guests at a wedding.*

EASTERN EUROPE

## Ukrainians

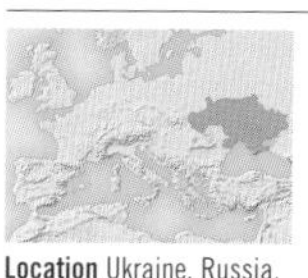

**Population** 36 million
**Language** Ukrainian (a Slavic language very closely related to Russian and Belarussian), Russian
**Beliefs** Eastern Orthodox, Uniate Church (also called Eastern Catholic Church)
**Location** Ukraine, Russia, Moldova, Poland, Romania

Ukraine (which means "land on the frontier") was without a state history of its own until 1991. During the 1920s, Ukrainians living under Soviet rule saw their language and culture flourish. A committee was even set up to create Ukrainian terminology for matters of administration and science. However, with Stalin's collectivization programme, launched in 1929, came mass famines. Due to this and his purges of intellectuals, millions of Ukrainians perished. After 1945, many survivors left their homeland for western Europe and the Americas.

Music is integral to Ukrainian life, and its traditions can be traced back to epic narrative poetry and lyrical ballads. The *Bandura Chorus (*a group from Kiev) and Mykola Lysenko (a classical composer) are two famous exponents of Ukrainian music.

**BANDURA**
*A bandura player plucks the bass strings with his left hand and strums with his right. The* bandura *is Ukraine's national instrument.*

**UKRAINIAN HOME**
*A woman at home is surrounded by floral fabric typical of Ukrainian textile design. Textiles are a principle product of Ukrainian light industry.*

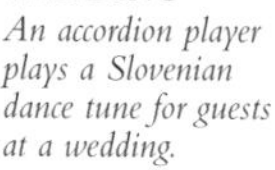

EASTERN EUROPE

## Belarussians

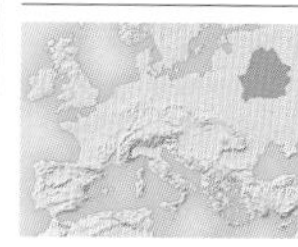

**Population** 8 million
**Language** Belarussian, Russian (both part of the eastern branch of the Slavic group of Indo-European languages)
**Beliefs** Eastern Orthodox, Roman Catholic
**Location** Belarus, Russia, Latvia, Lithuania, Poland

The Belarussians inhabit a land that has been, in turn, under Lithuanian, Polish, Russian, and Soviet control. At last, the Belarussians are now in control of their own independent state. They have revived traditional celebrations such as *Kypally* (summer solstice) and *Dzyed* (All Souls' Day), dedicated to family ancestors. Weaving and embroidery are traditional crafts.

**FOREST FESTIVITY**
*Several young Belarussians gather in the forest in January to celebrate the Orthodox Church's Epiphany, the baptism of Christ.*

EUROPE

## Ashkenazim

**Population** 10 million
**Language** Traditionally Yiddish (a blend of Hebrew with German, Slavic, and Aramaic), but most now speak the main language of their country
**Beliefs** Judaism
**Location** Europe, Israel, North America, worldwide

Ashkenazim (northern European Jews) have their own distinct language (Yiddish), vibrant music style (*klezmer*), a proliferation of stories and jokes, and strong family values. Today, the culture of most Ashkenazim is largely embodied in the humour of Woody Allen, the writing of Primo Levi, and culinary delicacies such as salt beef, chicken soup, and bagels.

History

### Shtetls

Ashkenazi culture developed in the *shtetls* of Central Europe before World War II. *Shtetls* were small towns or villages that were inhabited mainly or only by Jews and were often separate from the communities around them. Life in the *shtetl* was tough – Jews were not allowed to own land and they were banned from most colleges and universities. Most worked as merchants and craftsmen and endured much persecution from their neighbours.

**SABBATH SERVICE**
*Ukrainian Jewish men read the Torah during a Sabbath service. The Ukraine has Jewish communities totalling some 400,000 people.*

BALKANS

## Romanians

**Population** 20 million
**Language** Romanian (a Romance language related to Italian and French)
**Beliefs** Mostly Eastern Orthodox, small Roman Catholic and Protestant minorities
**Location** Romania, Moldova, Ukraine

During the 2nd and 3rd centuries BC, the land today known as Romania was the Roman province of Dacia. The presence of the Romans is still strongly felt in the Romanian language, on which Latin grammar and inflection have left their mark. During the 15th and 16th centuries, Renaissance-influenced warriors and the literate classes emerged as the Romanian elite. They termed themselves Romans and had a great influence on Romanian national identity.

At different periods in history, Romanians have come under the control of other nations, notably the Hungarians (from the 9th century), Turks (from 1526), and the Soviet Union (from 1945 until 1991). During World War II, Romania fought on the side of Nazi Germany and suffered greatly on the eastern front. As a result of their involvement, the Romanians also lost territory to the Soviet Union.

During the latter half of the 20th century, under what was in effect the last hardline Stalinist regime, the Romanians endured extreme poverty and severe repression. The country remains one of the poorest nations in Europe. Its main industries today are still those developed under communist rule, notably steel and chemical production, as well as machinery and vehicle manufacture. More than a quarter of the current Ukrainian workforce is employed in some form of agriculture.

**RURAL WOODEN HOUSE**
*Distinctive, traditionally built old wooden houses with porches and tiled roofs survive in remote, rural regions of Romania.*

**BEARSKIN COAT**
*A shepherd wraps himself in a bearskin to keep warm. Winters are cold in the wild Apuseni mountains, where wolves still roam.*

BALKANS

## Bulgarians

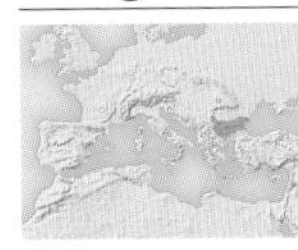

**Location** Bulgaria
**Population** 7.4 million
**Language** Bulgarian (a South Slavic language written in the Cyrillic script)
**Beliefs** Predominantly Bulgarian Orthodox Christianity; also Islam (about 12%)

Between the 12th and 14th centuries, Bulgaria was an empire, but it then spent many years under Turkish rule. In 1878, however, with assistance from Russia, Bulgaria gained its independence. The result was a long-standing friendship between the two countries. During World War II, although Bulgaria allied itself to Nazi Germany it was not able to declare war on the Soviet Union because of this allegiance.

The country's past continues to play an important role in the lives of Bulgarians – historical theatre is a popular pastime, although it tends to focus only on the high points of the nation's history.

After World War II, Bulgaria underwent much industrialization, yet 20 per cent of its economy is still agricultural. The main products include barley, wheat, and tobacco.

Roses have been grown here for 300 years, a fact that is celebrated with an annual rose festival at which a Rose Queen is elected.

**ROSE HARVEST**
*The Valley of the Roses in Bulgaria is one of the biggest producers of rose oil in the world and supplies top perfumeries.*

BALKANS

## Macedonians

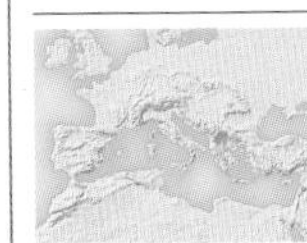

**Population** 1.5 million (including emigré communities elsewhere)
**Language** Macedonian (a Slavic language similar to Bulgarian)
**Location** Bulgaria, Greece, Albania, Macedonia, Serbia
**Beliefs** Predominantly Orthodox Christianity

The historic region of Macedonia stretches from the Aegean Sea to the Balkans. It is divided between the Republic of Macedonia and Greece, with small numbers of people in Bulgaria, Albania, and Serbia.

There has traditionally been large-scale emigration from Macedonia. It is often reflected in folk songs, which describe the nostalgic feelings of those who had to go abroad in search of jobs and better living conditions. Similar emotions have inspired rich decoration on clothing and carpets.

The language is closely related to Bulgarian. There is debate in both countries as to whether Macedonian qualifies as a distinct language or is merely a strong dialect of Bulgarian.

**TRADITIONAL DRESS**
*A woman from the Greek part of Macedonia wears a traditional headdress. Macedonian people are still proud of their folk culture.*

**LIFE IN THE CAPITAL**
*A barber tends to a client's needs in Macedonia's capital city, Skopje. The city has about 600,000 people, but it is not prosperous.*

BALKANS

## Greeks

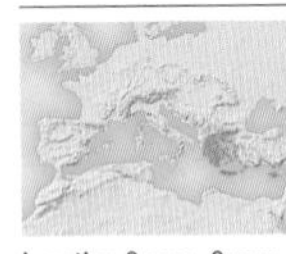

**Location** Greece, Cyprus
**Population** 11.5 million (including emigré communities elsewhere)
**Language** Greek (an Indo-European language), English, Turkish, French
**Beliefs** Predominantly Greek Orthodox Christianity

The ancient Greeks were one of history's greatest civilizations. They made a significant contribution to the development of such disciplines as philosophy, mathematics, and drama. Their concept of *demokratia* (from *demos* meaning people, and *kratein* meaning rule) was the form of government that inspired the French and the American revolutions. Also, although the original system differs in many respects from the democratic systems in place today, the ancient Greek concept underpins Western contemporary political life.

Today, Greek people proudly view themselves as Westerners despite their geographic location in the Balkans.

Education and learning continue to be held in high regard, while participation in domestic political life is widespread. There are close political, linguistic, and cultural ties between Greeks and Greek Cypriots, (who live on the island of Cyprus). Both peoples attach great importance to familial bonds and are renowned for pride in Hellenic (Greek) culture.

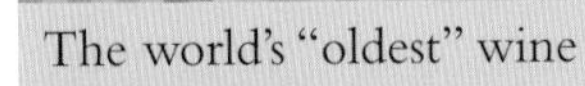

History

### The world's "oldest" wine

The national drink of Cyprus is a sweet red wine called Commandaria. It has the longest history of any wine: the poet Hesiod mentioned it as long ago as 800BC, when it was known as "Cypriot Nama". Later, in the 13th century, the King of France crowned it the "Apostle of Wines". It is still produced using traditional methods.

**ANCIENT WINE POURER**
*Ancient Greeks may have served Cypriot Nama wine from Griffin-headed jugs (left). Nama is made from the Mavro and Xynistery varieties of grapes.*

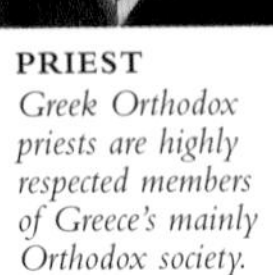

**PRIEST**
*Greek Orthodox priests are highly respected members of Greece's mainly Orthodox society.*

**WREATH**
*Wreaths and sprays of gold leaves were often placed on the heads of gods' statues during processions.*

**CELEBRATION**
*Dancers wear traditional costume as they perform in their town square. In Greece, dancing is an essential part of any celebration.*

PEOPLES

BALKANS

## Albanians

**Population** 3.3 million (including communities in Croatia and Greece)
**Language** Albanian (an Indo-European language)
**Beliefs** Sunni Islam (70%), Roman Catholic, Orthodox Christianity
**Location** Albania, Kosovo Macedonia, Montenegro

There are two major groups of Albanians – the Ghegs to the north of the Shkumbin River and the southern Tosks. They differ in dialect, dress, music, and customs. Tosk is recognized as the official dialect.

After initial resistance, Albania became part of the Ottoman Empire in the 15th century. The first Albanian state was established only in 1913, a year after the Albanian uprising of 1912. Decades later, in 1946, Albania – or Shqiperia (Land of the Eagles), as it was known to its native population – became a communist state under the leadership of Enver Hoxha, who ruled until his death in 1985. Just 4 years later, in 1989, Communist rule in Eastern Europe collapsed and, in 1991, Albania established a multiparty democracy. However, the legacy of years of corrupt rule persists, and Albania has the lowest standard of living in Europe. It has a rundown infrastructure and high unemployment.

As a result of emigration during the 19th and 20th centuries, more Albanians now live in North America and Australia than in Europe.

**RUFAIS MUSLIMS**
*The Rufais or "Howling Dervishes" are Sufis (Muslim mystics). They may inflict pain on themselves during worship.*

**TRADITIONAL DRESS**
*Muslim Albanian women wear traditional dress to a market.*

Issue

### Kosovo

Since the mid-20th century, the region of Kosovo in Serbia has been dominated by ethnic Albanians. These people were politically oppressed under the regime of Slobodan Milosevic in the 1990s. Attempts by the Albanians to retain autonomy led to the breakup of Yugoslavia and the NATO bombing of Serbia in 1999. The province is now administered by the UN. Many Kosovo Albanians hope for full independence.

BALKANS

## Croats

**Population** 4.5 million
**Language** Croatian (a South Slavic language that uses the Roman alphabet)
**Beliefs** Mainly Roman Catholic, with a minority of Orthodox Christians
**Location** Croatia, Bosnia, Herzegovina, Slovenia

From 1102, the Croats were dominated first by Hungary and then Austria. These circumstances separated them from ethnically similar neighbours and introduced Roman Catholicism. To strengthen their identity against foreign influences during the 19th and 20th centuries, the Croats turned to fellow southern Slavs (Yugoslavs). This association led to closer cultural and political ties with those who share their language: the Serbs and Bosniaks.

**CROAT MARKET**
*The south coast and islands of Croatia are Mediterranean in character, as is shown by the produce on sale at this market in Split.*

BALKANS

## Serbs/Montenegrins

**Population** Serbs: 8.5 million; Montenegrins: 450,000 (including some in Serbia)
**Language** Serbian (a South Slavic language)
**Beliefs** Orthodox Christianity
**Location** Serbia, Montenegro

The Serbs and Montenegrins are two closely related groups of southern Slavs, who share a common origin, religion, and language.

In Vojvodina, the northern part of Serbia, the architecture, lifestyle, cuisine, and mentality reflect strong central European influences. Southern Serbia is characterized by the specific Balkan mixture of indigenous and oriental cultural elements. The capital, Belgrade, is a blend of the two.

Serbs like to think of their country as "something in between". Most Serb families, even atheist ones, adhere to the tradition of celebrating *slava* (thanksgiving or glory-giving) on the feast day of the family patron saint.

"Montenegro" means "Black Mountain". The inaccessibility of the terrain enabled the state to remain independent of the Ottoman Empire (15th–19th centuries), unlike their Serb neighbours, but in modern times it has hindered economic development. A notable feature of Montenegrin culture is great respect for the power of the word, both oral and written.

**SERBIAN MUSIC**
*As a result of long domination by Ottoman Turks, Serbian music, especially in the south, has European, Roma, and Turkish influences.*

**LESKOVA FAIR**
*The annual fair in this Serbian village attracts 10,000 people. New brides and women who are ready to marry attend to show themselves.*

BALKANS

## Bosniaks

**Population** 2 million (including small groups in the Sandzak region – Serbia and Montenegro)
**Language** Bosnian (a Central South Slavic language)
**Beliefs** Sunni Islam
**Location** Bosnia, Herzegovina

Bosniaks are southern Slavs whose ancestors were converted to Islam during the period of Ottoman rule in the Balkans (15th–19th centuries). The culture has a lot in common with that of neighbouring Serbs and Croats. The architecture, national dress, and music show oriental influences, while the literature, education, and customs exhibit Western ones.

*Sevdalinka* is a style of Bosnian song that is traditionally accompanied by a guitar, an accordion, and a violin. It represents a melancholic longing for a love that cannot be realized.

**REFUGEES' PRAYER**
*Women pray at a refugee camp. During the war with the Serbs in the early 1990s, many thousands of Bosniaks had to flee their homes.*

RUSSIA

## Russians

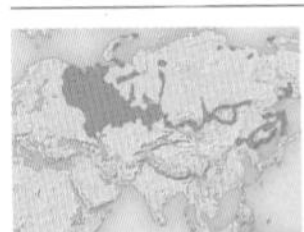

**Location** Russian Federation

**Population** 120 million (in Russia, Ukraine, Belarus, Kazakhstan, and other former Soviet Republics)
**Language** Russian
**Beliefs** Russian Orthodox Christianity

Russians are descended from the Slavic people who lived north of the Black Sea. During the era of the tsarist aristocracy, most Russians were rural peasants employed by wealthy landowners. The ruling class was overthrown during the communist revolution of 1917, after which the state took control of all farmland and industry and made itself responsible for the allocation of housing. Today most Russians live in towns, many in vast apartment blocks built during the communist period (1917–1991).

Russians endured centuries of authoritarian rule, from the tsars to communists, but have been developing a new democratic society since the collapse of communism in the early 1990s. A class of wealthy entrepreneurs is emerging, but the majority of the population are still to feel the benefits of economic reform and the free market. With the burden of a perceived lost empire behind them, modern Russians are trying to rediscover their national identity. Many have returned to Russian Orthodox Christianity, which was suppressed but never eliminated during communist rule. The church takes part in uniquely Russian rituals; on Easter Sunday, the priest blesses the breakfast meal, which traditionally consists of *kulich* (a sweet cylindrical bread) served with *pashka* (a pyramid of cheese). Distinctive pre-Christian festivals also survive, such as *Maslenitsa*, which marks the rebirth of nature in spring.

In their spare time, Russians enjoy a range of outdoor activities, regardless of the weather. Winter sports include cross-country skiing, ice-skating, ice-fishing, and swimming in frozen waters. In summer, many Russians spend weekends or holidays in *dachas* (wooden houses) in the forest. *Shashliki* (barbecues) are popular, as are chess games in the park and holidays on the coast of the Black Sea.

**WINTER FURS**
*Luxurious fur clothes remain a popular solution to Russia's bitter winter weather.*

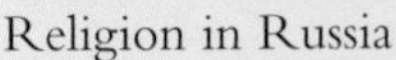

Fact

### Religion in Russia

Russians were pagan until AD987, when the Kievan Rus, the first Russian state, converted to Christianity. Today an estimated 40–80 million people belong to the Russian Orthodox branch of Christianity. The church was repressed in the early Soviet era because of its links with the pre-revolutionary tsarist regime, but since the fall of the Soviet Union in the 1980s, many Russian churches have been restored, and there has been a resurgence of open faith and worship.

**ICY PLUNGE**
*Many Russians believe that plunging the body into icy water helps build resistance to cold weather and disease.*

**KREMLIN**
*A gilded hammer and sickle, symbols of the former USSR, adorn the Kremlin in Moscow.*

**POLLING DAY**
*Russian sailors leave polling booths after voting in a 2003 election. Russians first took part in an election in 1989.*

EUROPEAN RUSSIA

## Komi

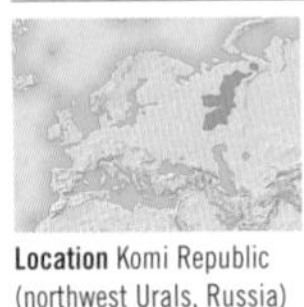

**Location** Komi Republic (northwest Urals, Russia)

**Population** 400,000
**Language** Russian, Komi (a language belonging to the Finno-Ugric language family, related to Finnish)
**Beliefs** Nominally Russian Orthodox Christianity; also animist

The Komi live in the icy far north of Russia. In the past most Komi led nomadic lives, following reindeer herds on their seasonal migrations as well as hunting and fishing. Like other reindeer herders, they obtained most of the resources needed for survival from their semi-tame animals. The Komi were converted to Christianity in the 14th century but held on to many of their animist beliefs. They still have special rituals and ceremonies for births, deaths, and marriages, as well as elaborate superstitions about luck.

The Komi's homeland is rich in minerals and timber that Russians have been exploiting since the 18th century. In the 20th century, Komi lands received an influx of outsiders, and the Soviet government instigated a series of modernization efforts that largely wiped out the Komi's traditional way of life, especially in the south. Many Komi moved to towns, and those who stayed in the country were forced to collectivize. In recent years the area has been designated the Komi Autonomous Republic, but the Komi remain a poor minority.

**KOMI GIRL**
*Reindeer provide the Komi with clothing, shoes, meat, and milk.*

**REINDEER HERDER**
*The traditional way of life continues in northern parts of the Komi's homeland.*

EUROPEAN RUSSIA

## Udmurts

**Location** Mainly Republic of Udmurtia (Russia)

**Population** 750,000 (most live near the Volga, Kama, and Vyatka Rivers)
**Language** Russian, Udmurt (a Finno-Ugric language related to Mari and Finnish)
**Beliefs** Eastern Orthodox, Protestant

Udmurts ("meadowmen") are related to the Mari and the Komi, and more remotely to the Nenets (*see* p428, and shown as the northerly area on the map above). They lived as seminomadic forest dwellers in *kars* (riverside settlements), but most Udmurts now live in towns. Historians claim there are 70 ancient clans, and clan feelings remain strong, although the clan-based social structure no longer exists.

Udmurtia became a republic in 1934 and was one of the first places in Russia to industrialize, with sawmills, ironworks, and shipyards, built in the early 18th century. Much of the marshy pine forest that once covered the area has been cleared to make way for farmland. Udmurtia is known for its distinctive honey, which is derived from lime blossom.

EUROPEAN RUSSIA

# Mari

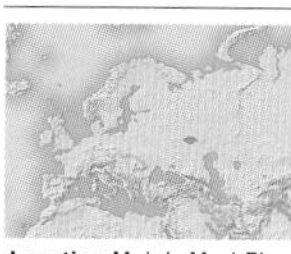

**Population** 650,000
**Language** Russian, Mari (a Finno-Ugric language consisting of two mutually unintelligible dialects)
**Beliefs** Orthodox Christianity, pagan cult of Kuga Sorta ("big candle")
**Location** Mainly Mari El (Russian Federation)

The Mari are a Finno-Ugric people who have lived in the middle Volga river valley for thousands of years. About 1,000 years ago they split into meadow and mountain dwellers, a split reflected in two main branches of the Mari language, both of which now use the Cyrillic script. Despite centuries of Russian influence, the Mari have kept their ethnic identity. In the 19th century they invented the pagan cult of *Kuga Sorta*, which combines beliefs in the sanctity of nature with elements of Christianity. Rituals take place in forest glades around a giant candle.

**MARI DRESS**
*Necklaces of coins and beads form part of a Mari woman's traditional dress.*

EUROPEAN RUSSIA

# Chuvash

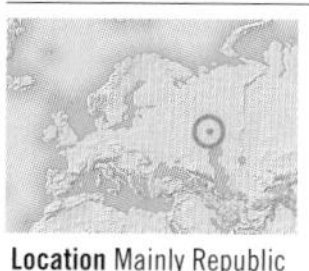

**Population** 1,774,000
**Language** Russian and Chuvash (an ancient and unusual branch of the Turkic language family)
**Beliefs** Russian Orthodox Christianity, incorporating aspects of paganism
**Location** Mainly Republic of Chuvashia (Russia)

Chuvashia is an area of low, rolling hills to the west of the Volga River, where the forests of Russia begin to give way to the grassy steppes of Asia. The region was annexed by the Russian empire in 1551 and became a separate republic within the Russian Federation in 1992. Chuvash believe they are descended from the Bulgars, who migrated from Central Asia to the Volga valley in the 4th century AD. Their unusual language is thought to be the only surviving form of Bulgar Turkic, one of the two main branches of the Turkic language group.

Although nominally Russian Orthodox Christians, the Chuvash adhere to elements of paganism and include aspects of Zoroastrianism, Judaism, and Islam in their religion.

Chuvash people account for about 70 per cent of the population of the Chuvash Republic and are divided into two groups: lower and upper Chuvash. They have close relations with the neighbouring Mari, especially through the tradition of bridal exchange.

Industries involving wood have been practised in Chuvashia for centuries, with many Chuvash being involved in lumbering, sawing, and carving. The republic is mainly agricultural and exports a wide range of crops and other goods, including grain, hemp, tobacco, hops, potatoes, textiles, and leather.

**CHUVASH DRESS**
*Chuvashia is famous for its beautiful embroidery and finely woven fabrics.*

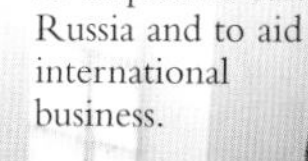

**CHUVASH ART**
*Chuvashia has a strong artistic tradition, expressed through weaving, woodcarving, ceramics, and embroidery. This artist uses traditional Chuvash motifs in his paintings.*

EUROPEAN RUSSIA

# Tatars

**Population** 6,650,000. (most in west-central Russia; also in Central Asia, Siberia, and Crimea in the Ukraine)
**Language** Russian, Tatar (a Turkic language)
**Beliefs** Sunni Islam
**Location** Tatarstan and Siberia (Russia), Ukraine

Tatarstan, where the majority of Tatars live, is a republic in the Volga river region of the Russian Federation. It has many self-governing rights, but does not have full sovereignty. The Tatars are a Turkic people who are thought to have come to Europe with the Mongol invasion led by Genghis Khan in the 13th century. Their genetic ancestry is mixed, and they include people with blond hair and blue eyes, as well as people who appear Mongolian. Their Asian ancestors were probably nomadic shepherds, but after settling in Russia the Tatars became largely sedentary farmers, and their society has since lost its former tribal structure.

The Tatars lived in several khanates (kingdoms), notably the powerful Khanate of Kazan (now the capital of Tatarstan). The Russians, under Ivan the Terrible, conquered most of the khanates in the 16th century. Tatars soon became valued as traders and intermediates in Russia's dealings with newly acquired Central Asian territories. They also had a reputation for craftsmanship for their high-quality goods of wood, leather, metal, ceramics, and cloth. Tatars are the second-largest ethnic group in Russia and a large minority in many cities. Since the collapse of communism, Islam has become a powerful force in Tatar society. Most Tatars have a liberal attitude towards women's rights and education.

The Tatar script has changed over the years from Arabic to Roman and finally Cyrillic. In recent years there have been moves to revert to Roman to emphasize cultural differences with Russia and to aid international business.

**TATAR DRESS**
*Though Muslim, Tatar women seldom cover their faces with veils.*

**COUNTRY LIFE**
*In the rural heartland of Tatarstan, most people are farmers. The water supply in this village comes from a well.*

**CRIMEAN MOSQUE**
*Crimean Tatars of the Ukraine pray in their mosque. Tatars make up 6 per cent of the population in the Crimea.*

RUSSIA

## Kalmyks

**Population** 177,000
**Language** Russian, Kalmyk (a language of the western branch of the Mongolian language group)
**Beliefs** Tibetan Buddhism; also Islam
**Location** Republic of Kalmykia (SW Russia)

The Kalmyk people are descended from Mongolian nomads who migrated to the lower Volga region, near the northwest shore of the Caspian Sea, in the 17th century, in search of grazing land. The area is now the Republic of Kalmykia and a part of the Russian Federation.

Like the ancestral Mongolians, Kalmyks are nomadic herders, driving flocks of sheep, cattle, horses, and goats between seasonal grazing grounds. They live in *yurts* – circular felt tents that are easily dismantled and carried on trips. Some Kalmyks have become sedentary farmers living in permanent houses. Only about a fifth of the population lives in towns.

By tradition, Kalmyks live in large, extended families. Married sons and unmarried adults stay with their parents, but daughters move home after marrying. Most Kalmyks are Tibetan Buddhists, practising a form of Buddhism that also incorporates strong elements of shamanism. A small number of people have converted to Islam.

The Republic of Kalmykia has an elected president, the first elections having being held in 1993. As head of state, the president upholds the "Code of the Steppe", inspired by the idealized conduct of 17th-century Kalmyk nobility.

**KALMYK COUPLE**
*Despite their east Asian looks, the Kalmyks have been living in Europe for about 400 years.*

CAUCASUS

## Circassians (Adygei)

**Population** 1,600,000
**Language** West Adyghe, East Adyghe, Kabardian (all languages of the North Caucasian family, closely related to Abkhaz), Russian, Turkish, Arabic
**Beliefs** Mostly Islam
**Location** SW Russia, Turkey, Syria, Jordan

The Circassians (also called Adygei) are a Muslim people native to the area around the northeastern shore of the Black Sea. When their homeland was colonized by Russia in the 19th century, most Circassians fled, forming a diaspora in Turkey and other parts of the Middle East. Circassian tradition includes a code of knightly conduct, demanding respect for clan leaders and elders, protection of family honour, and blood feuds. Circassians perceive themselves as an embattled people. The national dress for men, the *cherkesska*, incorporates a row of small pockets for shotgun cartridges on each breast. The *cherkesska* is worn with a sheepskin hat during the incredibly athletic folk dance that Circassians perform at social events. Women traditionally wear ornate, full-length dresses, a style shared with related peoples of the Caucasus.

**CIRCASSIAN PRINCE**
*Jordan's Prince Ali, who is part Circassian, led a group of Circassians on a horseback trek from Jordan to Russia in 1998.*

CAUCASUS

## Chechens and Ingush

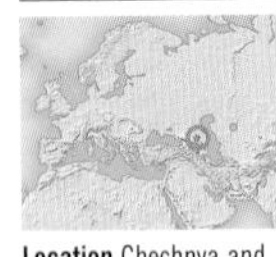

**Population** 1,063,000 (Chechens 899,000, Ingush 164,000)
**Language** Chechen, Ingush (both North Caucasian language family), Russian
**Beliefs** Sunni Islam, Sufism (mystical branch of Islam)
**Location** Chechnya and Ingushetia (SW Russia)

The Chechens and Ingush live in the Caucasus Mountains of southwestern Russia. They share similar traditions and understand each other's language. They are Muslim and traditionally lived by farming the lowlands and herding cattle in the mountains. In recent years, petroleum reserves in the region have become an important source of both income and conflict.

Chechen and Ingush village life is organized around a system of *taips* (clans), which cut across the territorial boundaries. Like other Caucasian peoples, Chechens and Ingush are known for their horsemanship and love of the mountains.

The Chechens are also known for being fiercely independent and have had troubled relations with Russia for nearly 200 years. Under the Avar leader Imam Shamil, Chechen tribes resisted Russian annexation in the 19th century until 1859, when Shamil was finally captured. However, the Chechens never really accepted defeat. Antagonism towards Russia intensified in World War II, when Chechens and Ingush were deported en masse to Central Asia and Siberia by Stalin; they were only allowed to return in 1957. Calls for Chechen independence in the 1990s led to a prolonged guerilla war in which thousands died. In 1992, the Russian republic of Checheno-Ingushetia was divided into the two republics of Chechnya and Ingushetia, which remain under Russian control.

**TRADITIONAL DRESS**
*Chechen men wear the* papakha *(traditional sheepskin hat) at all times. The hat is regarded as a powerful symbol of personal dignity.*

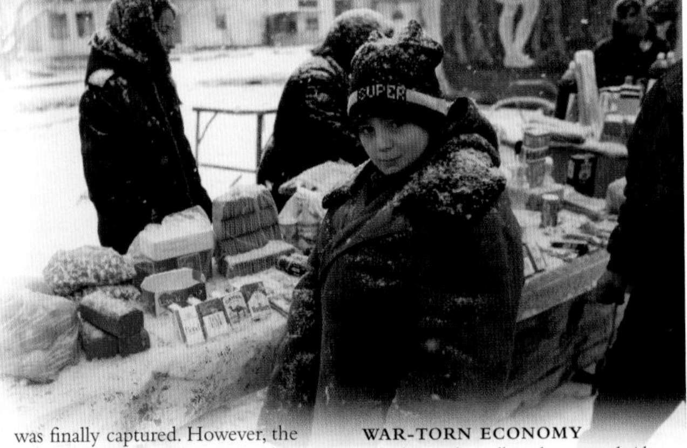

**WAR-TORN ECONOMY**
*Chechen traders sell goods at a road-side stall. Chechnya's economy and infrastructure were left in ruins by the war of the 1990s.*

Issue

### Chechen refugees

Up to 100,000 people died during conflict in Chechnya in the 1990s, and 400,000 were forced to leave their homes, some crowded on to trucks (left). Many Chechen refugees now live in Ingushetia and Moscow. For those that remain, reconstruction is likely to be difficult because of poor infrastructure and distrust of the pro-Russian authorities.

CAUCASUS

## Avars

**Population** 604,000
**Language** Avar, Andi, Karata, Dido, or Tsez, Bezhta, and several other languages of the North Caucasian language family
**Beliefs** Sunni Islam
**Location** Republic of Dagestan (SW Russia)

The Avars are a Caucasian people who make up the largest ethnic group in the Russian republic of Dagestan. They traditionally lived by herding animals in the mountains, where their villages, consisting of multistorey stone houses and fortress-like towers, were built on the slopes. Avar villages are made up of *tl'ibils* (clans) and were historically administered by village elders and Muslim leaders.

**AVAR HEADDRESS**
*Avar women traditionally wear a headdress called a* chikhta. *Men wear a felt cloak with a quilted coat, like other Caucasians.*

CAUCASUS

## Abkhaz

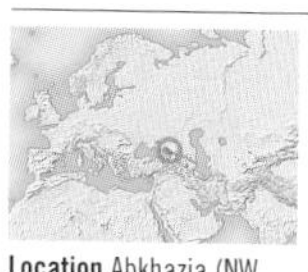

**Population** 500,000
**Language** Abkhaz (a North Caucasian language related to Circassian languages), Russian
**Beliefs** Nominally Orthodox Christianity, Sunni Islam; also traditional beliefs
**Location** Abkhazia (NW Georgia), Turkey, Syria

The Abkhaz are closely related to the Circassians and have a similar culture, with a rural economy traditionally based on maize, honey, and farming sheep. In keeping with the Caucasian tradition of hospitality, the Abkhaz hold elaborate feasts and entertain guests with particularly eloquent toasts. Among traditional entertainments are horseback games, folk dancing, and singing. The Abkhaz are famed for their longevity, although experts now call into question the age of alleged Abkhaz centenarians. People attribute their long lifespan to the wholesome mountain air and a healthy diet of pulses, maize-flower, yoghurt, and honey. Abkhaz food is often flavoured with *ajika*, a paste made of red peppers, herbs, and salt.

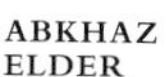

**ABKHAZ ELDER**
*Some believe the age of the famous Abkhaz elders to be exaggerated.*

**ABKHAZ FAMILY**
*An Abkhaz mother comforts her child after being forced out of her country by fighting between Abkhaz and Georgians.*

**MARKET TRADERS**
*Merchants sell grain and spices at Tbilisi market. Crops such as maize and citrus form the backbone of Georgia's rural economy.*

CAUCASUS

## Georgians

**Population** 5 million
**Language** Georgian (South Caucasian family), Russian
**Beliefs** Georgian Orthodox Christianity (one of the first nations to adopt Christianity); also Islam (small minority)
**Location** Georgia, Turkey, southwestern Russia

The Georgians are an ancient people who have lived in the Caucasus region since the dawn of history. They have a distinctive culture, architecture, language, and their own alphabet, which bears little resemblance to any other. Meals are the focus of family life and often begin with a toast. Georgians pride themselves on their food, which was famous throughout the Soviet Union. Delicacies include cheese-filled bread, walnut preserve, and strings of walnuts coated in grape jelly. During Soviet rule, Georgian restaurants spread to Moscow and St Petersburg, where they remain popular today. Georgia also has a rich literary and artistic tradition stretching back 15 centuries to the invention of the Georgian script.

**DIFFERING ENVIRONMENTS**
*Georgians occupy environments ranging from the sunny Black Sea coast to the cold northern mountains.*

**ARMENIAN ARTS**
*Performing arts have a long history in Armenia, from plays in the Byzantine era to modern ballet and film.*

CAUCASUS

## Armenians

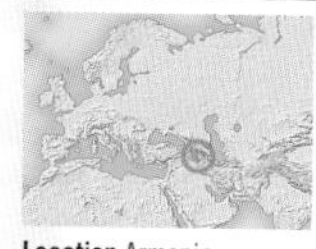

**Population** 6 million (around 1 million in the US)
**Language** Armenian (an Indo-European language)
**Beliefs** Armenian Apostolic Church (Orthodox Christianity)
**Location** Armenia, Azerbaijan, US

The Armenians are one of the world's most ancient peoples, and Yerevan, the capital of Armenia, is one of the oldest continuously inhabited places on Earth. Armenians call themselves the *Hayer* because they believe they descend from Haik, the great-grandson of biblical Noah, whose Ark came to rest on Mount Ararat in the Caucasus mountains. The ancient land of Armenia remains divided between the modern country, Turkey, and the unrecognized republic of Nagorno-Karabakh, an exclave in Azerbaijan. Armenia became Christian in AD301, becoming the first nation to do so. The relationship between the resulting Apostolic Church and the nation remains strong. The Bible was the first text produced in the unique Armenian script, in the 5th century. Armenian folk music, which sounds strongly Middle Eastern, is an important part of daily life. Food also has a Middle Eastern flavour and includes boiled lamb, chickpeas, beans, aubergine, and yoghurt. Among Armenian festivals is *Vardavar* (water festival) in which people fling buckets of water at each other at the peak of Armenia's stifling summer.

**ARMENIAN PATRIARCH**
*A patriarch inspects a student's work at an Armenian school. Since independence from the Soviet Union in 1991, Armenian history and religion play a bigger role in education.*

**BOARD GAMES**
*Armenian men often meet in cafes after work to exchange news and play games, such as this traditional Armenian game played with tiles.*

**DEFINITE IDENTITIES**
*Moroccan horsemen demonstrate the strong sense of tradition that remains in North Africa, despite the fact that European influence is spreading.*

# MIDDLE EAST AND NORTH AFRICA

In this region, great tracts of inhospitable deserts intersperse areas watered by the Nile in Egypt, and the Euphrates and Tigris in Mesopotamia (now Turkey, Syria, and Iraq). The "Fertile Crescent" created by these rivers enabled sophisticated agriculture and led to the building of the first cities (4000–2500BC), and to three of the world's great religions.

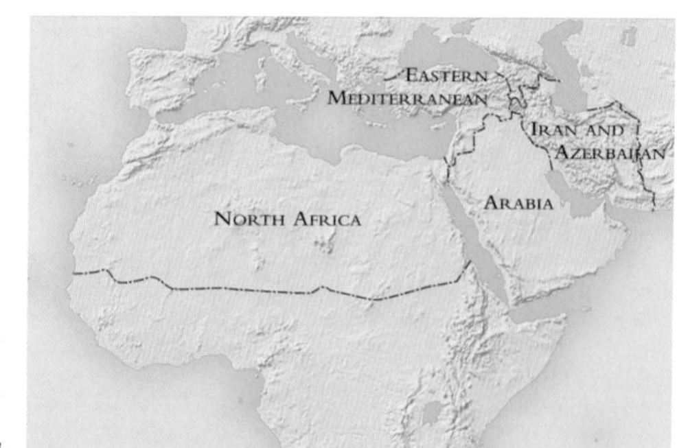


**THE REGION**
*The Middle East here includes Arabia and extends to Iran in the northeast, and northwest to Turkey. North Africa stretches west from Egypt to Mauritania.*

The Middle East and the coast of North Africa are at the point where Africa, Europe, and Asia meet. This is a region of ancient peoples and cultures, and of great empires. First the Babylonian, then the Assyrian, Persian and Arabian, and then the Ottoman Turkish cultures spread their influence over the region.

## MIGRATION

The first arrivals came from the south, from sub-Saharan Africa. They established communities in the fertile strip along the north African coast and in the great Saharan desert it fringed. The fertility of the terrain led to settlement – the area has some of the world's oldest continuously inhabited cities, from Luxor in Egypt to Jericho on the West Bank. Some of the oldest records of agricultural activity are from the so-called "Fertile Crescent" centred on great river valleys: the Nile, Tigris, and Euphrates.

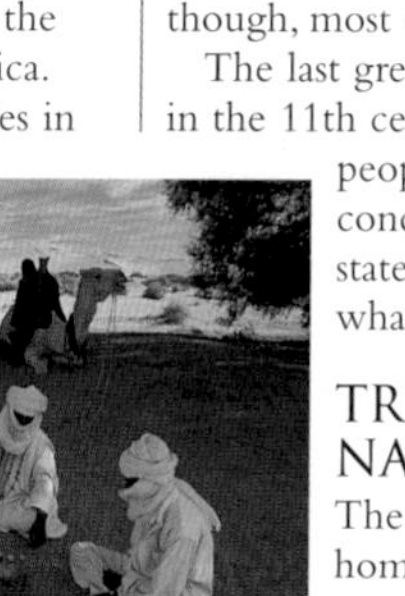

**IN THE STEPS OF THE TUAREG**
*The Tuareg live in the Sahara – a desert bissected by international borders. As these are enforced, the nomads cannot migrate.*

Thousands of years later, in the 7th century AD, what was perhaps the most significant migration occurred when Arabian peoples spread from their home in southern Arabia to conquer the region and held it in the name of their new religion, Islam. The Arabian conquest forced native peoples such as the Berbers of North Africa either to migrate or adopt Arabic language and customs. Some moved south and became nomads in the Sahara. Eventually, though, most converted to Islam.

The last great migration came in the 11th century when Turkic peoples of Central Asia conquered the Arabian states and settled in what is now Turkey.

**ASSYRIAN DEITY**
*The ancient Assyrians, centred in present-day Iraq, built an empire in the region and left carvings of deities still visible today.*

## TRIBES AND NATIONS

The region has been home to some of the greatest empires of the ancient world. These included Babylonia (18th–14th centuries BC) and Assyria (14th–7th centuries BC), both now parts of Iraq, Syria, and Turkey. They stood between the Tigris and Euphrates rivers, while Egypt dominated the Nile.

Many foreign rulers conquered the lands, beginning with the Persians, followed by the Macedonian Greeks, the Romans, the Mongols, the Christian Crusaders, and finishing with the Turks – the Ottoman dynasty lasted from the 13th century to the end of World War I.

The collapse of Arab domination from the 11th century onwards (under pressure from the Ottoman Turks), led to the rise of competing mini-states. Distance divided Arabian culture into groups with distinct traditions and practices. Today, there are still strong regional identities: dialects and customs vary between diverse Arabian peoples from the Jalayin and Juhaynah Arabs of the Sudan in the south to the Marsh Arabs of Iraq in the east and the Moroccan Arabs in the west. Nevertheless, the idea of a Muslim brotherhood and the widespread use of Arabic promote desire for Arab unity.

**PROTECTING THEIR EXISTENCE**
*The Imagruen, an endangered Berber minority of Mauritania, celebrate the launch of a patrol boat to protect their fish stocks.*

In the modern era, the peoples of the region still feel the influence of outside powers: peoples such as those of Iraq, Syria, and Jordan, are divided by arbitrary borders drawn up by colonizing powers such as Britain. Perhaps the most significant modern creation is the state of Israel. In recompense for the years of persecution of Jews in Europe that culminated in the Holocaust, the Western powers created a Jewish state in the "Holy Land". The competing interests of the peoples in the region are still a source of global tension. The plight of the Arabian Palestinians, who were displaced, has mobilized Islamic extremists.

**OIL PROSPERITY**
*The discovery of oil in the Gulf states has dramatically changed people's aspirations and lifestyles.*

## CHANGE

A catalyst for change in the region has been the balance between the power of Islamist teaching and the increasing global dominance of Western social expectations and culture. The stand-off could have consequences for long-held ideals of the Arabian, Iranian, and other peoples of the region. Liberalism, capitalism, and the wealth made from oil have provoked a desire for change, felt mostly by the young; traditionalists balk at the influx of Western ideals and the perceived erosion of Islamic principles; while the secular governments struggle against the popularity of Islamist teaching among the poor.

Issue

### One city, many faiths

Judaism, Christianity, and Islam centre on Jerusalem (right). For Jews, it is the capital of the Promised Land, for Christians it is the place their messiah, Jesus, was crucified, while Muslims believe their prophet, Muhammad, was taken there on the night of his death to ascend into heaven. The city has become a focus of tension between the region's peoples.

MIDDLE EAST, NORTH AFRICA

# Arabs

**Location** North Africa, the Arabian peninsula

**Population** 287 million
**Language** Arabic; also English, French
**Beliefs** Sunni and Shia Islam; also Greek Orthodox, Coptic Christianity (an ancient Egyptian religion), Armenian Apostolic Church

Prior to the Islamic (Muhammadan) era in the 7th century AD, the term Arab described the Semitic-speaking tribes and the nomadic Bedouin inhabitants of the Arabian Peninsula. These tribes went on to spread Islam across the Byzantine and Persian empires, from the mountain ranges of the Pyrenees to the Himalayas. As their influence grew, the term Arab came to describe those communities in the Middle East sharing the Islamic religion and the Arabic language. Arabic became a language of high culture and commerce, as well as the language of the Koran (the holy book of Islam). The Arab people also contributed significantly to scientific discovery, technological innovation, and geographic exploration in the medieval era.

The fortunes of the Arabic countries of the Arabian peninsula were dramatically changed in the 20th century with the beginning of large-scale exploitation of their abundant natural oil reserves. Some of the Arab microstates in the Persian Gulf have among the highest standards of living in the world as a direct consequence of the oil industry.

Arab identity is inseparable from Islamic civilization and religion (*see* p291), and the Arab people are at the heart of the *dar al-Islam* (Islamic world). Even today it is not uncommon for Arabs to give their nationality as Muslim rather than the nation of their birth. There are two dominant Muslim groups in the Arab world: Sunnis and Shias. The former stress their adherence to the *sunna* (path) of the Prophet, while the latter emphasize allegiance to the *shia* (party) of Ali, the Prophet's son-in-law. Fasting, pilgrimage, and the rites of Muslim prayer – performed five times a day facing towards Mecca (in western Saudi Arabia) – are central to all Arab societies. Islam also prescribes a strict code governing relations between the sexes. Despite increasing occupational mobility and educational opportunities, women have traditionally held a subordinate position in Arab society. In many conservative countries, women wear the dark-coloured Muslim *hijab*, which leaves their bodies (and sometimes their faces) covered up to varying degrees.

There is no one notion of kinship or family in the Arab world, but Arab hospitality is widely noted. So, too, is the Arab love of the art of storytelling; the Arabic word *qara'a*, meaning to recite, is the root for the word Koran.

**CALLIGRAPHY**
*The decorative art of calligraphy has become important in the Arab world. This vase bears an inscription from the Koran.*

**HEADDRESS**
*This Saudi Arabian man displays the headgear typical of people of the Arabian peninsula.*

**TIME TO RELAX**
*In the privacy of a modern Saudi home, an Arabian family may relax together and watch the news on television.*

History

## The Arab League

First proposed at an Arab summit in Alexandria, Egypt, in 1944, the Arab League was established by charter in March 1945. It is comprised of representatives from all the Arab nations and provides a forum for inter-Arab discussions on social, defence, economic, and political issues. Its senior official, the secretary-general, also represents Arab political interests in the international arena.

**MOSQUE VISIT**
*A man stands at a Shia mosque in Dubai, United Arab Emirates – one of the richest Gulf states.*

**AFTER A DAY IN THE FIELDS**
*Parts of Syria are more fertile than other areas in the Arabian peninsula and North Africa, and some Syrian Arabs, such as these women, are agricultural workers.*

NORTH AFRICA

## Berbers

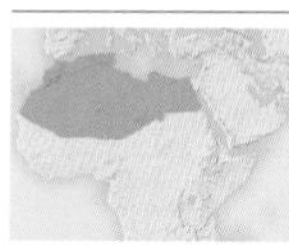

**Population** Estimated 12–18 million in Morocco and 3–4 million in northern Algeria
**Language** Amazigh languages (some 300 dialects), Arabic, French
**Beliefs** Islam
**Location** Mainly Morocco, Algeria, and Libya

Berber peoples often prefer to call themselves Amazigh rather than Berber, a term thought to derive from an Arabic word meaning "wild" or "uncivilized", or "barbaric". They refer to their homeland as Tamazgha, which includes parts of several modern North African countries. The largest number of Amazigh people is found in Morocco, Algeria, and Libya; other communities live in Tunisia, Burkina Faso, Egypt, and Mauritania.

Early Roman and Greek accounts mention nomadic tribes who controlled the five trade routesacross the Sahara, and they were collectively known as Berbers at the time of the Arab invasion in the 7th century. With the decline of trans-Saharan trade routes, the majority of Berber people became cultivators or settled traders, but they preserve a strong sense of their separate cultural heritage. There are three main groups of Berbers in Morocco alone: the Rif, Berraber, and Schluh. All of them are associated with making fine jewellery, metalwork, woven goods, and embroidery, which reflect their earlier role in the trade of easily portable luxury goods.

The Berber language has about 300 different dialects, some of which are distinct enough to be called separate languages. In France, the Amazigh language is now a subject for the national Baccalaureate examination, and Departments of Berber Studies have been created in Morocco and France.

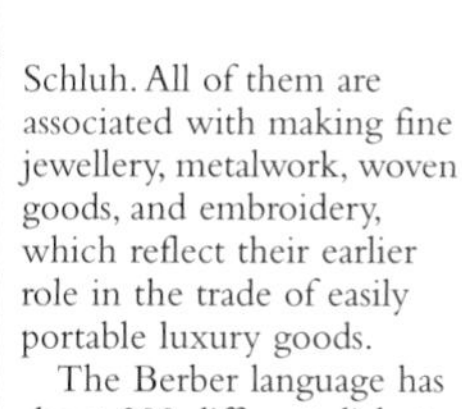

**WEDDING FESTIVAL**
*A young bride, with the pale skin and green eyes of many Berbers, is presented at the Imilchil Wedding Festival in Morocco.*

**BERBER METALWORK**
*The Berbers are still known for their beautiful metalwork, one of the many goods they traded in earlier times.*

**ON THE WAY TO A FESTIVAL**
*These Berber women are attending the Sahara Festival in Douz, Tunisia. The festival features dance, acrobatics, and horsemanship.*

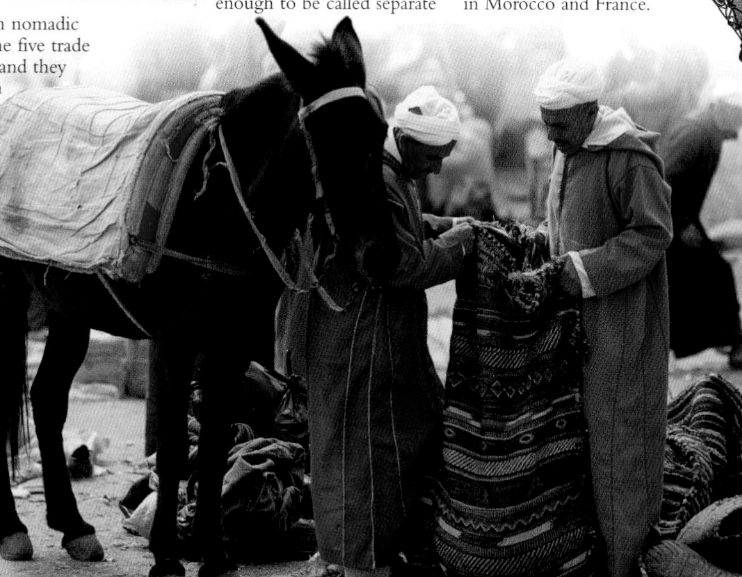

**AT THE MARKET**
*Berber men in Morocco examine goods for sale at a market.*

NORTH AFRICA

## Moroccans

**Population** 31 million
**Language** Officially Arabic, but 40–50% of people speak Berber, or Tamazight; also French, Spanish
**Beliefs** Islam
**Location** Morocco

Moroccan identity is firmly bound up with national identity and religion: "God, nation, king", as the national anthem says. Although of Berber, or Amazigh, ancestry, the native peoples of Morocco accepted Islam when Arabs invaded the country in the 7th century and fully adopted Arab culture. Many Arabs who settled here ended up intermarrying with Berbers, producing a largely mixed population, which sees itself as both Moroccan and Muslim.

Traditional Moroccan weddings are a reflection of the nation's love of dressing up, eating well, and enjoying music and dancing. Moroccan music has a variety of styles, from complex, sophisticated orchestral music to a simpler style involving only the voice and drums. The modern version of traditional dress, especially the *jellaba* (a long tunic with a hood), is widely worn by both men and women, and for special occasions women wear the traditional kaftan, or variations on this. Moroccan cuisine – such as couscous, tajine, and mint tea – is now increasingly appreciated in other countries, and is a source of pride to Moroccans.

**SILVER EARRING**
*Moroccan craftsmanship has a rich heritage in the production of jewellery, leatherware, pottery, and woodwork.*

**DANCING AT THE SOUK**
*The vibrant heart of old Moroccan towns is the souk, or market, where these performers are dancing.*

NORTH AFRICA

## Maures

**Population** 950,000
**Language** Hassaniya (a form of Arabic that includes a high proportion of Berber vocabulary); also French, and Arabic
**Beliefs** Islam (Sunni with membership of Sufi orders)
**Location** Mauritania, Gambia, Mali, Morocco

The Maures, or Moors, have mixed Berber and Arab origins; many also have African ancestors. Maures have a highly stratified social structure, in which distinctions between people are based on occupation, and ideas of rank, race, and heritage. The history of slavery in the area has left its mark: importance is given to descent from people of free birth rather than slaves. Distinctions are also made between Black and White Maures (*Bidan*), and between nobles, artisans, craftspeople, and herders. Each "caste" has preserved its rank and way of life by intermarriage and remaining closed to outsiders.

**MAURE FACE**
*There is no typical Maure appearance, because the people have diverse origins. Subgroups remain separate due to inherited status.*

**HEAVENLY HAIRSTYLE**
*According to the Haratin Maures, their topknot hairstyle helps Allah pull them to heaven.*

NORTH AFRICA

## Tuareg

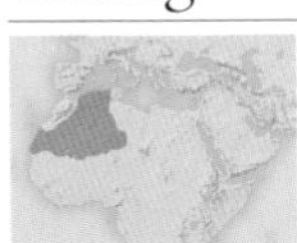

**Population** Between 100,000 and 300,000
**Language** Tamashek (a Berber language that has an alphabet called Tifinagh), also Arabic
**Location** Sahara desert regions in northwest Africa
**Beliefs** Islam (mixed with pre-Islamic beliefs)

Tuareg are people of the Sahara who speak one of the Berber languages, known as Tamashek. Berber languages are written in a script called Tifinagh; among the Tuareg, the Tifinagh script was traditionally used by women, often for writing letters. Tuareg consider themselves descended from the nomadic tribes whose camel caravans played a main part in trade across the Sahara for thousands of years. They traded in luxury goods, which could be transported easily and sold for a large profit. Modern Tuareg art still includes elaborate jewellery, leather saddles, swords, and metalwork.

**UNVEILED**
*In Tuareg society, people trace descent though women. This gives Tuareg women high status and some do not wear face veils.*

**LEATHERWORK**
*Among the Tuareg's finely crafted leather goods are decorated men's pouches.*

Today, many Tuareg live in communities at the edge of Saharan cities. Others have replaced camels with trucks and continue a nomadic trading existence. Most Tuareg are Muslim, although Islam often coexists with practices and beliefs associated with their ancient pre-Islamic past. Many men wear protective amulets, which contain verses from the Koran.

**BIANOU FESTIVAL**
*Bianou commemorates Muhammad's flight from Mecca to Medina in AD622. It features dancers and musicians (left), as well as a parade of warriors.*

Fact

### Veiled men

Tuareg male dress is among the most distinctive of any North African people. Men, rather than women, begin to veil and conceal their faces in early adulthood. At full manhood, a turban-veil almost completely covers the face. Face-coverings have great practical value as protection from the harsh desert climate, but are traditionally never removed, even when with close family.

**INDIGO CLOTH**
*A Tuareg man is swathed in yards of costly indigo-coloured cloth. He has attached a protective silver Koran-holder to his turban.*

NORTH AFRICA

## Copts

**Population** 4–6 million
**Language** Arabic, Coptic (the ancestral Coptic language survives only in religious worship and is thought to be related to Ancient Egyptian)
**Location** Egypt
**Beliefs** Coptic Christianity

The Copts, who form a large minority community in Egypt, are considered to be descendants of the native non-Arab peoples of the region. Many Copts regard themselves as historically linked to the first North African converts to Christianity, and define their cultural distinctiveness mainly in religious terms as people who resisted conversion to Islam.

In the centuries following Egypt's emergence as an Arab and Muslim state, Copts have maintained close-knit, family-centred communities across the country. They tend to be relatively well-educated members of the professional and business classes. In recent years, tension between Muslim and Coptic peoples has intensified, and has resulted in violence in urban centres.

**HEAD COVER**
*Copts follow some of the customs of their Muslim neighbours, such as covering the head with a scarf.*

NORTH AFRICA

## Egyptians

**Population** 74 million
**Language** Arabic; also English and French (among educated people)
**Location** Egypt, Kuwait, Oman, Qatar, Saudi Arabia
**Beliefs** Sunni Islam (huge majority), Coptic and Orthodox Christianity (small communities)

The Egyptian people are heirs to one of the oldest civilizations in the world, dating back 5,500 years. In the time of the Pharaohs (3200–341BC), the ancient Egyptians built some of history's great architectural works, including the Pyramids and the Valley of the Kings outside Luxor. They also introduced the use of embalming for mummification, papyrus for writing, hieroglyphics, and the solar calendar.

Egypt's geographic location as a sea link between the Indian Ocean and Mediterranean Sea, and the only land bridge between Africa and the East, explains the cosmopolitan nature of Egyptian society, which can claim Jewish, Nubian, Greek, Armenian, and European influences. Since gaining complete independence in the 1950s, Egypt – the most populous Arab state – has continued to play a pivotal role in regional affairs. The Egyptian people have a reputation as hospitable hosts. They feel a social obligation to be generous to guests, but place a high premium on mutual respect, and it is inappropriate for a guest to enter an Egyptian home uninvited.

History

### Hieroglyphics

The Ancient Egyptians used a form of picture writing called hieroglyphics. There were around 700 different symbols, some of which represented whole words – a series of wavy lines represented water, for example. Because the system was so intricate, most people could not read or write, and eventually a simplified form of the script was developed for everyday use.

**BOATMAN**
*A boatman waits to ferry tourists across the Nile. Many Egyptian people earn their living from tourism.*

**SWORD DANCE**
*Ritual sword play takes place during a festival in Gezira, near Luxor.*

**FRUITS OF THE NILE**
*Egypt owes its five millennia of civilization to the River Nile, which irrigates crops such as olives.*

NORTH AFRICA

# Nubians

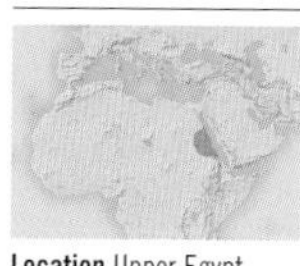

**Location** Upper Egypt, northern Sudan

**Population** 1.5 million
**Language** Nobin, Kenuzi-Dongola, other Nubian languages of the Nilotic family, also Arabic
**Beliefs** Islam, traditional beliefs (which feature rainmaking ceremonies)

Today, in Upper Egypt, especially in Aswan, dark-skinned Arabic-speaking people are often called Nubians. They are thought to be descended from people who migrated north or were bought as slaves during wars between the ancient Nubian kingdoms and Egypt. A number of unique Nubian temples and sculptures, more than 3,500 years old, are now housed in the Nubian Museum in Aswan.

In recent decades, there has been an increase in interest in Nubian history, folklore, and traditional culture. Nubian artists in Aswan have developed a contemporary school of art, which blends African and Egyptian styles.

**HOOKAH**
*Many Nubian men enjoy smoking the hookah – the traditional, often elaborate water-pipe used across the Middle East.*

EASTERN MEDITERRANEAN

# Turks

**Location** Turkey, northern Cyprus, Bulgaria

**Population** 68 million
**Language** Turkish (a Turkic language of the Altaic family, related to Turkmen and Azeri)
**Beliefs** Overwhelmingly Sunni Islam; also Shia Islam, Christianity, Judaism

Turks have lived in what is now Turkey since AD800, when Muslim tribes started arriving from Central Asia to fight the ailing Byzantine empire. One family in particular, the Osmans (or Ottomans), consolidated its grip and built an empire based in Istanbul. Although no longer the power it was in the period of the Ottoman empire, modern Turkey boasts a unique position in both Europe and Asia. Turks view themselves as the most Westernized of all Muslim peoples, and women have equal rights in divorce and marriage. This secular character is largely due to a programme started by Mustafa Kemal Ataturk, president of Turkey between 1923 and 1938. During this period, Islam was disestablished, the Arabic alphabet was replaced by Roman (Western) script, and the Sultanate was abolished. Ataturk's revolution, which emphasized the Turkish rather than the Islamic roots of the people, has influenced Turkish identity more than any other social, political, or cultural development.

Turkish cuisine, which is world famous, is probably best known for its kebabs and *baklava* (sweet pastries).

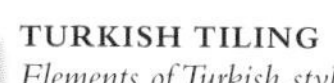

**TURKISH TILING**
*Elements of Turkish style, such as decorated tiles (often used on buildings), link the Turks with their relatives in Central Asia.*

**GRAND BAZAAR**
*Istanbul's grand bazaar is festooned with icons of Turkish culture, from lamps and tea sets to belly-dancing outfits.*

**MOUSTACHE**
*Many Turks grow moustaches after national service, during which they are clean-shaven.*

EASTERN MEDITERRANEAN

# Dom

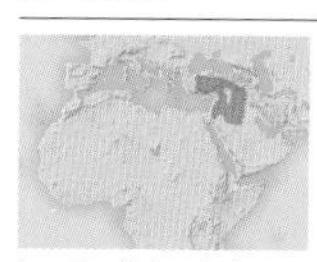

**Location** Turkey, Jordan, Israel, Lebanon, Iraq, Iran

**Population** Estimated 2–3 million (but some Dom are excluded from national census and others label themselves in national terms rather than as Dom)
**Language** Domari, Arabic
**Beliefs** Islam

The Dom (or Domi) of the Middle East are part of the larger ethnic and linguistic group often called Roma, or Gypsies. Dom means "man", and later this developed into the terms Roma and Romani (*see* p392). The Dom have a rich oral tradition and express their culture and history through music, poetry, story, and dance. Some groups believe they migrated to the Middle East when Persia invaded their original homeland in India.

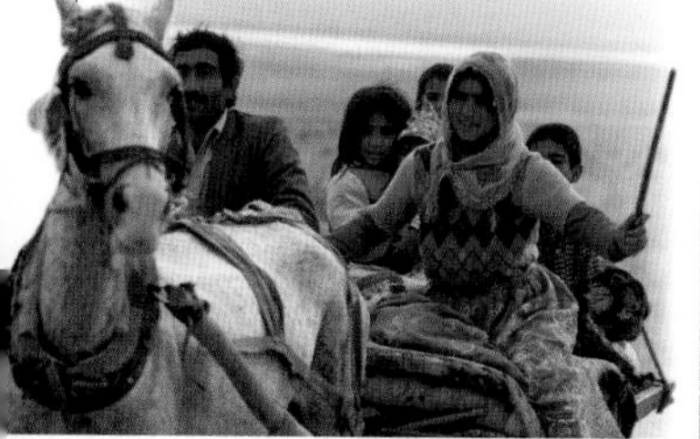

**TRAVELLING DOM**
*Some Dom still live as travelling entertainers, moving in camel caravans or in carts drawn by mules or (as here) horses.*

EASTERN MEDITERRANEAN

# Kurds

**Location** Kurdistan (parts of Iraq, Turkey, Iran, Syria)

**Population** Estimated 16–20 million (Iraq: 2–3 million, Turkey: 4–10 million, Iran: 5–6 million, Syria: 1.5 million)
**Language** Kurdish, Farsi, Arabic, Turkish
**Beliefs** Sunni Islam

The Kurds trace their history back several thousand years. The Silk Road (the trade route linking Europe with Asia) passed through Kurdistan, and Kurds traded fine rugs and textiles. There are some 800 different tribes within Kurdistan, and a person's last name often identifies his or her tribe.

Many aspects of Kurdish history and culture have been shaped by the dramatic setting of their homeland. Kurdish tales are often set against a backdrop of high peaks and lush valleys, while traditional forms of dress reflect the natural landscape in their use of bright colours. Kurds also have a rich tradition of poetry, dance, and music, which often involves types of *dohol* (drum) and *zornah* (flute). Some Kurds work in mountain areas as herders and cereal cultivators. Elsewhere, they have adopted urban trades and professions, following migrations to cities as a consequence of urbanization, war, or natural disasters.

Issue

## Struggle for independence

Despite being the fourth largest ethnic group in the Middle East, the Kurdish people have no separate nation-state, and for centuries they have struggled against subjugation by neighbouring peoples. Sometimes loyalty to tribe and historic rivalries between leaders have weakened the development of a unified national Kurdish movement. Many Kurds have now organized along separate political lines in their fight for independence and economic rights.

**KURDISH WEDDING**
*Guests at a wedding throw shaving cream at the bride and groom to celebrate their marriage.*

EASTERN MEDITERRANEAN

# Druze

**Location** Syria, Lebanon, Israel, Jordan

**Population** 700,000
**Language** Levantine Arabic (a dialect of Arabic)
**Beliefs** Secret religion related to Islam (involves belief in reincarnation; heaven is an escape from the cycle of rebirths)

Forming closed communities in parts of the Middle East, the Druze follow a secret religion that split from Islam during the 11th century. Conversion to the Druze religion and marriage to non-Druze people are forbidden.

Druze communities may live in their own villages, often in secluded mountain valleys. Elsewhere, they blend in with urban societies, in part to avoid unwanted attention. They have no separatist national aspirations and are loyal to the states in which they reside. This loyalty may include service in the army. Within the context of the Arab–Israel conflict, it has meant that they often find themselves serving in armies that are in confrontation.

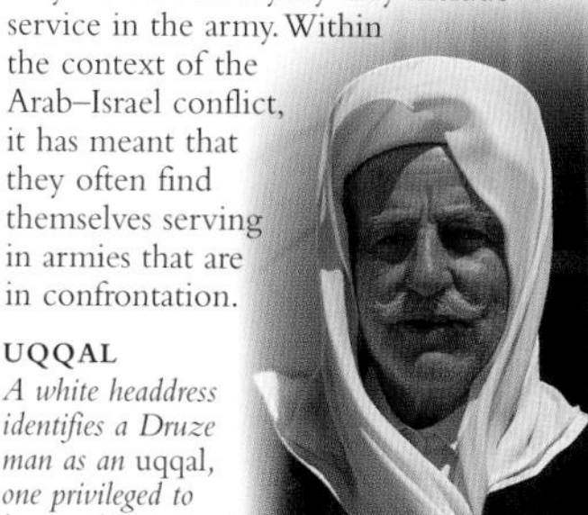

**UQQAL**
*A white headdress identifies a Druze man as an uqqal, one privileged to know religious secrets.*

EASTERN MEDITERRANEAN

## Jews

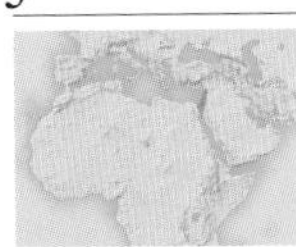

**Population** 5.3 million
**Language** Hebrew (revived as new vernacular language in late 19th century); Arabic, English, and Russian also commonly spoken
**Location** Israel; also in the US, Canada, and Europe
**Religion** Judaism

There has been a continuous Jewish presence in what is now Israel since the Abrahamic era (around 4,000 years ago). Despite mass expulsion of the Jews from Palestine at the hands of the Babylonians in 586BC and the Romans in AD70, a Jewish presence continued under Roman, Byzantine, Ottoman, and British rule until the establishment of Israel in 1948. Following the restoration of Hebrew as a vibrant language, Israelis could claim to be the only Middle Eastern people speaking the same language as their forefathers in the 1st century AD.

Israel is home to Jews from more than 100 nations. The Right of Return, enshrined in Israeli law since 1950, stipulates that all Jews have a right to settle permanently in Israel, and Israeli society has been influenced by the absorption of Jews facing hardship or persecution elsewhere in the world. Jewish rabbinical law plays a role in matters of marriage and divorce, while *Kashrut* (Jewish dietary) laws and Sabbath observance in some public and commercial institutions also underline the distinctively Jewish nature of the state.

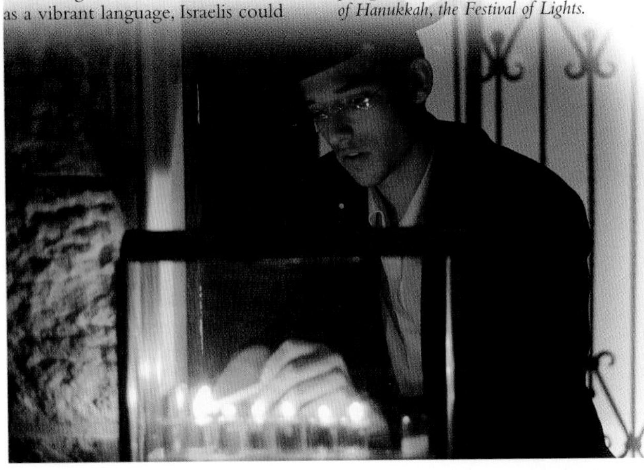

**FESTIVAL OF LIGHTS**
*The Jewish people are defined by their unique religion and its many festivals. A young Orthodox Jew observes a ritual of Hanukkah, the Festival of Lights.*

EASTERN MEDITERRANEAN

## Palestinians

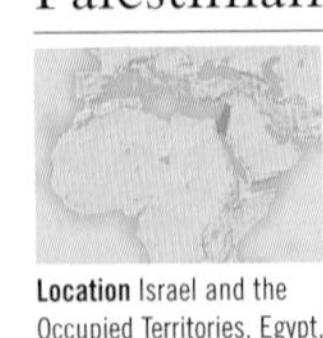

**Population** 8.8 million worldwide, of whom 4.3 million live in Israel and the Occupied Territories
**Language** Levantine Arabic (a dialect of Arabic)
**Location** Israel and the Occupied Territories, Egypt, Jordan, Lebanon, Syria
**Religion** Islam (Sunni); a few people practise Catholic Christianity

The historical region of Palestine includes parts of Israel, the Occupied Territories, Egypt, and Jordan. It has been invaded and occupied by numerous peoples over the last 3,000 years. However, the people who now call themselves Palestinians draw their culture from the arrival of the Arabic people in the 7th century.

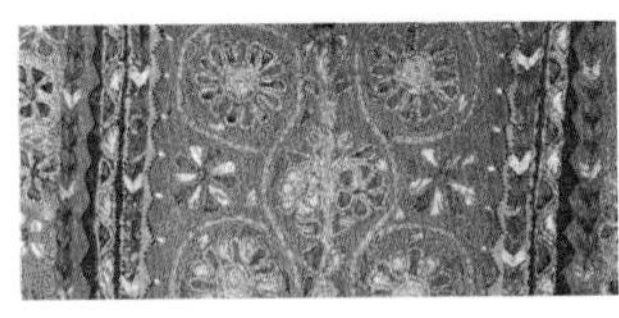

**TRADITIONAL EMBROIDERY**
*The Palestinians have been prominent in textile production since medieval times. The word "gauze" probably comes from the town Gaza in the Occupied Territories.*

Palestinians are Arabs, and in accordance with Arabic culture, Palestinian culture places high value on poetry and the decorative arts. The recent troubled history of the region associated with the creation of the state of Israel has made it difficult for Palestinians to hold onto their cultural identity. Many still live in the refugee camps to which they were displaced decades ago. Displays of Palestinian nationalism meet with a hostile reaction. Nevertheless, even in the refugee camps, the people conserve their culture by continuing to produce textiles in the traditional way.

**REFUGEES**
*Many Palestinians live as virtual prisoners in a land they regard as their own.*

ARABIA AND NORTH AFRICA

## Bedouin

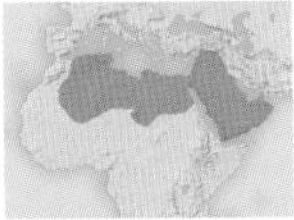

**Population** 11 million worldwide (3 million in the Middle East, 8 million in North Africa); perhaps 10 per cent remain nomadic
**Language** Arabic dialects including Bedawi, Najdi, and Hassani
**Location** North Africa, Israel, Syria, Jordan, Arabian Peninsula
**Religion** Islam (Sunni)

Bedouin are regarded by many Arabs as the original or ideal representation of Arab culture. Like other Arabs, Bedouin consider their heartland to be the Arabian desert, the Negev and Sinai deserts, and the deserts of Jordan and Syria. Since the Middle Ages, however, they have been dispersed across the northern Sahara to Algeria and Mali. Today, a nomadic lifestyle probably accounts for less than one per cent of the world's Arabs. Some Bedouin still make few concessions to modern life other than a pick-up truck to transport animals, plastic water containers, and perhaps a kerosene stove.

**HEADGEAR**
*A Bedouin man wears a* kufiyya *cloth secured by the symbolically important* 'agal *rope.*

**SPINNING WHEEL**
*This Bedouin wife and mother sits beside her spinning wheel in a tent. The yarn from the wheel may be sold or used to make clothing.*

History

### Arabian diaspora

Beginning their expansion in the 7th and 8th centuries, the ancestral Arabian people spread from their home in southern Arabia north to Syria and west into North Africa. Their familiarity with the difficulties of desert travel surely aided them, but they were also spurred by their mission to spread the message of Islam. Today, nomadic Bedouin still travel as far as Tunisia (below).

The harsh environment of the desert has profound effects on Bedouin society and culture. It is the custom to turn no traveller away. Instead, Bedouin often regard strangers as guests and take them into their black woven goat-hair tents for shelter and offer them refreshment in the form of coffee spiced with cardamom. The first cup of coffee is usually tasted by the Bedouin, in order to make the guest feel safe. A typical tent is divided into two sections, one for the men, and the other for the women – in which they cook, and receive female guests. Desert survival also required that Bedouin develop extensive kinship networks, which provide them with the community support they need. This is coupled with a strict code of honour, infringements of which are not easily forgiven or forgotten.

Emotions are guarded in daily life. Bedouin believe that such reserve ensures group solidarity. They have a rich tradition of emotional oral poetry, however, which they recite from memory in classical Arabic.

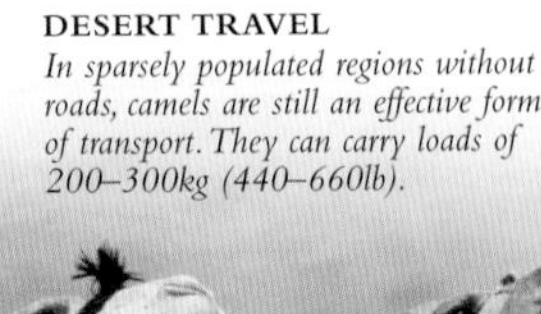

**DESERT TRAVEL**
*In sparsely populated regions without roads, camels are still an effective form of transport. They can carry loads of 200–300kg (440–660lb).*

EASTERN MEDITERRANEAN

## Assyrians

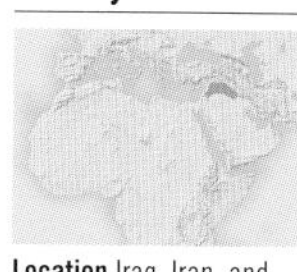

**Location** Iraq, Iran, and Syria

**Population** Estimated 200,000–250,000
**Language** Aramaic Arabic (an ancient Semitic language)
**Religion** Christianity (mostly Orthodox or Syrian Catholic)

Assyrians are often considered to be descendants of traders and invaders of Ancient Mesopotamian times. Language and religious affiliation both play a key part in defining Assyrian people's ethnic and cultural identity, although they may not be recognized as a separate ethnic group in their country of residence. They represent a religious minority as Christians in their communities, and a particular feature of importance to them is their sense of historic connection to the early Christian communities who did not convert to Islam during the rapid expansion of Islam in the region from the 12th century onward.

Today most Assyrians live in major cities and towns, where they tend to be members of the educated business or professional classes. Assyrians in rural areas of northeastern Iraq are mostly engaged in agriculture as fairly prosperous independent farmers.

ARABIA

## Marsh Arabs

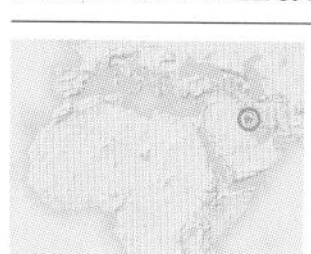

**Location** Euphrates delta wetlands in Iraq and Iran

**Population** 40,000 remaining in the ancestral homeland of the marshes, plus an estimated 180,000 displaced within Iraq and to Iran
**Language** Arabic
**Religion** Shi'a Islam

Until the 1990s, the Marsh Arabs, or Ma'dan, had a unique, independent lifestyle, fishing and farming, and living largely by floating on the marshes. They built astonishing floating structures from reeds – not only small dwellings but also halls, mosques, and meeting places. Their lifestyle had apparently evolved through 5,000 years' continuous occupation of the marshes. Today, only a handful of Ma'dan remain on the dwindling wetlands, and they cannot sustain their way of life. Most have fled to refugee camps or to find jobs in cities. There are plans to restore the marshes, but some newly urbanized Ma'dan would be reluctant to return to occupation of the marshes.

**LIFE ON WATER**
*Before the marshes were drained, Marsh Arabs lived entirely on water. Here, they hunt fish with spears.*

**REED BUILDINGS**
*Reeds from the marsh can be used to weave houses that float on the water.*

Issue

### End of the marshes

When Saddam Hussein's Iraqi regime drained the water from the marshes of the Euphrates delta during the 1980s and 1990s, the centuries-old lifestyle of the Marsh Arabs became impossible. As a result, most people fled to refugee camps in Iraq and Iran, where the children pictured below now live.

ARABIA

## Yemenis

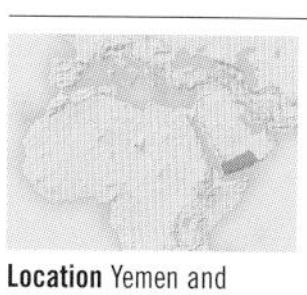

**Location** Yemen and adjacent countries

**Population** 22 million
**Language** Arabic dialects including Ta'izzi-Adeni and Sanaani
**Religion** Islam; mainly Zaydi Shi'a Islam and Shafi'i Sunni Islam, plus Ismaili Shi'a Islam

Most Yemenis are Arabic people, but as with the Omanis, their culture is distinct from that of Arabs elsewhere, perhaps due to the isolation of high mountains and rugged terrain. Yemen has strong links with both Southeast Asia and Africa, and many people calling themselves Yemeni Arabs are African in appearance.

A large proportion of the Yemeni population is organized into tribes, particularly in the mountainous north, and cultural differences abound. Tribal identity among the Hashid and Bakil, for example, can take priority over Yemeni or Arab identity. Yemeni society was once layered into caste-like groups, but these distinctions are now blurring and vanishing, allowing people more social mobility.

Yemenis are deeply committed to their local communities, but Yemen is not without its social problems. The country has a high rate of illiteracy.

Men and women lead separate social lives, but the social institution of chewing *qat* plays a central role in each case. *Qat* is a plant with mildly narcotic leaves, which Yemenis enjoy during afternoons of socializing.

**VIBRANT TEXTILES**
*Bedouin influence is apparent in this bright red shawl and hat from the northern region.*

**JAMBIYYA**
*Like the Omanis, Yemeni men carry an ornate dagger on a belt at their waist. The Yemeni dagger is called the* jambiyya.

Yemeni building is uniquely accomplished. In some Yemeni towns, people construct mud-brick skyscrapers up to 14 storeys high. In these buildings, the animals sleep at ground level, and the people live above them. Some people may adorn their dwellings with stained glasswork.

**TENDING FIELDS IN THE DESERT**
*Yemen is the most fertile country in the Arabian Peninsula, and many Yemenis are settled farmers. In the Wadi Hadramaut region, the desert sun is nonetheless fierce, and women keep cool with* madhalla *straw hats.*

PEOPLES

ARABIA

# Omanis

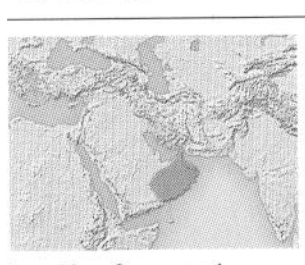

**Population** 2.7 million in Oman plus small numbers in nearby countries
**Language** Arabic (Omani, Dhofari, and Gulf Arabic)
**Religion** Mainly Ibadi Islam (a sect distinct from both Sunni and Shia Islam)
**Location** Oman and neighbouring countries

Oman has been a distinct nation for centuries (although, during the Middle Ages, it came under the influence of Arab dynasties ruling from Baghdad, Cairo, and Tunisia). It has long had seafaring and merchant traditions, and until 1856, had a trading empire controlling Zanzibar and ports in Iran and India. Omanis are mainly Arabs, but most are Ibadi Muslims, members of a sect distinct from both Sunni and Shia Islam since the 8th century. The Ibadi sect was one of Islam's earliest fundamentalist movements, and its conservative doctrines are still preached. Therefore, despite rapid economic changes since oil was discovered in the 1960s, life for many Omanis remains traditional.

**BOAT-BUILDING**
*Dhows, the boats traditionally used in the Persian Gulf, are still built by hand today.*

It is still common for men to wear the traditional *dishdasha*, a plain, ankle-length robe, usually in white. A tassel worn around the neckline may be impregnated with perfume. At public engagements and feasts, a man would feel naked without his symbolic silver dagger, the *khanjar*, at his waist. Women wear more colourful robes and dresses, under which they wear ankle-length trousers. Omani women are active in the economy and may work in offices and in the police force alongside men. Omanis in the broader sense includes other citizens of Oman, such as Sunni and Shia Muslims, and many Arabian tribes, including the Dhofari and the seminomadic Shihuh, each speaking its own dialect.

**WOMEN'S DRESS**
*Some Omani women wear the traditional* burqa *(face mask) and* abaya *(black robe).*

**OMANI DANCERS**
*Omani dress includes a* khanjar *(dagger) and a* muzzar *(a square of fabric wrapped and folded into a turban).*



IRAN AND AZERBAIJAN

# Persians

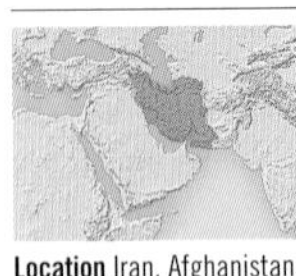

**Population** 40 million
**Language** Persian, or Farsi (Iranian branch of the Indo-Iranian languages)
**Beliefs** Mainly Shia Islam; also Sunni Islam, Judaism, Zoroastrianism, Baha'ism, Christianity
**Location** Iran, Afghanistan, United Arab Emirates

The historic Persian empire covered much of Central Asia, from the shores of the Mediterranean in the West to the borders of China in the East. The first of the major Persian dynasties, the Achaemenid, dates back more than 2,500 years (550–330BC). Since 1935, the country at the centre of the Persian empire has been called Iran.

In the contemporary era, Iran was ruled by the Pahlavi dynasty. After the revolution of 1979, the Persian people established a unique political entity – an Islamic republic. Iran is the home of Zoroastrianism, the state religion of the Persian empire prior to the Islamic era. Founded by Zarathustra some time between 700 and 600BC, it is one of the world's oldest religions.

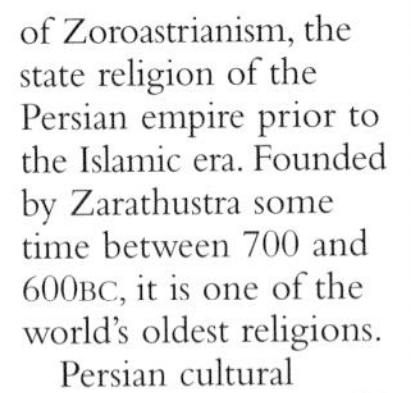

Persian cultural influence extends well beyond the borders of present-day Iran and has had a profound and lasting influence on the literature and art of the region. Persians speak an Indo-European language, Farsi, which is written in Arabic characters and includes much Arabic vocabulary. They are also distinctive in their adherence to the Shia branch of Islam; they are the world's largest group of Shia Muslims.

**SHIA MAN**
*Most Persians follow the Shia interpretation of Islam. This Persian man stands in front of a Shia mosque.*

History

## A national epic

The *Shah-nameh*, or Book of Kings, is the most famous work of Persian poetry. Written by the poet Ferdowsi around AD1000, it tells of ancient heroes such as Rustam and their battles against evil. Although Arabic was the most commonly used written language at the time, this work was written in Persian, similar to the language the Iranians use today. Iranians regard the *Shah-nameh* as a central part of their culture.

**MEDIEVAL TILE PAINTING**
*Persian artists excelled in painting miniatures and crafting painted ceramics. Unlike other Muslims, they painted people and natural forms as well as patterns.*

**IMAM'S SHRINE**
*A woman smokes a hookah pipe while gathering with others to pay respect beside an Imam's tomb. Shia Muslims regard Imams not only as leaders, but divinely appointed and preserved from sin.*

IRAN AND AZERBAIJAN

## Azeris

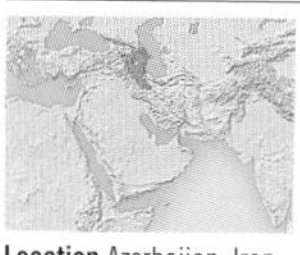

**Location** Azerbaijan, Iran, Russian Federation

**Population** 23 million
**Language** Azerbaijani, also known as Azeri (a Turkic language), Russian, Farsi
**Beliefs** Predominantly Shia Islam; also Russian Orthodox and Armenian Apostolic Christianity

The Azeris are native to Azerbaijan, but there are more ethnic Azeris (15 million) living in Iran than anywhere else; there they are known as Azaris.

Azeris were under Russian control from the early 19th century until 1991. Russian influence was responsible for Azeris introducing European culture into the Turkic world, but traditional Azeri theatre, music, and dance have Persian and Turkish roots. Since their independence in 1991, there has been a noticeable revival in Islamic practice among the Azeris of Azerbaijan.

AZERI WOMAN

IRAN AND AZERBAIJAN

## Qashqai

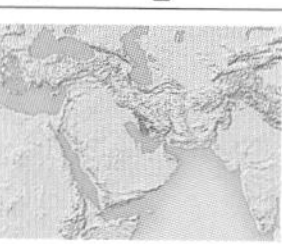

**Location** Fars province (western Iran)

**Population** Estimated 300,000–400,000
**Language** Qashqai (closely related to Azeri; Qashqai are the second largest Turkic-speaking people group in Iran), Farsi
**Beliefs** Shia Islam

Traditionally, the Qashqai live as pastoral nomads and depend on a mix of cultivation and shepherding, according to the season. Their route through the mountain passes is considered the most treacherous in all Iran. Suspicious of the encroachment of outsiders, the Qashqai have a deep attachment to their nomadic past and a tradition of self-reliance. They value independence, and the headmen of village communities are revered for their bravery and fearlessness.

The colourful Qashqai form of dress appears to have changed little across millennia. Qashqai people wear a highly decorated short tunic with wide-legged trousers, the voluminous folds of wool and cotton offering protection from the elements in this harsh climate. Unlike the round, head-hugging skull-cap of other nomadic mountain peoples in Iran, the Qashqai hat resembles Napoleonic military headwear.

**SPINNING WOOL**
*Many Qashqai people are shepherds. The Qashqai use the wool, which is prized for its sheen and white colour, to make their famous, richly coloured carpets.*

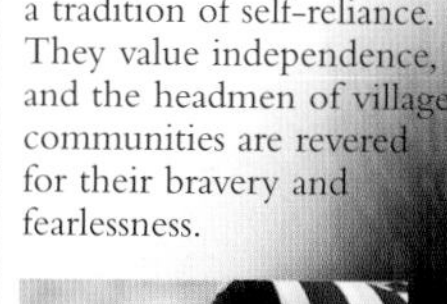

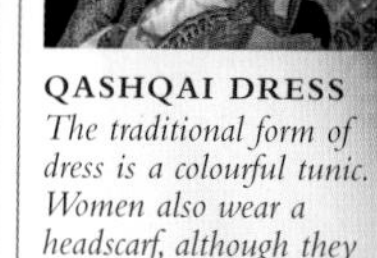

**QASHQAI DRESS**
*The traditional form of dress is a colourful tunic. Women also wear a headscarf, although they do not cover their faces.*

**FOLKLORE SONG**
*A blind musician, wearing the typical Qashqai felt hat and holding a bamboo flute, sings a folk song.*

IRAN AND AZERBAIJAN

## Lur

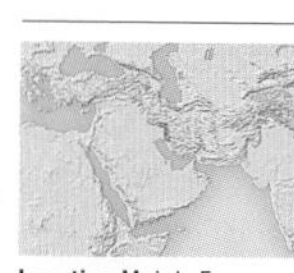

**Location** Mainly Fars province (western Iran)

**Population** Estimated 500,000–600,000 (second largest tribal group in Iran)
**Language** Lur Kucklik, Lur Buzurg (both Iranian languages related to Farsi); also Farsi (Persian)
**Beliefs** Mostly Shia Islam

Most Lur people live in the central and southern ranges of the Zagros Mountains in western Iran. They are thought to have developed as a separate tribe from the ancient Kurds of Iran, and some claim descent from invaders in the first millenium BC, who settled in the area during their migration from Central Asia. In the 18th and 19th centuries, the Lur were known as one of the fiercest tribes in Persia, and they still prize their reputation for strength and endurance as mountain dwellers. They are often seen as distinct from other peoples of the area because of their tall, lean stature.

The Lur are a culturally distinctive group who have remained relatively unaffected by outside influences. The two main branches speak different languages: Lur Kucklik and Lur Buzurg. The latter is also spoken by the nearby Bakhtiari people. The Lur are further subdivided into more than 60 tribal units. Most live by small-scale cultivation and shepherding.

Lur poetry and songs celebrate the virtues of being tough and fearless in the face of an ongoing struggle for survival, as well as the beauty and harshness of their homeland. The Lur are known for their rich material culture, especially the intricate weaving and colourful, highly decorated textiles.

**NOMAD FAMILY**
*The Lur have a long history of earning their living as travelling musicians. This family continues the tradition.*

**BAKHTIARI MAN**
*The Bakhtiari people are closely related to the Lur. They have similar customs, including forms of dress.*

Profile

### Shahpour Bakhtiar

A Lur born in the Bakhtiari region in 1914, Dr Shahpour Bakhtiar was one of Iran's most prominent politicians. In the 1950s, he served in the Iranian cabinet. He was critical of the Shah's administration; however, late in 1978 the Shah made him Prime Minister. Dr Bakhtiar was in office for just 37 days before the Iranian revolution took place. He fled to France, where he founded the National Resistance Front, but was assassinated in 1991.

**NDEBELE HOUSE PAINTING**

*It is part of many African peoples' cultures to decorate their houses. This Ndebele woman in South Africa adds the finishing touches to the bright colours and geometric patterns.*

# SUB-SAHARAN AFRICA

The broad expanse of the Sahara desert separates Africa into two distinct areas. The peoples living north of the desert have much in common with Middle Eastern peoples, while those living south of it boast vastly different cultures. This more southerly area, known as sub-Saharan Africa, is generally viewed as the birthplace of humanity. Fossil evidence of our evolutionary ancestors has been found there dating back millions of years, and the first *Homo sapiens* (modern man as we know it) evolved here around 150–200,000 years ago.

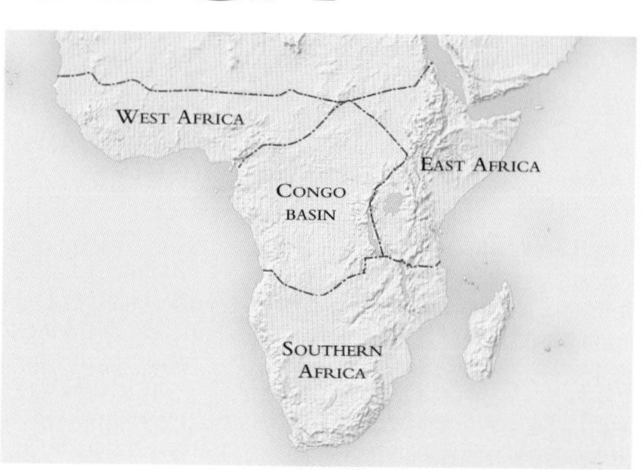


**THE REGION**
*Sub-Saharan Africa includes all the African countries south of the immense Sahara desert (which is situated in the north).*

The fact that people have been living in this part of Africa for much longer than anywhere else in the world means that the level of genetic and ethnic diversity is very high. The people with the greatest diversity are believed to be the San (or "Bushmen", *see* p424), Africa's earliest known inhabitants.

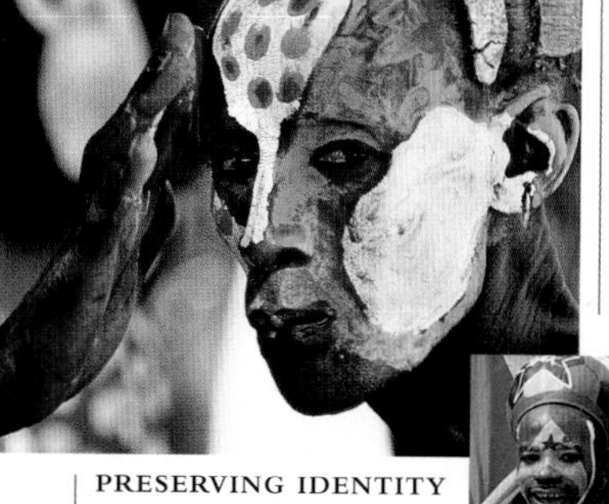

**PRESERVING IDENTITY**
*The Karo are a small but unique community. They represent just one of some 200 peoples living within Ethiopia.*

## MIGRATION AND SETTLEMENT

It is thought that the ancestors of the hunter-gatherer San people lived mainly in southern, eastern, and central Africa. However, around 1000BC another people, known as the Bantu, living in West Africa – who today include groups such as the Ganda (*see* p419) and Zulu (*see* p425) – started making iron tools and practising agriculture. This efficient means of food production led to a rapid increase in the Bantu population, which started migrating east and south, changing Africa's linguistic and cultural landscape along the way. Many native peoples were displaced or assimilated by them, with only a few small groups of San people retaining their identity by moving to more inhospitable lands. Amazingly, the Bantu people were still migrating south towards South Africa's Cape region when the Europeans began to settle the area in the 17th century.

12TH-CENTURY BENIN BRONZE HEAD

## TRIBES AND NATIONS

Most African states were founded by European colonial powers, rather than by local ethnic groups, which means it is rare for state boundaries to match ethnic ones in Africa. Instead, each state includes a variety of ethnic groups (often up to two or three hundred); some ethnic groups span several states; and most states have hundreds of regional languages as well as one or two national ones.

This diversity has led to a host of problems in contemporary Africa as some people feel a stronger tie to their ethnic group than to their country. Extreme cases have led to violence between groups, and even to genocide. One shocking example was the animosity between the Hutu and Tutsi peoples in Rwanda, which resulted in the deaths of some 800,000 people within just 100 days in 1994. Over a longer period, the ethnic separatism enforced in South Africa during the Apartheid era led to deep unrest and many deaths.

Nevertheless, there are also many Africans today who prefer to identify with a nation state rather than an ethnic group. Urban-educated Africans generally prefer to think of themselves as "Kenyan" or "Nigerian" rather than "Kikuyu" or "Igbo", and competition between national sports teams is as strong in Africa as it is elsewhere in the world.

**FOOTBALL FAN**
*Modern contests such as the African Nations Cup give people a chance to express their national pride.*

## CHANGE

Rural African areas are generally very slow to embrace change, mainly as a result of their cultural isolation. The trend for urbanization, however, has led to more rapid change in towns and cities, and urban migration has caused an increasing divide between the rich and the poor. Along with social changes come cultural changes. As people from a range of ethnic backgrounds flock to cities, old cultural distinctions and ways of life are being absorbed within the new, shared urban lifestyle. However, industrialization is taking place at a slow rate, and the majority of Africans continue to make a living either as herders or subsistence farmers.

Issue

### Shanty towns

One stark feature of modern Africa is the rapid growth of shanty towns near urban centres. People migrate to these areas in search of work but often find no jobs and are forced to make houses from materials such as corrugated iron. Few shanty towns have proper sanitation, hospitals, or schools, and inhabitants face disease and abject poverty on a daily basis.

**TRADING INFLUENCES**
*Zanzibar – a trading centre for centuries – has gradually absorbed east African, Arab, Persian, British, and Portuguese elements.*

HORN OF AFRICA

## Tigray

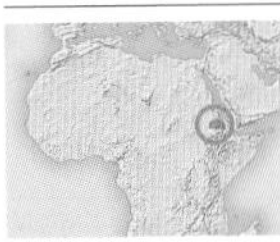

**Population** 4 million
**Language** Tigrinya (a Semitic language)
**Beliefs** Ethiopian Orthodox Christianity (arose in the 4th century AD and is separate from European Christianity)
**Location** Northern Ethiopia, Eritrea

The Tigray, who are descended from Semitic people, live in Eritrea and northern Ethiopia. Their language, Tigrinya, is one of the few written languages in Africa, and it is closely related to Amharic. Like the Amhara, the Tigray traditionally cultivate grain, using ox-driven ploughs. In 1991, the Tigray people were instrumental in bringing about the change of government that led to the creation of Eritrea as a separate country and to the formation of many ethnic "mini-states" throughout Ethiopia.

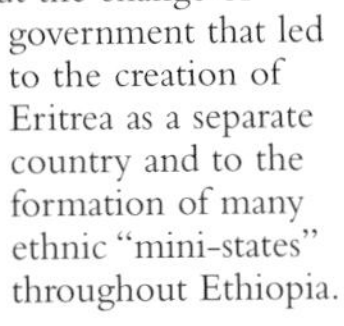

**TATTOOS**
*Many Tigray women have circles and crosses tattooed on their neck, chest, and forehead.*

HORN OF AFRICA

## Somalis

**Population** 10 million (including almost 8 million within Somalia)
**Language** Somali (a Cushitic tongue related to Oromo and Afar), Arabic
**Beliefs** Sunni Islam; also tariqa (related to Sufism)
**Location** Somalia, Ethiopia, Kenya, Djibouti

The majority of Somali people live in Somalia, but sizable numbers are also found in Djibouti, in the Ogaden region of Ethiopia, and in Kenya. Most Somalis are fully nomadic or seminomadic and inhabit harsh, arid environments. Men and boys herd the prized camels; women look after sheep or goats. The Somali diet is dominated by milk and dairy produce, although today Somali people also eat maize-meal and rice.

Nomadic families live in portable huts made of bent saplings and woven mats. Villages consist of a group of huts inhabited by related families. The huts are arranged in a circle or semicircle. Home-building and home-making are the women's responsibility.

Somalis belong to a number of different clans, and there has never been one political system to unify all Somali people. Competition over scarce resources has often led to violent conflicts between the clans, many of which are still raging in southern Somalia.

Most Somalis are Sunni Muslims. There is also a strong tradition of *tariqa*, a religious system associated with Sufism, a mystical strand in Islam. The orders of *tariqa* are social and religious brotherhoods that serve as centres of learning and religious leadership.

**SOMALI HEADDRESS**
*A Somali woman wears a flamboyant red-and-yellow headdress. Married Somali women usually do not wear veils (unlike other Muslim women).*

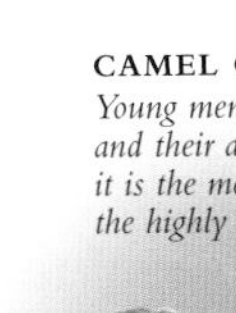

**CAMEL CARE**
*Young men draw water for themselves and their animals. In Somali culture, it is the men and boys who look after the highly valued camels.*

**ORTHODOX PRIEST**
*Holding an ornate cross and an illustrated Bible, an Amhara priest stands among painted angels and biblical scenes.*

**EPIPHANY**
*The rock-hewn, 12th-century church of Bet Giorgis is the setting for a Timkat (epiphany) ceremony.*

HORN OF AFRICA

## Amhara

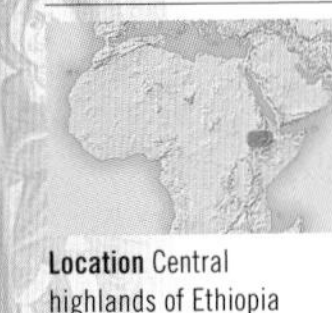

**Population** 14 million
**Language** Amharic (a Semitic tongue also spoken by many other peoples in Ethiopia as a second language)
**Beliefs** Ethiopian Orthodox Christianity
**Location** Central highlands of Ethiopia

The Amhara people live in the central highlands of Ethiopia and make up around one-quarter of the country's population. Although not as numerous within Ethiopia as the Oromo (*see opposite page*), their language, Amharic, is the country's official language. At one time, the Amhara formed the core of the Abyssinian empire, which was powerful in the Horn of Africa for many centuries before the formation of modern Ethiopia in 1855. Today, some feel alienated by Ethiopia's Tigray-dominated government.

Today, many Amhara people live in the countryside and farm a local grain called *t'eff*. Others work in Ethiopian towns, employed in business, education and government. Most Amhara people practise Christianity and belong to the Ethiopian Orthodox Church. Their form of Christianity was established in the 4th century AD and since then has developed separately from the European Christian tradition. Inside any Ethiopian Orthodox Church is a *tabot* (ark), which is kept hidden for most of the year but is paraded around the church on special occasions. The orthodox priests, known as *k'essotch*, wear colourful robes and carry a large cross for followers to kiss.

*Injera*, a spongy, pancake-like bread made from *t'eff*, forms the base of most Amhara dishes. A tasty stew, known as *wat*, is placed on top. On Wednesdays and Fridays, most Amhara people, along with all Ethiopian Orthodox Christians, refrain from eating animal products. On these days, they eat special vegetable meals, known as "fasting food".

History

### An African script

Amharic is one of the few African written languages. It has its own script, known as *fidel*, seen below on a shop sign. The system developed from the Ge'ez script of the medieval Abyssinian empire. The basic letters are all consonants, and their shape is subtly changed to indicate which vowel should be added. Amharic is now the main language used in the urban areas of Ethiopia.

HORN OF AFRICA

## Karo

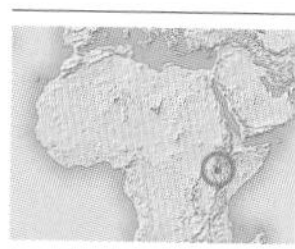

**Location** Lower Omo Valley (southern Ethiopia)

**Population** 500
**Language** Karo (a member of the Omotic language group; related to Gamo)
**Beliefs** Animism (a system of beliefs based on the idea that natural objects have a soul)

The Karo are an agricultural people, and mainly raise cattle. The people live in the Omo Valley, located in the hot, dry lowlands of Ethiopia's far south. With only around 500 Karo people surviving today, they are one of the smallest and most endangered ethnic groups in the area.

Elaborate body decorations and face-painting distinguish the Karo from their neighbours, such as the Hamar and the Bume. The Karo paint themselves daily with coloured ochre, white chalk, yellow mineral rock, charcoal, and pulverized iron ore, all of which are found naturally in the area. Men use clay to construct elaborate hairstyles and headdresses that signify status, beauty, and bravery. For example, if a man wears his hair in a grey-and-red bun with an ostrich feather sticking out, it indicates that he has killed an enemy or a dangerous animal, such as a lion or a leopard.

The Karo people, like the Hamar, initiate young males into adulthood with a ritual in which the boy must leap over rows of cattle six times, without once falling. If successful, the initiate is deemed ready for marriage. He is also accorded the honour of being allowed to appear publicly with the elders in the community's sacred places.

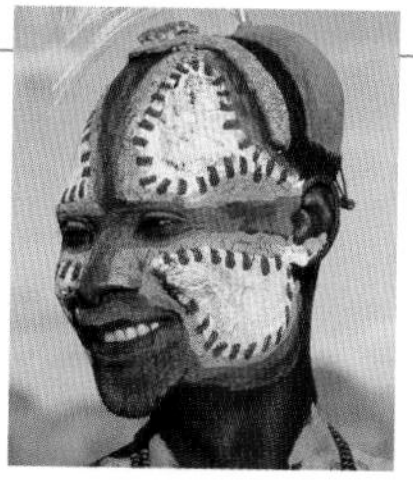

**FACE-PAINTING**
*This young man's ornate facial paintings are typical of the Karo people.*

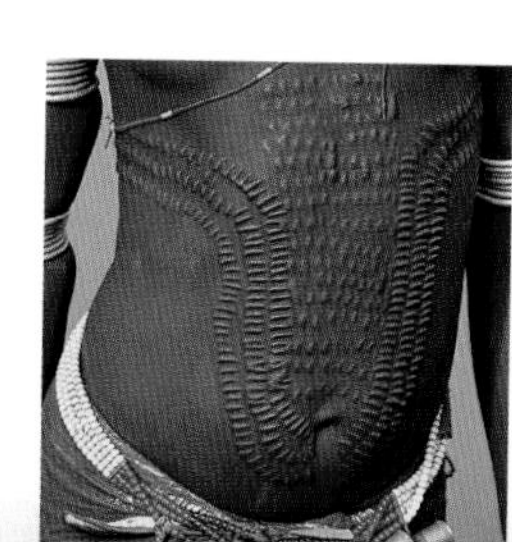

**SCARIFICATION**
*Karo girls' abdomens are ritually scarred to enhance beauty and sensual appeal (left). The scarring of a man's chest, on the other hand, shows that he has killed enemies from other ethnic groups, which wins him respect.*

**KARO VILLAGE**
*The few remaining Karo people live in small grass huts clustered together on the eastern bank of the Omo River.*

HORN OF AFRICA

## Gamo

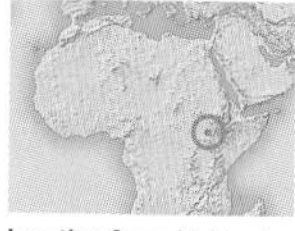

**Location** Gamo highlands (southwestern Ethiopia)

**Population** 700,000
**Language** Gamo (a language of the Omotic group, related to Karo)
**Beliefs** Mainly animism, Orthodox Christianity; also Protestant (since recent missionary visits)

Gamo *dere* (communities) are scattered over the mountainous terrain of the Gamo highlands in southwestern Ethiopia. Some are situated more than 3,000m (9,800ft) above sea level. Each *dere* consists of a collection of bamboo huts. The family dwellings are tall buildings, often more than 6m (20ft) high. Frequently, the Gamo also build simple round huts alongside the family home. They are not as tall as the main hut, and they either house the family's son and his wife or are used for cooking and storing grain.

Most Gamo are subsistence farmers. Men who own plots of land rarely tend them alone; instead, work-groups of 10–15 people spend a day or two on each individual's plot. The Gamo grow barley and wheat in the higher altitudes and maize and sorghum on the lower slopes. They also grow *enset*, a relative of the banana that has edible roots and stems. *Enset* ensures that the Gamo avoid famine, because when other crops fail in years of poor rainfall, this virtually indestructible plant usually survives.

The main festival in the Gamo calendar is the New Year celebration, which takes place in September, after the heavy rains. Sheep are sacrificed, and a large bonfire is prepared. Men carrying smoking bamboo poles from the fire greet each other with the exclamation "Yo!". There are also special rituals for those who have been initiated or married in the previous year. In addition, New Year is a time for reconciliation within the *dere*, and many assemblies take place to resolve arguments. It is believed that peace between people and the fertility of the soil are closely linked. Resolving disputes is therefore thought to be crucial for successful agriculture and general wellbeing.

Whenever a person dies, the men, together with the rest of the community, run around a mourning field in battle dress, chanting war songs. This expresses the warriorlike strength it takes to confront death.

**WATERPROOF HOMES**
*Gamo dwellings are made of woven bamboo, which allows cooking smoke to escape from the interior but keeps out the rain.*

**NEW YEAR FESTIVITIES**
*During the New Year celebrations, men parade in line, holding smouldering bamboo poles taken from the bonfire.*

HORN OF AFRICA

## Oromo

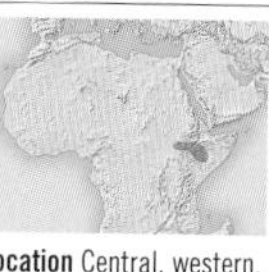

**Location** Central, western, and southeastern Ethiopia; northern Kenya

**Population** 18–30 million
**Language** Oromifa (a Cushitic language of the Afro-Asiatic family)
**Beliefs** Mainly Islam and Orthodox Christianity, generally coupled with animist beliefs; also Protestant Christianity

The largest ethnic group in Ethiopia is the Oromo people. Traditionally they are a pastoral people, herding cattle and goats. The Boran Oromo, one of the southern, pagan groups, retain this herding lifestyle. Other branches of the Oromo have taken up different forms of agriculture, while others have moved to urban centres. At the core of Oromo culture is the *gada* system, which divides men into age grades (*see* p260), each with a role in society. Rituals govern the transfer of men between age grades. The *gada* system structures the political and ritual life of the community.

**COFFEE TRADE**
*An Oromo woman sells raw coffee beans at the local market. Coffee is a popular drink among the Oromo.*

HORN OF AFRICA

## Dinka

**Location** Southeast Sudan, either side of the White Nile; west Ethiopia

**Population** 2 million
**Language** Dinka, a language of the Nilotic group; many of the people also speak Arabic
**Beliefs** Animism; there are also a small number of Dinka who are Protestant Christians

The Dinka are a semi-nomadic people who live in southern Sudan, on the savannas near the swamps of the Nile basin. For most of the year they live in small riverside camps, tending and herding their highly valued long-horned cattle. However, during the rainy season (June to August), the flooding forces the people to move into villages on higher ground. This means living in close proximity to each other, which often leads to arguments.

**HERD SECURITY**
*Dinka peoples' wealth is in their cattle. During the fighting in Sudan, many cattle have been killed. Herdsmen must arm themselves to protect their animals.*

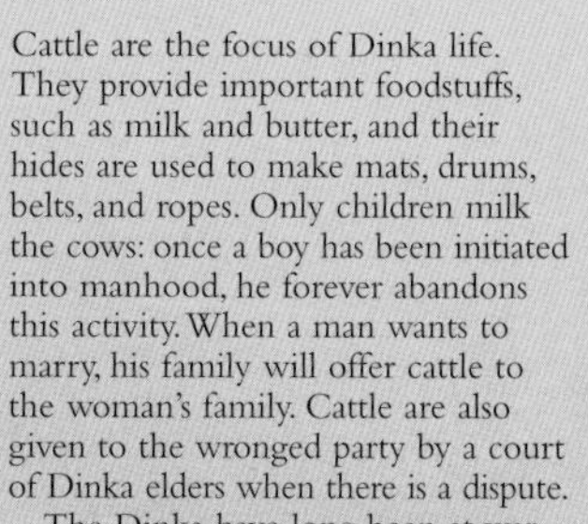

Cattle are the focus of Dinka life. They provide important foodstuffs, such as milk and butter, and their hides are used to make mats, drums, belts, and ropes. Only children milk the cows: once a boy has been initiated into manhood, he forever abandons this activity. When a man wants to marry, his family will offer cattle to the woman's family. Cattle are also given to the wronged party by a court of Dinka elders when there is a dispute.

The Dinka have long been at war with their neighbours, the Nuer. Since 1983, a civil war has been raging in the southern Sudan, which has led to a huge escalation in the violence and increasing casualties.

**COW HORN SCARS**
*The Dinka sometimes make scars in the shape of the horns of their beloved cattle, as seen on the face of this young woman. Their cattle are a representation of the Dinka's wealth.*

**DRY SEASON ACCOMMODATION**
*During the dry season, which runs from September to May, the Dinka live in simple grass huts and are constantly on the move in search of pasture and water for their cattle.*

Fact

## Insect repellent

Keeping flies and mosquitoes at bay is an important concern for the Dinka. This is not only because these insects are extremely annoying but also because insects spread diseases, including serious ailments such as malaria. The Dinka cover themselves, and often their animals, from head to toe with an effective natural insect repellent made from the white ash of burned cow dung. Cow urine, which is sterile, is used as an antiseptic for any bites received.

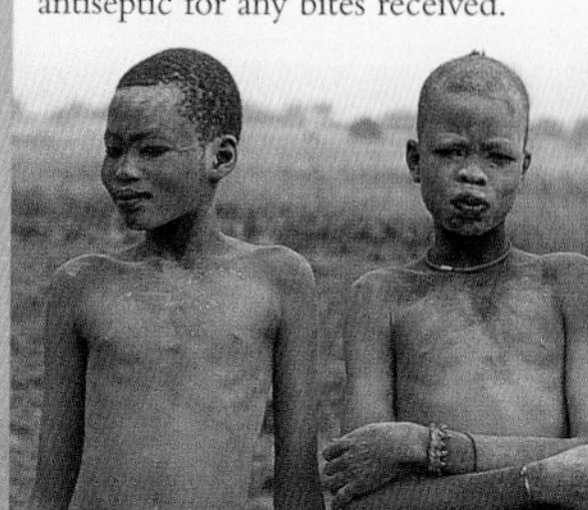

WEST AFRICA

## Hausa

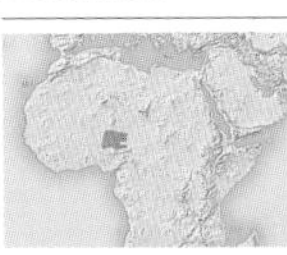

**Population** 20 million
**Language** Hausa (a language of the Chadic group of the Afro-Asiatic family; many other West African peoples use it as a second language)
**Location** Northwestern Nigeria, southwestern Niger
**Beliefs** Islam

The Hausa people live principally in northwestern Nigeria and southwestern Niger. They make up the largest ethnic group in the area and are traditionally farmers. They belong to seven large city states that first rose up around 600, taking control of the formerly decentralized villages of the region in approximately 1200. In the early 19th century, the Hausa lost power to the Fulani, who waged a holy war against them. The Fulani later adopted a settled lifestyle and assimilated into the Hausa way of life.

The Hausa are great traders and often travel long distances to markets. They are accomplished craftspeople, known for a range of skills, including weaving, thatching, silversmithing, and leatherworking. Their language (also called Hausa) is one of the most widely spoken in Africa. It is the mother tongue of about 20 million people, and it is also spoken by another 25 million as a lingua franca. Hausa includes several Arabic words owing to the influence of Islam. There was Islamic presence in the Hausaland region as long ago as the 12th century, although large-scale conversion to Islam did not take place until the 1800s. Today, most Hausa are Muslim.

**MARKET LIFE**
*Market trading has long been a key aspect of Hausa life, both in the towns and in the countryside. Hausa regularly travel long distances to market.*

**COURTLY MUSIC**
*Brightly dressed musicians herald the local ruler, called an emir, by playing long, thin metal trumpets known as* kakaki.

WEST AFRICA

## Fulani

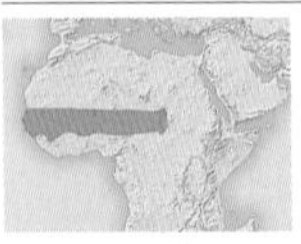

**Population** 18 million (spread across 10 countries including The Gambia, Mali, Mauritania, Burkina Faso, and Guinea)
**Language** Fulani (also known as Fula, Fulbe, Peul, and Pulaar)
**Location** West African countries from Senegal east to Nigeria and Niger
**Beliefs** Islam

The Fulani are traditionally nomadic herders who roam across a large area of West Africa. They are credited with spreading Islam across much of the region in the early 19th century. Today, many Fulani have agreements with other nearby agricultural people, according to which they exchange milk and other cattle products for grains and vegetables.

There are many Fulani customs, but what unites most Fulani is first their love of cattle and second the concept of *pulaaku*, or "pastoral chivalry". *Pulaaku* involves disciplines such as *munyal* (patience and self-control), *semteende* (respect for others), and *hakillo* (wisdom, forethought, and hospitality). Fulani men are trained not to show their feelings and to maintain respect for others by keeping their distance. They can therefore appear aloof to outsiders.

The Fulani celebrate a number of Muslim and traditional festivals. One branch of the Fulani, known as the Wodaabe, observe a colourful festival called the *yaake*. Young men dress up in their finery, wearing face paint, jewellery, and headdresses, and try to charm a group of female judges.

**TATTOOS**
*Lip tattoos are very fashionable among Fulani women. Some even have tattoos on their gums.*

**HEAVY LOAD**
*Cattle are central to Fulani life. This animal is carrying storage gourds, called calabashes, to market.*

**CHARM DANCE**
*The* yaake *celebrations are held at the end of the dry season. If one of the female judges is impressed by a young man's extravagant costume and face-pulling, she may consent to marry him.*

PEOPLES

WEST AFRICA

## Buduma

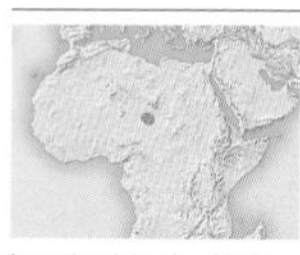

**Population** 60,000 (mainly in Chad)
**Language** Buduma, or Yidena (a member of the Chadic group of the Afro-Asiatic language family)
**Beliefs** Islam
**Location** Islands of Lake Chad, Cameroon, Nigeria

The Buduma, or Yidena, a fiercely independent people who strongly believe in preserving their distinctive culture, live on the northern islands and shores of Lake Chad. They are mainly herders and fishermen. Buduma literally means "people of the reeds", and they use papyrus reeds for a variety of purposes, such as building fishing boats and constructing lightweight huts that can be readily shifted to higher ground when the lake rises. The Buduma are closely related to the Kouri people, who speak the same language and live on the lake's southern shores.

The Buduma people raise cattle that have very large, hollow horns, which help the animals float when they are transported across the lake. The Buduma eat mainly cow's milk and fish, but they also collect the roots of water lilies and grind them into a flour to supplement their diet. They do not kill their cattle for food.

**FISHING ON LAKE CHAD**
*Buduma men fish from a papyrus-reed boat. There is some commercial fishing, although Buduma mainly fish to feed their families.*

WEST AFRICA

## Wolof

**Population** 3 million
**Language** Wolof (a member of the Niger-Congo language family; many other peoples in Africa use it as a second language)
**Beliefs** Islam
**Location** West Senegal, west Gambia, Mauritania

The Wolof people, most of whom live in The Gambia and Senegal, have a social system that is stratified into three castes. The *geer* consists of farmers. Below the *geer* are the *nyenyoo*, or occupational specialists, which includes blacksmiths, weavers, and musicians. At the very bottom of the hierarchy are the *jaam*, who are the descendants of slaves. This system still largely guides social behaviour today.

**KORA PLAYER**
*A man plays the kora, an instrument with 21 strings and a large calabash (gourd) resonator. It is very popular among the Wolof in Senegal, and is sometimes amplified to produce a more wide-reaching sound.*

Profile

### Youssou N'Dour

A singer and songwriter, Youssou N'Dour was born in Dakar, Senegal, in 1959. Of Wolof heritage, he has developed a musical style that successfully blends traditional Wolof percussion with jazz, hip hop, soul, and Afro-Cuban influences. He is known as the "King of mbalax", *mbalax* being the Wolof name for Senegalese pop music. He has recorded and performed worldwide with his band, The Super Etoile. N'Dour is also politically active and has given benefit concerts for causes such as Amnesty International and children with AIDS.

WEST AFRICA

## Dogon

**Population** 100,000–400,000, (including a small number in Burkina Faso)
**Language** Dogon (a member of the Niger-Congo language family)
**Beliefs** Animism
**Location** Bandiagara area (southeastern Mali)

**DOGON MAN**
*Dogon men and women mostly associate with their people of their own gender. Men often congregate in the* toguna, *or "house of words", a special men's house built at the centre of most villages.*

The Dogon are an agricultural people who live in southeastern Mali, mainly around the Bandiagara cliffs. They grow millet, sorghum, and corn as subsistence crops, and also produce onions as a cash crop.

The Dogon are best known for their fantastical myths and rituals, which represent an extremely intricate and complex cosmology. There are a number of all-male religious societies that conduct rituals throughout the year. One of the most important is the Awa Society, which performs masked dances at funerals, leading the souls of the dead into the spirit world. Another of their societies is Lebe, the members of which are responsible for agricultural spirits and build altars from clay and dirt.

Dogon masks are known as *inima* and are believed to possess the life force, or *nyama*. There are more than 65 masks used during Dogon rituals, and many of these tower high above the head of the wearer. Red, white, black, and brown are the most common colours.

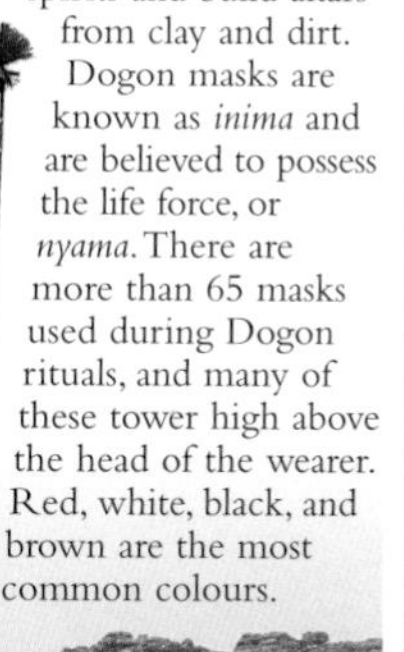

**MASK DANCE**
*Dogon men dance at an important ritual. Each extravagant mask is made from the branch of a single tree.*

WEST AFRICA

## Igbo

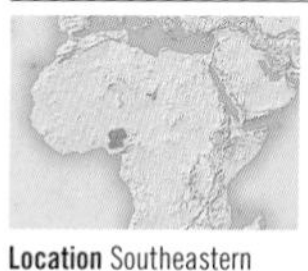

**Population** 20 million
**Language** Igbo, or Ibo (a member of the Niger-Congo family of languages)
**Beliefs** Christianity (both Roman Catholic and Protestant)
**Location** Southeastern Nigeria

The Igbo are a farming people living on either side of the Niger River in southeastern Nigeria. Their main crop is the yam, and its harvesting is the time for celebration. In August, at the end of the agricultural cycle, the Igbo hold the *Iri Ji (*New Yam Festival). During the celebrations, the first yams are eaten by the village head, then everyone feasts on a variety of yam dishes. The Igbo are well-known for their fantastical arts and crafts. They have a particularly strong tradition of metalwork and woodwork and use several different figurines and masks in their ritual practices.

**SMALL STATUE**
*This wooden figurine of a mother nursing a baby represents the power and importance of female fertility.*

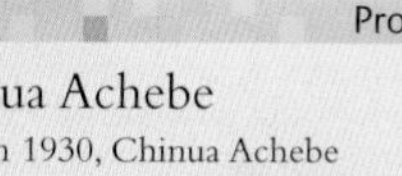

Profile

### Chinua Achebe

Born in 1930, Chinua Achebe became an internationally acclaimed author in 1958 on the publication of his first book, *Things Fall Apart*, which vividly portrays the traditional life and culture of Igbo people. His later books are concerned with the impact of colonialism on the Igbo.

**KEEPING COOL**
*Igbo women in a Nigerian town use an umbrella to shade them from the intense heat of the sun. Their lightweight skirts and light-coloured tops also help keep them cool.*

WEST AFRICA

## Ashanti

**Location** Southern Ghana

**Population** 1 million
**Language** Ashanti, or Asante (a dialect of Twi, part of the Niger-Congo language family);
**Beliefs** Animism (involving ancestor worship), Christianity

The largest ethnic group in modern-day Ghana is the Ashanti people, sometimes spelled Asante. Members of the Ashanti are mostly farmers, growing yams, cassava, maize, and a variety of fruits for subsistence, and cocoa and kola nuts as cash crops. The Ashanti have a complex system of rituals that involves elaborate ceremonies, witchcraft, sorcery, ancestor worship, divination, and belief in spirits. Traditionally, Ashanti people used divination to try to find the source of problems. One divination method involved giving a rooster a little poison and then asking a question that could be answered either "yes" or "no". The rooster's fate, whether it lived or died, indicated the answer. The Ashanti are a matrilineal society. They believe that children inherit flesh and blood from their mother and therefore belong to their mother's clan. From their father they inherit *ntoro* (spirit) and while this is not unimportant, it does not confer clan ties. Bearing children is therefore very highly valued, and young women sometimes own an *akua'ba* (fertility doll), to help them conceive a child. A diviner or priest prays over the doll, then the woman carries it like a real baby, hoping that she will soon give birth.

**DANCING FOR THE KING**
*A young woman, dressed in a gold costume and wearing abundant jewellery, dances for the asantehene (king).*

**COCOA HARVEST**
*Cocoa is a very important source of income for the Ashanti people.*

**STATUETTE**
*A fertility statue such as this one is carried by a young woman when she is hoping to conceive a child.*

History

### Ashanti kingdom

In the 18th and early 19th centuries, the Ashanti kingdom developed from a number of smaller, previously autonomous chiefdoms. During the installation of an *asantehene* (king), the royal person is lowered and raised three times over the Golden Stool, the Ashanti symbol of unity.

**ROYAL SYMBOLS**
*The traditional stool and the brightly coloured kente cloth are two of the most important symbols of Ashanti royalty.*

WEST AFRICA

## Senoufo

**Location** Ivory Coast, Mali, Burkina Faso

**Population** 1 million (mainly living in northern Ivory Coast; small numbers in southern Mali and Burkina Faso)
**Language** Senoufo (Niger-Congo family)
**Beliefs** Animism, Islam

The Senoufo are a farming people who mostly live in small villages in the northern savannah region of the Ivory Coast. They have a complex ritual life that incorporates secret associations. These are divided into the *Poro* cult for boys and the *Sakrobundi* cult for girls. The associations prepare children for adulthood by teaching them the people's traditions and instilling self-control through a variety of strenuous tests held in the forest. After several years, the youths "graduate" with an initiation ceremony. This involves circumcision, a period of isolation, and finally a ritual in which they are permitted to use masks for the first time.

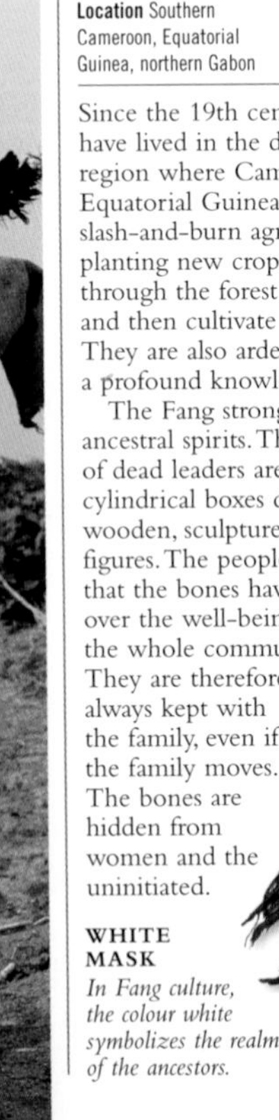

**MASKED SPIRIT DANCE**
*Masked Senoufo dance around their ancestral burial grounds, chasing away the spirits of the dead. These masks have walrus tusks, crocodile teeth, and porcupine-quill plumes.*

**SENOUFO WOMAN**
*In Senoufo society, men value women as wives and mothers, but often exclude them from rituals.*

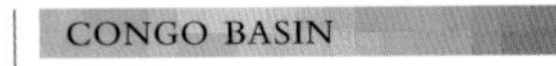

CONGO BASIN

## Fang

**Location** Southern Cameroon, Equatorial Guinea, northern Gabon

**Population** 900,000
**Language** Fang, also known as Pamue or Pahouin (a member of the Bantu branch of the Niger-Congo language family)
**Beliefs** Animism, Christianity

Since the 19th century, the Fang have lived in the dense forest in the region where Cameroon, Gabon, and Equatorial Guinea meet. They are slash-and-burn agriculturists. When planting new crops, the Fang cut through the forest using huge knives and then cultivate the land with hoes. They are also ardent hunters and have a profound knowledge of the forest.

The Fang strongly believe in ancestral spirits. The skull and bones of dead leaders are kept in special cylindrical boxes decorated with wooden, sculptured guardian figures. The people believe that the bones have power over the well-being of the whole community. They are therefore always kept with the family, even if the family moves. The bones are hidden from women and the uninitiated.

**WHITE MASK**
*In Fang culture, the colour white symbolizes the realm of the ancestors.*

CONGO BASIN

## Kongo

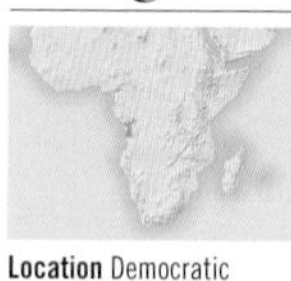

**Location** Democratic Republic of Congo, Congo, northern Angola

**Population** 3 million, living either side of lower Congo River
**Language** Kongo (a member of the Bantu branch of the Niger-Congo language family)
**Beliefs** Animism, Roman Catholic Christianity

The kingdom of Kongo was one of the great dominions of central Africa. It was at its peak in the 15th century, but by the 17th century it had split into a number of smaller realms due to the impact of colonialism. Now, after decades of civil war, much of the traditional Kongo culture has disappeared as people have left their villages to go to war or to work or trade in the towns and cities. During the days of the kingdom, there was a strong trade in ivory, copper, and slaves, and the majority of the people worked the land as farmers. Today, many continue these agricultural ways, cultivating crops such as bananas, maize, and taro. Traditional Kongo beliefs centre around spirit and ancestor cults.

EAST AFRICA

## Ganda

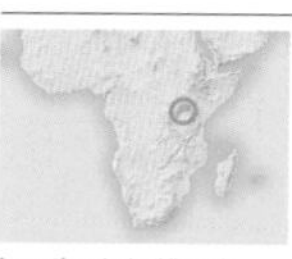

**Location** Lake Victoria area (southeast Uganda)

**Population** 3 million
**Language** Ganda, also known as Luganda (a member of the Bantu branch of the Niger-Congo language family)
**Beliefs** Animism, Christianity, Islam

The Ganda, or Baganda, the largest ethnic group in Uganda, live on the fertile lands around Lake Victoria in the southeast of the country. They are traditionally farmers, growing bananas and a variety of other crops including beans, millet, maize, sweet potato, peanuts, and cassava. They also grow coffee and tea as cash crops for export. The Ganda are among the best-educated people in the country, and many of them live in the capital city Kampala and other towns in Uganda, often holding government positions. They also have a better standard of living and are more modernized than other Ugandan peoples.

The Ganda have lived in the region of southeastern Uganda since around AD1000. By the 18th or 19th century they had developed a complex system of central government, courts, and taxation. At the head of this system was the *kabaka* (king) who was not only the supreme judge but also the high priest. Furthermore, the king was in charge of the all-important task of land allocation. In the villages today, the traditional culture continues to thrive. Respect for the *kabaka* is strong, and the hierarchical social organisation of chiefs, clan heads, and lineage heads continues. The society is patrilineal, with clan membership reckoned through the male line. Most people can name their ancestors back through four or even five generations.

**BANANA HARVEST**
*Bananas are the most important crop for the Ganda people.*

**GANDA ROYALTY**
*This extremely valuable leopard skin is a symbol of the king's royalty. Ganda kings still function as regional administrators.*

EAST AFRICA

## Swahili

**Location** Coastal regions of Kenya and Tanzania

**Population** 700,000
**Language** Swahili (more than 30 million other people also speak Swahili as either a first or second language)
**Beliefs** Islam (very strict observance)

Swahili people are a mixture of Bantu Africans (who migrated to the coast from central areas in the early part of the 1st millennium AD) and merchants and traders from the Persian Gulf (who arrived around AD900). The two peoples and cultures mixed to form the Swahili, whose language has become the most widely spoken tongue in East Africa. The term "Swahili" comes from the Arabic *sawahil*, meaning coast.

The Swahili people have always been traders. Their trading has prompted the emergence of large city-states, such as Lamu, Zanzibar, and Mombasa, on the coast. Swahili merchants transported goods in sailing boats known as dhows, which would cross the Indian Ocean to Arabia (now generally known as the Middle East) and India. On occasion, Swahili have also acted as intermediaries between inland ethnic peoples and colonial governments.

The Swahili are strict Muslims, and most people pray regularly in the mosques. Many wear amulets that contain verses from the Koran to protect them from *djinns* (harmful spirits). The Koran is also used for divination and for treating disease. A teacher of Islam may instruct a sufferer to dip into water paper that has verses from the Koran written on it. This water, which is now thought to contain the word of Allah, is either drunk or used to wash the patient's body to cure the affliction. The prosperity of the Swahili has inspired some neighbouring peoples to adopt Islam.

**TRADING DHOW**
*Large-sailed Swahili dhows are still common on the Kenyan and Tanzanian coast. They follow monsoon winds to India.*

**HERALD**
*In coastal town of Lamu, the siwa horn is blown to announce events, such as marriages, circumcisions, and religious festivals.*

**FOLLOWERS OF ISLAM**
*The way these young Swahili women are dressed, particularly their head coverings, readily identifies them as Muslims.*

GATHERING HONEY

*Honey is a highly valued food source, and Pygmy men will climb extremely tall trees in search of beehives. A person returning to camp with honeycomb is always greeted with terrific excitement.*

CONGO BASIN

# Pygmy peoples

**Location** Forest areas in central Africa

**Population** 250,000 (living in Democratic Republic of Congo, Cameroon, Gabon, Rwanda, Burundi, Uganda)
**Language** All groups adopt the languages of their non-Pygmy neighbours
**Beliefs** Animism

Many Pygmy peoples live in the rainforests of central Africa. They include groups such as the Mbuti, the Twa, the Efe, and the Baka, or Bayaka. Pygmy people all have in common an average height of less than 1.5m (5ft). They live mostly by hunting and gathering and prefer a nomadic existence in the forest to a settled life in villages, although they trade with their agricultural neighbours.

Forest hunting requires great skill and a profound knowledge of the environment and the animals that live there. Different Pygmy groups use different hunting techniques. Baka men, for example, hunt with spears, while Efe men prefer bows and arrows. Among the Mbuti, most hunting is done with nets; men, women, and children help by chasing the animals into the net. Many Pygmies use a sign language that allows silent communication when hunting. All Pygmy peoples share the meat from a large kill among the whole camp, irrespective of who participated in the hunt or who has successfully hunted previously.

Pygmy women mostly gather edible products from the forest, but they also raise the children and cook.

Pygmy peoples have an egalitarian social structure (*see* p260), with no leader and no procedures for solving disputes or punishing misbehaviour. If two people argue, it is likely that one will simply move to a different camp.

**THE PRIZE**
*A Pygmy gatherer places fresh honeycomb into a pan. He will take it back to camp to share with the whole community. This man is one of the Ba-benzele people living in the Republic of Congo.*

Issue

## Changing environment

The increase in logging activities by international companies is seriously threatening the lives of many Pygmy peoples. As roads are built into the forest and trees are felled, the wildlife is diminishing and the natural balance of the forest is crumbling. Although most Pygmy people continue a mainly traditional life, as food sources dwindle, some are seeking work with the logging companies.

**LEARNING NEW SKILLS**
*Pygmy children must get the chance to attend school, so that they are equipped to cope with economic changes in their forest.*

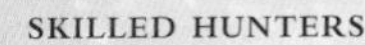

**SKILLED HUNTERS**
*Pygmies search the forest, looking for the next kill. They will use the nets they are carrying to ensnare their prey.*

PEOPLES

EAST AFRICA

## Kikuyu

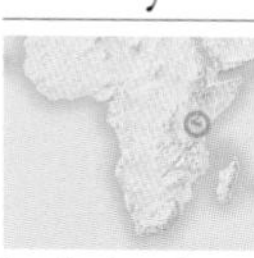

**Population** 6.5 million
**Language** Kikuyu (a member of the Bantu branch of the Niger-Congo family); Swahili, English (as second languages)
**Location** Central highlands area (Kenya)
**Beliefs** Protestant Christianity

The Kikuyu, who live in the highland areas northeast of Nairobi, are Kenya's largest ethnic group – they make up around 20 per cent of the country's population. Traditionally they are farmers, growing mainly millet, maize, yams, beans, bananas, and sugar cane. Today, the most important cash crops include wattle, coffee, tea, fruits, vegetables, and maize. While many Kikuyu continue an agricultural way of life, others have moved to towns and become successful in business and politics. Whatever they do, Kikuyu people have a reputation for being extremely industrious.

**PICKING TEA LEAVES**
*Many Kikuyu people work harvesting tea by hand on the one of the many large plantations found in Kenya.*

History

### Mau Mau rebellion

The Mau Mau was an insurgent organization consisting mainly of Kikuyu people. It formed in the 1940s with the aim of expelling white settlers from Kenya. Although the British crushed the uprising, Jomo Kenyatta, a Mau Mau leader, became the first president of Kenya in 1964.

JOMO KENYATTA

EAST AFRICA

## Hadza

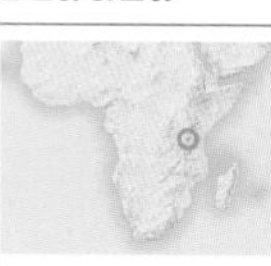

**Population** 1,000
**Language** Hadza (a "click" language of the Khoisan family; related to San, or Bushman, languages; a set of click sounds serve as consonants)
**Location** Lake Eyasi area (northern Tanzania)
**Beliefs** Animism

The Hadza are a small group of nomadic hunter-gatherers who live in the dry savanna near Lake Eyasi, northern Tanzania. They live in camps of 20–30 people and move to another site every 2 or 3 weeks. The men hunt animals that roam in the area, including wildebeest, waterbucks, buffalo, and impala, as well as smaller animals such as hares, hyraxes, and porcupines. The women gather foods such as vegetables, roots, berries, and the fruit of the baobab tree.

As is typical of hunter-gatherers, the Hadza do not have a strong notion of private property. The people are free to hunt or gather wherever they please, and there are no individual- or group-owned lands. When large animals are killed, the meat is divided among the community. The Hadza own most things communally, including arrows, pipes, and clothes, and those who have more than they need are required to share. The Hadza therefore have one of the most egalitarian societies in the world.

**FORAGED FOOD**
*A Hadza family poses with the bow and arrows that the man uses for hunting. Although the Hadza prize meat above all, most of their diet comes from roots, berries, and fruits.*

EAST AFRICA

## Masai

**Population** 900,000
**Language** Masai (a member of the Nilotic language group also known as Maa; it is related to Dinka)
**Location** Southern Kenya, northern Tanzania
**Beliefs** Animism, Christianity

The Masai are a seminomadic pastoral people who live in southern Kenya and northern Tanzania. Herding cattle, and also sheep and goats, is at the centre of their way of life. The Masai have resisted intense external pressure to become settled, and remain seminomadic. They eat mainly milk, meat, and blood from their cattle, which they value far more highly than the crops raised by their neighbours.

Masai men are grouped into age grades (*see* p260), in which they remain for their entire lives. Around the time of puberty, Masai boys are circumcised in preparation for their initiation into warriorhood. As *moran* (warriors), the young men live together in an *emanyatta* (warriors' camp) and spend their time hunting, tending the cattle, and protecting the community. During this period, they cover themselves in red ochre and observe several cultural restrictions, the most crucial being that they do not marry. Only after they have been *moran* for 10 years can the men be initiated into junior elderhood and finally marry.

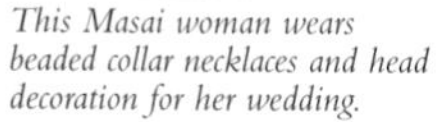

**MASAI BRIDE**
*This Masai woman wears beaded collar necklaces and head decoration for her wedding.*

**A WARRIOR'S DUTIES**
*A Masai of the* moran *(warrior) age grade brings cattle back to the village for the night. There, the cattle are protected from predators inside a stockade of thorny bushes.*

**WARRIOR INITIATION**
*Wearing red, and with their heads covered with bright red ochre, Masai men take part in the ceremony that initiates them all into the warrior grade.*

EAST AFRICA

## Tutsi

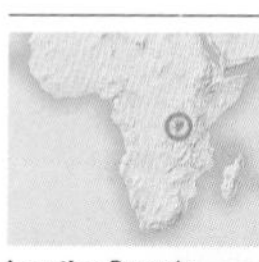

**Location** Rwanda, Burundi

**Population** 1.4 million (1.9 million before the genocide in 1994)
**Language** Kirundi, Kinyarwanda (both Bantu languages)
**Beliefs** Roman Catholic Christianity, animism

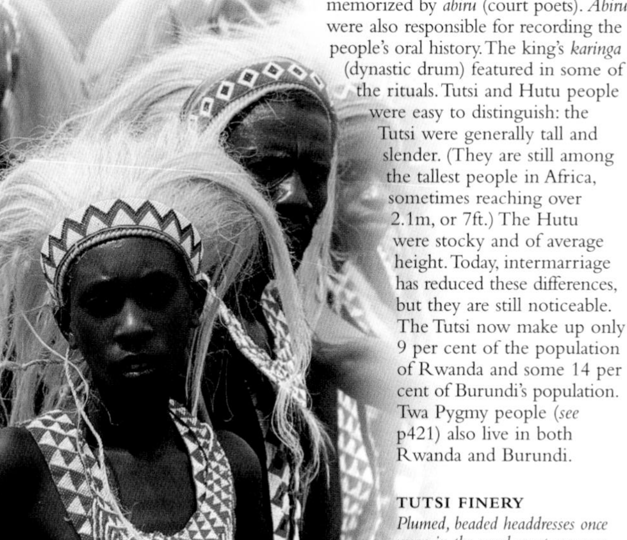

A cattle-herding people with a fierce warrior tradition, the Tutsi arrived in Burundi and Rwanda in the late 14th or 15th century. They quickly dominated the local Hutu then set up a stratified caste-like society in which the minority Tutsi people were the aristocratic rulers, while the majority Hutu people were the subjugated commoners. The Tutsi court was known for its rituals, whose form was memorized by *abiru* (court poets). *Abiru* were also responsible for recording the people's oral history. The king's *karinga* (dynastic drum) featured in some of the rituals. Tutsi and Hutu people were easy to distinguish: the Tutsi were generally tall and slender. (They are still among the tallest people in Africa, sometimes reaching over 2.1m, or 7ft.) The Hutu were stocky and of average height. Today, intermarriage has reduced these differences, but they are still noticeable. The Tutsi now make up only 9 per cent of the population of Rwanda and some 14 per cent of Burundi's population. Twa Pygmy people (*see* p421) also live in both Rwanda and Burundi.

**TUTSI FINERY**
*Plumed, beaded headdresses once worn in the royal court are now the costume of Tutsi dancers.*

EAST AFRICA

## Hutu

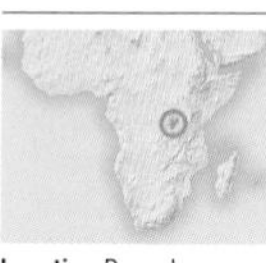

**Location** Rwanda, Burundi

**Population** 12 million (forming the majority in both Rwanda and Burundi)
**Language** Kirundi, Kinyarwanda (both Bantu languages)
**Beliefs** Roman Catholic Christianity, animism

**HUTU FACES**
*Many Hutu have fairly dark skin and a round face, distinguishing them from the Tutsi, who have paler skin and a long face.*

The Hutu, who traditionally have a clan-based society, have lived in Burundi and Rwanda since around the 1st century AD. When the Hutu arrived, they set up a number of small kingdoms and ruled over the Twa Pygmies, who had been living in the area previously. Life revolved around small-scale agriculture. In the 14th or 15th century, however, the Tutsi subjugated the Hutu. They set up a caste system, whereby the Hutu were subordinate to the Tutsi. The Hutu now constitute more than 80 per cent of the population of Burundi and Rwanda.

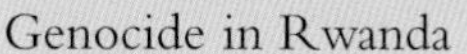

Issue

### Genocide in Rwanda

In 1962, Rwanda became an independent republic, with the Hutu people in power, after a civil war between the Hutu and the Tutsi. In 1994, Hutu extremists murdered the (Hutu) Rwandan president, sparking another civil war. In the next 100 days about 800,000 Rwandans, mostly Tutsi, were killed, and the killings were condemned as genocide. Many Hutu were also killed, however, unless they fled to refugee camps (right).

SOUTHERN AFRICA

## Bemba

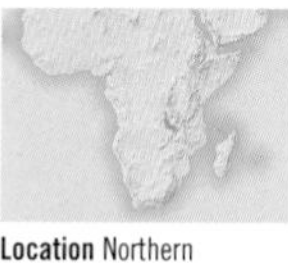

**Location** Northern Province (Zambia)

**Population** 2 million
**Language** Bemba (a member of the Bantu branch of the Niger-Congo family, and the lingua franca of Zambia)
**Beliefs** Christianity, animism

Bemba are slash-and-burn cultivators of finger millet and manioc. They also keep sheep and goats. Many Bemba have migrated to southern Zambia to work in mines. The Bemba are one of the few African societies in which offspring belong to the mother's clan, not the father's. Married couples live with the wife's parents, maintaining a bond between mother and daughter.

**FEMALE KINSHIP**
*Bemba women together pound cassava with pestle sticks to make flour. Female bonding is important in this matrilineal society.*

SOUTHERN AFRICA

## Chewa

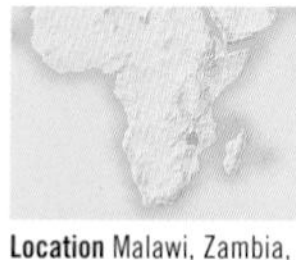

**Location** Malawi, Zambia, Mozambique, Zimbabwe

**Population** 4 million
**Language** Chichewa (a Bantu language spoken by about two-thirds of the population of Malawi)
**Beliefs** Christianity, animism

The largest ethnic group in Malawi is the Chewa. Chewa people live in compact villages and are traditionally slash-and-burn horticulturists. Their main crops are corn and sorghum. They are thought to have come from Zaire (now the Democratic Republic of Congo) in the 14th or 15th century. The Chewa have a complex mythology surrounding ancestral spirits. They consider that women govern regeneration and the realm of the living, while men govern death and the domain of the spirits. Chewa venerate ancestral spirits through sacrifices, ceremonies, music, and dance. They believe that spirits can intercede between people and God, helping to overcome problems.

The Chewa people have a men's secret society, called *nyau*, which performs a number of masked rituals. There are also community-wide rituals, such as the *Kulamba* annual thanksgiving ceremony. Other rituals are performed at funerals, exorcisms, and initiations.

**MASK RITUAL**
*During this ceremony, participants light a fire on the masked dancer's head and roast maize over the flames.*

SOUTHERN AFRICA

## Shona

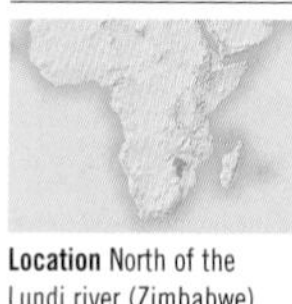

**Location** North of the Lundi river (Zimbabwe)

**Population** 7 million
**Language** Shona (a Bantu language, and the dominant African language in Zimbabwe); many also speak English
**Beliefs** Christianity, animism

The ancestors of today's Shona people are thought to have built "Great Zimbabwe", a walled city whose ruins have survived to this day (and from which the nation of Zimbabwe took its name). Great Zimbabwe is the largest ancient stone construction south of the Sahara, and in the 14th century it was probably home to around 18,000 people.

The Shona are now an agricultural people. Their main crop is maize, but they also grow millet, beans, pumpkins, and sweet potatoes, as well as keeping cattle, sheep, and chickens. Shona people consider cattle taboo for women, so men do the milking. Shona are known for their beautiful stone sculptures with flowing curves and for their music, played on *mbiras* (thumb pianos).

**SHONA WOMAN**
*A shona women takes a break from working among the crops. In Shona society, women carry out most of the agricultural work.*

SOUTHERN AFRICA

## San

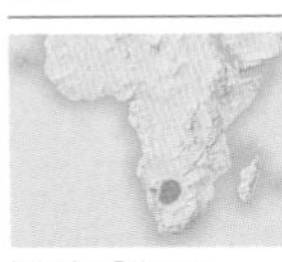

**Population** 60,000
**Language** Several Khoisan "click" tongues; many people also speak Tswana
**Beliefs** Animism (including beliefs in supernatural beings and in the spirits of the dead)
**Location** Botswana, Namibia, Angola

Often called "Bushmen", the San are the original inhabitants of the Kalahari desert, which stretches across Botswana and Namibia. They are thought to be direct descendants of the original ancestor of all humans. One piece of evidence for this idea is their genetic diversity – a sign of ancient origins. Another is their "click" languages, which have over 100 sounds (most languages have 20–40); linguists believe this range of sounds indicates the great age of the languages. The San are not one homogeneous people but consist of several groups, such as the !Kung and the /Xam (the ! and / represent different click sounds). They are nomadic hunter-gatherers, constantly searching the desert for food and water.

By dating rock art found in the area, scientists have discovered that the San people have inhabited the Kalahari for at least 27,000 years. During most of this time (until the intrusion of national governments in recent years), the San's way of life has probably changed very little. Most San today are still hunter-gatherers. The men hunt with bows and arrows, the arrowheads dipped in poison from beetle larvae. They may silently follow an animal (usually a gemsbok, a wildebeest, a giraffe, or an antelope) for several days before successfully bringing home a kill. Smaller animals, such as reptiles and birds, are caught using traps. The women mainly gather roots, tubers, berries, and nuts.

The San have an egalitarian social system whereby most things are shared, especially the meat of hunted animals. They also have a complex ritual life, which includes trance dances and ceremonies to mark a girl's first period and a boy's first kill.

Issue

### Ancestral lands

The government of Botswana is forcing the San out of the Kalahari, wanting them to live in permanent camps and take up agriculture. San living in the Central Kalahari Game Reserve are set limits to the number of animals they can kill each year, making them dependent on government rations. While some San have resisted, many have been moved and resettled. However, they have not taken to a settled life, and many social problems have arisen.

**FACIAL FEATURES**
*The San have unusual facial features – smaller and sharper than those of the Bantu Africans, who occupy most of the continent.*

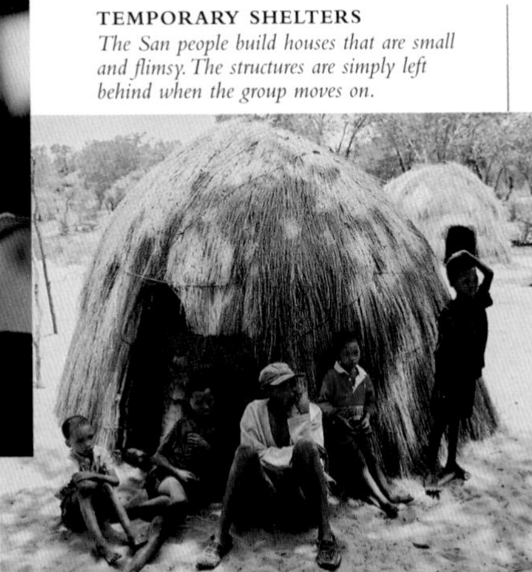

**TEMPORARY SHELTERS**
*The San people build houses that are small and flimsy. The structures are simply left behind when the group moves on.*

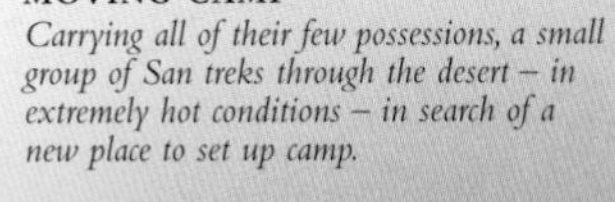

**MOVING CAMP**
*Carrying all of their few possessions, a small group of San treks through the desert – in extremely hot conditions – in search of a new place to set up camp.*

SOUTHERN AFRICA

## Himba

**Population** 15,000 (less than 1 per cent of the population of Namibia)
**Language** Himba
**Beliefs** Animism (the ancestors and their burial grounds are vital elements in Himba beliefs)
**Location** Kunene region (northwestern Namibia)

The Himba are a semi-nomadic, pastoral people who live in the hot, arid Kunene region of northwestern Namibia. They are particularly proud of their cattle, but also herd sheep and goats. Occasionally, the people also set up small-scale farms on the banks of the Kunene River.

The Himba have resolutely stuck to their traditions and, partly due to the inhospitable environment that they inhabit, have been relatively undisturbed by outsiders.

Himba women cover their faces, bodies, and hair daily with red ochre and butter fat, which gives them a distinctive, rich red appearance and provides some protection against the harsh desert sun. They also wear ornate iron and shell jewellery. Most women wear metal rings around their ankles. Married women also wear a heavy necklace with a conch shell attached to it.

**MARRIED WOMEN**
*Himba women have elaborate hairstyles and, if married, wear conch-shell necklaces.*

SOUTHERN AFRICA

## Ovimbundu

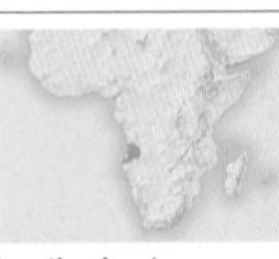

**Population** 4 million (traditionally found in the Benguela highlands; now scattered through Angola due to the civil war)
**Language** Umbundu
**Beliefs** Animism, Roman Catholic Christianity
**Location** Angola

The Ovimbundu, who were originally organized into a number of kingdoms, form the largest ethnic group in Angola. In previous centuries they were successful traders, particularly in the rubber business, and controlled many local trade routes. They were also farmers, mainly growing maize.

Today, most of the Ovimbundu work on coffee plantations or in towns. The civil war, which began in the 1970s, has heavily affected their lives. Most support the National Union for the Total Independence of Angola, which opposes the Popular Liberation Movement of Angola.

SOUTHERN AFRICA

## Xhosa

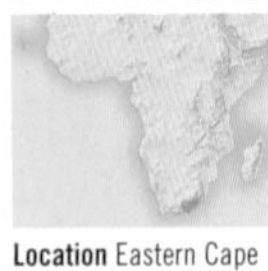

**Population** 7 million
**Language** Xhosa (the most widely distributed African language in the country – although there are more speakers of Zulu); also English
**Beliefs** Christianity
**Location** Eastern Cape (South Africa)

The Xhosa people were traditionally farmers and herders, living in the Eastern Cape area of South Africa. Today, the people live throughout South Africa and have a variety of occupations. Many of them have salaried jobs in South Africa's major cities. Several of the business people and lawyers who opposed apartheid were Xhosa. Nelson Mandela is Xhosa by origin.

XHOSA BEADWORK

PEOPLES

SOUTHERN AFRICA

## Zulu

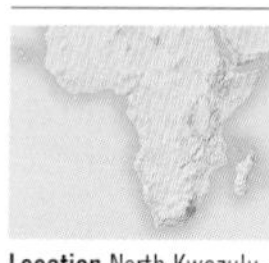

**Population** 9 million (the largest ethnic group in South Africa)
**Language** Zulu, English
**Beliefs** Traditional ancestor worship; also Christianity (which has become important in recent times)
**Location** North Kwazulu-Natal (South Africa)

The Zulu mostly live in Kwazulu-Natal province in the eastern part of South Africa. The name Zulu means "the people of heaven". They are traditionally farmers and herders, but nowadays many Zulu men have taken mining jobs. They travel long distances from their homes and live in basic dormitory accommodation near the mines for much of the year, only returning occasionally to see their wives and children.

The Zulu have retained many of their traditional ideologies and healing ceremonies. Some believe that illness can be caused by sorcery or bad spirits and will often consult a traditional healer, called an *isangoma*.

**ISICHOLO HATS**
*The traditional headdress of married women is the* isicholo *hat. It consists of dyed grass fibres over a basket frame.*

**DIFFICULT WORK**
*Many Zulu men work for very low wages in harsh, cramped conditions in the gold and diamond mines of South Africa.*

History

### Commemorating Shaka

In the 19th century, the Zulu people first came together as a great nation under the leadership of Shaka Zulu. Shaka was a highly respected warrior who led his people to conquer many neighbouring peoples and form a truly formidable nation. The time of Shaka Zulu is considered to be the heyday of the Zulu people, and ever since then the Zulus have held colourful celebrations every year to commemorate him.

**SHAKA CELEBRATIONS**
*The yearly Shaka celebrations include dancing and the wearing of traditional warrior dress, as seen below.*

MADAGASCAR

## Merina

**Population** 4 million
**Language** Merina and Malagasy (both Malayo-Polynesian languages related to the languages of Borneo)
**Beliefs** Animism, Protestant Christianity
**Location** Central highlands (Madagascar)

The Merina historically constituted the largest and strongest kingdom in Madagascar. Today, they are mainly farmers, and the major crop is rice.

The Merina intern their dead in huge tombs and periodically remove the remains in order to "turn the bones". This ceremony, known as *famadihana*, is the occasion for huge family reunions and lively festivities.

**MODERN TOMB**

MADAGASCAR

## Zafimaniry

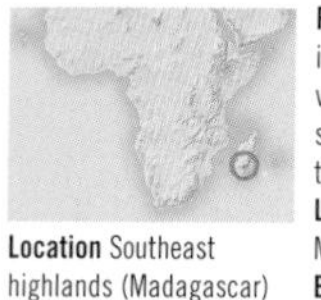

**Population** 25,000 (living in approximately 100 villages and hamlets scattered throughout the highlands)
**Language** Zafimaniry, Malagasy
**Beliefs** Animism
**Location** Southeast highlands (Madagascar)

The Zafimaniry are a group of slash-and-burn agriculturists who live in the mountainous forest region of southeastern Madagascar. They are well known for their fine woodwork and intricate carving. Zafimaniry houses are made from vegetable fibres, wood, and bamboo. When a couple marries, their first task is to build a simple house, which represents the newness of their relationship. As years pass and their relationship strengthens, the pair add to their house until it is a solid and permanent building.

**FAMILY LIFE**
*A girl holds her younger sibling as she prepares food. Over the years, as children are born, a couple add new parts to their house.*

SOUTHERN AFRICA

## Afrikaners

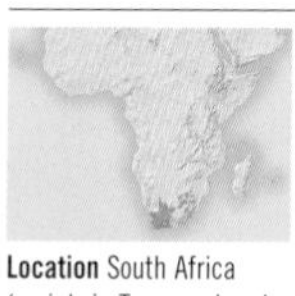

**Population** 3 million
**Language** Afrikaans (which was declared an official language in 1925)
**Beliefs** Calvinism (a form of Protestant Christianity) blended with nationalist belief that Afrikaners are God's chosen people
**Location** South Africa (mainly in Transvaal and Orange Free State)

The Afrikaners, or Boers ("farmers"), have disparate origins. Many were from the Netherlands, while others came from France, Germany, and Scotland. They came together in the Cape of Good Hope in the 17th century. The people were bound by a strongly held, common faith in Calvinism and by the sharing of a new life in the Cape. A strong national identity soon arose.

History

### Apartheid

The apartheid system was introduced by the Afrikaner government in 1948. It gave different races different rights and turned black people into second-class citizens. Many people feel that apartheid was based in the Boers' Calvinist beliefs, whereby they saw themselves as a special people with special rights, chosen by God above others. While an all-race election in 1994 produced a government with a black majority, many of the social and economic effects of apartheid remain.

**REMEMBERING THE VOORTREK**
*Members of the former Afrikaner Resistance Movement (AWB), committed supporters of apartheid, re-enact the* Voortrek.

When the British took control of the Cape in 1814, the Afrikaners decided to find their own land in the interior. From 1835 to the early 1840s, an estimated 12,000 Afrikaners travelled inland on ox-drawn carts, fighting off Xhosa and Zulu attacks as they went. This journey became known as the *Voortrek* and is a defining point in Boer history. The travellers set up three states, but the British later annexed the areas. This action led to two Anglo-Boer wars and then, eventually, to the establishment of the Union of South Africa in 1910. Today, the Afrikaners are mostly farmers. They are deeply religious and have fiercely maintained their traditions.

**HUNTING IN THE BUSH**
*Many Afrikaners hold dear the freedom to hunt wild animals. Extensive private game reserves occupy much land in South Africa.*

**MAMPOER LIQUOR**
*Mampoer, or "fire water", is a traditional Afrikaner liquor. It is a very strong, homemade brandy and is popular with locals and tourists alike.*

**THE ART OF FALCONRY**
*Hunting with birds of prey is a tradition that originated with Central Asian nomads and is still alive today. This man is attending the Golden Eagle Festival in Mongolia.*

# NORTH AND CENTRAL ASIA

The few nations in northern and Central Asia combine to offer a rich and ancient tapestry of peoples and cultures. They have been affected by centuries of migration and invasion from all sides, including Persians, Turks, Chinese, and Russians. Both cultural assimilation and the struggle to uphold cultural distinctiveness have been inevitable. Today's political divisions rarely correspond to the history of the regions' peoples, but they do serve to offer them a newfound sense of national identity alongside their original ethnic ones.

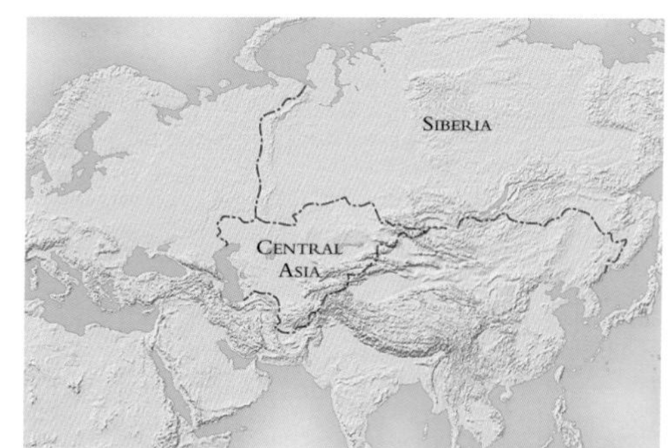


**THE REGION**
*The area covered in this section covers Siberia and Central Asia (which includes northwest Pakistan, Afghanistan, and Mongolia).*

Some experts believe that threads of some northern Asian cultures can be traced back as far as the ancient Sumerians, inventors of the wheel and writing. Despite this common history, there is now a wide variety of peoples in northern and Central Asia. Many ethnic groups live in northwestern Pakistan, Mongolia, and Afghanistan, as well as in the "-stans" – states that belonged to the Soviet Union until 1991: Kazakhstan, Uzbekistan, Tajikistan, Turkmenistan, and Kyrgyzstan.

Issue

### Death of the Aral Sea

The Aral Sea, between Kazakhstan and Uzbekistan, is fed by two large rivers: the Amu Darya and Syr Darya. From the 1930s, these rivers were diverted for irrigation. This has caused over 60 per cent of what was once the world's fourth largest lake to dry up in just 35 years. Much of the fishing industry has now ceased and many coastal dwellers have no access to fresh water. Polluted water and dust storms on the dried sea bed also cause health problems.

## MIGRATION AND SETTLEMENT

The majority of people in the region were originally nomads. The Persians (Indo-Iranian peoples) among them used the wealth made from the Silk Route to create grand Central Asian cities such as Bukhara, which attracted other tribes, including the Baluchi. Turkic tribes also settled in the area: Mongolians headed into the land that now bears their name, while Turks headed south into Central Asia and beyond. Another migration came from the west when the Russian empire expanded from the 16th century.

## TRIBES AND NATIONS

The Persians settled into varied communities, as reflected in the diversity of modern Afghanistan, which still has over 45 ethnic languages. However, descendants of Turkic warriors, initially employed by Persian rulers, gained power in the 11th century, and it is they who now occupy most of the "-stans", with only Tajikistan home to a mainly Persian people, the Tajiks. The Mongolians, led by Genghis Khan, then conquered much of northern and Central Asia during the 12th and 13th centuries, but their empire dissolved only a century later. Later still came Russian invasion, which rigorously stamped its culture across the "-stans" to the detriment of indigenous cultures. Even hunter-gatherers of the far north, such as the Evenk, were touched by Soviet and Chinese modernization and industrialization, including the forced decline in nomadism as a result of urbanization. When the Soviet empire fragmented in 1991, the "-stans" gained their independence, but the political boundaries now in place by no means match ethnic groupings. Most peoples of the region span several national borders.

**PREHISTORIC ARTEFACTS**
*Tombs found in the Altai mountains contain artefacts like this decorative deer (once part of a headdress), which dates back to a nomadic culture of the 4th and 5th centuries* BC.

**SIBERIAN CHURCH**
*A Russian Orthodox church in Siberia bears witness to the Russian colonization of the area in the 17th century.*

## CHANGE

The effects of industrialization on the "-stans" and Siberia have been huge. Former Soviet states now have the challenge of overcoming the remnants of communism in their quest for democracy and self-determination. Since their creation as independent states in 1991, the dominance of national languages and populations (as opposed to Russian) has often been reasserted. However, tribal dialects and cultures still struggle to survive, as the emphasis is often now on national identity, rather than traditional clans. Conflict even broke out in the Tajikistan mountains in the early 1990s as a result of the existence of some 20 dialects (and clans) that are unrecognized in official Tajik culture. Many states have formed groups to preserve such cultural diversity, like the Association of Peoples of the North set up by Nenets. Such groups have meant that many people have upheld age-old traditions, like allegiance to clans among the Kazhaks, the building of *gers* (tents) by Mongolians, and the resurgence of Buddhist culture, also among the Mongolians.

**DRILLING FOR OIL**
*Parts of the region are rich in oil, gas, and minerals. These resources play a key role in the area's continued industrialization.*

**REACHING INTO SPACE**
*The Baikonur cosmodrome is a fine example of modernization in Kazakhstan (a former Soviet state). It is the launch site for International Space Station missions.*

SIBERIA

## Evenk

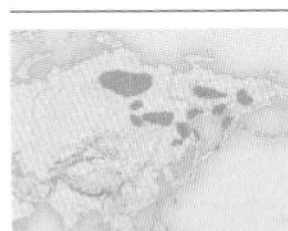

**Population** 65,000
**Language** Evenki (Tungusic branch of the Altaic family, related to the Manchu tongue of China), Russian, Chinese, Yakut, Mongolian
**Location** Siberia (Russia); Mongolia, NE China
**Beliefs** Shamanism

Evenk, or Ewenki, are widespread across the taiga (northern coniferous forest) of Siberia, Mongolia, and Heilongjiang in China. Both Russia and China have named autonomous regions for the Evenk, but only a minority of Evenk live within their borders. Chinese Evenk are settled pastoralists and farmers, but almost half of the northerly Russian Evenk still have a traditional economy of reindeer herding, hunting for food and fur, and fishing. They rear completely domesticated reindeer, which they milk, ride, and use as pack animals. Their reindeer are not large, and must be ridden on the shoulders instead of the back. The Evenk will not eat their reindeer, except at times of desperation or for ceremonies. The animals are otherwise too valuable. Shamanism had a powerful effect on Evenk life – the word shaman has Evenk origins. A shaman would enter a trance by eating hallucinogenic mushrooms and would reputedly tell the future, predict the weather, and foretell success in hunting or herding.

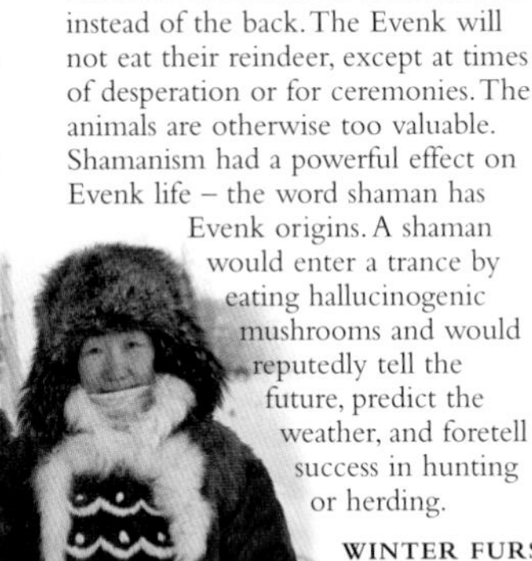

**WINTER FURS**
*Warm clothes such as these are made with reindeer skin and trapped sable, otter, and ermine.*

SIBERIA

## Yakut (Sakha)

**Population** 380,000
**Language** Yakut, or Sakha (a member of the northern Turkic branch of the Altaic language family), Russian
**Location** Republic of Yakutia (Sakha), Russia
**Beliefs** Russian Orthodox Christianity since conversion in 17th century

The Yakut, or the Sakha as they prefer to be called, are the largest indigenous group in Siberia. They inhabit the largest administrative division of Russia – the Republic of Yakutia, a vast swathe of eastern Siberia that includes the Verkhoyansk region, the "Pole of Cold", where temperatures can drop below –60°C (–76°F). They arrived in the 13th century, probably as a northerly expansion of the Turkic peoples from Central Asia. They brought with them techniques of breeding cattle and horses, which they managed to adapt to the cold climate. Yakut culture flourished with the Russian fur trade, but under Stalin their economy was brutally collectivized. President Yeltsin officially apologized in 1994.

**SERVING TEA**
*Tea being served from a samovar (urn) is a sign of Russian influence on Yakut culture, while the woman's clothes and coin headdress reveal connections to a wider Turkic culture.*

**HARDY HORSES**
*Although they live in some of the coldest inhabited places on Earth, Yakut breed horses, not reindeer.*

SIBERIA

## Nenets

**Population** 34,000
**Language** Russian, Nenets (a language remotely related to Lapp and Finnish)
**Location** Arctic Russia, either side of the Urals
**Beliefs** Nenets religion, including elements of animism and shamanism

Nenets are a polar people, living on the Arctic tundra, a land of permafrost and marshes. They live near the Arctic coast of Siberia, and are the most extreme northerly people of Europe, since they also live on the European side of the Ural mountains. Their home includes the Yamal peninsula (Yamal means "the end of the Earth"), which extends far into the Arctic Ocean. Nenets is the name the people give themselves and it means "true, genuine man". Together with the Nganasan to the east and Enets to the south (whose names also mean "true, genuine man"), Nenets are part of a wider grouping called the Samoyeds. The Samoyeds are connected by having related languages and cultures.

After arriving in the region early in the 1st millennium AD, the Nenets became renowned as the best breeders of reindeer and reared animals strong enough to be ridden like a horse.

Owing to the remoteness of their home, the Nenets culture has to some extent survived the attentions of first the Russian Christian missionaries, then the collectivizing ambitions of the Soviet authorities. Today, their traditional religion is still practised, and many people remain true nomads. They have no one permanent living area and travel with the reindeer on their migrations twice a year – south from the coastal regions to the forest in the autumn, and back to the coastal tundra grazing grounds in the spring.

The Nenets' ancient religion is both animist (they attribute natural objects, both living and inanimate, with spirits) and shamanistic (a holy person called a shaman carries out rituals). On their migrations they drag a sacred sleigh, which carries religious idols, bear skins, coins, and other sacred possessions. The sleigh is only unpacked at times of sacrifice or on special occasions, and only then by a respected elder.

**BABY CRADLE**
*Babies are kept insulated from the frozen ground in hanging cradles. They wear clothes of baby reindeer skin and nappies lined with wood shavings.*

**YEARLING REINDEER**
*A young reindeer is fed boiled fish. The woman wears a yagushka – a reindeer-skin coat with fur facing inside and out.*

**REINDEER-SKIN DWELLINGS**
*Nenets lands are remote, so it is practical to build nomadic dwellings from local materials: reindeer skins and larch poles.*

**ALMOND EYES**
*Like most Siberian peoples, Nenets have almond-shaped eyes – common in Far East peoples and some Native Americans.*

SIBERIA

## Chukchi

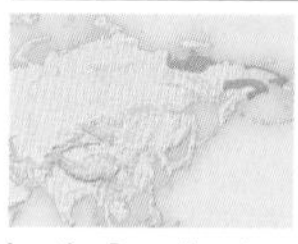

**Population** 15,000
**Language** Chukchi (one of the Chukotko-Kamchatkan languages of the Russian Far East), Russian
**Beliefs** Shamanism, animism, Russian Orthodox Christianity
**Location** Far northeastern Siberia (Russia)

Many of the Chukchi live in their own autonomous region of the Russian Federation, occupying the Chukotka peninsula between the Bering and Chukchi seas. The coastal Chukchi and the inland Chukchi have contrasting traditional lifestyles. While the inland people are partly nomadic reindeer herders, the coastal people are settled fishers and hunters of sea mammals. The coastal Chukchi have a similar traditional lifestyle to Alaskan peoples across the Bering Straits, like the Yupik, and some use similar walrus-skin boats, called *umiak*. In the 20th century, the Chukchi put up armed resistance to being settled by the Soviet regime and have endured the encroachment of polluting industries and nuclear testing on their land.

Issue

### Collectivization

Like all indigenous Arctic peoples of Asia, the Chukchi suffered from Soviet collectivization in the 1930s, which sought to settle and organize nomads. Authorities forced nomadic families into villages. The hunting and herding men eventually had to join them, while children were sent to boarding schools. The Chukchi's future is a balancing act between Russian-style education (below) and the preservation of Chukchi culture.

**CHUKCHI PEOPLE**
*In facial appearance, clothing, and culture, the Chukchi look a lot like indigenous Arctic peoples in North America. Experts are looking for links in their genes and languages.*

**INLAND CHUKCHI**
*An inland Chukchi herder prepares a rope for capturing a reindeer.*

**IVORY CARVING**
*Coastal Chukchi people carve sculptures from walrus ivory. Their subjects, such as this polar bear, are often figures taken from myths and legends.*

**CARE FOR ANIMAL SPIRITS**
*A Chukchi child gives a drink to the spirit of a dead seal. Chukchi believe a hunted animal gives itself to the hunter and must be respected.*

SIBERIA

## Tuvinians

**Population** 240,000
**Language** Tuvin (a Turkic language), Russian, Mongolian, Chinese
**Beliefs** Mostly Buddhism allied to the lamas of Tibet, some ancestral shamanism
**Location** Republic of Tuva (Russia), Mongolia

The Republic of Tuva lies at the edge of Siberia, bordering Mongolia and lying close to China. For centuries, Tuvinians have been tossed between powerful neighbours such as Turkish khanates, Chinese empires, the Mongols of Genghis Khan, and the Russian Empire. Tuva finally declared independence from China in 1912, only to become a protectorate of Russia, then an autonomous republic within the Soviet Union.

Tuvinians are related to Mongolians, and their language, although mainly Turkic, has many Mongolian words. Since the 17th century, Tuva has come under increasing Russian cultural influence, and the region underwent Soviet-style urbanization and industrialization during the 20th century. During this period, Tuvinians had autonomy in name only. Since 1991, however, a nationalist revival has called for increased Tuvin-language education and help in rebuilding Buddhist monasteries, which were abandoned under the Soviets.

Traditionally, the Tuvinian people are seminomadic herders, rearing sheep, cattle, horses, reindeer, or yaks. Their twin religions, both suppressed by the Soviet regime in the 20th century, are Buddhism and shamanism. Both have enjoyed revivals since 1991, and many people follow both belief systems, using each in different parts of their lives.

Tuvinians have a strong musical tradition, and have developed a unique technique of overtone singing, also called throat singing. A single vocalist produces two, or even three, notes at once, sequencing them into melodies. Overtone singing combines with instruments such as the horsehead fiddle to represent the natural sounds and landscapes of traditional Tuvinian life on the steppe – a life revolving around horses and herding.

**THROAT SINGERS**
*Huun Huur-Tu are leading exponents of the Tuvinian overtone-singing tradition, also called throat singing.*

**SEMINOMADIC LIFE**
*Some Tuvinians maintain a seminomadic lifestyle in conical hide tents in the forest (left) or in circular yurts on the dry steppe.*

SIBERIA

## Altay

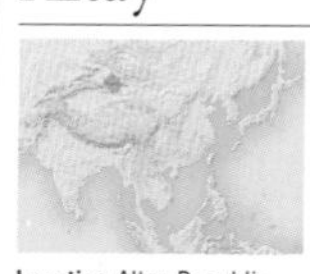

**Population** 69,000
**Language** Altay, or Oirot (a Turkic language related to Tuvin and Yakut), Russian
**Beliefs** Burkhanism (a blend of Buddhism and shamanism), Christianity, Islam
**Location** Altay Republic and elsewhere in Russia

The Altay region, home to today's Altay people, may be the source of the flowering of languages that gave us all of today's Altaic languages, spoken by all Turkic and Mongolian peoples.

Today's Altay people have survived attempts in past centuries by both China and Russia to extinguish their culture. Ethnic Russians now form the majority in Altay territory. The Altay were a seminomadic society of herdspeople until most were settled by the Soviets. Even settled families sometimes keep a *yurt* (traditional nomadic tent) in the garden as an extra room or summer kitchen. In around 1900, a unique local religion called Burkhanism emerged when an Altay shepherd learned in a vision that a messianic figure would lead his people to freedom. Burkhanism combined elements of Buddhism with ancestral shamanistic nature religion, but it has been discouraged by the authorities.

CENTRAL ASIA

## Tajiks

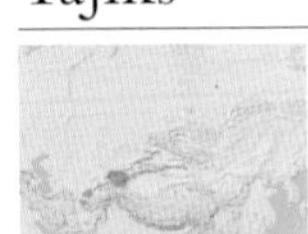

**Population** 8.6 million
**Language** Tajik (and various Persian-related languages), Russian
**Beliefs** Mainly Sunni Islam, Ismaili Islam (a breakaway sect of Shia Islam followed by Pamiri Tajiks)
**Location** Tajikistan, Afghanistan, Uzbekistan, Kazakhstan, China

The Tajiks can trace their origins back as early as the emergence of the Bactrians and the Sogdians in the 1st century BC. They differ from other central Asian groups in that they speak a Persian rather than a Turkic language. Tajiks from the mountainous Pamir region use dialects that differ markedly from those spoken in the lowland areas. In the Soviet era, the Tajiks were granted their own union republic (in 1929), and this became the independent state of Tajikistan after the dissolution of the Soviet Union in 1991. From 1991 to 1996, Tajikistan endured a clan-based civil war that ultimately cost approximately 50,000 lives. Many Tajiks live in small villages and grow crops such as cotton, wheat, and barley. Those living in lowland regions have a culture similar to the neighbouring Uzbeks (for example, much respect is afforded to elders, known as *mui-safed* – "white haired"). Tajik dress for men traditionally includes a *tapan* (a heavy quilted coat) and a *tupi* (a black embroidered cap). Tajik women, meanwhile, wear strikingly colourful *kurta* (dresses) with matching *rumol* (headscarves).

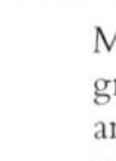

**MOUNTAIN HORSEMEN**
*Tajiks live in expansive steppe and rugged mountains. Horsemanship is a necessary skill in remote areas.*

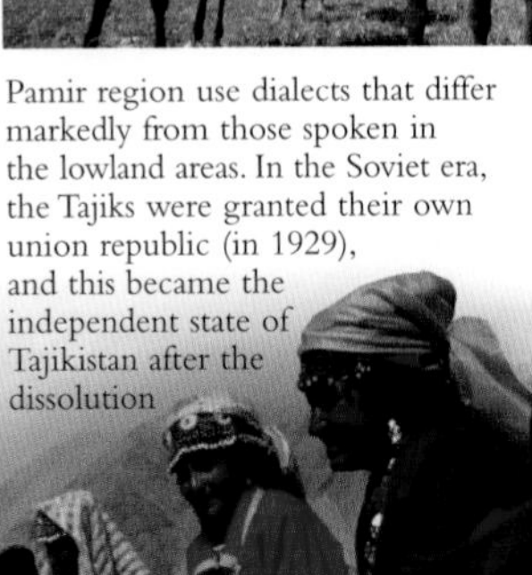

**BANQUET**
*Tajiks form minorities in all Tajikistan's neighbouring countries. These Tajiks enjoying a feast live in the Xinjiang region of China.*

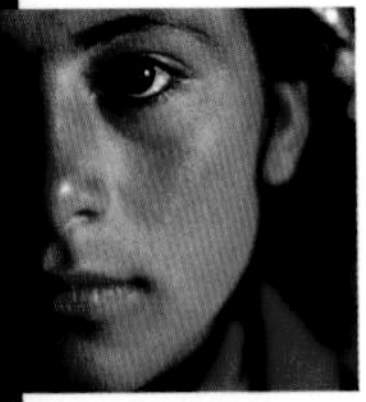

**TAJIK APPEARANCE**
*Tajik people tend to have light brown skin, dark hair, and brown eyes. The hair is usually covered with a headscarf for women or a cap for men.*

CENTRAL ASIA

## Pamiris

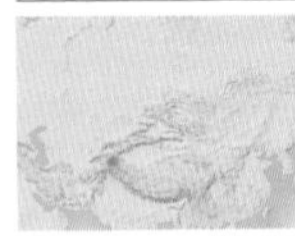

**Population** 200,000
**Language** Pamiri Farsi
**Beliefs** Ismaili Islam (followers believe that only a direct descendant of Muhammad can understand the Islamic teachings), Sunni Islam
**Location** Tajikistan, China, Afghanistan, Pakistan

Most Pamiris live in villages and small towns in the Pamir mountain range (often known as the "roof of the world"). They speak a distinct dialect of Farsi, yet there are many regional variations. They also identify themselves not only as Pamiris but also as people from specific valleys and regions. Most Pamiris, however, are Ismaili Muslims, and this is a focus for national identity.

The Pamir region suffers harsh winters, land shortages, and occasional droughts, and some people have traditionally taken opium to relieve stress. The Pamiri area is also very volatile politically, and people's lives have been deeply affected by the civil wars in Tajikistan in the 1990s and in Afghanistan since 1979. In Afghanistan, Sunni governments and organizations have targeted Pamiri people because they are followers of Ismaili Islam. As a result, many Pamiris have left the region.

CENTRAL ASIA

## Uzbeks

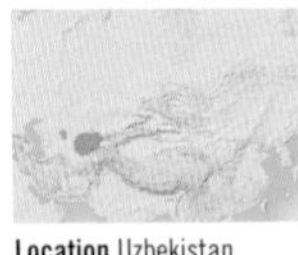

**Population** 22 million (the largest population of all the Central Asian republics)
**Language** Uzbek (a Turkic language with two dialects), Russian
**Beliefs** Islam
**Location** Uzbekistan, Afghanistan, Kyrgyzstan

**THREE-PART WEDDING**
*Female guests attend the afternoon ceremony of an Uzbek wedding. Men are invited to the morning ceremony, then everyone celebrates together in the evening.*

Uzbeks emerged with their own, independent state of Uzbekistan in 1991 following the dissolution of the Soviet Union. Uzbek communities have strong ties to the *mahalla* (neighbourhood), and the elders, known as *aksakal* ("white beards"), enjoy considerable respect and authority. The Uzbeks also attach great importance to hospitality: they relish entertaining guests. Traditional Uzbek foods include *plov* (pilau), nan (flat bread), and noodles.

CENTRAL ASIA

## Kazakhs

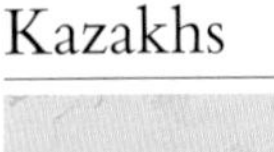

**Population** 11 million (Kazakhs make up only 50% of the population of Kazakhstan, the world's ninth largest country)
**Language** Kazakh (a Turkic language), Russian
**Beliefs** Islam
**Location** Kazakhstan, China, Uzbekistan, Russia

Traditionally nomadic pastoralists, the Kazakhs dominated the steppe region between the southern Urals and the Altai Mountains until large-scale Russian settlement in the 19th century deprived them of much of their grazing land.

Over the last 100 years, the Kazakhs have become a more settled people. The main shift away from nomadism took place during the era of early Soviet collectivization, from the 1920s onwards. Nonetheless, allegiance to clans (a common feature of many nomadic peoples) is still a vital feature of Kazakh society, and Kazakhs divide themselves into three main groups: the Great, Middle, and Little Hordes (*zhuz*). Ancestral ties also remain important, and many introductory conversations begin with, "What *zhuz* do you belong to?" Furthermore, the Kazakh's nomadic heritage plays an important part in cultural identity – for example, the *yurt* (a type of tent used by nomads) is a national symbol.

Kazakh oral literature is based on a body of heroic epics. Many of these describe the 16th-century clashes between the Kazakhs and Kalmyks and the *batyrs* (warriors) of the age.

Issue

### Soviet secularism

During the Soviet era, the traditional role of Islam – which is central to the culture and identity of Kazakhs (and other central Asian cultures) – was suppressed. One way in which the Soviets did this was to introduce secular "wedding palaces", such as the one shown here. Some of these palaces are still used today.

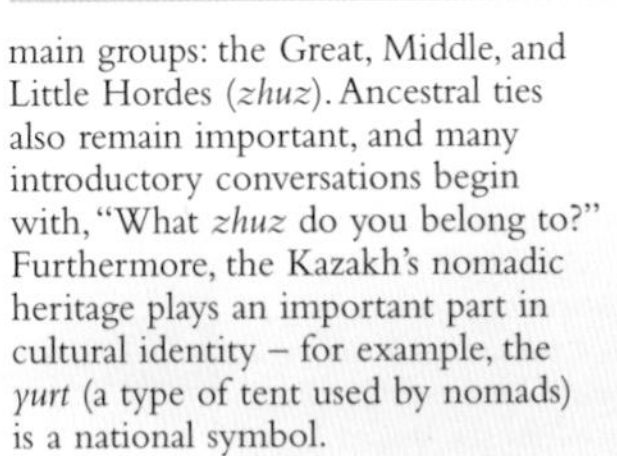

**COLOURFUL CLOTHING**
*Although some Kazakhs prefer "Western" clothes, women traditionally wear colourful handmade dresses and headscarves.*

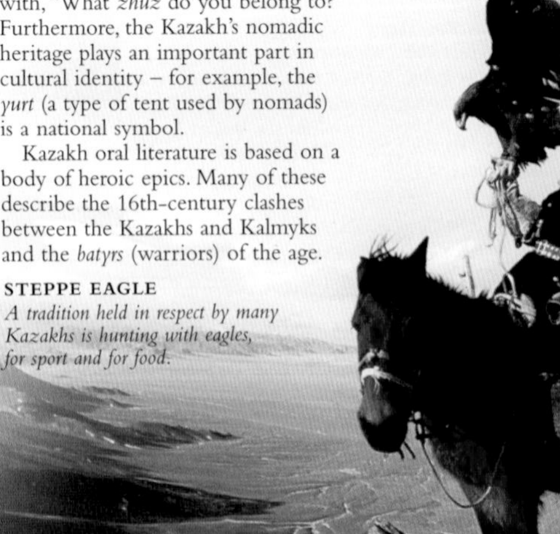

**STEPPE EAGLE**
*A tradition held in respect by many Kazakhs is hunting with eagles, for sport and for food.*

CENTRAL ASIA

# Kyrgyz

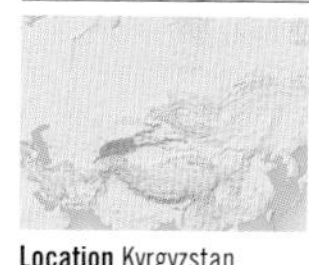

**Population** 3.35 million
**Language** Kyrgyz, Russian
**Beliefs** A mix of shamanism and Islam (natural features such as mountains are regarded as holy, and shamans, *bakhshi*, are seen as guides and healers)
**Location** Kyrgyzstan, Uzbekistan, China, Afghanistan

The Kyrgyz traditionally engaged in pastoral nomadism, but under Soviet rule the majority of Kyrgyz became settled. Following the break-up of the Soviet Union, the independent state of Kyrgyzstan emerged.

The Kyrgyz people have a rich oral literature in the form of the *Manas*, a collection of tales describing the heroic deeds of Manas (the legendary founder of the Kyrgyz) and his men.

**KYRGYZ CLOTHING**
*This pair are wearing typically colourful clothing, with bold patterns. Such features can also be seen in the local embroidery.*

Kyrgyz males are easily recognizable by their large, white felt hats known as *ak Kalpak*. The Kyrgyz diet is mainly based on meat (including entrails), milk products, and bread, reflecting their nomadic life. *Kumis*, fermented mare's milk, is also popular.

**KYRGYZ YURT**
*Made of layered felt over a collapsible wooden frame, yurts are light and portable. A small number of Kyrgyz still live in yurts.*

CENTRAL ASIA

# Turkmen

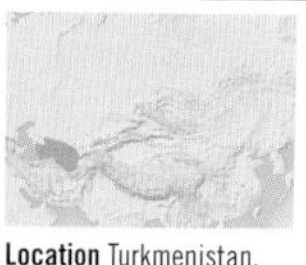

**Population** 5.25 million
**Language** Turkmen (which is most closely related to the languages of other Oghuz descendants, such as the Turks and the Azeris), Russian
**Beliefs** Islam
**Location** Turkmenistan, Iran, Afghanistan

The Turkmen people are descended from Oghuz Turks, who in the 11th century remained in Central Asia when most of their number moved into southwestern Asia.

Turkmen engaged in nomadic pastoralism around the oases fringing the Karakum desert. They are split into over 100 clans, each distinguished by dialect, clothing, jewellery, and the patterns in their gillams (carpets).

Following the fall of the Soviet Union in 1991, the Turkmen Soviet Socialist Republic, founded in 1924, became an independent state. Since that time, President Niyazov, the self-styled *Turkmenbashi* (father of all Turkmen), has ruled the nation of Turkmenistan. He has installed a n autocratic government and a cult that depicts him as the embodiment of Turkmen national identity.

Along with their distinctive *telpek* hats (*see below*), Turkmenistan men traditionally wear baggy trousers tucked into knee-length boots and a red jacket, known as a *khalat*. Turkmen women wear heavy velvet or silk dresses, often decorated with red and maroon colours.

**TILE DESIGN**
*Patterns and motifs on the tiled panels of Turkmen mosques inspire the design of the people's famous decorated carpets.*

**TELPEK**
*Turkmen males wear a* telpek *– a huge sheepskin hat that is either white or made up of a dense set of black ringlets.*

CENTRAL ASIA

# Mongolians

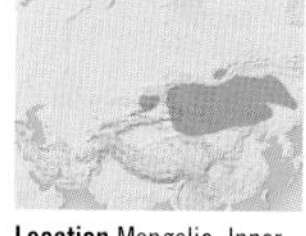

**Population** 2.7 million
**Language** Khalkha Mongol (spoken by 90% of the population of Mongolia), Russian, Chinese
**Beliefs** Tibetan Lamaist Buddhism, Islam, shamanism
**Location** Mongolia, Inner Mongolia (China), Russia

The Mongolians created one of the largest empires in history. At its peak, it stretched from Korea to Hungary and as far south as Vietnam. Mongols now live not only in today's Mongolia, but also through Inner Mongolia in China and Buryatia in the Russian Federation. Until the end of the 12th century, the Mongols were little more than a loose alliance of rival clans inhabiting the Mongolian grasslands. Under Genghis Khan (1162–1227), however, the clans united and set off on a path of conquest throughout much of Eurasia. Khan's descendants, especially Kublai Khan, consolidated and extended the empire, establishing the Yuan Dynasty (1271–1368) in China. From the end of the 14th century onwards, the Empire dissolved as the Mongols reverted to a collection of small tribal domains.

Following their incorporation into the Soviet Union, the Mongolians were the victims of extreme cultural intolerance: by 1939, 27,000 people had been executed. Nevertheless, national traditions and institutions thrive in post-Soviet Mongolia. A number of Mongolians still exist by pastoral nomadism, and almost half live in tents known as *gers* (similar to the *yurts* used by Kazakhs and Kyrgyz). *Gers* are erected throughout Mongolia, including in the suburbs of the capital, Ulaanbaatar. However, recent dry summers and harsh winters have impacted on the nomadic way of life. Hundreds of thousands of the Mongolians' animals died in 1999 and 2000, and as a result, many people have since settled in urban areas.

History

## Genghis Khan

After uniting the Mongol clans, the great leader originally called Temujin became known as Genghis Khan ("universal king"). Khan's ruthlessness became evident when, as a teenager, he killed his half-brother in cold blood. In modern Mongolia, he has become a national symbol, and his face appears on a diverse range of items, such as bottles of vodka and currency notes.

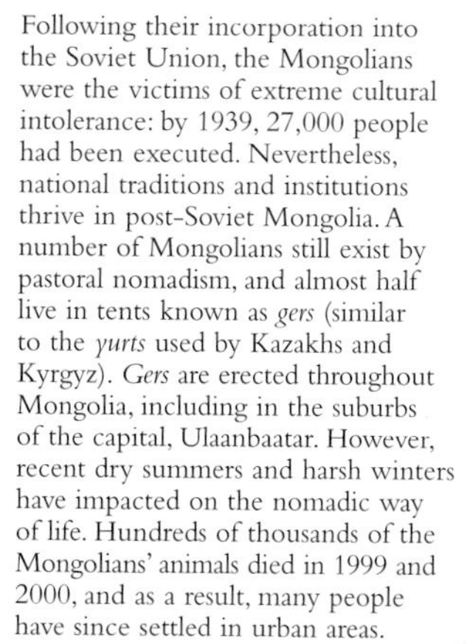

**ARCHERY**
*One of Mongolia's national pursuits, archery is often an ingredient of festivals, including this, the Naadam Festival near Ulaanbaatar.*

**GER INTERIOR**
*The traditional, circular nomadic tent is called a ger in Mongolia, but it is similar to the yurt of the Kyrgyz and the Kazakhs.*

**HORSEMANSHIP**
*The horse is central to traditional, nomadic Mongolian life, and skill on horseback is still highly valued. Children are often taught to ride from an early age.*

**FESTIVE WRESTLING**
*Wrestlers take part in a summer festival. There are neither weight classes nor time limits in Mongolian wrestling.*

CENTRAL ASIA

## Hazaras

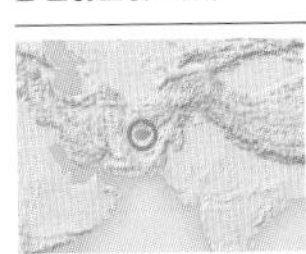

**Population** 1 million
**Language** Hazaragi Farsi (closely related to Farsi, or Persian)
**Beliefs** Mostly Shia Islam; also Ismaili Islam (a breakaway sect of Shia Islam)
**Location** Central Afghanistan

The high, barren Hazarajat plateau of central Afghanistan is home to the Hazara people, who speak Hazaragi Farsi. The Hazara have a rich, poetic heritage, and the poems they write and recite are often about the delights and pains of love. They are mostly followers of the Shia branch of Islam, in contrast to the predominantly Sunni Pashtun and Tajik people, who dominate neighbouring areas.

Because the Hazarajat is such an isolated region, it was relatively free of violence during the Soviet invasion of Afghanistan. However, drought and famine drove many of the Hazaras to migrate to other parts of the country.

Religious differences, as well as the Hazaras' distinctly Mongol-like faces, have often made these people the target of racial discrimination by other Afghans. In the past, they were often used as slaves and servants, and more recently they were persecuted by the Taliban regime. As a result, many Hazaras fled the country and now live as refugees in Pakistan and Iran, but even in exile they have been victims of prejudice. In the Pakistani city of Quetta, their mosques have been attacked by Sunni extremists.

History

### The Bamiyan Buddhas

Although Muslim, the Hazaras have long been regarded as custodians of the giant, 1,800-year-old Buddhas carved out of a mountain at Bamiyan, in the Hindu Kush of northern Afghanistan. The Taliban regime destroyed the statues in 2001, and forced Hazara men to help. A Hazara man (right) prays in front of a recess once filled by a Buddha.

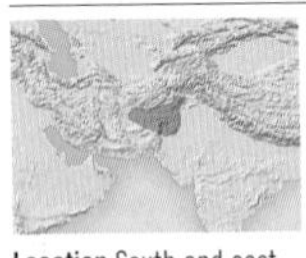

**REFUGEES**
*During the period of Taliban rule, thousands of Hazara people were victimized and hundreds were massacred. Many Hazaras fled from Afghanistan, most going to Pakistan.*

**CAVE DWELLERS**
*Some Hazara people have been displaced within Afghanistan. A few hundred in Bamiyan have set up homes in caves.*

CENTRAL ASIA

## Aimaq

**Population** 400,000 (more than half in Afghanistan)
**Language** Aimaq (related to Farsi, a member of the Iranian branch of the Indo-European language family)
**Beliefs** Sunni Islam
**Location** Afghanistan, Iran, Tajikistan

The Aimaq people live in central northwestern Afghanistan, to the west of the Hazara. Unlike their Hazara neighbours, who are Shia Muslims, the Aimaq are Sunni. They belong to one of four major Aimaq tribes, all of which speak a language related to Farsi (Persian), albeit with Turkic and Mongolian vocabulary mixed in. Many have Mongolian facial features and claim descent from the armies of Ghenghis Khan (*see* p431).

Many of the Aimaq live either nomadic or seminomadic lifestyles. Some live in distinctive canonical or rectangular *yurts* (felt tents); others live in black tents similar to those used by nomadic Pashtun tribes.

The Aimaq people are mainly goat and sheep herders. They move up and down the valleys and mountains searching for grass for their flocks. Some of the Aimaq trade with village farmers during their journeys between central Afghanistan's pastures. They offer milk, dairy products, skins, and carpets in exchange for the grains, nuts, fruit, and vegetables grown by the settled farmers.

CENTRAL ASIA

## Pashtuns

**Population** 27 million in Pakistan and Afghanistan
**Language** Pashto (also known as Pukhto), Dari (in Afghanistan); also Urdu (in Pakistan)
**Beliefs** Mostly Sunni Islam; also Shia Islam
**Location** South and east Afghanistan, Pakistan

Pashtuns live in a wide area straddling the rugged, mountainous region along the border between Afghanistan and Pakistan. Much of the population is organized into tribes, with tribal affiliations being more important than national or even Pashtun identity.

Pashtuns pride themselves on being devout Muslims, and they are well known for their code of honour, or *Pashtuwali*, which encourages them to be hospitable, to take revenge, and to provide refuge to those in trouble.

In rural areas, life is centred on the men's *hujera* (guest-house) and the mosque. The *hujera* is the place where elders gather in the evenings and where visitors are received. Praying five times a day in the mosque is an important expression of faith for most Pashtun people.

Historically, many Pashtuns were traders and merchants, selling cloth, food, horses, and other goods as far afield as Central Asia, Iran, and India. Today, city life (which has long been a part of Pashtun culture) is increasingly important. War and poverty have driven many to work in towns as bus drivers, butchers, traders, and scrap collectors. Others are doctors, bankers, and construction workers, both in Pakistan's biggest city, Karachi, and in the Persian Gulf.

Family life is important, and Pashtun migrants maintain close contacts with home. Every year, thousands return to Afghanistan and Pakistan to celebrate the Muslim festivals of *Eid al-Fitr* and *Eid al-Azha* with their relatives.

**HEADDRESS**
*This woman's headdress is decorated with a fine example of Pashtun jewellery, featuring carnelian and lapis lazuli.*

**SOCIAL GATHERING**
*Pashtun men chat over cups of green tea at a tea shop in Peshawar, Pakistan.*

**WARRIOR'S DANCE**
*A dancer wearing a Pakistani badge performs the Khatak sword dance, which was once a prelude to battle.*

CENTRAL ASIA

## Chitralis

**Population** 300,000
**Language** Khowar (also called Chitrali); also Urdu, Pashto, Dari
**Beliefs** Sunni Islam, Ismaili Islam (a branch of Shia Islam)
**Location** Himalayan valleys (northern Pakistan)

The Chitralis live in a remote part of northern Pakistan that is accessible only by high mountain passes. They describe their culture as different from that of their Pashtun neighbours, saying their language is "sweeter" and their customs more polite and sophisticated. Chitralis are known for their love of sports, especially polo. Chitrali polo is said to have no rules, with players often wielding their sticks at each other. Music and dance are a major part of Chitrali culture. Love songs are sung to the accompaniment of sitars and petrol-can drums.

**CHITRALI HAT**
*A woollen* pakol *is made in one size, then rolled up until it fits.*

**FAMILY GROUP**
*Members of this family wear the* salwar kameez *(trousers and long tunic), popular throughout the region.*

CENTRAL ASIA

## Burushos

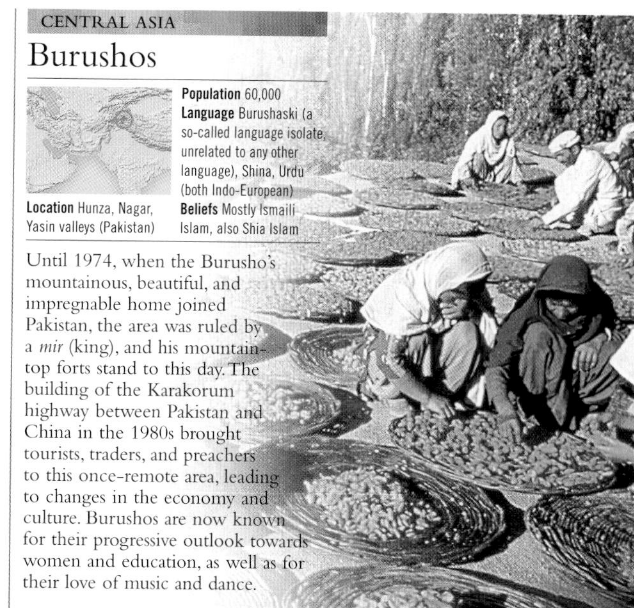

**Population** 60,000
**Language** Burushaski (a so-called language isolate, unrelated to any other language), Shina, Urdu (both Indo-European)
**Beliefs** Mostly Ismaili Islam, also Shia Islam
**Location** Hunza, Nagar, Yasin valleys (Pakistan)

Until 1974, when the Burusho's mountainous, beautiful, and impregnable home joined Pakistan, the area was ruled by a *mir* (king), and his mountain-top forts stand to this day. The building of the Karakorum highway between Pakistan and China in the 1980s brought tourists, traders, and preachers to this once-remote area, leading to changes in the economy and culture. Burushos are now known for their progressive outlook towards women and education, as well as for their love of music and dance.

**APRICOT HARVEST**
*Apricots are an important food crop for Burusho people. The fruit is dried for use during the long winter.*

**JOSHI FESTIVAL**
*Kalash welcome spring with a festival called* Joshi, *during which they add a second headdress on top of their everyday one.*

CENTRAL ASIA

## Kalash

**Population** 3,000
**Language** Kalasha; also Urdu, Khowar
**Beliefs** Ancient Kalasha religion (animist), Islam (some Kalash converted either in the 16th century or in recent years)
**Location** Chitral (northwestern Pakistan)

The Kalash are mainly non-Muslims who live in three remote valleys in the mountains of northern Pakistan, where they grow crops and herd goats in high mountain pastures. The Kalash ruled a wide area until the Islamic conquests of the 16th century, after which many converted or became slaves and serfs to Chitral's new rulers. They were still respected, however, for their engineering skills, especially in the building of bridges across the region's treacherous mountain rivers.

Festivals are a central part of the Kalash people's animist religion and involve feasts of meat and milk from their herds, ancestral praise songs, poetry, and group dancing. The Kalash face problems in modern Pakistan, including the erosion of their culture by tourism and Islamic missionaries. In recent years, they have begun to reassert their ethnic identity and revive their language.

**LONG HEADDRESS**
*Kalash women wear long, finely worked headdresses made with beads, buttons, and cowrie shells.*

CENTRAL ASIA

## Baluch

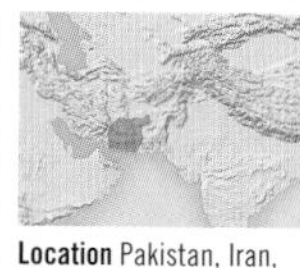

**Population** 6 million
**Language** Baloch, also known as Baluch or Baluchi (an Iranian language related to Farsi and Kurdish)
**Beliefs** Sunni Islam (vast majority)
**Location** Pakistan, Iran, Afghanistan; also Oman

The Baluch live in the mountainous, arid region called Baluchistan. The hostile land and climate in this part of Asia has shaped the lifestyle of the Baluch; many of the people lead a seminomadic lifestyle, herding goats and sheep as well as raising crops. Tribal affiliations are important in Baluch identity, and the diversity of their culture is enhanced by the isolated desert communities in which they live. The shared Baluchi language and folklore provide a common focus of identity nevertheless.

Baluchistan is a strategically sensitive area and has substantial deposits of minerals and oil. These factors make the region economically important to both Iran and Pakistan. As a result, the governments of both countries are anxious about the activities of the Baluch. Tensions peaked in the 1970s, when the Baluch fought a guerrilla war with the Pakistani army. Many Pakistani Baluch still feel marginalized, and sporadic armed conflicts sometimes erupt, often over disputes about the region's oil.

**FINE FABRIC**
*Baluchi clothing, such as this man's shirt and skullcap, is often adorned by the women with fine needlework.*

**BALUCH NOMAD**
*A shepherd pours tea outside his tent. A nomadic way of life allows herders to follow the rains and find grazing.*

**SHEPHERD GIRL**
*A Baluch girl tends a young goat. The inhabitants of Baluchistan have been herding goats for some 9,000 years.*

**ELEPHANT FESTIVAL**

*An elephant festival in the state of Kerala, southern India, honours the place of the elephant in Indian culture.*

# INDIAN SUBCONTINENT

The peninsula of southern Asia, now occupied by India, Pakistan, and Bangladesh, is home to one of the world's oldest and most influential civilizations. Although the peoples of this region have been united by various empires during different periods of history, they remain extraordinarily diverse, with 850 languages still in daily use within India alone. Furthermore, countless distinct cultures occupy the valleys in the mountainous fringes – in Nepal, Bhutan, northern Pakistan, and eastern Bangladesh.



**THE REGION**
*The region is bounded by the Himalayas, the hill tracts on the border with Myanmar, and by the river Indus, Pakistan's main artery. India is divided between Dravidian speakers in southern India and Indo-Aryan speakers in the centre.*

The hearts of this region are the Indus and Ganges river valleys. The Indus valley of Pakistan separates peoples of the east, in the Indian subcontinent from peoples influenced by the Persian empire to the west.

## MIGRATION

The recorded history of the Indian subcontinent dates from at least 2500BC, the early days of the Indus Valley civilization. Between 1500BC and 200BC, Aryan peoples from Central Asia invaded the north. These people spoke Indo-European languages similar to Sanskrit and were probably related to the same peoples, such as the ancestors of Celts and Greeks, who migrated to Europe before written records began there. Their descendants, from Europe to Sri Lanka, still speak related languages. In India, the Aryans forced the original Dravidian inhabitants, such as Tamils and Telugus, to the South.

By the 8th century BC, the region's dominant Hindu society was becoming consolidated into the caste system, a social and ritual hierarchy that is still prevalent today. Ever since this time, the region's peoples have absorbed successive religious movements and invading forces. Buddhism became influential in the region in the 3rd century BC. The 10th and 11th century AD saw the spread of Islam in the north, with the arrival of Muslims from Afghanistan. The Mogul emperors invaded and founded a Muslim dynasty in the 16th century, but the Maratha Hindu empire grew in influence in the 17th century and kept control until the British empire, in the form of the British East India Company, took over in the early 19th century. The Portuguese, French, Danes, and Dutch also all had trading centres and controlled parts of India at one time.

**MYRIAD NATIONS**
*The mountainous fringes of the region are home to a diversity of peoples, such as these Ladakhis wearing traditional costume.*

Issue

### Caste and identity

The ethnic identity of people in India's Hindu caste system is complex because each person also has a caste identity. Traditionally there is no intermarriage between castes, so castes form closed communities with their own defining occupations and customs. The Indian authorities now campaign to loosen the grip of the caste system, giving more freedom to *Dalit* people (right).

**PARTITIONED PEOPLES**
*Indian and Pakistani border guards perform an elaborate ritual at their common border. The ceremony expresses, in ritual form, true animosity between the two nations.*

## TRIBES AND NATIONS

The peoples of the region are now split into only six countries, which fact hides endless cultural diversity. Even the old kingdoms of Nepal and Bhutan are each home to many distinct peoples. The political entities of India, Pakistan, and Bangladesh (then East Pakistan), were created by Partition in 1947, and are not equivalent to ethnic territories at all. Each hosts dozens or hundreds of peoples, and several prominent ones, such as Punjabis, Bengalis, and Kashmiris, live in more than one country. The populations are a mix of people who subscribe to national identity (Bangladeshis, Indians, Pakistanis), those with strong regional ethnicity (Bengalis, Tamils) and finally, some 56 million people in India who belong to tribal (*adivasi*) communities. *Adivasis*, the "original inhabitants" of India, are officially called the "Scheduled Tribes" and are divided between at least 4,635 communities scattered throughout India. These communities tend to live in densely forested and hill regions. Many still speak languages completely different from the country's main regional or national languages, and follow traditions that have no relation to Hinduism, Islam, or other Indian religions.

**CULTURAL HISTORY**
*A Mogul-era shield represents one of India's more recent cultural periods (1526–1748).*

## CHANGE

The region's peoples are left with the legacies of the caste system, and of Partition. The caste system is becoming less dominant in Hindu society, but the *Dalit* (the "Untouchable" people below the base of the caste system) must still fight to shake off the stigma of their inherited status. Partition of the region into Pakistan, India, and Bangladesh was carried out hastily in 1947 to set aside separate states for Muslims. The political borders created have been a source of tension and bloodshed ever since, but in the new countries of Pakistan and Bangladesh, pride in the people's new national identity has gradually emerged.

**EMERGING IDENTITIES**
*Pakistan was founded as recently as 1947, yet consciousness of identity is strong among these cricket supporters in national colours.*

INDIA

## Indians

**Location** India
**Population** 1.05 billion
**Language** Hindi, Urdu, Tamil, Bengali, Kashmiri, numerous national and regional languages
**Beliefs** Hinduism, Islam, Christianity, Sikhism, Buddhism, Jainism, numerous tribal religions

Few countries in the world are as extraordinarily diverse as India. From east to west and north to south, India boasts distinct peoples, languages, customs, and traditions.

The country is home to all of Earth's great religions. Around 80 per cent of Indians practise Hinduism, but Islam and Sikhism are also major cultural forces. Ethnic identity for most Indians is a mixture of Indian nationhood and factors such as religion, regional identity, caste membership for Hindus (*see* p262), and tribal identity for *adivasi* (tribal) peoples. Despite their great diversity, many Indians regard themselves first and foremost as Indians, and look to those cultural traits all Indians have in common. India's cuisine, for instance, although regionally varied, has common themes. The basis for most meals is either rice or wheat, which is eaten in the form of *chapatti*, *nan*, or *roti* (various flattened breads). Regardless of region, most Indians serve *dahl* (lentils), vegetables, and chutneys with nearly every meal and all use many spices. *Mithai* (Indian sweets) and desserts are also distinctive.

The graceful sari, worn by women throughout India, is one of the most enduring images of the country. Some Indian women also wear the *salwar kameez*, a loose tunic top over drawstring trousers, and a scarf called a *dupatta*, which is draped over the shoulders or head. Although most Indian men wear Western-style shirts and trousers, they also commonly wear the white *dhoti*, a loose, thin cloth pulled between the legs; the *kurta*, a long tunic-like shirt; and the coloured *lungi*, a type of sarong tied around the waist.

Indians have a wealth of artistic traditions. Among classical Indian dances are Bharata Natyam and Kathakali, both of which have southern Indian origins, along with Kathak, a northern Indian dance tradition. India also has a rich variety of folk dance forms. Classical music in India has two main forms, Hindustani from the north and Carnatic from the south. India also has a long history of literature, dating back to Sanskrit writings of 1500BC. Painting in India has been a major art form since at least AD500, when scenes from Buddha's life were painted onto the ceilings and walls of the Ajanta caves in western India.

**MUSLIMS AT THE TAJ MAHAL**
*Emperor Shah Jahan completed this Islamic mausoleum in 1648 in honour of his wife.*

**SACRED COWS IN A HOLY CITY**
*Varanasi is one of the seven sacred cities of the Hindus. Cows, too, are venerated and so are tolerated when they cause an obstruction.*

**BOLLYWOOD**
*Posters in Mumbai (Bombay) advertise recent film releases. The city's studios are the most productive in the world. The film industry has been nicknamed "Bollywood".*

**TYPICAL SADDHU**
*The clothing and face paint show this man has given up his home and possessions to concentrate on spiritual life.*

Health

### Traditional medicine

India has varied medical traditions. Among the popular ones are ayurveda, which means "knowledge of life" and includes massage techniques (below); homeopathy, which is based on the principle of "like cures like"; and accupressure, an ancient Chinese technique that stimulates the body's own recuperative powers. *Adivasi* communities also use herbs, roots, and other forest remedies.

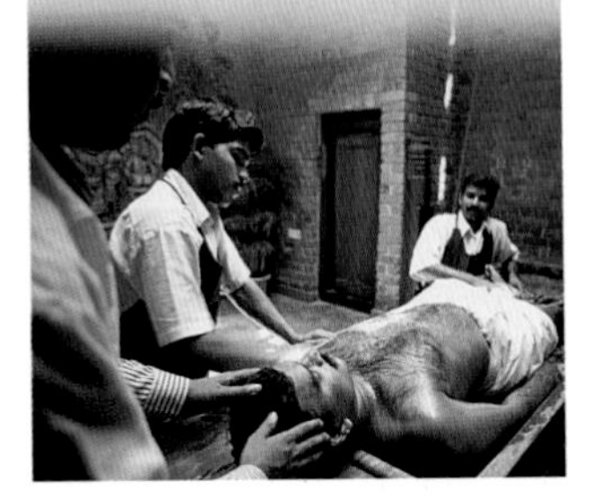

HIMALAYAS

## Nepalis

**Location** Nepal
**Population** 26 million
**Language** Nepali (official language), Bhojpuri, Gurung, Maithili, Newari, Sherpa, Tamang, Tharu, some 100 other languages
**Beliefs** Mainly Hinduism (including Tantrism); also Buddhism, Islam

Nepal is officially a Hindu country, but in practice the religion is a mix of Hindu and Buddhist beliefs with Tantrism (a mystical movement of Hinduism) sprinkled in. Most larger towns and cities sport ancient Hindu temples and Buddhist *stupas* (monuments housing sacred relics), some of which are thought to be more than 2,500 years old.

Nepal is a complex mix of more than 40 ethnic groups with distinctive cultures and lifestyles. These include the Sherpa, a Buddhist group from the High Himalayas, who are believed to be migrants from eastern Tibet. The Gurung, a group from the Middle Himalayas and Valleys, form part of the famous Gurkha military contingent. Groups from the Terai belt (the subtropical region of the southern Nepal border) include the Tharus, the oldest ethnic group in the Terai belt, who are Hindus and practise farming.

Considering the country's position between the two great culinary giants of India and China, Nepali food is relatively simple. Meals largely consist of *dhal bhat tarkari* – a combination of lentils, rice, and curried vegetables.

**ELDERS**
*This Nepali couple have achieved great age. In most countries women live longer than men do. In Nepal, however, the men outlive the women.*

**SHERPAS**
*The Sherpas of the Himalayas are famous for their trekking and mountaineering skills. They use yaks to carry heavy goods.*

HIMALAYAS

## Baltis

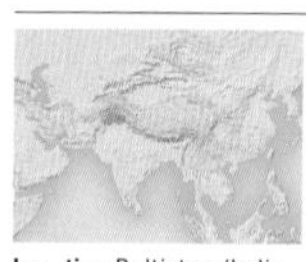

**Location** Baltistan (India and Pakistan)

**Population** 400,000, split between India and Pakistan

**Language** Balti (a Sino-Tibetan language; the Balti script is now extinct); also Shina and Urdu as second languages

**Beliefs** Islam

Baltis are of mixed ethnic origin and maintain close relations with both Muslim Kashmiris and Buddhists. Balti people are divided into groups who each trace their descent to a common ancestor. At the birth of a child, the sacred *sheikh* or *akhun* is called to perform *azaan* (a prayer or chant) in the ears of the newborn.

Balti cuisine is world famous. Spicy, predominantly non-vegetarian curries are cooked in a *karahi* – a heavy cast-iron dish with two handles – and eaten with thick *nan* bread.

TRAVELLING THE KARAKORAM PASS

HIMALAYAS

## Kashmiris

**Location** Jammu and Kashmir (India), Pakistan

**Population** 4 million, mainly in the Kashmir Valley (central Jammu and Kashmir)

**Language** Kashmiri (an Indo-Aryan language with Persian script); also Urdu

**Beliefs** Hinduism, Islam

Nestled in the Himalayas, Kashmir is known throughout the world for its scenic beauty and bracing climate. The Kashmiri people have light skin and eyes and are said to descend from members of Alexander the Great's army, which invaded India in the 4th century BC. The many occupations, invasions, and religious movements of Kashmir's rich history have left a palpable impression on Kashmiri arts and crafts, which have been influenced by Hindu, Buddhist, and Islamic cultures for centuries.

One of the most celebrated Kashmiri textiles is the pashmina shawl, which is traditionally made from the wool of a *kel* goat usually by female handloom artisans. Kashmiri carpets and rugs, made from beaten wool or silk, and the *kasida* embroidery, used to decorate shawls and saris, are also world famous. Kashmiri dress also includes the long cloak known as the *pheran*, which is worn by both men and women alike.

Issue

### Conflict over Kashmir

There are many issues surrounding the disputed status of Kashmir. Claims are disparate and driven by various groups, whose demands range from full independence to a merger with Pakistan or full integration of the state with India. Kashmiri-speaking Sunni Muslims have dominated Kashmir State politics since 1947.

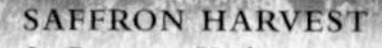

SAFFRON HARVEST

*In Pampore, Kashmiri women pick the* Crocus sativus *blossoms from which saffron, the world's most expensive spice, is made.*

HOUSEBOATS ON DAL LAKE

*Dal Lake, Srinigar, is famous for its setting and its waterborne culture. The people who live here buy their supplies from merchants in boats.*

FACE OF KASHMIR

*Kashmiris are taller than most other Indian peoples, usually with green eyes, Aryan features, olive complexions, and black hair.*

HIMALAYAS

## Bhutanese

**Location** Bhutan

**Population** 500,000

**Language** Dzongkha (official language, related to Tibetan), Tshangla, Khengkha (Tibetan languages), Nepali

**Beliefs** Drukpa Kagyu Buddhism (official religion of Bhutan)

The kingdom of Bhutan remained in self-imposed isolation for centuries. Only in 1959 did it forge its first diplomatic ties, and its doors were not opened to visitors until 1974. Bhutan's society is made up of four broad ethnic groups: the Ngalop and the Sharchop (together known as the Drukpas), several indigenous aboriginal peoples, and the ethnic Nepalese. In 1987, the Bhutanese government introduced a code of traditional Drukpa dress and etiquette called *Driglam Namzhag*. The edict required men to wear the *gho*, a knee-length robe, and women the *kira*, an ankle-length dress, when outside administration premises and government offices, at schools, in monasteries, or at public functions. People who did not abide by these orders were fined or imprisoned. Many aspects of Bhutanese life are guided by the ethics of its official religion, Drukpa Kagyu Buddhism. The government has long supported the state religion through subsidies given to shrines and monasteries. Bhutanese artistic traditions are steeped in Buddhism and almost all art, music, and dance represents the eternal struggle between good and evil.

BHUTANESE WOMAN

*Most Bhutanese people have features similar to Tibetans and related peoples to the north and east.*

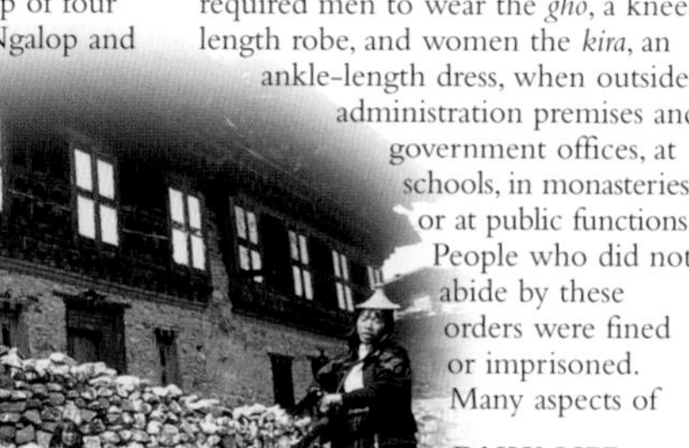

DAILY LIFE

*Bhutanese clothes are often made from yak's wool. Here, two village women wash and wring out a yak's wool dress.*

Fact

### Into the internet age

Not until 1998 was Bhutan's official ban on television loosened. The football-mad kingdom demanded, and was permitted, to watch the World Cup final on a giant screen erected temporarily in the national stadium. In 1999, Bhutan became the last nation in the world to begin broadcasting television and to join the worldwide web. Today, Bhutanese monks choose videos to rent (below).

HIMALAYAS

## Lepchas

**Location** Sikkim and West Bengal (India), Bhutan

**Population** 50,000

**Language** Rong-ke, Rong, or Lepcha (Sino-Tibetan language with its own script), Nepali and Sikkim as second languages

**Beliefs** Animism, Buddhism, Christianity

Lepchas, thought to be the earliest inhabitants of the mountainous region of Sikkim, call themselves *Rongkup*, meaning "dwellers of a rocky land", or "ravine folk". A *li* (Lepcha house) is made of bamboo and rests 1.2–1.5m (4–5ft) above ground on stilts.

Lepchas are mainly agriculturalists, cultivating rice, oranges, and cardamom. They carry out weaving and basketry and have a rich heritage of folk dancing, songs, and tales. They practise endogamy, only marrying with each other.

LEPCHA LUTE

*In the Darjeeling district of West Bengal, India, a Lepcha man uses a bow to play his four-stringed lute.*

PAKISTAN

## Muhajirs

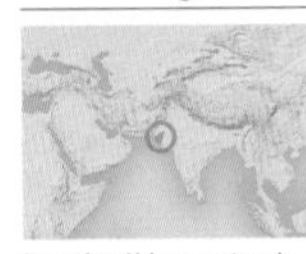

**Location** Urban centres in Sindh (Pakistan)

**Population** 12 million (approximately 8% of Pakistan's population, most of whom reside in Karachi and Hyderabad)
**Language** Urdu (related to Hindi; uses Persian script)
**Beliefs** Islam

The term Muhajir is the Urdu and Arabic word for migrant and refers traditionally to the Islamic exodus that marked the beginning of the Islamic calendar. After Partition, in 1947, the term was attributed to the highly educated group of mostly Urdu-speaking Muslims who emigrated mainly from Uttar Pradesh and Bihar (northern India) to the newly formed nation of Pakistan.

In the mid-1980s, the Muhajir identity took on greater political importance with the establishment of the Muhajir Quami Movement (MQM), a political party that has demanded the recognition of Muhajirs as a separate and distinct nationality and ethnic identity.

**KARACHI WEDDING**
*Muhajir wedding ceremonies can last several days. On Shadi, the wedding day proper, the bride and groom are adorned in red and gold.*

PAKISTAN AND INDIA

## Sindhis

**Location** Sindh (Pakistan), India

**Population** 18 million (a small proportion of which are Hindu, Sikh, and Sufi Sindhis, predominantly residing in India)
**Language** Sindhi
**Beliefs** Islam, Hinduism, Sikhism, Sufism

The river Sindh (Indus) is one of the most important sources of livelihood for the Sindhis, who dwell, for the most part, in rural regions and who are largey agriculturalists. Despite the Sindh province's boundaries, with the desert to the east, mountains to the west, the Arabian sea to the south, and the Indus river to the north, it has been subject to conquests. Persian and Arab Muslim, as well as Buddhist and Hindu invaders have influenced the culture, economy, and linguistic features of Sindhi people, who are known today principally for their beautiful pottery, lacquer and leather work, and embroidery and textile design.

**SINDHI WOMAN**
*This young woman's colourful, patterned clothing is typical of Sindhi textiles.*

Sindhis also have their own rich cultural traditions, including Sufism, a religious philosophy associated with Islamic mysticism. Brought to Sindh by Arab Muslim conquerors, Sufism is portrayed in a large body of Sindhi poetry and literature.

PAKISTAN AND INDIA

## Punjabis

**Location** Punjab (India and Pakistan)

**Population** 80 million
**Language** Panjabi (related to Rajasthani)
**Beliefs** Islam (in the Province of Punjab, Pakistan), Sikhism (in the State of Punjab, India); also Hinduism

Punjab means "the land of the five rivers", and these have contributed to the Punjab's status as the most affluent state in India. Farming is the occupation of most Punjabis, who supply the country with a large proportion of its wheat and rice.

Most Pakistani Punjabis are Muslim, and with 37 per cent of the population, Punjabis are the largest ethnic group in Pakistan. A large majority of Indian Punjabis follow Sikhism, a monotheistic religion that began as a reaction against both the caste system and the brahman domination of ritual.

The exuberance of Punjabi people is displayed in their folk dances, for which they are famous. One of the more renowned dances, the bhangra, traditionally celebrates the harvest and is associated with the festival of *Baisakhi*.

**SIKH LEADER**
*A Nihang (warrior) wearing a damala (full folded) turban. Iron rings protect the head from the blow of a blade.*

Punjabis have also lent their name to a clothing style found all over India called the Punjabi suit or *salwar kameez*. This is a two-piece outfit that consists of a long shirt with a drawstring and billowy trousers, worn with a flowing scarf called the *dupatta*.

**MAKING MUSIC**
*Punjabi dancers and musicians get ready to perform. The two-headed dhol (drum) is worn around the neck.*

CENTRAL INDIA

## Warli

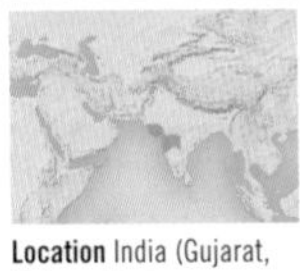

**Location** India (Gujarat, Maharashtra)

**Population** 550,000 (mainly residing in the Dahanu Taluka, or Thane District)
**Language** Warli (which some classify as a dialect of Gujarati or Bhili)
**Beliefs** Hinduism

The Warli (or Varli) are a Scheduled Tribal group (that is, one recognized by the Indian government, with specific rights and resources devoted to its development). Traditionally, the Warli were seminomadic, living in small groups under a headman. Today, they are primarily agriculturalists, growing crops such as rice and wheat. The Warli occasionally practise polygyny (marriage to more than one woman at the same time). Necklaces and toe-rings are worn by women as symbols of marriage.

Warli people are world famous for their detailed pictograph work (pictures or heiroglyphs expressing certain ideas). Called *chawk* in Warli, the pictographs are traditionally done by women and adorn the majority of Warli houses. They are drawn on the walls with a rice-powder paste and are painted in white with touches of brilliant saffron red. They depict Warli life and nature and include landscapes and pictures of the harvest, along with people, animals, and trees. They usually convey a time and space where humans live sustainably with nature.

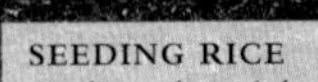

**SEEDING RICE**
*Warli people work hard, planting rice in a field that is irrigated by water taken from a community well.*

CENTRAL INDIA

## Gond

**Location** Broad area of central India

**Population** 8 million (from Uttar Pradesh to Andra Pradesh and Orissa to Maharashtra)
**Language** Gondi, Hindi, Marathi
**Beliefs** Hinduism, Sarna (a form of animism)

Gonds are numerically the most dominant tribal group in India. Most earn their livelihood through paddy and pulse farming, but many are also migrant workers in forestry, mining and quarrying, and on tea plantations.

The Gond people have a rich art and craft tradition, which includes *rangoli* (floor painting), woodcarving, pottery, and basketry. Their pictographs (pictures or hieroglyphs expressing ideas), which they paint in red and black on a white background, are made to celebrate festivals and are also used for aphrodisiac purposes.

**GOND SCULPTURE**
*This terracotta horse is a typical example of the craftsmanship of the Gond people.*

Gond men often wear a choker around their necks made up of two or three rows of cowrie shells stitched to a cloth band. These shells are believed to possess magical powers.

**WEDDING DANCE**
*Men with headdresses of horns and feathers simulate a bullfight at a Gond wedding.*

CENTRAL INDIA

## Oraons

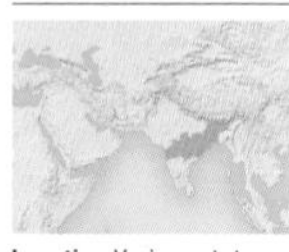

**Location** Various states in Central India

**Population** 3 million (in Bihar, Madhya Pradesh, Orissa, Chhattisgarh, Jharkhand)
**Language** Kurukh
**Beliefs** Hinduism, Christianity, Sarna (a form of animism)

Oraon (or Uraon) people call themselves *Kurukh*, which is also the name of their language. Traditionally, the Oraons depended on the forest for their ritual and economic livelihood, but in recent times they have become mainly settled agriculturalists. Small numbers have emigrated to parts of northeastern India, where they are employed as migrant labourers on tea estates.

Women traditionally wear a thick, white, cotton sari with borders stitched in detailed designs of red or purple thread, and tattoo their bodies with intricate symmetrical patterns around their chests, forearms, and ankles. Men wear a smaller thick cloth with similarly embroidered borders as a *lungi* or *dhoti* (*see* India, p436).

Oraon people have a vast range of folk songs, dances, and tales, as well as traditional musical instruments.

**BATHING THE KARMA**
*The Oraon venerate the* karma *("king") tree, which represents growth and regeneration. Here, women bathe its branches in milk.*

Both men and women participate in the dances, which are performed at seasonal and life-cycle events. Both consume rice beer and *arkhi*, an alcoholic drink made from the mahua flower, on special occasions.

The sacred *ojha* or *mati* look after the general social and ritual health of the Oraon community and cure diseases by appeasing evil spirits.

CENTRAL INDIA

## Bhils

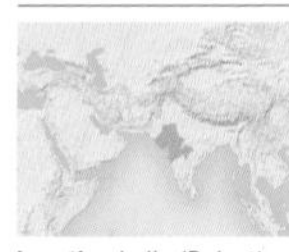

**Location** India (Rajasthan, Udaipur, Chittaurgarh)

**Population** 8 million (mainly in the southern districts of Rajasthan)
**Language** Bhili (related to Gujarati and Rajasthani), Hindi, Marathi
**Beliefs** Hinduism, Bhagat (Bhakti)

Bhil villages are under the leadership and control of a *gameti*, the village headman, who has jurisdiction and ultimate decision-making powers over local disputes. Although traditionally Bhils are monogamous, *gameti* often have more than one wife.

**STRIKING TURBAN**
*The colour of a turban, and the way it is tied, usually reflect the wearer's caste, religion, and place of origin.*

Bhil pictographs, in red over white-washed walls, are supposed to promote fertility, avert disease, propitiate the dead, and fulfil the demands of the ghost spirits.

**SOLID RAINCOAT**
*A Bhil woman uses a bamboo "coat" to shelter from the rain.*

CENTRAL INDIA

## Rajasthanis

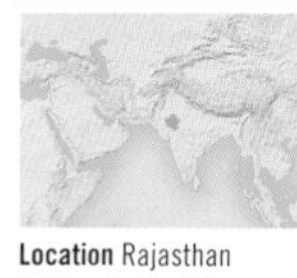

**Location** Rajasthan (NW central India)

**Population** 44 million
**Language** Rajasthani (consists of five main dialects: Marwari, Dhundhari, Mewari, Mewati, Hadauti)
**Beliefs** Hinduism, Jainism, Islam

Rajasthani clothing is some of the most colourful in India: bright red, yellow, green, and orange are the preferred colours. The traditional dress is influenced by the hot, desert climate. Bright tie-dye saris for women and turbans for men are popular. There are about 1,000 different styles and types of turbans, each denoting the class, caste, and region of the wearer. The use of henna originated here, as did the custom of applying *mehndi* (intricate body decorations in henna) by women on auspicious occasions.

Rajasthan's rich cultural tradition is captured in the vibrant and evocative music of the desert. There are many professional performers, who hand down their skills from generation to generation. Some of the more famous folk musical instruments include the *sarangi*, a haunting stringed instrument; the *jantar*, made from two gourds, four strings and 14 frets; the *ektaara*, with its single string; and the *morchang* and *ghoralio*, which resemble the Jewish harp.

History

### Rajput legacy

From the 7th to the 12th century, the Rajputs were one of the most influential of the Rajasthani groups. Although they never constituted more than a tenth of the population, they held leadership and governmental power in Rajasthan. They left behind many buildings of archaeological significance, including the City Palace, Jaipur (detail below).

**JEWELLERY**
*The Rajasthanis' bright clothing is complemented by jewellery, such as these colourful gold bangles.*

**HEADWEAR**
*Rajasthani women wear colourful, decorated headscarves, which can act as veils and reach to below the waist.*

**BRIGHT COLOURS**
*Rajasthanis favour bright colours, as typified by this man's turban. The turban is the most important part of a man's attire.*

**CAMEL FAIR**
*The annual Pushkar Camel Fair is the world's largest. It attracts thousands of people and camels.*

CENTRAL INDIA

## Gujjars

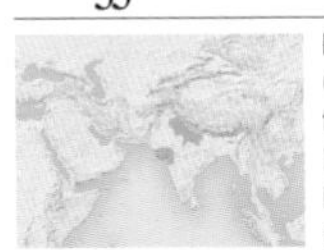

**Location** NW central India, northern Pakistan

**Population** 840,000 (scattered due to traditionally nomadic lifestyle)
**Language** Gujjari
**Beliefs** Predominantly Islam; also Hinduism

Gujjars probably originated in Central Asia before migrating to northwestern India. Today, an increasing number are settled agriculturalists. Others trek with their sheep, goats, cattle, and buffalo from the lowland plains, where they stay in winter, to the higher regions in search of green pastures in summer. They subsist by selling milk and its products and enjoy grazing rights with a formal permit in the forests.

The Gujjar's growing contact with the urban population and the government's continuing efforts to rehabilitate them have resulted in changes to their lifestyle.

Gujjar men wear a long shirt and trousers, with a Rajasthani-style turban or headcap called the *tamba*. Women usually wear a *salwar* or *churidar kameez* (long shirt and trousers), with a cap or scarf on their head.

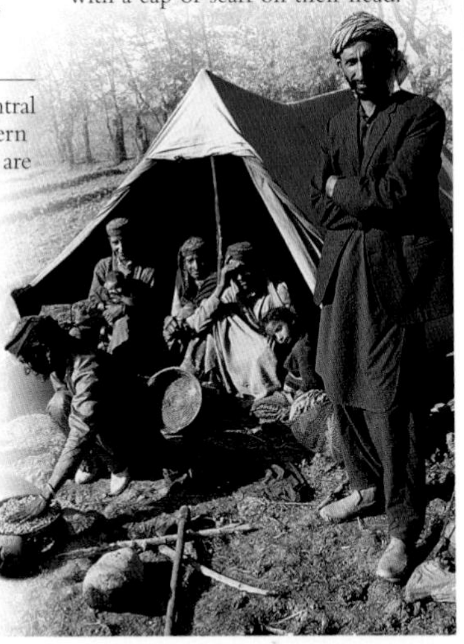

**FAMILY ON THE MOVE**
*The traditionally nomadic Gujjars keep sheep, goats, and cattle. When on migration, they live in tents called* tok.

NORTHEAST HILL TRACTS

## Nagas

**Location** Nagaland, Assam, Manipur (India)

**Population** 3 million
**Language** Naga (a Tibeto-Burman language of which there are many variations between tribes)
**Beliefs** Christianity, traditional animism

The term Naga is a generic name given to a cluster of 16 major and 12 minor tribal communities, each of which has a distinctive character and identity in terms of its traditions, language, and customs.

Traditionally, many Nagas practised head-hunting, believing that the enemy's *yaha* (animated soul) could only be freed through beheading. The *mio* (spiritual soul) was also thought to reside in the head, and the Nagas would carry home the heads of their dead comrades in an attempt to acquire prosperity and good fortune.

Today, Naga people are a highly educated group. While some still engage in agriculture, others are employed in service-oriented professions. They are excellent craftsmen, too, and their artistic talent is expressed in various ceremonial dresses, shawls, woodcarvings, engravings, and bamboo and cane basketry. Each Naga community has a unique colourful shawl design woven skilfully by women. Shawls are embroidered with raised designs of stylized and symbolic horses, butterflies, elephants, and flowers in white, red, yellow, and green on black backgrounds.

Nagas observe and participate in festivals and dances related to harvest, fertility, and other occasions.

**BASKET CARRIER**
*A Naga woman transports her harvest in a basket suspended from her head.*

**GATE-PULLING CEREMONY**
*The most important part of the Angami tribe's 10-day Sekrenyi festival is the gate-pulling ceremony, performed on the eighth day.*

INDIA AND BANGLADESH

## Bengalis

**Location** West Bengal (India), Bangladesh

**Population** 60 million in West Bengal, 120 million in Bangladesh
**Language** Bengali
**Beliefs** Mainly Hinduism in West Bengal, mainly Islam in Bangladesh

Home to political activists, writers, artists, and social reformers, Bengal has long been considered the cultural centre of India. In 1949, Bengal split into Muslim-dominated East Bengal (East Pakistan) and the principally Hindu West Bengal. In 1971, East Bengal won its independence from Pakistan to become Bangladesh.

Bengali people are known across the subcontinent and beyond for the rich, milk-based sweets they make, which include *misthi dhoi* (sweetened curd) and *rasgulla* (sweet cream-cheese balls flavoured with rose-water).

**FISH SELLER**
*A man proudly holds his fish, for sale at the market. Rice and fish (usually curried or fried) are the mainstays of Bengali cuisine.*

Bengali folk heritage is enriched by its ancient animist, Hindu, Buddhist, and Muslim traditions. Folk theatre is common at the village level and usually takes place during harvest time or at *melas* (village fairs).

**FLOODING**
*A woman of West Bengal wades through one of the region's frequent devastating floods to collect fresh water.*

INDIA AND BANGLADESH

## Santals

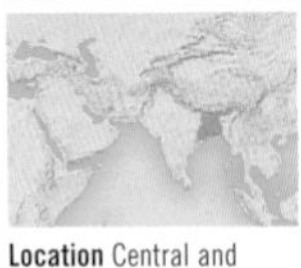

**Location** Central and eastern India, Bangladesh

**Population** 4.5 million (the third largest tribal community in India)
**Language** Santali (an Austro-Asiatic language)
**Beliefs** Hinduism, animism

The Santals were among the first of India's tribal communitues to engage in what could be described as a peasant war in the tribal India of 1855–1856. The uprising was directed against moneylenders and middlemen and was waged in defence of the native people's rights to the land.

The Santals were traditionally hunter-gatherers, but their primary occupation is now settled agriculture. The village council is headed by a person known as a *manjhi*, who is assisted by other council members in looking after village affairs.

The Santal's wooden bridal palanquins, used by people of various religious and ethnic groups across India, bear carvings of riders on horseback, elephants, and men in fishing boats. Carved birds and animals, hunting and dancing scenes, and geometric forms are also used to ornament the walls of the Santals' homes. Sadly, however, the rich artistic traditions of the Santal are fast disappearing.

**THE ART OF WEAVING**
*A woman weaves palm fronds to make items such as mats. Some Santals weave to supplement their income from agriculture.*

NORTHEAST HILL TRACTS

## Khasis

**Location** Meghalaya and Assam

**Population** 650,000 (in the Khasi and Jainta hills)
**Language** Khasi (a Mon-Khmer language)
**Beliefs** Christianity, traditional Khasi animism and ancestor worship

The Khasi people are divided into a number of clans, bound together by strict ties of religion, ancestor worship, and funeral rites. The Khasis are what is known as exogamous: marriage within the clan is prohibited.

Unlike most tribal and caste groups within the Indian subcontinent, descent among the Khasi people is matrilineal: children take on their mother's clan name, and the *ka khadduh* (youngest daughter) is the child who inherits the ancestral home. She is also the one responsible for looking after her parents in their old age. Khasi women have always played an important role in economic and social structures: they run family businesses, dominate their households, and make all key family decisions. Khasis now only tend to wear traditional clothing for special ceremonies due to Western influence.

**TOBACCO TRADERS**
*Having collected their tobacco leaves, Khasi women are ready to trade. Women play an important role in most aspects of Khasi life.*

History

### Nongkrem dance

The Khasis have held an annual festival since ancient times. Dancers perform the Nongkrem dance to thank the gods for a plentiful harvest, and for peace, and prosperity. In earlier times, the festival was held in midsummer; now it is usually held every November.

**DANCING IN THE RAIN**
*Traditional Khasi dancers are assigned an umbrella carrier each to shade them and protect them from rain.*

**TEA BASKET**
*A conical basket, made of woven bamboo, is used to collect tea leaves.*

**ARCHERY COMPETITION**
*The city of Shillong hosts a traditional archery competition on a daily basis. Some of the many spectators place bets on the contest.*

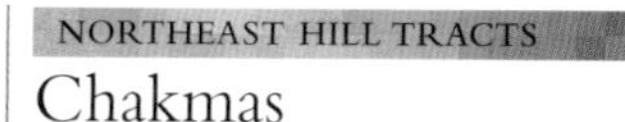

NORTHEAST HILL TRACTS

## Chakmas

**Location** Tripura, Mizoram, Assam (India), Bangladesh

**Population** 300,000 (mainly in hilly areas)
**Language** Chakma (of which there are several dialects)
**Beliefs** Mostly Buddhism; also Hinduism

Slash-and-burn cultivation is the traditional occupation of Chakma people, but many have adopted plough cultivation, and poultry farming often provides a subsidiary source of income for Chakma families.

Their temporary houses, built with bamboo and thatch and supported on wooden logs, are called *monoghara*, a word that figures prominently in traditional Chakma folk songs.

**OPEN-FIRE COOKING**
*The Chakmas mainly cook over an open fire. Most are meat-eaters, although they do not eat beef. Rice is the staple food.*

NORTHEAST HILL TRACTS

## Meiteis

**Location** Manipur (India)

**Population** 1.2 million (the dominant ethnic group in Manipur)
**Language** Meiteilon (Manipuri)
**Beliefs** Hinduism, Christianity, Islam, traditional Meitei animism

The Meiteis were the original valley inhabitants of Manipur and are sometimes referred to as "Manipuris". The Meitei people are made up of seven tribes or clans, which trace their history back to before the 1st century AD. The Meitei Pangals (Muslims) form a minority community in Manipur.

Meitei women have always enjoyed an equal status with men, and today they work in nearly every social and economic sphere of society. In particular, they control the markets and trade in vegetables and in traditional clothing. Weaving and basketry are traditional crafts and most households will have a loom.

The Meitei people are primarily agriculturalists, cultivating rice as their staple food. They also grow fruit such as oranges, pineapples, mangoes, lemons, and guavas. Fishing is popular as a hobby as well as a profession.

Additionally, the Meitei people are well-known for their sporting prowess: hockey and polo are traditional, and the Meitei martial art form, *thang ta*, has recently been recognized as one of the official forms of international martial arts.

Sadly, Manipur has one of the highest rates of HIV in India. This is thought to be linked to increasing levels of heroin addiction among the unemployed youth of Manipur.

**TRADITIONAL GAME**
Kang *is a traditional game in Manipur. Players throw a flat, buffalo-horn puck at fixed targets to see who can get closest.*

**TEXTILE MERCHANT**
*Meitei women have equal status with their men. They often control markets in clothing and textiles.*

PEOPLES

SOUTHERN INDIA

# Tamils

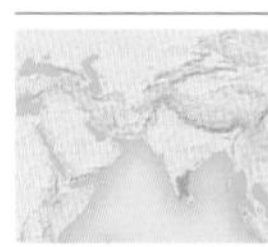

**Location** Tamil Nadu (India), Sri Lanka

**Population** 66 million (3.5 million of which are in Sri Lanka)
**Language** Tamil (a Dravidian language)
**Beliefs** Hinduism, Islam, Christianity, Jainism

Tamil Nadu in southern India is the traditional home of the Tamil people, but there is also a substantial number of Tamils living in Sri Lanka, as well as in communities in Malaysia, Mauritius, Singapore, and other countries. Tamil people are known by other Indians (and often themselves) as "Tamilians".

Tamil Nadu is sometimes referred to as the centre of Dravidian culture. Dravidians are said to have migrated to the area from northern India around 1500BC. Tamil legends claim that their history is about 10,000 years old and that the sea twice swallowed up their lands. The Tamil language is one of the Dravidian tongues, once spoken widely across the Indian subcontinent but now largely confined to its southern quarter.

**BRICKS**
*A Tamil woman takes handmade bricks out of their moulds to dry in the sun.*

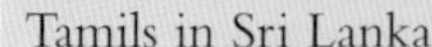

**HARVESTING TEA LEAVES**
*Traditionally dressed female workers collect tea leaves on a plantation in Tamil Nadu. Tea is an important cash crop in the region.*

**CLASSICAL DANCE**
Bharata Natyam *is a dance that originated in Tamil Nadu but is popular all over India.*

Issue

## Tamils in Sri Lanka

In the 14th century, a south Indian dynasty seized power in northern Sri Lanka and established a Tamil kingdom there. Since the 1950s, there have been long periods of ethnic violence between the minority Tamils and the dominant Sinhalese in Sri Lanka.

**HONOURING THE DEAD**
*These Tamils are taking part in a ceremony at a cemetery for people killed fighting for a separate Tamil state in Sri Lanka.*

The state of Tamil Nadu is well known for its plethora of Hindu temples and festivals, and the caste system (*see* p262) is still very strong among Tamils.

Although Tamil Nadu is one of the most urbanized states in India, most Tamil people continue to live in rural areas and to earn their livelihood through some kind of agricultural work. Agriculture is also the principal occupation of Tamils in Sri Lanka. Tamils also have a rich artistic culture. The popular Indian classical dance, *Bharata Natyam*, and the southern Indian classical music, *Carnatic*, are very important for the Tamil people. Tamils are also known for their handicraft traditions, which include hand-loomed silk, metalwork, leatherwork, and *kalamkari* – fabric painted by hand using natural dyes.

Tamil cuisine is traditional vegetarian and includes *dosas* (a lentil-flour fried pancake) and *idlis* (steamed rice dumplings), snacks that are favourites all across India.

**TANJORE PAINTING**
*An intricate painting from the town of Tanjore depicts the Hindu god Shiva and his bull, Nandi.*

**HARVEST FESTIVAL**
*As part of the harvest festival in Tamil Nadu, bundles of money are tied around the neck or horns of a bull. The people then attempt to snatch the money.*

INDIAN OCEAN HARVEST
*Sinhalese men and boys work together to pull in a huge fishing net. Fish forms an important part of the Sinhalese diet.*

SRI LANKA

## Sinhalese

**Location** All of Sri Lanka apart from the far north

**Population** 14.6 million
**Language** Sinhala (an Indo-European language and the national language of Sri Lanka)
**Beliefs** Theravada Buddhism, Hinduism, Christianity

Legend has it that Sinhalese people are descended from a lion (*sinha*), whose blood (*le*) continues to flow through their veins. Despite the fact that the Sinhalese mainly follow Theravada Buddhism (the entire population embraced the religion in the 3rd century BC), Sinhalese culture has been heavily influenced by the caste system of the Hindus. There is no sizable priestly or warrior caste, so the highest caste group is the *goviya* (farmer). The Sinhalese social system is further divided between the "hill-country" (Kandyan) and "low-country" branches, due to the legacy left when the Kandyan hill kingdom remained independent while the Dutch and Portuguese controlled the coastal areas.

Sinhalese dancing is renowned and is similar to South Indian dance forms but more acrobatic. The Sinhalese are also famous for their woodcarving, weaving, pottery, and metalwork.

Sinhalese people cherish a belief in their common ancestry. Tensions with Sri Lanka's Tamil community serve to unify the Sinhalese, as does the use of Sinhala as the national language.

TRADITIONAL DANCE
*The Kandyan dance is the national dance of Sri Lanka. The performer wears decorative silverware, which rattles during the dance.*

SOUTHERN INDIA

## Telugus

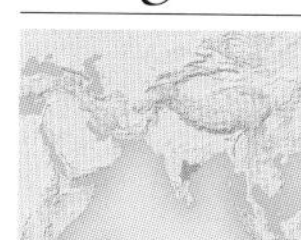

**Location** Andhra Pradesh (India's fifth largest state)

**Population** 70 million
**Language** Telugu (a Dravidian language)
**Beliefs** Predominantly Islam; also Buddhism (dating back to the 3rd century BC), Hinduism

Like other Indian peoples, Telugus celebrate many festivals. March is the time of *Ugadi*, the Telugu New Year, which people celebrate through *puja* (worship) and by making sweets and other special food. During the month of Ramadan, the men congregate around clay ovens known as *bhattis*, taking turns to vigorously pound and prepare the *haleem* (a mixture of meat and wheat) for the evening meal.

Telugus are largely agriculturists. They are also well-known for their exquisite craftsmanship, particularly the traditional *bidriware*, in which silver is inlayed onto black metal.

ANDAMAN ISLANDS

## Jarawa

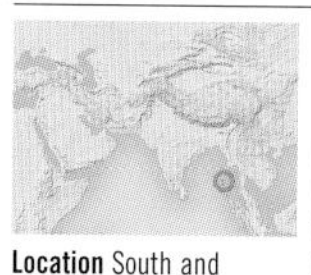

**Location** South and Middle Andaman Islands

**Population** 250–400, all within a 750sq km (290sq mile) reserve
**Language** Jarawa (a language unrelated to any other)
**Beliefs** Animist

The Jarawa people are hunter-gatherers who live a seminomadic existence. Hunting, fishing, and collecting honey are viewed as jobs for men only, and it is also the men who handle bows, arrows, and spears. Women do small-scale fishing with baskets and help collect roots and tubers, yet seem to enjoy a status at least equal to men.

Their diet consists mainly of pork, turtle meat, eggs, fish, roots, tubers, and honey. They also eat rice and banana, supplied by groups who are trying to establish contact with them.

Issue

### Foreign Settlement

Attempts to resettle many minority tribal groups resulted in their virtual extinction, so the Government of India decided against resettling the Jarawa people. In 2002, in an effort to protect the Jarawa and their way of life, the Indian Supreme Court ordered the closure of a trunk road through their reserve, banned logging and poaching in the area, and moved Indian settlements from there to avoid further non-Jarawa contact.

SOUTHERN INDIA

## Syrian Christians

**Location** Kerala (southern India)

**Population** 5 million (approximately 18% of Kerala's population)
**Language** Malayalam
**Beliefs** Orthodox Syrian Christianity (allegedly the first church established outside Palestine)

Christianity in India predates that in most of Europe, thanks to the Syrian Christians, who trace their origin to the advent of St Thomas the Apostle in AD52. With strong connections to their place of origin, they have kept their identity for nearly 2,000 years.

Syrian Christians form one of the most affluent communities in Kerala. They own more land, are better educated, and have lower fertility rates than any other group.

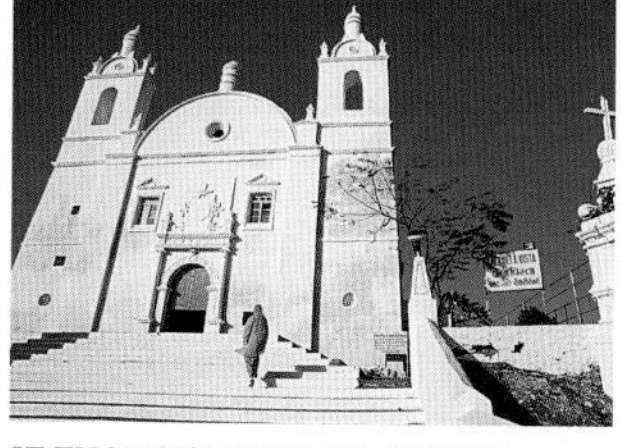

ST THOMAS'S CHURCH, KERALA

SRI LANKA

## Veddha

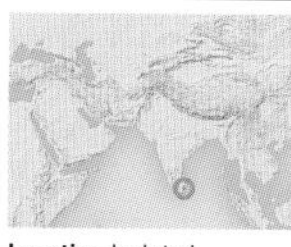

**Location** Isolated communities in Sri Lanka

**Population** 380
**Language** Veddha, Sinhala
**Beliefs** Traditional Veddha religion (centred around a cult of ancestor spirits called "Ne Yaku"), Buddhism

Veddha people, Sri Lanka's aboriginal inhabitants, call themselves *wanniya-laeto* (forest dwellers). Their Sinhalese neighbours call them Veddha, which comes from *vyadha* – a Sanskrit word meaning hunter with bow and arrow.

The Veddha are hunter-gatherers and have a rich diet of venison, rabbit, turtle, lizard, monkey, and wild boar. Since the creation of National Parks, where hunting is banned, access to hunting grounds is restricted, so some Veddha have become farmers. Veddha people traditionally preserve meat in the hollow of a tree, enclosing it with clay. Dried meat preserve soaked in honey is a Veddha delicacy.

Until recently, Veddha men wore only a loincloth suspended with a string at the waist, while women dressed only in a piece of cloth that extended from navel to the knees. Nowadays, both men and women wear Sinhalese-style sarongs. In a traditional Veddha marriage, the bride tied a bark rope around the waist of her bridegroom to symbolize her acceptance of him as her partner.

SUBSISTENCE HUNTERS
*Veddha men, wearing sarongs, string a bow. Although the shotgun is now widely used, handmade bows and arrows are still important to the Veddha way of life.*

PEOPLES

**DRAGON BOATS**
*As a centre of trade and commerce, Hong Kong has long been a meeting place for the Far East's peoples. The dragon boat race is an event that is centuries old.*

# FAR EAST

From China's teeming cities to the virgin forests of Indonesia's myriad islands and the diversity of the tiny city-state of Singapore, the Far East presents a dazzling array of cultures and languages. Prehistoric migrations, the rise and fall of ancient kingdoms, movements of people in search of trade or conquest, and – more recently – colonialism and independent nation-building have all played their own part in making the region into the cultural mosaic we see today.

**THE REGION**
*The Far East includes China, Indochina (Thailand, Vietnam, Cambodia, Myanmar, and Laos), Pacific Asia (Japan, North and South Korea, and Taiwan), and Indo-Malaysia (Malaysia, Indonesia, and the Philippines).*

Although the modern political map of the Far East is composed of scarcely more than a dozen nations, a closer look at any one of them quickly reveals a rich tapestry of both indigenous and immigrant populations.

## MIGRATION

Among the most ancient movements of people through the Far East were migrations along the coast of East Asia into the Japanese islands, Korea, and the Russian Far East, and south into Southeast Asia. The Ainu people of Japan are thought to be descended from such early migrants, as are the short, dark-skinned people of the Philippine mountains. Population centres grew up in the fertile lowlands. In China, a distinct people known as the Han came to dominate through military conquest and cultural assimilation.

Over two millennia, traders and conquerors from India, Arabia and Europe brought successive waves of cultural, political, and religious influence to bear. Indian practices are still found today in places such as Bali, where the main religion is a form of Hinduism; Islam has had a profound effect on hundreds of millions of people; and Christianity has made inroads into societies such as the Bidayuh of Malaysia.

Seeking rare luxuries such as tea, silks, and spices, Portuguese and later Spanish, Dutch, and British travellers established colonies across the East, from Macau to Singapore.

**PAGODAS**
*Buddhism is a major religion in the Far East. Pagodas (sacred buildings), like these Chinese ones dating from the 9th century AD, are a feature of many Asian countries.*

**CHINESE ARTISTRY**
*This ceramic Bactrian camel dates from the Tang dynasty (AD618–960), a time of great artistic and technical innovation in China.*

## TRIBES AND NATIONS

European influence came to affect ancient ethnic identities: the diverse peoples of Southeast Asia had foreign colonial identities imposed on them. After 1945, Southeast Asia's colonies gained independence as modern nation states. The borders this created did not match ethnic boundaries and divided peoples with distinct geographies and cultures. The Thai peoples, for example, live in southern China, Myanmar, Laos, and Vietnam, as well as in Thailand. Disparate ethnicities have been subsumed under the flags of large nation states. Ethnic confrontations in Indonesia were sparked by government-sponsored moves from the crowded island of Java to more spacious Indonesian islands. The migrations took little account of the land rights of culturally distinct indigenous islanders.

Within China's borders, many minority groups have been granted limited autonomy since the 1950s. The government recognizes 55 of the country's hundreds of peoples. However, 90 per cent of Chinese people view themselves as Han.

## CHANGE

There was an explosion of urban populations in the Far East during the late 20th century. As dynamic economies developed, farmers left villages to find work in expanding cities. The distinct identities of many minorities, such as Taiwan's indigenous peoples (*see* p458), are rapidly changing as young people adopt the majority way of life. Rapid economic growth and industrialization is inevitably leading to the loss of ethnically unique ways of life in favour of a shared, urbanized lifestyle.

**Fact**

### East Timor

The republic of East Timor is one of the world's newest nations. It is made up of 20 peoples, some related to New Guineans, others related to peoples in Malaysia and Indonesia. All these peoples were ruled by Indonesia, but after a fierce struggle and vigorous campaigning (below), they won independence. They elected their first president and joined the United Nations in 2002.

**TRADITIONAL SAILING VESSEL**
*A traditional Chinese sailing "junk" sails into modern Hong Kong. Vessels like this were used as long ago as the 14th century.*

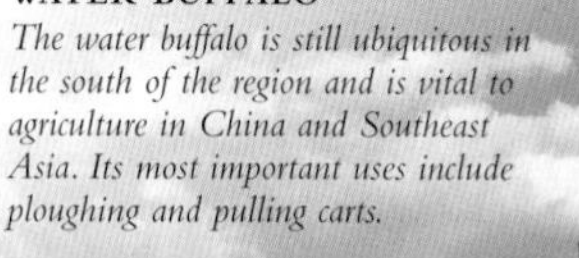

**WATER BUFFALO**
*The water buffalo is still ubiquitous in the south of the region and is vital to agriculture in China and Southeast Asia. Its most important uses include ploughing and pulling carts.*

CHINA

# Han

**Population** 900 million
**Language** Mandarin Chinese, various dialects (written Mandarin is based on the Beijing dialect)
**Beliefs** Confucianism, Taoism, Christianity, Islam, atheism
**Location** Mainland China, the island of Taiwan

The Han represent 15 per cent of the global population. The name Han is derived from the ancient dynasty that first united the northern Chinese and was used after 1644 to distinguish the *Guanhua* (Mandarin) speakers from their new Manchu rulers. It is often applied to all people who use the same written characters (but whose language is very different), but such "Han chauvinism" is frowned upon. The Mandarin class of bureaucrats were traditionally predominantly Han.

**LANTERN SELLER**
*This seller of paper lanterns cycles through the old capital of the Chinese empire, Xi'an.*

Confucianism, which governs social interaction, was founded by the Han 2,500 years ago, alongside Taoism (meaning "the Way"), a mystic religion that incorporates many aspects of traditional Han Chinese culture, including meditation, herbal remedies, and acupuncture. After the nationalist revolution of 1911 many Taoist shrines were destroyed. The communist regime, which came to power in 1949, then repressed religious worship, and atheism is now prevalent, particularly among urban Han people. The largest population of Han outside mainland China is the 22 million who live on the island of Taiwan (*see* p458). Those who migrated from the mainland after the ascendancy of the Manchu are viewed separately from the descendants of the 100,000 or so nationalists who set up government in Taipei in 1949 after communist victory on mainland China. With greater connections to the West, Taiwanese Han live in a more liberalized society.

Han cuisine is traditionally based on wheat noodles and mutton. Wheat is the regional staple in the north, as opposed to rice in the south of modern China. Steamed pancakes are also popular, as is bean-curd, a soft food made from soya-bean milk. There are similarities with Mongolian food, and Han cooking is different from the Cantonese and spicy Sichuan styles, which are the popular types of Chinese food in the West.

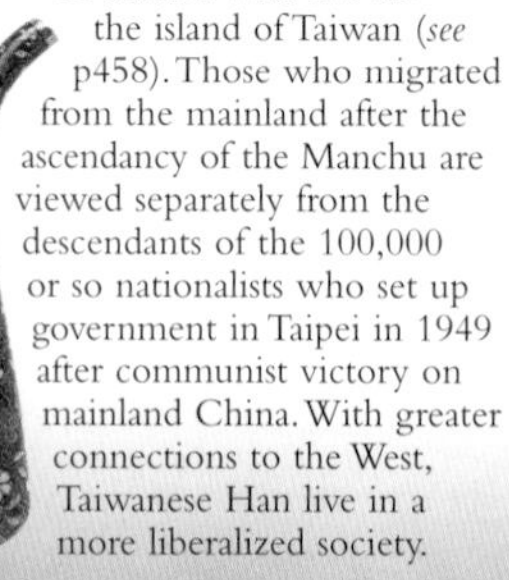

**NEW YEAR**
*Chinese communities all over the world celebrate New Year with songs and Dragon dances.*

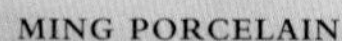

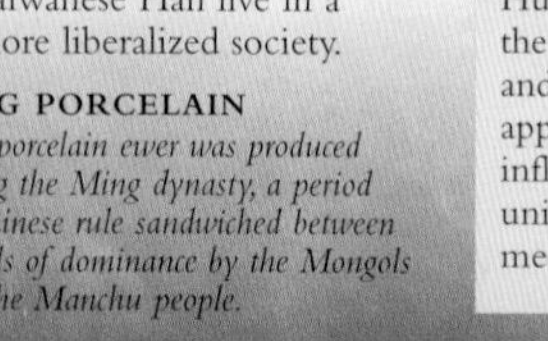

**MING PORCELAIN**
*This porcelain ewer was produced during the Ming dynasty, a period of Chinese rule sandwiched between periods of dominance by the Mongols and the Manchu people.*

Health

## Chinese medicine

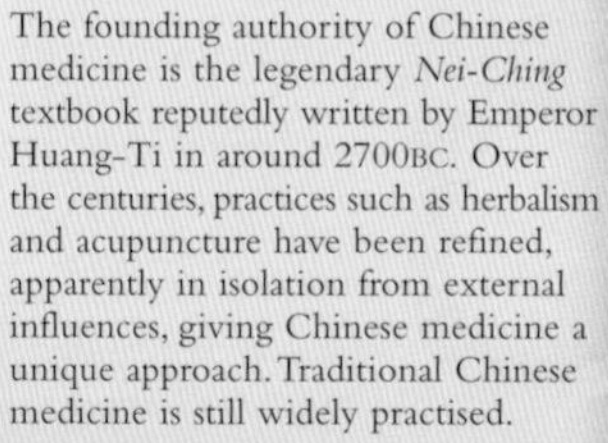

The founding authority of Chinese medicine is the legendary *Nei-Ching* textbook reputedly written by Emperor Huang-Ti in around 2700BC. Over the centuries, practices such as herbalism and acupuncture have been refined, apparently in isolation from external influences, giving Chinese medicine a unique approach. Traditional Chinese medicine is still widely practised.

**TAOIST TEMPLE**
*In Taoist temples, people leave offerings such as incense to gods. Taoism is a philosophical and religious tradition that permeates Chinese culture.*

CHINA

## Uighurs

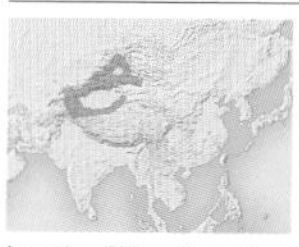

**Population** 7.2 million **Language** Uighur (a member of the Turkic branch of the Altaic language family, and closely related to Turkish) **Location** China, Kyrgyzstan, Uzbekistan, Kazakhstan **Beliefs** Islam; also Buddhism, Christianity

The Uighurs are descended from Turkic-speaking tribes of ancient Central Asia. Although they are now exclusively Muslims, the Uighurs' Central Asian geography gave them a complex religious history, including belief in Buddhism and Christianity.

Today, most Uighurs are farmers, living in the towns and villages at the foot of the Tianshan and Kunlun mountains. Farming in such an arid region has made them experts in irrigation: for centuries they have constructed remarkable underground channels called *karez* to bring water from the mountains to their fields. Using such resources, they have become famous for their agricultural products – notably sweet melons from Kumul, apples from Gulja, and grapes from the oasis of Turpan, often called Grape City. The Uighurs also work in the production of cotton and in modern petrochemical, mining, and manufacturing industries. Their exquisite rug-weaving and jade-carving form part of what are still traditional cottage industries.

Uighur food is very different from that in the rest of China. Uighurs eat large amounts of sheep meat – *plov* is a popular dish of rice, vegetables, and mutton. *Samsas* (baked parcels filled with ground lamb and spices) and kebabs are cooked by street vendors, while across China, Uighurs sell blocks of *matang* – nougat mixed with dried fruit and nuts. The Uighur staple is *nan* (baked flatbread).

**SILK SPINNER**
*A Uighur woman is spinning silk in the age-old way. Much of the ancient trade route, the Silk Road, runs through Uighur homeland.*

**STRIKING A DEAL**
*In the trading centre of Kashgar, two merchants negotiate in the customary way – head-to-head.*

CHINA

## Hui

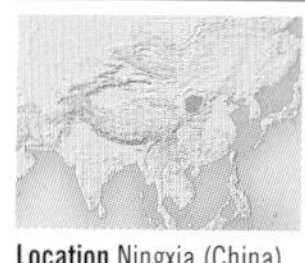

**Population** 8.6 million (most in the Ningxia Hui Autonomous Region) **Language** Chinese **Beliefs** Islam (Hui communities in cities across China are focused around a local mosque) **Location** Ningxia (China) and throughout China

The Hui are descendants of merchants, soldiers, and artisans from Persia and Central Asia, who settled in China from the 7th to the 13th century. Travelling vast distances along the Silk Road or sailing across the Indian Ocean to trade, the Hui's ancestors brought with them the Islamic beliefs that are their defining characteristic today. Islam is central to Hui society, and an *imam* (teacher) presides over weddings. Children are given a Hui name in addition to their Chinese one. Most Hui live in the countryside, raising sheep or farming. Many city dwellers are traders, especially in jewellery, jades, spices, and tea. Others are restaurateurs: their distinctive lamb hotpots and spiced noodle dishes are popular throughout China.

**HUI MAN**
*Most Hui are almost indistinguishable in appearance from the Han. The skullcap worn by Hui men is one distinguishing feature.*

**PRAYERS AT HUI MOSQUE**
*The Hui live throughout China. Men pray on Muhammad's birthday in the Hui mosque in Ili, a town in the Xinjiang Autonomous Region of China.*

CHINA

## Manchu

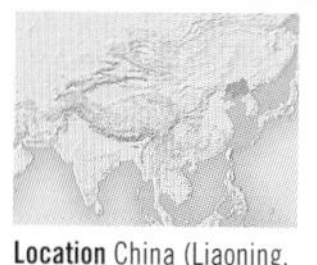

**Population** 9.9 million **Language** Mandarin Chinese; formerly Manchu (a Mongolian-related language) **Beliefs** Shamanism, although Manchus are now mostly atheist **Location** China (Liaoning, Jilin, Heilongjiang)

Modern Manchus are the descendants of ancient tribes in what is now northeastern China. The distinctive hairstyle of Manchu men – the head shaved but for a long pigtail at the back – was compulsory throughout China until 1911 and is a familiar symbol of China's imperial past.

Traditional Manchu culture was based on a rugged outdoor life of riding and hunting. Although the Manchus once practised shamanism, scarcely a trace remains of traditional Manchu customs and religion, and their language is nearly extinct.

Today, most Manchu are farmers: the soil of their temperate homeland is suited to soybeans, maize, sorghum, corn, and tobacco. Apple orchards and silkworm-breeding provide many jobs, as does the gathering of ginseng and wild fungi.

History

### Imperial past

In the 17th century, Manchu tribes conquered the Ming dynasty and established the Manchu Qing, China's last dynasty. The Qing dynasty at first closed its homeland (Manchuria) to keep their Manchu culture pure, but the Manchu slowly adopted the Chinese way of life. The boy below is dressed as the emperor for tourists.

**ROYAL TOMBS**
*Shenyang in northeastern China still boasts the tombs of Manchu Qing-dynasty rulers (below) and a palace of the Manchu kings.*

CHINA

# Tibetans

**Location** China (Tibet, Qinghai, Sichuan) and adjacent countries

**Population** 6.5 million, (4.5 million in Tibet, the rest in neighbouring countries)
**Language** Tibetan (various levels of speech reflect the complex, traditional social hierarchy)
**Beliefs** Tibetan Buddhism

Until 1950, Tibet communicated relatively little with the rest of the world. It's cultural identity was most strongly defined by the Tibetan language and Tibetan Buddhism. However, in 1950, China invaded and occupied Tibet, later forming the "Tibet Autonomous Region". Tibetan Buddhist religious structures disappeared during this period, but many have now been reestablished.

The Tibetans occupy the "roof of the world", Asia's highest plateau. Most are nomadic pastoralists or settled farmers, their practices and culture well adapted to the harsh environment. The staple diet is *zamba* (roasted barley); mutton; and meat, milk, and cheese from the yak. Tibetans drink tea mixed with yak butter and salt in a wooden tub; they also enjoy *chang*, a barley wine.

The Tibetan calendar follows the phases of the moon. It observes a cycle of 60 years, each 354 days long. The years are designated with reference to five natural elements, to the forces of yin and yang, and to 12 symbolic animals. *Losar* (the Tibetan New Year) is the most important date in the calendar, and involves an elaborate festival.

The traditional Tibetan method of parting with the dead is called *jator*, "feeding the birds". It involves taking the body to the mountainside, where it is laid out to be consumed by vultures.

History

## Enduring Buddhist faith

After some relaxation of Chinese rule since the mid-1980s, Tibetans can once again express their Buddhist beliefs in private and in public. One way of doing this is to sound the *dung-chen*, a horn, 3m (10ft) long, used in Buddhist rituals. Its bass notes are said to resemble the call of an elephant.

**TENT LIFE**
*While settled Tibetans live in stone houses, herders shelter in long, rectangular tents made from yak hair. This colourful tent, situated in the Qinghai Province, is for use in summer.*

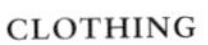

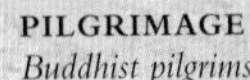

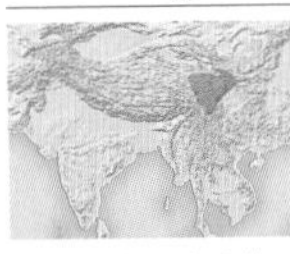

**CLOTHING**
*Both Tibetan men and women typically wear a felt or woollen hat to help protect against the cold. Women often carry small children in colourful slings arranged around the shoulders and upper back.*

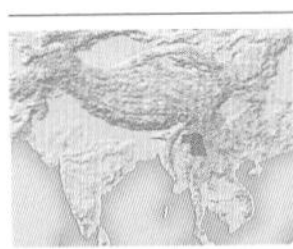

**PILGRIMAGE**
*Buddhist pilgrims circle Mount Kailas in western Tibet. The mountain is sacred to both Buddhists and Hindus.*

CHINA

# Khampa

**Location** China, including southeastern Tibet

**Population** 1.5 million (in southeastern Tibet and in adjacent parts of the Chinese provinces of Qinghai and Sichuan)
**Language** Tibetan
**Beliefs** Buddhism

The Khampa are skilled horse riders, animal herders, and farmers inhabiting the remote, inhospitable mountains of the eastern Tibetan plateau. Although they are not an official national minority within China, the Khampa are generally taller and more rugged than their Tibetan and Chinese counterparts.

Khampa clothing is of animal skin, wool, leather, and silk, and is well-adapted to their rugged way of life in a cold environment. Khampa grow their black hair very long and wind silk thread around it. The hair is then worn up, with red tassels hanging down. Young women often wear several kilograms of gems, jades, and precious metal in their hair.

**XINGLONG FESTIVAL**
*These Tibetan Khampa women are attending the Xinglong Festival, held in Sichuan Sheng, China. Their traditional dress is typically colourful and flamboyant.*

CHINA AND INDOCHINA

# Dai

**Location** China, Myanmar, Laos, Thailand, Vietnam

**Population** 1.5 million
**Language** Dai (a Tai language), Chinese
**Beliefs** Therevada Buddhism (introduced from India 1,000 years ago), animism (worship of nature and ancestor spirits)

The term "Dai" covers a large number of linguistically similar groups living in northern Indochina and southern China (concentrated along the upper reaches of the Mekong). Most Dai work as wet-rice farmers. However, tourism is creating new economic opportunities for many Dai living in mountain forests, as these areas are full of exotic wildlife that attract visitors to the region.

The Dai are renowned for their cleanliness. They have a special reverence for water as a symbol of purity and may wash themselves many times daily. During the annual Water Splashing Festival, the Dai throw buckets of water over members of the opposite sex. The same festival includes dragon boat races that chase away the bad elements of the past year and welcome a good harvest in the coming one.

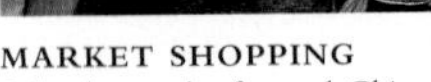

**MARKET SHOPPING**
*Like the people of several Chinese minorities in the Yunnan region, Dai women wear their distinctive ethnic dress every day.*

PEOPLES

CHINA

# Bai

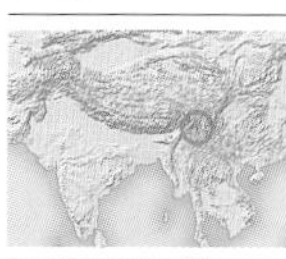

**Population** 1.6 million
**Language** Bai (part of the Yi group of Tibeto-Burman tongues, Bai only recently became a written language), Chinese
**Location** China (Yunnan, Sichuan, Guizhou)
**Beliefs** Buddhism, deity- and ancestor-worship

Indigenous to Yunnan, the Bai were assimilated into the cultural sphere of the Han Chinese (*see* p446) early in the dynasty's rule (206BC–AD220). The name Bai means "white", a reference to the Bai's veneration of that colour, and white materials feature in their clothing. Many Bai women wear their hair in a long braid wrapped in a headcloth, a style called "the phoenix bows its head". Most Bai retain an age-old lifestyle, farming rice and vegetables, and fishing. Today, some sell handicrafts to tourists.

The Bai are remarkable for their achievements in the natural sciences, especially astronomy, and in literature and the arts. They have their own operatic form, a mixture of folk music and dance called *chuichui*.

HARVESTING CORN
*An elderly Bai man carries corn cobs in his wicker backpack. Other crops grown by the Bai include rice, wheat, beans, oil-seed rape, sugar cane, tobacco, millet, and cotton.*

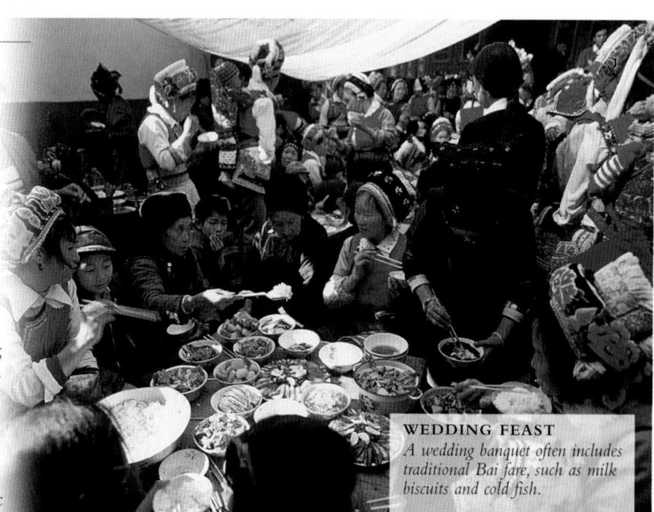

WEDDING FEAST
*A wedding banquet often includes traditional Bai fare, such as milk biscuits and cold fish.*

CHINA

# Zhuang

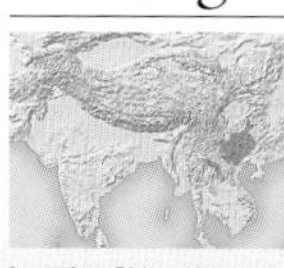

**Population** 15.5 million (most populous minority in China)
**Language** Zhuang (a Tai language related to Dong)
**Location** China (Guangxi and adjacent provinces)
**Beliefs** Animism (natural oddities may be venerated), polytheism, Buddhism

The Zhuang's predecessors were known to the Han Chinese (*see* p446) more than 2,000 years ago, and historical and archaeological evidence reveal a rich culture dating from antiquity. Ancient Zhuang frescoes still exist in southwestern Guangxi, the largest being over 100m (330ft) long and depicting thousands of figures. Hundreds of Zhuang bronze drums have also been excavated.

The Zhuang are Tai-speaking people and share their homeland with closely related Tai-speakers such as the Dong. Zhuang musical tradition is thriving, and songs are the main method by which cultural information has been handed down. Song contests are also popular. In a major annual festival, the Zhuang compete in "antiphonal singing" – a woman sings a line of song and a man must hastily improvise a suitable sung response.

Although the Zhuang mostly wear mainstream Chinese clothing, some elderly women still wear traditional garments, including embroidered collarless jackets and black headscarves.

Although the Zhuang have to a large extent adopted Han Chinese culture, animism and polytheistic beliefs remain. Ancestral spirits are offered fowls and glutinous rice at the Ghost Festival. During the Ox Soul Festival, the Zhuang offer extra grass and a basket of rice to entice back the souls of their water buffalo. The souls are believed to leave the animals' bodies during the harsh work of the spring ploughing season.

INTRICATE EMBROIDERY
*The Zhuang and their neighbours, the Dong, are known for their intricate and colourful embroidery work. This child's head wrapper was made by the Dong people.*

CORMORANT FISHING
*Fishermen in Guangxi Zhuang region use tame cormorants, leashed to the boats, to catch fish. When a fish is captured, the man forces the bird to disgorge its prey.*

CHINA AND INDOCHINA

## Yao (Mien)

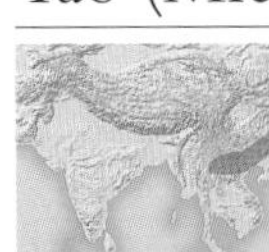

**Population** 2.2 million in China, 700,000 elsewhere
**Language** Iu Mien, Kim Mun, Hmong Njua (and several other languages in the Miao-Yao, or Hmong-Mien, family)
**Beliefs** Animism, ancestor worship
**Location** Guangxi and nearby provinces (China), Vietnam, Laos, Thailand

The Yao are geographically widespread and culturally diverse. Economically, they range from settled rice farmers in southern China to slash-and-burn cultivators and opium growers in neighbouring countries, where they are often called the Mien. The majority of Yao people are still mountain-dwelling farmers whose political system focuses on the home village.

Like the Zhuang people (*see* p449), the Yao are adept at antiphonal singing, using it in courtship. Singing also plays a role in farming: communal ploughing and sowing in neighbours' fields are directed by a man who leads many households in song to the sound of a drum strung by his waist.

A Yao legend tells of an ancient Chinese emperor who had long fought a rebel general and vowed to give his daughter's hand in marriage to whomever brought him the rebel's head. When a dog called Panhu appeared, carrying the rebel's head in his jaws, the emperor kept his word. The dog brought the emperor's daughter to the southern mountains and the Yao people were born of the union. Some Yao groups still honour this animal as a totem.

The Yao in northern Guangxi enjoy eating *youcha*, or "oily tea". Tea leaves are fried in oil, boiled in salted water, and cooked with rice or beans into a thick, savoury soup. Some Yao preserve small birds and other meats by sealing them in airtight pots with salt and riceflour. Left long enough, such salted preserves are considered a delicacy.

**YAO DRESS**
*Colourful and elaborate costumes are everyday clothes for many Yao people, such as this group in Thailand.*

**PADDY TERRACES**
*Like several other minority peoples in south China, the Yao live in hilly regions and cultivate rice in terraced paddies.*

CHINA AND INDOCHINA

## Hani (Akha)

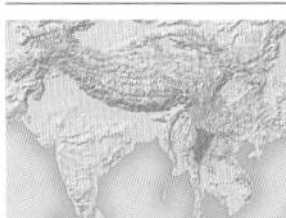

**Population** 1.3 million
**Language** Hani, or Akha, dialects (Akha is a member of the Sino-Tibetan language family)
**Beliefs** Ancestor worship, polytheism (worship of many gods)
**Location** Yunnan (China), Myanmar, Laos, Thailand

The remoteness of the Hani's homes has resulted in distinct local customs. In some areas, the parents of a couple who wish to marry will walk out together into the countryside in a custom called "treading the path". If they meet with no wild hares or wolves on their way, the engagement may go ahead. The Hani possess a strong storytelling tradition, but recent attempts to introduce an alphabet to their language have met with limited success.

Hani areas of China are slowly opening up to modern industries: the beautiful rainforests of Xishuangbanna especially are a popular destination for tourists. Hani sell their traditional handicrafts to supplement their slash-and-burn agriculture.

**WEARING ONE'S WEALTH**
*Hani women display their wealth by wearing headdresses decorated with silver coins and rings and with colourful feathers and beads.*



CHINA AND INDOCHINA

## Lisu

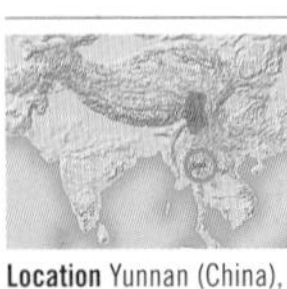

**Population** 600,000
**Language** Lisu
**Beliefs** Animistic polytheism (based around the worship of natural phenomena and the spirits thought to reside in all natural objects)
**Location** Yunnan (China), Myanmar, Laos, Thailand

The Lisu build distinctive bamboo houses, notable for the many wooden supporting piles, which allow a flat floor to be constructed on a hillside. A large firepit at the centre of each room is the focus of family activity.

Most Lisu people are farmers, their main crops rice, maize, and wheat. The opium poppy is a major economic crop for many of the Lisu in northern Thailand. The ability to hunt is greatly admired, and Lisu men tend to be adept at making powerful crossbows.

**FLOWERY LISU**
*Dress varies between Lisu subgroups, from the rough hemp clothes of the Black Lisu to the bright costumes of the Flowery Lisu (left).*

CHINA AND INDOCHINA

## Miao (Hmong)

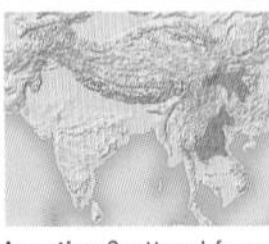

**Population** 7.4 million
**Language** Miao, or Hmong (a group of many dialects or languages of the Miao-Yao, or Hmong-Mien, family, including Hmong Njua, Kim Mun)
**Beliefs** Shamanism, ancestor worship
**Location** Scattered from Hubei to Hainan Island (China), Laos, Thailand

The Miao are known in Indochina as Meo, Hmong, and H'moong. With distant cultural roots and little social organization beyond the village, centuries of dispersed development have resulted in many dozens of distinct Miao identities, with varying clothing, dialects, and traditions.

Antiphonal singing, practised among many peoples of southwestern China and Indochina, forms the basis of courtship rituals in Miao villages. Mass courtship festivals take place, when the young women of a particular village will play host to young men from others. Couples whose singing attracts them to one another will exchange love tokens and go on to try to win the approval of their parents.

Agriculture, principally of rice and maize, is the main economic activity for all Miao people. Some also sell their beautiful embroidery, batik, and woven fabrics to supplement their income.

**EMBROIDERY**
*Miao girls are taught traditional handicraft skills from a very young age. The exquisite embroidery is used to decorate their clothing.*

**HEADDRESS OF HAIR**
*The long-horned Miao are just one of many distinctive Miao subgroups. They weave human hair around buffalo horns to create their headdresses.*

INDOCHINA

## Shan

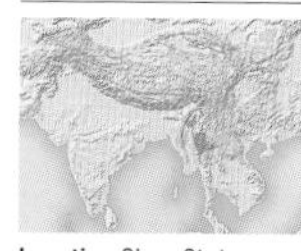

**Location** Shan State (northern Myanmar)

**Population** 3.9 million
**Language** Shan (a language of the Tai family; some Shan champion a greater Tai identity that includes all minority Tai-speaking peoples)
**Beliefs** Buddhism, animism

The Shan empire dominated most of Myanmar (Burma) from the 13th to the 16th century and the Shan are still Myanmar's largest minority. Today the Shan living under the rule of the Myanmar government are actively seeking complete autonomy.

Traditionally, Shan people live in fertile river valleys where they cultivate wet-rice and grow other crops such as corn, tobacco, and cotton. Buddhism is a key element of Shan identity, and the temple and its many rituals form the focal point of village life.

**WOMEN'S WORK**
*A group of Shan women enjoy a moment of leisure. Women traditionally do most of the cooking, cleaning and the caring for children. They also control the day-to-day finances.*

INDOCHINA

## Karen

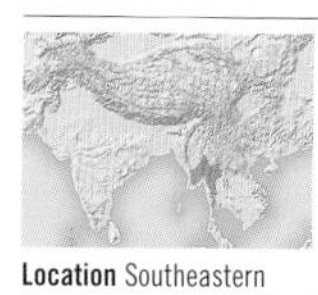

**Location** Southeastern Myanmar, western Thailand

**Population** 3.2 million
**Language** Karen (some 20 languages of the Karen branch of the Sino-Tibetan family), Burmese, Thai
**Beliefs** Theravada Buddhism, animism, Christianity

Karen people comprise 7 per cent of Myanmar's population, making them the second largest minority group after the Shan. Political tensions in Myanmar have caused many Karen people to flee to Thailand. Like other ethnic minorities, they seek autonomy from the Myanmar government and the right to return to their homelands.

**ORNAMENTS**
*Women in one group of Padaung Karen ornament themselves by pushing heavy earrings through their ear lobes, which gradually elongate the ears.*

Karen people are not one single people, but a collection of subgroups, the largest of which are the Sgaw and the Pwo, and the most well-known of which is the Padaung (*see* right). The groups share a common language, but there is a high degree of variation in regional dialects. Some authorities consider the dialects as including up to 20 separate languages. Originally all Karen practised animism, but many later adopted Buddhism, and there is now a growing minority of Christian converts.

Karen women are famed for their weaving skills. Traditional clothing often uses black, white, blue, and red. White represents purity. Unmarried women traditionally wear long white dresses. Married women, on the other hand, wear two-piece outfits consisting of a *longyi* (sarong sewn as a tube) – often in red – and a shirt, which is usually blue or black. The colour red symbolizes bravery; blue is said to denote loyalty. Although there is significant variation in embroidery and other decorations among Karen peoples, the basic form of the clothing remains distinct from that of non-Karen tribes. Many Karen now save their traditional attire to wear on special occasions, and opt to wear contemporary Western fashion for everyday activities.

**VILLAGE CHORES**
*The Padaung are an originally Mongolian tribe who have been assimilated into the Karen group. The women wear elaborate jewellery; however, it seems not to impede them in their daily chores.*

Issue

### Padaung ethnotourism

Certain women of the Padaung – a Karen tribe numbering only about 7,000 – wear brass rings around their legs, arms, and necks, which confer status on the wearer's family. The weight of the neck rings collapses a woman's collar bone, giving her a long-necked appearance. The custom was dying out until tourist interest sparked it off again. Some see this "ethnotourism" as exploitation.

INDOCHINA

## Burmese

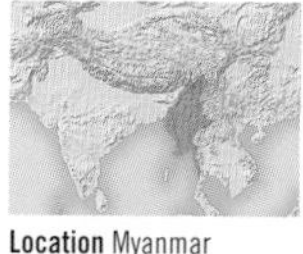

**Location** Myanmar (Burma), Bangladesh

**Population** 28.9 million, and 230,000 in Bangladesh
**Language** Burmese (a Sino-Tibetan language related to Akha and Lisu)
**Beliefs** Mainly Buddhism; also Christianity, Islam, animism

The Burmese people derive their name from their native homeland, Burma, a country that was renamed Myanmar in 1989 by the newly installed military government. The Burmese live alongside numerous minority ethnic groups, including the Shan, the Karen, the Mon, and the Rakhine. Each of these groups has its own cultural and linguistic identity and culinary traditions.

The Burmese have a rich artistic culture, which is closely related to their history and religion. Pagodas, temples, and palaces are full of carvings and paintings by renowned Burmese artists. Drama is one of the primary artistic forms, and nearly all celebrations are cause for a *pwe* (show), which recounts Buddhist legends in comedy, in dance, and with giant puppets. Burmese music, which often accompanies these performances, originated in Thailand.

Rice is the food staple for all Burmese people. The curried dishes are nearly all flavoured with *ngapi*, a dried and fermented shrimp paste. Sugar-cane juice is a popular drink.

Despite the fact that Myanmar is a country rich in natural resources, there is abject rural poverty. The nation's politics have been dominated by the ongoing struggle between opposing political parties and the military government, which means that Myanmar has been unable to achieve financial stability. The black market and border trade are often estimated to be one or two times the size of the official economy.

**TEASHOP**
*Burmese men relax with tea on a boat in the floating market at Inle lake. Much socializing in Myanmar (Burma) is done around the teashop table.*

**BURMESE BREAKFAST**
*A dish of rice-noodle soup called* mohinga *constitutes a typical first meal of the day for Burmese people.*

**WATER HIGHWAYS**
*The Irrawaddy, Myanmar's main waterway, provides a route through regions that have few roads.*

PEOPLES

INDOCHINA

## Thais

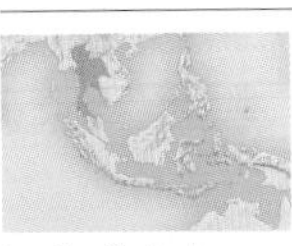

**Location** Thailand

**Population** 65.6 million
**Language** Thai, Chinese dialects, Malay
**Beliefs** Predominantly Theravada Buddhism (about 95%); also Islam, Hinduism, Sikhism, Christianity

Of all the states of Southeast Asia, Thailand alone avoided European colonial penetration. An inclination to "bend with the wind", or fit in with more powerful influences, has long characterized the country's foreign policy. During the period of the Cold War, the anticommunist military regime in Bangkok (now Krung Thep) developed close links with the US against Vietnam and China. This independence has allowed the Thai people to pursue modernization on their own terms and is a source of pride for them.

Buddhism is the official religion in Thailand and there are some 24,000 temples and 200,000 monks; all Thai boys are expected to spend some time in a monastery. The World Fellowship of Buddhists has its headquarters in Krung Thep.

Thais use two systems for telling the time: the 24-hour clock and the Thai 6-hour clock. The latter system is the one used in colloquial speech and divides the day into four parts.

**DURIAN FRUIT**
*The fleshy fruit of the durian is popular in the Far East. Durian fruit taste delicious; however, they are so smelly that they are banned from many public places.*

**BUDDHIST WORSHIPPER**
*A nun prays by the Buddha of Sukothai. About 95 per cent of Thais are Buddhists. Thousands are monks or nuns.*

INDOCHINA

## Khmer

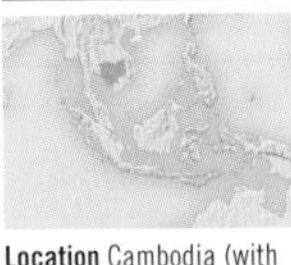

**Location** Cambodia (with expatriates elsewhere)

**Population** 16 million (95% of Cambodia's population; the Muslim Cham people is the only large ethnic minority)
**Language** Khmer
**Beliefs** Buddhism, Islam, folk religion

The Cambodians, or Khmer, have inhabited their present homeland since the beginning of recorded history. The brutal regime of the Cambodian communists, the *Khmers Rouges*, resulted in the deaths of possibly millions of Cambodians from brutal treatment, starvation, and disease from 1975–1978.

Khmer celebrate holidays and religious days with traditional dances. At Khmer New Year, an ox-cart race attracts thousands of spectators and their cows. Khmers value these animals greatly, and during the Cow New Year, farmers give their cattle gifts, bathe them, and put perfume on them.

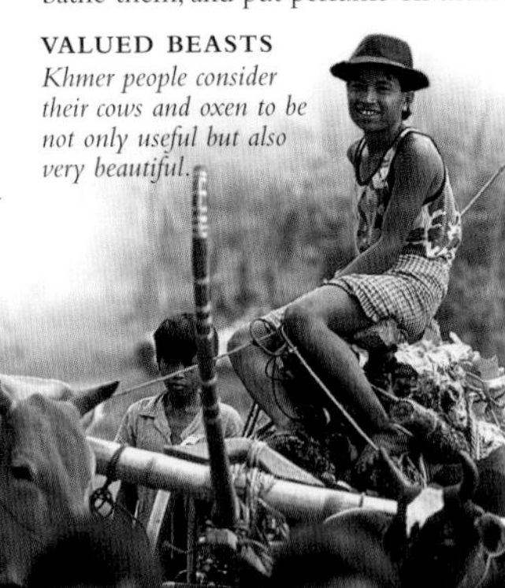

**VALUED BEASTS**
*Khmer people consider their cows and oxen to be not only useful but also very beautiful.*

History

### Angkor Wat

The stunning Angkor Wat pyramid temple in Cambodia is regarded as the supreme masterpiece of Khmer architecture. Built by Suryavarman II in 1113–1150, it is dedicated to the Hindu god Vishnu and boasts some of the world's most beautiful pieces of Khmer and Hindu art. The temple is one of over 100 on the Angkor site. In 1992, the United Nations declared Angkor a World Heritage Site.

INDOCHINA

## Laotians

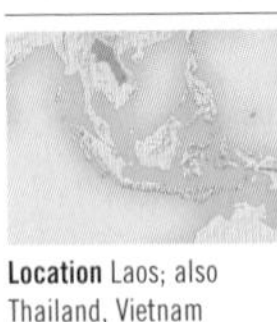

**Location** Laos; also Thailand, Vietnam

**Population** 6.7 million (only 50% of Laos' population are Lao Lum)
**Language** Lao, Tai dialects, Hmong (or Yao), Lao Theung dialects
**Beliefs** Theravada Buddhism, animism

The presence of ethnic minorities in the upland areas of Laos (notably the Tai, Kha, and Hmong, or Yao) means that, inside the country, ethnic Lao Lum, or Valley Lao, make up barely half of the total population. However, the term "Lao" includes a substantial number of people living outside Laos, mainly in northeastern Thailand.

In 1975, the Laotian Communist movement took control of the nation. It changed its name to Lao People's Revolutionary Party and abolished the monarchy. Since 1975, it has made education a high priority and aims to establish schools at the primary level in every village in the country. Many Buddhist *wats* (temples) serve as schools. In Laos, the Buddhist New Year festival is called *Pi Mai*. The festivities include the ceremonial washing away of the old year's sins with water, which often results in water fights that may continue for several days.

A Laotian delicacy is the meat from snakes such as the python. People in rural areas also cook and eat the flesh of wild animals such as squirrels, deer, civets, frogs and lizards.

**TRADE AND TRANSPORT HUB**
*Street vendors sell vegetables in Muang Xai, a market town lying at the crossroads of several major roads in northern Laos.*

**WILD FOOD**
*Like many people who live in rural areas, Lao eat wild animals like these frogs cooking on an open-air grill.*

**RIVER TRAFFIC**
*Traders meet at a floating market. Laos has about 9,000km (5,600 miles) of navigable rivers; boats are widely used.*

INDOCHINA

## Vietnamese

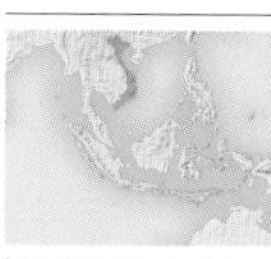

**Population** 83 million
**Language** Vietnamese, French, Chinese, Khmer
**Beliefs** Buddhism; also Taoism, Confucianism, Hoa Hao (a faith based on Buddhism), Cao Dai (a new faith), Islam, Christianity
**Location** Vietnam; also US, Canada, France

Vietnam has a great many traditional and religious holidays; however, none can compare to the New Year festival. This event, *Tet Nguyen-Dan*, or *Tet*, is held in late January or early February. Vietnamese believe that the first week of the New Year will determine the fortunes or misfortunes for the rest of the year. It is a time to pay debts, forgive others, and correct one's faults. It is auspicious if fruit tree branches bloom on the first morning of *Tet*: apricot blossoms are reputed to keep demons out of the homes at this time. Some families buy entire apricot trees and decorate them with greeting cards from well-wishers. Families gather together, making the traditional *banh chung*, a cake of sticky rice, and *mut* (fruit candies).

**NEW YEAR BLOOMS**
*Branches of plum blossom are among the decorations used during* Tet *(New Year).*

Vietnam has made great economic progress in recent years, changing from a significant importer of rice in the early 1980s to the world's second-largest rice exporter in 2003. Poverty has been reduced from over 70 per cent in the mid 1980s to about 37 per cent in 2000, and average per capita income has more than doubled over that period. The government has also made progress in promoting the country for tourism.

**LIFE ON THE RIVER**
*Many Vietnamese people live on houseboats. Some people spend almost all of their time on rivers.*

INDO-MALAYSIA

## Filipinos

**Population** 84 million
**Language** Tagalog, or Filipino (including Visaya dialect); Ilocano, Cebuano
**Beliefs** Roman Catholic Christianity, Sunni Islam, Protestant Christianity, Buddhism, animism
**Location** Philippines; also US, Middle East, Europe

The Filipinos are of Malay origin with some Chinese, US, Spanish, and Arab features. Most Filipinos live in the lowland areas of the Philippines and are Catholics. This shared religion has tended to promote a common Filipino culture; however, it alienates the large Muslim minority.

Some 4 million Filipinos work abroad, in around 140 countries. Regardless of background, the great majority are employed as domestic workers, although some are nurses, midwives, and medical technicians.

**CATHOLIC DEVOTION**
*Despite having Malay origins, the majority of Filipinos are devout Roman Catholics. The Philippines has the largest Catholic population of any Asian country.*

The fiesta is an important part of Filipino culture. Each city and *barrio* (quarter or village) has at least one festival of its own, usually on the feast of its patron saint, which means that there is always a fiesta under way somewhere in the country. The biggest and most elaborate of the festivals is Christmas, a season that is always celebrated with much pomp and pageantry.

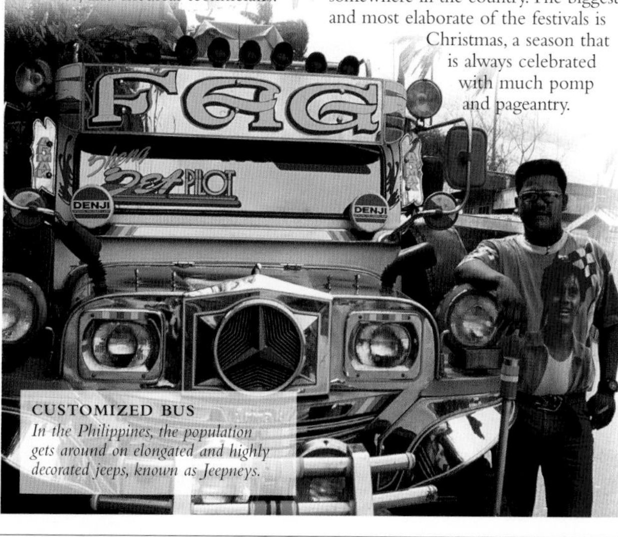


**CUSTOMIZED BUS**
*In the Philippines, the population gets around on elongated and highly decorated jeeps, known as Jeepneys.*

INDO-MALAYSIA

## Igorots

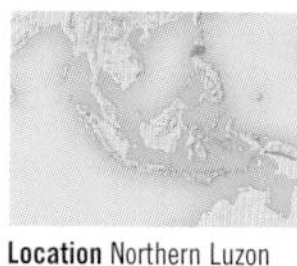

**Population** 600,000
**Language** Numerous related languages, with many dialects, of Malayo-Polynesian origin
**Beliefs** Animism with some Christianity
**Location** Northern Luzon (Philippines)

Igorot is the collective name for several tribes of people living in the Cordillera Central mountains of northern Luzon in the Philippines. They include the Bontoc, Ibaloi (Benguet), Ifugao, Isneg (Apayao), Kankanay, and Kalinga peoples. Each tribe speaks a different language, and some of them even have different dialects.

The remoteness of the Igorots' mountain homes means that they have remained largely beyond the influence of both the Spanish people who colonized Luzon, and the Muslim *Moros*. They have therefore retained many of their own beliefs and customs.

**JEWELLERY**
*Igorots make necklaces from coloured coconut shell. They are also known for their silverwork.*

Igorots are mainly agriculturalists and have adapted to their difficult terrain by building dramatic stonewalled rice terraces high on the steep mountainsides. Many terraces are falling into disrepair, however, as farmers leave to try their luck at gold mining or woodcarving. Ifugao also practise the slash-and-burn cultivation of sweet potatoes and maize. Their houses are in small groups at the edges of their fields.

Igorot customs include many feasts, and elaborate dances are also performed. Some Igorot villages (notably of the Bontoc and Kankanay tribes) have dormitories called *afong* or *dap-ay* for men only and *ulog* or *ebgan* for unmarried women only.

**PRESERVING THE DEAD**
*An Ifugao man holds the bones of his dead father that have been carefully wrapped in fabric. The Ifugao honour their deceased family members by keeping their bones.*

**HILL PEOPLE**
*An Ifugao woman stands on a hilltop in Banaue with the characteristic terraces on which rice is grown in the background.*

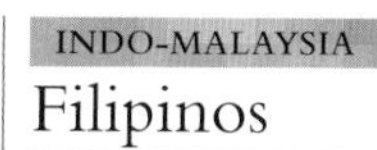

PEOPLES

INDO-MALAYSIA

# Mangyan

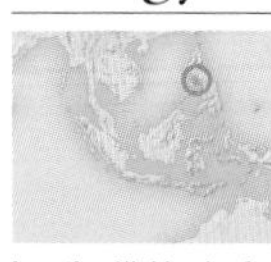

**Location** Highlands of Mindoro (Philippines)

**Population** 111,500
**Language** Buhid, Tabuhid, Hanunoo, (Buhid people), Iraya, Mindoro, Alangan, Tadyawan (Iraya people); unique syllabic writing system (Surat Mangyan)
**Beliefs** Animism

The Mangyan comprise six different ethnolinguistic groups: the Iraya, Alangan, and Tadyawan of the north, and the Hanunoo, Buhid, and Tabuhid peoples of the south. Many Mangyan are subsistence farmers, practising slash-and-burn cultivation. The Buhid are famous pot makers and the Iraya skilled weavers. Alangan people are known for chewing betel nut, while Tabuhid start smoking pipes from an early age.

The Mangyan are famous for their colourful clothes and beaded jewellery. Successive arrivals of colonists forced the Mangyan to retreat from the coastal regions into their highland home.

**WILD FOOD**
*In the moist highland forests of Mondoro in the Philippines, wild food may come in the form of a log covered in edible mushrooms.*

**MANGYAN CLOTHING**
*The cloth used to make Mangyan clothing is often woven from bark or from the leaves of palm trees.*

INDO-MALAYSIA

# Mindanao highlanders

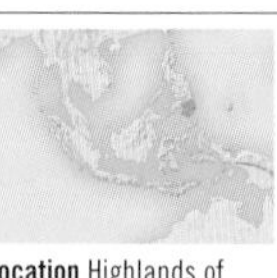

**Location** Highlands of Mindanao (Philippines)

**Population** 1 million
**Language** Several dialects of Manobo (a member of the Malayo-Polynesian family), only some of which are mutually intelligible
**Beliefs** Islam, animism

Mindanao island is home to many groups of indigenous Filipinos, including the Bagobo, Bukidnon, Bilaan, Mandaya, Manobo, Subanun, T'boli, and Tiruray peoples. Mindanao villages are governed by *datus*, village leaders who uphold the laws and settle disputes. Many groups have a warrior caste known as *bagani*. Mindanaos are mostly slash-and-burn agriculturalists, who move each year to a new plot of land within the forest. Their way of life is threatened by outsiders who purchase land and by logging of the forest.

**MANOBO**
*The Agusanon Manobo usually wear red. These highlanders predate Malay-related Filipinos.*

**HOUSE-WARMING**
*Stages in the Manobo's lives are accompanied by rituals. Here, Manobos sacrifice a chicken after building a new house.*

INDO-MALAYSIA

# Bajau

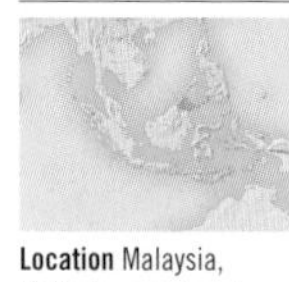

**Location** Malaysia, Philippines, Indonesia

**Population** Estimated 260,000 on Sabah (northern Borneo), in Malaysia; 47,000 in the Philippines
**Language** Bajau, or Moken
**Beliefs** Sunni Islam (Shafite branch)

The Bajau, or Sama Dilaut, peoples are divided into the East Coast Bajau (known as *Orang Laut*), often called "sea gypsies", and the West Coast Bajau (*Orang Sama*), who are known as the "cowboys of Sabah".

The East Coast Bajau of Sabah, the Philippines, and Indonesia are traditionally nomadic boat-dwellers. At one time their range was centred on Tawitawi island off the east coast of Sabah, but they are now more widespread. This is in part due to burgeoning trade in *trepang*, a sea slug that is considered a speciality by the Chinese, but the Bajau also exploit more than 200 different species of fish. These people used to be without land, except for access to freshwater and a small patch of land for burial use. Increasingly though, the sea gypsies are abandoning their seafaring lives for homes on the land.

The West Coast Bajau live around Kota Kinabalu in western Sabah. They are cattle-rearers and formidable horsemen. Just as the sea gypsies have a regatta to celebrate their heritage on the water, the *Orang Sama* hold an annual gymkhana, called *Tamu Besar,* to show off their skill at horsemanship. Both riders and horses are elaborately dressed in embroidered cloth. Tamu is also the name given to the weekly markets where the Bajau trade their produce and crafts.

The dominant religion of the Bajau is Sunni Islam, but some pre-Islamic traditions – such as giving thanks to Omboh Dilaut (the god of the sea) after a particularly good fishing catch – are still followed.

**PERFORMER**
*Some Bajau people present their unique culture to tourists in perfomances of dance.*

Fact

## Lepa

The traditional wooden boat of the Bajau is a *lepa*. In the Philippines, these boats resemble dugout canoes, but in Malaysia they are larger. Five or six people might live on each craft, and in shallow water, several lepas are linked together to form a living platform. The design of the sail allows the boat to travel almost straight into the wind.

**LIFE ON THE WATER**
*A traditional East Coast Bajau person may still live entirely at sea, returning for fresh water and trade.*

INDO-MALAYSIA

## Malays

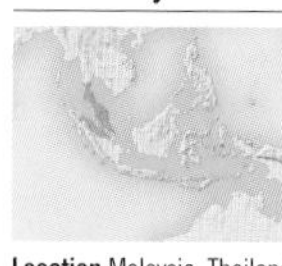

**Population** 25 million
**Language** Bahasa Melayu (Malay), Bahasa Indonesia, English, Chinese languages
**Beliefs** Principally Islam; also Christianity, Hinduism, and Buddhism
**Location** Malaysia, Thailand, Singapore, Indonesia

The name Malay can be used to describe all the numerous related groups in the Malay archipelago, however the term more usually refers to the subgroup that is native to the Malay peninsula, southern Thailand, Singapore, the Riau Islands, and the eastern part of Sumatra.

Traditional Malay customs and arts are practised in Kelantan – the state in northeastern Malaysia known as the "cradle of Malay culture". Here, Malays play *Rebana Ubi* (giant drums) during weddings and rice harvests. They practise *Mak Yong*, an ancient form of dance-theatre; *Wau*, the flying of kites after the rice harvest; *kertok*, coconut-husk percussion; *silat*, stylized sequences of self defence; and *gasing uri*, the spinning of giant tops.

Malay names are written with a person's given name first, plus *bin* (son of) or *binti* (daughter of) and then their father's given name (the *bin* and *binti* component is often omitted). The tracing of family trees is difficult.

**MALAY WOMAN**
*The Malay people are a widespread and influential people in the region. They power one of the "Tiger Economies" of eastern Asia.*

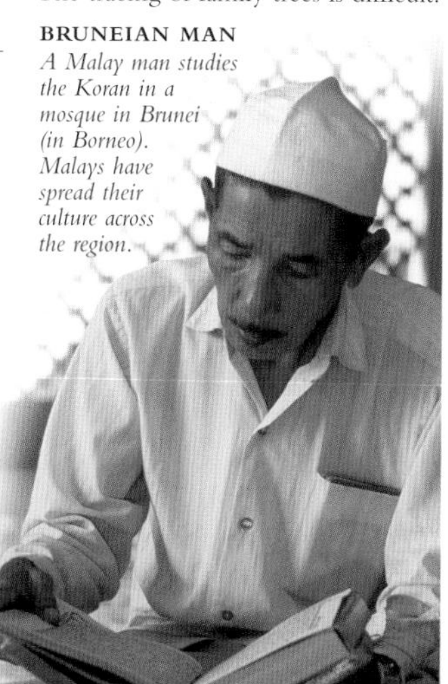

**BRUNEIAN MAN**
*A Malay man studies the Koran in a mosque in Brunei (in Borneo). Malays have spread their culture across the region.*

INDO-MALAYSIA

## Singaporeans

**Population** 4.2 million (Singapore's population density is one of the highest in the world)
**Language** Malay, English, Chinese, Tamil
**Beliefs** Buddhism, Islam, Christianity, Hinduism
**Location** Singapore

Singapore's strategic position at the tip of the Malay peninsula made it an economic success, under British supervision, in the 19th century. Its riches attracted immigrants from all over the region, leaving Singapore a cultural and ethnic mix. Three-quarters are Chinese, one-sixth Malay, and one-twentieth Indian in origin.

Since 1959, the government has stressed racial tolerance and discipline, and has introduced laws aimed at controlling personal behaviour. Chewing-gum was banned until recently and anyone caught littering or smoking in public is still liable to be fined. Government control of the citizens' life also extends as far as reproduction: cash bonuses are paid to families with more than one child. Efforts are also being made to stamp out the use of Singlish, the Singaporean pidgin English.

**HIGH-RISE LIVING**
*A Singaporean construction worker works on a high-rise apartment building. With land in high demand, many residents live in such housing.*

**OPERA HOUSE**
*In Singapore, a healthy economy and the fusion of cultures have helped create distinctive buildings such as this opera house.*

INDO-MALAYSIA

## Semang

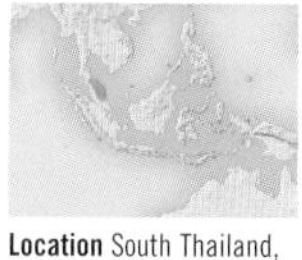

**Population** 2,500
**Language** Semang (several tribal dialects)
**Beliefs** Animism (involving communication with forest spirits through a shamanic trance); also Sunni Islam
**Location** South Thailand, highlands of Malaysia

The Semang are the oldest of the *Orang Asli*, the nomadic aboriginal peoples of Thailand and the Malay peninsula. The seven main tribal groups are the Kensiu, Jahai, Batek, Chewong, Kintak, Mendriq, and Lanoh, each of which has its own dialect. The tribes, which live in extended family groups of up to 30, have elected representatives to negotiate with the outside world but are otherwise equals.

Semang people are traditionally hunter-gatherers, foraging for fruit such as durian and wild yam, and hunting monkeys, birds, and other wildlife. They have long used a system of "silent trade" with the Malays; they deposit forest products near settlements and later go back to collect goods left in return.

They are renowned for their abhorrence of violence, but in recent times there have been clashes with Malays, who have encouraged them to settle. This pressure, combined with massive deforestation by logging companies, has led to many settling and turning to agriculture.

INDO-MALAYSIA

## Iban

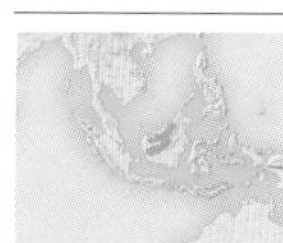

**Population** 510,000 (mostly in Sarawak in Malaysian Borneo)
**Language** Iban (a dialect of Malay, and a Malayo-Polynesian language)
**Beliefs** Christianity, animism
**Location** Borneo (Malaysia and Indonesia)

Although also known as "Sea Dayak", the Iban usually inhabit the fringes of the main rivers of Sabah. They traditionally live in hardwood and bamboo longhouses. Longhouses are complete settlements of between 4 and 50 family units (*see* p220). Tattoos were once a compulsory part of Iban culture for both sexes. The *uker degok*, a traditional neck tattoo for males, signified headhunting skill. Young men made a *bejalah*, a journey to gain life experience. Women did not travel, but had equal status. The Iban primarily grow rice in dry paddy fields called *ladang*. Rituals, such as the *ngajat* dance accompanied by gong music, focus on the success of the rice crop. Some Iban are now involved with exporting rubber and timber, or fishing.

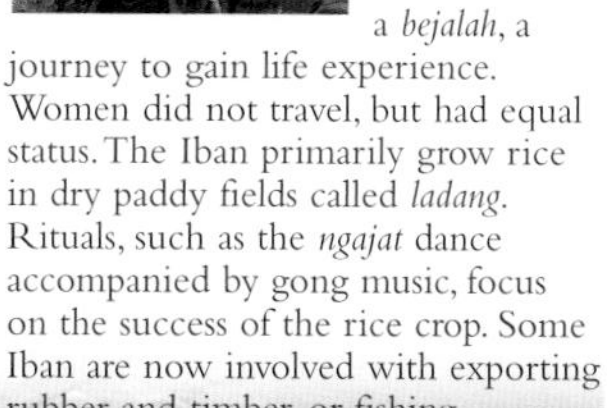

**ELDER**
*An elder displays tattoos on her forearms, once routinely applied to Iban women.*

**WARRIOR HISTORY**
*An Iban man poses with a feathered headdress, a mark of the people's pride in their historically warlike nature.*

INDO-MALAYSIA

## Bidayuh

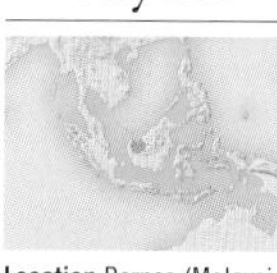

**Population** 160,000
**Language** Biatah, Salakau, Bukar Sadong, Bau Jagoi (all members of the "Land Dayak" group of Malayo-Polynesian tongues)
**Beliefs** Christianity, animism
**Location** Borneo (Malaysia and Indonesia)

The various Bidayuh tribes were labelled "Land Dayaks" by the British in the 1840s to distinguish them from the "Sea Dayak" Iban people. Like the Iban, they were formerly headhunters, and skulls can still be seen stored in the *baruk* – a village roundhouse used for meetings and ritual ceremonies. Most Bidayuh now live in villages, but some traditional Dayak longhouses and *baruk* still exist and form a focus, along with the Bidayuh's basketry, for the region's tourist industry.

The Bidayuh cultivate both dry and wet rice on the hills, clearing new fields each year with curved machetes called *parang*. They are renowned for their fruit and for their rice wine.

**CUSTOMARY DRESS**
*Men's traditional dress includes a loin cloth,* burang sumba *headgear, and in the Bukar-Sadung subgroup, heavy necklaces.*

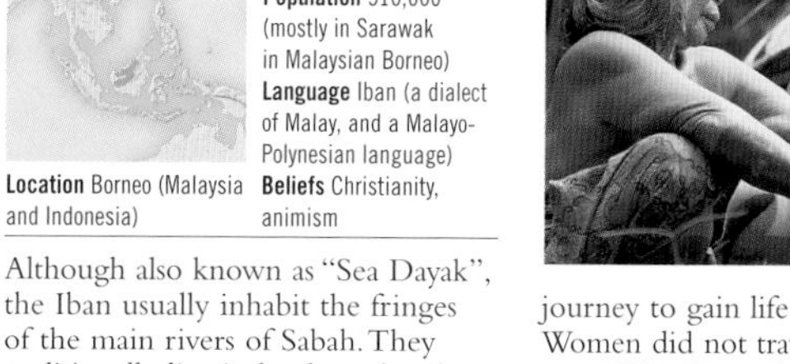

INDO-MALAYSIA

## Madurese

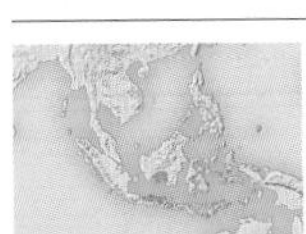

**Location** Madura; Kangean, Sapudi (Java); Kalimantan **Population** 13.7 million, which includes 4 million on Madura and another 9 million on Java **Language** Madurese (Western and Sumenep dialects), Kangean **Beliefs** Sunni Islam

The Madurese are the third largest ethnic group in Indonesia. Crowding and poor soil on Madura have led to large numbers of Madurese settling on Java, and in the 1970s, Indonesian authorities began resettling around 100,000 in Kalimantan (Indonesian Borneo). The policy proved highly controversial: the Madurese people thrived in the timber trade to the disadvantage of Kalimantan's native peoples, who killed 3,000 Madurese in 1999.

The Madurese are a proud people, united by powerful family bonds and a code of honour called *carok*, which traditionally entails killing an enemy by slitting the throat or stomach from behind. This custom, and a tendency to plain speaking, have fostered an overstated reputation for violence.

Bull-racing (*karapan sapi*) is a hugely popular sport with the Madurese, who are also known for their striking ritual masks, which are carved out of wood and painted with natural pigments.

**RETURN TO MADURA**
*Madurese refugees from massacres in Kalimantan listen as the Indonesian president warns against ethnic hatred.*

INDO-MALAYSIA

## Acehnese

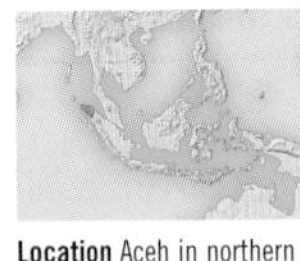

**Location** Aceh in northern Sumatra in Indonesia **Population** 4 million **Language** Acehnese; Bahasa Indonesia (second language for most) **Beliefs** Orthodox Islam (the Acehnese were the first Indonesians to embrace Islam and are still devout)

In the 13th century, the Acehnese were the first people in Indonesia to embrace Islam. The sultanate of Aceh remained independent after the rest of Indonesia had fallen under Dutch control. It was finally defeated in 1903, but the independent Acehnese spirit survives. Rebels in the partly autonomous state of Aceh still fight with the Indonesian government, partly over the control of Aceh's natural gas and oil.

If an Acehnese woman marries, her parents provide a house or offer the couple their own home. Women are the homeowners and spend most of their time in the house. Men, however, spend most of their time in public spaces doing business.

**MEN'S WORK**
*An Acehnese man fills in a form to apply for a teaching job. Most Acehnese men earn a living outside the home.*

**WOMEN'S DOMAIN**
*Acehnese women do most of the work the fields, such as planting rice, as well as tending the house.*

INDO-MALAYSIA

## Javanese

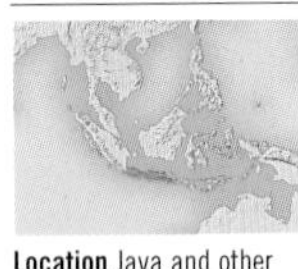

**Location** Java and other main islands of Indonesia **Population** 86 million **Language** Bahasa Indonesian and Javanese; Dutch is understood by many Javanese **Beliefs** Predominantly Islam, with animism and mystical beliefs

As Indonesia's most populous island and the site of the national capital, Jakarta, Java and its people have long been dominant in the country's affairs, and this has created resentment in other parts of Indonesia.

The Javanese mostly live in small, compact villages, in homes structured like huts (*kampongs*). Cultural traits include an intense spirituality and intimate family relationships. Special periods of the year are marked with a religious meal known as *slametans*.

**GAMELAN GONGS**
*The typically contemplative mood of Javanese music is created by a gamelan orchestra. The orchestra is made up of bronze instruments including gongs.*

Fact

### Javanese shadow plays

Javanese shadow puppet theatre (*wayang kulit*) is one of the oldest forms of storytelling in the world. The traditional *gamelan* orchestra accompanies the *dalan* (puppetmaster). The heroes and heroines of these plays – deeply embedded in Javanese culture – are drawn from Hindu epic literature.

**MUSLIM GIRL**
*The majority of Javanese are Muslims. This young girl wears a traditional Muslim headdress, which completely covers the forehead and is decorated with simple beads.*

**TEA PICKER**
*Along with coffee, rubber, and tobacco, tea is one of Java's major exports. Two-thirds of the island is cultivated.*

INDO-MALAYSIA

## Minangkabau

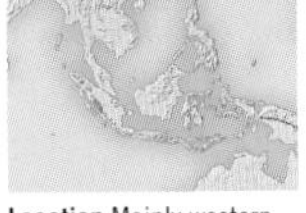

**Location** Mainly western Sumatra around Padang **Population** 7 million; 4 million in western Sumatra, the remainder in major cities of Indonesia **Language** Minangkabau (Minang), thought to be the parent language to Malay **Beliefs** Islam

The Minangkabau are the world's largest matrilineal society. Men must wait for a marriage proposal from a woman and, when married, will live with the bride's mother and married sisters in a *rumah gagang* family unit. Men practise *merantau*: they leave the community and must thrive on their own in the world of commerce.

Buffalo imagery permeates much of Minangkabau culture (Minangkabau means "victory of the water buffalo"); for example, their houses and hats both have distinctive "horns".

MINANGKABAU HEADDRESS

INDO-MALAYSIA

## Batak

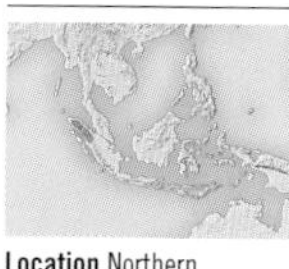

**Population** 4 million
**Language** Toba and Dairi, Malayo-Polynesian family
**Beliefs** Mostly Christianity and Islam; also Batak ancestral animism, which has become blended with Buddhism
**Location** Northern Sumatra in Indonesia

The Batak are divided into two main groups according to language. The Toba, Mandailang, and Angkola (Siporok) peoples speak Toba family dialects; and the Dairi, Karo, and PakPak speak dialects of Dairi. A third group, called the Simalungen, use an intermediate language.

**TOBA HOUSE**
*Antlers decorate the thatched gables on this typical house in Lake Toba. Batak peoples also live in communal longhouses.*

The Batak live in large family groups, known as *margas*. Marriages are only permitted between people from different *margas*, and it is always the women who move from a group to be with their husbands.

The first contact the Batak people had with the world beyond the Toba region was through trading benzoin and camphor, both aromatic products from trees. Most Batak subsist on rice and other crops.

**KARO BATAKS**
*These Karo Batak men are attending a festival. Many of the Karo Batak still live by* adat *customs, which specify how they should interact with different family members.*

INDO-MALAYSIA

## Kubu

**Population** 15,000 total, including 2,700 Orang Rimba and several other subgroups
**Language** Kubu, a Malayan tongue of the Malayo-Polynesian family
**Beliefs** Animism
**Location** Southern Sumatra in Indonesia

Kubu is a name used to represent several tribes living in the rainforests of southern Sumatra. The Indonesian government has given the name Orang Dalat, or "people of the land", to the Kubu, but they call themselves by the name of their specific tribe.

The Kubu people are hunter-gatherers and also tend forest gardens. Their culture prohibits the raising of domestic animals and the hunting of those they consider sacred, including tigers, elephants, and gibbons. The Kubu generally live in groups of 20–40 family members. When a member of the group dies, the others must move far away from the area.

Destruction of their forest habitat and a state-sponsored transmigration programme have resulted in many Kubu integrating into Indonesian society and adopting its ways.

**SHELTERING IN THE FOREST**
*When on hunting expeditions in the forest, Kubu construct makeshift temporary shelters. This family sit under a tarpaulin.*

INDO-MALAYSIA

## Sundanese

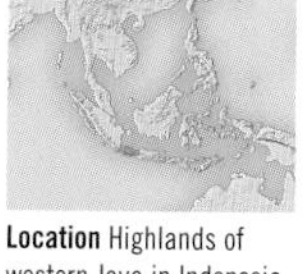

**Population** Estimated 33 million
**Language** Sundanese (also called Sunda or Priangnan)
**Beliefs** Islam and animism, with limited Hindu and Buddhist influences
**Location** Highlands of western Java in Indonesia

The Sundanese are one of three main ethnic groups found on the island of Java, Indonesia. They view their homeland region of western Java as their own cultural island – Sunda.

Although the Sundanese share many cultural, political, and economic similarities with the Javanese, there are many differences. Sundanese language is unintelligible to the Javanese, and Sundanese culture is less marked by Hindu and Buddhist influences.

The majority of Sundanese have adopted the Islamic faith. Many took part in the Darul Islam Muslim rebellion between 1948 and 1962, although it has been suggested that some did so more out of allegiance to village leaders than because of their own religious convictions. The traditional ancestral religion of the Sundanese, which still influences their cultural traditions, is animist in origin and includes the belief that spirits reside in rocks, trees, and streams.

The Sundanese have a strong literary tradition, and they perform elaborate puppet dramas using wooden dolls called *wayang golek*.

INDO-MALAYSIA

## Balinese

**Population** 2.86 million
**Language** Bahasa Indonesian, Balinese
**Beliefs** Balinese Hinduism (Bali is the only island in Indonesia where a variation of Hinduism is the primary religion)
**Location** The Indonesian island of Bali

Bali is the only island in Indonesia whose population is primarily Hindu. In addition to their Hindu faith, the Balinese believe in a vast number of spirits, which are present everywhere. They take their religious beliefs very seriously and hold elaborate and frequent rituals, for which they are famous. These rituals include dance and spirit possession.

**SURFER**
*Many Balinese, such as this local surfer, enjoy the island's renowned white beaches. Bali is also a popular holiday destination with foreign visitors.*

Graceful poise in composure and behaviour are highly valued by the Balinese, who see it as a means of maintaining self-respect and conforming to standards of social etiquette. Public display of emotion is considered improper and an indication that a person is unable to control his or her heart.

The Balinese maintain a caste system similar to that of Hindus in India but in a culturally distinct form, and they also belong to individual clans. These clans share the same customs and caste, and it is therefore traditional for people to marry a member of the same clan. Balinese people trace their ancestry along the father's side.

Bali came under Dutch control in the early 20th century. Remarkably, each of the ruling royal families, located in various parts of the island, committed suicide upon imminent or actual capture by the Dutch.

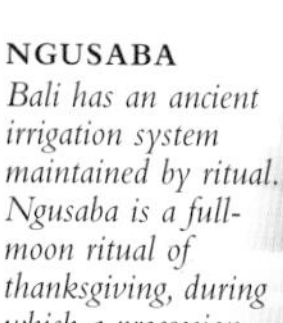

**NGUSABA**
*Bali has an ancient irrigation system maintained by ritual. Ngusaba is a full-moon ritual of thanksgiving, during which a procession takes an offering to an irrigation shrine.*

**GRACEFUL POISE**
*Aesthetic perfection and elegance are important not only in Balinese ritual practices (above), but also in daily life.*

INDO-MALAYSIA

## Toraja

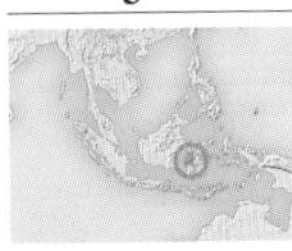

**Population** 2 million
**Language** Toraja-Sa'dan
**Beliefs** Christianity (around 80%, although traditional animist beliefs influence the culture and customs)
**Location** Highlands of central Sulawesi (Indonesia)

Although most Toraja people claim to follow Christianity, beliefs in spirits of the dead and the underworld influence their culture and customs. At the centre of the Toraja's *aluk to dolo* (law of the ancestors) is respect for the spirits of the dead. Lavish funerals sometimes take place several years after death in order to give relatives sufficient time to save money for the elaborate ceremonies. Temporary buildings to house the coffin may be constructed on burial grounds, and many tens, or even hundreds, of water buffalo may be slaughtered. The Toraja's dead are buried in caves hewn in cliff faces, often very high up – a practice that began when relatives sought to prevent grave robbers from stealing the expensive goods that are interred with the body. The caves are closed with doors and guarded by effigies of the dead called *tau-tau*. The water buffalo, which is the Toraja's most highly prized animal, features prominently in their building. The Toraja build ceremonial houses called *tongkonan*, the roofs of which curve up at either end like a water buffalo's horns. Houses are often decorated with water buffalo motifs, carvings, and sets of horns collected during funeral ceremonies.

**HARVEST TIME**
*Toraja men mark a harvest festival by wearing festive dress.*

**ELABORATE FUNERAL**
*A line of Toraja people follow a funeral procession in a village. They place the coffin in a decorative wooden structure, which is carried through the streets.*

INDO-MALAYSIA

## Weyewa

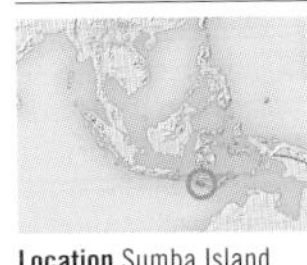

**Population** Estimated 100,000
**Language** Weyewa and regional dialects
**Beliefs** Marapu (Weyewa ancestral religion); many are now Christians, some follow Islam
**Location** Sumba Island (Indonesia)

For years the Weyewa were isolated because their island was not on the many spice-trading routes of the region, but in the 18th century they were forced to build defensive hill-top villages against Muslim slave traders.

Ritual speeches, poems, and songs formed a central part of the Weyewa's cultural identity and were the way in which they communicated with ancestral spirits. Weyewa society was traditionally headed by a *raja* or "big man". Recent attempts to unify the Indonesian people have resulted in many of the *raja* being sidelined; and with government officials making decisions concerning the land and the people, ritual feasts to influence the spirits are rendered unimportant.

Weyewa are mostly slash-and-burn agriculturalists, with some rice paddy fields and livestock-rearing. They produce colourful woven textiles.

INDO-MALAYSIA

## Tanimbarese

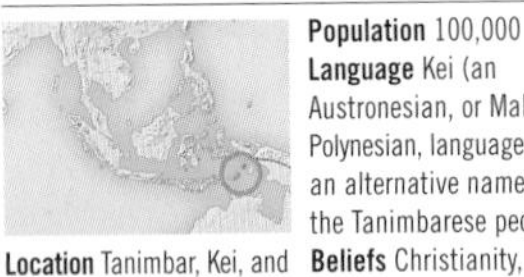

**Population** 100,000
**Language** Kei (an Austronesian, or Malayo-Polynesian, language; also an alternative name for the Tanimbarese people)
**Beliefs** Christianity, Sunni Islam
**Location** Tanimbar, Kei, and Maluku Islands (Indonesia)

Tanimbarese villages number between 300 and 1,000 people. Village heads, known as *rat* or *orang kaja*, belong to a ruling lineage called *mel mel*. People of different villages marry, forming alliances to maintain peace.

The Tanimbarese engage in slash-and-burn farming, fishing, and wild boar hunting. According to custom, they fear the spirits of people who died violently or during childbirth.

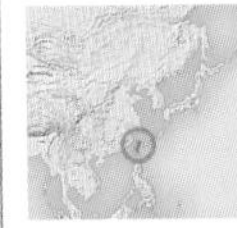

ANCESTOR STATUE

**DANCER**
*A Tanimbarese dancer performs in Portuguese style, proof of the lasting infuence of Portuguese colonization.*

PACIFIC ASIA

## Native Taiwanese

**Population** 300,000
**Language** Chinese, various Austronesian (Malayo-Polynesian) languages
**Beliefs** Polytheism, pantheism, ancestor worship
**Location** Taiwan

Although the people in Taiwan are mainly Han (*see below*), the island also has many colourful indigenous cultures belonging to some 10 distinct peoples. Most numerous are the Ami, notable for their matrilineal kinship structure. The next largest, the Atayal, are skilled hunters and weavers. Some elderly Atayal have distinct facial tattoos. In the Paiwan festival of *maleveq*, held every 5 years, ancestral spirits are invited down from a mountain to take part in days of celebrations, including head-hunting rituals.

The Ceremony of the Pygmies is an event held by the Saisiat: a legend tells of how the tribe was once helped by its neighbours, who were clearly smaller in stature, and whom they later massacred. The spirits of the murdered "pygmies" are invited to a week-long dance festival, before being appeased in the hope that they will not take revenge.

The Yami (or Tao) number just 4,000. Their isolated existence on Orchid Island, 40 miles off the Taiwanese coast, has preserved their distinctive culture. They are expert boat-carvers, fishermen, and potters.

Fact

### Taiwanese Han

Today, 98 per cent of Taiwan's people are Han Chinese (*see* p346). They first arrived in the 16th century, escaping war and famine in China. A second wave followed in 1949, when a further 1.5 million Chinese people arrived after the Communist victory on the mainland. China still regards Taiwan as a rebel province.

ETHNIC CHINESE MAN

**INDIGENOUS PEOPLE**
*Although many Taiwanese people are becoming assimilated into mainstream Chinese–Taiwanese culture, some retain old customs, such as this group wearing colourful ethnic dress.*

**LIVING TRADITIONS**
*An indigenous Taiwanese man wears a wet-weather outfit once common throughout China, made of woven bamboo and reeds.*

PACIFIC ASIA

# Koreans

**Location** Korea, northeastern China

**Population** 75 million
**Language** Korean (possibly related to Mongolian languages, but with much Chinese vocabulary)
**Beliefs** Confucianism, Taoism, Buddhism, Christianity, Chondogyo

The majority of Koreans – around 70 million people – live on the Korean peninsula. It is widely accepted that modern Koreans are descended from Central Asian tribes, and many people can trace their ancestry back thousands of years. Regional origin is important in determining blood heritage, since there are only 270 Korean surnames and, in South Korea, half the population is named Kim, Lee, Park, or Choi. Personal names are used only by very close friends, and even young children address their elder siblings as *eonni* (sisters) and *oppa* (brothers). Displays of emotion are generally avoided in Korean society.

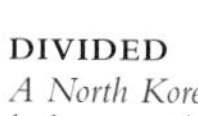

**DIVIDED**
*A North Korean soldier looks across the border that divides the Korean lands and people.*

Korean literature has a long history, and the world's first metal type for printing was produced in Korea in the 12th century. The unique Korean script, *Hangul*, was created in 1443 to avoid the use of Chinese characters. The lack of significant regional variations in the Korean language has helped to forge a strong sense of national identity. Since 1948, Koreans have been divided between the capitalist South of the country and the communist North. In the South, society has developed along Western lines, with strong influences from the US. The North has been controlled by a single-party state, and opposition to the government there is severely repressed. From an early age, children in North Korea are looked after by an extensive system of state-run crèches.

**KOREAN BUDDHISTS**
*Believers balance printing blocks from the* Tripitaka Koreana, *an ancient Buddhist text, to attain enlightenment.*

PACIFIC ASIA

# Ainu

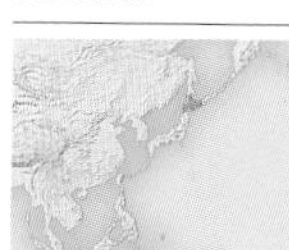

**Location** Hokkaido island (Japan), Russian Far East

**Population** 25,000
**Language** Ainu (a unique language with no known relationship to any other tongue), Japanese
**Beliefs** Shamanism (although this tradition has largely died out)

Originally seafaring hunters and trappers throughout Japan, the Ainu were restricted to the extreme north as Japan expanded as a modern nation in the 19th century. Many who claim Ainu heritage are now employed in labouring or construction. Although Ainu language and culture have all but died out, ancient art forms are beginning to be revived.

**ENIGMATIC PEOPLE**
*The Ainu's origins are a mystery: they often look different from neighbouring peoples. Here, an Ainu couple wear traditional attire.*

**GEISHA GIRL**
*Geisha women entertain men with song, dance, and chat. This one is wearing the clothes of a 12th-century noblewoman.*

PACIFIC ASIA

# Japanese

**Location** Japanese islands, expatriates worldwide

**Population** 128 million
**Language** Japanese (various dialects)
**Beliefs** Shinto (the indigenous Japanese religion), Buddhism; also Confucian influence

Japan has become a leading economy since 1945, with supremacy in high-tech industries. Japanese management culture promotes strong social ties between employee and company, even extending to sanctioning marriages. Employment status is often very closely related to personal identity, unemployment being socially stigmatized. Education is pressurized, and a university degree is essential for attaining higher management roles.

The Japanese regard for formal politeness can be seen in their stylized *kabuki* theatre, the tea ceremonies performed by geisha, and even in the national sport, sumo, all of which involve a high degree of ritual. Modern culture is less formal: baseball is a popular import and karaoke (singing in public to recorded accompaniment) a famous export. Japanese comics (manga) account for 40 per cent of all published material in Japan. The stories play up to popular Japanese ideas of honour and challenging adversity. Manga has a characteristic artwork that has influenced design and youth culture in the West as well as Japan.

The cultivation of rice has long been a central part of Japanese life and is heavily subsidized by the state, despite the abundance of cheaper imports. Other Japanese specialities include noodles, miso soup, the fiery wasabi paste, sushi, and sashimi (raw fish).

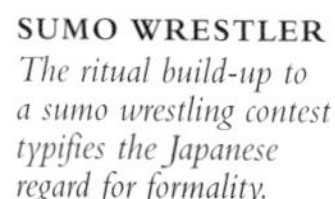

**SUMO WRESTLER**
*The ritual build-up to a sumo wrestling contest typifies the Japanese regard for formality.*

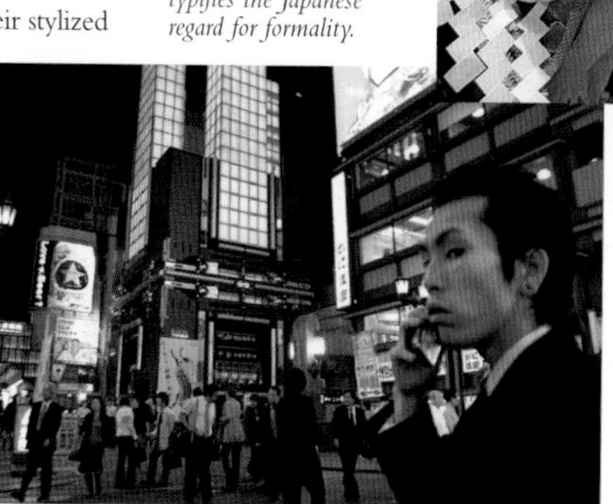

**HIGH-TECH CITY**
*Urban life in Japan is fast-paced, hi-tech, and crowded. This businessman talks on his mobile phone amid the city hustle and bustle.*

**FRUITS OF THE SEA**
*Oceania's islanders often live off the sea. Here, a coral reef diver is searching for seafood in the National Park of American Samoa, Ofu Island.*

# OCEANIA

Despite being the smallest continent in terms of land mass, Oceania boasts thousands of islands across the huge expanse of the Pacific Ocean, which means it is home to wide ethnic diversity. There are 32 million people living in this region, two-thirds of them in Australia. European scholars in the 19th century recognized three geographical and cultural areas outside Australia: Micronesia, Melanesia, and Polynesia. These divisions are still accepted because they encapsulate broad trends of culture across the ocean.

**THE REGION**
*Oceania consists of Australia, as well as the island groups of Micronesia, Melanesia (including New Guinea), and Polynesia (including New Zealand).*

Each island group encompasses myriad islands and peoples. Even within islands, particularly New Guinea, the ethnic diversity is breathtaking. The 5 million people of Papua New Guinea, the western half of the island, is split evenly among speakers of more than 800 languages, each associated with a distinct ethnic identity.

## MIGRATION AND SETTLEMENT

The earliest inhabitants of Oceania were ancestors of the Aboriginal people of Australia and New Guinea, who settled there at least 50,000 years ago, after migrating from Southeast Asia or southern China. The Melanesians and Micronesians are believed to be related to these Aborigines – based on both linguistic and genetic evidence. Islands here, however, were only settled around 3,500–4,000 years ago, also from the direction of Southeast Asia. The settlement of Polynesia occurred later still. Hawaii was settled in AD200–600; and Kiribati and New Zealand were not settled before AD1000.

European influence in Oceania only began in the 16th century, with Portuguese explorer Ferdinand Magellan finding the Mariana Islands in 1521. The arrival of the Portuguese began a profound transformation of many native cultures of the region. Christian missionaries discouraged a host of cultural practices in the Pacific, and European settlers dispossessed the indigenous peoples of their land in Australia. European interest in the Pacific region soared in the 19th century with the arrival of traders. Britain, France, Germany, and Spain often fought to control island groups. After World War II, many island groups were given US "trust territory" status or gained full independence.

**BOOMERANG**
*The boomerang is now recognized worldwide as a symbol of Aboriginal culture in Australia.*

**UNITING PEOPLES**
*A Papua New Guinea native dresses as a "Mudman" to help celebrate the island's ethnic diversity at a cultural show.*

## TRIBES AND NATIONS

Most of the peoples of Oceania have been hugely affected by colonialism over the past four centuries. The British even claimed Australia as *terra nullius* ("land without people") at the time of Captain Cook's landing in 1770, despite the presense of over 500 native peoples. On the smaller islands, colonial rulers varied: some were unintrusive, others impinged on ethnic customs and rights. Some indigenous peoples maintain a strong sense of their own identity. The Maori in New Zealand even seem to have upheld their cultural integrity partly as a result of their united opposition to colonial rule. Their resistance to European settlement led to the creation of a treaty of coexistence in 1840.

Some Pacific islands are still colonies or ruled by international agreements. In places such as New Caledonia, which is allied to France, the indigenous people are now in a notable minority as waves of Europeans, Japanese, Americans, and Asians have settled, often in pursuit of an idealized tropical lifestyle. However, most islands have broken away from colonialism in the latter half of the 20th century to form small, independent nations.

**DECORATIVE DIVERSITY**
*The Huli people, one of hundreds in New Guinea, assert their particular ethnic identity by wearing wigs and face paint for ceremonial dances known as sing-sings.*

## CHANGE

New-found autonomy can cause new problems. States often foster a sense of shared national identity rather than of cultural diversity within their boundaries. When Fiji gained independence, there was conflict between native Fijians and Indians (descended from labourers brought by the British), with many of the latter leaving in the 1990s. However, Aboriginal reconciliation in Australia and the revival of ethnic consciousness in the rest of Oceania tend to be increasingly high on national agendas. Another positive aspect is the evolution of new composite cultures in countries, such as Australia and New Zealand, whose populations mainly arrived relatively recently.

Many indigenous peoples now make money from their traditional customs by displaying them to a growing influx of tourists to the region. Papua New Guinea – the most ethnically diverse region in the world – now hosts annual shows to bring together its numerous and varied peoples.

Issue

### Land rights

Aboriginal peoples in Australia have preserved key elements of their culture, particularly in relation to their land, upon which they rely. Australian colonial powers have also laid claim to this land owing to what they have always viewed as a lack of a land ownership system on the part of the Aborigines. In fact, it was not until 1992 that Aboriginal land rights in Australia and New Zealand were recognized, allowing the Aboriginal peoples to reclaim their homelands, such as the Queensland forest below.

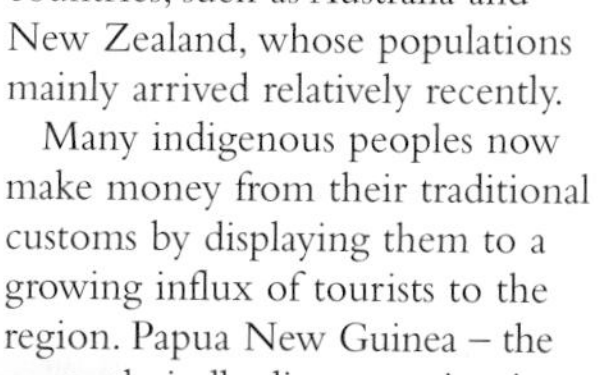

**MIGRATION BY CANOE**
*The peopling of Pacific islands was achieved by outrigger canoes, such as this one off Hawaii.*

MELANESIA

## Motu

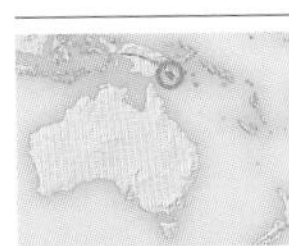

**Population** 30,000
**Language** Motu (an Austronesian language), Hiri Motu (a trade language based on Motu), Tok Pisin, English
**Location** Port Moresby area (Papua New Guinea)
**Beliefs** Christianity, traditional beliefs

The Motu are a group of navigators and fishermen who live around Port Moresby, the capital of Papua New Guinea. They traditionally engaged in *hiri* (trade expeditions) along the coast to obtain food because the area was prone to drought. *Hiri* also involved ritual, religious, and political exchanges, and gave rise to a trade language called *Hiri motu*. The introduction of Western practices, such as cash employment, has led to many changes in Motu society. However, their traditional institutions remain, such as bridewealth (a sum paid from a groom's to a bride's family). The Motu people have also retained most of the land rights in the Port Moresby area.

**MUSICIANS**
*Motu people now share their home of Port Moresby with travelling musicians from elsewhere in the country.*

MELANESIA

## Dani

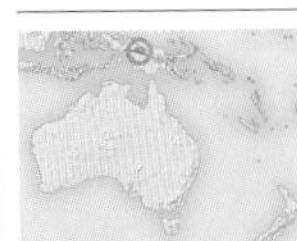

**Population** 200,000
**Language** Dani (a non-Austronesian language), Indonesian
**Location** Irian Jaya, or West Papua (Indonesia)
**Beliefs** Roman Catholic and Protestant Christianity; also traditional beliefs

**DANI MEN**
*The adornments traditionally worn by Dani men include braided bark-string necklaces, face paint, and feather headdresses (above). Some also wear nose pieces made from pig tusks.*

The Dani live in Irian Jaya, the western part of the island of New Guinea. It was not until the 1950s that that they had sustained contact with outsiders. At this time, the Dani, who grew sweet potatoes, herded pigs, and made stone tools, were considered one of the last Stone-Age societies. Their use of penis sheaths and mummification of corpses also indicated an ancient culture.

The Dani social system, however, was highly complex. The people were divided into a series of village-based political units. These communities engaged in ritualized wars involving hundreds of men. The conflicts were formal events with several functions (mostly based on ritual exchanges). There were few casualties.

The Indonesians, who annexed Irian Jaya in 1963, stopped these wars, but they also encouraged migration from Java, claimed local land, and introduced logging and mining. In the late 1970s, the Dani, armed with bows and arrows, united and revolted. The conflict lasted for two years, but the Indonesians' modern weaponry (including napalm) was too powerful.

**PIT OVEN**
*The Dani cook their sweet potatoes in a pit oven. They wrap the vegetables in leaves and place them on hot stones in the earth.*

MELANESIA

## Abelam

**Population** 40,000
**Language** Abules (a non-Austronesian language: part of the Ndu language family), Tok Pisin, English
**Location** East Sepik area (Papua New Guinea)
**Beliefs** Roman Catholic and Protestant Christianity; also traditional beliefs

The Abelam people are traditionally subsistence cultivators who mainly produce yams. During annual harvest rituals, decorated "long yams" are exchanged by ceremonial partners or relatives. Abelam authority is held by "big men", whose power is based on success in growing these ceremonial yams, along with skills in oration and diplomacy. Today, although many traditional ways have been abandoned under the influence of Western missionaries, "long yam" ceremonies are still held.

**LONG-YAM HOUSE**
*"A"-shaped ceremonial houses, built up to 15m (50ft) high, are used to teach the secret magic used to grow long yams.*

LINTEL WITH ANCESTRAL FIGURES

MELANESIA

## Huli

**Population** 65,000
**Language** Huli (a non-Austronesian language), Tok Pisin (New Guinea pidgin English), English
**Location** S Highlands (Papua New Guinea)
**Beliefs** Roman Catholic and Protestant Christianity; also traditional beliefs

Living in the Southern Highlands province of Papua New Guinea, the Huli occupy an area mostly comprising dense rainforest and swampy valleys. They are principally horticulturists, typically growing sweet potatoes; they also hunt pig and other game (such as cassowaries and possums). Like many other New Guinea societies (including the Abelam, Dani, and Amungme peoples), the Huli practise slash-and-burn cultivation (*see* p253). Under this system, patches of rainforest are burnt, cultivated for a couple of years, then left fallow for two or three generations – a period protected by customary beliefs and prohibitions.

Authority is based on the "big man" system, whereby prestige and power in the community and surroundings is determined by a person's talents, skills, and knowledge. Huli society, in common with several Papua New Guinea peoples, is marked by a strongly gender-based division of daily tasks. Men prepare the gardens and plant the food, for example, whereas women gather the harvest and take care of the pigs (which are used in rituals). The people divide their tasks according to principles of separation of sexes, not because the roles complement each other.

The Huli are well known for their striking wigs, made from human hair and adorned with bird-of-paradise and parrot feathers. They also wear intricate, colourful body adornments and face paintings. These decorations are associated with complex beliefs and ceremonies, notably the long *haroli* initiation process for young men and the *sing-sing* festivals, which involve the whole community. The traditional religion includes belief in several classes of supernatural beings, which control the climate and the fertility of the land.

**MOURNING RITUAL**
*A dancer, her body covered in ash, performs a mourning ritual at a sing-sing. The festival may be an occasion for mourning as well as celebration.*

**SING-SING FESTIVAL**
*Huli people line up to perform dances during a sing-sing, an annual festival celebrating clan pride.*

**COLOURFUL FACE**
*The ceremonial dress of a Huli man includes a wig with colourful plumage, bright face paint, and a nose feather.*

PEOPLES

MELANESIA

# Korowai

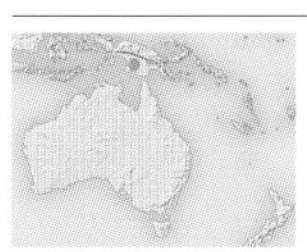

**Location** Near southern Irian Jaya coast (Indonesia)

**Population** Estimates vary from 700 to 4,000; mainly living in lowland, swampy rainforest areas
**Language** Korowai (an unclassified, non-Austronesian language)
**Beliefs** Animism

The Korowai have only been known to the West since the 1970s. They live in family groups in *khaim* (houses) built high in the forest canopy. They dwell in the trees because they believe that the ground is inhabited by evil spirits, and to protect themselves from hostile neighbours, insects, and the heat; those under threat from other tribes tend to build highest.

The Korowai have few possessions other than stone or wooden tools and bows and arrows for hunting. Women dress in a skirt made of leaves and wear a flying-fox bone through the nose. Men wear a penis sheath; some also wear pig-tooth necklaces and have cassowary feathers in the nose.

Headhunting occurs in tribal wars, and rumours persist that the Korowai people practise cannibalism. Women are seen as inferior to men and are often kidnapped from nearby villages. The men have several partners. Until recently, the Korowai had little contact with other peoples, but tourism and Indonesian logging companies now threaten their lifestyle.

**FEAST OF SAGO GRUBS**
*Sago grubs – the larvae of the capricorn beetle – are an important source of nutrition. The grubs are either steamed or eaten raw.*

**CANOPY HOUSE**
*Korowai homes are usually about 25m (80ft) high but can be up to 50m (165ft) tall.*

MELANESIA

# Amungme

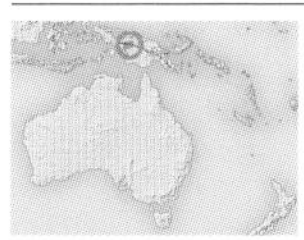

**Location** Tsinga Valley, south central highlands, Irian Jaya (Indonesia)

**Population** 12,000
**Language** Amung (non-Austronesian), Indonesian
**Beliefs** Roman Catholic and Protestant Christianity; traditional beliefs (in which natural features, such as mountains, are sacred and linked to family spirits)

Like other peoples in Irian Jaya, the Amungme are shifting cultivators, who use land for a while and then leave it fallow. They grow sweet potatoes and raise pigs. Amungme society is based on clans whose members trace kinship through the male line. They follow a belief system in which they revere mountains and glaciers associated with the spirits of their ancestors.

The Amungme have a traditional Melanesian system of land ownership. Land cannot be sold or bought; only temporary rights of use are granted (with accompanying compensations), but these rights are generally not given to people outside the community.

Issue

## The impact of mining

Mining has deeply affected the Amungme people. Vast geographical areas have been granted to mining concerns (with little regard for local claims to ancestral lands), and in the 1970s, the Amungme were relocated to make way for the "Copper City" settlement. These factors, combined with the environmental impact of huge open-pit mines (below), have created tensions and conflict.

**DANCERS IN COSTUME**
*Ni-Vanuatu dancers wear colourful costumes to perform at one of the many rituals that takes place on the islands of Vanuatu.*

MELANESIA

# Ni-Vanuatu

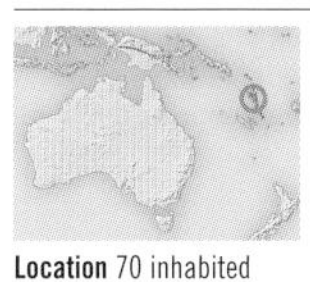

**Location** 70 inhabited islands of Vanuatu

**Population** 160,000
**Language** 113 Austronesian languages, Bislama (Melanesian pidgin), English, French
**Beliefs** Roman Catholic and Protestant Christianity; also traditional beliefs

The inhabited territory of Vanuatu comprises 70 islands. The people, known as Ni-Vanuatu, have a wide range of cultures and languages. Some Ni-Vanuatu societies have distinctive features; two such elements are the "land-diving" of Pentecost Island (the origin of bungee jumping; *see* p204), which is associated with initiations for young men, and the "grades societies" of Malekula and Ambrym.

The grades societies are a hierarchy through which men seeking political power must progress. The men attain a new rank by gathering wealth, usually in the form of pigs; the pigs are then slaughtered and offered to members of the desired grade. As an individual passes through the levels, he acquires new rights and duties, which are symbolized by insignias of colourful ornaments and paraphernalia, such as curved pig-tusks. (The length of these teeth determines the material and sacred value of the pig.) More than a hundred pigs are required to reach the highest ranks; these levels are only accessible to elders who have managed to acquire ritual knowledge and establish partnerships with neighbouring islands. Over time, the grades societies have successfully integrated Christianity and modernity into their traditional system.

**BODY DECORATION**
*A man from Malekula island wears a headband and armband made from local plants.*

MELANESIA

## Trobrianders

**Location** Trobriand Islands (Papua New Guinea)

**Population** 15,000
**Language** Kiriwinian (part of the Austronesian language family), Tok Pisin, English
**Beliefs** Roman Catholic and Protestant Christianity, traditional belief systems

The Trobriand archipelago is a group of low-lying coral islands situated on the south west of the Papua New Guinea mainland. The Trobrianders are horticulturists, growing food in the islands' fertile soil; they are also fishermen. Both activities involve the use of specialized magic. The society is based on hereditary clan chiefdoms, and on matrilineal descent, in which people's rights are inherited along the female line. Its most distinctive feature is the ritualized *kula* exchange system (*see right*), which comes into many aspects of life.

**DISTINCTIVE EARRINGS**
*A Trobiander woman wears colourful home-made earrings.*

Fact

## Kula

The *Kula* system is a complex cycle in which ceremonial goods are exchanged by partners from different islands. At first *kula* was based on shell armbands and necklaces, but it came to involve many other goods as well. Decorated canoe prows, body ornaments, carved shields, and songs and dances were used to entice the trading partner and facilitate a swap. It now mobilizes the whole society. The exchanges do not just have an economic purpose; success in *kula* also determines prestige and influence.

**KULA GOODS**
*Goods used for* kula *exchanges include yams (below), bananas, pigs, artefacts, and even the "copyright" of songs.*

MELANESIA

## Fijians

**Location** Fiji, New Zealand

**Population** 835,000 (many Indo-Fijians left Fiji after a failed coup in 2000)
**Language** Fijian (a Malayo-Polynesian language), Hindustani, English
**Beliefs** Hinduism, Islam, Christianity

Early Melanesians and Polynesians occupied Fiji from around 1500BC. There have since been many waves of migration, but the most important was when Indians were brought in to work as labourers between 1879 and 1916. Now, almost half the population is Indian in origin. Although many Indo-Fijians are adopting Fijian ways, some Fijian nationalists object to them.

Fijian villages are communally arranged, with families living in single-room *bures* and sharing community chores. Society has a firm hierarchical structure, with the village chief the unquestionable head at local level, and a Great Council of Chiefs playing an important political role. Fijian culture features a strong oral tradition and crafts such as mat-weaving and woodcarving. Food is cooked in the *lovo* oven, half buried in the ground.

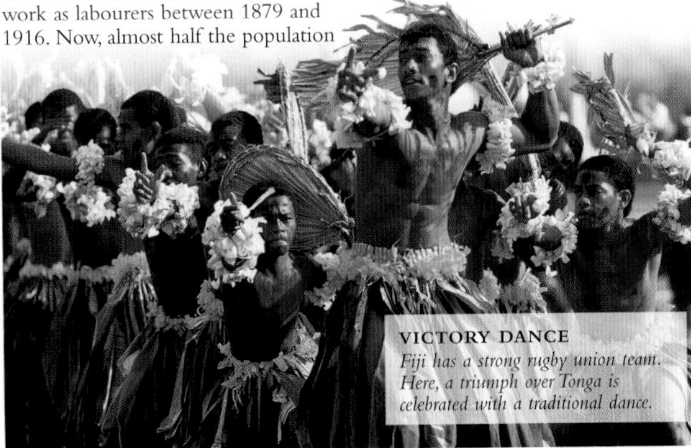

**VICTORY DANCE**
*Fiji has a strong rugby union team. Here, a triumph over Tonga is celebrated with a traditional dance.*

AUSTRALIA

## Australians

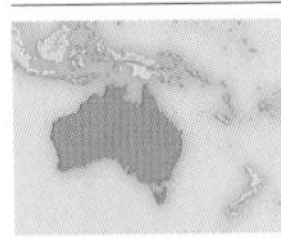

**Location** Australia

**Population** 19.1 million
**Language** English, Italian, Cantonese, Greek, Arabic, Vietnamese, various other immigrant languages
**Beliefs** Christianity (especially Roman Catholic, Anglican), nonreligious

The earliest European migrants to Australia were British and Irish people who arrived in the late 18th and early 19th centuries. Many were convicted criminals, who were sent to penal colonies run by the British imperial administration. Numbers increased during the gold rushes of the 1850s. At this time, many immigrants also arrived from China. During the 20th century, Australian governments started viewing immigration as a way to bolster the economy, and since 1945 the population has doubled. The governments gradually wound down the exclusive "white Australia" policy, enabling more Asians to join the nation. Australian identity is therefore increasingly multicultural. However, indigenous Australians were harshly treated over the years by colonists, and Aboriginal reconciliation now ranks high on the national agenda.

**SYDNEY OPERA HOUSE**
*The distinctive design of the Sydney Opera has become a symbol of Australia. In the foreground, diners enjoy an outdoor meal.*

**LIFEGUARD COMPETITION**
*New South Wales lifeguards start an ocean-swim competition. Lifeguards are essential on many of Australia's beaches.*

Health

## The bronzed Australian

Australians are often seen by foreigners as athletes who divide their time between the surfboard and the Australian-rules football ground. This image tends to be promoted worldwide by the widespread success of Australian sports competitors. However, a large part of the population suffers from the same problems of inactivity and unhealthy diet as people in other parts of the developed world.

With over 90 per cent of people living in towns and cities, urban life is an important part of Australian culture. Most of the cities are characterized by small centres surrounded by vast stretches of suburbs, filled with large houses, back yards, and pools.

Popular culture is influenced by Europe and the US, but Australians also export their music and television programmes worldwide. The cuisine is eclectic, using seafood and other natural resources and heavily influenced by nearby Southeast Asia. The winemakers are also globally successful. Just 9 per cent of Australians live in the interior, but the iconic figures of the bushman and jackaroo (ranch hand) are still important to the national self-image.

**BARBECUE**
*The "barbie" is a popular way to grill food, often washed down with a "tinny" (can of beer).*

**SMALL TOWN**
*A man transports his sheep through the local town. Rural communities still encapsulate a significant part of the Australian self-image.*

AUSTRALIA

## Aboriginal peoples

**Location** Australia

**Population** 410,000
**Language** English, more than 100 Australian languages (a distinct group that are largely unrelated to any other languages)
**Beliefs** Animism

The Aboriginal peoples, or indigenous peoples of Australia (although regional peoples prefer to be known by their own names) migrated from Southeast Asia at least 50,000 years ago. When Europeans arrived in the late 18th century, the Aboriginal population was at least 300,000. Within a century, this figure had dropped by as much as 90 per cent, and the population has only recently started recovering.

**CLAP STICKS**
*Clap sticks are percussion instruments played at ceremonial dances.*

Traditional Aboriginal culture is mainly nomadic. Hundreds of peoples rove across huge territories. The lands are defined not so much by area as by the relationship of each people with its totemic plants and animals. It is also defined by the pathways, or "songlines", of its gods. The peoples believe that their ancestral gods walked the primordial Earth and gave it form: many features of the landscape commemorate these paths.

Other distinctive elements of the culture include the boomerang and the didgeridoo (*yidaki*), a resonant pipe used to help induce a trance and enter a state of "dreaming".

Early settlers were often welcomed by Aboriginal tribes, but their land "ownership" claims – a concept alien to Aboriginal philosophy – removed much of the Aboriginal peoples' traditional hunting and gathering areas and exposed them to new diseases that often destroyed whole tribes. Since colonization, Aboriginal peoples have fought, occasionally violently but mostly with patience, to have their sacred sites and traditional lands recognized. They have had mixed success: Uluru (known to Europeans as Ayer's Rock) was returned to the Anangu in 1985; however, in 2003, many Nyungah were evicted from the Swan Valley near Perth.

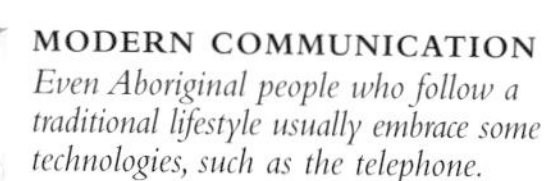

**MODERN COMMUNICATION**
*Even Aboriginal people who follow a traditional lifestyle usually embrace some technologies, such as the telephone.*

**HEALTH CLINIC**
*A nurse works at one of the many local clinics that have been set up in Aboriginal communities to raise standards of healthcare.*

**INITIATION CEREMONY**
*Singing and didgeridoo playing are parts of an initiation ceremony for boys, held at Nangalala, Northern Territory.*

Fact

### Dreamtime

Dreamtime is central to Aboriginal culture. It is a "beginning that never ended", when the gods made the land and all living things (and the relationships between them). Each tribe has sacred animals, plants, and landforms; the "dreamings" (echoes of creation) of these elements can be accessed through rituals. Dot paintings like the one below represent aspects of Dreamtime.

AUSTRALIA

## Warlpiri

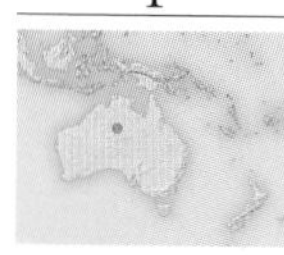

**Location** Tanami region (Northern Territory)

**Population** 2,700
**Language** Warlpiri, English
**Beliefs** Animism (dreamings are central, notably the fire dreaming, in which pregnant women predict the conception spirit of their unborn child)

The Warlpiri (who usually refer to themselves as "yapa") were originally seminomadic, following sources of food and water across their largely desert homeland. Contact with European settlers was limited until the 1920s and 1930s, when a brief gold rush attracted miners and missionaries to the region. The Warlpiri people resisted invasion; they suffered the last notable massacre of indigenous Australians, at Coniston in 1928. The Warlpiri then withdrew from contact with Europeans, but the Australian government forced them to settle in the town of Yuendumu in 1946 and even placed restrictions on their movements.

Warlpiri boys are officially initiated into manhood at the age of 12. This is celebrated by ritual dances: the men dance the *purlapa*, while the women perform a dance that expresses grievance at the "loss" of the child.

The Warlpiri have embraced modern technology since the 1970s, especially the use of video conferencing, which is popular both for communicating and for sharing Warlpiri traditional art.

**WARLPIRI PORTRAIT**
*A Warlpiri man wears a nose-bone for decoration. The septum of his nose was probably pierced when he was a boy.*

AUSTRALIA

## Arnhem Land peoples

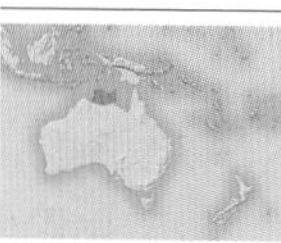

**Location** Arhem Land (Northern Territory)

**Population** 11,000
**Language** Yolngu family, many others (several not related to Pama-Nyungan, the main Aboriginal language family), English
**Beliefs** Animism, Christianity

Arnhem Land is home to more than a dozen peoples, and these groups exhibit the greatest linguistic diversity in Australia. Long before Europeans arrived, the Arnhem Land peoples had established a highly successful system of trade with the Macassans, annual visitors from Indonesia. Permanent European settlers came to the area as late as the 1930s, so the peoples of Arnhem Land have preserved their cultures more vigorously than those in other areas. Arnhem Land art is painted on bark and employs distinctive cross-hatching as opposed to the dot paintings of other Aboriginal peoples.

The Yolngu, the dominant Arnhem Land people, comprise approximately 20 tribes divided into two moieties: the Dhuwa and the Yirritja. The two groups depend on each other, but their rituals, dreamings, and artistic styles are separate. The Yolngu have been at the forefront of land rights campaigns.

**NATURAL HUNTERS**
*Many Arnhem Land people acquire skills at finding natural food sources early in life. A home-made spear is sufficient to capture a sizeable crab.*

AUSTRALIA

## Arrernte

**Population** 3,800
**Language** Various dialects of Arandic (a subgroup of the Pama-Nyungan language family, the main Aboriginal language group), English
**Beliefs** Animism
**Location** Around Alice Springs (Northern Territory)

The Arrernte people (also known as the Aranda) are among the most well known Aboriginal peoples outside Australia. Anthropological studies on the Arrernte conducted from the start of the 20th century predicted that Aboriginal cultures were doomed to die out. However, the Arrernte have proven otherwise, and theirs is one of the largest indigenous language groups in Australia. The language is widely taught, and a major Arrernte/English dictionary was published in 1994. There are many dialects, and these are intimately connected to family kinships. Of these dialects, two have come to dominate: the Western and the Eastern dialect. However, now that several clans live together in Mparntwe (Alice Springs), many dialectal subtleties have been confused. The Arrernte are seminomadic, but they build circular shelters out of branches during the cold months. They inhabit the Macdonnell Ranges area, and although the region is served by the Finke river, it is often subject to drought in summer. The Arrernte are hunters (catching emus, kangaroos, bandicoots, and birds) and gatherers (collecting honey ants, grubs, and lizards). Vital to hunting is the *merra*: a hooked, wooden sleeve used to increase the spear's throwing distance.

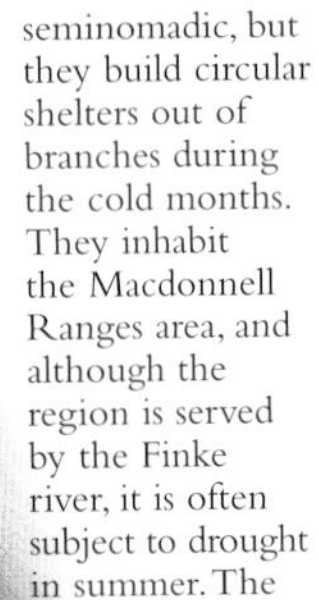

**TOUR GUIDES**
*Some Arrernte people act as guides at the sacred site of Uluru.*

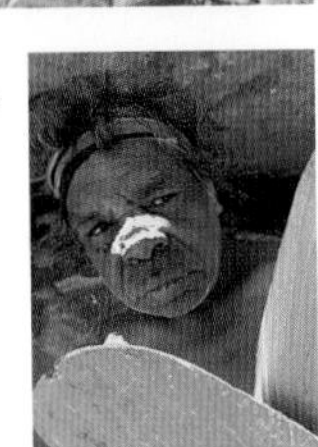

**SACRED SITE**
*An Arrernte man poses at a sacred site for adult males. Women and children are allowed only restricted access.*

Fact

### Yeperenye festival

In 2001, the Yeperenye Dreaming festival in Alice Springs recognized and celebrated the centenary of Australia becoming a federation. The festival, which began with a welcoming ceremony by Arrernte people, was attended by Aboriginal performers from all over Australia.

**TRADITIONAL DANCE**
*Exotically decorated Arrernte men start off proceedings at the 2001 Yeperenye Dreaming festival with a traditional Arrernte dance.*

AUSTRALIA

## Torres Strait Islanders

**Population** 28,700
**Language** Torres Strait Creole, Meriam Mir, Kalau Lagau Ya, English
**Beliefs** Ailan Kastom (a fusion of traditional island totemic religion with Christianity)
**Location** Torres Strait Islands, Queensland

Although the Torres Strait Islanders are related to the peoples of Papua New Guinea and the Australian Aboriginals, they retain their own distinct culture. They have lived on the islands, of which there are more than 100, for at least 10,000 years. The islands, 17 of which are permanently occupied, became part of Queensland in 1879, but limited autonomy was restored in the 1930s. The Islanders have had their own flag since 1995.

Torres Strait Islanders are seafarers, hunting turtles and dugong, and fishing for pearls. They also have a system of adoption, whereby children are moved from one extended family to another.

**COSTUME**
*A Torres Strait dancer, wearing an elaborate headdress (*dari*), performs at a festival.*

POLYNESIA

## New Zealanders

**Population** 4 million
**Language** English, Maori
**Beliefs** Anglican most common, also Presbyterian, other Protestant denominations, Roman Catholic, nonreligious
**Location** New Zealand

New Zealanders are all descended from recent migrants. The first arrivals, the Maoris, came from Polynesian islands and settled in New Zealand some 1,200 years ago. They were then joined by mainly British settlers from the mid-19th century onwards. The majority of modern New Zealanders are European in origin, and relations between the ethnic Maoris and more recent Pacific migrants – who retain a distinct cultural heritage and separate languages – can sometimes be tense. In recent years, there has been a distinct revival of Maori cultural traditions and language. There has also been a great influx of Asian people over the past 20 years: more than 200,000 have settled here.

The "Kiwis", named after the small flightless bird native to the islands, pride themselves on their egalitarian society. For example, New Zealand's women were the first in the world to gain the vote (in 1893) and are well represented in the professions and in politics (in 2001, the prime minister, the leader of the opposition, and the governor-general were all women). New Zealanders also have access to a long-established welfare system, although this is not as strong as it once was.

Although 86 per cent of the New Zealand population live in cities (and most of these on the smaller North Island; the South Island is sparsely populated), the rural tradition is still strong. Sheep rearing was the foundation of the country's wealth, and sheep still outnumber people ten-to-one. Modern farming in New Zealand is a highly scientific operation, and agricultural goods make up the country's main exports.

In recent times, tourism has become a key part of the economy. Visitors are attracted by the clean, green image of New Zealand, and appreciation of the outdoors is a central part of the culture. Hiking (known as tramping locally), boating, and other water sports are popular pastimes, as are rugby and cricket, at which the New Zealand teams are among the world's best. Both tourists and locals often indulge in adventure sports such as bungee jumping and skydiving.

**SPORTING AGGRESSION**
*The All Blacks – New Zealand's national rugby union team – perform a haka before every match. This is a ceremonial Maori war dance accompanied by loud, repeated chants.*

**SHEEP ROUNDUP**
*Some New Zealand farms have steep hills that are easier for farmers to tackle by horse.*

**WELLINGTON**
*Many cities, including Wellington (right), are built on the fault line that runs through New Zealand. Earthquakes are not uncommon.*

POLYNESIA

## Maori

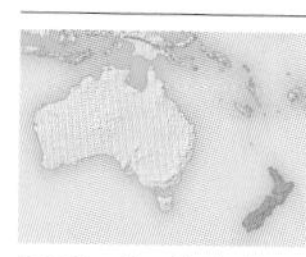

**Population** 310,000 (50–70,000 of whom speak the Maori language)
**Language** English, Maori (a Polynesian language closely related to Tahitian)
**Beliefs** Protestant Christianity
**Location** New Zealand (mainly in the North Island)

The Maori are a Polynesian people who first explored New Zealand, a place they called *Aotearoa*, or "land of the long, white cloud", no more than 1,200 years ago. According to a popular account, they did not settle in numbers until an expedition of seven canoes, each carrying a different tribe, landed in around 1350. They brought seeds of their familiar food plants. Not all of them thrived, so the Maoris had to adapt their culture to the colder climate.

Later, the Maori came into conflict with European settlers, but the 1840 Treaty of Waitangi, although not always honoured, allowed coexistence of the peoples.

Today, the Maori form an active and prominent part of New Zealand society. Owning and running many of the country's tourism enterprises, they greet visitors with the *powhiri* (welcoming ceremony), followed by cultural displays and *hangi* (traditional feasts cooked in a pit oven). The *powhiri* takes place on the *marae* (the sacred courtyard in front of the meeting house of a Maori village). Through several courteous steps, mutual trust and respect is built until visitors and hosts are united in peace.

**CHALLENGE**
*A tattooed Maori uses facial expressions to issue a challenge to visitors arriving at his village. It is the first part of an elaborate ceremonial welcome called a* powhiri.

**HISTORY IN WOOD**
*The Maori are master carvers, as shown by this figure with seashell eyes. Carvings often tell stories or signify events in Maori history.*

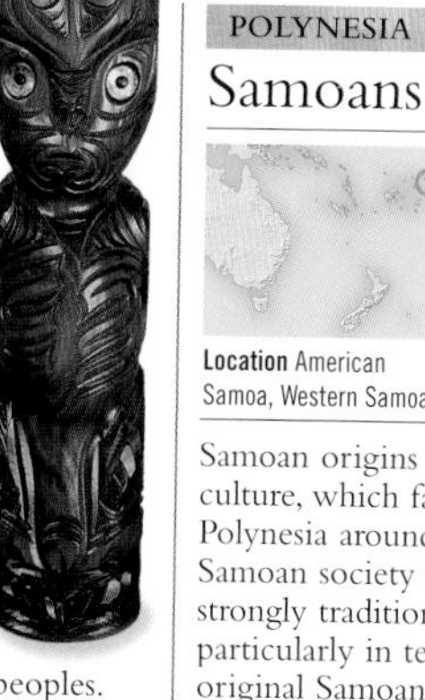

**TANGI**
*At a tangi (the Maori funeral rites) friends and relatives gather round the body of the deceased until the burial.*

POLYNESIA

## Samoans

**Population** 480,000
**Language** Samoan (a Polynesian tongue, part of the large Austronesian family), English
**Beliefs** Adapted Christianity: Methodist, Congregational
**Location** American Samoa, Western Samoa

Samoan origins lie in the Lapita culture, which fanned out through Polynesia around 3,000 years ago. Samoan society is known for being strongly traditional and conservative, particularly in terms of religion. The original Samoan religion included ancestor worship coupled with belief in a pantheon of human and non-human gods. Today, although Samoa is often referred to as the Pacific "Bible Belt", Christianity there has been heavily adapted by the Samoans.

The ritual tattoo is still commonly administered to young Samoan boys. The month-long procedure – performed by the *tufuga* (tattooist) using traditional shark's teeth implements – is a test of endurance. At the end, the boys are adorned from waist to knees. Though women are sometimes tattooed, female initiation traditionally consists of weaving a fine mat for her future dowry.

Rugby football is a popular sport, and the Samoan national squad – Manu Samoa – is among the best in the world. Before kick-off, Manu Samoa perform *Ma'ulu'ulu Moa*, a traditional dance with many similarities to the Maori's "haka".

In recent times, many young Samoans have emigrated in search of better education and employment opportunities. Most ethnic Samoans now live outside the islands. However, the conflict between traditional Samoan ways and imported Western cultures has led to serious social problems for returning emigrants.

**ISLAND CHIEF**
*This man is one of the local chiefs that make up* Fa'a matai, *the Samoan government system.*

**INDEPENDENCE DAY**
*Dancing women perform at the Western Samoan Independence Day celebrations. The country gained independence in 1962.*

**COOLING OFF**
*A group of Samoans enjoys a swim in one of the picturesque bays of American Samoa.*

POLYNESIA

## Tongans

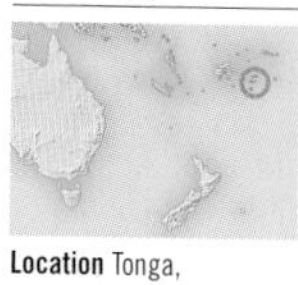

**Population** 120,000
**Language** Tongan (a Polynesian language, part of the large Austronesian family), English
**Beliefs** Free Wesleyan, Roman Catholic, other Christian denominations
**Location** Tonga, New Zealand, US

Having arrived in the Tongan islands in around 1000BC, the Tongans went on to dominate the South Pacific region as traders. Along with Samoa and Fiji, Tonga is often described as the cradle of Polynesian culture. Tongans are very proud that they were never fully colonized (merely "protected" by the UK) and that they have the Pacific's only monarchy.

Tongans have adopted Christianity wholeheartedly. Consequently, Tongan society is very conservative. However, while some Polynesian practices, such as tattooing, have gradually become less important, traditional Tongan ways – such as living in an extended family, leading a communal lifestyle, and respect for social hierarchy – remain central to the modern culture.

As in other regions of Polynesia, the oral tradition is still strong, and dancing almost always accompanies song. Tongan crafts, such as mats made of woven palm leaves, are considered the most beautiful in the Pacific.

**TAPA**
*A woman paints on bark from a mulberry tree, which has been beaten into a paper known as* tapa.

POLYNESIA

## Hawaiians

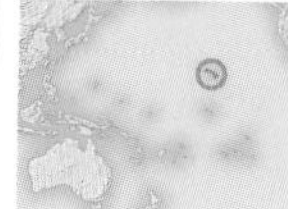

**Population** 340,000 (including 100,000 on US mainland)
**Language** English, Creole English, Hawaiian (a Polynesian tongue related to Tahitian and Maori)
**Beliefs** Mainly Christianity
**Location** Hawaiian island chain and US mainland

Native Hawaiian people, or *Kanaka Maoli*, are descendants of a strand of the Polynesian culture that spread across the Pacific by outrigger canoe, reaching Hawaii by around AD300. In isolation, on one of the world's most remote island groups, Hawaiian people developed a sophisticated political system and a society featuring an array of specialist classes. Among them were priests, chiefs, and craftsmen. Today, native Hawaiians are mostly of mixed ancestry and share their islands with settlers from the US and eastern Asia. They have not always enjoyed high status relative to these newcomers. The 1970s, however, saw a renaissance of Hawaiian language and customs, including revival of the hula dance and traditional healing. A source of ethnic pride was provided by the 1976 voyages of the sailing canoe *Hokule'a*, guided only by stars, wind, and waves.

**OUTRIGGER**
*Hawaiian men propel a canoe fitted with a traditional outrigger.*

PEOPLES

POLYNESIA

# Tahitians

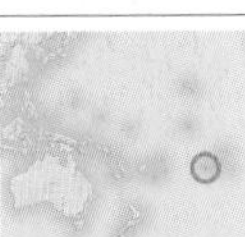

**Location** Tahiti and other parts of French Polynesia

**Population** 200,000 (109,000 of which are on Tahiti)
**Language** Tahitian (a Polynesian language related to Maori), French
**Religion** Protestant and Catholic Christianity

The Tahitians are a Polynesian people who first settled on the island of Tahiti between AD300 and 600, after long canoe journeys. Many Tahitians identify with a greater Polynesian nationality, which includes all the islands in the triangle between Hawaii, Easter Island, and New Zealand. The culture of these eastern Polynesians is called *Ma'ohi*, from which comes the word *Maori*, the name of the New Zealand Polynesians. The name of the ancient sacred buildings of Polynesia, *marae*, also stems from the same word.

The appearance of most Tahitians reflects their *Ma'ohi* roots. Many are tall, broad, and muscular people. Genes from elsewhere are now mixed in, however, and today's Tahitians may also claim Chinese or French ancestry.

Tahiti became a French colony in 1842, but decades earlier, Christian missionaries had begun stamping out Tahitian culture and imposing Christian values. *Maraes*, where Tahitians once worshipped many gods, were destroyed. *Heiva*, or festivals, were banned because they featured not only frenetic drumming, but also supposedly erotic, hip-wiggling dancing.

Throughout the 20th century, Tahitian cultural identity slowly emerged again. *Heiva* returned, and the distinctive hip-wiggling dance is now a mainstay of both tourist entertainment and cultural identity.

Tahitian society today includes a third sex called *mahu*. They say that someone is *mahu* just as someone is a man or a woman. *Mahu* are born male, but their families raise them as girls. They take on the tastes and appearance of Tahitian women. They help in the house, raise children, and in tourist areas, they tend to work in hotels and restaurants.

Flowers are common in Tahitian customs. Tahitians weave *tiare* flowers, which are abundant on high ground, into garlands. They bestow these garlands on visitors and those leaving the islands. Women can also signal that they are attached with a *tiare* flower behind the ear.

Before trade with Europeans began, Tahitians exploited local raw materials. They made drums with shark skin, and their cloth came from the bark of the breadfruit tree and paper mulberry. For sustenance, they fished and tended gardens of taro, yams, and sweet potatoes. Today, most goods are instead imported at great cost. People tend to work in commercial fishing, mainly for tuna, in culturing black pearls for export, or in the tourism industry. The French military was also a major employer until nuclear testing in the region ended, amid protests, in 1996. Tahitians now receive subsidies and compensation money from the French government for damage done by the tests, which first began in 1962.

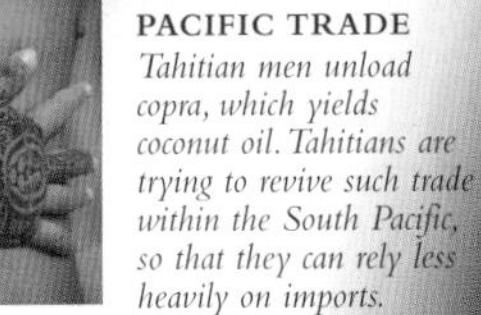

**TATTOOS**
*Throughout Polynesia, tattooing was demanded by social custom. Its revival in Tahiti fulfils a desire for cultural identity.*

**PACIFIC TRADE**
*Tahitian men unload copra, which yields coconut oil. Tahitians are trying to revive such trade within the South Pacific, so that they can rely less heavily on imports.*

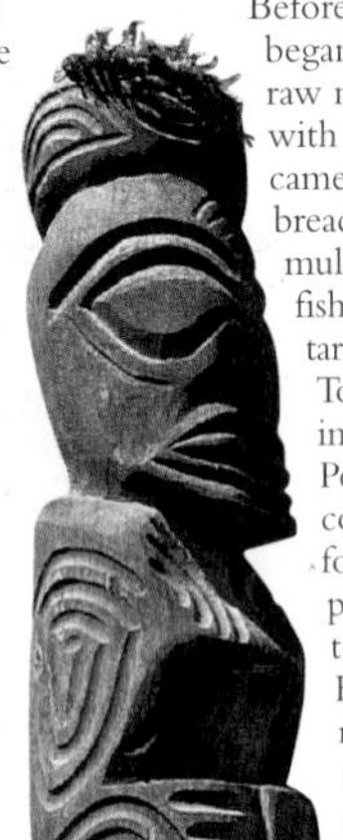

**TIKI CARVINGS**
Tiki *figures represent gods or ancestors. Missionaries outlawed such carving, but many figures survive.*

History

## Gauguin's Tahiti

The French painter Paul Gauguin settled in Tahiti in 1891 to "renew himself in the purity of nature". The primitive clarity of his pictures influenced art worldwide in the following century. In Tahiti, his impact is still apparent in the profusion of galleries and in the many painters who have followed him.

PEOPLES

**FIRE DANCE**
*Tahitians may use traditional dances such as the fire dance to entertain tourists, but for many, they are also an opportunity to assert and celebrate their cultural identity.*

POLYNESIA

## Rapa Nui

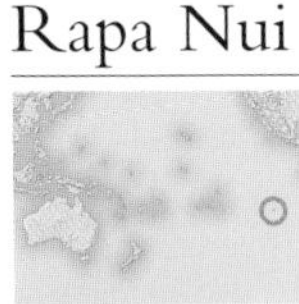

**Population** 2,500 (2,200 on Easter Island, 300 on the Chilean mainland)
**Language** Spanish, Rapa Nui (a Polynesian language of the Malayo-Polynesian family)
**Beliefs** Christianity
**Location** Easter Island and Chilean mainland

The people of Easter Island call their land, their language, and themselves Rapa Nui. The island has been part of Chile since 1888 but was first settled by Polynesian people in about AD300. At that time, experts think it was called *Te-Pito-Te-Henua*, or "the navel of the world". Isolated on their remote island, the Rapa Nui developed a rich culture that featured Oceania's only pre-20th-century writing system – a hieroglyphic script called *rongorongo*. Among their beliefs was the Bird-Man cult, which centred on the quest for the first sea-swallow egg of spring. The winner of the quest became Bird-Man, the reincarnation of the bird-god *Make Make*, until the following spring.

The Rapa Nui exhausted the island's resources and became marooned, with no trees left to make canoes. Poverty, civil war, and tragic encounters with Europeans reduced their numbers and divorced the people from their oral history. Today, the Rapa Nui have revived their culture and hold the annual Tapati festival honouring their heritage.

**LIVING TRADITION**
Kai kai – *telling stories with string figures – is a skill passed down the generations.*

History

### Moai

The Rapa Nui people once carved gigantic stone figures called Moai. In their full glory, the volcanic basalt carvings had eyes of shell and coral and were topped with red stone headdresses. The Moai's secrets were lost when most of the priestly class of the island died during South American slaving expeditions.

MOAI STONE FIGURES

MICRONESIA

## Palauans

**Population** 16,000
**Language** Palauan, English, Sonsorolese, Tobi, Angaur
**Beliefs** Roman Catholic and other forms of Christianity, Jehovah's Witness, Modekngei, Baha'i
**Location** Palau (western Pacific Ocean)

Palau – an archipelago of more than 200 islands – was originally settled by a mix of Malays, Melanesians, and Micronesians. From the 16th century, Palau was owned by Spain; in 1899, Germany took over; in 1914, the Japanese occupied; and finally (in 1944), the US assumed control. Palau became a sovereign state in 1994. Modern Palauans therefore lead a mostly Western lifestyle, and since Japanese occupation, property has been held by individuals instead of being owned communally, as was traditional. Western diseases and the introduction of Christianity greatly changed the Palauans. Diseases killed 90 per cent of the native Palauans and the population is yet to return to its pre-colonial numbers. Missionaries, who had been visiting Palau from before the 19th century, brought with them conservative sensibilities concerning social behaviour; these attitudes are now shared across the Pacific islands. Over two thirds of Palauans are strict churchgoers. *Modekngei*, the Palauan traditional religion, was banned under German colonial administration but has recently undergone a revival. Based on indigenous polytheistic traditions, the religion adapted Christian beliefs and involves the worship of a supreme god called Ngirchomkuuk. Religious teaching is passed from generation to generation through *keske* chants. Traditional *Modekngei* rites are still important elements of key events such as funerals.

**MEN'S HOUSE**
*The elaborately carved bai is the focus of village life. It is used exclusively by male chiefs.*

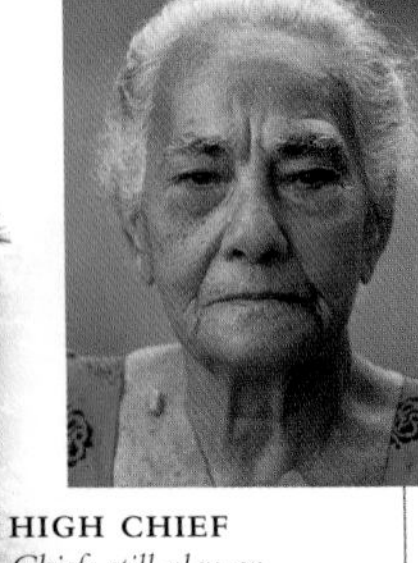

**HIGH CHIEF**
*Chiefs still play an important role in Palau. This woman is one of the high chiefs.*

MICRONESIA

## Chamorro

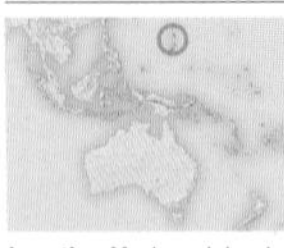

**Population** 78,000
**Language** Chamorro, English, Japanese
**Beliefs** Christianity; traditional beliefs (involving the worship of *taotaomona*, or ancestor spirits)
**Location** Mariana Islands (western Pacific Ocean)

The Chamorro are descended from Indo-Malaysian and Polynesian seafarers that arrived on the Mariana Islands around 3,500 years ago. When Magellan visited in 1521, he dismissed the Marianas as "islands of thieves" and burned a village. The first permanent Spanish colony was established in 1668 and lasted until 1899. During this time, war and disease caused a drop in the population from about 50,000 to less than 5,000 (by 1741). It is thanks to the women (traditional Chamorro society is matrilineal) that the culture survived this period to thrive today. After the Spanish, brief German and Japanese occupations followed. The islands are now all US territories.

The Chamorro were described by early visitors as a powerful and muscular people. They sailed long distances in swift *proa* (canoes) to trade, and subsisted on fishing. Worship of *taotaomona* (ancestor spirits) was important, and the bones of the dead were often fashioned into spears, imbued with the ancestors' power. There were three traditional classes: the *matua* nobility, the arbitrators of disputes; the semi-noble *atchaot*; and the *mangatchang* lower orders. Today, Chamorro culture retains the philosophy of reciprocity and respect, with people greeting elders by kissing their hands. Many traditional ceremonies are also still observed. However, one custom has had a severe impact on people's health; recent research has attributed the unusually high incidence of neurological disease to the eating of flying-fox bats on ceremonial occasions.

**PERFORMER**
*A Chamorro dancer wears a coconut-leaf cloak and a flower headdress to take part in a local festival.*

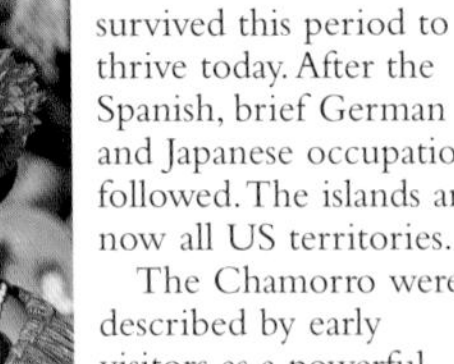

**RELIGIOUS PROCESSION**
*Young girls carrying bouquets take part in a procession during a Christian festival in Guam. The festival honours Our Lady of Camarin, the patron saint of Guam.*

MICRONESIA

## Chuukese

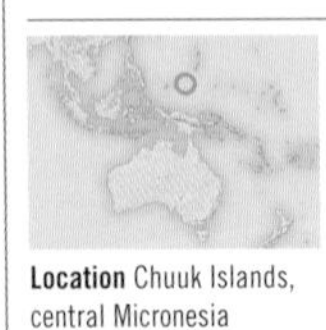

**Population** 50,000
**Language** Chuukese (Trukese), English
**Beliefs** Christianity; also spirit beliefs (many women still undergo a trancelike possession by the spirits of ancestors and nature)
**Location** Chuuk Islands, central Micronesia

Around 2,000 years ago, Malay or Polynesian seafarers settled the Chuuk Islands. Europeans first appeared in 1565, when a Spanish ship was repelled by islanders throwing spears from their canoes. Serious Western interest came in the late 19th century with the arrival of US missionaries, and the islands have been closely allied to the US ever since. Japanese occupation during World War II ended in 1944 following an enormous battle with US forces, which left more than 100 wrecks in the lagoon at the centre of the major Chuuk islands.

**LOCAL FACE**
*The Chuukese, such as this islander, have largely Micronesian facial features.*

**FEAST FROM THE SEA**
*Men cook fresh turtle meat over an open fire after a successful fishing expedition to a nearby island.*

MICRONESIA

## Yapese

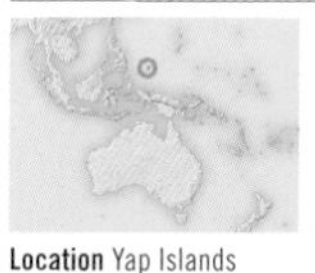

**Population** 12,000
**Language** Yapese (part of the Austronesian language family), English, Ulithian, Woleaian, Satawalese, Japanese
**Location** Yap Islands (western Pacific Ocean)
**Beliefs** Christianity; traditional beliefs

The Yapese are descended from Indo-Malaysian migrants who came to the islands around 3,000–4,000 years ago. In the 16th century, Portuguese invaders arrived, followed by the Spanish and Germans in the 1800s and then the Japanese between the World Wars. In 1945, US forces seized control. The islands have been independent for around 20 years but remain closely tied to the US.

In traditional Yap culture, women remain in the family dwelling, and men fish or spend time in the men's *faluw* (hut). Until the 20th century, it was common for women to be taken to the *faluw* for young men's sexual initiation. Yapese society is caste-based, and until recently, members of different clans would eat from separate clay dishes. Although US culture is influential, the Yapese still often wear traditional dress, such as the *thuu* loincloth for men and skirts woven from leaves for women.

**GIANT STONE MONEY**
*The Yapese traditionally used big stone discs as money. Discs were valued according to age, type of stone, and how difficult they were to obtain.*

MICRONESIA

## Marshallese

**Population** 56,000
**Language** Marshallese (an Austronesian language; has two intelligible dialects), English
**Location** Marshall Islands (central Pacific Ocean)
**Beliefs** Protestant (mainly United Church of Christ) and Catholic Christianity

The Marshallese arrived on their current atolls around 2,000 years ago. Already proficient sailors, they refined their navigation and canoe-building skills to become the most highly regarded seafarers in the Pacific.

Living in close proximity on small land areas, the Marshallese have created a uniquely benevolent culture. Anger is seen as an inappropriate response to conflict situations and is often met with laughter in an effort to dispel tensions. An ability to remain calm and friendly is highly praised.

Marshallese society has a strict class structure that is still observed today. Each person belongs to a *jowi* (clan). There are three strata: the *Iroij* (chief) administers land and settles disputes, while the *Alap* (*jowi* head) oversees the daily tasks carried out by the *Rijerbal* (workers). Land inheritance is matrilineal.

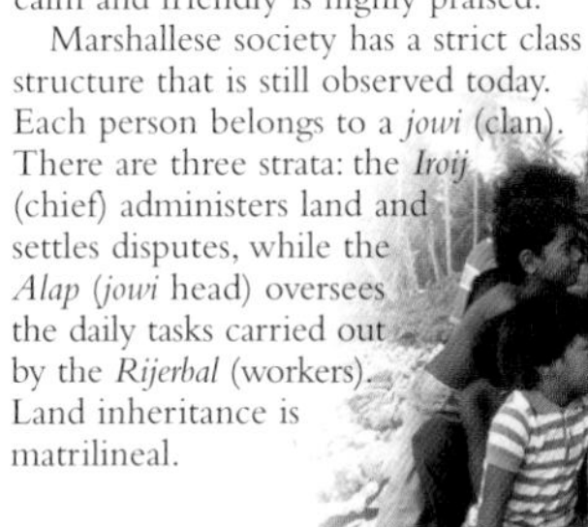

**ORAL TRADITION**
*Community historians keep the past alive by passing on remembrances to the next generation.*

History

### Atomic testing at Bikini

Bikini Atoll – a ring of about 20 coral outcrops that are part of the Marshall Islands – was used by the US during the 1940s and '50s as the base for experiments to determine the effect of atomic weapons on naval ships. The natives of Bikini – such as these people at Kili Atoll – were relocated throughout the Marshall Islands. From the late 1960s, some returned to Bikini but were unable to remain due to dangerously high levels of radioactivity.

The Chuukese people have a practical outlook on life, and land – which indicates status and provides income – is the most valued possession. Houses are built of palm thatch, and families maintain close bonds. The culture is patrilineal, with women having notably lower status than men. The Chuukese people are also known for a courtship tradition involving *fanai* love sticks. Young men push these elaborately carved, dagger-shaped wooden rods through the thatch of a female's dwelling at night. The woman, who will recognize the carvings, then either accepts or rejects his advances.

**FISHING A CORAL LAGOON**
*The clear waters of the Chuuk Islands are home to an abundance of fish. Women as well as men are involved in fishing.*

# FUTURE

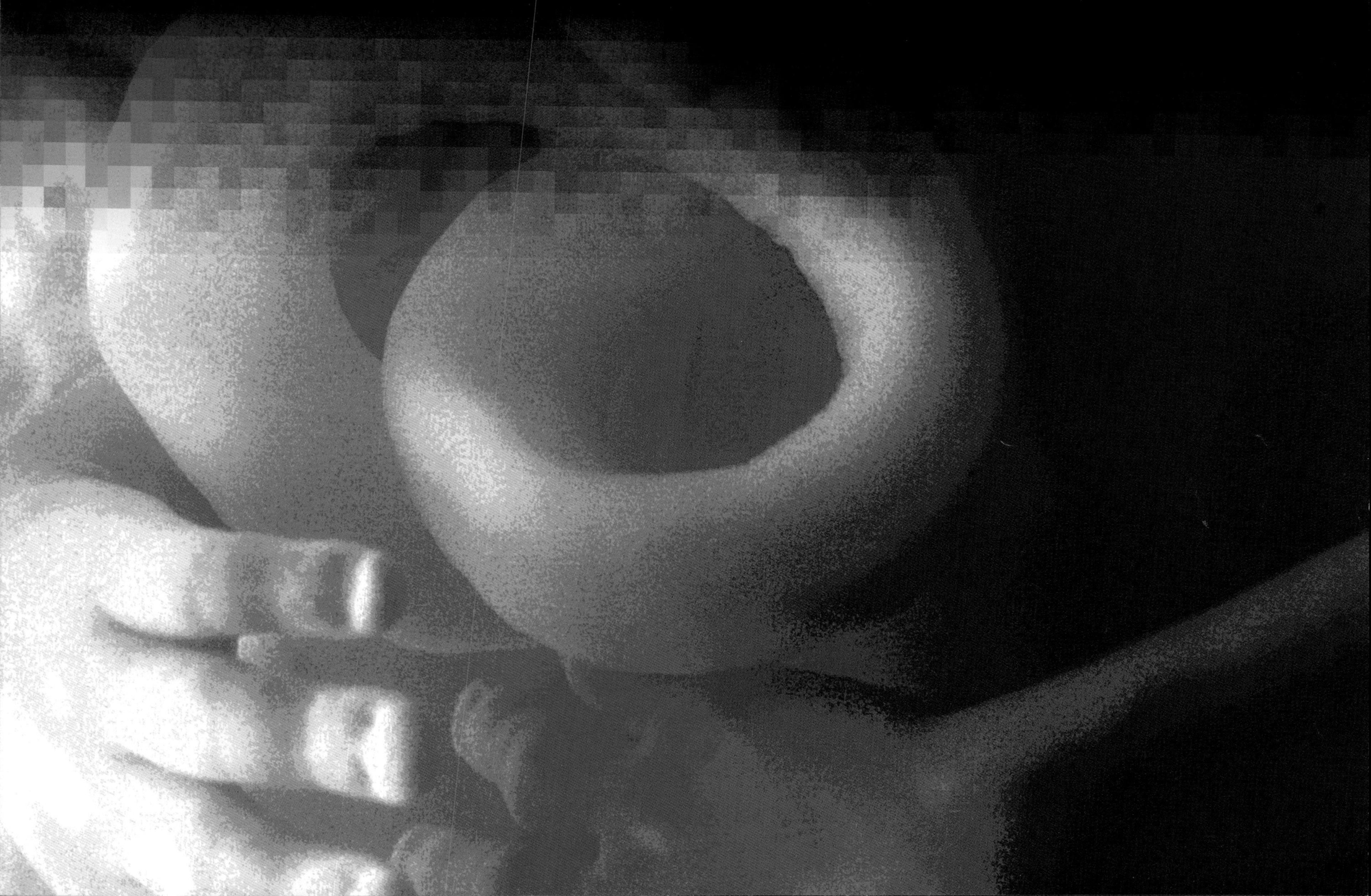

# FUTURE

Our species has existed for well over 100,000 years, but in just the last few centuries, there has been an explosion of technology and an extraordinary expansion of human activity. In a fraction of the time that *Homo sapiens* has been living, we have invented printing, harnessed electricity, created computers, and flown men to the Moon. The pace of change is now increasing so fast that it is hard to predict with any certainty what the future will hold for us.

**SEX CHROMOSOMES**
*By selecting sperm that carry X (left) or Y (right) chromosomes, doctors may be able to inseminate a woman with either "male" or "female" sperm and thus guarantee the gender of a baby.*

Many scientists state that the Darwinian process of evolution by natural selection, in which only the "fittest" survive, now operates more weakly on our species than it did in the past, mostly because we can control our environment by artificial means. We take pride in our mastery of nature, our modern standards of living, and our ability to lead longer, healthier lives than ever before. However, the advantages that technology and medicine bring are not available to all. One third of the world's population are still malnourished and prone to preventable diseases, and in many poor countries, large numbers of children still die from starvation and infectious disease. Indeed, infection with bacteria and many other parasites is a growing problem worldwide, a threat to rich as well as poor people. Germs have a very short generation time (some can reproduce every 15 minutes) and so can evolve alarmingly quickly. They can get round any defence that the human immune system can construct and overcome any drug that our ingenuity might devise.

So survival of the fittest still matters. But perhaps survival of the species is becoming an even greater concern. More than ever, we need to be aware of our global responsibilities. We need to remember that the Earth is a small planet and a fragile environment, and that conditions afflicting people in poor parts of the world are of real concern to luckier people in richer countries.

## FUTURE MEDICINE

Our understanding of the human body and mind improved enormously in the 20th century, in part due to a plethora of new technologies from magnetic resonance imaging (MRI) to DNA sequencing. The coming decades promise even more exciting developments. The rapid accumulation of knowledge in such fields as reproductive biology, genetics, pharmacology, and tissue engineering will almost certainly enable us, in the future, to choose the gender and possibly some of the characteristics of our children. Genetics will help us to predict and possibly avoid the diseases to which we are most susceptible. We may enhance our physical abilities with machines implanted in the body. In addition, computer technology may be used to increase our mental power.

**MRI BODY SCAN**
*Techniques such as magnetic resonance imaging (MRI) give astonishingly clear, detailed views of the body's interior.*

## FERTILITY ON DEMAND

It is a reasonable prediction that, within 50 years, nearly every infertile person in the developed world will be able to have a genetically related child. Various forms of assisted conception already exist. One is in vitro fertilization (IVF), in which eggs and sperm are incubated together in a laboratory. Another, used for male infertility, is intracytoplasmic sperm injection (ICSI), in which sperm are introduced into an egg with a microscopically small needle. These techniques will become more widely available as they grow cheaper.

One likely development will be the ability to mature eggs from tiny slivers of tissue taken painlessly from a woman's ovary. This approach could extend childbearing life, lessen the emotional trauma associated with IVF, and dramatically reduce the cost of treatment. Embryos will be chemically screened so that doctors can choose those most likely to implant successfully in the mother's uterus. The success rate of IVF treatment will therefore increase, and the risk of multiple pregnancy will be reduced.

It may become possible to create "artificial" eggs or sperm from other body cells, to treat people who cannot produce viable sex cells. It may also become possible to treat an infertile woman by transplanting her genetic material into an empty egg cell from a donor woman, creating a viable egg – an extension of cloning technology.

Certain other innovations, however, are less likely. Uterus transplants are unlikely to happen in coming decades. Doctors have also tried growing animal fetuses outside the body in an artificial womb, but the technology looks unlikely to be of great benefit, largely because the placenta – the baby's life support system – needs a natural environment in order to develop. Although conventional childbirth may be painful, inconvenient, and occasionally hazardous, it is unlikely that there would ever be much demand for artificial wombs. Mothers have a powerful instinct to conceive and deliver their babies naturally.

## REPRODUCING BY CLONING

Cloning has been suggested as a way of helping infertile people who cannot produce sperm or eggs. It involves transferring a person's genetic material into a donated egg that has had its nucleus removed, producing an embryo. The

Fact

### Artificial uterus

A few researchers are working to create an artificial uterus that might liberate women from the need to carry babies. In 1999, a Japanese scientist kept a goat fetus alive for 10 days in a clear acrylic tank filled with amniotic fluid. However, the technology entails huge technical difficulties, and it seems unlikely that artificial uteruses will be of any use to humans in the foreseeable future.

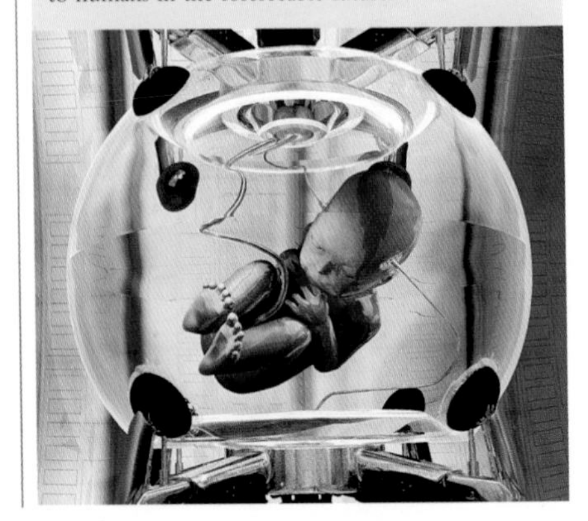

FUTURE

**INJECTING SPERM**
*Sperm are injected into an egg. This form of assisted conception, known as ICSI, is highly effective and is expected to improve even more in the future.*

embryo would become a child that is genetically almost identical to the parent (although a tiny fraction of the child's DNA, less than 0.00005 per cent, would come from the egg).

Cloning is likely to remain too risky to use in humans. The technique has been used experimentally in animals, and it tends to cause severe genetic abnormalities in nearly every species. Even if these problems are overcome, cloning may remain controversial because so many people believe that such manipulation devalues human life. However, some people argue that cloning is not as unnatural as it might appear – many natural clones already exist, in the form of identical twins.

## ASSESSING EMBRYOS

People have always wanted to influence their children's characteristics. In recent years, however, altering the inherent attributes of offspring has started to become a possibility.

Preimplantation genetic diagnosis (PGD) involves selecting embryos created by in vitro fertilization. PGD is usually done a few days after fertilization, when one or two cells can be removed from the embryo. DNA tests and chromosome tests are then carried out on the cells to reveal the embryo's sex and certain other characteristics. PGD is already used to screen embryos for genetic diseases. It is also increasingly employed to select embryos that give the best chance of a successful IVF treatment, and to reduce the risk of miscarriage in patients who carry a chromosomal abnormality. Some people have speculated that PGD could be used to test embryos for "desirable" characteristics – such as high intelligence, mathematical or musical aptitude, strength, and sporting ability. However, testing for such characteristics might prove to be very difficult. These complex elements of human personality are the result of an interaction between many different genes and other factors, and PGD seems too crude a process to have much of an impact in this area.

PGD could, in theory, be used to select a baby's gender, but simpler methods (such as sperm selection) are being developed for that purpose. The extent to which gender selection will be practised in future societies will be shaped partly by cultural influences and partly by legal constraints. Nevertheless, it is likely to become commonplace within a few decades at most. It remains to be seen whether it will affect the proportion of boys and girls that are born.

## DESIGNER CHILDREN

Germ-line engineering (also called germ-line therapy) is an even more radical idea than preimplantation diagnosis (*see above*), giving future parents a real prospect of having a child with a specific set of desirable characteristics.

The therapy is not yet technically feasible, but would involve changing genes in sperm, eggs, or embryos to create genetically modified people. The altered genes would exist in every cell in a person's body and would be passed on to his or her descendants. Germ-line engineering in humans has already been advocated as a future means of preventing genetic disorders, but if it does become possible, it is more likely to be used to "improve" babies – by giving them genes to make them intelligent or attractive, for instance.

Research into germ-line engineering is outlawed in many countries, largely because the risks are considered too great; there are serious worries that genetic changes could disrupt the development of the fetus in unexpected ways or lead to tragic results in later generations, for example. Because this technology could have such dangerous and unpredictable consequences, it seems very unlikely that it will be used on any scale for many years to come.

## CONTROLLING FERTILITY

To date there is no such thing as an ideal contraceptive. None is 100 per cent effective, reversible, and free from side-effects, and it is likely that research will continue until better contraceptives are created. One long sought-after goal is a safe hormonal contraceptive for men – the so-called "male pill".

Issue

### Striving for perfection

Many people have deep concerns about how genetic science could change the way we have children in the future. There are fears that parents will one day be able to design the personality and appearance of their children from a menu of genetic options. Valuable genes would be highly sought after, and people would pay a premium for the genes that could make their children intelligent, successful, popular, or attractive. If such technology does become possible (which some scientists doubt), it would probably be available only to the wealthy few. In time, society would gradually become divided into a genetically enhanced elite and a biologically inferior underclass.

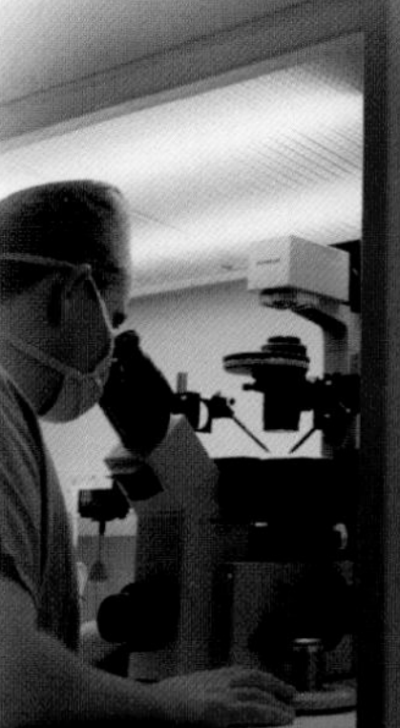

**SCREENING EMBRYOS**
*Doctors can already screen embryos for genetic defects by removing cells under the microscope for DNA tests.*

The male pill would work by reducing a man's sperm count to such a low level that he would become unable to father a child. The problems with current male contraceptive pills are that they work slowly and their effects last for varying amounts of time and are not immediately reversible. Developing a more fast-acting and short-lived alternative has proved difficult, but it seems likely that a safe and effective male pill will be available by about 2010. This contraceptive would be based on an injection of progestogen (a synthetic form of the sex hormone progesterone) every few months, accompanied by testosterone-releasing implants.

Future research is likely to focus on contraceptives that work by targeting specific genes or proteins that play a key role in reproduction. It is hoped that such contraceptives would cause fewer side-effects than conventional hormonal drugs, but it will be at least 10 years before any become widely available.

## USING GENETIC PROFILES

One major benefit expected to come from the Human Genome Project (which has deciphered the genetic instructions for making a human being) is genetic profiling. A genetic profile would reveal the diseases from which a person is most likely to suffer. Such profiles would enable people to avoid activities or substances that might trigger their diseases. Profiles would also enable doctors to tailor medical examinations and treatments to individual needs. A person's response to drugs is currently influenced in unpredictable ways by his or her genes. As we learn more about how genes and drugs interact, doctors will be able to prescribe only drugs that are compatible with a person's genetic profile.

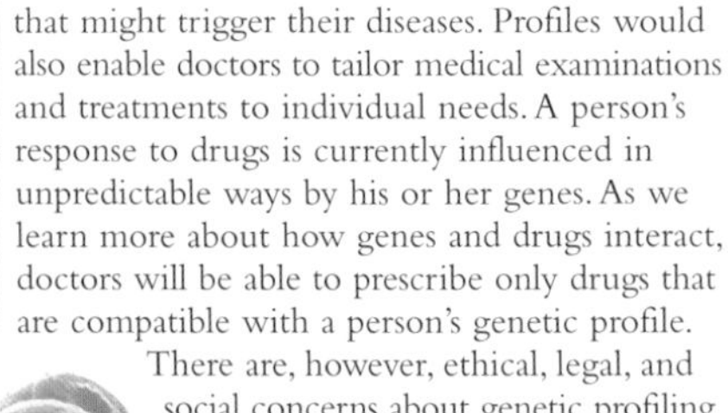

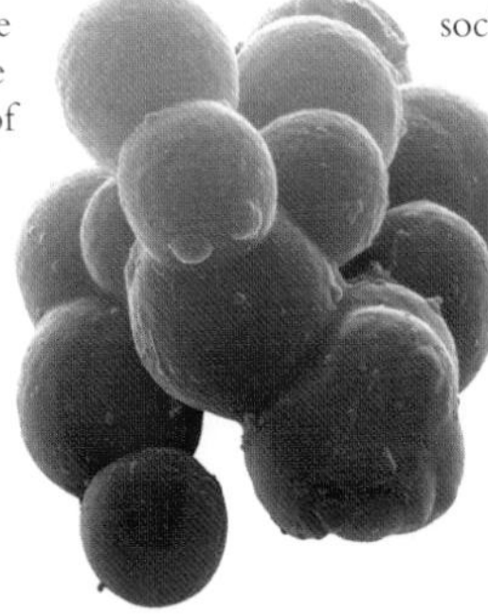

**STEM CELLS**
*Embryonic stem cells can be collected from the blood in a baby's umbilical cord and could be used to generate many types of body tissue.*

There are, however, ethical, legal, and social concerns about genetic profiling. Some people might feel doomed by a discouraging profile and choose not to marry or pursue particular careers. Insurance companies or employers might demand access to genetic profiles and discriminate against people seen as having problem genes.

## STEM CELLS

A promising area of medical research involves the use of primitive body cells known as "stem cells". These cells have the unique ability to develop – to differentiate – into many kinds of mature cell, such as skin, blood, liver, brain, or muscle cells. Most body cells are already fully differentiated and specialized for one particular function. When cells become specialized they often lose the ability to divide and generate new cells, and those that can still divide can only produce cells of the same type. Stem cells, in contrast, can renew themselves and give rise to a wide range of tissues. The cells in embryos are supreme examples of stem cells. As an embryo develops and grows into a baby, its stem cells differentiate to form all the tissues of the human body. In babies, children, and adults, stem cells can currently be obtained from only a few parts in the body, such as bone marrow.

Stem cells taken from bone marrow seem to be able to form only a limited range of body tissues, but the evidence suggests that stem cells from embryos can be manipulated artificially to give rise to any type of tissue. Embryonic stem cells, therefore, seem to have most potential for treating disease. It is thought that the blood in a baby's umbilical cord, which can be collected shortly after birth, might be a good source of stem cells, but isolating and growing them from this source has proved to be difficult.

Stem cells could be used to regrow damaged tissues or organs for transplant operations. One of the current problems with transplants is that foreign tissues, such as donated organs, are likely to be rejected by the immune system unless the patient takes powerful immunosuppressant drugs. Stem cell research may lead to several ways around this problem. One strategy is to transplant tissues that have been derived from a patient's own stem cells. Another is to modify embryonic stem cells genetically so that tissues derived from them are less likely to trigger an immune reaction even if the cells are implanted into another person. A third option is therapeutic cloning. This technique would involve fusing the patient's genetic material with an empty egg cell from a donor. Any embryo that developed thereafter would be stimulated to produce stem cells, and these cells would be immunologically compatible with the patient.

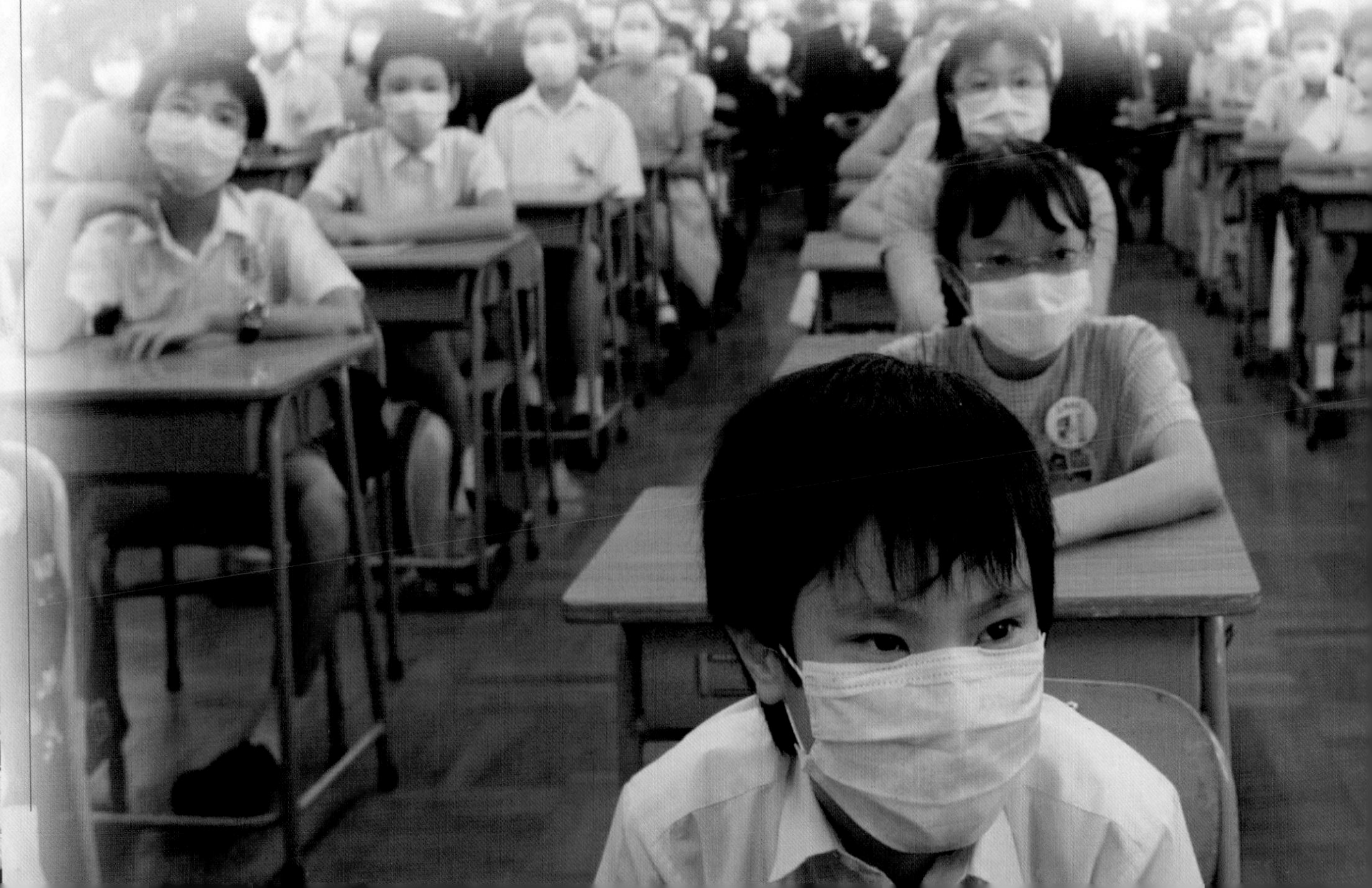

If research into therapeutic cloning continues, the fruits of the technology might be seen as soon as 2020. However, some people oppose the research on ethical grounds because it involves the utilization of human embryos.

## TREATING GENETIC DISEASE

In coming decades, scientists are likely to discover many of the genes that underlie a wide range of diseases. The next step will be to study the proteins produced by both normal and abnormal genes – a field known as proteomics. Scientists hope that proteomic studies will lead to the development of drugs that act on these proteins. The drugs could be used to treat currently incurable genetic diseases, and would cause fewer side-effects than existing drugs. By the year 2020, thousands of these new drugs may be available.

Another technique, a likely by-product of the Human Genome Project (*see* Using genetic profiles, *opposite page*), is gene therapy. This technology aims to cure people suffering from genetic disorders caused by a single faulty gene. A corrected copy of the faulty gene is introduced into body cells, where it becomes active and produces a protein that was previously missing or abnormal. There are few ethical objections to gene therapy; it is akin to giving the patient a transplant, although the transplant is a gene rather than an organ.

Researchers have been trying to develop effective gene therapies for more than a decade, but the various methods have so far failed to live up to the much-publicized expectations. As yet, there are no approved gene therapy products on sale, and the technology remains purely experimental. One of the stumbling blocks is integrating the inserted gene with the patient's DNA so that body cells reproduce the gene when they divide (*see* Making new body cells, p53). Unless this copying can be achieved, the therapy can have only short-term benefits. Researchers are currently experimenting with viruses that can insert genes into DNA, but there are serious concerns that such viruses might cause novel diseases. Other techniques under development include using liposomes (tiny droplets of fat) to carry genes into cells, and adding an extra chromosome to a patient's cells.

If the hurdles can be overcome, gene therapy could become increasingly important as more disease-causing genes are discovered. However, most common diseases involve lots of genes working together rather than a single faulty gene, and gene therapy is unlikely to be helpful for these diseases for many decades.

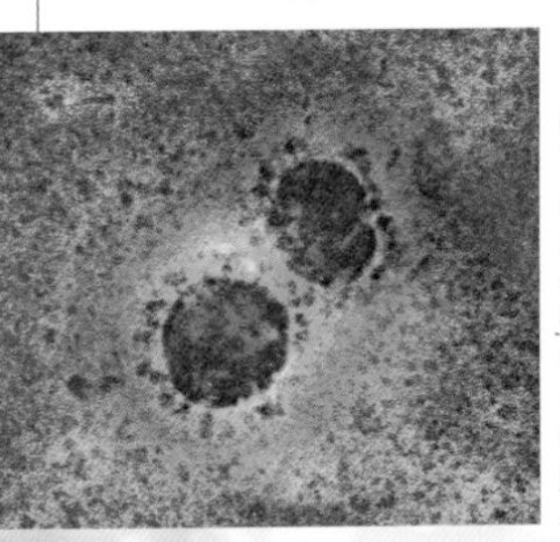

**SARS VIRUS**
*The SARS virus is a mutant form of the virus that normally causes colds. The future will see many more viruses mutating and becoming more dangerous, with potentially devastating consequences.*

**A NEW THREAT**
*Chinese schoolchildren had to wear masks to ward off the SARS disease in 2003. Even deadlier viruses are likely to appear in the future.*

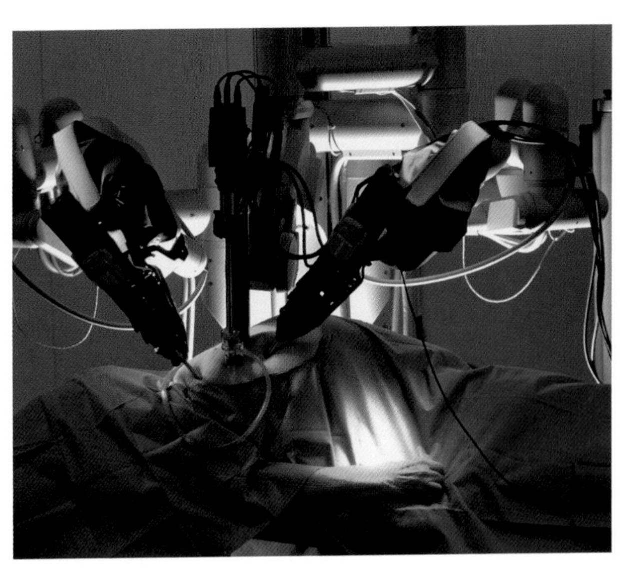

**ROBOTIC SURGERY**
*Surgical robots could become widespread in the future. As well as allowing surgeons to work from a distance, the robots can make much tinier and more precise cuts than a human hand, reducing healing times.*

## NEW DISEASES

The AIDS epidemic, and the 2003 outbreak of SARS (severe acute respiratory syndrome) and bird flu (avian influenza), have all alerted the world to the dangers posed by new diseases. AIDS is thought to have first infected humans in equatorial Africa, where a closely related virus (simian immunodeficiency virus) infects chimpanzees. SARS probably originated from a cold or flu virus that mutated and became much more virulent, and bird flu is caused by a potentially lethal form of flu virus that can jump from chickens to humans.

In the future, outbreaks of new viral diseases and the evolution of antibiotic-resistant bacteria will pose grave threats to health. As the human population expands, cities become larger, and international travel increases, the potential for new bugs to evolve and spread will grow.

## FUTURE SURGERY

Computer technology is likely to play an increasing part in future surgical procedures, particularly in the planning of complex, novel, or delicate operations. Many procedures will be rehearsed on a deformable, three-dimensional computer model of a patient's body before being performed in reality. Such rehearsals will enable surgeons to predict the potential complications of a particular procedure and to change their surgical strategy accordingly.

Many surgical operations will be performed with the surgeon at a distance from the patient – perhaps in a different hospital or a different city, or even on a different continent. As a result, highly specialized surgeons will be able to treat people in widely scattered locations. The surgeon will view the operation using a high-bandwidth communications link and carry it out by manipulating robotic instruments remotely.

Issue

### Nanotechnology

An area of general technology that is expected to have widespread applications in future medicine is nanotechnology. This field involves constructing tiny machines and materials atom by atom, such as the "nanotube" below. In theory, nanomachines smaller than body cells could be guided to remove obstructions in blood vessels or kill cancer cells. Drugs could be carried to specific target sites in the body by "nanoshells" – tiny, medicine-packed particles whose movement and action could be controlled by light or infrared radiation.

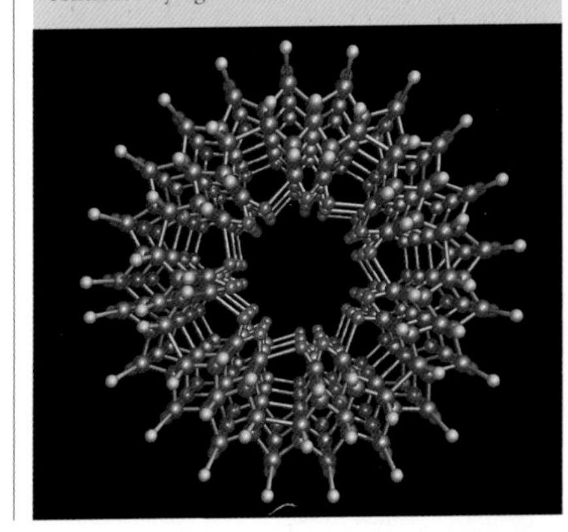

**ARTIFICIAL SKIN**
*Skin transplants are increasingly performed with artificial skin, which is grown in culture from living skin cells. It takes about 3 weeks to grow about 1 sq m (11 sq ft) of skin.*

**ARTIFICIAL CORNEA**
*Transplant operations to replace the front part of the eye (the cornea) are already routine. The future will see more use of artificial rather than donated corneas.*

Until it becomes possible to grow replacement organs from stem cells or make artificial organs, the demand for donor organs for transplant operations will remain very high. The shortfall might be made up with organs from animals, such as pigs. Transplants from animals, known as xenotransplants, have the potential to save many lives, but they could also provoke scares about the spread of new viral diseases.

Transplants of specific types of tissue will become more common, and could lead to cures for some diseases that are currently treated with drugs. For example, it might be possible to cure insulin-dependent diabetes by transplanting the pancreatic cells that make insulin. Transplants of muscle cells could be used to treat heart disease and muscular dystrophy.

## ARTIFICIAL BODY PARTS

There are already millions of people who have "prosthetic" body parts, including artificial joints, blood vessels, teeth, eye lenses, breast implants, heart valves, and so on. Parts that will soon be available include prosthetic blood, skin, tendons, and "biosynthetic" organs, such as an implantable liver that is partly machine and partly living tissue. In addition, we are likely to see implantable gadgets with computer chips that monitor body fluids and release drugs or hormones on demand. People with diabetes, for example, might receive an implanted chip that monitors blood glucose levels and releases insulin when levels are high.

Some experts think technology will go further and allow the human nervous system to interface with computers. It has already been shown that people can operate computers with the power of thought alone, using electrodes on the scalp or wires linked to their nerves. Such technology might lead to improvements in artificial limbs or hands, allowing increasingly complex manoeuvres. Artificial eyes could be made for blind people; they would rely on implanted light sensors and computer chips connected by means of thousands of tiny wires to the brain's visual areas. Crude devices of this type already exist.

Profile

### Kevin Warwick

Professor Kevin Warwick is a cybernetics researcher at Reading University in the UK. He has had a series of computer chips implanted in his body, allowing him to control electric devices in his home, connect his nervous system to a computer, and send crude "telepathic" signals to a chip in his wife's body. Many scientists question the value of Warwick's work and accuse him of courting publicity. Warwick, who says he wants to be a cyborg, claims the research is vital to protect the human race from being taken over by robots.

## ENHANCING OUR BODIES

Prosthetic body parts are purely restorative at present; they give back their owners a body function that was lost due to disease or injury. At some point, technology will give us the ability to enhance our bodies as well as repair them. If a fully functional artificial eye can be developed, why not make one that can see at night or sense wavelengths outside the visible spectrum? If artificial limbs work as well as real ones, why not make them bigger and stronger? And if the nervous system can interface with computers, why not add radio transmitters so that people can communicate telepathically?

Films such as the Hollywood blockbusters *Robocop* and *Terminator* depict a future society populated by cyborgs – entities that are partly machine and partly human. Some commentators think that the evolution of cyborgs is not just inevitable but necessary (*see* Kevin Warwick, *left*). They see a future in which people must either merge with machines or accept domination by robots. Others envisage future humans extending their abilities with external gadgetry rather than implants. Perhaps we will become surrounded by ever-increasing numbers of miniaturized gizmos, such as handheld computers and intelligent phones, with broadband wireless networks putting us in constant, universal contact with the internet and with each other.

## LONGER LIVES

Average life expectancy will increase at only a subdued rate in the immediate future. A low rate is expected because life expectancy is falling in much of Africa as a result of AIDS, although it is increasing in other developing countries thanks to improvements in nutrition and medical care. In the developed world, life spans may be approaching a peak. In the US, for instance, average life expectancy reached a record high of 77.2 years in 2003, but scientists expect life spans to fall in coming decades because of a surge in the incidence of obesity.

As medical technology improves, increasing numbers of people will live to be 110–120 years old, which scientists think is the upper limit for the human life span. Living beyond this age will become possible only when scientists identify the genes that control the aging process and find ways of manipulating them. Some scientists think there may be just a handful of key genes controlling the speed of aging, but increasing numbers suspect that aging is caused by the harmful side-effects of many genes that are vital to health during early life. If this is the case, extending the human life span will prove very difficult and might lead to genetic abnormalities and other health problems during the first decades of life.

Some people try to cheat death by having themselves frozen shortly after dying, in the hope that they can be revived in the future when medical science has become sufficiently advanced to cure them. In some cases the whole body is frozen, but many "cryogenics"

**CYBORG**
*In the film* Robocop, *a murdered policeman is resurrected in robotic form. Such technology, if it ever becomes possible, is many decades away at best.*

enthusiasts choose to have just the head preserved. In either case, large amounts of antifreeze chemicals are introduced into the blood to reduce tissue damage caused by ice crystals. Even so, mainstream scientists think the delicate tissue in the brain would be irreparably damaged by freezing, making revival impossible.

## FUTURE MIND

Human consciousness is one of the biggest conceptual stumbling blocks in science and is likely to be a focus of research for some time to come. It is almost certain that we will learn increasing amounts about human behaviour and psychology in the near future thanks to advances in imaging techniques and neuroscience. Research into the human mind will go in two directions: we will continue trying to uncover the causes of mental illness and find new ways of improving mental health; and we will search for ways of enhancing the intellectual capacity of the human mind.

## BRAIN IMAGING

We are currently able to produce sophisticated scans of the brain using imaging techniques such as positron emission tomography (PET) scanning and magnetic resonance imaging (MRI). The main purpose of these scans is to diagnose problems such as tumours or strokes, but they can also give us an insight into how our brains function. In the future, brain scanners and machines that monitor brainwave patterns could become small enough to mass-produce and sell as household items, allowing us to look into our brains and check our mental state whenever we want. This kind of instant access to the inner workings of the brain will give us, among other things, valuable information about our daily cycle of alertness. At any moment, we will be able to find out where we are on an "ultradian alertness wave" – a daily biological rhythm characterized by troughs of fatigue every 90 minutes and corresponding peaks in which we feel brighter and more awake. If we can tap into this and other biorhythms, we will be able to plan our activities more carefully and determine when work will be most productive and when rest will be most restorative for us.

Personal brain scanners will also allow us to observe the prefrontal cortex – the "intelligent" part of the human brain, which is responsible for conscious thought, planning, and making decisions. The prefrontal cortex is essential for careful thought and concentration, but it tires more quickly than other parts of the brain. With a scanner, we could quickly find out when the prefrontal cortex is at its sharpest and tailor activities accordingly.

**PET SCAN**
*A PET scan of the brain shows which parts are most active (red and yellow). In the future, personal PET scanners might allow us to monitor our mental state.*

## BOOSTING IQ

Is human intelligence limited or can we boost it? Research suggests that average IQ is on the increase in many parts of the world. In Japan, for instance, IQ has gone up by an average of 12 points in the last 50 years or so. Some psychologists think this increase merely reflects an improvement in the ability to complete IQ tests, thanks to improvements in education. Others say humans are actually becoming more intelligent, a trend that might continue.

In the future, research is likely to focus on ways in which intelligence can be artificially increased. There is some evidence that newborn babies given extra oxygen through a lung machine become more intelligent than other children. Such evidence suggests that newborns have a developmental capacity that is restricted under normal circumstances and can be artificially enhanced, in this case with extra oxygen. Another technique claimed to have the power to boost intelligence is transcranial magnetic stimulation (TMS). A TMS machine creates a powerful magnetic field that can "switch off" activity in specific parts of the brain when held close to the head. The device is currently used to predict the effects of brain surgery, but a few scientists believe that it can unlock certain hidden skills that are normally suppressed but sometimes unmasked in people who have autism or brain damage.

Fact

### Artificial intelligence

In the 1970s, great hopes existed for artificial intelligence. During the 1980s, these hopes faded as it became clear that computers have great difficulty grasping context, which the human brain does easily and intuitively. Although computers are a long way from replicating human thought, they excel at logical processing – as Gary Kasparov found during his 2003 chess battle with computer "Deep Junior" (below).

## LEARNING BRAINS

Although all human brains are built to the same general plan, the size of different parts of the brain and the number of connections between the two halves of the brain differ from person to person. This variation has important implications for learning. In the future, schoolchildren could

**MRI BRAIN SCANS**
*An MRI scanner can see soft brain tissues in great detail. By monitoring blood flow, "functional MRI scans" can also show areas of activity, giving a window into our thoughts.*

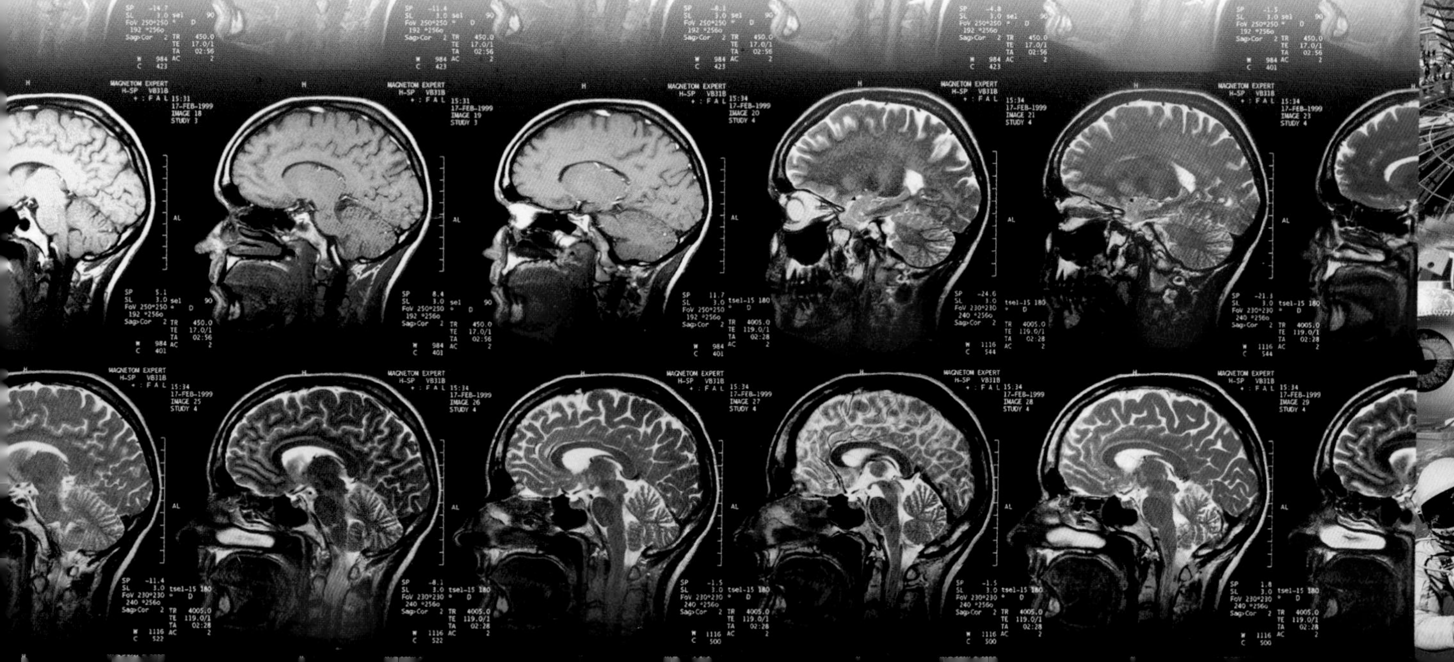

**IN SAFE HANDS**
*Pilots have to stay alert for long periods on international flights. Personal brain scanners would tell them when they would most benefit from taking a rest.*

**FIT TO FLY?**
*This brain scan shows activity in the front of the brain (red and yellow) while a person is concentrating. Such scans could one day be used to assess whether pilots have sufficient mental alertness to fly.*



be taught in ways that are tailored to the specific structure of their brains. Perhaps children will wear electronic headsets incorporating brain scanners so that a brain consultant, resident in every school, can make observations about how well each child is responding to teaching. If the brain scans showed that children were having problems, the teacher could try a different technique or simplify lessons – learning is most effective when the level of difficulty is just right. The scans would help to reveal which children were falling behind or finding lessons too easy, and they would help teachers to work out the times of day when children would be most attentive and receptive.

The nature of school reports would change. In addition to receiving a teacher's personal observations about a child's progress, parents would receive a readout showing how well different parts of their child's brain had coped with the learning process, revealing weaknesses in need of further attention or talents that might prove useful in later career choices.

## FIT FOR THE JOB

The widespread availability of brain-scanning equipment in the future will have an impact on the way that we do our jobs. Using brain scans to detect periods of alertness and fatigue will not only tell us when we are most productive but will also help prevent accidents caused by human error. Brain scans could become vital in screening people for occupations in which safety is paramount, such as the jobs of airline pilot or air traffic controller.

Testing whether someone is mentally fit for work could also be applied to those in powerful positions. The US president, for example, is obliged to have an annual medical examination in which any health issues are detailed. Such health checks take place because past presidents were dishonest about their health and held office when they were not fit to govern. In the future, the president's medical examination might include a check on the integrity and function of the brain, particularly the prefrontal cortex – the part of the brain most needed for making thoughtful decisions.

In addition, brain scans could be used to detect people with specific mental disorders. For example, people who apply for jobs working with children might have a scan to assess their likelihood of committing sex abuse.

## STRESS-BUSTERS

We accept stress as an inevitable part of life, just as we accepted illness as a fact of life in the past. However, stress evolved as a reflex to help us survive in the wild savannas of prehistoric Africa; in the modern urban world, stress is often redundant and sometimes harmful.

The future will bring new ways of managing stress. We will be able to monitor stress levels with "stressometers" worn on the body. Such devices already exist in the form of electronic armbands that monitor the body's sympathetic nervous system, which is activated by stress. Stressometer armbands are currently used only for research, but miniaturization and mass production could make them available to all. Once people can monitor stress, they will be able to use stress-busting techniques to calm themselves more effectively. Instead of feeling stressed while waiting for a delayed plane, for instance, people would look at a small screen on a wristwatch showing breathing rate, pulse, levels of the stress hormone cortisol, and other indicators of stress, and then use relaxation techniques to bring these elements under control. Used over long periods, stressometers would reveal an individual's stress patterns.

Over time, stress-monitoring devices will become smaller and smaller, until they are small enough to be implanted in the body. As well as giving much more precise data on internal physical states, these implanted chips might be combined with chemicals that counter the effect of hormones released during stress.

Antistress implants could be used to treat people with anxiety disorders such as phobias or panic attacks. They could also help to treat serious mental illness; it is well known that intense stress can trigger depressive or schizophrenic breakdowns in vulnerable people. People who work in stressful situations, such as soldiers and rescue workers, would also benefit from antistress implants, which would protect them from post-traumatic stress disorder.

Long-term stress can damage a part of the brain called the hippocampus, which plays a key role in memory. Research suggests that people with anxious personalities are at greater risk of memory impairment and dementia in later life, perhaps because of hippocampus damage brought on by chronic stress. Stressometers will identify such people, and antistress implants would protect them.

## VIRTUAL FRIENDS

For everyday stress, people in the future may turn to electronic friends or robotic pets for comfort and tension relief. Just as children report having imaginary friends, it is possible that adults will increasingly share their lives

Issue

### Virtual pets

Scientists have long known that pets have a calming effect on owners, lowering blood pressure and other symptoms of stress. Makers of robotic pets hope that their inventions will eventually provide just the same benefit, but without the inconvenience of having to be fed, walked, or allowed out of the house to answer calls of nature. Robotic dogs that can learn tricks and respond to voice commands are already available. Future advances in artificial intelligence may allow for more lifelike interactions with owners.

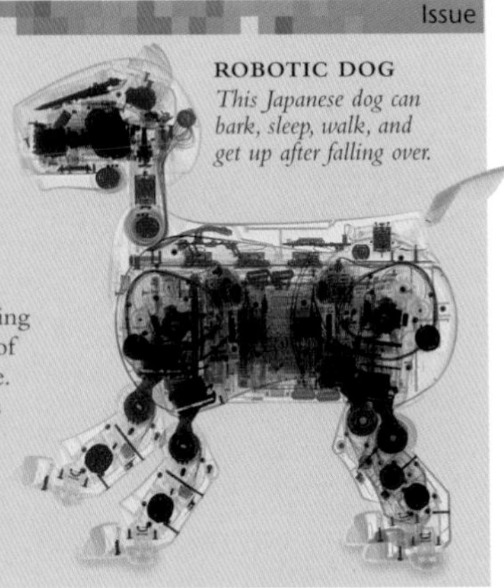

**ROBOTIC DOG**
*This Japanese dog can bark, sleep, walk, and get up after falling over.*

**FIGHTING TRAUMA**
*Anti-stress implants might help people such as rescue workers keep a level head during a crisis and protect them against post-traumatic stress disorder.*

with virtual ones, although some people think there are moral issues involved in encouraging lifelike interaction with machines. Instead of coming back to an empty home, people who live alone would be greeted by a synthetic but perfectly realistic voice, accompanied perhaps by an equally realistic computer-generated face, capable of the full range of human facial expressions. A computer-generated friend could adopt any number of personalities according to a user's mood or requirements. It might be talkative and nosey, or as quiet and discreet as a butler. Alternatively, it might play the role of sympathetic therapist, asking personal questions and giving meaningful advice.

## CRIMINALS AND LIARS

Research suggests that some crimes are associated with particular brain abnormalities. Acts of unprovoked violence, for example, are sometimes linked with impairment of the prefrontal cortex, the front part of the brain's cortex (outer layer). In the future, as well as having a DNA sample taken, someone suspected of an unpremeditated rape or murder might have a brain scan to detect any prefrontal impairment.

Brain scans might also be used to aid lie detection in police interviews. Studies show that taking longer than usual to answer a question is a common sign of lying; such delays occur because the prefrontal cortex takes time to invent and check the plausibility of a lie. In a brain scan, the prefrontal cortex would show unusually high activity. Other technologies that could assist lie detection include stressometers, which reveal telltale activity in the nervous system when an interviewee is faced with a difficult question; and transcranial magnetic stimulation (*see* p479), which impairs the ability of the prefrontal cortex to organize lies. (Sleep deprivation, sometimes used during interrogation, has the same effect.)

## VISUAL MUSIC

One of the aims of future entertainment will be to make experiences as multidimensional as possible. For example, when we listen to a music CD at home, we hear the music but we do not "feel" it in the same way as when we sit in front of a live band or a live orchestra. This difference occurs partly because the sight of musicians playing instruments stimulates brain cells called mirror neurons (*see* Learning by imitation, p155), creating a richer sensory experience.

Mirror neurons respond to the sight of body movement, and this movement could be digitally encoded and incorporated in music CDs to augment a listener's experience, creating "mirror neuron music". The technique used to encode body movement is known as motion capture and is already widely used in the film industry to create computer-generated characters, such as Gollum in *Lord of the Rings*. The performer wears a special suit containing light diodes fixed to key parts of the skeleton, and the movement of these lights is captured by a camera and used to animate a 3-D "wireframe" model in a computer. In mirror neuron music, the listener would simultaneously listen to a music CD while watching a graphic animation of the captured movement – either in two dimensions on a computer screen or TV or in three dimensions in virtual-reality goggles.

## COMPUTER GAMING

On-line computer games are likely to become increasingly popular in the future. Multiplayer games already allow people to adopt on-line personalities and interact in virtual environments. Such games are enormously popular in the Far East and are quickly catching on in North America and Europe.

At present, the virtual characters in on-line games are not directly linked to the emotional states of the players, but this situation could change. Smart software and webcams could read players' facial expressions and body language and replicate them in the on-line characters, making their behaviour richer and more human. Information from stressometers (*see* Stress-busters, *opposite page*) or implanted chips could also be incorporated, allowing physiological states such as sexual arousal and stress to factor in the game. As a result, virtual interactions will increasingly have the quality of real-life interactions.

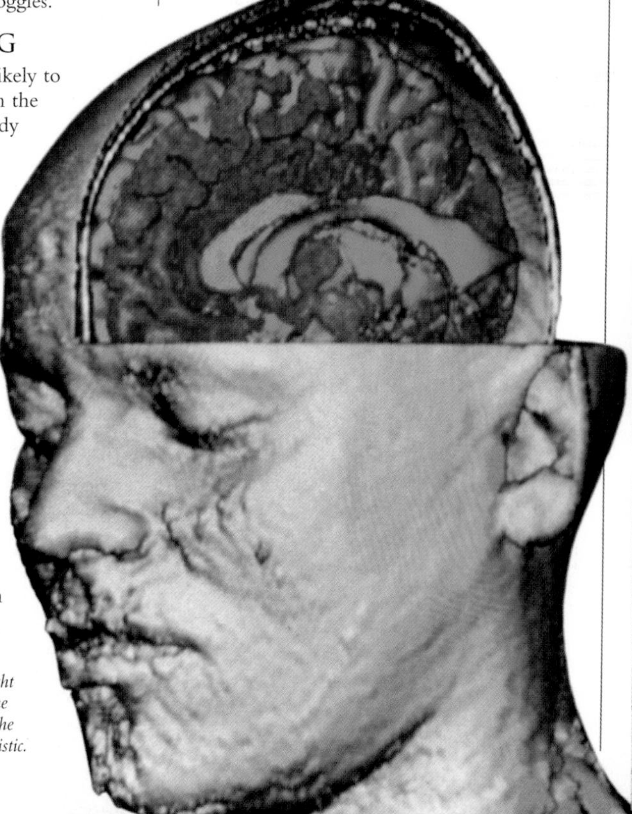

**TAPPING INTO EMOTIONS**
*Virtual-reality games of the future might tap into the brain's emotional core – the limbic system (green areas) – making the virtual world more compelling and realistic.*

## FUTURE DRUGS

Until the development of the impotence drug Viagra (sildenafil), a working aphrodisiac was thought by many experts to be impossible. Future drugs could prove to have an even bigger impact on sex lives than Viagra. Most men are currently unable to experience multiple orgasms, possibly because of a surge in the hormone prolactin after orgasm. (Men who experience multiple orgasms apparently have no prolactin surge.) Perhaps by using implants that detect the moment of orgasm and make a synchronized release of prolactin inhibitors, men would not experience a loss of libido after orgasm. Loss of libido is also linked to the release of natural opiates, which cause pleasure. A drug that blocked the dampening action of opiates on libido without removing the pleasurable effects might also help men have multiple orgasms.

Political and legal attitudes to recreational drugs are difficult to predict, but the drugs themselves will probably never disappear. Future research might enable us to replace unhealthy or addictive drugs with safer alternatives. Advanced delivery systems, such as the use of heat rather than fire to vaporize drugs that are smoked, could make cannabis safer. Pharmaceutical companies are already hunting for synthetic compounds that trigger the same receptors as cannabis, and their discoveries will probably be among the recreational drugs of the future.

In recent years there has been a surge of interest in "smart drugs". These drugs are said to boost intelligence or improve memory by stimulating blood flow to the brain or by increasing the level of brain chemicals. Some of the compounds suggested as smart drugs are currently used for medical reasons, including vasopressin (a natural hormone involved in the control of learning and memory); hydergine (which increases blood flow in the brain); and piracetam (used to treat memory loss in Alzheimer's disease). Although such compounds may be helpful in dementia, there is no clear evidence that they can improve memory in a normal brain.

Could working smart drugs become available in the future? Opinion is divided. Many scientists think that healthy brains already have optimum levels of blood flow and neurotransmitters, neither of which can therefore be improved. Others believe that the drugs of the future might boost memory and intelligence just as easily as current drugs can elevate mood and alertness. They see a future dominated by increasingly sophisticated chemicals that can fine-tune our mental abilities and and adjust our moods.

## TOMORROW'S PEOPLE

How human society develops in the future depends to a large extent on how population grows, but will the global population spiral out of control or eventually start to decline? After quadrupling in less than a century, the world's population has now reached more than 6 billion. Experts expect the number to rise to 9 billion by the middle of the 21st century, after which time the global population will begin to fall. One of the main reasons for the decline is that people are increasingly choosing to have fewer children. In the 1950s, women had an average of 5 children each, but today they have an average of 2.7 each. In order to sustain a population at a stable level, women need to have an average of 2 children each. In many European countries, however, the average has fallen below 1.5 children. As a result, population is declining in much of Europe.

Families are becoming smaller principally because of economic changes. In poor rural economies, people benefit from having large families because children provide labour. In wealthy industrial societies, however, children are an economic burden, so parents choose to have fewer. Economists believe that birth rates will drop in many developing countries as they become wealthier, mirroring the drop in birth rates in 20th-century Europe.

The combination of falling birth rates and increased life expectancy in the developed world will lead to a large increase in the proportion of older people within the population. This change in the age structure of society could cause economic stresses – eventually there may be too few people of working age to financially support the growing number of retired people. Many people will be forced to continue working into their 70s to build up a pension fund large enough to support them in retirement.

Almost all of the population growth in the next 50 years will occur in developing countries. Eventually, however, as these countries become more industrialized, falling birth rates and increasing longevity will take effect there too, and their population structures will begin to resemble those in other parts of the world. However, in the short term, it seems inevitable that large-scale migrations will occur, with people moving from poor to richer countries to make up for the shortfall in labour.

## URBAN LIVING

By 2030, most Chinese, Indonesians, Nigerians, Egyptians, Saudis, South Koreans, and Brazilians will live in cities. All over the world, cities will continue to be centres of trade, technology, tourism, investment, and transport.

Cities in the developing world will expand horizontally to accommodate an increasing number of people, but the growth of cities in the developed world will be more restrained. In developed countries, government programmes

Fact

### Future cities

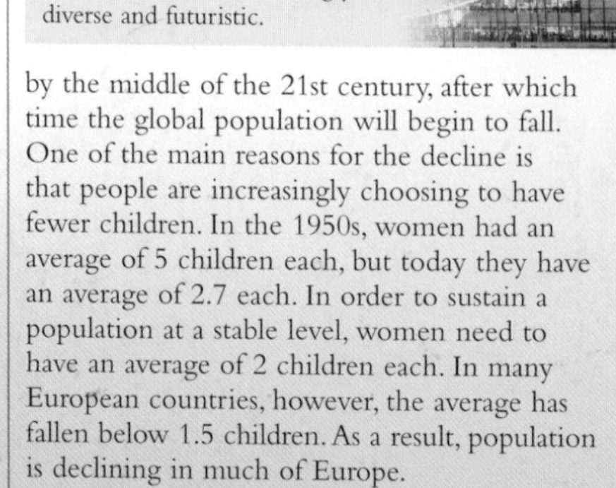

Central Shanghai, in China, might be a vision of the future, with its space-age glass domes and towers. As in many of the world's major cities, growth is likely to continue skywards as the government strives to limit urban sprawl while still providing space for homes and businesses. The architectural trend for designing innovative and sculptural new buildings looks set to continue, making our cities increasingly diverse and futuristic.

**CAPSULE HOTEL**
*City workers who have missed their train home prepare for sleep in a capsule hotel in Osaka, Japan. Compact living spaces, from microflats to capsule hotels, will become widespread if city populations and land prices keep rising.*

**ELECTRIC CAR**
*Battery-powered vehicles like this three-wheeled car, which is recharging at a roadside station, could help relieve pollution in increasingly crowded cities.*

**SEGWAY**
*A possible alternative to cars, the Segway can carry a standing rider 24km (15 miles) on one charge.*

to preserve the countryside will keep cities within strict physical boundaries. City-dwellers will increasingly live in regenerated brownfield sites, and tiny living spaces such as microflats will become more common. Governments will insist on low energy usage, local energy generation, and local waste disposal.

In developing nations, cities could become uncontrollably crowded. Many cities will grow to more than 10 million people, which could lead to fears of disease, unrest, and terrorism.

Governments throughout the world will become increasingly concerned about urban infrastructure. New buildings will need to house high numbers of people while also being safe and providing a sense of community. Key buildings might be erected in the countryside, where terrorist attacks are less likely to happen.

## TRANSPORT

Rather than driving to work in a car, future commuters might travel standing up on a self-balancing, two-wheel device such as the Segway, which is already available in the US. Personal hovering machines are being developed in several countries, and researchers are looking into ways of using liquid hydrogen as an alternative to petrol. Rechargeable electric cars have been in development for years, but they have yet to become a commercial success.

As the developing world becomes able to mass-produce cars, the transition from travelling by bike or train to travelling by car will accelerate. The developing world may lead the way in new technologies designed to replace fossil fuels. Brazil is already a world pioneer in cars that run on alcohol derived from fermented sugar-cane; and China may become the first major user of maglev (magnetic levitation) trains, which use powerful magnets to hover above the track. For longer journeys, we could fly on superjumbo jets that can carry up to 800 passengers.

Concerns about congestion, safety, and security of transport will increase throughout the world, especially in cities. There will be more tolls for road use, more pedestrianization, more controversy over new airports, and more environmental campaigns about aircraft noise and emissions. Costs and fares on most modes of transport will rise, and there are fears that new high-tech means of transport may become the preserve of rich people only.

## WORKING WORLD

High-speed internet connections have had a profound effect on the working world and will continue to do so. People who work in financial services, information technology (IT), design, and the media can already supply much of their work electronically and have less need to visit the office. As this trend continues, career aspirations will change – more people will choose to become freelance or develop skills that allow them to work from home in exotic or rural locations, away from the corporate environment. As a result, lifestyle will become a more important sign of career success than position in the corporate hierarchy.

As the internet becomes more ubiquitous, work will encroach on more aspects of people's lives, spreading from the office to homes, streets, cars and trains, and plaguing people's weekends, evenings, and holidays.

At work there will be more bureaucracy and meticulous measuring of performance and productivity. Governments will insist that large organizations offer financial transparency and social and environmental responsibility. Having an ethical reputation will be an important way of maintaining shareholder value. Expanding areas of future employment will include healthcare, education, care for elderly people, and general domestic service. Young immigrants from developing economies will take many of

Issue

### Cashless society

Cash could soon be a thing of the past. Credit and debit cards have already replaced it for many transactions. In years to come, digital cash, stored on smart cards or gadgets that can "beam" money, may replace hard currency. Proponents claim digital cash will reduce violent crimes such as muggings and attacks on shopkeepers. Electronic transactions would also be easier to trace, reducing tax evasion through illicit "cash in hand" payments for work.

**URBAN SPRAWL**
*Seen from the air at night, Chicago looks like a sea of light. During the 21st century, cities in the developing world will outgrow even the biggest US cities.*

these jobs. Meanwhile, young people in the developed world will continue the trend of delaying full-time work until they have spent several years studying at university and travelling around the world.

Multinational corporations will extend their global reach and continue the current trend of relocating operations to countries with cheap labour. Manufacturing industries will be transferred to China, while service industries will be relocated to India. In developing countries, the female workforce will grow and the IT industry will expand.

## INFORMATION REVOLUTION

There is much debate about the long-term impact of information technology (IT) on human ways of living and working. Some people say that we are in the middle of an information revolution, comparable in impact to the agrarian and industrial revolutions. Others say that the ability of technology to change human lives has been exaggerated.

Computer power has increased in leaps and bounds in recent decades as the process of miniaturization has made computer chips ever smaller and faster. Handheld computers the size of a pocket calculator can now do almost anything that desktop machines of a decade ago could do, and some computerized gadgets are small enough to be sewn into "smart clothes".

According to Moore's Law – an observation made by computer pioneer Gordon Moore in the 1960s – the number of transistors that an integrated circuit can hold doubles every two years or so. This law has held true over the last two or three decades, but many computer scientists think the process of miniaturization could reach a dead end when microcircuits shrink to the smallest size that silicon chips can physically support without becoming unstable. With the world economy increasingly reliant on advances in computing, an end to the increase in computer processing speeds could destabilize stockmarkets and cause a global economic crisis. Many computer scientists are investigating new ways of making computer chips that would allow the trend of miniaturization to continue. Some people think nanotechnology offers the best hope of keeping Moore's Law alive. For example, "nanotubes" constructed from carbon atoms could replace copper connections in computer chips. It may eventually become possible to construct transistors from single organic molecules and grow microscopic computers as living cells, but such technology remains in the realm of science fiction for now.

Fact

### Computerized clothing

The trend of miniaturization will eventually make computers small enough to be sewn invisibly into clothes. Researchers are even developing washable computers made entirely from fabrics that conduct electricity. These fabrics will contain a microchip, and be powered by solar energy or by devices that create electricity from body movements.

**SMART JACKET**
*This prototype jacket features a microchip, a fabric keyboard on one sleeve, and speakers built into the hood.*

In the meantime, people will continue to use and depend on IT in their everyday lives. Fears that the need for travel or face-to-face contact will be made redundant by computers are likely to be unfounded. Nevertheless, the concern that IT will make human interaction a less personal business is valid.

## GLOBAL SOCIETY

Some people think that national identity might disappear in the future, with the world's people becoming global citizens united by their shared humanity. They believe there will be a global economy, a global media, a single global currency, and a global democracy headed by one world leader. However, critics of this idea say that although some aspects of society, such as the media, could quickly become globalized, for others globalization is unlikely. The hopes for a global democracy, for example, look bleak as long as inequalities of wealth and conflicts of interest exist between nations.

There is agreement, however, that people will become more cosmopolitan and "global-thinking" in their attitude. Our increasing contact with foreign cultures and common experiences with other people will expand our horizons. Foreign events will seem increasingly significant and relevant to us.

**GLOBAL BRANDING**
*In the global free market, brands will become increasingly international – a trend led by fast foods, fizzy drinks, and designer clothes labels.*

**ONLINE ANYWHERE**
*A Buddhist monk in Cambodia works on a laptop computer. The increasing availability of computers and the spread of the internet are creating a global culture.*

In recent decades, many aspects of Western culture have spread across the planet, from designer labels to pop music and blue jeans. Some people fear this trend could lead to a culturally homogeneous world in which everyone eats the same food, drinks the same drinks, wears the same clothes, and watches the same films and TV programmes. The counter-argument is that, although Westernization is a domineering force, the "culture" that is being exported is a narrow one marketed primarily at the young. As people grow older, they come to value the individuality of their own culture. In Muslim countries there has been a resurgence in local customs and a rejection of Western values, a trend that looks likely to continue.

## WAR AND POWER

Wars will continue to occur in many parts of the world as a result of ethnic and religious conflicts and as people struggle to free themselves from corrupt regimes, dictators, and foreign occupation. At the same time, developing nations may attempt to extend their power through war. More than any country, the US will continue to exercise power through the use of technologically advanced weaponry, electronic surveillance, armies, and space science.

Power isn't always military – in the future the East will rival the US in both economic and demographic terms. China especially has a rapidly expanding manufacturing industry and plays an increasingly important role in the world economy. With a population of more than 1 billion and a colossal workforce, China looks likely to become a future economic superpower, in spite of its Marxist history. In the short term, however, China's growth is constrained by poor infrastructure and draconian laws imposed by the country's centrally planned economy.

## A HEALTHIER WORLD?

Differences in life expectancy in the developed and developing world will decline in the near future, but the health divide between rich and poor will persist. While people in many wealthy countries will reap the rewards of advances in medicine, those in parts of Africa and elsewhere will continue to suffer from poor sanitation and a lack of basic medical care.

Scientific research is likely to lead to major health improvements in the developing world in the next few decades. AIDS, tuberculosis, and malaria could all become diseases with which we will deal differently. Vaccines against AIDS and tuberculosis will probably be developed, and genetically-modified strains of mosquito that cannot carry malaria parasites might replace their

**OZONE HEALING**
*The ozone hole over Antarctica shows as a white area in this satellite image. Some scientists think the hole will disappear within about 50 years owing to an international ban on CFCs.*

wild counterparts. Programmes are also under way to bring reliable supplies of clean drinking water and better sanitation to millions of people in Asia, Africa, and Central and South America.

In rich countries, changes in the age structure of society (*see* Tomorrow's people, p482) and in lifestyle will lead to changes in the pattern of disease. As average age increases, so too will the incidence of intractable diseases associated with aging, such as heart disease, cancer, and dementia. Healthcare costs will rise, putting a financial strain on national health systems and leading to increases in insurance premiums and taxes.

**RISING WATER**
*Thousands of homes were submerged when the Mississippi River burst its banks in 1993. Such disastrous floods could become much more common as world temperatures rise.*

## THE ENVIRONMENT

People are stripping the Earth of its natural resources. Unless action is taken soon, nearly all of the world's remaining tropical rain forests will be destroyed, as well as most of the coral reefs, wetlands, and mangrove forests. Several million species of animals and plants will be wiped out in the process, including many plants that are potentially useful sources of medicines. Studies of global warming predict that average global temperatures will rise by 1.4–5.8°C (2.5–10°F) in the 21st century, mainly as a result of carbon dioxide emissions from the burning of forests and fossil fuels. Thermal expansion of the oceans and melting of glaciers could push up sea levels by 20–70cm (8–28in), inundating low-lying islands and many coastal areas, including densely populated parts of Bangladesh and Egypt. In the US, vast areas of land could be lost, including much of Florida. Global warming is also expected to produce more destructive weather, with increased numbers of hurricanes, floods, and forest fires. The change could also disrupt ocean currents, making some parts of the world much colder. For example, northwest Europe would experience a return to ice-age conditions if the Gulf Stream stopped flowing northeast across the Atlantic. Up to 70 per cent of the world's dry land could be damaged by soil erosion, with disastrous consequences for farming and food production. By 2100, famines could be a regular occurrence in many areas.

**DIGGING FOR WATER**
*Farmers dig a well in Gambia to obtain water. Clean water supplies and effective sanitation are among the basic elements that will enable people in the developing world to enjoy longer, healthier lives and higher living standards.*

## A BETTER WORLD?

Despite all the gloomy forecasts, there are some grounds for optimism. Apocalyptic predictions made in the 1970s about overpopulation and the fuel crisis proved to be exaggerated, and some people even think forecasts about global warming and environmental damage could be inaccurate as well. The hole in the atmosphere's ozone layer over Antarctica, which threatened to expand and cause levels of dangerous solar radiation to rise in the southern hemisphere, now appears likely to shrink owing to an international ban on CFCs. Nevertheless, the burning of fossil fuels may present a real risk of changing the Earth's climate irreversibly.

Whether human standards of living will rise depends on the extent to which the undesirable aspects of capitalism, such as economic instability and inequality, can be softened, and the positive aspects, such as technological innovation, can be exploited. For example, a minimum wage could be introduced internationally to help eliminate sweatshops and child labour – although such a policy might make it harder for developing countries to compete in the global economy.

Technical change and uncertainty about the future give a window for optimism. In years to come, we will be able to communicate more quickly and easily than ever before, we will have more control over our physical and mental wellbeing, and most of us will lead longer, healthier, and perhaps happier lives.

**SUSTAINABLE ENERGY**
*Although fossil fuel supplies could last for centuries, renewable energy, such as power from wind turbines, will be vital to prevent a dangerous rise in global temperature.*

# GLOSSARY

In this glossary, references to terms defined in other entries are identified with *italic* text.

## A

**ABORIGINAL** An indigenous inhabitant of a region; specifically, a member of Australia's indigenous population.

**ACCOMMODATION** The process by which the eyes adjust to focus on nearby or distant objects.

**ADENOIDS** A collection of lymphatic tissue on each side of the upper throat, at the back; part of the immune system.

**ADOLESCENCE** The period of life between childhood and adulthood, approximately equivalent to the teenage years in Western societies.

**ADRENALINE** see *epinephrine*.

**AIDS (ACQUIRED IMMUNE DEFICIENCY SYNDROME)** A syndrome that occurs after infection with HIV (the human immunodeficiency virus). AIDS is spread by sexual intercourse or infected blood. It results in loss of resistance to infections and some cancers.

**ALGONKIAN** A family of native American languages that is spoken in eastern North America.

**ALLERGEN** Any substance causing an allergic reaction in a person previously exposed to it.

**ALVEOLI** Tiny air sacs in the lungs, through the walls of which gases diffuse into and out of blood during respiration.

**AMYGDALA** An almond-shaped structure within the *limbic system* of the brain that plays a key role in triggering human emotions.

**ANGLICAN CHURCH** One of the main branches of the *Protestant Church*. The Anglican Church is a global organization but follows the traditions of the Church of England.

**ANIMISM** A religion characterized by the belief that ordinary objects, such as plants or animals, are inhabited by souls.

**ANTIBODY** A soluble protein that attaches itself to harmful microorganisms in the body, such as bacteria, and helps to destroy them.

**AORTA** The central and largest artery of the body, arising from the heart's left ventricle and supplying oxygenated blood to all other arteries except the pulmonary artery.

**APE** A type of primate that differs from a monkey in lacking a tail and having large, powerful arms and mobile shoulders. The ape superfamily (Hominoidea) includes human beings and our closest relatives, such as chimpanzees and gorillas.

**APPENDIX** The wormlike structure attached to the initial part of the large intestine. The appendix has no known function.

**AQUEOUS HUMOUR** The fluid that fills the front chamber of the eye, between the back of the cornea and the front of the iris and lens.

**ARTERIOLE** Small terminal branch of an *artery*, leading to even smaller *capillaries*.

**ARTERY** An elastic, muscular-walled tube that transports blood away from the heart to all other parts of the body.

**ARTICULATION** A joint, or the way in which jointed parts are connected.

**ARTIFICIAL INTELLIGENCE** A branch of computer science concerned with machines or computer programs that are designed to perform tasks requiring humanlike intelligence, such as the ability to reason and generalize.

**ASSIMILATION** The incorporation of people of a different ethnic or cultural background into the dominant culture of a society, usually with some loss of the minority's distinctive heritage.

**ATHEISM** The belief that there is no God.

**ATRIUM** One of the two thin-walled upper chambers of the heart.

**AUSTRALOPITHECUS** The generic name given to several extinct African *hominid* species that had an upright posture and small brains, and lived some 2–5 million years ago.

**AUSTRONESIAN** A family of languages spoken in much of Southeast Asia, Madagascar, New Zealand, and many Pacific islands.

**AUTONOMIC NERVOUS SYSTEM** The part of the nervous system that controls unconscious functions such as heart rate. It has two divisions: the *sympathetic nervous system* and the *parasympathetic nervous system*.

**AXON** The long, fibrelike process, or projection, of a nerve cell that conducts nerve impulses to or from the cell body. Bundles of many axons form nerves.

## B

**BACTERIA** Types of microorganism consisting of one cell. There are many species but only a few cause disease.

**BAND** A small community of hunter-gatherers, such as Arctic Inuit or Aboriginal Australians, with informal leadership and minimal political or hierarchical organization.

**BANTU** A term referring to several hundred ethnic groups of central, eastern, and southern Africa and their associated languages, which include Swahili, Zulu, and Xhosa.

**BAPTIST CHURCH** A branch of the *Protestant Church* whose members are initiated into the religion by profession of their belief as adults, prior to baptism (ritual immersion in water).

**BASAL GANGLIA** Paired masses of nerve cell bodies, or nuclei, that lie deep in the brain and are concerned with the control of movement.

**BILE** A greenish-brown fluid produced by the liver and stored in a sac called the gallbladder. Bile aids the digestion of fats.

**BILIARY SYSTEM** The network of *bile* vessels formed by the ducts from the *liver* and the gallbladder (the sac that stores bile) and the gallbladder itself.

**BIPEDAL** Two-footed. The emergence of bipedal walking was an important stage in early human evolution.

**BLOOD CLOT** A mesh of fibrin (strands of protein), *platelets*, and blood cells that forms when a blood vessel is damaged.

**BOLUS** A chewed-up quantity of food that is swallowed and passes into the stomach.

**BONE MARROW** The fatty tissue within bone cavities, which may be red or yellow. Red bone marrow produces red blood cells.

**BRAIN STEM** The lower part of the brain, which houses the centres controlling vital functions such as the breathing and heartbeat.

**BRIDEWEALTH** Money or property given by the groom or his family to the bride's family upon marriage. Payment of bridewealth is especially common in African societies. See also *dowry*.

**BRONCHI** Air tubes that lead from the *trachea* to the lungs and divide into smaller bronchioles.

**BRONCHIAL TREE** The *trachea* and branching system of air tubes in the lungs, consisting of progressively smaller *bronchi* and bronchioles.

**BROW RIDGE** A ridge of bone above the eyes, characteristic of certain extinct *hominid* species including *Neanderthals*.

**BUDDHISM** A religion and philosophy, inspired by the Indian mystic Gautama Buddha (6th century BC), that teaches followers to strive for enlightenment and liberation from suffering through mental training and good behaviour.

## C

**CALLIGRAPHY** A form of art based on decorative writing.

**CAPILLARY** One of the tiny blood vessels that link the smallest arteries and the smallest veins.

**CAPITALISM** An economic system that permits the personal accumulation of wealth (capital) and the private ownership of means of production. Capitalism is based on the concept of a free market, but many capitalist governments impose regulations that intervene to some extent in the workings of the market.

**CARTILAGE** A tough, fibrous connective tissue, also known as gristle.

**CASTE SYSTEM** A hierarchical social system in Hindu India, in which a person's socioeconomic status and choice of profession depend on family background.

**CATHOLIC** See *Roman Catholic Church*.

**CENTRAL NERVOUS SYSTEM** The brain and spinal cord. The central nervous system receives and analyses sensory data and then initiates a response.

**CEREBELLUM** The region of the brain behind the *brain stem*. The cerebellum is concerned with balance and the control of fine movement.

**CEREBRAL CORTEX** The wrinkly exterior of the main part of the human brain (the *cerebrum*). The cerebral cortex is divided into prominent lobes in which specialized mental functions appear to be localized.

**CEREBRUM** The largest part of the brain, which is made up of two hemispheres. The cerebrum contains the nerve centres for thought, personality, the senses, and voluntary movement.

**CHIEFDOM** A settled community with greater political complexity than a tribe but less than a nation state. Chiefdoms have a leadership, a permanent political structure, and some social stratification (organization into different ranks).

**CHRISTIANITY** A monotheistic religion that developed from Judaism but is based on the teachings of Jesus Christ, who is believed to be the Son of God.

**CHROMOSOMES** Threadlike structures present in all body cells that have a nucleus. They carry the genetic code for the formation of the body. A normal human body cell carries 46 chromosomes arranged in 23 pairs.

**CIRCUMCISION** An operation in which genital tissue is removed, usually the foreskin in males and the clitoris in females. Circumcision is carried out more often as a social custom than for medical reasons.

**CIVILIZATION** A large, complex society characterized by urban settlements, extensive trade, and the development of advanced culture and technology.

**CLAN** A kinship group defined by descent from a common ancestor. Membership of a clan may be inherited through the maternal or paternal line, and marriage between members of the same clan is usually forbidden.

**CLASS** A social group whose members have a similar socioeconomic status. The term "class" originally referred to the sharply defined social categories of 19th century Europe, but it is often used more loosely to refer to any ranking in a hierarchical society.

**COCHLEA** The coiled structure in the inner ear in which sound vibrations are converted into nerve impulses for transmission to the brain.

**COLD WAR** A period of intense rivalry that developed between the US and the

Soviet Union in the 20th century but did not lead to outright conflict.

**COLLAGEN** An important structural protein present in bones, tendons, ligaments, and other connective tissues. Collagen fibres are twisted into bundles.

**COLON** The major part of the large intestine, extending from the caecum (first part of the large intestine) to the rectum. Its main function is to conserve water in the body by absorbing it from the bowel contents.

**COMMUNISM** An economic and political system advocated by German philosopher Karl Marx, in which private ownership is abolished and the means of production is owned by a whole community, with the aim of achieving a fairer, more equal society.

**CONCEALED OVULATION** Release of a mature egg from a female animal's ovary without accompanying physical or behavioural changes that display fertility (such as the swelling of genital tissue that occurs in chimpanzees).

**CONDITIONING** A form of learning in which a process of reinforcement makes a behavioural response more likely. In classical conditioning, an animal learns to associate two stimuli (such as the sound of a bell and the arrival of food). In operant conditioning, a spontaneous behaviour becomes more or less frequent as a result of reinforcement with a reward or a punishment, respectively.

**CONFUCIANISM** An east Asian ethical tradition, based on the teachings of the Chinese philosopher Confucius (6th–5th century BC), that emphasizes humane, honest, and gentlemanly behaviour. Unlike a true religion, Confucianism does not involve spiritual beliefs.

**CONSCIOUSNESS** The psychological state of wakeful awareness. Some definitions of consciousness refer only to sensory awareness (sentience); others include self-awareness and introspection.

**CORNEA** The transparent dome at the front of the eyeball that is the eye's main focusing lens.

**CORONARY** A term meaning "crown"; it refers to the arteries that encircle and supply the heart with blood.

**CORPUS CALLOSUM** The wide, curved band in the brain of about 20 million nerve fibres that connects the two hemispheres of the *cerebrum*.

**CORTEX** See *cerebral cortex*.

**COUP D'ÉTAT** A sudden and often violent overthrow of the upper echelons of government by a small group. Unlike a revolution, which changes the whole political system, a coup d'état merely replaces those at the top.

**CREOLE** A complex, grammatically advanced hybrid language that forms, over more than one generation, from a *pidgin* language that is used as a native language.

**CRITICAL PERIOD** A period in life during which the brain is receptive to a specific type of learning. The development of normal vision, for example, involves a critical period during the first year of life.

**CRO-MAGNON** A race of anatomically modern *Homo sapiens* who lived between 35,000 and 10,000 years ago in southern France and are associated with the emergence of art and highly sophisticated tools.

**CT SCAN (COMPUTERIZED TOMOGRAPHY SCAN)** A detailed, cross-sectional image of soft body tissues, produced by a machine that fires short pulses of X-rays through the body at multiple angles.

**CYRILLIC** The alphabet used for Slavic languages including Russian, Bulgarian, and Serbian. Cyrillic developed in the 9th–10th century AD. The modern form of Cyrillic script arose in 1708, owing to the reforms of Peter the Great.

# D

**DEMOCRACY** A form of government in which power is exercised by the public, usually indirectly through elected representatives.

**DERMIS** The inner layer of skin, made of connective tissue and containing various structures such as blood vessels, nerve fibres, hair follicles, and sweat glands.

**DIAPHRAGM** The dome-shaped muscular sheet that separates the chest from the abdomen. It alters the volume of the chest cavity to enable breathing.

**DICTATORSHIP** A form of government in which absolute power is exercised by one person (a dictator) or a small group of people, with no limitation on power by constitution, law, or recognized opposition.

**DNA** Deoxyribonucleic acid, a complex organic molecule that carries genes in the form of a chemical code. DNA is stored in the nucleus of cells and is a chief component of *chromosomes*.

**DOMESTICATION** The process by which wild animals or plants are brought under human control. Long-term domestication can lead to genetic change as human breeders strive to create increasingly useful progeny.

**DOWRY** Money or property given by the bride's family to her husband upon marriage. See also *bridewealth*.

**DRAVIDIAN** A term referring to either the indigenous peoples of southern India or the family of languages spoken by those peoples.

**DUODENUM** The first part of the small intestine, into which stomach contents empty. Ducts from the gallbladder, liver, and pancreas enter the duodenum.

**DYNASTY** A succession of rulers of the same family, or the period during which they rule.

# E

**EASTERN ORTHODOX CHURCH** One of the main branches of *Christianity*. The Eastern Orthodox Church grew out of the Eastern Roman (Byzantine) Empire. It gave rise to the Churches of Russia, Greece, Romania, and Serbia.

**EGALITARIAN** Advocating or relating to the political, social, and economic equality of all people.

**ELECTORATE** The body of people qualified to vote in an election.

**EMBRYO** An unborn, developing baby from conception until the 8th week of pregnancy.

**ENDORPHIN** A substance produced within the body that relieves pain.

**ENZYME** A protein acting as a catalyst to accelerate a chemical reaction.

**EPIDERMIS** The outer layer of the skin; its cells flatten and become scale-like towards the surface.

**EPIGLOTTIS** A flap of cartilage located at the entrance of the larynx, which covers the opening during swallowing and prevents food or liquid from entering the airways.

**EPINEPHRINE (ADRENALINE)** A hormone that is released by the adrenal glands during stress or excitement, increasing alertness and preparing the body for rapid physical exertion.

**EVANGELICAL CHURCH** A branch of the *Protestant Church* that stresses salvation by conversion to Christianity and literal belief in the gospels. Evangelical Protestantism is associated with Christian *fundamentalism*.

**EVOLUTION** The gradual change of biological species, driven mainly by the process of natural selection.

**EXECUTIVE** One of the three main branches of government, concerned with implementing laws and decisions. See also *judiciary* and *legislature*.

**EXTENDED FAMILY** A social unit made up of a family and close relatives spanning several generations.

# F

**FARSI** The official language of Iran (also called Persian).

**FEUDALISM** A medieval European social system in which a wealthy landowner (a lord) granted property to peasants (vassals) and gave them protection, in return for loyalty in war.

**FINNO-UGRIC** A family of languages spoken in pockets of Europe that includes Finnish, Estonian, and Hungarian. The Finno-Ugric languages are not part of the *Indo-European* language family.

**FLINT** A type of quartz rock used by stone-age people to make small blades and other tools.

**FREE-MARKET CAPITALISM** See *capitalism*.

**FRONTAL LOBE** The foremost of the four lobes that make up each hemisphere of the *cerebrum*. The frontal lobes play an important role in higher mental faculties such as planning and decision-making.

**FUNDAMENTALISM** Advocacy of strict adherence to the fundamental principles of a religion. Fundamentalist *Christians* believe in the literal truth of the Bible. *Muslim* fundamentalists believe in strict application of Islamic law.

**GASTRIC JUICE** A mixture containing digestive enzymes and hydrochloric acid, which is produced by the cells of the stomach lining.

**GENE** A section of a *chromosome* that is the basic unit of inheritance. Genes contain information for growth and development.

**GERMANIC** A branch of the *Indo-European* language family that includes German, English, Dutch, Flemish, Afrikaans, and Scandinavian languages.

**GLIAL CELL** A type of cell that provides support for *neurons*.

**GLOBALIZATION** The tendency towards greater economic, technological, and social exchange between countries. Globalization often leads to cultural convergence.

**GREEK ORTHODOX CHURCH** The Greek branch of the *Eastern Orthodox Church*.

**GREY MATTER** The regions of the brain and spinal cord that are composed mainly of *neuron* cell bodies, as opposed to their projecting fibres (which form white matter).

**GUERRILLA** A member of a small, irregular fighting force that uses tactics such as sabotage and terrorism against conventional forces for political aims.

# H

**HAEMOGLOBIN** The protein that fills red blood cells and combines with oxygen, which it transports from the lungs to all parts of the body.

**HAREM** An Arabic term referring to the part of a house set aside for women. Zoologists use the term "harem" to refer to a group of females that share a single mate.

**HIJAB** A head scarf worn by *Muslim* women.

**HINDUISM** The dominant religion in the Indian subcontinent. Hinduism is a composite of various doctrines and cults, involving *polytheism* and belief in cycles of reincarnation.

**HIPPOCAMPUS** Part of the brain dealing with learning and long-term memory.

**HIV (HUMAN IMMUNODEFICIENCY VIRUS)** The virus that causes *AIDS*. HIV destroys certain cells of the body's immune system, seriously undermining its efficiency.

**HOMINID** A term often used of prehistoric human ancestors, such as *Homo erectus* and *Australopithecus*; more accurately applied to any member of the great ape family Hominidae – including gorillas, chimpanzees, orangutans, and humans.

**HOMININ** A member of the Homininae, a subfamily of great apes made up of humans and extinct relatives such as *Australopithecus* but excluding all living relatives such as chimpanzees.

**HOMO** The genus (group of closely related species) to which humans belong.

**HOMO ERECTUS** An extinct, humanlike species thought to be a direct ancestor of modern humans. *Homo erectus* appeared around 1.8 million years ago and may have survived until as late as 100,000 years ago.

**HOMO SAPIENS** The scientific name for modern humans.

**HORMONES** Chemicals, released into the blood from some glands and tissues, that act on sites elsewhere in the body.

**HUMAN GENOME PROJECT** An international scientific research effort to locate and sequence every *gene* in human *DNA*. It was completed in 2003.

**HUMANISM** A doctrine or attitude based on the belief in the intrinsic worth of human beings and rejecting religion.

**HUMANITARIANISM** A doctrine asserting the intrinsic worth of all people and promoting human welfare and social reform.

**HUNTER-GATHERER** A person who lives on animal and plant foods gathered from the wild. Before the invention of agriculture, all the world's people were hunter-gatherers.

**HYPOTHALAMUS** A small structure at the base of the brain where the nervous and hormonal body systems interact.

# I

**ICE AGE** A period in Earth's history when the climate is cool and polar ice sheets expand, covering large areas of land. The most recent ice age ended 10,000 years ago.

**IDEOGRAPHIC SCRIPT** A form of writing consisting of small pictures.

**ILEUM** The final part of the small intestine, in which absorption of nutrients is mainly completed.

**IMAGING TECHNIQUES** Medical procedures used to produce images of internal structures in the human body, usually to help diagnose disease. See also *MRI scan, CT scan (computerized tomography scan), PET scan (positron emission tomography scan).*

**IMPERIALISM** The extension of a nation's power and dominion by acquiring new territories, whether by military, economic, or political means.

**INCEST TABOO** Prohibition of sexual intercourse between closely related people.

**INDO-EUROPEAN** A large family of languages spoken in most of Europe and large parts of Asia. Indo-European languages include Celtic, Germanic, Italic, and Slavic languages.

**INDUSTRIAL REVOLUTION** The transformation of 18th- and 19th-century European agricultural economies into economies based on industrial production. The Industrial Revolution led to massive urbanization and population increases.

**INITIATION** A ceremony in which a person is formally accepted as a member of adult society or a sect.

**ISLAM** A major world religion founded in Arabia in the 7th century AD by the prophet Muhammad. Islam belongs to the same monotheistic tradition as *Judaism* and *Christianity*.

**ISMAILI** The largest branch of *Shia Islam*, whose followers believe that Ismail, son of the sixth imam, was the rightful seventh imam.

# JK

**JAINISM** An Indian religion founded in the 6th century BC, whose followers seek spiritual perfection through self-denial and non-violence towards people and animals.

**JEHOVAH'S WITNESS** A member of an offshoot of *Christianity* founded in the late 19th century and known for door-to-door evangelizing and belief in an apocalyptic "second coming" of Christ.

**JUDAISM** The religion followed by Jews, characterized by belief in a single God (monotheism) and observance of commandments and teachings recorded in the Old Testament and Talmud.

**JUDICIARY** One of the three main branches of government, concerned with administration of justice. See also *executive* and *legislature*.

**KORAN** The holy book of *Islam*, written by the prophet Muhammad and believed by Muslims to record the divine revelations of Allah (God), made through the angel Gabriel.

# L

**LEGISLATURE** One of the three main branches of government, concerned with creating and amending laws. See also executive and judiciary.

**LIGAMENT** A band of tissue consisting of *collagen*. Ligaments support bones, mainly at joints.

**LIMBIC SYSTEM** A part of the brain that plays a role in automatic body functions, emotions, and the sense of smell.

**LINGUA FRANCA** A common language used for communication between people of different mother tongues.

**LUTHERANISM** One of the main branches of the *Protestant Church*, founded in the 16th century by German theologian Martin Luther.

**LYMPHATIC SYSTEM** A network of lymph vessels and lymph nodes. The lymphatic system returns excess tissue fluid to the circulation and helps to combat infections and tumour cells.

**LYMPHOCYTE** A type of small white blood cell that is a vital part of the immune system, giving protection against viral infections and cancer.

# M

**MACHIAVELLIAN INTELLIGENCE** A term used by psychologists to refer to the apparently sophisticated ability of many *primates* to understand and manipulate social relationships.

**MAMMAL** A type of animal that is characterized by the ability to produce milk, warm-bloodedness, hairy skin, and giving birth to live young.

**MANUAL DEXTERITY** Agility of the hands. Humans have much better manual dexterity than other animals.

**MARXISM** See *communism*.

**MATRILINEAL** Tracing descent through the female line. In a matrilineal society, titles and sometimes property are inherited from mothers rather than fathers. See also *patrilineal*.

**MEIOSIS** The stage in the formation of sperm and egg cells when chromosomal material is randomly redistributed and the number of chromosomes is reduced to 23 (instead of the 46 found in other body cells).

**MENARCHE** The first occurrence of *menstruation* in a woman's life.

**MENINGES** Three membranes (the pia mater, the arachnoid, and the dura mater) that surround and protect the brain and spinal cord.

**MENINGITIS** Inflammation of the *meninges*, usually resulting from an infection.

**MENOPAUSE** The end of the reproductive period in a woman, when egg development in the ovaries and *menstruation* have ceased.

**MENSTRUATION** The approximately monthly discharge of blood and uterine tissue from a woman's body that occurs as part of the reproductive cycle.

**MERITOCRACY** A society in which people's success depends on ability rather than birthright.

**METABOLISM** The sum of all the physical and chemical processes that take place in the body.

**METHODISM** A branch of the *Protestant Church* founded in the 18th century by English theologian John Wesley, who believed in methodical study of the Bible.

**MIDDLE CLASS** One of the three main socioeconomic groups of traditional European society, made up of working professionals such as businessmen, teachers, lawyers, and doctors.

**MILITARY DICTATORSHIP** A form of government in which absolute power is held by unelected military leaders. Such dictatorships often form by *coup d'état*.

**MIRROR NEURON** A type of brain cell that becomes active in a monkey (and hence possibly in humans) when it watches another individual performing a task.

**MITOCHONDRIA** A class of microscopic cell organ concerned with provision of energy for the various cell functions and containing genetic material.

**MITOCHONDRIAL EVE** The most recent common ancestor of all the world's women, who lived about 150,000 years ago. Her existence is deduced from studies of *mitochondria*, which are maternally inherited.

**MITOSIS** The process of division of most body cells. In the process, one cell produces two daughter cells, each with the identical genetic make-up of the parent cell.

**MONARCHY** A form of government ruled by a single hereditary leader. In a constitutional monarchy the monarch is merely head of state and government is run by an elected body.

**MONOGAMY** The habit of having only one mate.

**MONOTHEISM** Belief in a single God. See also *polytheism*.

**MORMONISM** An offshoot of *Christianity* that combines biblical tradition with divine revelations made to founder Joseph Smith in the 19th century, recorded in his Book of Mormon.

**MOTOR CORTEX** The part of the surface layer of each hemisphere of the *cerebrum* in which voluntary movement is initiated. It can be mapped into various areas that are linked to specific parts of the body.

**MOTOR NEURON** A type of neuron (nerve cell) that carries impulses to muscles to produce movement.

**MRI (MAGNETIC RESONANCE IMAGING) SCAN** A detailed, cross-sectional image of soft body tissues produced by a machine that detects radio waves emitted by the body in response to powerful magnetism.

**MUCOUS MEMBRANE** The soft, skinlike, mucus-secreting layer lining the tubes and cavities, such as the respiratory tract, in the body.

**MULTICULTURAL** Consisting of several distinct cultures that coexist. Many modern states are multicultural.

**MUSLIM** A follower of Islam.

**MUTATION** An accidental change in an organism's *genes*, usually with harmful effects.

**MYOCARDIUM** The heart muscle, in which the fibres make up a network that will contract spontaneously.

**MYOFIBRILS** Cylindrical elements within muscle cells (fibres). Each myofibril consists of thinner filaments, which move to produce muscle contraction.

# N

**NATION STATE** A large group of people that live in the same region, are united by cultural and linguistic similarities, and are governed by a political structure that exercises complete authority within the region's borders.

**NEANDERTHAL** An extinct species of human being that inhabited ice-age Europe. Although the Neanderthals were closely related to modern humans, they are not thought to be our ancestors.

**NEURON** A single nerve cell that transmits electrical impulses.

**NEUROTRANSMITTER** A chemical released from a nerve fibre, creating an electrical "message" that passes from one nerve to another or to a muscle.

**NOMAD** A person who moves from place to place without settling permanently. Many nomads travel with herds of animals in continual search of pasture.

**NOREPINEPHRINE (NORADRENALINE)** A *neurotransmitter* released from nerve endings in the *sympathetic nervous system*. It has similar effects to *epinephrine*, preparing the body for action.

**NUCLEAR FAMILY** A family made up of husband, wife, and children. See also *extended family*.

**NUCLEUS** The central part of a cell, holding almost all of the genetic material.

# O

**OCCIPITAL LOBE** The hindmost of the four lobes that make up each hemisphere of the *cerebrum*. The occipital lobe is important in vision.

**OESTROGEN** A female sex hormone, produced mainly in the ovaries, that stimulates female development at puberty and regulates the menstrual cycle.

**OESTRUS** The regularly occurring period of sexual receptivity in the reproductive cycle of a female mammal. Humans are unusual in being sexually receptive throughout the reproductive cycle.

**OLIGARCHY** Government by a small group of people, or concentration of power in the hands of a small group. Members of an oligarchy often abuse power to further their own interests.

**OMNIVORE** An animal that eats both plant and animal foods.

**OPPOSABLE THUMB** A thumb that can swing around to face and press the tips of the fingers, giving a precision grip. Most primates have opposable thumbs but those of humans are particularly highly developed.

**ORTHODOX CHRISTIANITY** See *Eastern Orthodox Church*.

**OSSICLE** One of three tiny bones of the middle ear. These are called the malleus, incus, and stapes, and convey vibrations from the eardrum to the inner ear.

**OSTEON** The rod-shaped unit, also called a Haversian system, that is the building block of hard bone.

**OVARIES** The two structures, on each side of the uterus, that produce ova (eggs) and female sex hormones.

**OVULATION** The release of an ovum (egg) each month from a follicle that has matured within the ovary. Ovulation usually occurs in the middle of the menstrual cycle.

# P

**PAGAN** A person who follows a polytheistic or animist religion. The term "pagan" has traditionally been used in a pejorative way to refer to anyone who is not *Christian*, *Muslim*, or *Jewish*.

**PARASYMPATHETIC NERVOUS SYSTEM** One of the two divisions of the autonomic nervous system. It maintains and restores energy, for example by slowing the heart rate during sleep.

**PARIETAL LOBE** One of the four lobes that make up each hemisphere of the *cerebrum*. The parietal lobe is involved in the interpretation of touch, pain, and temperature.

**PASTORALISM** A way of life based on keeping herds of grazing animals. Many pastoralists are nomads, especially in arid grassland areas.

**PATRILINEAL** Tracing descent through the male line. In a patrilineal society, titles and usually property are inherited from fathers. See also *matrilineal*.

**PELVIS** The basinlike ring of bones to which the lower spine is attached and with which the thigh bones articulate.

**PERICARDIUM** The tough, fibrous, two-layered sac that encloses the heart and the roots of the major blood vessels that emerge from it.

**PERIPHERAL NERVOUS SYSTEM** All the nerves and their coverings that fan out from the brain and spinal cord, linking them with the rest of the body.

**PERISTALSIS** A coordinated succession of contractions and relaxations of the muscular wall of a tubular structure, such as the intestines, that moves the contents along.

**PERITONEUM** A double-layered membrane that lines the inner wall of the abdomen, covering the abdominal organs and secreting fluid to lubricate them.

**PET SCAN (POSITRON EMISSION TOMOGRAPHY SCAN)** A coloured image showing areas of high activity in the brain, produced by a machine that detects radioactive compounds injected into the body prior to scanning.

**PHEROMONE** A chemical substance that is secreted by an animal, often as a scent, and changes the behaviour of other animals. Many animals produce pheromones to attract mates.

**PHOBIA** An intense and irrational attack of fear in response to something harmless. Many phobias are triggered by animals.

**PHONEME** A speech sound. A phoneme is the smallest unit of speech that can be used to distinguish between words, such as the sound "p" that makes "pin" different from "sin". In many languages, phonemes correspond to letters of the alphabet, but in English, phonemes correspond only approximately to the alphabet.

**PIDGIN** A crude hybrid language, with poorly developed grammar, that forms in a community made up of people of different native languages who often speak together. See also *Creole*.

**PLASMA** The fluid part of blood from which all cells have been removed. Plasma contains proteins, salts, and various nutrients that regulate the volume of blood.

**PLATELETS** Fragments of large cells, called megakaryocytes, that are present in large numbers in blood and are needed for blood clotting.

**PLEURA** A double-layered membrane, the inner layer of which covers the lungs, and the outer layer of which lines the chest cavity. Fluid lubricates and enables movement between the two.

**POLYANDRY** Marriage of a woman to more than one husband at once.

**POLYGAMY** Marriage of a man or woman to more than one spouse at once.

**POLYGYNY** Marriage of a man to more than one wife at once.

**POLYTHEISM** Belief in more than one god. See also *monotheism*.

**PREDATOR** An animal that lives by hunting and killing other animals.

**PREFRONTAL CORTEX** The front part of the frontal lobe in the brain. It is important in motivation, critical thinking, and planning.

**PRESBYTERIANISM** A branch of the *Protestant Church* in which church leaders (elders or presbyters) are elected and considered equal. Presbyterianism developed in reaction to episcopalian churches, which are governed by a hierarchy of bishops.

**PRIMATE** A type of mammal characterized by grasping hands, nails instead of claws, large, forward-facing eyes, and a large brain.

**PROGESTERONE** A female sex hormone secreted by the ovaries and placenta. Progesterone prepares the uterine lining to receive and retain a fertilized egg.

**PROSTAGLANDINS** A group of fatty acids made naturally in the body that act much like hormones.

**PROSTATE GLAND** The structure at the base of the male bladder that secretes some of the fluid in semen.

**PROTEIN** A complex biological molecule formed from a chain of units called amino acids. Proteins called enzymes mediate chemical processes in cells. Other proteins form structural tissues such as bone, hair, skin, and collagen.

**PROTESTANT CHURCH** One of the main branches of *Christianity*. Protestantism grew out of a 16th-century reform movement in northern Europe that criticised corruption and Medieval doctrine in the Roman Catholic Church and rejected the authority of the pope.

**PUBERTY** The period of development when a person becomes capable of reproducing sexually.

**PULMONARY ARTERY** The artery conveying blood from the heart to the lungs for reoxygenating.

**PULSE** The rhythmic expansion and contraction of an artery as blood is forced through it.

# R

**RACE** A group of people of common ancestry, distinguished by physical characteristics such as skin colour or eye shape.

**RED BLOOD CELLS** Small, biconcave discs, also known as erythrocytes, that are filled with haemoglobin. Each cubic millimetre of blood contains about five million red blood cells.

**RENAISSANCE** A European cultural movement that took place between the 14th and 17th centuries and involved a revival of classical scholarship and a flowering of arts, architecture, literature, and science. The Renaissance marks the end of the Middle Ages and the beginning of the modern era.

**REPUBLIC** A form of government in which the head of state is appointed or elected, in contrast to a monarchy.

**RESPIRATION** The process by which oxygen is conveyed to, and carbon dioxide is removed from, the body cells.

**RETINA** A light-sensitive layer lining the inside of the back of the eye. The cells of the retina convert optical images to nerve impulses, which travel to the brain via the optic nerve.

**ROMAN CATHOLIC CHURCH** One of the main branches of *Christianity*, headed by the pope in Rome and governed by a hierarchy of bishops. The Roman Catholic Church is known for its uniform doctrine and centralized structure. It is the largest single religious denomination in the world.

**RUSSIAN ORTHODOX CHURCH** The Russian branch of the Eastern Orthodox Church. See also *Eastern Orthodox Church*.

# S

**SALIVA** A watery fluid secreted into the mouth, via ducts from the salivary glands, to aid chewing, tasting, and digestion.

**SAVANNA** Tropical grassland with scattered trees and bushes.

**SCARIFICATION** The process of making small cuts in the skin to produce scar tissue in a decorative pattern.

**SCAVENGER** An animal that feeds on carrion or refuse.

**SECULAR** Not religious or bound by religious law.

**SEMITIC** A family of languages or peoples native to the Middle East. Semitic languages include Arabic, Hebrew, Aramaic, and Amharic. Semitic peoples include the Arabs and the Jews. However, the term "Semitic" is sometimes used to refer to specifically Jewish people and culture.

**SEROTONIN** A neurotransmitter that plays an important role in mood, among other functions.

**SEX HORMONES** Steroid substances that bring about the development of bodily sex characteristics. Sex hormones also regulate sperm and egg production and the menstrual cycle.

**SEXUAL SELECTION** An evolutionary process driven by mate choice that can lead to the development of sexual adornments, such as bright colours in male birds and antlers in deer.

**SHAFII ISLAM** One of the four schools of religious law in *Sunni Islam*, founded in the 8th century by the Arab scholar Imam Shafii. Shafii Islam is followed by 15 per cent of Muslims, including many in Southeast Asia, East Africa, and southern Arabia.

**SHAMANISM** A religion centred on a shaman, a person with the power to communicate with spirits and heal the sick. Elements of shamanism are found in various hunter-gatherer societies, but the term applies primarily to the religion of certain North Asian peoples.

**SHARIA** The body of law that governs the life of Muslims. Sharia is based on both the Koran and the way the prophet Muhammad lived.

**SHIA (SHIITE) ISLAM** The smaller of the two main branches of Islam, accounting for 10–15 per cent of Muslims. Unlike the majority Sunni Muslims, Shias believe that Muhammad's son-in-law and cousin Ali was the rightful legal successor to Muhammad. See also *Sunni Islam*.

**SHINTO** The indigenous animist religion of Japan.

**SIKHISM** A monotheistic religion founded in 16th-century India and incorporating elements of Hinduism and Islam. Most Sikhs live in the state of Punjab.

**SINOATRIAL NODE** A cluster of specialized muscle cells in the heart's right atrium that act as its natural pacemaker.

**SLAVIC (SLAVONIC)** A family of Indo-European languages and peoples native to eastern Europe. Slavic languages include Bulgarian, Serbo-Croatian, Czech, Polish, Ukrainian, and Russian.

**SOCIALISM** A term referring to a broad range of ideologies ranging from communism to forms of capitalism involving state intervention. All socialist ideologies advocate some form of social control to redistribute wealth.

**SPECIES** A category of animals, plants, or other organisms whose members can interbreed and produce fertile offspring.

**SPHINCTER** A muscle ring, or thickening of the muscle coat, surrounding an opening in the body, such as the outlet between the stomach and the duodenum.

**SPINE** The column of 33 ring-like bones, called vertebrae, that is divided into seven cervical, 12 thoracic, and five lumbar vertebrae, and five fused vertebrae making up the sacrum and four fused vertebrae of the coccyx.

**STATE** See *nation state*.

**STEROID** A fat-soluble organic molecule made up of rings of carbon atoms. The sex hormones testosterone and oestrogen are steroids. The term steroid is used informally to refer to drugs called corticosteroids, which reduce inflammation by mimicking natural corticosteroid hormones.

**STONE AGE** The period in human history characterized by the use of stone tools and lasting from around 2.5 million years ago to 6,000 years ago.

**SUBCONSCIOUS** Below the level of conscious awareness.

**SUBSISTENCE FARMING** A type of farming that produces sufficient food only for the farmer's household.

**SUFISM** A mystical branch of Islam that emphasizes personal contact with God through experience of his divine love.

**SUNNI ISLAM** The mainstream branch of Islam, accounting for about 85 per cent of Muslims. Sunni Islam split from Shia Islam in a dispute over leadership following Muhammad's death. Sunnis believe Muhammad's father-in-law Abu Bakr was the rightful Caliph (leader). See also *Shia Islam*.

**SYMPATHETIC NERVOUS SYSTEM** One of the two divisions of the autonomic nervous system. In conjunction with the *parasympathetic nervous system*, it controls many involuntary activities of the glands, organs, and other parts of the body.

**SYNTAX** The arrangement of words in sentences according to grammatical rules that govern meaning.

**SYSTOLE** The period in the heartbeat cycle when first the atria and then the ventricles contract to squeeze blood out of the heart; it alternates with the relaxation period (diastole).

# T

**TEMPORAL LOBE** One of the four lobes that make up each cerebral hemisphere of the brain. The temporal lobe is important in hearing, speech, and memory.

**TENDON** A strong band of collagen fibres that joins muscle to bone and transmits the pull caused by muscle contraction.

**TESTOSTERONE** The principal sex hormone, produced in the testes (in males) and in small amounts in the adrenal cortex and the ovaries.

**THALAMUS** A mass of grey matter that lies deep within the brain. The thalamus receives and coordinates sensory information.

**THE ENLIGHTENMENT** An 18th-century intellectual movement that challenged religious authority and sought to advance human progress through the power of reason.

**THEORY OF MIND** The ability to understand another person's thought processes. Theory of mind develops when children are 3–4 years old and seems to be a uniquely human ability.

**THORAX** The part of the trunk between the neck and the abdomen that contains the heart and lungs.

**TONSILS** Oval masses of lymphatic tissue, on the back of the throat, either side of the soft palate. The tonsils help to protect against childhood infections.

**TRACHEA** The windpipe. A tube lined with a mucous membrane and reinforced with cartilage rings.

**TRADING BLOC** A group of countries that coordinate trading policies and have strong economic links. There are three major trading blocs: Europe, North America, and Asia.

**TRIBE** A social structure typically made up of agricultural villages united by economic, military, and kinship ties and sharing a common culture and usually a common language.

**TURKIC** A family of languages and peoples native to Central and western Asia. Turkic people live mostly in Turkey and the states of Central Asia. Turkic languages include Turkish, Tatar, Turkmen, and Kyrgyz.

**TYRANNY** A government that is cruel, oppressive, and usually dictatorial.

# U

**UNITED NATIONS** An international organization of nation states formed in 1945 to promote peace and cooperation. Nearly all countries of the world are members.

**UPPER CLASS** One of the three main socioeconomic groups of traditional European society, made up of the wealthiest and most powerful members of society.

**UREA** A waste product of protein breakdown that contains nitrogen and is included in urine.

**URINARY TRACT** The waste system, consisting of the kidneys, ureters, bladder, and urethra, that forms urine and excretes it from the body.

**UTERUS** The hollow muscular organ of the female reproductive system in which an unborn baby develops.

# V

**VEIN** A thin-walled blood vessel that returns blood from body organs and tissues to the heart.

**VENTRICLES** The two lower chambers of the heart. The term is also used for the four cavities in the brain that are filled with cerebrospinal fluid.

**VERTEBRATE** An animal with a backbone, such as a fish, amphibian, reptile, bird, or mammal.

**VIRUS** A small infectious agent capable of invading and damaging cells and reproducing within them.

**VODOU** An animist religion common in Haiti and the French Caribbean and characterized by witchcraft and communication by trance with spirits. Vodou combines elements of Roman Catholic ritual with West African animism. Vodou is also known as Vodun and also popularly misnamed Voodoo.

# W

**WARM-BLOODED** Having a constant body temperature, independent of environmental fluctuations. Mammals and birds are warm-blooded; all other animals are cold-blooded.

**WERNICKE'S AREA** A part of the temporal lobe that plays a key role in speech. Damage to Wernicke's area causes Wernicke's aphasia, a disorder in which speech is fluent but unintelligible and meaningless.

**WHITE BLOOD CELLS** The colourless blood cells that play various roles in the body's immune system.

**WHITE MATTER** The regions of the brain and spinal cord made up of nerve fibres.

**WORKING CLASS** One of the three main socioeconomic groups of traditional European society, made up of the poorest members of society, such as factory workers.

**WORLD BANK** An international organization established by the United Nations that finances projects intended to enhance economic development in poor countries.

# XYZ

**X CHROMOSOME** A sex chromosome. All the body cells of females have two X chromosomes.

**Y CHROMOSOME** A sex chromosome, necessary for the development of male characteristics. Male cells have one Y and one X chromosome.

**YOGA** An Indian philosophy whose followers seek to attain a higher spiritual state through physical and mental exercises.

**ZOROASTRIANISM** The pre-Islamic monotheistic religion of Iran, founded by the prophet Zoroaster in the 6th century BC and still practised in parts of Iran and India. Zoroastrianism was a major religion in biblical times and probably had a significant influence on Judaism, Christianity, and Islam.

# INDEX

Page numbers in **bold** indicate main references. *Italic* numbers indicate pages on which the topic is illustrated.

## A

# C

# D

# E

# F



# G

# I

# M

# N

# O

# U



# V

# ACKNOWLEDGMENTS


**Dorling Kindersley** would like to thank the following people for their help in the preparation of this book: Dr Daniel Carter for additional medical consultancy; Martyn Page for developmental work; Simon Adams, Anne Charlish, Kesta Desmond, Jolyon Goddard, Rebecca Lack, Anne McDowall, Lisa Magloff, Susannah Marriott, Sean O'Connor, Teresa Pritlove, Hazel Richardson, Francis Ritter, and Kelly Thompson for additional editorial assistance; Polly Boyd, Lee Wilson, and May Corfield for proofreading; Chloe Alexander, Isabel de Cordova, and Vimit Punater for additional design assistance; Mark Bracey, Gemma Casajuana, Sonia Charbonnier, Jon Goldsmid, and Jackie Plant for additional DTP assistance; Michelle Thomas and Liz Cherry in Production; Richard Tibbitts for additional Illustrations; Rhydian Lewis, Lisa Mason, and Lee Stafford for modelling; John Fenn for assistance with photography; Jeremy Crane, Jacqueline McGee, Erin Richards, and Kathryn Wilkinson for administrative support; and Dr Enkhmandakh of the Mongolian Embassy, London for supplying an example of Mongolian script.

**PICTURE CREDITS**

Dorling Kindersley would like to thank the following for their help with images: John Rutter, Debbie Li, and Rob Henry of the National Geographic Society; Bryan and Cherry Alexander Photography; South American Pictures; and Giovanni Cafagna at Corbis.

**Key:** a = above, b = below, c = centre, l = left, r = right, t = top, f = far, s = sidebar

**Abbreviations: NGS:** National Geographic Society Image Collection. **DK:** DK picture library www.dkimages.com. **Getty:** Getty Images. **B & C Alexander:** Bryan and Cherry Alexander Photography **Alamy:** Alamy images

**1 DK**/American Museum of Natural History (c) **2-3 NGS**/Jodi Cobb. **4 NGS**/Frans Lanting (tl). **4 SPL**/Neil Bromhall (br). **4 Getty**/Randy Wells (cl). **5 Alamy**/ Marc Hill (cl). **5 Corbis**/Charles O'Rear (tc); Torleif Svensson (b). **5 Robert Estall Photo Library**/Carol Beckwith/Angela Fisher (tr). **5 Shehzad Noorani** (cbr). **5 Photonica**/dreamtime (tl). **6 Alamy**/Iain Masterton (bl). **6 Al Deane** (tc). **6 Corbis**/Bob Krist (br); D.Turnley (tr). **6 Robert Estall Photo Library**/Angela Fisher/ Carol Beckwith (tr). **6 Getty**/Karan Kapoor (tcl). **7 B & C Alexander** (tr). **7 Corbis**/Anthony Bannister (br); Michael Yamashita (bc). **7 Robert Estall Photo Library**/Carol Beckwith/Angela Fisher (bl). **7 NGS**/ Pablo Corral Vega (tl). **8 SPL**/D. Roberts (cl); Simon Frazer (cl). **8-9 Getty**/Steve Bloom. **10-11 NGS**/Frans Lanting. **12-13 Alamy**/Frank Whitney; Jan Stromme (s2). **12 Corbis**/Craig Lovell (s3). **12 Lonely Planet Images**/ Richard I'Anson (s1). **Alamy**/Benno de Wilde (s6); **12 Alamy**/Image Source (s5). **12 SPL**/A. B. Dowsett (bcl); Laguna design (bl); Martin Dohrn (s4); Sovereign. ISM (tl). **13 DK**/Marwell Zoological Park (bc); Twycloss Zoo (bl). **13 SPL**/James King-Holmes (cr). **14 SPL**/Mehau Kulyk (br). **14 Getty**/Anup Shah (tr); /The Mooks (tl). **15 Corbis**/Gallo Images (tc); Lindsay Hebberd (br). **16 Alamy**/B & C Alexander (bc); Pictor (br). **16 Toni Angermayer** (ca). **16 Corbis**/Cydney Conger (t). **17 Corbis**/Bohemian Nomad Picturemakers (br). **17 SPL**/ Dr Arthur Tucker (tl); Martin Dohrn (tc). **18 Corbis**/Bob Krist (bl); Gianni Dagli Orti (br); Sygma (tc). **18 SPL**/D. Roberts (tr).**19 Alamy**/Robert Harding Picture Library Ltd (bc). **20-21 Getty**/Randy Wells. **22 Alamy**/A.C Waltham (s1); Steve Bicknell (s5). **22 Corbis** (s6); Archivo Iconografico, S.A. (s2); Luca I. Tettoni (s4); Penny Tweedie (tr); Wolfgang Kaehler (s3). **22 DK**/Natural History Museum (cr). **22 The Natural History Museum, London** (cr). **22 SPL** (cbr). **23 Corbis**/Bryan Allen (bl); David Aubrey (tr); Elio Ciol (tl); Hulton-Deutch (c); Stocktrek (br). **24 Corbis**/Abrecht G Schaefer (br); Galen Rowell (tl). **24 SPL**/David Gifford (bc). **25 DK**/Natural History Museum (cl), (tc). **25 Getty**/AFP (b). **25 John Gurche** (cr). **25 SPL**/John Reader (c). **26 Corbis**/ Jonathon Blair (b). **26 DK**/Natural History Museum (c), (cl); Pitt Rivers Museum (bcl), (clb). **27 Madrid Scientifc Films** (bc). **27 NGS**/Kenneth Garrett (tr). **27 SPL**/John Reader (bl); Volker Steger/Nordstar - 4 million years of man (tl). **28 DK**/Natural History Museum (cb). **28 Madrid Scientifc Films**(cal), (cbl), (cl). **28 The Natural History Museum, London** (bl); Maurice Wilson (tr). **28 SPL**/Tom McHugh (br). **29 DK**/Natural History Museum (br). **29 The Natural History Museum, London** (bl). **29 NGS**/Ira Block (tc); Smithsonian Institution, Washington, DC (tl). **29 SPL**/ Martin Land (cr); Volker Steger (crb); Volker Steger/ Nordstar - 4 Milllion years, of man (cl). **30 Corbis**/ Charles and Josette Lenars (tr). **30 The Natural History Museum, London** (bl). **30 SPL**/Biophoto Associate (cr); BSIP (cl); Pascal Goetgheluck (br). **31 Corbis**/Bettmann (tc); Bob Krist (crb); Michael S Lewis (br). **31 Hutchison Library**/Bernard Gerard (ctr). **31-32 NGS**/Chris Johns. **32 Corbis**/Gianni Dagli Orti (bc); Pierre Vauthey (cfr). **32 C. M. Dixon** (tr). **32 NGS**/Kenneth Garrett (cla); Sisse Brimberg (clb), (cra), (crb). **33 Corbis**/Hubert Stadler (bl); O. Alamany & E. Vicens (clb); Thom Lang (br). **33 Werner Forman Archive**/Tanzania National Museum (cra). **33 NGS**/Kenneth Garrett (ca). **34 Corbis**/Bojan Brecelj (clb); Lowell Georgia (cal); Sakamoto Photo Research Laboratory (cbr). **34 DK**/British Museum (tr); (c); (cla); Museum of London (c). **34 NGS**/Kenneth Garrett (cbl). **35 DK**/Alistair Duncan (crb), (l); National Museum of New Delhi (cbr). **35 Sonia Halliday Photographs** (tr); Jane Taylor (tl). **36 Corbis**/Alinari Archives (cr); Araldo de Luca (ca); Gianni Dagli Orti (bl). **36 DK**/National Maritime Museum (cbr); Pitt Rivers Museum (bl); Science Museum (c). **37 Ancient Art & Architecture Collection** (br), (cr). **37 Corbis**/Jon Hicks (tr); Michael Nicholson (bc); Vanni Archive (cl). **37 DK**/British Museum (car). **38 Corbis**/Gianni Dagli Orti (tr); Stapleton Colllection (bl). **38 DK**/Edinburgh University (car). **39 Ancient Art & Architecture Collection** (bc), (cal), (tr). **39 Corbis**/Christel Gerstenberg (c); Owen Franklin (cl). **39 DK**/Mary Rose Trust (br); Science Museum (cla). **40 Corbis**/Bettman (cr); (bcr); Gustavo Tomsich (cr); Joseph Sohm (clb); Michael Nicholson (br); Stefano Bianchetti. **41 Ancient Art & Architecture Collection** (tl). **41 Corbis**/Araldo de Luca(c); Archivo Iconografico (tr); (cl); Pizzoli Alberto (br). **42-43 Corbis**/ Bettman (cr), (tr); Bettmann (bl); Historical picture Archive. **43 Corbis**/Christie's Images (tr); Hulton-Deutch (br), (cr). **43 DK**/National railway museum (tl). **44 Corbis** (car); Bettmann (c), (tr); David Brauchli/Reuters (br); Hulton-Deutsch Collection (cla); Peter Turnley (bl). **44 DK**/Imperial War Museum (cfl). **45 Alamy**/Bill Barksdale (cbl). **45 Corbis**/Chris Rainier (cl); Najlah Feanny (cla); Thomas Hartwell (tr). **45 Reuters**/Anton Meres (bl); Gregg Newton (cfr). **45 Rex Features** (tc). **48 Alamy**/ Mike Hill (s6). **48 SPL** (c), (bc), (s3), (s5); CNRI (s4); D.Phillips (tl); Mike Devlin (s2); Robert Daly (bcr); Susumu Nishinaga (s1). **49 SPL**/CNRI (tl); Hank Morgan (cbr); Omikron (crb); Simon Frazer (bc), (bcr); VVG (cr). **50 SPL**/CNRI(c). **51 SPL**/Andrew Syred (bcr); Don Fawcett (bl); Geoff Tompkinson (tr); Mauro Fermariello (bl); P.Motta & T. Naguro(c). **52 SPL**/A. Barrington Brown (bl); CNRI (cbr). **53 SPL**/Lawrence Livermore Laboratory (br). **54 DK**/Spike Walker (cb). **54 SPL**/Astrid & Hans-Frieder Michler (bc); Biophoto Associates (br); Dr Gary Gaugler (clb); Dr Tony Brain (bl); Eric Grave (bcr); John Burbidge (cla); Manfred Kage (cbl); Prof. P. Motta (bcl); VVG (cbr). **55 SPL**/Dr Jeremy Burgess (tr); Professors P. Motta & T. Naguro (br). **56 SPL** (car); Eye Of Science (tcr); ISM (ca); J. L. Martra, Publiphoto Diffusion (crb); Prof. P. Motta/Dept. of Anatomy/ University "La Sapienza", Rome (tc); VVG (cla). **60 SPL** (bcl); Astrid & Hanns-Frieder Michler (cal). **61 SPL** (cr). **61 The Untitled Organisation Ltd** (br). **62 SPL**/Dr Jeremy Burgess (br). **64 SPL** (bl), (cl), (cr), (tr). **65 SPL** (bc), (cr), (tr). **66 Corbis**/Roger Ressmeyer (br). **66 SPL** (car); CNRI (bcl), (cfl); GCa (cl); Mehau Kulyk (cbl); Simon Brown (cra); Simon Fraser (bl); Stammers/ Thompson. **67 SPL**/Francoise Sauze (br); GJLP (cbr). **68 SPL** (bl), (clb); Innerspace Imaging (cbl); Manfred Kage (cfl). **69 Corbis** (crb). **70 Alamy**/Kolvenbach (br). **70 SPL** (cl), (tl). **71 Getty**/Shaun Botterill (br). **71 Photolibrary.com** (bl). **71 SPL** (bl), (tc), (tl). **72 Getty**/ Karan Kapoor (tl). **73 Corbis**/Duomo (br). **73 SPL**/ CNRI (bl); National Cancer Instititute (cr). **74 SPL**/Dr Gary Settles (bcl). **75 Corbis**/Amos Nachoum (tr). **75 SPL**/CNRI (bcl), (cb); Eye Of Science (bl). **76 SPL**/St Bartholomew's Hospital (bcl); VVG (cr). **77 SPL**/BSIP, S & I (tl); Manfred Kage (cra); Zephyr (crb). **78 SPL** (cr); BSIP, James Cavallini (tl); CNRI (bcl). **79 DK**/British Museum (tr). **79 SPL**/Frieder Michler (br); Manfred Kage (tr).**80 SPL**/CNRI (cra), (tr); D. Phillips (tl); DU Cane Medical Imaging Ltd (br). **81 SPL**/Alfred Pasieka (br). **82 SPL**/Eye of Science (bl); Martyn F. Chillmaid (tl). **83 SPL** (bl); BSIP Amar (car); Prof. P. Motta (cfr); Scott Camzine, Sue Trainor (tr). **84 Corbis**/Kevin Christopher Ou Photography (c) **85 Getty**/AFP (br). **85 SPL** (clb); CNRI (cfr), (cl); Professors P. Motta & T. Naguro(c) **86 SPL**/ BSIP VEM (cb), (tr). **87 SPL**/Prof P. Motta/Dept. Of Anatomy/University, "La Sapienza", Rome (clb); Prof. P. Motta/Dept. Of Anatomy/University, "La Sapienza", Rome (tl); Professors P. M. Motta, K. R. Porter & P. M. Andrews (cra). **88 SPL**/David M. Martin, M.D (tr); Dr K.F.R. Schiller (cr). **89 SPL** (br); Eye Of Science (cra); Prof. P. Motta/Dept. of Anatomy/University, "La Sapienza", Rome (tl). **90 SPL**/Biophoto Associates (cl); BSIP Dr Pichard (tr); David M. Martin M.D (cr); Dr Karl Lounatmaa (bc). **91 SPL**/David M. Martin M. D (bc); Prof P. Motta (br); Scott Camazine (cr). **92 SPL**/Astrid & Hann-Freider Michler (cr); Astrid & Hanns-Freider Michler (br); Prof. P. Motta (bl); Zephyr (tr). **93 SPL**/ Susumu Nishinaga (bc). **94 SPL** (bl). **95 SPL**/Susumu Nishinaga (cl). **96 Alamy**/Microworks (c). **97 SPL** (cr); Andrew Syred (ca); Custom Medical Stock Photo (br); Dr Linda Stannard, UCT (cb), (cl). **98 SPL** (tr); Astrid and Hanns-Freider Michler (tl). **99 Corbis** (br). **99 SPL** (cfr), (cr); Andrew Syred (cbr); Richard Wehr (tl). **100 SPL**/Ed Reschke, Peter Arnold Inc. (tr); Francis Leroy, Biocosmos (cfl). **100 The Wellcome Institute Library, London** (bl). **101 SPL**/CNRI (crb). **102 SPL**/A.B. Dowsett (cl); Alfred Pasieka (br); Biollogy Media (cal); Dr Andejs leipins (cr); NIBSC (br). **103 SPL**/David Scharf (c); K.H. Kjeldsen (tl); National Library of Medecine (bl). **103 Getty**/Barros & Barros (cr). **104 SPL**/J De Mey, ISM (cl), (clb), (cr), (crb). **105 SPL** (cr); Eye of Science (b). **106 Corbis**/David Jett (c). **107 Alamy**/Dennis Kunkel (crb). **107 SPL** (bl), (br); CNRI (ca). **108 SPL** (cla); J.(c) Revy, ISM (clb). **109 SPL** (br); BSIP, Cavallini James (tl); CNRI (bcl); Michael Abbey (crb). **110 SPL** (cbl); Manfred Kage (tr); Welcome dept of Cognative Neurology (tl); Zephyr (br). **111 SPL**/Pascal Goetgheluck (cr); Zephyr (br). **112 SPL**/Biophoto Associates (cfr); Wellcome Dept. of Cognitive Neurology (bcl). **114 Corbis** (tcr). **114 SPL**/ Adam Hart-Davis (tl); Martin Dohrn (cla); Roseman, Custom Medical Stock Photography (cra); Tek Image (car). **115 Corbis**/Charles Gupton (br). **115 SPL**/Hank Morgan (bcl), (bl); Prof. P. Motta/Dept. Of Anatomy/ University, "La Sapienza", Rome (cfl); Scott. **116 SPL**/ Astrid and Hanns-Frieder Michler (tl). **117 SPL**/Dr John Zajicek (car); /Nancy Kedersha/UCLA (tr); Synaptek (br). **118 SPL**/Astrid and Hanns -Freider Michler (car); Satrun Stills (br); Sovereign, ISM (cl). **119 SPL**/Bsip, Laurent/ Laeticia (cra); CNRI (br); Dee Bruger (tl). **120 Corbis**/ Randy Faris(c). **121 Alamy**/Sandii McDonald (crb). **121 Corbis**/Layne Kennedy (br). **121 Robert Harding Picture Library**/T V Goubergen (tr). **121 SPL**/Ralph Eagle (cbl); Susumu Nishinaga (cl). **122 SPL**/Prof. P. Motta/Dept of anatomy/University,"La Sapenzia" Rome (tr); Ralph Eagle (tl). **123 SPL** (bl); CNRI (tcr); Geoff Bryant (cr); James King-Holmes (br); Martin Dohn (cr); Martin Dohrn (cr); Prof. P. Motta/Dept of Anatomy/ University, "La Sapenzia" Rome (tl). **124 Getty**/ Photodisc Green (bc). **124 SPL**/VVG (tc). **125 SPL** (bl), (tl); Mauro Fermariello (blc); Mehau Kulyk (tcl). **128 SPL**/BSIP, Bussy/Leheutre (bc); Eric Grave(c); Eye Of Science (tr); J. C. Revy (cbl); Prof. P. Motta/G. Franchitto/ University, "La Sapienza", Rome (crb). **129 Corbis**/Javier Pierini (br). **129 SPL**/Pascal Goetgheluck (tl); VVG (bl). **130 Corbis**/Dennis Wilson(c). **131 Corbis**/Andrew Brookes (br). **131 Corbis** (bl). **131 SPL**/Christian Darkin (clb); Josh Sher(c). **132 SPL**/Manfred Kage (tl); P. M. Motta & J Van Blerkom (cr). **133 DK**/Mr T.Hillard (br). **133 SPL**/Christina Pedrazzini (cr); Proff P. M. Motta & E. Vizza (bl); Proff P. M. Motta & E. Vizza (tc). **134 SPL**/ CNRI (cbr); Manfred Kage (tl); National Library of medicine (bc). **135 SPL**/James King-Holmes (cr), (tr); John Burbridge (bc); Parviz M. Pour (tl); Susumu Nishinaga (bl). **136 SPL**/Alfred Pasieka (b). **137 SPL**/(b); Dr Yorgos Nikas(c); Mehau Kulyk (t). **138 SPL**/CNRI (cla).**139 SPL**/Annabella Bluesky (bcl). **140-141 SPL**/ Edelmann (cbl); Neil Bromhall ; Simon Frazer (bl). **140 The Wellcome Institute Library, London**/Yorgos nikas (tr). **142 SPL**/BSIP, Keene (tr); Ruth Jenkinson/ Midirs (bcl). **143 Corbis**/Bettmann (br). **144-145 Photonica**/Dreamtime. **146 Alamy**/Aleksandar Dragutinovic (s3); BCA&D Photo Illustration (s5). **146 DK**/Dave King (br); Denoyer-Geppert (s1); John Lepine (cb); Science Museum (cbr). **146 Alamy**/Imagestate (s2); Alamy/Justin Yeung (s4). **146 SPL**/Nancy Kedersha/ UCLA (s6); Sovereign,ISM (tr); sovereign,ISM (tr). **146 Getty**/David Sacks (bl); /The Bridgeman Art Library/ Joos van Gent (ca). **147 Al Deane** (br). **147 Corbis**/ Roger Ressmeyer (cr). **147 DK**/Susanna Price (tr). **148 SPL**/GJLP (l). **149 Getty**/Terry Vine (bl); /Valder-Tormey (br). **150-151 Corbis**/David H. Wells ; Roger Ressmeyer (cb), (cl); Stewart Tilger (cl). **150 Getty**/ Dennis Novak (cl); /Image Bank Film (cb). **151 DK**/ Conaculta-Inah-Mexico (tr). **151 SPL**/Dr Robert Spicer (c). **152 DK**/David Murray and Jules Selmes (cl). **152 SPL**/Peter Menzel (tr). **152-153 Getty**/David Noton; Photomondo (c). **153 Corbis**/David Leeson(c); William Gottlieb (cr). **153 Getty**/Robert Daly (ct). **154 Corbis**/ Jose Luis Pelaez, Inc. (cl); Liba Taylor. **154-155**; Paul Almasy (bl). **154 SPL**/Hank Morgan (bl). **154 Getty**/ Christoph Wilhelm (ct). **155 Corbis**/Bettmann (tr); Bob Krist (bl); Jose Luis Pelaez (cr); Tom Stewart (cr). **155 DK** John Bulmer (br). **156 Corbis**/Jennie Woodcock; Reflections Photolibrary (bl); Michael St. Maur Sheil. **156-157 DK**/Guy Ryecart (c). **156 Getty**/Patti McConville (br). **157 Alamy**/D. Hurst (tr). **157 Getty**/ Steven Weiberg (cb); Lori Adamski Peek (c). **158-159 Corbis**/Anthony Bannister; Gallo Images; Sygma (tr). **158 DK**/Commissioner for the City of London Police (car). **158 Getty**/Walter Hodges (cl). **159 Corbis**/Alison Wright (tl); David Turnley (cr). **159 Dan Newman** (tl). **159 DK**/Brian Cosgrove (br). **159 Topfoto**/The British Museum/HIP (bl). **160-161 Corbis**/Dan Lamont. **160 SPL**/Mauro Fermariello (b); Wellcome dept. of Cognitive Neurology (ca); Zephyr (cl). **161 Alamy**/Elmtree Images (cb). **161 Corbis** (ca); LWA - Dan Tardiff (br); Ralph A. Clevenger (t). **161 Getty**/Chris Hondros (bl). **162 Alamy**/Bill Bachmann (ca). **162 Corbis**/Seth Joel (tr). **162 DK**/National Maritime Museum (cb). **162 SPL** (cl); /Tony Craddock (b). **163 Corbis**/Ariel Skelley (tc); Christian Blackman (bl). **163 DK**/Frederick & Laurence Arvidsson (bc). **163 SPL**/Dr Jeremy Burgess (cal), (cla). **163 Getty**/Ed Honowitz (cb); Nick Dolding (ca). **164-165 PA Photos**/Corbis/Jonathan Cavendish (cr). **165 Getty**/Allsport Concepts (br); Bruce Herman (tr). **166 Alamy**/ Bill Bachman (tl). **166 Corbis**/Chris Jones (bl). **166 Reuters** (tl). **166 SPL** (ct); Lauren Shear (tl). **166 Getty**/Alan Danaher (bl); Micheal Nemeth (tc); Patrick Molnar (c); Peter Beavis (tl). **167 Alamy**/Stock Connection, Inc. (tr). **167-166 Corbis**/Kevin Fleming (cl); Robert Patrick (br); Ted Spiegel. **167 DK**/Guy Ryecart (tl). **167 Getty** (cr). **168 Corbis**/Kelly-Mooney Photography (br); Rick Friedman (bl). **168 SPL**/Klaus Guldbrandsen (bl). **168-169 Getty**/Martin Barraud; Renee Lynn (c). **169 Corbis** (br); Claudia Kunin (cb). **169 Getty**/Ted Russell (br). **170 Corbis**/Charles Gupton (br); Jeremy Horner (tr); Stephanie Maze (l). **170 Getty**/ J P Fruchet (cr). **171 Corbis**/Trinette Reed (r). **171 DK**/Magnus Rew (tr). **171 Katz**/FSP/Aurora (cr). **171 SPL**/Dolphin Institute (bl). **171 Still Pictures**/Ron Giling (cl). **171 Getty**/ Andy Sacks (tl). **172 Corbis**/J B Russell Sygma (c) **173 Corbis**/Francois Carrel (br); Owen Francken (bl). **173 Getty**/Photodisc (cr). **173 SPL**/Alfred Pasieka (ca). **174-175 Alamy**/Guy Moberly. **174 Corbis**/Eric K K Yu (clb2); Jon Feingersh (cbl2); Richard Hutchings (bcl). **174 Getty**/Cherie Steinberg Cote (cfl2); Don Klumpp (cfl1); Ralf Gerrard (cbl1); Romilly Lockyer (cl2). **174 DK**/Alistair Hughes (cl1). **175 AKG-images** (tl). **175 Alamy**/John Powell Photographer (cr); **175 The Photolibrary Wales** (br). **175 Corbis** (bl). **175 Corbis** (cl). **175 Getty**/Adri Berger (tr). **176 Alamy**/Liam Bailey (cbr). **176 Corbis**/ Annie Griffiths Belt (br); David Turnley (cb); Laura Dwight (cl); Micheal Yamashita (tl); Micheal Yamashita (bl); Steve Raymer (cr). **176 Getty**/UHB Trust (tr). **177 Alamy**/Cosmo Condina (bc). **177 DK**/Simon Brown (tl). **177 Getty**/Larry Dale Gordon (br). **178-179 Corbis**/J. Hawkes. **178 Corbis**/H. Davies (cl); **178 Corbis** (br). **178 Getty**/Andy Bullock (bl); Erik Dreyer (ca). **179 Corbis**/C. Gupton (cr). **179 Corbis** (br). **179 Getty**/Marc Romanelli (bc); Paul Chesley (bcr). **180 Alamy**/Alois Weber (c). **180 Corbis**/David Reed (bc); Keren Su (bl); Randy Faris (tr). **181 Corbis**/Alain Nouges (bcr); Earl & Nazima Kowall (c). **181 DK**/British Museum (crb). **181 Getty**/Brendan Beirne (br); Daly & Newton (tr); David Sacks (bl). **182-183 Alamy**/Marc Hill. **184 Alamy**/bobo (s4); Imagestate (s3). **184 Corbis**/ Anthony Bannister (cr); Ken Redding (s5); Lawrence Manning (s6); Lindsay Hebberd (bl); Will & Deni McIntyre. **184 Alamy**/William McKellar (s1). **185 Corbis**/Annie Griffiths Belt (tl); Chris Lisle(c); Chris Rainer (br). **186 Still Pictures**/Gil Moti (c) **187 Corbis**/ Peter Poby (br). **187 DK**/British Museum (cra), (tr). **188 Corbis**/Annie Griffiths Belt (car); Ariel Skelley (bcl). **188 Sue Cunningham Photographic** (cbl). **188 Nancy Durrell Mckenna** (cl). **188-189 Robert Estall Photo Library**/Angela Fisher/Carol Beckwith. **188 Eddie Lawrence** (bl). **189 Corbis**/Jeremy Horner (car); Stephanie Maze (br). **189 DK**/Johnson/Johnson (tc). **189 SPL**/CNRI (cfr). **190 Corbis**/Norbert Schaefer (tl); Owen Franken (cr), (crb); Roy Morsh (bc); Smailes Alex (cfr). **191 Corbis**/Cat Gwynn (cr); Darwin Wiggett (crb); Owen Franken (br). **191 DK**/Judith Miller (clb). **191 Nancy Durrell Mckenna** (bl). **191 NGS**/Tim Laman (tr). **191 Getty**/Darren Rob (cla). **192 Corbis**/Christine Osborne (cfl); Dustko Despotovic (br). **192 Nancy Durrell Mckenna** (tcl), (cra). **192 Robert Estall Photo Library**/Carol Beckwith/Angela Fisher (bl). **193 Corbis**/ A Ryman (bl); D.H.Wells (t); F. Graham (cra); L.Hebbard (br). **194 B & C Alexander** (br). **194 Associated Press**/ A. P. Lovetsky (cra). **194 Corbis**/D. Conger (cfr), (cr); J. Woodcock (cb). **194 Nancy Durrell Mckenna** (bl). **194 NGS**/Paul Chesley (tr). **195 Corbis**/Bohemian Nomad (bl); D. Degnan (bcl); Yang Liu (br). **195 Panos Pictures** (ca). **195 Corbis** (cla). **196 Corbis**/Earl & Nazima Kowall (crb); Tom Stewart (c); Vince Streano (bcr). **196 NGS**/ Beall, Cynthia M. DR. & Goldstein, Melvyn M.DR (bl). **196 SPL**/BSIP VEM (cla). **196 Getty**/Donna Daly (cra). **197 Corbis**/Ariel Skelley (cbr); Jon Feingersh (t). **197 Nancy Durrell Mckenna** (br). **197 Still Pictures**/Hartmut Schwarzbach (bcl). **198 Alamy**/Joe Sohm (crb). **198 Corbis**/Peter Johnson (cl); Richard Powers (b). **198 Still Pictures**/Jorgen Schytte (t). **198 Getty**/Renzo Mancini (cr); Robert Daly(c). **199 Corbis**/

Jon Feininger (ca); Michael Yamashita (tr). **199 Jo De Berry** (cb). **199 DK**/Janeanne Gilchist (cl). **199 Getty**/ Jim Cummins (br); Robert E Daemmrich (br);Yellow Dog Production (bc). **200 Axiom**/Gordon D R Clements (cl); Jim Homes (cr). **200-210 Corbis**/Ed Kashi (tl); Jeremy Horner (tr); Michael S Lewis; Terres du Sud (b). **201 Still Pictures**/T. Srinivasa Reddy (tc). **201 Getty**/Inc Archive Holdings (tr); Mark Segal (t). **202-203 Corbis**/Duomo (br); Peter M fisher (tl); Peter M Wilson (tl); Steve McDonough . **203 Corbis**/Joe Bator (bl); Richard T Nowitz (bc); Ronnie Kaufman (tl); Roy McMahon(c). **203 NGS**/Jodi Cobb (tc); Pablo Corrall Vega (tr). **204 Corbis**/Anders Ryman (tl), (tl). **204 Robert Estall Photo Library**/Angela Fisher/Carol Beckwith (tr). **204 Panos Pictures**/Penny Tweedie (bl). **205 Corbis**/Arne Hodalic (bl); David. H.Wells (tc); Mark Peterson (ca); Richard T. Nowitz (cl). **205 Robert Estall Photo Library**/Carol Beckwith/Angela Fisher (br), (cr). **205 NGS**/Bruce A. Dale (br); Bruce Dale (bc); Chris Johns (cr); Paul Chesley (cb). **206 Corbis**/Otto Lang (tr). **206 Robert Estall Photo Library**/Angela Fisher/Carol Beckwith (bl); Carol Beckwith/Angela Fisher (br), (clb). **206 Eye Ubiquitous**/Hutchison/Michael Macintyre (tcl). **207 Alamy**/David Levenson (cra); Sylvain Grandadam (tr). **207 Corbis**/Haruyoshi Yamaguchi (crb). **207 NGS**/Joel Sartore (br); Karan Kasmauski (cal). **208 Corbis**/Michael Pole(c). **209 Corbis**/Paul A.Souders (tr); Peter Stone (cl). **209 DK**/Gables (br); Guy Ryecart (cr). **209 SPL**/CNRI (bl). **210 Corbis**/Archivo Iconografico, S.A. (br); Lawrence Manning (cl). **210-211 Robert Estall Photo Library**/Angela Fisher/Carol Beckwith (cb). **210 Rex Features**/Richard Saker (tc). **211 Corbis**/Bettman (tr); Peter Turnley (br). **211 DK**/Stephen Oliver (bl). **211 Robert Estall Photo Library**/Angela Fisher/Carol Beckwith (tl). **212 Corbis**/ Dean Conger (cr); The Cover Story (cb). **212 Eye Ubiquitous**/John Hulme (bl). **212 Hutchison Library**/ Liba Taylor (car); Michael Macintyre (cl). **213 Corbis**/ Francoise de Mulder (tr). **213 Dena Freeman** (bl). **213 Still Pictures**/Peter Schickert (l). **214 Alamy**/Paula Solloway (tl). **214 Corbis**/Bob Krist (tr); Chuck Savage (bl); Paul A Souders (bc); Raymond Gahman (br). **214 Eye Ubiquitous**/John Hulme (cal). **214 NGS**/William Albert Allard (cr). **215 Corbis**/Bettmann (crb); Dave Bartruff (tl); Earl & Nazima Kowall (tr); Koren Ziv/Sygma (br); Lindsay Hebberd (car); Michael Yamashita (c). **215 NGS**/Sam Abell (bl). **216 Alamy**/Camerapix (tcl). **216 Corbis**/Michelle Garrett (cra); Sergio Dorantes (tcr). **216 Robert Estall Photo Library**/Angela Fisher/ Carol Beckwith (b), (cl). **217 Alamy**/David Hoffman Photo Library (cr). **217 Corbis**/Jeffrey L Rotman (tcr); Roger de la Harpe/Gallo Images (c), (cla). **217 NGS**/Jodi Cobb (bl). **217 Reuters**/Rafiqur Rahman (br). **218 Corbis**/Daving Rubinger (bl). **218 Nancy Durrell Mckenna** (ca). **218 Lonely Planet Images**/Noboru Komine (bc). **218 Corbis** (c) **219 Alamy**/Doug Houghton (crb). **219 Corbis**/Caroline Penn (cla); Kevin Cozad (tcr); Lynsey Addario (br); Nik Wheeler (cb). **219 Corbis** (tr). **220 B & C Alexander** (tr). **220 Corbis**/ Chris Hellier (cb); Dean Conger (bl); Lui Liqun (tl). **220 NGS**/Joel Sartore (br). **220 Getty**/Wayne R Bilenduke (ca). **221 Corbis**/ Alison Wright (tl); S.P.Gillette (tc). **221 Robert Estall Photo Library**/Carol Beckwith/Angela Fisher (tr). **221 Hutchison Library**/Nigel Howard (bl). **221 Still Pictures**/Peter Schickert (bl). **222-223 Corbis** ML Sinibaldi. **222 Corbis**/Chuck Savage (cr); Jeremy Horner (t); Jim Craigmyle (b); Nik Wheeler (cl). **223 Corbis**/Chuck Savage (b); Helen King (cl); S P Gillette (t); Tom Stewart (cr). **223 Panos Pictures**/Jean-Leo Dugast (crb). **224 Alamy**/Ami Vitale (bl). **224 Graham Barratt** (bc). **224 Corbis**/Caroline Penn (br); David Turnley (cr); Earl & Nazima Kowall (tl); K M Westermann (cr). **224 Panos** Pictures/Crispin Hughes (c); Mark Henley (ct). **225 Graham Barratt** (tl). **225 Corbis**/ Lester Lefkowitz (b); Lindsay Hubbard (c); Michael Busselle (cr); Setboun(c). **225 Panos Pictures**/Chris Stowers (tr); Fernando Moleres (cbr). **226-227 Getty**/ Derek Berwin. **227 Alamy**/Janine Wiedel Photolibrary (cl); Robert Mullan (bl). **227 Corbis**/Chin Allan (tr); Gabe Palmer (tc); Sigma/Jerome Favre (br). **228 Corbis**/ Bob Krist (br); David H. Wells (cr); Dex Images (tr); Liba Taylor (bl). **228 Panos Pictures**/Mikkel Ostergaard (cl). **228 Still Pictures**/Mike Schroeder (cr). **229 Alamy**/ Comstock Images (cl); Image Source (tr); Nicholas Pitt (tl);Vittorio Sciosia (bl). **229 Corbis**/Christophe Paucellier/Photo & Co. (cr); Larry Lee Photography (br); Todd A. Gipstein (tr). **230 The Art Archive**/Jean Walter & Paul Guillaume Coll/Dagli Orti (A) (bc). **230 SPL** (br); Dr Gopal Murti (cr). **230 Still Pictures**/Achinto/ Christian Aid (tl). **231 Corbis**/Chris Lyle (tr); Dean Conger (tr). **231 NGS**/Karen Kasmauski (bl). **231 PA Photos** (tr). **232 SPL**/FK Photo (c); Mehau Kulyk (tl). **233 Corbis**/Annie Griffiths Belt (br); Tim Page (tr). **233 NGS**/James L. Stanfeild (bl). **233 Getty**/Andrea Pistolesi (cr). **234 Corbis**/Chris Lisle (cfl); Craig Lovell (ca); Jonathon Blair (cbr); Setboun (cr). **234 NGS**/Randy Olson (cra). **234 Panos Pictures**/Mark Henley (tc). **234 Getty**/David Hiser (br). **235 Robert Estall Photo Library**/Angela Fisher/Carol Beckwith (cra); Carol Beckwith/Angela Fisher (bl), (bcl), (cr), (cfr), (tr). **235 Hutchison Library**/Mary Jelliffe (tr). **236 Corbis**/ David Turnley (tl); Ear & Nazima Kowall (cl); Joseph Khakshouri (bl); Paul Almasy (tr). **236-237 Robert Harding Picture Library**/Syvian Granadam. **238 Corbis**/Chase Swift (cr); Lindsay Hebard (tr). **238 DK**/ Imperial War Museum (cr). **238 Dena Freeman** (tl). **238 Hutchison Library**/Jeremony Horner (bl). **238 NGS**/ Steve Raymer (br), (br). **239 Corbis**/Les Stone (cr), (cra); Michael Yamashita (cla); Micheal Yamashita (tr). **239 Eye Ubiquitous**/John Davies (tr). **239 Getty**/ AFP/Omar torres (cb); Demetrio Carrasco (br); Demitrio Carrasco (br). **240-241 Corbis**/Charles O'Rear. **242 Corbis**/ Caroline Penn (tc); Charles O'Rear (s5); Craig Aurness (s2); Danny Lehman (s3); Jerome Sessini (bc); Kat Wade (tr); Nik Wheeler (s1); Richard T. Nowitz (s4). **242 Corbis** (s6). **243 Corbis**/Najlah Feanny Hicks (tl); Owen Franken (b); Pablo Corral (bl). **243 DK**/Musee D'Orsay Paris (tr). **244 Robert Harding Picture Library**/ Advertasia (c). **245 Corbis**/Bettmann (cl); Hashimoto Noboru (br);Vernier Jean Bernard (cr). **245 Still Pictures**/Mark Edwards (clb). **246 Corbis**/Joel W. Rogers (cra); Massimo Mastrorillo (cfr). **246 Robert Harding Picture Library** (b). **246 Still Pictures**/Mark Edwards (cfl); /Ron Giling (cr). **247 Corbis**/Jeremy Horner (cla); Jose Luis Pelaez, Inc. (cra). **247 Panos Pictures**/Belinda Lawley (tcr). **247 Still Pictures**/Argus (ca). **248 Corbis**/Historical Picture Archive (bcl); Jean Miele (clb); Keith Dannemiller (cla); Paul Cadenhead/ Reuters (br). **248 Still Pictures**/Shehzad Noorani (tr). **249 Alamy**/Greg Pease (tcl). **249 Corbis**/Charles O'Rear (cra); David Turnley (cbl); Reinhard Krause/ Reuters (br); Sergio Dorantes (cfr). **249 DK**/Erik Svensson/Jeppe Wilkstrom (c). **250 Robert Harding Picture Library** (tc). **250 Panos Pictures**/Caroline Penn (cfl). **250 SPL**/Chris Knapton (bl). **250-251 Getty**/Denis Waugh;Yann Layma (tr). **252 Alamy**/ B & C Alexander (br). **252 Corbis**/Penny Tweedie (cl). **252 Robert Estall Photo Library**/David Coulson (cr). **252 Hutchison Library**/Tim Beddow (br). **252 NGS**/ Sam Abell (tc). **252 N.H.P.A.** (cr). **253 Darvell Bruderhof** (bl). **253 Corbis**/Richard T. Nowitz (bl). **253 Findhorn Foundation** (bcr). **253 Hutchison Library**/ Jeremy Horner (tr). **253 Lonely Planet Images**/Dennis Wisken (tr). **253 South American Pictures** (tl). **253 Still Pictures**/Francois Gilson (cl). **254 Corbis**/ Bettmann (br); Jeremy Horner (bc); Juan Carlos Ulate/ Reuters(c); Paul A. Souders (tr). **254 Hutchison Library**/Trevor Page (cbl). **254 Still Pictures**/Friedrich Stark (tc). **255 Alamy**/Panoramic Stock Photos Co Ltd (br). **255 Corbis**/Earl & Nazima Kowall (cl); Lester Lefkowitz (bcl). **255 Eye Ubiquitous**/Bennett Dean (cla). **255 Robert Harding Picture Library**/D. Lomax (cr). **255 Lonely Planet Images**/Patrick Horton (tr). **255 Panos Pictures**/Mark Henley (tcl). **255 Alamy**/ Robert Mullan(cb). **256 Corbis**/Peter Turnley (c) **257 Corbis**/Charles & Josette Lenars (cl); Graham Tim (bl); Hans Halberstadt (br). **257 DK**/Nijo Castle (cra). **258 Alamy**/David Tothil (clb). **258-259 Corbis**/Nik Wheeler; Reuters/Paul Hanna (cfl); Setboun (bc); Ted Spiegel (cra). **258 Panos Pictures**/Mark Henley (car). **259 Corbis**/David Lucas (cla); David Turnley (cfr); Gallo Images (br); Roger Ressmeyer (cra). **259 Katz**/FSP/ Contrasto (tcl). **260 Alamy**/B & C Alexander (tr). **260 Corbis**/Peter Johnson (cfl). **260 Robert Estall Photo Library**/Carol Beckwith/Angela Fisher c, (bl). **260 Hutchison Library**/Crispin Hughes (bcr). **260 Rex Features** (cal). **261 Corbis**/Ariel Skelley (tr); Bisson Bernard (cr); Christopher Cormack (bl); Hulton-Deutsch Collection (cfl); John B. Boykin (cb); Michael S. Lewis (tc); Steve Raymer (br). **262 Corbis**/Charles & Josette Lenars (tc); Robert Holmes (bl). **262 Katz**/FSP/Robb Kendrick/Aurora Photos (cfl). **262-263 NGS**/ William Albert Allard. **262 Reuters**/Savita Kirloskar (tcl). **264 Corbis**/Bettmann (cfr), (tl), (tr); Caroline Penn (bl); Massimo Mastrorillo (br); Pascal Parrot (bc). **264 Topfoto**/Jason Laure/The Imageworks (tcr). **264-265 Zefa Visual Media**/Scot Gilchrist. **265 Corbis**/ Commander John Leenhouts (tc); David Turnley (tr); IPA/B.Slatensek (tl); Pete Turnley (br); Reuters (cr). **266 Corbis**/Anders Ryman (bl); Guy Stubbs; Gallo Images (br); Penny Tweedie (cl). **266 Still Pictures**/Friedrich Stark (cfr). **266 Topfoto**/Alison Wright/The Image Works (tcl); John Maier, Jr./The Image Works (cra). **267 Corbis**/Athar Hussain (br); Attar Maher (car); Bettmann (bl); David Turnley (cra); Langevin Jacques (cla); Reuters/ Dadang Tri (cr); Reuters/Juda Ngwenya (cl); Ted Streshinsky (tc). **268 Corbis**/Ahmad Masood/Reuters (cbr); Anthony Njuguna (br); Christophe Calais (br); Finbarr O'Reilly/Stringer (cbr); Gong Yidong/Xinhua Photo (cl); Reuters (cr); Sygma/Sven Creutzmann (tr); Thom Lang (tl). **269 Corbis**/Antoine Serra (br); Fredrik Naumann (cra); Peter Turnley (bl); Reuters (bl), (cbr), (tr); Uwe Schmid (tl). **270-271 Corbis**/Henry Romero/ Reuters. **271 Corbis**/Erik Freeland (bc); Michael Samojeden/Reuters (cra);Yiorgos Karahalis/Reuters (ca). **271 Panos Pictures**/Sven Torfinn (cr). **272-273 Corbis**/ Dave Bartruff (cfl); David Turnley (cra); Mark Peterson. **272 NGS**/Jodi Cobb (bc). **273 Corbis**/Bettmann (cb); Christophe Calais/In Visu (cal); Greg Smith (tl); Reuters/ Paul McErlane (bcr); Tim Wright (cra). **274 Corbis**/Dave Bartruff (cl); Paul Vreeker/Reuters (br); Reuters/Antony Njuguna (cla); Reuters/China Photo (crb); Reuters/ Robert Elliott/Feature Matcher/Leisure-Guiana-Papillon (br); Reuters/Zohra Bensemra (tr); Van Parys. **275 Corbis** (tcl); Gyori Antoine (tr); Hulton-Deutsch Collection (cr); Neville Elder (br); Reuters/Dan Chung (cl); Yomiuri. **276 Corbis**/Antoine Gyori(c); Bernard Bisson (cbl); Reuters/ Faleh Kheiber (br); Jim Bourg/Reuters (tr); Patrick Robert (cal), (cb); Polak Matthew (tr); Reuters/Shamil Zhumatov (bcl). **276 Alamy**/Comstock Images (cbr). **277 Corbis**/Hulton-Deutsch Collection (cfl); Jorge Silva/ Reuters (br); Peter Yates (bl); Reuters/Stringer (car). **278 Alamy**/Sam Tanner (clb). **278 Corbis** (b); Howard Davies (cla); Setboun (cra). **279 Corbis**/Caroline Penn (crb); Chris Lisle (cbr); Dean Conger (br); Owen Franken (cra). **280 Corbis**/David Turnley (tr); Ed Kashi (clb); Howard Davies (tl); Tom Wagner (c);Viviane Moos (bcl). **281 Alamy**/Popperfoto (cfr). **281 Corbis**/Jose Luis Pelaez, Inc. (crb); Mark Peterson (bl); Peter Turnley(c) **281 Getty**/Sean Murphy (tr). **282-283 Shehzad Noorani**. **284 Corbis**/Anthony Bannister; Gallo Images (cl); Keren Su (s1); Lindsay Hebberd (tr); Nathan Benn (br); Robert Semeniuk (s3); Roger Ressmeyer (bl); Rose Hartman (s5). **284 Alamy**/David Jones (s6); /Steve Allen (s4); Corbis (s2). **285 Corbis**/Edward Bock (cra); Gianni Dagli Orti (cal); Hans Georg Roth (tr). **285 Panos Pictures**/Penny Tweedie (bl). **286-287 Lonely Planet Images**/Paul Beinssen. **287 Corbis**/Adam Woolfitt (bc); Charles & Josette Lenars (cl); Craig Lovell (br). **287 DK** Pitt Rivers Museum (tr). **287 Zefa Visual Media**/A.Inden (c) **288-289 Corbis**/Carl & Ann Purcell (bl); Chris Rainier (cb); Vittoriano Rastelli. **288 DK**/Ashmolean Museum (tl). **Corbis** (c). **289 Corbis**/Lindsay Hebberd (tl); Setboun (cr). **289 DK**/British Museum (c). **289 Reuters** (bl). **290 Corbis**/ Bernard and Catherine Desjeux (br); Kevin Fleming (b), (tr); Reuters NewsMedia Inc (ca). **290 DK**/Florence Nightingale Museum (cl). **290 Lonely Planet Images**/Rick Gerharter (tc). **291 Corbis**/ Christine Osborne (cr); Lawrence Manning (br). **291 DK**/National Maritime Museum (bl). **291 Robert Harding Picture Library**/ M.Amin (bcl). **291 NGS**/ Reza (t). **292 Corbis**/Robbie Jack (b). **292 DK**/Judith Miller/Sloan's (cl). **292 NGS**/Maria Stenzel (bl), (clb). **292 Shehzad Noorani** (tc). **292-293 Getty**/Grant Faint. **294 Corbis**/Lindsay Hebbard (b). **294 DK**/British Museum (cr); Glasgow Museum (ca). **294 Robert Harding Picture Library**/Jeremy Bright (tc). **295 Corbis**/Dallas & John Heaton (bl); Hanan Isachar (cl); Nathan Benn (tr); Phil Schermeister (bc); R Ressmeyer (cr). **295 DK**/Glasgow Museum, Andy Crawford (tc). **295 Hutchison Library**/ Felix Greene (cr). **296 B & C Alexander** (tl). **296 Corbis**/Chris Rainier (cr); Lindsay Hebberd (blc); Robert Essel (br); Robert Holmes (cr); Tim Thompson (cr). **296 Eye Ubiquitous**/Dean Bennett (cl). **296 Michael Freeman** (ct). **297 Corbis**/Christine Kolisch (cr); Robert van der Hilst (bl). **297 DK**/American Museum of Natural History (tc). **297 Robert Estall Photo Library**/Carol Beckwith/Angela Fisher (bc). **297 Getty**/AFP/Rajesh Jantilal (tc); AFP/Thony Belezaire (bl). **297 Corbis** (br). **298 Corbis**/Adam Woolfitt (cl); Daniel Laine (br); Paul A.Souders (t); Penny Tweedie (tr); Ralph A.Clevenger (ct); Rebecca McEntee (cb). **299 British Humanist Association**/Maria MacLachlan (bl). **299 Corbis**/Bettmann (bl); Owen Franken (ct); Salt Lake Tribune (cl); Stone Les (tr). **300-301 Corbis**/Walter Hodges. **301 Corbis**/Jean Miele (b); Kennan Ward(c) **301 Photonica** (tr). **302 Corbis**/Archivo Iconografico (tl); Bradley Smith (bl); Charles & Josette Lenars (cr); James Marshall (tr); Stephanie Maze **302 DK**/Jerry Young (br). **303 Corbis**/Alison Wright (t); Jacques Langevin(c) **303 Katz**/FSP/Aurora (b). **304 Corbis**/Jaques Pavlovsky (cbr); Peter Turnley (bl); Robert Patrick (br); Roger Ressemeyer (tr). **304 Corbis** (bc). **305 Corbis**/Randy Francis (br). **305 DK**/Kelly Walsh (ca). **305 NGS**/Annie Griffiths Belt (tl). **306 Alamy**/Wolff, M. (tl). **306 Corbis**/ David Butow (tr); David Lawrence (bl); Jed & Kaoru Share (cla); Ludovic Maisant (car). **306 Panos Pictures**/ Andy Johnstone (cl). **307 Alamy**/Molly Cooper (cfr); Popperfoto (br). **307 Corbis**/David Samuel Robbins (cl); Ellis Richard (cra); Howard Davies (cbr); Ricardo Azoury (cal);Viviane Moos (bl). **307 Alamy**/Image Source (tr). **308 Alamy**/Lebrecht Music Collection (c) **308 Corbis**/ Christophe Calais/In Visu (cra); Collart Herve/Sygma (bcr); Dean Conger (bl); Gordon R Gainer (crb); Lawrence Manning (cla). **308 DK**/British Library (cal), (tl). **308** Corbis (tr). **309 Alamy**/Classic Image (tr); Iain Masterton (tcl); Kim Karpeles (cfr); Photo Japan (cfl). **309 Corbis**/Austrian Archives (car); Wolfgang Kaehler (cb). **309 Panos Pictures**/Giacomo Pirozzi (br). **309** Corbis (bcl). **310 Corbis**/Ariel Skelley (ca); Sandy Felsenthal (b). **310-311 Reuters**/CJ Gunther (tr); Savita Kirloskar. **310 Getty**/Laurence Monneret (cb). **311 Corbis**/Alan Schein Photography (cl); David Pollack (t). **311 Corbis** (cr). **312 Corbis**/Dave G. Houser (br); Mark M Lawrence (tr). **312 Getty**/AFP EPA/Yuri Kochetkov-STF (crb). **312 ImageState/Pictor** (tl). **312 NGS**/Annie Griffiths Belt (bl); James P Blair (clb). **312 Getty**/Bettina Salomon (cfr); Debora Jaffe (cra); Martin Riedl (cal). **313 Corbis**/ Reuters (cra). **313 Robert Harding Picture Library**/ D Webster (crb). **313 NGS**/James L. Stanfield (tr); Steve Raymer (bl); Tomasz Tomaszewski (br). **313 Corbis** (cla). **313 Getty**/Holos (cb). **314 Alamy**/J Horton (tr). **314 Corbis**/Earl & Nazima Kowall (tc); John Henley (crb); Layne Kennedy (b); Lynn Goldsmith (cb); Patrick Robert (clb). **315 John Birdsall Photo Library** (bl). **315 Corbis**/Dimitri Lundt (cl); James Marshall (tr); Reka Hazir/Sygma (br) Earl & Nazima Kowall (tc). **315 DK**/ Brian Pitkin (bc); Gables (cbr). **315 NGS**/Jodi Cobb (cr). **315 Panos Pictures**/Penny Tweedie (tl). **316 Corbis**/ Bettmann (cl); Ed Bock (t); M Eliason/Santa Barbara News/Zuma (br). **316 Panos Pictures**/Mark Henley (cr). **317 Coral Planet Photography**/Patrick Robert/ Sygma (br). **317 Corbis**/Carl & Ann Purcell (ca); Catherine Karnow (cl); Gail Mooney (cb); Gideon Mendel (bl); Herrmann/Starke (cr); Swim Ink (tc); Tom Wagner (br). **318-319 NGS**/James L.Stanfeild. **319 Alamy**/James Montgomery (tr). **319 Corbis**/Buddy Mays (bl), (cl). **319 Reuters**/J.T.Lovette (br). **320 Alamy**/ Brett Jorgensen (bl); R.Warburtons (br). **320 Corbis**/ Ainaco (cfr); Francesco Venturi (crb); Kevin R Morris (t); Robert Weight (cl). **320 DK**/Judith Miller (ca). **320-321 Robert Estall Photo Library**/Angela Fisher/Carol Beckwith. **321 Alamy**/Chuck Pefley (tl); thislife pictures (cb). **321 Corbis**/Gideon Mendel (b); Michael Pole (tr); Penny Tweedie (tc). **322 B & C Alexander** (cl), (tl). **322 Corbis**/ Archivo Iconographica (bl); Charles O'Rear (cr); Joe Baraban (b); W.Perry (tc). **322 Robert Estall Photo Library**/Angela Fisher/Carol Beckwith (bc). **323 Alamy**/David R Frazier Photolibrary (tl). **323 Corbis**/ Bob Krist (br); Craig Aurness (tr); D. Turnley (bl); Franco Vogt (cr); Neil Rabinowitz (cr); Sygma S R Frank (bl); Todd A Gipstein (tc). **324-325 Robert Estall Photo Library**/Carol Beckwith/Angela Fisher. **325 Corbis**/ Catherine Karnow (cbr); Craig Lovell (c); Graham Tim (b). **325 Corbis** (tr). **326 Alamy**/Martin Dalton (tl). **326 Corbis**/Joe Malone (bc); Kevin Flemming (bl); Margaret Courtney Clarke(c) **326 Robert Harding Picture Library** (bl). **326 Katz**/FSP/Aurora (tc). **326 Reuters**/ Dan Chung (tr). **327 Corbis**/Bob Krist (bc); Charles O'Rear (br); Fulvio Roiter (tc); Graham Tim (cr); Ludovic Maisant (cl); Mark Peterson (br). **327 Robert Estall Photo Library**/Angela Fisher/Carol Beckwith (c). **328 Corbis**/Jose Fuste Raga (c) **329 Corbis**/Bettmann (bc); Gianni Dagli Orti (cl). **329 SPL**/Chris Butler (cbr); Professors P M Motta & S Correr (cr). **330 Corbis**/ Archivo Iconografico (tl); Chuck Savage (br); Keren Su (tc); Robbie Jack (bl). **330 DK**/British Museum (tr). **330 Panos Pictures**/Penny Tweedie (cr). **331 Corbis**/ Archivo Iconografico/Le Corbusier FLC/ADAGP, Paris and Dacs, London 2004.(c); Danny Lenham (t); Douglas Kirkland (bc); Tim Graham (bl). **332 Corbis**/Alan Schein Photography (br); Bettmann (bl); Christie's Images (crb); Farrell Grehan (tr); John Van Hasselt (cla); Massimo Mastrorillo (cfr); Tim Graham (tc). **333 Corbis** (c) Jefferey L. Rotman (bl); Jim Zuckerman (cra); Leonard de Selva (crb); Lindsay Hebberd (br); Nik Wheeler (tcr). **333 Robert Aigen Oriental Carpets and Textiles** (bc). **333 Rex Features** (cla). **334 Corbis**/Dallas and John Heaton (bl); Hulton-Deutsch Collection (cr); Julia Waterlow; Eye Ubiquitous (cal); Michael Freeman (br); Mimmo Jodice (cbl); Rune Hellestad (cra). **335 Corbis**/ Bettmann (bl); Bob Krist (c); Henry Diltz (tc); Lynsey Addario (cla); Swift/Vanuga Images (tr); The Cover Story (br). **336-337 Corbis**/Peter Johnson (car); Roger Ressmeyer. **336 DK**/Natural History Museum (cbl). **337 Corbis**/Firefly productions (cb); Roger Du Buisson (tr). **337 SPL**/Makoto Iwafuji/Eurelios (crb); Mehau Kulyk (cl). **338 Corbis**/Massimo Mastrorillo (tr); NASA/Roger Ressmeyer (br); Paul A Souders (crb); Richard Hamilton (c); Steve Chenn. **338 SPL**/Novosti Photo Library (cla); US Department of Defence (ca). **339 Corbis**/Joseph Sohm, ChromoSohm Inc (tr); Larry Lee (cla); Lester Lefkowitz (cra); NASA/Roger Ressmeyer (br); RM (cfr); Roger Ressmeyer (tcl). **340 Corbis**/Astier Frederik/ Sygma (tc); Liba Taylor (br); Tom Stewart (bl); W Perry Conway (ca). **341 Corbis**/Ann Kaplan (cal); Charles O'Rear (cr); David Turnley (cra); Dewitt Jones (cbl); Gianni Dagli Orti (bl). **341 DK**/Guy Ryecart (bcr). **342-343 Robert Estall Photo Library**/Carol Beckwith/ Angela Fisher. **344 Alamy**/B & C Alexander (s1), (s2). **344 Corbis**/Cat Gwynn (tr); David Stoecklein (s3); Dean Conger (bl4); Jacques Langevin (bl3); James Davis (bl3); James Nelson/The Cover Story (s6); John Van Hasselt (br6); Lawrence Manning (cal); Leland Bobbe (bc); Morton Beebe (br5); Pat O'hara (br3); Reuters/Darren Whiteside (bl4); Richard T. Nowitz (s5);Yann Arthus-Bertrand (bl). **344 Hutchison Library**/John Wright (br2). **344 NGS**/Nicolas Reynard (br1); Sam Abell (bl2). **344 Panos Pictures**/Penny Tweedie (bl1). **344 Alamy**/ Jack Hollingsworth (s4). **344 Corbis** (bl). **345 B & C Alexander** (bl2), (bl3). **345 Corbis**/Anthony Bannister; Gallo Images (br2); Jacques Langevin (cbr); Michael Yamashita (bl); Nevada Wier (tr); Owen Franken (bl5); Patrick Ward (bl8); William Manning (br); Wolfgang Kaehler (cal). **345 Eye Ubiquitous**/Bennett Dean (bl6). **345 Hutchison Library**/Nancy Durrell McKenna (br4); Sarah Errington (br1), (br3) Edward Parker (br5). **345 Impact Photos**/Daniel White (bl4). **345 Katz**/FSP/ Gamma (bl1), **345 NGS**/Annie Griffiths Belt (bl7); Michael Nichols (br6). **346 Corbis**/Barry Lewis (bc);

George Tiedemann (tc); Peter Guttman (tcr). **346 Robert Estall Photo Library**/Angela Fisher/Carol Beckwith (cfl). **347 Corbis**/Peter Turnley (bcr); Stephanie Maze (clb); Strauss/Curtis (cra); Wally McNamee (cr). **347 Robert Estall Photo Library**/Carol Beckwith/Angela Fisher (tl). **348 Corbis**/Darrell Gulin (c). **349 Corbis**/Bettmann (bl); Bob Daemmrich (crb); Craig Aurness (br). **349 DK**/Scott Polar Institute(c). **349 Corbis** (cal). **350 Corbis**/Bernd Obermann (tc); Bettman (tr); Bob krist (tr); Francis. G. Mayer (bl); James. L. Amos (clb); Jim Erikson(c). **350-351 Photorush**. **352 Corbis**/Bettmann (bc); Morton Beebe (cl). **352 Robert Harding Picture Library**/Roy Rainford (tr). **352 John Hyde** (br). **352 Impact Photos**/Steve Parry (c). **352 Zefa Visual Media**/Alec Pytlowany (cr). **353 B & C Alexander** (clb). **353 Corbis**/Pat O'hara (tl); Wolfgang Kaehler (car). **353 DK**/American Museum of Natural History (bl). **353 NGS**/Nick Caloyianis (cla); Richard Olsenius (cr). **353 Still Pictures**/B&C Alexander (br). **354 B & C Alexander** (cl), (tcl). **354 Corbis**/Dewitt Jones (bl); Richard A. Cooke (cfr), (tr). **354 Lonely Planet Images**/John Elk III(c). **354 Still Pictures** (br). **355 B & C Alexander** (cla), (tc). **355 Corbis**/Christie's Images (cfr); Earl & Nazima Kowall (br); Phil Schermeister (clb). **355 NGS**/Joel Sartore (bc). **356 B & C Alexander** (bc), (br), (crb). **356 Corbis**/David Turnley (cfr), (tr). **356 Lonely Planet Images**/Richard I'Anson (tr). **356 NGS**/ Randy Olson (cla). **356 Zefa Visual Media**/John de Visser (cfl). **357 Corbis** (ca), (cfr); Brian A. Vikander (bc); Macduff Everton (crb). **357 Katz**/FSP/Jeff Jacobson (tr). **357 NGS**/Nicolas Reynard (cla). **358 Corbis**/Peter Turnley (tl). **358 DK**/American Museum of Natural History (cla), (tr). **358-359 Magnum**/David Alan Harvey. **359 Corbis**/Bettmann (t). **360 Corbis**/Bob Krist (bl); Burstein Collection (t); Ed Eckstein (br); Lee White (cr); Philip Gould (bc); Raymond Gehman (cl); Richard A. Cooke (cfl). **360 DK**/American Museum of Natural History (cr). **361 Corbis**/Catherine Karnow (b); Christie's Images/Edward S. Curtis (cr); George H H Huey (cl); Mark Peterson (tl); Philip Gould (tc). **361 DK**/American Museum of Natural History (cr). **361 NGS**/Paul Chesley (cb). **362 Corbis**/Ben Wittick (cl); Bettmann (cbr); Bob Rowan (cb); Bowers Museum of Cultural Art (b); Gunter Marx (bl); Raymaond Gehman (cra). **362 DK**/American Museum of Natural History (br), (tc). **362 Still Pictures**/Martha Cooper (tr). **363 Corbis**/Bob Krist (cb); Christopher J Morris (cbr). **363 DK**/American Museum of Natural History (t); Andre Leyerle (br). **363 NGS**/Bruce Dale (ca); Kenneth Garrett (bl). **364 Corbis**/Charles & Josette Lenars (cfr); Danny Lehman (c), (bc), (cb), (cra); Robert van der Hilst (bl); Sergio Dorantes (cfl), (tcl); Wolfgang Kaehler (br). **365 Corbis**/Danny Lehman (bl), (cfr); Dave G. Houser (cla); Morton Beebe (cra). **365 Katz**/FSP/Russell Gordon/ Aurora Photos (br). **365 NGS**/Maria Stenzel (ca); Nicholas Reynard (tcr). **366 Liba Taylor** (c). **367 Corbis**/Charles & Josette Lenars (bl); Macduff Everton (br), (cbl). **367 Katz**/FSP/Kactus Foto/Gamma (cbr). **367 NGS**/Michael Nichols (car). **368 Corbis**/Alain Le Garsmeur (cfr); Daniel Laine (bcl); Jeremy Horner (tc); Nik Wheeler (br); Robert van der Hilst (cra). **368 Lonely Planet Images**/Martin Fojtek (cla). **368 Reuters**/Joe Skipper (clb). **369 Corbis**/Alison Wright (tl); Bob Krist (bl); David Turnley (cfr); Morton Beebe (tr); Stephanie Maze (cl). **369 NGS**/Michael Melford (cbr); Steve Winter (br). **370 James Davis Travel Photography** (cfr). **370 Robert Harding Picture Library**/A.Woolfitt (br). **370 Hutchison Library**/Jeremy Horner (tcr); Jon Fuller (cl). **370 South American Pictures** (bcr). **370 Mireille Vautier** (bl). **371 Corbis**/Bill Gentile (cfr); Jay Dickman (cla), (tc). **371 South American Pictures** (bc), (bl). **372 Corbis**/Colita (cr). **372 Hutchison Library**/Jeremy Horner (c), (cl); Teremy Horner (tl). **372 Lonely Planet Images**/Krzysztof Dydynski (tr). **372 NGS**/Pablo Corral Vega (cb), (clr). **372 South American Pictures**/Tony Morrison (br). **372 Still Pictures**/Roland Sietre (br). **373 Corbis**/Paulo Fridman (cr). **373 NGS**/Micheal Nichols (tc), (tr). **373 South American Pictures**/Tony Morrison (br), (cr), (crb). **373 Still Pictures**/Mark Edwards. **374-375 Katz**/FSP/Gamma Ribeiro Antonio. **375 Mary Evans Picture Library** (bl). **375 Hutchison Library**/ Jeremy Horner (cb). **375 Still Pictures**/Mark Edwards (tr); /Roland Seitre (cbr). **376 Alamy**/Sergio Pitamitz (bc). **376 South American Pictures**/Luke Peters (bl). **376 Still Pictures**/Herbert Giradet (c), (tc), (tr);/ Marcella Hugard-Christian Aid (tc). **376 Mireille Vautier** (br). **377 South American Pictures**/Frank Nowikowski (br); Luke Peters (bl); Marion Morrison(c); Tony Morrison (tc). **377 Still Pictures**/Marcella Hugard-Christian Aid (tl). **377 Mireille Vautier** (br), (tr). **378 Corbis**/Firefly Productions (cr). **378 Hutchison Library**/Eric Lawrie (cra); John Wright (cb). **378 Katz**/FSP/Gonzalez/Laif (tc). **378 Lonely Planet Images**/Eric L Wheater (cla). **378 Still Pictures**/Julio Etchart/Reportage (br). **378 Mireille Vautier** (bcl). **379 Hutchison Library**/Eric Lawrie (cra); Nancy Drew Mckenna (crb). **379 NGS**/Joel Sartore (bcl). **379 South American Pictures** (cal), (cla); Fiona Good (clb). **379 Mireille Vautier** (br). **380 Corbis**/ Mark L Stephenson (c). **381 Corbis**/Howard Davies (br); Kit Houghton (bcl); McPherson Colin(c). **382 B & C Alexander** (cl). **382 Corbis**/Bo Zaunders (cbl); Hans Strand (bl); Staffan Widstrand (tr). **382 Lonely Planet Images**/Martin Moos. (cfr). **382 NGS**/Sisse Brimberg (cla). **383 B & C Alexander** (cb). **383 Corbis**/Dave G. Houser (br); Nik Wheeler (cl), (tcl). **383 Hutchison Library**/Nick Haslam (cra). **383 NGS**/George F. Mobley (bl). **384 Associated Press** (bl). **384 Bridgeman Art Library**/The Board of Trinity College, Dublin, Ireland (cbr). **384 Hutchison Library**/Jeremy Horner (br). **384 Lonely Planet Images**/Jonathan Smith (cla), (tc). **384 Getty**/Alan Becker (tr); Mike Powell (car). **385 Corbis** (cfr); Annebicque Bernard (br). **385 Empics Ltd**/Tony Marshall (bl). **385 Eye Ubiquitous**/Kevin Wilton (tc). **385 Robert Harding Picture Library**/Roy Rainford (ca). **385 Impact Photos**/Tony Page (cla). **385 Photolibrary Wales** (bcl), (cb). **386 Corbis**/David Turnley (tr); Michel Setboun (br). **386 Robert Harding Picture Library**/J. Lightfoot (cl). **386 Katz** /FSP/ Bruno De Hodges/Gamma (cr); Ogando/Laif (tl); Specht/Laif (tc). **386 Lonely Planet Images**/Guy Moberly (bl). **387 Axiom**/Alberto Arzoz (cr). **387 Corbis**/Julio Donoso (tc); Owen Franken (cbl), (cl). **387 Robert Harding Picture Library**/M.Short (cr). **387 Katz**/FSP/Osang/Laif (bl). **388 Corbis**/ABC Press-Hofstee (cr); Marc Garanger (bl); Tim De Waele (cl); Tiziana and Gianni Baldizzone (cb). **388 Robert Harding Picture Library**/Explorer (br). **389 Corbis**/ David Turnley (c); Uwe Walz (c); Wally McNamee (tr). **389 Katz**/FSP/Adenis/GAFF(c); H & D Zielske (bl); Jung (cl). **389 PA Photos**/DPA (br). **390-391 NGS**/Sisse Brimberg . **391 Katz**/FSP/Laif/Ogando (tr). **391 NGS**/ William Albert Allard (ca). **391 Corbis**/Francesco Talenti (br). **392 Corbis**/David Ball (ca); Peter turnley (bl). **392 Hutchison Library**/James Brabazon (tr); Liba Taylor (bc); Yuri Shpagin (bl). **392 Lonely Planet Images**/ Richard Nebesky (tl). **392 NGS**/Sarah Leen (cb). **392 Still Pictures** (tc). **393 Corbis**/Gavriel Jecan (br); Peter Turnley (ca). **393 Hutchison Library**/Liba Taylor (crb); Peter Moszynski (cl). **393 Lonely Planet Images**/ Christina Dameneya (tr). **393 NGS**/Steve Raymer (bl). **393 Reuters**/Vasily Fedosenko (cra). **394 Corbis**/ Charles Lenars (tr); Daniel Laine (b). **394 DK**/British Museum (bl), (c). **394 Hutchison Library**/Melanie Friend (tl); Simon McBride (bl). **394 NGS**/Pritt Vesiland (tr). **395 Corbis**/Jonathon Blair (cl); Michel Setboun (tr); Pascal Le Segretain (br); Peter Turnley (cra). **395 Katz**/ FSP/Gamma/Art Zamur (bc). **395 Lonely Planet Images**/Lee Foster (ca). **395 NGS**/James L. Stanfield (bl). **396 B & C Alexander** (bc), (tl). **396 Corbis**/Alexander Natruskin (cl); Brian A Vikander (ca); David Cumming (tr); Gleb Garanich (cr); Gregor Schmid (tr); Peter Turnley (tc). **397 Corbis**/Ed Kash (br); Gregor Schmid (bl), (cbr), (tc), (tcr). **397 Novosti (London)** (cl). **398 Corbis**/Attar Maher (tr); David Turnley (cb); Gregor M. Smith (tl); Pete Turnley (bc), (bl). **398 Magnum**/Thomas Dworzak (br). **399 Corbis**/Annie Griffiths Belt (br); Dave Bartruff (br); David Turnley (bl), (cr), (tc), (tr); Patrick Robert (tr). **400 Corbis**/Penny Tweedie(c). **401 Corbis**/Georgina Bowater (cr). **401 DK**/British Museum (bl). **401 Corbis** (br). **401 Still Pictures** (c);/Frans Lemmens (cl). **402 Corbis**/Thomas Hartwell (tr). **402 DK**/British Museum (ct). **402 Impact Photos**/Robin Laurance (cr). **402 NGS**/Annie Griffiths Belt (b) Reza (cl); Steve Raymer (tl). **403 Corbis**/Jonathon Blair (br); Patrick Ward (cl); Paul A Souders (cr); Paul Almasy (cl). **403 DK**/Pitt Rivers Museum (clb). **403 Impact Photos**/Alan Keohane (ct). **403 Still Pictures**/Sean Sprague (bl). **404 DK**/British Museum (cr). **404 Robert Estall Photo Library**/ Angela Fisher/Carol Beckwith (cl); Carol Beckwith/ Angela Fisher (cr). **404 Eye Ubiquitous**/Julia Waterlow (bl). **404 NGS**/Kenneth Garratt (bc), (cbr); Kenneth Garrett (br). **404 Still Pictures** (t). **405 Corbis**/Nik Wheeler (bl); Owen Franken (cr); Richard Hamilton Smith (c); Richard T Nowitz (br). **405 DK**/British Library (t). **405 NGS**/ Michael Yamashita (bc); Stephen St John (cl). **406 Corbis**/Charles & Josette Lenars (cbl); Dave Bartruff (cbr); Jeffrey L Rotman (br); Peter Turnley (bl); Richard T Nowitz(c). **406 Hutchison Library**/ Billie Rafaeli (cl). **406 Getty**/Lorne Resnick (cr). **407 Corbis**/Nik Wheeler (cl), (t). **407 Ffotograff** (cbr); Patricia Aithie (cbl). **407 NGS**/Steve McCurry (b). **407 Getty**/Mario Tama (cr). **408 Corbis**/Arthur Thevenart (c); Dave G. Houser (tc); Diego Lezama Orezzoli (bl). **408 DK**/British Library (crb). **408 Eye Ubiquitous**/Bennett Dean (c) **408 Hutchison Library** (br). **408 NGS**/James L. Stanfield (tr). **409 Corbis**/Alain Nogues (br); Arthur Thevenart (tr); Earl & Nazima Kowall (cl), (cr); Paul Almasy (bl); Roger Wood (bc). **409 Impact Photos**/ Caroline Penn (cfl). **410 Robert Estall Photo Library**/ Carol Beckwith/Angela Fisher(c) **411 Corbis**/Christie's Images (cbl); Gideon Mendel (br); Habans Patrice (cfr). **411 Robert Estall Photo Library**/Angela Fisher/Carol Beckwith (cal). **411 Reuters**/Mike Hutchings (c). **412 Al Deane** (br). **412 Robert Estall Photo Library**/Carol Beckwith/Angela Fisher (bl), (tc), (tr); David Coulston (bc). **412 Hutchison Library**/Sarah Errington (tl). **413 Robert Estall Photo Library**/Carol Beckwith/Angela Fisher (br), (cr), (tc), (tr). **413 Dena Freeman** (bc), (bl). **414-415 Robert Estall Photo Library**/Fabby Nielsen. **415 Robert Estall Photo Library**/Angela Fisher (tcr). **415 Hutchison Library**/Sarah Errington (tc), (tr). **416 Robert Estall Photo Library**/Carol Beckwith (cr); Carol Beckwith/Angela Fisher (tr), (b). **416 Hutchison Library**/Mary Jelliffe (cr). **416 Impact Photos**/Giles Moberly (tl). **417 Corbis**/Neal Preston (tr); Paul Almasy (br); Werner Forman (cr). **417 Robert Estall Photo Library**/Carol Beckwith/Angela Fisher (bl). **417 Hutchison Library**/Edward Parker (c); Juliet Highet (cr); Sarah Errington (tl). **417 PA Photos**/DPA (br). **418 Corbis**/Archivo Iconografico (car); Brian A.Vikander (bl); Margaret Courtney-Clarke (cra); Owen Franken (cla); Werner Forman (br). **418 Robert Estall Photo Library**/Carol Beckwith/Angela Fisher (cb). **418 NGS**/ Fabby Nielsen (ct). **419 James Davis Travel Photography** (c) **419 Robert Estall Photo Library**/ Carol Beckwith/Angela Fisher (br), (tcr). **419 Images Of Africa Photobank**/Carla Signorini Jones (cal). **419 Lonely Planet Images**/Ariadne Van Zandbergen (bl). **420-421 NGS**/Micheal Nichols. **421 NGS**/Michael Nichols (br); Micheal Nichols (cl), (tr). **422 Corbis**/Carl & Ann Purcell (cal); W. Perry Conway (cfr). **422 Robert Estall Photo Library**/Angela Fisher/Carol Beckwith (b); Ariadne Van Zandbergen (cra); Carol Beckwith/ Angela Fisher (c). **422 Images Of Africa Photobank**/ Charlotte Thege (cla). **423 Sue Cunningham Photographic** (bl). **423 Douglas Curran** (cb). **423 Hutchison Library**/Bernard Gerard (cla); Liba Taylor (cr); Trevor Page (tr). **423 Mark Read Photography** (br). **424 Robert Estall Photo Library**/Angela Fisher/CarolBeckwith (br). **424 Mark Read Photography** (c), (tc), (tr). **424 Still Pictures**/Roger De La Harpe (bl). **425 Corbis**/David Turnley (cb); Guy Stubbs (bc); Peter Johnson (bl). **425 Eye Ubiquitous**/ M.Wilson (ca). **425 Hutchison Library**/Ingrid Hudson (cl); Nancy Durrell McKenna (tc). **425 Images Of Africa Photobank**/Brian Charlesworth (tr). **425 Lonely Planet Images**/Oliver Cirendini (br). **426 Corbis**/ Michael Setboun (c). **427 Corbis**/Archivo Iconografico, S.A. (cbl); Brian A. Vikander (bcl); David Turnley (car); Roger Ressmeyer (br). **427 Reuters**/Shamil Zhumatov (cbr). **428 B & C Alexander** (bcl), (br), (cfr), (cl), (cla), (cra). **428 Katz**/FSP/Naundorf/laif (tr). **429 B & C Alexander** (bl), (ca), (cfl), (cl), (tr). **429 Jak Kilby** (cbr). **429 Mark Read Photography** (cr). **430 Corbis**/Dean Conger (cfr); Nevada Wier (tc); REZA (cla); Reza/ Webistan (ca). **430 Katz**/FSP/Gamma (cbl). **430 Mark Read Photography** (clb). **430 Getty**/NGS (br). **431 Corbis**/Galen Rowell (cla); Zylberman Laurent (cra). **431 Katz**/FSP/Gamma (car). **431 Mark Read Photography** (tcl). **431 NGS**/James L. Stanfeild (bcl), (cl); James L. Stanfeild (br). **431 Getty**/ Paul Harris (cfr); The Bridgeman Art Library (bl). **432 Hutchison Library**/ Sarah Errington (bc). **432 Katz**/FSP/Aventurier Patrick/ Gamma (tc); Noel Olivier/Gamma (cr). **432 Reuters**/ Kamal Kishore (cl). **432 Still Pictures**/Shehzad Noorani (bl), (br). **433 Hutchison Library**/Sarah Eddington (bc); Sarah Errington (br); Sarah Murray (cl). **433 Lonely Planet Images** Richard L'Anson (tc). **433 NGS**/James L. Stanfeild (cl); James. L. Stanfeild (br); Jonathan Blair (tr). **433 Still Pictures**/Shehzad Noorani (cr). **434 Lonely Planet Images**/Eddie Gerald (c). **435 Corbis**/Antoine Serra/In Visu (bcl). **435 DK**/City Palace Museum, Jaipur (cr). **435 Impact Photos**/Colin Jones (cfl). **435 Reuters**/Mohsin Raza (br); Munish Sharma (car). **436 Corbis**/Catherine Karnow (cl); Floris Leeuwenberg (cfr). **436 Eye Ubiquitous**/J. Burke (br); Patrick Field (bc). **436 Impact Photos**/Christophe Blunzer (tc); Daniel White (cb); Mark Henley (tr). **437 Corbis**/Arthur Thevenart (cr); Earl & Nazima Kowall (br), (tr); Jacques Langevin (cl); Lindsay Hebberd (cfr). **437 NGS**/Bobby Model (cl); James L. Stanfeild (cbl). **437 Still Pictures**/ Mark Edwards (bc); Mark Edwards (bl); Roland Seitre (bl). **438 Corbis**/Charles & Josette Lenars (cr); Charles Rotkin (bl); Chris Lisle (tl); Hulton-Deutch collection (cr). **438 Hutchison Library**/Stephen Pern (bl). **438 Lonely Planet Images**/ Anders Blomqvist (cbr). **438 Still Pictures**/Shehzad Noorani(c). **439 Corbis**/Jeremy Horner (bc); Lindsay Hebbard (tc); Tiziana Baldizzione (tr). **439 DK**/National Museum Delhi (bcl). **439 Peggy Froerer**(tl). **439 Hutchison Library**/William Holtby (bl). **439 Impact Photos**/Christophe Bluntzer (bl). **439 Still Pictures**/Chris Caldicott (cr). **440 Corbis**/Earl & Nazima Kowall (cr); Lindsay Hebberd (cal), (cl), (tr). **440 Shehzad Noorani** (bl), (br), (cl). **441 Corbis**/Earl & Nazima Kowall (car), (tcr); Lindsay Hebberd (clb); Lindsay Jebberd (br). **441 DK**/National Museum-New Delhi (ca). **441 Shehzad Noorani** (cra). **442 Axiom**/Vicki Couchman (tl). **442 DK**/National Museum Delhi (cr). **442 Eye Ubiquitous**/Bennett Dean (cl). **442 Reuters**/ Anuruddha Lokuhapuarachchi (tl). **442 Still Pictures**/ John Issac (b), (ca). **443 Eye Ubiquitous**/Bennett Dean (bl). **443 Impact Photos**/David Palmer (tl); Dominic Sansoni (cr). **443 Reuters**/Anuruddha Lokuhapuarachchi (bc). **443 Zefa Visual Media**/R.Ian Lloyd (bl). **444 Corbis**/James Marshall (c) **445 Corbis**/Liu Liqun (ca); Setboun (clb). **445 DK**/David Gower (cl). **445 Lonely Planet Images**/Bernard Napthine (br). **445 Reuters**/ Lirio da Fonseca (cfr). **446 Corbis**/Carl & Ann Purcell (tr); Keren Su (cbl); Michel Setboun (cr). **446 DK**/British Museum (cl). **446 Independent Photographers Network**/Michael Yamashita (b). **447 Corbis**/Todd A. Gipstein (br). **447 Independent Photographers Network**/Micheal Yamashita (cr), (tr). **447 NGS**/Reza (c), (bl); Sidney Hastings (cr). **448 Corbis**/Corbis/Owen Franken (cl); Galen Rowell (cr); Tiziana and Gianni Baldizzone (bl). **448 Eye Ubiquitous**/Bennett Dean (bl). **448 Robert Harding Picture Library**/G.Corrigan (tc). **448 Katz**/FSP/Aurora/Micheal Yashamita(c). **449 Corbis**/Bennett Dean (tl); David Lawrence (b); Keren Su (cr). **449 SPL**/Noboru Komine (tr). **450 Corbis**/Keren Su (bc), (cbr), (cl); Michael Yamashita (tc). **450 Michael Freeman** (cr). **450 Colin Prior** (bl). **451 Corbis**/Alison Wright (cr); Chris Lisle (b); Jeremy Horner (tr); Kevin Morris(c); Kevin R Morris (tc); Owen Franken (cbr). **451 Impact Photos**/Alain Evrad (cl). **452 Corbis**/ Catherine Karnow (cla); Chris Lisle (cbl); Jacques Langevin (tr); Kevin R. Morris (cr); Lindsay Hebberd (cal); Owen Franken (bcl). **452 NGS**/Paul Chesley (br). **453 Corbis**/Abbie Enock (bl), (cb); Alison Wright (cl); Cheryl Ravelo/Stringer/Reuters (tr); Jan Butchofsky-Houser (cr); Paul A. Souders (br); Steve Raymer (tcl). **454 Corbis**/Charles & Josette Lenars (cbl); Christophe Loviny (cla), (tcl); Dean Conger (br); Michael Freeman (cra), (tr); Paul Almasy (bcl). **455 Bryn Walls** (car). **455 Corbis**/ Charles & Josette Lenars (crb); Dean Conger (cal), (cfl), (tr); Peter Guttman (bc); Reinhard Eisele (c). **456 Corbis**/ Charles & Josette Lenars (bl), (cl); Michael Freeman (bc); Reuters/Beawiharta (cal); Reuters/Darren Whiteside (cra); Tarmizy Harva/Reuters (tr); Wolfgang Kaehler (br). **456 DK**/Museum of Moving Image (cbr). **457 Corbis**/ Catherine Karnow (cbl); Dennis Degnan (br); Lindsay Hebberd (cla); Paul Almasy (tcl); Roger Ressmeyer (bc); The Cover Story (cra). **458 Corbis**/Charles & Josette Lenars (ca), (tc); Dave Bartruff (bc); Lindsay Hebberd (clb); Paul Almasy (br), (cr); Wolfgang Kaehler (bl). **459 Corbis**/Bisson Bernard (crb); John Van Hasselt (ca); Morton Beebe (cra); Reuters/Kim Kyung-Hoon (tc). **459 Katz**/FSP/ Aerts Layla-Labrun Didier/Photo News/ Gamma (br). **459 NGS**/Jodi Cobb (bl). **460-461 NGS**/ Randy Olson. **461 Corbis**/Bob Abraham (br); Yann Arthus-Bertrand (cl). **461 NGS**/Jodi Cobb (car); Sam Abell (cfr). **462 Corbis**/Albrecht G. Schaefer (cra); Bowers Museum of Cultural Art (cfr); Charles & Josette Lenars (c), (tcr); Wendy Stone. **462 NGS**/Jodi Cobb (bl), (crb). **462 Still Pictures**/Art Wolfe (bcr). **463 Corbis**/ Michael Freeman (cra); W. Perry Conway (cal). **463 Hutchison Library**/Michael Macintyre (cfr). **463 Katz**/ FSP/Dutilleux Jean-Pierre/Gamma (bl). **463 NGS**/Sam Abell (bcr). **464 Corbis**/Albrecht G. Schaefer (br); Caroline Penn (ca); Dallas and John Heaton (c); Paul A. Souders (bl). **464 Hutchison Library**/Michael Macintyre (cla). **464 Lonely Planet Images**/Peter Hines (cbr). **464 NGS**/James L. Stanfield (cra); Sarah Leen (cfr). **465 Panos Pictures**/John Miles (bl); Penny Tweedie (br), (cfr), (cl), (tcr), (tr). **466 Corbis**/Anders Ryman (cra); Catherine Karnow (cal); John Van Hasselt (car), (tcr); Kit Houghton (bl). **466 Getty**/Chris McGrath (cfr). **466 Zefa Visual Media**/George Simhoni (bcr). **467 Corbis**/ ML Sinibaldi (cla). **467 Robert Harding Picture Library**/James Strachan (bcl), (cfr). **467 Hutchison Library**/Michael Macintyre (tr). **467 NGS**/Chris Johns (br); John Eastcott and Yva Momatiuk (cfl); Randy Olsen (cr). **468 Corbis** Charles & Josette Lenars (lc); D. G. Houser (tc); Jack Fields (tc); Paul Gauguin (bl). **468-469 NGS**/Jodi Cobb. **470 Corbis**/Jack fields (br), (tr); James Davis(c); Michael Yamashita (bl); Peter Gutterman (tc); Wolfgang Kaehler (cr). **470 NGS**/Richaed Nowitz (cl). **471 Corbis**/Anders Ryman (tl); Jack Fields (b),(cl), (tr).**471 NGS**/James P. Blair (cr). **472-473 SPL**/Neil Bromhall. **474 Alamy**/Pat Behnke (s4). **474 Corbis**/ Annie Griffiths Belt (s3); Lester Lefkowitz (s1). **474 Alamy**/NASA/Index Stock (s6); **474 Corbis**; (s2). **474 SPL** (bl), (br), (tl), (tl); Sam Ogden (s5). **475 Corbis**/ Lester Lefkowitz (tc). **475 SPL** (br), (tr). **476 Corbis**/ Reuters (b). **476 SPL** (tc). **477 SPL**/ (bl), (cl), (tr). **478 Kobal Collection** (bl). **478 SPL** (b), (tc), (tl). **479 Corbis**/Reuters (tr). **479 SPL** (b), (cl). **480 Corbis**/ Roger Ressmeyer (tr). **480 SPL** (bl), (tl). **481 Reuters** (tl). **481 SPL** (bl). **482-483 Corbis**/Jose Fuste Raga; Tibor Bognar (tcr). **482 Getty**/Paul Chesley (clb). **483 Corbis**/Michael Macor/San Francisco Chronicle (cla). **483 Corbis** (cfr). **483 SPL**/Mehau Kulyk (tcl). **484 Corbis**/Richard T. Nowitz (bl). **484 Infineon Technologies AG** (tc). **484 Panos Pictures**/Mark Henley (cr). **485 Alamy**/Cameron Davidson (tc). **485 Corbis**/Liba Taylor (cfl); ML Sinibaldi (bc). **485 SPL**/ Laboratory for Atmospheres, NASA Goddard, Space Flight Centre (tcl).

**Endpapers Getty**/Jonathon Kirn.